U0220915

The Living Record of Science

《自然》学科经典系列

总顾问：李政道（Tsung-Dao Lee）

英方总主编：Sir John Maddox
Sir Philip Campbell

中方总主编：路甬祥

生命科学的进程 III

PROGRESS IN LIFE SCIENCES III

（英汉对照）

主编：许智宏

外语教学与研究出版社 · 麦克米伦教育 · 《自然》旗下期刊与服务集合

FOREIGN LANGUAGE TEACHING AND RESEARCH PRESS · MACMILLAN EDUCATION · NATURE PORTFOLIO

北京 BEIJING

图书在版编目（CIP）数据

生命科学的进程. III ：英汉对照 / 许智宏主编. -- 北京 ：外语教学与研究出版社，2021.9
（《自然》学科经典系列 / 路甬祥等总主编）
ISBN 978-7-5213-2929-2

Ⅰ. ①生… Ⅱ. ①许… Ⅲ. ①生命科学－文集－英、汉 Ⅳ. ①Q1-53

中国版本图书馆 CIP 数据核字（2021）第 176594 号

地图审图号：GS（2021）3947 号

出 版 人　徐建忠
项目统筹　章思英
项目负责　刘晓楠　顾海成
责任编辑　刘晓楠
责任校对　王　菲　夏洁媛
封面设计　孙莉明　高　蕾
版式设计　孙莉明
出版发行　外语教学与研究出版社
社　　址　北京市西三环北路 19 号（100089）
网　　址　http://www.fltrp.com
印　　刷　北京华联印刷有限公司
开　　本　787×1092　1/16
印　　张　61.5
版　　次　2021 年 9 月第 1 版　2021 年 9 月第 1 次印刷
书　　号　ISBN 978-7-5213-2929-2
定　　价　568.00 元

购书咨询：（010）88819926　电子邮箱：club@fltrp.com
外研书店：https://waiyants.tmall.com
凡印刷、装订质量问题，请联系我社印制部
联系电话：（010）61207896　电子邮箱：zhijian@fltrp.com
凡侵权、盗版书籍线索，请联系我社法律事务部
举报电话：（010）88817519　电子邮箱：banquan@fltrp.com
物料号：329290001

《自然》学科经典系列

（英汉对照）

生命科学的进程

（英汉对照）

主编：许智宏

审稿专家（以姓氏笔画为序）

于天源　王　昕　王晓晨　王敏康　邢　松　同号文　刘　力
刘　武　刘佳佳　刘京国　孙　军　李芝芬　李典谟　李素霞
吴新智　沈　杰　张健旭　陈建国　陈继征　陈新文　林圣龙
昌增益　金　城　周　江　周筠梅　郑家驹　赵凌霞　胡卓伟
秦志海　袁　峥　顾孝诚　黄晓航　曹文广　崔　巍　崔娅铭
梁前进　董　为　曾少举　曾长青

翻译工作组稿人（以姓氏笔画为序）

王耀杨　刘　明　关秀清　李　琦　何　铭　沈乃澂　蔡　迪

翻译人员（以姓氏笔画为序）

王耀杨	毛晨晖	邓铭瑞	田晓阳	冯　琛	吕　静	刘　霞
刘振明	刘皓芳	李　梅	吴　彦	沈乃澂	张锦彬	周志华
郑建全	荆玉祥	姜　薇	董培智			

校对人员（以姓氏笔画为序）

于平蓉	马　荣	马晨晨	王　羽	王　菲	王　敏	王帅帅
王阳兰	王珊珊	王晓敏	王晓蕾	王海纳	王赛儿	化　印
公　晗	孔凌楠	代　娟	丛　岚	吕秋莎	乔萌萌	任　奕
任嵛铭	刘　伟	刘　明	刘　佩	刘　婷	刘若青	闫　妍
许梅梅	阮玉辉	孙　瑶	李　四	李　娟	李　梅	李　琦
李　景	李　静	李红菊	李盎然	邱彩玉	何　铭	邹伯夏
张玉光	陈思婧	陈露芸	范艳璇	周少贞	周玉凤	周平博
周晓明	郑建全	郑娇娇	赵广宇	赵凤轩	侯彦婕	姜　薇
夏洁媛	顾海成	黄　欢	黄　璞	黄元耕	黄晓东	第文龙
葛　越	董　为	焦晓林	曾芃斐	谢周丽	蔡军茹	潮兴娟
潘卫东						

Contents
目　　录

1972

1973

1974

1975

1976

1977

Volume III

"Self-recognition" in Colonial Marine Forms and Flowering Plants in Relation to the Evolution of Immunity

F. M. Burnet

Editor's Note

The Australian F. MacFarlane Burnet was one of the pioneers of the modern understanding of the immune response of animals (including people) to infectious pathogens. For his pioneering work he was awarded a Nobel Prize for medicine in 1960. In this article, he raises questions about the similarities between immune reactions in plants as well as animals and speculates about the relationship between immunity and the genetic code.

THERE is a growing tendency to regard the evolutionary origin of adaptive immunity, as manifested in ourselves and in the standard experimental mammals of the laboratory, as being related to something other than defence against pathogenic microorganisms. In 1959 Thomas[1] suggested that there were two mammalian phenomena which might be relevant to the understanding of immunity and its evolution—pregnancy and malignant disease. Both suggestions have subsequently been discussed extensively and the relationship of the foetus as a mass of antigenically foreign cells embedded in the maternal tissues to the mother's immune system has provoked much interest and experimentation[2,3]. Similarly, the antigenic quality of a large proportion of autochthonous tumours became one of the central themes of recent oncological research[4-7]. I have written extensively[8] on the concept of immunological surveillance, which postulates that antigenic patterns abnormal to the genotype which arise by somatic mutation may provoke an antigenic response and play a part in preventing or delaying the appearance of malignant disease.

All metazoan forms must have a capacity to inhibit the multiplication of bacteria and other potential pathogens in their tissues and in that sense they are possessors of an immune system. For obvious anatomical and functional reasons, if we are to consider the evolution of the mammalian immune system, we must confine ourselves to the vertebrates and their immediate ancestral forms. Recently there have been intensive studies of sea lamprey[9,10] and hagfish[11-13] as representative modern cyclostomes, the most primitive extant vertebrates. It is probably a fair summary of the results to say that homograft rejection is well shown; there are other T-type (cell-mediated) responses and restricted and weak antibody responses. The immunoglobulins have typical vertebrate characteristics but they are in very low concentration[10].

群体海洋生物和开花植物中的“自我识别”与免疫进化的关系

伯内特

编者按

澳大利亚的麦克法兰·伯内特作为先驱之一，建立了动物（包括人）对传染性病原体的免疫反应的现代解释。由于所作的开创性工作，他在 1960 年被授予诺贝尔医学奖。在这篇文章中，他就动植物间免疫反应的相似性提出了一些问题，并思考了免疫和遗传密码间的关系。

根据在我们自己体内以及在实验室标准实验哺乳动物体内所发现的情况，人们越来越倾向于将适应性免疫的进化起源与除抵御病原微生物外的某些因素关联起来。1959 年，托马斯[1]提出，哺乳动物中有两种现象可以用免疫和免疫的进化来解释——怀孕和恶性疾病。这些建议后来被广泛讨论。胎儿可以被看作是一团嵌入到母体组织中的抗原性外源细胞，其与母体免疫系统的关系激起了科学家们的极大兴趣，他们对此进行了大量实验[2,3]。类似地，大部分原发肿瘤的抗原特性已经成为近来肿瘤学研究的中心议题之一[4-7]。我在一本专著中[8]广泛讨论了免疫监视的概念，认为由体细胞突变引起的基因型异常的抗原模式可能会诱发抗原反应，并在阻止或延缓恶性疾病的发生中起着一定的作用。

所有后生动物（译者注：此处指多细胞动物）都一定具有抑制细菌和其他潜在病原体在自身组织内繁殖的能力。从这个意义上讲，它们要拥有一套免疫系统。基于显而易见的解剖学和功能方面的原因，如果我们要考虑哺乳动物免疫系统的进化，就必须将我们的注意力集中在脊椎动物和它们最直接的祖先上。最近已经对海生七鳃鳗[9,10]和盲鳗进行过深入研究[11-13]，它们是现代圆口纲脊椎动物的代表，也是现存最原始的脊椎动物。对这些结果的合理总结很可能可以说明同种移植排斥反应已得到很好的体现；还有其他 T 细胞型的（细胞介导）反应和一些受限且微弱的抗体反应。其免疫球蛋白虽然具有典型的脊椎动物特征，但浓度很低[10]。

Studies of Colonial Tunicates

No immunological studies have yet been made to my knowledge of the protochordates and the work of Oka[14] on the genetics of compatibility in colonial ascidians has not been considered previously for its immunological implications. Here I shall explore the possibility that Oka's studies of colonial tunicates (*Botryllus*) and the less extensive studies of Theoder[15] of the anthozoan (*Eunicella*), also a colonial form, may throw light on primitive types of "self and not-self" recognition from which adaptive immunity may have evolved.

The colonial ascidian *Botryllus* produces star shaped colonies, each ray being an individual with its own inlet and outlet pores for seawater circulation and feeding, but with a common vascular system ending in a series of peripheral ampullae. The colony is initiated by a larva that develops from a fertilized ovum, but once the larva has settled down it buds off new individuals to form the colony. Each colony is therefore essentially a clone of genetically identical units. As might be expected, therefore, if colony I is divided into two and the parts allowed to grow separately and then brought into contact, the two daughter colonies fuse together and reconstitute a single colony. If, however, an unrelated colony II of the same species is brought into contact with colony I there is a positive rejection and a barrier of necrotic material develops between the two colonies. Finally, if another compound ascidian of the same general type, *Botrylloides*, comes into contact with *Botryllus* I or II, nothing happens. One will grow over the other as if it were just an inert attachment surface. Clearly, therefore, there is recognition of foreignness within the species and recognition must always be the basic phenomenon of immunity.

Sexual reproduction of *Botryllus* is dominated by mechanisms to avoid self-fertilization since the organisms are hermaphrodite, liberating both sperm and ova into the environment. To summarize Oka's work, it is convenient to accept his assumption, which is validated by much preliminary work, that fusion or rejection between colonies depends on a single locus with many alleles which can be referred to as recognition genes. All natural colonies are heterozygous and can be represented as AB, CD and so on. At the margin of each colony or regenerated sub-colony there is a row of ampullae (with a single cell wall) the contents of which are part of the common blood circulation of the colony. It is the reaction of contact between these ampullae which determines the gross character of the interaction between two colonies. When two parts of the same AB colony are brought together again, the walls of adjacent ampullae fuse and a lumen is developed to extend the common circulation. When an F1 generation is produced, the progeny from an AB× CD cross will be of four different types: AC, AD, BC and BD. Any two sets of these which have a gene in common—AC/BC, AC/AD etc.—will show the same type of fusion and integration of the circulation. Where there is no common gene—AC/BD, AD/BC—the ampullae in contact undergo necrosis, producing a barrier between the two colonies.

The other feature of interest concerns cross-fertilization. These organisms are hermaphrodite and there are special arrangements to prevent self-fertilization. The gametes which are, of course, haploid are A or B from AB and C or D from DC. Self-

对群体被囊动物的研究

据我所知，目前还没有针对原索动物的免疫学研究，此前人们并没有把奥卡[14]关于群体海鞘相容性的遗传学研究看作是免疫学方面的工作。这里我将探究以下可能性：利用奥卡对群体被囊动物（菊海鞘）的研究以及西奥多[15]对另一种群体动物——珊瑚虫（网柳珊瑚）的小范围研究，或许可以揭示出“自我和非我”识别的原始类型，适应性免疫可能就是由此演化而来的。

一种群体海鞘——菊海鞘形成星形群落，每一射线都是一个个体，各自用自己的进水孔和出水孔进行海水的循环和进食。但是它们共用一套末端位于一系列外周壶腹的脉管系统。群落始于由一个受精卵发育而来的幼体，但是一旦这个幼体定居下来，就会出芽产生新的个体，最终形成群落。因此，每个群落本质上是一个由相同遗传单元组成的克隆体。正如我们所料，如果群落 I 一分为二，并且两个部分都能独立生长，然后使它们相互接触，那么这两个子群落就会融合在一起，重组为一个单一的群落。然而，如果将一个属于同一物种但不相关的群落 II 与群落 I 放在一起互相接触，就会发生主动排斥，并且两个群落间会形成一道由坏死物质组成的屏障。最后，如果用另一种大致相同类型的混合海鞘——拟菊海鞘，与菊海鞘 I 或 II 放在一起互相接触，将什么都不发生。一种会覆盖在另一种上生长，就好像下面只是一种非生物的附着表面。因此，很明显，菊海鞘能够识别同一物种内的外来物质，而识别通常是免疫的基本现象。

由于菊海鞘是雌雄同体，且同时向环境中释放精子和卵子，因而这种生物体的有性繁殖由一些能够避免自体受精的机制所支配。要总结奥卡的工作，方便的方法是接受他的假设，这些假设已经被很多初步工作证实，那就是群落间的融合或排斥取决于具有很多等位基因的某一单基因座，可以认为这些等位基因就是识别基因。所有自然的群落都是杂合的，可以用 AB、CD 等来表示。在每一群落或者再生亚群落的边缘存在一排壶腹（具有单一的细胞壁），壶腹的内容物是群落共同血液循环的一部分。两个群落间相互作用的总体特征由这些壶腹间的接触反应决定。当同属于 AB 群落的两部分再次放在一起时，邻近壶腹的壁相互融合，发育成一个共同的内腔，将二者独立的循环扩展为一个共同的循环。当 F1 代产生时，AB 与 CD 杂交的后代会有四种不同的类型：AC、AD、BC 和 BD。其中具有一个共同基因的任何两个组合，如 AC/BC、AC/AD 等，会表现出同种类型的融合和循环整合。如果不存在共同基因，如 AC/BD、AD/BC，则接触的壶腹会发生坏死，从而在两个群落间形成一道屏障。

值得关注的另一个特征是异体受精。这些生物体都是雌雄同体，具有阻止自体受精的特殊办法。它们的配子必然是单倍体，其中 A 或 B 来自 AB，C 或 D 来自

fertilization does not occur presumably because of the layer of follicular cells which surrounds the ovum. The nature of this structure is not elaborated in Oka's review[14] but the cells surrounding an A ovum from an AB individual are presumably diploid and AB, and prevent the entry of either an A or a B spermatozoon. Any other type of spermatozoon, such as C or D, encounters no obstruction.

Other Colonial Forms

Both phenomena seem "designed" to ensure the greatest degree of heterozygosity in all viable colonies. It is virtually impossible for a homozygous colony to be formed and, at the same time, the situation cannot be muddied by the fusion of unrelated colonies. Both seem to produce colonies which are at a disadvantage for survival, although it is not easy to see just what the disadvantage is. Oka points out that the situation is directly analogous to the self-incompatibility of angiosperms such as plums, but it is even more analogous to the findings of Theodor[15] on some colonial coelenterates (Anthozoa). It may well be of importance for survival that colonial marine forms should be able to distinguish their own colony from another. In the sea-fan *Eunicella* the colony has a branching form supported by a firm sclero-protein axis on which there is a continuous sheath of living cells from which the polyps derive a dense efflorescence. Taking 25–30 mm segments of the branches, Theodor denuded a central 2 or 3 mm portion to expose the rigid non-living axis. Two such preparations were combined so that the two denuded areas formed the meeting point of a right-angled cross. When such preparations were maintained in suitable aquaria, regrowth eventually covered the denuded regions. If both segments were from the same colony all four growing surfaces fused without any sign of discontinuity. When one segment was from colony A and the other from B, A and B tissues failed to fuse and a clear line of demarcation was visible. When two different genera, *Eunicella* and *Lophogorgia*, were placed in contact there was a damaging interaction in the form of the development of "blistery" tissue along the surface of contact.

Self-recognition in *Botryllus*

As far as the evolution of adaptive immunity is concerned, it can be assumed that the phenomena Oka described for colonial ascidians, and probably also those of Theodor in the Anthozoa, facilitate the achievement and maintenance of heterozygosis. In the better studied system of Oka, the aspect of immediate interest to the immunologist is the destructive interaction between cells lacking appropriate common alleles of the recognition genes. This, however, has no more than a superficial resemblance to the cytotoxic action of sensitized lymphocytes on a larger cell. There must be three distinct sets of recognition phenomena. First, a somatic cell in contact with another recognizes that they possess at least one common recognition allele and as a result the necrotic response[2] is inhibited. Second, a somatic cell in contact with another which has no common specific allele but is otherwise genetically similar (of the same species) can recognize this conspecificity in the sense that it induces mutual necrosis. This is not seen when wholly unrelated forms make contact. Third, a sperm (haploid) recognizes the presence of the same allele by a gene product present in the periovular cells and cannot fertilize such an egg.

DC。没有发生自体受精的原因可能是有一层卵泡细胞包围着卵子。奥卡的综述[14]并没有详尽阐述这种结构的性质，只提到围绕在来自AB个体的A卵子周围的细胞有可能是二倍体并且是AB，这样就阻止了A精子或B精子的进入。而其他类型的精子，例如C或D，将可以毫无阻碍地进入。

其他群体生物

以上两种现象看上去是为了确保在所有可繁殖后代群落中的杂合程度最大化而“设计”的。现实中不太可能形成纯合子群落；与此同时，情况也不会因不相关群落的融合而被搅乱。这两种情况看上去都产生了不利于生存的群落，虽然不容易发现不利之处到底是什么。奥卡指出这种情况恰好与李子等被子植物的自交不亲和性类似，但是它更类似于西奥多[15]在一些群体腔肠动物（珊瑚虫）中的发现。群体海洋生物应该能区别自身群落与其他群落，这对于生存来说很可能具有重要意义。海扇形网柳珊瑚的群落具有分支结构，其中的支撑体是坚固的硬蛋白轴，轴上有一层由活细胞组成的连续鞘状结构，珊瑚虫致密的骨骼就是由这些活细胞分泌的。西奥多取分支上25 mm~30 mm的断片，剥开中间的一段2 mm或3 mm的部分，以暴露坚硬的非生物轴。准备这样的两段，将它们结合在一起，使两个剥离的区域形成直角交叉的接触点。将这样的制备物长期置于合适的鱼缸中，再生结构会最终覆盖剥离区域。如果两个断片来自同一群落，那么所有四个生长表面就会完全连续地融合在一起。如果一段源自群落A，另一段源自群落B，A和B的组织就不能融合，可以看到两者之间有一条清晰的分界线。如果两个不同的属（网柳珊瑚和柔枝柳珊瑚）被放在一起互相接触，就会发生破坏性相互作用，表现为沿接触表面形成“起泡”组织。

菊海鞘的自我识别

就适应性免疫的进化而言，可以假定：是奥卡所描述的群体海鞘现象，也可能是西奥多所描述的珊瑚虫现象，促进了杂合性的实现和维持。在奥卡的更为完善的研究系统中，最令免疫学家感兴趣的方面是：细胞间因为缺乏合适的作为识别基因的共同等位基因，从而产生破坏性相互作用。然而，这与致敏淋巴细胞对较大细胞的细胞毒作用只具有表面上的相似性。这里必然存在三种不同情况下的识别现象。第一，一个体细胞与另一个体细胞接触，认出它们至少拥有一个共同的识别用等位基因，因而坏死反应[2]被抑制。第二，一个体细胞与另一个虽不具有共同的特定等位基因但在其他方面有类似遗传学特征（属于同一物种）的体细胞接触，会识别出这种同种性，从而引起相互坏死。如果是完全不相关的生物互相接触，则不会出现这一现象。第三，精子（单倍体）通过卵周细胞内的基因产物识别出相同等位基因的存在，从而不能使这样的卵子受精。

The concept of recognition is so important that these three examples justify discussion in some detail. The essence of the matter is that organism tissue or cell A comes into contact or other definable relationship with two or more generally similar entities, X^1, X^2, X^3 and so on, and in one instance (say, with X^1) a demonstrable and reproducible reaction occurs which is not seen with X^2, X^3 etc. A is then said to be capable of positive recognition of X^1. Sometimes, as in the case of *Botryllus*, the individual reaction by which recognition of one character as against "any others" is shown takes the form of an inhibition of necrosis or some otherwise inevitable response. The crux of the matter is that recognition is only meaningful when it is a recognition of one out of many alternative candidates. An antibody reacts visibly with only one of a thousand antigens. In the *Botryllus* examples we must postulate that there is a positive recognition of the presence of a common allele in the somatic cell genome, a positive recognition of conspecificity by the mutually damaging reaction and a positive recognition by the sperm that the follicular cells of the ovum have the same R allele, since it can fertilize "all other" combinations which lack that allele.

Nature of Recognition

All those positive recognitions are conventionally and almost certainly correctly interpreted as representing specific union, reversible or irreversible, between chemical groupings on the surfaces of the interacting cells. There are possible ways by which like chemical groupings can recognize each other as exemplified in the growth of any crystal, but where proteins are concerned this seems to be a cumbersome approach inappropriate for the extremely heterogeneous micro-environment where all recognition reactions occur. All immunologists, every biochemist concerned with enzyme-substrate relationships and every pharmacologist interested in drug-receptors implicitly accept the axiom that all such specific ("recognizing") reactions are based on sterically complementary (+/–) chemical patterns. We are bound to do the same here, but to do so brings us immediately up against a genetic difficulty whose solution may be of first rate importance for immunological theory. Two of the examples of recognition seem to be recognition by cell X that cell Y has an allele identical to its own. If we exclude, as we must, any suggestion that recognition involves interaction between nucleic acids, this seems that any mutation to a new allele must in some way produce corresponding but complementary changes in both + and – patterns of the mutual recognition sites. It is extremely difficult to conceive any way by which information in a DNA sequence (A) can provoke the formation of another DNA sequence which will code for a complementary three-dimensional structure that will react specifically with the product of A. It seems more promising to consider the situation where the + and – complementary receptors are controlled by distinct genes. There is much to be said for an *ad hoc* assumption that both + and – types of recognition gene are derived by duplication from a common progenitor gene, but they would require to be capable of independent mutation. We have to assume that a mutation in, say, the + gene cannot be expressed until a corresponding mutation arises in the – gene. If such a process were to function there would have to be a high mutation rate in genes, a situation which has obvious relevance to the genetics of immune pattern and to Jerne's ideas[16] of a special relationship between the potential antibody patterns transmissible by the germ line and histocompatibility antigens.

识别的概念太重要了，以至于值得详细讨论以上三个例子。识别问题的关键是生物组织或细胞 A 与两个或更多大致类似的实体 X^1、X^2、X^3 等发生接触或发生其他可定义的关系。如果在一个例子中（比如与 X^1）发生了可论证和可重复的反应，但与 X^2 和 X^3 等则未见此类现象，那么就可以说 A 能够主动识别 X^1。有时，例如就菊海鞘来说，赖以识别一种特征有别于"所有其他"特征的个体反应表现为抑制坏死或其他一些不可避免的反应。此问题的关键是，识别只有在能从很多可供选择的候选项中识别出一个时才是有意义的。一个抗体显然只会与上千抗原中的一个发生反应。在菊海鞘的例子中，我们必须假定存在三种主动识别：一是通过体细胞基因组中的一个共同等位基因的主动识别，二是通过发生相互破坏反应的同物种间的主动识别，三是通过精子的主动识别——精子可以识别出卵子的卵泡细胞是否含有相同的 R 等位基因，因为它可以使"所有其他"缺乏该等位基因的组合受精。

识别的本质

所有主动识别通常都被几乎确定无疑地解释为代表了相互作用细胞表面上化学基团之间的可逆或不可逆的特异性结合。比如化学基团可以通过一些可能的方式相互识别，这已被任一晶体的生长过程所证实；但是对蛋白质而言，这看上去是一种很繁复的方式，不太容易与发生每种识别反应的极端异质的微环境相适应。所有免疫学家、每一位关注酶–底物关系的生化学家和每一位对药物–受体感兴趣的药理学家都绝对接受如下公理，即所有这些特异（"识别"）反应都基于空间互补（+/–）的化学模式。我们注定要采用空间互补的处理方法，但是这样做会使我们立刻面临遗传学上的难题，其解决方案对于免疫学理论来说可能具有第一位的重要性。两个识别的例子看上去是：细胞 X 发现细胞 Y 具有同一等位基因，从而实现识别。如果正如我们必须要做的那样，我们排除了识别涉及核酸间相互作用的任何建议，那么似乎任何产生某种新等位基因的突变必然会以某种方式在相互识别位点的 + 和 – 模式中产生相应但互补的变化。很难想象，DNA 序列（A）中的信息可以通过任何一种方式引起另一 DNA 序列的形成，后者将编码某一互补的三维结构，这一结构将与 A 的产物发生特异性反应。看上去考虑以下情况更有前景，即 + 和 – 这对互补受体由不同的基因控制。识别基因的 + 型和 – 型均源自某共同祖先基因的复制，但要求它们能够独立突变，对于这一特殊假说还有很多需要交代。举例来说，我们不得不假设：在 + 基因中的一个突变将不被表达，直到 – 基因中发生相应的突变。如果这一过程具有功能，那基因中的高突变率也就是必然的了，这种情况显然与免疫模式的遗传学以及杰尼的想法 [16] 相关联，杰尼认为在可通过生殖细胞遗传的潜在抗体模式和组织相容性抗原之间存在着一种特殊的关系。

All that I have been discussing in regard to the colonial ascidians is solely concerned with germ line genetics. Each cell is regarded as expressing appropriately the instructions of the germ line genome. Somatic genetic changes come later in the story.

The situation in *Botryllus* is represented schematically in Fig. 1. Here it is assumed that at each surface of contact with an adjacent cell there are + and − receptors for the two relevant alleles of each heterozygous diploid colony. A positive receptor is shown as a knob, a negative one as a depression. By hypothesis, +/− union stabilizes the two adjacent cell surfaces, and in the absence of any +/− unions mutual disorganization with liberation of damaging products takes place. It is inexpedient to attempt any biological explanation of this phenomenon except to point out that in all larval–adult metamorphoses there is extensive cell destruction in which some such processes must be involved and also that similar cell damage is seen in all severe T-D immune reactions in mammals.

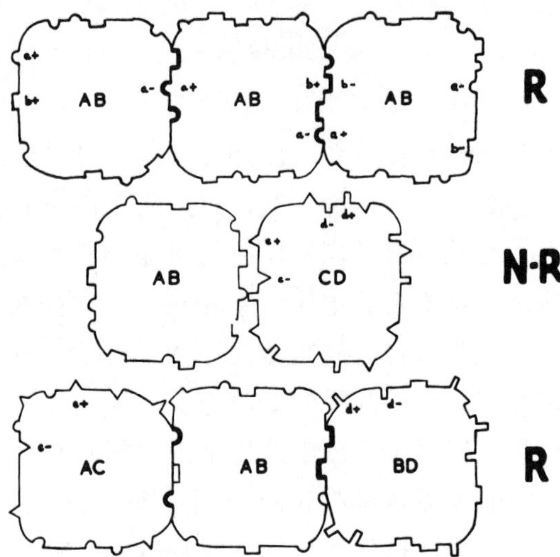

Fig. 1. A diagram to suggest the type of +/− relationship between recognition units that is postulated for *Botryllus*. A positive receptor is shown as a projection, the corresponding complementary negative receptor by a depression of the same shape. Where there is a reciprocal relationship (R) allowing receptor union (heavy line), the relation is stabilized. In its absence (N-R), fusion is impossible.

Relation to Immune Response

At most this can only be a foreshadowing of the adaptive immune system of vertebrates. There is, for example, no hint as to how the capacity to recognize foreign qualities of conspecific cells evolved and there are manifest differences of the tunicate system from that subserving tissue integrity (the T-immune system) in vertebrates. Like every biological invention the utilization of steric complementarity involves both the informational store in the genome and the protein molecules needed for its phenotypic expression. There arises immediately the outstanding problem of how mutation and other informationally random genetic processes can produce complementary structure in the products of two

我一直在讨论的关于群体海鞘的所有内容都只与生殖细胞遗传学有关。每个细胞都被认为是适当地表达了生殖细胞基因组的指令。体细胞的遗传学改变将在本文稍后部分进行描述。

菊海鞘中的情况如图 1 所示。这里假设：相邻细胞的每一接触表面都存在 + 受体和 − 受体，分别代表每一杂合二倍体群落的两个相关等位基因。如图所示，球形凸起表示正受体，凹陷表示负受体。根据假设，+/− 结合使两个相邻细胞的表面稳定，在 +/− 结合完全缺失的情况下，会发生相互解体，并释放出破坏性产物。对于这一现象，任何用生物学解释的尝试都是不明智的，除非可以说明以下两点：一是在所有幼体 – 成体的变态过程中存在着大面积的细胞破坏，在细胞破坏时还必须涉及一些识别过程；二是哺乳动物中所有严重的 T–D 免疫反应也都可见类似的细胞破坏。

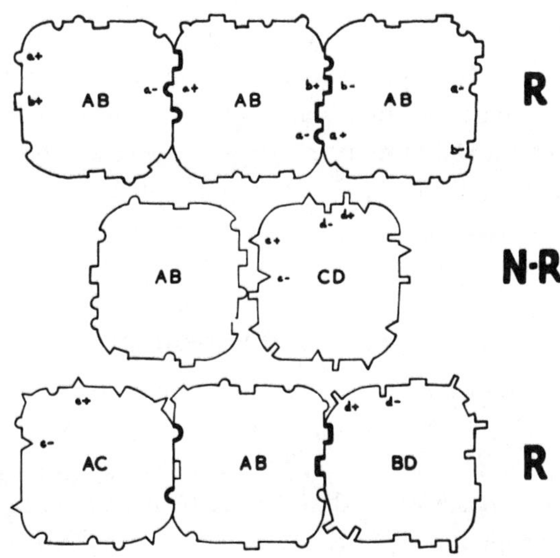

图 1. 以菊海鞘为例表示识别单元间 +/– 关系类型的示意图。凸起代表正受体，与凸起形状匹配的凹陷代表相对应的互补负受体。一旦出现允许受体结合（粗线）的相互关系（R），这种关系就会是稳定的。在不满足这种关系（N-R）的情况下，融合是不可能的。

与免疫反应的关系

这充其量只能算是脊椎动物适应性免疫系统的铺垫。例如：没有迹象表明识别外源同种细胞的能力是如何进化的，而且在被囊系统与脊椎动物中促进组织整体性的系统（T– 免疫系统）之间存在着明显差异。与每一生物学创造一样，对空间互补性的利用牵涉到基因组中的信息贮存和表型表达所需要的蛋白质分子。这样就立刻引发了一个难解决的问题：在两种不同的基因产物中，突变和其他能使信息发生随机遗传的过程如何能产生互补结构。然而，一旦发现了这种进化过程，它就能够提

distinct genes. Yet once that evolutionary discovery had been made it could provide the raw material from which in colonial organisms means could have evolved to ensure the genetic integrity of the colony and to avoid the long term dangers of self-fertilization and inbreeding. A continuation of biological invention along the same general path could in principle lead to the construction of a vertebrate immune system. For this to happen, there are three requirements: first, mutual recognition systems would have to be elaborated and certain cells differentiated to specialize in recognition and in the local damaging response; second, recognition functions of such cells would have to become progressively specialized until there existed a wide diversity of immune patterns, each limited to a single clone, and third, such "immunocytes", in certain conditions of specific stimulation, would have to be able to take on the blast form and proliferate.

It is probably unwise to attempt to imagine the various steps by which such changes could be made. One can foresee a period of great research activity in these fields of tissue fusion and rejection in invertebrates and protochordates during the next decade. Undoubtedly a variety of intriguing phenomena will be uncovered, differing from group to group. Some may be further along the road toward adaptive immunity than the colonial ascidians. Much more extensive comparative studies are called for and in due course analysis of the results should allow a clear evolutionary history to emerge. Whatever form that history eventually takes we can be certain that gene duplication (gene expansion) plays a major part, and that progressive specialization of cell function and phenotypic restriction will be as conspicuous as it is in all other organs and functions.

Viviparity and Parasitism

Evolution is much more than an expansion and specialization of rudimentary functions present in the form we choose as a starting point. It is a process of modification for more effective survival of populations. Major evolutionary changes probably always mean that a major danger to survival has appeared and must be overcome or that a new ecological niche has arisen to be occupied by species which can first make the necessary adaptations. Often, of course, the two reasons overlap.

A year or two ago I suggested that when free swimming marine protovertebrates ancestral to cyclostomes first appeared it is conceivable that one of the early results was the emergence of a new ecological niche, survival by parasitism on one's own kind. The stimulus to develop this point of view was the immense ecological effect of the entry of the sea lamprey into the great lakes of North America around 1930[17]. Lampreys are cyclostomes that live by blood sucking—one lamprey can kill 14 kg of fish per annum—and there were countless streams running into the lakes which were suitable for larval development. Lamprey proliferation reduced the economically significant fish species to such a level that the lakes' fishing industry was destroyed. The subsequent ecological history of the Great Lakes is of great interest but not relevant to my theme, which is that with the entry of an effective predator-parasite in the form of a cyclostome a major ecosystem was wholly changed. If free swimming protovertebrates in Cambrian seas represented the current evolutionary success, a wide "radiation" into new ecological

供一些原始材料以使群体生物体的进化能保证群落的遗传整体性以及避免自体受精与同系繁殖的长期危险性。大致沿同一方向进化的连续生物学创造原则上会导致脊椎动物免疫系统的形成。要实现这一点，需要三个前提条件：第一，相互识别的系统必须十分精细，并且分化出了一些专门用于识别和用于局部破坏反应的细胞；第二，这些细胞的识别功能必须逐渐特化，直到能够产生多种多样的免疫模式，每一免疫模式限制在单克隆内；第三，在某些特定的刺激条件下，这样的“免疫细胞”必须能以爆发形式大量增殖。

试图想象出实现这些变化所需的多步过程很可能是不明智的。可以预见未来十年将是大量研究活动集中在无脊椎动物和原索动物的组织融合与排斥领域的时期。毫无疑问，很多有趣现象将会被一批一批地揭示出来。一些研究可能会进一步沿适应性免疫方向进行，而不是沿群体海鞘方向发展。当务之急是进行更加广泛的比较研究，从对结果的正确分析中应该能够得到一段清晰的进化史。无论进化史最终将以何种形式呈现，我们都可以断定基因复制（基因扩展）在其中起着主要的作用，并且细胞功能的不断特化和表型的限制都与在所有其他器官和功能中一样明显。

胎生和寄生

进化的意义远远超过基本功能的扩展和特化，我们选择基本功能的存在形式作为起点。进化是为使种群更有效生存而进行的修正过程。主要的进化变化通常意味着生存遇到了一个巨大的危机，必须战胜这种危机；或者意味着出现新生态位，并被第一个能作出必要适应性改变的物种占领。当然，这两种原因经常是交织在一起的。

一两年前我曾提出：当在海洋中自由游泳的原始脊椎动物——圆口类的祖先开始现身时，可以想象早期结果之一是出现了一个新生态位，即通过在同类身上寄生来实现生存。产生这种观点的诱因源自 1930 年前后海洋七鳃鳗进入北美五大湖所造成的巨大生态效应[17]。七鳃鳗是靠吸食其他鱼类血液生存的圆口纲脊椎动物——一条七鳃鳗每年能杀死 14 千克鱼——适于七鳃鳗幼体发育的湖泊有数不清的河流汇入。七鳃鳗数目的激增使重要经济鱼种的减少达到这些湖泊的渔业蒙受损失的地步。五大湖随后的生态史非常引人注目，但与我的主题无关，我的主题是：随着圆口纲脊椎动物这类强大捕食性寄生动物的引入，主要生态系统发生了彻底的变化。如果在寒武纪海洋中自由游泳的原始脊椎动物与当前的进化产物一样具有很强的繁殖力，那么进入新生态位的广泛“辐射”必然会随之很快开始。现有证据表明，类似七鳃

niches must have begun immediately. There is evidence that lamprey-like forms were present in the Silurian[18], presumably living like modern lampreys but necessarily on other cyclostomes until fishes had evolved.

Against this background I asked[19] whether parasitism of a cyclostome by its own young or by related species might not provide an urgent demand for the development of a more refined capacity to recognize the difference between self and not-self. This could be the evolutionary stimulus needed to set the construction of an adaptive immune system on its way. In the light of what has been published since, I should now be inclined to state the possibilities slightly differently. What I failed to comment on was the significance of viviparity which appears in many lower organisms including tunicates and fish as well as in mammals. When a living embryo is nourished in the tissues of the parent the situation is barely distinguishable from parasitism, and oscillation between a controlled situation and uncontrolled parasitism must have occurred not infrequently. There may well have been a special predilection for viviparous larvae kept under control in their own parental environment to make an early switch to parasitism on a similar type of organism.

Comparison with Homograft Reactions

At this stage it is necessary to attempt a formal comparison at the genetic level of what is observed in the colonial ascidians and what holds for homograft reactions in mammals. If in both the simplifying assumption is made that histocompatibility is determined by alleles of a single gene, we can consider the results of matings between heterozygous animals whose genotypes are AB and CD, AB being used for the female in each case. For both types of animal the F1 generation will contain the genotypes AC, AD, BC and BD. In *Botryllus* all possible offspring will fuse (be compatible) with the parent. In a mammal, skin grafts from any of the offspring will be rejected. The evolutionary problem is how to pass from the first condition to the second.

First, it is necessary to establish a process by which recognition can have a second type of sequel. Instead of serving only as a means of stabilizing cell interaction at contact surfaces, recognition must under other circumstances be followed by a damaging liberation of cell products. Second, complementary patterns must be provided which can recognize not only gene products which can be produced by the individual itself but also those characteristic of all the patterns (histocompatibility antigens) which it is within the capacity of the species to produce.

The requirements can be discussed in relation to Fig. 2, which is a simple diagram of an offspring in a situation in which it is potentially parasitic on the mother's tissues. On a simple extension of the *Botryllus* situation, the cells should tolerate each other and the parasitic relationship should develop easily. In the situation of incipient parasitism that we have pictured, the biological requirement, if the parasite is to be expelled, is that the parent AB should produce a specialized cell with what in *Botryllus* would be a – type C receptor analogous to an immune receptor (cell-fixed antibody) in a vertebrate. Such a cell must be able to reach the region of contact between AB and AC and, following

鳗的生物出现于志留纪[18]，据推测，它们的生活方式类似于现在的七鳃鳗，但是必须寄生在其他圆口纲脊椎动物身上，直到进化出鱼类。

以此为背景，我提出以下问题[19]：一种被同类幼体或相关物种寄生的圆口纲动物，会不会并不急迫需要发育出更佳识别自我和非我之间差异的能力。对这种能力的需求可以成为进化的刺激因素，适应性免疫系统的构建在确立时需要这种刺激。基于迄今为止已经发表的结果，我现在倾向于以稍有不同的方式阐述一些可能性。对于胎生在包括被囊动物和鱼类在内的众多低等生物体以及哺乳动物中的意义，我还无法作出评价。一个活的胚胎在母体组织中被滋养时的情况与寄生几乎没有什么差别，在可控的胎生和不可控的寄生之间的摇摆必然会经常发生。母体很可能对生活在其环境中的受控胎生幼体具有一种特殊的偏好，正是这种偏好实现了寄生在类似类型生物体中的早期转化。

与同种移植反应的比较

在此阶段，有必要尝试对群体海鞘中所观察到的现象和造成哺乳动物同种移植反应的原因进行基因水平上的正式比较。在两种情况中，如果都简单假设组织相容性由单个基因的等位基因决定，那么我们可以考虑基因型分别为 AB 和 CD 的杂合动物之间进行交配的结果，每次都选择 AB 作为母本。两种类型动物的 F1 代将包含以下基因型：AC、AD、BC 和 BD。在菊海鞘中，所有可能的子代都会与亲代融合（是相容的）。而在哺乳动物中，移植自任何子代的皮肤都会被亲代排斥。如何从第一种情况发展到第二种情况是进化学上的一个难题。

首先，需要建立一个过程，识别能够通过这一过程产生第二种类型的结果。识别的后果不仅仅是使细胞在接触表面的相互作用稳定化，在另一些情况下，紧随识别之后的必然是细胞产物的破坏性释放。第二，必须提供互补模式，这种互补模式不仅能够识别个体自身产生的基因产物，而且能识别所有该物种可接受的模式所特有的基因产物（组织相容性抗原）。

根据图 2 可以讨论一些必要的条件，图 2 是关于子代在某种情况下具有寄生在母体组织上的潜能的一个简单示意图。基于对菊海鞘这种情况的简单拓展，细胞间应该能够相互容忍，并且应该很容易发展寄生关系。在我们已经描绘出的初期寄生的情况下，如果要阻止寄生，那么生物学前提是：亲本 AB 应产生一种特化的细胞，并且在菊海鞘体内的这种细胞应具有一种与脊椎动物中的免疫受体（细胞结合抗体）类似的–型受体 C。这种细胞必须能够到达 AB 和 AC 间的接触区域，随着对 C 的识别，

recognition of C, provoke a destructive local reaction. The parasitic embryonic form and the adjacent host cells will be killed, thrown off and the region repaired. The result will be wholly equivalent to the repair of a traumatic wound or of a patch of localized bacterial or other type of parasitic intrusion on a surface. Every aspect of such a reaction except what was due to the specialized anti-C cell is within the capacity of any invertebrate. Survival of any metazoan requires that it can deal with incipient infection by bacteria and repair minor injuries. Highly complex activities may be required but they do not require an adaptive immune system. To complete the sequence shown in Fig. 2 we need only to add an immunological barrier between the two sets of tissue, something with the same function as is usually ascribed to the trophoblast in placental mammals.

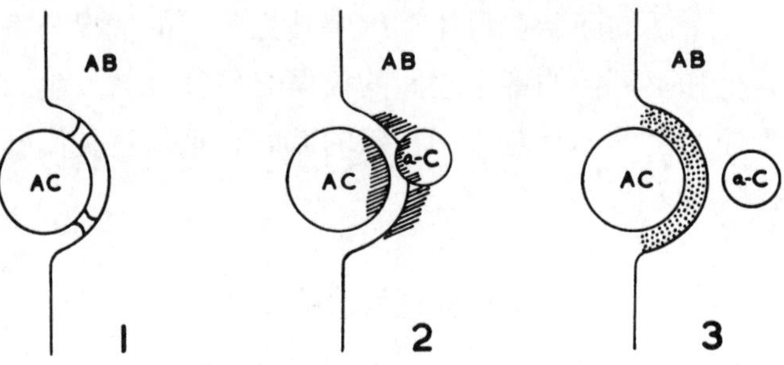

Fig. 2. A schematic diagram to indicate possible relationships of viviparous offspring (AC) to maternal tissues (AB) (see text).

So far I have been considering the danger that a viviparously produced embryo in the wrong place might become a damaging parasite. Essentially the same considerations apply if we take the broader possibility of a primitive cyclostome becoming a parasite of any other available species of cyclostomes that were not able to develop a means of protection. On this view the first evolutionary task of the free-swimming progenitors of the vertebrates was to devise a way of recognizing whether a group of living cells within its body was self or not-self. If "not-self", then in some way it must be eliminated. In retrospect one must assume that there must have developed "markers"—what we now call histocompatibility antigens by which cells genetically proper to the individual can be distinguished from foreign ones even if these are essentially similar in other respects.

Evolution of the Vertebrate Pattern

The requirement for segregation of the communities of colonial forms has sufficient similarity to suggest that the basic recognition mechanism was adapted to this new "immunological" requirement. But in the tunicates the recognition factors conduced to stable relationships between cells. As long as our potential parasite has one such "marker" in common it could, on the basis of pre-vertebrate analogies, expect to be tolerated. There are plenty of analogies, both in the history of evolution and of human war, where an initial advantage to one contestant can eventually provide the means by which it is defeated. Again thinking retrospectively, the essential change that must have been made was to produce specialized wandering cells or immunocyte prototypes in

引起该处的破坏性反应。寄生性胚胎和邻近的寄主细胞将被杀死，然后被抛出，该区域得以修复。这种结果完全等同于外伤修复以及局部细菌斑块或表面上其他类型的寄生性侵入的修复。除了由特化的抗 C 细胞引起的部分之外，这种反应的每一方面都在任何无脊椎动物的能力范围之内。所有后生动物为了生存都必须能够处理由细菌引起的初期感染和修复细小伤口。这可能是一种很复杂的行为，但并不需要具备适应性免疫系统。为了达到图 2 显示的结果，我们仅需要在两个系列的组织间加上一道免疫屏障，它的功能通常被认为与有胎盘类哺乳动物胚胎滋养层的功能相同。

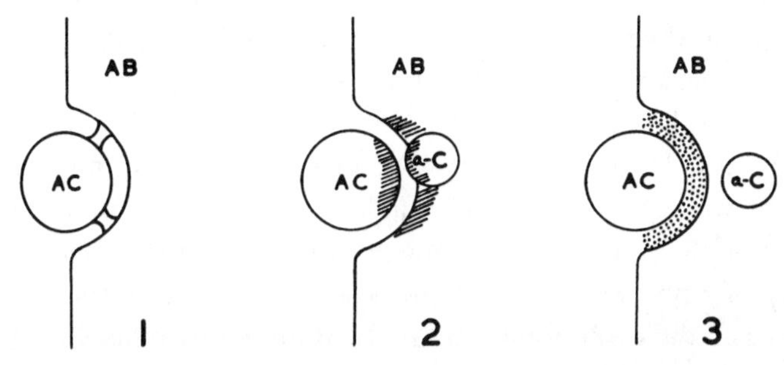

图 2. 表明胎生子代（AC）与母体组织（AB）的可能关系的示意图（见正文）。

到目前为止，我一直在考虑以下这种危险，那就是胎生动物在错误位置产生的胚胎可能会变成一种破坏性的寄生物。当我们把可能性拓展至原始圆口类脊椎动物成为不能发展保护手段的任意一种现存圆口类物种的寄生物时，也可以采用本质上相同的处理。按照这种观点，自由游泳的脊椎动物祖先的第一进化要务是，设计出一种方式以识别自身体内的一组活细胞是自我还是非我。如果是“非我”，那就必须以某种方式消除它。现在想起来应该假设之前已经发展出“标记”——我们现在称之为组织相容性抗原，通过这种抗原可以将对于个体来说基因正常的细胞与外源细胞区别开来，即使两者在其他方面基本一致。

脊椎动物模式的进化

各种群体生物的隔离都要求非常类似的条件，这说明基本识别机制可以适应这一新的“免疫学”的需要。但是在被囊动物中，识别因子导致了细胞间的稳定关系。只要我们说的潜在寄生物有这样一个共同的“标记”，那么根据与原脊椎动物的类比，就可以认为潜在寄生物会被容忍。在进化史和人类的战争史中有大量这样的例子，某一竞争者最初的优势可能最终转变为导致其失败的因素。再回想一下，必然发生的本质变化是产生专门的游走细胞或者免疫细胞原型。在后者中 +/– 接触不会使邻近细胞表面稳定，而是发生潜在的坏死相互作用。这样可能会导致出现自动强

which +/− contact led not to stabilization of adjacent cell surfaces but to a potentially necrotic interaction. This would have the automatic capacity to enforce "suicide" of any immunocytes whose immune receptors (−) corresponded to (+) groupings on accessible cell surfaces and therefore ensure that only not-self (−) groupings were represented on the surviving immunocytes. It requires only a minor extension of the intra-genomic mechanism discussed in regard to *Botryllus* to ensure that (−) receptors corresponding to all the (+) receptors likely to be produced as a result of somatic (or germinal) mutation within the species should be those most commonly produced. In other words, the dominant (−) (that is, immune pattern) receptors in the primitive immunocyte population correspond to the other (+) receptors (histocompatibility antigens) that are characteristic of the species. Once again this brings us very close to the Jerne concept. At this speculative level of discussion it is immaterial whether the new patterns arise by somatic mutation or by germ-line mutation. Since there would be advantages to be gained if on the basis of what was transmitted in the germ line it was easy to ensure that all the "not-self" histocompatibility patterns of the species should be represented in the complementary patterns (−) of the immunocytes, there would be a drive to establish the necessary information in the germ line. Perhaps it is not altogether naive to suggest that when the genetic mechanism which generates diversity of immune pattern is properly understood it will be relatively easy to rewrite this concept of the evolutionary origin of immune specificity in the relevant terms. For the present, nothing would be gained by going beyond the present very general statement.

If this general approach is to serve as a stimulus to further comparative study of the evolution of immune responses, it is important to make one final point. In the lamprey and in primitive fishes there are active immune responses to homografts and delayed necrotic response to Freund's complete adjuvant, but antibody production is poor. In current immunological terminology T-responses are much better developed than B, and in line with many other indications this points to the T system—concerned with homograft immunity, delayed hypersensitivity and so on—being the more primitive. In any investigation of invertebrates or pre-vertebrate chordates for evolutionary precursors of an adaptive immune system, conventional tests for antibody production are likely to prove futile. The best approaches are probably (1) to test for differences between the responses to auto and homografts of skin or other tissue, and (2) to compare the histological response where a region has been damaged by physical means with what results from injection of material likely to induce a response with an immunological component, such as that seen in the lamprey given Freund's complete adjuvant.

More Research Needed

The capacity to recognize the difference between self and not-self appears early in animal evolution and is well marked in at least two colonial marine forms: *Eunicella*, an anthozoan (Coelenterata), and *Botryllus*, a colonial ascidian (Tunicata). In the ascidian, Oka's work has allowed analysis of the genetics of recognition in terms not wholly remote from the histocompatibility antigens of mammalian immunology. If the current opinion is correct that vertebrates arose from the chordate larvae of some primitive intermediate between echinoderm and tunicate, it is legitimate to look at what is necessary to derive primitive

制免疫细胞“自杀”的能力，这些被强制“自杀”的免疫细胞的免疫受体（–）对应于可接触细胞表面上的（+）型,因而确保在存活的免疫细胞上只有非我（–）型存在。就菊海鞘而论，它只需要基因内机制的微小延伸就可以确保大量生成与所有可能由体细胞（或生殖细胞）突变产生的（+）受体相对应的（–）受体。换句话说，在原始免疫细胞群中的显性（–）（也就是免疫模式）受体对应的是其他一些能代表物种特征的（+）受体（组织相容性抗原）。这又一次使我们非常接近于杰尼的概念。就这种推测水平上的讨论而言，是通过体细胞突变还是通过生殖细胞突变引发新模式这一问题并不重要。如果以生殖细胞系中所传递的物质为基础，就很容易确保物种的所有“非我”组织相容性模式都能被免疫细胞的互补模式（–）所代表，这样会有助于在生殖细胞中产生一种建立必需信息的推进力。或许不能完全把以下看法当成异想天开：当能正确理解产生免疫模式多样性的遗传学机制时，用相关术语重写免疫特异性的进化起源这一概念就变得相对容易。就目前的水平而言，想超出现有的通用化陈述是不可能取得成功的。

如果这种通用的方法能促进对免疫反应进化的深层对比研究，那么制定一个终点是十分重要的。七鳃鳗和原始鱼类会对同种移植物产生积极的免疫反应，对弗氏完全佐剂会产生延迟坏死反应，但是生成的抗体很少。在现在的免疫学术语中，T 反应发展得远强于 B 反应，并且与许多其他现象一致，这表明 T 系统——涉及同种移植免疫、迟发型超敏反应等——是更原始的系统。在研究各种无脊椎动物或原脊椎脊索动物的适应性免疫系统的进化前体时，常规的抗体产物检测方法很可能会失效。最佳方法可能是：(1) 检验皮肤或其他组织在自体移植和同种移植后所产生的反应间的差异，(2) 比较遭受物理破坏区域的组织学反应与注入可能会和一种免疫成分产生反应的物质后所导致的结果，例如给七鳃鳗注射弗氏完全佐剂后所观察到的结果。

深入研究的必要性

识别自我与非我间差异的能力出现于动物进化的早期，且在至少两种群体海洋生物中有很好的体现：网柳珊瑚，一种珊瑚虫（腔肠动物）；以及菊海鞘，一种群体海鞘（被囊动物）。在对海鞘的研究中，奥卡已经能够运用与哺乳动物免疫学中的组织相容性抗原有一点儿关系的术语来分析识别遗传学了。如果目前的看法是正确的，即认为脊椎动物是由棘皮动物和被囊动物之间的某种作为原始中间体的脊索动物幼

vertebrate immune responses as seen in the extant cyclostomes from the recognition processes in *Botryllus*. A far-reaching reorganization is clearly necessary and the "invention" of the immunocyte as a mobile cell specialized to carry receptors for recognition of foreign pattern must have been the important step. The borderline between viviparity and parasitism may have been important as well as the very early differentiation of some cyclostomes to a semi-parasitic way of life. Without much more factual information it is impossible even to sketch the early stages. Intuitively, however, one can feel that much more extensive study of invertebrates and protochordates, particularly at the genetic level, may be specially enlightening in regard to the nature of the processes by which diversity of immune pattern arises and the evident importance of histocompatibility genes in determining the nature of that diversity, as Benacerraf, Jerne, McDevitt and others have emphasized.

"Self-recognition" in Flowering Plants

As Oka has pointed out, there are important similarities in the self-sterility of *Botryllus* colonies and self-incompatibility in flowering plants. Under the guidance of Professor J. S. Turner and Miss Mary Ellis, I have consulted enough of the recent literature on self-incompatibility in plants to feel that it might be helpful for an immunologist to offer some comments on self-incompatibility in plants in the light of the foregoing discussion of the evolution of self-recognition in animals.

Self-incompatibility is very common in flowering plants and has been well known since Darwin's time. There is much to suggest that by enforcing heterozygosity it has been of great importance in the "explosive success" of angiosperm evolution. There are a number of different mechanisms, but the commonest and most widely studied example is that shown by failure of the haploid pollen tube to grow through the diploid stylar substance and to effect fertilization[20]. Many but not all such examples are controlled by a single locus with a very large number of possible alleles. In order to keep the topic as simple as possible I shall confine myself wholly to this class. It has been studied in *Petunia, Lilium, Oenothera, Trifolium* and *Nicotiana* among others.

In general the phenomenon of self-incompatibility takes the following form. Pollen from an individual plant is incapable of setting seed in flowers of that plant or in other plants of the same incompatibility group, that is, with a common, S^1 say, self-incompatibility allele. S^1 pollen, however, is fully active in fertilizing any other plant of the species. Similar findings hold for other types of pollen. The number of distinct S alleles in a species can be very high: seventy-eight are mentioned in red clover and more than twenty-four in cherries. The condition is usually symmetrical, but in some species the incompatibility is one-sided in the sense that pollen from plants of strain A will fail to fertilize strain B, but B pollen can fertilize strain A.

To take a classical example. Crane and Lawrence[21] tabulated fifty-two varieties of cherry to show that they fell into twenty-four compatibility groups, the three largest containing nine, eight and four varieties respectively. All the incompatibilities shown in the table

体进化而来，那么就会合理地看待从菊海鞘的识别作用衍生出原始脊椎动物免疫反应所需要的条件，原始脊椎动物的免疫反应可从现存圆口类动物中看到。影响深远的重组明显是必需的；而“创造”出可以流动的免疫细胞，使其专门用于携带能识别外源模式的受体，也一定会是很重要的一步。胎生和寄生之间的界限可能非常重要，其重要性与一些圆口类动物在很早以前分化为半寄生生活方式等同。在没有更多确凿证据的情况下，即使想对早期阶段的大概情况进行描述也是不可能的。但还是能够直观地感觉到通过拓展对无脊椎动物和原索动物的研究，特别是在基因水平上的研究，有可能在以下两方面受到特别的启示：一是关于免疫模式多样性的产生进程的本质，二是关于组织相容性基因在决定免疫模式多样性的本质中的显著重要性，正如贝纳塞拉夫、杰尼、麦克德维特和其他人所强调的那样。

开花植物中的“自我识别”

像奥卡已经指出的那样，菊海鞘群落的自交不育性和开花植物的自交不亲和性之间存在着重要的相似性。在特纳教授和玛丽·埃利斯小姐的指导下，我已经查阅了足够多的关于植物自交不亲和性的最新文献。我的感觉是：对于一名免疫学家来说，用动物中自我识别进化的前述讨论来评论植物中的自交不亲和性可能会有助于自己的研究。

自交不亲和性在开花植物中是很普遍的，自达尔文时代起就广为人知了。有很多结果表明，强制杂合性在被子植物进化的“爆发式成功”中具有非常重要的意义。自交不亲和性有多种不同的作用机理，但是最普通的也是研究最广泛的例子是，单倍体花粉管不能生长穿过二倍体花柱内物质和实现受精[20]。尽管数量众多，但并非所有这样的例子都是由单个基因座上的大量可能的等位基因控制的。为了使话题尽可能简化，我会把讨论内容完全限制在这类例子中。其中已经研究过的有矮牵牛花、百合、月见草、三叶草和烟草等。

自交不亲和性现象的表现通常是：某一株植物的花粉不能使自身或属于同一自交不亲和性组中的其他植株的花结籽，即这些植株都具有某一共同的自交不亲和性等位基因，比如说是 S^1。虽然如此，S^1 花粉具有完全活性，能够使此物种内任意其他植株受精。类似的发现也适用于其他种类的花粉。在某一物种内，不同的 S 等位基因的数量可以非常高：在红三叶草中据称有 78 个 S 基因，比樱桃中的 24 个要多。不亲和性通常是对称的，但是在有些物种中，不亲和性是一边倒的，也就是说，品种 A 植株的花粉不能使品种 B 受精，但是品种 B 的花粉能使品种 A 受精。

举一个经典的例子。克兰和劳伦斯[21]将 52 个樱桃品种列成表格以说明可以将它们分为 24 个相容性组，3 个最大的相容性组分别包括 9、8 和 4 个品种。表中显

are symmetrical. This is, however, by no means universal among cultivated fruits; there are some strikingly one-sided examples among plums, and apple varieties in general show quantitative differences in the capacity to set seed instead of the relatively clear cut results + or − of cherries. Confining ourselves to the latter, the failure of the pollen tube to develop in styles of self type must be looked on as a positive recognition mediated by components in pollen tube surface and stylar tissue, both of which are coded for by a common allele. The ability for the pollen tube to grow and reach the ovule in any other strain is at this level a negative finding.

In a recent discussion of the general problem, Lundqvist[22] has described three essential genetic features. First, the S allele has two sites (pollen tube and stylar tissue) for phenotypic expression. Second, these two gene products are exactly coordinated to bring about incompatibility by their interaction. Third, a changed allele may lose its characteristic phenotypic expression at one site while fully active at the other.

If only the usual symmetrical form of self-incompatibility is considered, plants seem formally similar to *Botryllus*, as Oka had pointed out. If instead of using S^1, S^2 and so on for the alleles we use the A, B, C and D nomenclature, as for the colonial ascidians, the rules governing the behaviour of heterozygote forms are as follows for two plants AB and CD. Both are self-incompatible, that is, when either A or B pollen grains lodge on the stigma of an AB flower, growth of the pollen tube is inhibited. When A or B pollen lodges on CD, however, the combinations are all fertile and AC, AD, BC and BD offspring can be obtained. When these are mated, the types of offspring are:

♀AD×AC♂	AC, CD	♀AC×AD♂	AD, CD
BC×AC	AB, AC	AC×BC	AB, BC
BD×AC	AB, AD, BC, CD		

These are essentially the same rules as obtain in *Botryllus*. The symmetry of the reaction is not always complete and it is well known that irradiation of pollen will often allow fertilization of the normally self-incompatible strain from which it was obtained. This has made it possible to produce homozygous plants which still cannot be fertilized by normal pollen of the original strain. There are also a number of natural examples of one-sided incompatibility as well as various degrees of partial compatibility. One gains the impression that as in so many other genetic phenomena an intensely complex set of genetic factors is concerned and that specially suitable material must be chosen to provide a system expressible in simple terms. The change from self-incompatibility to compatibility is operationally extremely easy to detect and measure and it has therefore attracted many workers. Perhaps one of the most striking of their results is that irradiation is more likely to delete an S allele than to produce any other type of mutant.

At the physiological and biochemical level there is much on record about specific situations, but few generalizations have been established from the several examples of

示的所有不亲和性都是对称的。虽然如此，在栽植的水果品种中，这绝不是普遍现象；在李子中存在一些显著一边倒的例子；苹果品种通常会表现出在结籽能力方面的数量差异，而不是樱桃中 + 或 – 的相对清楚的界线。就后者而言，花粉管在自身类型的花柱上不能发育，这必然被看作是一种由花粉管表面和花柱组织中的成分介导的主动识别，这两种成分均由同一个等位基因编码。在任意其他品种中，花粉管能够生长并到达胚珠即说明不存在自交不亲和性。

在最近一次对这一普遍问题的讨论中，伦德奎斯特[22]描述了三个基本的遗传学特征。第一，S 等位基因有两个表型表达位点（花粉管和花柱组织）。第二，这两个基因产物能通过相互作用协调一致地导致不亲和性。第三，一个已发生变化的等位基因可能在一个位点失去它的表型表达特征，而在其他位点却具有全部活性。

像奥卡所指出的那样，如果只考虑自交不亲和性中常见的对称形式，那么植物就会与菊海鞘具有形式上的相似性。如果对于这些等位基因，我们不使用 S^1、S^2 等名称，而是和研究群体海鞘时一样采用 A、B、C、D 等来命名，那么对于两株植物 AB 和 CD，其杂合子的形式应遵循如下规则。两者都是自交不亲和，也就是说，当 A 或 B 的花粉颗粒落到一朵 AB 花的柱头上时，花粉管的生长会受到抑制。虽然如此，当 A 或 B 的花粉落到 CD 的柱头上时，所有组合方式都能结果，由此得到的子代是 AC、AD、BC 和 BD。当这些子代杂交时，产生的下一代的类型是：

♀AD×AC♂	AC, CD	♀AC×AD♂	AD, CD
BC×AC	AB, AC	AC×BC	AB, BC
BD×AC	AB, AD, BC, CD		

以上这些与从菊海鞘中总结出的规则基本相同。反应的对称性并不一定总能实现，大家都知道，正常自交不亲和品种的花粉经辐射后通常能使自体受精。这使产生纯合植物成为可能，这种纯合植物仍然不能被原始品种的正常花粉受精。自然界中还存在许多一边倒的不亲和性的例子，以及程度不同的部分亲和性的例子。由此得到的印象是：正如研究许多其他遗传现象时一样，这项研究也要涉及极其复杂的一系列遗传因子，必须选择特别合适的材料以提供一个可用简单方式体现的系统。从操作上来讲，发现和检测从自交不亲和性到亲和性的变化是很容易做到的，因此吸引了众多研究者。在他们的研究成果中，最为显著的一个也许是：相较于产生任意其他类型的突变体，辐射处理更可能的后果是去除 S 等位基因。

在生理学和生物化学水平上有许多关于具体情形的记载，但是根据不同物种中自交不亲和性的几个例子，几乎没能概括出普适的结论。很多作者，特别是林斯肯

self-incompatibility in different species. A number of writers, notably Linskens[23], have spoken of antigen–antibody reactions and applied serological methods. Lewis[24] identified individual antigens in pollen extracts of *Oenothera* by Ouchterlony methods and correlated each with a specific incompatibility allelotype. He obtained similar findings in the stylar tissue of *Petunia*, while Linskens, using radioisotopic labelling, was able to show that a complex of pollen protein and style protein was formed in the course of incompatible pollination. Knox *et al.*[25], using immunofluorescence techniques, showed that most of the antigens capable of provoking antibody in rabbits immunized with simple saline extracts of ragweed pollen were located in the inner (intine) part of the cell wall of the pollen grain. Studies by immuno-electrophoresis showed that their pollen extracts contained as least seven antigens, some of which were likely to be enzymes known to be present in the intine. They noted, however, the possibility that some of the wall-associated antigenic material could be concerned with incompatibility reactions. One might add that it would be even more interesting to look for a possible relationship in some appropriate species between the incompatibility protein coded by the S allele and the protein responsible for allergic reactions to the pollen in human subjects.

The nature of the process by which the pollen tube fails to grow effectively is of no obvious importance in the present context. It is said that there are various morphological changes—the generative nucleus fails to divide and the vegetative nucleus disappears—and, as would be expected, there are a variety of enzymatic changes. From our point of view, the nature of the specific interaction between a pollen substance and a style substance is the important matter. What happens after that has no bearing on vertebrate immunology.

There are many reasons for avoiding the use of the terms antigen and antibody, but we can legitimately use the concept of mutual recognition between cell surfaces. The facts of self-incompatibility in plants demand that there is a positive recognition by which a pollen tube carrying allele A has a surface protein which on "recognizing" another protein in the style also carrying allele A interacts with it, so triggering the progressive degeneration of the pollen tube. Strictly speaking, some substance other than protein might be concerned, but from every point of view the concept that the substance is a protein directly coded for by the allele seems to be the only one worth consideration. All the current hypotheses make this assumption.

Probably only two basic hypotheses are possible. Both agree that one allele codes for proteins in two different sites. Both have been formulated by Lewis, the second being the one he currently favours. In the first, the allele codes for two distinct proteins, one for the pollen tube, one for the style. The proteins S^{P} and S^{st} have a complementary configuration analogous to antigen–antibody or enzyme–substrate. In the second, the allele is a complex which codes for (*a*) specific protein pattern common to both sites; (*b*) activator for protein production in pollen tube, and (*c*) activator for protein production in style. For reasons associated with the fact that grasses have two loci, S and Z, concerned with self-incompatibility, Lewis postulates that the primary gene product takes the form

斯[23]曾谈到过一些抗原抗体反应以及适用的血清学方法。刘易斯[24]使用乌赫特朗尼法鉴别了月见草花粉提取物中的抗原，并使每一个抗原都与特定的不亲和性等位基因型联系在一起。他在矮牵牛花的花柱组织中也有类似的发现；而林斯肯斯用放射性同位素标记法显示，花粉蛋白和花柱蛋白的复合体是在不相容的授粉过程中形成的。诺克斯等人[25]利用免疫荧光技术证实：用豚草花粉的简单生理盐水提取物使兔子免疫后，大多数能激活兔子体内抗体的抗原位于花粉颗粒的细胞壁内部（内壁）。免疫电泳的研究结果表明：在花粉提取物中至少含有 7 种抗原，其中的一些很可能是已知存在于内壁中的酶。然而他们特别提到，一些与细胞壁关联的抗原物质有可能参与了不亲和反应。也许还应补充一点，更有趣的研究方向是：在一些合适的物种中，寻找 S 等位基因编码的不亲和蛋白与使人类被试者产生花粉过敏反应的蛋白之间的可能关系。

从本文的研究角度考虑，花粉管不能有效生长这一过程的本质并不具有明显的重要性。据说其中存在各种各样的形态学变化——生殖核不能分裂并且营养核消失，以及像预期的那样，多种酶发生变化。依我们之见，花粉物质和花柱物质间特异性相互作用的本质才是关键问题。之后发生的事与脊椎动物免疫学没有关系。

尽管有很多理由阻止我们使用抗原和抗体这两个术语，但我们可以正当地使用细胞表面间的相互识别概念。在植物中发现的众多自交不亲和性现象表明存在一种主动识别：携带 A 等位基因的花粉管中具有一种表面蛋白，一旦这种蛋白“识别”出同样携带等位基因 A 的花柱中的另一蛋白，就会与之发生相互作用，从而引发花粉管的逐步退化。严格地说，除了蛋白，可能还会有其他一些物质与识别有关，但是从各个角度看，似乎只有以下概念才是唯一值得考虑的事情，即这种物质是一种由等位基因直接编码的蛋白。目前所有的假说都支持这一假定。

或许只有两种基本假说是可行的。两者都认为一个等位基因编码两个不同位点的蛋白。刘易斯已经系统地阐述过这两种假说，目前他更推崇的是第二种。在第一种假说中，等位基因编码两种不同的蛋白，一种在花粉管，一种在花柱。S^{p} 和 S^{st} 蛋白具有互补的结构，类似于抗原 – 抗体或酶 – 底物。在第二种假说中，等位基因是一个复合体，编码：(*a*) 两个位点共有的特异蛋白模式；(*b*) 在花粉管中产生蛋白所需的激活因子；(*c*) 在花柱中产生蛋白所需的激活因子。草类具有两个基因座，分别是 S 和 Z，都与自交不亲和性相关。为了给出与此相关的原因，刘易斯假设，最初

of a dimer. When this makes contact with an identical dimer on the interacting surface, union, presumably involving an allosteric molecule, occurs and the tetramer functions as a growth inhibitor. The idea is admittedly based on the structure of haemoglobin or immunoglobulin G, and, as Lewis[26] himself points out, cannot even be tested experimentally until the dimer and tetramer molecules are available for physical and chemical evaluation.

No one seems to have claimed that a decision can yet be made between the two possibilities: interaction of two identical proteins (or polypeptide determinants) or of sterically complementary patterns.

Simply because I am primarily concerned only in looking for analogies with immunological processes that involve the recognition of self from not-self, I should like to elaborate Lewis's first hypothesis of sterically complementary proteins along the lines we have adopted for *Botryllus*. The concept then becomes for S alleles A and B that as the pollen grain germinates and the tube begins to grow, a specific interaction between chemical groupings A+ on pollen tube, A– in the cells of the style and/or A– on pollen tube, A+ in the cells of the style results in cessation of growth. In view of the symmetry of the relationship, we assume that both + and – patterns are present on both sides and that the cells of style, being diploid, will carry + and – patterns of each gene A + B.

Because there are many plants which are self-fertilizing and more which show minor degrees of self-incompatibility, it is probable that + – contact is merely an initiating signal which, depending on other evolved cellular mechanisms, may be neutral, stimulatory or inhibitory as in self-incompatible species and varieties. Without having any detailed knowledge of the process of morphogenesis in plants, one can postulate that, as in animals, there must be some interchange of information between adjacent cells, part at least of which will take the form of + – union between specific patterns. The point to be made is simply that the possibility and the need for intercellular information exchange has provided an opportunity for the evolution of means of recognizing the difference between self and not-self when this was required for evolutionary purposes, whether to prevent self-fertilization or to produce an adaptive immune system.

The still totally unresolved problem of how a single allele can produce two mutually reactive proteins in the two relevant sites again draws attention to the currently teasing question in animal immunology as to how the capacity to produce a certain type of antibody is firmly linked to a specific histocompatibility antigen. Obviously the answer must come from the isolation of the reactive S proteins in pollen tube and stylar tissue, and assuming that they are polypeptides to determine the difference in amino-acid sequence corresponding to allelic change.

Reverting once more to Jerne's[16] concept that the organism "knows" what are the structures of the histocompatibility antigens which the species can produce, we can hardly avoid pondering—heretically, no doubt—whether there is any genetic way by which

的基因产物表现为二聚体形式。当这一二聚体在相互作用表面上与同样的二聚体相接触时，两者发生相互结合，可以假定能形成一种变构分子，这种四聚体的功能是抑制生长。诚然，上述假设是基于血红蛋白或免疫球蛋白 G 的结构而建立的，而且正如刘易斯[26]自己所指出的那样：在获得可用物理法和化学法鉴定的二聚体和四聚体分子之前，根本无法用实验来验证上述想法。

目前还未曾有人在以下这两种可能性的取舍上表明过自己的态度：是由于两个同样蛋白（或多肽决定因子）的相互作用，还是由于空间互补模式的相互作用。

仅仅因为我主要关心的只是寻找对识别自我与非我的免疫过程的模拟，所以我倾向于按照我们研究菊海鞘的方式详尽说明刘易斯阐述过的第一种假说——关于空间互补蛋白的假说。于是 S 等位基因概念就化为 A 和 B，当花粉颗粒萌发、花粉管开始生长时，花粉管上的化学基团 A + 与花柱细胞中的化学基团 A– 之间的特异性相互作用，和 / 或花粉管上的化学基团 A– 与花柱细胞中的化学基团 A + 之间的特异性相互作用，会导致花粉管生长中止。考虑到这种关系的对称性，我们假定两者都存在 + 模式和 – 模式，花柱细胞是二倍体，因而会携带每一 A+B 基因的 + 和 – 模式。

因为有很多植物是自体受精，而更多植物只表现出很小程度的自交不亲和性，所以 + – 接触可能仅仅是一种启动信号，依赖于进化而来的其他细胞机制。这些信号可以是中性的、刺激的或抑制的，后者正如自交不亲和性物种和变种中所表现的那样。在不详细了解植物中形态发生过程的情况下，我们可以假定：像动物中那样，植物的邻近细胞间一定也存在某种信息交换，至少有一部分会采用特殊模式间的 + – 结合形式。我只不过想说明：当为了进化的目的需要识别自我和非我间的差异时，细胞间信息交换的可能性和需要早已为识别方式的产生准备好了一个机会，不管进化的目的是为了防止自体受精还是为了产生适应性免疫系统。

仍然完全未解决的问题是，单个等位基因如何能在两个相关位点中产生两个互相反应的蛋白，这再一次把人们的注意力吸引到目前极具挑战性的问题上，即：在动物免疫学中，产生某种类型抗体的能力是如何与特殊的组织相容性抗原紧密地联系在一起的。很明显，答案必然来自花粉管和花柱组织中参与反应的 S 蛋白的分离，如果它们就是能根据等位基因变化决定氨基酸序列差异的多肽。

再一次重提杰尼[16]的概念，即生物体"知道"本物种所能产生的组织相容性抗原的结构，我们不得不考虑——毫无疑问，这与常规思路背道而驰——是否存在一

a nucleotide sequence coding for an amino-acid group A–B–C–D can automatically produce another sequence P–Q–R–S (say) which can "recognize" it. One has an intuitive feeling that the "choice" of the twenty biological amino-acids and the form of the genetic code might in the last analysis depend on some such requirement for mutual recognition.

(**232**, 230-235; 1971)

F. M. Burnet: School of Microbiology, University of Melbourne, Parkville, Victoria 3052.

Received June 22, 1971.

References:

1. Thomas, L., in *Cellular and Humoral Aspects of the Hypersensitive States* (edit. by Lawrence, H. S.), 529 (Hoeber-Harper, New York, 1959).
2. Billingham, R. E., *New Engl. J. Med.*, **270**, 667 and 720 (1964).
3. Currie, G. A., van Doorninck, W., and Bagshawe, K. D., *Nature*, **219**, 191 (1968).
4. Burnet, F. M., *Brit. Med. Bull.*, **20**, 154 (1964).
5. Hellström, K. E., and Möller, G., *Prog. Allergy* (*Karger*), **9**, 158 (1965).
6. Klein, G., *Ann. Rev. Microbiol.*, **20**, 223 (1966).
7. Prehn, R. T., *Ann. NY Acad. Sci.*, **164**, 449 (1969).
8. Burnet, F. M., *Immunological Surveillance* (Pergamon, Oxford, London, New York, Toronto, Sydney, 1970).
9. Finstad, J., and Good, R. A., *J. Exp. Med.*, **120**, 1151 (1964).
10. Marchalonis, J. J., and Edelman, G. M., *J. Exp. Med.*, **127**, 891 (1968).
11. Papermaster, B. W., Condie, R. M., and Good, R. A., *Nature*, **196**, 355 (1962).
12. Acton, R. T., Weinheimer, P. F., Hildemann, W. H., and Evans, E. E., *J. Bact.*, **99**, 626 (1969).
13. Hildemann, W. H., and Thoenes, G. H., *Transplantation*, **7**, 506 (1969).
14. Oka, H., in *Profiles of Japanese Science and Scientists* (edit. by Yukawa, H.), 198 (Kodansha, Tokyo, 1970).
15. Theodor, J. L., *Nature*, **227**, 690 (1970).
16. Jerne, N. K., in *Immune Surveillance* (edit. by Smith, R. T., and Landy, M.), 343 (Academic Press, New York and London, 1970).
17. Howell, J. H., in *Phylogeny of Immunity* (edit. by Smith, R. T., Miescher, P. A., and Good, R. A.), 263 (University of Florida, Gainesville, 1966).
18. Ritchie, A., *Palaeontology*, **11**, 21 (1968).
19. Burnet, F. M., *Acta Pathol. Microbiol. Scand.*, **76**, 1 (1969).
20. East, E. M., and Mangelsdorf, A. J., *Proc. US Nat. Acad. Sci.*, **11**, 166 (1925).
21. Crane, M. B., and Lawrence, W. J. C., *The Genetics of Garden Plants*, third ed., 185 (London).
22. Lundqvist, A., *Genetics Today* (edit. by Geerts, S. J.), **3**, 637 (Pergamon, Oxford, 1965).
23. Linskens, H. F., *Genetics Today* (edit. by Geerts, S. J.), **3**, 629 (Pergamon, Oxford, 1965).
24. Lewis, D., *Proc. Roy. Soc.*, B, **140**, 127 (1952).
25. Knox, R. B., Heslop-Harrison, J., and Reed, C., *Nature*, **225**, 1066 (1970).
26. Lewis, D., *Genetics Today* (edit. by Geerts, S. J.), **3**, 657 (Pergamon, Oxford, 1965).

种遗传学方式，通过它，编码一个氨基酸组 A–B–C–D 的核苷酸序列能自动产生另一氨基酸序列，（比如说是）P–Q–R–S，使 P–Q–R–S 能“识别”A–B–C–D。给人的一种直观感觉是，可能在最后的分析中要根据某种相互识别需求来确定对生物体内 20 种氨基酸的“选择”以及遗传密码的形式。

（吕静 翻译；刘京国 审稿）

Sex Pheromone Mimics of the American Cockroach

W. S. Bowers and W. G. Bodenstein

Editor's Note

The specificity of insect sex pheromones is questioned here with the discovery of a number of plant extracts shown to drive male American cockroaches "wild with desire". Male cockroaches (*Periplaneta americana*) respond to a female-produced airborne pheromone by fluttering their wings, extending their abdomen and attempting to copulate. But identical behaviour is seen when males are exposed to certain conifer-derived essential oils and flowing plant extracts, William Bowers and William Bodenstein report. The plant products differ chemically from the natural pheromone, which the researchers extracted from the midguts of virgin females. This is the first documented report of a sex pheromone-like substance found in plants.

DURING an investigation of several essential oil preparations for juvenile hormone activity[1], we discovered that a volatile component(s) in several coniferyl needle distillates stimulated apparent sexual excitement in male American cockroaches, *Periplaneta americana* (L.). The response, consisting of movement, wing flutter, extended abdomen and attempted copulation, was behaviourally identical to that elicited by the natural sex pheromone of this insect[2] and in marked contrast to the behaviour (hyperactivity) induced by exposure to highly irritating chemicals such as amyl acetate.

Needle and cone oil distillates of spruce, *Picea rubra* Link, and fir, *Abies siberica* Ledeb., *Abies alba* Mill., were fractionated by column chromatography over "Florisil", and the active fractions were examined by gas–liquid chromatography. (A column 4 mm × 2 m was used, packed with 0.75% methyl silicone on 200 mesh gas chrom P.) The effluent of the column was directed into a 2 l. jar containing ten male cockroaches which had been isolated from female cockroaches for 6 weeks. Sexual display by the cockroaches was associated with only one peak in these fractions. Final purification of the pheromonal substance was carried out by preparative gas chromatography. (An F and M Model 775 Prepmaster equipped with gold plated injector block and stainless steel columns 0.75 inches ×16 feet was used. It contained 4% methyl silicone on 150 mesh "Chromosorb W".)

By its pleasing and characteristic odour the active compound was tentatively characterized as bornyl acetate, a known constituent of the source trees. Comparison with standards by gas–liquid chromatography and by infrared, nuclear magnetic resonance and mass spectroscopy confirmed its identity.

美洲蜚蠊的性外激素类似物

鲍尔斯，博登施泰因

编者按

因为作者发现完全可以从植物中提取出一些能使雄性美洲蜚蠊“神魂颠倒”的物质，所以昆虫性外激素具有特异性的提法在本文中受到了质疑。雄性蜚蠊（美洲大蠊）对一种由雌性蜚蠊产生并能在空气中传播的激素所作出的反应是：扇动翅膀、扩张腹部和试图交配。然而，当雄性蜚蠊暴露于某些松类植物精油和开花植物的提取物中时也会表现出同样的行为，威廉·鲍尔斯和威廉·博登施泰因在本文中讨论了这一现象。这些植物源物质的化学成分与研究人员从蜚蠊处女虫中肠中分离出来的天然性外激素并不相同。这篇文章是第一篇报道来源于植物的性外激素类似物的文献。

在研究激活保幼激素所必需的几种精油制品时[1]，我们发现在多种松柏针叶的馏分中有一种易挥发的成分能够刺激雄性美洲蜚蠊（拉丁学名为美洲大蠊）产生明显的性兴奋。性冲动反应包括飞来飞去、扇动翅膀、扩张腹部以及试图交配，与在这种昆虫天然性外激素作用下所产生的行为[2]完全一样，但与由暴露于醋酸戊酯等强刺激性化学物质中所诱发的行为（极度活跃）相去甚远。

用以硅酸镁为载体的柱层析法分离由云杉（红果云杉）和冷杉（西伯利亚冷杉和欧洲冷杉）的针叶和球果蒸馏出来的油分，并用气 – 液色谱法鉴定其中的活性成分（所用色谱柱的尺寸为 4 mm × 2 m，在 200 目的硅藻土型色谱载体 P 上涂渍 0.75% 的甲基硅酮）。将色谱柱的流出物直接导入一只装有 10 只雄性蜚蠊的容积为 2 升的育虫瓶里，这些雄性蜚蠊已经和雌性蜚蠊隔离了 6 周。在洗脱组分中只有一个峰与蜚蠊性兴奋的产生相对应。用制备气相色谱法对性外激素组分进行最后的分离纯化（使用的是配备有镀金注射模块和 0.75 英寸 × 16 英尺不锈钢柱的 F 和 M 型 775 制备母机。在 150 目的“硅烷化白色硅藻土”担体上含有 4% 的甲基硅酮）。

根据这种活性化合物所散发出来的令人愉悦的特征性气味，我们把它初步鉴定为乙酸冰片酯，它也是这些原料树中的一种已知成分。通过与气 – 液色谱、红外光谱、核磁共振谱以及质谱的标准物进行比较，可以确认该活性化合物的成分就是乙酸冰片酯。

Bioassays were performed according to Bodenstein[3]. Thus, the compound in hexane solution, was absorbed on a piece of Whatman No. 1 filter paper (1 cm^2) and held within the 2 l. rearing jar with forceps. The sensitivity of male cockroaches to the natural sex pheromone and to the pheromone mimics is greatly magnified by maintaining the males in isolation from females for at least 4–6 weeks before attempting bioassays. The response of newly segregated males is erratic or indifferent except to extremely high concentrations of natural pheromone, whereas isolation for several weeks makes them highly sensitive and uniform in response.

Related compounds such as camphor and borneol were inactive, although the propionate ester prepared from borneol was very active. Other plant-derived essential oils were examined for activity and, of these, α- and β-santalol were found to be equivalent to bornyl acetate in activity with *Periplaneta* (Table 1). Although the L and D bornyl acetate standards were of the highest optical purity obtainable it is possible that the slight activity of the L isomer was due to minute contamination with the D isomer. Alternatively, partial racemization may take place during the period of exposure or at the receptor site of the insect.

Table 1. Compounds that Elicit Sexual Display in Isolated Male *Periplaneta americana*

Compound	Active concentration* (mg/cm^2)
L-Bornyl acetate	7.0
D-Bornyl acetate	0.07
α-Santalol	0.07
β-Santalol	0.07
$C_{15}H_{24}$ plant hydrocarbon	0.14

* Threshold concentration of compound on 1 cm^2 of Whatman No.1 filter paper which induces sexual display in at least eight of ten groups of male cockroaches containing ten males per group.

After these studies we examined several flowering plant species for sex pheromone-like activity with *Periplaneta* by briefly soaking various parts of the plants in hexane or methanol and offering the extract on filter paper to the isolated groups of male cockroaches by the usual bioassay. Of one hundred common plants, eighteen were found to produce behaviour characteristic of exposure to the natural sex pheromone (Table 2). The freshly cut stems of several species, notably *Aralia spinosa* L. and several Compositae produced strong responses from the male cockroaches without solvent extraction. The active material seemed to be present mostly in the fruits and in the stems just beneath the outer cuticle, and varied with the season or stage of growth of the plants. In addition to these plants a very active hydrocarbon fraction has been obtained from the fruits and buds of *Liquidambar styraciflua* L. and *Liriodendron tulipifera* L. The pheromonal principle was readily obtained by rinsing the plants (or steeping for a few minutes) in hexane. Column chromatography over Florisil gave a highly active hydrocarbon fraction. Final purification was achieved by column chromatography on silver nitrate coated silica gel. The active hydrocarbon constituent from several of these plants was found to be identical in a comparison by gas–liquid chromatography, infrared spectroscopy, nuclear magnetic resonance and mass

按照博登施泰因所用的方法进行生物测定[3]。首先用一片沃特曼1号滤纸（1 cm²）吸收己烷溶液中的该化合物，然后用镊子将其放入容积为2升的育虫瓶中。在进行生物测定前将雄性蜚蠊与雌性蜚蠊隔开至少4~6周，雄性蜚蠊对天然性外激素和上述性外激素类似物的敏感性会因此而得以放大。隔离时间不长的雄性蜚蠊对天然性外激素的反应不是很稳定或者十分微弱，除非使用浓度极高的天然性外激素，然而长达数周的隔离则可以使蜚蠊的敏感性和反应均一性大大提高。

相关化合物，比如樟脑和冰片，是无活性的，但是由冰片制备的丙酸酯却具有很高的活性。我们还检测了其他植物精油的活性，结果发现α–和β–檀香醇对于大蠊的活性与乙酸冰片酯相当（表1）。尽管L－和D－乙酸冰片酯标准物的光学纯度已经达到了目前能够达到的最高水平，但L异构体中仍存在弱活性，这可能源自微量D异构体的污染。也有可能是L－乙酸冰片酯在暴露过程中或者在昆虫的受体部位发生了部分外消旋化。

表1. 能使经过隔离的雄性美洲大蠊产生性兴奋的化合物

化合物	活性浓度*（mg/cm²）
L–乙酸冰片酯	7.0
D–乙酸冰片酯	0.07
α–檀香醇	0.07
β–檀香醇	0.07
$C_{15}H_{24}$ 植物烃	0.14

* 每cm²沃特曼1号滤纸上的该化合物能够诱导共10组每组10只雄性蜚蠊中至少8组产生性兴奋的阈值浓度。

在这些研究结束之后，我们又检测了几种开花植物作为大蠊的性外激素类似物的活性，方法是：将植物的不同部分短时间地浸泡在己烷或者甲醇中，然后用滤纸吸取并按照常规的生物测定方法将其作用于几组经过隔离的雄性蜚蠊。在100种常见植物中，我们发现有18种能使动物表现出类似于接触天然性外激素的行为（表2）。有几种植物（尤其是多刺楤木和数种菊科植物）新切下来的茎能够在不用溶剂提取的情况下使雄性蜚蠊产生强烈的反应。有活性的成分似乎主要存在于果实和茎中紧挨着表皮的部分，并会随着季节或植物生长阶段的不同而变化。除了这18种植物以外，我们还从美国枫香和北美鹅掌楸的果实和芽中提取到一种活性很强的烃类组分。通过用己烷漂洗（或者浸泡数分钟）的方法很容易获得其中的性外激素成分。我们用以硅酸镁为载体的柱层析法分离出了一种活性很强的烃类组分。最后的分离纯化采用的是以硝酸银包覆的硅胶为载体的柱层析法。在比较了气－液色谱、红外光谱、核磁共振谱和质谱的分析结果之后，我们发现从几种植物中分离得到的活性烃类组

spectroscopy. In the mass spectrum, the parent ion gave a molecular weight of 204, consistent with the empirical formula $C_{15}H_{24}$. This assignment was fully supported by the fragmentation pattern. Approximately 0.1 μg of this material was capable of exciting eight out of ten groups of isolated male cockroaches. A study of its structure has been undertaken.

Table 2. Flowering Plants[3] Containing Hydrocarbons Causing Sex Pheromone-like Behaviour in *Periplaneta americana* Male Cockroaches

Simaroubaceae
Ailanthus altissima (Mill.) Swingle
Araliaceae
Aralia hispida Vent., *nudicaulis* L., *racemosa* L., *spinosa* L.
Labiatae
Perilla frutescens (L.) Britton
Compositae
Achillea millefolium L.
Ambrosia artemisiifolia L., *trifida* L.
Chrysanthemum cinerariaefolium (Trev.) Vis., *leucanthemum* L., *morifolium* (Ramat.), Hemsl.
Erigeron annuus (L.) Pers., *canadensis* L.
Eupatorium hyssopifolium L., *purpureum* L.
Rudbeckia hirta L.
Solidago juncea Ait.

The discovery of a diversity of structures which elicit this characteristic sexual display in *Periplaneta* is very disconcerting in view of the oft reported specificity of sex pheromones. Although a relationship between these plant products and the natural cockroach sex pheromone may eventually be established, we are certain that none are identical with the natural attractant. To investigate this possibility we extracted virgin female midguts as described by Bodenstein[3], obtained a highly active lipid extract. The active principle withstood refluxing in 5% methanolic potassium hydroxide but was destroyed by treatment with 10% methanolic base. Pheromonal activity was destroyed by brief treatment with sodium borohydride in methanol and then quantitatively regenerated by oxidation with chromic acid in acetone. Treatment of the crude extract in hexane solution with a drop of bromine eliminated activity, but debromination with zinc dust in 5% acetic acid in ether restored about one-half the original activity. This brief study leads us to speculate that the natural attractant is an unsaturated ketone with a hindered ester moiety or containing other base labile groups. By trapping the effluent from gas chromatographic columns we have obtained unweighable samples (<50 μg) of the natural pheromone which can be diluted 1 billion times and still elicit the characteristic mating display from male cockroaches. Thus, none of the plant derived pheromonal mimics are as active as the natural pheromone.

We hope that the information derived from structural characterization of the active plant hydrocarbons taken together with the known active compounds such as bornyl acetate and santalol will provide leads in the elucidation of the structure of the natural sex pheromone.

分均属于同一类物质。由质谱分析得出母离子的分子量是 204，这与实验式 $C_{15}H_{24}$ 相吻合。其裂解后的质量分布也完全支持这一分子式。大约 0.1 μg 该物质就能使 10 组经过隔离的雄性蜚蠊中的 8 组产生性兴奋。对其结构的研究已经开始进行。

表 2. 含有能使雄性美洲大蠊产生类似性冲动反应的烃类物质的开花植物 [3]

苦木科
臭椿
五加科
硬毛楤木、裸茎楤木、甘松、多刺楤木
唇形科
紫苏
菊科
千叶蓍
美洲豚草、三裂叶豚草
除虫菊、滨菊、白菊花
一年篷、加拿大飞篷
泽兰、紫苞泽兰
黑心菊
一枝黄花

性外激素的特异性常常见诸报道，而现在却发现有好几种结构都能使大蠊产生性兴奋，这使人感到十分尴尬。虽然我们最终也许能在这些植物源物质与蜚蠊天然性外激素之间建立起某种关系，但我们可以肯定，没有一种植物源物质在结构上与天然引诱剂完全相同。为了研究这种可能性，我们按照博登施泰因所述的方法 [3] 分离出了蜚蠊处女虫的中肠，得到了一种活性很高的脂质提取物。这种活性成分在 5% 氢氧化钾甲醇溶液中回流后仍能保持活性，但用 10% 氢氧化钾甲醇溶液处理则会被破坏。用硼氢化钠甲醇溶液短暂处理性外激素可使其失去活性，然后通过铬酸的丙酮溶液将其氧化从而使活性得以定量恢复。在粗提物的已烷溶液中加入一滴溴能使其灭活，但用锌粉在 5% 乙酸的醚溶液中进行脱溴反应即可恢复大约一半的原初活性。我们根据这些简单的研究过程推测，天然引诱剂是带有一个阻碍酯基或者含有其他不稳定碱性基团的不饱和酮。通过收集从气相色谱柱中流出的物质，我们可以得到一些不可称量的天然性外激素样品（<50 μg），它们在被稀释 10 亿倍之后仍能诱导雄性蜚蠊出现特征性的交配表现。因此，任何一种来源于植物的性外激素类似物都不可能具有和天然性外激素同样强的活性。

我们希望由这些有活性的植物源烃类物质以及由乙酸冰片酯和檀香醇等已知活性化合物的结构特征得到的信息能够为我们提供一些解释天然性外激素结构的线索。

Although it has been shown that a plant substance is necessary for sex pheromone release in an insect[4], this is the first report of a sex pheromone-like substance taken directly from plants.

We thank Dr. William L. Ackerman, Dr. Theodore R. Dudley, Dr. E. E. Terrell, Dr. Elbert L. Little and Dr. Melvin L. Brown for assistance in the identification of the plant materials. Coniferyl oil distillates were given by Ungerer and Co. and Fritsche Bros. and santalol by Givaudan Inc.

(**232**, 259-261; 1971)

William S. Bowers and William G. Bodenstein: Entomology Research Division, Agricultural Research Service, US Department of Agriculture, Beltsville, Maryland 20705.

References:

1. Bowers, W. S., Fales, H. M., Thompson, M. J., and Uebel, E. C., *Science*, **154**, 1020 (1966).
2. Roth, L. M., and Willis, E. R., *Amer. Midland Naturalist*, 47, 61 (1952).
3. Bodenstein, W. G., *Ann. Entomol. Soc. Amer.*, **63**, 336 (1970).
4. Riddiford, L. M., *Science*, **158**, 139 (1967).

尽管已经有人报道过昆虫释放性外激素需要一种植物成分[4]，但是本文首次报道了直接从植物中提取的性外激素类似物。

感谢威廉·阿克曼博士、西奥多·达德利博士、特雷尔博士、埃尔伯特·利特尔博士和梅尔文·布朗博士帮助我们鉴别各种植物材料。松柏油馏分由恩格乐公司和弗里切兄弟公司提供；檀香醇由奇华顿公司提供。

（毛晨晖 翻译；沈杰 审稿）

Monosodium Glutamate Induces Convulsive Disorders in Rats

H. N. Bhagavan *et al.*

Editor's Note

Fanning the flames of the debate over whether monosodium glutamate (MSG) causes "Chinese-restaurant syndrome", H. N. Bhagavan and colleagues here report tremors and seizures in rats injected with the food additive. Three years earlier, reports of adverse reactions to Chinese food had been described, with MSG mooted as the culprit. But subsequent research produced conflicting data. In 1995, a report prepared for the US Food and Drug Administration concluded that MSG and related substances are "safe food ingredients for most people when eaten at customary levels," but cautions that some people may be intolerant to large quantities of this flavour enhancer.

THE physiological and pharmacological effects of L-glutamate have received much attention since its implication in the aetiology of the "Chinese restaurant syndrome"[1-3]. Although an excess of L-glutamic acid (GA) as the monosodium salt (MSG) was known to cause retinopathy in experimental animals[4-7], brain lesions have been noticed only recently[8-10]. There seems to be no agreement about its pharmacological and neurophysiological effects in humans[11] and experimental animals[12].

While investigating glucose metabolism[13-15], we used L-glutamate (as MSG) as a gluconeogenic precursor *in vivo* (unpublished), and observed that rats became somnolent after intraperitoneal administration of carrier doses. In a study of the effect of L-glutamate on the *in vivo* incorporation of amino-acids into cerebral proteins (unpublished), several animals experienced severe tonic-clonic convulsions, usually between 30 and 120 min after intraperitoneal injection of a large dose of MSG (2 mmol/100 g body weight). Some convulsions terminated in death. We have now made a detailed study of the incidence of seizures following treatment with MSG. As well as taking an electroencephalogram (EEG) during the convulsions, we studied the influence of pyridoxine on the incidence of these convulsive disorders, for pyridoxine is directly involved in the metabolism of GA and gamma-amino-butyric acid (GABA).

We used weanling Sprague–Dawley male rats maintained on either laboratory chow (Teklad), or pyridoxine-control or deficient diets of the following composition: vitamin-free casein, 22%; sucrose, 69%; corn oil, 4%; salt mixture (Northern Regional Research Laboratories), No. 446, 4%; vitamin-fortification mixture (pyridoxine-free), 1%. This basal diet supplemented with 30 mg of pyridoxine HC1/kg served as the pyridoxine-control diet. Controls weighed 380–550 g (10–20 weeks old) and the deficient animals 80–120 g

谷氨酸一钠诱发大鼠痉挛性障碍

巴加万等

编者按

巴加万及其同事的这篇文章挑起了大家对谷氨酸一钠（MSG）是否会引起“中国餐馆综合征”的争论（译者注：MSG是味精的主要成分），他们报告称大鼠在被注射食品添加剂后会出现颤抖和痉挛症状。三年前，有人列举了食用中国菜后出现不良反应的诸多报道，指出罪魁祸首是MSG。但后来的研究得到了相反的结论。1995年，一份为美国食品和药品监督管理局提供的报告所显示的结论是：MSG及其相关物质“在食用常规水平时，对绝大多数人来说是安全的食品添加剂”，但也要警惕在大量食用这种调味剂后某些人可能会无法耐受。

由于牵涉到“中国餐馆综合征”的病因[1-3]，L–谷氨酸盐的生理和药理作用受到了广泛关注。尽管人们知道过量L–谷氨酸（GA）的单钠盐（MSG）会导致实验动物视网膜病变[4-7]，但其对大脑的损害直到最近才被注意到[8-10]。看似人们在MSG对人类[11]以及实验动物[12]的药理学和神经生理学作用上还没有达成一致的意见。

在研究糖代谢时[13-15]，我们使用L–谷氨酸盐（和MSG一样）作为体内糖异生的前体（尚未发表），结果观察到大鼠在腹腔内注射负荷量L–谷氨酸盐之后变得嗜睡。在体内研究L–谷氨酸盐对氨基酸合成大脑蛋白质的影响时（尚未发表），数只动物出现了严重的强直阵挛性抽搐，这种症状通常在腹腔内注射大剂量MSG（2 mmol/100 g体重）后30 min ~ 120 min内发生。一些痉挛最终导致死亡。现在我们已经对注射MSG后出现的痉挛进行了详细研究。除了在痉挛过程中采集脑电图之外，我们还研究了维生素B_6对这些痉挛性障碍的影响，因为维生素B_6直接参与GA和γ–氨基丁酸（GABA）的代谢。

我们使用刚断乳的SD雄性大鼠作为实验动物，给予实验室动物饲料（特克拉德实验室）或维生素B_6对照饲料或如下成分的缺陷型饲料：不含维生素的酪蛋白，22%；蔗糖，69%；玉米油，4%；盐混合物（北部地区研究实验室），第446号，4%；维生素加强的混合物（不含维生素B_6），1%。这个基础饲料再加上每千克体重30 mg的盐酸维生素B_6作为维生素B_6对照饲料。对照组动物的体重为

(9–12 weeks old). MSG (2.5 mmol/ml.) was injected intraperitoneally (2.0 mmol/100 g body weight). Pyridoxine HC1 (5.0 mg per 100 g body weight) was administered intraperitoneally to some animals 30 min before MSG treatment. For EEG recording, animals were prepared under pentobarbital and chlorprothixene anaesthesia with chronically implanted stainless steel electrodes and then allowed to recover for 1 week before recordings were taken.

Table 1 summarizes the incidence of the various symptoms observed during 2 h after injection of MSG. Marked somnolence was observed in 99% of the animals within 5–20 min. After this (1 h after injection), 52% of the animals salivated copiously, and 31% then displayed what we have described as spastic tremors, varying in intensity from mild to severe myoclonic jerking sometimes followed by vigorous running about the cage and stereotyped biting. This behaviour occurred significantly more often (χ^2=5.25; d.f.=1; P<0.05) in rats on the pyridoxine-control diet pretreated with pyridoxine than in animals on the same diet pretreated with 0.9% NaCl. Seizures, which occurred 30–120 min after MSG injection, were invariably preceded by spastic tremors, and terminated in death in some cases, although the animals usually recovered within 2–3 min and resumed the spastic tremors until the next seizure. The largest proportion of seizures was observed in the animals fed pyridoxine-control diet and pretreated with pyridoxine; the lowest proportion was found in the pyridoxine-deficient and laboratory chow animals, both pretreated with pyridoxine.

Table 1. Incidence of Somnolence, Hypersalivation, Spastic Tremors, Seizures and Death in MSG*-treated Rats

Diet	Pretreatment	Somnolence		Hypersalivation		Spastic tremors		Seizures		Death†	
		N	%	*N*	%	*N*	%	*N*	%	*N*	%
Laboratory chow (21)	Pyridoxine HCl‡	20	95	6	29	5	24	1	5	2	10
Laboratory chow (22)	0.9% NaCl	22	100	16	73	7	32	5	23	6	27
Pyridoxine-control (25)	Pyridoxine HCl	25	100	15	60	15	60	7	28	9	36
Pyridoxine-control (25)	0.9% NaCl	25	100	17	68	6	24	5	20	8	32
Pyridoxine-deficient (15)	Pyridoxine HCl	15	100	4	27	2	13	1	7	1	7
Pyridoxine-deficient (11)	0.9% NaCl	11	100	4	36	2	18	1	9	3	27
Total (119)		118	99	62	52	37	31	20	17	29	24

Figures in parentheses indicate the number of animals.
* 2 mmol/100 g body weight, intraperitoneal.
† During 24 h after injection.
‡ 5 mg/100 g body weight.

To ascertain whether the effects of MSG could be attributed to the administration of excess sodium *per se* and/or to the tonicity of the solution injected, two groups (eight and four rats respectively) were injected with either NaCl or sodium DL-lactate (2.5 mmol/ml., 2 mmol/100 g body weight). None of the animals developed any symptoms characteristic of MSG treatment.

Of the eleven MSG-treated animals from which EEG recordings were taken, five displayed tonic-clonic convulsions during the 2 h recording session. The seizures were characterized

380 g ~ 550 g（10~20 周龄），缺陷饲料组动物的体重为 80 g ~ 120 g（9~12 周龄）。MSG（2.5 mmol/ml）经腹腔注射给予（2.0 mmol/100 g 体重）。盐酸维生素 B_6（5.0 mg/100 g 体重）则在一些动物注射 MSG 前 30 min 经腹腔注射给予。为了记录脑电图，动物在戊巴比妥和氯普噻吨麻醉下长期埋植不锈钢电极，经过一周的恢复后开始记录数据。

表 1 总结了在注射 MSG 之后 2 h 内观察到的各种症状的发生率。在 5 min ~ 20 min 内 99% 的动物出现了明显的嗜睡。之后（注射后 1 h），52% 的动物大量地流涎，随后 31% 的动物出现了我们曾描述过的那种痉挛性颤抖，严重程度轻重不一，最严重的出现肌阵挛性抽动，有时随后出现在笼子里狂乱地跑动和刻板地咀嚼。维生素 B_6 对照饲料组用维生素 B_6 预处理的大鼠，表现出的这种行为明显多于相同饲料组用 0.9% NaCl 预处理的大鼠（$\chi^2=5.25$；自由度 =1；$P<0.05$）。注射 MSG 后 30 min ~ 120 min 内出现的痉挛毫无例外都发生在痉挛性颤抖之后，有时以死亡终结，不过动物通常会在 2 min ~ 3 min 内恢复并重新进入痉挛性颤抖直至下一次发作。观察到动物痉挛发生比例最高的是维生素 B_6 对照饲料组并用维生素 B_6 预处理的动物；而发生比例最低的则是用维生素 B_6 缺乏饲料和实验室动物饲料喂养的动物，两者都用维生素 B_6 预处理。

表 1. MSG* 处理大鼠的嗜睡、唾液分泌增多、痉挛性颤抖、痉挛和死亡的发生率

饲料	预处理	嗜睡		唾液分泌增多		痉挛性颤抖		痉挛		死亡 †	
		N	%	*N*	%	*N*	%	*N*	%	*N*	%
实验室动物饲料（21）	盐酸维生素 B_6‡	20	95	6	29	5	24	1	5	2	10
实验室动物饲料（22）	0.9% NaCl	22	100	16	73	7	32	5	23	6	27
维生素 B_6 对照饲料（25）	盐酸维生素 B_6	25	100	15	60	15	60	7	28	9	36
维生素 B_6 对照饲料（25）	0.9% NaCl	25	100	17	68	6	24	5	20	8	32
维生素 B_6 缺乏饲料（15）	盐酸维生素 B_6	15	100	4	27	2	13	1	7	1	7
维生素 B_6 缺乏饲料（11）	0.9% NaCl	11	100	4	36	2	18	1	9	3	27
总计（119）		118	99	62	52	37	31	20	17	29	24

圆括号内数字代表实验动物的数量。

* 2 mmol/100 g 体重，腹腔内注射。

† 注射后 24 h 内。

‡ 5 mg/100 g 体重。

为了确定 MSG 的作用是否可以归因于给予过量的钠和 / 或注射液的张力，我们给两组动物（8 只大鼠和 4 只大鼠）分别注射了 NaCl 或者 DL– 乳酸钠（2.5 mmol/ml, 2 mmol/100 g 体重）。没有一只动物发生注射 MSG 后出现的症状特征。

MSG 处理的动物中有 11 只动物记录了脑电图，有 5 只动物在 2 h 的记录时间内出现了强直阵挛性抽搐。这种形式的痉挛发作具有高幅尖波的典型特征。

by high amplitude spikes typical of this form of convulsion.

Thus MSG in a dose smaller than that used by Olney[8] in adult mice, and by Adamo and Ratner[12] in infant rats, can induce tonic-clonic seizures in adult rats. Administration of pyridoxine to animals which already had an excess of this vitamin (30 mg/kg diet) significantly increased the occurrence of spastic tremors. It should be noted that we used intra-peritoneal administration rather than the subcutaneous method of the previous authors.

The types of seizure seen in the rat are remarkably similar to those noted by Wiechert *et al.*[16-18] in dogs and rats after intracisternal injection of L-glutamate (but not other amino-acids), and pyridoxal-5′-phosphate (PLP). Wiechert *et al.*[16-18] and Elliot and van Gelder[19] have suggested that the balance between GA and GABA but not the absolute amount of GABA *per se*[20] is a decisive factor in the control of cerebral neuronal activity. Unphysiological shifts in this balance have been implicated in the onset of seizures. Our findings suggest that intraperitoneally administered L-glutamate in the dose used may cross the blood-brain barrier, which would affect the steady state levels of GA and GABA, thus inducing seizures. The effect of pyridoxine, on the other hand, may be a consequence of differential stimulation of the two enzyme systems—GA decarboxylase and GABA aminotransferase—the latter having a higher affinity for PLP[21]. The resulting change in the relative concentrations of GA and GABA may determine the seizure activity. Other possibilities such as a disturbance in the electrolyte balance do exist and a detailed neurochemical study of L-glutamate-induced convulsions is necessary.

We thank George Strutt, Robert Brackbill and John Koogler, jun., for technical assistance. This investigation was supported in part by the National Institute of Neurological Diseases and Stroke and general research support, US Public Health Service.

(**232**, 275-276; 1971)

H. N. Bhagavan and D. B. Coursin: Research Institute, St Joseph Hospital, Lancaster, Pennsylvania 17604.
C. N. Stewart: Department of Psychology, Franklin and Marshall College, Lancaster, Pennsylvania 17604.

Received October 27, 1970; revised January 18, 1971.

References:

1. Schaumburg, H. H., and Byck, R., *New Engl. J. Med.*, **279**, 105 (1968).
2. Ambos, M., Leavitt, N. R., Marmorek, L., and Wolschina, S. B., *New Engl. J. Med.*, **279**, 105 (1968).
3. Schaumburg, H. H., Byck, R., Gerstl, R., and Mashman, J. H., *Science*, **163**, 826 (1969).
4. Lucas, D. R., and Newhouse, J. P., *Amer. Med. Assoc. Arch. Ophthalmol.*, **58**, 193 (1957).
5. Potts, A. M., Modrell, R. W., and Kingsbury, C., *Amer. J. Ophthalmol.*, **50**, 900 (1960).
6. Freedman, J. K., and Potts, A. M., *Invest. Ophthalmol.*, **1**, 118 (1962).
7. Olney, J. W., *J. Neuropathol. Exp. Neurol.*, **28**, 455 (1969).
8. Olney, J. W., *Science*, **164**, 719 (1969).
9. Olney, J. W., and Sharpe, L. G., *Science*, **166**, 386 (1969).
10. Olney, J. W., and Ho, O., *Nature*, **227**, 609 (1970).

这样，MSG 的剂量比奥尔尼[8]在成年小鼠中以及阿达莫和拉特纳[12]在幼年大鼠中所用的剂量更小就能够导致成年大鼠的强直阵挛性发作。将维生素 B_6 用于已经是维生素 B_6 过量（30 mg/kg 饲料）的动物，能显著增加痉挛性颤抖的发生率。应当注意，我们使用的是腹腔内注射的方式而不是上述文献作者所用的皮下注射方式。

在大鼠中看到的发作类型与维歇特等人[16-18]在狗和大鼠脑池内注射 L– 谷氨酸盐（而不是其他氨基酸）和 5′– 磷酸吡哆醛（PLP）后发现的发作类型非常相似。维歇特等人[16-18]以及埃利奥特和范格尔德[19]曾提示，控制大脑神经元活性的决定性因素是 GA 和 GABA 之间的平衡而非 GABA 本身的绝对含量[20]。这种平衡的非生理性偏移被认为与痉挛发作有关。我们的研究结果提示：腹腔内注射的所使用剂量的 L– 谷氨酸盐有可能穿过了血脑屏障，这会影响到 GA 和 GABA 水平的稳定状态进而诱发痉挛。另一方面，维生素 B_6 的作用可能源自其对两个酶系统（GA 脱羧酶和 GABA 氨基转移酶）的不同刺激所导致的结果，后者对 PLP 有更高的亲和力[21]。GA 和 GABA 相对浓度的变化可能决定了痉挛发作。其他可能性（如电解质平衡的紊乱）确实存在，因此有必要对 L– 谷氨酸盐诱发的抽搐进行细致的神经化学研究。

感谢乔治 · 斯特拉特、罗伯特 · 布拉克比尔和小约翰 · 库格勒为我们提供了技术支持。本研究的部分经费来自美国国立神经疾病与中风研究所，此外美国公共卫生署还为本项目提供了一般性研究支持。

（毛晨晖 翻译；李素霞 审稿）

11. Morselli, P. L., and Garattini, S., *Nature*, **227**, 611 (1970).
12. Adamo, N. J., and Ratner, A., *Science*, **169**, 673 (1970).
13. Bhagavan, H. N., Coursin, D. B., and Dakshinamurti, K., *Arch. Biochem. Biophys.*, **110**, 422 (1965).
14. Bhagavan, H. N., Maruyama, H., and Coursin, D. B., *Abstr. Seventh Intern. Cong. Biochem., Tokyo*, 802 (1967).
15. Bhagavan, H. N., Coursin, D. B., and Stewart, C. N., *Life Sci.*, **8**, 299 (1969).
16. Wiechert, P., and Herbst, A., *J. Neurochem.*, **13**, 59 (1966).
17. Wiechert, P., and Göllnitz, G., *J. Neurochem.*, **15**, 1265 (1968).
18. Wiechert, P., and Göllnitz, G., *J. Neurochem.*, **17**, 137 (1970).
19. Elliot, K. A. C., and van Gelder, N. M., *J. Neurochem.*, **3**, 28 (1958).
20. Kamrin, R. P., and Kamrin, A. A., *J. Neurochem.*, **6**, 219 (1961).
21. Roberts, E., Rothstein, M., and Baxter, C. F., *Proc. Soc. Exp. Biol. Med.*, **97**, 796 (1958).

New Hominid Skull from Bed I, Olduvai Gorge, Tanzania

M. D. Leakey *et al.*

Editor's Note

The problems of Louis Leakey's perhaps over-bold definition of the genus *Homo*, consequent on the description of *Homo habilis* in 1964, came back to haunt him. This paper is a description of a crushed skull from the lower levels at Olduvai that resembled *Homo* from the more recent Bed II — but which differed in various ways. Leakey saw no reason to modify his bold claims in his 1964 paper, but neither that it was necessary to suppose that the new skull represented a female *Homo habilis*. The first author of this paper is Louis' archaeologist wife Mary, who continued to live and work at Olduvai until her death in 1996.

THE cranium described by L. S. B. L. later in this article (Olduvai Hominid 24) was found by P. Nzube in October 1968 (ref. 1) at a site in Olduvai Gorge that lies in the eastern part of the gullies known as DK, approximately 300 m east of the living site in lower Bed I that was excavated during 1962 and 1963.

The Site

The cranium was embedded in a mass of calcareous matrix measuring approximately 20×12×8 cm. All the bones were covered by matrix with the exception of a small portion of the right supra-orbital region and the posterior part of the palate. Indeed, so few identifiable parts were visible that it is remarkable it was recognized as hominid. The deposit on which it was found consists of a tuffaceous clay that can be traced over a wide area in Bed I, extending from the eastern part of the Gorge as far west as the FLK sites. This horizon overlies the basalt and is in turn overlain by the marker tuff IB for which an apparently reliable average date of 1.8 m.y. has been obtained by K/Ar dating[2]. At the hominid site (DK East), the tuffaceous clay is 3 m thick but the thickness varies considerably at different localities, depending on the surface configuration of the underlying lava. The clay at several other sites has yielded artefacts and fossil bones, including the molar tooth of Olduvai H. 4, found at MK during 1959, that has been attributed to *Homo habilis*[3] as well as the only remains of a chalicothere known from Olduvai.

After the discovery of the cranium the surface deposit was sieved extensively and an area of approximately 400 m² was eventually worked over. Three complete and some broken teeth were recovered as well as substantial portions of the maxillae, part of the occipital,

在坦桑尼亚奥杜威峡谷第I层中新发现的原始人类头骨

利基等

编者按

路易斯·利基的麻烦可能在于对人属的定义过于大胆，人属的定义来自他在 1964 年对能人的描述，这个定义后来又反过来困扰着他。这篇论文描述了一件在奥杜威遗址靠下层位发现的已被压碎的头骨，它类似于在更晚的第 II 层中发现的人属头骨，但又在很多方面有所不同。利基认为没有必要修正他在 1964 年的论文中所作的大胆定义，但又不能完全肯定新发现的头骨代表的是女性能人。这篇文章的第一作者是利基的太太考古学家玛丽，她一直在奥杜威峡谷生活和工作，直到 1996 年去世。

路易斯·利基在下文中所描述的头骨（第 24 号奥杜威原始人类）是由尼祖毕于 1968 年 10 月（参考文献 1）在奥杜威峡谷的一处遗址发现的，该遗址坐落于通常所说的 DK 冲沟的东部，位于第 I 层下部的生活遗址以东约 300 m 处，这个生活遗址是在 1962 年至 1963 年间被发掘出来的。

关于该遗址

该头骨被包裹在大块的石灰质围岩中，经测量该石灰质围岩的大小约为 20 cm × 12 cm × 8 cm。除小部分右眶上区和上腭后部外，其余所有骨头都被围岩所覆盖。的确，可鉴别的部分实在太少，以至于人们对能将它判定为原始人类的头骨而感到惊诧。我们发现该头骨之下的沉积物是由凝灰质黏土组成的，这种黏土分布在第 I 层中从峡谷的东部向西一直延伸到 FLK 遗址的广大区域中。该层位覆盖在玄武岩之上，其自身又被标志层凝灰岩 IB 所覆盖，由 K/Ar 年代测定法 [2] 得到这种标志层凝灰岩的平均年代为 180 万年，这个数据看来还是比较可靠的。在该原始人类遗址处（DK 东部），凝灰质黏土有 3 m 厚，但在不同地点，凝灰质黏土的厚度差别很大，这取决于地下熔岩的地表形态。在其他几处遗址的黏土中发现了人工制品和骨头化石，包括第 4 号奥杜威原始人类的臼齿，这是 1959 年间在 MK 发现的，该臼齿被归入能人 [3] 名下，还有从奥杜威发掘出来的唯一一个爪兽类动物的遗骸。

在该头骨被发现之后，通过对地表堆积物的广泛筛查，最终确定了约 400 m^2 的搜索区域。从中又找到了三枚完整的牙齿以及一些破碎的牙齿，还发现了绝大部分上颌骨、部分枕骨、几块顶骨破片和额骨的左眶上区。有些破片是在侵蚀斜坡下

several fragments of the parietals and the left supra-orbital region of the frontal bone. Some fragments were found low down the erosion slope and at a depth of just under 1 m from the present ground level: others lay on the surface, close to the block of matrix containing the cranium. It is evident that the specimen had been exposed for many years and that, as fragments broke off, some rolled down the slope and were buried under hill wash, while others became detached more recently and remained on the surface near the parent block. No trace of the canine and incisor teeth was found and it is likely that they broke off first and were transported downhill by erosion, possibly as far as the present river course.

The matrix in which the cranium was embedded is indistinguishable from the limestone concretions that frequently occur within the tuffaceous clay that overlies the basalt at DK and elsewhere (the mandibular fragment containing the molar of Olduvai H. 4 was similarly embedded). These concretions often form round fossil bones that can be seen as central cores when the blocks are broken open.

Crocodile remains predominate among the faunal material from this site and more than 2,000 teeth were found. Tortoise plates, shells of Urocyclid slugs, fish vertebrae and scales, bird bones and pieces of ostrich eggshell were also relatively common. Mammalian bones and teeth included those of primates, rodents, carnivores, equids, suids, giraffids and bovids. There were also more that 37,000 unidentifiable bone fragments. The state of preservation of these remains is very similar to that of the cranium and many were also found embedded in limestone concretions.

Associated Stone Industry

Apart from a few quartzite flakes and chips, no artefacts were found, but the living floor previously excavated at DK—which included the circle of lava blocks considered to represent an artificial structure—has yielded a large series of artefacts that represent the contemporary stone industry[4]. This site consists of a living floor within the same tuffaceous clay that occurs at DK East and is similarly overlain by Tuff IB. In some places the accumulation of occupation debris rests on the surface of the lava, so that the living floor is at a slightly lower level than the layer of limestone concretions from which the cranium is believed to have eroded. Artefacts recovered from the living floor and from the clay above amount to 1,198 specimens, consisting of 154 tools, 187 utilized pieces and 857 debitage. The industry is typical of the Oldowan as it occurs throughout Bed I. It includes choppers of various forms, generally made on water-worn lava cobbles, polyhedrons, discoids, sub-spheroids, heavy and light duty scrapers and burins. Although the tools are of the usual Oldowan types, a proportion of the choppers, polyhedrons and discoids are smaller than those from higher levels of Bed I.

The associated faunal material includes all the species noted at DK East with the addition of *Elephas* and *Deinotherium*. Crocodile remains are similarly the most numerous.

的低处找到的，深度仅为现在的地面水平以下 1 m：其余样本位于地表，就在包含头骨的围岩块附近。很明显，该标本已暴露多年，并且因为骨头破片都折断了，有些破片就沿着斜坡滚下而被埋在了坡滑之下，另外一些破片则是后来分解开的，因而仍留在离母体块很近的地表上。没有发现犬齿和门齿的半点痕迹，很可能它们是最先发生断裂的，然后由于侵蚀作用沿着坡向下滑，可能一直滑进了目前的河道中。

很难将包裹该头骨的围岩与石灰岩结核体区分开来，后者通常存在于覆盖在 DK 和其他地方的玄武岩之上的凝灰质黏土之内（包含第 4 号奥杜威原始人类的臼齿的下颌骨断片也被相似的围岩包裹）。这些结核体常常会在骨头化石周围形成，当块状物被打开时，就可以看到中心的核是骨头化石。

在该遗址发现的动物群材料中，鳄鱼的遗骸占主导地位，有 2,000 多枚鳄鱼牙齿被发现。龟甲、蛞蝓壳、鱼脊椎和鱼鳞、鸟骨和鸵鸟蛋壳碎片也是相对常见的遗骸。找到的哺乳动物骨骼和牙齿包括灵长类的、啮齿类的、食肉类的、马科的、猪科的、长颈鹿科的和牛科的。还有 37,000 多块无法鉴定的骨骼碎片。这些遗骸的保存状况与头骨的保存状况很相似，其中有很多碎片也被发现包裹于石灰岩结核体中。

共生的石器业

除了一些石英岩片和碎屑之外，没有发现任何人工制品，但早先在 DK 挖掘出的居住地面——所包含的熔岩块围成的圈被认为是一种人造结构——曾出土过一系列代表当时石器工业的人工制品[4]。该遗址由一个居住地面构成，和 DK 东部的遗址一样都处于同样的凝灰质黏土中，而且也同样被凝灰岩 IB 所覆盖。在有些地方，人类活动留下的生活垃圾积聚在熔岩的表面，所以居住地面所在的层位略微低于人们认为头骨曾在此被侵蚀出来的石灰岩结核层。从居住地面以及覆盖于其上的黏土中发现的人工制品总计达到了 1,198 件，包括 154 件工具、187 件使用器具和 857 件废片。这种工业是奥杜威文化的典型特征，它广泛分布于第 I 层之中。包括通常由用水磨蚀的熔岩卵石制造而成的各式砍砸器、多面体、盘状器、亚球状器、重型和轻型刮削器以及雕刻器等。尽管这些工具都属于奥杜威文化常见的类型，但一部分砍砸器、多面体和盘状器要比从第 I 层较高层位出土的同类物体小。

共生的动物群材料包括了在 DK 东部发现的所有物种，此外还有亚洲象属和恐象属。鳄鱼遗骸在这两处都是数量最多的。

Restoration of Olduvai H. 24

The tools used by R. J. C. in the removal of matrix from Olduvai H. 24 were a chisel-ended dental pick and a small hammer, a diamond drill and an S. S. White Industrial Air-Abrasive machine. Acetic acid could not be used because although the exposed pieces of bone had become hardened by weathering, those not exposed were softer than the matrix.

Parts of the matrix contained impressions of bone fragments that had fallen out during weathering. Some of these were found during excavation and it was possible to fit them back into position and thus build up the cranium into the form it had been before they became detached. It was apparent that the specimen was reasonably complete but had been badly crushed and distorted, the chief pressure having been downward and backward from the brow region, with the result that the nasal bones, although complete and perfectly preserved, were lodged down behind the infra-orbital region. The glabella, still attached to the nasals, was crushed, and the brow ridges were squeezed close together on top of the glabella.

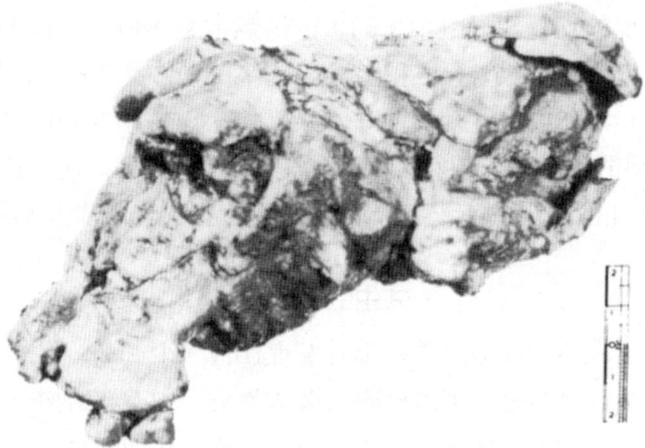

Fig. 1. Cast of Olduvai H. 24 after partial cleaning and before reconstruction.

The vault fragments were first separated and then completely cleaned of matrix under a binocular microscope. The facial fragments were more difficult to separate because they were interlocked and rammed against each other, but once they were free of matrix there were perfect joins between the nasals and the rest of the face and between the right brow ridge and the right orbital margin, just above the fronto-zygomatic suture.

The base of the cranium had been depressed into the brain cavity and was rammed against the right petrous, and the basi-occipital had been pushed under the vomer.

After thorough cleaning, consideration of sutures, bone thickness and curvature it was found that all the vault fragments but one joined together and that there was contact from the foramen magnum almost to the glabella. The sides of the vault, however, were not

第24号奥杜威原始人类的复原

克拉克用来剔除第24号奥杜威原始人类上的围岩的工具包括：一件牙科医生用的末端类似凿子的工具、一把小榔头、一个金刚石钻头和一台怀特工业公司生产的气压喷砂磨光器。不能使用醋酸的原因是：尽管被暴露的骨头破片已经由于风化作用而变硬，但是未暴露的那些部位要比围岩软。

部分围岩中含有风化期间脱落的骨头断片的印痕。只要在挖掘时找到一部分骨头断片，就有可能把它们拼接到原来的位置，由此将头骨复原成断片分离之前的样子。显而易见的是：这个标本相当完整，只是在严重的挤压下发生了变形，主要压力来自眉区的下面和后面，这使得虽然得以完整保存下来的鼻骨被卡到了眶下区之后。眉间区仍与鼻部相连，但已被压碎，眉脊在眉间区顶部被挤压在了一起。

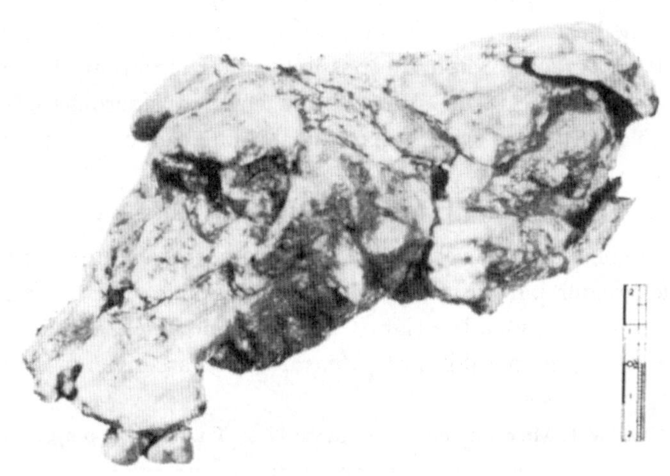

图1. 部分清理后、进行复原前的第24号奥杜威原始人类的模型。

先分离出颅顶破片，然后在双目显微镜下将围岩完全去除。由于面部破片相互纠结、彼此挤压，因而比较难于分离，但是一旦将它们与围岩分离开来，就会看到鼻部与面部其他部分之间以及右眉脊和额颧缝之上的右眶缘之间的完美连接。

头骨底部被压进脑腔，与右侧颞骨岩部挤在一起，而枕骨底部被推至犁骨之下。

彻底进行清洗之后，从骨缝、骨骼厚度和曲率的情况来看，除一块骨头之外，其余所有颅顶破片都被挤压在了一起，并且从枕骨大孔差不多到靠近眉间处。然而，并没有发现颅顶的侧面。后部已经严重变形，为了将该区域正确地连接到一起，有

present. The posterior part was considerably crushed and in order for the joins in this region to lock properly it was necessary to straighten out the buckled lambdoid region by breaking through the bone along one crushed bend and by partially cutting and breaking the bone just above the suture.

The crowns of the left P^4, M^1 and right M^1 that were found during sieving fitted back directly on to the palate, but the left P^3 and the fragment of right M^2 had to be placed in position with plaster supports. A small but perfect contact between the mesiolingual fragment of the right M^3 and the major portion of this tooth could be seen under the microscope. The two pieces were fitted together and were placed in position on the palate by means of an impression that had been left in the matrix.

The cranium is still warped slightly and it is doubtful whether it can be completely straightened out, particularly as the sides of the vault are missing. The right petrous and tympanic plate are still in their crushed position and it would be a major operation to attempt to restore these bones to their proper relationship. The base of the cranium is also still depressed into the brain cavity and its thinness in places and the fact that it is covered with a network of fine cracks inhibit the possibility of restoration. The crushing of the whole cranium has also to be taken into account when considering the cranial capacity, which must inevitably have been greater than the absolute capacity as measured now.

Description of Olduvai H. 24

This specimen now consists of the following parts: almost the entire supra-orbital region of the frontal, the greater part of the vault, a considerable part of the occipital bone, nearly all the sphenoid, nearly all the right temporal bone and a considerable part of the face and palate, including the nasal bones. (Measurements are given in Tables 1 and 2.)

Table 1. Measurements of Olduvai H. 24 Cranium (in mm)

Glabella to inion cord	147.0
Glabella to bregma cord	88.5
Bregma to inion cord	78.5
Bimastoid crest breadth	122.0
External orbital angle width	107.0
Minimum frontal width	75.0
Foramen magnum length	28.5
Foramen magnum width	25.0
Length nasal bones	19.5
Nasion to prosthion	60.0
Nasion to orale	61.0
Nasion to left side of pyriform aperture	38.5
Nasion to naso-spinale	36.0
Naso-spinale to prosthion	25.0
Maximum breadth of pyriform aperture	25.0
Palatal width at alveolae of M^1	37.0

必要沿着一个破碎的弯曲部分将骨头拆开并局部切割和打断位于骨缝正上方的骨头来弄直已变形的人字形区域。

筛查时发现：左 P^4、M^1 和右 M^1 的齿冠可以直接安回到上腭上去，但左 P^3 和右 M^2 的破片就只能借助石膏来回复原位了。在显微镜下可以看到：右 M^3 近中舌侧的破片与这颗牙齿的主体部分之间存在着一个虽小但很完美的衔接。将这两部分拼在一起，并参照留在围岩上的印痕把它们安装到上腭上的适当位置处。

这件头骨还是稍微有点弯曲，尚不确定是否可以将其完全弄直，特别是在颅顶侧面缺失的情况下。右侧颞骨岩部和鼓板仍然处于它们被压碎时的位置，尝试将这些骨头按原有的关系复位将是一项很重要的工作。头骨基底部仍处于被挤压进脑腔的位置，它在某些地方厚度很薄以及被细裂纹的网络所覆盖的现状使复原工作难以完成。当衡量颅容量时，也应该把整个头骨的破碎程度纳入考虑当中，实际容量必定大于现在测量出来的绝对容量。

对第 24 号奥杜威原始人类的描述

当前，该标本由以下几部分组成：额部的几乎整个眶上区、绝大部分颅顶、大部分枕骨、几乎整个蝶骨、几乎全部右颞骨以及大部分面部和上腭，包括鼻骨。(具体测量结果见表 1 和表 2。)

表 1. 对第 24 号奥杜威原始人类头骨的测量结果（单位：mm）

项目	数值
眉间－枕骨隆突索	147.0
眉间－前囟索	88.5
前囟－枕骨隆突索	78.5
两乳突脊间的宽度	122.0
眶外角宽度	107.0
前额最小宽度	75.0
枕骨大孔长度	28.5
枕骨大孔宽度	25.0
鼻骨长度	19.5
鼻根到齿槽中点	60.0
鼻根到切牙颌内缝终点	61.0
鼻根到梨状孔左侧	38.5
鼻根到鼻棘点	36.0
鼻棘点到牙槽中点	25.0
梨状孔的最大宽度	25.0
M^1 齿槽处的上腭宽度	37.0

Table 2. Measurements of Teeth of Olduvai H. 24 (in mm)

Left side	Bucco-lingual breadth	Mesio-distal length
P^3	(12.0)	9.0
P^4	(12.5)	9.1
M^1	(13.0)	12.6
M^3	(14.5)	12.0

(1) The frontal bone. The supra-orbital region on the right side is almost intact as well as most of the glabella region. A large part of the left supra-orbital region is also preserved but it does not extend quite as far as the external orbital angle. There is also a small piece missing between the left supra-orbital fragment and the glabella region. On the right side the external orbital angle is intact and in normal articulation with the orbital process of the zygomatic. The supra-orbital region as a whole, while forward projecting, is not very massive. The glabella contact with the nasal bones at nasion is preserved and nasion lies well behind the overhanging glabella. Behind the supra-orbital region there is a depressed valley that is more pronounced above the glabella than on either side. In such parts as are preserved the frontal bone rises from this valley in a low curve towards bregma. There is sufficient of the frontal bone to be reasonably certain of the glabella–bregma length, which is approximately 90 mm.

On the right side of the cranium the region immediately behind the external orbital angle and the supra-orbital torus is well preserved. Part of this area is also present on the left side so that the minimum frontal width can be closely estimated. The figure is 75 mm.

(2) The parietals. Although neither of the two parietal bones is intact, the whole of the sagittal suture is preserved so that the bregma–lambda cord and arc can be measured with reasonable accuracy: these are 76.5 mm and 81.0 mm respectively. On either side of the sagittal suture the parietal bones extend laterally with only a slight downward curve for about 50 mm. Although the lateral parts of both parietals are missing, these bones must then have turned abruptly downwards if they were to meet the squamous parts of the temporal bones. This can clearly be seen on the more complete right side of the cranium. The sagittal suture is very largely fused but can still be traced. This condition is somewhat surprising in an individual whose third molars were barely erupted and not yet in occlusion. It suggests that the eruption of the third molars in early hominids based on analogy with modern man may not be a reliable indication of age.

(3) The occipital bone. A large part of this bone is well preserved, especially on the right side. There is enough of the nuchal area intact as well as continuous contact from lambda to opisthion to make a reasonable reconstruction of the upper half of the occipital and it is possible to estimate, with a fair degree of accuracy, the lambda–opisthion arc and cord: these measure 87.5 mm and 71.0 mm respectively. The arc/cord index (Martin's index 125)[5] is 81.14. By doubling the measurement from the right asterion to the midline of the occipital it is also possible to estimate the bi-asterionic width, which is approximately 99 mm. The occipital index (Martin's index 129) is approximately 72.0.

表 2. 对第 24 号奥杜威原始人类牙齿的测量结果（单位：mm）

左侧	颊–舌宽	近中 – 远中长
P^3	(12.0)	9.0
P^4	(12.5)	9.1
M^1	(13.0)	12.6
M^3	(14.5)	12.0

（1）额骨。右侧的眶上区和大部分眉间区都几乎完好无损。左眶上区的大部分也被保存了下来，但并未一直延伸到眶外角那么远。在左眶上区破片和眉间区之间还有一小块缺失的部分。右侧眶外角是完整无缺的，并且与颧骨眶突的连接也是正常的。作为一个整体，眶上区虽然向前突出，但并不算很粗大。眉间与鼻骨在鼻根处的衔接部位被保存了下来，鼻根恰好位于突出的眉间之后。在眶上区后面有一个凹陷处，该处在眉间之上比在两侧更明显。在被保存下来的这些部分中，额骨以一条低曲线从上述的凹陷处升向前囟。由现有的额骨足以合理地确定出眉间 – 前囟的长度，大约为 90 mm。

头骨右侧紧挨着眶外角和眶上圆枕后面的区域保存状况很好。这一区域的左侧也被保存了下来，因而可以比较准确地估计出最小前额宽度。该数值为 75 mm。

（2）顶骨。尽管两块顶骨都不完整，但是整条矢状缝被保存了下来，因此可以相当准确地测量出前后囟索和前后囟弓来：它们分别是 76.5 mm 和 81.0 mm。在矢状缝的两侧，顶骨以一条稍微有一点向下弯曲的曲线向侧面横向延伸了约 50 mm。尽管两块顶骨的侧面部分都丢失了，但是如果这些骨头与颞骨的鳞状部分会合的话，那么它们的走向肯定是突然向下的。从该头骨保存得比较完整的右侧部位可以清楚地看到这一点。大部分矢状缝融合在了一起，但仍然可以看到它的痕迹。在第三臼齿几乎没有萌出且尚不能咬合的个体中出现这种情况的确有点令人吃惊。这说明按照与现代人相似的方法用第三臼齿的萌出来推断早期原始人类的年龄可能是不太可靠的。

（3）枕骨。大部分枕骨保存完好，尤其是右侧部分。完整的项区以及从后囟点到枕后点间的连续衔接部位的长度已经足够，由此可以正确地复原枕骨的上半部分，而且可以相当精确地估计出后囟 – 枕后弓和后囟 – 后索的长度：测量结果分别是 87.5 mm 和 71.0 mm。弓 / 索指数（马丁指数 125）[5] 为 81.14。采用将从枕骨右星点到中线的测量尺寸加倍的方法还有可能估计出两侧星点间的宽度，所得结果约为 99mm。枕骨指数（马丁指数 129）约为 72.0。

Viewed in profile, when the skull is orientated in the Frankfurt plane, the occipital region is seen to project far beyond a vertical line through the external auditory meatus, a feature seen in *Homo* but not in *Australopithecus*. The occipital condyles lie between the external auditory meati and the foramen magnum is elongate. The basi-occipital projection is short.

(4) The sphenoid bone. Nearly all the sphenoid is preserved including the pterygoid plates but not the pterygoid hamuli. In general appearance this bone is very similar to that seen in *Homo*, but this area requires detailed study in order to elucidate its significance taxonomically.

(5) The temporal bones. In the right temporal only the superior portion of the squamosal and part of the zygomatic process are missing. The mandibular fossa is perfectly preserved and is deep and narrow. The external auditory meatus is large and rather parallel-sided. The posterior part of the tympanic plate rises steeply to separate the meatus from the mastoid region. The tip of the mastoid process is broken off and the original size can only be estimated approximately from the preserved part. The zygomatic process is slenderly built and lacks the wide lateral shelf and outward flaring generally seen in *Australopithecus*.

The greater part of the left temporal bone is missing except for a small fragment in the medial region of the mandibular fossa.

(6) The face and palate. (*a*) The nasal bones. Both nasal bones are preserved with only the tips of the lateral processes on either side slightly damaged. The contact of the two bones in the midline is marked by a distinct keel which continues almost to the end of the nasal area. The margins of the pyriform aperture are not quite intact, but at its widest point it measures approximately 25 mm. The nasal length from nasion to the left lateral lip of the nasal aperture is 42.5 mm and nasion to nasospinale 36.0 mm. The nasal index (Martin's index 148) is 69.4, that is hyperchamaerrhine. There is a well developed anterior nasal spine at the center of the pyriform aperture and from this point a very distinct keel runs downwards in the mid-line on the naso-alveolar clivus. (*b*) The zygomatic bones. The right zygomatic bone is reasonably well preserved; only the zygomatic process and a chip from the frontal process are missing. The root of the zygomatic arch lies approximately above a point between the first and second molars. The left zygomatic bone is missing. (*c*) The maxillae. The right maxilla is almost complete; parts of the alveolar process are missing, particularly in the region of M^2, and at the canine jugum the bone is also missing. The left maxillary bone is less complete, the zygomatic process, part of the nasal surface, and much of the posterior part of the alveolar process are missing. At the incisor sockets the alveolar margin seems to be preserved lingually but is broken labially. At present the margin runs straight across between the two canine alveoli and is not curved; but it is difficult to assess the original form on account of the damage. The whole of the median palatal suture is preserved. The right side of the bony palate has been displaced posteriorly by approximately 2 mm. The palate is seen to be reasonably deep (depth of alveolar processes at the first molars is approximately 15.0 mm). (*d*) The palatine bones.

从侧面看来，当头骨被定位在法兰克福平面上时，就可以看到枕骨区明显突出于穿过外耳道的垂直线之外，这是一个存在于人属中的特征，南方古猿并不具备。枕髁位于外耳道之间，枕骨大孔呈细长形状。枕骨底部的突出很短。

（4）蝶骨。除翼钩外，包括翼板在内的绝大部分蝶骨被保存了下来。从整体外观来看，这个蝶骨与在人属中见到的很相似，但是要想阐明它在分类学上的意义，尚需要详细的研究。

（5）颞骨。在右侧颞骨中，只缺少鳞部的上部和颧突的一部分。深且窄的下颌窝部分保存完好。外耳道很大，并且侧边几乎是平行的。鼓板后部急剧上升将耳道与乳突区分开。乳突顶端断裂，只能根据保存下来的部分对其原来的尺寸进行粗略的估计。颧突细长而缺少宽阔的侧面支架和外部扩口，这种特征通常是南方古猿所具有的。

左侧颞骨的绝大部分丢失，仅剩下下颌窝中间区的一小块破片。

（6）面部和上腭。(*a*) 鼻骨。两块鼻骨都保存了下来，只是每一侧外侧突起的尖端稍微有点损伤。中线上两块骨头衔接处以一条明显的龙骨状突起为标志，龙骨状突起几乎延伸到接近于鼻区的末端。梨状孔的边缘并不十分完整，但是在最宽处测量得到的值大约为 25 mm。鼻部从鼻根到鼻孔的左外侧唇的长度为 42.5 mm，从鼻根到鼻棘点的长度为 36.0 mm。鼻指数（马丁指数 148）是 69.4，属于超阔鼻型。在梨状孔中部有一个很发达的鼻前棘，另一个非常明显的龙骨状突起从此处开始沿处于中线的鼻齿槽斜坡向下延伸。(*b*) 颧骨。右颧骨保存得相当好；只缺少颧突和额突上的一个薄片。颧弓根部大致位于第一和第二臼齿间的某点之上。左侧颧骨缺失。(*c*) 上颌骨。右上颌骨基本上是完整的；齿槽突有部分丢失，尤其是 M^2 区，犬齿隆突处的骨头也找不到了。左上颌骨不太完整，缺少颧突、部分鼻表面以及齿槽突后部的很大一部分。门齿槽上的近舌侧齿槽边缘看似被保存了下来，但是近唇侧部分断掉了。目前齿槽边缘笔直地从两颗犬齿的齿槽之间穿过，毫不弯曲；但是由于发生了损坏，因而很难估计出其原来的形式。整个腭中缝都保存了下来。硬腭右侧向后错位了约 2 mm。上腭看起来非常深（第一臼齿处的齿槽突深度约为 15.0 mm）。(*d*) 腭骨。腭骨是完整的，它们与上颌粗隆间以及与蝶骨间的衔接部位都完好地保存了下来。

The palatine bones are intact and the areas of their contact with the maxillary tuberosities and with the sphenoid are well preserved.

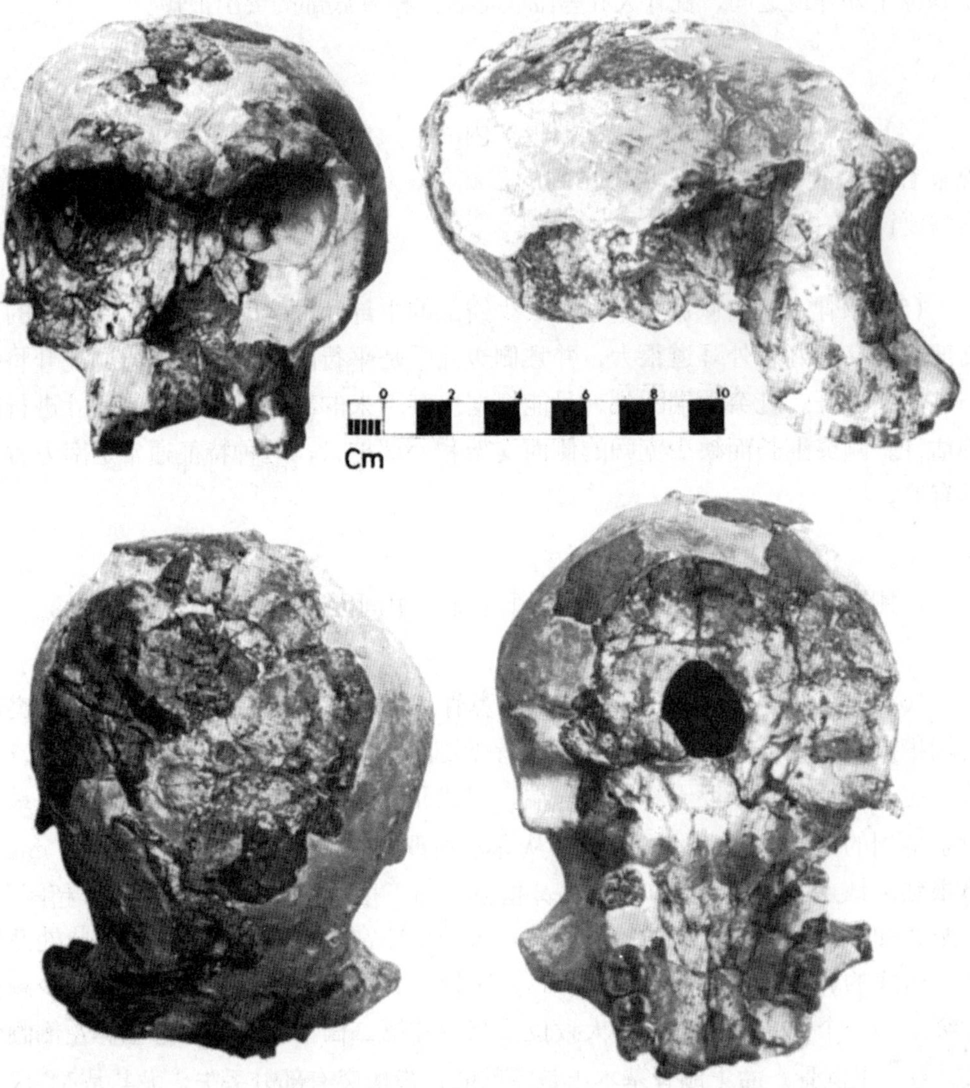

(7) The teeth (see Table 2) present in the cranium as now reconstructed are as follows; (*a*) The crown of the left M^3. This is slightly broken on the distal aspect. The occlusal surface is undamaged and is very wrinkled with no sign of wear. (*b*) The left M^1 is intact and is firmly rooted in the maxilla. The enamel of the crown shows a degree of flat wear and the dentine is only exposed in a very small area on the protocone. (*c*) The left P^4 is complete. Its bucco-lingual width is less than that of M^1 and it is rather rectangular in form. (*d*) The left P^3 is partially preserved, the mesio-lingual area is broken away. (*e*) the right M^3 is partially preserved, the paracone and part of the protocone are broken away. The tooth compares closely with the intact left M^3. (*f*) The right M^2 is incomplete; it is broken

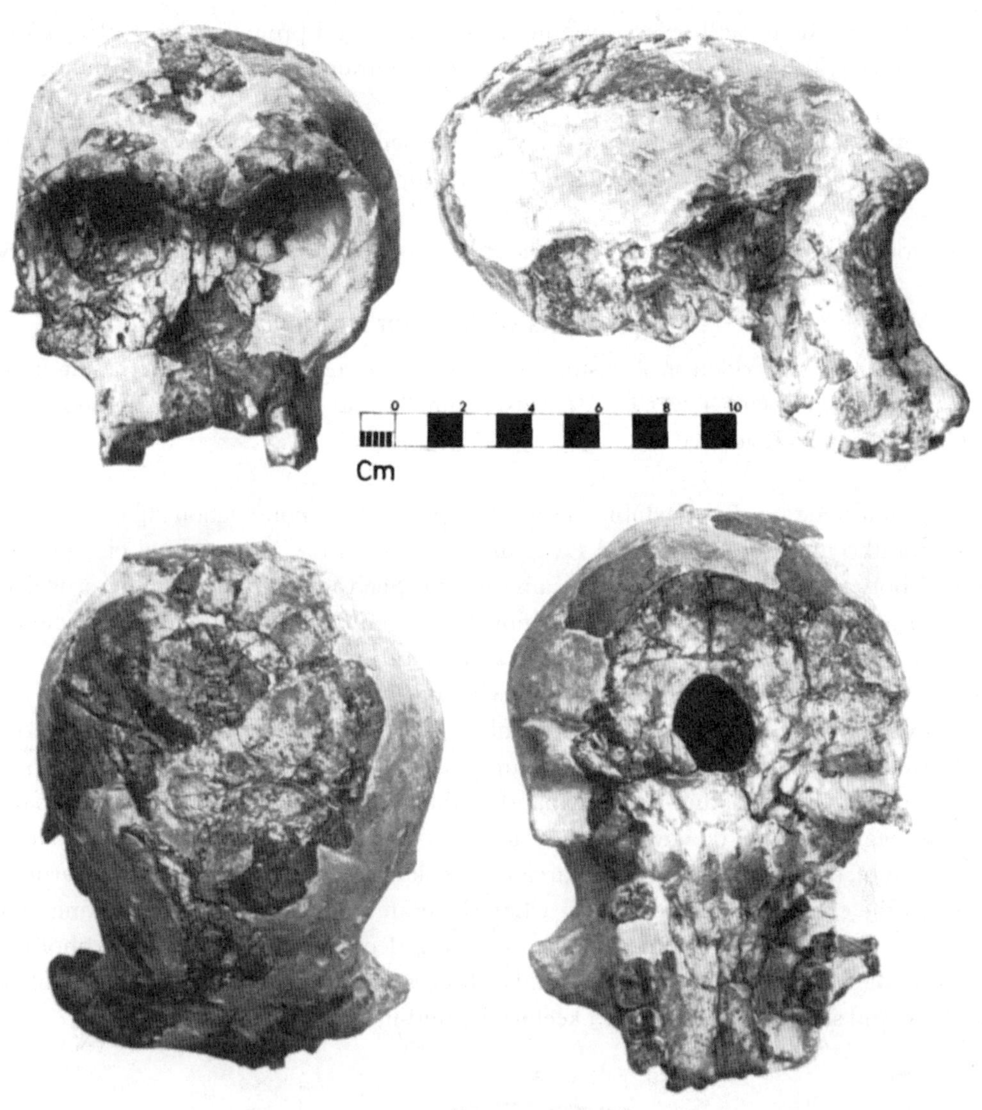

（7）现存于复原头骨上的牙齿（见表 2）情况如下：(*a*) 左 M^3 的齿冠。其远端面有轻微的破裂。咬合面没有破损，因为没有磨损，所以齿冠面上褶皱得很厉害。(*b*) 左 M^1 完整，牢固地植根于上颌骨中。齿冠的珐琅质出现了某种程度的磨平，齿质仅仅暴露于上原尖上的一个很小的区域内。(*c*) 左 P^4 完整。其颊舌宽度比 M^1 的要小，且外形非常接近于矩形。(*d*) 左 P^3 仅有部分被保存了下来，近中舌侧区断掉了。(*e*) 右 M^3 仅有部分被保存了下来，上前尖和部分上原尖已经断掉。该牙与保存完整的左 M^3 非常相像。(*f*) 右 M^2 不完整；它在沿对角线方向发生了断裂，所以近舌侧

diagonally so that the lingual and distal area is missing. (*g*) The right M^1 is nearly complete, only the protocone is missing. Its wear and structure are like those of the left first molar.

In addition to these teeth several fragments of molars and premolars were also recovered during sieving. It has not yet been possible to fit these fragments on to the cranium.

(8) Cranial capacity. It is not possible at this stage to give a close estimate of cranial capacity but preliminary determinations give a figure in the order of 560 cm^3.

(9) Endocranial region. Considerable parts of the endocranial morphology are preserved.

Comparisons with Other Hominids

There is still some evidence of distortion in the reconstruction. This has resulted in the vault of the skull being lower than it was originally and the backward projection of the occipital is now exaggerated.

The frontal bone does not exhibit the marked post-orbital constriction that is seen in the australopithecines from South and East Africa, nor does the supra-orbital element of the frontal bone curve backwards to the same extent. Seen in both lateral and vertical views the supra-orbital torus and glabellar region form a more or less straight line whereas in all known species of *Australopithecus* the glabella tends to project further forward with the lateral edges of the supra-orbital torus swinging backward. The parietal bones extend outwards from the sagittal suture with only a slight downward curve at first, and then bend steeply downwards to meet the squamosal parts of the temporal bones. The parietal region of the brain is thus more expanded than in *Australopithecus*, where the parietals usually begin to curve downwards close to the mid-line. Even allowing for the backward distortion of the occipital region this area of the skull extends further behind a vertical line through the external auditory meatus, when the cranium is in the Frankfurt plane, than in the australopithecines. The mandibular fossae are deep and very similar morphologically to those of *Homo sapiens* and wholly unlike those seen in *Australopithecus*. The nasal bones are long and slender with a marked keel in the mid-line.

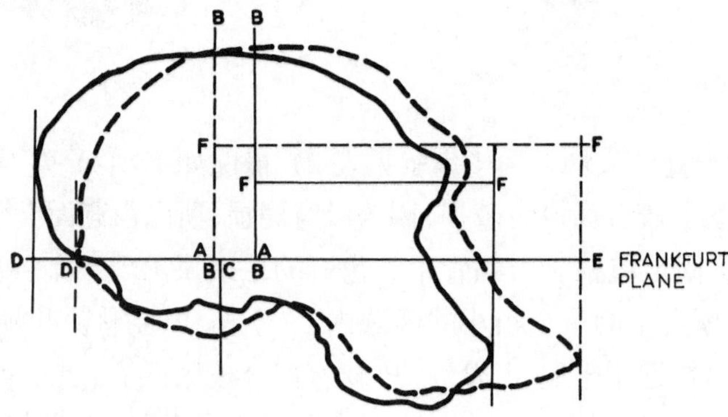

Fig. 3. Profile of Olduvai Hominid 24 (solid line) superimposed on that of *Australopithecus africanus* ST. 5 (broken line).

和远端区域都不见了。(*g*) 右 M^1 基本上是完整的，只有上原尖缺失。其磨损程度和结构都与左侧的第一臼齿类似。

除这些牙齿外，在筛查过程中还发现了一些臼齿和前臼齿的破片。现在想把这些破片复原到头骨上还不太可能。

（8）颅容量。现阶段还不可能准确地估计出颅容量的大小，不过，初步确定的结果给出的数量级为 560 cm^3。

（9）颅腔区。颅腔区的大部分形态仍保持原状。

与其他原始人类的比较

在复原过程中仍能发现一些变形的迹象。这导致头骨顶部低于原来的位置，同时夸大了枕骨的向后突出。

额骨并未表现出向眶后明显收缩的现象，在南非和东非的南方古猿类中可以看到这种现象，额骨的眶上区部分也没有发生同等程度地向后弯曲。从侧面和垂直角度来看，眶上圆枕和眉间区域几乎形成了一条直线；但在所有已知的南方古猿种中，眉间都倾向于更加向前突出，而眶上圆枕的侧面边缘则向后摆动。顶骨从矢状缝处向外延伸，起初只是稍微有一点儿向下弯曲，而后则急剧地向下弯曲而与颞骨的鳞部会合。因此，第 24 号奥杜威原始人类脑子的顶骨区比南方古猿的要大，南方古猿的顶骨通常在接近中线处开始向下弯曲。当头骨位于法兰克福平面时，即便考虑到枕骨区的向后变形，第 24 号奥杜威原始人类头骨的这部分区域也比南方古猿类在更大程度上延伸到了穿过外耳道的垂直线之后。下颌窝很深，在形态上非常类似于智人，而完全不同于在南方古猿中看到的情形。鼻骨又长又细，在中线处有一个明显的龙骨状突起。

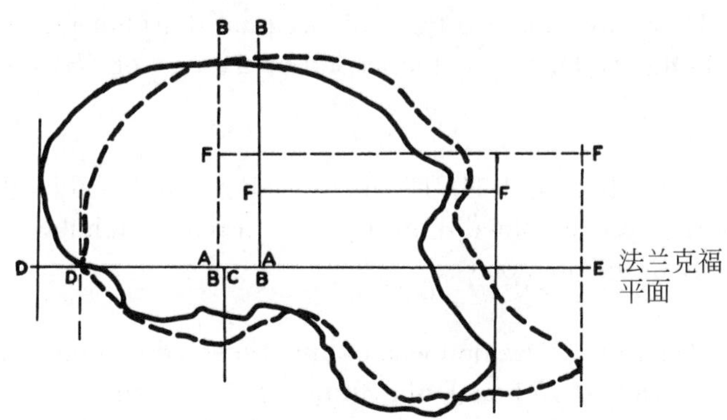

图 3. 第 24 号奥杜威原始人类的头骨侧面图（实线）与 ST. 5 号南方古猿非洲种的头骨侧面图（虚线）叠加在一起的效果图。

The discovery of this skull, which is clearly a member of the genus *Homo* as defined in 1964[6], although it has certain primitive morphological features, raises an important issue. Although it is clearly not an australopithecine it nevertheless differs in certain important respects from the incomplete type of *Homo habilis* (Olduvai H. 7). Yet it resembles the paratype from Bed II (Olduvai H. 13) in both cranial and dental characters. At the same time, the dentition of the type of *Homo habilis* shows considerable resemblances to that of Olduvai H. 16, from lower Bed II, Maiko Gully, although differing in some respects. The problem therefore is whether or not there are two forms of *Homo* in Beds I and II or whether the differences between the two groups can be accounted for on the basis of individual variation and sexual dimorphism.

Although only part of the upper dentition is preserved in the cranium from DK, there are other specimens from Beds I and II which have a comparably small or even smaller tooth size and are morphologically similar. These clearly belong within the same group. They include a complete upper molar (Olduvai H. 6) from the site of *Australopithecus boisei* and an isolated molar (Olduvai H. 21), also probably from Bed I.

In Olduvai H. 24 from DK and in H. 13 from MNK certain important diagnostic characters of the skull are preserved, including the area of the temporal bones around the external auditory meatus and most of the occipital bones. These features, together with the morphology of the teeth, clearly indicate that both specimens belong within the genus *Homo*.

Although the occipital bone of the type specimen of *Homo habilis* (Olduvai H. 7) is not preserved it is evident from the structure of the parietal bones, and especially the form of the lambdoid suture, that the occipital bone must have been morphologically similar to that of *Homo erectus*, that is to say, the lambdoid suture is in the form of a low, wide arch and not V-shaped. This form is also seen in the skull of Olduvai H. 16 from lower Bed II. In contrast, the form of this suture in both Olduvai H. 24 from DK and Olduvai H. 13 from MNK and, indeed, the whole form of the occipital bone, is very similar to that seen in *Homo sapiens* today: the line of the suture descends sharply from lambda in the form of an inverted V. There are thus two types of occipital morphology, one represented by Olduvai H. 24 and Olduvai H. 13 and the second by the type of *Homo habilis* and Olduvai H. 16.

The teeth of Olduvai H. 7 and H. 16 are megadont and within the size range of *Australopithecus africanus*; yet, in other respects, they display a number of morphological differences, especially in the premolars.

Distinct differences from the australopithecines are also evident in the cranial morphology, particularly in the parietals. Although the frontal bone of the type of *Homo habilis* is missing, a large part of this bone is preserved in Olduvai H. 16. Both the minimum frontal width and the width between the temporal crests is greater than in any known australopithecine.

尽管该头骨具有一定的原始形态特征，但其显然是 1964 年被定名的人属的一个成员[6]，这一发现引出了一个重要问题。虽然它显然不是南方古猿类中的一种，但是在某些重要方面，它又与能人的不完整类型（第 7 号奥杜威原始人类）有所区别。而它的头骨和牙齿特征类似于在第 II 层中发现的能人副型（第 13 号奥杜威原始人类）。与此同时，虽然能人类型的齿系与在第 II 层下部舞子沟中发现的第 16 号奥杜威原始人类在某些方面有差异，但二者之间毕竟存在着很多相似之处。因此，问题就变成了在第 I 层和第 II 层中发现的是否是两种形式的人属，或者，这两个组群之间的差异是否可以用个体差异和两性异形来解释？

尽管 DK 遗址出土的头骨中只有部分上齿系保存了下来，但在第 I 层和第 II 层中还发现了其他标本，这些标本有相对较小或更小的牙齿，并且在形态上是相似的。它们显然属于同一组群。其中包括一枚完整的上臼齿（第 6 号奥杜威原始人类）和一枚单独的臼齿（第 21 号奥杜威原始人类），前者是从南方古猿鲍氏种遗址得到的，后者也可能来自第 I 层。

在 DK 遗址出土的第 24 号奥杜威原始人类和 MNK 遗址出土的第 13 号奥杜威原始人类的头骨中，某些可用于鉴定的重要特征仍保存至今，包括外耳道周围的颞骨区和大部分枕骨。这些特征以及牙齿的形态都明确显示上述两个标本可归入人属。

尽管能人模式标本（第 7 号奥杜威原始人类）的枕骨没有被保存下来，但是从顶骨的结构，尤其是人字缝的形式就可以清楚地得到以下结论：能人的枕骨肯定在形态上与直立人的枕骨相似，也就是说，人字缝的形式是低而宽的弓形而非 V 字形。这种形式也见于在第 II 层下部发现的第 16 号奥杜威原始人类的头骨。相比之下，DK 第 24 号奥杜威原始人类和 MNK 第 13 号奥杜威原始人类的人字缝的形式，确切地说，是枕骨的整体形式，都与在今天智人中看到的情况非常相似：缝线从人字点处以倒 V 字形急剧下降。因此，枕骨有两种类型的形态：一种以第 24 号奥杜威原始人类和第 13 号奥杜威原始人类为代表；另一种则以能人类型和第 16 号奥杜威原始人类为代表。

第 7 号奥杜威原始人类和第 16 号奥杜威原始人类的牙齿属于巨型齿，其牙齿的大小在南方古猿非洲种的牙齿尺寸变化范围之内；然而，另一方面，它们在形态上又存在着很多差异，尤其是前臼齿的特征。

与南方古猿类的显著差异也明显地表现在头骨的形态上，尤其是顶骨。尽管找不到能人类型的额骨，但在第 16 号奥杜威原始人类中还保存着大部分的额骨。其最小额宽和颞嵴之间的间距都要大于任何已知的南方古猿。

Until the discovery of Olduvai H. 24 it was considered possible that the difference in morphology and in size between the cranial parts and dentition of the type of *Homo habilis* from Bed I and those of the paratype from Bed II (Olduvai H. 13) might have resulted from the elapse of a prolonged time interval. Because Olduvai H. 24 is chronologically near to the type of *Homo habilis* and yet more closely resembles the paratype from Bed II this explanation is no longer tenable. There remains the question of sexual dimorphism and it is possible that Olduvai H. 13 and H. 24 represent females of *Homo habilis* with the type and H. 16 representing the males. The difference in the morphology of the occipital region in the two groups, however, cannot be disregarded and suggests the possibility of taxonomic variation.

Beyond doubt, the new specimen represents the genus *Homo* as defined by Leakey, Tobias and Napier and differs fundamentally from the australopithecines. At the same time it is not entirely certain that it necessarily represents a female of *Homo habilis*.

We thank Dr. A. Walker of the University of Nairobi for checking the text and measurements and for most valuable comments.

(**232**, 308-312; 1971)

M. D. Leakey, R. J. Clarke and L. S. B. Leakey: Centre for Prehistory and Palaeontology, PO Box 30239, Nairobi.

Received April 2, 1971.

References:

1. Leakey, M. D., *Nature*, **223**, 754 (1969).
2. Curtis, G. H., and Hay, R. L., in *Calibration of Hominoid Evolution* (edit. by Bishop, W. W., and Miller, J.) (Scottish Academic Press for Wenner Gren Foundation for Anthropological Research, in the press).
3. Leakey, L. S. B., and Leakey, M. D., *Nature*, **202**, 3 (1964).
4. Leakey, M. D., in *Background to Evolution in Africa* (edit. by Bishop, W. W., and Clark, J. D.) (University of Chicago Press, Chicago and London, 1967).
5. Martin, R., *Lehrbuch der Anthropologie* (Fischer, Stuttgart, 1959).
6. Leakey, L. S. B., Tobias, P. V., and Napier, J. R., *Nature*, **202**, 5 (1964).

在发现第 24 号奥杜威原始人类以前，以下这一解释被认为是有可能成立的，即来自第 I 层的能人类型和来自第 II 层的副型（第 13 号奥杜威原始人类）在头骨部分和齿系方面存在的形态和尺寸差异是由于两者之间跨越了漫长的岁月。因为第 24 号奥杜威原始人类在年代上接近于能人类型，但却与来自第 II 层的副型更加相像，因而上述解释再也站不住脚了。那么就只剩下两性异形的问题了，也许第 13 号和第 24 号奥杜威原始人类代表的是女性能人，而第 16 号奥杜威原始人类则代表了男性能人。但不能忽视这两个组群在枕骨区形态上存在的差异，这种差异暗示了分类上变异的可能性。

毫无疑问，这个新标本属于利基、托拜厄斯和内皮尔所定义的人属，而与南方古猿类有着本质上的不同。同时，并不能完全确定它一定代表着一个女性的能人。

内罗毕大学的沃克博士对文章内容和测量结果进行了核查，并提出了非常有价值的意见，我们对此表示衷心感谢。

（刘皓芳 翻译；林圣龙 审稿）

Non-random X Chromosome Expression in Female Mules and Hinnies

J. L. Hamerton *et al.*

Editor's Note

Equine hybrids, such as mules (produced from a male donkey and a female horse) and hinnies (produced from a male horse and a female donkey), have proved useful in the study of developmental genetics. Here John L. Hamerton and colleagues use female hybrids to lend support to English geneticist Mary Lyon's hypothesis that one of the X chromosome copies in female mammals is inactivated early in development. The team used expression of species-specific glucose-6-phosphate dehydrogenase to show that, in any given cell, only one of two X chromosomes is functional. Today, the principle of X chromosome inactivation is well accepted.

THERE have been several studies of X chromosome DNA replication and the behaviour of X-linked genes in the female mule[1-6] but the results are not in agreement on the frequency with which the two parental X chromosomes (X^D and X^H) show late DNA replication. Here we present additional data on the expression of the X-linked glucose-6-phosphate dehydrogenase (*Gd*, E.C.1.1.1.49) locus in a female mule and two female hinnies. (The hinny is the reciprocal cross *Equus caballus* ♂ × *E. asinus* ♀.) The mule and one hinny were less than one year old and the third animal was a 35 day old hinny foetus obtained by hysterotomy. All were bred at the Veterinary School and ARC Unit of Reproductive Physiology and Biochemistry, Cambridge, England.

Fibroblast cultures were established from the three animals by standard methods. The medium used was McCoy's 5a (modified) supplemented with 15% foetal calf serum (McFC15). Cloning of fibroblasts was carried out as follows[7]. Healthy exponentially growing fibroblast cultures were trypsinized, and the cell suspension was diluted as necessary to a final dilution of about thirty cells per ml. One drop of this suspension (0.03 ml.) was placed in each well of an appropriate number of Falcon Microtest II culture plates. (Each plate contains ninety-six wells of 0.5 ml. capacity. For each experiment five or six plates were seeded.) Then 0.2 ml. of McFC15 was added to each well and the plates were covered and incubated in a 100% humidity incubator in an atmosphere of 5% CO_2 in air. They were examined after 3 and 5 days and wells containing one colony were marked. The plating efficiency varied from 20–40%, which was similar to that obtained in conventional plating experiments using the same cell suspensions.

After 10–14 days, wells containing single colonies were trypsinized and the cell suspension transferred to 60 mm Falcon Petri dishes. When these dishes were confluent, the cells were either transferred to culture bottles to allow more extensive growth or, alternatively, the

母马骡和母驴骡的X染色体非随机表达

哈默顿等

编者按

马的杂交后代，如马骡（由公驴和母马杂交产生）和驴骡（由公马和母驴杂交产生），已被证实在发育遗传学研究方面是很有用处的。在本文中，约翰·哈默顿及其同事用雌性杂交动物证实了英国遗传学家玛丽·莱昂的假说，即雌性哺乳动物的两条X染色体副本中的一条会在发育的早期失活。该研究小组用物种特异的葡萄糖–6–磷酸脱氢酶的表达来说明：在任意给定的细胞中，两条X染色体中只有一条在发挥作用。时至今日，X染色体失活的理论已经得到了大家的普遍认可。

目前我们已经见到了几篇关于母马骡中X染色体DNA复制和X连锁基因行为方式的研究报告[1-6]，但是在关于两条亲本X染色体（X^D和X^H）上的DNA发生复制延迟的频率方面，所得结果之间并不一致。在本文中，我们将给出关于一匹母马骡和两匹母驴骡中X连锁的葡萄糖–6–磷酸脱氢酶（*Gd*, E.C.1.1.1.49）基因位点表达的一些额外数据。（驴骡是由公马和母驴互交得到的。）这匹马骡和其中一匹驴骡的年龄都不到一岁，另外一匹驴骡是通过子宫切开术得到的35天大的胚胎。所有这些动物都是由英国剑桥大学的兽医学院以及农业研究委员会生殖生理与生物化学研究室饲养的。

采用标准方法培养建立这三只动物的成纤维细胞系。所使用的培养基是添加了15%胎牛血清（McFC15）的（改良）麦科伊5a。成纤维细胞的克隆按以下步骤进行[7]。将健康的、指数生长的成纤维细胞培养物进行胰蛋白酶消化，首先根据需要将细胞悬浮液稀释至每ml约含30个细胞的终浓度。将这样的悬浮液加入孔数适当的猎鹰牌微量测定II型培养板的各个孔中，每个孔中加一滴（0.03 ml）。（每个培养板含有96个孔，每个孔的容量为0.5 ml。每次实验需要用5到6块培养板来进行接种。）接着在每个孔中加入0.2 ml McFC15，然后盖上培养板，并在含5% CO_2的大气条件下于湿度为100%的恒温箱中进行培育。分别在3天和5天后检查培育情况，并对含有一个克隆体的孔进行标记。培养板的接种效率在20%~40%之间变化，这与用同一细胞悬浮液在常规平板接种实验中所得的结果类似。

10天~14天后，我们对含有一个克隆体的孔内的细胞进行胰蛋白酶消化，然后将细胞悬浮液转移至直径为60 mm的猎鹰牌皮氏培养皿中。当细胞铺满这些培养皿时，可以将它们转移至培养瓶中扩大培养，或者停止进一步培养而去检测皮氏培

G6PD types were determined on the cells in the Petri dishes without further passage. Cells from each original culture and from a number of clones were stored at −70 or −196°C in a cryoprotective medium containing 15% glycerol. Lysates were prepared from both original cultures and clones by freezing and thawing in 0.015 M Tris-borate buffer (*p*H 7.8) containing 0.005 M EDTA, 0.02 g% "Triton X–100", 2 mg% NADP and 0.05% (v/v) 2-mercaptoethanol. Electrophoresis ("Cellogel", Chemetron, Milan) was carried out for 1.25 h at 4°C at a potential gradient of 45 V/cm and the gels were stained for glucose-6-phosphate dehydrogenase[5].

Gd is seen as a single band in both parental species. Gd^{H} moving slower than the Gd^{D} (refs. 2 and 8). The faint slower bands seen in both parental species and the hybrids after starch gel electrophoresis represent breakdown products. The earlier samples from animals 100 −196 were typed on starch as described earlier[2]. A number of other hybrid animals of both sexes as well as the parental species have been studied using both erythrocytes and fibroblasts without cloning. The *Gd* types are summarized in Table 1, and clearly confirm X-linkage of *Gd* in the Equidae. The parents of the three hybrids that were cloned are included in this table and show the expected *Gd* patterns. The interrelationship of the various hybrids which we have studied in detail is shown in Fig. 1. The hybrids, with one exception, develop a strong Gd^{H} band and either a faint or no Gd^{D} band. The exception (animal 192) has a stronger Gd^{D} band than Gd^{H} band. This inter-animal variation may explain the lack of agreement between different workers[3,4].

Table 1. Expression of Glucose-6-phosphate Dehydrogenase in Horses and Donkeys and Their Hybrids

Animal No.	Type	Material	Gd^{D}	Gd^{H}	Source
Female hybrids					
100	Mule	E	±	+	Cambridge*
		F	−	+	
138	Hinny	E	±	+	Portugal
139	Hinny	E	−	+	Portugal
143	Mule	E	−	+	Portugal
187	Hinny	E	±	+	Portugal
188	Hinny	E	±	+	Portugal
189	Hinny	E	±	+	Portugal
190	Hinny	E	±	+	Portugal
		F	−	+	
191	Hinny	E	−	+	Portugal
192	Hinny	E	+	±	Portugal
195	Mule (foetus)	F	−	+	Cambridge†
196	Mule (foetus)	F	−	+	Cambridge†
200	Hinny (foetus)	F	−	+	Cambridge†
202	Mule	E	−	+	Cambridge
		F	±	+	Cambridge

养皿中细胞的葡萄糖–6–磷酸脱氢酶（G6PD）的类型。将来自每个原始培养物和来自许多克隆体的细胞在–70℃或者–196℃下储存于含15%甘油的冷冻保护液中。通过对原始培养物和克隆体进行冷冻和融解来制备溶菌液，融解介质为0.015 M三羟甲基硼酸缓冲液（pH值7.8），其中含0.005 M乙二胺四乙酸、0.02 g%“曲拉通X–100”（译者注：成分为聚乙二醇辛基苯基醚）、2 mg%烟酰胺腺嘌呤二核苷酸磷酸和0.05%（体积比）的2–巯基乙醇。以45 V/cm为电位梯度在4℃下电泳1.25 h（电泳介质为“醋酸纤维素凝胶”，米兰凯美创公司产品），然后对凝胶进行葡萄糖–6–磷酸脱氢酶染色[5]。

在两个亲代物种中都只能看到一条单一的*Gd*电泳条带，并且Gd^{H}比Gd^{D}移动得要慢（参考文献2和8）。经过淀粉凝胶电泳之后，在亲代物种和杂交体中都能看到一些移动较慢的微弱条带，它们代表的是裂解的产物。我们对来自100号至196号实验动物的早期样品进行了淀粉凝胶电泳分析结果的归类，这已在以前发表的文章中描述过[2]。现在我们用未经克隆的红细胞和成纤维细胞对其他许多物种进行研究——不仅研究了其亲代，而且研究了包括雌雄两性的杂种。*Gd*类型总结于表1中，由此显然可以证明马科动物中*Gd*的X连锁特性。三只被克隆的杂交动物的亲本都包括在这张表中，其结果都与预期的*Gd*类型一致。我们曾经仔细研究过的各个杂交动物之间的相互关系示于图1。除一只以外，其他杂交动物都显示出一条很强的Gd^{H}条带和一条或者很弱或者消失的Gd^{D}条带。不过，也有例外者（编号为192的动物），其Gd^{D}条带要比Gd^{H}条带强。动物间的这种变异也许可以用来解释不同工作者的研究结果之间为什么并非完全一致[3,4]。

表1. 葡萄糖–6–磷酸脱氢酶在马、驴和它们的杂交后代中的表达

动物编号	类型	原料	Gd^{D}	Gd^{H}	样品来源
雌性杂交体					
100	马骡	E	±	+	剑桥＊
		F	−	+	
138	驴骡	E	±	+	葡萄牙
139	驴骡	E	−	+	葡萄牙
143	马骡	E	−	+	葡萄牙
187	驴骡	E	±	+	葡萄牙
188	驴骡	E	±	+	葡萄牙
189	驴骡	E	±	+	葡萄牙
190	驴骡	E	±	+	葡萄牙
		F	−	+	
191	驴骡	E	−	+	葡萄牙
192	驴骡	E	+	±	葡萄牙
195	马骡（胚胎）	F	−	+	剑桥†
196	马骡（胚胎）	F	−	+	剑桥†
200	驴骡（胚胎）	F	−	+	剑桥†
202	马骡	E	−	+	剑桥
		F	±	+	剑桥

Continued

Animal No.	Type	Material	Gd^D	Gd^H	Source
203	Hinny	F	±	+	Cambridge
		E	±	+	
217	Mule	E	±	+	Cambridge
Male hybrids					
	Mule (n=1 foetus)	F	–	+	Cambridge
	Mules (n=4)	E	–	+	Portugal and Cambridge
	Hinnies (n=7)	E	+	–	Portugal and Cambridge
Parental species					
Males	Donkeys (n=7)	E	+	–	Cambridge
	Horses (n=2)	E	–	+	Cambridge
Females	Donkeys (n=12)	E	+	–	Cambridge
	Horses (n=8)	E	–	+	Cambridge

* This animal (No. 100) reported by Hamerton *et al.*, 1969.

† 195 45D gestation
196 60D gestation
200 35D gestation
201 ♂ mule 35D gestation } All obtained by hysterotomy

+ Strong; ±, faint; –, absent; E, erythrocytes; F, fibroblasts.

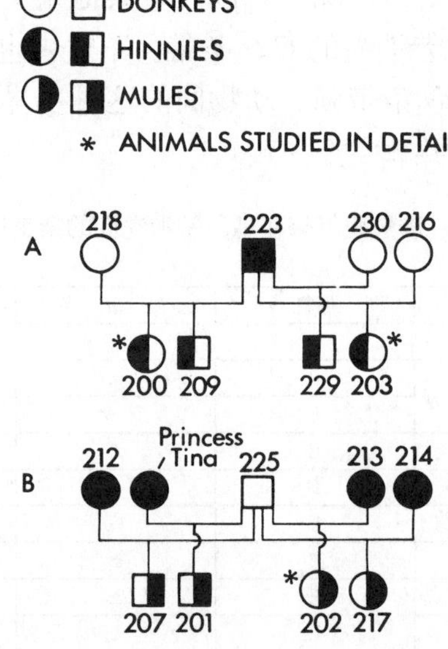

Fig. 1. Pedigree charts of the mules and hinnies studied. The numbers refer to numbers in Table 1. A, Hinnies; B, mules.

The *Gd* results obtained on 303 clones derived from three hybrids are given in Table 2. Two hundred and ninety-seven of these clones showed either the Gd^H or Gd^D band. Six clones were mixed (Gd^H/Gd^D), two of these were recloned, and 106 out of 107 of these sub-clones showed either Gd^H or Gd^D while one was again mixed (Table 3). In all animals,

续表

动物编号	类型	原料	Gd^D	Gd^H	样品来源
203	驴骡	F	±	+	剑桥
		E	±	+	
217	马骡	E	±	+	剑桥
雄性杂交体					
	马骡（n=1，胚胎）	F	−	+	剑桥
	马骡（n=4）	E	−	+	葡萄牙和剑桥
	驴骡（n=7）	E	+	−	葡萄牙和剑桥
亲代物种					
雄性	驴（n=7）	E	+	−	剑桥
	马（n=2）	E	−	+	剑桥
雌性	驴（n=12）	E	+	−	剑桥
	马（n=8）	E	−	+	剑桥

* 这只动物（编号为 100）是由哈默顿等人在 1969 年报道的。

† 195：怀孕 45 天；196：怀孕 60 天；200：怀孕 35 天；201 公马骡：怀孕 35 天——均通过子宫切开术获得

+ 代表强，± 代表弱，− 代表无；E 代表红细胞，F 代表成纤维细胞。

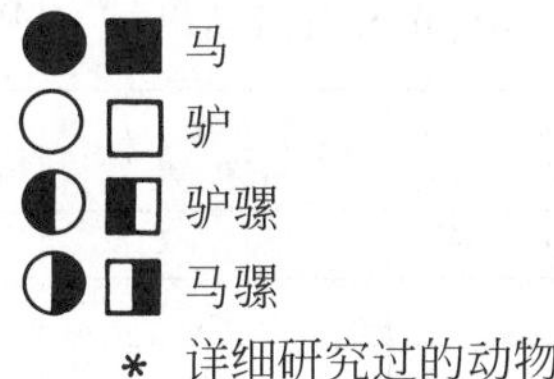

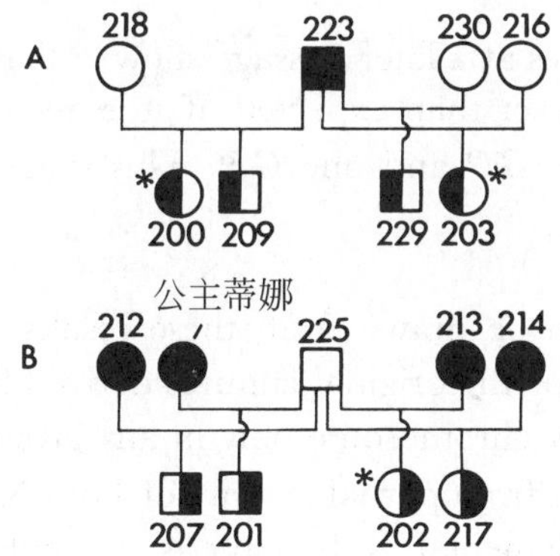

图 1. 所研究的马骡和驴骡的谱系图。图中的编号指的是表 1 中的编号。A 为驴骡；B 为马骡。

由来自 3 只杂交动物的 303 个细胞克隆得到的 *Gd* 类型列于表 2 中。其中有 297 个克隆表现出 Gd^H 或者 Gd^D 条带。有 6 个克隆表现出混合条带（Gd^H/Gd^D），将其中两个再次进行克隆，结果在 107 个亚克隆中，有 106 个表现出 Gd^H 条带或者 Gd^D 条带，剩下的那一个仍然表现为混合条带（表 3）。在所有实验动物中，不论杂交与否，这

irrespective of the cross, the frequency of the two types is significantly different from 50% (P>0.0005). This results from a deficiency of clones showing Gd^D.

Table 2. *Gd* Expression in Clones Derived from Fibroblasts from Three Equine Hybrids

Animal No.	Expt.	No. Gd^D	No. Gd^DGd^H	Total	Proportion Gd^D
200 (Hinny ♀)	VIII	4	–	40	0.100
35 D Foetus	X	6	1	68	0.088
Total		10	1	108	0.093
202 (Mule ♀)	II	5	3	11	0.454
	IV	3	–	20	0.150
	VI	7	–	41	0.171
	XII	14	1	42	0.333
Total		29	4	114	0.254
203 (Hinny ♀)	VII	1	–	3	0.333
	IX	3	–	26	0.115
	XIII	8	1	52	0.154
Total		12	1	81	0.148
Grand total		51	6	303	0.168

Table 3. Expression of *Gd* Locus in Sub-clones from Two Mixed (Gd^DGd^H) Clones from Animal 202

Clone	Expt.	No. Gd^D	No. Gd^DGd^H	Total sub-clones	Proportion Gd^D
C1	XIV	9	1	58	0.155
C8	XV	8	–	49	0.163
Total		17	1	107	0.159

Recloning the two mixed clones at a later passage showed that the proportion of Gd^D sub-clones (0.16) is significantly lower than expected, if it is assumed that these mixed clones originated from two cells, one Gd^D and one Gd^H. This suggests powerful selective forces acting *in vitro*.

A number of conclusions can be drawn from these results. First, the mixed expression seen in the hybrids and some of the original cultures derived from them can be accounted for by the expression of one X chromosome only in any given cell and its descendants, as postulated by Lyon[9,10], and not from partial activity of both X chromosomes as postulated by Grüneberg[11] in his complemental X hypothesis. It could be argued, however, that the minor allele is expressed in less than 10% of the cells, in which case it would not be detected by the methods used here without sub-cloning. It seems unlikely that in a system where if the expression of an allele is equated to 100, all relative activities from 100:0 to 0:100 can reversibly occur for two alleles as suggested by Grüneberg[11], we would expect to find 81.7% of one type, 16.6% of the other and only 1.7% mixed, which are the results obtained here. Second, it can be concluded that active cell selection over a fairly short period of time occurs during *in vitro* fibroblast culture and that this selection

两种类型的发生频率明显偏离 50%（P>0.0005）。这一结果归因于表现为 Gd^D 的克隆数量不足。

表 2. *Gd* 在源于三匹马科动物杂交后代的成纤维细胞克隆中的表达

动物编号	实验	Gd^D 的数量	Gd^DGd^H 的数量	总数	Gd^D 所占比例
200（母驴骡）35 天的胚胎	VIII	4	–	40	0.100
	X	6	1	68	0.088
共计		10	1	108	0.093
202（母马骡）	II	5	3	11	0.454
	IV	3	–	20	0.150
	VI	7	–	41	0.171
	XII	14	1	42	0.333
共计		29	4	114	0.254
203（母驴骡）	VII	1	–	3	0.333
	IX	3	–	26	0.115
	XIII	8	1	52	0.154
共计		12	1	81	0.148
合计		51	6	303	0.168

表 3. *Gd* 在源于 202 号动物两个混合（Gd^DGd^H）克隆的亚克隆中的表达

克隆	实验	Gd^D 的数量	Gd^DGd^H 的数量	亚克隆总数	Gd^D 所占比例
C1	XIV	9	1	58	0.155
C8	XV	8	–	49	0.163
共计		17	1	107	0.159

在后面的传代中再次克隆那两个表现出混合条带的克隆，结果 Gd^D 亚克隆的比例（0.16）明显低于预期，如果假设这些表现出混合条带的克隆是来自两个细胞（一个是 Gd^D，另一个是 Gd^H）的话。这表明有一种在体外发挥作用的强大选择力。

从这些结果中可以推出很多结论。首先，在杂交体和源于它们的一些原始培养物中观察到的混合表达可以用莱昂假说 [9,10] 来解释，即在任意一个给定的细胞及其后代中只表达了一条 X 染色体；而非格吕内贝格在他的 X 染色体补偿假说 [11] 中所提出的混合表达是来自两条 X 染色体的部分活性。然而，有人可能认为少数等位基因的表达发生在不超过 10% 的细胞中，在这种情况下，不采用亚克隆是无法用上述方法检测出来的。似乎不会像格吕内贝格所指出的那样 [11]：在一个体系中，如果一个等位基因的表达量为 100，那么从 100:0 到 0:100 之间的所有相对活性都可以在两个等位基因之间可逆地发生。我们预期能够发现一种类型为 81.7%，另一种类型为 16.6%，而混合类型只有 1.7%。这正是我们在本文中得到的结果。第二个可以得出结论的是：在体外成纤维细胞培养过程中，活性细胞的选择所经历的时间非常短，

would usually seem to be against those cells expressing Gd^D irrespective of whether the X^D is maternal or paternal in origin. This indicates that the maternal cytoplasm does not preferentially activate the maternal X chromosome or effect subsequent cell selection in these hybrids. Third, studies of X-linked loci using electrophoretic methods seem to demonstrate activity of a given locus only when it is expressed in more than 10% of the cell population. Original cultures from animals 202 and 203 both showed a minor fast band (Gd^D) at an early passage; this was subsequently lost during culture. In both these animals the frequency of Gd^D cells as demonstrated by the typing of clones was between 0.1 and 0.3 in different experiments. The original cultures from the foetus (200) never showed a fast Gd^D band; in spite of this, 9.0% of clones showed expression of Gd^D. These results are interesting in the light of the results reported by Sharman and his co-workers[12-14] on X-inactivation in marsupial hybrids in which they demonstrated preferential late replication of the paternal X chromosome and inactivation of the paternal *Gd* and phosphoglycerokinase (*PGK*) loci. Finally, the results obtained here agree with other data showing that the proportion of late replicating X^D ranged from 0.67–1.00 in six female mules and eight female hinnies[6]. If late replication of a given chromosome represents its genetic inactivation, then the autoradiographic results suggest that the proportion of cells expressing Gd^D should vary between 0.33 and zero which fits very closely to the results obtained here. Similar results have recently been reported for four mules by Cohen and Rattazzi[5].

These findings support the single active X hypothesis originally put forward by Lyon[9,10]. They demonstrate clearly activity of only one *Gd* allele in any given cell, which has also been shown for man[15,16]. Unfortunately, the other X-linked enzyme systems *PGK* and α-galactosidase which we have tested have failed as yet to show different mobilities in horse and donkey and so cannot be tested. Furthermore, these data demonstrate *in vitro* cell selection in favour of those cells carrying an active X^H. Examination of uncloned fibroblasts and erythrocytes from these animals also shows predominant expression of Gd^H (Table 1) suggesting that the proportion of cells with an active X^H *in vivo* are reflected in our *in vitro* results and that these are not therefore due to differential plating efficiencies of the two cell types. The one animal showing equal expression of the two types is unavailable for further study. Lyon[9,10] has postulated *in vivo* cell selection to account for variation in expression of sex linked loci in the mouse and man and the demonstration of such strong selective forces *in vitro* due to the activity or inactivity of a single X chromosome lends support to the hypothesis that similar selective forces may be acting *in vivo*. Finally, the close agreement between the autoradiographic data[2,4-6] and the genetic data presented here adds further support to the hypothesis that late replication is the cytological manifestation of genetic inactivity. A final proof of this would be provided by autoradiographic studies on the two clonal types derived here. These studies are at present in progress.

We thank Mrs. G. Anderson, Mrs. E. Cameron and Miss V. Niewczas-Late for technical assistance. This work was supported by a grant to J. L. H. from the Medical Research Council of Canada. Financial support from the Children's Hospital Research Foundation of Winnipeg is also gratefully acknowledged. W. R. A. is supported by the Thoroughbred

并且这种选择似乎总是不利于那些表达了 Gd^{D} 基因的细胞，不管 X^{D} 是来自母本或是父本。这表明：在这些杂交体中，母本细胞质并没有优先激活母本 X 染色体，也没有影响随后的细胞选择。第三，利用电泳法对 X 连锁基因位点的研究似乎说明：只有当一个给定基因位点在超过 10% 的细胞群中被表达时才能用电泳方法给出该位点的活性。来自 202 号和 203 号实验动物的原始培养物在早期阶段都表现出一条较小的快速条带（Gd^{D}）；在随后的培养过程中又消失了。根据克隆类型，这两只动物 Gd^{D} 细胞的发生频率在不同实验中得到的结果介于 0.1~0.3 之间。来自动物胚胎（编号 200）的原始培养物从未表现出快速的 Gd^{D} 条带；尽管如此，其克隆中仍有 9.0% 显示出 Gd^{D} 表达。这些结果在解释沙曼及其同事所报道的关于有袋动物杂交体中 X 染色体失活方面的研究成果 [12-14] 上很有意义，他们发现这些杂交体表现出父本 X 染色体的优先复制延迟以及父本的 *Gd* 和磷酸甘油酸激酶(*PGK*)位点的优先失活。最后，有数据显示：在 6 匹母马骡和 8 匹母驴骡中，X^{D} 复制延迟的比例在 0.67~1.00 之间 [6]，这与本文中得到的结果相符。如果一个给定染色体的复制延迟能反映出它的遗传失活，那么由放射自显影结果就可以给出表达 Gd^{D} 的细胞比例应该在 0~0.33 之间，这与本文中得到的结果非常吻合。最近，科亨和拉塔齐报道的有关 4 匹母马骡的结果 [5] 也与此类似。

上述发现支持了由莱昂最先提出的一条活性 X 染色体的假说 [9,10]。这些结果清楚地表明，在任一给定的细胞中只有一个 *Gd* 等位基因是具有活性的，人类中的情况也是如此 [15,16]。不幸的是，我们在对马和驴的其他 X 连锁酶系——*PGK* 系和 α– 半乳糖苷酶系进行测试时，并没有发现它们的迁移率有何不同，因而无法进行验证。此外，这些数据还表明：体外的细胞选择倾向于那些带有一个活性染色体 X^{H} 的细胞。我们对这些动物中未克隆的成纤维细胞和红细胞进行检测的结果也表明，占主导地位的是 Gd^{H} 基因的表达（表 1），这意味着体内带有一个活性染色体 X^{H} 的细胞比例与我们在体外的测试结果相符，因而在体外得到的结果不能用培养板对这两种细胞类型具有不同的接种效率来解释。进一步研究所需的、能等量表达这两种基因类型的单个动物是很难得到的。莱昂 [9,10] 曾提出，用体内细胞的选择可以解释小鼠和人类性别连锁基因位点在表达上的变化。发现由单一 X 染色体的活性或者失活在体外引起的强烈选择力也支持了如下假设：在体内可能同样存在着类似的选择力。最后，放射自显影数据 [2,4-6] 与本文提供的基因数据之间的严格一致进一步支持了复制延迟是基因失活的一种细胞学表现的假说。支持上述假说的最终证据将通过对源于本实验的两种克隆类型进行放射自显影研究得到。目前这些研究还在进行之中。

感谢安德森夫人、卡梅伦夫人和纽蔡斯 – 雷特小姐在技术上对我们的帮助。本工作是由加拿大医学研究委员会提供给哈默顿的一项经费支持的；温尼伯儿童医院研究基金会在经济上给予我们慷慨援助，谨此一并致谢。艾伦有幸受到了良种马饲

Breeders Association. Collection of material from some of the animals in Portugal was supported by a grant to J. L. H. and R. V. S. from the Wellcome Trust. We thank Dr. M. J. Freire and his veterinary colleagues for Portuguese samples.

(**232**, 312-315; 1971)

J. L. Hamerton, B. J. Richardson and Phyllis A. Gee: Department of Medical Genetics, Children's Hospital, Winnipeg, and Departments of Paediatrics and Anatomy, University of Manitoba.

W. R. Allen and R. V. Short: ARC Unit of Reproductive Physiology and Biochemistry, Animal Research Station, Huntingdon Road, Cambridge, and Department of Veterinary Clinical Studies, School of Veterinary Medicine, Madingley Road, Cambridge.

Received March 29; revised June 7, 1971.

References:

1. Mukherjee, B. B., and Sinha, A. K., *Proc. US Nat. Acad. Sci.*, **51**, 252 (1964).
2. Hamerton, J. L., Giannelli, F., Collins, F., Hallett, J., Fryer, A., McGuire, V. M., and Short, R. V., *Nature*, **222**, 1277 (1969).
3. Mukherjee, B. B., Mukherjee, A. B., and Mukherjee, A. B., *Nature*, **228**, 1321 (1970).
4. Hamerton, J. L., and Giannelli, F., *Nature*, **228**, 1323 (1970).
5. Cohen, M. M., and Rattazzi, M. C., *Proc. US Nat. Acad. Sci.*, **68**, 544 (1971).
6. Giannelli, F., and Hamerton, J. L., *Nature*, **232**, 315 (1971).
7. Cooper, J. E. K., *Texas Rep. Biol. Med.*, **28**, 29 (1970).
8. Trujillo, J. M., Walden, B., O'Niel, P., and Anstall, H. B., *Science*, **148**, 1603 (1965).
9. Lyon, M. F., *Ann. Rev. Genet.*, **2**, 31 (1968).
10. Lyon, M. F., *Phil. Trans. Roy. Soc.*, B, **259**, 41 (1970).
11. Grüneberg, H., *J. Embryol. Exp. Morphol.*, **22**, 145 (1969).
12. Sharman, G. B., *Nature*, **230**, 231 (1971).
13. Richardson, B. J., Czuppon, A. B., and Sharman, G. B., *Nature*, **230**, 154 (1971).
14. Cooper, D. W., Vandeberg, J. L., Sharman, G. B., and Poole, W. E., *Nature*, **230**, 155 (1971).
15. Davidson, R. G., Nitowsky, H. M., and Childs, B., *Proc. US Nat. Acad. Sci.*, **50**, 481 (1963).
16. Linder, D., and Gartler, S. M., *Amer. J. Human Genet.*, **17**, 212 (1965).

养者协会的资助，在葡萄牙搜集动物样本的工作则得益于维康基金会授予哈默顿和肖特的一项经费。感谢弗莱雷博士和他的兽医团队为我们提供了葡萄牙的动物样品。

（刘振明 翻译；梁前进 审稿）

Non-random Late Replication of X Chromosomes in Mules and Hinnies

F. Giannelli and J. L. Hameron

Editor's Note

X chromosome inactivation—where one of the two copies of the X chromosome present in female mammals is silenced—had been proposed by geneticist Mary Lyon's several years earlier. But it was unclear which of the parental X chromosomes lost its function, and if the process was random. Here F. Giannelli and John L. Hamerton use autoradiographs of labelled cultured cells from female mules and hinnies to propose that donkey X chromosomes are more commonly inactivated than horse X chromosomes, regardless of the direction of the cross. This non-random inactivation was subsequently questioned by others who put the finding down to artefact, and today it is thought that X inactivation exists in two forms: random and imprinted.

THE process of DNA replication and genetic activity in X chromosomes of female mules and hinnies has been studied[1-6] without any agreement on the relative frequency with which the two parental X chromosomes (X^D and X^H) become late replicating and genetically inactive. Additional autoradiographic results on six female mules (*E. asinus* ♂ × *E. caballus* ♀) and eight female hinnies (*E. caballus* ♂ × *E. asinus* ♀) are presented here. One of the hinnies (No. 203) was bred in Cambridge[5], three of the mules (Nos. 27, 33, 100), subject of a preliminary communication[2], were obtained from various sources in Great Britain, while the remainder of the hybrids were from Portugal.

Lymphocyte and fibroblast cultures were established by standard methods[7,8]. Mixed lymphocyte cultures, using female donkey and horse blood were set up as controls, for the hybrid cultures. ^{3}H-Tdr (specific activity 5 Ci/mmol, Amersham) in a concentration 0.3–0.5 μCi/ml. was added to the leucocyte cultures 3–4 h before collection and colcemid was added 1.5 h before termination of the culture. Fibroblast cultures were labelled with 0.3 μCi/ml. 6 h before termination and colcemid was added 3 h later. Slides were prepared by air drying, stained lightly with lacto-acetic orcein and suitably spread metaphase plates were photographed. Autoradiographs were prepared by standard methods[2,9]. The cells which had previously been photographed were relocated and examined for a differentially labelled X chromosome. (This was defined as a medium sized chromosome which had at least twice as many autoradiographic grains as any other chromosome of similar size. This definition was adoped because the donkey has a submetacentric autosome much larger than the X and with a large heterochromatic paracentric region which is often densely labelled in late S.) The morphology of the differentially labelled X chromosome (DLX) was then determined by examination of the photographs taken before autoradiography. The two X chromosomes in the female hybrids are quite different, one (X^D) being

马骡和驴骡中X染色体的非随机复制延迟

詹内利，哈默顿

编者按

遗传学家玛丽·莱昂在几年之前就已经提出了X染色体失活假说——雌性哺乳动物体内两条X染色体中有一条发生沉默。但当时尚不清楚失活的是哪一条亲本X染色体，也不知道失活过程是否是随机的。在本文中，詹内利和约翰·哈默顿根据母马骡和母驴骡标记细胞培养物的放射自显影照片提出：驴X染色体要比马X染色体更经常发生失活的情况，而这与杂交的方向并无关系。这种非随机性失活的说法随即遭到了其他人的质疑，他们认为这一发现是人为因素造成的。而现在人们的看法是：X染色体的失活存在两种形式——随机性的和印记性的。

目前已经有不少人研究过母马骡和母驴骡X染色体中的DNA复制过程和遗传活性[1-6]，但在两条亲本X染色体（X^D 和 X^H）发生复制延迟和遗传失活的相对频率上还没有得到统一的结果。我们在本文中将会补充介绍一些有关6匹母马骡（公驴×母马）和8匹母驴骡（公马×母驴）的放射自显影结果。其中一匹驴骡（编号为203）是在剑桥大学饲养的[5]；3匹马骡（编号分别为27、33和100）是我们在先前的一篇文章中提到过的[2]，它们源自英国的不同地方；而其余的杂交动物则来自葡萄牙。

淋巴细胞和成纤维细胞的培养物通过标准方法[7,8]制备。杂交体培养物的对照样品是用母驴和母马的血液制备的混合淋巴细胞培养物。在采样前3 h~4 h，将浓度为0.3 μCi/ml ~ 0.5 μCi/ml的氚–胸腺嘧啶脱氧核苷（^{3}H–Tdr）（比活度为5 Ci/mmol，安玛西亚公司产品）加入白细胞培养物中；在终止培养前1.5 h加入秋水仙酰胺。成纤维细胞培养物在终止培养前6 h用0.3 μCi/ml的^{3}H–Tdr进行标记，3 h后再加入秋水仙酰胺。采用空气干燥的方式制备载玻片样本，并用乳酸–醋酸地衣红进行轻微染色，然后对那些载有适当分散的分裂中期细胞的载玻片进行拍照。按照标准方法[2,9]制作放射自显影图。重新定位并检查先前被拍过照的细胞以搜索一条差异标记的X染色体。（对这个“差异”的界定是：中型大小染色体的放射自显影颗粒计数至少是任意其他相似大小染色体的2倍。采用上述定义的原因是：驴含有的近中着丝粒常染色体要比X染色体大很多，并且这个常染色体上有一大片异染色质着丝粒旁区域，在S期后期经常会被标记得很致密。）然后通过查看在放射自显影之前拍摄的照片来确定这个差异标记的X染色体（DLX）的形态。在雌性杂交体中的两条X染色体明显不相同：一条（X^D）是近中着丝粒的，而另一条（X^H）则是

submetacentric, while the other (X^H) is metacentric[1,2] (Fig. 1).

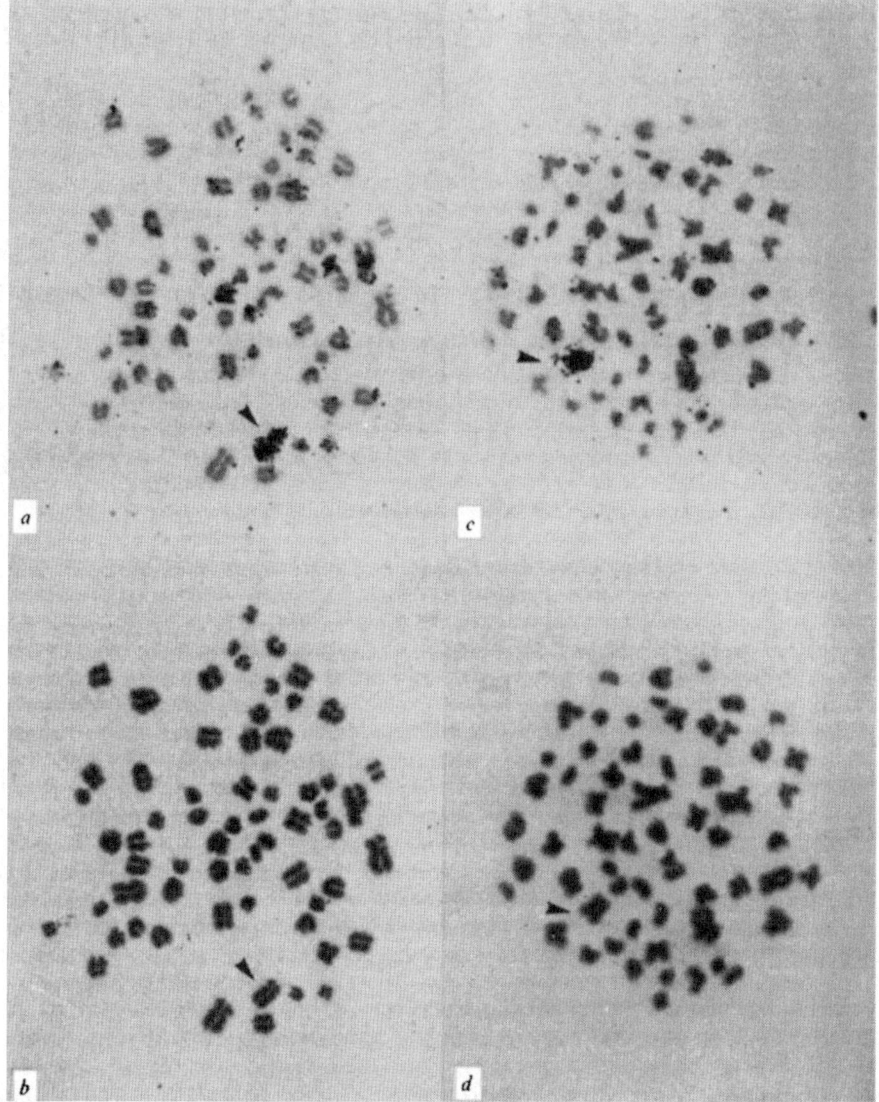

Fig. 1. Two metaphase plates from the same hybrid, the first showing (*a* and *b*) a submetacentric chromosome (arrowed) and the second (*c* and *d*) a metacentric chromosome (arrowed) differentially labelled. These chromosomes presumably represent, in order, the donkey and horse-derived X chromosomes.

Twenty-one suitable metaphase plates from each of two hinnies and one mule (Nos. 100, 188, 203) were selected for quantitative autoradiographic studies, and karyotypes prepared. The total autoradiographic grains for each cell and for each of the five longest medium sized submetacentric and metacentric chromosomes were then counted. Grain counts were also made on samples of cells from the mixed leucocyte cultures, selected either at random or for the presence of a differentially labelled X chromosome.

中央着丝粒的[1,2]（图 1）。

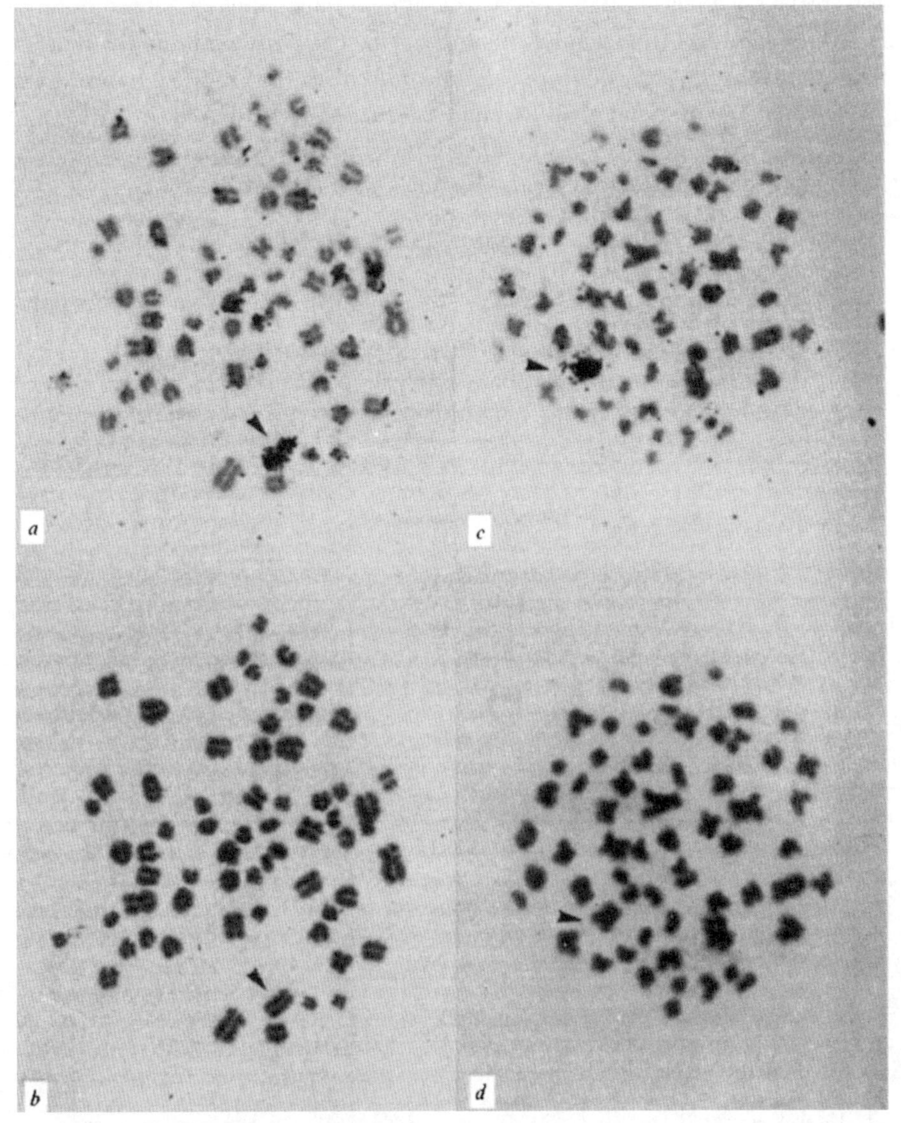

图 1. 两块来自同一杂交体的分裂中期细胞载玻片。第一块中（a 和 b）显示出一条近中着丝粒的染色体（见箭头所指）；第二块中（c 和 d）显示出一条差异标记的中央着丝粒染色体。这两条染色体可能依次代表了源自驴和马的 X 染色体。

将来源于两匹驴骡和一匹马骡（编号分别为 100、188 和 203）中每一个个体的细胞培养物制成载玻片样本，从中选取 21 块适当分散的分裂中期细胞载玻片以进行定量的放射自显影研究以及核型的制备。然后清点每个细胞中的放射自显影颗粒总数以及中型大小近中着丝粒染色体和中央着丝粒染色体中 5 条最长的染色体的放射自显影颗粒总数。此外还清点了来自混合白细胞培养物的细胞样本的放射自显影颗粒数目，选择样本的方式或者是随机性的或者是针对那些存在差异标记 X 染色体的样本。

The blood and skin cultures from the fourteen animals studied contained varying proportions of labelled cells and of cells with a differentially labelled X chromosome (Table 1). This did not influence the relative frequency of cells with a differentially labelled X^D or X^H. The relative frequencies differed significantly from 0.5 in every hybrid from which enough cells could be analysed and in each case there was an excess of DLX^D (Table 4). In the mixed leucocyte cultures, on the other hand, the number of horse cells which has a DLX was greater than the number of donkey cells with a DLX (Table 2).

Table 1. Autoradiographic Date from Six Mules and Eight Hinnies

Animal No.	Tissue cultured	No. of cells scored	Total No. of labelled cells	No. of cells with a DLX		No. of cells with a "very hot" DLX‡	
				Total	No. DLX^D	Total	No. DLX^D
Mules							
27	Skin	70	65	14	13	8	8
33	Blood*	1,000	35	27	24	19	18
	Skin	105	77	45	43	33	33
100	Blood*	755	453	68	60	55	50
	Blood†	15	9	3	3	2	2
	Skin	45	40	8	8	8	8
143	Blood*	12	2	1	1	1	1
	Blood†	18	16	7	7	5	5
144	Blood*	22	13	8	7	7	6
	Blood†	142	124	33	28	28	24
145	Blood*	12	7	1	1	0	0
	Blood†	17	14	2	2	2	2
Totals	Blood*	1,801	510	105	93	82	75
	Blood†	192	163	45	40	37	33
	Skin	220	182	67	64	49	49
Hinnies							
138	Blood*	9	5	2	2	1	1
	Blood†	3	2	0	0	0	0
139	Blood*	27	12	6	6	5	5
187	Blood*	172	155	28	22	21	18
	Blood†	17	16	0	0	0	0
188	Blood*	168	120	13	11	8	8
	Blood†	23	22	0	0	0	0
189	Blood*	326	133	18	12	14	10
190	Blood*	152	127	12	8	9	6
	Blood†	37	32	1	1	0	0
191	Blood*	159	135	5	5	3	3
	Blood†	32	30	4	3	3	2
203	Blood*	165	105	23	18	17	13
	Blood†	131	121	211	17	19	16
Totals	Blood*	1,178	792	107	84	78	64
	Blood†	243	223	26	21	22	18

* 3 h incubation with ^{3}H-Tdr.

† 4 h incubation with ^{3}H-Tdr.

‡ DLX with at least 2.2 times the grains counted over any other chromosome of similar size.

来自所有 14 匹马骡或驴骡的血液和皮肤培养物各自含有不同比例的标记细胞以及含有不同比例的含差异标记 X 染色体的细胞（表 1）。这并不影响含一个差异标记染色体 X^D 或者 X^H 的细胞的相对频率。在每一只有足够多细胞可供分析的杂交动物中都出现了相对频率明显偏离 0.5 的情况，并且每次都是 DLX^D 过量（表 4）。而另一方面，在混合的白细胞培养物中，含有一个 DLX 的马细胞数量要多于含有一个 DLX 的驴细胞数量（表 2）。

表 1. 6 匹马骡和 8 匹驴骡的放射自显影数据

动物编号	培养的组织	记录的细胞数量	标记细胞的总数	含一个 DLX 染色体的细胞数量		含一个“过热”DLX 染色体‡的细胞数量	
				总数	含 DLX^D 的细胞数量	总数	含 DLX^D 的细胞数量
马骡							
27	皮肤	70	65	14	13	8	8
33	血液 *	1,000	35	27	24	19	18
	皮肤	105	77	45	43	33	33
100	血液 *	755	453	68	60	55	50
	血液 †	15	9	3	3	2	2
	皮肤	45	40	8	8	8	8
143	血液 *	12	2	1	1	1	1
	血液 †	18	16	7	7	5	5
144	血液 *	22	13	8	7	7	6
	血液 †	142	124	33	28	28	24
145	血液 *	12	7	1	1	0	0
	血液 †	17	14	2	2	2	2
总计	血液 *	1,801	510	105	93	82	75
	血液 †	192	163	45	40	37	33
	皮肤	220	182	67	64	49	49
驴骡							
138	血液 *	9	5	2	2	1	1
	血液 †	3	2	0	0	0	0
139	血液 *	27	12	6	6	5	5
187	血液 *	172	155	28	22	21	18
	血液 †	17	16	0	0	0	0
188	血液 *	168	120	13	11	8	8
	血液 †	23	22	0	0	0	0
189	血液 *	326	133	18	12	14	10
190	血液 *	152	127	12	8	9	6
	血液 †	37	32	1	1	0	0
191	血液 *	159	135	5	5	3	3
	血液 †	32	30	4	3	3	2
203	血液 *	165	105	23	18	17	13
	血液 †	131	121	211	17	19	16
总计	血液 *	1,178	792	107	84	78	64
	血液 †	243	223	26	21	22	18

* 用 ^{3}H–Tdr 孵育 3 h。

† 用 ^{3}H–Tdr 孵育 4 h。

‡ 显影颗粒数量至少是任何其他相似大小染色体的 2.2 倍的 DLX 染色体。

Table 2. Autoradiographic Data from Mixed Lymphocyte Cultures

Cell type	No. of cells scored	No. of labelled cells	Cells with a DLX		Difference in proportions	s.e.	P
			No.	Proportion			
Horse	158	144	33	0.2291	0.1205	0.0515	0.05>P>0.025
Donkey	96	92	10	0.1086			

The relative frequencies of DLX^D and DLX^H in cultures from individual mules or hinnies did not differ significantly from the overall frequencies for their own group (Tables 1 and 3). The hinnies behave in a similar fashion to the mules, but are possibly less homogeneous and seem to show a slightly higher frequency of DLX^H (Table 3).

Table 3. Observed Proportion of Cells with an X^D Differentially Labelled and Their 95% Confidence Limits

Hybrids		Tissue	Proportion DLX^D	95% Confidence limits
Type	No.			
Mules	27	S	0.928	0.661–0.998
	33	B	0.889	0.708–0.976
		S	0.955	0.848–0.995
	100	B	0.887	0.790–0.950
		S	1.00	0.631–1.00
	143	B	1.00	0.631–1.00
	144	B	0.854	0.708–0.944
	145	B	1.00	0.292–1.00
	Total	B	0.887	0.824–0.932
		S	0.955	0.875–0.991
Hinnies	138	B	1.00	0.158–1.00
	139	B	1.00	0.541–1.00
	187	B	0.786	0.590–0.917
	188	B	0.846	0.545–0.981
	189	B	0.667	0.410–0.867
	190	B	0.692	0.386–0.909
	191	B	0.889	0.517–0.997
	203	B	0.795	0.647–0.902
	Total	B	0.789	0.747–0.885
Total		B	0.841	0.801–0.853

S, Skin fibroblasts; B, lymphocytes.

In the sample of twenty cells selected at random from a mule and each of two hinnies (Table 4) the grain counts over the most densely labelled medium sized submetacentric and metacentric chromosomes showed the same regression line on the total cell grain counts, but the former chromosome had clearly higher grain counts than the latter. If these chromosomes are taken to represent respectively the X^D and X^H chromosome, these data seem to agree with the qualitative observation that the X^D chromosome is more frequently differentially labelled. Furthermore, the proportion of X chromosome grains counted over X^D seems to be very similar in the mule and the two hinnies.

表 2. 混合淋巴细胞培养物的放射自显影数据

细胞类型	记录的细胞数量	标记细胞的数量	含有一个 DLX 的细胞		比例之差	标准误差	概率
			数量	比例			
马	158	144	33	0.2291	0.1205	0.0515	0.05>P>0.025
驴	96	92	10	0.1086			

DLX^D 和 DLX^H 在每匹马骡或者驴骡细胞培养物中的相对频率与其在马骡或者驴骡所属群体中的总体频率并没有明显的差异（表 1 和表 3）。驴骡的表现与马骡很类似，但在均一性上可能要稍差一些，而且驴骡的 DLX^H 出现频率似乎比马骡的略高（表 3）。

表 3. 含有一个差异标记的 X^D 染色体的细胞比例与它们的 95% 置信区间

杂交体		组织	DLX^D 的比例	95% 置信区间
类型	编号			
马骡	27	S	0.928	0.661 ~ 0.998
	33	B	0.889	0.708 ~ 0.976
		S	0.955	0.848 ~ 0.995
	100	B	0.887	0.790 ~ 0.950
		S	1.00	0.631 ~ 1.00
	143	B	1.00	0.631 ~ 1.00
	144	B	0.854	0.708 ~ 0.944
	145	B	1.00	0.292 ~ 1.00
	总计	B	0.887	0.824 ~ 0.932
		S	0.955	0.875 ~ 0.991
驴骡	138	B	1.00	0.158 ~ 1.00
	139	B	1.00	0.541 ~ 1.00
	187	B	0.786	0.590 ~ 0.917
	188	B	0.846	0.545 ~ 0.981
	189	B	0.667	0.410 ~ 0.867
	190	B	0.692	0.386 ~ 0.909
	191	B	0.889	0.517 ~ 0.997
	203	B	0.795	0.647 ~ 0.902
	总计	B	0.789	0.747 ~ 0.885
总计		B	0.841	0.801 ~ 0.853

S：皮肤成纤维细胞；B：淋巴细胞。

在一匹马骡和两匹驴骡中各随机抽取 20 个细胞样本（表 4），对标记最致密的中型大小近中着丝粒染色体和中央着丝粒染色体的放射自显影颗粒进行计数，在用计数结果对全部细胞显影颗粒计数进行回归分析时得到了相同的回归线，但前者染色体的显影颗粒计数明显高于后者。如果认为这些染色体分别代表了 X^D 和 X^H 的话，那么上述数据似乎与定性观察结果相吻合，即 X^D 染色体在差异标记中出现得更频繁。此外，这匹马骡和这两匹驴骡的 X^D 染色体显影颗粒所占的比例似乎非常接近。

Table 4. Comparisons between the Grain Counts over the Most Densely Labelled X^D and X^H-like Chromosomes) $?X^D$, $?X^H$) in a Random Sample of Cells from Mule 100 and the Hinnies 188 and 203

Hybrid No.	No. of cells analysed	Comparison of mean grain counts over $?X^D$ and $?X^H$	Comparison of the regression coefficients of the logarithm of $?X^D$ and $?X^H$ grain counts on TGC*	Comparison of the contribution (D) of $?X^D$ to the grains counted over the presumptive X chromosomes in the mule and the two hinnies
100	21	$\bar{Y}_1 = 12.85$ $\bar{Y}_2 = 9.00$ $\bar{Y}_1 - \bar{Y}_2 = 3.85$ s.e. $(\bar{Y}_1 - \bar{Y}_2) = 1.35$ $P<0.02$	$b_1 = 0.000956$ $b_2 = 0.001864$ $b_1 - b_2 = -0.000908$ s.e.$(b_1 - b_2) = 0.000869$ $P>0.3$	$D_m=0.5882$
188	20	$\bar{Y}_1 = 8.40$ $\bar{Y}_2 = 6.10$ $\bar{Y}_1 - \bar{Y}_2 = 2.30$ s.e.$(\bar{Y}_1 - \bar{Y}_2) = 0.964$ $P<0.05$	$b_1 = 0.00209$ $b_2 = 0.00199$ $b_1 - b_2 = 0.00010$ s.e.$(b_1 - b_2) = 0.00077$ $P>0.9$	$D_h=0.5892$ $D_m - D_h = -0.0010$
203	18	$\bar{Y}_1 = 15.61$ $\bar{Y}_2 = 10.61$ $\bar{Y}_1 - \bar{Y}_2 = 5.00$ s.e. $(\bar{Y}_1 - \bar{Y}_2) = 2.25$ $P<0.05$	$b_1 = 0.00102$ $b_2 = 0.00205$ $b_1 - b_2 = -0.00103$ s.e.$(b_1 - b_2) = 0.000613$ $P>0.1$	s.e.$(D_m - D_h)=0.023$ $P>0.8$

* Scatter diagrams have shown that the regressions of the logarithms of the ?X chromosomes grain counts on the total complement grain counts (TGC) are linear.

1, $?X^D$; 2, X^H; m, mule; h, hinny.

Finally, in the mixed leucocyte cultures the regression of the grains counted over the differentially labelled X chromosomes, or over the most densely labelled X-like chromosomes, on the total cell grain counts was very similar in the horse and donkey cells. The covariance analysis (Table 5) showed that the grain counts over the X^H and X^D were very similar after allowance had been made for the labelling over the rest of the complement.

Table 5. Comparison by Covariance Analysis of the Grain Counts over Differentially Labelled X Chromosomes (A) or the Most Densely Labelled X-like Chromosomes (B) in Horse–donkey Mixed Blood Cultures

Analysis	A chromosome		B chromosome	
	X^D	X^H	$?X^D$	$?X^H$
No. cells analysed	10	33	20	20
Mean grain counts	$\bar{Y}_1 = 30.21$	$\bar{Y}_2 = 26.00$	$\bar{Y}_1 = 25.65$	$\bar{Y}_2 = 19.95$
Adjusted mean	$\bar{Y}_1^1 = 29.55$	$\bar{Y}_2^1 = 28.17$	$\bar{Y}_1^1 = 22.90$	$\bar{Y}_2^1 = 22.70$
Difference	$\bar{Y}_1^1 - \bar{Y}_2^1 = 1.38$		$\bar{Y}_1^1 - \bar{Y}_2^1 = 0.20$	
s.e. of difference	1.88		2.55	
Probability	$0.5>P>0.4$		$0.95>P>0.90$	

Scatter diagrams were made before these analyses. These have shown a linear relationship between the presumptive X chromosome grain counts and the counts over the rest of the complement. The variance of *Y* was fairly constant over the range of total grain counts used and this was similar in the two samples compared. The regression coefficients of the individual samples used in this comparison were significantly different from zero, but did not differ significantly from each other.

These results show that in both crosses irrespective of the direction of the cross, X^D is

表 4. 在 100 号马骡以及 188 号和 203 号驴骡的细胞随机抽样中标记最致密的类 X^D 和类 X^H 染色体（$?X^D$ 和 $?X^H$）之间的颗粒计数对比

杂交动物编号	被分析细胞的数量	对比 $?X^D$ 和 $?X^H$ 的平均颗粒计数	对比 $?X^D$ 和 $?X^H$ 的颗粒计数 log 值对 TGC* 的回归系数	对比 $?X^D$ 在这匹马骡和这两匹驴骡中类 X 染色体的颗粒计数上的贡献（D）
100	21	$\overline{Y}_1 = 12.85$ $\overline{Y}_2 = 9.00$ $\overline{Y}_1 - \overline{Y}_2 = 3.85$ 标准误差 = 1.35 $P<0.02$	$b_1 = 0.000956$ $b_2 = 0.001864$ $b_1 – b_2 = –0.000908$ 标准误差 = 0.000869 $P>0.3$	$D_m=0.5882$ $D_h=0.5892$ $D_m–D_h=–0.0010$ 标准误差=0.023 $P > 0.8$
188	20	$\overline{Y}_1 = 8.40$ $\overline{Y}_2 = 6.10$ $\overline{Y}_1 - \overline{Y}_2 = 2.30$ 标准误差 = 0.964 $P<0.05$	$b_1 = 0.00209$ $b_2 = 0.00199$ $b_1 – b_2 = 0.00010$ 标准误差 = 0.00077 $P>0.9$	
203	18	$\overline{Y}_1 = 15.61$ $\overline{Y}_2 = 10.61$ $\overline{Y}_1 - \overline{Y}_2 = 5.00$ 标准误差 = 2.25 $P<0.05$	$b_1 = 0.00102$ $b_2 = 0.00205$ $b_1 – b_2 = –0.00103$ 标准误差 = 0.000613 $P>0.1$	

* 散点图显示 ?X 染色体颗粒计数的 log 值对全部标记物颗粒计数（TGC）的回归是呈线性的。
1 代表 $?X^D$；2 代表 X^H；m 代表马骡；h 代表驴骡。

最后，在混合白细胞培养物中，用差异标记的 X 染色体或者标记最致密的类 X 染色体的颗粒计数对全部细胞显影颗粒计数进行回归分析的结果显示：马的细胞与驴的细胞非常相近。在考虑了剩余标记物对标记的影响后，协方差分析结果（表 5）显示染色体 X^H 和 X^D 的显影颗粒计数非常接近。

表 5. 马 – 驴混合血液培养物中差异标记 X 染色体（A）和标记最致密的类 X 染色体（B）的颗粒计数协方差分析结果对比

分析结果	A 染色体		B 染色体	
	X^D	X^H	$?X^D$	$?X^H$
被分析细胞的数量	10	33	20	20
平均颗粒计数	$\overline{Y}_1 = 30.21$	$\overline{Y}_2 = 26.00$	$\overline{Y}_1 = 25.65$	$\overline{Y}_2 = 19.95$
校正后的平均计数	$\overline{Y}_1^1 = 29.55$	$\overline{Y}_2^1 = 28.17$	$\overline{Y}_1^1 = 22.90$	$\overline{Y}_2^1 = 22.70$
偏差	$\overline{Y}_1^1 - \overline{Y}_2^1 = 1.38$		$\overline{Y}_1^1 - \overline{Y}_2^1 = 0.20$	
标准误差	1.88		2.55	
概率	$0.5>P>0.4$		$0.95>P>0.90$	

在进行这些分析之前先制作散点图。散点图表明了类 X 染色体的颗粒计数与剩余标记物颗粒计数之间的线性关系。方差 Y 在所用的全部颗粒计数范围内是相当稳定的，并且这种情况在用于对比的两个样品中都是类似的。在上述对比分析中用到的单个样品的回归系数显然不是 0，但各个回归系数之间并没有明显的差异。

上述结果表明：在两种杂交体中，不论杂交的方向如何，X^D 在差异标记中的出

consistently more often differentially labelled than X^H. This differs from the findings of Mukherjee and his co-workers[1,3] but is in broad agreement with the observation of Cohen and Rattazzi[6].

Mukherjee and Sinha[1] used both longer labelling and colchicine times for blood cultures, while for skin cultures Mukherjee *et al.*[3] used a shorter labelling time than ours. Their results, however, differ in the same direction in both culture systems from those reported here; this suggests that the discrepancy cannot be explained by a variation in labelling procedure.

In the present study the DLX was defined as the chromosome of compatible morphology which had at least twice as many grains as any other chromosome of a similar size. This is in agreement with the illustration of Mukherjee and Sinha[1] who did not discuss their criteria in detail, but is less strict than that used in their later article[3] in which the DLX is defined as the chromosome with a grain count of at least 50% the total grain count over the whole complement.

In our opinion, this definition is too strict because of the difficulty of counting fifty grains or more over an X-like chromosome, which means that all cells with a total grain count of 100 or more over the remainder of the complement would have to be rejected. To see whether a more strict definition of DLX could affect the results presented here, however, all cells in which the grain counts over the DLX were not greatly in excess of twice that over any chromosome of similar size were rejected, leaving only those cells with a "very hot" DLX available for analysis. This procedure did not alter the ratio of DLX^H to DLX^D (Table 1). It seems reasonable to conclude from this that the way in which the DLX is defined is unlikely to account for the differences observed. Could any peculiarity or intrinsic difference in the labelling of X^D and X^H, not associated with the process of facultative heterochromatinization of the X chromosome, for example, a higher rate of DNA synthesis in X^D than in X^H at some stage of late S, be responsible for the difference which we have observed between the frequency of differentially labelled X^D and X^H in the hybrids' cultures?

Data from the control cultures suggest that this is not so. The mixed lymphocyte cultures showed that the horse lymphocytes have a G_2 similar to that of the donkey and have a differentially labelled X chromosome at least as frequently. Admittedly a mixed lymphocyte culture is not a perfect control for cultures from hybrids in which each chromosome must be compared with the haploid sets from both parental species. But covariance analysis of grain counts over the DLX in the mixed lymphocyte cultures shows that, when account is taken of the labelling over the rest of the complement, DLX^D and DLX^H label in a similar fashion during the part of S studied. This is confirmed by the similarity of the regression of grain counts over the presumptive X^D and X^H on total complement grain counts

现频率始终比 X^H 高。这与慕克吉及其同事的发现[1,3]不相符，但与科恩和拉塔齐的观察结果[6]在很大程度上一致。

在对血液培养物进行标记和用秋水仙碱处理时，慕克吉和辛哈[1]用的时间比我们用的时间长；但在皮肤培养物中，慕克吉等人[3]在标记时用的时间比我们的短。然而，他们在这两个培养系统上得到的结果与本文中报道的结果之间存在着同一方向上的偏差；这表明结果上的这种差异并不能用标记过程的不一致来解释。

在本研究中，DLX 被定义为符合下述形态特征的染色体：显影的颗粒计数是具有相似大小的任何其他染色体的至少 2 倍。这与慕克吉和辛哈的描述[1]一致，他们没有对自己的这个标准进行详细的论述。但这个标准不及他们在后续文章[3]中所用的标准那么严格。在后面的文章中，他们把 DLX 定义为颗粒计数占整个标记物全部颗粒计数的至少 50% 的染色体。

我们认为，这种定义过于严格，因为在一个类 X 染色体上清点 50 个或者更多的颗粒数目是有一定困难的，这意味着我们将不得不放弃那些剩余标记物的总颗粒计数达到 100 或 100 以上的所有细胞。不管怎样，为了了解对 DLX 染色体的一个更严格的定义是否会对本文报道的研究结果产生影响，我们将不考虑那些在 DLX 染色体上的颗粒计数没有远远大于任何相似大小染色体上的颗粒计数 2 倍的所有细胞，那么就只剩下那些含有一个“过热”DLX 染色体的细胞可供分析了。这种研究方式并没有改变 DLX^H 与 DLX^D 的比值（表 1）。由此似乎可以很合理地得到如下结论：观察到的差异不可能用定义 DLX 染色体的方式不同来解释。标记染色体 X^D 和 X^H 时存在的独特性或者固有差异——这种独特性或者固有差异与 X 染色体的功能异染色质化过程无关，例如：在 S 期后期的某一阶段，X^D 染色体中的 DNA 合成速率比 X^H 染色体的要高——能否解释我们在杂交动物培养物中观察到的差异标记 X^D 和 X^H 染色体的出现频率之间的差异？

从对照培养物中得到的数据表明答案是否定的。混合淋巴细胞培养物的结果显示：马的淋巴细胞具有与驴的淋巴细胞相似的 G_2 期，并且至少经常会含有一条差异标记的 X 染色体。诚然，混合淋巴细胞培养物并不是杂交体培养物的最佳对照物，因为我们需要将杂交体中的每一条染色体与来自两个亲本物种的单倍体组进行对比。但是对混合淋巴细胞培养物中 DLX 染色体的颗粒计数的协方差分析结果显示：在考虑了剩余标记物对标记的影响后，染色体 DLX^D 和 DLX^H 在 S 期的部分阶段表现出相似的标记方式。这一观点得到了以下事实的证实：在我们用于定量研究的 3 匹杂交骡子上，对类 X^D 和类 X^H 染色体的颗粒计数进行基于全部标记物颗粒

(TGC) in the three animals studied quantitatively (Table 4). Furthermore, an analysis of lymphocyte cultures from the hybrids shows that with an increase in labelling time the proportion of labelled cells increases and that of cells with a DLX decreases, while the relative frequency of DLX^D and DLX^H remains the same (Tables 1 and 6).

Table 6. Analysis of the Tendency for the Proportion of DLX^D to Vary According to the Proportion of Labelled Cells in Blood Cultures of Mules and Hinnies

Type of hybrid	Regression coefficient	SS due to regression	χ^2	*P*
Mules	−0.5341	352.4	0.429	>0.5
Hinnies	−1.033	1,531	1.86	>0.1
All*	−1.189	3,777	4.6	<0.05

This analysis was conducted using the angular transformation method[11].
* Since the proportion of DLX^D in the hinnies is lower than that of mules and their cell samples contribute chiefly to the class with a relatively high proportion of labelled cells the data regarding both types of hybrid together are not very meaningful.

Finally, can these results be explained on the basis of an error in the identification of the X^D due to the presence of a morphologically similar late replicating autosome which mimics the behaviour of the DLX? Our definition of the DLX precludes this explanation unless the DLX^H had less than half the grains counted over this autosome much more often than the DLX^D. This could occur either if the DLX^H usually had lower grain counts than the DLX^D or if it was in the differentially labelled state less often. The data presented do not fit the first of these alternatives so that it seems reasonable to accept the second.

A further possibility which might account for our results is biased selection of cells or chromosome identification. To avoid this, the analysis of autoradiographs was conducted as independently as possible of the morphological chromosome analysis (a chromosome was classified as differentially labelled before analysis of the unlabelled photographs, and karyotypes were made by an individual who was unaware of the autoradiographic results). It is difficult, however, completely to exclude bias in an analysis of this type. The most serious cause of observer bias is the *a priori* expectations. In the study described here our intention was to investigate the relationship between differential labelling and genetic inactivity of one X chromosome, and it was expected that a 1:1 relationship would be found between the DLX^H and DLX^D in accordance with earlier findings[1]. Preliminary observations on differential labelling in cells with three mules[2] showed this was not so and it was considered a tentative hypothesis that the reciprocal cross might show the reverse, namely a greater frequency of DLX^H than DLX^D. On both counts therefore the results reported here are contrary to *a priori* expectations and lead to the conclusion that X^D is significantly more often differentially labelled than X^H in both mules and hinnies.

These results must now be considered in relation to the Lyon hypothesis (LH). In contrast to the suggestion of others[1,3] the mule may not provide a simple cytological test of random inactivation for reasons which have already been discussed[4]. These hybrids are, however, useful in studying the relationship between differential labelling of one X chromosome

计数（TGC）的回归分析，所得的结果是相似的（表 4）。此外，关于杂交体的淋巴细胞培养物的一项分析结果显示：随着标记时间的增长，被标记细胞的比例增加，而含 DLX 染色体的细胞比例却降低，尽管 DLX^D 和 DLX^H 的相对频率保持不变（表 1 和表 6）。

表 6. 马骡和驴骡血液培养物中 DLX^D 所占比例随标记细胞所占比例的变化而变化的趋势分析

杂交动物类型	回归系数	回归的离均差平方和	χ^2	P
马骡	–0.5341	352.4	0.429	>0.5
驴骡	–1.033	1,531	1.86	>0.1
所有 *	–1.189	3,777	4.6	<0.05

这项分析是采用角变换法进行的 [11]。

* 因为驴骡中 DLX^D 染色体的比例要低于马骡，并且它们的细胞样品主要是对标记细胞所占比例比较高的类别有贡献，所以合并两种杂交类型的数据并没有太大的意义。

最后，能否用在识别 X^D 染色体时的一个错误来解释这些结果呢？出现错误的原因是：存在形态上类似且延迟复制的常染色体，它可以表现出类似于 DLX 染色体的行为。我们对 DLX 染色体的定义即可排除上述解释，除非 DLX^H 染色体出现颗粒计数少于这种常染色体的一半的情况总是多于 DLX^D 染色体。如果 DLX^H 染色体的颗粒计数通常比 DLX^D 的低，或者 DLX^H 染色体在差异标记中的出现频率低，那么上述情况就有可能出现。本文提供的数据并不符合两种假设中的第一种，因而第二种似乎可以被合理地接受。

还有一种可能性，即细胞或者染色体识别上的偏差性选择，或许可以解释我们的分析结果。为避免发生这种情况，我们在进行放射自显影分析的时候尽可能做到不依赖于染色体的形态学分析（先把一个染色体归入差异标记类，然后再对未标示的照片进行分析；并让一位不知道放射自显影分析结果的研究人员来确定核型）。然而，要想在这类分析中完全不出现偏差是十分困难的。观察者产生偏差的最重要的原因是先验的预期。我们进行这项研究的目的是检查染色体的差异标记与一条 X 染色体在遗传性状上表达失活之间的联系，根据早先的报道 [1]，我们预计染色体 DLX^D 和 DLX^H 之间应存在 1:1 的关系。在 3 匹马骡的细胞中观察差异标记的初步结果 [2] 显示情况并非如此，于是我们便提出了一个尝试性假说，认为反交也许会表现出结果上的扭转，也就是说染色体 DLX^H 的出现频率要比染色体 DLX^D 的高。因此，在两者的计数上，本文报道的结果与先前的预期是相反的，由本文中的结果可以推出：在马骡和驴骡中，染色体 X^D 在差异标记中的出现频率显然高于染色体 X^H。

现在必须认为这些结果影响到了莱昂假说。与其他人的结论 [1,3] 相反，由于先前已经讨论过的原因 [4]，由马骡给出的简单细胞学检测结果可能说明不了随机失活。然而，这些杂交体在研究单一 X 染色体的差异标记与遗传失活之间的联系方面

and genetic inactivation. Data on glucose-6-phosphate dehydrogenase (*Gd*)[5] show that the frequency with which Gd^{D} is expressed in fibroblast clones derived from three hybrids is in almost exact inverse proportion to the frequency with which X^{D} is observed to be differentially labelled in both lymphocytes and fibroblasts. Furthermore, qualitative data on expression of Gd^{D} show that, with the exception of one animal, it is expressed far more weakly than Gd^{H} in all the hybrids studied[10]. Cohen and Rattazzi[6] have reported a close correlation between the frequency of DLX^{D} and the expression of Gd^{D} in four further female mules. In each animal the X^{D} was more frequently late replicating and, correlated with this, Gd^{D} was the minor component. Hook and Brustman[10] have studied *Gd* expression in organs from 37 female mules. They found that in most tissues (blood, pancreas, brain, cervical cord, kidney and parotid gland) Gd^{H} is preferentially expressed. In thyroid and lung there appeared to be random expression while in the liver Gd^{D} seemed to be preferentially expressed.

The excellent correlation observed in our studies and those of Cohen and Rattazzi[6] between differential labelling and the expression of an X-linked locus implies a definite relationship between these two phenomena; proof of such a relationship will be provided by autoradiographic studies on the clones discussed in the previous article[5]. These are in progress. A comparison of X chromosome behaviour and *Gd* espression in both the mule and the hinny suggests that neither the maternal cytoplasm nor the maternal environment is important in determining which X chromosome becomes genetically inactive and differentially labelled in the female hybrids or in determining which cell type proliferates preferentially.

The significant excess of cells with an inactive and differentially labelled X^{D} is most easily explained by a difference between the two X chromosomes either in respect to the mechanism of "induction" of facultative heterochromatin or to the increased fitness which an active X^{H} confers on the hybrid cells resulting in selection in their favour. There are at least two alternatives which might account for a difference between X^{D} and X^{H} in regard to the mechanism of heterochromatinization; first, that X^{D} is simply more prone to become heterochromatic, or, second, that heterochromatinization commences earlier in the donkey than in the horse and that this is retained in the hybrids and so leads to an earlier involvement of X^{D} compared with X^{H}. The studies reported in the previous article[5] on fibroblast clones suggest *in vitro* cell selection in favour of cells with an active X^{H}; studies by Hook and Brustman[10] suggest that *in vivo* cell selection occurs in different organs.

Our finding of one animal (No. 192)[5] with a preponderance of Gd^{D}, on which unfortunately autoradiographic studies were not possible, suggests that there may be inter-animal variation in the expression of *Gd*. Inter-animal variation could perhaps account also for the discrepancy between our autoradiographic results and those of Mukherjee and his colleagues[1,3].

Our data and those of Cohen and Rattazzi[6], however, clearly show that X^{D} tends to be more often heterochromatic than X^{H} in the tissues of female horse-donkey hybrids which have been available for study.

还是有用处的。有关葡萄糖–6–磷酸脱氢酶（*Gd*）的数据[5]显示，在源于3个杂交体的成纤维细胞克隆中 Gd^D 的表达频率与在淋巴细胞和成纤维细胞中观察到的染色体 X^D 差异标记的频率几乎精确成反比。而且，关于 Gd^D 表达的定性数据表明：除了一只动物以外，Gd^D 在所有研究过的杂交动物中的表达远远弱于 Gd^H 的表达[10]。科恩和拉塔齐[6]曾报道过在另外4匹母马骡中染色体 DLX^D 的频率与 Gd^D 表达之间的密切关系。在每只动物中，染色体 X^D 都表现出更频繁的延迟复制，与此相关的是，Gd^D 成为了一种较少的成分。胡克和布鲁斯特曼[10]已经在37匹母马骡的器官中研究了 *Gd* 的表达。他们发现：在绝大多数组织（血液、胰腺、脑、颈髓、肾脏和腮腺）中，Gd^H 是优先表达的。在甲状腺和肺中看似为随机表达，而在肝脏中优先表达的似乎是 Gd^D。

我们与科恩、拉塔齐[6]在研究中观察到的差异标记与X连锁基因表达之间的良好相关性暗示着这两种现象之间存在着一定的联系；证明这种联系的证据可以通过对在先前文章中讨论过的克隆体[5]进行放射自显影研究得到。这些实验仍在进行中。对比马骡和驴骡中X染色体的行为与 *Gd* 的表达的结果表明：在决定雌性杂交体中哪条X染色体会发生遗传失活和差异标记，或者在决定哪种细胞类型会优先增殖上，母系的细胞质或者母系的环境都不重要。

那些含有失活和差异标记的 X^D 染色体的细胞出现显著过量现象很容易用两条X染色体之间的差异来解释，或者是因为功能异染色质的“诱导”机制，或者是因为活性 X^H 染色体赋予了杂交细胞更强的适应度，从而产生了有利于自身的选择。从异染色质化机制的角度来看，至少有两种解释可以说明染色体 X^D 和 X^H 之间的差异。第一，染色体 X^D 更容易发生异染色质化；或第二，异染色质化发生在驴身上比发生在马身上更早，并在杂交体上保留下来，于是就产生了染色体 X^D 发生异染色质化要早于 X^H 的结果。我们在以前文章[5]中报道的关于成纤维细胞克隆体的研究结果表明：体外细胞的选择偏向于含活性染色体 X^H 的细胞；胡克和布鲁斯特曼的研究结果[10]表明：体内的细胞选择会发生在各种器官中。

我们发现有一只动物（编号为192）[5]中 Gd^D 占多数，这表明 *Gd* 的表达也许存在动物的个体间差异，可惜我们无法对其进行放射自显影研究。动物的个体间差异或许还可以用于解释我们的放射自显影结果与慕克吉及其同事的结果[1,3]之间为什么有差别。

然而，我们的数据与科恩、拉塔齐的数据[6]显然说明：在曾被用于研究的母马–驴杂交体的组织中，染色体 X^D 倾向于比 X^H 更经常出现异染色质化。

We thank Drs. R. V. Short, W. R. Allen and M. J. Freire, with his colleagues in Portugal, for samples of material. Collection of Portuguese samples was supported by grants to J. L. H. from the Wellcome Trust. During this work J. L. H. and F. G. were supported by the Spastics Society. J. L. H. also acknowledges support from ARC and the Canadian Medical Research Council. We also thank Mrs. F. F. Collins, Mrs. J. Hallett, Miss A. Fryer, Miss V. M. McGuire and Mr. L. Grixti for technical assistance.

(**232**, 315-319; 1971)

F. Giannelli: Paediatric Research Unit, Guy's Hospital Medical School, London SEI.

J. L. Hamerton: Department of Genetics, Children's Hospital, Winnipeg, and Departments of Paediatrics and Anatomy, University of Manitoba.

Received May 7, 1971.

References:

1. Mukherjee, B. B., and Sinha, A. K., *Proc. US Nat. Acad. Sci.*, **51**, 252 (1964).
2. Hamerton, J. L., Giannelli, F., Collins, F., Hallett, J., Fryer, A., McGuire, V. M., and Short, R. V., *Nature*, **222**, 1277 (1969).
3. Mukherjee, B. B., Mukherjee, A. B., and Mukherjee, A. B., *Nature*, **228**, 1321 (1970).
4. Hamerton, J. L., and Giannelli, F., *Nature*, **228**, 1322 (1970).
5. Hamerton, J. L., Richardson, B. J., Gee, P. A., Allen, W. R., and Short, R. V., *Nature*, **232**, 312 (1971).
6. Cohen, M. M., and Rattazzi, M. C., *Proc. US Nat. Acad. Sci.*, **68**, 544 (1971).
7. Hsu, T. C., and Kellogg, jun., D. S., *J. Nat. Cancer Inst.*, **25**, 221 (1960).
8. Moorhead, P. S., Howell, P. C., Mellman, W. J., Battips, D. M., and Hungerford, D. A., *Exp. Cell Res.*, **20**, 613 (1960).
9. Giannelli, F., *Human Chromosomes DNA Synthesis, Monographs in Human Genetics*, 5 (edit. by Beckman, L., and Hauge, M.) (Karger, Basel and New York, 1970).
10. Hook, E. B., and Brustman, L. D., *Genetics*, **64**, 2, Part 2 (abstract, 1970).
11. Fisher, R. A., and Yates, F., *Statistical Tables for Biological, Agricultural and Medical Research*, fifth ed. (Oliver and Boyd, Edinburgh, 1957).

感谢肖特博士、艾伦博士、弗莱雷博士及其葡萄牙同事为我们提供了动物样本。在葡萄牙搜集动物样本的工作得益于维康基金会对哈默顿的资助。在这项研究中，哈默顿和詹内利得到了痉挛协会的支持，哈默顿还得到了英国农业研究委员会和加拿大医学研究委员会的支持。柯林斯夫人、哈利特夫人、弗吕耶小姐、麦圭尔小姐和格里克斯蒂先生为我们提供了技术上的帮助，在此一并表示感谢。

（刘振明 翻译；梁前进 审稿）

Evidence for Selective Differences between Cells with an Active Horse X Chromosome and Cells with an Active Donkey X Chromosome in the Female Mule

E. B. Hook and L. D. Brustman

Editor's Note

Conflicting evidence meant that researchers were unsure whether X-inactivation, the process in females where one copy of the X chromosome is silenced during embryonic development, was random or not. Here Ernest B. Hook and Loretta D. Brustman look at the contribution of paternal and maternal X chromosomes in different tissues from female mules and find there are two organ types: "those in which the horse phenotype was usually predominant, and those in which expression was not significantly different from what would be expected on a random basis." If X-inactivation were non-random, they argue, the favoured allele should be found in all tissues. Any discrepancies between previous studies, they conclude, probably reflect sampling differences.

THE glucose-6-phosphate dehydrogenase (G6PD) locus is known to be X-linked in the horse and donkey[1,2]. The female mule, an interspecific hybrid of these two species, is an obligatory heterozygote for two electrophoretically distinguishable G6PD alleles[1,2].

Studies of the DNA replication patterns of the chromosomes of two female mules by Mukherjee *et al.*[3,4] have shown that, in about half the cells investigated in each animal, the X chromosome derived from the female parent (the horse X) was late replicating and the X chromosome derived from the male parent (the donkey X), was early replicating; the reverse pattern obtained in the other half of the cells. Hamerton *et al.*[5] found in three female mules, however, that most lymphocytes and fibroblasts had a late replicating donkey X chromosome and early replicating horse X chromosome. The distribution was significantly different from the roughly equal representation of cell types noted by Mukherjee *et al.* In the only cell line studied electrophoretically by Hamerton *et al.* horse G6PD was preferentially expressed, consistent with the hypothesis that the early replicating mammalian X chromosome in females is the active one[6]. The observations of the latter workers suggest that in some female mules, either (1) the donkey X chromosome is preferentially inactivated early in development, or (and) (2) cells with an active horse X chromosome have a selective advantage in at least some tissues during ontogeny.

We have investigated this question by studying the G6PD phenotypes of samples from tissues and organs of fifty-four female mules. The results for representative organs are

母马骡中含马活性X染色体的细胞和含驴活性X染色体的细胞之间选择差异的证据

胡克，布鲁斯特曼

编者按

由于所获得的证据间互相矛盾，研究人员难以确定在雌性动物中发生的X染色体失活（X染色体的一个拷贝在胚胎发育过程中被沉默）过程是否是随机的。在本文中，欧内斯特·胡克和洛蕾塔·布鲁斯特曼考察了母马骡不同组织中父本X染色体和母本X染色体的贡献并发现存在两种器官类型："在一种类型中马的表型通常占主要地位，而在另一种类型中表型表达却与随机抽样所得的结果没有明显区别。"他们认为，如果X染色体失活是非随机的，那么应该在所有组织中都能发现优势等位基因。他们推断：在先前的研究结论之间所存在的任何偏差有可能都只是采样差异的反映。

众所周知，马和驴的葡萄糖–6–磷酸脱氢酶（G6PD）位点都是X连锁的[1,2]。作为这两个物种种间杂交体的母马骡必然是一个有两条电泳清晰可辨的G6PD等位基因的杂合子[1,2]。

慕克吉等人[3,4]曾对两匹母马骡染色体的DNA复制模式进行过研究，结果显示：在每匹母马骡的细胞中，大约有一半会出现母本X染色体（马X染色体）发生延迟复制而父本染色体（驴X染色体）发生早期复制的现象；而在另一半细胞中会出现相反的复制方式。然而，哈默顿等人[5]发现在三匹母马骡中绝大多数淋巴细胞和成纤维细胞会含有一条发生延迟复制的驴X染色体和一条发生早期复制的马X染色体。这种分布情况与慕克吉等人提出的两个细胞类型表现大致相等的说法显然不符。在哈默顿等人用电泳研究的唯一一个细胞系中，优先表达的是马的G6PD，这与在雌性哺乳动物中发生早期复制的X染色体具有活性的假说相符[6]。后期的研究者通过观察发现，在一些母马骡中：(1)在发育初期，优先失活的是驴的X染色体；或者（并且）(2)在个体发育中，至少某些组织内带有马活性X染色体的细胞会具有选择上的优势。

为了探索这一问题，我们对54匹母马骡的组织和器官样品中的G6PD表型进行了研究。表1中列出的是几种典型器官的结果。表型结果的实例以及对应的解释示

listed in Table 1. Examples of patterns and their interpretation are illustrated in Fig. 1.

Table 1. G6PD Phenotypes of Female Mules

Predominant phenotype	Cervical cord	Pancreas	Parotid	Liver	Lung	Thyroid
H>D	20 (18)	19 (12)	23 (15)	9 (6)	8 (8)	8 (6)
H≌D	1 (1)	4 (4)	4 (3)	12 (7)	11 (5)	12 (9)
D>H	2 (2)	4 (3)	3 (3)	13 (8)	6 (4)	9 (5)

The number of animals with the observed phenotypes are listed. Not all organs from every animal could be investigated, but in twenty-two of the fifty-four animals at least five of the six organs listed in the table were studied. The results for these twenty-two animals are listed in parentheses. The patterns were scored independently by E. B. H. and L. D. B. and the sample rerun if there was a discrepancy or uncertainty (see Fig. 1). Examination of blood samples revealed that the horse band was the usual predominant one here also. Mixing experiments with variable proportions of donkey and horse blood samples with known activity indicated that, when the bands of approximately equal intensity were observed, there was at least 35% activity of the major component in the mixture. In a mule in which expression of donkey and horse phenotype was close to equality, there was no observed difference in the proportion of parental phenotype between reticulocyte-rich and reticulocyte-poor preparations, although there was less activity in the latter. Similarly, there was no observed difference between fresh and stored cells in the proportion of parental phenotype, although activity was lower in stored cells. The specific activities of G6PD in peripheral blood of a young donkey and of a young horse were within 10% of each other. Earlier studies[8] suggested that the donkey phenotype was preferentially expressed in liver, but this no longer seems likely.

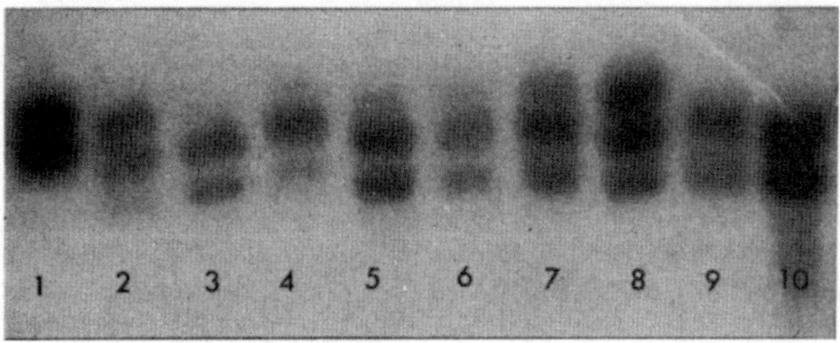

Fig. 1. The bands in the slots at extreme left (1) and right (10) illustrate donkey (D) and horse (H) patterns respectively. The other samples, from pancreas and parotid, were interpreted from left to right as follows: (2) D>H, (3) H>D, (4) ?, (repeat), (5) H>D, (6) H>D, (7) ?H ≌ D, (repeat), (8) D>H, (9) H>D. All organs were studied using starch gel electrophoresis in EBT buffer with standard methods[9] at *p*H 8.6. Almost all animals were relatively old. Organ samples were collected at the abattoir on dry ice about 60–120 min after death and exsanguination. Blood was obtained in citrate about 5–15 min after death and stored at 4°C.

There were two organ types: those in which the horse phenotype was usually predominant, and those in which expression was not significantly different from what would be expected on a random basis. In none of the organs studied was the donkey phenotype the usual predominant one. If X inactivation were non-random, and occurred at the early stage usually hypothesized in the mammalian female embryo, representation of the favoured allele should be relatively uniform in all derived tissues, unless cell selection in particular organs had also occurred. Since X inactivation is thought to occur at about the blastocyst stage[7], before observable organ differentiation, it seems unlikely that preferential

于图 1。

表 1. 母马骡的 G6PD 表型

主要表型	颈髓	胰腺	腮腺	肝脏	肺脏	甲状腺
马 > 驴	20 (18)	19 (12)	23 (15)	9 (6)	8 (8)	8 (6)
马 ≌ 驴	1 (1)	4 (4)	4 (3)	12 (7)	11 (5)	12 (9)
驴 > 马	2 (2)	4 (3)	3 (3)	13 (8)	6 (4)	9 (5)

表中列出的是与观察到的表型相对应的动物数量。不能保证每只动物的全部器官都能用于研究，不过在 54 只动物中，我们对其中 22 只动物列于表中的 6 个器官中的至少 5 个进行了研究。这 22 只动物的研究结果列在圆括号中。表型结果（见图 1）是由欧内斯特 · 胡克和洛蕾塔 · 布鲁斯特曼分别得到的，如果两个人的结果之间存在不一致或者不确定的情况，那么样品实验就需要重复进行。对血液样品的检测结果显示：在这里马的显带仍然是普遍显著的。对已知活性的驴和马的血液样品进行不同比例混合实验，结果表明，当观察到显带的强度大致相同时，混合物中主要成分的活性至少可达 35%。在一匹驴和马的表型接近相等的马骡中，并没有观察到在富含和缺乏网织红细胞的培养物之间存在亲本表型比例上的差异，尽管在后者中活性会更低一些。同样地，在新鲜的细胞和储存的细胞之间也没有观测到亲本表型比例上的差别，尽管在储存的细胞中活性更低。在一匹幼驴和一匹幼马的外周血中，两者的 G6PD 比活性相差不到 10%。我们早先的研究结果 [8] 表明：在肝脏中优先表达的是驴的表型，但现在看来这是不可能的。

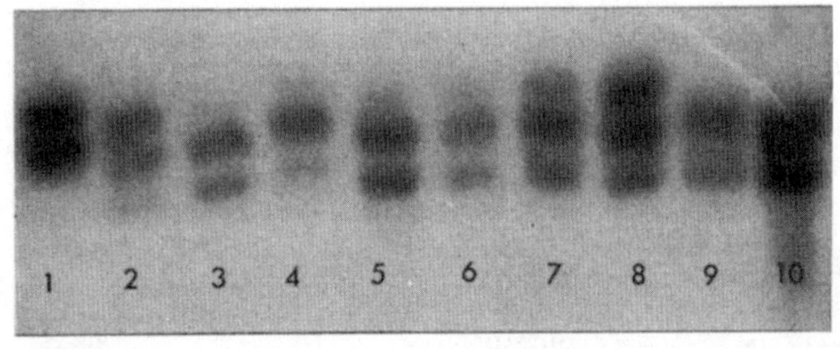

图 1. 电泳槽中的显带：最左边（1）和最右边（10）的显带分别代表驴（D）和马（H）的表型。其他样品均来自胰腺和腮腺，从左到右依次为：(2) D>H，(3) H>D，(4) ?，(重复)，(5) H>D，(6) H>D，(7) ?H ≌ D，(重复)，(8) D>H，(9) H>D。对所有器官的研究都是按照标准方法 [9] 在 pH 值 8.6 的 EBT 缓冲液（译者注：即 EDTA– 硼酸 –Tris 缓冲系统）中利用淀粉凝胶电泳进行的。几乎所有的动物年龄都相当大。在屠宰场收集已经死亡并被放血约 60 分钟 ~120 分钟后的动物的器官样本，放置在干冰上；在动物死亡约 5 分钟 ~ 15 分钟后取血并溶于柠檬酸盐中，然后 4℃储存。

我们发现有两种器官类型：一种是马的表型通常占主要地位的类型，另一种是表型表达与随机抽样所得结果没有明显区别的类型。在研究过的所有器官中，从未发现过任何一个驴的表型通常会占主要地位的器官。如果 X 染色体的失活是非随机的，并且按照通常的假设是发生在哺乳动物雌性胚胎生长的早期阶段，那么那些优势等位基因应该在所有的衍生组织中都表现出相对的一致性，除非在某些特殊器官中还会发生细胞选择。因为 X 染色体的失活大约发生在囊胚阶段 [7]，即在可观察到

inactivation has occurred just in the cell precursors of particular organs. It seems more probable that X-linked alleles of the donkey and horse have had different effects during development and maturation, with a relative selective advantage for cells with an active horse X chromosome in organs such as the parotid gland. But preferential expression of the horse X chromosome in these latter organs did not occur in all animals studied, indicating that the observed phenomenon represents a trend rather than a uniform developmental event in the female mule. Thus discrepancies between studies such as those of Hamerton *et al.* and Mukherjee *et al.* seem likely to reflect sampling differences in animals studied.

An alternative explanation for the observed distribution should be mentioned. Unknown organ specific factors may have differentially suppressed (or enhanced) expression of a particular parental G6PD phenotype (for example, perhaps inhibiting expression of the donkey band in cervical cord, pancreas and so on). Although this seems unlikely, it cannot be completely excluded. But when samples of pancreas with horse phenotype were mixed with samples of other organs which expressed the donkey phenotype, there was no evidence for an inhibiting factor.

(**232**, 349-350; 1971)

Ernest B. Hook and Loretta D. Brustman: Birth Defects Institute, New York State Department of Health, and Department of Pediatrics, Albany Medical College of Union University, Albany, New York 12208.

Received February 18, 1971.

References:

1. Trujillo, J. M., Walden, B., O'Neil, P., and Anstall, H. B., *Science*, **148**, 1603 (1965).
2. Mathai, C. K., Ohno, S., and Beutler, E., *Nature*, **210**, 115 (1966).
3. Mukherjee, B. B., Mukherjee, A. B., and Mukherjee, A. B., *Nature*, **228**, 1321 (1970).
4. Mukherjee, B. B., and Sinha, A. K., *Proc. US Nat. Acad. Sci.*, **51**, 252 (1964).
5. Hamerton, J. L., Gianelli, F., Collins, F., Hallett, J., Fryer, A., McGuire, V. M., and Short, R. V., *Nature*, **222**, 1277 (1969).
6. Lyon, M. F., in *Adv. Teratol.*, **1** (edit. by Woolam, D. H. M.) (Logos Press, London, 1966).
7. Kinsey, J. D., *Genetics*, **55**, 337 (1967).
8. Hook, E. B., and Brustman, L. D., *Genetics*, **64**, 530 (1970).
9. Motulsky, A. G., and Yoshida, A., in *Biochemical Methods in Red Cell Genetics* (edit. by Yunis, J. Y.), 51 (Academic Press, New York, 1969).

的器官分化出现之前，所以染色体优先失活刚好发生在某些特殊器官的细胞前体中似乎是不大可能的。可能性更大的似乎是：驴和马的 X 连锁等位基因在胚胎发育阶段和成熟阶段所起的作用不同，它们对腮腺等器官中含有马活性 X 染色体的细胞具有相对的选择性优势。但是在腮腺等器官中，马 X 染色体被优先表达的现象并没有发生在所有研究过的母马骡上，这说明观察到的现象所代表的只是一种倾向，而不是母马骡中始终如一的发育事件。因此，诸如哈默顿等人和慕克吉等人研究结果之间的差异，看起来很可能是被研究动物的采样差异的反映。

应该指出，还有一种能说明所观测到的分布规律的解释。器官中一些未知的特殊因子也许会差异性地抑制（或者增强）某种特定的亲本 G6PD 表型的表达（例如，也许会在颈髓、胰腺等器官中抑制驴的显带的表达）。尽管这看上去可能性不大，但也不能被完全排除。然而，当把带有马的表型的胰腺样品与带有驴的表型的其他器官样品混合在一起时，就找不到任何有关抑制因子的证据了。

（刘振明 翻译；梁前进 审稿）

Bomb ^{14}C in the Human Population

R. Nydal *et al.*

Editor's Note

Understanding the cycling of carbon between the atmosphere, oceans and biosphere (owing mostly to the uptake of carbon dioxide in photosynthesis) is essential for predicting how anthropogenic greenhouse gases affect climate. This paper from scientists at the Norwegian Institute of Technology shows how interest in the carbon cycle was already burgeoning in the early 1970s. It is expressed within the context of its time: it focuses on how the radioactive isotope carbon-14 (^{14}C), released into the atmosphere by nuclear weapons tests, will increase in the human body, which was considered a potential health risk. Strikingly, this increase is very short-lived, because the bomb ^{14}C is rapidly diluted by carbon dioxide released from fossil-fuel burning—today considered a much greater hazard.

IN the atmosphere ^{14}C occurs principally as $^{14}CO_2$ and is usually produced by nuclear reactions between cosmic ray neutrons and the nitrogen atoms of the air. The natural equilibrium between production and disintegration of ^{14}C determines a part of the natural background radiation to the human population. From 1955 there has been a gradual increase of ^{14}C in the atmosphere, the land biosphere and the ocean, as a result of nuclear tests. Although ^{14}C was initially not regarded as an important hazard to man[1], it was later pointed out[2-4] that ^{14}C could be a source of appreciable genetic hazard in the world's population, because of its long half life (5,700 yr).

At this laboratory we have studied the ^{14}C concentration in the human body[5]. The correspondence between ^{14}C in the atmosphere and in the human body, mediated as it is by photosynthesis, has been confirmed in 6 yr of measurements[6-8].

Since 1955, about two-thirds of the total nuclear energy liberated in nuclear tests has resulted from tests carried out in the atmosphere at high northern latitudes in 1961 and 1962 (ref. 9). The subsequent transfer of ^{14}C down to the troposphere, the biosphere and the ocean has been followed in detail[10-16]. The ^{14}C excess[16] ($\delta^{14}C$) in the northern troposphere (Fig. 1) is representative chiefly of the region between 30°N and 90°N, although the curve for the southern troposphere is representative of the region from the equator to 90°S.

进入人体的核弹^{14}C

尼达尔等

编者按

了解碳在大气、海洋和生物圈中的循环（主要归因于生物体在光合作用中对 CO_2 的吸收）对于预测人类活动所产生的温室气体将怎样影响气候相当重要。这篇由挪威技术研究所的科学家们所撰写的文章使我们了解到，在 20 世纪 70 年代早期人们是如何开始注意到碳循环的。本文反映了那个年代的焦点问题：由核武器试验释放到大气中的放射性同位素碳–14（^{14}C）在人体中的含量将会增加，这在当时被认为是一种潜在的健康威胁。显然，^{14}C 含量的增加是非常短暂的，因为核弹产生的 ^{14}C 很快就会被矿物燃料燃烧释放的 CO_2 所稀释，现在人们认为后者才是人类将要面临的更大威胁。

大气中 ^{14}C 主要以 $^{14}CO_2$ 的形式存在，通常由宇宙射线中子和大气中的氮原子通过核反应而产生。^{14}C 在产生和衰变之间的自然平衡决定了人类所受到的一部分天然本底辐射的大小。由于核试验的原因，从 1955 年起，大气层、陆地生物圈和海洋中的 ^{14}C 开始逐步增加。虽然起初人们并不认为 ^{14}C 会对人类造成重大危害 [1]，但后来有人指出 [2-4]：由于 ^{14}C 的半衰期很长（5,700 年），因而它对人类基因造成一定程度危害的可能性还是存在的。

在本实验室中，我们研究了人体内的 ^{14}C 浓度 [5]。经过 6 年时间的测试，我们已证实大气中的 ^{14}C 与人体中的 ^{14}C 是相互关联的，它们之间的媒介是光合作用 [6-8]。

自 1955 年起到现在，从核试验中释放出来的全部核能的 2/3 左右来自 1961 年和 1962 年人们在北纬高纬度地区大气层中所进行的试验（参考文献 9）。有不少人对 ^{14}C 接下来向对流层、生物圈和海洋中的转移进行了详细的报道 [10-16]。北半球对流层中 ^{14}C 的过量值 [16]（$\delta^{14}C$）主要针对从北纬 30° 到北纬 90° 之间的区域（图 1），而南半球对流层曲线所代表的区域则是从赤道到南纬 90° 之间。

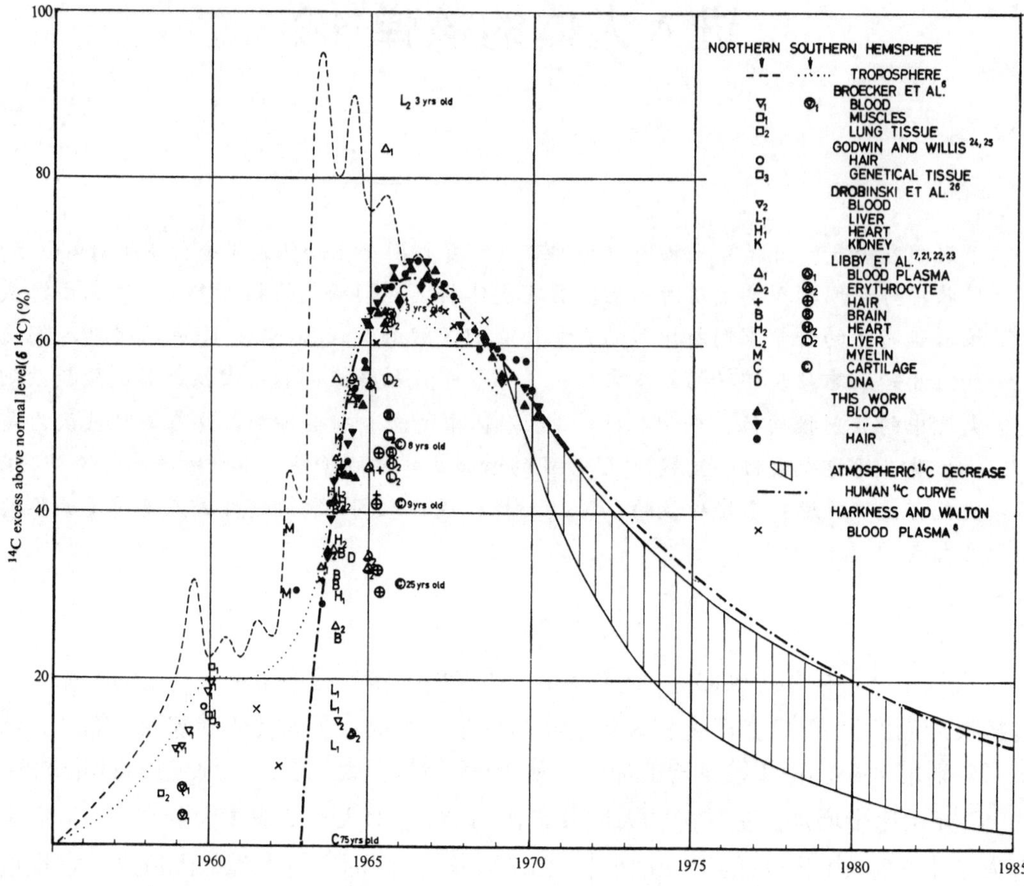

Fig. 1. Radiocarbon in the troposphere and human body.

The model for CO_2 exchange between the various reservoirs is shown in Fig. 2, in which the CO_2 in the troposphere is in exchange with CO_2 in the stratosphere, the land biosphere and the ocean. For CO_2 exchange between the troposphere and the biosphere we share the view of Münnich (for discussion, see ref. 17), who divided the biosphere into two parts. The first (b_1), which consists of leaves, grass, branches, and so on, is in rapid exchange with the troposphere and is combined with this reservoir, but the larger part of the vegetation (b_2) has a much slower exchange rate and is combined with the humus layer.

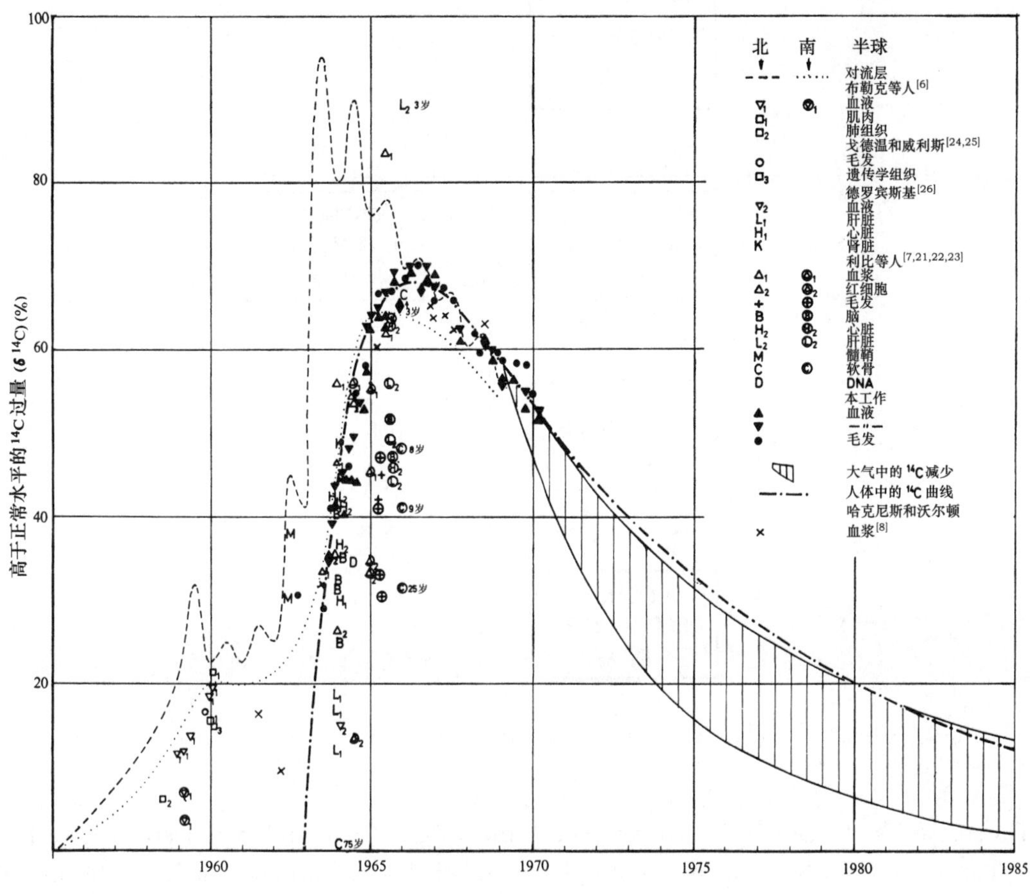

图 1. 在对流层和人体中的放射性碳。

CO_2 在不同碳库之间的交换模型如图 2 所示。从图中可以看出，对流层中的 CO_2 会与平流层、陆地生物圈和海洋之中的 CO_2 发生交换。就 CO_2 在对流层和生物圈之间的交换而言，我们赞同明尼希的观点，即认为生物圈可以被分为两个部分（讨论过程见参考文献 17）。第一部分（b_1）由树叶、草、树枝等组成，这部分与对流层之间发生着快速的交换并且与之结合；但相当多的植物属于第二部分（b_2），这部分与对流层之间的交换速率很慢并且是和腐殖质层结合在一起的。

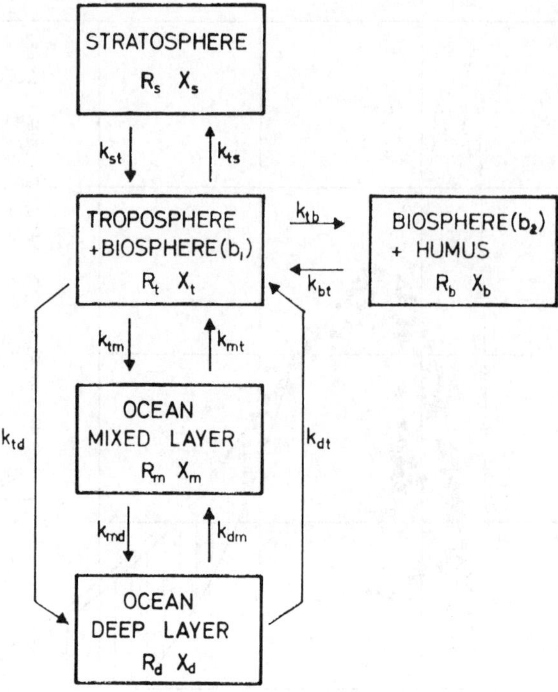

Fig. 2. Exchangeable carbon reservoirs, where: R_i, total carbon amount; X_i, ^{14}C excess; k_{ij}, exchange coefficients; i, j, t, b, m, d $(i \neq j)$.

The ocean is divided into two reservoirs, the mixed layer and the deep ocean. The exchange of CO_2 between the troposphere and the ocean occurs chiefly in the mixed layer, but according to Craig[18] there is also the possibility of a direct exchange with the deep ocean. The ^{14}C concentration in the mixed layer of the ocean is now 10 to 15% above normal[19,20].

Using the model of Fig. 2, we have treated the decrease of $\delta^{14}C$ in the troposphere in a previous article[20]. We showed that the measured variation of $\delta^{14}C$ (x_t) in the troposphere is approximately reproduced by the following two-term exponential function:

$$x_t = A_1 e^{-k_1 t} + A_2 e^{-k_2 t} \qquad (1)$$

in which the parameters A_1, A_2, k_1 and k_2 depend on the various exchange coefficients. Because the errors on some of these coefficients are large, the extrapolation is uncertain and is given in the shaded area of Fig. 1.

The amount of bomb-produced ^{14}C in the atmosphere has increased the total natural amount of ^{14}C in nature by about 3%. According to Harkness *et al.*[8], the production of CO_2 from fossil fuel would lower the natural ^{14}C concentration in the atmosphere to about 16% below normal at the end of this century. It is thus reasonable that the ^{14}C excess caused by the atomic bomb will be more than compensated for by the dilution of inactive

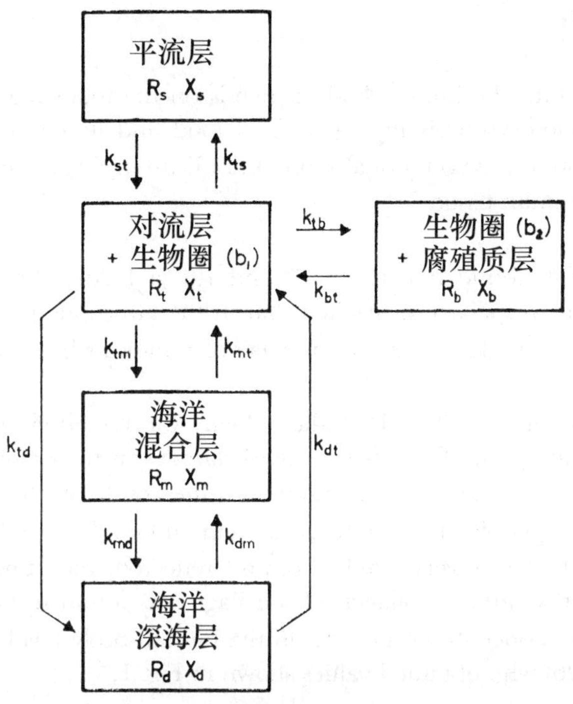

图 2. 可进行交换的碳库，其中：R_t 为总碳量；X_t 为 ^{14}C 过量；k_{ij} 为交换系数；i、j 代表 t、b、m、d $(i \neq j)$。

海洋被分成两个碳库，分别是混合层和深海层。CO_2 在对流层和海洋之间的交换主要发生在混合层，但根据克雷格的说法 [18]，直接交换也可能会发生在对流层和深海层之间。目前在海洋混合层中的 ^{14}C 浓度要高出正常值 10% ~ 15% [19,20]。

在之前发表的一篇文章中 [20]，我们利用图 2 的模型讨论了对流层中 $\delta^{14}C$ 的下降。我们认为，对流层中 $\delta^{14}C$（x_t）测量值的变化可大致由以下这个包含两项指数函数的式子来模拟：

$$x_t = A_1 e^{-k_1 t} + A_2 e^{-k_2 t} \tag{1}$$

其中，参数 A_1、A_2、k_1 和 k_2 由不同的交换系数决定。因为其中有一些系数误差很大，所以外推结果有一定的变化范围，在图 1 中用阴影部分表示。

核弹在大气中产生的 ^{14}C 已使自然界中 ^{14}C 的天然总含量增加了约 3%。根据哈克尼斯等人的说法 [8]：到本世纪末，由矿物燃料产生的 CO_2 将使大气中的天然 ^{14}C 浓度下降到比正常值低大约 16% 的水平。因此，我们可以合理地推出，非放射性碳的稀释作用完全可以补偿由核弹引起的 ^{14}C 过量（苏斯效应）。

carbon (the Suess effect).

The transfer of ^{14}C into the human body depends on the following three factors: (1) the time between the photosynthesis in vegetational food and its consumption; (2) the diet, particularly the amount of vegetational food, and (3) the residence time of the carbon in the constituents of human tissue.

Broecker *et al.*[6] (Fig. 1) found that it took 1 and 1.8 yr before the ^{14}C concentration in blood and lung tissue, respectively, reached that in the atmosphere. They also found that the $\delta^{14}C$ value of blood had a maximum time lag of 6 months behind food.

Berger and collaborators[7,21-23] (Fig. 1) studied chiefly the metabolic turnover time of the constituents of human tissue. For this they used samples from persons who had travelled from the southern to the northern hemisphere. One result of their work was that the incorporation in these people of ^{14}C in brain protein and lipids, liver, heart, plasma protein and erythrocyte protein was very similar to, and reflected, the atmospheric ^{14}C content present several months earlier. Collagen of cartilage was found to be metabolically inert in older persons. The concentration of ^{14}C in the human body has been studied by other workers (refs. 8, 24-26) who obtained values shown in Fig. 1.

At our laboratory the transfer of ^{14}C into the human body was studied by following the time-variation of $\delta^{14}C$ in blood and hair for three persons[5,27] (Tables 1–3). No separation between blood plasma and erythrocyte protein was performed, and the measured ^{14}C activity is thus a mean value for the total blood samples. Fig. 1 shows that there is excellent agreement between data obtained for the blood and for the hair samples. The values of the ^{14}C concentration in blood plasma obtained by Harkness and Walton[8] are slightly lower than ours.

Table 1. Carbon in Neck Hair, from a Boy (K.N.) Born in 1962

Time of collection		$\delta^{14}C$ %	$\delta^{13}C$/‰	$\Delta^{14}C$ %
November	1962	30.7±1.3		
June	1963	29.2±0.8	−19.7	26.7
October	1963	41.0±1.0	−19.2†	39.3
March	1964	46.0±1.0	−18.9	44.2
July	1964	54.8±1.0	−19.2†	52.9
October	1964	58.0±0.8	−18.6	55.9
February	1965	66.9±1.0	−19.2†	64.9
July	1965	67.2±1.0	−18.6	64.5
February	1966	68.7±1.0	−19.2	66.7
May	1966	70.0±0.9	−19.2†	68.0
December	1966	65.9±0.9	−19.2†	63.9
February	1967	67.5±1.1	−19.2†	65.4
July	1967	65.9±0.9	−19.2†	63.9
February	1968	61.5±0.9	−19.2†	59.6

^{14}C 向人体内的迁移过程取决于以下三个因素：(1) 从植物性食物进行光合作用到它被人类食用之间的时间；(2) 饮食结构，尤其是植物性食物所占的比重；(3) 碳在人体组织各组成成分中的滞留时间。

布勒克等人[6]（图 1）发现：要使血液和肺组织中的 ^{14}C 浓度达到大气中的浓度分别需要 1 年和 1.8 年的时间。他们还发现，从进食到血液中出现 $\delta^{14}C$ 之间的最大时间滞后可达 6 个月。

伯杰及其合作者[7,21-23]（图 1）重点研究了人体组织各组成成分的新陈代谢周转时间。他们为此选用的样本均来自那些有过从南半球到北半球旅行经历的人。他们获得的一项研究成果显示：^{14}C 与这些人的脑蛋白以及脂类、肝、心、血浆蛋白和红细胞蛋白的结合情况非常接近于数月前大气中的 ^{14}C 含量，可以认为它能够反映数月前大气中的 ^{14}C 含量。他们发现老年人软骨组织中的胶原蛋白具有新陈代谢惰性。图 1 中还显示出了一些由其他研究者（参考文献 8 及 24~26）得到的人体中的 ^{14}C 浓度值。

在我们的实验室，对 ^{14}C 进入人体的研究是通过跟踪 3 个人血液和毛发中 $\delta^{14}C$ 随时间的变化来进行的[5,27]（表 1 ~ 表 3）。因为没有对血液中的血浆和红细胞进行分离，所以测得的 ^{14}C 放射性是整个血样的平均值。从图 1 中可以看出，由血样得到的数据和由毛发样本得到的数据吻合得非常好。哈克尼斯和沃尔顿测得的血浆中的 ^{14}C 浓度数据[8] 略微低于我们测得的值。

表 1. 颈部毛发中的碳，来自一个 1962 年出生的男孩（K.N.）

采集时间	$\delta^{14}C$ %	$\delta^{13}C$ /‰	$\Delta^{14}C$ %
1962 年 11 月	30.7 ± 1.3		
1963 年 6 月	29.2 ± 0.8	−19.7	26.7
1963 年 10 月	41.0 ± 1.0	−19.2†	39.3
1964 年 3 月	46.0 ± 1.0	−18.9	44.2
1964 年 7 月	54.8 ± 1.0	−19.2†	52.9
1964 年 10 月	58.0 ± 0.8	−18.6	55.9
1965 年 2 月	66.9 ± 1.0	−19.2†	64.9
1965 年 7 月	67.2 ± 1.0	−18.6	64.5
1966 年 2 月	68.7 ± 1.0	−19.2	66.7
1966 年 5 月	70.0 ± 0.9	−19.2†	68.0
1966 年 12 月	65.9 ± 0.9	−19.2†	63.9
1967 年 2 月	67.5 ± 1.1	−19.2†	65.4
1967 年 7 月	65.9 ± 0.9	−19.2†	63.9
1968 年 2 月	61.5 ± 0.9	−19.2†	59.6

Continued

Time of collection		$\delta^{14}C$ %	$\delta^{13}C$/‰	$\Delta^{14}C$ %
May	1968	60.2±1.1	−19.2†	58.2
August	1968	61.8±1.2	−19.2†	59.9
November	1968	59.7±1.1	−19.2†	57.7
January	1969	58.7±1.2	−19.2†	56.7
April	1969	58.2±1.2	−19.2†	56.5
May	1969	58.3±1.2	−19.2†	56.6
July	1969	58.1±1.1	−19.2†	56.4
October	1969	54.9±0.9	−19.2†	53.1
December	1969	54.7±1.1	−19.2†	52.9

† Not measured (mean value).

Table 2. Carbon in Blood Samples, from a Woman (I.N.), 26 Yr Old in 1963

Time of collection		$\delta^{14}C$ %	$\delta^{13}C$/‰	$\Delta^{14}C$ %
September	1963	34.9±1.1	−22.0	34.0±1.1
October	1963	39.3±0.8	−22.2	38.5±0.8
November	1963	43.6±0.8	−22.7	42.9±0.9
February	1964	45.6±1.1	−22.0	44.7±1.2
April	1964	48.1±1.1	−21.7	47.1±1.1
June	1964	49.6±1.1	−21.7†	48.6±1.1
July	1964	53.7±1.0	−22.1	52.8±1.1
November	1964	62.9±0.9	−19.8	61.2±0.9
January	1965	64.2±0.9	−19.7	62.4±0.9
March	1965	65.0±0.9	−20.5	63.5±1.0
May	1965	67.1±1.0	−21.1	65.6±1.0
September	1965	69.1±0.9	−22.0	67.9±1.0
November	1965	65.6±0.9	−21.4	64.2±1.0
March	1966	69.7±0.9	−22.3	68.7±0.9
July	1966	67.6±1.1	−21.2	66.2±1.1
September	1966	69.8±1.0	−21.5	68.6±1.1
December	1966	67.4±1.0	−21.9	66.3±1.0
September	1967	62.5±0.9	−21.7†	61.4±0.9
June	1968	60.6±0.9	−21.7†	59.5±1.0
August	1968	59.9±0.9	−21.7†	58.8±1.0
January	1969	56.0±0.7	−21.7†	55.0±0.7
September	1969	55.5±1.2	−21.7†	54.5±1.2
March	1970	52.6±1.2	−21.7†	51.6±1.2

† Not measured (mean value).

Table 3. Carbon in Blood Samples, from a Woman (A.L.), 26 Yr Old in 1963

Time of collection		$\delta^{14}C$ %	$\delta^{13}C$/‰	$\Delta^{14}C$ %
September	1963	35.2±0.8	−21.6	34.2±0.8
November	1963	41.5±0.6	−21.7†	40.6±0.7

续表

采集时间	$\delta^{14}C$ %	$\delta^{13}C$ /‰	$\Delta^{14}C$ %
1968 年 5 月	60.2 ± 1.1	−19.2†	58.2
1968 年 8 月	61.8 ± 1.2	−19.2†	59.9
1968 年 11 月	59.7 ± 1.1	−19.2†	57.7
1969 年 1 月	58.7 ± 1.2	−19.2†	56.7
1969 年 4 月	58.2 ± 1.2	−19.2†	56.5
1969 年 5 月	58.3 ± 1.2	−19.2†	56.6
1969 年 7 月	58.1 ± 1.1	−19.2†	56.4
1969 年 10 月	54.9 ± 0.9	−19.2†	53.1
1969 年 12 月	54.7 ± 1.1	−19.2†	52.9

† 未测量（平均值）。

表 2. 血样中的碳，来自一位在 1963 年时年龄为 26 岁的妇女（I.N.）

采集时间	$\delta^{14}C$ %	$\delta^{13}C$/ ‰	$\Delta^{14}C$ %
1963 年 9 月	34.9 ± 1.1	−22.0	34.0 ± 1.1
1963 年 10 月	39.3 ± 0.8	−22.2	38.5 ± 0.8
1963 年 11 月	43.6 ± 0.8	−22.7	42.9 ± 0.9
1964 年 2 月	45.6 ± 1.1	−22.0	44.7 ± 1.2
1964 年 4 月	48.1 ± 1.1	−21.7	47.1 ± 1.1
1964 年 6 月	49.6 ± 1.1	−21.7†	48.6 ± 1.1
1964 年 7 月	53.7 ± 1.0	−22.1	52.8 ± 1.1
1964 年 11 月	62.9 ± 0.9	−19.8	61.2 ± 0.9
1965 年 1 月	64.2 ± 0.9	−19.7	62.4 ± 0.9
1965 年 3 月	65.0 ± 0.9	−20.5	63.5 ± 1.0
1965 年 5 月	67.1 ± 1.0	−21.1	65.6 ± 1.0
1965 年 9 月	69.1 ± 0.9	−22.0	67.9 ± 1.0
1965 年 11 月	65.6 ± 0.9	−21.4	64.2 ± 1.0
1966 年 3 月	69.7 ± 0.9	−22.3	68.7 ± 0.9
1966 年 7 月	67.6 ± 1.1	−21.2	66.2 ± 1.1
1966 年 9 月	69.8 ± 1.0	−21.5	68.6 ± 1.1
1966 年 12 月	67.4 ± 1.0	−21.9	66.3 ± 1.0
1967 年 9 月	62.5 ± 0.9	−21.7†	61.4 ± 0.9
1968 年 6 月	60.6 ± 0.9	−21.7†	59.5 ± 1.0
1968 年 8 月	59.9 ± 0.9	−21.7†	58.8 ± 1.0
1969 年 1 月	56.0 ± 0.7	−21.7†	55.0 ± 0.7
1969 年 9 月	55.5 ± 1.2	−21.7†	54.5 ± 1.2
1970 年 3 月	52.6 ± 1.2	−21.7†	51.6 ± 1.2

† 未测量（平均值）。

表 3. 血样中的碳，来自一位在 1963 年时年龄为 26 岁的妇女（A.L.）

采集时间	$\delta^{14}C$ %	$\delta^{13}C$ /‰	$\Delta^{14}C$ %
1963 年 9 月	35.2 ± 0.8	−21.6	34.2 ± 0.8
1963 年 11 月	41.5 ± 0.6	−21.7†	40.6 ± 0.7

Continued

Time of collection		$\delta^{14}C$ %	$\delta^{13}C$/‰	$\Delta^{14}C$ %
February	1964	40.5±0.6	−20.9	39.3±0.7
March	1964	44.4±0.8	−22.5	43.5±0.9
April	1964	44.2±0.7	−23.2	43.6±0.7
June	1964	44.2±1.0	−22.1	43.3±1.0
August	1964	53.4±0.8	−21.9	52.4±0.9
November	1964	57.6±0.9	−20.7	56.2±0.9
January	1965	62.4±0.9	−19.1	60.5±0.9
March	1965	64.3±0.7	−20.1	62.6±0.7
May	1965	63.9±0.9	−21.0	62.5±1.0
September	1965	68.4±0.9	−22.0	67.3±1.0
November	1965	65.9±0.7	−21.4	64.4±0.8
March	1966	69.6±1.0	−22.0	68.5±1.1
July	1966	67.5±1.0	−21.2	66.2±1.1
September	1966	68.3±1.0	−20.9	66.9±1.0
December	1966	69.0±0.9	−20.8	67.6±1.0
September	1967	60.9±0.9	−21.7†	59.8±0.9
June	1968	61.4±0.9	−21.7†	60.3±0.9
August	1968	58.8±0.9	−21.7†	57.7±0.9
January	1969	56.6±0.7	−21.7†	55.6±0.7
May	1969	56.4±1.0	−21.7†	55.4±1.0
September	1969	53.0±1.0	−21.7†	52.0±1.0
March	1970	51.6±1.1	−21.7†	50.6±1.1

† Not measured (mean value).

The blood and hair data are almost representative for persons living in the northern hemisphere. Because $\delta^{14}C$ in the southern troposphere at present lags behind that of the northern troposphere by about 1 yr, there should also be a similar lag for $\delta^{14}C$ in the human populations of the respective hemispheres. After about 1970, the ^{14}C concentrations in people in the northern and southern hemispheres will be similar, and equal to those in the troposphere. Fig. 1 shows that $\delta^{14}C$ (x_H) in the human body appears as a pulse, delayed with respect to that of the atmosphere. The observed values for x_H can be fitted reasonably well by the following two-term exponential function:

$$x_H = 108\,(e^{-0.1t} - e^{-0.75t}) \tag{2}$$

The coefficients in this function were determined by a least mean squares method, using the upper limits of the measured values for ^{14}C in the human body in the period from 1963 to 1970, and extrapolated values in the troposphere in the period from 1970 to 2000. The upper limit values were chosen because there was some excess ^{14}C before 1963 which should also be considered. There is also a tendency for the blood and hair data during the last 2 yr to correspond with the upper limit of the shaded area in Fig. 1. Function (2) was simplified by assuming that all previous nuclear tests occurred within a short time interval, and that the ^{14}C increase in the human population started in about January 1963.

续表

采集时间	$\delta^{14}C$ %	$\delta^{13}C$ /‰	$\Delta^{14}C$ %
1964 年 2 月	40.5±0.6	−20.9	39.3±0.7
1964 年 3 月	44.4±0.8	−22.5	43.5±0.9
1964 年 4 月	44.2±0.7	−23.2	43.6±0.7
1964 年 6 月	44.2±1.0	−22.1	43.3±1.0
1964 年 8 月	53.4±0.8	−21.9	52.4±0.9
1964 年 11 月	57.6±0.9	−20.7	56.2±0.9
1965 年 1 月	62.4±0.9	−19.1	60.5±0.9
1965 年 3 月	64.3±0.7	−20.1	62.6±0.7
1965 年 5 月	63.9±0.9	−21.0	62.5±1.0
1965 年 9 月	68.4±0.9	−22.0	67.3±1.0
1965 年 11 月	65.9±0.7	−21.4	64.4±0.8
1966 年 3 月	69.6±1.0	−22.0	68.5±1.1
1966 年 7 月	67.5±1.0	−21.2	66.2±1.1
1966 年 9 月	68.3±1.0	−20.9	66.9±1.0
1966 年 12 月	69.0±0.9	−20.8	67.6±1.0
1967 年 9 月	60.9±0.9	−21.7†	59.8±0.9
1968 年 6 月	61.4±0.9	−21.7†	60.3±0.9
1968 年 8 月	58.8±0.9	−21.7†	57.7±0.9
1969 年 1 月	56.6±0.7	−21.7†	55.6±0.7
1969 年 5 月	56.4±1.0	−21.7†	55.4±1.0
1969 年 9 月	53.0±1.0	−21.7†	52.0±1.0
1970 年 3 月	51.6±1.1	−21.7†	50.6±1.1

† 未测量（平均值）。

来自血液和毛发的数据基本上代表了生活在北半球的所有人。因为南半球对流层的 $\delta^{14}C$ 值比北半球对流层滞后大约 1 年，所以居住在两个半球的人体内的 $\delta^{14}C$ 值也应该存在类似的滞后。大致到 1970 年之后，^{14}C 在南北半球的人体内的浓度将会趋于持平，并等于对流层中的 ^{14}C 浓度。从图 1 中可以看出：人体内部的 $\delta^{14}C$（x_H）值就像一个落后于大气中 $\delta^{14}C$ 值的脉冲。用以下这个包含两项指数函数的式子可以很好地拟合 x_H 的观测值：

$$x_H = 108\left(e^{-0.1t} - e^{-0.75t}\right) \tag{2}$$

函数式中的系数是通过最小均方法得到的，用到了在 1963 年～1970 年间测得的人体中 ^{14}C 的上限值，以及将对流层的数据外推到 1970 年～2000 年时的数据。选择上限值是因为还应当考虑到 1963 年以前的一些 ^{14}C 过量。在最近两年内得到的血液和毛发数据趋向于与图 1 阴影部分的上限值相符。化简式 (2) 的依据是以下两个假设：假设之前所有的核试验都集中发生在一个很短的时间间隔内，并且假设人体中 ^{14}C 含量开始增加的时间大约在 1963 年 1 月。

The first term in the brackets of equation (2) indicates that the excess ^{14}C in the human body has a mean lifetime of about 10 yr and the second term that ^{14}C enters the human body with a mean delay time of about 1.4 yr after production in the atmosphere. The latter value is probably accurate to within 30% and agrees with previous estimates[6,7].

The hazard to the human population from artificial radiocarbon arises largely from inventory radiation of the body. Natural ^{14}C contributes with a certain dose rate, r_0, as a result of its decay rate of about fourteen disintegrations per min per gram of carbon. The average value of r_0 is about 1.06 mrad/yr[9]. The dose from natural ^{14}C is distributed as follows: 1.64 mrad/yr in the bones, 1.15 mrad/yr in the cells lining bone surfaces and 0.71 mrad/yr in bone marrow and soft tissue. Applying function (2), the dose D_1 absorbed in the human body during a time t can be calculated from the formula:

$$D_1 = 1.08\, r_0 \int_0^t (e^{-0.1t} - e^{-0.75t})\, dt \qquad (3)$$

For a period of about 30 yr the total radiation dose from this source will be $9r_0$. We thus obtain a total radiation dose of 16 mrad to bone, 11 mrad to cells lining bone surfaces and 7 mrad to bone marrow and soft tissue. The genetic hazard is caused by the latter. That dose constitutes about 10% of the total gonad dose from all radioactive fallout. Purdom[4] pointed out, however, that the actual gonad dose is somewhat larger because of a transmutation process in the DNA molecule, in which the decaying ^{14}C atoms are replaced by nitrogen atoms. Purdom assumed that the biological damage from the transmutation was equal to that from β-radiation.

The amount of artificial ^{14}C in the human body at about the year AD 2000 will constitute about 3% of the total amount of ^{14}C. This isotope has a half-life of 5,700 yr, and several scientists[2,3,9,28] think that it would therefore cause a most serious genetic threat. The long term radiation dose can be calculated from the formula

$$D_2 = 0.03\, r_0 \int_0^t e^{-0.000125t}\, dt \qquad (4)$$

The total radiation doses (D_1+D_2) which will be received by the bone cells, the cells lining bone surfaces, and the bone marrow and soft tissue in the next 10,000 yr will be 410, 290 and 180 mrad, respectively. These doses, which are in agreement with values given in a United Nations report[9] (page 45), are more important than those from all other radioactive fallout. We question, however, the value of the long term radiation dose (D_2) because, as previously mentioned, the use of fossil fuel might reduce the ^{14}C concentration in man below normal. We are of the opinion that the only ^{14}C hazard from previous tests which should be taken into account is attributable to a total genetic dose (D) of the order of 10 mrad, received in a period of about 30 yr. This dose is, however, negligible compared with the dose received from natural sources, which constitutes about 100 mrad per yr.

式（2）括号内的第一项表明，人体内过量 ^{14}C 的平均寿命大约为 10 年；第二项表明，^{14}C 从在大气中产生到进入人体之间会有平均 1.4 年左右的时间滞后。滞后值的精确度大约可以达到 30% 以内，并与以前文献中的估计值相符 [6,7]。

人造放射性碳对人类造成的危害主要来源于人体内的辐射量。天然 ^{14}C 的剂量率 r_0 是一个固定值，由它的衰变速率——每克碳每分钟衰变约 14 次——决定。r_0 的平均值大致是 1.06 毫拉德 / 年 [9]。天然 ^{14}C 的剂量是这样分布的：在骨骼中为 1.64 毫拉德 / 年，在骨骼表层细胞中为 1.15 毫拉德 / 年，在骨髓和软组织中为 0.71 毫拉德 / 年。根据函数式（2），人体在 t 时间内吸收的剂量 D_1 可由下式计算：

$$D_1 = 1.08\, r_0 \int_0^t (e^{-0.1t} - e^{-0.75t})\, dt \tag{3}$$

在大约 30 年内，来自该辐射源的总辐射剂量将为 $9r_0$。因此我们的骨骼所接受的辐射剂量为 16 毫拉德，骨骼表层细胞为 11 毫拉德，骨髓和软组织为 7 毫拉德。7 毫拉德的剂量会对基因造成危害。这一剂量大致等于生殖腺从所有放射性尘埃中所吸收的总剂量的 10%。但珀德姆 [4] 指出：实际上生殖腺所接收到的辐射剂量还要更大一些，因为 DNA 分子会发生嬗变，在嬗变过程中产生衰变的 ^{14}C 原子被替换成了 N 原子。珀德姆认为该嬗变过程对生物造成的伤害就相当于 β 辐射对生物的伤害。

到大约公元 2000 年时，人体中的人造 ^{14}C 量将占到 ^{14}C 总量的 3% 左右。这种同位素的半衰期为 5,700 年，因而有几位科学家 [2,3,9,28] 认为它将对人类基因造成严重的威胁。长期的辐射剂量可由下式计算：

$$D_2 = 0.03\, r_0 \int_0^t e^{-0.000125t}\, dt \tag{4}$$

在未来的 10,000 年里，由骨骼细胞、骨骼表层细胞以及骨髓和软组织所吸收的总辐射剂量（D_1+D_2）分别为 410、290 和 180 毫拉德。这些剂量与一份联合国报告 [9]（第 45 页）中给出的值一致，它们比所有其他的放射性尘埃都更重要。然而，我们对长期辐射剂量（D_2）有所怀疑，因为正如我们之前曾经提到的，矿物燃料的使用可能会导致人体中的 ^{14}C 浓度降低至正常水平以下。我们认为：在大约 30 年内，由以前有足够规模的核试验所造成的 ^{14}C 危害对总遗传剂量（D）的贡献只有 10 毫拉德数量级。而从天然源中吸收的剂量约为 100 毫拉德 / 年，可见，与天然源相比，人工源的剂量是不值一提的。

We thank Norges Almenvitenskapelige Forskningsråd for financial support.

(**232**, 418-421; 1971)

Reidar Nydal, Knut Lövseth and Oddveig Syrstad: Radiological Dating Laboratory, Norwegian Institute of Technology, Trondheim, Norway.

Received November 10, 1969; revised August 24, 1970.

References:

1. Libby, W. F., *Science*, **123**, 657 (1956).
2. Pauling, L., *Science*, **128**, 3333 (1958).
3. Sakharow, A. D., *Soviet Scientists on the Danger of Nuclear Tests*, 39 (Foreign Languages Publishing House, Moscow, 1960).
4. Purdom, C. E., *New Scientist*, **298**, 255 (1962).
5. Nydal, R., *Nature*, **200**, 212 (1963).
6. Broecker, W. A., Schulert, A., and Olson, E. A., *Science*, **130**, 331 (1959).
7. Libby, W. F., Berger, R., Mead, J. F., Alexander, G. V., and Ross, J. F., *Science*, **146**, 1170 (1964).
8. Harkness, D. D., and Walton, A., *Nature*, **223**, 1216 (1969).
9. *Rep. UN Sci. Comm. Effect of Atomic Radiation*, No. 14 (A/5814) (United Nations, New York, 1964).
10. Münnich, K. O., and Roether, W., *Proc. Monaco Symp.*, 93 (Vienna, 1967).
11. Bien, G., and Suess, H., *Proc. Monaco Symp.*, 105 (Vienna, 1967).
12. Rafter, T. A., *NZ J. Sci.*, **8**, 4, 472 (1965).
13. Rafter, T. A., *NZ J. Sci.*, **11**, 4, 551 (1968).
14. Young, J. A., and Fairhall, A. W., *J. Geophys. Res.*, **73**, 1185 (1968).
15. Lal, D., and Rama, *J. Geophys. Res.*, **71**, 2865 (1966).
16. Nydal, R., *J. Geophys. Res.*, **73**, 3617 (1968).
17. Nydal, R., *Symp. on Radioactive Dating and Methods of Low-Level Counting, UN Doc. SM 87/29* (International Atomic Energy Agency, Vienna, 1967).
18. Craig, H., *Second UN Intern. Conf. on the Peaceful Uses of Atomic Energy*, A/CONF. 15/P/1979 (June 1958).
19. Rafter, T. A., and O'Brien, B. J., *Proc. 12th Nobel Symp.* (Almquist and Wiksell, Uppsala, 1970).
20. Nydal, R., and Lövseth, K., *CACR Symp.* (Heidelberg, 1969); *J. Geophys. Res.*, **75**, 2271 (1970).
21. Berger, R., Fergusson, G. J., and Libby, W. F., *Amer. J. Sci., Radiocarbon Suppl.*, 7, 336 (1965).
22. Berger, R., and Libby, W. F., *Amer. J. Sci., Radiocarbon Suppl.*, **8**, 467 (1966).
23. Berger, R., and Libby, W. F., *Amer. J. Sci., Radiocarbon Suppl.*, **9**, 477 (1967).
24. Godwin, H., and Willis, E. H., *Amer. J. Sci., Radiocarbon Suppl.*, **2**, 62 (1960).
25. Godwin, H., and Willis, E. H., *Amer. J. Sci., Radiocarbon Suppl.*, **3**, 77 (1961).
26. Drobinski, jun., J. C., La Gotta, D. P., Goldin, A. S., and Terril, jun., J. G., *Health Phys.*, **11**, 385 (1965).
27. Nydal, R., and Lövseth, K., *Nature*, **206**, 1029 (1965).
28. Pauling, L., *Les Prix Nobel*, 296 (1963) (Nobelstiftelsen, Stockholm, 1964).

感谢挪威自然和人文科学研究理事会为我们提供了经费上的支持。

（邓铭瑞 翻译；刘京国 审稿）

Lithium in Psychiatry

I. B. Pearson and F. A. Jenner

Editor's Note

The medicinal uses of lithium salts far predate their current application to treat the psychiatric illness manic depression (bipolar disorder). But as psychiatrists I. B. Pearson and F. A. Jenner say in this overview, such uses were often over-optimistic and heedless of side-effects. The modern psychotropic application stemmed from a chance finding of mood alteration in 1949 by Australian doctor John Cade. Perhaps it is understandable that only two decades of use had not yet shed much light on the mode of action, but little more is known today. Pearson and Jenner hint that lithium might interfere with some aspect of brain metabolism; it now seems likely that it regulates production of a key neutrotransmitter, but the details remain unclear.

LITHIUM salts have been used medicinally in various ways for more than a century, and their history has been one of mistaken enthusiasm. Garrod recommended them for the treatment of kidney stones, gout and other rheumatic conditions in 1859[1], but without appreciable success. Subsequently lithium bromide was given as an anticonvulsant in the treatment of epilepsy, but it was no more effective than other bromides. At the end of the nineteenth century lithium salts were first tried in the treatment of mental illness, Lange[2] considered that lithium was helpful in mental depression, but fifty years were to elapse before further and more extensive psychiatric uses developed. More recently lithium salts have been offered as a substitute for sodium chloride in patients requiring a restricted sodium intake, but there were fatalities, and it is now recognized that cardiac and renal decompensation are both indications that the use of lithium should be avoided. Psychiatric interest was revived after work reported by Cade in 1949[3] when, following an accidental observation of a sedative effect on guinea-pigs, lithium was found to be helpful in controlling mania. The modern psychotropic use of lithium dates from that time, but dissemination has been slow and probably reflects a natural caution after the earlier fatalities, combined with the introduction of many new and effective psychoactive compounds.

During the past two decades lithium has been used increasingly in the treatment of manic-depressive illnesses in which the mood of the patient is elevated in mania and lowered in depressions. Results so far suggest that lithium salts have considerable potential in the treatment of manic depressive illnesses, although it is far from clear how they exert their effects.

Very occasionally the patient's mood changes in a regular clock-like manner, but it is more usual to find only occasional brief episodes of mania interspersed in a series of depressive episodes. The remarkably regular patients, though very rare, offer unusual opportunities to prove that lithium salts have a psychopharmacological effect. This has been done, but the relevance of such findings to the more usual clinical studies requires a more complicated

锂在精神病学中的应用

皮尔逊，詹纳

编者按

锂盐在被人们用于治疗躁狂抑郁症类精神病（双向障碍）之前早已作为药物使用。但正如精神病学家皮尔逊和詹纳在这篇综述中所指出的，人们过去对锂的应用总是过于乐观，毫不在意副作用。1949 年，澳大利亚医生约翰·凯德偶然发现了锂在改变情绪上的作用，从此开创了锂在现代精神类药物中的应用。也许 20 年的使用时间还不足以让人们认识锂的作用机制，但直到现在我们也几乎不比那个时候知道得更多。皮尔逊和詹纳指出锂可能干扰了脑内部分新陈代谢；现在看来它也可能控制着一种重要神经递质的产生，但具体细节尚不清楚。

一个多世纪以来，锂盐一直被人们用作各种形式的药物，其发展史已属于一种错误的狂热。1859 年，加罗德曾推荐用锂盐治疗肾结石、痛风以及其他风湿性疾病，可是并没有获得令人称道的成功 [1]。随后，溴化锂被用作一种治疗癫痫的抗惊厥药，但其疗效仅与其他溴化物相当。19 世纪末期，锂盐第一次被尝试用于治疗精神类疾病。朗厄 [2] 提出锂有助于治疗精神抑郁，但直到 50 年后，人们才把它更多、更广泛地用于治疗精神类疾病。最近，有人提出锂盐可以作为氯化钠的替代物用于那些需要限制钠摄入量的病人，但这种做法是致命的；因而现在人们认识到，为了防止心脏代偿和肾代偿失调，应该避免使用锂。在 1949 年凯德偶然观察到锂对豚鼠具有镇定作用 [3] 之后，人们发现锂有助于控制躁狂症，随后对锂在精神病治疗方面的研究兴趣又开始恢复。从那时起到现在，锂一直被用作治疗精神病的药物，但普及过程十分缓慢，可能是早期致死事件之后人类本能产生的谨慎以及许多高效精神类新药物的出现所致。

在过去的 20 年里，锂越来越多地被用于治疗躁狂抑郁性精神病，患有这种病的病人在躁狂时情绪激动，在抑郁时情绪又低落。目前的研究结果表明，锂盐在治疗躁狂抑郁性精神病方面具有很可观的疗效，但很难说清产生这种作用的机理到底是什么。

病人情绪的变化极少遵守严格的规律性，更常见的情况是，在一系列抑郁周期之中只会偶尔出现短暂的躁狂阶段。虽然定期发作的病人很少，但我们仍然从他们身上获得了一些能证明锂盐具有精神药理学效用的难得机会。这一工作已经完成，但上述结果与更普遍的临床研究之间的相关性还需要通过更为复杂的统计学方法来

statistical approach, taking proper account of the variable natural history of the condition.

It is difficult to classify these illnesses since unipolar affective disorders are much commoner that the bipolar type, and patients who have recurrent episodes of depression for several years may then have a manic episode. Depressions also merge into anxiety and other psychiatric states. The natural history and heterogeneity of the clinical group studied present all sorts of obvious difficulties to the investigator, and it is these types of problem which have influenced much of the debate about lithium in psychiatry. Schou, who produced a useful guide to the literature in 1968[4], has been the most active and assiduous student in this field.

The effectiveness of lithium in the treatment of mania is now well recognized, and has been confirmed by controlled studies. But this treatment requires daily doses of the order of 1.5–2 g of lithium carbonate, when therapeutic and toxic levels closely approximate. In recent years the prophylactic use of lithium has been advocated, using rather smaller doses in the range 0.75 to 1 g daily of lithium carbonate, when toxic effects are reduced. The efficacy of lithium salts in the prophylaxis of manic depressive illness has been the subject of some debate. It has been claimed that when effective, lithium salts diminish the change in mood and in addition reduce the frequency of relapse in affective illness. Angst *et al.*[5] have demonstrated a natural tendency in manic depressive illness for the frequency of episodes to increase, but the pattern of relapse in affective illness is variable, although it does tend to show a progression of increasing frequency which is greater in bipolar illness than in recurrent depressive illness.

Blackwell and Shepherd[6] properly challenged some of the earlier favourable reports, pointing out that with such patients seen at a time of frequently recurring illness, a reduced rate of relapse might be accounted for solely by chance, and certainly this could be so with patients having frequent episodes, even if there is an overall tendency for the cycles to shorten. They suggested that the prophylactic value of lithium could only be determined by a comparative trial. Subsequently several clinical studies have been reported, including double blind trials, and most of them support the original conclusion that prophylactic lithium is of value. In general it seems that the frequency of relapse is appreciably reduced and that the natural trend of increasing frequency is reversed.

Lithium belongs to the group of alkali metals which includes sodium and potassium and it shares several properties with the other members of this group, but Birch has pointed out that in certain respects it resembles magnesium and calcium to which it bears a diagonal relationship in the periodic table[7]. It is absorbed promptly and virtually completely from the gastro-intestinal tract, and this is independent of whether lithium is given as carbonate, citrate or acetate. Lithium is not bound to plasma proteins, but is distributed freely in the tissues, and equilibration occurs, although this is somewhat slower in bone and brain. Distribution in the tissue differs sharply from that of sodium and potassium since lithium is more evenly divided between intracellular and extracellular compartments. Concentrations in the serum reflect those in tissue and can be used satisfactorily to monitor treatment. Lithium salts are largely excreted by the kidneys, and only small amounts are lost in faeces and sweat. Renal lithium excretion is proportional to the plasma concentration over a wide range; renal clearance is of

确定，同时要适当考虑这种疾病的病程极易发生变化。

很难给这些精神病归类，因为单向情感性疾病比双向情感性障碍更为普遍，在几年内反复出现抑郁期的病人随后可能会表现出一个躁狂发作期。抑郁症也会合并焦虑和其他精神病症状。临床上的病程和异质性给研究者造成了各种各样显而易见的困难，也正是这些困难左右了很多关于锂用于精神病治疗的讨论。舒尔在1968年发表了针对这些文献的指南[4]，他堪称这个领域内最活跃和最勤奋的研究者。

现在人们已经承认了锂在治疗躁狂症方面的有效性，这一点已经被对照实验所证实。但治疗需要每天服用1.5克～2克碳酸锂，治疗所需剂量与中毒剂量非常接近。近年来，大家都倡导将锂作为预防用药，当每天服用碳酸锂的量降低至0.75克～1克时，其毒性就会降低。锂盐在预防躁狂抑郁症方面的效果已经成为众多争论的焦点。有人指出：当锂盐发挥功效时，它能减少情绪上的变化并且能降低情感性疾病的复发频率。昂斯特等人[5]已经证明躁狂抑郁症的一个自然趋势是发病频率不断增加，而情感性疾病的复发模式却是变化无常的，不过后者也会表现出一种发病频率逐渐增加的趋势，但双向性疾病的发病频率比复发性抑郁症的发病频率增加得更快。

布莱克韦尔和谢泼德[6]彻底挑战了早期的一些支持锂具有预防作用的报道。他们指出：对于那些处于疾病反复发作时期的病人，复发频率的下降可能只是偶然出现的现象，频繁发病的病人显然也会发生类似的情况，即使整体趋势是周期在缩短。他们认为，锂的预防作用只能用对比试验来确定。随后，又有几项临床研究的成果被发表出来，包括一些双盲试验，这些结果大多支持最初的结论，即锂作为预防性药物是有效的。在通常情况下，锂似乎会使复发频率明显下降，而发病频率自然增加的趋势也会被逆转。

锂和钠、钾一样都属于碱金属，它的一些性质与碱金属中的其他成员相同，但伯奇指出，在某些方面，锂也类似于在元素周期表中与它呈对角线关系的镁和钙[7]。无论锂的存在形式是碳酸锂、柠檬酸锂还是醋酸锂，它都会立刻被胃肠道完全吸收。锂没有与血浆蛋白结合，而是自由地分布于组织中，并会达到分布平衡，不过这种平衡在骨和脑中出现的速度比较慢。锂在组织中的分布与钠和钾完全不同，因为锂在细胞内结构和细胞外结构中的分布更平均。血清中的含量可以反映出组织中的含量，因而能够满意地用于监测治疗效果。大部分锂盐由肾排泄，只有很少量从粪便和汗液排出。从肾排出的锂在很宽的范围内正比于血浆浓度，肾清除率的数量级为15毫升 / 分钟～30毫升 / 分钟，但会随着年龄的增长而下降，在老年人中低至

the order of 15–30 ml./min, but decreases with age, and values of 10–15 ml./min are not uncommon in elderly people. Lithium excretion decreases when sodium intake is restricted.

The principal disadvantage of lithium therapy lies in the narrow safety margin between therapeutic and toxic doses. Therapeutic plasma concentrations are usually considered to be between 0.8 and 1.6 mequiv./1.; at these concentrations only minor side effects occur. The commonest of these effects are gastro-intestinal disturbances, muscular weakness, tremor, drowsiness and a dazed feeling. When plasma concentrations are higher, serious toxic effects are liable to ensue, and poisoning results in coma, hyperreflexia, muscle tremor, attacks of hyperextension of the limbs and convulsions; fatalities have occurred. Haemodialysis is the most effective treatment if available but where this is not possible patients should be treated by forced diuresis. Long term treatment may give rise to small goitres and hypothyroidism but these usually respond to modest doses of thyroxine. Teratogenic effects have now been reported in two out of forty patients who received lithium during pregnancy[8]. Clearly lithium should be used with care, and concentrations in the blood should be monitored regularly.

Mode of Action

The mode of action of lithium in affective illness remains obscure, and most theories have related to interference in electrolyte metabolism or to effects on the metabolism of brain amines. More recently evidence has been adduced to suggest a specific effect of lithium on the limbic system of the brain, an area known to be associated with emotion[9]. These uncertainties about the action of lithium reflect the unsatisfactory state of understanding of the biological disturbance associated with manic-depressive illness.

When the fundamental importance of sodium and potassium for neuronal activity is considered together with the similarity of their ions to those of lithium, the suggestion that lithium interferes with the metabolism of these ions seems to be reasonable. Evidence is accumulating that disturbance of electrolyte and water balance accompanies manic-depressive illness although it is not clear whether this is causal or consequential. It has been shown that the administration of lithium leads to changes in the body water spaces, and this observation is supported by the natriuresis following the intake of lithium, which is more striking in normal people than in manic-depressive patients. More lithium is retained by patients than by normal people, and the retention is greater in mania and in patients who respond to lithium therapy. In addition a proportion of patients taking lithium develop persisting polyuria and thirst. This may be explained by the fact that lithium ions specifically and reversibly inhibit the action of the antidiuretic hormone on the kidney. Conceivably the consequences of this inhibition are relevant to the therapeutic effects.

Further work on this particular action has demonstrated that lithium probably blocks the action of the cyclic AMP released by vasopressin[10]. This raises the possibility that inhibition of the effect of vasopressin is just one example of an action on the effects of all hormones releasing cyclic AMP in target cells. In addition to its action on cyclic AMP, lithium also inhibits adenyl cyclase necessary for the production of cyclic AMP from ATP, but in the kidney this only occurs with high doses[11].

10 毫升 / 分钟～15 毫升 / 分钟的情况并不少见。当钠的摄入量受限时，锂的排泄也会下降。

锂作为治疗药物的主要缺点在于治疗剂量和中毒剂量之间的安全范围太窄。一般认为血浆治疗的浓度在 0.8 毫当量 / 升～1.6 毫当量 / 升之间，在这样的浓度范围内，只会出现很有限的副作用。其中最常见的是胃肠道障碍、肌肉无力、颤抖、嗜睡和晕眩。当血浆中的药物浓度更高时，很可能会随即出现严重的中毒反应，中毒的后果是昏迷、反射亢进、肌肉颤动、肢体伸展过度发作和痉挛，以至于致命。如果有条件，血液透析是最有效的治疗手段；但在条件不允许的情况下，必须对病人采用强迫利尿的治疗方法。长期治疗可能会引起甲状腺腺体缩小和甲状腺功能减退，但在通常情况下保留了对适度的甲状腺素的反应。在 40 例怀孕期间接受过锂治疗的病人中，目前已报道出现 2 例致畸 [8]。显然在使用锂时应格外小心，并且要定期检测其在血液中的浓度。

作用机制

现在还不清楚锂在治疗情感性疾病方面的作用机制，大多数理论认为它与干扰电解质的代谢或者与影响脑胺的代谢有关。最近有人证明，锂可以对与情感相关的区域——大脑边缘系统产生特定的影响 [9]。关于锂作用机制的这些不确定性说明，人们在理解与躁狂抑郁症有关的生物失调方面还不能令人满意。

当把钠和钾对神经元活动的重要性同它们的离子类似于锂离子联系起来考虑的时候，就会认为锂干扰这些离子新陈代谢的观点是有道理的。不断积累的证据表明电解质和水平衡的紊乱与躁狂抑郁症有关联，但现在还不清楚两者之间是因果关系还是继发关系。有人指出服用锂会导致体内水区发生变化，而且这一发现也被摄入锂后出现的尿钠排泄所支持，这种症状在正常人身上的表现比在躁狂抑郁症患者身上更明显。在病人体内锂潴留比正常人更多，尤其是躁狂症病人和对锂治疗有反应的病人。另外，有一部分服用锂的病人会持续出现多尿和口渴的症状。这也许可以用以下论点来解释：锂离子通过特异性和可逆的方式抑制了肾脏内抗利尿激素的作用。或许可以认为这种抑制作用与疗效有关。

有人对这种特异性作用进行了深入的研究。结果表明，锂可能抑制了由抗利尿激素引起的环磷酸腺苷的释放 [10]。这就引出了这样一种可能性，即抗利尿激素的作用受到抑制仅仅是锂对靶细胞中所有能导致释放环磷酸腺苷的激素的其中一种作用。除了能对环磷酸腺苷产生影响之外，锂还能抑制由三磷酸腺苷生成环磷酸腺苷所需的腺苷酸环化酶，但在肾脏中这种抑制只有在高剂量的情况下才会出现 [11]。

Perhaps a more exciting observation is an increase of α-ketoglutaric acid (shown by Bond *et al.*) in the urine of patients receiving lithium. This observation, although unequivocally established, still requires explanation. Dose lithium, for example, inhibit enzymes in the tricarboxylic acid cycle and so cause a build-up of α-ketoglutaric acid? Is it a response to alkalosis and does it occur in brain? This, like many other questions, is being actively explored. It could also be relevant to Delgado and DeFeudis's observation[9] that lithium leads to an increased level of brain glutamate.

Inhibition of several enzyme systems could be the result of effects of lithium on tissue magnesium and calcium which have now been observed. As these changes may also occur in brain, possible hypotheses of the consequences are legion.

The work of Smith, Balagura and Lubran[12] demonstrates an apparently direct action of lithium on the lateral nuclei of the hypothalamus. There is stimulates drinking behaviour and presumably thirst. Application of lithium ions directly to the hippocampus and amygdala of monkeys[9] evoked rather specific EEG effects which could be associated with the reduction of aggression reported in man, rodents and fish[13-15]. These are among the few reported direct effects of lithium on nervous tissue, and it is not surprising that neural function can be altered in this way, but the significance of these observations in relation to the therapeutic actions of lithium is not entirely clear.

Lithium does, however, have very clear effects on the electroencephalogram in man. In particular, it increases the amplitude of the alpha rhythm and causes slow waves to appear. After quite usual doses of lithium carbonate have been discontinued, the effects on the electroencephalogram persist for up to 6 weeks.

Lithium, like so many other agents which have therapeutic uses, is like a heavy spanner in a delicate piece of machinery. Like electroconvulsive therapy, it changes so many things that clues to as what is relevant are difficult to assess. Among other effects which require further study, one must include inhibition of carbohydrate transport[16-18] across various membranes including the inhibition of the accumulation of myoinositol by the lens[19] and kidney[20]. The latter substance may well have a special role in the central nervous system.

The most popular current theories suggest that brain monoamine metabolism is altered in manic-depressive illness, but again, as with electrolytes, it is uncertain whether this is cause or effect; moreover, it is not clear which of the amines is related to changes in mood. Lithium has effects on brain monoamine metabolism, but these are confined to noradrenaline and the function of noradrenergic neurones. Noradrenaline metabolism is shifted from O-methylation to intraneuronal deamination with an increase in noradrenaline destruction; noradrenaline uptake by synaptosomes is increased and the release of synaptic neurotransmitters may be inhibited. Unfortunately these findings do not easily fit into models developed to account for the actions of psychotropic drugs, since such drugs change mood in one direction only, whereas lithium restricts mood swing in both directions. One school of thought, however, considers mania as a more profound form of depression[21].

一个更令人兴奋的发现也许是，在病人服用锂后其尿液中的α– 酮戊二酸含量会有所增加（邦德等人的研究成果）。尽管没有人怀疑这个结果的真实性，但仍需要对其进行解释。例如：锂是否会在三羧酸循环中抑制酶的作用从而使 α– 酮戊二酸得以积聚？这是碱中毒的一个反应吗？会不会在大脑中出现？就像对许多其他问题一样，人们一直在积极探索这个问题的答案。德尔加多和德费乌迪斯的观察结果也与此相关[9]，他们发现锂会导致脑谷氨酸水平的提高。

几个酶系统之所以被抑制可能是由于最近观察到的锂对组织镁和组织钙所起的作用。因为这些变化也可能会出现在脑中，所以对后果的假设不胜枚举。

史密斯、巴拉古拉和卢布兰[12]通过研究证明：锂会直接作用于下丘脑的外侧核。这样就会刺激人去饮水，可能是因为感到了口渴。将锂离子直接作用于猴的海马区和杏仁核[9]会引发特异性很强的脑电效应，这种脑电效应可能与人类、啮齿动物及鱼类的攻击性下降有关[13-15]。关于锂可以对神经组织产生直接影响的报道寥寥无几，以上列举了其中的几个，它们无疑会使神经功能发生变化，但目前还不完全清楚这些发现对于解释锂的疗效有什么帮助。

然而，锂确实对人类的脑电图有很明显的影响，尤其是它可以增加α节律的波幅，进而导致慢波的出现。在停止使用常规剂量的碳酸锂之后，它对脑电图的作用还能持续长达 6 周。

如同许多其他用于治疗的药物一样，锂所起的作用也类似于一台精密机器中的重型扳手。和电痉挛治疗一样，锂改变的要素太多以至于很难评估到底哪一个与它的疗效相关。在其他一些需要深入研究的效应中，我们应该把抑制碳水化合物穿过不同生物膜的转运[16-18]考虑在内，其中包括晶状体[19]和肾脏[20]对肌醇堆积的抑制。因为肌醇在中枢神经系统中很可能起着特殊的作用。

当前最流行的理论认为，躁狂抑郁症病人脑内的单胺代谢会发生变化，但和电解质代谢一样，我们还是不清楚这到底是致病原因还是患病后的结果，而且也不清楚哪一种胺与情绪改变有关。锂对脑内的单胺代谢确实有影响，但影响范围只限于去甲肾上腺素和去甲肾上腺素能神经细胞的功能。随着去甲肾上腺素被破坏程度的增加，去甲肾上腺素代谢从 O–甲基化转变到神经细胞内的脱氨基作用；突触体对去甲肾上腺素的摄取也在增加，而突触神经递质的释放可能会减少。不幸的是，这些发现不太容易适用于那些解释神经药物作用机制的模型，因为这些药物只在一个方向上改变了情绪，而锂对情绪波动的限制是双向的。然而，有一个学派认为可以把躁狂症看作是抑郁症的一种深度发展的形式[21]。

The pharmacological actions of lithium salts are curiously complex, and knowledge about them is far from complete, Indeed, there is ample room for further research designed to increase our understanding of the effects of lithium, which may in turn shed more light on the biological mechanisms possibly involved in manic-depressive illnesses.

(**232**, 532-533; 1971)

I. B. Pearson: University Department of Psychiatry, Sheffield S10 3TL.
F. A. Jenner: MRC Unit for Metabolic Studies in Psychiatry, University Department of Psychiatry, Sheffield S10 3TL.

References:

1. Garrod, A. B., *Gout and Rheumatic Gout* (Walton and Maberly, London, 1859), cited by Ottosson, J. O., *Acta Psychiat. Scand.*, Suppl. 207 (1969).
2. Lange, C., *Bidrag til urinsyrediates ens klinik. Hospitalstidende*, **5**, 1 (1897), cited by Ban, T., *Psychopharmacology* (Williams and Wilkins, 1969).
3. Cade, J. F. J., *Med. J. Austral.*, **36**, 349 (1949).
4. Schou, M., *J. Psychiat. Res.*, **6**, 67 (1968).
5. Angst, J., Weis, P., Grof, P., Baastrup, P. C., and Schou, M., *Brit. J. Psychiat.*, **116**, 604 (1970).
6. Blackwell, B., and Shepherd, M., *Lancet*, **i**, 968 (1968).
7. Birch, N. J., *Brit. J. Psychiat.*, **116**, 461 (1970).
8. Schou, M., and Amdisen, A., *Lancet*, **i**, 1391 (1970).
9. Delgado, J. M. R., and DeFeudis, F. V., *Exp. Neurol.*, **25**, 255 (1969).
10. Harris, C. A., and Jenner, F. A., *J. Pharmacol.* (in the press, 1971).
11. Dousa, T., and Hechter, O., *Life Sci.*, **9**, 1, 765 (1970).
12. Smith, D. F., Balagura, S., and Lubran, M., *Physiol. Behav.*, **6**, 209 (1971).
13. Sheard, M. H., *Nature*, **230**, 113 (1971).
14. Sheard, M. H., *Nature*, **228**, 284 (1970).
15. Weischer, M. Z., *Psychopharmacologia*, **15**, 245 (1969).
16. Bhattacharya, G., *Biochim. Biophys. Acta*, **93**, 644 (1964).
17. Bihler, I., and Adamic, S., *Biochim. Biophys. Acta*, **135**, 466 (1967).
18. Bosackova, J., and Crane, B. K., *Biochim. Biophys. Acta*, **102**, 423 (1965).
19. Varma, S. D., Chakrapani, B., and Reddy, V. N., *Invest. Ophthal.*, **9**, 794 (1970).
20. Margolis, R. U., and Heller, A., *Biochim. Biophys. Acta*, **98**, 438 (1965).
21. Coppen, A., and Shaw, D. M., *Lancet*, **ii**, 805 (1967).

锂盐的药理作用极其复杂，目前人们在这方面的知识还远远不够。确实，为了增加对锂作用机制的了解，我们还需要进行广泛的研究，这也许会有助于揭开隐藏在躁狂抑郁症背后的生物学机制。

（刘霞 翻译；陈建国 审稿）

Formation of New Connexions in Adult Rat Brains after Partial Deafferentation

P. D. Wall and M. D. Egger

Editor's Note

Here Patrick D. Wall and M. David Egger provide dogma-challenging evidence that the adult mammalian central nervous system contains some capacity to reorganize itself functionally after injury. The researchers severed part of a neuronal pathway linking leg to brain in the rat, and found to their surprise that "leg-responsive" brain cells began responding to "arm" stimulation after the surgery. The most likely explanation, they conclude, is that undamaged neurons grew new projections that formed connextions with their damaged neighbours. Such "plasticity", where the brain reorganizes neural pathways based on new experiences, is now well documented after brain injury, and is known to play an important role in learning and memory.

BRAIN damage in adult mammals may be followed by remarkable recovery or by a depressingly permanent defect. Any understanding of this sometime plasticity would be important not only for therapy but also for understanding how the central nervous system adjusts to its environment during development and learning. During experiments designed for a different purpose, we came across evidence that a group of cells which had been deprived of their major input began to respond to quite novel stimuli after a few days. In these experiments, the cell bodies of the normal major input had been destroyed. The one class of explanation which cannot be invoked is true regeneration in which cells had divided and replaced the destroyed cells. In the mammalian adult central nervous system (CNS) there is no evidence for nerve cells which are still capable of division, growth and differentiation: following destruction of nerve cells, non-neural elements divide and migrate into the region of damage, phagocytosing the wreckage and replacing it with an impenetrable scar. Even if the cell bodies are intact but their axons are severed, there is no evidence of successful regeneration of axons within the CNS.

Our experiments involved the dorsal columns, through which cutaneous sensory afferents travel from the hind limbs to terminate on cells of nucleus gracilis. From this nucleus, axons decussate and project to the ventral posterior nucleus of the thalamus (VPL). Similar projection systems pass from the forelimb by way of the cuneate nucleus and from the face through the trigeminal nucleus. Within this region of the thalamus, the entire contralateral body surface is represented in an exact somatotopic map (Fig. 1). Recent experiments have suggested that discriminative cutaneous sensation survives the destruction of the dorsal columns[1]. One explanation for this lack of effect is that other systems converge onto the thalamic nucleus and transmit the necessary afferent information by alternative pathways. To test this hypothesis, experiments were in progress to examine the response of cells

部分传入神经阻滞后成年大鼠脑内新联系的形成

沃尔，埃格

编者按

在本文中，帕特里克·沃尔和戴维·埃格提供了挑战传统观点的证据，该证据证明成年哺乳动物的中枢神经系统在受损后，具有部分功能重组的能力。研究者切断了大鼠身上连接后肢和脑的那部分神经通路，他们惊奇地发现：原先“对后肢产生应答”的脑细胞在术后开始对“前肢”所受刺激作出反应。他们总结出的最可能的解释是：未损坏的神经元长出新投射，与已受损的相邻神经元之间产生联系。如今，脑在受损后基于新经验重组神经通路的这种“可塑性”已被很好地证明，这种“可塑性”也在学习和记忆中扮演重要的角色，此观点已成为共识。

成年哺乳动物脑损伤后可能会奇迹般地恢复，或者成为悲剧性的永久缺陷。对这一偶然发生的可塑性的任何理解不仅仅对治疗来说非常重要，而且对于理解中枢神经系统在发育和学习过程中如何适应它周围的环境也同样重要。在为另一个目的而设计的实验中，我们偶然发现了以下现象：一组已经失去自身主要输入的细胞，在几天后开始对全新的刺激产生应答。在这些实验中，具有正常主要输入的细胞体已被破坏掉。有一类解释是我们不能接受的，即发生了真正的再生，在再生过程中细胞进行分裂并取代了被破坏的细胞。在成年哺乳动物的中枢神经系统（CNS）中，没有证据表明神经细胞仍然能够分裂、生长和分化：在神经细胞遭到破坏后，非神经细胞就会分裂并移入受损区域，吞噬神经细胞残片并用一个难以穿过的伤疤替代它。即使细胞体未受损而只是它们的轴突被切断，也没有证据能证明中枢神经系统中的轴突可以成功再生。

我们的实验涉及背柱，来自后肢的皮肤感觉传入经过它到达终端薄束核的细胞。轴突从这一神经核起交叉成十字形并投射到丘脑腹后外侧核（VPL）。类似的投射系统转为前肢的话，将经过楔束核；而来自面部的类似投射系统将经过三叉神经核。在丘脑的这一区域内，全部对侧身体表面可反映在一张准确的躯体位置图上（图 1）。最近的实验表明，具有辨别能力的皮肤感觉在背柱破坏后仍能幸存[1]。对于这种免受影响现象的一种解释是：其他系统汇聚到丘脑神经核团并通过替代的通路来传输必要的传入信息。为了检验这种假说，我们正在进行一系列实验，以检测背柱内侧丘系系统被破坏之前和之后丘脑腹后外侧核细胞的反应。分别在切断 11 只大鼠胸背

in VPL before and after part of the dorsal column-medial lemniscus system had been destroyed. Responses had been recorded in VPL immediately before and after section of thoracic dorsal columns in eleven rats, and after destruction of one nucleus gracilis in eight rats. We suspected that the full potentiality of the cells might not be revealed immediately after deafferentation and therefore decided to leave the animals for various times after the lesion before examining the thalamus. We encountered the unexpected result that cells in VPL which normally respond to leg stimulation began to respond to arm stimuli some time after destruction of nucleus gracilis. We shall report elsewhere the effect of acute and chronic deafferentation on the response of VPL cells to leg and body stimulation. Here we concentrate on those cells whose receptive fields switched from leg to arm.

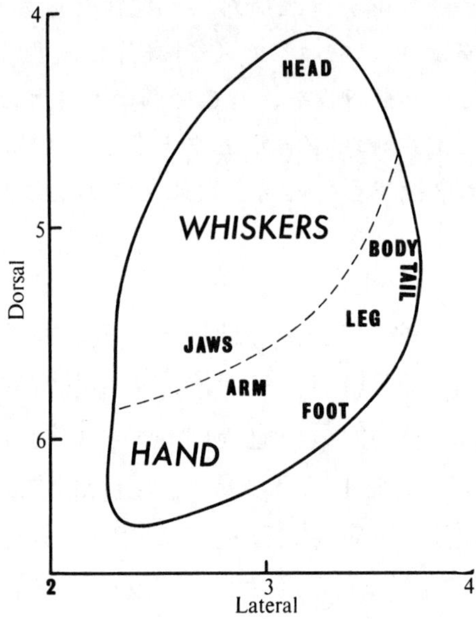

Fig. 1. Diagram of the representation in rat thalamus of the contralateral body surface. The region in which arm, trunk, leg and tail is represented in nucleus ventralis-posterior-lateralis. Maps are produced by placing microelectrodes successively at many locations in a single transverse plane in the region. For each recording point, the face and body surface is mechanically stimulated. Cells close to the recording electrode respond to a small area of skin. A combination of the receptive fields for all the points provides a map of the entire skin surface. The transverse plane is 4.5 mm rostral to external auditory meatus and the areas show distances in mm lateral to the midline and vertically below the cortical surface.

Recordings were made in seventy-seven adult rats, ASH–CSE strain, weighing 200–400 g, under urethane (25%) anaesthesia, 0.65 ml./100 g. The head was held in a stereotactic machine with the lambda vertically above the ear bars in which position the centre of VPL is A 4.5, L 3.0 and 5.0 mm below cortical surface. The nucleus is crescent shaped in cross section and extends 1.4 mm rostrocaudally and mediolaterally. Thalamic recordings were made with glass-covered platinum microelectrodes whose impedance was 600–800 kohms at 1,000 Hz. When marking of electrode sites was required, recording was first carried out with silver plated steel microelectrodes followed by electrolytic deposition of iron which was later located by the prussian blue reaction. Cortical unit recordings were made with

柱之前和之后，以及在破坏 8 只大鼠的一个薄束核后，立刻记录丘脑腹后外侧核的反应。我们怀疑在传入神经刚刚被阻滞之后细胞可能还没有发挥出全部的潜能，于是决定在损伤后把实验动物放养不同的时间然后再检验其丘脑。我们得到了出乎意料的结果：在薄束核受损一段时间之后，通常应该对来自后肢的刺激产生反应的丘脑腹后外侧核细胞开始对来自前肢的刺激有反应了。我们将在别处报道在急性和慢性传入神经阻滞后丘脑腹后外侧核细胞对来自后肢和躯干的刺激的反应。这里我们将把精力集中在那些感受野从后肢转换到前肢的细胞上。

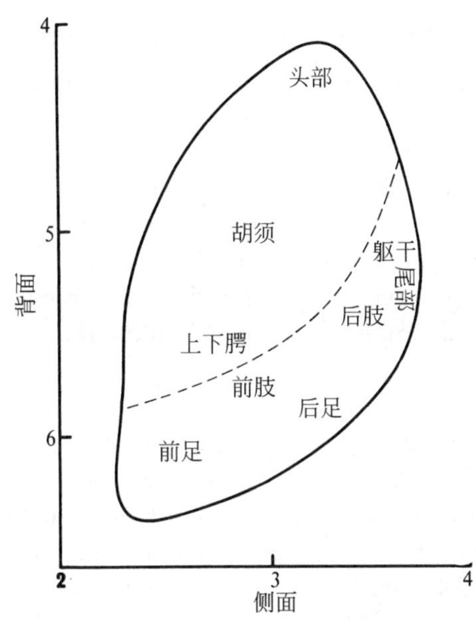

图 1. 对侧身体表面在大鼠丘脑中的映射图。前肢、躯干、后肢和尾部传入投射到丘脑腹后外侧核上的对应代表区域。这类映射图是通过在该区域单一横断面上多个位置陆续放置微电极得到的。每个记录点都是对动物面部和身体表面进行机械刺激的结果。记录电极周边的细胞对皮肤的一小块区域作出反应。所有这些点提供的感受野组合在一起就拼成了整个皮肤表面的映射图。横断面位于外耳道喙侧 4.5 mm 处，该区域到中线的距离为 mm 级，并垂直地位于皮层下方，距离也是 mm 级。

取 77 只 ASH–CSE 品种的成年大鼠，每只体重 200 g ~ 400 g，每 100 g 体重注射 0.65 ml 乌拉坦（25%）来实现麻醉，然后进行记录。将头部固定在脑立体定位仪中，使人字缝尖垂直高于耳夹，在这种定位中，丘脑腹后外侧核的中心在皮层表面之下，距离腹面 4.5 mm，距离侧面分别为 3.0 mm 和 5.0 mm。该神经核在横断面上为新月形，并向喙尾侧和中间外侧延伸 1.4 mm。用玻璃包裹的在 1,000 Hz 时电阻为 600 kΩ ~ 800 kΩ 的铂微电极记录丘脑信号。当电极位点需要标记时，先用镀银的钢微电极记录信号，接下来进行铁电解沉积，稍后用普鲁士蓝反应定位。皮层单元信号用充满

glass microelectrodes filled with 3 M KCl with impedance of 2–4 M ohms. When latencies were required, electrical stimuli were applied to the cutaneous receptive field of the unit through intradermal 30 gauge needles. Chronic lesions of nucleus gracilis or dorsal columns were made in sterile conditions with the animal anaesthetized with "Equithesin". With the head fully flexed and the cisterna magna open, nucleus gracilis can be observed directly without further dissection. The nucleus was destroyed by repeated maceration with honed jeweller's forceps. At various times after this lesion, thalamus or cortex was mapped and then the extent of the lesion was determined from serial frozen sections stained with darrow red–luxol fast blue. Only those animals with the lesion restricted to gracile nucleus are reported. In some, the extreme caudal end of the nucleus was spared.

Distribution of Receptive Fields

The distribution of receptive fields of cells in VPL was mapped in sixteen intact animals. In a further thirty-five animals maps were made of the intact VPL on one side so that these maps could be compared with those from the opposite VPL which had been partially deafferented by nucleus gracilis destruction. Successive recording points were distributed in a regular grid separated by 100 μm vertically and 200 or 250 μm rostrocaudally and mediolaterally. For each station on this three dimensional grid, the entire body surface was brushed with medium pressure and the position was noted of any area of skin which evoked unit responses with an amplitude greater than 50 μV.

In the normal VPL individual cells had small discrete receptive fields. Neighbouring cells recorded at a single electrode site had overlapping receptive fields restricted to some small area of the contralateral body surface. In 35% of the recording sites, cells responded to stimulation of trunk or proximal portions of the limbs or tail and these cells lie in the dorsolateral part of the nucleus. Twenty-four percent responded to lower hind leg or feet or toes and lie ventrolaterally. Forty-one percent responded to lower forearm or hands or fingers and are located ventromedially. The ventral surface of the nucleus is made up of a lamina about 200–400 μm thick which contains cells that respond to passive movement of the limbs. Ventral to the arm area these cells respond to arm movements and the cutaneous leg zone has leg movement cells ventral to it.

Cross-sectional maps were made first of one side and then continued across the midline to investigate the opposite nucleus. Fifty-seven pairs of maps were made in twenty animals at various stages after destruction of n. gracilis. Five variables were compared in the normal nucleus and its partner which had lost afferents: (1) the stereotactic coordinates of areas responding to particular parts of the body; (2) the ratio of the number of points on each side which responded to certain stimulus sites; (3) the monopoly by hind limb cells of the rostral pole of the normal nucleus; (4) the map in the ventral lamina of movement detection cells, which was unaffected by gracilis lesions and therefore provided another marker; (5) the receptive fields and latencies of selected single cells.

In rats mapped 1 or 2 days after destruction of the n. gracilis the picture is the same as immediately after the lesion (Fig. 2). The lower arm–hand–finger area is unchanged. In the leg–foot–toe region, there is an almost complete disappearance of sites where stimuli evoked activity. We shall describe the nature of these scattered responses to leg stimuli and

3 M KCl的电阻为2 MΩ～4 MΩ 的玻璃微电极记录。当需要测量反应时间时，可以用30号皮内针对与该单元对应的皮肤感受野进行电刺激。腹腔注射“Equithesin”(译者注：0.6% 戊巴比妥钠和水合氯醛混合物）麻醉动物，在无菌条件下慢性损坏薄束核或背柱。使头部充分弯曲，打开小脑延髓池，不用进一步解剖就能直接观察到薄束核。用精巧珠宝镊反复浸离，破坏该神经核。在这种损伤后的不同时间点，绘制丘脑或皮层映射图，接下来用达罗红–罗克沙尔固蓝对连续冷冻切片进行染色，以确定损伤范围。我们只报道了那些损伤限制在薄束核内的动物的实验结果。在一些实验动物中，该神经核的极尾端没有被破坏。

感受野的分布

作16只正常动物的丘脑腹后外侧核细胞的感受野分布图。再作另外35只动物未受损一侧的丘脑腹后外侧核的感受野分布图，在此基础上，可以与对侧的因薄束核损伤导致部分传入神经阻滞的丘脑腹后外侧核感受野分布图进行比较。连续记录点分布在规则网格上，垂直间距为100 μm，喙尾侧和中间外侧间距为200 μm或250 μm。对于这一三维网格的每一记录点，用中等压力刷遍体表，标出能够引起单元反应幅度超过50 μV的任意皮肤区域。

在正常的腹后外侧核中，个别细胞具有小的离散感受野。由单电极位点记录的邻近细胞具有重叠的感受野，这些感受野都局限在对侧身体表面的小区域内。在35% 的记录位点中，细胞对来自躯干或四肢最接近躯干部分或尾部的刺激有反应，这些细胞分布在该神经核的背外侧部分。在24% 的记录位点中，细胞对来自后肢下部或后足或后爪的刺激有反应，这些细胞分布在该神经核的腹外侧部分。在剩下41% 的记录位点中，细胞对来自前肢下部或前足或前爪的刺激有反应，这些细胞位于该神经核腹正中。该神经核的腹部表面由200 μm～400 μm厚的单一薄层组成，薄层里含有对四肢的被动运动有反应的细胞。在前肢反应区腹侧的这些细胞对前肢运动有反应，而对后肢运动有反应的细胞位于皮肤的后肢映射带腹侧。

先完成一侧的横断面映射图，然后继续越过中线研究对侧的神经核。取20只动物，在破坏薄束核后的不同阶段作出了57对映射图。比较正常神经核和它对侧失去传入功能的神经核的5个变量：(1）对身体特定部位作出反应的区域的立体定位坐标；(2）两侧对确定刺激位点能产生反应的点的数量比；(3）正常神经核喙极对应后肢的细胞单独支配的区域；(4）由运动感知细胞构成的腹侧薄层中的对应图，它不受薄束损伤的影响，因此提供了另一标记；(5）被选单细胞的感受野和反应时间。

薄束核破坏1到2天后再对大鼠的反应作映射图，得到的图和损伤后立即作的图相同(图2)。前肢下部–前足–前爪区域没有发生改变。在后肢–后足–后爪区域，由刺激能够引起活性的位点几乎完全消失。仅存在一些对后肢和躯干区域刺激产生

those in the body region in another paper. In eight of nine animals mapped 3–17 weeks after the lesion, responses to stimuli on trunk, tail and hind leg remained essentially the same as in the acute and short survival animals but there was an expansion of the arm area into regions normally responding to leg stimuli. Sites responding to body and proximal limb stimuli averaged 25% fewer than normal, those for leg averaged 87.5% fewer but the number of arm points increased by 24%. Examination of the maps in detail shows that this expansion is produced by a lateral movement of the lateral edge while the medial part remains unchanged (Fig. 3). This lateral zone subserves the lower arm and wrist and these sites increased by an average of 132%. The change was most marked at the rostral pole where the zone normally monopolized by leg cells was now occupied by arm cells. Arm cells now lay dorsal to cells which responded to leg movement. The 16 and 17 week survival animals showed no greater expansion than the 3 week survivors. In no animal was the entire area which normally responded to hind limb invaded by the forelimb area. In a ninth animal with 3 week survival there were signs of expansion of the body area.

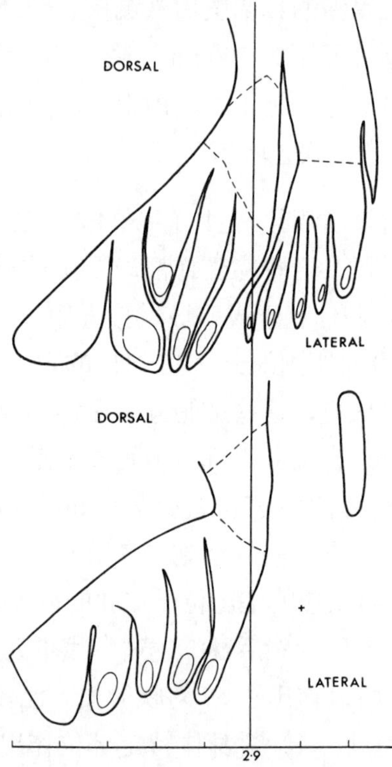

Fig. 2. Transverse map of distribution of receptive fields in VPL in rat 1 day after destruction of nucleus gracilis on one side. The map above shows the distribution in the thalamus supplied by the intact dorsal column nuclei with the forelimb representation medial, leg lateral, and body dorsolateral. The dotted lines mark the elbow and wrist on arm and the ankle on leg. The face area is not mapped. The map below shows the result of continuing the transverse search plane directly across the midline to the opposite thalamus which is not receiving an input from nucleus gracilis. The arm–hand–finger area is very similar in both maps. The leg area contained no responding cells in this plane with one exception marked by a cross. The horizontal axis marks 200 μm intervals in the mediolateral direction. The vertical line marks electrode tracks 2.9 mm from the midline penetrating both left and right thalamic maps. These tracks passed through the lateral arm region in each thalamus and show how similar the two arm regions are.

的零散反应，我们将在其他文章中描述这些反应的性质。在损伤 3 周 ~ 17 周后作映射图，9 只动物中有 8 只对于来自躯干、尾部和后肢的刺激的反应本质上与急性短期存活的动物一样，但是前肢区域会扩展到通常对后肢刺激有反应的区域。对来自躯干和四肢接近躯干部分的刺激产生反应的位点平均比正常动物少 25%，对后肢产生反应的位点平均比正常动物少 87.5%，但对前肢产生反应的位点数量增加了 24%。仔细检查映射图，发现这一扩展是由侧面边缘的侧向运动产生的，而中间部分仍维持原样（图 3）。这个侧面带促进了前肢下部和腕部，这些位点平均增长了 132%。这种变化大多集中在喙极，这一区域通常只分布着对后肢产生反应的细胞，现在却被对前肢产生反应的细胞所占据。对前肢产生反应的细胞现在位于对后肢运动产生反应的细胞的背侧。破坏 16 周和 17 周后依然存活的动物与破坏 3 周后的存活动物相比，扩展区没有进一步扩大。没有一只动物表现出通常对后肢产生反应的区域完全被对前肢产生反应的区域所占领的行为。在存活 3 周的动物中，有九分之一存在躯体区域扩展的迹象。

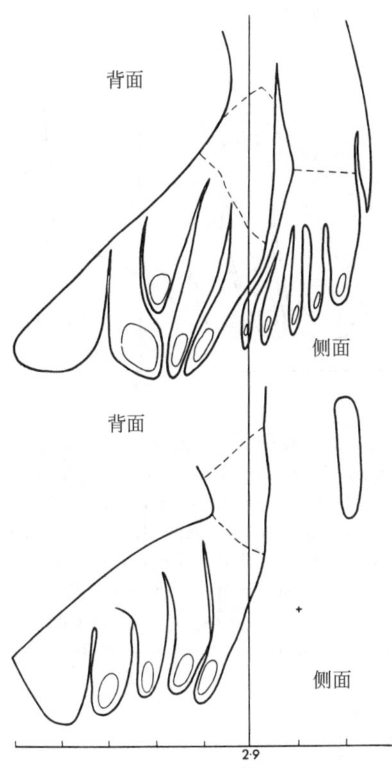

图 2. 单侧薄束核破坏 1 天后大鼠丘脑腹后外侧核中感受野的分布横断面图。上图显示了由未受损伤的背柱神经核提供的丘脑分布，中间代表前肢、侧面代表后肢以及背侧代表躯干。虚线标出了前肢的肘部和腕部所在处以及后肢的踝关节所在处。面部区域没有标示到图上。下图显示的是直接越过中线，对没有接收来自薄束核输入的对侧丘脑继续横向搜索断面的结果。前肢 – 前足 – 前爪区域在两幅图中非常相似。在这一断面上后肢区域不含有反应的细胞，有一个例外，用十字表示。水平轴沿中间外侧方向，刻度间隔 200 μm。垂线标出了电极轨迹，距离中线 2.9 mm，贯穿丘脑图的左右两侧。这些轨迹穿过每一丘脑的前肢区域侧面，并显示出两个前肢区域有多相似。

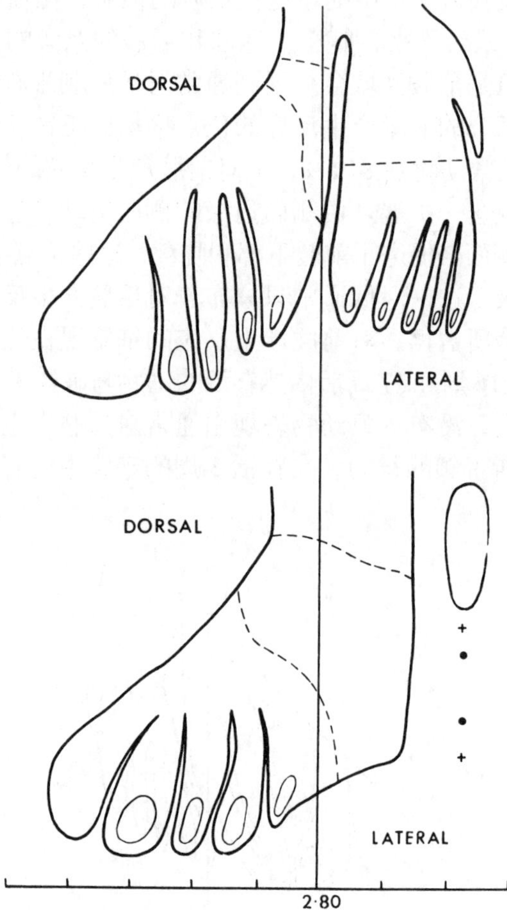

Fig. 3. Transverse map of distribution of receptive fields in intact VPL (above) and the map produced by continuing the mapping plane across the midline into the opposite VPL studied 7 weeks after destruction of the n. gracilis which projected to this nucleus (below). The vertical line marks an electrode track 2.8 mm from the midline which samples a similar region of the thalamus on the intact side to the vertical line shown in Fig. 2. The thalamus on the medial side of the line contains a similar map on both the intact and deafferented side. But the region representing the arm, especially the lower arm, has expanded on the operated side to invade a region which responds to leg on the intact side. At the lateral edge of the nucleus four cells were encountered which responded to body or leg stimulation but most cells in this region failed to respond to any peripheral stimuli.

Time Course and Nature of the Change

The simplest explanation for the expansion would be the shrinking of deafferented cells with a simple moving over of the neighbours. In this case there would be no new connexions and the cortical somatotopic map would not be distorted. Successful maps were completed of the arm and leg area of main sensory-motor cortex on both sides in animals 2, 5 and 6 week after destruction of one n. gracilis. The animals were anaesthetized with urethane and the body surface stimulated as in the thalamic recording experiments. In order to obtain a sufficiently detailed map it was necessary to record the

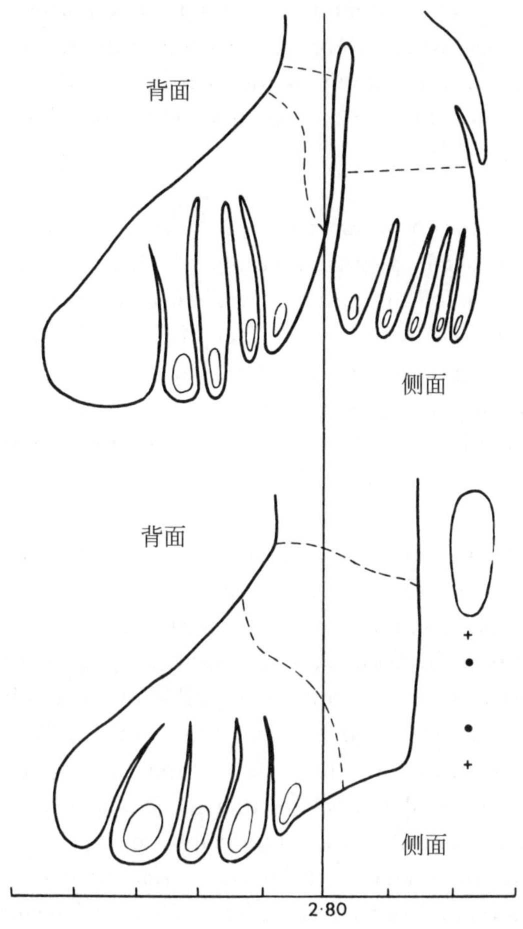

图 3. 未受损伤的丘脑腹后外侧核中感受野的分布横断面图（上图）；在投射到这一核团的薄束核被破坏7 周后进行研究，将断面继续绘至越过中线到对侧丘脑腹后外侧核，得到对应的映射图（下图）。垂线标出了电极轨迹，距离中线 2.8 mm，在丘脑未受损一侧垂线上的取样与图 2 中的垂线相同。垂线内侧的丘脑在未受损一侧和传入神经阻滞一侧的映射图类似。但是代表前肢的区域，尤其是代表前肢下部的区域，在手术一侧发生扩展，侵入到未受损一侧对后肢产生反应的区域。在该神经核侧面边缘找到 4 个对来自躯干或后肢的刺激产生反应的细胞，但这一区域中的大多数细胞对外周刺激不产生反应。

变化的时间进程和本质

对于这种扩展，最简单的解释可能是：传入阻滞的细胞发生萎缩，于是相邻细胞移动过来，简单地覆盖了萎缩细胞。在这种情况下，不会产生新联系，而且皮层躯体位置图也不会变形。分别在破坏一个薄束核 2 周、5 周和 6 周后，我们成功地绘制出动物两侧主要感觉–运动皮层前后肢对应区的映射图。用乌拉坦麻醉动物，并以在丘脑记录实验中所用的方式刺激体表。为了得到足够详细的映射图，需要记

receptive fields of cortical units by penetrating 500 μm into cortex with glass, 2–4 Mohms, KCl filled microelectrodes. Maps were constructed from a grid of recording points separated by 500 μm. In all three animals, the arm area had expanded into the leg area and it must therefore be concluded that thalamic cells projecting to cortex had changed their peripheral receptive fields.

The simplest explanation for the appearance of new receptive fields would be that the deafferentation had unmasked existing afferents which are normally inhibited. This is unlikely because in this type of preparation even massive electrical stimuli to arm in the intact animal fail to excite cells with discrete receptive fields on leg. Furthermore, there is a time delay in the switching of receptive fields. During this period when the former leg cells fail to respond to peripheral stimuli, ongoing activity continues in them at at least the frequency of normally connected cells showing that the deafferented cells themselves do not go through a period of lowered excitability.

To postulate the formation of new connexions, it is necessary first to discover the time course of their development. Eleven animals were examined 1–7 days after the lesion. At 1 and 2 days there was no significant change. At 3 days, two of three animals showed some expansion but without invasion of the rostral pole. At 4 days one animal showed both expansion and rostral invasion while a second showed no change. Of two animals at 5 days, one showed the full picture of expansion and the other some expansion. A 6-day survivor was inconclusive but the 7-day animal showed the same full picture as those examined after 3 weeks. We conclude that the changeover begins at 3 days and is fully established between 1 and 3 weeks. The cell bodies of the changed cells lay 200–400 μm from the original boundary of the normal area but one must remember that the dendrites of the innervated and deafferented cells might be intermingled so that any shift of connexion might be over very small distances.

It is conceivable that the stimulus for the change of connectivity is not the degeneration of the terminals but the absence of the normal afferent impulses which bombard the thalamic cells. This possibility was ruled out by sectioning the dorsal columns just caudal to n. gracilis leaving the nucleus itself intact. The section of the fibres was made to spare the more lateral component of dorsal columns which supplies the cuneate nucleus. Thalamic maps from this animal taken 7 days after the lesion showed no signs of expansion of the arm area into the silent region which no longer responded to leg stimuli. It is true that the thalamus would still be bombarded by the ongoing activity of the deafferented nucleus gracilis but this is presumably slight by comparison with that transmitted in the normal freely moving animal.

A possible explanation for the newly effective connexion is that a background of ineffective fibres from the cuneate nucleus lie scattered through the area normally innervated by n. gracilis and that these fibres mature and successfully stimulate the cells vacated by degenerating axons. The change could be either presynaptic or postsynaptic. If such a diffuse system of fibres existed, one might expect the receptive fields of the newly

录皮层单元的感受野，方法是用电阻为 2 MΩ ~ 4 MΩ 的充满 KCl 的玻璃微电极插入皮层下 500 μm。用间隔 500 μm 的记录点组成的网格构建映射图。在所有 3 只动物中，前肢区域都扩展到了后肢区域，因而我们只能认为丘脑投射到皮层的细胞已经改变了它们的外周感受野。

对于新感受野的出现，最简单的解释就是：传入阻滞使现存的、正常情况下被抑制的传入暴露出来。这种解释不太可能，因为在这类实验中，即使对未损伤动物的前肢施加强大电刺激，也无法激发后肢的离散感受野细胞。此外，感受野的转换还有一段时间延迟。在此前对后肢刺激有反应的细胞无法响应外周刺激的这段时间里，这些细胞中延续着持续的活性，至少维持正常联系细胞的频率，这说明传入阻滞的细胞本身并没有经历一个兴奋性降低的时期。

为了假定新联系的形成，首先需要揭示它们发育的时间进程。取 11 只动物，分别在损伤 1 天 ~ 7 天后加以检验。第 1 天和第 2 天没有明显的变化。在第 3 天，三分之二的动物显现出一些扩展，但是没有喙极的侵入。到第 4 天，有一只动物同时出现了扩展和喙极侵入，另一只仍没有显现出变化。在第 5 天，两只动物中的一只显现出完整的扩展图，另一只出现了一些扩展。一只损伤 6 天后的幸存动物结果不明确，但是损伤 7 天后的动物显示出了与那些损伤 3 周后才用于检查的动物一样完整的图。我们得出的结论是：转换过程在损伤后 3 天内开始，在损伤后 1 周到 3 周之间完全建立。转变后细胞的细胞体位于距离正常区域的原始边界 200 μm ~ 400 μm 处，但是必须注意到，受神经支配的和传入阻滞的细胞的树突有可能会混在一起，因此联系的任何改变也许仅跨越了非常小的距离。

有人认为：使连通性发生改变的刺激并不是终端的退化，而是冲击丘脑细胞的正常传入脉冲的缺失。通过切断紧靠薄束核尾侧的背柱同时保证其神经核完好无损的方法，可排除这种可能性。将纤维切断时需保留给楔束核提供信号的背柱更外侧部分。从损伤 7 天后的动物得出的丘脑映射图来看，不存在前肢区域扩展进入已不再对后肢刺激产生反应的沉默区的迹象。事实上，丘脑仍然会被传入阻滞的薄束核的持续活性所冲击，但是比起能够自由活动的正常动物的传入，大概变得很轻微。

关于新的有效联系的一个可能解释是：来自楔束核的无效纤维的背景分布零星穿过了通常由薄束核神经所支配的区域，且这些纤维发育成熟并成功地对由于轴突不断退化而空出的细胞产生刺激。这一变化可能发生在突触前，也可能发生在突触后。如果存在这样的一个纤维扩散系统，就可以预期新联系起来的细胞的感受野将不同

connected cells to differ from those in the established part of the nucleus. In fact we were unable to differentiate the receptive fields of cells in the presumably newly invaded region from those medial to them either in terms of field size or threshold. It is a particularly suitable region for examining receptive fields because many cells respond to movement of one or more of the long whisker-like hairs protruding from the rat's volar wrist. Cells with precisely this type of receptive field seemed to be duplicated in the new territory. Another possible mechanism is that the cells respond to an indirect input perhaps by way of cortex rather than by way of axons originating in cuneate nucleus. This seems unlikely because the latency of response of the switched cells was identical to that in the unchanged zone.

Collateral Sprouting

Probably the most likely explanation is that sprouts have grown from the terminal arborization of intact axons from cuneate nucleus and have established successful contact with deafferented neighbours. Collateral sprouting and new contact formation have been shown to occur in the periphery both in denervated muscle[2] and in sympathetic ganglia[3]. In the central nervous system, light microscope evidence of sprouting into denervated regions has been provided for dorsal root fibres[4] and for optic nerve fibres[5]. Raisman[6] has shown clear electron microscope evidence for sprouting and end knob formation in the partially denervated septal nuclei of adult rat. But in the case of collateral sprouting, one must still explain the small receptive fields of the newly connected cells. The distance over which the sprouts would have to extend is not known, because although the cell bodies of the switched cells can be as much as 400 μm from the intact area it may be that the dendritic fields of the denervated cells overlap with those of the intact cells. If a sprout had only to grow a few μm to occupy a new cell, a pioneer sprout might lay down a pathway along which its neighbours could grow. Another way in which a restricted receptive field might be formed would be if the fibres which fire together grow together. This would require a mechanism by which the presence of more or less coincident impulse traffic in terminals would exaggerate the growth of terminals which sensed impulses in themselves and their neighbours.

Clearly an explanation of the phenomenon reported here will have to await an electron microscope examination of morphological changes in the region of changed response. In assessing the significance of the phenomenon of new connexions, one must give certain warnings. These observations were made in the rat, an animal which continues to grow throughout adult life, perhaps therefore retaining certain embryological characteristics. The distance over which new connexions could be established is evidently limited since there were no signs of occupation of vacant cells in the lateral edge of the nucleus. The new connexions do not represent recovery from the original lesions because information from arms is now channelled into a system presumably specialized to handle leg information. However, one might speculate that if scattered single cells were destroyed, as

于神经核已建立部分的那些细胞的感受野。事实上，我们无法区分假设新被入侵区域中的细胞的感受野与位于它们内侧的细胞的感受野，无论是在感受野的大小方面还是在阈值方面。有一个特别适合作感受野检查的区域，因为很多细胞对大鼠腕掌部突出来的一根或更多胡须状长毛的运动有反应。恰好具有这种类型感受野的细胞看上去在这个新领域中发生了复制。另一可能的机制是：细胞可能对经由皮层的间接输入有反应，而不是对经由源于楔束核的轴突的间接输入有反应。这一机制似乎不太可能，因为已转换细胞的反应时间与那些未变化带中细胞的反应时间完全相同。

侧索生芽

也许最有可能的解释是：源自楔束核的未受损轴突的分支终端长出新芽，并与传入阻滞的相邻细胞之间成功地建立起联系。侧索生芽和新联系的形成已被证明可以发生在外周的去神经肌肉[2]和交感神经节[3]中。光学显微镜提供的证据表明：在中枢神经系统中，生芽进入去神经区发生在背根神经纤维[4]和视神经纤维[5]。雷斯曼[6]已经给出明确的电子显微镜证据，该证据证明在成年大鼠体内，部分去神经的隔核中存在生芽和形成尾结的现象。但是，在侧索生芽的例子中，还需要解释新联系起来的细胞的小感受野。芽不得不延伸跨越的距离仍属未知，因为虽然已转换细胞的细胞体距离未受损区域可达 400 μm，但是去神经细胞的树突野也有可能和那些未受损细胞的树突野发生重叠。如果一个芽为了占领新细胞，只需长几 μm，那么一个先导芽可能会铺好一条通路，它邻近的细胞可以沿着这条通路生长。形成受限感受野的另外一种可能方式是，同时产生神经冲动的纤维一起生长。这就要求有一种机制能使终端有几乎同时产生的神经冲动发生，这会促进能感受到自身和相邻细胞冲动的终端的生长。

显然，要想给出此现象的一个解释，还需要等待用电子显微镜对反应发生改变区域中的形态学变化的检查结果。在评价新联系现象的意义时，必须给出某些提醒。这些观察结果是从大鼠身上得到的，大鼠是一种在整个成年期中持续生长的动物，也许因此保留了一些胚胎学特征。可以建立新联系的跨越距离显然有限，因为没有发现该核团侧面边缘中空闲细胞被占领的迹象。新联系并不代表从原来的损伤中康复，因为来自前肢的信息现在被导入到一个被认为专门处理来自后肢的信息的系统。虽然如此，我们可以推测，如果零星分布的单细胞被破坏，就像患脑膜炎时那样，

in an encephalitis, sprouting from associated collaborating neighbours might provide some functional advantage.

(**232**, 542-545; 1971)

P. D. Wall and M. D. Egger: MRC Cerebral Functions Research Group, Department of Anatomy, University College, Gower Street, London WCIE 6BT.

Received April 8; revised July 30, 1971.

References:

1. Wall, P. D., *Brain*, **93**, 505 (1970).
2. Edds, jun., Mac V., *Quart. Rev. Biol.*, **28**, 260 (1953).
3. Guth, L., and Bernstein, J., *J. Exp. Neurol.*, **4**, 59 (1961).
4. Liu, C. N., and Chambers, W. W., *Arch. Neurol.*, **79**, 46 (1958).
5. Goodman, D. C., and Horel, J. A., *J. Comp. Neurol.*, **127**, 71 (1966).
6. Raisman, G., *Brain. Res.*, **14**, 25 (1969).

那么从相邻的有关合作细胞上生芽也许会带来某些功能上的益处。

（毛晨晖 邓铭瑞 翻译；刘力 审稿）

Change in Methylation of 16S Ribosomal RNA Associated with Mutation to Kasugamycin Resistance in *Escherichia coli*

T. L. Helser *et al.*

Editor's Note

The ribosome is the central component of the protein-manufacturing machinery found in all cells, and is made up of a large and small subunit. In this paper, James E. Dahlberg and colleagues suggest that a change in chemical modification (methylation) of a small subunit constituent (16S rRNA) causes resistance to the antibiotic kasugamycin in *Escherichia coli* bacteria. The mutation, they suggest, is linked with a change in methylation of a specific nucleotide sequence close to one end of the 16S rRNA. Several years later it was found that such a sequence participates directly in the initiation of protein synthesis by forming base pairs with messenger RNA.

GENES which code for proteins of both the 50S and 30S ribosomal subunits of *E. coli* seem to be clustered on the linkage map between minutes 62 and 64[1]. Mutations which affect proteins of the 30S subunit (streptomycin[2] and spectinomycin[3] resistances), those which affect proteins of the 50S subunit (erythromycin resistance[4]), and those in the translocation factor G (fusidic acid resistance[5-7]), all map in this region, as do several other 30S and 50S proteins[8,9]. In contrast, resistance to the aminoglycoside antibiotic kasugamycin (ksg), which is known to interact with the 30S subunit[10], maps close to minute 1[10-12]. Kasugamycin also inhibits the binding of fMet-tRNA to the mRNA-ribosome initiation complex in the presence of β,γ-methylene-guanosine triphosphate (GMPPcP), which indicates a different mode of action from the other aminoglycosides[11,13]. In this article we show that resistance to kasugamycin is determined by the properties of the 16S ribosomal RNA and that the mutation is associated with a change in the methylation of a specific nucleotide sequence close to the 3′ end of the 16S RNA[14,15].

Ribosomal Component Conferring Kasugamycin Sensitivity

The work of Sparling[10] has determined the site of action of kasugamycin as the 30S ribosome subunit. We have further localized its action using 30S subunits reconstituted from "core particles" and "split proteins" (obtained from cesium chloride density centrifugation of 30S subunits[16]) and find that sensitivity or resistance to kasugamycin is a property of the "core particles" (Table 1). 30S subunits reconstituted from pure 16S RNA and total protein[17] were then tested for sensitivity or resistance to kasugamycin in an *in vitro* polypeptide synthesizing system. The results with several different preparations of 16S RNA and total 30S proteins were all essentially the same: the particles containing 16S RNA from ksg^s strains were always more susceptible than particles containing 16S RNA from ksg^r strains (Table 2). Furthermore, we have analysed the proteins obtained from 30S

大肠杆菌中春日霉素抗性突变与16S核糖体RNA的甲基化变化相关

赫尔泽等

编者按

核糖体在所有细胞中都是蛋白质合成体系中的核心成分，它由一大一小两个亚基构成。在这篇论文中，詹姆斯·达尔伯格和他的同事们指出，小亚基中成分（16S核糖体RNA）的化学修饰的变化（甲基化）会引起大肠杆菌对一种抗生素——春日霉素的抗性。他们还指出，突变是与靠近16S核糖体RNA一端的某一特定核苷酸序列的甲基化变化联系在一起的。几年后人们发现，这些序列通过与信使RNA形成碱基对直接参与了蛋白质合成的启动。

编码大肠杆菌50S和30S核糖体亚基蛋白质的基因看上去聚集在连锁图上62 min和64 min之间[1]。影响30S亚基蛋白的突变（链霉素[2]和壮观霉素[3]抗性），以及那些影响50S亚基蛋白的突变（红霉素抗性[4]）和那些易位因子G中的突变（夫西地酸抗性[5-7]），都位于这一区域，其他几种30S和50S蛋白也是如此[8,9]。相反，对氨基糖苷类抗生素春日霉素（ksg）的抗性，已知是和30S亚基[10]发生相互作用，却位于靠近1 min的位置[10-12]。在β,γ–亚甲基–鸟苷三磷酸盐（GMPPcP）存在的条件下，春日霉素也会抑制甲酰甲硫氨酸–转运RNA与信使RNA–核糖体启动复合物的结合，这显示出一种和其他氨基糖苷类不同的作用方式[11,13]。在本文中，我们将证明：对春日霉素的抗性是由16S核糖体RNA的性质决定的，并且突变与靠近16S RNA 3′末端的某一特定核苷酸序列的甲基化变化有关[14,15]。

赋予春日霉素敏感性的核糖体成分

斯帕林[10]的工作已经确定春日霉素的作用位点是30S核糖体亚基。我们用由“核心粒子”和“脱落蛋白”（由30S亚基的CsCl密度离心获得[16]）重组的30S亚基进一步定位了春日霉素的作用位点，并发现对春日霉素的敏感性或抗性是“核心粒子”的性质（表1）。用纯16S RNA和全蛋白重组30S亚基[17]，然后在体外多肽合成系统中测试它对春日霉素的敏感性或抗性。由16S RNA和全30S蛋白制备的几种不同产物都能得到基本一致的结果：含有来自春日霉素敏感性（ksg^{s}）品系的16S RNA的粒子通常比含有来自春日霉素抗性（ksg^{r}）品系的16S RNA的颗粒更易受影

subunits of *ksg*^{s} and *ksg*^{r} strains for chemical differences by one dimensional polyacrylamide gel electrophoresis, and by co-chromatography of ^{3}H-labelled "sensitive" proteins and ^{14}C-labelled "resistant" proteins on carboxymethylcellulose columns. We have found that the sensitive and resistant strains seem to have complete and identical complements of 30S ribosomal proteins.

Table 1. Reconstitution of 30S Ribosomes from Split Proteins and Core Particles

Reconstituted 30S ribosomes		Incorporation—c.p.m. (% of control c.p.m.)		
Split protein	Core particle	Concentration of ksg (μg/ml.)		
		0	20	100
S	S	19,230 (100)	12,100 (63)	5,800 (30)
R	S	16,200 (100)	10,430 (64)	5,700 (35)
S	R	17,900 (100)	16,400 (92)	13,260 (74)
R	R	16,100 (100)	13,240 (82)	10,800 (67)

Analysis of CsCl "split" and "core" fractions for ksg sensitivity and resistance. 30S subunits were reconstituted from the appropriate split protein and core particle preparations using the methods of Traub and Nomura[16] and assayed for poly UG-directed ^{14}C-valine incorporation. In a volume of 100 μl., the reactions contained from 1.1 to 1.3 A_{260} units of reconstituted 30S and 1.5 A_{260} units of 50S from the resistant strain, 10 μg of poly UG (2:1–U:G input), and magnesium acetate at a final concentration of 7.5 mM. ^{14}C-valine (25 μCi/μmol) and other components were added as described by Nirenberg[27]. Reactions were incubated for 30 min at 37°C and terminated by the addition of 1 ml. of 10% trichloroacetic acid. After collection, the precipitates were counted in a Packard scintillation counter using a toluene-based fluid.

Table 2. Reconstitution of 30S Ribosomes from Total Protein and 16S RNA

Reconstituted 30S ribosomes		Incorporation—c.p.m. (% of control c.p.m.)		
Total protein	16S RNA	Concentration of ksg (μg/ml.)		
		0	20	100
S	S	681 (100)	346 (51)	188 (28)
R	S	761 (100)	389 (51)	191 (25)
S	R	634 (100)	400 (63)	260 (41)
R	R	471 (100)	329 (70)	300 (64)

30S subunits were reconstituted from the appropriate total protein preparation (urea–lithium chloride extracted) and 16S RNA (phenol extracted) using the methods of Traub and Nomura[17]. They were assayed as in Table 1 using from 0.86 to 1.0 A_{260} units of reconstituted 30S and 0.9 A_{260} units of 50S from the resistant strain.

16S Ribosomal RNA Structures

With this evidence implicating 16S ribosomal RNA as the determinant of kasugamycin sensitivity or resistance we have compared the RNA sequences by the "fingerprint" analysis technique of Sanger and co-workers[18,19] using radioactively labelled ribosomal RNA obtained from *ksg*^{s} and *ksg*^{r} strains.

Initial experiments were performed on ^{32}P-labelled 16S RNA obtained from resistant and sensitive cells. The RNAs were digested with RNAase T_1 plus alkaline phosphatase and the resulting oligonucleotides were separated, as described in Fig. 1. The fingerprints of these RNAs are shown in Fig. 1*a* and Fig. 1*b*.

响（表 2）。此外，我们还使用一维聚丙烯酰胺凝胶电泳法和联合色谱法（用羟甲基纤维素色谱柱分离 ^{3}H 标记的“敏感”蛋白和 ^{14}C 标记的“抗性”蛋白）来分析从 *ksg*^{s} 品系和 *ksg*^{r} 品系的 30S 亚基获得的蛋白，以寻找它们间的化学差异。我们发现，敏感性品系和抗性品系的 30S 核糖体蛋白成分似乎是完全相同的。

表 1. 来自脱落蛋白和核心粒子的 30S 核糖体的重组

重组的 30S 核糖体		重组后的 c.p.m.（占对照 c.p.m. 的百分比）		
脱落蛋白	核心粒子	春日霉素浓度（μg/ml）		
		0	20	100
敏感性	敏感性	19,230 (100)	12,100 (63)	5,800 (30)
抗性	敏感性	16,200 (100)	10,430 (64)	5,700 (35)
敏感性	抗性	17,900 (100)	16,400 (92)	13,260 (74)
抗性	抗性	16,100 (100)	13,240 (82)	10,800 (67)

春日霉素敏感性和抗性的“脱落”和“核心”片段的 CsCl 离心法分析。采用特劳布和野村法 [16] 用合适的脱落蛋白和核心粒子制剂重组 30S 亚基，并化验多聚 UG 指导的 ^{14}C– 缬氨酸的掺入。在 100 μl 体积内，反应物含有 1.1 A_{260} ~ 1.3 A_{260} 单位的重组 30S 和 1.5 A_{260} 单位来自抗性品系的 50S、10 μg 多聚 UG（输入 U:G 为 2:1）和终浓度为 7.5 mM 的醋酸镁。^{14}C– 缬氨酸（25 μCi/μmol）和其他成分按尼伦伯格所述添加 [27]。反应在 37℃下进行 30 min，加入 1 ml 10% 的三氯乙酸终止反应。收集后，沉淀物使用测量溶液为甲苯的帕卡德闪烁计数器计数。

表 2. 来自全蛋白和 16S RNA 的 30S 核糖体的重组

重组的 30S 核糖体		重组后的 c.p.m.（占对照 c.p.m. 的百分比）		
全蛋白	16S RNA	春日霉素浓度（μg/ml）		
		0	20	100
敏感性	敏感性	681 (100)	346 (51)	188 (28)
抗性	敏感性	761 (100)	389 (51)	191 (25)
敏感性	抗性	634 (100)	400 (63)	260 (41)
抗性	抗性	471 (100)	329 (70)	300 (64)

采用特劳布和野村法 [17] 由 16S RNA（苯酚提取）与合适的全蛋白制剂（尿素 – 氯化锂提取）重组得到 30S 亚基。用如表 1 注所述的方式分析它们，使用 0.86 A_{260} ~ 1.0 A_{260} 单位的重组 30S 和 0.9 A_{260} 单位来自抗性品系的 50S。

16S 核糖体 RNA 的结构

上述证据表明，16S 核糖体 RNA 是春日霉素敏感性和抗性的决定因素。采用桑格及其同事 [18,19] 发明的“指纹”分析技术，用放射性标记取自 *ksg*^{s} 和 *ksg*^{r} 品系的核糖体 RNA，我们比较了 RNA 序列。

最初的实验是在 ^{32}P 标记的取自抗性和敏感性细胞的 16S RNA 上进行的。用加入碱性磷酸酶的 T_1 核糖核酸酶消化这些 RNA，分离得到寡核苷酸，具体过程参见图 1 注中的描述。这些 RNA 的指纹如图 1*a* 和图 1*b* 所示。

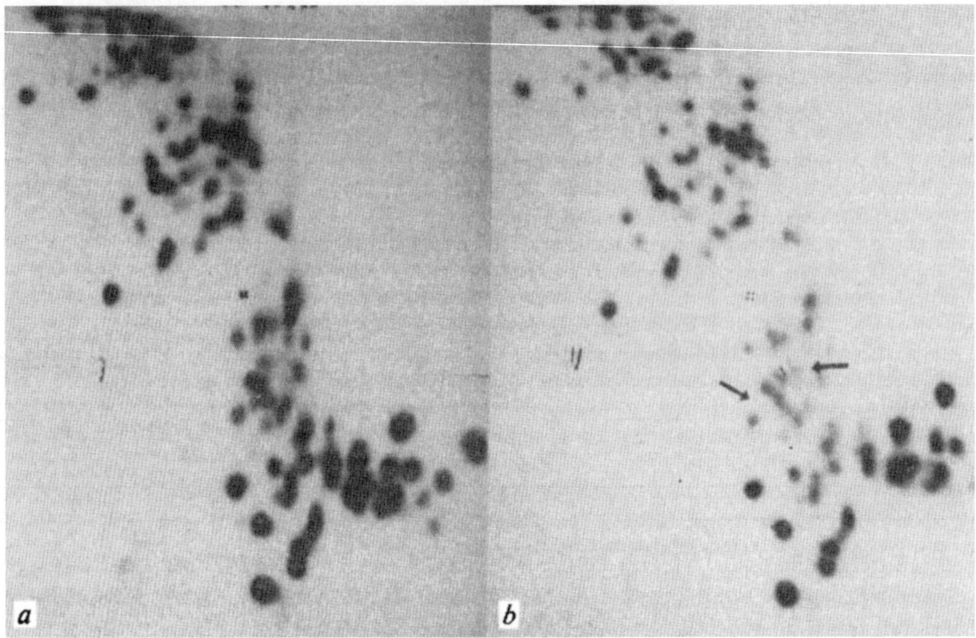

Fig. 1. RNAase T_1 plus phosphatase fingerprints of ^{32}P-labelled 16S RNA isolated from *ksg* sensitive (*a*) and resistant (*b*) *E. coli.* The arrows in *b* indicate the positions of oligonucleotides which are present in only one or the other digest. The strains were labelled by growth in 1% "Bactotryptone", 0.5% yeast extract, 0.3% glucose medium from which inorganic phosphate had been removed by $MgCl_2$ precipitation at *p*H 9 before sterilization. ^{32}P-Phosphate (1 mCi) was added to 25 ml. of the above medium (*p*H 7.8). Cells were grown to late log phase, collected, washed by centrifugation and lysed by lysozyme/ EDTA treatment. Total RNA was extracted with phenol and purified on a 5–20% sucrose gradient. The 16S RNA peaks were collected and ethanol-precipitated to remove EDTA. 20 μg (~10^6 c.p.m.) of each RNA was digested with 1 μg RNAase T_1 (Sankyo) plus 2 μg bacterial alkaline phosphatase (Worthington, BAP–F) in 2 μl. of 0.05 M Tris Cl, *p*H 7.7, 0.0001 M $ZnCl_2$ for 30 min at 37°C. Electrophoresis from right to left was on a 90 cm Oxoid cellulose acetate strip in *p*H 3.5 pyridine acetate containing 8 M urea for 3 h at 5 kV. Ionophoresis in the second dimension (from top to bottom) was on a 40×80 cm DEAE cellulose paper (Whatman) in 7% formic acid for 16 h at 1 kV. The autoradiogram was exposed for 1 day.

A comparison of the two fingerprints reveals only two differences, which are indicated by the arrows in Fig. 1*b*. Oligonucleotide No. 71 (using the numbering system of Fellner *et al.*[14,20]) is absent from the digest of RNA from the resistant cells, and a different oligonucleotide, No. p-71, is present. Oligonucleotide 71 has the sequence $m_2^6Am_2^6$ACCUG, while oligonucleotide p-71 is probably its unmethylated precursor and has the sequence AACCUG[21]. Thus, mutation from kasugamycin sensitivity to resistance could be accompanied by the inability of the cells to convert adenine in oligonucleotide p-71 to dimethyl adenine.

This inference was confirmed by fingerprint analysis of ^{14}C-methyl-labelled 16S RNAs. These fingerprints (Fig. 2) show that an oligonucleotide with the mobility of $m_2^6Am_2^6$ACCUG (No. 71) is present only in the RNA from the sensitive cells (2*a*). It is clear that no other methylation changes in the 16S RNA fragments can be detected (the

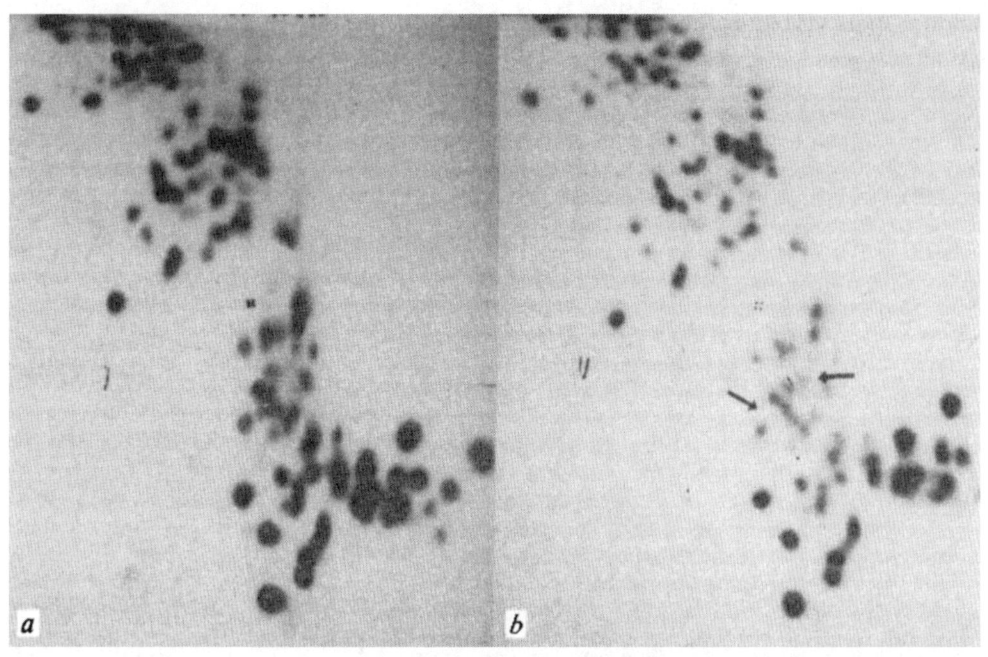

图 1. 从春日霉素敏感性（*a*）和抗性（*b*）大肠杆菌中分离 ^{32}P 标记的 16S RNA，并经加入磷酸酶的 T_1 核糖核酸酶消化后的指纹图。图 *b* 中箭头指示存在只被一种（或另一种）酶消化的寡核苷酸的位置。配制含 1%“细菌用胰蛋白胨”、0.5% 酵母提取物和 0.3% 葡萄糖的培养基，灭菌前在 pH 9 条件下用 $MgCl_2$ 沉淀除去无机磷酸盐。用这种培养基培养大肠杆菌品系，从而标记它。加入 ^{32}P 标记的磷酸盐（1 mCi），使上述培养基（pH 7.8）的体积达到 25 ml。细胞在这种培养基中生长到对数期后期，收集，用离心法洗涤，再用溶菌酶 / EDTA 溶液处理以溶解细胞。用苯酚提取全 RNA，并经 5% ~ 20% 蔗糖梯度纯化。在 16S RNA 峰值收集，并用乙醇沉淀除去 EDTA。每 20 μg（~10^6 c.p.m.）RNA 在 2 μl pH 值为 7.7 的 0.05 M Tris-HCl 和 0.0001 M $ZnCl_2$ 中，用 1 μg T_1 核糖核酸酶（三共）加上 2 μg 细菌碱性磷酸酶（沃辛顿，BAP–F）消化，在 37℃ 下反应 30 min。电泳从右到左是在一个长 90 cm 的奥克欧德牌醋酸纤维素带上进行的，电泳液为 pH 3.5 含有 8 M 尿素的吡啶醋酸盐溶液，在 5 kV 条件下电泳 3 h。离子电泳作用在第二维上，（从上到下）是在一个 40 cm × 80 cm 的二乙氨基乙基纤维素纸（沃特曼）上进行的，电泳液为 7% 的甲酸，在 1 kV 条件下电泳 16 h。放射自显影照片的曝光时间为 1 天。

两幅指纹图的比较只显示出两处不同，如图 1*b* 中的箭头所示。当消化的是来自抗性细胞的 RNA 提取物时，71号寡核苷酸（采用费尔纳等人制订的编号系统 [14,20]）没有出现，但出现了一个与之不同的寡核苷酸 p–71号。71号寡核苷酸具有序列 $m_2^6Am_2^6ACCUG$，而 p–71号寡核苷酸很有可能是它的非甲基化前体，具有序列 AACCUG[21]。因此，从春日霉素敏感性到抗性的突变可能伴随着细胞能力的缺失——不能将 p–71号寡核苷酸中的腺嘌呤转化为二甲基腺嘌呤。

对 ^{14}C– 甲基标记的 16S RNA 进行指纹分析的结果证实了这一推论。这些指纹图（图 2）显示，一种具有 $m_2^6Am_2^6ACCUG$ 迁移率的寡核苷酸（71 号）只存在于敏感性细胞的 RNA 中（图 2*a*）。目前已清楚的是，在 16S RNA 片段中没有检测到其

additional spots visible in the resistant fingerprint (2*b*) are due to contamination with 23S RNA).

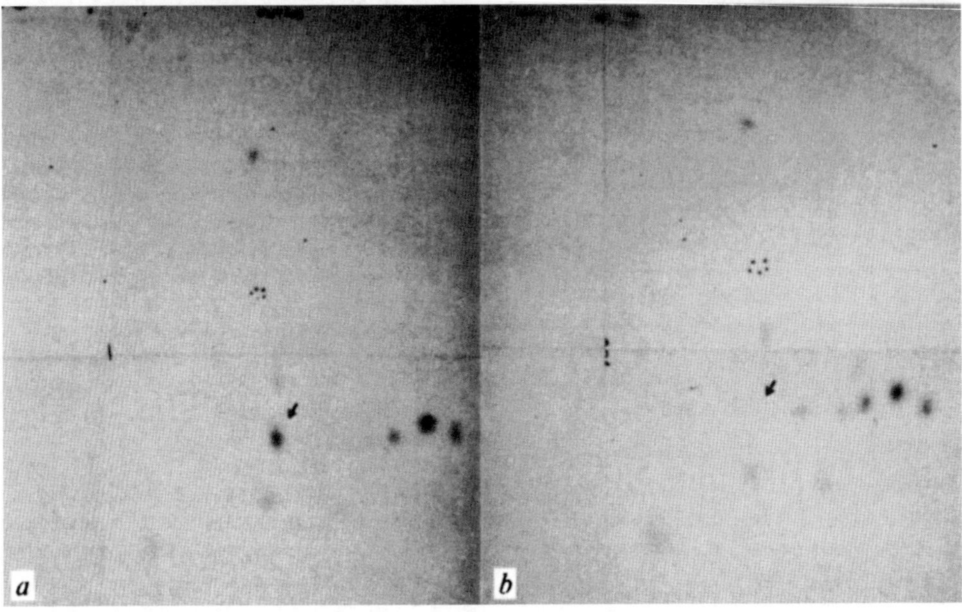

Fig. 2. RNAase T_1 plus phosphatase fingerprints of ^{14}C-methyl labelled 16S RNA isolated from *ksg* sensitive (*a*) and resistant (*b*) *E. coli*. The RNAs were isolated as described in the legend to Fig. 1 from cells which had been grown on Tris minimal medium[28] supplemented with 50 μg/ml. each of threonine, leucine, cytidine, thymidine, and uridine. ^{14}C-methyl-L-methionine was added (25 μCi/10.5 μmol in 25 ml. of medium) and the cells were allowed to grow for one generation, when unlabelled methionine (2 μg/ml.) was added. The cells were collected in late log phase. The enzymatic digestion and fingerprinting were performed as described in the legend to Fig. 1 except that 40 μg of each RNA ($\sim 2\times 10^4$ c.p.m. ^{14}C) were digested with 2 μg RNAase T_1 and 4 μg phosphatase in a volume of 4 μl. The autoradiogram was exposed for 12 days.

Significance of Methylation

Until recently, the only mutations known to affect ribosome structure and function have been those in which a ribosomal protein was altered. The phenotypes of these mutations have been antibiotic resistance, ribosome ambiguity, or ribosome assembly[22,23]. Lai and Weisblum, however, have reported that the induced appearance of erythromycin resistance in *Staphylococcus aureus* is accompanied by methylation of the 23S ribosomal RNA, consistent with the fact that erythromycin resistance is a property of the 50S subunit of the ribosome[24]. Kasugamycin resistance also seems to be an example of a change in ribosomal RNA, since we find that the kasugamycin sensitive and resistant phenotypes are determined by the source of the 16S RNA in reconstitution experiments. The change from kasugamycin-sensitivity to resistance is apparently associated with a failure to methylate a specific sequence in this RNA. We have yet to establish, however, a direct cause and effect relationship between the presence of the $m_2^6Am_2^6ACCUG$ sequence and the sensitive phenotype, or between the AACCUG sequence and the resistant phenotype.

他甲基化变化（抗性指纹图（2*b*）中的一些额外的可视点是由 23S RNA 的污染所致）。

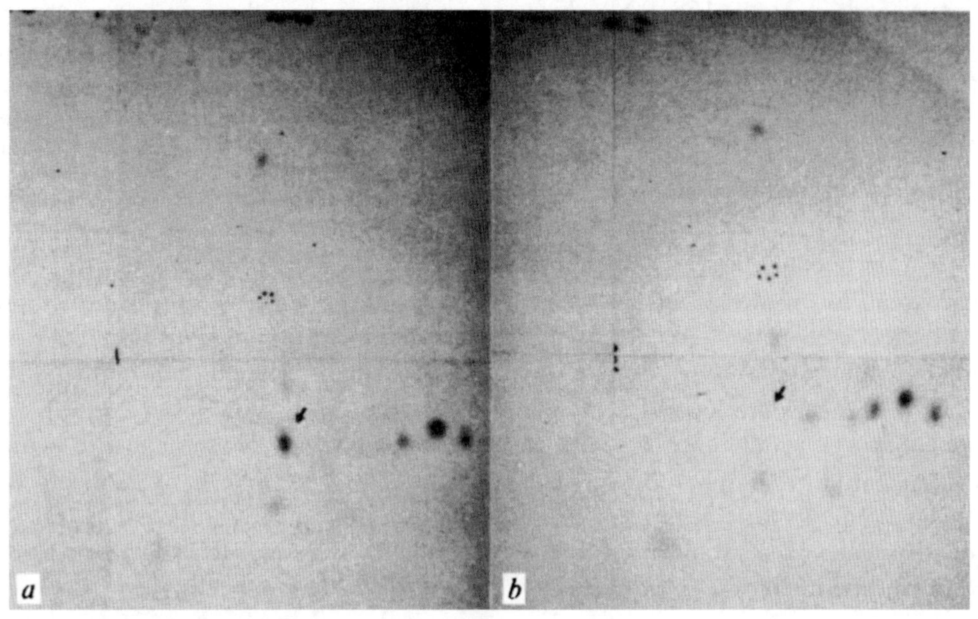

图 2. 从春日霉素敏感性（*a*）和抗性（*b*）大肠杆菌中分离 ^{14}C– 甲基标记的 16S RNA，并经加入磷酸酶的 T_1 核糖核酸酶消化后所得的指纹图。向三羟甲基氨基甲烷（Tris）基础培养基 [28] 中分别加入 50 μg/ml 的苏氨酸、亮氨酸、胞嘧啶、胸腺嘧啶和尿嘧啶，以此培养细胞，从这些细胞中分离 RNA 的方法见图 1 注。加入 ^{14}C– 甲基 –L– 甲硫氨酸（每 25 ml 培养基中加入 25 μCi/10.5 μmol），允许这些细胞生长一代，然后加入未标记的甲硫氨酸（2 μg/ml）。在对数期后期收集这些细胞。按照图 1 注所述进行酶消化和指纹实验，不同之处是用每份 40 μg RNA（$\sim 2\times10^4$ c.p.m. ^{14}C）在体积为 4 μl 含 2 μg T_1 核糖核酸酶和 4 μg 磷酸酶的体系中消化。放射自显影照片的曝光时间为 12 天。

甲基化的意义

截止到最近，已知能影响核糖体结构和功能的突变都是那些使核糖体蛋白发生改变的突变。这些突变的表型具有抗生素抗性、核糖体双关性或核糖体组装 [22,23] 的特点。虽然如此，莱和韦斯布卢姆曾报道称：伴随 23S 核糖体 RNA 的甲基化，金黄色葡萄球菌出现被引发的红霉素抗性表象，这与红霉素抗性是核糖体 50S 亚基性质的事实一致 [24]。春日霉素抗性看上去也是核糖体 RNA 变化的一个例子，因为我们发现，春日霉素敏感性和抗性表型是由重组实验中 16S RNA 的来源决定的。从春日霉素敏感性到抗性的转化显然和该 RNA 中特定序列的甲基化失败有关。虽然如此，我们至今也还没有在 $m_2^6Am_2^6ACCUG$ 序列的存在和敏感性表型之间，或在 AACCUG 序列和抗性表型之间，建立起直接的因果关系。

Assuming that a cause and effect relationship does exist, there are at least three possible mechanisms which could result in failure to dimethylate the adenylic acid residues in oligonucleotide p-71 of 16S RNA. First, a minor change in the primary sequence of the RNA, which may not be detectable by fingerprint analysis, could modify the RNA conformation so that it would no longer be a good substrate for a methylase. Second, an as yet undetected change in the structure of one of the ribosomal proteins could prevent methylation of oligonucleotide p-71 during maturation. Third, a specific RNA methylase could be altered by mutation.

No clear distinction between these possibilities can be made at present. In diploids heterozygous for the ksg_A locus, the sensitive phenotype is dominant, which might support any of these explanations (ref. 12 and Helser, unpublished observations). Studies on the mapping of the ribosomal RNA genes by DNA–RNA hybridization indicate that at least half of these genes are located near minute 74 of the *E. coli* chromosome[25,26]. Since we cannot detect any difference between the ribosomal proteins of kasugamycin sensitive and kasugamycin resistant strains, we suggest that the ksg_A locus is the gene for an RNA methylase. Support for this proposal would come from the demonstration that 16S RNA from a kasugamycin resistant strain can be converted to "sensitive" RNA by methylation *in vitro*. Such experiments are in progress.

Fingerprinting of the 23S RNA and total soluble RNA fractions (unpublished observations) has provided two other points of interest. First is the fact that, in the kasugamycin-resistant strain in which dimethylation of adenine does not occur, monomethylation of the N-6 position of adenine in 23S RNA does occur. Thus it is very likely that different enzymes are responsible for monomethylation and dimethylation of adenine residues in RNA. Second, one oligonucleotide from a RNAase T_1 and phosphatase digest of low molecular weight RNA (approximately 4S) is unmethylated in the resistant strain. If this sequence is from a distinct tRNA species, it would indicate the absence of dimethylated adenine in the only two species of RNA in *E. coli* that normally contain this base. But because this oligonucleotide seems to have the same sequence as oligonucleotide 71 from 16S RNA, this apparent tRNA fragment could have come from degradation of the 16S RNA during isolation.

This work was supported by a grant from the National Institutes of Health and the National Science Foundation. The Graduate School, University of Wisconsin, provided funds for the purchase of zonal centrifugation equipment. J. E. D. holds an NIH research career development award. T. L. H. was a recipient of NIH training grant support, awarded to the Department of Biochemistry. Kasugamycin was provided by Dr. M. J. Cron of Bristol Laboratories. We thank Dr. B. Weisblum for interest and comments.

(*Nature New Biology*, **233**, 12-14; 1971)

假设因果关系确实存在，那么至少存在三种可能的机制，均会使 16S RNA 的 p–71 号寡核苷酸中的腺苷酸残基二甲基化失败。第一，RNA 初级序列的一个小变化，可能用指纹分析法检测不到，但会改变 RNA 的构象，使它不再是甲基化酶的有效底物。第二，某一核糖体蛋白可能发生了到目前为止未能检测到的结构变化，这种变化会阻止 p–71 号寡核苷酸在成熟过程中的甲基化。第三，一种特异的 RNA 甲基化酶通过突变发生了改变。

目前还不知道如何明确区分这几种可能性。在具有 *ksg*$_A$ 基因座的二倍体杂合子中，敏感性表型为显性，这可能支持上述解释中的任何一条（参考文献 12 和赫尔泽尚未发表的观察结果）。通过 DNA–RNA 杂交研究核糖体 RNA 基因的图谱，结果表明这些基因中至少有一半位于大肠杆菌染色体 74 min 附近 [25,26]。由于我们未能检测到春日霉素敏感性和春日霉素抗性品系的核糖体蛋白间的任何区别，我们提议，*ksg*$_A$ 基因座是对 RNA 甲基化酶产生反应的基因。对这一提议的支持将来自下述证据：在体外能够通过甲基化将来自春日霉素抗性品系的 16S RNA 转化为“敏感性”RNA。这样的实验正在进行中。

23S RNA 和全部可溶 RNA 片段的指纹图（尚未发表的观察结果）提供了另外两点值得关注之处。第一点，在没有发生腺嘌呤二甲基化的春日霉素抗性品系中，的确存在 23S RNA 中腺嘌呤的 N–6 位置被单甲基化的现象。因此非常有可能由不同的酶负责 RNA 中腺嘌呤残基的单甲基化和二甲基化。第二，在抗性品系中发现，小分子量（大约 4S）RNA 经 T_1 核糖核酸酶和磷酸酶消化后得到的一个寡核苷酸产物是未甲基化的。如果该序列来自另一种转运 RNA 类型，那么将说明：在大肠杆菌中，所有两种通常含有这一碱基的 RNA 类型都存在二甲基化腺嘌呤缺失的现象。但因为该寡核苷酸看上去和来自 16S RNA 的 71 号寡核苷酸具有相同的序列，所以这一貌似转运 RNA 的片段可能来自分离时降解的 16S RNA。

本工作得到了美国国立卫生研究院和美国国家科学基金会的拨款支持。用于购买区带离心设备的资金是由威斯康星大学研究生院提供的。朱利安 · 戴维斯获得了美国国立卫生研究院的研究事业进步奖金。特里 · 赫尔泽是美国国立卫生研究院授给生化系的培训基金资助的获得者。春日霉素由布里斯托尔实验室的克龙博士提供。感谢韦斯布卢姆博士对我们的关注和评价。

（邓铭瑞 翻译；陈新文 陈继征 审稿）

Change in Methylation of 16S Ribosomal RNA Associated with Mutation to Kasugamycin Resistance in *Escherichia coli*

Terry L. Helser and Julian E. Davies: Department of Biochemistry, College of Agricultural and Life Sciences, University of Wisconsin, Madison, Wisconsin 53706.
James E. Dahlberg: Department of Physiological Chemistry, University of Wisconsin, Madison, Wisconsin 53706.

Received June 4; revised July 12, 1971.

References:

1. Flaks, J. G., Leboy, P. S., Birge, E. A., and Kurland, C. G., *Cold Spring Harbor Symp. Quant. Biol.*, **31**, 623 (1966).
2. Ozaki, M., Mizushima, S., and Nomura, M., *Nature*, **222**, 333 (1969).
3. Bollen, A., Davies, J., Ozaki, M., and Mizushima, S., *Science*, **165**, 85 (1969).
4. Takata, R., Osawa, S., Tanaka, K., Teraoka, H., and Tamaki, M., *Mol. Gen. Genet.*, **109**, 123 (1970).
5. Bernardi, A., and Leder, P., *J. Biol. Chem.*, **245**, 4263 (1970).
6. Kuwano, M., Schlessinger, D., Rinaldi, G., Felicetti, L., and Tocchini-Valentini, G. P., *Biochem. Biophys. Res. Commun.*, **42**, 441 (1971).
7. Tanaka, N., Kawano, G., and Kinoshita, T., *Biochem. Biophys. Res. Commun.*, **42**, 564 (1971).
8. Dekio, S., Takata, R., and Osawa, S., *Mol. Gen. Genet.*, **109**, 131 (1970).
9. O'Neil, D. M., Baron, L. S., and Sypherd, P. S., *J. Bact.*, **99**, 242 (1969).
10. Sparling, P. F., *Science*, **167**, 56 (1968).
11. Helser, T. L., and Davies, J. E., *Bact. Proc.*, P82 (1971).
12. Sparling, P. F., *Bact. Proc.*, P81 (1971).
13. Okuyama, A., Machiyama, N., Kinoshita, T., and Tanaka, N., *Biochem. Biophys. Res. Commun.*, **43**, 196 (1971).
14. Ehresman, C., Fellner, P., and Ebel, J. P., *FEBS Lett.*, **13**, 325 (1971).
15. Bowman, C. M., Dahlberg, J. E., Ikemura, T., Konisky, J., and Nomura, M., *Proc. US Nat. Acad. Sci.*, **68**, 964 (1971).
16. Traub, P., and Nomura, M., *J. Mol. Biol.*, **34**, 575 (1968).
17. Traub, P., and Nomura, M., *Proc. US Nat. Acad. Sci.*, **59**, 777 (1968).
18. Sanger, F., Brownlee, G. G., and Barsell, B. G., *J. Mol. Biol.*, **13**, 373 (1965).
19. Brownlee, G. G., and Sanger, F., *J. Mol. Biol.*, **23**, 337 (1967).
20. Fellner, P., Ehresman, C., and Ebel, J. P., *Nature*, **225**, 26 (1970).
21. Lowry, C. V., and Dahlberg, J. E., *Nature New Biology*, **232**, 52 (1971).
22. Nomura, M., *Bact. Revs.*, **34**, 228 (1970).
23. Weisblum, B., and Davies, J., *Bact. Revs.*, **32**, 493 (1968).
24. Lai, C. J., and Weisblum, B., *Proc. US Nat. Acad. Sci.*, **68**, 856 (1971).
25. Yu, M. T., Vermeulen, C. W., and Atwood, K. C., *Proc. US Nat. Acad. Sci.*, **67**, 26 (1970).
26. Birnbaum, L. S., and Kaplan, S., *Proc. US Nat. Acad. Sci.*, **68**, 925 (1971).
27. Nirenberg, M. W., in *Methods in Enzymology* (edit. by Colowick, S. P., and Kaplan, N. O.), **6**, 17 (Academic Press, New York, 1964).
28. Grossman, L., in *Methods in Enzymology* (edit. by Grossman, L., and Moldave, K.), **12A**, 700 (Academic Press, New York, 1967).

Problems of Artificial Fertilization

R. G. Edwards

Editor's Note

With *in vitro* fertilization (IVF) capable of producing 6-day old human embryos, IVF pioneer Robert Edwards describes the practical and social problems related to his technique. He highlights the need to use pre-ovulatory eggs, and suggests that ovulation-inducing compounds such as clomiphene may prove superior and less costly than gonadotrophin injections—a debate that continues today. Successful IVF, he argues, could give some infertile couples their own children so "comments about overpopulation seem to be highly unjust to such an underprivileged minority." Much of the debate so far, with its constant references to nuclear cloning and survival of the family unit, are "unreal", and Edwards urges for rational discussion of the emerging social issues related to IVF.

UNTIL recently, opportunities for scientific and clinical studies on human conception have been extremely limited, largely because of the inaccessibility of the ovary and the difficulty in obtaining preovulatory oocytes. One egg per month was a discouraging target and experience had shown that even this single egg was difficult to recover because ovulation could not be predicted with any degree of accuracy. In the past two or three years, considerable progress has been made with the clinical and scientific problems associated with the study of human fertilization and cleavage. The opening came with the discovery that human oocytes would undergo their final stages of maturation—those occurring just before ovulation—when removed from their follicles and placed in a suitable culture medium. These stages could be timed with considerable accuracy, and the availability of this human material provided opportunities for studies on meiosis, ovulation and fertilization[1]. Oocytes matured *in vitro* thus provided the material for studies on human fertilization *in vitro*, which was achieved two years ago.

Patients Involved

When fertilization had been accomplished using these oocytes, the emphasis of the work changed to include studies on patients in order to realize the potential of these studies for human infertility. There were various reasons for this extension into clinical studies. While oocytes matured *in vitro* provide good material for fertilization, they are unsuitable for studies on cleavage because observations in animals has shown that the resulting embryos fail to develop normally[2]. For normal development to full term, oocytes must be collected after completing, or almost completing, their maturation in the ovary (see Fig. 1). Endocrine methods were thus needed to induce follicular growth and oocyte maturation in the patients, the preovulatory growth of oocytes had to be timed exactly from the estimates made from cultures and surgical methods had to be developed for the collection of the oocytes just before ovulation. The achievement of these objectives has opened

关于人工受精的几个难题

爱德华兹

编者按

在本文中，体外受精技术（IVF）的先驱者罗伯特·爱德华兹描述了与这项技术相关的实际操作问题和社会问题，那时他已经能够运用该项技术获得6天大的人类胚胎。他强调了使用排卵前卵子的必要性，并指出可诱导排卵的化合物如克罗米酚等可能比注射促性腺激素更有效而且更便宜——关于两者效用对比的讨论至今仍在继续。他指出，体外受精技术的成功可使一些不孕夫妇拥有自己的孩子，所以"和这些生育能力低下的少数群体谈论人口过剩问题似乎是很不公平的"。迄今为止，那些不断提及核克隆和家庭维持问题的大量讨论是"毫无意义"的，并且爱德华兹极力主张人们要对由IVF技术引发的社会问题进行理性的讨论。

直到最近，进行人类受孕的科学和临床研究一直受到很大的限制，主要是因为很难从卵巢中获得排卵前的卵母细胞。人每月只能排出一个卵子已经够令人沮丧的了，但以往的经验告诉我们：即使是这样一个卵子也很难取得，因为我们无法准确地预测出排卵时间。最近两三年间，人们已经在与人类受精和卵裂研究相关的临床和科学难题上取得了重要突破。首先是发现将人类卵母细胞从卵泡内取出并置于适当的培养基中之后，它们能够发育到成熟的最后阶段——与即将排卵时相对应的阶段。这样就可以非常精确地确定取出卵子的时间，而人体中卵子的获得为研究减数分裂、排卵和受精提供了可能[1]。因此，体外成熟的卵母细胞就为两年前已取得成功的人类体外受精研究提供了原材料。

临床研究

在实现对这些卵母细胞的受精后，工作重点转移到了包括对患者的研究方面，以便了解将这些研究成果运用于治疗人类不孕症的可能性。将研究扩展到临床领域的原因有很多。虽然体外成熟的卵母细胞为受精实验提供了良好的材料，但它们却不适合用于研究卵裂，因为在动物实验中发现由此得到的胚胎不能正常发育[2]。为了使胚胎能正常发育到妊娠足月，就必须等到卵母细胞在卵巢内发育成熟或接近发育成熟时再进行收集（见图1）。因而需要用内分泌法来诱导患者体内的卵泡生长和卵母细胞成熟，并且需要根据体外培养得出的预测结果精确地确定排卵前卵母细胞的生长进程，同时还必须发展外科技术以保证刚好在即将排卵之前收集到卵母细胞。

the possibility of growing and analysing cleaving embryos in culture, replacing them in the mother, and studying the processes of implantation. Each of these projects opens fundamental opportunities for the study of human conception. What progress has been made in these various areas of investigation?

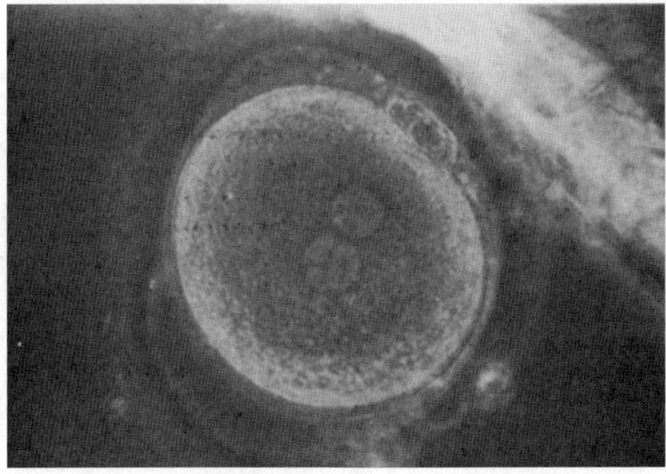

Fig. 1. Human pronucleate stage.

The surgical methods for collecting human oocytes are perhaps almost as refined by now as they ever will be. Laparoscopy has proved to be a simple and very safe procedure permitting the necessary manipulations to be carried out on the ovary[3]. Oocytes are collected from more than one-half of the available Graafian follicles, and although the proportion could be improved the loss of one-half of the oocytes in this way is not holding up further developments in any serious manner. The timing of oocyte maturation and ovulation are also under fine control. None of the many patients so far examined has ovulated before the expected time, and many of the oocytes were fertilized soon after collection, showing that they were certainly preovulatory. The major problems in the initiation of follicular growth and oocyte maturation concern the response of the patients to the endocrine treatments, which are based on the use of human menopausal gonadotrophins on amenorrhoeic women[4]. Some of the patients have few or no preovulatory follicles after the treatment.

There are well known difficulties with the method—for example, the great variation between patients or even between successive treatments of the same patient[5]. Three or four preovulatory follicles are the minimum required to provide sufficient oocytes to work with. Measurements of the excretion of oestrogen in the urine of the patients during treatment provides some indication of their response to the treatment, particularly as a safeguard against overstimulation, but its usefulness as a measure of the growth of preovulatory follicles remains to be decided. There is some scope for improvements in the hormone treatment. The amount of gonadotrophin being administered to the patients is modest in comparison with that use for the treatment of amenorrhoea, partly because the patients have a normal menstrual cycle which might heighten their response. Simply raising the

这些目标实现后，我们就有可能在培养基中培养和分析分裂中的胚胎，然后将其置回母体内，并对植入过程进行研究。所有这些科研项目都为研究人类受孕带来了重要的机遇。那么，在这些不同的研究领域中我们取得了哪些进展呢？

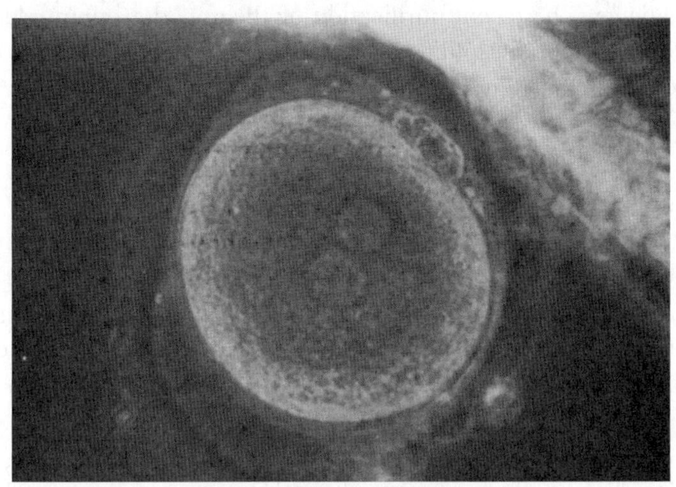

图 1. 原核期的人类胚胎

现在我们用于收集人类卵母细胞的外科方法或许已经几近完善了。腹腔镜手术已被证明是一种既简单又十分安全的处理方式，我们可以利用它在卵巢内进行必要的操作[3]。卵母细胞是从超过一半的可利用格拉夫卵泡中收集到的，并且尽管这个比例还可以再提高，但是以此种方式损失一半的卵母细胞并不会严重影响进一步的发育。卵母细胞成熟和排卵的时间点也在我们的精确控制之中。迄今为止，在参与研究的众多患者中没有一人在预期时间之前排卵，并且相当多的卵母细胞在收集后不久便受精成功，这表明它们的确处于排卵前的状态。对于接受内分泌治疗的患者——这种内分泌治疗是基于给闭经患者使用人绝经期促性腺激素[4]，需担心的主要问题是：卵泡生长初期和卵母细胞成熟过程中患者对内分泌治疗的反应。治疗后，有些患者只产生了很少或者根本不产生排卵前的卵泡。

此种方法遇到了一些众所周知的难题——例如：在不同患者之间，甚或同一患者的连续治疗期间也存在着很大的差异[5]。为给人工受精提供足够的卵母细胞，至少需要 3 ~ 4 个排卵前的卵泡。在治疗期间检测患者尿液中的雌激素浓度可提供了解患者对治疗的反应的某些指征，特别是可以作为一种安全指征，以预防过度刺激；但将这一检测作为排卵前卵泡发育情况的量度的有效性仍需进一步确证。激素治疗法还有进一步改善的余地。给患者服用的促性腺激素的剂量与用于治疗闭经的剂量相比并不算大，在一定程度上是因为月经周期正常的患者可能会对药物有更强烈的反应。简单地加大促性腺激素的剂量将使大多数患者产生更重大的药物反应。

dosage of gonadotrophins should induce a greater response in most patients. Alternatively, the use of compounds such as clomiphene, to induce the release of endogenous hormones, might prove superior, cheaper, and more acceptable than the injection of gonadotrophins. Clomiphene is widely used in the treatment of female hypopituitarism, and should work well with cyclic patients. The recent purification of the gonadotrophin releasing hormones of the hypothalamus could well offer another alternative, provided that the dosage and response can be controlled.

Human Ova

Studies on the fertilization of mammalian ova *in vitro* have moved quickly in the past two or three years after a decade of uncertainty. Belief in the necessity of uterine spermatozoa had become almost dogmatic as a result of experiments involving mostly rabbit ova. Even the demonstration that hamster ova could be fertilized *in vitro* by epididymal spermatozoa[6] had scarcely shaken this fixed belief. When the conditions needed in the culture medium for hamster fertilization were defined, the work was extended to human material, using oocytes matured *in vitro* and washed, ejaculated spermatozoa. Fertilization was established using the same criteria defined for studies with animal ova[7]. Moreover, similar studies are now being extended to other species such as the mouse and cow, and even in rabbits spermatozoa have now been found to penetrate through the membranes of the ova. The culture media are simple: a basic Tyrode's solution is supplemented with bovine serum albumin, pyruvate, and bicarbonate to a *p*H of 7.6. Pyruvate is obviously an energy source, and the albumin presumably serves as a source of nitrogen or to stabilize cell membranes. The significance of the high *p*H, or of the contributions of the cumulus cells surrounding the ova, remain unexplained.

Media suitable for fertilization and for cleavage are rarely identical. The development of media to support the cleavage of human ova has not relied on experiments with mammalian ova to nearly the same degree. Simple defined media had been developed for the cleavage of mouse ova, following the recognition of the importance of pyruvate as an energy source during these early embryonic stages[8]. The initial attempts to culture human embryos utilized these media, but cleavage progressed only to the 8-celled stage[9] (Fig. 2). More complex media developed for routine tissue culture have proved much more successful. A medium composed of Ham's F10 supplemented with calf or human serum, at a *p*H of 7.3 under a gas phase of 5% O_2, 5% CO_2 and 90% N_2, has supported development to the blastocyst stage[10]. Several blastocysts, each having 100 or more nuclei and several mitoses, have been grown under these conditions (Fig. 3). The inner cell mass, trophoblast and blastocoelic cavity were well defined, and some blastocysts expanded in culture. The blastocoel appeared to form from the secretions of large cells that accumulated at one pole of the morula. Modified body fluids are also reported to support the cleavage of human embryos[11]. According to one report, oocytes matured *in vitro* had to be fertilized by human spermatozoa previously incubated in the uterus of a monkey[12]; the abortive cleavage of the embryos might have been a consequence of maturation *in vitro* or some other cause.

另一种方案是：或许可以证明使用诸如克罗米酚这样的化合物来诱导内源性激素的释放会比注射促性腺激素的效果更好、更经济和更容易被接受。克罗米酚被广泛用于治疗女性垂体功能减退，并且对周期性患者应该有很好的疗效。最近从下丘脑中得到了提纯的促性腺激素释放激素，如果在临床上可对这种激素的剂量和用药后的效应加以调控，则很可能会为治疗提供另一种选择。

人类卵子

经过十年的困惑期之后，关于哺乳动物卵子体外受精的研究在近两三年内发展得很迅速。根据主要基于兔卵子的实验结果，人们曾认为在实验中必须使用从子宫内取出的精子，这一结论几乎成了金科玉律。即使发现仓鼠卵子可与附睾精子在体外受精[6]这样的证据也几乎没能动摇这一固有的观念。在使仓鼠受精所需的培养基条件确定之后，研究工作随即扩展到包括体外成熟的卵母细胞和清洗过的射出精子在内的人类材料。在设定人类体外受精条件时采用了从动物卵子的研究中得到的标准[7]。此外，类似的研究如今已经扩展到了其他物种，如小鼠和牛，现在甚至发现兔精子也可以穿透卵子的细胞膜。所用的培养基非常简单：在一种碱性台罗德氏溶液中加入牛血清白蛋白、丙酮酸和碳酸氢盐并将 pH 值调至 7.6。丙酮酸显然是一种能量来源，而白蛋白可能作为氮源或者起稳定细胞膜的作用。现在我们还不能解释这一高 pH 值的重要性以及卵子周围的卵丘细胞所起的作用。

适合于受精和适合于卵裂的培养基各不相同。研制能促进人卵子分裂的培养基并不完全依赖于哺乳动物的卵子实验。在认识到丙酮酸作为胚胎早期阶段能量来源的重要性之后[8]，目前已研制出了用于小鼠卵子分裂的、配方明确的简单培养基。在最初尝试培养人类胚胎时曾使用过这些培养基，可惜分裂只能进行到 8 细胞期[9]（图 2）。现已证明，更为复杂的用于常规组织培养的培养基可更成功地用于人类胚胎培养。使用由哈姆氏 F 10 组成，并添加小牛或人血清及校正 pH 值为 7.3 的培养基，在 5% O_2、5% CO_2 和 90% N_2 的气相条件下，可使细胞发育到囊胚阶段[10]。在这种条件下培养出了好几个囊胚，每个囊胚都有 100 或更多个细胞核和若干个有丝分裂（图 3）。内细胞团、滋养层和囊胚腔之间界限分明，有些囊胚还在培养过程中发生了膨胀。囊胚腔似乎是由累积在桑葚胚一极的大细胞分泌形成的。另有报道称改良的体液也能维持人类胚胎的分裂[11]。据报道，体外成熟的卵母细胞只能用事先在猴子子宫中孵育过的人类精子受精[12]；胚胎分裂的失败或许是由体外成熟或其他一些原因造成的。

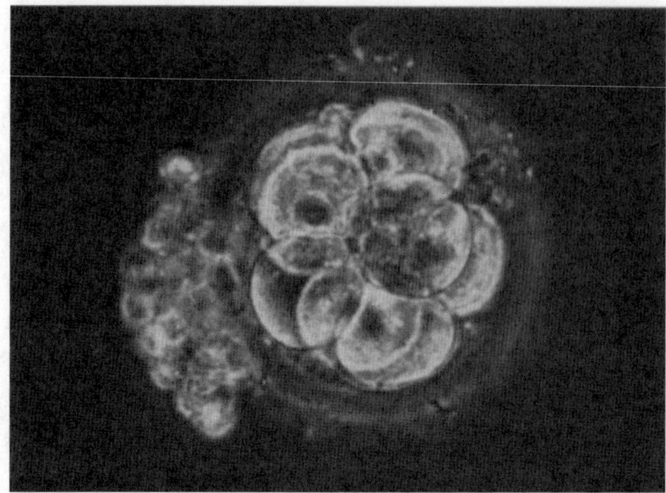

Fig. 2. Human eight-celled embryo.

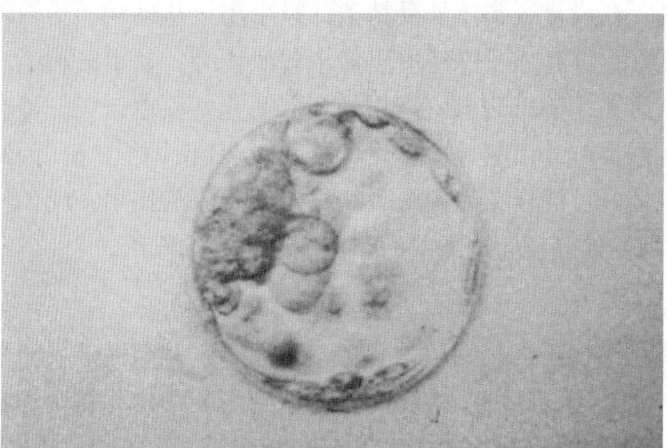

Fig. 3. Human blastocyst.

Where Next?

Further development beyond the blastocyst stage must surely be the next study. This raises the question of whether the embryos will be able to implant in the mother, that is, if the uterus has been sufficiently stimulated by the endocrine treatments used to induce follicular maturation. Another question concerns the capacity of the embryos to implant. The indications are good: the pronounced differentiation of many embryos in culture augurs well for their continued development. Conditions in the mother also seem to be satisfactory: most patients have had a secretory endometrium, excreted suitable amounts of pregnanediol, and their induced cycle has resembled a natural cycle[3].

What is known of the capacity of the embryos for normal development? The criteria for cleavage are now well recognized, and provide the first safeguard. Several of the embryos have possessed a chromosome complement that appeared to be diploid, thus excluding

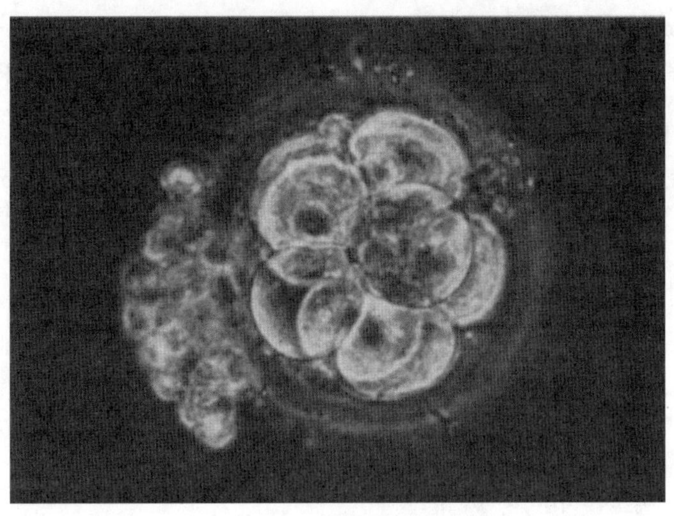

图 2. 人类的 8 细胞期胚胎

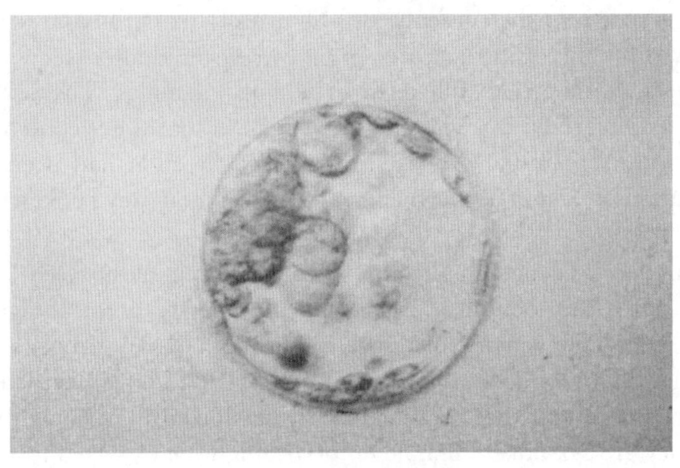

图 3. 人类的囊胚

下一步的目标?

下一步要研究的当然是囊胚阶段以后的发育过程。这就引出了这样一个问题：这些胚胎能否被移植到母体中？更确切地说，子宫能否被用以诱导卵泡成熟的内分泌治疗充分刺激。另一个问题是要考虑胚胎的植入能力。目前的迹象是好的：许多培养的胚胎出现了显著的分化，这预示着它们将会有良好的进一步发育。母体中的情况似乎也很令人满意：大多数患者的子宫内膜能够分泌适量的孕二醇，并且它们的诱导周期接近于自然周期[3]。

人们对胚胎正常发育能力的认识都有哪些呢？现在，人们对分裂的标准已经有了很深入的了解，这就提供了第一道安全保障。有几个胚胎拥有一套看似是二倍体

the possibility of triploidy. Trisomy has not been excluded. Yet the more that is learnt about natural human pregnancy, the more obvious becomes the conclusion that many embryos conceived naturally are trisomics, triploids or mosaics destined to perish in the first trimester. The possibility of trisomy in cultured embryos thus becomes less important, and especially since Down's syndrome can be identified in foetuses by amniocentesis. Nor should the problem of teratogenic development present undue difficulties, because widespread experience in animals and man has shown that agents known to be teratogenic in later stages of pregnancy are lethal to preimplantation embryos.

The immediate social consequences of the successful reimplantation of human embryos would be the cure of some forms of infertility, notably where the wife has occluded oviducts or the husband is oligospermic. There should be no criticism in giving these couples their own children: comments about overpopulation seem to be highly unjust to such an underprivileged minority.

The next development could well be the sexing of embryos before transfer. Rabbit blastocysts were sexed from the presence or absence of sex chromatin in small pieces of trophoblast excised by microsurgery[13]; the embryos recovered after a few hours in culture and some of them developed to full term in recipient females. The sex of all foetuses had been predicted correctly; this was the first occasion that the secondary sex ratio had been controlled in any mammal. Most efforts in this direction had been devoted to the attempted separation of X and Y spermatozoa, so far without any convincing success.

Improvements are now needed in the methods for sexing blastocysts. Two approaches are currently feasible: the identification of enzyme activity or of antigens determined by sex-linked genes. The activity of glucose-6-phosphate dehydrogenase (G6PD) and hypoxanthine-guanine phosphoribosyl transferase (HGPRT) have been studied in mouse embryos; it is not yet certain that these genes are sex-linked in mice as they are in man. Assuming that they are sex-linked, female embryos might be expected to possess twice the activity of male embryos. Unfortunately, levels of G6PD were dominated by the massive amounts of maternal enzyme "inherited" from the oocyte, and which declined steadily during cleavage[8,14]. The activity of HGPRT in embryos increased after fertilization, indicating that this enzyme was being synthesized during these early stages and, perhaps more important, that the sex chromosomes are active at this time[15]. This enzyme might thus be present in different amounts in male and female embryos. Evidence for sex-linked antigens is restricted to one report showing that the sex ratio of mouse offspring can be altered by interfering with the normal immunological processes of pregnancy[16]. Sexing blastocysts would be an excellent approach to the control of sex-linked mutant genes in man by avoiding the birth of affected boys. Strict control would have to be exerted over sexing if it could be demonstrated that widespread parental choice would seriously disturb the sex ratio in human populations.

Studies on human conception open new approaches to human problems including preventive medicine and family planning. More knowledge about the growth of ovulatory

的染色体组，这样就排除了三倍性的可能性。但三体性并没有被排除。人们对人类自然受孕的了解越多，就越清楚地看到以下这一点，即许多自然受孕的胚胎是在前三个月内注定要死亡的三体、三倍体或嵌合体。这样，在培养的胚胎中可能会出现三体性就变得不那么重要了，尤其是在利用羊水诊断就能知道胎儿是否患有唐氏综合征之后。解释畸形发育问题也不会有太大的难度，因为从动物和人身上得到的大量经验表明，怀孕后期能致畸的化学物质对于植入子宫前的胚胎来说是致命的。

成功再植入人类胚胎的直接社会效应将是可以治疗某些类型的不育症，尤其是女方输卵管堵塞或者男方精子太少的情况。让这样的夫妻拥有自己的子女不该受到人们的指责：和这些生育能力低下的少数群体谈论人口过剩问题似乎是很不公平的。

下一步的研究方向很可能是移植之前的性别判断。兔囊胚的性别可以通过检查由显微外科手术切除得到的小片滋养层细胞是否具有性染色质来判断[13]；胚胎在培养几小时后回收，有一些胚胎可在雌性受体内发育到足月。所有胎儿的性别都已被正确地预测出来；这是人们对哺乳动物出生性别比进行控制的首次尝试。以前朝这个方向的努力大多集中于尝试分离 X 精子和 Y 精子，不过至今尚未取得令人信服的成果。

目前我们还需要改进判断囊胚性别的方法。现在有两种方法是可行的：一种是鉴定酶的活性，另一种是鉴定由性连锁基因决定的抗原。我们已经在小鼠胚胎中研究了葡萄糖–6–磷酸脱氢酶（G6PD）和次黄嘌呤–鸟嘌呤磷酸核糖转移酶（HGPRT）的活性；目前还不能确定这些在人身上表现为性连锁的基因是否在小鼠身上也是性连锁的。假设这些基因是性连锁的，那么雌性胚胎中的酶活性应该是雄性胚胎中的两倍。但遗憾的是，G6PD 的水平主要以卵母细胞“遗传”的大量母源酶为主，而这些母源酶会在分裂过程中不断减少[8,14]。胚胎中 HGPRT 的活性在受精后有所增加，这表明此酶是在早期阶段合成的，也许更重要的是可以表明性染色体在这时是活跃的[15]。因此，HGPRT 在雄性胚胎和雌性胚胎中可能有不同的量。目前仅有一篇文献与性连锁抗原有关，该文指出，小鼠后代的性别比例可以通过干扰怀孕时的自然免疫过程来改变[16]。囊胚性别鉴定将是一种通过避免患病男婴出生从而控制性连锁突变基因的绝好方法。如果有证据表明父母们的普遍意愿会严重干扰人口的性别比例，那么就必须对性别鉴定进行严格的控制。

人类受孕研究开辟了解决包括预防医学和计划生育在内的人类问题的新途径。更多地了解关于排卵卵泡生长、囊胚分泌以及精子获能成分方面的知识也许会使我

follicles, the secretions of the blastocyst or the components of capacitation could lead to the development of novel methods of contraception. Rational discussion is needed about the emerging social issues involved in these studies. Much of the debate so far has been unreal in the sense of being far removed from the actual clinical problems involved: constant reference to nuclear cloning is an example. Earlier remonstrations about the disposal of tiny blastocysts appear almost irrelevant when foetuses of four months' gestation or older are being aborted for social reasons. Alleviating infertility has even been criticized because more consideration is needed about the survival of the family unit in our society! Realization of the benefits that could accrue to many people should lead to more rational discussion and conclusions.

(**233**, 23-25; 1971)

References:

1. Edwards, R. G., *Lancet*, ii, 926 (1965).
2. Chang, M. C., *J. Exp. Zool.*, **128**, 379 (1955).
3. Steptoe, P. C., and Edwards, R. G., *Lancet*, i, 683 (1970).
4. Hack, M., Brish, M., Serr, D. M., Insler, V., and Lunenfeld, B., *J. Amer. Med. Assoc.*, **211**, 791 (1970).
5. Crooke, A. C., *Brit. Med. Bull.*, **26**, 17 (1970).
6. Yanagimachi, R., and Chang, M. C., *J. Exp. Zool.*, **156**, 361 (1970).
7. Edwards, R. G., Bavister, B. D., and Steptoe, P. C., *Nature*, **221**, 632 (1969).
8. Brinster, R. L., Chap 17 in *Biology of the Blastocyst* (edit. by Blandau R. J.) (University of Chicago Press, 1971).
9. Edwards, R. G., Steptoe, P. C., and Purdy, J. M., *Nature*, **227**, 1307 (1970).
10. Steptoe, P. C., Edwards, R. G., and Purdy, J. M., *Nature*, **229**, 132 (1971).
11. Shettles, L. B., *Nature*, **229**, 343 (1971).
13. Gardner, R. L., and Edwards, R. G., *Nature*, **218**, 346 (1968).
14. Brinster, R. L., *Biochem. J.*, **101**, 161 (1966).
15. Epstein, C. J., *J. Biol. Chem.*, **245**, 3289 (1970).
16. Lappe, O. O., and Schalk, M. J., *Transplantation*, **11**, 491 (1971).

们找到避孕的新方法。关于由这些研究引发的社会问题尚需要通过理性的讨论来解决。迄今为止的讨论大多因为远远脱离了实际临床问题而显得毫无意义：经常被引用的核克隆就是一个例子。当怀孕四个月或者更大的胎儿因社会原因而被流产时，早先人们对丢掉小囊胚的抗议也就显得无关紧要了。缓解不孕症一度曾受到了批评，因为当前需要更多考虑的是家庭单元在现实生活中是否能维持下去！在认识到人工受精会使众多的民众受益之后，人们才会进行更理性的讨论，并作出结论。

（冯琛 翻译；郑家驹 审稿）

Tyrosine tRNA Precursor Molecule Polynucleotide Sequence

S. Altman* and J. D. Smith

Editor's Note

Biologist Sidney Altman had previously isolated the transcript of the gene for tyrosine transfer RNA (tRNA), the small RNA molecule that transfers the amino acid tyrosine to the growing polypeptide chain during protein translation. Here Altman and his colleague John D. Smith report the discovery of extra nucleotides at both ends of the transcribed precursor tRNA molecule. When the precursor tRNA was mixed with an extract of *Escherichia coli*, it became apparent that enzymes in the cell extract could cleave off the "extra" nucleotides yielding mature tRNA. The researchers demonstrated that the RNA itself had catalytic properties: it is a so-called ribozyme. Altman went on to isolate the enzyme responsible, ribonuclease P, earning himself a Nobel Prize.

RIBOSOMAL RNAs in bacterial and mammalian cells are transcribed in units longer than the functional final product[1-4]. Transfer RNA in mammalian cells is also thought to be made through precursor molecules[4,5]. Although the biological role of the extra pieces in these transcription products is not known, they may contain sequences affecting the level of production of the final product. The isolation of precursor molecules to one species of *E. coli* tyrosine transfer RNA, which is amenable to total nucleotide sequence analysis and specific mutant selection, was recently described[6]. We now report (*a*) the nucleotide sequence of the new segment of one of these precursor molecules and verify the absence of the usual nucleotide modifications from this molecule; (*b*) that a mutation[7] (A2P) that increases the amount of mature tRNA produced by tyrosine tRNA-mutant A2 is located in the precursor segment close to the usual 5′ end of the tRNA sequence, and (*c*) that a nuclease activity, present in crude extracts of *E. coli*, specifically cleaves the mature tRNA sequence from the precursor *in vitro*.

The sequence that we describe is characteristic of a class of single base change mutants of tyrosine tRNA and differs from the wild type precursor in a few nucleotides at the 3′ end of the molecule. The newly defined segment at the 5′ end of the precursor molecule can, in principle, assume different configurations. We will briefly discuss a kinetic model of tRNA biosynthesis in which secondary structure of precursor plays an important role.

* Present address: Department of Biology, Yale University, New Haven, Connecticut.

酪氨酸转运RNA前体分子的多聚核苷酸序列

奥尔特曼*，史密斯

编者按

生物学家西德尼·奥尔特曼之前曾分离出了酪氨酸转运RNA（tRNA）基因的转录本。tRNA是一种小RNA分子，在蛋白质翻译过程中可以将酪氨酸转运到延伸中的多肽链上。在本文中，奥尔特曼和他的同事约翰·史密斯报道称，他们发现在转录的tRNA前体分子两端存在额外的核苷酸。当tRNA前体与大肠杆菌提取物混合时，很明显地，细胞提取物中的酶可以切掉"多余"的核苷酸，产生成熟的tRNA。研究人员证明：RNA自身具有催化活性，它是一种核酶。奥尔特曼接着分离出了这个起作用的酶——核糖核酸酶P，并因此获得了诺贝尔奖。

细菌和哺乳动物细胞中的核糖体RNA在转录时的长度比功能性终产物更长[1-4]。哺乳动物细胞中的tRNA也被认为是经过前体分子而形成的[4,5]。尽管现在还不清楚这些转录产物多余片段的生物学功能，但猜测它们可能含有影响终产物产生水平的序列。前不久曾报道过已分离得到一种大肠杆菌酪氨酸tRNA的前体分子[6]，用它可以进行总核苷酸序列分析和特异性突变筛选。我们现在要报道的是：(*a*) 这些前体分子之一的新片段核苷酸序列，并证实这一分子中缺少常见的核苷酸修饰；(*b*) 有一个突变[7]（A2P）能增加由酪氨酸tRNA–突变体A2产生的成熟tRNA的数量，这个突变在前体片段中的位置很靠近tRNA序列的常见5′端；(*c*) 大肠杆菌粗提物的核酸酶活性可以在体外特异性地将成熟tRNA从前体中切割出来。

我们所描述的核苷酸序列是酪氨酸tRNA的一类单碱基改变突变体所特有的，它与野生型前体相比在分子3′端存在少数几个核苷酸的区别。原则上讲，新确定的前体分子的5′末端片段能够呈现不同的立体结构。我们将简要讨论一种有关tRNA生物合成的动力学模型，在该模型中，前体的二级结构起着重要的作用。

* 现在的地址：康涅狄格州纽黑文市耶鲁大学生物系。

Isolation and Nucleotide Sequence of Precursor

^{32}P-labelled precursor tRNA was isolated after pulse labelling of *E. coli* infected with φ80 phage carrying a mutant su_{III} gene specifying tyrosine tRNA with a single base substitution[6]. This type of mutant gives a much higher yield of precursor than the original su^{+}_{III} gene. We used mutant A25 whose tRNA has a G→A substitution at residue 25[7]. All the phage mutants used here yielded a precursor band identical to band Y of ref. 6, migrating just behind 5S RNA in acrylamide gels.

The T_1 and pancreatic ribonuclease fingerprints of A25 precursor tRNA are shown in Fig. 1. The unlabelled nucleotides correspond to those found in digests of mature tRNA and the numbered nucleotides are derived from the additional segments of precursor tRNA. Their sequences are given in Table 1. All the nucleotides derived from the tRNA portion of the precursor were identified by further enzymatic or alkaline digestion as previously described[8]. This analysis showed that the precursor contained the complete primary tRNA sequence. Comparison with mature tRNA, however, showed two important differences: (1) those bases which are normally modified were almost completely unmodified, and (2) the T_1 and pancreatic digestion products derived from the 5′ and 3′ termini of mature tRNA were absent in digests of precursor.

Table 1. Additional Products of Enzymatic Digestion of Precursor tRNA

T_1 ribonuclease No. in Fig. 1*a*	Pancreatic ribonuclease No. in Fig. 1*b*
1 CCAG	1 AGGC
2 AUAAG	2 AAAAGC
3 UAAAAG	3 GAU
4 CUUCCCG	4 AGU
5 CAUUACCCG	5 GGU
6 pppGp	6 AAGGGAGC
7 $AAUCCUUCCCCCACCACCAUCU_{OH}$	7 pppGC
CAG	AC
AG	AU (2 moles)
UG	GU

All products are in molar yield except where indicated. Sequences were derived from the products of T_1, pancreatic or U_2 ribonuclease digestion. CUUCCCG was determined by partial digestion of the dephosphorylated nucleotide with snake venom diesterase. Overlap data from larger fragments were used for AAGGGAGC.

前体分离及其核苷酸序列

su_{III} 突变基因可特异性地使酪氨酸 tRNA 的单个碱基发生替换，利用携带 su_{III} 突变基因的 φ80 噬菌体感染大肠杆菌，并对感染后的大肠杆菌进行脉冲标记，最后将 ^{32}P 标记的 tRNA 前体分离出来[6]。与原始的 su^{+}_{III} 基因相比，由这类突变基因得到的前体产量要高很多。我们采用了 A25 突变体，该突变体的 tRNA 在第 25 位残基处有一个 G → A 的替换[7]。在此使用的所有噬菌体突变都能产生与参考文献 6 中的 Y 条带完全一致的前体条带，这个条带在丙烯酰胺凝胶电泳中的迁移距离紧跟在 5S RNA 后面。

图 1 是 A25 tRNA 前体的 T_1 核糖核酸酶和胰核糖核酸酶指纹图。未标记的核苷酸与那些在成熟 tRNA 酶解物中发现的核苷酸一致，有数字编号的核苷酸是由 tRNA 前体的额外片段产生的。表 1 给出了相应核苷酸的序列。前体中 tRNA 部分的所有核苷酸都用以前描述过的酶解或碱溶法进行了鉴定[8]。分析结果表明前体具有完整的初级 tRNA 序列。但与成熟 tRNA 相比，有以下两点重要的不同：(1) 那些通常情况下被修饰的碱基几乎完全未被修饰；(2) 源于成熟 tRNA 5′和 3′末端的 T_1 核糖核酸酶和胰核糖核酸酶酶解产物在前体的酶解物中不存在。

表 1. tRNA 前体酶解的额外产物

T_1 核糖核酸酶 按图 1*a* 中的编号	胰核糖核酸酶 按图 1*b* 中的编号
1 CCAG	1 AGGC
2 AUAAG	2 AAAAGC
3 UAAAAG	3 GAU
4 CUUCCCG	4 AGU
5 CAUUACCCG	5 GGU
6 pppGp	6 AAGGGAGC
7 $AAUCCUUCCCCCACCACCAUCU_{OH}$	7 pppGC
CAG	AC
AG	AU (2 mol)
UG	GU

除非特别标注，所有产物的产量均以摩尔为单位。这些序列均源于 T_1 核糖核酸酶、胰核糖核酸酶或 U_2 核糖核酸酶的酶解产物。CUUCCCG 由蛇毒二酯酶对去磷酸化的核苷酸的部分酶解作用所确定。AAGGGAGC 由较大片段的重叠数据所确定。

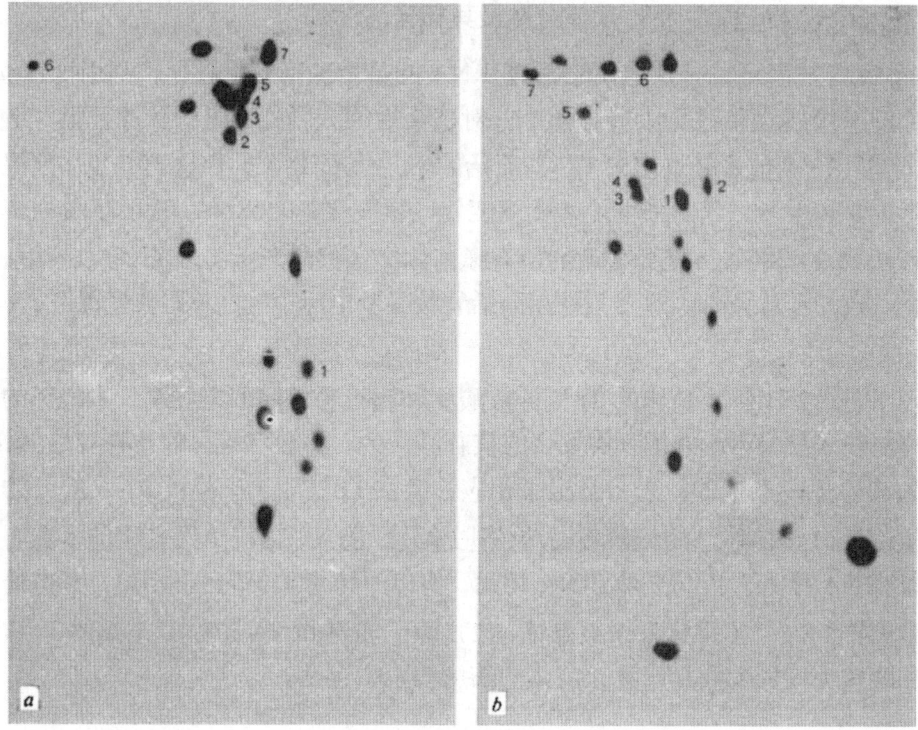

Fig. 1. Fingerprints of products of (*a*) T_1 ribonuclease and (*b*) pancreatic ribonuclease digests of A25 precursor tRNA. The numbered spots are those not found in digests of A25 tRNA (Table 1). Autoradiographs of two dimensional separations: right to left, electrophoresis on cellulose acetate in pyridine-acetate *p*H 3.5 containing 7 M urea and 0.002 M EDTA; top to bottom, electrophoresis on DEAE paper in 7% v/v formic acid. In *b* nucleotide 1 (AGGC) is not separated from GAGC.

Su_{III} tyrosine tRNA contains seven modified nucleoside residues[8], which are (numbered from the 5′ end) 4-thiouridines, 8 and 9; 2′O-methyl guanosine, 17; 2-thiomethyl, 6-isopentenyl adenosine, 38; pseudouridines, 40 and 64; and thymine riboside, 63. In the precursor tRNA G17 and A38 are unmodified and uridine replaces ψ40 and T63. Residue 64 (present as ψ in the TψCG sequence of the mature tRNA) is partly modified in the precursor; about 10% was isolated as ψ and 90% as U. We have not measured the extent of modification of the thiolated residues 8 and 9.

It has been argued from hybridization experiments that the common CCA_{OH} terminal sequence of mature tRNA is not part of the tRNA gene[9]. So it is important to know whether this sequence is present in the precursor tRNA.

The 3′ terminus of tyrosine tRNA gives the T_1 ribonuclease product $AAUCCUUCCCCCACCACCA_{OH}$. This is absent in digests of precursor tRNA but the entire sequence is present in nucleotide 7 occupying a different position on the fingerprint (Fig. 1). Nucleotide 7 contains the 3′ terminus of the tRNA with three additional nucleotides at the 3′ end (Table 1). Since it does not contain G this is the 3′ terminus of the precursor and all the other extra nucleotides in the digests must be part of the 5′ additional segment.

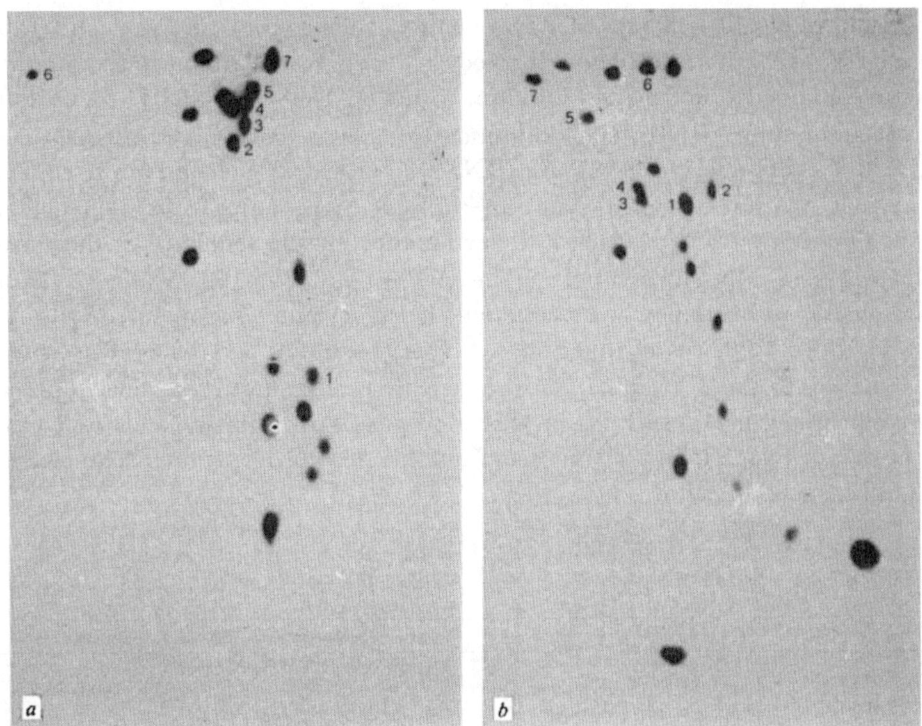

图 1. A25 tRNA 前体的（*a*）T_1 核糖核酸酶酶解产物和（*b*）胰核糖核酸酶酶解产物的指纹图。图中有数字编号的点是在 A25 tRNA 酶解物中找不到的点（表 1）。二维分离的放射自显影图：从右至左，电泳液为含有 7 M 尿素和 0.002 M EDTA 的 pH 3.5 的醋酸哌啶溶液，在醋酸纤维素膜上进行电泳检测；从上至下，电泳液为体积比为 7% 的甲酸溶液，在二乙氨基乙基纤维素纸上进行电泳检测。图 *b* 中的第 1 号核苷酸（AGGC）与 GAGC 没有分开。

su_{III} 酪氨酸 tRNA 含有 7 个修饰的核苷酸残基 [8]，这些残基分别是（从 5′端开始编号）：4–硫尿核苷，8 和 9；2′–O–甲基鸟苷，17；2–甲硫基–6–异戊烯基腺苷，38；假尿苷，40 和 64；胸腺嘧啶核苷，63。在 tRNA 前体中，G17 和 A38 未发生修饰，并且尿苷代替了 ψ40 和 T63。第 64 位残基（成熟 tRNA 的 TψCG 序列中出现的 ψ）在前体中被部分修饰：分离出来的残基约有 10% 为 ψ，90% 为 U。我们还没有测定硫醇化残基 8 和 9 的修饰情况。

从杂交实验角度看，通常情况下成熟 tRNA 的 CCA_{OH} 末端序列并不是 tRNA 基因的一部分 [9]。因此了解该序列是否存在于 tRNA 前体中很重要。

由酪氨酸 tRNA 3′末端给出的 T_1 核糖核酸酶酶解产物的序列为 $AAUCCUUCCCCCACCACCA_{OH}$。在 tRNA 前体的酶解物中不存在这一序列，但整个序列存在于第 7 号核苷酸中，在指纹图里出现在另一个不同的位置上（图 1）。第 7 号核苷酸包含 tRNA 的 3′末端，在 3′端有 3 个额外的核苷酸（表 1）。因为第 7 号核苷酸不包含 G，所以是前体的 3′末端；酶解物中所有其他多余核苷酸一定是 5′端额外片段的一部分。

Comparisons of the 3′ terminus of A25 precursor (nucleotide 7 in Fig. 1 and Table 1) with that of A25 mature tRNA showed: (1) pancreatic ribonuclease digests of precursor contained 1 mol AU in addition to the AAU, AC, C and U found both in precursor and mature tRNA; (2) U_2 ribonuclease digests[10] gave CCUUCCCCCA, AA, A, CCA and as expected from mature tRNA and an extra nucleotide only containing U and C. CCA_{OH} was absent. Because of the repetition of the CCA sequence at the terminus of tyrosine tRNA (Table 2) determination of the sequence at the terminus of A25 precursor-tRNA depended on the yield of CCA in the U_2 digest. To avoid reliance on this measurement we examined the 3′ terminus of precursor tRNA in a mutant in which C80 was changed to U. (This mutant A2 U41 U80 P is described below.) Analysis of the 3′ terminal T_1 ribonuclease product from this mutant precursor tRNA is illustrated in Table 2. As U_2 digestion gives equal molar yields of UCA and CCA we can conclude that the precursor contains the tRNA terminal sequence CCA. This is linked to a U residue.

Table 2. Sequence of the 3′ Terminus of Precursor tRNA

	3′ terminus isolated after T_1 ribonuclease digestion
A25 tRNA	$AAUCCUUCCCCCACCACCA_{OH}$
A2 U80 U41 P tRNA	$AAUCCUUCCCCCAUCACCA_{OH}$
A2 U80 U41 P precursor	$AAUCCUUCCCCCAUCACCAUCU_{OH}$
A25 precursor	$AAUCCUUCCCCCACCACCAUCU_{OH}$

Identified products from U_2 ribonuclease digestion are defined by a line above the sequence, those from pancreatic ribonuclease digests by a line below.

The 5′ terminus of tyrosine tRNA is pGGU—which gives pGp with T_1 ribonuclease and pGGU with pancreatic ribonuclease. Both of these products are absent from digests of precursor. The additional pancreatic ribonuclease product GGU is the only sequence which could contain the 5′ terminus of the tRNA itself, suggesting that this is joined to the additional 5′ segment as pyr.GGU. This was confirmed by the sequence analysis.

We have examined the precursor from four different base substitution mutants A25[7], A15, A17 and A31[11] and all give the same additional T_1 and pancreatic ribonuclease digestion products shown in Table 1. Quantitative analysis of the nucleotides from the digests showed that those unique to the precursor tRNA sequence are present in approximately molar yield and that the dinucleotides and trinucleotides listed are also present in the additional segment. The identification of pppGp and pppGC defines the 5′ terminus of the precursor and we conclude that the initial transcript is intact at this end.

The sequence of the 5′ segment was determined by analysis of products of partial digestion with T_1 ribonuclease[8,12] and is shown in Fig. 2. Examination of T_1 ribonuclease partial digestion products which include portions of the tRNA shows that they all correspond to the fragments isolated from similar digests of tyrosine tRNA[8]. This suggests that the secondary structure of the tRNA portion of precursor is similar to that of mature

比较 A25 前体的 3′ 末端序列（图 1 和表 1 中的第 7 号核苷酸）与 A25 成熟 tRNA 的 3′ 末端序列，结果如下：（1）前体的胰核糖核酸酶酶解物除了含有在前体和成熟 tRNA 中都能找到的 AAU、AC、C 和 U 之外，还含有 1 mol AU；（2）前体 U_2 核糖核酸酶酶解物 [10] 中含有 CCUUCCCCCA、AA、A 和 CCA（成熟 tRNA 产物中也有这些序列），而且还含有一个仅由 U 和 C 组成的多余核苷酸；不存在 CCA_{OH}。因为酪氨酸 tRNA 末端具有重复的 CCA 序列（表 2），所以 A25 tRNA 前体末端的序列是不是 CCA 取决于 U_2 核糖核酸酶酶解物中 CCA 的生成量。为了避免单独依靠这种测定方法，我们检测了一个 tRNA 前体突变体的 3′ 末端序列，在这个突变体中，第 80 位的 C 转变成了 U（关于该突变体 A2 U41 U80 P 的描述见下文）。对此 tRNA 前体突变体的 3′ 末端的 T_1 核糖核酸酶酶解物进行分析，结果见表 2。由于 U_2 核糖核酸酶的酶解物中 UCA 和 CCA 的摩尔数相等，因此我们可以推断该前体含有 tRNA 的末端序列 CCA。这个 CCA 序列与一个 U 残基相连。

表 2. tRNA 前体的 3′ 末端序列

	T_1 核糖核酸酶酶解后分离出的 3′ 末端序列
A25 tRNA	$AAUCCUUCCCCCACCACCA_{OH}$
A2 U80 U41 P tRNA	$AAUCCUUCCCCCAUCACCA_{OH}$
A2 U80 U41 P 前体	$AAUCCUUCCCCCAUCACCAUCU_{OH}$
A25 前体	$AAUCCUUCCCCCACCACCAUCU_{OH}$

序列上方有线代表是由 U_2 核糖核酸酶酶解鉴定出的结果，序列下方有线代表是由胰核糖核酸酶酶解鉴定出的结果。

酪氨酸 tRNA 的 5′ 末端是 pGGU——使得 T_1 核糖核酸酶酶解物具有 pGp，胰核糖核酸酶酶解物具有 pGGU。这两种产物在前体酶解物中都不存在。胰核糖核酸酶酶解的额外产物 GGU 是唯一可能含有 tRNA 自身 5′ 末端的序列，这说明该序列可以以吡啶 GGU 的形式连接到 5′ 端的额外片段上。上述结论已通过基因序列分析确认。

我们检测了来自 4 个不同碱基替换突变体 A25[7]、A15、A17 和 A31[11] 的前体，所有这些前体在 T_1 核糖核酸酶和胰核糖核酸酶酶解后都生成了完全相同的额外产物，结果见表 1。对酶解物中核苷酸的定量分析结果显示：那些为 tRNA 前体所特有的核苷酸序列的产量都达到了摩尔水平，表 1 中列出的二核苷酸和三核苷酸在额外片段中也存在。通过鉴别 pppGp 和 pppGC 可以确定前体的 5′ 末端，而且我们断定此末端的初期转录本是完整的。

通过对 T_1 核糖核酸酶部分酶解的产物进行分析确定 5′ 端片段的序列 [8,12]，结果如图 2 所示。对含有 tRNA 部分的 T_1 核糖核酸酶的部分酶解产物进行检测，结果发现它们与从酪氨酸 tRNA 的类似酶解物中分离出来的片段完全一致 [8]。这表明前体分子中 tRNA 部分的二级结构与成熟 tRNA 中的类似。假设 tRNA 部分具有三叶草形二

tRNA. Assuming the tRNA portion has the clover leaf secondary structure and is not interacting with the 5′ segment, a possible secondary structure for the latter is shown as structure II (Fig. 2).

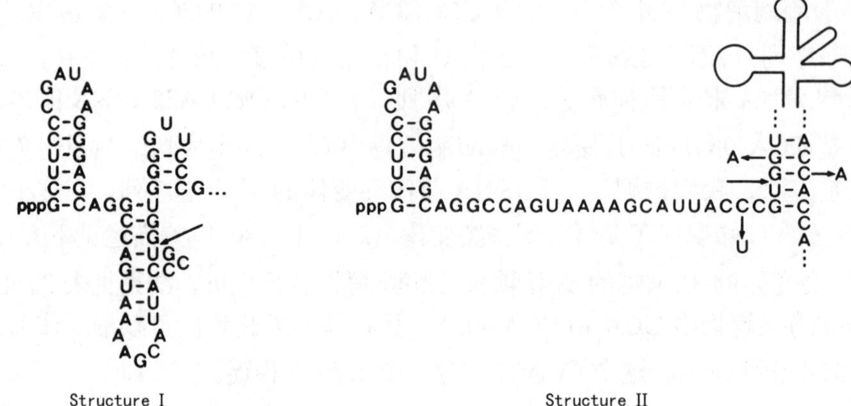

Fig. 2. Possible configuration for tyrosine precursor. The hydrogen-bonded loops shown are stable according to the approximate criteria cited in ref. 20. The arrows pointing towards the sequence indicate the beginning of the 5′ end of the tRNA moiety and a cleavage point. In the mature tRNA, none of the last four nucleotides shown at the 3′ end of structure II are involved in hydrogen bonding. The arrows pointing outward indicate the positions of certain mutants: P (C → U); A2 (G → A); and A81 (C → A), all mentioned in the text.

Mutant Sequence in Precursor Segment

Mutation A2P increases the production of a particular mutant (A2) tyrosine tRNA[7]. As P does not change the sequence of the mature tRNA it must lie outside this portion of the gene, so we examined the possibility that it produced a sequence change in the additional parts of the gene transcribed as part of the precursor. This could not be determined directly because both A2 and A2P, unlike base substitution tRNA mutants such as A25, do not lead to accumulation of precursor. Several hydroxylamine induced su^- mutants of A2 and A2P also failed to accumulate precursor, so it seemed that the apparent rapid *in vivo* degradation of precursor might be due to the change in tRNA structure resulting from the A2 mutation. We therefore examined derivatives of A2P in which the base pairing between residues 2 and 80 was restored by the mutation C80→U.

These derivatives also had a third base substitution within the tRNA so as to be analogous to the base substitution mutants of the type A25 which accumulate precursor tRNA. The su_{III} multiple mutant A2 U80 U41 P was isolated starting with *E. coli* MB93 su_{III} A2P and using selection techniques previously described[7]. This was obtained as φ80 psu A2 U80 U41 P. Two other completely independent isolations yielded φ80 psu A2 U80 U54 P and φ80 psu A2 U80 A46 P. These were identified by tRNA sequence analysis. Cells infected with each of these phages accumulated precursor tRNA in ^{32}P pulse-label experiments.

The T_1 ribonuclease fingerprint of precursor from cells infected with φ80 psu A2 U80 U41 P showed the altered products resulting from the three mutations in the tRNA portion of

级结构，并且与 5′ 端的序列不发生相互作用，则后者可能的二级结构如结构 II 所示（图 2）。

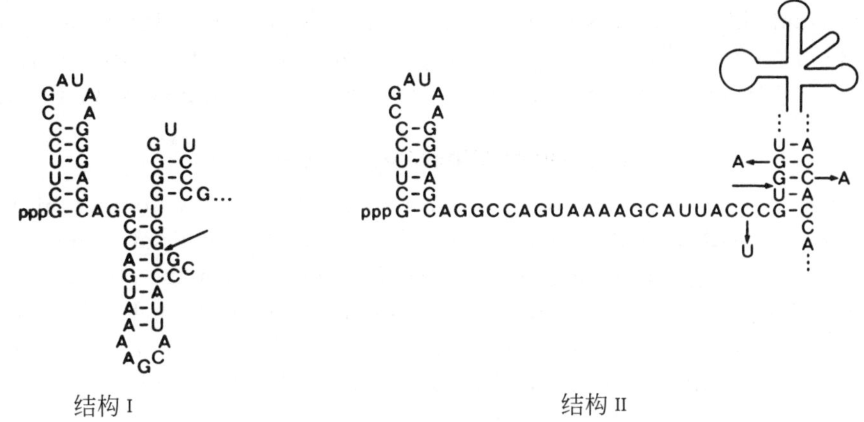

图 2. 酪氨酸前体的可能结构。根据参考文献 20 中的大概标准，图中氢键环是稳定的。指向序列的箭头表示 tRNA 组分 5′ 端的起始点，同时也是一个切割点。在成熟 tRNA 中，结构 II 中 3′ 端的最后 4 个核苷酸都不含氢键。指向序列外的箭头表示发生突变的位置：P (C → U)；A2 (G → A)；A81 (C → A)，这些突变都在文中被提到过。

前体片段的突变序列

突变 A2P 增加了一种特定突变体（A2）酪氨酸 tRNA 的产量[7]。既然 P 不改变成熟 tRNA 的序列，它一定存在于该段基因的外部，因而我们检测了如下可能性，即 P 会在前体部分转录的基因的额外部分产生序列改变。这一点不能直接进行检测，因为 A2 和 A2P 与 tRNA 的碱基替换突变体（例如 A25 突变体）不同，它们都不能导致前体积累。A2 和 A2P 的几个羟胺诱导的 su⁻ 突变体同样不能积累前体，因此前体在体内迅速降解可能是由于 A2 突变引起 tRNA 的结构变化所致。所以我们检测了 A2P 的几种衍生物，在这些衍生物中，第 2 位残基和第 80 位残基之间的碱基配对由于第 80 位的 C 突变为 U 而得到了恢复。

这些衍生物在 tRNA 内部还有第三个碱基发生替换，以便与能够积累 tRNA 前体的 A25 型碱基替换突变体类似。从大肠杆菌 MB93 su_{III} A2P 步开始，采用以前描述过的筛选技术[7]，将 su_{III} 多重突变体 A2 U80 U41 P 分离出来。得到了 φ80 psu A2 U80 U41 P。另外两种完全独立的分离过程则产生了 φ80 psu A2 U80 U54 P 和 φ80 psu A2 U80 A46 P。这些产物通过 tRNA 序列分析得以鉴别。在 ^{32}P 脉冲标记的实验中，被所有这些噬菌体感染的细胞都能积累 tRNA 前体。

从感染 φ80 psu A2 U80 U41 P 的细胞中提取的前体 T_1 核糖核酸酶指纹图显示：前体的 tRNA 部分的 3 个突变造成了酶解产物的改变；A2 产生 AUG 而不是 UG；

the precursor; A2 gives AUG instead of UG; U80 gives a 3′ terminus product with altered mobility; U41 changes the mobility of the anticodon-containing fragment. In addition nucleotide 5 (CAUUACCCG) (Fig. l*a*), part of the precursor segment, is absent and replaced by a nucleotide in a position corresponding to a C→U change. Sequence analysis of this nucleotide identified it as CAUUACUCG. In this precursor tRNA the 5′ segment is altered by a C→U change at the position indicated in Fig. 2. As all three su_{III} mutants derived from A2P showed the same change we consider this is specified by the P mutation.

Precursor Cleavage *in vitro*

We have detected a nucleolytic activity, which may produce tRNA from its precursor, in extracts of *E. coli* MRE 600. After incubation of purified A25 precursor with cell extracts, there are three major products (Fig. 3): (*a*) a band migrating near the position of mature tRNA (band B in Fig. 3); (*b*) a much faster moving band (band C in Fig. 3); and (*c*) two diffuse bands moving with the salt front, which consist primarily of 5′ mononucleotides but also containing a small percentage of dinucleotides and trinucleotides.

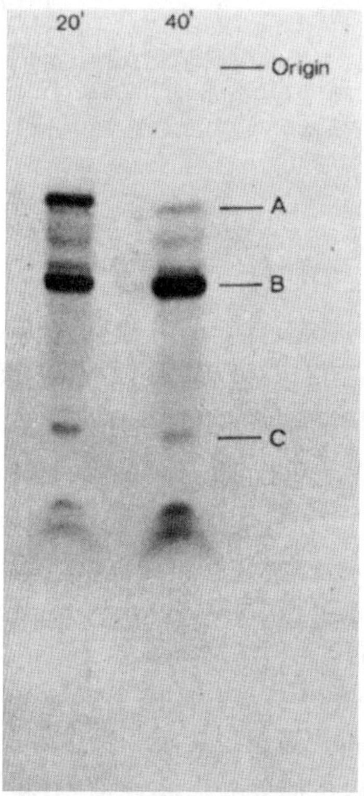

Fig. 3. Autoradiogram of acrylamide gel separation of precursor cleavage products. Purified ^{32}P-labelled A25 precursor was incubated at 37°C in siliconized glass tubes containing 5 mM $MgCl_2$, 0.1 mM beta-mercaptoethanol, 0.1 mM EDTA, 10 mM Tris (hydroxymethyl) amine, *p*H 8, an excess (approximately 1.5 A_{260} units) of cold carrier *E. coli* tRNA and 5 A_{260} units of a 30,000*g* supernatant of *E. coli* MRE 600 extract (a gift of Dr. H. D. Robertson) in a final volume of 0.4 ml. At the times indicated, 20 μl. of 0.4 M EDTA, *p*H 9.4 was added to 0.2 ml. of the reaction mixture which was then evaporated to dryness. Each sample was resuspended in 35 μl. 40% sucrose made 0.04 M in EDTA, and containing 2 μl. of a saturated bromphenol blue solution, and was then layered on a 10% acrylamide gel. Electrophoresis was for approximately 16 h at 400 V and 4°C. In separate experiments we have observed that mature tRNA marker runs in the same position as band B.

U80 产生的是迁移率发生变化的 3′ 末端产物；U41 则改变了含反密码子片段的迁移率。此外，作为前体片段一部分的第 5 号核苷酸（CAUUACCCG）（图 1*a*）缺失了，取而代之的是一个在相应位置由 C 变为 U 的核苷酸。这个核苷酸经过序列分析鉴定被确定为 CAUUACUCG。在这个 tRNA 前体中，5′ 端片段因图 2 所示位置上 C → U 的变化而发生了改变。因为源于 A2P 的所有 3 种 su_{III} 突变体都表现出相同的改变，所以我们认为这种特异性是由 P 突变引起的。

前体的体外切割

我们已经从大肠杆菌 MRE 600 的提取液中检测到了核溶解活性，该活性可能使前体产生 tRNA。用细胞提取液孵育纯化的 A25 前体之后，得到 3 种主要产物（图 3）：（*a*）成熟 tRNA 位置附近的条带（图 3 中的 B 带）；（*b*）迁移速度非常快的条带（图 3 中的 C 带）；以及（*c*）两条随着盐缘移动而移动的弥散带，这两条弥散带主要由 5′ 端的单核苷酸组成，但也含有很小比例的二核苷酸和三核苷酸。

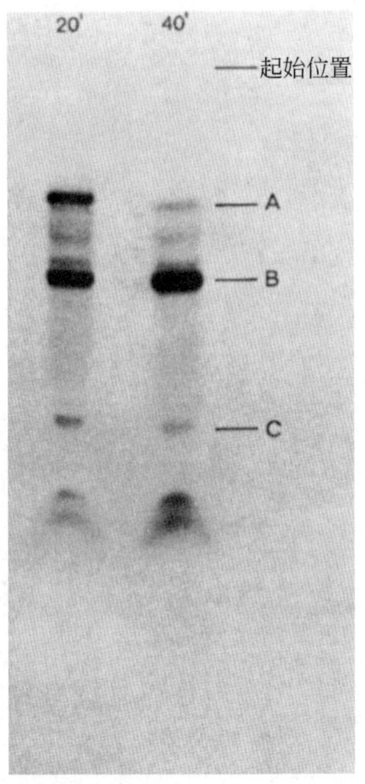

图 3. 前体切割产物的丙烯酰胺凝胶电泳分离放射自显影图。将纯化了的 ^{32}P 标记的 A25 前体置于硅玻璃管中，并在 37℃ 下孵育，硅玻璃管中含有 5 mM $MgCl_2$、0.1 mM β– 巯基乙醇、0.1 mM EDTA、pH 值为 8 的 10 mM Tris、过量（大约 1.5 个 A_{260} 单位）的冷大肠杆菌 tRNA 载体和 5 个 A_{260} 单位的经 30,000*g* 离心的大肠杆菌 MRE 600 提取液（罗伯逊博士捐赠）的上清液，最终混合溶液的体积为 0.4 ml。在图中所标的时间内，将 20 μl、pH 值为 9.4 的 0.4 M EDTA 加入 0.2 ml 的反应混合物中，然后将混合物蒸发至干燥。每个样品被重悬于 35 μl、40% 含有 0.04 M EDTA 和 2 μl 饱和溴酚蓝的蔗糖溶液中，然后加样于 10% 的丙烯酰胺凝胶中。在 400 V 电压和 4℃ 下持续电泳约 16 h。在另外几个独立的实验中，我们观察到成熟 tRNA 标准品迁移到了与 B 条带相同的位置。

Specific *in vitro* cleavage of the precursor is established because a two dimensional fingerprint of a pancreatic ribonuclease digest of band B shows that the 5′ end is pGGU—as expected for the mature tRNA and the additional 5′ segment of the precursor has been completely removed. A fingerprint of the T_1 ribonuclease digest of band B (Fig. 4) is also identical to that of mature A25 tRNA except that two spots derived from the 3′ terminus are present. One of these has the same mobility as the precursor 3′ oligonucleotide. The other, nearby in the fingerprint, is present in about 20% yield and lacks at least the two terminal nucleotides, presumably due to the action of another nuclease in the cell extract. The yield of the modified 3′ end is dependent upon the time of incubation of the reaction mixture. Prolonged electrophoresis can resolve band B into two bands, the faster moving of which contains exclusively the modified 3′ end.

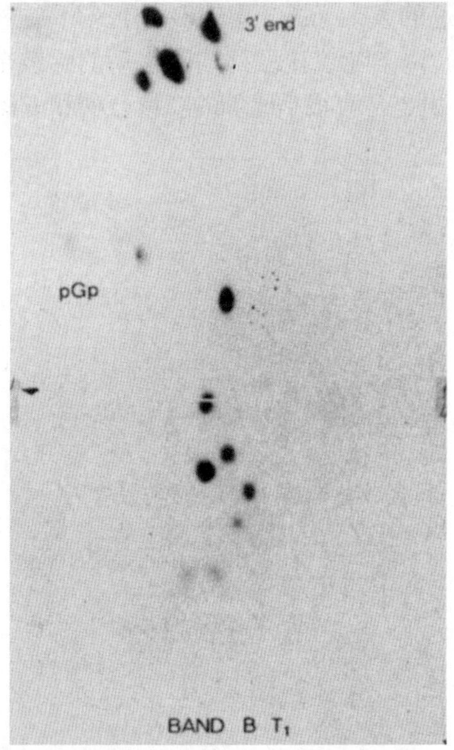

Fig. 4. Autoradiogram of a two dimensional fingerprint of the T_1 ribonuclease digestion products of A25 precursor cleavage product. Band B of Fig. 3 was eluted from the gel, digested with T_1 ribonuclease and the products separated by two dimensional electrophoresis. The newly generated 5′ end is indicated on the autoradiogram as is the position of the precursor 3′ terminal oligonucleotide and a nearby spot in molar yield near the position of the usual 3′ terminal oligonucleotide of mature tyrosine tRNA. These products are discussed further in the text.

Band C contains the extreme 5′ end of the precursor molecule reading in to the sequence...AGGCC(A). Presumably the rest of the precursor sequence at this end of the molecule has been degraded to mononucleotides.

Precursor exposed to cell extract (band A in Fig. 3), but still retaining its original mobility

前体的特异性体外切割已得到确认，因为B条带的胰核糖核酸酶酶解物的二维指纹图显示5′末端是pGGU——与成熟tRNA一样，且前体的5′端额外片段已被完全移除。除了源于3′末端的两个点以外，B条带的T_1核糖核酸酶酶解物的指纹图（图4）也与成熟A25 tRNA的指纹图相同。其中一个点的迁移率与前体3′端寡核苷酸的迁移率相同。而另一个点，在指纹图上该点的附近，呈现出大约20%的产量，并且至少缺失了两个终端核苷酸，我们猜测是因为细胞提取液中其他核酸酶的作用所致。修饰的3′端的产量依反应混合物的孵育时间而定。延长电泳时间能使B条带分解成两个条带，迁移较快的条带仅含有修饰的3′端。

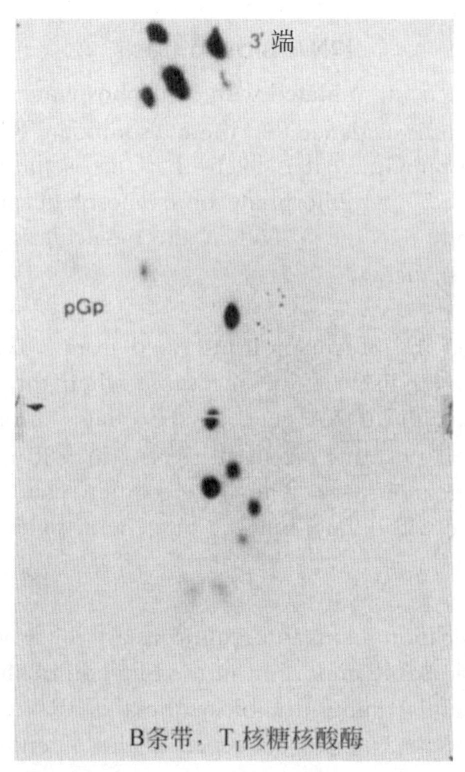

图4. A25前体切割产物T_1核糖核酸酶酶解物的二维电泳分离放射自显影图。从凝胶中洗脱出图3中的B条带，然后用T_1核糖核酸酶进行酶解，最后用二维电泳法分离产物。放射自显影图中显示出了新产生的5′端以及前体3′端寡核苷酸的位置，在成熟酪氨酸tRNA的常见3′端寡核苷酸的位置附近还有一个产量为摩尔水平的斑点。对这些产物的讨论详见正文。

C条带含有前体分子5′端的序列，其序列为…AGGCC(A)。我们猜测分子这一端的前体序列的其余部分已经被降解成单核苷酸了。

虽然前体被浸没在细胞提取液中（图3中的A条带），但孵育后它仍能在凝胶

in gels after incubation, does not have any interruptions in the nucleotide sequence.

Residual endonucleolytic activity on ribosomes washed twice in 2 M NH_4Cl (given by D. Ish-Horowicz) can carry out the maturation reaction in a standard mixture lacking requirements for nucleotide modification[13-15]. It seems likely, therefore, that addition of isopentyl, methyl or thio groups or pseudouri-dylation is not necessary for cleavage *in vitro* (One indication of the lack of dependence on modification is the absence of spots containing 2′O methyl G from the two dimensional fingerprints of band B in Fig. 3) The enzyme does not degrade mature tRNA. A more complete description of the novel endonucleolytic activity will be given (S. A., H. D. Robertson, and J. D. S., in preparation).

tRNA Biosynthesis

In vitro transcription products are initiated with 5′-triphosphate purine nucleotide residues at their 5′ end[16,17] and there is evidence that these also initiate RNA *in vivo*[18]. As the tRNA precursor begins with 5′ pppGp…, it is likely that this is the start of a transcriptional unit unless some complex 5′ pyrophosphate or triphosphate addition reaction is active *in vivo*. We cannot conclude that the 3′ end of the molecule is the appropriate bacterial transcription termination sequence.

Although we have studied the precursor transcribed from a tRNA inserted in a phage chromosome, transcription is likely to be the same when the gene is on the bacterial chromosome. In particular, the tRNA gene on the phage chromosome still responds to relaxed-stringent control as do other bacterial tRNAs in φ80 infected cells[19] and in this respect differs from phage messages. Therefore, certain characteristics of the tyrosine tRNA precursor which we will discuss may be of general significance for other bacterial tRNAs.

First, as nucleotide modification seems to be unnecessary for *in vitro* cleavage of precursor, the rare nucleotides are probably important in the function of the individual tRNA in the translational machinery rather than in the biosynthesis of tRNA itself. Second, our results regarding the presence of the—CCA sequence at the 3′ end of the molecule suggest that this triplet is part of the transcriptional unit, in contrast to earlier published work[9]. We have not yet determined the sequence of the 3′ terminal oligonucleotide of wild type tRNA precursor. Preliminary experiments indicate that this precursor is a longer molecule and differs from mutant precursor only at the 3′ end.

Lastly, we consider the significance of the lengthy 5′ segment of the precursor. While the specificity for cleavage may be entirely governed by structural factors, it seems more likely that recognition of a specific nucleotide sequence is also very important. This is apparent from considering the mutation P, four nucleotides from the cleavage point, which alters greatly the level of production of mature tRNA *in vivo* when compared with its parent. It is possible that this mutation simply changes the binding constant of the enzyme to a

中保持原有的电泳迁移率，没有对核苷酸序列造成任何干扰。

将核糖体用 2 M NH_4Cl（由伊什 – 霍罗威茨提供）清洗两次后，残留在其上的核酸内切酶活性能在一个缺少核苷酸修饰所需条件的标准混合物中完成成熟反应 [13-15]。因此，添加异戊基、甲基、硫代基团或者假尿苷化可能对于体外切割是不必要的（不依赖修饰的一个迹象是：在图 3 中 B 条带的二维指纹图中缺少含有 2′–O– 甲基鸟苷的斑点）。这种酶不降解成熟 tRNA。今后我们将对这种奇特的核酸内切酶活性进行更加完整的描述（奥尔特曼、罗伯逊和史密斯，完稿中）。

tRNA 的生物合成

体外的转录产物是从 5′ 端以 5′– 三磷酸嘌呤核苷酸残基开始的 [16,17]，有证据证明体内 RNA 也是从这个位置起始的 [18]。由于 tRNA 前体起始于 5′ pppGp…，很可能这就是一个转录单位的起点，除非体内发生了某种复杂的 5′ 焦磷酸或者三磷酸加成反应。我们不能断定分子的 3′ 端就是细菌合适的转录终止序列。

虽然我们研究的是由插入到噬菌体染色体中的 tRNA 转录得到的前体，但是细菌染色体上的基因的转录过程很可能是一样的。特别是，噬菌体染色体上的 tRNA 基因仍然对松弛 – 紧密控制有反应，就像其他被 φ80 感染的细胞里的细菌 tRNA 一样 [19]，在这方面不同于噬菌体信息。因此，我们将要讨论的酪氨酸 tRNA 前体的某些特点有可能对于其他细菌 tRNA 来说是有普遍意义的。

首先，因为核苷酸修饰对于前体的体外切割而言看起来是不必要的，所以稀有核苷酸对单个 tRNA 的功能的重要意义很可能体现在翻译机制而非 tRNA 自身的生物合成方面。其次，与前人发表的研究结果 [9] 相反，我们关于前体分子 3′ 端存在—CCA 序列的结果暗示这一三联体是转录单位的一部分。我们尚未确定野生型 tRNA 前体的 3′ 端寡核苷酸序列。初步实验表明：野生型前体有较长的分子，而且仅在 3′ 端与前体突变体不同。

最后，我们来考虑冗长的前体 5′ 端片段的意义。尽管切割的特异性可能完全由结构因素决定，但似乎更有可能的是对特定核苷酸序列的识别也同样非常重要。显然，同其母体相比，与切割点距离 4 个核苷酸的突变体 P 极大地改变了体内成熟 tRNA 的产量。有可能这种突变仅仅改变了某种特定 RNA 酶的结合常数。突变体 A2 和 A81 都能产生少量的 tRNA[7]，并且都不能在体内积累前体。这些突变位于（在序列

particular stretch of RNA. Mutants A2 and A81 both produce low levels of tRNA[7] and do not accumulate precursor *in vivo*. These mutations are located in the finished tRNA molecule very close (in the sequence or in space) to the cleavage point (Fig. 2) and may also effect "precursor-endonuclease" binding in an undetermined fashion such that breakdown of precursor to products other than tRNA is enhanced. The phenotype of other mutants (A15, A25, A17 and A 31) distant from the cleavage point in the molecule suggests that a structural factor is also important in precursor endonuclease activity.

We think it plausible that, as precursor synthesis proceeds from the 5′ to the 3′ end of the molecule, it may first assume a structure similar to structure I (Fig. 2) in which the total sequence near the cleavage point is involved in a hydrogen-bonded structure. In this model, when transcription of the entire molecule is completed, the clover leaf configuration is assumed, as shown in structure II (Fig. 2), since it is an energetically favoured configuration[20]. Structure II, we propose, may be more susceptible than structure I to precursor-endonuclease attack. If the transition from I to II is inhibited by the inability to form certain hydrogen bonds in the tRNA moiety, precursor effectively has a longer half-life and should accumulate *in vivo*. This is what we observe for A15, A17, A25 and A31, single base change mutants all involved in breaking hydrogen bonds normally present in the wild type tRNA. An alternative kinetic scheme, in which the rate of transition of precursor from I to II is unchanged in the mutants, but the rate of production of tRNA from mutant structure II is decreased relative to wild type, is also feasible. In either case, secondary as well as primary structure of the precursor is likely to be important in determining the kinetics of tRNA biosynthesis.

It should be noted that precursor we isolate is susceptible to precursor endonuclease attack *in vitro* suggesting, according to our model, that if it accumulates as structure I *in vivo*, it can rapidly assume structure II *in vitro*. Furthermore, to explain the low level of mature tRNA produced by these mutants, it is necessary to postulate an additional pathway of degradation not producing tRNA *in vivo*.

Our experiments with φ80 carrying a single tRNA gene have yielded the first sequence analysis of a tRNA precursor. Similar data regarding T4-induced tRNAs in *E. coli* will be presented (W. H. McClain, personal communication). No direct evidence is available yet regarding the nature of transcription of tRNAs from the *E. coli* chromosome.

We thank our colleagues, especially Dr. H. D. Robertson for helpful discussions, and Miss E. Higgins and Mr. T. V. Smith for technical assistance.

Note added in proof. Although the last residue of fragment 7 in Table 1 has been written as U_{OH}, its identity has not been conclusively established.

(*Nature New Biology*, **233**, 35-39; 1971)

或者空间上）距切割点非常近的已形成的 tRNA 分子中（图 2），它们也可能以一种未知的方式影响“前体－内切酶”的结合，以使前体分解成产物，但产物并不是成熟 tRNA。分子中远离切割点的其他突变体（A15、A25、A17 和 A31）的表型说明，结构因素对于前体的内切酶活性也很重要。

因为前体的生物合成是从分子的 5′ 端开始向 3′ 端延伸，我们认为以下模型似乎是合理的，即可以首先假设一个与结构 I（图 2）相类似的结构，其中靠近切割点的全部序列都以某种氢键结构相结合。在这个模型中，当整个分子完成转录时，就可以假设形成了如结构 II 所示的三叶草结构（图 2），因为三叶草结构是能量最低的结构[20]。我们建议，结构 II 可能比结构 I 更容易受到前体内切酶的攻击。如果因为在 tRNA 部分不能形成某种氢键而阻碍了从结构 I 转换到结构 II 的过程，那么前体就会拥有更长的半衰期，并在体内可以积累。这正是我们观察 A15、A17、A25 和 A31 时所发现的，单碱基改变突变体都会使通常存在于野生型 tRNA 中的氢键断裂。在另一种动力学模型中，突变体前体从结构 I 转换到结构 II 的比例没有变化，但从突变体的结构 II 生成 tRNA 量的比例相对野生型来说减少了，这也是有可能的。在以上两种模型中，前体的二级结构和一级结构一样，都可能对确定 tRNA 生物合成的动力学起着重要作用。

需要注意的是：根据我们的模型，我们分离出的前体在体外很容易受到前体内切酶的攻击这一点表明，如果前体以结构 I 的形式在体内积累，那么在体外它就会快速转换为结构 II 的形式。此外，为了解释由这些突变体产生的成熟 tRNA 产量低的原因，还需要假设在体内存在着另外一条降解而非生成 tRNA 的途径。

我们用只携带一个 tRNA 基因的 φ80 噬菌体进行实验，得到了 tRNA 前体的第一个序列分析结果。关于大肠杆菌中 T4 诱导的 tRNA 的类似数据（麦克莱恩，个人交流）也将被报道。到目前为止还没有直接的证据能证明大肠杆菌染色体中 tRNA 的转录本质。

感谢我们的同事，特别是罗伯逊博士与我们进行了多次有益的讨论，还要感谢希金斯小姐和史密斯先生给我们提供了技术支持。

附加说明：尽管表 1 中第 7 号片段的最后一个残基被写作 U_{OH}，但现在还没有完全证实这一点。

（邓铭瑞 翻译；李素霞 审稿）

S. Altman and J. D. Smith: MRC Laboratory of Molecular Biology, Hills Road, Cambridge CB2 2QH.

Received June 21; revised July 26, 1971.

References:

1. Hecht, N. B., and Woese, C. R., *J. Bact.*, **95**, 986 (1968).
2. Adesnik, M., and Levinthal, C., *J. Mol. Biol.*, **46**, 281 (1969).
3. Forget, B. G., and Jordan, B., *Science*, **167**, 382 (1970).
4. Darnell, jun., J. E., *Bact. Rev.*, **32**, 262 (1968).
5. Burdon, R. H., Martin, B. T., and Lal, B. M., *J. Mol. Biol.*, **28**, 357 (1967).
6. Altman, S., *Nature New Biology*, **229**, 19 (1971).
7. Smith, J. D., Barnett, L., Brenner, S., and Russell, R. L., *J. Mol. Biol.*, **54**, 1 (1970).
8. Goodman, H. M., Abelson, J., Landy, A., Zadrazil, S., and Smith, J. D., *Europ. J. Biochem.*, **13**, 461 (1970).
9. Daniel, V., Sarid, S., and Littauer, U. Z., *Science*, **167**, 1682 (1970).
10. Arima, T., Uchida, T., and Egami, F., *Biochem. J.*, **106**, 609 (1968).
11. Abelson, J. N., Gefter, M. L., Barnett, L., Landy, A., Russell, R. L., and Smith, J. D., *J. Mol. Biol.*, **47**, 15 (1970).
12. Brownlee, G. G., and Sanger, F., *Europ. J. Biochem.*, **11**, 395 (1969).
13. Abrell, J. W., Kaufman, E. E., and Lipsett, M. N., *J. Biol. Chem.*, **246**, 294 (1971).
14. Bartz, J. K., Kline, L. K., and Söll, D., *Biochem. Biophys. Res. Commun.*, **40**, 1481 (1970).
15. Johnson, L., and Söll, D., *Proc. US Nat. Acad. Sci.*, **67**, 943 (1970).
16. Maitra, U., Novogrodsky, A., Baltimore, D., and Hurwitz, J., *Biochem. Biophys. Res. Commun.*, **18**, 801 (1965).
17. Bremer, H., Konrad, M. W., Gaines, K., and Stent, G. S., *J. Mol. Biol.*, **13**, 540 (1965).
18. Jorgensen, S. E., Buch, L. B., and Nierlich, D. P., *Science*, **164**, 1067 (1969).
19. Primakoff, P., and Berg, P., *Cold Spring Harbor Symp. Quant. Biol.*, **35**, 391 (1970).
20. Tinoco, jun., I., Uhlenbeck, O. C., and Levine, M. D., *Nature*, **230**, 362 (1971).

Poly A Sequences at the 3′ Termini of Rabbit Globin mRNAs

H. Burr and J. B. Lingrel

Editor's Note

Here Henry Burr and Jerry B. Lingrel report the presence of polyadenylic acid (poly A) "tails" at one end of two different rabbit globin messenger RNAs. They speculate that the sequences may have been conserved during evolution, and may play a role in messenger RNA maturation after transcription. Conserved, non-translated regions in RNAs which direct peptide synthesis had already been observed in certain RNA viruses, raising the possibility that they may be acting as regulators of gene expression. It is now accepted that polyadenylation (the addition of a poly A tail to an RNA molecule) plays a role in gene expression, forming part of the process that yields mature messenger RNA for translation into protein.

THE 9S RNA fraction of reticulocyte polyribosomes includes the mRNAs for the α and β-chains of haemoglobin and rabbit 9S RNA directs the synthesis of rabbit α and β-chains when added to either an *E. coli*[1] or guinea-pig reticulocyte cell-free system (Jones and Lingrel, unpublished results). Similarly, mouse 9S RNA directs the synthesis of mouse β-chains in the rabbit cell-free systems[2] as well as α and β-globin chains in the guinea-pig and duck reticulocyte cell-free systems[3].

We report the isolation of 3′ terminal fragments of rabbit globin mRNAs produced by T_1 and pancreatic RNAases, and the base sequence of the pancreatic RNAase fragments.

T_1 RNAase Fragments

Milligram quantities of the haemoglobin mRNA were prepared using procedures described previously[4,5] except that RNAs were separated by zonal rotor. The mRNA is well resolved from the other RNAs and gives a single band in acrylamide gel electrophoresis.

The 3′ terminal nucleoside of the RNA was labelled using the periodate oxidation-tritiated borohydride reduction technique[6,7] and, as tritium is incorporated only at the 3′ terminus, this method assured identification of the terminal fragments. T_1 and pancreatic RNAase digestion products of ^{3}H-mRNA and carrier RNA were fractionated by "DEAE-Sephadex" chromatography according to the number of their phosphate groups[6].

Two major 3′ terminal fragments are resolved when the mRNA is digested with T_1 RNAase (Fig. 1). Considerable radioactivity in early fractions is not derived from the RNA because it was present in undigested samples. Furthermore, when digests are appropriately chromatographed on "DEAE-Sephadex A-25", no radioactivity is present in nucleoside

兔球蛋白信使RNA 3′末端的多聚腺苷酸序列

伯尔，林格里尔

编者按

在本文中，亨利·伯尔和杰里·林格里尔报道了在两种不同的兔球蛋白信使 RNA 分子同一末端存在多聚腺苷酸“尾巴”。他们推测这类序列在进化过程中可能是保守的，并且可能在信使 RNA 转录后的成熟过程中发挥着某种作用。在某些 RNA 病毒中已经观察到指导肽合成的 RNA 非翻译保守区域，这就增加了它们是基因表达调控者的可能性。多聚腺苷酸化（将多聚腺苷酸尾巴加到一个 RNA 分子上）在基因表达中的作用现在已经被普遍接受，它构成产生蛋白质翻译所需成熟信使 RNA 的过程的一部分。

网状细胞多聚核糖体的 9S RNA 部分包含血红蛋白 α 链和 β 链的信使 RNA (mRNA)。将兔 9S RNA 加入大肠杆菌 [1] 或豚鼠网状细胞的无细胞体系中时，它能指导兔 α 链和 β 链的合成（琼斯和林格里尔，尚未发表的结果)。与此类似，小鼠 9S RNA 既能在豚鼠和鸭网状细胞的无细胞体系中指导小鼠 α 球蛋白链和 β 球蛋白链的合成 [3]，也能在兔的无细胞体系中指导小鼠 β 链的合成 [2]。

我们报告的是通过 T_1 核糖核酸酶和胰核糖核酸酶产生的兔球蛋白 mRNA 3′ 末端片段的分离，以及胰核糖核酸酶片段的碱基序列。

T_1 核糖核酸酶片段

除了 RNA 是通过区带转头离心机分离外，毫克级血红蛋白 mRNA 是按以前描述过的程序 [4,5] 制备的。mRNA 与其他 RNA 的分离很彻底，在丙烯酰胺凝胶电泳中只有单一条带。

因为氚元素只与 3′ 末端结合，所以我们使用过碘酸盐氧化–氚标记的硼氢化物还原技术来标记 RNA 的 3′ 末端核苷 [6,7]，这种方法能确保识别出末端片段。利用“二乙氨基乙基–葡聚糖”色谱柱将 ^{3}H–mRNA 和载体 RNA 的 T_1 核糖核酸酶和胰核糖核酸酶酶解产物按照磷酸基团的数量分离成不同的组分 [6]。

当 mRNA 被 T_1 核糖核酸酶酶解时，两个主要的 3′ 末端片段被分离出来（图 1)。因为在未酶解的样品中出现放射现象，所以早期洗脱组分的大量放射性并非来自 RNA。此外，当用“二乙氨基乙基–葡聚糖 A-25”色谱柱对酶解物进行适当分离时，

derivatives or mononucleotides. This non-RNA radioactivity has been observed by other workers[6,8,9] and is probably labelled side-reaction products which are not completely removed from the RNA by purification[7]. Approximately 90% of this tritium label could be removed by an MAK column, but this procedure was not used because the RNA appeared to be somewhat resistant to enzymatic digestion.

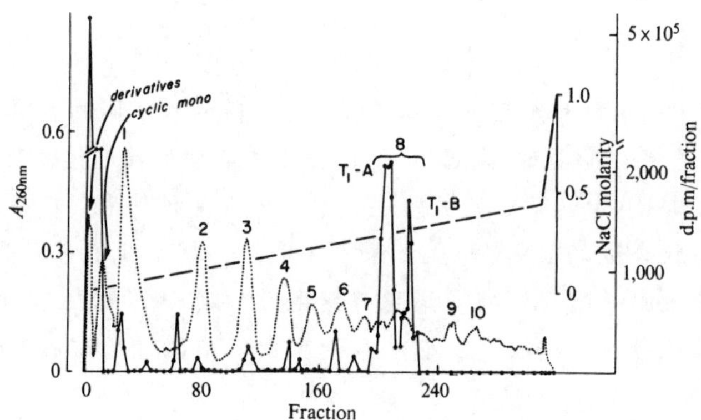

Fig. 1. T_1 RNAase digest of terminally labelled globin mRNA. 200 μg of globin mRNA, with rabbit reticulocyte 18S RNA as ultraviolet-carrier, was digested with T_1 RNAase (E.C. 2.7.7.26, Sigma) for 1 h at 37°C in 0.02 M Tris buffer (*p*H 7.0) containing 2 mM EDTA. These conditions were essentially identical to those used by other workers[34,35]. Carrier nucleoside derivatives[6] and monophosphates were added as markers for numbering the peaks and the digest was made 7 M in urea and applied to a 0.9×23 cm "DEAE Sephadex A-25" column previously equilibrated with 7 M urea in 0.02 M Tris (*p*H 7.6). Elution was carried out using a 3 litre gradient of 0.0–0.45 M NaCl in Tris (*p*H 7.6) and 7 M urea. The column was stripped with a 100 ml. gradient of 0.45–1.0 NaCl in the same buffer. Samples of 3 ml. of each fraction with 1 ml. H_2O were counted in 10 ml. of "Insta-Gel" (Packard) at approximately 12% efficiency. Globin mRNA was labelled using a procedure similar to that of DeWachter and Fiers[6]. RNA was dissolved in 1 ml. of 0.01 M sodium phosphate buffer (*p*H 6.0) and 0.05 ml. of 0.1 M $NaIO_4$ was added and allowed to stand in the dark. After 1 h, 0.05 ml. of 1 M ethylene glycol was added and allowed to stand 30 min. Next, 1 ml. of 0.5 M sodium phosphate buffer (*p*H 7.0) was added an mixed. NaB^3H_4 (502 mCi/mM, Amersham-Searle, crystalline solid) was dissolved in cold 1 N NaOH so that 0.1 ml. contained 10 mCi; this level of radioactivity was used routinely. Reduction was allowed to proceed for 2 h at room temperature. The RNA was precipitated two or three times with cold ethanol and further purified from excess ^3H by sedimentation through one or two sucrose gradients 5–20% (w/w) in the Spinco SW-27 rotor for 24 h at 2°C with additional precipitation between sedimentations. The 3′-terminal nucleoside after oxidation and reduction is referred to as a nucleoside derivative in this paper. These steps removed much of the extraneous label; however, some counts not derived from RNA can be seen in the first fractions of this figure and Fig. 2. (See text.) Peaks are designated according to the number of negative charges in the fragments; beyond seven the numbering is somewhat arbitrary. ——, Radioactivity; . . . . ., absorbance at 260 nm; - - - - -, NaCl gradient.

The T_1 fragment elutes in the region of eight phosphate groups inferring that it comprises nine bases as the terminal fragment lacks the phosphate on the 3′ terminus, but this is only tentative as base composition begins to influence the elution pattern of longer oligonucleotides[6,10]. The first guanosine nucleotide would be at position ten, counting from the 3′ end. This may be an underestimate as DeWachter and Fiers have isolated a fragment from MS2 bacteriophage RNA which elutes as though it contained eight

核苷衍生物或单核苷酸中都没有表现出放射性。这种不是来自 RNA 的放射性已被其他研究者观察到[6,8,9]，很可能是被标记的副反应产物在提纯时没有完全从 RNA 中移走产生的[7]。大约 90% 的氚标记物能通过一个 MAK 柱去除，但是这样做似乎会使 RNA 对酶解作用有一定程度的抗性，因此我们没有使用这个步骤。

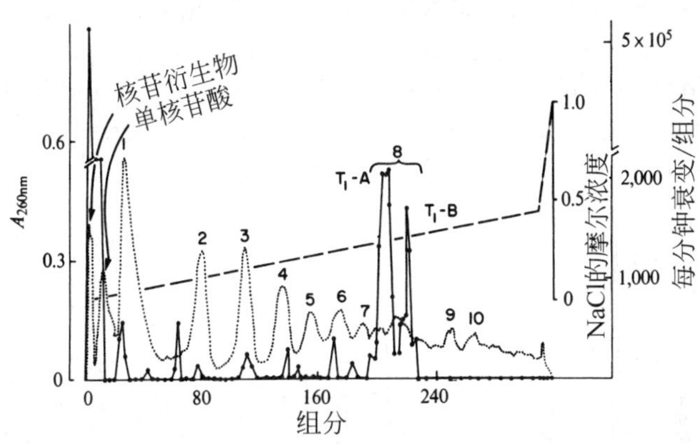

图 1. 末端标记的球蛋白 mRNA 的 T_1 核糖核酸酶酶解产物。用兔网状细胞 18S RNA 作为紫外辐射载体，200 μg 球蛋白 mRNA 在含 2 mM EDTA 的 0.02 M Tris（pH 值为 7.0）中用 T_1 核糖核酸酶（E.C. 2.7.7.26，西格马公司）在 37℃下分解 1 h。这些条件与其他研究者采用的条件[34,35]基本一致。为了给这些峰编号，我们加入核苷衍生物载体[6]和单磷酸盐作为标准物，在尿素中将酶解物调至 7 M，然后使用被 7 M 尿素的 0.02 M Tris 溶液（pH 值为 7.6）预先平衡过的 0.9 cm × 23 cm"二乙氨基乙基 – 葡聚糖 A-25"色谱柱进行分离。用 Tris（pH 值为 7.6）和 7 M 尿素溶液配制浓度梯度为 0.0 M ~ 0.45 M 的 3 L NaCl 溶液以进行洗脱。色谱柱用同一缓冲液配制的浓度梯度为 0.45 M ~ 1.0 M 的 100 ml NaCl 溶液彻底洗脱。从每个组分中取 3 ml 样品加入 1 ml H_2O，然后置于 10 ml 大约 12% 效能的"英斯达凝胶"（帕卡德公司）中进行计数。用与德瓦赫特和菲耶尔所采用的类似步骤[6]来标记球蛋白 mRNA。将 RNA 溶解于 1 ml 0.01 M 的磷酸钠缓冲液（pH 值为 6.0）中，再加入 0.05 ml 0.1 M $NaIO_4$，在黑暗中静置 1 h，然后加入 0.05 ml 1 M 的乙二醇并静置 30 min。接下来，加入 1 ml 0.5 M 的磷酸钠缓冲液（pH 值为 7.0）并混合均匀。将 NaB^3H_4（502 mCi/mM，安玛西亚 – 瑟尔公司，结晶固体）溶解于冷的 1 N NaOH 中，这样每 0.1 ml 含有 10 mCi；这就是常规情况下使用的放射性水平。还原反应在室温下持续进行 2 h。用冷乙醇沉淀 RNA 2 到 3 次，然后按照下述方法从过量的 ^3H 中进一步提纯：在斯平科 SW-27 离心机中，通过 1 到 2 个 5% ~ 20%（重量比）蔗糖浓度梯度在 2℃条件下离心 24 h，得到从两次沉淀中提纯的新沉淀。经过氧化和还原的 3′ 末端核苷在本文中被认为是核苷衍生物。上述步骤能除去大量外来的标记物；然而，在本图和图 2 的第一个组分中可以看到一些并非来自 RNA 的计数（见正文）。根据片段中负电荷的数量命名波峰；编号超过 7 则有一定的主观性。——：放射性；……：在 260 nm 处的吸光度；- - -：NaCl 梯度。

由于末端片段缺少 3′ 末端的磷酸基团，根据 T_1 片段在 8 个磷酸基团区域被洗脱下来可以推断出它包含 9 个碱基。但上述推断只是我们的假设，因为碱基组成开始影响较长寡聚核苷酸的洗脱曲线[6,10]。从 3′ 端开始计数，第一个鸟苷酸将出现在第 10 个核苷酸的位置。这个数字也许被低估了，因为德瓦赫特和菲耶尔曾经从 MS2 噬菌体 RNA 中分离出一个片段，洗脱出来它似乎含有 8 个碱基，而最终测序

bases while the final sequence revealed ten[6]. As a 3′ terminal G would not be cleaved enzymatically, the terminal base of the T_1-A fragment was determined by paper chromatography[6] of a KOH digest. Radioactivity was found only in adenine residue.

Sequence of Pancreatic RNAase Fragments

Chromatography of the pancreatic RNAase digest (Fig. 2) shows a pattern similar to that with T_1 RNAase. These fragments are five and six bases long, indicating that there are no pyrimidine bases in the terminal five or six nucleotides of the α and β-chain mRNAs with the exception of the 3′ terminal base. End-group analysis carried out as before on both pancreatic RNAase fragments indicated only A again.

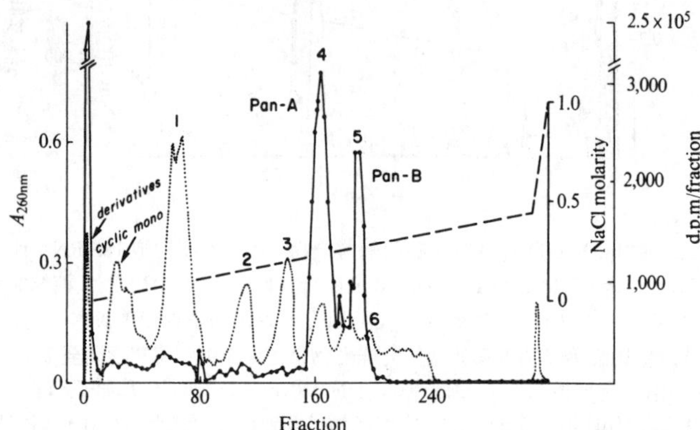

Fig. 2. Pancreatic RNAase digest of terminally labelled globin mRNA. Conditions of digestion and fragment separation were identical to those described in the legend to Fig. 1. Bovine pancreatic RNAase (E.C. 2.7.7.16) was obtained from Worthington Biochemical Corp. ——, Radioactivity; . . . . ., absorbance at 260 nm; - - - - -, NaCl gradient.

Therefore, there is no guanosine in the last nine bases, and no pyrimidines in the last five to six bases. Consequently, the α and β-chain mRNAs must terminate in a sequence of five to six adenine nucleotides. Consistent with this conclusion is the finding that chromatography of each fragment on the basis of base composition (DEAE cellulose, *p*H 3.5) produced only one peak. Electrophoretic immobility of T_1 and pancreatic RNAase fragments at *p*H 3.0 suggested to Hunt and Laycock that the 3′ terminal fragments might be enriched in adenylic acid[11].

The fragments from both RNAase digests occur in a 2:1 ratio with the smaller one predominating. This would not be anticipated if each fragment originated from a different globin chain mRNA; indeed, one would expect, *a priori*, that the quantities of α and β-chain mRNAs would be reflected by the amount of each chain found in a reticulocyte. The unequal amounts of the two fragments could be the result of several anomalies: (1) labelling difficulties due to secondary structure in the RNA, base preference in reaction, addition of ^3H label elsewhere in the RNA, esterification of 3′-OH, (2) incomplete digestion, (3) contamination with ribosomal RNAs.

显示的碱基数是 10 个[6]。由于 3′ 末端鸟苷酸不能用酶切开，因此 T_1–A 片段的末端碱基用 KOH 分解的纸层析法[6]测定。放射性只在腺嘌呤残基中被发现。

胰核糖核酸酶片段的序列

胰核糖核酸酶酶解产物的色谱图（图 2）所显示的洗脱曲线类似于 T_1 核糖核酸酶酶解产物的色谱图。酶解产物片段有 5 到 6 个碱基长，这表明：除了 3′ 末端碱基外，在 α 链和 β 链 mRNA 末端的 5 个或 6 个核苷酸中没有嘧啶碱基。跟之前一样，对两个胰核糖核酸酶片段进行末端基团分析，结果显示还是只有腺苷酸。

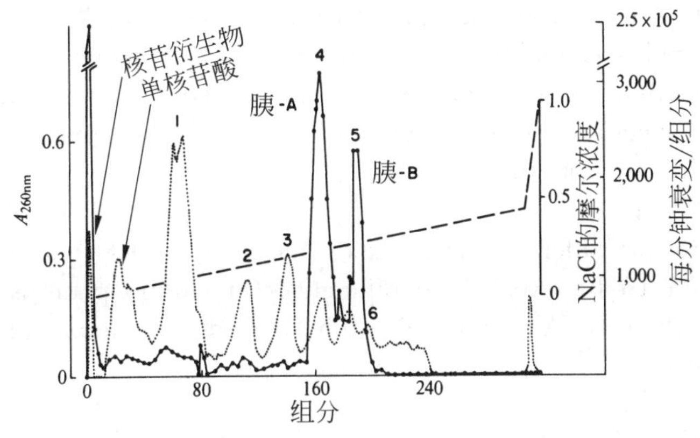

图 2. 末端标记的球蛋白 mRNA 的胰核糖核酸酶酶解产物。分解条件和片段分离条件与图 1 注中所描述的完全一样。牛胰核糖核酸酶（E.C.2.7.7.16）来自沃辛顿生物化学公司。——：放射性；……：在 260 nm 处的吸光度；- - -：NaCl 梯度。

因此，在最后 9 个碱基中没有鸟苷，并且在最后 5 到 6 个碱基中也不会有嘧啶。所以，α 链和 β 链 mRNA 一定是以 5 到 6 个腺嘌呤核苷酸序列结束的。每个片段用色谱法（二乙氨基乙基纤维素，pH 值为 3.5）分析碱基组成都只产生一个峰，该分析结果符合上述结论。T_1 核糖核酸酶和胰核糖核酸酶片段在 pH 值为 3.0 时的电泳停滞使亨特和莱科克认为 3′ 末端片段可能富含腺苷酸[11]。

两种核糖核酸酶酶解产物的片段以 2∶1 的比率产生，其中较小片段占多数。无法判断两者是否分别来自球蛋白链不同的 mRNA；的确，我们会先验地认为，α 链和 β 链 mRNA 的量可以通过在一个网状细胞中发现的每种链的数量来反映。两种片段在数量上的不相等可能由以下几种异常造成：(1) 因 RNA 二级结构、碱基在反应上的偏好性、RNA 的其他位置被 ^3H 标记和 3′–OH 的酯化作用而导致的标记困难；(2) 分解不完全；(3) 被核糖体 RNA 污染。

The efficiency of periodate oxidation-borohydride reduction of RNA is known to be variable and lower as the size of the RNA increases[6,8,12], because of the increased secondary structure of the larger molecules[9]. Our labelled preparations have had specific activities indicating a range of 5 to 44% of ends labelled if each end carries two tritium atoms. As both preparations digested gave essentially the same fragment patterns, secondary structure does not seem to be influencing the type of end available for labelling. The chemical specificity and apparent absence of base specificity have been shown by others[7,13,14]. Separation of KOH hydrolysis products revealed tritium in only the derivatives, showing that label is not in the rings or ribose of bases other than the one at the 3′ terminus.

Addition of label to the 3′ end does not occur unless the mRNA is previously oxidized, indicating that there is a *cis* diol on the terminal ribose. The known variability of labelling efficiency and requirement for previous oxidation make it doubtful that a substantial proportion of molecules are esterified. We cannot conclude, however, that all molecules in the 9S RNA fraction are not esterified.

Enzymatic digestion at different enzyme substrate ratios (1:100 and 1:20) for different times (30 min and 16 h) revealed no differences in elution patterns. Acrylamide gel electrophoresis of the mRNA used in this study showed no contamination by any of the intact ribosomal RNAs.

Consequently, we feel that these results are a realistic reflexion of the molecular composition of the globin mRNA fraction and that both globin mRNAs terminate in a region of poly A which does not correlate with the known amino-acid sequence of either of the globin chains. These results are consistent with the presence of a UAA chain terminating triplet.

Terminal Fragments and Globin Synthesis

Each terminal fragment is derived from one of the messengers. A pool of α-chains in rabbit reticulocytes[15-17] and the finding that nascent α-chain appear on smaller polysomes[18] are compatible with an excess of α-chain mRNA. It seems reasonable, then, the smaller fragment might be derived from the α-messenger and the larger fragment from the β-mRNA. Hunt and co-workers, however, have found that α-chain synthesis is substantially faster[19], and at present the data do not permit assignment of α or β-mRNA to either fragment. Both globin mRNAs may be represented in only one of the fragments while the other fragment is derived from some other mRNA. The fact that the two fragments differ by only one adenine nucleotide might be explained by the addition of a nucleotide as has been found elsewhere[20].

Terminal Sequences and mRNA Synthesis

The current models of eukaryotic gene expression view mRNA synthesis in terms of a high molecular weight, rapidly labelled, nuclear RNA (HnRNA)[21,22]. Synthesis of a large

已经知道RNA的过碘酸盐氧化–硼氢化物还原的效能是可变的——随着RNA分子的增大而降低[6,8,12]，这是因为较大分子的二级结构更复杂[9]。如果每个末端都能带上2个氚原子，那么由我们标记的制剂的放射性比活度大小可以说明5%～44%的末端被标记。由于两种酶解产物制剂给出了基本相同的片段图谱，因此二级结构似乎不影响标记的末端类型。其他研究者已证明存在化学特异性但显然不存在碱基特异性[7,13,14]。KOH水解产物的分离结果表明氚只存在于衍生物中，这说明不是在碱基环或核糖上而是在3′末端标记。

除非预先对mRNA进行氧化处理，否则不能在3′端添加标记，这表明在末端核糖中存在着顺式二醇。前文已提及的标记效能的可变性和对前期氧化的需要使得有多少比例的分子发生酯化变得不确定。然而，我们不能断定9S RNA部分的所有分子都没有被酯化。

不同酶底物比（1∶100和1∶20）在不同作用时间（30 min和16 h）下得到的酶解产物在洗脱曲线上没有显示出任何差别。丙烯酰胺凝胶电泳结果显示，本研究中所用的mRNA没有被任何完整的核糖体RNA所污染。

因此我们认为，这些结果真实地反映了球蛋白mRNA组分的分子组成，并认为两种球蛋白mRNA都终止于多聚腺苷酸区域，该多聚腺苷酸区域与两种球蛋白链上的已知氨基酸序列都不相关。这些结果与UAA链终止三联体的存在相符合。

末端片段和球蛋白合成

每一个末端片段都源自一个mRNA。兔网状细胞中的α链库[15-17]以及新生α链在较小的多聚核糖体上出现的发现[18]都与α链mRNA的过量相一致。因此，较小片段可能源于α链mRNA而较大片段源于β链mRNA的推断似乎是合理的。然而，亨特和他的同事们曾发现，α链的合成实际上更快一些[19]，而且目前的数据没有将α链mRNA或β链mRNA与两个片段中的任何一个对应起来。两种球蛋白mRNA也许都仅对应于两个片段中的一个，而另一个片段则源于其他的mRNA。仅由一个腺嘌呤核苷酸造成两个片段不同的事实可以通过添加一个核苷酸得到解释，这种情况曾在别处发现过[20]。

末端序列和mRNA的合成

真核生物基因表达的当前模型是从高分子量、快速标记、核RNA（HnRNA）方面来考虑mRNA的合成[21,22]。大前体mRNA的合成意味着要在切除以及可能的后

pre-mRNA implies recognition of the messenger region for excision and perhaps further processing. The possible precursor-product relationship between HnRNA and polysomal mRNA has been indicated recently by the finding that both RNAs contain A-rich sequences[23-25].

Lim and Canellakis have isolated an A-rich fragment approximately fifty to seventy bases long (70% A) from the purified rabbit globin mRNA fraction[26]. Our calculations of the maximum percentage of A which could be present in a fragment of this size, if it were derived from within the globin coding region, indicate that the large A-rich fragment must have originated from outside the peptide-coding region and therefore must be located in one of the termini. To relate our results to those mentioned above, it seems that the small poly A sequence at the 3′ terminus of both globin mRNAs represents a small part of a longer, untranslated A-rich region, which may play a role in post-transcriptional processing.

Untranslated A-rich regions may be common to many or all mRNAs with the result that their dT-rich complements are reiterated in the DNA. Williamson *et al.*[27] have hybridized purified mouse globin mRNA fraction and obtained very interesting results. The DNA was not saturated until the RNA/DNA input ratio was three; yet at each RNA input, the time needed for maximum binding was only 10 min. These results are compatible with the presence of a component in the RNA which is hybridizing to reiterated DNA sequences. Kedes and Birnstiel, however, have hybridized sea urchin histone mRNA fraction with a large excess of DNA and failed to demonstrate a rapidly hybridizing RNA component which would be expected if binding were due to a sequence common to all messengers and were uniformly distributed throughout the genome[28]. It is not clear whether the A-rich regions found in crude mRNA fractions and globin mRNA are actually transcribed from DNA or are added later by another mechanism[29].

An alternative interpretation of our results has as its basis the idea that different proteins have emerged during evolution through gene duplication, followed by independent mutation of each gene[30]. Ingram[31] used this idea to explain the remarkable similarity of the different globin chains. This would produce identical regions in the mRNAs coding for the proteins, and we cannot exclude the possibility that the results reported here are due to such a process. Should this be the explanation of our results, the terminal sequences seem to have been conserved during evolution at least as well as the non-variable globin-coding regions.

Conserved, non-translated regions in RNAs which direct peptide synthesis have also been observed in RNA bacteriophages[32,33] and may be a general feature of informational RNAs. Immunity from mutational pressure implies a rather strict requirement for the region. Since these regions are apparently not translated, they become likely candidates for regulators in the complicated, highly specific process of gene expression.

续处理过程中对 mRNA 区域进行识别。HnRNA 和多聚核糖体 mRNA 之间可能具有的母核–产物关系最近已经通过两种 RNA 都富含腺苷酸序列这一发现得到了证明 [23-25]。

利姆和卡耐尔拉基斯曾从纯化的兔球蛋白 mRNA 组分中分离出一个大约长 50 ~ 70 个碱基（70% 腺苷酸）的富含腺苷酸的片段 [26]。假设富含腺苷酸的片段源于球蛋白编码区域内部，我们所计算出的腺苷酸可能存在于这一大片段上的最大百分率表明富含腺苷酸的大片段一定源于多肽编码区域之外，因此必定位于其中一个末端。将我们的结果和上述内容联系起来，两种球蛋白 mRNA 3′ 末端的小多聚腺苷酸序列似乎代表了一个较长的、非翻译的富含腺苷酸区域的一小部分，这个区域在转录后的加工过程中可能起作用。

对许多或者全部 mRNA 来说，非翻译的富含腺苷酸区域也许很常见，其结果是：与它们互补的富含胸腺嘧啶脱氧核苷酸的区域在 DNA 中重复出现。威廉森等人 [27] 曾将纯化的小鼠球蛋白 mRNA 组分进行杂交并得到了一些非常有趣的结果。在 RNA/DNA 的加入比例等于 3 之前，DNA 是不饱和的；然而对于每个加入的 RNA，达到最大键联所需的时间只有 10 min。这些结果与 RNA 中存在一个与重复 DNA 序列杂交的组分是一致的。然而，基德斯和比恩施蒂尔曾将海胆组蛋白 mRNA 组分与大量过量的 DNA 杂交，却没有发现一种能快速杂交的 RNA 组分。如果键联是基于所有 mRNA 所共有的序列，且均匀地分布在整个基因组中，这种 RNA 组分应该能找得到 [28]。在天然 mRNA 组分和球蛋白 mRNA 中发现的富含腺苷酸区域确实是由 DNA 转录而来还是由其他机制后来加上去的，至今还不是很清楚 [29]。

我们所得结果的另一种解释基于以下理论，即在通过基因复制的进化过程中出现了不同的蛋白质，而后每个基因都发生了独立的突变 [30]。英格拉姆 [31] 用这个理论解释了不同球蛋白链间惊人的相似性。这将在编码蛋白质的 mRNA 中产生完全一样的区域，我们不能排除本文报告的结果源于这一过程的可能性。如果这就是对我们所得结果的解释，那么末端序列在进化过程中被保留下来的程度至少会和未发生变化的球蛋白编码区域一样完好。

在 RNA 噬菌体中也发现了指导肽合成的 RNA 非翻译保守区域 [32,33]，这或许是带有信息的 RNA 的一个普遍特征。未受突变压力的影响意味着对该区域有更加严格的要求。这些区域显然并没有被翻译，因而它们很可能在复杂的、高度特异的基因表达过程中成为候选的调控者。

This work was supported by research grants from the American Cancer Society, USPHS National Institutes of Health, and the National Science Foundation. J. B. L. is a career development awardee of the USPHS National Institutes of Health.

(*Nature New Biology*, **233**, 41-43; 1971)

Henry Burr and Jerry B. Lingrel: Department of Biological Chemistry, University of Cincinnati College of Medicine, Cincinnati, Ohio 45219.

Received May 24; revised July 26, 1971.

References:

1. Laycock, D. G., and Hunt, J. A., *Nature*, **211**, 1118 (1969).
2. Lockard, R. E., and Lingrel, J. B., *Biochim. Biophys. Res. Commun.*, **37**, 204 (1969).
3. Lingrel, J. B., in *Methods in Protein Biosynthesis*, Methods in Molecular Biology Series (edit. by Laskin, A. E., and Last, J. A.), **2** (Dekker, New York, in the press).
4. Evans, M. J., and Lingrel, J. B., *Biochemistry*, **8**, 829 (1969).
5. Evans, M. J., and Lingrel, J. B., *Biochemistry*, **8**, 3000 (1969).
6. DeWachter, R., and Fiers, W., *J. Mol. Biol.*, **30**, 507 (1967).
7. RajBhandary, U. L., *J. Biol. Chem.*, **243**, 556 (1968).
8. Glitz, D. G., and Sigman, D. S., *Biochemistry*, **9**, 3433 (1970).
9. Glitz, D. G., Bradley, A., and Fraenkel-Conrat, H., *Biochim. Biophys. Acta*, **161**, 1 (1968).
10. Robinson, W. E., Tessman, I., and Gilham, P. T., *Biochemistry*, **8**, 483 (1969).
11. Hunt, J. A., and Laycock, D. G., *Cold Spring Harbor Symp. Quant. Biol.*, **34**, 579 (1969).
12. Leppla, S. H., Bjoraker, B., and Bock, R. M., in *Methods in Enzymology* (edit. by Grossman, L., and Moldave, K.), **12B**, 236 (Academic Press, New York, 1968).
13. Schmidt, G., in *Methods in Enzymology* (edit. by Grossman, L., and Moldave, K.), **12B**, 230 (Academic Press, New York, 1968).
14. Khym, J. X., and Cohn, W. E., *J. Amer. Chem. Soc.*, **82**, 6380 (1960).
15. Baglioni, C., and Campana, T., *Europ. J. Biochem.*, **2**, 480 (1967).
16. Shaeffer, J. R., Trostle, P. K., and Evans, R. F., *Science*, **158**, 488 (1967).
17. Tavill, A. S., Grayzel, A. I., London, I. M., Williams, M. K., and Vanderhoff, G. A., *J. Biol. Chem.*, **243**, 4987 (1968).
18. Hunt, R. T., Hunter, A. R., and Munro, A. J., *Nature*, **220**, 481 (1968).
19. Hunt, T., Hunter, A., and Munro, A., *J. Mol. Biol.*, **43**, 123 (1969).
20. Kamen, R., *Nature*, **221**, 321 (1969).
21. Scherrer, K., and Marcaud, L., *J. Cell. Physiol.*, **72**, supp. **1**, **181** (1968).
22. Georgiev, G. P., *J. Theoret. Biol.*, **25**, 473 (1969).
23. Darnell, J. E., Wall, R., and Tushinski, R. J., *Proc. US Nat. Acad. Sci.*, **68**, 1321 (1971).
24. Lee, S. Y., Mendecki, J., and Brawerman, G., *Proc. US Nat. Acad. Sci.*, **68**, 1331 (1971).
25. Edmonds, M., Vaughan, M. H., and Nakazato, H., *Proc. US Nat. Acad. Sci.*, **68**, 1336 (1971).
26. Lim, L., and Canellakis, E. S., *Nature*, **227**, 710 (1970).
27. Williamson, R., Morrison, M., and Paul, J., *Biochem. Biophys. Res. Commun.*, **40**, 740 (1970).
28. Kedes, L. H., and Birnstiel, M. L., *Nature New Biology*, **230**, 165 (1971).
29. Twu, J. S., and Bretthauer, R. K., *Biochemistry*, **10**, 1576 (1971).
30. Lewis, E. B., *Cold Spring Harbor Symp. Quant. Biol.*, **16**, 159 (1951).
31. Ingram, V. B., *Nature*, **189**, 704 (1961).
32. DeWachter, R., Vanbenberghe, A., Merregaert, J., Contreras, R., and Fiers, W., *Proc. US Nat. Acad. Sci.*, **68**, 585 (1971).
33. Cory, S., Spahr, P. F., and Adams, J. M., *Cold Spring Harbor Symp. Quant. Biol.*, **35**, 1 (1970).
34. Sanger, F., Brownlee, G. G., and Barrell, B. G., *J. Mol. Biol.*, **13**, 373 (1965).
35. Brownlee, G. G., and Sanger, F., *J. Mol. Biol.*, **23**, 337 (1967).

这项工作的研究经费来自美国癌症协会、美国公共卫生署国立卫生研究院和美国国家科学基金会。杰里·林格里尔是美国公共卫生署国立卫生研究院颁发的职业发展奖的获得者。

（邓铭瑞 翻译；李素霞 审稿）

On the Mechanism of Action of *lac* Repressor

B. Chen *et al.*

Editor's Note

The *E. coli lac* operon is a set of genes responsible for the breakdown of lactose into sugars used for metabolism. In the cell's default state, a repressor molecule prevents gene expression by binding to a control region of bacterial DNA. But when lactose enters the cell and binds to the repressor, the repressor is released and RNA polymerase can begin transcription of the operon. Here Robert R. Perlman and colleagues show that the *lac* repressor and RNA polymerase bind independently to *lac* DNA *in vitro*, but that a mutation in the promoter can lead to competitive binding between repressor and polymerase. The *lac* operon was the first genetic regulatory mechanism to be described in detail, and remains a textbook classic.

THE expression of the *lac* operon is controlled by a specific repressor which prevents transcription of the operon, as well as by cyclic AMP and a cyclic AMP receptor protein (CRP) which stimulate transcription[1,2]. The *lac* repressor is a protein and binds to *lac* DNA at the operator locus[2]. It could repress transcription by interfering with one of several steps in DNA transcription: (*a*) it may prevent binding of RNA polymerase to *lac* DNA; (*b*) it may act subsequent to the binding of RNA polymerase, but before the formation of the first nucleotide bond; or (*c*) it may act after the formation of the first nucleotide bond. We have described a purified system capable of transcribing the *lac* operon and controlled by *lac* repressor[3]. We have used this system and rifampicin, which inhibits RNA polymerase, to establish whether repressor prevents binding of RNA polymerase to the *lac* promoter or acts at a subsequent step in the transcription process. We find that for the wild type *lac* operon, *lac* repressor does not inhibit the binding of RNA polymerase. A mutation in the *lac* promoter which increases the expression of the operon both *in vivo* and *in vitro* (*lac* p^s)[7], however, also alters the operon so that the polymerase and repressor compete for binding.

Wild Type *lac* DNA

When λh80d*lac* DNA containing a normal *lac* promoter is preincubated with CRP and RNA polymerase in the presence of cyclic AMP, a complex is formed which is resistant to inhibition by rifampicin. In the presence of the drug, transcription of the *lac* operon is thought to begin at this "preinitiation complex" and terminate after a single round of transcription is completed[3,4]. *Lac* transcription is repressed by adding the *lac* repressor to the preincubation mixture; repression is observed whether repressor is added after or before the addition of CRP and RNA polymerase (Table 1, lines 1, 2 and 4). In each case, repression is overcome by the later addition of the inducer, IPTG, together with rifampicin (lines 3 and 5). These results indicate that (*a*) repressor will bind effectively and

关于乳糖操纵子阻遏物的作用机理

陈等

编者按

大肠杆菌乳糖操纵子是一套基因，负责将乳糖降解成能用于新陈代谢的糖。细胞处于默认状态时，阻遏物分子可以通过结合到细菌 DNA 的控制区域上而阻遏基因的表达。但在乳糖进入细胞并与阻遏物结合之后，如果阻遏物被释放，RNA 聚合酶就能开始操纵子的转录。在本文中，罗伯特·帕尔曼及其同事们证明：在体外，乳糖操纵子阻遏物和 RNA 聚合酶可以独立地结合到乳糖操纵子 DNA 上，但启动子上发生的某一突变可以导致阻遏物和聚合酶之间出现竞争性结合。乳糖操纵子是第一个被详尽描述的遗传调控机制，至今仍是教科书中的经典。

乳糖操纵子的表达被一个能阻止操纵子转录的特异阻遏物所控制，同时也受到能促进转录的环磷酸腺苷（AMP）和环磷酸腺苷受体蛋白（CRP）的影响[1,2]。乳糖操纵子阻遏物是一种蛋白质，并且结合在乳糖操纵子 DNA 的操纵基因位点上[2]。它能通过干扰 DNA 转录过程的几个步骤中的一个来抑制转录：(*a*) 它可能阻止 RNA 聚合酶和乳糖操纵子 DNA 的结合；(*b*) 它可能在 RNA 聚合酶结合之后、第一个核苷酸键形成之前发挥作用；或者 (*c*) 它可能在第一个核苷酸键形成之后对其进行干扰。我们曾描述过一个能使乳糖操纵子进行转录的纯化系统，这一系统受乳糖操纵子阻遏物的调控[3]。我们使用这套系统和能够抑制 RNA 聚合酶的利福平来证实：阻遏物到底是通过阻止 RNA 聚合酶与乳糖启动子之间的结合发挥作用，还是在转录过程的后续步骤中发挥作用。我们发现，对于野生型乳糖操纵子，阻遏物并未抑制 RNA 聚合酶的结合。然而，乳糖启动子中的一个无论在体内还是在体外都能增强操纵子表达的突变（*lac* p^s）[7]，也同时造成了聚合酶与阻遏物之间对操纵子的竞争性结合。

野生型乳糖操纵子 DNA

在环 AMP 存在的情况下将含有一个正常乳糖启动子的 λh80d 乳糖操纵子 DNA 与 CRP 和 RNA 聚合酶预孵育后，形成了一个能抵抗利福平抑制作用的复合体。在有这种药物存在的条件下，乳糖操纵子的转录过程被认为是从这个“起始前复合体”开始，并在一轮转录完成后终止[3,4]。通过往预孵育混合物中加入阻遏物，乳糖操纵子转录受到抑制；不管加入阻遏物是在添加 CRP 和 RNA 聚合酶之前还是之后，都能观察到阻遏现象（表 1，第 1、2 和 4 行）。在每种情况中，阻遏都能被之后与利福平一起加入的诱导剂——异丙基 –β–D– 硫代半乳糖苷（IPTG）所逆转（第 3 行和

reversibly to the *lac* operator when a preinitiation complex exists, and (*b*) RNA polymerase will bind to a *lac* DNA–repressor complex. This last conclusion assumes that free RNA polymerase does not bind to *lac* DNA and form a preinitiation complex in the presence of rifampicin. We have determined this directly by showing that free RNA polymerase is inactivated by rifampicin in less than 10 s, whereas the binding of RNA polymerase to the *lac* promoter has a half-life of 80 s. In the presence of 10^{-2} M IPTG the *lac* operator–*lac* repressor complex has a half-life too fast to be measured by existing techniques[5], whereas rifampicin inactivates RNA polymerase bound to the *lac* promoter with a half-life of 90 s. Our experimental conditions therefore prevent the binding of active RNA polymerase to *lac* DNA after the addition of rifampicin, while permitting the dissociation of the repressor DNA complex at a rate much faster than the inactivation of the preinitiation complex by rifampicin. We conclude therefore that with λh80d*lac* DNA containing a normal *lac* promoter, RNA polymerase and *lac* repressor bind independently.

Table 1. Repressor Action with λh80d*lac* DNA

Time of addition (min)			Experiment	
0	5	10	I	II
CRP+RNP	—	Rif+XTP	4.0	3.6
CRP+RNP	Repressor	Rif+XTP	1.6	1.6
CRP+RNP	Repressor	Rif+XTP+IPTG	4.5	3.5
Repressor	CRP+RNP	Rif+XTP	1.9	1.2
Repressor	CRP+RNP	Rif+XTP+IPTG	5.3	—

Each reaction mixture of 0.15 ml. contained 0.02 M Tris-HCl, *p*H 7.9, 0.01 M $MgCl_2$, 0.06 M KCl, 1×10^{-4} M dithiothreitol, 4.6 μg λh80d*lac* DNA, 0.75 μg rho, 1×10^{-4} M cyclic AMP, 1.2 μg repressor, 1.7 μg CRP, 3.6 μg RNA polymerase (350 U/mg), 0.15 mM each of ATP, GTP and UTP and 0.075 mM CTP (20 Ci/mmol). DNA, cyclic AMP and the basic salt solution were mixed together at 0°C and then brought to 30°C (zero time). Then CRP together with RNA polymerase or *lac* repressor were added at 5 min intervals as indicated. Five minutes after the addition of the last component, rifampicin and nucleotides were added and the samples incubated for another 15 min before the reaction was stopped and samples prepared for hybridization as previously described[3]. IPTG at 10^{-2} M was added as indicated with rifampicin and nucleotides. In the absence of cyclic AMP and CRP a background of 1.3–1.7% *lac* mRNA was obtained and subtracted from the values shown. In a typical experiment with λh80d*lac* or λh80d*lac* p^s DNA, 1.6×10^6 c.p.m. of RNA was made. This value was not affected by the presence of cyclic AMP, CRP or repressor.

Preparation of the components of the *in vitro* transcription system for the *lac* operon has already been described. DNA from λh80t68d*lac* or λh80t68d*lac* p^s was used as template. *Lac*-specific transcription was detected by prehybridization of the RNA to λ*c*I*857Sam7* DNA followed by hybridization of the unannealed RNA to the separated DNA strands of λ*c*I*857Sam7*p*lac*5[3]. The latter phage was prepared from λ*c*I*857* p*lac*5 (gift of J. Beckwith).

Mutant *lac* p^s DNA

We described a mutation of the *lac* operon, p^s, which results in increased synthesis of β-galactosidase both *in vivo* and *in vitro*[6]. DNA containing the *lac* p^s mutation serves as a

第 5 行）。这些结果说明：(*a*) 当起始前复合体存在时，阻遏物能有效并可逆地同乳糖操纵子的操纵基因结合；(*b*) RNA 聚合酶将同乳糖操纵子 DNA– 阻遏物复合体结合。后一个结论假设游离的 RNA 聚合酶在利福平存在的条件下不能与乳糖操纵子 DNA 结合而形成起始前复合体。我们通过下列结果直接证实了上述假设，即：利福平在不到 10 秒的时间内就能使游离 RNA 聚合酶灭活，而 RNA 聚合酶与乳糖启动子的结合则具有 80 秒的半衰期。在浓度为 10^{-2} M IPTG 存在的条件下，乳糖操纵基因 – 阻遏物复合体的半衰期非常短以至于用现有的技术还无法对其进行测量 [5]，而经利福平灭活的 RNA 聚合酶结合乳糖启动子的半衰期则为 90 秒。因此在加入利福平之后，我们的实验条件阻止了活性 RNA 聚合酶与乳糖操纵子 DNA 之间的结合，而允许阻遏物 –DNA 复合体发生分离，其速度远远超过利福平使起始前复合体灭活的速度。因此我们的结论是：对于含一个正常乳糖启动子的 λh80d 乳糖操纵子 DNA 来说，RNA 聚合酶和乳糖操纵子阻遏物在与它的结合上是独立的。

表 1. 阻遏物与 λh80d 乳糖操纵子 DNA 的作用

加入时间（分钟）			实验	
0	5	10	I	II
CRP+RNA 聚合酶	—	利福平 + XTP	4.0	3.6
CRP+RNA 聚合酶	阻遏物	利福平 + XTP	1.6	1.6
CRP+RNA 聚合酶	阻遏物	利福平 + XTP + IPTG	4.5	3.5
阻遏物	CRP+RNA 聚合酶	利福平 + XTP	1.9	1.2
阻遏物	CRP+RNA 聚合酶	利福平 + XTP + IPTG	5.3	—

每 0.15 ml 反应混合物包含：pH 值为 7.9 的 0.02 M Tris-HCl，0.01 M 氯化镁，0.06 M 氯化钾，1×10^{-4} M 二硫苏糖醇，4.6 μg λh80d 乳糖操纵子 DNA，0.75 μg rho，1×10^{-4} M 环 AMP，1.2 μg 阻遏物，1.7 μg CRP，3.6 μg RNA 聚合酶（三磷酸胞苷，350 U/mg），ATP（三磷酸腺苷）、GTP（三磷酸鸟苷）、UTP（三磷酸尿苷）各 0.15 mM 以及 0.075 mM CTP（20 Ci/mmol）。将 DNA、环 AMP 和基础盐溶液在 0℃下混合后升温至 30℃（记为 0 时）。接下来以 5 分钟为间隔按表中所述加入 CRP 和 RNA 聚合酶或者乳糖操纵子阻遏物。加入最后成分后再过 5 分钟，加入利福平和核苷酸（XTP），在反应停止前再孵育样品 15 分钟，按以前描述过的方法准备样品以便进行杂交 [3]。如表中所述在利福平和核苷酸中加入 10^{-2} M IPTG。在缺乏环 AMP 和 CRP 的条件下，获得背景为 1.3% ~ 1.7% 的乳糖信使 RNA 并从显示值中扣除。在 λh80d 乳糖操纵子或 λh80d 乳糖操纵子 p^s DNA 的典型实验中，生成了 1.6×10^6 放射性计数 / 分钟的 RNA。该值不受环 AMP、CRP 或阻遏物影响。

乳糖操纵子体外转录系统的成分制备已经叙述。用 λh80t68d 乳糖操纵子 DNA 或 λh80t68d 乳糖操纵子 p^s DNA 作为模板。通过将 RNA 和 λ*cI857Sam7* DNA 预杂交后接着对未退火的 RNA 和分离自 λ*cI857Sam7*p*lac*5[3] 的 DNA 条带进行杂交，来对乳糖操纵子特异转录进行检测。后一种噬菌体是由 λ*cI857*p*lac*5（由贝克威思赠送）制备的。

乳糖操纵子突变型 p^s DNA

我们曾描述过乳糖操纵子的一种突变 p^s，它在体内体外都能导致 β– 半乳糖苷酶合成加速 [6]。在体外纯化系统中，与母本乳糖操纵子 DNA 相比，含有乳糖操纵子

better template for *lac* RNA synthesis in a purified *in vitro* system than the parental *lac* DNA[3]. As some single mutations in the *lac* operon affect both promoter and operator function (Smith and Sadler, personal communication), it was of particular interest to test the action of repressor on this mutant. Our first experiments failed to demonstrate any difference between wild type and the mutant. The *in vivo* synthesis of β-galactosidase from the *lac* p^s operon is fully repressible as is the *in vitro lac* RNA synthesis. Furthermore, the wild type and mutant operons seem to be equally sensitive to repressor; the repressor concentration required for 50% inhibition of *lac* transcription is approximately the same for both templates (data not shown). Nevertheless, the action of repressor on the mutant operon is, at least in part, different from its action on the wild type operon (Table 2). These results show that (*a*) *lac* p^s DNA is a more effective template for *lac* transcription than normal *lac* DNA (compare Table 2, line 1, with Table 1, line 1); (*b*) the addition of repressor after formation of the preinitiation complex produces only partial repression, which is removed by the addition of IPTG (lines 2 and 3); (*c*) the addition of repressor before RNA polymerase produces complete repression (line 4), and partially interferes with the binding of RNA polymerase since subsequent addition of IPTG produces only a partial reversal of inhibition (lines 1 and 5).

Table 2. Repressor Action with λh80d*lac* p^s DNA

Time of addition (min)			Experiment	
0	5	10	I	II
CRP+RNP	—	Rif+XTP	7.7	6.4
CRP+RNP	Repressor	Rif+XTP	4.6	4.4
CRP+RNP	Repressor	Rif+XTP+IPTG	6.6	6.1
Repressor	CRP+RNP	Rif+XTP	1.6	1.4
Repressor	CRP+RNP	Rif+XTP+IPTG	3.4	2.6

The experiment was performed as in Table 1. In the absence of cyclic AMP and CRP, a background of 1– 2% *lac* mRNA was made which was subtracted from the values shown.

Although the addition of IPTG and rifampicin does not restore *lac* transcription to unrepressed levels, prolonged incubation with inducer before addition of rifampicin does (Fig. 1). RNA polymerase was added to a repressor–DNA complex in the presence of CRP and cyclic AMP. After 5 min of incubation, IPTG was added and the incubation continued. At various times, rifampicin and the four ribonucleoside triphosphates were added, and the amount of *lac* transcription measured after additional incubation. Following an initial lag of about 1 min, transcription gradually returns to unrepressed levels in 3–4 min, probably because of the binding of free RNA polymerase. Independent experiments have shown that RNA polymerase required 5 min to reach maximal binding to available *lac* promoters.

p^s 突变的 DNA 是合成乳糖操纵子 RNA 的更好的模板 [3]。由于乳糖操纵子中的一些单突变能同时影响启动子和操纵基因的功能（史密斯和萨德勒，个人交流），因此我们对测试阻遏物对这种突变体的作用特别感兴趣。我们最初的实验结果显示野生型和突变体之间没有任何区别。来自 p^s 乳糖操纵子的 β– 半乳糖苷酶的体内合成同体外发生的乳糖操纵子的 RNA 合成一样受到完全阻遏。此外，野生型操纵子和突变体操纵子似乎对阻遏物具有同等的敏感度；对于两种模板来说，要使对乳糖操纵子转录的抑制率达到 50% 所需要的阻遏物浓度是近乎一样的（数据未给出）。尽管如此，阻遏物对突变体操纵子的作用，至少在某种程度上，与它对野生型操纵子的作用是不同的（表 2）。这些结果表明：(*a*) 乳糖操纵子 p^s DNA 是比普通乳糖操纵子 DNA 更有效的乳糖转录模板（对比表 2 的第 1 行和表 1 的第 1 行）；(*b*) 在起始前复合体形成之后加入阻遏物只能产生部分阻遏，其阻遏作用可在加入 IPTG 后去除（第 2 和第 3 行）；(*c*) 在 RNA 聚合酶之前加入阻遏物能产生完全的阻遏（第 4 行），并对 RNA 聚合酶的结合造成部分干扰，因为随后加入的 IPTG 只部分逆转了抑制效应（第 1 行和第 5 行）。

表 2. 阻遏物与 λh80d 乳糖操纵子 p^s DNA 的作用

加入时间（分钟）			实验	
0	5	10	I	II
CRP+RNA 聚合酶	—	利福平 + XTP	7.7	6.4
CRP+RNA 聚合酶	阻遏物	利福平 + XTP	4.6	4.4
CRP+RNA 聚合酶	阻遏物	利福平 + XTP + IPTG	6.6	6.1
阻遏物	CRP+RNA 聚合酶	利福平 + XTP	1.6	1.4
阻遏物	CRP+RNA 聚合酶	利福平 + XTP + IPTG	3.4	2.6

实验按表 1 中描述的步骤进行。在缺乏环 AMP 和 CRP 的条件下，获得背景为 1%～2% 的乳糖信使 RNA，该值已经从显示值中扣除。

尽管加入 IPTG 和利福平并没有使乳糖操纵子转录恢复到未受阻遏时的水平，但在加入利福平之前延长与诱导剂一起孵育的时间却能做到（图 1）。在 CRP 和环 AMP 存在的条件下将 RNA 聚合酶加入阻遏物 –DNA 复合体中。孵育 5 分钟后，加入 IPTG 继续孵育。选择不同的时间将利福平和 4 种核苷三磷酸盐加入，在附加的孵育过程结束之后测量乳糖操纵子转录的数量。在大约 1 分钟的起始滞后期之后，大概是因为游离 RNA 聚合酶结合的原因，转录在 3 分钟～4 分钟之内逐渐恢复到未受阻遏时的水平。另有一些独立的实验显示，RNA 聚合酶需要 5 分钟时间才能与可结合的乳糖启动子达到最大程度的结合。

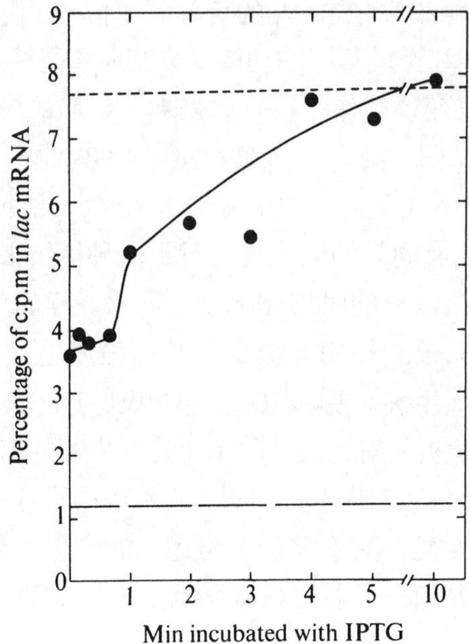

Fig. 1. Kinetics of repressor dissociation by IPTG. Incubation medium was the same as in Table 1. DNA, CRP and *lac* repressor were mixed at 0°C. After RNA polymerase was added the temperature was brought to 30°C for 5 min. Then IPTG (10^{-2} M) was added followed at the times indicated in the figure by rifampicin and nucleotides. — — —, *lac* RNA synthesis in the absence of IPTG; - - -, *lac* RNA synthesis in the absence of repressor.

Both the preinitiation and repressor–DNA complexes are stable (Fig. 2). When repressor is added after formation of the preinitiation complex, the level of inhibition does not increase beyond the initial (~50%) level even if incubation is continued for 30 min. A five-fold increase in the concentration of repressor also does not produce further inhibition (data not shown). Similarly, only background levels of *lac* RNA are made when CRP and RNA polymerase are added after formation of a repressor–DNA complex, even when the mixture is incubated for 30 min. The addition of a five-fold excess of CRP and RNA polymerase does not overcome repression (data not shown). The stability of the *lac* repressor–*lac* DNA complex has already been described by Bourgeois and Riggs[7], who measured the binding of repressor to radioactively labelled *lac* DNA.

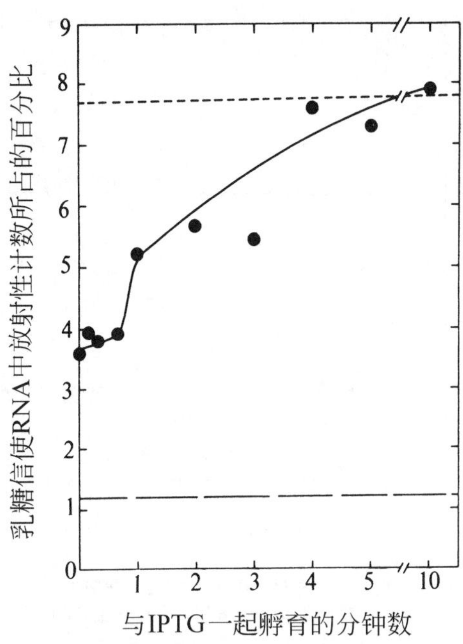

图 1. IPTG 使阻遏物发生解离的动力学曲线。孵育环境和表 1 中的一样。在 0℃ 下混合 DNA、CRP 和乳糖操纵子阻遏物。加入 RNA 聚合酶后升温至 30℃ 并持续 5 分钟。此时添加 IPTG（10^{-2} M），随后按照图中所标示的时间加入利福平和核苷酸。— — — 表示在缺少 IPTG 条件下的乳糖操纵子 RNA 合成；-------- 表示在缺少阻遏物条件下的乳糖操纵子 RNA 合成。

起始前复合体和阻遏物–DNA 复合体都很稳定（图 2）。如果在起始前复合体形成后加入阻遏物，那么即使孵育时间持续 30 分钟，抑制水平的增加也未能超过刚开始时的水平(约 50%)。阻遏物的浓度增至 5 倍亦不能产生更强的抑制(数据未给出)。同样，如果在阻遏物–DNA 复合体形成后才加入 CRP 和 RNA 聚合酶，即使混合物孵育时间长达 30 分钟，也只会产生背景水平的乳糖操纵子 RNA。加入 5 倍过量的 CRP 和 RNA 聚合酶也不能消除阻遏（数据未给出）。布儒瓦和里格斯已经对乳糖操纵子阻遏物–乳糖操纵子 DNA 复合体的稳定性进行过描述 [7], 他们曾经测量过阻遏物与放射性标记的乳糖操纵子 DNA 的结合情况。

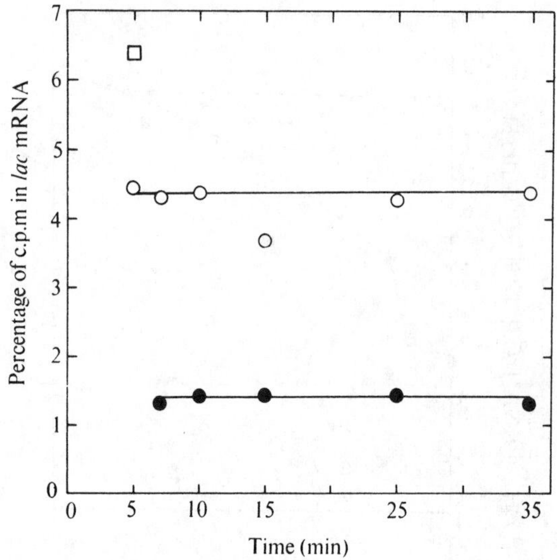

Fig. 2. Stability of preinitiation complex. The experiments were performed as described in Table 1. ○, CRP and polymerase added before repressor; ●, r epressor added before CRP and polymerase; □, control, no repressor added.

CRP has an affinity for DNA which is greatly increased by cyclic AMP[8]. We believe that CRP acts by binding to the *lac* promoter, altering the promoter in a manner which allows RNA polymerase to bind. Unlike CRP plus RNA polymerase, CRP alone does not compete with repressor for binding to the *lac* p^s template. Previous incubation of the *lac* p^s DNA with CRP and cyclic AMP does not prevent repressor from exerting maximal inhibition of *lac* transcription (Table 3).

Table 3. Effect of the Order of Addition of CRP, RNA Polymerase and Repressor on *lac* RNA Synthesis with *lac* p^s DNA

Time of addition (min)			Experiment
0	5	10	% *lac* mRNA
CRP	RNP	Rep	6.0
CRP	Rep	RNP	2.8
CRP	RNP	—	9.4
RNP	—	—	1.5
Repressor	CRP	RNP	2.8

Conditions were similar to those of Table 1 except that *lac* p^s DNA was used. All the initial components were preincubated at 0°C, the temperature raised to 30°C and the various additions made at 5 min intervals. Five minutes after the last addition, rifampicin and nucleotides were added. The background *lac* mRNA synthesis was not subtracted.

We have presented evidence based on *in vitro* transcription studies which indicates that the mechanism by which the *lac* repressor inhibits transcription in a wild type *lac* operon is different from its mechanism of action in a *lac* "promoter" mutant, *lac* p^s. With normal

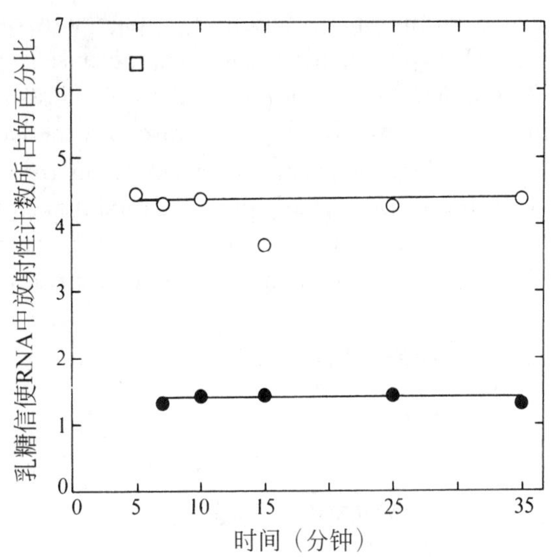

图 2. 起始前复合体的稳定性。实验按表 1 描述的步骤进行。○：CRP 和聚合酶在阻遏物之前加入；●：CRP 和聚合酶在阻遏物之后加入；□：对照，未加入阻遏物。

CRP 对 DNA 具有亲和力，且这种亲和力在环 AMP 存在的情况下显著增强[8]。我们认为 CRP 的作用方式是通过结合乳糖操纵子的启动子，从而在某种程度上改变启动子以使 RNA 聚合酶能结合到操纵子上。与 CRP 和 RNA 聚合酶共同作用时不同，单独的 CRP 并不与阻遏物竞争与乳糖操纵子 p^s 模板的结合。用 CRP 和环 AMP 与乳糖操纵子 p^s DNA 的预孵育也不能阻止阻遏物最大程度地发挥对乳糖操纵子转录的抑制（表 3）。

表 3. CRP、RNA 聚合酶和阻遏物的加入顺序对用乳糖操纵子 p^s DNA 的 RNA 合成的影响

加入时间（分钟）			实验
0	5	10	% 乳糖信使 RNA
CRP	RNA 聚合酶	阻遏物	6.0
CRP	阻遏物	RNA 聚合酶	2.8
CRP	RNA 聚合酶	—	9.4
RNA 聚合酶	—	—	1.5
阻遏物	CRP	RNA 聚合酶	2.8

除了使用的是乳糖操纵子 p^s DNA 外，其他条件和表 1 所述相似。所有初始成分都在 0℃下预孵育，然后升温至 30℃，各种添加成分以 5 分钟为间隔加入。在加入最后一个成分后再过 5 分钟加入利福平和核苷酸。背景的乳糖信使 RNA 合成没有被扣除。

我们已经提出了基于体外转录研究的证据，该证据说明：在野生型乳糖操纵子中阻遏物抑制转录的机理和它在乳糖操纵子的“启动子”突变体（*lac* p^s）中的作用机理不同。在正常乳糖操纵子 DNA 里，阻遏物在 RNA 聚合酶－乳糖操纵子 DNA *lac* DNA, *lac* repressor exerts its inhibitory action at a stage subsequent to formation of the RNA polymerase–*lac* DNA preinitiation complex, because repressor is unable to prevent formation of this complex, but is able to prevent polymerase once bound to DNA from transcribing the *lac* operon. Thus, RNA polymerase and the *lac* repressor must bind independently of one another (Fig. 3*a*). With *lac* p^s DNA, on the other hand, previous incubation of the DNA with repressor partially prevents the formation of the preinitiation complex, and transcription from preinitiation complexes already formed in the absence of repressor is partially inhibited by the subsequent addition of repressor.

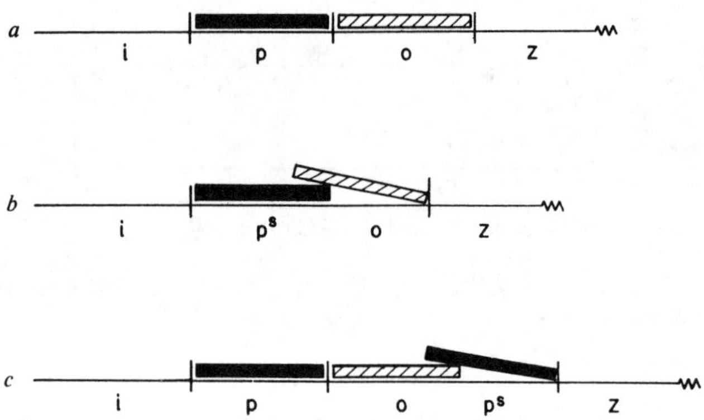

Fig. 3. Mechanism of action of *lac* repressor. RNA polymerase is represented by solid bar and *lac* repressor by striped bar. *a*, Normal *lac* DNA; *b* and *c*, *lac* p^s DNA.

Previous experiments on the mechanism of action of the *lac*, *gal*, *bio* and λ repressors do not present a conclusive or consistent picture. Reznikoff *et al.*[9] found that *lac* repressor binding *in vivo* to the *lac* operator could partially block transcription initiated at a distal promoter. They suggested that repression could occur subsequent to RNA chain initiation. Buttin (personal communication) found that induction of a λ lysogen in which λ DNA synthesis cannot occur results in synthesis of *gal* enzymes in the presence of the *gal* repressor. This "escape synthesis" seems to be caused by transcriptional read-through into *gal* by an RNA polymerase initiating at a λ promoter. Escape synthesis of the dethiobiotin synthetase on induction of a λ prophage can also be explained by a similar mechanism (K. Krell, M. Gottesman, J. Parks, M. Eisenberg, in preparation; A. Campbell, personal communication). The λ repressor seems to inhibit the binding of RNA polymerase to λ DNA (Hayward and Green[10]; Wu *et al.*[11]). Wu *et al.* found also that λ repressor could inhibit λ transcription even when added after RNA polymerase. This is in apparent contradiction to the findings of Chadwick *et al.*[12] who found that the binding of labelled λ repressor was inhibited by previous binding of RNA polymerase to DNA.

Our findings, in agreement with previous genetic studies, suggest that the promoter and operator binding sites may overlap in the *lac* p^s mutant. Ordal[13] finds that some virulent (o^c) mutants map on each side of a promoter mutant in the x region of λ. Smith and Sadler (personal communication), who have studied a large number of *lac* operator and promoter mutants, concluded that a single mutation could show both operator–constitutive and

起始前复合体形成之后的某个阶段发挥它的抑制作用，原因是阻遏物无法阻止这种复合体的形成，但能阻止已结合到 DNA 上的聚合酶转录乳糖操纵子。因此，RNA 聚合酶和乳糖操纵子阻遏物一定是独立地与 DNA 结合的（图 3*a*）。另一方面，在乳糖操纵子 p^s DNA 中，阻遏物与 DNA 的预孵育可以部分阻止起始前复合体的形成，而由形成于无阻遏物存在条件下的起始前复合体开始的转录也会被随后加入的阻遏物部分抑制。

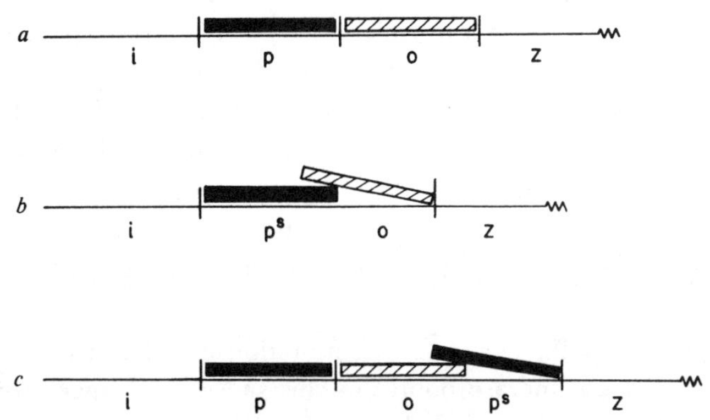

图 3. 乳糖操纵子阻遏物的作用机理。RNA 聚合酶用实心条表示，乳糖操纵子阻遏物用斜线条表示。*a*，正常乳糖操纵子 DNA；*b* 和 *c*，乳糖操纵子 p^s DNA。

以前关于乳糖、半乳糖、生物素和 λ 阻遏物作用机理的实验都没有给出一个结论性或一致性的描述。列兹尼科夫等人 [9] 发现：在体内，阻遏物与乳糖操纵子操纵基因的结合能够部分地抑制开始于一个远端启动子的转录。他们认为阻遏作用能够在 RNA 链起始后发生。比坦（个人交流）发现：在 λ DNA 无法合成的情况中，λ 溶素原的诱导将导致半乳糖酶的合成发生在半乳糖操纵子阻遏物存在时。这种“逃避合成”似乎是由起始于一个 λ 启动子的 RNA 聚合酶引起的转录通读到半乳糖操纵子所致。由 λ 前噬菌体诱导的脱硫生物素合成酶的逃避合成也能用相似的机理来解释（克雷尔、戈特斯曼、帕克斯、海森伯格，完稿中；坎贝尔，个人交流）。λ 阻遏物似乎能抑制 RNA 聚合酶与 λ DNA 的结合（海沃德和格林 [10]，以及吴等 [11]）。吴等人还发现，甚至在 RNA 聚合酶之后加入 λ 阻遏物也能抑制 λ 转录。这明显与查德威克等人 [12] 的发现相冲突，后者发现 RNA 聚合酶和 DNA 结合后会对标记的 λ 阻遏物的结合产生抑制。

与以前基于基因水平的研究一致，我们的研究结果表明，在乳糖操纵子 p^s 突变体中启动子和操纵基因的结合位点可能是部分重叠的。奥德尔 [13] 发现一些毒性突变体（o^c）出现在 λ 的 x 区域中某个启动子突变的两边。史密斯和萨德勒曾研究过大量乳糖操纵子的操纵基因和启动子突变体（个人交流），他们得出的结论是：一个单

promoter phenotypes; the predominance of one phenotype depends on the location of the mutation.

Our biochemical data for transcription from a wild type *lac* DNA template fail to demonstrate any overlapping of operator and promoter regions (Fig. 3*a*). With *lac* p^s DNA, however, the non-competitive situation characteristic of the wild type *lac* operon has been converted to a partially competitive one. Two models for the molecular basis of this conversion are presented below.

(*a*) The *lac* p^s mutation is a deletion which brings the promoter and operator closer together, causing partial overlapping of the repressor and RNA polymerase binding sites (Fig. 3*b*). Similarly, a point mutation might produce functional interference between the two sites. This hypothesis, however, does not explain the biochemical observation that inhibition of transcription from a *lac* p^s–RNA polymerase preinitiation complex by the subsequent addition of repressor is only 50% effective and is independent of repressor and CRP and RNA polymerase concentration over a five-fold range.

(*b*) The new promoter hypothesis, *lac* p^s, is a mutation creating a promoter site near the operator (Fig. 3*c*). RNA polymerase bound to this site is no longer repressible, although repressor-binding to the operator and inhibition of transcription at the normal promoter still occur. If repressor is bound first to the promoter site, RNA polymerase can no longer bind to the new promoter, although binding at the normal site is unaffected. This model explains the observed partial inhibition and competition by invoking two promoter sites. It also supposes that a cyclic AMP–CRP-dependent promoter has been translocated close to the *lac* operator or that a mutation can create a new promoter in a very small region of the *lac* operon which is still repressible by *lac* repressor and still cyclic AMP–CRP-dependent. No examples of such a mutation are known.

Studies are now in progress to resolve these models by fine structure genetic mapping.

We thank T. Platt and W. Gilbert for *lac* repressor, J. Beckwith for λp*lac*, P. Nissley and J. Parks for helpful discussion, and F. Herder for technical assistance.

(*Nature New Biology*, **233**, 67-69; 1971)

Beatrice Chen, Benoit De Crombrugghe, Wayne B. Anderson, Max E. Gottesman and Ira Pastan: Laboratory of Molecular Biology, National Cancer Institute, National Institutes of Health, Bethesda, Maryland 20014.

Robert L. Perlman: Clinical Endocrinology Branch, National Institute of Arthritis and Metabolic Diseases, National Institutes of Health, Bethesda, Maryland 20014.

Received July 13, 1971.

突变能同时显示出操纵基因组成型表型和启动子表型；突变发生的位置决定了有一种表型会占主导。

我们以野生型乳糖操纵子 DNA 为模板进行转录而得到的生化数据不能证明操纵基因区域和启动子区域之间有任何重叠（图 3*a*）。然而，在乳糖操纵子 p^s DNA 中，野生型乳糖操纵子所特有的无竞争环境转化成了部分竞争环境。描述这一转化的两种分子基础的模型如下所述。

（*a*）乳糖操纵子 p^s 突变是导致启动子和操纵基因挨得更近的缺失突变，这使得阻遏物和 RNA 聚合酶的结合位点部分重叠（图 3*b*）。与此类似，一个点突变也可能会导致两个位点间的功能性干扰。然而，这个假设并不能解释如下的生化观察结果：后来加入的阻遏物对源自乳糖操纵子 p^s–RNA 聚合酶起始前复合体的转录的抑制效用只有 50%，而且在 5 倍的浓度范围内，这种效应与阻遏物、CRP 和 RNA 聚合酶的浓度无关。

（*b*）新的启动子假说认为：乳糖操纵子 p^s 突变在操纵基因附近产生了一个启动子位点（图 3*c*）。尽管阻遏物与操纵基因的结合以及对正常启动子转录的抑制依然会发生，但是 RNA 聚合酶对新启动子位点的结合不再受到阻遏。如果是阻遏物首先与这种启动子位点结合，那么 RNA 聚合酶就不能再结合这种新型启动子了，但在正常位点的结合不受影响。这个模型通过引入两个启动子位点解释了观察到的部分抑制和竞争现象。它还假定环 AMP–CRP 依赖的启动子被转移到靠近乳糖操纵基因的位置；或者说一个突变在乳糖操纵子的一个非常小的区域创造了一个新启动子，这个新启动子同样受到乳糖操纵子阻遏物的阻遏，并且仍为环 AMP–CRP 依赖。目前还没有发现一例这样的突变。

通过精细结构遗传图来解析这些模型的研究目前正在进行中。

感谢普拉特和吉尔伯特为我们提供乳糖操纵子阻遏物，感谢贝克威思提供 λp*lac*，感谢尼斯利和帕克斯与我们进行了有益的讨论，还要感谢赫德的技术支持。

（邓铭瑞 翻译；金城 审稿）

References:

1. Pastan, I., and Perlman, R. L., *Science*, **169**, 339 (1970).
2. Gilbert, W., and Muller-Hill, B., *Proc. US Nat. Acad. Sci.*, **56**, 1891 (1966).
3. de Crombrugghe, B., Chen, B., Anderson, W., Nissley, P., Gottesman, M., Perlman, R., and Pastan, I., *Nature New Biology*, **231**, 139 (1971).
4. Bautz, E. F. K., and Bautz, F. A., *Nature*, **226**, 1219 (1970).
5. Riggs, A. D., Newby, R. F., and Bourgeois, S., *J. Mol. Biol.*, **51**, 303 (1970).
6. de Crombrugghe, B., Chen, B., Gottesman, M., Pastan, I., Varmus, H. E., Emmer, M., and Perlman, R. L., *Nature New Biology*, **230**, 37 (1971).
7. Riggs, A. D., Suzugi, H., and Bourgeois, S., *J. Mol. Biol.*, **48**, 67 (1970).
8. Pastan, I., de Crombrugghe, B., Chen, B., Anderson, W., Parks, J., Nissley, P., Straub, M., Gottesman, M., and Perlman R., *Proc. Miami Winter Symp.* (North Holland, Amsterdam, 1971).
9. Reznikoff, W. S., Miller, J. H., Scaife, M. J., and Beckwith, J. R., *J. Mol. Biol.*, **43**, 201 (1969).
10. Hayward, W. S., and Green, M. H., *Proc. US Nat. Acad. Sci.*, **64**, 962 (1969).
11. Wu, A., Ghosh, S., and Echols, H., *Fed. Proc.*, **30**, 1529 (1971).
12. Chadwick, P., Pirrotta, V., Steinberg, R., Hopkins, N., and Ptashne, M., *Cold Spring Harbor Symp. Quant. Biol.*, **35**, 283 (1970).
13. Ordal, G., in *The Bacteriophage Lambda* (edit. by the Cold Spring Harbor. Lab., Cold Spring Harbor, New York) (in the press).

Functional Organization of Genetic Material as a Product of Molecular Evolution

T. Ohta and M. Kimura

Editor's Note

Motoo Kimura's influential neutral theory of molecular evolution states that, at the molecular level, evolutionary changes are caused by the random drift of selectively neutral mutants. The theory was based on constancy in rates of amino-acid change seen in proteins seen over time, most of which, Kimura argued, had no influence on individual fitness. Here Tomoko Ohta and Kimura calculate that an average of 8 amino-acid substitutions occur in the human genome every year, and conclude that, since base substitutions have no effect, a large part of an organism's DNA must be non-essential. Although some viewed Kimura's theory as an argument against Darwinian evolution, Kimura argued that the two theories could coexist, with natural selection and genetic drift both influencing evolution.

MOLECULAR evolution consists of a sequence of events in which originally rare molecular mutants (DNA changes) spread into the species. Two important classes of mutations are nucleotide (or amino-acid) replacement and gene duplication.

From comparative studies of the protein sequences of related organisms, we now have some information on the rate of amino-acid substitution in evolution. The rate is different from protein to protein, but for each particular molecule, such as haemoglobin α or cytochrome c, it is remarkably uniform per year in diverse lines over geological times[1]. This constitutes very strong evidence for the hypothesis[1-3] that most nucleotide substitutions in evolution are the result of random fixation of selectively neutral or near neutral mutants. Furthermore, the rate at which neutral mutant genes are substituted in the population per generation is equal to the rate of occurrence of such mutants per gamete[5].

There are significant differences in evolutionary rates among proteins. The highest rate (fibrinopeptides) is some 1,500 times greater than the lowest rate (histones)[4]. This means that the rate depends on the functional requirement of the molecule (refs. 1 and 3 and unpublished work of M. K. and T. O.). The greater the chance that the mutations are deleterious, the lower the evolutionary rate. The uniformity of evolutionary rate for each molecule implies that the fraction of neutral mutations among all mutations in each cistron remains constant per year.

The average rate of amino-acid substitutions estimated using nine proteins is about 1.6×10^{-9}/amino-acid site/yr[3]. If the difference of evolutionary rates among proteins

功能性遗传物质是分子进化过程中的产物

太田朋子，木村资生

编者按

木村资生颇具影响力的分子进化中性理论认为：在分子水平上，进化的改变是由选择性中性突变的随机漂移导致的。该理论是基于观察到蛋白质中氨基酸的变化速率在很长时间内保持恒定，木村资生认为大多数这种变化对个体适应度并无影响。在本文中，太田朋子和木村资生计算出人类每个基因组平均每年会发生 8 次氨基酸替换，他们得出结论：因为碱基替换没有产生任何影响，所以生物体中的 DNA 一定有很大一部分是非必需的。虽然有人提出木村资生的理论有悖于达尔文的进化论，但木村本人认为这两种理论是可以共存的，自然选择和遗传漂变都对进化过程有影响。

分子进化由一连串事件构成，在这些事件中，最初稀少的分子突变（DNA 变化）逐渐扩散到各物种中。其中有两类重要的突变，即核苷酸（或氨基酸）替换和基因重复。

通过对相关生物体蛋白质序列的对比研究，现在我们已经得到了关于进化过程中氨基酸替换速率方面的一些信息。每一种蛋白质都有不同的氨基酸替换速率，但是对于每个特定的分子而言，例如血红蛋白 α 或细胞色素 c，处于不同地质时代的各种生物世系在每一年中的变化速率却非常一致 [1]。这为以下假说 [1-3] 提供了非常有力的证据：进化过程中的大部分核苷酸替换是选择性中性突变或近中性突变随机固定的结果。此外，每一代群体的中性突变基因发生替换的速率都与每个生殖细胞发生这种突变的速率相同 [5]。

各种蛋白质之间的进化速率存在显著差异。最高速率（血纤维蛋白肽）是最低速率（组蛋白）的 1,500 倍左右 [4]。这意味着该速率取决于蛋白质分子的功能性需求（参考文献 1、3 以及木村资生和太田朋子尚未发表的研究结果）。有害突变发生的机会越多，进化速率就越慢。每个分子都具有同样的变化速率说明：在每个顺反子的所有突变中，中性突变的比例每年都保持恒定。

有人用 9 种蛋白质估算出氨基酸替换的平均速率——每个氨基酸位点每年约为 1.6×10^{-9}[3]。如果蛋白质进化速率的差异是由于功能性需求上的差异而非它们的

is due to the difference in functional requirement rather than the difference in their mutation rate, we should expect that for any cistron the frequency of mutation at the time of occurrence is equal to that of the fibrinopeptides showing the highest substitution rate. This allows us to estimate the true mutation rate at the molecular level.

From the atlas of Dayhoff[6] we sampled seven organisms for which the complete amino-acid sequences of fibrinopeptides A, B and cytochrome c are given, from which to calculate the regression of substitution rate of the fibrinopeptides on that of cytochrome c for which the rate of substitution is well known. We therefore chose the rapidly evolving parts corresponding to the amino-acid positions four to seven in fibrinopeptide A and nine to twenty in fibrinopeptide B (alignment 9 of Dayhoff[6]). The remaining amino-acid positions not only evolve much more slowly but also show some non-randomness with respect to the pattern of substitution. We have therefore excluded these slowly evolving parts from our calculation. Fig. 1 illustrates the regression based on all the pairs of comparisons between seven organisms (human, rhesus monkey, rabbit, dog, horse, pig and kangaroo) with observed values shown by dots. The abscissa gives the number of substitutions in terms of $-\log_e(1-p_d)$ for cytochrome c while the ordinate gives that for fibrinopeptides, where p_d is the fraction of different amino-acids. (For the rationale of using such a logarithmic scale, see ref. 1.) The regression is about 22.4, suggesting that the true mutation rate is 22.4 times the substitution rate of cytochrome c, which is about 0.37×10^{-9}/amino-acid site/yr in mammalian evolution. The resulting mutation rate becomes

$$0.37\times10^{-9}\times22.4\approx8.3\times10^{-9}$$

per amino-acid site/yr. This is roughly five times the average rate of substitutions of the nine proteins and agrees with the conclusion of Corbin and Uzzell[7], and with the conventional mutation rate per locus per generation of $10^{-5}\sim10^{-6}$ if the average cistron codes for the protein of several hundred amino-acids long and if the average generation time is not much different from 1 yr.

突变率差异，那么我们就可以认为任何顺反子发生突变的频率都与替换速率最高的血纤维蛋白肽的突变频率相等。这使我们能够估计出分子水平上的真实突变率。

从戴霍夫[6]的图集中，我们抽取了7种生物，这7种生物的血纤维蛋白肽A、B和细胞色素c的完整氨基酸序列均已给出，由这些序列出发，根据细胞色素c的已知替换速率对血纤维蛋白肽替换速率的回归系数进行计算。为此我们选择了一些快速进化的部分，即与血纤维蛋白肽A中氨基酸位点4到7及血纤维蛋白肽B中氨基酸位点9到20对应的部分（戴霍夫比对9[6]）。其余的氨基酸位点不仅在进化速率上要慢很多，而且替换模式还具有一些非随机性。因此我们在计算中没有将这些进化缓慢的部分考虑进去。图1中绘出了对7种生物（人类、恒河猴、兔、狗、马、猪和袋鼠）的所有配对进行比较得到的回归结果，观察值用点表示。横坐标给出了以 $-\log_e(1-p_d)$ 表示的细胞色素c的替换数目，而纵坐标则给出了血纤维蛋白肽的替换数目，式中 p_d 代表不同氨基酸所占的比例。（关于使用这种对数标尺的基本原理，请参见参考文献1。）得出的回归系数约为22.4，这说明真实突变率是细胞色素c的替换速率的22.4倍，在哺乳动物的进化中，细胞色素c的替换速率大致为每个氨基酸位点每年 0.37×10^{-9}。于是真实突变率为每个氨基酸位点每年：

$$0.37 \times 10^{-9} \times 22.4 \approx 8.3 \times 10^{-9}$$

这大概是那9种蛋白质的平均替换速率的5倍，与科尔宾和尤泽尔[7]得到的结果一致。如果平均而言顺反子编码的蛋白质长度达到几百个氨基酸，并且如果平均世代时间大约等于一年，那么这个值也与每个基因座每世代 $10^{-5} \sim 10^{-6}$ 的公认突变率一致。

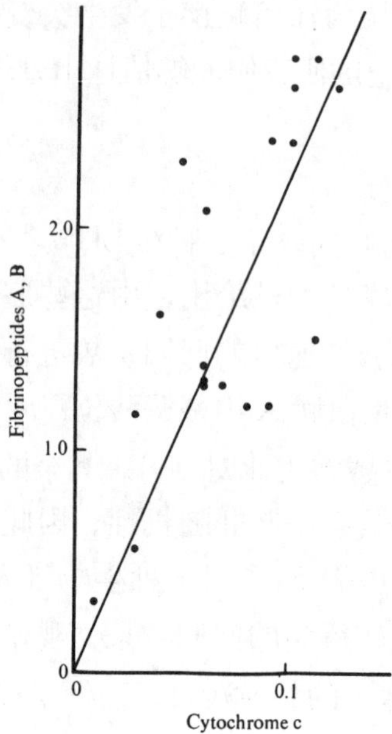

Fig. 1. The regression of the rate of amino-acid substitution of fibrinopeptides (actually their rapidly evolving parts) on that of cytochrome c. In each coordinate, the number of different amino-acid sites between two homologous proteins is expressed in terms of $-\log_e(1-p_d)$ where p_d is the fraction of differing sites. Dots represent twenty-one observed values (of which two sets of values coincide) from paired comparisons involving seven organisms (human, rhesus monkey, rabbit, dog, horse, pig and kangaroo). The estimated regression coefficient is 22.4.

If a large fraction of mutations are deleterious as discussed above, every higher organism must suffer a heavy genetic load. According to the Haldane–Muller principle[5] the mutation load of equilibrium populations is equal to one to two times the total deleterious mutation rate depending on the degree of dominance of mutations[8]. Recent investigations in *Drosophila* have shown that the detrimental mutations have a fairly high degree of dominance[9,10]. The mutation load must therefore be nearly twice the total mutation rate, although epistatic (synergistic) interaction in fitness among the loci may reduce the load somewhat[10,11].

Various species of mammals have about the same amount of DNA. The human haploid genome consists of about 3×10^9 nucleotide sites[12,13], equivalent to 10^9 amino-acid sites: thus the total mutation rate due to base replacement per genome per year amounts to

$$8.3\times10^{-9}\times10^9\approx8$$

If the mutation rate is strictly proportional to chronological time, the rate per generation is probably twenty times this figure in man but half as large in mouse. In the main course of mammalian evolution, the average generation time was probably 2~3 yr, so we tentatively equate 1 yr with one generation in mammals. If most of these mutations are deleterious, any mammalian species must suffer an intolerable mutation load: Muller argued[14] that

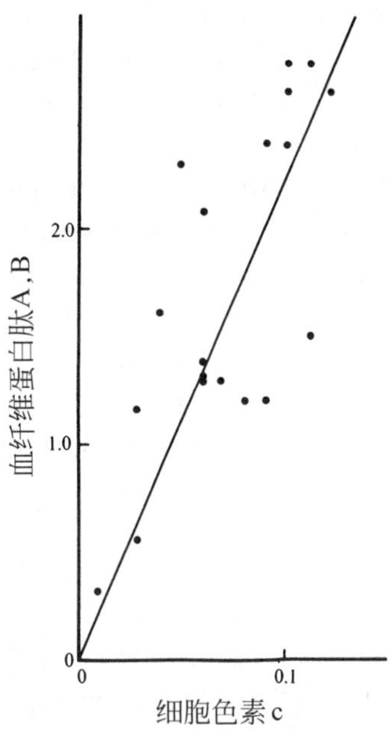

图 1. 根据细胞色素 c 的氨基酸替换速率对血纤维蛋白肽（实际上是其中快速进化的部分）的氨基酸替换速率进行回归分析得到的结果。在每个坐标上，两个同源蛋白质之间的不同氨基酸位点的数目以 $-\log_e(1-p_d)$ 来表示，式中 p_d 表示不同位点所占的比例。图中的点代表 7 种生物（人类、恒河猴、兔、狗、马、猪和袋鼠）配对比较的 21 个观察值（其中有两组数据是重叠的）。回归系数的估计值为 22.4。

如上所述，如果大部分突变是有害变异的话，那么每种高等生物都必须承受非常大的遗传负荷。根据霍尔丹 – 马勒原理 [5]，平衡群体的突变负荷依突变显性程度的不同而相当于总有害突变率的 1 ~ 2 倍 [8]。对果蝇的最新研究结果显示，有害突变的显性度相当高 [9,10]。因此，尽管基因座间存在适当大小的上位（协同）相互作用也许会在某种程度上减轻这种负荷 [10,11]，但突变负荷肯定接近于总突变率的 2 倍。

不同种类哺乳动物所含的 DNA 数目大致相同。人类的单倍体基因组由大约 3×10^9 个核苷酸位点构成 [12,13]，相当于 10^9 个氨基酸位点，因此由每个基因组每年的碱基替换速率导致的人类基因组的总突变率为：

$$8.3 \times 10^{-9} \times 10^9 \approx 8$$

如果突变率严格与年代时间成正比的话，那么人类每世代的突变率可能是这一数字的 20 倍，而小鼠的突变率则仅为这一数字的 2 倍。在哺乳动物的主要进化路线中，平均世代时间可能为 2 年 ~ 3 年，所以我们试探性地将哺乳动物的一个世代定为 1 年。如果这些突变中的大部分是有害的，那么任何一个哺乳动物物种肯定都承受着无法

the total rate for detrimental mutations in man is at most 0.5 per generation. What does such great discrepancy signify? We conclude that a large part of DNA in the genome is not essential for the life of the organisms in that base substitutions have no effects[3]. If we compare the above estimate of the total mutation rate with Muller's maximum detrimental mutation rate we are led to the conclusion that only 0.5/8 or 6% of the total DNA are as important in function as the cistrons such as those coding for cytochrome c or haemoglobins. If the estimated total mutation rate per generation is twenty times that per year and the total detrimental mutation rate is 1.0 (taking into account the "viability polygenes" of Mukai)[15], then the fraction becomes 0.6%. Although we conclude, with Muller[16], that the total number of "genes" in man is about 3×10^4, whereas he assumes a large gene size (corresponding to 30,000 amino-acids), we assume that the genes are much smaller in size (corresponding to several hundred amino-acids on average) but a large fraction of DNA is not "informational".

Mammalian DNA is about 1,000 times larger than bacterial DNA, so that the genetic material duplicated some ten times on average (2^{10}=1,024) in the course of evolution from a unicellular organism to the mammals[17]. After each fixation of duplication in the species, irrespective of whether the duplicated part is a whole chromosome or a tiny fraction, the duplicated genes must have differentiated from the original genes through nucleotide substitution. It is possible that in this process many mutations which would have been deleterious before duplication become harmless (selectively neutral) if they occurred after duplication, for one set of genes provides the essential function of the organism. The originally deleterious mutants will spread into the population by random drift, so that there will be degeneration in many duplicated genes. The possibility of DNA degeneration after gene duplication has been pointed out by Ohno *et al.*[18] and Nei[17]. The probability of such degenerated parts acquiring new functions might be quite small. Some duplicated genes will acquire new functions, establishing themselves as important genes of the organism. But mutation is a random event, so that the chance that duplicated genes will acquire a new function must be much smaller than that they will become inert. Thus, we should expect that higher organisms have much non-informational (inert) DNA in their genome.

In this case, the rate of nucleotide substitution must be very rapid within those parts, for all the mutations are neutral. We estimate this rate approximately as $1/3 \times 8.3 \times 10^{-9}$/ yr/nucleotide site, which is one-third of the estimated rate per amino-acid sites in rapidly evolving portions of the fibrinopeptides. We further multiply this by 1.2 to give total nucleotide substitution rate by taking account of the synonymous mutations[19]. The resulting rate is 0.33×10^{-8}/yr/nucleotide site. This is about five times as high as the rate estimated from the average proteins.

Recent investigations using DNA hybridization techniques indicate that the differentiation of DNA in mammalian evolution is very rapid. Laird *et al.*[20] estimated that the rate of DNA evolution is about three times as fast in artiodactyls and thirty times in rodents as the corresponding rate inferred from haemoglobins. Walker[21] estimated that the differentiation

忍受的突变负荷：马勒认为[14]人类的有害突变率每世代总计至多为0.5。如此大的差异意味着什么？我们推测基因组中的大部分DNA并不是生物体维持生命所必需的，因而碱基替换没有产生任何影响[3]。如果我们将上述总突变率的估计值与马勒的最大有害突变率相比较，就可以得到以下结论：在全部DNA中，只有0.5/8或6%与顺反子（如编码细胞色素c或血红蛋白的顺反子）在功能上具有同等的重要性。如果估计出的每世代总突变率是每年突变率的20倍，且总有害突变率是1.0（考虑到向井辉美的“活性多基因”）[15]，那么上述比例就变成了0.6%。尽管我们和马勒[16]都确信人类“基因”的总数目大概是3×10^4，不过他认为基因很大（含30,000个氨基酸），而我们认为基因的大小要小很多（平均只含几百个氨基酸），但大部分DNA是非“编码”的。

哺乳动物的DNA大约是细菌DNA的1,000倍，所以在从单细胞生物进化到哺乳动物的过程中，遗传物质平均要复制10次左右（2^{10}=1,024）[17]。在每次对物种中的复制进行固定之后，不论复制的部分是一整条染色体还是一个微小的片段，复制的基因都必然是通过核苷酸替换从最初的基因分化而来的。在这一过程中，许多复制之前是有害的突变在发生复制之后可能反而变成无害了（选择中性），这是因为其中有一套基因为生物体提供了必要的功能。原来有害的突变体将通过随机漂移扩散到整个群体之中，因此在许多基因副本中会出现退化现象。大野乾等人[18]和根井正利[17]曾经指出，基因复制之后可能会出现DNA退化现象。这些退化部分获取新功能的概率可能会非常小。有些基因副本获得了新功能，成为了生物体的重要基因。但突变是一个随机事件，因而基因副本获取新功能的概率一定会比它们变成无义突变的概率小很多。因此，我们应当认为高等生物的基因组中有很多非编码（无义）的DNA。

在这种情况下，由于非编码DNA部分的所有突变都是中性的，所以它们的核苷酸替换速率一定很快。我们估计这个速率大致为每个核苷酸位点每年$1/3 \times 8.3 \times 10^{-9}$，该值是血纤维蛋白肽快速进化部分每个氨基酸位点的替换速率估计值的1/3。考虑到同义突变[19]，我们再将该值乘以1.2以给出总的核苷酸替换速率。最终得到的速率为每个核苷酸位点每年0.33×10^{-8}。这大概是从几种蛋白质估计出的平均速率的5倍。

最近使用DNA杂交技术进行的研究表明，在哺乳动物的进化过程中DNA发生分化的速率很快。莱尔德等人[20]根据从血红蛋白突变速率推导出的相应速率估计出：哺乳动物DNA的进化速率大概是偶蹄动物的3倍，是啮齿类动物的30倍。沃

of DNA of rodents is fifteen times as fast as that of known proteins. It is possible that the evolutionary rate of total DNA is more nearly proportional to generations rather than to chronological time. More recently, Kohn[22], who has summarized these results, emphasizes that the substitution rate in DNA of mammalian genome is proportional to generations. He concludes that the number of nucleotide substitutions in non-repeated DNA is roughly five per genome per generation, and the total rate of substitutions is probably two to four times greater. It is interesting that his figure is roughly equal to our estimate, that is, eight substitutions per genome per year.

We thank Dr. Kazutoshi Mayeda for reading out manuscript and correcting the English.

(**233**, 118-119; 1971)

Tomoko Ohta and Motoo Kimura: National Institute of Genetics, Mishima, Shizuoka-ken, 411.

Received December 2, 1970; revised March 17, 1971.

References:

1. Kimura, M., *Proc. US Nat. Acad. Sci.*, **63**, 1181 (1969).
2. Kimura, M., *Nature*, **217**, 624 (1968).
3. King, J. L., and Jukes, T. H., *Science*, **164**, 788 (1969).
4. McLaughlin, P. J., and Dayhoff, M. O., in *Atlas of Protein Sequence and Structure 1969* (edit. by Dayhoff, M. O.), 39 (National Biomedical Research Foundation, Silver Spring, Maryland, 1969).
5. Crow, J. F., and Kimura, M., *An Introduction to Population Genetics Theory* (Harper and Row, New York, 1970).
6. Dayhoff, M. O., *Atlas of Protein Sequence and Structure* (National Biomedical Research Foundation, Silver Spring, Maryland, 1969).
7. Corbin, K. W., and Uzzell, T., *Amer. Nat.*, **104**, 37 (1970).
8. Kimura, M., *Jap. J. Genet.*, **37** (Suppl.), 179 (1961).
9. Crow, J. F., in *Population Biology and Evolution* (edit. by Lewontin, R.), 71 (Syracuse University Press, New York, 1968).
10. Mukai, T., *Genetics*, **61**, 749 (1969).
11. Kimura, M., and Maruyama, T., *Genetics*, **54**, 1337 (1966).
12. Muller, H. J., *Bull. Amer. Math. Soc.*, **64**, 137 (1958).
13. Vogel, F., *Nature*, **201**, 847 (1964).
14. Muller, H. J., *Amer. J. Human Genet.*, **2**, 111 (1950).
15. Mukai, T., *Genetics*, **50**, 1 (1964).
16. Muller, H. J., in *Heritage from Mendel* (edit. by Brink, R. A.), 419 (University of Wisconsin Press, Madison, 1967).
17. Nei, M., *Nature*, **221**, 40 (1969).
18. Ohno, S., Wolf, U., and Atkin, N. B., *Hereditas*, **59**, 169 (1968).
19. Kimura, M., *Genet. Res.*, **11**, 247 (1968).
20. Laird, C. D., McConaughy, B. L., and McCarthy, B. J., *Nature*, **224**, 149 (1969).
21. Walker, P. M. B., *Nature*, **219**, 228 (1968).
22. Kohn, D. E., *Quart. Rev. Biophys.*, **3** (3), 327 (1970).

克[21]估计啮齿类动物的DNA分化速率是已知蛋白质的15倍。总DNA的进化速率可能更接近于与世代时间成正比，而非与年代时间成正比。就在前不久，总结出上述结果的科恩[22]强调指出：哺乳动物基因组的DNA替换速率与世代成正比。他推测：在非重复DNA中，核苷酸替换的数目大概为每个基因组每世代5次，总的替换速率可能是这一数值的2～4倍。有趣的是，他得到的数值与我们的估计值大致相等，即每个基因组每年发生8次替换。

感谢前田和俊（译者注：音译）博士通读了我们的手稿并对英文进行了修改。

（刘皓芳 翻译；陈新文 陈继征 审稿）

Covalently Linked RNA–DNA Molecule as Initial Product of RNA Tumour Virus DNA Polymerase

I. M. Verma *et al.*

Editor's Note

A year earlier, Nobel laureate David Baltimore co-discovered the enzyme reverse transcriptase, which converts messenger RNA into DNA and thus enables RNA-encoded viruses to integrate with the host genome. Baltimore and his coworkers show here that the same enzyme from an avian tumour virus converts RNA into a chemically bound DNA–RNA species, which they suggest contains a small RNA "primer" molecule. The primer, they correctly speculate, is a host-encoded transfer RNA. The study hinted that primers could be used to kick start the production of genes from messenger RNAs in the laboratory, a reaction that was to prove important to the development of the biotechnology industry.

THE DNA polymerase found in virions of the RNA tumour viruses[1,2] can be assayed in two ways. If disrupted virions are incubated without addition of a template, the endogenous viral RNA is copied by the DNA polymerase[3-5]. If exogenous templates are provided, often these are copied at a much higher rate than the endogenous RNA[6-12]. With exogenous templates, however, the DNA polymerase requires the presence of a homologous polynucleotide primer to initiate polymerization of nucleotides[10]. The primer, which can be as short as a tetranucleotide[11,12], is physically incorporated into the product (Smoler, Molineaux and Baltimore, submitted for publication). The primer requirement of the enzyme, demonstrated with exogenous polynucleotide templates, suggests that in the absence of such templates, when the enzyme copies the endogenous 60–70S viral RNA, a primer might also be present to initiate polymerization.

The initial reaction product formed when the virion DNA polymerase copies the endogenous viral RNA consists of small pieces of DNA attached to the 60–70S RNA[3,4,13-15]. Analysis by buoyant density in Cs_2SO_4 indicated that the DNA product could be released from the viral RNA by procedures which disrupt hydrogen bonds[3].

Further analysis of the product released from the viral RNA by heat treatment has now revealed that the material is not free DNA. After 30 min or less of reaction the product contains molecules which behave like DNA–RNA duplexes even after heat denaturation. One interpretation of this result is that the initial product of the reaction might be a covalently linked DNA–RNA molecule and that the primer for initiation of synthesis might be an RNA species.

The initial observation was with Moloney mouse leukaemia virus (MLV). After 20 min of incubation, the product of the virion DNA polymerase reaction was extracted with

RNA肿瘤病毒DNA聚合酶的初产物是通过共价键连接的RNA–DNA分子

维尔马等

编者按

一年前，诺贝尔奖获得者戴维·巴尔的摩和其他研究者各自独立发现了逆转录酶，这种酶能将信使RNA转化为DNA，由此使得RNA病毒基因能够整合到宿主基因组。巴尔的摩和他的同事在本文中展示了一种来自禽肿瘤病毒的逆转录酶，这种酶可以将RNA转化为通过化学键连接的DNA–RNA分子，他们认为此DNA–RNA分子含有一个小的RNA“引物”分子。巴尔的摩等人正确地推测出此引物是一个宿主编码的转运RNA。本项研究暗示，在实验室中这些引物可以用于启动从信使RNA开始的基因生产，此反应在生物科技产业发展中的重要性现已被证实。

可以通过两种途径检测出在RNA肿瘤病毒的病毒体中发现的DNA聚合酶[1,2]。如果破裂的病毒体在没有额外模板的条件下培养，那么内源病毒RNA就会通过这种DNA聚合酶进行复制[3-5]。如果提供外源模板，通常这些模板的复制速率会远高于内源RNA[6-12]。虽然如此，在提供外源模板时，DNA聚合酶仍需在同源多聚核苷酸引物存在的条件下才能启动核苷酸聚合[10]。该引物最小为4个核苷酸[11,12]，它会与聚合产物物理混合在一起（斯莫勒、莫里纽克斯和巴尔的摩，已提请出版）。此酶对引物的需求表明：在没有外源多聚核苷酸模板的条件下，当酶复制内源60–70S病毒RNA时，可能也存在引物以启动聚合反应。

当病毒体DNA聚合酶复制这些病毒内源性的RNA时，会形成许多附在60–70S RNA上的小片段DNA，这就是最初的反应产物[3,4,13-15]。Cs_2SO_4溶液浮力密度分析结果显示，DNA产物可以通过切断氢键的过程从病毒RNA里释放出来[3]。

通过热处理对从病毒RNA中释放的产物做进一步分析，现已揭示该物质不是自由DNA。经过30 min或者更短时间的反应，产物中含有表现得很像DNA–RNA双链体的分子，即使经过热变性处理，这种情况仍然存在。对这个结果的一种解释是：此反应的初产物也许是一个通过共价键连接的DNA–RNA分子，用于启动该合成的引物可能是一种RNA。

最初的观测基于莫洛尼氏小鼠白血病病毒（MLV）。培养20 min后，用“肌氨酰”提取病毒体DNA聚合酶反应的产物，再用“葡聚糖凝胶G-50”柱提纯，煮沸使二

"Sarkosyl", purified on a "Sephadex G-50" column, boiled to denature the secondary structure and centrifuged to equilibrium in a Cs_2SO_4 gradient. The principal component of the boiled product banded at a density slightly greater than denatured P22 phage DNA (Fig. 1*a*). Some material of greater density was also evident. After treatment with alkali (Fig. 1*b*), ribonuclease in low salt (Fig. 1*c*) or ribonuclease in 0.3 M NaCl plus 0.3 M Na citrate (Fig. 1*d*), the product DNA banded coincidentally with the denatured P22 DNA although the polymerase product formed a much wider band than the marker. The effects of alkali and ribonuclease indicate that RNA was attached to the product DNA even after boiling and was causing the DNA to band at a higher density than the marker phage DNA.

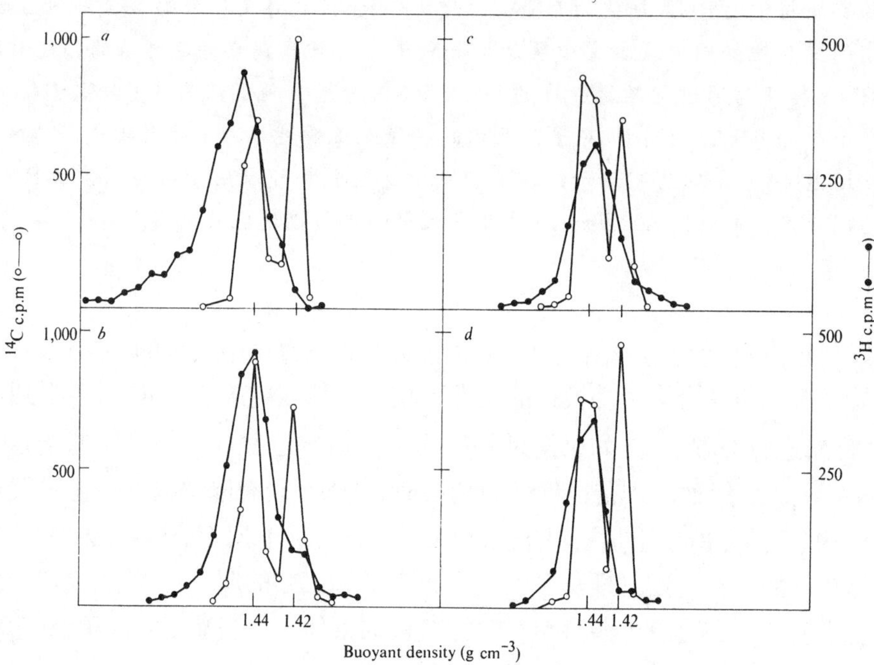

Fig. 1. Analysis in Cs_2SO_4 gradients of the product of the mouse leukaemia virus DNA polymerase. A 20 min reaction was carried out in standard conditions[13] with virions of Moloney mouse leukaemia virus and ^{3}H-TTP. The fast sedimenting fraction was separated on sucrose gradients, collected by precipitation with ethanol, and dissolved in 0.01 M Tris-HCl (*p*H 7.6), 0.01 M NaCl. Portions were placed in a boiling water bath for 5 min and chilled. Parts of the heated samples were treated for 10 min at 37°C in 0.4 ml. with 16 μg of pancreatic ribonuclease (Worthington) plus 3 μg of T_1 ribonuclease (Calbiochem, 5,000 U/mg) contained either in 0.01 M Tris-HCl (*p*H 7.6), 0.01 M NaCl (low salt) or 0.3 M NaCl, 0.03 M Na citrate (high salt). A separate portion of the product was adjusted to 0.3 M NaOH and placed in a boiling water bath for 5 min, chilled and neutralized with HCl. For analysis, the treated samples were mixed with 1.40 ml. of 0.01 M Tris-HCl (*p*H 7.4), 0.001 M EDTA saturated at 24°C with Cs_2SO_4 and enough Tris-EDTA buffer to make 3.0 ml. The samples were placed in polyallomer tubes, a cushion of 0.15 ml. of saturated Cs_2SO_4 was carefully placed at the bottom and the top was covered with "Nujol". The solutions were centrifuged for 65–70 h at 33,000 r.p.m. and 23°C in a Spinco SW 50.1 rotor in a model L ultracentrifuge. All the gradients contained ^{14}C-labelled heat-denatured and native P22 DNA as standard markers with buoyant densities of 1.44 and 1.42 respectively[13]. Three to five drop fractions were collected from the bottom of the tube, acid-precipitated, collected on filters as previously described[13] and counted in a mixture of 5 g PPO and 100 g naphthalene in 1 l. of p-dioxane. *a*, Boiled product; *b*, alkali-treated product; *c*, boiled product after ribonuclease treatment in low salt; *d*, boiled product after ribonuclease treatment in high salt.

级结构变性，然后用 Cs_2SO_4 密度梯度离心达到平衡。煮沸后产物的主要成分在密度略高于变性后 P22 噬菌体 DNA 的地方形成条带（图 1*a*）。存在某种密度更高的物质也是显而易见的。用碱液处理（图 1*b*）、用溶于低浓度盐溶液的核糖核酸酶处理（图 1*c*）或用溶于 0.3 M NaCl 外加 0.3 M 柠檬酸钠溶液的核糖核酸酶处理之后（图 1*d*），产物 DNA 恰好与变性 P22 噬菌体 DNA 聚集在一起，不过聚合酶产物形成的条带要比标志基因宽很多。碱液和核糖核酸酶的作用结果表明：即使在煮沸后，RNA 仍与产物 DNA 结合在一起，并使 DNA 在比标志物噬菌体 DNA 密度更高的地方形成条带。

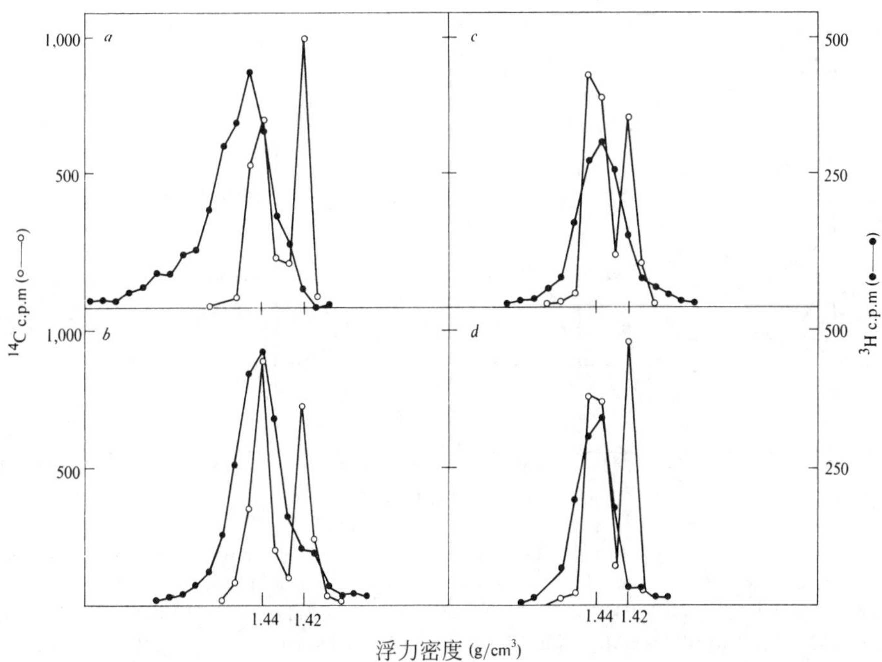

图 1. 用 Cs_2SO_4 梯度分析小鼠白血病病毒 DNA 聚合酶的反应产物。在标准条件下[13]，莫洛尼氏小鼠白血病病毒的病毒体和 ^{3}H– 胸腺嘧啶核苷三磷酸反应 20 min。用蔗糖梯度法分离快速沉降级分，再用乙醇沉淀法收集这些级分，然后溶解在 0.01 M Tris-HCl（pH 7.6）和 0.01 M NaCl 溶液中。取部分置于沸水浴中 5 min，随后快速冷却。在 37℃条件下，将两份加热后的样品分别加入 0.4 ml 含有 16 μg 胰核糖核酸酶（沃辛顿）外加 3 μg T_1 核糖核酸酶（卡尔生物化学公司，5,000 U/mg）的 0.01 M Tris-HCl（pH 7.6）、0.01 M NaCl（低盐）溶液或 0.3 M NaCl、0.03 M 柠檬酸钠（高盐）溶液中处理 10 min。将独立的一份产物调整到 NaOH 浓度为 0.3 M 并置于沸水浴中 5 min，然后快速冷却并用 HCl 中和。为了分析，将处理过的样品同 1.40 ml 0.01 M Tris-HCl（pH 7.4）、24℃下 Cs_2SO_4 饱和的 0.001 M EDTA 溶液以及足量 Tris-EDTA 缓冲液混合，以使溶液体积达到 3.0 ml。将样品放置在异质同晶聚合物管里，小心地将 0.15 ml 饱和 Cs_2SO_4 垫在管底部，顶部用“纽加尔”美国药典矿物油覆盖。将溶液置于 L 型超速离心机的斯平科 SW 50.1 转子中，在 23℃下以每 min 转数 33,000 的速度离心 65 h ~ 70 h。所有梯度中都含有 ^{14}C 标记的经热变性处理的 P22 噬菌体 DNA 和原生 P22 噬菌体 DNA，这两种 DNA 作为标准标志，浮力密度分别为 1.44 和 1.42[13]。从管底部收集 3 到 5 滴级分，用酸沉淀，以之前发表的文章中描述过的过滤器[13]进行收集，然后在含 5 g 多酚氧化酶和 100 g 萘混合物的 1 L 对二氧杂环己烷中计数。*a*，煮沸过的产物；*b*，碱液处理过的产物；*c*，在低盐条件下用核糖核酸酶处理的煮沸产物；*d*，在高盐条件下用核糖核酸酶处理的煮沸产物。

To confirm and extend these data the reaction product of the avian myeloblastosis virus (AMV) DNA polymerase was investigated because larger amounts of this virus were available. The endogenous product formed after 2, 5, 10, 15 and 30 min of reaction was banded in Cs_2SO_4 in its native state, after boiling and after alkali treatment.

The native products consisted of some DNA banding coincidentally with an RNA marker, at a density of 1.65 (Fig. 2*a*). Increasing amounts of DNA banding at lower density appeared as incubation time increased. These complexes have been described previously[3,4,13-15].

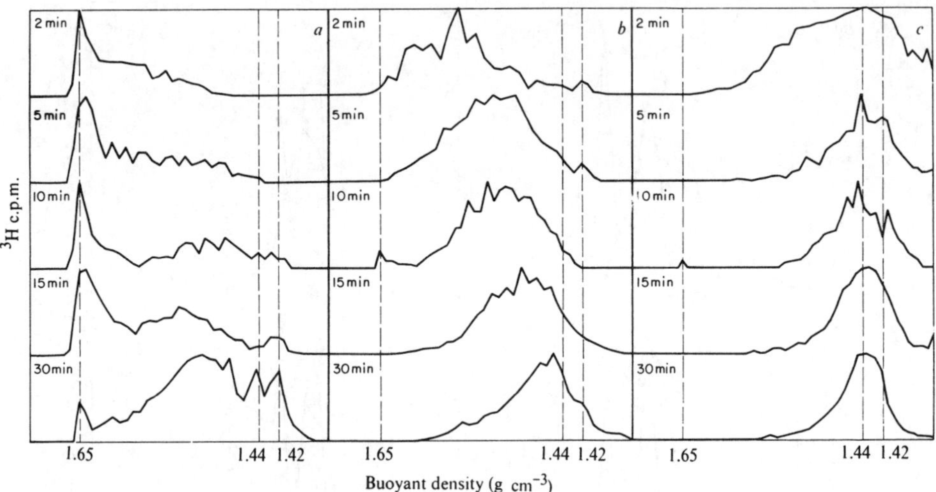

Fig. 2. Analysis in Cs_2SO_4 gradients of the product of the avian myeloblastosis virus DNA polymerase after various times of reaction. The products were prepared using the following 0.2 ml. reaction mixture: 50 mM Tris-HCl (*p*H 8.3), 6 mM magnesium acetate, 20 mM dithiothreitol, NaCl, 1 mM dATP, 1mM dGTP, 1 mM dCTP, 6.5 μM ^{3}H-dTTP (6,000 c.p.m./pmol), 0.2% "Nonidet P-40" and 1–2 mg of AMV protein in purified virions[10]. Samples were incubated at 37°C for the given length of time and the reaction terminated by the addition of an equal volume of 3% sodium dodecyl sarcosinate. The samples were further incubated for 5 min at 37°C. Portions were withdrawn from the samples to measure acid-insoluble material. The 2, 5, 10, 15 and 30 min products in a typical experiment had 51,000, 165,000, 280,000, 400,000 and 500,000 total c.p.m. respectively. The products were then separated from low molecular weight material by chromatography on "Sephadex G-50" columns using 0.1 M NH_4HCO_3 buffer, *p*H 8.0, for elution. The recovery of the purified product was over 70%. For heat denaturation, 0.01–0.05 ml. of the product was diluted to 0.5 ml. with 0.01 M Tris-HCl, *p*H 7.6, and 0.01 M NaCl, placed in a boiling water bath for 5 min and cooled rapidly. For alkaline hydrolysis, portions in the same buffer containing 0.3 M NaOH were placed in a boiling water bath for 5 min, chilled and neutralized with HCl to *p*H 7.0. Cs_2SO_4 gradients containing 8,000–12,000 c.p.m. of products were prepared and analysed as in Fig. 1. ^{14}C-18S ribosomal RNA was also included as a marker. Recoveries from the gradients containing native material were 20–50% and from the other gradients were 55–80%. Drops from the gradient tubes were collected directly into scintillation fluid. *a*, Native products; *b*, boiled products; *c*, alkali-treated products.

The heat-denatured product after 2 min of incubation (Fig. 2*b*) banded heterogeneously with most of the DNA heavier than 1.5 g/ml. which is the density of a 1:1 DNA · RNA hydrogen-bonded duplex[16]. With increasing time of incubation, the heterogeneous band

为了确认和扩展这些数据，我们对禽成髓细胞瘤病毒（AMV）DNA 聚合酶的反应产物进行了研究，因为可以很方便地获得更多量的这种病毒。将反应 2 min、5 min、10 min、15 min 和 30 min 后形成的内源产物，分别以原生、经过煮沸和经过碱液处理的状态放入 Cs_2SO_4 溶液中，聚集成条带。

原生产物中含有某种 DNA，这种 DNA 碰巧与一种 RNA 标志物聚集在一起，在密度为 1.65 处形成条带（图 2*a*）。随着培养时间的增加，聚集在低密度处的 DNA 的数量看上去也在增加。前人已经在文章中描述过这些复合物[3,4,13-15]。

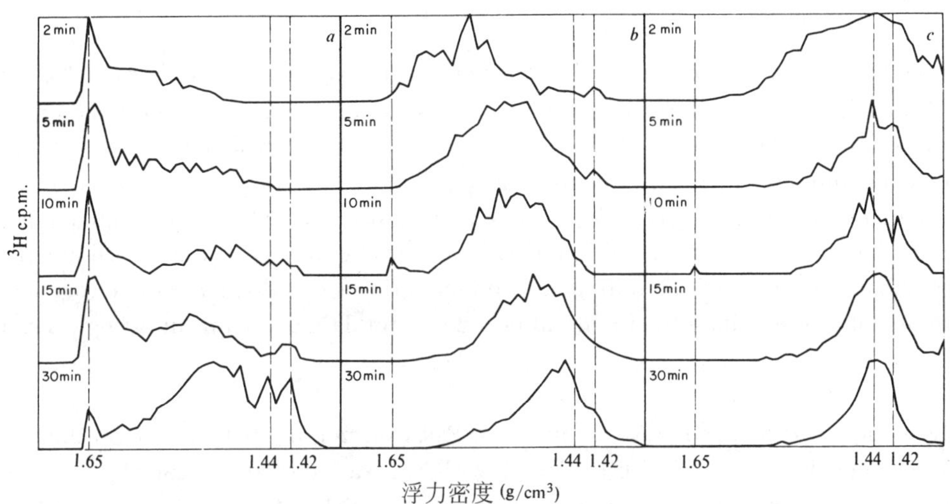

图 2. 用 Cs_2SO_4 梯度分析不同反应时间段禽成髓细胞瘤病毒 DNA 聚合酶的产物。使用 0.2 ml 的以下反应混合物制备产物：50 mM Tris-HCl（pH 8.3）、6 mM 醋酸镁、20 mM 二硫苏糖醇、NaCl、1 mM 脱氧腺苷三磷酸、1 mM 脱氧鸟苷三磷酸、1 mM 脱氧胞苷三磷酸、6.5 μM ^3H– 脱氧胸苷三磷酸（6,000 c.p.m./pmol）、0.2%“诺乃洗涤剂 P-40”和 1 mg ~ 2 mg 提纯病毒体中的 AMV 蛋白[10]。在 37℃条件下按给定时长培养样品，加入等体积的 3% 十二烷基肌氨酸钠终止反应。接下来样品在 37℃继续培养 5 min。从样品中取出一部分用于测量不溶于酸的物质。反应 2 min、5 min、10 min、15 min 和 30 min 后的产物在典型实验中的总 c.p.m. 值分别为 51,000、165,000、280,000、400,000 及 500,000。使用“葡聚糖凝胶 G-50”柱层析，用 0.1 M NH_4HCO_3 缓冲液（pH 8.0）洗脱，从而将产物从低分子量物质中分离出来。纯化产物的回收率超过 70%。热变性处理：取 0.01 ml ~ 0.05 ml 产物用 0.01 M pH 为 7.6 的 Tris-HCl 和 0.01 M NaCl 稀释至 0.5 ml，置于沸水浴中 5 min，然后迅速冷却。碱水解：将部分产物置于含 0.3 M NaOH 的同种缓冲液中，沸水浴 5 min，快速冷却后用 HCl 中和至 pH 为 7.0。按图 1 中的方法制备和分析含 c.p.m. 为 8,000 ~ 12,000 的产物的 Cs_2SO_4 梯度。^{14}C 标记的 18S 核糖体 RNA 也被包含在其中作为标志物。含有原生物质的梯度的回收率为 20% ~ 50%，其他梯度的回收率为 55% ~ 80%。将梯度管中的滴液直接收集到闪液中。*a*，原生产物；*b*，煮沸过的产物；*c*，碱液处理过的产物。

培养 2 min 后，热变性产物（图 2*b*）与大多数密度比 1.5 g/ml 高的 DNA 聚集在一起形成异质性条带，其密度与 DNA 和 RNA 按 1:1 比例以氢键结合的双链体相同[16]。随着培养时间的增加，这个异质性条带移动到梯度中更轻的位置。培养 30 min 的产

moved to lighter positions in the gradient. By 30 min the product resembled that seen after 20 min of reaction by the MLV DNA polymerase.

Treatment of the DNA polymerase products with alkali in conditions which hydrolyse all RNA caused them to band in Cs_2SO_4 at a lower average density than the boiled product (Fig. 2*c*). The bands were quite broad, indicating a low molecular weight. The samples from later times of incubation gave a sharper band which centred at a density slightly lighter than the denatured P22 phage DNA. The earliest sample gave an especially board band but still the alkali-treated product was on the average less dense than the boiled product.

To show that the boiled product did not spontaneously renature, the effect of the single-strand-specific nuclease from *Neurospora** was investigated. A portion of the boiled product from a 15 min incubation of AMV was treated with *Neurospora* nuclease in previously specified conditions which only degrade single stranded regions of nucleic acids[13]. The treated material was then centrifuged to equilibrium in Cs_2SO_4 as was an equivalent amount of boiled but undigested material. The nuclease-treated product contained only 4% of the radioactivity of the untreated product and this was distributed heterogeneously. Therefore, after the boiling treatment, almost all of the DNA remains in a single stranded state.

To see if the 60–70S viral RNA contained an RNA primer, the reaction of purified AMV DNA polymerase with isolated 60–70S AMV RNA was studied. A reaction was carried out for 15 min and portions of the product were centrifuged to equilibrium in Cs_2SO_4 either without any treatment, after boiling, or after alkaline hydrolysis. As with the early endogenous product, most of this product banded with RNA at $\rho=1.65$ g cm^{-3} before treatment (Fig. 3*a*). Boiling converted it to material banding between RNA and DNA (Fig. 3*b*) and alkaline hydrolysis caused it to form a wide band centring slightly lighter than the P22 denatured DNA (Fig. 3*c*). The product formed when purified enzyme is allowed to copy 60–70S viral RNA therefore behaves similarly to the product of the endogenous reaction, supporting the idea that these two systems are comparable.

* In a previous paper[13] this enzyme was called the *Neurospora* endonuclease and was thought to be identical to the enzyme prepared by Linn and Lehman[20]. Dr. M. Fraser, who prepared the nuclease used in the previous studies, has informed us that the nuclease is a different enzyme from the Linn and Lehman enzyme. It is an exonuclease combined with some endonuclease activity but has very high specificity for single stranded regions of nucleic acid like the Linn and Lehman enzyme. Using a modification of the procedure of Rabin, Preiss and Fraser[21], we have isolated the enzyme used in the present experiments and have confirmed its high specificity for single stranded nucleic acid as previously reported[13].

物类似于用 MLV DNA 聚合酶反应 20 min 后观察到的产物。

经碱液处理水解掉所有 RNA 后，DNA 聚合酶产物在 Cs_2SO_4 中平均密度比煮沸产物低的位置形成条带（图 2*c*）。这些条带非常宽，说明分子量很小。培养后期的样品给出了一个更窄的条带，其中心在密度比变性 P22 噬菌体 DNA 略低的位置。培养最早期的样品所给出的条带特别宽，但平均看来，碱液处理产物的密度依然低于煮沸产物。

为了说明煮沸产物并没有自发复性，我们对来自脉孢菌 * 的单链特异性核酸酶的效应进行了研究。取一份培养了 15 min 的 AMV 的煮沸产物，在此前只降解核酸单链区的指定条件下 [13] 用脉胞菌核酸酶处理。之后将这些处理过的物质和一份等量的煮沸过但未经消化的物质分别在 Cs_2SO_4 中离心至达到平衡。核酸酶处理过的产物只具有未处理产物 4% 的放射性，并且分布也是异质的。因此，在煮沸处理后，几乎所有的 DNA 仍保持在单链状态。

为了检验 60–70S 的病毒 RNA 中是否包含 RNA 引物，我们对提纯的 AMV DNA 聚合酶与分离的 60–70S AMV RNA 的反应进行了研究。反应 15 min 后，将未经过任何处理的、煮沸后的或经过碱液水解的各份产物分别在 Cs_2SO_4 中离心达到平衡。因为混有早期内源产物，所以在处理前，大多数这种产物都与 RNA 一起在密度为 1.65 g/cm^3 处形成条带（图 3*a*）。煮沸使得这种产物转变为条带位于 RNA 和 DNA 之间的物质（图 3*b*），碱液水解使它形成一条宽带，其中心在比 P22 变性 DNA 略轻的位置上（图 3*c*）。用提纯过的酶复制 60–70S 病毒 RNA 时所形成的产物表现得类似于内源反应产物，该事实支持了外源系统与内源系统具有相似性的观点。

* 在之前的一篇论文中 [13]，这种酶被称为脉孢菌核酸内切酶，并被认为与林和莱曼 [20] 制备的酶相同。此前研究中使用的这种核酸酶是由弗雷泽博士提供的，他告诉我们这种核酸酶与林和莱曼制备的酶并不相同。这是一种带有一些核酸内切酶活性的核酸外切酶，但对核酸单链区具有非常高的特异性，这一点与林和莱曼制备的酶类似。通过改进雷宾、普赖斯和弗雷泽提出的步骤 [21]，我们已经分离出现有实验中使用的这种酶，并已证实它与以前所报道的 [13] 一样，都对单链核酸具有很高的特异性。

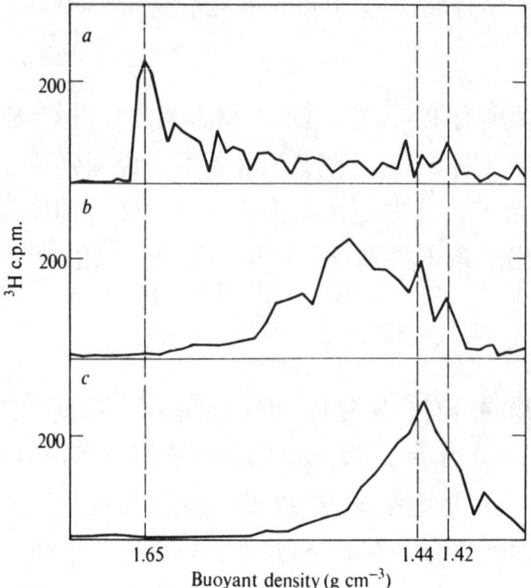

Fig. 3. Analysis in Cs_2SO_4 gradients of the product of purified avian myeloblastosis virus DNA polymerase copying 60–70S viral RNA. The polymerase was purified by disruption of the virions with 0.2% "Nonidet-P40", chromatography on "DEAE–Sephadex" and then chromatography on phosphocellulose. A yield of 44% of the initial activity was obtained using poly(C) · $(dG)_{14}$ as template-primer[10] and a purification of fifty times was achieved (I. M. V. and D. B., unpublished results). The 60–70S RNA was isolated by centrifugation through a sucrose gradient of purified avian myeloblastosis virus[10] disrupted with sodium dodecyl sulphate. The RNA was identified by absorbance at 260 nm and ethanol precipitated. Two μg of DNA polymerase and 5 μg of viral RNA were incubated at 37°C for 15 min in the reaction mixture described in Fig. 2, lacking detergent and virus. The product was purified as in Fig. 2 and contained 25,000 c.p.m. of ^{3}H label. Portions were treated as in Fig. 2 and analysed as in Fig. 1. *a*, Native product; *b*, heat-denatured product; *c*, alkali-treated product.

These date indicate that the initial product of the endogenous DNA polymerase reaction of both MLV and AMV and of the reaction of purified AMV polymerase with AMV 60–70S RNA is a covalently bonded DNA–RNA molecule. Covalent bonding is indicated by the resistance of the product to boiling. The sensitivity of the boiled product to *Neurospora* nuclease and the change of density of the product after treatment with ribonuclease in 0.3 M NaCl indicate that little or no RNA remains hydrogen-bonded to the product DNA after heat denaturation. The ability of alkali and ribonuclease to convert the boiled product to material banding coincidentally with DNA indicates that RNA and not some other material was responsible for the density of the boiled product being greater than that of free DNA. The decreasing density of the boiled product with time of incubation (Fig. 2*b*) coupled with the increasing size of the DNA product indicated by its decreasing band width (Fig. 2*c*) suggest that a constant size piece of RNA is attached to a growing DNA strand.

The RNA in the product presumably acts as a primer for the endogenous reaction. Experiments utilizing the inhibition of the polymerase reaction by dideoxythymidine

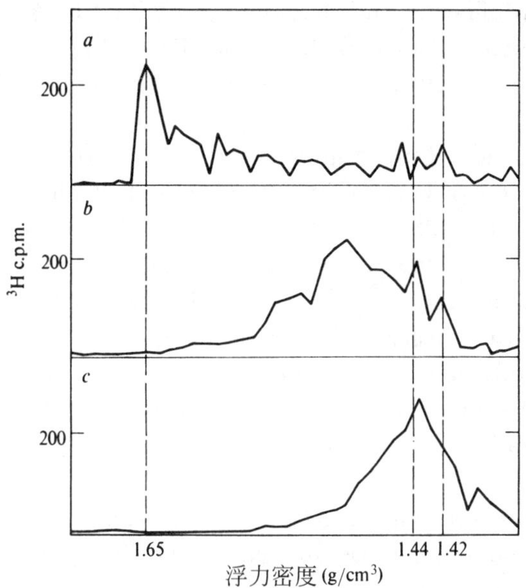

图 3. 用 Cs_2SO_4 梯度分析经提纯的禽成髓细胞瘤病毒 DNA 聚合酶复制 60–70S 病毒 RNA 的产物。先用 0.2%“诺乃洗涤剂 –P40”破坏病毒体，再经“DEAE– 葡聚糖”层析，最后用磷酸纤维素层析，以提纯聚合酶。使用聚（C)·(dG)$_{14}$ 作为模板引物 [10]，获得的活性为初活性的 44%，并达到 50 倍的提纯（英德尔 · 维尔马和戴维 · 巴尔的摩，尚未发表的结果）。先用十二烷基硫酸钠破坏经提纯的禽成髓细胞瘤病毒 [10]，再通过蔗糖梯度离心分离 60–70S RNA。通过 260 nm 紫外吸光率鉴别 RNA，再用乙醇使 RNA 沉淀。2 μg DNA 聚合酶和 5 μg 病毒 RNA 在 37℃条件下培养 15 min，反应混合物中的成分如图 2 注中所述，只是缺少洗涤剂和病毒。产物用图 2 描述的方法提纯，其中含有的 ^{3}H 标记为 25,000 c.p.m.。按图 2 中的方式处理各份产物，并按图 1 的方法进行分析。*a*，原生产物；*b*，热变性的产物；*c*，碱液处理的产物。

这些数据表明：MLV 和 AMV 与内源 DNA 聚合酶反应的初产物，以及经提纯的 AMV 聚合酶与 AMV 60–70S RNA 反应的初产物，都是一种以共价键结合的 DNA–RNA 分子。产物对煮沸的耐受力说明它们是共价结合的。煮沸后产物对脉孢菌核酸酶的敏感性，以及用溶于 0.3 M NaCl 的核糖核酸酶处理后产物密度的变化，都说明：在热变性处理后，很少或几乎没有 RNA 能保持以氢键与产物中的 DNA 相结合。碱液和核糖核酸酶能使煮沸产物变成碰巧与 DNA 共同形成条带的物质，这说明：使煮沸产物的密度大于自由 DNA 的是 RNA，而不是其他物质。随着培养时间的增加，煮沸产物的密度不断减小（图 2*b*)，加之不断变窄的带宽反映出 DNA 产物的体积在不断增加（图 2*c*)，这些都说明有一条大小不变的 RNA 附着在一条持续增长的 DNA 链上。

产物中的 RNA 很可能在内源反应中作为引物。用双脱氧胸苷三磷酸抑制聚合酶反应的实验（斯莫勒、莫里纽克斯和巴尔的摩，已提请出版）表明，DNA 附着在

triphosphate (Smoler, Molineaux and Baltimore, submitted for publication) indicate that the DNA is attached to the 3′-OH of the primer RNA. Experiments are in progress to determine whether unique nucleotides in the DNA and RNA are linked.

The nature of the attached RNA is under investigation. Two models seem most likely. One is that the 60–70S bends back on itself to form a "hair pin" and that DNA synthesis is initiated on an end of 60–70S RNA. The sensitivity of boiled product to the *Neurospora* nuclease argues against this model because a "hair pin" structure should renature spontaneously. The second model invokes a small RNA primer and we favour this idea at present. The existence of transfer RNA species in the 60–70S RNA complex[17] suggests that they might serve this function. Previous evidence that multiple DNA molecules are formed from a single region of viral RNA[13] argues that multiple primer molecules may exist attached to the viral RNA.

The use of an RNA primer for initiating DNA synthesis might occur in systems other than the RNA tumour viruses. For example, an RNA molecule might serve the function of the postulated initiator in bacterial DNA synthesis[18,19]. Such an initiator molecule could be synthesized from a DNA template or could be formed by specific enzymes in the absence of a template.

This work was supported by a grant from the American Cancer Society and a contract from the Special Virus Cancer Program of the National Cancer Institute. I. M. V. was a fellow of the Jane Coffin Childs Memorial Fund for Medical Research, K. F. M. was a fellow of, and D. B. was a faculty research awardee of, the American Cancer Society.

(*Nature New Biology*, **233**, 131-134; 1971)

Inder M. Verma, Nora L. Meuth, Esther Bromfeld, Kenneth F. Manly and David Baltimore: Department of Biology, Massachusetts Institute of Technology, 77 Massachusetts Avenue, Cambridge, Massachusetts 02139.

Received August 19, 1971.

References:

1. Baltimore, D., *Nature*, **226**, 1209 (1970).
2. Temin, H., and Mizutani, S., *Nature*, **226**, 1211 (1970).
3. Spiegelman, S., Burny, A., Das, M. R., Keydar, J., Schlom, J., Travnicek, M., and Waston, K., *Nature*, **227**, 563 (1970).
4. Rokutanda, M., Rokutanda, H., Green, M., Fujinaga, K., Ray, R. K., and Gurgo, C., *Nature*, **227**, 1026 (1970).
5. Duesberg, P. H., and Canaani, E., *Virology*, **42**, 783 (1970).
6. Spiegelman, S., Burny, A., Das, M. R., Keydar, J., Schlom, J., Travnicek, M., and Watson, K., *Nature*, **227**, 1029 (1970).
7. Spiegelman, S., Burny, A., Das, M. R., Keydar, J., Schlom, J., Travnicek, M., and Waston, K., *Nature*, **228**, 430 (1970).
8. Mizutani, S., Boettiger, D., and Temin, H. M., *Nature*, **228**, 424 (1970).
9. Riman, J., and Beaudreau, G. S., *Nature*, **228**, 427 (1970).
10. Baltimore, D., and Smoler, D., *Proc. US Nat. Acad. Sci.*, **68**, 1507 (1971).
11. Baltimore, D., and Smoler, D., *Proc. Third Annual Miami Winter Biochemistry Symp.* (in the press).

引物 RNA 的 3′–OH 端。用以测定 DNA 和 RNA 中相连接的是否是某些特定核苷酸的实验正在进行中。

目前正在研究附着的 RNA 的性质。看起来有两个模型可能性最大。一个模型认为 60–70S 自身会弯曲，形成“发卡”，并且 DNA 合成始于 60–70S RNA 的一端。煮沸产物对脉孢菌核酸酶的敏感性不支持这一模型，因为“发卡”结构应该自发复性。第二个模型引入了一个小的 RNA 引物，目前我们更倾向于这种提法。在 60–70S RNA 复合体中存在转运 RNA 类型 [17] 说明转运 RNA 也许具有这一功能。此前有人发现多种 DNA 分子从病毒 RNA 的某一单区形成 [13]，这一证据表明，也许存在多种附着在病毒 RNA 上的引物分子。

使用 RNA 引物来启动 DNA 合成也许还会发生在除 RNA 肿瘤病毒以外的系统内。比如，在细菌的 DNA 合成中 [18,19]，或许就可以认为 RNA 分子具有启动子的功能。这种启动子分子能由 DNA 模板合成，在模板缺失的条件下也可以通过特异的酶合成。

这项工作得到了美国癌症协会的拨款并与国家癌症研究所特殊癌症病毒项目签订了合同。英德尔·维尔马是简·科芬·蔡尔兹医学研究纪念基金学院的成员，肯尼思·曼利是美国癌症协会的会员，戴维·巴尔的摩是美国癌症协会的一名教授研究奖的获得者。

（邓铭瑞 翻译；陈新文 陈继征 审稿）

12. Baltimore, D., Smoler, D., Manly, K. F., and Bromfeld, E., in Ciba Foundation Symp. *The Strategy of the Viral Genome* (Churchill, London, in the press).
13. Manly, K., Smoler, D. F., Bromfeld, E., and Baltimore, D., *J. Virol.*, 7, 106 (1971).
14. Fujinaga, K., Parsons, J. T., Beard, J. W., Beard, D., and Green, M., *Proc. US Nat. Acad. Sci.*, 67, 1432 (1970).
15. Garapin, A.-C., McDonnell, J. P., Levinson, W. E., Quintrell, N., Fanshier, L., and Bishop, J. M., *J. Virol.*, **6**, 589 (1970).
16. Chamberlain, M., and Berg, P., *J. Mol. Biol.*, **8**, 297 (1964).
17. Erikson, E., and Erikson, R. L., *J. Virol.* (in the press).
18. Jacob, F., and Brenner, S., *CR Acad. Sci.*, **256**, 298 (1963).
19. Jacob, F., Brenner, S., and Cuzin, F., *Cold Spring Harbor Symp. Quant. Biol.*, **28**, 329 (1963).
20. Linn, S., and Lehman, I. R., *J. Biol. Chem.*, **240**, 1287 and 1294 (1965).
21. Rabin, Preiss and Fraser, *Preparative Biochemistry* (in the press).

Titration of Oral Nicotine Intake with Smoking Behaviour in Monkeys

S. D. Glick *et al.*

Editor's Note

By the 1970s, research had hinted that orally administered nicotine could decrease the desire to smoke, and Stanley D. Glick and colleagues had shown that monkeys trained to inhale cigarette smoke puffed less when treated with the nicotinic antagonist mecamylamine. Here Glick's team confirm this effect, and go on to show that combining oral nicotine with mecamylamine induces a profound drop in smoking behaviour. The results raised interesting questions about the role of nicotine in smoking, and the authors suggest nicotinic blockade rather than activation makes large doses of oral nicotine unpleasant. Blockage of nicotinic activating, presumably rewarding, actions of nicotine by mecamylamine would then allow non-rewarding blocking actions to predominate.

SEVERAL studies have indicated that nicotine is an incentive in smoking. Johnston[1] found that heavy smokers reported a pleasurable sensation when given nicotine, whereas nonsmokers reported an unpleasant effect. Lucchesi *et al.*[2] showed that intravenous administration of nicotine diminished the number of cigarettes smoked by heavy smokers. Jarvik *et al.*[3] found that oral nicotine administration produced a very small but significant decrease in the number of cigarettes smoked but because humans were used, they[3] were not able to use large doses of nicotine which might have produced larger decrements in smoking. Previous work in this laboratory with monkeys trained to puff cigarette smoke showed that mecamylamine, a nicotinic-blocking agent, reduced their smoking[4], and the work reported here was conducted to establish more definitely that orally administered nicotine could substitute for nicotine derived from smoking.

Four mature rhesus monkeys were trained to puff cigarette smoke. Only one (Phoebe) had served as a subject in the previous study with mecamylamine. Puffing behaviour was instilled initially by making the monkeys suck on a tube in order to drink. The mouthpiece of a cigarette smoking apparatus[5] was then substituted for the water tube. The smoking apparatus allowed a monkey to smoke lit cigarettes by automatically lighting each cigarette, spacing the cigarettes over time, and sensing changes in pressure as the monkey puffed. When the monkeys had learned to puff, they were so trained that they had to puff but were allowed to choose between smoke and air[6]. This procedure was developed to reduce the large variability from subject to subject inherent in free puffing.

A smoking apparatus was attached to a "Plexiglas" panel covering a 36 cm square opening in the door of each monkey's home cage. Two tubes, one delivering cigarette smoke and a "dummy" tube with access only to air, were also mounted on the panel. A water solenoid

口服尼古丁的摄入量与猴子的吸烟行为

格利克等

编者按

到20世纪70年代时，研究工作已经暗示，口服尼古丁有可能会降低烟瘾。斯坦利·格利克及其同事曾指出：在给予四甲双环庚胺这种尼古丁拮抗剂之后，受过香烟烟雾吸入训练的猴子会减少吸烟。在本文中，格利克的研究小组确认了上述效应，并进一步证实，口服尼古丁与注射四甲双环庚胺相结合会导致吸烟行为的显著下降。这些结果提出了关于尼古丁在吸烟中所起作用的有趣问题，本文作者认为是尼古丁的阻滞作用而非激活作用使口服大剂量尼古丁时产生不快的感受。尼古丁的激活作用可能是有快感的，在它被四甲双环庚胺阻断之后，就会让没有快感的阻滞作用占了上风。

一些研究指出，尼古丁是吸烟的一种诱因性刺激。约翰斯顿[1]发现：重度吸烟者在接受了尼古丁之后会产生一种愉悦的感觉，而不吸烟者则报告说有不舒服的效果。卢凯西等人[2]指出，静脉注射尼古丁可以减少重度吸烟者的吸烟量。亚尔维克等人[3]发现，口服尼古丁可以使吸烟的支数发生很小但有意义的减少。尽管加大尼古丁剂量可能会造成更大的吸烟量减少，但因为是以人为实验对象，所以他们[3]不能使用大剂量的尼古丁。本实验室以前用受过香烟烟雾吸入训练的猴子进行过实验，结果表明，一种尼古丁阻断剂——四甲双环庚胺可以减少猴子的吸烟量[4]。本文所报告的工作旨在更确切地证实口服尼古丁能够替代由吸烟摄入的尼古丁。

通过训练使四只成年恒河猴学会吸香烟的烟雾。其中只有一只（菲比）曾作为实验对象用于之前的四甲双环庚胺研究。这些猴子之所以养成了吸烟的习惯，是因为起初它们只能通过吸管才能喝到水。然后用吸烟装置的烟嘴[5]来替代吸水管。为了让猴子吸到点着的烟，吸烟装置每间隔一定时间就自动点燃一支香烟，并能感知猴子吸气时的压力变化。如果猴子已经学过吸气，就要训练它们必须去吸，但可以选择是吸烟还是吸空气[6]。发展这种方法是为了减少个体与个体之间生来就具有的在自主吸气上的巨大差异。

在每只猴笼的门上开一个面积为36 cm^2的洞口，吸烟装置就装在覆盖于洞口上的一个“普列克斯玻璃”面板上。面板上还装有两根管子，一根管子释放香烟产生的烟雾，另一根为“虚构”管，仅与空气连通。在吸烟管和空气管之间有一个控制

was mounted between the smoke and air tubes. A puff on either the smoke or air tubes preset the solenoid and would release a small amount of water when the monkey touched the solenoid spout. A light mounted above the solenoid signalled the availability of water. A fixed ratio (FR) contingency was added, so that a monkey could be required to puff a specific number of times to obtain water.

The monkeys received all their water during a daily 4 h puffing session. Thirty cigarettes were loaded into the smoking apparatus each morning and the test was started at 1030 h. A new cigarette was automatically lit every 7.5 min. The positions of the smoke and air tubes were interchanged each morning so that side preferences would not develop. Initially, water could be won with a single puff on either the smoke or air tube. The number of puffs for a reward of water was then increased to five, ten, twenty and finally to thirty. The FR 30 schedule was then maintained. Puffs were recorded on Sodeco counters.

All four monkeys preferred smoke to air; that is, although they could get water by puffing on either the smoke or the air tubes, they reliably took more puffs on the smoke tube than on the air tube. When puffing rates on the FR 30 schedule stabilized, the nicotine experiment was begun (except for the monkey used in the previous mecamylamine experiment[4]; the present experiment was started 2 months after completion of the previous experiment).

Various quantities (50 mg to 200 mg) of nicotine tartrate were dissolved in the water which each monkey obtained through puffing. The amount of nicotine on a given day was always dissolved in 700 ml. of water although the monkeys would usually consume only 300–400 ml. during a puffing session. Increasing dosages of nicotine were self-administered in this way with at least 4 days between successive nicotine trials. Each dose was repeated two or three times before starting a higher dose.

The results obtained from such oral nicotine administration are shown in Table 1. The data on days immediately preceding oral nicotine days were used as control data in paired *t* tests. For each monkey, doses of oral nicotine were found which significantly decreased and in some cases reversed the smoke–air preference without significantly affecting the overall puffing rate.

Table 1. Changes in Smoke Preference Induced by Oral Nicotine

	Phoebe		Alex		Ivan		Dot	
Dose (mg)	Total puffs	% pref.	Total puffs	% pref.	Total puffs	% pref.	Total puffs	% pref.
Control	6,103	88.7	4,844	86.4	2,786	87.2	4,322	91.1
50	6,509	87.5			2,594	87.0	3,666	88.3
75	6,466	91.0	5,191	90.0	3,465	87.0	4,437	86.0
100	6,400	92.0			3,453	75.0	4,754	81.5
125	5,784	64.1*	4,798	91.9	2,589	41.7*	4,308	14.8*
175			4,973	75.0				
225			4,623	56.7*				

* Significantly less than control at $P<0.05$.

进水的螺线管。不管是用吸烟管吸气还是用空气管吸气，都会使螺线管具有预设值，并在猴子触及螺线管的喷水口时释放出一小股水。螺线管上方的指示灯可以显示是否有水。还要引入事件发生的某个固定比值（FR），要求猴子需要吸特定的次数才能喝到水。

每天安排一个 4 小时的吸水时段，让猴子获取它们一天所需的所有水量。每天早上将 30 支香烟装入吸烟装置中，测试开始的时间是上午 10 点 30 分。吸烟装置每隔 7.5 分钟会自动点燃一支新的香烟。每天早上都要交换吸烟管和空气管的位置以防形成对某一侧的偏好。起先，吸一口吸烟管或空气管后都能得到水。随后为了喝到水需要吸的次数逐渐增加至 5 次、10 次、20 次，最后达到 30 次。然后维持 FR 比值在 30 次。用索迪科计数器记录吸的次数。

所有四只猴子都更愿意吸烟而不是吸空气；也就是说，尽管它们既可以通过吸吸烟管也可以通过吸空气管来获得水，但它们确实会多吸吸烟管，而不是空气管。一旦为喝到水猴子需要吸的次数稳定在 FR 值 30 次，尼古丁实验就可以开始了（除了那只先前曾用于四甲双环庚胺实验的猴子[4]；本项实验在四甲双环庚胺实验结束 2 个月后开始）。

让每只猴子通过吸气获得溶解于水中的不同量的尼古丁酒石酸盐（50 mg ~ 200 mg）。在某一天所给予的一定剂量的尼古丁总是被溶于 700 ml 水中，不过猴子在一个吸水时段内通常只能消耗 300 ml ~ 400 ml 水。用这种方式让猴子自主服用剂量逐渐增加的尼古丁，两次尼古丁实验之间至少间隔 4 天。针对每个剂量都会重复 2 ~ 3 次实验，然后才开始使用更高的剂量。

由这样的口服尼古丁实验得到的结果见表 1。在配对 t 检验中，用口服尼古丁之前一天的数据作为对照数据。从表中的数据可以看出：对每只猴子来说，都存在一个口服尼古丁的剂量，这个剂量能明显降低甚至在某些情况下会逆转猴子对香烟 - 空气的偏好，而不显著影响猴子的总吸入量。

表 1. 口服尼古丁引起的吸烟偏好的变化

	菲比		亚历克斯		伊万		多特	
剂量	总吸入量	偏好率 %	总吸入量	偏好率 %	总吸入量	偏好率 %	总吸入量	偏好率 %
对照	6,103	88.7	4,844	86.4	2,786	87.2	4,322	91.1
50	6,509	87.5			2,594	87.0	3,666	88.3
75	6,466	91.0	5,191	90.0	3,465	87.0	4,437	86.0
100	6,400	92.0			3,453	75.0	4,754	81.5
125	5,784	64.1*	4,798	91.9	2,589	41.7*	4,308	14.8*
175			4,973	75.0				
225			4,623	56.7*				

* 显著低于对照组，$P<0.05$。

One additional drug trial was conducted with one of the monkeys. Because previous work had shown that mecamylamine inhibited monkeys' smoking, it was of interest to determine whether mecamylamine would antagonize the inhibition of smoking by orally administered nicotine, that is, whether the combination would depress smoking behaviour less than either treatment alone. For this purpose, the monkey used in the previous mecamylamine experiment as well as the more recent oral nicotine experiment was used again. A dose of 125 mg of nicotine tartrate was administered orally as already described. In addition, a dose of 0.8 mg/kg of mecamylamine was given intramuscularly 15 min before the beginning of the same puffing session. This combination was repeated twice. Thereafter, two administrations of the mecamylamine alone were conducted. The results are shown in Table 2. All three treatments—oral nicotine alone, mecamylamine alone and the combination—produced significant inhibition of smoking behaviour. Only mecamylamine alone and the combination significantly decreased overall puffing. Most surprising, however, was the finding that the inhibition of smoking produced by the combination was significantly greater ($P<0.05$) than that produced by either drug treatment alone.

Table 2. Changes in Smoke Preference of Phoebe induced by Oral Nicotine (125 mg) and/or Mecamylamine (0.8 mg/kg)

	Total puff	% pref.
Control	6,055	88.2
Nicotine	5,784	64.1*
Mecamylamine	4,842*	65.5*
Nicotine+mecamylamine	3,358*	33.5*

* Significantly less than control at $P<0.05$.

These results raise interesting questions about the role of nicotine in smoking. If oral nicotine depressed smoking by providing a greater than optimal level of nicotine, such that additional nicotine derived from smoking was aversive, then mecamylamine should have antagonized rather than potentiated this effect. However, different central actions of nicotine may have been responsible. Perhaps the aversiveness produced by the large oral nicotine doses was a function of nicotinic blockade rather than of activation.[7] Blockage of nicotinic activating, presumably rewarding, actions of nicotine by mecamylamine would then allow non-rewarding blocking actions to predominate. Clinical studies are being conducted in this laboratory to evaluate the therapeutic significance of this hypothesis.

This work was supported by grants from the US National Institute of Mental Health and the American Cancer Society.

(**233**, 207-208; 1971)

S. D. Glick*, B. Zimmerberg and M. E. Jarvik: Department of Pharmacology and Psychiatry, Albert Einstein College of Medicine, 1300 Morris Park Avenue, Bronx, New York 10461.

Received June 21, 1971.

* Present address: Department of Pharmacology, Mount Sinai School of Medicine, New York, NY, 10029.

我们还对其中一只猴子进行了额外的药物实验。因为之前的实验结果已经表明四甲双环庚胺会抑制猴子吸烟，所以我们很想知道四甲双环庚胺会不会拮抗口服尼古丁对吸烟的抑制，也就是说，这两种处理相结合对吸烟行为的抑制作用是否会比单独给予任意一种时更小。出于这个目的，那只在以前的实验中接受过四甲双环庚胺并在最近的实验中口服过尼古丁的猴子被再次使用。按照前述方法，让这只猴子口服 125 mg 的尼古丁酒石酸盐。另外，在开始同一吸烟实验之前 15 分钟时，给猴子肌肉注射剂量为 0.8 mg/kg 的四甲双环庚胺。这种联合用药的方式被重复了两次。然后再让猴子只注射四甲双环庚胺两次。实验结果见表 2。三种处理方式——仅口服尼古丁、仅注射四甲双环庚胺和两者同时使用都产生了对吸烟行为的显著抑制作用。只有在单独使用四甲双环庚胺和同时使用两者时会显著减少总吸入量。然而，最令人惊讶的是，我们发现在两者合用时所产生的对吸烟的抑制作用要远远大于（$P<0.05$）仅使用其中某一种处理时的效果。

表 2. 口服尼古丁（125 mg）和 / 或四甲双环庚胺（0.8 mg/kg）引起的猴子菲比的吸烟偏好变化

	总吸入量	偏好率 %
对照	6,055	88.2
尼古丁	5,784	64.1*
四甲双环庚胺	4,842*	65.5*
尼古丁 + 四甲双环庚胺	3,358*	33.5*

* 显著低于对照组，$P<0.05$。

这些结果提出了关于尼古丁在吸烟中所起作用的有趣问题。如果口服尼古丁对吸烟的抑制作用是由于提供了超过最佳水平的尼古丁，因而通过吸烟得到的额外的尼古丁就会使人产生不快的感觉，那么四甲双环庚胺应该是拮抗而不是加强了这种作用。然而，尼古丁的其他重要功用也有可能是产生原因。或许由口服大剂量尼古丁所产生的不快的感觉代表的是尼古丁的阻滞作用而不是激活作用[7]。尼古丁的激活作用可能是有快感的，在它被四甲双环庚胺阻断之后，就会让没有快感的阻滞作用占了上风。为了评估这个假设对治疗烟瘾的意义，本实验室正在进行临床研究。

美国国立精神健康研究所和美国癌症协会为这项研究提供了经费上的支持。

（董培智 翻译；顾孝诚 审稿）

References:

1. Johnston, L. M., *Lancet*, ii, 742 (1942).
2. Lucchesi, B. R., Schuster, C. R., and Emley, G. S., *Clin. Pharmacol. Ther.*, **8**, 789 (1967).
3. Jarvik, M. E., Glick, S. D., and Nakamura, R. K., *Clin. Pharmacol. Ther.*, **11**, 574 (1970).
4. Glick, S. D., Jarvik, M. E., and Nakamura, R. K., *Nature*, **227**, 969 (1970).
5. Pybus, R., Goldfarb, T., and Jarvik, M. E., *Exp. Anal. Behav.*, **12**, 88 (1969).
6. Glick, S. D., Canfield, J. L., and Jarvik, M. E., *Psychol. Rep.*, **26**, 707 (1970).
7. Goodman, L. S., and Gilman, A., *The Pharmacological Basis of Therapeutics*, 585 (Macmillan, London, 1965).

Redistribution and Pinocytosis of Lymphocyte Surface Immunoglobulin Molecules Induced by Anti-Immunoglobulin Antibody

R. B. Taylor *et al.*

Editor's Note

Here Roger Taylor and colleagues use fluorescent-labelled antibodies to visualize immunoglobulin molecules on the surface of lymphocytes. In a normal cell the molecules are diffusely scattered, but when the antibody binds, the immunoglobulin molecules gather over one pole of the cell and are then brought inside. This rapid clearing of immunoglobulin from the cell surface provided an explanation for "antigenic modulation", where cell-surface antigens "disappear" following incubation with antibodies. A similar mechanism has since been implicated in the control of many cell-surface receptors following the binding of their extracellular ligand. The paper also suggested that cell membranes are two-dimensional fluids that can traffic molecules around, changing the way researchers thought about membrane structure and function.

IMMUNOGLOBULIN (Ig) determinants, presumably functioning as antigen receptors, can be demonstrated on the surface of lymphocytes in various ways[1]. The distribution of Ig determinants on the cell surface, when living lymphocytes have been studied in suspension with fluoresceinated anti-Ig antibodies, has been variably reported as diffuse and patchy, giving a spotted or ring-like appearance[2] (Fig. 1*a*), or polar, giving a crescent or cap-like appearance over one pole of the cell[3,4] (Fig. 1*b*). Immunoferritin electron microscopy showed that the difference in distribution was temperature dependent and that the former pattern was seen when cells were studied at 0°C, while the latter was seen at 20°C (S. de P. and M. C. R., unpublished work). Using immunofluorescence, we demonstrate here that the distribution on resting lymphocytes is diffuse, while the polar distribution is induced by the interaction of the anti-Ig antibodies with the Ig molecules of the cell membrane. Following this redistribution the Ig molecules are pinocytosed. This sequence of events is temperature dependent and suggests an explanation for antigenic modulation, a possible mechanism for lymphocyte triggering by antibody, mitogens and antigen, and raises important questions about the structure of mammalian cell membranes.

抗免疫球蛋白抗体诱导的淋巴细胞表面免疫球蛋白分子的重分布和胞饮现象

泰勒等

编者按

这里，罗杰·泰勒和同事们利用荧光素标记的抗体来观察淋巴细胞表面的免疫球蛋白分子。在正常细胞中，这类分子呈现为弥散分布；然而当与抗体结合时，免疫球蛋白分子就会聚集到细胞的一极，并被带到细胞内部。这种免疫球蛋白分子从细胞表面快速清除的现象为“抗原调变”提供了解释。“抗原调变”指的是：在细胞与抗体孵育之后，细胞表面的抗原会“消失”。在结合细胞外配体后，很多细胞表面受体的调制也牵涉到了与之相似的机制。这篇论文还提示细胞膜是可以转运分子使之移动的二维液体，这改变了以往研究者考虑膜结构及其功能的思维方式。

免疫球蛋白决定簇，其功能很可能如抗原受体一样，能在淋巴细胞表面以多种方式表现出来[1]。当用荧光素标记的抗免疫球蛋白抗体研究悬浮液中的活淋巴细胞时，所报道的关于免疫球蛋白决定簇在细胞表面的分布并不一致。一种是弥散或斑块分布，显示为点状或环状[2]（图 1*a*）；另一种是极性的，显示为集中在细胞一极的新月状或冠状[3,4]（图 1*b*）。采用免疫铁蛋白技术在电子显微镜下观察，结果显示分布的不同是由温度决定的，前一种分布是在 0℃下研究的细胞中被发现的，而后一种是在 20℃下被观察到的（斯特凡内洛·德·彼得里斯和马丁·拉夫，尚未发表的研究结果）。利用免疫荧光的方法，我们在此展示了它在静止淋巴细胞上的分布是弥散的，而极性分布是由抗免疫球蛋白抗体与细胞膜上的免疫球蛋白分子之间的相互作用引起的。随着这种重分布，免疫球蛋白分子被胞饮掉。这一系列依赖于温度的现象暗示了对抗原调变的一种解释（抗原调变是淋巴细胞在抗体、促细胞分裂剂和抗原作用下被激发出的一种可能的机制），并提出了若干关于哺乳动物细胞膜结构的重要问题。

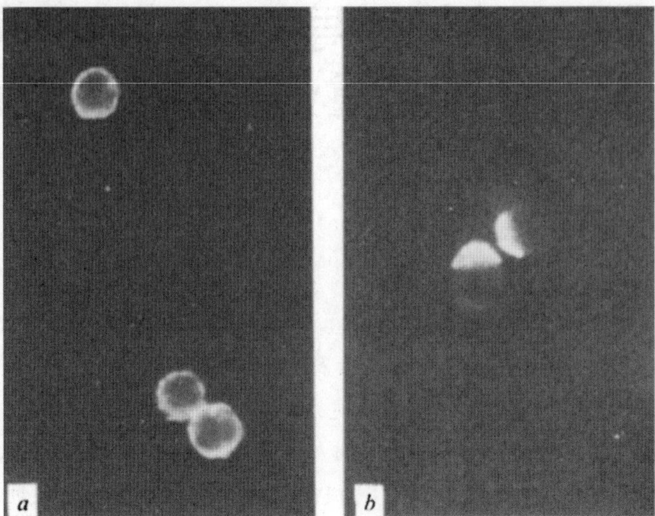

Fig. 1. Pattern of immunofluorescence in mouse spleen cells incubated in R anti-MIg-Fl in VBS for 30 min, (*a*) at 0°C ("ring" pattern), (*b*) at room temperature ("cap" pattern). The cells were washed and examined at room temperature in the presence of 3×10^{-3} M sodium azide.

Most of these studies were carried out independently in two laboratories. Since the results were in complete agreement we publish them here together.

CBA spleen cell suspensions were prepared by teasing in veronal buffered saline with 0.1% bovine serum albumin (VBS), Leibovitz medium[5] with 10% foetal calf serum (L-15) or phosphate-buffered balanced salt solution (BSS)[6]. Anti-Ig sera were prepared, fractionated, conjugated with fluorescein and sometimes purified, as previously described[3,4]. The sera used and the preparation of univalent Fab fragments are described in Table 1. Anti-θC3H (anti-θ), anti-H-2^k and anti-lymphocyte serum (ALS) were prepared as previously described[7,8]. Direct and indirect immunofluorescent staining was carried out, using 25 μl. of live spleen cells ($10–20\times10^6$ cells/ml.) and 25 μl. of antiserum at concentrations indicated in Table 1, unless otherwise specified[3]. Cells were examined in suspension under coverslip with ultraviolet or blue-violet light, and the percentage of leucocytes stained and those showing "caps" and "ring" staining was determined. Cells were counted as "caps" if the fluorescence was at one pole and occupied one half or less of the surface. Cells were counted as "rings" when fluorescence occupied the entire surface or was broken up into patches without obvious polar distribution.

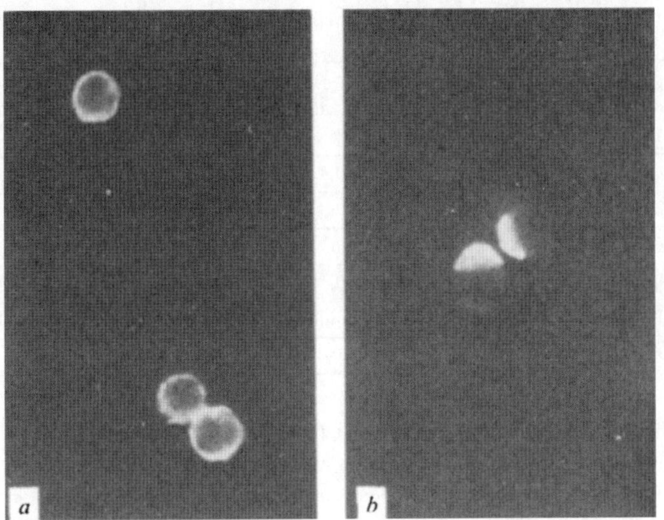

图 1. 在 VBS 中与兔抗鼠免疫球蛋白荧光标记物共孵育 30 min 后小鼠脾细胞的免疫荧光检测结果，(*a*) 在 0℃时（呈“环状”分布），(*b*) 在室温下（呈“冠状”分布）。对细胞的冲洗和检测是在室温下、有浓度为 3×10^{-3} M 叠氮化钠存在的情况下进行的。

这些研究大多是在两个实验室中独立进行的。由于结果完全一致，我们在此将其一并公布。

将撕碎的脾浸泡在由巴比妥缓冲的含 0.1% 牛血清白蛋白的生理盐水（VBS）或含 10% 胎牛血清的莱博维茨培养基（L-15）[5] 或由磷酸盐缓冲的平衡盐溶液（BSS）[6] 中，制得 CBA 脾细胞悬浮液（译者注：CBA 是一种小鼠品系，这种品系具有较低的肿瘤发生率）。按已描述过的方法 [3,4] 制备抗免疫球蛋白血清、层析、与荧光素结合，有时还需要提纯。使用的血清和单价抗原结合片段的制备方法如表 1 所述。按照以前描述过的方法 [7,8] 准备好抗 θC3H（抗 θ）、抗 $H\text{-}2^k$ 和抗淋巴细胞血清（ALS）。如果没有特别说明，用 25 μl 活脾细胞（每 ml 含 $10 \times 10^6 \sim 20 \times 10^6$ 个细胞）和 25 μl 如表 1 所示浓度的抗血清直接或间接进行免疫荧光染色 [3]。盖上盖玻片用紫外光或蓝紫光照射。对处于悬浮状态的细胞进行检验，分别测定染上色的、显现出“冠状”和“环状”染色的白细胞的百分比。如果荧光集中在一极并占据了一半或更少的表面，那么对应的细胞就被计入“冠状”。当荧光占据了整个细胞表面或破散成没有明显极性分布的斑点时，则被计入“环状”。

Table 1. Properties of Antisera

Antiserum	Fluoresceinated	Abbreviation	Protein used	Usual concentration used (mg/ml.)	
				Protein	Antibody
Rabbit anti-mouse Ig	+	R anti-MIg-Fl	γG*	5	–
Rabbit anti-mouse Ig	–	R anti-MIg	γG*	6.1	0.21
Rabbit anti-mouse Ig	–	Fab R anti-MIg	Fab- γG†	5.7	0.19
Goat anti-mouse Ig	+	G anti-MIg-Fl	Purified antibody‡	0.08	~0.05
Goat anti-mouse Ig	+	Fab G anti-MIg-Fl	Fab-purified antibody§	0.08	~0.05
Goat anti-rabbit Ig	+	G anti-RIg-Fl	γG*	1	–

* Prepared by fractionation on DEAE–cellulose (in 0.02 M phosphate buffer and 0.01 M NaCl at *p*H 7.5) after precipitation with 50% saturated ammonium sulphate.

† The R anti-MIg γG was digested by papain (one part papain to 100 parts γG (w/w)) in the presence of 0.01 M mercaptoethanol for 18 h at 37°C and the Fab fragments were separated on "Sephadex G-100".

‡ G anti-MIg or G anti-MIg-Fl was purified by elution from "Sepharose" immunoabsorbent columns[33].

§ Purified G anti-MIg was digested as above in the presence of 0.005 M dithiothreitol for 4 h. The Fab fragments were separated as above and conjugated as previously described[3].

Direct and indirect immunofluorescence testing gave similar results except where indicated, and results using various anti-MIg sera and various media were not significantly different. Although we only illustrate representative experiments in the tables and figures, each type of experiment was done at least twice and gave consistent results.

Redistribution of Surface Ig Determinants

As previously reported[3], approximately 30–50% of spleen cells could be demonstrated to have surface Ig by immunofluorescence. When cells were treated at 0°C, surface fluorescence was entirely ring-like (Fig. 1*a*). If the procedure was carried out at room temperature, or 37°C, the percentage of stained cells was the same, but most showed cap fluorescence (Fib. 1*b*). Cap staining frequently overlaid a foot-like projection of the cell; the uropod[9]. Ring cells which had been stained at 0°C transformed into cap cells when warmed (Fig. 2). The transformation from ring-staining to cap-staining was rapid at 37°C (50% by 1.5 min) and considerably slower at 24°C (50% by 4 min). Increasing the concentration of BSA to 5% slowed cap formation to some extent (50% by 5 min at 24°C), but even in neat foetal calf serum, cap formation occurred rapidly. Transformation to caps could be observed in individual ring-staining cells when they were warmed under a coverslip.

表 1. 抗血清的性质

抗血清	是否荧光染色	缩写	使用的蛋白质	通常使用浓度 (mg/ml)	
				蛋白质	抗体
兔抗鼠免疫球蛋白	+	R 抗 –MIg–Fl	γ 球蛋白 *	5	—
兔抗鼠免疫球蛋白	–	R 抗 –MIg	γ 球蛋白 *	6.1	0.21
兔抗鼠免疫球蛋白	–	Fab R 抗 –MIg	抗原结合片段 –γ 球蛋白 †	5.7	0.19
山羊抗鼠免疫球蛋白	+	G 抗 –MIg–Fl	纯化抗体 ‡	0.08	~ 0.05
山羊抗鼠免疫球蛋白	+	Fab G 抗 –MIg–Fl	纯化抗体的抗原结合片段 §	0.08	~ 0.05
山羊抗兔免疫球蛋白	+	G 抗 –RIg–Fl	γ 球蛋白 *	1	—

* 用 50% 饱和硫酸铵沉淀后，经二乙氨基乙基纤维素（洗脱剂为 pH 值 7.5 的 0.02 M 磷酸盐缓冲液和 0.01 M 氯化钠）层析制得。

† 将兔抗鼠免疫球蛋白 γ 球蛋白在 37℃、0.01 M 巯基乙醇存在的条件下用木瓜蛋白酶（1 份木瓜蛋白酶对应 100 份 γ 球蛋白（重量比））消化 18 h，得到的抗原结合片段在“葡聚糖 G-100”上分离。

‡ 山羊抗鼠免疫球蛋白或山羊抗鼠免疫球蛋白荧光标记物经“琼脂糖”免疫吸附柱洗脱提纯 [33]。

§ 提纯后的山羊抗鼠免疫球蛋白按如上条件在 0.005 M 二硫苏糖醇存在的条件下消化 4 h。抗原结合片段的分离方法如上所述并照以前描述过的方法结合 [3]。

除另有说明的以外，直接和间接的免疫荧光检验都给出了相似的结果；当使用多种抗鼠免疫球蛋白血清和多种培养基时，结果也没有表现出显著差异。虽然我们在图表中只举例说明了具有代表性的实验，但每种实验都至少重复过一次且都给出了一致的结果。

表面免疫球蛋白决定簇的重分布

根据之前发表的报告 [3]，大约有 30% ~ 50% 的脾细胞可以被免疫荧光检验法证实拥有表面免疫球蛋白。当细胞在 0℃ 下被处理时，表面荧光为完整的环状（图 1*a*）。如果这个过程是在室温或 37℃ 下进行的，则虽然被染色细胞的百分比没有变化，但大多数表现为冠状荧光（图 1*b*）。冠状染色经常覆盖在细胞的一个足状凸出物上，即尾足 [9]。0℃ 时被染色的环状细胞，在加热的情况下会转变为冠状细胞（图 2）。这种由环状染色变为冠状染色的转化在 37℃ 时相当迅速（50% 需要 1.5 min），而在 24℃ 时则要慢得多（50% 需要 4 min）。增加牛血清白蛋白的浓度至 5% 可以在一定程度上减缓冠状染色的形成（50% 在 24℃ 时需要 5 min），但即使是在纯胎牛血清中，冠状形成也会很快发生。在盖玻片下加热时，可以观察到单个的环状染色细胞转化为冠状细胞。

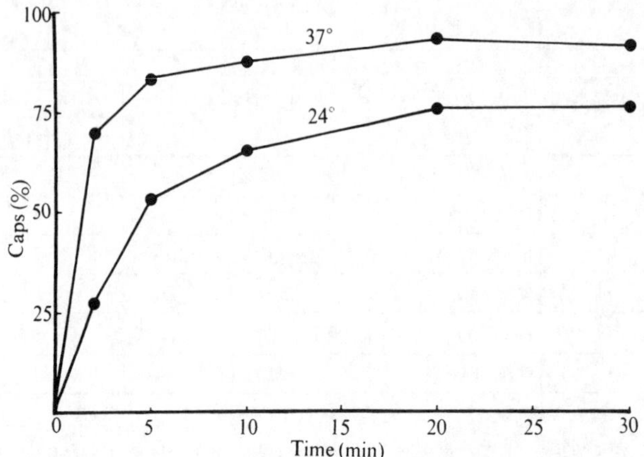

Fig. 2. Rate of cap formation at 37°C and 24°C. Cells were incubated in R anti-MIg-Fl in VBS for 30 min at 0°C, washed at 0°C and then warmed to 37°C or 24°C in a water bath. After the specified time the cells were cooled, 3×10^{-2} M sodium azide was added and the percentage of staining cells which showed cap formation was determined.

When spleen cells were treated with ferritin-labelled anti-MIg and examined by electron microscopy the same phenomenon was observed (unpublished work of S. de P. and M. C. R.). At 0°C the ferritin labelling was patchy and diffusely distributed over the cell surface. At 20°C the labelling was confluent and located over one pole of the cell, which contained the Golgi apparatus when this was visible in the section. As the total surface ferritin labelling was approximately the same at 0°C and at 20°C, cap formation must be the result of a redistribution of labelled Ig determinants which tend to gather over the Golgi pole of the cell.

Mechanism of Cap Formation

As Fig. 3 shows, cap formation was completely inhibited at 20°C by sodium azide; the inhibition was dose-dependent and readily reversible. At 37°C inhibition was 97% at 10^{-2} M and complete at 3×10^{-2} M. Dinitrophenol also inhibited cap formation and inhibition was complete at 10^{-3} M at 20°C, but was still reversible at 10^{-2} M. Thus, cap formation is a metabolically dependent active process. As cells warmed at 37°C and then treated with anti-MIg-Fl in the presence of azide showed only ring-staining, cap formation must be induced by the interaction of the anti-Ig antibody with the Ig molecules of the surface membrane, and not by increased temperature alone.

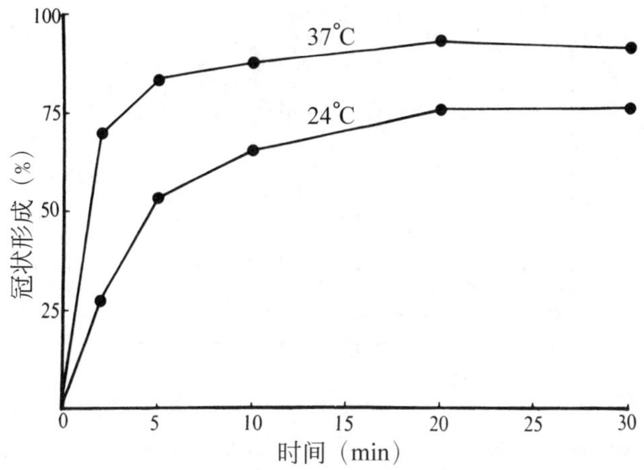

图 2. 37℃和 24℃下冠状形成的比例。将浸泡在 VBS 中的细胞在 0℃下与兔抗鼠免疫球蛋白荧光标记物共孵育 30 min，在 0℃下冲洗后，再用水浴加热到 37℃或 24℃。等细胞经过一定时间的自然冷却后，加入浓度为 3×10^{-2} M 的叠氮化钠并测定呈冠状染色的细胞的百分比。

当脾细胞用铁蛋白标记的抗鼠免疫球蛋白处理后，在电子显微镜下也能观察到同样的现象（斯特凡内洛 · 德 · 彼得里斯和马丁 · 拉夫，尚未发表的研究结果）。0℃时，铁蛋白标记在细胞表面表现为斑点和弥散分布。在 20℃时，标记汇合于细胞的一极，当这种情况出现时从剖面图可以看到这个位置包含有高尔基体。因为在 0℃和 20℃时表面上铁蛋白标记的总数大致相同，所以冠状形成必定是带有标记的免疫球蛋白决定簇重分布的结果，这种重分布趋向于集中在细胞含高尔基体的一极。

冠状形成的机理

如图 3 所示，冠状形成在 20℃时完全被叠氮化钠抑制；这种抑制是剂量依赖的并且很容易逆转。在 37℃时 10^{-2} M 叠氮化钠对冠状形成的抑制率达到 97%，浓度为 3×10^{-2} M 时完全抑制。二硝基酚也会抑制冠状形成，20℃时 10^{-3} M 就可以达到完全抑制，但在浓度为 10^{-2} M 时抑制依然是可逆的。因此，冠状形成是一个取决于新陈代谢的活性过程。因为细胞在 37℃加热后再在叠氮化物存在的条件下用抗鼠免疫球蛋白荧光标记物处理只显示出了环状染色，所以冠状形成必须由抗免疫球蛋白抗体和膜表面的免疫球蛋白分子之间的相互作用诱导，而不能仅由温度的增加诱导。

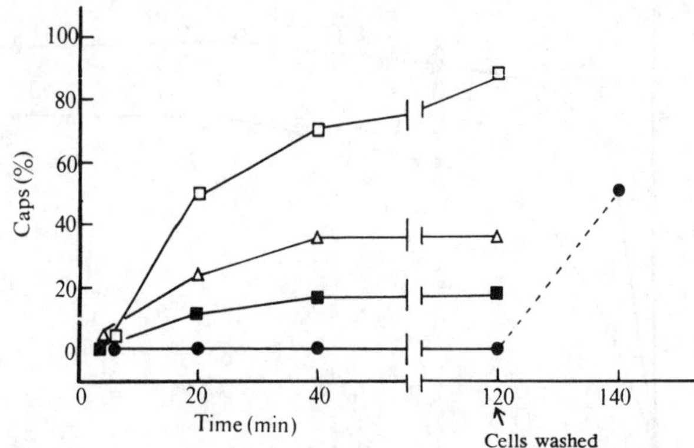

Fig. 3. Reversible inhibition of cap formation by sodium azide. Cells were incubated in sodium azide in BSS for 15 min at 20°C before G anti-MIg-Fl was added. Cells were sampled at specified times and the percentage of staining cells which showed cap formation was determined. After 2 h the cells in 10^{-3} and 10^{-2} M azide were washed and reincubated in BSS for another 20 min. Concentrations of sodium azide were 10^{-3} M (●), 10^{-4} M (■), 10^{-5} M (△), control (□). Experimental points at 10^{-2} M were coincident with those at 10^{-3} M.

Univalent Fab fragments (prepared by digesting G anti-MIg with papain[10] and labelling with fluorescein (Table 1)) stained the same percentage of spleen cells as undigested G anti-MIg-Fl but did not induce cap formation (Table 2). Also, ring fluorescence with the Fab antibody was invariably smooth and complete and never showed the granular or patchy distribution seen with bivalent antibody. When unlabelled Fab R anti-MIg was used as a middle layer in indirect testing with G anti-RIg-Fl, where the final staining with the latter reagent was done at 0°C in the presence of azide, no caps were formed, but the rings were generally granular. However, if the G anti-RIg-Fl was added without azide at 37°C, then caps were induced (Table 2). It seems likely that multivalent binding and lattice formation at the cell surface are required for cap formation. The lack of patchy fluorescence seen with Fab G anti-MIg-Fl suggests that the patchy distribution of surface Ig determinants may also be induced by multivalent binding of the anti-Ig antibody and demands caution in the interpretation of patchy distribution of other surface antigens such as H-2 when observed with multivalent labelled antibodies[11].

Table 2. Failure of Cap Formation with Fab-anti-Ig

Experiment No.	Type of immunofluorescence testing	Type of anti-MIg	% cells stained	% caps
1	Direct at 20°C*	G anti-MIg-Fl	25	75
		Fab G anti-MIg-Fl	23	0
2	Indirect with† development at 0°C	R anti-MIg	43	70
		Fab R anti-MIg	45	0
3	Indirect with† development at 37°C	Fab R anti-MIg	46	52

* Cells were incubated at 20°C for 40 min. The Fab-G anti-MIg-Fl was tested over a wide range of concentrations (0.001 to 1 mg/ml.) with similar results.

† Cells were incubated in R anti-MIg for 20 min at 37°C, washed and incubated with G anti-RIg-Fl at 0°C for 30 min or 37°C for 20 min. Undigested anti-MIg and Fab fragments were used at an antibody concentration of about 0.2 mg/ml.

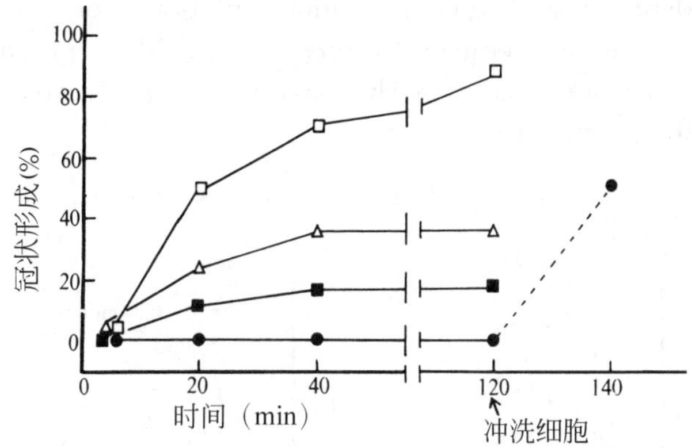

图 3. 叠氮化钠对冠状形成的可逆抑制。在加入山羊抗鼠免疫球蛋白荧光标记物之前将细胞置于 20℃下的含叠氮化钠的 BSS 中孵育 15 min。在特定时间抽取细胞样本并对染色细胞中出现冠状形成的百分比进行测定。2 h 后对用 10^{-3}M 和 10^{-2}M 叠氮化物处理的细胞进行冲洗并再一次在 BSS 中孵育 20 min。叠氮化钠的浓度为 10^{-3}M（●）、10^{-4} M（■）、10^{-5} M（△）和对照（□）。用 10^{-2} M 叠氮化钠处理得到的实验点与 10^{-3}M 的点重合。

与未经消化的山羊抗鼠免疫球蛋白荧光标记物相比，单价抗原结合片段（制备方法是：用木瓜蛋白酶消化山羊抗鼠免疫球蛋白 [10] 并进行荧光标记（表 1））可以使同样百分比的脾细胞染色，但没有引起冠状形成（表 2）。此外，由该抗原结合片段抗体染色导致的环状荧光总是很平滑，完全不会出现由二价抗体染色时表现出的颗粒状或斑点状分布。当在间接免疫荧光检验中用未标记的兔抗鼠免疫球蛋白的抗原结合片段作为中间层、然后用山羊抗兔免疫球蛋白荧光标记物作染色剂进行检测时，在 0℃且有叠氮化物存在的条件下，无冠状形成，但环状荧光一般为颗粒状。然而，如果山羊抗兔免疫球蛋白荧光标记物是在 37℃、没有叠氮化物的情况下加入的，则会引起冠状现象（表 2）。这似乎表明冠状形成需要多价结合和在细胞表面形成晶格结构。用山羊抗鼠免疫球蛋白抗原结合片段荧光标记物进行染色时未发现斑点荧光，这说明表面免疫球蛋白决定簇的斑状分布可能也是由抗免疫球蛋白抗体的多价结合引起的，在用多价标记抗体进行观察时，需要注意对诸如 H-2 等其他表面抗原引起的斑状分布的解释 [11]。

表 2. 用抗免疫球蛋白抗原结合片段不能导致冠状分布的形成

实验编号	免疫荧光检验法的类型	抗鼠免疫球蛋白的类型	染色的细胞%	冠状形成%
1	在 20℃下进行，直接 *	山羊抗鼠免疫球蛋白荧光标记物	25	75
		山羊抗鼠免疫球蛋白抗原结合片段荧光标记物	23	0
2	在 0℃下进行，间接 †	兔抗鼠免疫球蛋白	43	70
		兔抗鼠免疫球蛋白抗原结合片段	45	0
3	在 37℃下进行，间接 †	兔抗鼠免疫球蛋白抗原结合片段	46	52

* 将细胞在 20℃下孵育 40 min。用山羊抗鼠免疫球蛋白抗原结合片段荧光标记物在宽浓度范围内（0.001 mg/ml ~ 1 mg/ml）检验时所得的结果类似。

† 将细胞在 37℃下用兔抗鼠免疫球蛋白孵育 20 min，冲洗并在 0℃下用山羊抗兔免疫球蛋白荧光标记物孵育 30 min 或在 37℃下孵育 20 min。未经消化的抗鼠免疫球蛋白和其抗原结合片段的抗体使用浓度大约为 0.2 mg/ml。

If cap formation depends on lattice formation, inhibition of cap formation would be expected with increasing concentrations of anti-MIg (prozone effect), as in immunoprecipitation reactions. As Fig. 4 shows, cap formation decreased in the presence of high concentrations of anti-MIg.

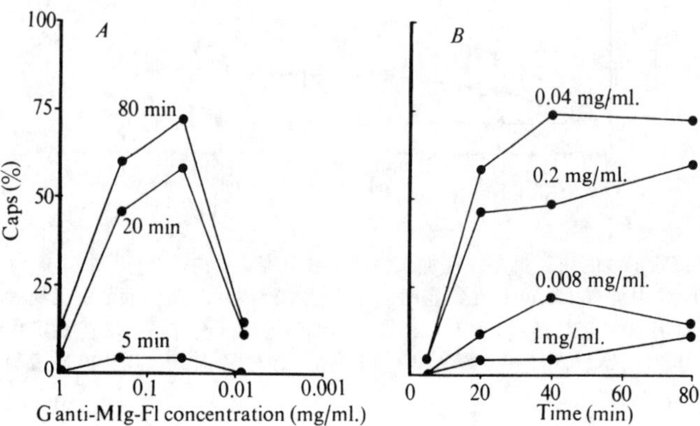

Fig. 4. Effect of anti-MIg concentration on the rate of cap formation. Cells were incubated at 20°C with various concentrations of G anti-MIg-Fl (purified antibody) for 5, 20, 40 and 80 min. Cells were washed at 4°C and examined immediately. The percentage of stained cells showing cap formation is plotted against antibody concentration (*A*) to emphasize the prozone effect with high antibody concentration, and against time (*B*) to illustrate the effect of protein concentration on the rate of cap formation. About 48% of the cells were stained at the antibody concentration of 1 mg/ml., 35% at 0.2 and 0.04 mg/ml., and 26% at 0.008 mg/ml.

In view of the frequent association of the caps with the lymphocyte uropod, it seemed possible that cap formation was somehow associated with membrane flow and cell motility. Cytochalasin B, which is claimed to specifically inhibit a contractile microfilament system responsible for cell movement[12], consistently inhibited cap formation, but only partially, even at high concentrations, and the effect was reversible (Table 3). After treatment with colcemid the cells looked very irregular but cap formation was not affected, suggesting that micro-tubular activity is not involved[13] (Table 3). To determine if cells free in suspension, without contact with substrate, could form caps, cells were kept in suspension at a concentration of 1.5×10^{6}/ml. by continuous stirring in the presence of R anti-MIg-Fl and then washed in the presence of azide. No significant inhibition of cap formation could be demonstrated. Pre-incubating cells at room temperature or at 37°C for 30 min in calcium-free medium in the presence of up to 4 mM of ethylene glycol-*bis* (2-amino ethyl ether)-tetra-acetic acid (EGTA), or calcium and magnesium-free medium in the presence of up to 4 mM ethylene diamino-tetra-acetic acid (EDTA) (at an estimated[14] free Ca^{2+} concentration of less than $1–5\times10^{-8}$), and treating and washing cells in the same conditions did not significantly inhibit cap formation.

如果冠状形成依靠晶格结构的形成，那么就会如同在免疫沉淀反应中一样，可以期望通过增加抗鼠免疫球蛋白的浓度而达到对冠状形成的抑制（前带效应）。如图4所示，在抗鼠免疫球蛋白浓度较高时冠状形成有所减少。

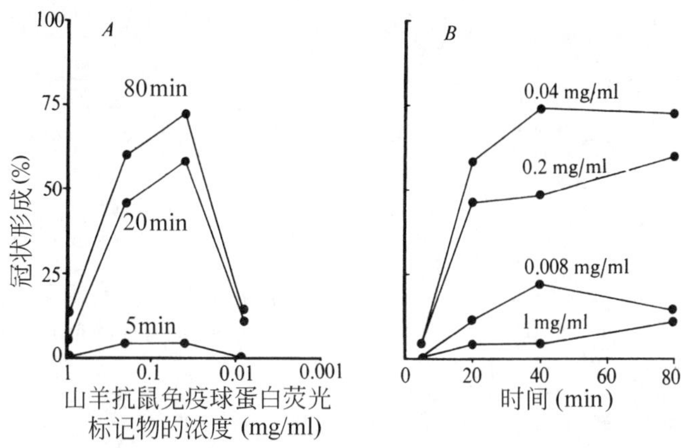

图4. 抗鼠免疫球蛋白浓度对冠状形成比例的影响。将细胞在20℃下用多种浓度的山羊抗鼠免疫球蛋白荧光标记物（纯化抗体）孵育5 min、20 min、40 min和80 min。在4℃下冲洗细胞并立即进行检验。将染色细胞中出现冠状形成的百分比对抗体浓度作图（*A*）以突出由高浓度抗体带来的前带效应，并对时间作图（*B*）以说明蛋白质浓度对冠状形成比例的影响。大约48%的细胞在抗体浓度为1 mg/ml、35%的细胞在0.2 mg/ml和0.04 mg/ml以及26%的细胞在0.008 mg/ml时被染色。

从冠状现象和淋巴细胞尾足经常会同时出现这一角度考虑，冠状形成看起来同膜流动性和细胞运动性之间可能存在着某种关系。细胞松弛素B被认为可以特异性破坏与细胞运动有关的、有收缩性的微丝系统[12]，它能持续地抑制冠状形成，但只是部分抑制，甚至在高浓度下这种影响也是可逆的（表3）。在用秋水仙酰胺处理后，细胞看起来十分不规则，但冠状形成并没有受到影响，说明这同微管活性无关[13]（表3）。为了弄清在悬浮液中自由存在、与培养基不接触的细胞是否会出现冠状形成，我们对细胞浓度为1.5×10^6/ml的悬浮液在有兔抗鼠免疫球蛋白荧光标记物存在的条件下不停搅拌，然后再在叠氮化物中冲洗，结果证明没有显著的冠状形成抑制现象发生。用含浓度可达4 mM的乙二醇双(2–氨基乙醚)四乙酸（EGTA）的无钙培养基或用含浓度可达4 mM的乙二胺四乙酸（EDTA）的无钙镁（估计[14]自由钙离子浓度低于$1\times10^{-8}\sim5\times10^{-8}$）培养基在室温或37℃下预孵育细胞30 min，然后在相同条件下处理和冲洗细胞，结果发现冠状形成并未受到显著的抑制。

Table 3. Effect of Cytochalasin B and Colcemid on Cap Formation

Experiment No.	Drug	Concentration (μg/ml.)	% caps	
			With drug	After recovery
1*	Cytochalasin B†	0	84	—
	Cytochalasin B†	10	67	—
	Cytochalasin B†	30	55	—
	Cytochalasin B†	50	52	75
2‡	Cytochalasin B†	0	90	—
	Cytochalasin B†	20	34	—
	Colcemid	20	91	—

* After preincubation at 2×10^6 cells/ml. in cytochalasin B in L-15 for 30 min at 37°C, cells were incubated with R anti-MIg for 15 min at 37°C washed in cytochalasin B and incubated with G anti-RIg-Fl at 0°C in 3×10^{-2} M azide for 30 min.

† Cytochalasin B (Imperial Chemical Industries, Alderley Park) was dissolved in dimethyl sulphoxide (DMSO) (10 μl./mg). The final DMSO concentration was the same in experimental and control samples and was 0.3% in experiment 1 and 1% in experiment 2.

‡ After preincubation at 10^6 cells/ml. in drug in VBS for 30 min at 37°C, cells were treated with R anti-MIg-Fl at 0°C for 15 min and then warmed up in drug to 37°C for 20 min.

Thus, Ig cap formation is an active, temperature-dependent process possibly involving contractile microfilament activity that is induced by multivalent binding of anti-Ig at the cell surface, and which seems not to require extracellular calcium or magnesium. The actual mechanism is unclear but we favour the possibility that surface Ig molecules are immobilized by cross-linking of the anti-Ig antibodies while the rest of the membrane can move or flow forward, resulting in the Ig molecules gathering over the trailing tail of the lymphocyte; the uropod. A more detailed discussion of possible mechanisms will be presented elsewhere (S. de P. and M. C. R. in preparation).

Antigenic Modulation

Cap formation is rapidly followed by pinocytosis of the Ig determinants. This could be seen when spleen cells were incubated with anti-MIg-Fl at 0°C, washed and then kept at 37°C. Cap formation occurred rapidly (Fig. 2) and, after 5 min, pinocytosis of the fluorescent label was seen and was marked by 10 min. By 30 min most of the label seemed to be inside the cells. No visible pinocytosis occurred, however, when cells were stained at 0°C in the presence of azide with univalent Fab anti-MIg-Fl. Spleen cells stained with ferritin-labelled anti-MIg at 20°C and examined by electron microscopy not only showed cap localization of the ferritin labelling but invariably contained ferritin-lined pinocytotic vesicles near the Golgi area, underlying the cap. Such vesicles were not seen when the cells were treated at 0°C.

To study further the disappearance of surface Ig, spleen cells were incubated with unlabelled R anti-MIg at 37°C. After 10 min or after 3 h, cells were washed and treated with G anti-RIg-Fl in azide at 0°C (Table 4). At 10 min, 40–50% of the cells were stained; most showed cap fluorescence. At 3 h, only 20–30% of the cells stained, and many of these had only a few fluorescent spots at one pole (Table 4). When the 3-h cells

表 3. 细胞松弛素 B 和秋水仙酰胺对冠状形成的影响

实验编号	药物	浓度 (μg/ml)	冠状形成%	
			用药	恢复后
1*	细胞松弛素 B†	0	84	—
	细胞松弛素 B†	10	67	—
	细胞松弛素 B†	30	55	—
	细胞松弛素 B†	50	52	75
2‡	细胞松弛素 B†	0	90	—
	细胞松弛素 B†	20	34	—
	秋水仙酰胺	20	91	—

* 37℃下将浓度为 2×10^6/ml 的细胞在 L-15 中与细胞松弛素 B 预孵育 30 min，再将其在 37℃下用兔抗鼠免疫球蛋白孵育 15 min，在细胞松弛素 B 中冲洗并在 0℃、有 3×10^{-2} M 叠氮化物存在的条件下用山羊抗兔免疫球蛋白荧光标记物孵育 30 min。

† 将细胞松弛素 B（帝国化学工业公司，奥尔德利公园）溶解在二甲基亚砜（DMSO）（10 μl/mg）中。实验组中和对照样本组中 DMSO 的最终浓度是一致的，在实验 1 中是 0.3%，在实验 2 中是 1%。

‡ 37℃下将浓度为 10^6/ml 的细胞在 VBS 中与药物预孵育 30 min，再将其在 0℃下用兔抗鼠免疫球蛋白荧光标记物处理 15 min，然后在药物中加热至 37℃并维持 20 min。

因此，免疫球蛋白冠状形成是一个活性的、依赖于温度的过程，可能牵涉到由细胞表面的抗免疫球蛋白的多价结合引起的微丝收缩活性，并且这个过程似乎不依赖于细胞外的钙和镁。明确的机理还不清楚，但我们支持如下这种可能性：表面免疫球蛋白分子被抗免疫球蛋白抗体的交联所固定，而膜的其余部位仍能向前移动或流动，导致免疫球蛋白分子集中在淋巴细胞拖后的尾部，即尾足。对几种可能机理的更详细的讨论将在其他地方发表（斯特凡内洛·德·彼得里斯和马丁·拉夫，完稿中）。

抗 原 调 变

冠状形成之后迅速发生免疫球蛋白决定簇的胞饮。在 0℃下将脾细胞用抗鼠免疫球蛋白荧光标记物孵育、冲洗然后保存于 37℃时，就能观察到这种现象。冠状形成迅速发生（图 2），5 min 后能观察到荧光标记的胞饮现象并在 10 min 时非常显著。到 30 min 时看起来大多数标记已处于细胞内。然而，如果细胞是在叠氮化物和抗鼠免疫球蛋白单价抗原结合片段荧光标记物存在的条件下于 0℃时染色的，则没有可见的胞饮现象发生。当用电子显微镜观察 20℃下用铁蛋白标记的抗鼠免疫球蛋白染色的脾细胞时，不仅能在冠状形成中找到铁蛋白标记物，还总能在冠状染色下方的高尔基区附近发现包含有铁蛋白的胞饮泡。在 0℃下处理细胞时观察不到这样的小泡。

为了进一步研究细胞表面免疫球蛋白的消失现象，将脾细胞在 37℃下用未标记的兔抗鼠免疫球蛋白孵育。分别在 10 min 和 3 h 后冲洗细胞，并于 0℃下用含叠氮化物的山羊抗兔免疫球蛋白荧光标记物处理这些细胞（表 4）。10 min 时，40% ~ 50% 的细胞被染上色；大多数显示出冠状荧光。3 h 时，只有 20% ~ 30% 的细胞被染色，许多被染色的细胞在细胞的一极只有少量荧光斑（表 4）。如果在 0℃的叠氮化物中用

were treated with anti-MIg-Fl in azide at 0°C instead of the G anti-RIg-Fl, only a few cells stained and all of these showed weak cap staining (Table 5). Thus, Ig determinants have been largely cleared off the cell surface by treatment with unlabelled anti-MIg for 3 h at 37°C, the process of removal appearing to be complete in 50% of Ig-bearing cells. However, when the 3-h cells were tested with ALS or anti-H-2 serum by indirect immunofluorescence in azide at 0°C, they showed normal intensity ring staining suggesting that only Ig determinants had been removed by the anti-Ig antibody (Table 5). When the spleen cells were incubated for 3 h at 37°C with unlabelled Fab-anti-MIg and then stained with G anti-RIg-Fl in azide at 0°C, 40–50% of the cells stained with ring fluorescence and there was no obvious loss of intensity of staining. Thus Fab fragments were unable to cause the disappearance of surface Ig molecules. Although cytochalasin B, at a concentration of 30 μg/ml., only partially inhibited cap formation (Table 3), it completely inhibited visible pinocytosis and markedly inhibited the disappearance of Ig determinants induced by anti-Ig. In the absence of extracellular calcium and in the presence of 3 mM EGTA, pinocytosis and Ig disappearance occurred normally.

Table 4. Disappearance of Ig Determinants*

Type of anti-MIg	Time (min)	% cells stained	% caps†		% rings
			Large	Small	
R anti-MIg	20	44	75	9	16
R anti-MIg	150	28	8	85	7
Fab-R anti-MIg	20	48	0	0	100
Fab-R anti-MIg	150	43	2	0	98

* 2×10^6 cells were incubated at 37°C in 0.25 ml. of R anti-MIg in L-15; cells were sampled at 20 and 150 min and stained with G anti-RIg-Fl for 30 min at 0°C in 3×10^{-2} M azide.

† Caps were considered "large" if estimated to occupy at least one-quarter of the cell surface and "small" if occupying less than this. Many of the small caps consisted of several fluorescent spots at one pole of the cell.

Table 5. Specificity of Ig Disappearance*

Anti-MIg-treated cells re-treated with		% cells stained	% caps	% rings
First layer	Second layer			
R anti-MIg-Fl	—	14 (faint)	0	100
R anti-MIg	G anti-RIg-Fl	25 (faint)	0	100
Anti-H-2^k (1:2)	R anti-MIg-Fl	100	0	100
ALS (1:100)	G anti-RIg-Fl	100	0	100

* Cells were from the same cell preparation as reported in Table 4, which were treated with R anti-MIg for 150 min at 37°C. They were then washed and subjected to direct or indirect testing as indicated at 0°C in 3×10^{-2} M sodium azide.

抗鼠免疫球蛋白荧光标记物替换山羊抗兔免疫球蛋白荧光标记物来处理孵育 3 h 的细胞，则只有少量细胞被染色，并且全都显示出微弱的冠状染色（表 5）（译者注：表 5 可能有误，前两行数据似应为冠状形成 100%）。因此，在 37℃下用未标记的抗鼠免疫球蛋白处理 3 h 就可以把大多数免疫球蛋白决定簇从细胞表面清除掉，这个清除过程看似在 50% 表面带有免疫球蛋白的细胞中是完全的。然而，当孵育了 3 h 的细胞在 0℃下的叠氮化物中通过间接免疫荧光法用 ALS 或抗 H-2 血清作测试时，它们会显示出正常强度的环状染色，这说明只有免疫球蛋白决定簇被抗免疫球蛋白抗体清除掉了（表 5）。将脾细胞在 37℃下用未标记的抗鼠免疫球蛋白抗原结合片段孵育 3 h，再在 0℃的叠氮化物中用山羊抗兔免疫球蛋白荧光标记物染色，这时 40%～50% 的细胞染色显示出环状荧光且未出现染色强度的明显降低。因此，抗原结合片段并不能引起表面免疫球蛋白分子的消失。虽然细胞松弛素 B 在浓度为 30 μg/ml 时只能部分抑制冠状形成（表 3），但它能完全抑制可见的胞饮现象并能显著抑制由抗免疫球蛋白引起的免疫球蛋白决定簇的消失。在细胞外缺少钙和有 3 mM EGTA 存在的条件下，胞饮和免疫球蛋白消失的现象仍能正常发生。

表 4. 免疫球蛋白决定簇的消失现象*

抗鼠免疫球蛋白的类型	时间（min）	染色的细胞%	冠状形成% †		环状形成%
			大	小	
兔抗鼠免疫球蛋白	20	44	75	9	16
兔抗鼠免疫球蛋白	150	28	8	85	7
兔抗鼠免疫球蛋白抗原结合片段	20	48	0	0	100
兔抗鼠免疫球蛋白抗原结合片段	150	43	2	0	98

* 37℃下将 2×10^6 个细胞在 L-15 中与 0.25 ml 兔抗鼠免疫球蛋白共孵育；分别在 20 min 和 150 min 时采集细胞样本并将采集的样本在 0℃、存在 3×10^{-2} M 叠氮化物的条件下用山羊抗兔免疫球蛋白荧光标记物染色 30 min（译者注：第 879 页最后一段中提到的采样时间分别是 10 min 和 3 h）。

† 如果冠状荧光染色区域至少占据了细胞表面的四分之一就被认为是“大”，如果小于四分之一则被认为是“小”。许多小冠状染色由细胞一极的若干荧光小斑点组成。

表 5. 免疫球蛋白消失现象的特异性*

经抗鼠免疫球蛋白处理的细胞的再处理		染色的细胞%	冠状形成%	环状形成%
第一层	第二层			
兔抗鼠免疫球蛋白荧光标记物	—	14（模糊）	0	100
兔抗鼠免疫球蛋白	山羊抗兔免疫球蛋白荧光标记物	25（模糊）	0	100
抗 $H-2^k$（1:2）	兔抗鼠免疫球蛋白荧光标记物	100	0	100
ALS（1:100）	山羊抗兔免疫球蛋白荧光标记物	100	0	100

* 细胞来源于表 4 中报道的细胞制备物，是在 37℃下用兔抗鼠免疫球蛋白处理 150 min 后得到的。随后进行冲洗并在 0℃、存在 3×10^{-2} M 叠氮化钠的条件下进行直接或间接的荧光检测。

The disappearance of surface Ig determinants induced by anti-Ig antibody is analogous to antigenic modulation described for the thymus-leukaemia (TL) antigens[15]. When TL-positive cells, particularly some lymphomas, are exposed to anti-TL antibody *in vivo* or *in vitro*, they rapidly lose their sensitivity to the cytotoxic action of anti-TL serum and complement. TL modulation does not occur at 0°C or in the presence of actinomycin D[15] but is said to occur with univalent Fab anti-TL[16]. Recently, Takahashi has observed similar modulation by anti-Ig antibody of Ig determinants on the surface of some murine myeloma cells and normal spleen cells[17]. The mechanism of antigenic modulation has remained a mystery, although the possibility of it being pinocytosis has recently been raised by Takahashi[17]. Our findings strongly suggest that modulation, of surface Ig determinants at least, is the result of antibody induced pinocytosis, preceded by the gathering of the determinants over the Golgi area. It is unclear why Fab fragments modulate TL but fail to modulate Ig. It will be important to see if cytochalasin B inhibits modulation of the TL antigens.

Anti-Ig sera are being widely used to inhibit the function and antigen-binding ability of lymphocytes[1]. The assumption has been that the anti-Ig inhibits lymphocyte antigen-binding by blocking or by steric hindrance[1]. Our findings, and those of Takahashi, suggest that modulation of the Ig receptors may sometimes be operating. If so, one might expect the inhibition due to modulation to be more effective than steric inhibition, which may explain the prozone effect that has been observed with increasing concentrations of anti-Ig[18], where cap formation would be inhibited.

Cap Formation with Anti-θ

The theta (θ) alloantigen is present on mouse thymocytes[19] and thymus-derived (T) lymphocytes[20,21]. One of us (M. C. R.) previously reported that anti-θ serum used in indirect immuno-fluorescence testing produced only ring-like staining of thymocytes and T cells, and that the morphology of the fluorescence could thus be used to distinguish θ-bearing T cells from non-θ-bearing B cells, since the latter showed mainly cap-like staining with anti-MIg-Fl[4]. To exclude the possibility that cap formation was unique for Ig determinants and/or B cells, we attempted to induce cap formation with thymocytes using various dilutions of anti-θ serum followed by a constant concentration of anti-MIg-Fl. When anti-θ was used at high concentrations (as was the case in the experiments previously reported), only rings were seen (Fig. 5), but when lower concentrations were used, cap formation occurred. In experiments where thymus cells were treated with dilute anti-θ at 37°C and anti-MIg-Fl used at 0°C, no cap formation occurred (Fig. 5), indicating that anti-θ alone did not induce caps during the 30 min incubation. When thymus cells were treated with dilute anti-θ, then with unlabelled R anti-MIg at 37°C for 20 min or 2 h, followed by G anti-RIg-Fl at 0°C, the caps were smaller after 2 h than at 20 min, suggesting that some disappearance had occurred. Our preliminary studies with anti-H-2^k serum gave similar results. Takahashi has recently found that anti-H-2 alone did not cause H-2 modulation, but anti-H-2 plus anti-Ig did modulate if the anti-H-2 was used in low concentrations[17].

由抗免疫球蛋白抗体引起的表面免疫球蛋白决定簇的消失现象与描述胸腺白血病（TL）抗原时的抗原调变[15]类似。当TL阳性细胞，尤其是一些淋巴瘤，在体内或体外暴露于抗TL抗体时，它们会迅速失去对抗TL血清和补体的细胞毒性作用的敏感性。尽管TL抗原调变在0℃时或者在有放线菌素D[15]存在的条件下不发生，但有报道称在与抗TL单价抗原结合片段在一起时会发生[16]。近来，高桥在一些鼠骨髓瘤细胞和正常脾细胞中观察到，由免疫球蛋白决定簇的抗免疫球蛋白抗体也能引起细胞表面发生类似的调变[17]。虽然高桥最近提出它由胞饮作用引起的可能性[17]，但抗原调变的机理至今仍是个谜。我们的发现显然表明：表面免疫球蛋白决定簇的调变现象，至少是由抗体引起的胞饮现象的结果，并发生在决定簇聚集在高尔基区之后。抗原结合片段能调变TL抗原但不能调变免疫球蛋白的原因尚不清楚。观察细胞松弛素B是否会抑制TL抗原的调变将是一件很重要的事。

抗免疫球蛋白血清被广泛用于抑制淋巴细胞的功能及其与抗原结合的能力[1]。之前的假说是，抗免疫球蛋白通过阻塞或位阻抑制了淋巴细胞与抗原的结合[1]。我们的发现以及高桥的发现都暗示，免疫球蛋白受体的调变可能会在有些情况下起作用。如果是这样的话，就可以预期由调变引起的抑制要比由位阻引起的抑制更有效，这样就可以解释在增加抗免疫球蛋白浓度从而使冠状形成受到抑制时观察到的前带效应[18]。

与抗θ有关的冠状形成

θ同种异体抗原出现在小鼠胸腺细胞[19]和源于胸腺的（T）淋巴细胞[20,21]中。我们中的一位作者（马丁·拉夫）之前曾报道过，抗θ血清在间接免疫荧光检验中只产生了胸腺细胞和T细胞的环状染色，因而可以用荧光形状来区分带有θ的T细胞和不带θ的B细胞，因为后者在用抗鼠免疫球蛋白荧光标记物染色时主要表现出冠状染色[4]。为了排除冠状形成是免疫球蛋白决定簇和（或）B细胞所特有的可能性，我们先用具有不同稀释度的抗θ血清，然后用某一恒定浓度的抗鼠免疫球蛋白荧光标记物来处理胸腺细胞，以期诱导出该细胞的冠状形成。当使用高浓度的抗θ时（如同之前报道的实验），只有环状染色出现（图5）；但当使用较低浓度的抗θ时，冠状形成发生。在37℃时用稀释的抗θ和0℃时用抗鼠免疫球蛋白荧光标记物处理胸腺细胞的实验中，冠状形成没有发生（图5），这表明在仅有抗θ的条件下孵育30 min并不能引起冠状形成。如果用稀释的抗θ处理胸腺细胞，再用未标记的兔抗鼠免疫球蛋白在37℃下孵育20 min或2 h，接着在0℃下用山羊抗兔免疫球蛋白荧光标记物处理，则孵育2 h的冠状染色要小于孵育了20 min的，这说明发生了一些消失现象。我们用抗$H\text{-}2^k$血清所作的初步研究也给出了相似的结果。高桥最近发现：仅用抗H-2不能引起H-2调变，但如果使用的是低浓度的抗H-2，则抗H-2加上抗免疫球蛋白就确实能导致调变[17]。

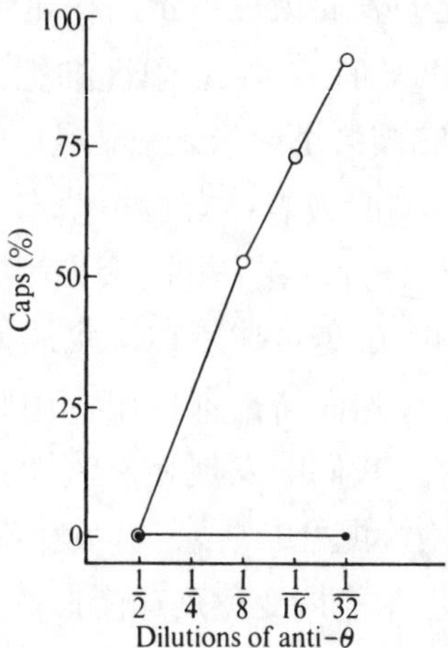

Fig. 5. Cap formation with anti-θ C3H in indirect immuno-fluorescence with CBA thymocytes. Cells were incubated in dilutions of anti-θ in VBS for 20 min at 37°C, washed and incubated with R anti-MIg-Fl for 30 min at 37°C (○) or 0°C (●). In all cases 100% of the cells stained.

The selective segregation and disappearance of specific surface determinants induced by specific antibody imply that at least some membrane components can move relative to one another. The significance of this in terms of membrane structure will be considered in more detail elsewhere and has recently been discussed by Frye and Edidin[22] who studied the movement of histocompatibility antigens on the surface of newly formed heterokaryons.

Immunological Significance

Are cap formation and pinocytosis involved in triggering lymphocyte transformation and/or tolerance induction? Although there is no direct evidence for this, there are a number of observations that are consistent with it. Our failure to induce cap formation or pinocytosis with univalent Fab anti-Ig mirrors the failure of Fab anti-Ig to transform rabbit lymphocytes or of Fab anti-mouse lymphocyte serum to activate mouse lymphocytes, where divalent $F(ab')_2$ did so[23,24]. Mitogens such as concanavalin A[25] and phytohaemagglutinin (PHA)[26] give cap staining localized over the uropod when surface binding was detected by immunofluorescence and concanavalin A binding was followed by pinocytosis[25]. In addition, pinocytosis in the area of the uropod has been observed during mixed lymphocyte reactions[27]. Cytochalasin B, which partially inhibits cap formation and markedly inhibits pinocytosis[28], has been found to inhibit PHA transformation, as well as the transformation of lymphocytes responding to PPD and allogeneic cells (unpublished work of D. Webster and A. C. Allison). Prozone effects observed when lymphocytes are stimulated with increasing concentrations of PHA or ALS[29] or anti-Ig[30] could possibly reflect prozones in inducing cap formation.

If cap formation or pinocytosis play an important physiological role in lymphocyte stimulation or tolerance induction, it should be possible to demonstrate that they can

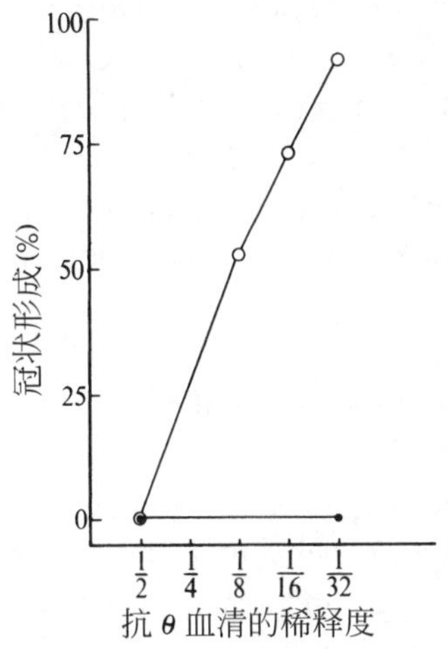

图 5. 用抗 θ C3H 对 CBA 胸腺细胞进行间接免疫荧光染色时的冠状形成。37℃下将细胞在 VBS 中与不同稀释度的抗 θ 共孵育 20 min，冲洗后在 37℃（○）或 0℃（●）下用兔抗鼠免疫球蛋白荧光染色物孵育 30 min。在各种稀释度下染色细胞所占的比例均为 100%。

选择性隔离和由特异抗体引起的特定表面决定簇的消失现象表明，至少有一些膜组分能够发生相互之间的移动。我们将另辟文章详细讨论这在膜结构方面的意义；最近，研究新形成异核体表面的组织相容性抗原运动的弗赖伊和埃迪登[22]也探讨过这个问题。

免疫学意义

冠状形成和胞饮现象是否牵涉到引起淋巴细胞转化和（或）耐受性诱导？虽然没有直接的证据表明如此，但是有一大堆观察结果与此相符。我们用抗免疫球蛋白单价抗原结合片段未能引起冠状形成或胞饮现象，这反映出抗免疫球蛋白抗原结合片段不能用来转化兔淋巴细胞或不能用抗鼠淋巴细胞血清中的抗原结合片段来激活鼠淋巴细胞,而二价抗原结合片段则能做到[23,24]。当通过免疫荧光法检测表面结合时，促分裂素原,如伴刀豆球蛋白 A[25] 和植物凝血素(PHA)[26],能在尾足处形成冠状染色，并且在伴刀豆球蛋白 A 结合后发生胞饮现象[25]。除此之外，在混合淋巴细胞反应[27]中也观察到了尾足区域的胞饮现象。有人发现，能部分抑制冠状形成和显著抑制胞饮现象的细胞松弛素 B[28] 也能抑制 PHA 的转化以及淋巴细胞为应对结核菌素(PPD)和同种异体细胞而发生的转化（韦伯斯特和艾利森，尚未发表的研究结果)。在用浓度不断增加的 PHA 或 ALS[29] 或抗免疫球蛋白[30] 刺激淋巴细胞时观察到的前带效应有可能反映了在诱导冠状形成时的前带。

如果冠状形成或胞饮现象在淋巴细胞刺激或耐受性诱导中起着重要的生理学作用，那就应该可以证明它们是通过与表面免疫球蛋白受体相结合的多价抗原来诱导

be induced by multivalent antigens binding to surface Ig receptors. Most lymphocyte antigen binding studies have been carried out at 0°C, often in the presence of azide, and cap-like polar binding has not been reported. In preliminary studies of antigen binding cells (demonstrated by exposing spleen cells to DNP_1BSA, DNP_8BSA, DNP_5BGG and $DNP_{23}BGG$ and staining with purified fluorescent antibodies against the carrier) we have found that at 37°C DNP_1BSA formed only rings, while all multivalent conjugates formed mainly caps.

The increasing evidence for the importance of matrix formation and multivalent binding at the lymphocyte surface for the induction of immunity (reviewed in refs. 18 and 31) and tolerance[32] could possibly reflect the need for multivalent binding in cap formation and pinocytosis. It should be possible to determine this by using *in vitro* models of induction and tolerance together with methods for purification of antigen-sensitive cells, to study the relationship between the pattern of antigen binding to lymphocytes and the induction-tolerance decision.

We thank Mr. M. R. Young for taking the micrographs used in Fig. 1, and Mr. J. Singh for technical assistance. M. C. R. was supported by a postdoctoral fellowship of the US National Multiple Sclerosis Society and S. de P. by a fellowship of the European Molecular Biology Organization (EMBO).

(*Nature New Biology*, **233**, 225-229; 1971)

Roger B. Taylor and W. Philip H. Duffus: MRC Immunobiology Group, University of Bristol, and Department of Pathology, Medical School, Bristol.

Martin C. Raff and Stefanello de Petris: National Institute for Medical Research, Mill Hill, London.

Received August 18; revised September 20, 1971.

References:

1. Greaves, M. F., *Transplant. Rev.*, **5**, 45 (1970).
2. Pernis, B., Forni, L., and Amante, L., *J. Exp. Med.*, **132**, 1001 (1970).
3. Raff, M. C., Sternberg, M., and Taylor, R. B., *Nature*, **225**, 553 (1970).
4. Raff, M. C., *Immunology*, **19**, 637 (1970).
5. Leibovitz, A., *Amer. J. Hyg.*, **78**, 173 (1963).
6. Mishell, R. I., and Dutton, R. W., *J. Exp. Med.*, **126**, 423 (1967).
7. Raff, M. C., *Nature New Biology*, **229**, 182 (1971).
8. Levey, R. H., and Medawar, P. B., *Ann. New York Acad. Sci.*, **129**, 164 (1966).
9. McFarland, W., Heilman, D. H., and Moorhead, J. P., *J. Exp. Med.*, **124**, 851 (1966).
10. Porter, R. R., *Biochem. J.*, **73**, 119 (1959).
11. Cerrottini, J.-C., and Brunner, K. T., *Immunology*, **13**, 395 (1967).
12. Wessells, N. K., Spooner, B. S., Ash, J. F., Bradley, M. O., Luduena, M. A., Taylor, E. L., Wrenn, J. T., and Yamada, K. M., *Science*, **171**, 135 (1971).
13. Tilney, L. G., *J. Cell Sci.*, **3**, 549 (1968).
14. Portzehl, H., Caldwell, P. C., and Rüegg, J. C., *Biochim. Biophys. Acta*, **79**, 591 (1964).
15. Old, L. J., Stockert, E., Boyse, E. A., and Kim, J. H., *J. Exp. Med.*, **127**, 523 (1968).
16. Lamm, M. E., Boyse, E. A., Old, L. J., Lisowska-Bernstein, B., and Stockert, E., *J. Immunol.*, **101**, 99 (1968).
17. Takahashi, T., *Transpl. Proc.* (in the press).

的。大多数有关淋巴细胞抗原结合的研究是在0℃、通常有叠氮化物存在的条件下进行的,从未有人报道过有冠状极性结合的现象。在对抗原结合细胞的初步研究中(通过将脾细胞暴露于 DNP_1BSA、DNP_8BSA、DNP_5BGG 和 $DNP_{23}BGG$ 并用提纯过的抗载体荧光抗体染色来证明),我们发现37℃时 DNP_1BSA 只形成了环状,而所有多价结合则主要形成冠状。

越来越多的证据表明淋巴细胞表面的基质建造和多价结合对诱导免疫(见参考文献18和31)和耐受性[32]具有重要意义,这有可能反映出冠状形成和胞饮现象是需要多价结合的。我们或许可以通过体外诱导模型和体外耐受性模型以及用提纯抗原敏感细胞的方法来确定上述结论,以便研究抗原结合淋巴细胞的方式和诱导–耐受决定之间的关系。

感谢扬先生为我们拍摄了图1中所用的显微照片,还要感谢辛格先生为我们提供技术支持。马丁·拉夫是美国国家多发性硬化症协会的博士后奖学金的获得者,斯特凡内洛·德·彼得里斯受到了欧洲分子生物学组织(EMBO)提供的奖学金的资助。

(邓铭瑞 翻译;秦志海 审稿)

18. Mitchison, N. A., *In vitro* (in the press).

19. Reif, A. E., and Allen, J. M. V., *J. Exp. Med.*, **120**, 413 (1964).

20. Schlesinger, M., and Yron, I., *Science*, **164**, 1412 (1969).

21. Raff, M. C., *Nature*, **224**, 378 (1969).

22. Frye, L. D., and Edidin, M., *J. Cell Sci.*, 7, 319 (1970).

23. Fanger, M. W., Hart, D. A., Wells, V. J., and Nisonoff, A., *J. Immunol.*, **105**, 1484 (1970).

24. Riethmüller, G., Riethmüller, D., Stein, H., and Hansen, P., *J. Immunol.*, **100**, 969 (1968).

25. Smith, C. W., and Hollers, J. C., *J. Reticuloendothel. Soc.*, **8**, 458 (1970).

26. Osunkoya, B. O., Williams, A. I. O., Adler, W. H., and Smith, R. T., *Afr. J. Med. Sci.*, **1**, 3 (1970).

27. McFarland, W., and Schechter, G. P., *Blood*, **34**, 832 (1969).

28. Allison, A. C., Davies, P., and dePetris, S., *Nature New Biology*, **232**, 153 (1971).

29. Foerster, J., Lamelin, J.-P., Green, I., and Benacerraf, B., *J. Exp. Med.*, **129**, 295 (1969).

30. Sell, S., and Gell, P. G. H., *J. Exp. Med.*, **122**, 423 (1965).

31. Mitchison, N. A., *Immunopathology*, **6**, 52 (1971).

32. Diener, E., and Feldmann, M., *J. Exp. Med.*, **132**, 31 (1970).

33. Wofsy, L., and Burr, B., *J. Immunol.*, **103**, 380 (1969).

Immunoglobulin E (Reagin) and Allergy

D. R. Stanworth

Editor's Note

In 1921, Carl Prausnitz showed it was possible to induce allergic sensitivity in non-allergic people by injecting them with sera from an allergic person. That landmark paper demonstrated that immediate hypersensitivity was mediated by a some tissue-sensitizing factor, which in 1966, Japanese couple Teruko and Kimishige Ishizaka identified as the antibody immunoglobulin E (IgE). Here British immunologist Denis R. Stanworth charts the progress from Prausnitz to present day, speculating that improved understanding of IgE's mode of action could lead to the development of new preventative therapies.

FIFTY years ago Prausnitz showed that it was possible to sensitize sites on his forearms by the intradermal injection of serum from one of his allergic patients (Küstner) who was hypersensitive to fish[1]. The injection of fish extract into the same sites 24 h later evoked immediate weal and erythema reactions similar to those shown by the patient himself (and by other fish-sensitive individuals) in response to the direct intradermal injection of fish extract.

This observation was a historic "landmark" in a field which has not always been noted for objectivity. Besides demonstrating that human hypersensitivity of this immediate-type could be transferred to normal tissue, outside the influence of the allergic individual, it meant that a passively induced reaction could be restricted to a localized area of the skin with minimal discomfort to the non-allergic recipient. Of even greater significance, however, was the inference that immediate hypersensitivity (unlike that of the delayed type, of which tuberculin sensitivity is an example) was mediated by a humoral factor.

There soon followed similar demonstrations of the local transfer of human hypersensitivity to many common inhalants (such as grass pollens, animal danders and house dusts); the Prausnitz–Küstner (P–K) test, as it is known, being used as an alternative to direct skin testing in the diagnosis of hay fever and allergic asthma. Its principal application, however, has been in the experimental study of immediate-type hypersensitivity; where, until a few years ago, it offered the only satisfactory means of assay. Consequently attempts to characterize the skin sensitizing factor more fully have been seriously handicapped.

Early investigations showed this serum constituent to resemble antibodies which appear in the circulation of humans and animals in response to conventional immunization procedures. But it seemed to lack certain properties commonly associated with immune antibodies, such as the capacity to form a precipitate and to fix complement when

免疫球蛋白E（反应素）与过敏反应

斯坦沃思

编者按

1921 年，卡尔·普劳斯尼茨指出向非过敏者体内注入过敏者的血清可能诱导其产生过敏性敏感。他在那篇具有里程碑意义的论文中证实，速发型超敏反应是由某种组织致敏性因子介导的。1966 年，日本夫妇石坂照子和石坂公成将其鉴定为抗体免疫球蛋白 E（IgE）。在本文中，英国免疫学家丹尼斯·斯坦沃思描绘了自普劳斯尼茨至今的研究进展，他推断对 IgE 作用方式的更好理解可能会引导人们发现新的预防疗法。

50 年前，普劳斯尼茨通过皮内注射将一个对鱼过敏的患者（屈斯特纳）的血清注入自己的前臂中，他发现注射部位可能会因此而变得敏感[1]。24 小时后向这个部位注射鱼的提取物可以立即引发斑痕和红斑反应，这些反应与该过敏患者本人（以及其他对鱼过敏的患者）直接接受皮内注射鱼提取物后产生的症状相似。

这一观察结果对这个一向缺乏客观性的领域具有历史性的“里程碑”意义。除了证明人的这种速发型超敏反应可以被转移到受过敏患者影响以外的正常组织中之外，这一结果还显示出这种被动导入的过敏反应仅限于皮肤的局部区域，且给非过敏的接受者带来的不适很小。然而，更重要的意义在于，可以由此推断出速发型超敏反应（与对结核菌素等过敏的迟发型超敏反应不同）是由体液中的一种因子介导的。

之后很快有类似的发现称，人类对许多常见吸入物（如草的花粉、动物的皮屑和房间里的灰尘）的超敏反应也可以被转移到局部区域；众所周知，在诊断枯草热和过敏性哮喘时，可以用普劳斯尼茨 – 屈斯特纳（P–K）试验取代直接的皮内试验。然而到目前为止，这种方法最重要的应用仍在于速发型超敏反应的实验研究，并且直到几年前它还是唯一令人满意的检测方法。随后在尝试对皮肤致敏因子进行更为详细的特征描述时却遇到了严重的障碍。

早期的研究显示这种血清中含有类似抗体的组分。在人和动物对常见的免疫过程做出应答时，其血液循环中会出现抗体。不过这一组分似乎并不具备免疫抗体的某些常见特性，例如：在体外实验中，它不能够在与特异性抗原结合时产生沉淀，

combined with specific antigens *in vitro*. For these reasons, some investigators[2] preferred to reserve judgment on its antibody status by terming it "atopic reagin" ("atopy" denoting a form of human hypersensitivity in which there was evidence of a hereditary predisposition, the sensitizing substances being referred to as "atopen"). The factor demonstrable in the serum of individuals with immediate hypersensitivity of the asthma–hay fever–urticaria type, has since been termed "reagin" and it has been customary to refer to the provoking agents (in inhalants, foods and so on) as "allergens".

P–K testing revealed that reagins become firmly "fixed" in normal isologous human skin where they are detectable (by subsequent allergen challenge) several weeks after the transfer of the allergic serum. In contrast, antibodies produced by the immunization of rabbits with allergenic substances (such as pollen extracts and egg protein) failed to fix to human skin; although unlike reagins, they could passively sensitize the skin of heterologous guinea-pigs for relatively short times. Human antibodies (γG type) against diphtheria toxoid likewise disappear rapidly when injected into isologous human skin (with a half-life of 12 h)[3], whereas reagins transferred isologously have been estimated[4] to have a "half-life" of 15 days.

Quantitative P–K testing[5] based on the accurate measurement of the areas of weals produced on the backs of normal recipients showed that a maximal skin response was achieved with a time interval of about 50 h between transfer of allergic serum and challenge with specific allergen. Once this was effected, however, a weal and erythema reaction developed rapidly with the weal approximately doubling in area during 10–20 min after introduction of the allergen. The susceptibility of reagins to relatively mild heat treatment was also confirmed[5]. Similar conditions are used in the destruction of the complement activity of immune sera, but the addition of fresh normal human serum to heated allergic serum fails to restore P–K activity[6] and other evidence indicates that reagins do not depend on the complement system for their activity.

Reagins were also found to differ from immune (7S type) antibodies in their inability to cross the placenta. This situation is obviously fortunate for the offspring of allergic women, as is the apparent absence of reagins from human colostrums. It has been attributed by some investigators to the macroglobulin (19S) antibody nature of the tissue sensitizing factor, which was thought to be of sufficiently large molecular size to be retained by the human placenta's supposed sieving mechanism. But extensive studies by Brambell and his associates[7], and others have indicated that the placental transmission of immunoglobulins is a highly selective process (controlled by sites located within their Fc regions).

The biological properties of reagins revealed by these early investigations, based principally on P–K testing, are summarized in Table 1. This was the position around 1940, when techniques of free-solution electrophoresis and ultracentrifugation were first applied to the fractionation of plasma proteins. Although such physico-chemical procedures offered better means of separating complex protein mixtures than had been

也不能结合补体。基于这些原因，一些研究人员[2]对它的抗体地位持保留态度，仍将其称为“特应性反应素”（“特应性”指的是人的一种超敏反应，有证据表明其具有遗传易感性，引起这种超敏反应的致敏物质被称为“特应原”）。在哮喘–枯草热–风疹类型的速发型超敏反应患者的血清中，与超敏有关的因子被称为“反应素”，而引发这些超敏反应的物质（在可吸入物或食物等中）则被习惯性地称为“过敏原”。

P–K 试验显示：反应素可以牢牢地“固定”到同源的正常人体的皮肤上，并且在过敏性血清移植实验后好几周仍能被检测到(即用过敏原刺激后会发生过敏反应)。与此相反，使用引起过敏的物质（如花粉提取物、鸡蛋蛋白）免疫兔子得到的抗体则不能固定到人的皮肤上；但与反应素不同，这种抗体可以使异源的豚鼠皮肤在相对较短的时间内被动致敏。人抗白喉类毒素抗体（γG 型）在以同样方式注射到同源人体的皮肤中之后很快便消失了（半衰期为 12 小时）[3]，但据估计[4]在同源间转移的反应素的“半衰期”可以达到 15 天。

定量 P–K 试验[5]是基于对正常接受者背部产生的斑痕面积的精确测量而进行的。其结果显示，在转移过敏性血清约 50 小时后再注入特定过敏原可引起最强烈的皮肤过敏反应。然而，即使马上接触过敏原，斑痕和红斑反应也会迅速发生，斑痕面积在引入过敏原之后 10 分钟～20 分钟内会增加一倍左右。此外，反应素对于比较温和的加热处理很敏感也得到了证实[5]。类似的实验条件同样可以破坏免疫血清中补体的活性，而向经加热处理的过敏性血清中加入正常人的新鲜血清并不能恢复该血清的 P–K 活性[6]，其他证据也证明反应素的活性并不依赖于补体系统。

人们还发现反应素与免疫（7S 型）抗体不同，它不能穿越胎盘。这一点对于过敏体质女性的后代来说显然是幸运的，在产妇的初乳中也明显不含有反应素。一些研究人员认为：组织中致敏因子的本质是巨球蛋白（19S），这些球蛋白的分子很大以至于不能够穿越血胎屏障而影响胎儿。不过布兰贝尔和他的同事们[7]以及其他一些研究者通过大量的研究证明，免疫球蛋白通过血胎屏障是一个高选择性的过程（由免疫球蛋白结晶片段（Fc）区上的某些位点控制）。

这些主要基于 P–K 试验的早期研究揭示出了反应素的生物学特性，表 1 中汇总了研究得到的所有生物学特性。在 1940 年前后，自由溶液电泳技术和超速离心技术首次被应用于分离血浆中的蛋白质组分。尽管这些物理化学方法在复杂蛋白质混合物的分离上比以前靠盐类和低温乙醇沉淀进行的分离更有效，但它们本质上仍是用

previously possible, by salt or low temperature-ethanol precipitation, they were essentially analytical tools. Nevertheless, electrophoretic analyses of the sera of animals before and after immunization provided the first evidence of the γ globulin nature of precipitating antibodies[8]; while corresponding ultracentrifugal studies indicated a size heterogeneity (antibody activity being associated with both 7S and 19S components).

Table 1. Main Biological and Physico-chemical Properties of Reagins (γE-Antibodies)

(1) Bind firmly to isologous, and closely related heterologous, tissues
(2) Heat labile (destroyed by 56°C for 1 h)
(3) Not dependent on complement
(4) Fail to cross the placenta
(5) Move in the "fast γ" region on electrophoresis
(6) Sediment near to 8 S

It was naturally hoped that a similar approach would provide more convincing evidence of the antibody nature of reagins; but this was thwarted by practical difficulties. It was established, however, that reagins were electrophoretically faster than immune γ globulins, moving in the β region (where 19S type immune antibodies were also found). But this conclusion was very much an extrapolation, and the minute amount of skin sensitizing factor present in allergic sera could have been moving in the electric field in combination with a major constituent. More refined preparative fractionation procedures were required.

Physico-chemical Characteristics

As in the characterization of many other minor active protein constituents of biological fluids, it was the advent of zone-fractionation procedures which gave the first definitive indication of the physico-chemical characteristics of the skin sensitizing factor. Thus, zone electrophoresis of allergic sera in starch blocks[9] confirmed that reagins moved in the "fast γ" (slow β) region; whilst zone-centrifugation in buffered sucrose gradients[10] showed clearly that they sedimented in the 7S region (and were not macroglobulins, as had been suggested to explain their failure to cross the placenta). There were, of course, limitations to this type of correlative approach because each of the major serum electrophoretic fractions, including the γ fraction, comprised several components. It was at this stage, however, that the application of highly-specific and sensitive immunological techniques began to compensate for the shortcomings of the physico-chemical procedures. Studies on the monoclonal γ globulins, isolated in large quantities from the sera of patients with multiple myelomatosis and Waldenström macroglobulinaemia, led to the production of specific antisera which distinguished three major immunoglobulin classes[11]: γG or IgG (previously referred to as 7S γ globulin): γM or IgM (previously referred to as 19S macroglobulin) and γA or IgA (a newly discovered third class[12], which when isolated from normal human serum by a zinc precipitation procedure was found to comprise a major 7S component, with minor amounts of polymerized material, and which migrated in the fast γ region on electrophoresis). It was established that the three major immunoglobulin classes then known possessed common light polypeptide chains, but distinctive heavy chains (Fig. 1).

于分析的手段。尽管如此，用电泳实验分析动物免疫前后的血清成分为证明沉淀出的抗体的本质是γ球蛋白提供了第一个证据[8]；而相应的超速离心研究则证明了抗体分子在大小上的不均一性（7S 和 19S 组分都与抗体活性有关）。

表 1. 反应素的主要生物学特性和物理化学特性（γE 型抗体）

（1）与同源组织紧密结合，与异源组织也密切相关
（2）热不稳定性（在 56℃下处理 1 小时后活性被破坏）
（3）不依赖于补体
（4）不能穿越血胎屏障
（5）在电泳实验中位于“快 γ”区
（6）沉降系数约为 8S

人们很自然地希望有一种类似的方法能提供更多令人信服的证据来证明反应素的本质是抗体；不过这种想法实行起来有难度。虽然反应素比γ免疫球蛋白的电泳速度更快，位于β区域（19S 型免疫抗体也位于这一区域），但这一结论在很大程度上只是一种推断：在电场作用下，过敏性血清中微量的皮肤致敏因子可能与血清中的某种主要成分结合而共同泳动。我们需要更精细的样品分离制备步骤。

物理化学特性

和表征生物液体样品中许多其他微量活性蛋白组分时一样，直到区带分离技术出现后人们才得以首次得到关于皮肤致敏因子的物理化学特性的决定性依据。例如：过敏性血清的淀粉区带电泳实验[9]证实，反应素在电泳中位于“快γ”（慢β）区；同时蔗糖密度梯度缓冲液的区带离心实验[10]显然说明，反应素的沉降系数位于 7S 区（也就是说它不是巨球蛋白，同时证明之前以反应素的分子太大为由来解释其不能穿越血胎屏障的说法是不成立的）。当然，这种关联法也有其局限性，因为各个主要的血清电泳组分，包括γ组分，都由数种成分组成。不过，在现阶段可以用高度特异性和敏感性的免疫学技术来弥补物理化学分析方法的不足。研究人员对从多发性骨髓瘤和瓦尔登斯特伦巨球蛋白血症患者血清中分离出来的大量单克隆γ球蛋白进行了分析，结果发现它是一种不同于三种主要免疫球蛋白类型的特殊抗血清[11]，即不同于γG（或 IgG，之前称为 7S γ球蛋白）、γM（或 IgM，之前称为 19S 巨球蛋白）和γA（或 IgA，最新发现的第三种类型[12]，它是人们利用锌离子沉淀法从正常人的血清中分离出来的，主要由 7S 组分组成，含有少量的多聚物质，在电泳中位于快γ区）。已经确认目前已知的这三种主要类型的免疫球蛋白都具有相同的轻多肽链，但重链有所不同（图 1）。

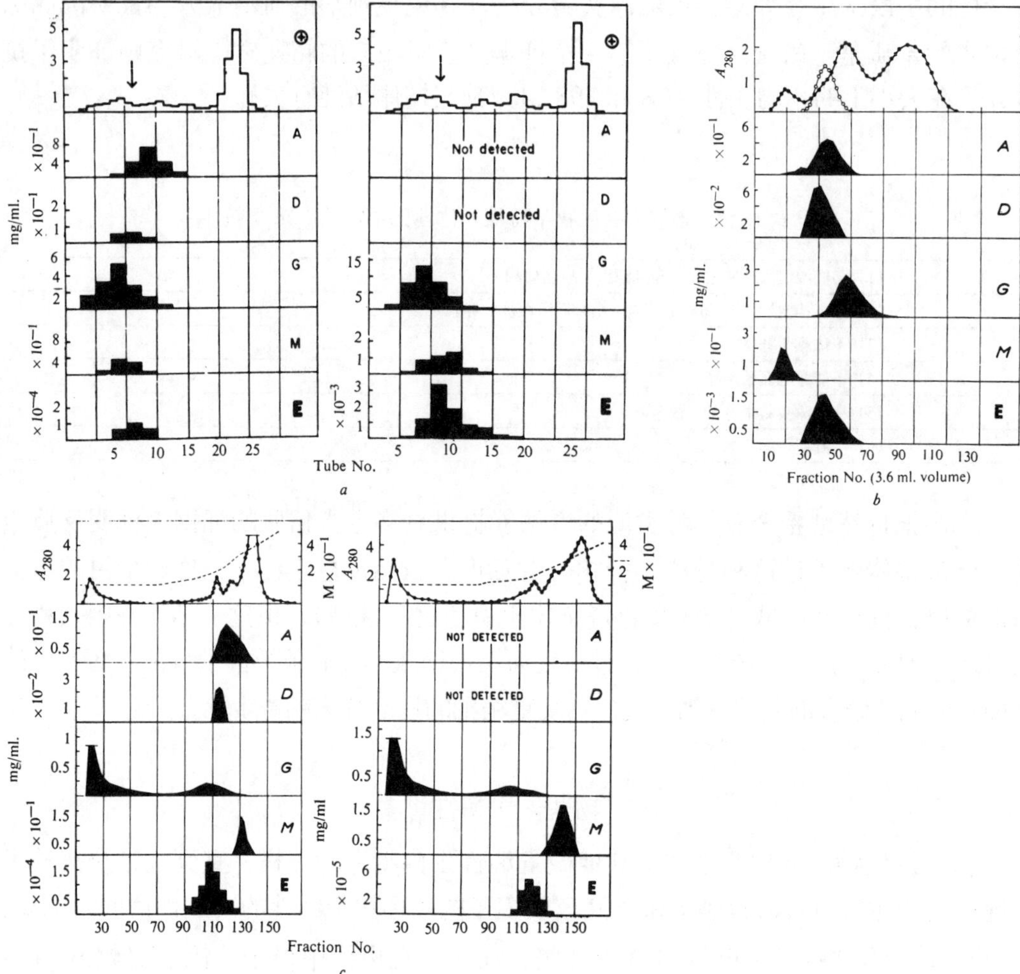

Fig. 1. A comparison of the physico-chemical properties of the five major human immunoglobulin classes. *a*, Zone electrophoresis; *b*, gel-filtration; *c*, ion-exchange chromatography on DEAE-"Sephadex" (reproduced from ref. 26, by courtesy of the authors).

The problem then was to show that skin sensitizing activity was a property of one (or perhaps more) of these antigenically distinguishable immunoglobulins. Antigenic and physico-chemical analysis of fractions of allergic human serum separated by chromatography on DEAE-cellulose (which provided a high degree of resolution of the various classes of immunoglobulin) revealed maximal P–K activity in a fraction in which only electrophoretically fast 7S γG globulin could be detected. As Fig. 1*c* shows, this was in an elution position well ahead of any detectable γM globulin; and even ahead of any γA globulin detectable by immunodiffusion analysis using specific antisera[13].

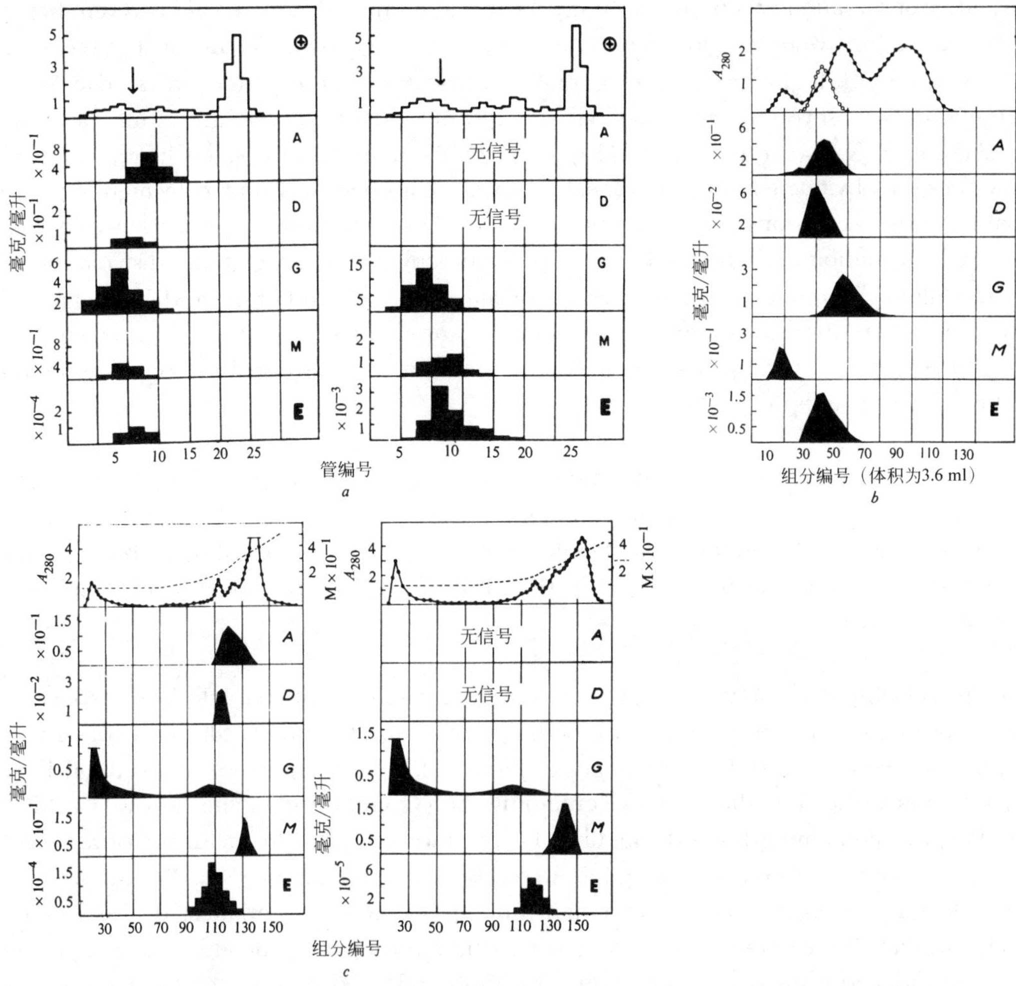

图 1. 五种主要类型人免疫球蛋白的物理化学性质比较。(*a*) 区带电泳实验；(*b*) 凝胶过滤实验；(*c*) 二乙氨基乙基－葡聚糖凝胶柱离子交换层析（经作者允许，从参考文献 26 中复制）。

接下来的问题是要证明，皮肤致敏活性是一种（也可能是多种）抗原不同的免疫球蛋白的特性。研究人员对经过二乙氨基乙基－纤维素层析（具有很高的分辨率，可以区分开不同类型的免疫球蛋白）分离的过敏患者血清各组分的抗原特性和物理化学性质进行了分析，发现在其中一个组分中能检测到最高的 P–K 活性。在电泳中，只有该组分能迁移到快 7S γG 球蛋白区。如图 1*c* 所示，该组分的洗脱峰位于所有能检测得到的 γM 球蛋白的前面；甚至还在用特异性抗血清进行免疫扩散分析时能检测得到的任一 γA 球蛋白的前面 [13]。

This latter observation was particularly significant, at a time when many claims were being made that reagins were γA globulins. These were based on such evidence as the recovery of P–K activity in highly purified γA globulin preparations[14], the removal of activity from allergic sera by absorption with specific sheep anti-human γA globulin antiserum[15] and the blockage of P–K reactions by normal human γA globulin or its isolated heavy chains[16]. They were of obvious teleological appeal as γA globulin had been found in high levels in nasal and other secretions, where it supposedly acted as a "front-line" defensive agent. But a lack of correlation was observed between reagin activity and γA globulin distribution in DEAE cellulose chromatorgraphic fractions of allergic sera[13]; and other studies[17] suggested that P–K activity was associated with a minor component of the sera of hypersensitive individuals, which had appeared as a contaminant in the supposedly pure γA globulin fractions investigated previously.

Were reagins a unique type of immunoglobulin (a prospect on which I had speculated in 1963 (ref. 18))? The discovery of a new human immunoglobulin class (IgD or γD)[19], which resembled γA globulin electrophoretically, seemed a possibility; but several independent investigations suggested this immunoglobulin did not possess skin sensitizing activity.

Reagins were found, however, to have unusual physico-chemical characteristics as well as distinctive biological properties (for antibodies). For example, the gel filtration of sera of people sensitive to ragweed (on "Sephadex G200"[20]) led to the recovery of maximal P–K activity slightly ahead of the 7S peak containing the γG and γA globulins, but well behind the 19S peak containing the γM globulin. Furthermore, reagins in such sera were found by density gradient ultracentrifugation[21] to possess a sedimentation coefficient of 7.7S (7.4–7.9S range)—significantly greater than human γG and γA globulins. It also seemed significant that the reagin-like antibodies, occurring naturally in some animals (such as the dog) and induced artificially in others (for example, rats and rabbits), had sedimentation coefficients greater than 7S but much less than 19S (Table 2).

Table 2. Sedimentation Characteristics of Reagin-type Antibodies Formed in Various Mammalian Species (as Determined by Density Gradient Ultracentrifugation)

Species	Mode of antibody formation	Sed. coeff. (S°)
Human	Spontaneously	>7.4S–7.9S
Dog	Spontaneously	6.8S–10.1S
Rat	Experimentally	>7S, ≪19S
Rabbit	Experimentally	>7S, ≪19S

Further evidence that reagins were probably a unique class of immunoglobulins was obtained by Ishizaka and his associates[22], who isolated active preparations by application of a multistep procedure to the fractionation of pools of sera from individuals showing marked hypersensitivity to ragweed pollen and who developed highly sensitive

后一个观察结果非常重要，因为当时有很多人认为反应素就是γA球蛋白。这一观点的提出主要基于以下证据：由高纯度γA球蛋白制备物能够恢复P–K活性[14]；使用特异的绵羊抗人γA球蛋白的抗血清除去过敏性血清中的γA球蛋白后，其P–K活性也随之消失[15]；此外，使用正常人血清中的γA球蛋白或其分离的重链都可以抑制过敏性血清的P–K活性[16]。γA球蛋白在鼻子和其他器官的分泌物中含量很高，这些分泌物被认为是抵御外来物质的"第一道防线"，基于此，人们很自然地把γA球蛋白和反应素联系在了一起。然而，在用二乙氨基乙基－纤维素层析分离得到的过敏性血清的各个组分中，人们发现含γA球蛋白的组分并不具有反应素的活性[13]；其他研究也表明P–K活性与超敏患者血清中的一种微量组分有关[17]，这说明在以前检测过的看似高纯度的γA球蛋白中很可能混入了这样的污染物。

反应素会不会是一种特殊类型的免疫球蛋白（在1963年我曾做过这样的推测，参考文献18）呢？一种新发现的人免疫球蛋白类型（IgD或γD）[19]具有与γA球蛋白相似的电泳特征，这似乎提供了一种可能性；不过，几项独立的研究结果显示，这种免疫球蛋白并不具有引起皮肤过敏的活性。

然而，人们发现（对于抗体来说）反应素具有不寻常的物理化学性质和与众不同的生物学特性。例如，用凝胶过滤（葡聚糖凝胶G200[20]）对豚草过敏的人的血清后，发现稍早于含γG和γA球蛋白的7S洗脱峰的组分具有最大的P–K活性，但该组分远远晚于含有γM球蛋白的19S洗脱峰。此外，密度梯度超速离心实验[21]的结果显示，该血清中反应素的沉降系数为7.7S（7.4S ~ 7.9S）——明显高于人γG和γA球蛋白。以下这一点显然也很重要：一些类似反应素的抗体在某些动物（例如狗）中会天然存在，而在另一些动物（如大鼠和兔子）中可以人为地诱导出来，它们的沉降系数都大于7S而远小于19S（表2）。

表2. 形成于几种哺乳动物中的类反应素抗体的沉降特性（用密度梯度超速离心法测量）

物种	抗体形成方式	沉降系数（S°）
人	自发	>7.4S~7.9S
狗	自发	6.8S~10.1S
大鼠	实验诱导	>7S, ≪19S
兔子	实验诱导	>7S, ≪19S

石坂公成及其同事发现了证明反应素很可能是一种独特的免疫球蛋白类型的新证据[22]。他们采用一个多步过程从对豚草花粉高度过敏的患者血清中分离得到了活性组分，并且发展出一种高敏感性的放射性免疫扩散分析方法，在体外实验中这种

radioimmune-diffusion assays as *in vitro* alternatives to the P–K test for the measurement of the reagin in the isolated fractions. Reagin preparations thus obtained were used to produce anti-reagin antisera, which were rendered specific by absorption with the major classes of immunoglobulin and which were used in radio-immunoelectrophoresis to provide evidence that a radio-labelled ragweed allergen (E) preparation combined with the reagin isolated from ragweed-sensitive individuals' sera.

Ishizaka and associates[23] suggested that reagin activity was carried by a unique immunoglobulin which they tentatively termed "γE" (in view of its specific binding capacity for ragweed allergen E). Despite the evidence presented in support of this idea, however, some investigators felt that such claims were perhaps a little premature until chemical evidence was available to suggest the existence of a unique type of polypeptide chain.

Myeloma Protein

The final evidence that reagins were a unique class of immunoglobulins came with the isolation of an atypical myeloma protein (originally designated "IgND"), from the serum of a patient ("ND") with myelomatosis and Bence Jones proteinuria[24], which lacked the antigenic determinants characteristic of the heavy polypeptide chains of the then known immunoglobulin classes (γG, γM, γA and γD) while possessing similar light chain determinants (of sub-type L). Like myeloma γA and γD globulins, this atypical protein moved in the "fast γ" region on zone electrophoresis (Fig. 1); but it was eluted slightly ahead of these classes of immunoglobulin during chromatography on DEAE-"Sephadex". Of greater significance, however, was its elution position during gel-filtration on "Sephadex G100", where it emerged after the γA globulin but just ahead of the γG globulin. Furthermore, free solution ultracentrifugal analysis[25] showed the atypical myeloma protein (IgND) to possess a sedimentation coefficient (S°20, W) of 7.92S (and molecular weight 196,000 from Archibald analysis); a value very similar to that which had been assigned to the reagins in sera of ragweed-sensitive individuals on the basis of the observed rate of sedimentation of the P–K activity in buffered sucrose gradients.

This similarity in sedimentation coefficient between the new type of myeloma protein and reagin could have been fortuitous, but it seemed significant that the myeloma protein isolated from the serum of patient "ND" was the first known type of immunoglobulin to have similar size characteristics to the skin sensitizing factor. Other observations were beginning to suggest a relationship between the new class of immunoglobulin and reagin. For example, by means of a radio-immunosorbent technique based on the use of "Sephadex"-coupled antibody directed specifically against the new myeloma protein[24], a significantly elevated concentration (5,900 ng/ml.) of an antigenically similar protein was detected in the serum of an individual with allergy to dog dander; in comparison with the much smaller amounts (100–700 ng/ml.) found in normal sera. Later comparative studies[26] of the antigenic characteristics of the new myeloma globulin and the reagin-rich fraction (designated γE) isolated from sera of ragweed-sensitive individuals were to provide evidence of a close structural relationship between the two proteins.

方法可以代替 P–K 试验用以检测分离出来的各组分中的反应素。这样得到的反应素被用来制备抗反应素的抗血清。他们特异地去除了抗血清中几种主要的免疫球蛋白类型，然后用这个抗血清进行放射性免疫电泳实验，以此证明放射性标记的豚草过敏原（E）可以结合从对豚草过敏的患者血清中分离出来的反应素。

石坂及其同事[23]认为反应素活性源自一种独特的免疫球蛋白，他们将其暂时命名为“γE”(因为它可以和豚草过敏原 E 特异地结合)。然而，尽管有上述证据支持石坂等人的观点，一些研究人员仍然认为，在找到化学证据证明存在这种独特的多肽链之前提出上述论断或许有点为时尚早。

骨髓瘤蛋白

最后一个能证明反应素是一种独特的免疫球蛋白类型的证据来自从患有骨髓瘤病和本周蛋白尿症的患者（“ND”）血清中分离出来的非典型骨髓瘤蛋白（起初被命名为“IgND”)[24]。这种蛋白缺乏四种已知免疫球蛋白（γG、γM、γA 和 γD）类型的重肽链所特有的抗原决定簇，但有与它们类似的轻链决定簇（L 亚型)。与骨髓瘤 γA 和 γD 球蛋白类似，这种非典型蛋白在区带电泳中位于“快 γ”区（图 1)；但是它在二乙氨基乙基 – 葡聚糖凝胶层析中的洗脱峰略早于这两类免疫球蛋白。然而，如果使用葡聚糖凝胶 G100 进行凝胶过滤层析，其洗脱位置的意义更重大：这种蛋白的洗脱峰晚于 γA 球蛋白但略早于 γG 球蛋白。此外，自由溶液超速离心分析[25]显示，这种非典型骨髓瘤蛋白（IgND）的沉降系数（S°20，W）为 7.92S（并且通过阿奇博尔德分析测得其分子量为 196,000)；这一数值与豚草过敏患者血清中反应素的沉降系数非常接近，后者是基于在蔗糖梯度缓冲液中测定 P–K 活性的沉降速度而得到的。

也许这种新发现的骨髓瘤蛋白与反应素在沉降系数上的相似性是一种巧合，不过似乎值得注意的一点是，从患者“ND”的血清中分离出来的这种骨髓瘤蛋白是人们发现的第一个与皮肤致敏因子具有相似尺寸特征的免疫球蛋白类型。其他观察结果正开始预示这种新型免疫球蛋白和反应素之间的某种联系。例如，使用葡聚糖凝胶偶联抗体进行放射免疫吸附试验，可以特异地识别这种新的骨髓瘤蛋白[24]，也可以显著提高一个对狗皮屑过敏的人的血清中一种有类似抗原性的蛋白的浓度（5,900 纳克 / 毫升)。在正常血清中这种蛋白的浓度要低得多（100 纳克 / 毫升 ~ 700 纳克 / 毫升)。后来，有人对这种新的骨髓瘤球蛋白的抗原特性和从豚草过敏患者血清中分离得到的富含反应素组分（命名为 γE）的抗原特性进行了比较研究[26]，从而为这两种蛋白在结构上的相近性提供了依据。

The most convincing evidence that the myeloma protein (IgND) was a pathological counterpart of reagin was obtained, however, from inhibition-P–K testing in a normal human recipient[27]. The intradermal injection of the myeloma protein mixed with, or 24 h before, the sensitizing serum (at a dosage of 5–50 times in excess of the level of reagin in the serum) completely inhibited a weal and erythema response on subsequent challenge with the specific allergen (isolated from horse dandruff). Thus not only was the myeloma protein (IgND) antigenically similar to reagin, it also seemed to have its striking affinity for isologous tissues. There was no evidence, however, that the myeloma protein could bind antigen (that is allergen). Thus it seemed to be a pathological counterpart of a new immunoglobulin class of which reagin was a normal representative, just as an over-production of monoclonal forms of the major immunoglobulin classes has often been seen in plasmocytic and lymphocytic neoplastic disorders.

Immunological and chemical studies[25] indicated that the myeloma protein IgND was a new class of immunoglobulin, with similar light chains to those in the other four classes, but distinctive heavy polypeptide chains somewhat larger (75,500 molecular weight) than those in the other immunoglobulins. Consequently at a meeting initiated by the World Health Organization in Lausanne in 1968 (ref. 28) it was decided that there was sufficient evidence to conclude that reagin was representative of this new class of immunoglobulin, which it was proposed should be designated "IgE" or "γE".

The availability of the myeloma protein offers a means of chemically characterizing reagins, and ultimately of explaining their unusual biological properties in structural terms. It has been calculated that the chance of finding a case of monoclonal production of γ globulin of the E type is about one in 50,000 that of finding a more common case of myelomatosis; so the protein isolated from the serum of the Swedish case has proved particularly valuable and highly sought after. Fortunately, however, a second case has been reported[29] in the United States, who has clinical manifestations similar to those seen in the case originally reported by Johansson and Bennich in Sweden.

Already the two γE myeloma proteins are providing important information about the role of the antibody (reagin) in immediate hypersensitivity reactions. For example, the myeloma IgE has been degraded into different types of polypeptide fragment by means of proteolytic and chemical cleavage procedures, and tested the ability of these to inhibit skin sensitization by the reagins (γE antibodies) in the sera of individuals sensitive to horse dandruff and grass pollen[30,31]. Of the various types of fragments tested (originating from the different parts of the IgE molecule as illustrated schematically in Fig. 2) only those incorporating the Fc region had the inhibitory activity of the parent molecule. Similar findings have resulted from inhibition studies in sub-human primates (for example, rhesus monkeys and baboons) which, like non-allergic humans, have proved receptive to sensitization by human γE antibodies but which (for obvious reasons) are now preferred as test recipients.

然而，有一个非常令人信服的证据能证明骨髓瘤蛋白（IgND）是反应素在病理上的对应物，该证据来自正常受试者血清的抑制性 P–K 试验[27]。将骨髓瘤蛋白与致敏性血清混合（以高于血清中反应素 5 倍 ~ 50 倍的剂量）或者在注射致敏性血清之前 24 小时进行皮内注射，都可以完全抑制随后由特定过敏原（从马的皮屑中分离得到）引起的斑痕和红斑反应。因此，这种骨髓瘤蛋白（IgND）不仅在抗原性上与反应素相似，它似乎还与同源组织具有很高的亲和力。然而，尚没有证据表明骨髓瘤蛋白可以与抗原（这里指过敏原）结合。如此看来，反应素似乎是这种新免疫球蛋白类型在体内的正常存在形式，而骨髓瘤蛋白则是其病理上的对应物，正如在患有浆细胞瘤和淋巴细胞瘤的人身上经常观察到的主要免疫球蛋白类型单克隆形式过度分泌的现象一样。

免疫学和化学研究[25]表明：骨髓瘤蛋白 IgND 是一种新型的免疫球蛋白，它具有与另外四种免疫球蛋白类型相似的轻链，但其重链（分子量为 75,500）却比其他类型的重链略大一些。因此，在 1968 年由世界卫生组织在瑞士洛桑举办的会议上（参考文献 28），大家一致认为有足够的证据可以证明反应素是这种新免疫球蛋白类型的代表，并建议把这种免疫球蛋白命名为“IgE”或者“γE”。

骨髓瘤蛋白的发现提供了一种用化学方法表征反应素的方式，这种方式可以使我们最终在结构水平上解释其独特的生物学特性。有人计算得出，找到一例 E 型单克隆 γ 球蛋白的概率大概是找到一例更典型的骨髓瘤蛋白的 1/50,000。因此，从这个瑞典患者的血清中分离出的这种蛋白是非常珍贵和难得的。然而，幸运的是，美国的研究人员已报道了第二个类似的例子[29]。这个患者的临床症状与约翰松和本宁最初在瑞典报道的病例相似。

这两例 γE 骨髓瘤蛋白现已为研究抗体（反应素）在速发型超敏反应中的作用提供了重要的信息。例如，使用蛋白酶解和化学消化的方法可以将骨髓瘤 IgE 降解成不同类型的多肽片段，然后检测这些片段对于由马皮屑和草花粉过敏患者血清中反应素引起的皮肤敏感化的抑制能力[30,31]。在检测的这些不同类型的片段中（如图 2 所示，来自 IgE 分子中的不同部分），只有那些包含 Fc 区的片段具有对亲本分子的抑制活性。类似的抑制性结果在非人灵长类（例如恒河猴和狒狒）中也被发现。和人类相似，原本不过敏的动物在注射人 γE 抗体后也会变得对某种过敏原过敏，（由于显而易见的原因）人们更倾向于用非人灵长类动物作为测试的受体。

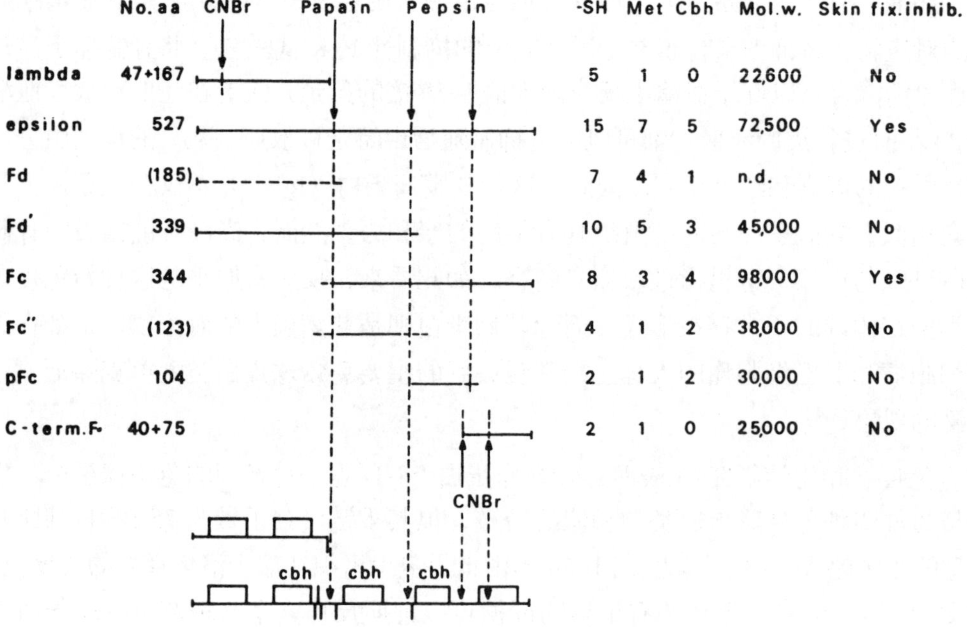

No. aa
CNBr
Papain
Pepsin
-SH
Met
Cbh
Mol.w.
Skin fix.inhib.
lambda
47+167
5 1 0 22,600 No
epsilon
527
15 7 5 72,500 Yes
Fd
(185)
7 4 1 n.d. No
Fd'
339
10 5 3 45,000 No
Fc
344
8 3 4 98,000 Yes
Fc''
(123)
4 1 2 38,000 No
pFc
104
2 1 2 30,000 No
C-term.F
40+75
2 1 0 25,000 No
CNBr
cbh
cbh
cbh

Fig. 2. Working model for molecular structure of IgE indicating the disposition of antigenically and biologically active fragments, and carbohydrate (cbh) prosthetic groups. (Reproduced from ref. 62, by permission of the authors.)

Other inhibition sensitization tests in baboons[32], have involved myeloma IgE preparations which have been reduced to various extents with 2-mercaptoethanol, to ascertain the importance of the relatively large number of disulphide bridges in maintaining the conformational integrity of that part of the IgE molecule involved in tissue binding.

Binding of IgE

Inhibition tests with proteolytic cleavage fragments of myeloma IgE have led to important conclusions about the binding of antibody IgE to isologous (or closely related heterologous) tissue. They suggest that, like the myeloma protein molecules, the antibody (reagin) molecules similarly bind to the tissue receptors through sites within their Fc regions (Fig. 3). Sites within the Fab region of the cell-bound antibody molecules are thus free to react subsequently with antigen (allergen) in the usual manner. In contrast, the myeloma IgE molecules (like most myeloma forms of the other immunoglobulin classes) cannot combine with antigen (Fig. 3). If, on the other hand, combination with the target cells had been found to occur through the Fab regions of the antibody IgE molecule, it would have been necessary to infer that immediate-type hypersensitivity is an auto-immune phenomenon involving the production of antibody against self-cell surface antigen (as occurs, for example, in autoimmune haemolytic anaemia).

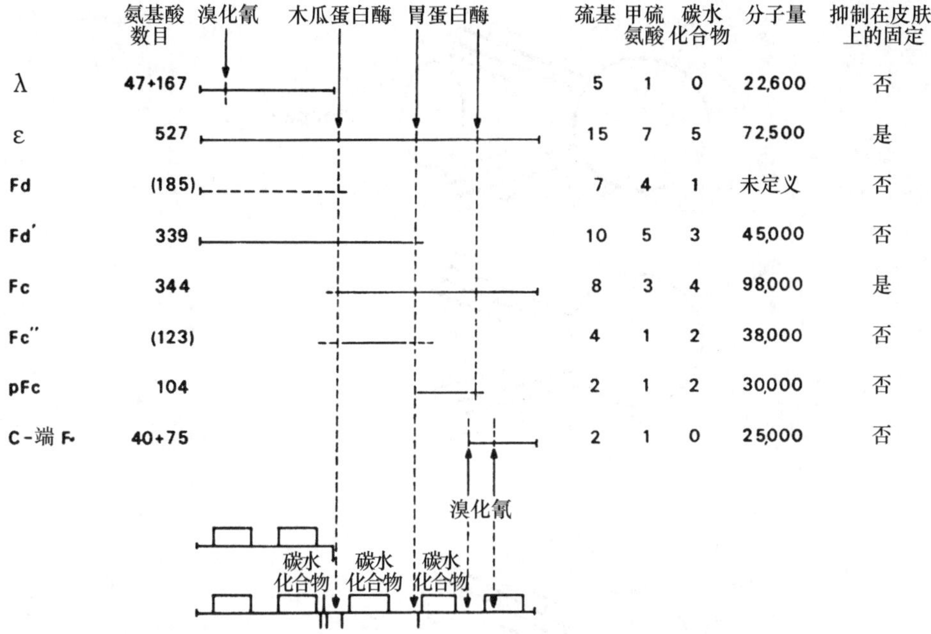

图 2. IgE 分子结构的实用模型，图中显示出抗原结合活性和生物学活性的排列，以及碳水化合物辅基的位置。（征得各位原作者许可后，从参考文献 62 中复制。）

另一些抑制皮肤敏感化的试验是在狒狒中进行的[32]，试验中使用的骨髓瘤 IgE 样品被 2– 巯基乙醇不同程度地还原，目的在于确定相对较多的二硫键在维持 IgE 分子中与组织结合部分的构象完整性方面的重要性。

IgE 的 结 合

使用经蛋白酶解的骨髓瘤 IgE 片段进行抑制试验，得出一些关于 IgE 抗体可与同源组织（或比较接近的异源组织）相结合的重要结论。结论认为：和骨髓瘤蛋白类似，抗体（反应素）分子同样可以通过其 Fc 区中的位点与组织上的受体发生结合（图 3）。随后细胞结合型抗体分子就可以通过其抗原结合片段（Fab）区中的位点自由地与抗原（过敏原）按照通常方式结合。与之相反，骨髓瘤 IgE 分子（与大多数其他免疫球蛋白类型的骨髓瘤蛋白类似）则不能与抗原结合（图 3）。另一方面，如果发现 IgE 抗体分子会通过其 Fab 区与靶细胞结合，那么就应该可以断定：速发型超敏反应是一种自体免疫现象，即身体产生了抗自身细胞表面抗原的抗体（正如在自体免疫性溶血性贫血症等中发生的一样）。

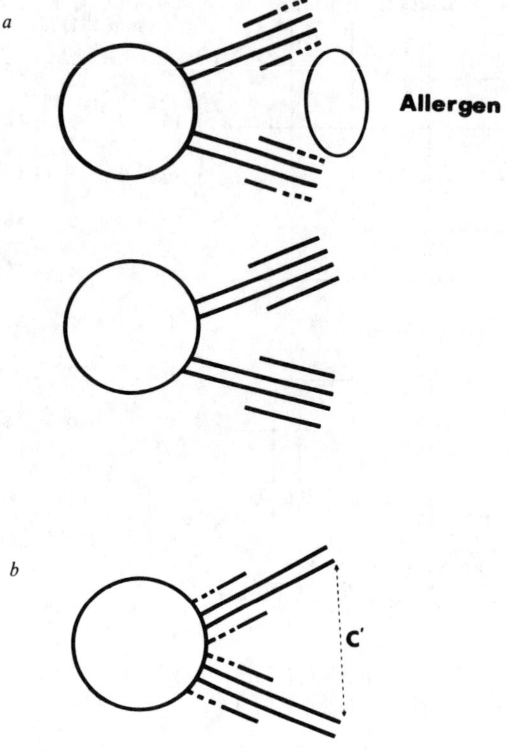

Fig. 3. *a*, Diagrammatic comparison of the properties of antibody and myeloma γE molecules, showing mode of attachment to target cells (this is by means of sites in the Fc regions). *b*, Mode of interaction of γG type cytolytic antibody molecules with target cells, showing subsequent interaction of complement (c′) with sites located in the Fc regions.

Thus it is now possible to start explaining how the immunological reactions occurring at tissue mast cell surfaces bring about the pharmacological manifestations characteristic of hypersensitivity reactions of the immediate-type. Whether such reactions are mediated by γE-type antibodies in passively sensitized human tissues (as, for example, in the classical P–K test), or by non-reaginic γG-type antibodies in the heterologous tissues of guinea-pigs passively sensitized with the serum of experimentally immunized animals, the sequence of events is remarkably similar. The antibody plays a crucial role, not only in the initial sensitization process (which, of course, occurs spontaneously in the allergic individual or the actively sensitized guinea-pig), but also in the manner in which it subsequently interacts with specific antigen (allergen) to activate the enzyme system supposedly responsible for effecting the release of histamine, 5-hydroxytryptamine and other pharmacologically active substances. The elucidation of the structural basis of these different, but complementary, function of tissue-sensitizing antibodies poses a problem of protein characterization. Moreover, it seems likely that studies along these lines on γE globulins will throw light on the mode of action of other classes of immunoglobulin fulfilling other roles on the surfaces of other types of cell (for example, receptor immunoglobulin or lymphocytes).

The emergence of *in vitro* alternatives to the P–K test in humans, and the PCA test in

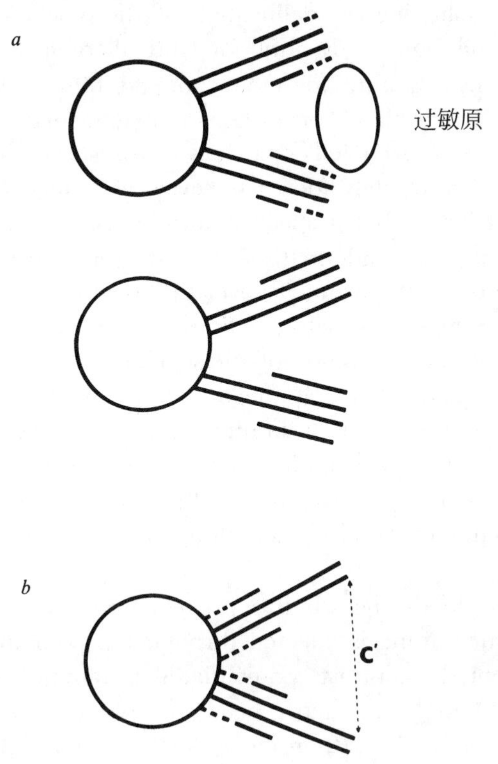

图 3. *a* 图，用示意图比较抗体分子和骨髓瘤 γE 分子的特性，画出了两者与靶细胞结合（通过 Fc 区的位点结合）的模式。*b* 图，溶细胞性 γG 抗体分子与靶细胞的相互作用模式，图中显示出补体（c′）随后与 Fc 区位点发生的相互作用。

这样我们就有可能来解释组织中肥大细胞表面发生的免疫反应是如何引起速发型超敏反应的药理学症状的。无论这些反应是由被动致敏的人体组织中的 γE 型抗体所介导（如在经典的 P–K 试验中），还是由豚鼠（由免疫动物的血清被动致敏）的异源性组织中的非反应素 γG 型抗体所介导，所发生的事件的顺序是非常相似的。不论是在致敏的起始阶段（当然，在过敏患者和主动致敏的豚鼠中会自发出现），还是在后续与特异的抗原（过敏原）结合并激活酶系统的过程中，抗体都起着关键性的作用。酶系统的激活被认为与组胺、5– 羟色胺以及其他药理活性物质的释放有关。要想从分子结构层面阐明这些组织致敏性抗体不尽相同但又互相补充的功能，目前还需要解决蛋白质表征方面的问题。此外，对 γE 型球蛋白的这些方面的研究可能会阐明其他类型的免疫球蛋白在其他类型细胞表面的作用方式（例如受体免疫球蛋白或淋巴细胞）。

人体外 P–K 试验的替代法的出现，以及非人灵长类中被动皮肤过敏反应（PCA）

sub-human primates, is facilitating the delineation of the biochemical events set in train by the combination of cell-bound γE antibody and allergen. Such systems are based on passive sensitization by γE antibodies of chopped human lung[33], normal human leucocytes[34], chopped monkey skin[35] or other suitable primate tissue. The histamine released from the washed sensitized cells or tissues as a result of interaction with specific allergen can be measured accurately (down to levels of a few nanograms) by bioassay using isolated guinea-pig ileum, but a spectrofluorometric procedure involving coupling with ophthalaldehyde[36] has been adopted as a satisfactory chemical alternative. These indirect procedures have obvious advantages over the previous pharmacological approach to the study of immediate hypersensitivity reactions, which involved the direct assay *in situ* of the histamine liberated on presentation of the specific antigen to the actively sensitized tissue set up in an organ bath. Their application has provided convincing evidence[37,38] that the allergen-induced release of histamine and other mediations of immediate-type hypersensitivity reactions is accomplished by a multi-step, energy-requiring process to which the glycolytic pathway is essential. There are precise *p*H and temperature requirements, and a dependence on calcium and magnesium ions.

It is important, however, to distinguish this process from cytolytic reactions in which destruction of the cell membrane is effected through the combined agency of anti-cell antibody and complement. In contrast, γE-mediated histamine release has been shown (from K^+ efflux studies[38]) to be an active secretory process which is not intrinsically injurious to the target cell. Thus, there is a fundamental difference in the release mechanism initiated by sensitizing antibodies (for example of the γE type), in the absence of complement, and those more drastic processes induced by cytolytic antibodies (for example of the γG type) with the aid of the complement system. As I have already implied (Fig. 3), the type of process evoked depends on the manner in which the antibody is presented to the cell. Antibodies of the γE type seem particularly suited to binding to histamine-containing cells through sites on their Fc regions, and to possess within their own structures sites capable of direct action on the target cell membrane (or other cell constituents). On the other hand, cytolytic antibodies (whether directed against cell membrane antigen, or coating antigen) react initially through sites within their Fab regions, a subsequent intermolecular interaction between their free Fc regions[39] supposedly leading to the activation of the complement system responsible for the ensuing lysis.

Tertiary and Quaternary Structure

The problem now is to establish how the interaction of the cell-bound γE antibody (reagin) with allergen triggers off the events outlined, which seem (from microscopic studies of viable cell preparations) to be associated with characteristic morphological changes[40] in the target cells, involving the release of their histamine-containing granules. The amino-acid sequencing in progress on the two myeloma IgE preparations should provide important information as might the characterization of the relatively large carbohydrate (11.7%) prosthetic group. Furthermore, studies of the inhibitory capacity of proteolytic cleavage fragments of the myeloma IgE (mentioned earlier) are being extended to smaller and smaller fragments, with the hope of isolating a readily characterizable low molecular weight peptide retaining tissue-binding activity. It is becoming increasingly obvious, however, that the peculiar biological properties of γE antibodies (as well as those of the

试验的建立，使得人们更容易描述细胞结合型 γE 抗体与过敏原结合时发生的生化过程。这些体系是基于由以下组织中的 γE 抗体引起的被动致敏而建立的，这些组织包括：捣碎的人肺组织[33]、正常人白细胞[34]、捣碎的猴皮肤组织[35]或其他合适的灵长类组织。当清洗后的致敏细胞或组织与特异性抗原发生相互作用后，便会释放出组胺。组胺的浓度可以通过生物测定法用分离的豚鼠回肠来精确测定（最小值在几纳克水平），不过后来人们采用邻苯二甲醛[36]和荧光光谱相结合的方法作为一种令人满意的化学替代检测法。与先前研究速发型超敏反应的药理学方法相比，这些间接的检测方法具有明显的优势。过去的方法需要在器官浴中使用特定的抗原激活组织，并对释放的组胺进行直接原位检测。新方法的应用为以下论点提供了具有说服力的证据[37,38]，即过敏原引起的组胺释放以及速发型超敏反应中其他中间产物的释放都是由一个多步骤的耗能过程完成的，在这个过程中糖酵解途径是必需的。这个过程对 pH 值和温度有着精确的要求，并且依赖于钙离子和镁离子。

不过，重要的是要把组胺的释放与细胞溶解反应区分开来。在细胞溶解反应中，由溶细胞性抗体和补体的结合中间体导致了细胞膜的破裂。与此相反，γE 抗体介导的组胺释放（通过钾离子外流实验[38]）是一个主动分泌的过程，对于靶细胞来讲也没有本质上的伤害。因此，不需要补体参与的、由致敏性抗体（如 γE 型抗体）引发的组胺释放过程和需要补体系统参与的、由溶细胞性抗体（如 γG 型抗体）引发的更为剧烈的过程之间存在着根本的区别。正如我在前面曾暗示过的那样（图 3），所激发过程的类型取决于细胞中存在的抗体种类。γE 抗体似乎特别适合通过它们的 Fc 区位点与含有组胺的细胞结合，并且其自身结构中含有能够与靶细胞膜（或其他细胞组分）发生直接作用的位点。另一方面，溶细胞性抗体（不管直接识别的是细胞膜表面抗原还是包被抗原）起初通过它们 Fab 区域的位点与细胞结合，之后在它们自由的 Fc 区之间会发生分子间相互作用[39]，这种相互作用可能激活了补体系统，从而导致细胞的溶解。

三级结构和四级结构

现在的问题在于细胞结合型 γE 抗体（反应素）与过敏原的相互作用是如何引发下游事件的，这似乎（根据用显微镜对活细胞进行的观察）与靶细胞的特征性形态改变有关[40]，同时伴随着含有组胺的颗粒物的释放。对两个骨髓瘤 IgE 的正在进行的氨基酸序列分析和对较大（占分子量 11.7%）的碳水化合物辅基的表征应该能够提供一些重要的信息。此外，用蛋白酶解的方法将骨髓瘤 IgE 抗体降解成越来越小的片段，并对这些片段的抑制活性进行分析（之前曾提到过），以期分离出一个具有组织结合活性且可进行表征的最小分子量的肽段。然而，日益明显的是，γE 抗体（以及在实验动物中产生的低效率致敏性 γG 抗体）的这种罕见的生物学功能是由其非

less efficient γG type sensitizing antibodies produced in experimental animals) will be explicable in terms of unusual tertiary and quaternary structural characteristics. It seems likely (as discussed fully elsewhere[41]) that the two structural features which will prove to distinguish sensitizing antibodies will be an ability to bind strongly to complementary tissue sites with no accompanying loss in the conformational integrity of their Fc region (which can soon occur as a result, for example, of relatively mild heat treatment); and the possession of sufficient flexibility within the "hinge regions" to permit the transmission of a critical allosteric change resulting from combination of allergen with sites in the Fab regions of adjacent cell-bound molecules.

There is evidence[42] that antigen-induced conformational changes can occur in free solution, in rabbit antibody directed against commonly used antigens such as bovine serum albumin and horse ferritin, and these lead to the exposure of new antigenic determinants within the Fc regions. It seems reasonable to conclude, therefore, that similar conformational changes could occur in γE-type antibody molecules in combination with specific allergen (as already proposed[43]). Moreover, if the γE antibody is already anchored to the target cell surfaces, through sites within its Fc region, newly exposed side chains could be brought into critical juxtaposition with points on the cell where activation of the appropriate enzyme system would occur (Fig. 4). Indeed there is evidence (from inhibition studies with aggregated IgE) that association of IgE molecules might lead to the formation of "tissue activation sites" which are distinct from the tissue binding sites in monomeric (antibody and myeloma) IgE molecules; and it is possible that activation sites are similarly formed in the Fc regions of adjacent cell-bound sensitizing antibody molecules as a result of the bridging by specific allergen, for which there is increasing evidence[45,46] and which presumable involves the cross-linking of sites within the antibody Fab regions (Fig. 4). The critical question then to be answered is whether groups exposed from the Fc regions of the cross-linked antibody molecules act cooperatively to form the activation site. In this event, the activation site could be linked to the active site of an enzyme, which of course is usually contained within a single polypeptide chain. Alternatively, the side-chains which are supposedly exposed in the Fc regions of the sensitizing antibody molecules might be proved to exert an independent effect on the target cell.

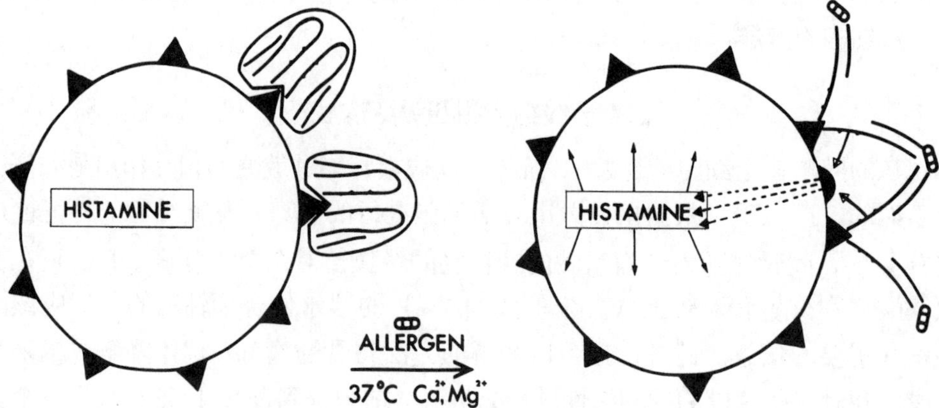

Fig. 4. Postulated mode of interaction between cell-bound γE antibodies and allergen leading to the triggering of release of vasoactive amines (modified from ref. 45).

同寻常的三级和四级结构所决定的。似乎是（正如我在另一篇文章中详细讨论的那样[41]）这两种结构特点使得致敏性抗体不同于其他抗体，因而致敏性抗体可以在与互补的组织位点紧密结合的同时不影响其 Fc 区的构象完整性（例如在相对温和的热处理中很快会出现的现象）；“铰链区”充分的结构柔性也使得相邻细胞结合型抗体 Fab 区的某些位点与过敏原结合所需的构象改变可以实现。

有证据表明[42]，由抗原引起的抗体构象改变可以在自由溶液中发生，兔抗体对牛血清白蛋白或马铁蛋白等常用抗原的识别会导致抗体 Fc 区暴露出新的抗原决定簇。因此我们似乎可以很合理地得出以下结论，当 γE 型抗体分子与特殊的过敏原结合时也会发生类似的构象改变（正如已经提出的那样[43]）。此外，如果 γE 型抗体事先已经通过其 Fc 区中的位点锚定在靶细胞的表面，那么新暴露的侧链就有可能与细胞上相应酶系统的激活点处于严格并行的位置（图 4）。事实上，的确有证据（来自对聚集的 IgE 分子的抑制活性研究）表明：多个 IgE 分子的结合可能会导致“组织激活位点”的形成，这与单体（抗体或骨髓瘤）IgE 分子的组织结合位点是不一样的；并且有越来越多的证据表明[45,46]，通过特异性过敏原的桥接作用，在相邻细胞结合型致敏抗体的 Fc 区有可能同样会形成激活位点，并且抗体 Fab 区的交联位点也可能参与其中（图 4）。于是，一个需要回答的重要问题是，从交联抗体分子 Fc 区暴露出的基团是否能互相协作形成激活位点？如果是的话，那么这个激活位点就可以连接到酶的活性位点了，当然这两个位点通常存在于同一条多肽链中。如果不是，也许我们能证明，致敏抗体分子 Fc 区中可能暴露出来的一些氨基酸侧链基团可以独立地对靶细胞施加某种作用。

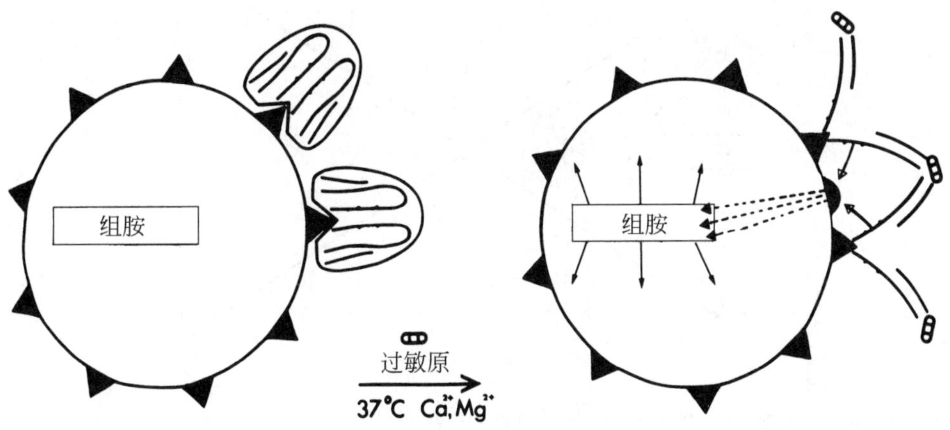

图 4. 细胞结合型 γE 抗体与过敏原相互作用引发血管活性胺释放的假想模式（根据参考文献 45 修改而成）。

In attempting to decide between these alternatives it is worth considering the results of studies designed to short-circuit the histamine release process initiated by reagin-allergen combination. These involve the use of alternative, artificial, ways of cross-linking the γE molecules, and seem to lead to pharmacological responses similar to those of classical immediate sensitivity reactions. Apart from the pre-aggregation of the γE molecule before presentation to the target cell (as already mentioned), it is possible to effect histamine release in human or monkey skin by bringing about the cross-linking of the cell-bound γE molecules with antibody directed specifically against determinants within their Fc regions. Thus, it can be demonstrated[46] that the intradermal injection of specific anti-human IgE antiserum induces an immediate oedematous and erythematous reaction similar to that observed in the P–K test.

Tissue Activation Sites

It is suggested that, as in the classical reagin-allergen mediated reaction, the exposure of tissue activation sites within the Fc region of the γE molecule triggers the release of the pharmacologically active mediators. But the cross-linking process which induces the required conformational change is less efficient than that effected by the cross-linking of the Fab regions of the γE antibody molecules by allergen. This might be expected by reference to the model shown in Fig. 5, which suggests that a force applied at position A (at maximum distance from the fulcrum) would be the most effective way of forming the activation site at position S; but application of a force at position B would accomplish the same effect in a less efficient manner. This analogy could offer an explanation of the skin reactions effected by the action of anti-IgE antibody, as well as that elicited by other agents such as protein A (an antigenic component of *Staphylococcus aureus*) which has an unusual capacity for combination with the Fc regions of immunoglobulins[47] and which presumably acts too by cross-linking cell-bound γE molecules[48].

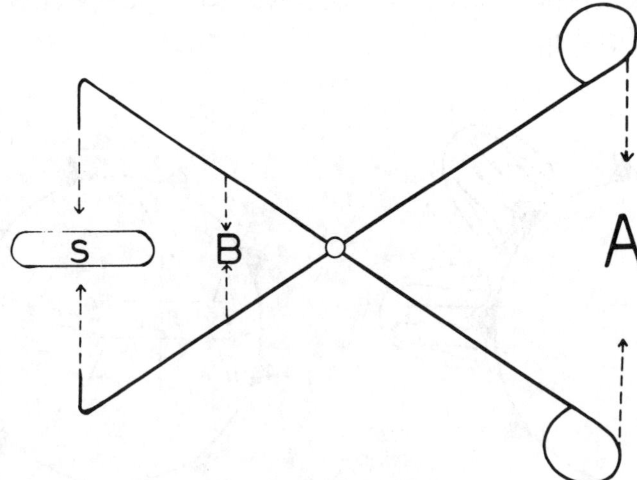

Fig. 5. Model illustrating various methods of inducing an allosteric transformation within cell-bound γE antibody molecules.

当我们试图从这些可能性中作出判断的时候，应该考虑旨在缩短由反应素–过敏原结合引起的组胺释放过程的研究结果。这些实验使用了一些使γE分子发生交联的人为替代方法，看似可以得到与经典的速发型过敏反应相似的药理反应。除了在暴露于靶细胞之前将γE抗体分子预聚集之外（如前所述），利用可以特异性识别Fc区中抗原决定簇的抗体与细胞结合型γE抗体分子的交联也有可能在人或猴子的皮肤中引起组胺的释放。由此可以证明[46]，皮内注射特异性抗人IgE抗血清所引发的速发型水肿和红斑反应与P–K试验中所观察到的现象类似。

组织激活位点

有人认为：和反应素–过敏原介导的经典反应一样，γE抗体Fc区中的组织激活位点一旦暴露出来就会引发药理活性因子的释放。不过，与引起必要构象改变的交联过程相比，由过敏原引起的γE抗体Fab区的交联过程效率更高。这一点可以参照图5中所示的模式推测得出。在这个模式图中，作用于A处（离支点最远的点）的力是在S处形成激活位点的最有效的方法，而如果在B处施加一个力，则要达到同样效果就会效率降低。以此类推，这一模式也可以解释由抗IgE抗体以及由蛋白A（金黄色葡萄球菌的抗原成分）等其他试剂引起的皮肤反应。蛋白A具有一种能与免疫球蛋白Fc区相结合的不同寻常的能力[47]，并且有可能也起着交联细胞结合型γE抗体的作用[48]。

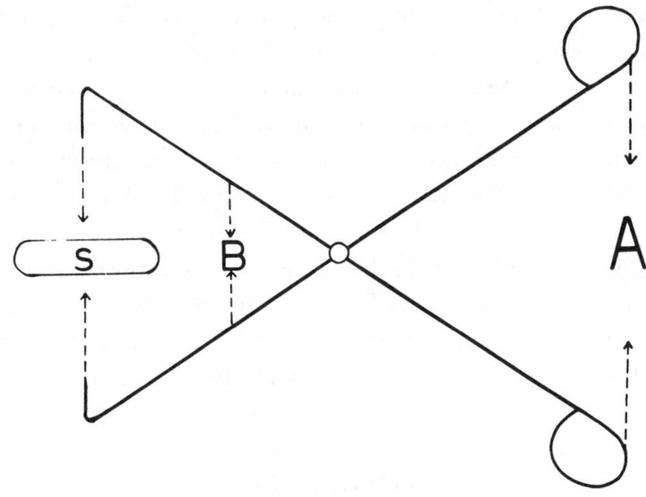

图5. 引起细胞结合型γE抗体分子构象改变的两种不同方式的模式图。

Alternatively, it seems likely that the conformational changes necessary for the formation of tissue activation sites can also occur in non-specifically aggregated IgE (brought about, for example, by lyophilization of monomeric myeloma IgE). As already mentioned, IgE treated in this way is also capable of effecting an immediate release of histamine on injection into human or monkey skin[46]. Thus, perhaps surprisingly, in spite of the non-specific nature of the aggregation process, a conformational change seems to be induced within the Fc regions similar to that resulting from antigen-mediated association of the γE molecules. This is directly analogous, however, to the situation with regard to the association of human (and rabbit) γG molecules where biological activities (for example heterologous skin reactivity and complement fixability), and new antigenic determinants (similar to those formed on specific antigen-antibody combination) are induced by non-specific denaturation treatments. Moreover, recent studies[49] on the nature of the structural changes shown by γG globulins as a result of physical or chemical treatments suggest that here too association is initiated by preliminary (non-covalent) interaction between the Fab regions of the molecule but in this case, of course, without the agency of the antigen. This, it seems, leads to an unfolding of the Fc regions, similar to that occurring in antibody γG molecules on combination with antigen, thereby providing scope for intermolecular Fc interaction with formation of the sites of the various biological activities shown by aggregated γG globulins. Thus what might be expected to be a random aggregation process seems to be a highly ordered polymerization, directed presumable by the initial intermolecular reactions of the Fab regions of the monomer molecules.

The complete characterization of these crucial tertiary and quarternary structural changes could have an important bearing on the explanation of other immunological phenomena. As far as immediate hypersensitivity is concerned, the need now is to characterize the "tissue activation sites" formed in γE globulin molecules on association, and to elucidate the nature of their "substrate" located on the target cells. Various possibilities are being considered. For example, could it be that reagin-allergen combination on the cell surface activates a phospholipase responsible for hydrolysis of membrane phospholipid[50]; or perhaps an esterase is activated, in a similar manner to that implicated in complement mediated cytolytic reactions[51]. Another possibility[52] is the stimulation of the adenyl-cyclase system, which is located on the inner surface of cell membranes (where it catalyses the conversion of ATP to cyclic AMP) and which has been shown to be involved in the action of certain hormones[53]. It could be more than fortuitous that certain polypeptide hormones (such as ACTH) which act in this manner are also potent histamine liberators[54]. It will, however, be important to exclude the simpler possibility that the primary triggering mechanism is a physical shearing of the cell membrane ("rigidification", as one investigator[44] has termed it), brought about by the association of surface-bound γE antibody molecules.

Future Tasks

Although the initiative is now passing to biochemists and biophysicists, many other aspects of the behaviour of γE antibodies require immunological investigation. For example, with the aid of specific antisera prepared against the myeloma proteins, it is possible to

或者，形成组织激活位点所需要的构象改变也可能在非特异聚集的 IgE（例如，由骨髓瘤 IgE 单体在冻干时导致）中发生。如前所述，当把这种 IgE 注射到人或猴的皮肤中时，也会立即引起组胺的释放[46]。因此，尽管这种聚集的过程是非特异的，但令人惊奇的是，在这些抗体的 Fc 区也发生了与结合了抗原的 γE 抗体类似的构象变化。这与人（或兔）γG 抗体分子的结合情况很相似，即非特异性的变性处理可以引起生物学活性（如异源皮肤反应和补体固定）和新抗原决定簇（与特异的抗原 – 抗体结合所形成的抗原决定簇类似）的产生。此外，最近对 γG 球蛋白在经物理或化学方法处理后发生结构变化的研究[49]也表明：这种结合是由抗体分子 Fab 区之间的初级（非共价结合）相互作用引起的，但在这种情况下当然不可能由抗原介导。这种结合似乎会导致 Fc 区的解折叠，这和抗体 γG 分子与抗原结合时所发生的变化相似。因此，在发生聚集的 γG 球蛋白中，这种结合为分子间 Fc 区发生相互作用从而形成各种生物活性位点提供了可能。因而看似应当为随机过程的分子聚集似乎是一个高度有序的聚合反应，这很可能是由单体分子 Fab 区的初始分子间反应引起的。

完整表征这些三级和四级结构的重要变化对于解释其他免疫学现象具有重大的意义。就速发型超敏反应来说，现在需要做的是表征 γE 球蛋白分子在结合时所形成的“组织激活位点”并阐明位于靶细胞上的“底物”的类别。需要考虑到各种可能性。例如，反应素 – 过敏原在细胞表面的结合能否激活可以水解细胞膜中磷脂的磷脂酶[50]；或者是否会通过与补体介导的细胞溶解反应相似的途径引起酯酶的激活[51]。另一种可能性[52]是，分布于细胞膜内表面的腺苷酸环化酶系统被激活（在这里它催化 ATP 转变为环化 AMP），该系统已被证明与某些激素的功能有关[53]。某些与腺苷酸环化酶系统相关的多肽类激素（例如 ACTH）同时又是有效的组胺释放者[54]，这一点并非偶然。然而，排除下面这种更为简单的可能性是很重要的，即认为最初的启动来源于表面结合型 γE 抗体分子在结合时所引起的细胞膜的物理剪切（有一位研究人员[44]称之为“硬化”）。

今后的任务

尽管目前研究的主动权已经转移到了生物化学家和生物物理学家的手里，但 γE 抗体的许多其他方面的行为仍需要免疫学角度的研究。例如，借助于抗骨髓瘤蛋白

begin to identify the sites of binding of γE antibodies on their target cells; as well as their sites of synthesis in human lymphoid tissues. Immunofluorescence studies[55] have provided evidence that γE antibodies are formed locally in the respiratory and gastro-intestinal tract where they probably become involved in allergic states affecting these organs. In any studies of γE antibody synthesis it will be important to answer the fundamental question about the definition of the factors which govern the production of sensitizing rather than immunizing antibodies. Here the use of animal models[56] is particularly revealing, because it is becoming apparent that the reagin-like antibodies produced in species like the rat, the rabbit and the mouse closely resemble the human γE antibodies. The chemical nature of the allergen, and its form and mode of administration are proving important factors in deciding the nature of the resultant antibody response, as is the influence of other classes of antibody produced concomitantly. Through such investigations it is becoming possible to define the immunological basis of "desensitization" (that is hyposensitization), a prophylactic measure which has been practiced by clinical allergists for several years and which is thought to encourage the production of γG-type antibodies which bind allergen preferentially (besides producing effects at the cellular level).

Other possible new forms of prophylaxis are suggested by studies[57] on the prevention of the passive sensitization of baboons with allergic human serum by the systemic administration of myeloma IgE, where a refractory state persisting for about a week has been induced by the intravenous injection of about 10 mg protein/kg body weight. The ultimate aim of this approach would be the isolation of a low molecular weight peptide fragment of the myeloma IgE, which retained tissue-binding activity and which could be made available in large amounts by application of modern methods of polypeptide synthesis.

Another advantage to the clinical allergist resulting from the discovery of the myeloma proteins has been the development of methods of measuring IgE concentration which do not rely on tissue or leucocyte sensitization. For example, radioimmunosorbent assay procedures involving the use of anti-IgE[24] or specific allergen[58] coupled chemically to dextran (or some other suitable insoluble carrier such as a cellulose carbonate derivative)[59] have been used to determine the total, or antibody, IgE levels in the sera of patients, and the use of radiolabelled anti-antibody[59] has rendered the single radial diffusion test sufficiently sensitive to detect the relatively low levels of IgE in allergic sera by autoradiography.

It is important to recognize, however, that γE antibodies need not necessarily be deleterious, as suggested, for example, by studies[60] of the immune reactions occurring in the mucous membranes of the rat during the expulsion of an infecting nematode (*Nippostrongylus brasiliensis*). There is evidence of an association between a sharp rise in intestinal mast cells, an alteration of mucosal permeability (effected by vasoactive amines released from the mast cells) and the expulsion of the parasites. It is tempting to conclude, therefore, in this situation that γE antibodies are fulfilling a protective role (the so called "self-cure" mechanism[61]); and it may be significant that one of the few conditions (other

的特异性抗血清，人们有可能开始进行γE抗体在其靶细胞上的结合位点以及在人淋巴组织中的合成位点的鉴定。免疫荧光实验[55]证明：γE抗体只在呼吸道以及胃肠道中形成，并且很可能与这些器官的过敏反应有关。在对γE抗体合成的研究中，对何种因子控制着产生致敏性抗体而不是免疫性抗体这一基本问题的解答很重要。在这一点上，基于多种动物模型的研究很有意义[56]，因为人们逐渐明确了在大鼠、兔子和小鼠等物种中所产生的类似反应素的抗体与人类的γE抗体非常相似。已证明过敏原的化学本质、形式和注射方式是决定所产生的抗体反应的重要因素，同时出现的其他类抗体所产生的影响也很重要。通过这些研究，人们也许可以弄清“脱敏反应”（即消除过敏）的免疫学基础，这种预防性的方法已经被临床过敏症专科医师应用了很多年，它被认为能够刺激产生可以优先与过敏原结合的γG抗体（除了产生细胞水平上的效应之外）。

通过全身注射骨髓瘤IgE阻止由人过敏性血清引起狒狒被动致敏反应的研究[57]为预防过敏提供了另外一些可能的新形式。每千克体重静脉注射约10毫克蛋白可使受试者在一周左右的时间内维持耐受状态。这种方法最终的发展趋势将是从骨髓瘤IgE中分离出一个低分子量的多肽片段，该片段能保持组织结合活性并且可以通过现代的多肽合成方法被大规模地制备出来。

骨髓瘤蛋白的发现带给临床过敏症专科医师的另一个好处是，得到了不依赖于组织或白细胞敏感化的测量IgE浓度的方法。例如：在放射免疫吸附试验中使用抗IgE抗体[24]或通过化学方式连接到葡聚糖（或其他合适的不溶性载体，例如纤维素碳酸盐衍生物）[59]上的特异过敏原[58]，并用这样的放射免疫吸附法测定患者血清中总的或抗体性的IgE水平；使用放射性标记的抗抗体[59]使得单向扩散试验灵敏到可以借助放射自显影术检测到过敏性血清中含量相对较低的IgE。

然而，意识到γE抗体并不一定都是有害的也很重要，正如在对大鼠黏膜排斥感染线虫（巴西日圆线虫）时产生的免疫反应的研究[60]中所揭示出的结果。有证据表明，在肠肥大细胞迅速增多、黏膜渗透性改变（由肥大细胞释放的血管活性胺引起）和排斥寄生虫之间存在着一种关联。因此，我们很容易得出这样一个结论，即γE抗体在这种情况下起到了一种保护作用（即所谓的“自愈”机制[61]）；并且，在血清中可以检测到很高浓度IgE抗体的少数几种情况（除了哮喘–枯草热型超敏反应和

than a hypersensitivity of the asthma-hay fever type, and atopic eczema) where very high serum IgE levels have been recorded is in human (as well as animal) parasitic infections.

(**233**, 310-316; 1971)

D. R. Stanworth: Department of Experimental Pathology, University of Birmingham.

References:

1. Frausnitz, C., and Köstner, H., *Zentr. Bakteriol., Parasiteuk.*, Abt. I, **86**, 160 (1921).
2. Coca, A. F., and Grove, E. F., *J. Immunol.*, **10**, 445 (1925).
3. Kuhns, W. J., *Proc. Soc. Exp. Biol. Med.*, **108**, 63 (1961).
4. Augustin, R., in *Handbook of Experimental Immunology* (edit. by Weir, D. M.), 1076 (Blackwell, Oxford, 1967).
5. Stanworth, D. R., and Kuhns, W. J., *Immunology*, **8**, 323 (1965).
6. Stanworth, D. R., and Kuhns, W. J., quoted in Stanworth, D. R., *Adv. Immunol.*, **3**, 181 (1963).
7. Hemmings, W. A., and Brambell, F. W. R., *Brit. Med. Bull.*, 17, 96 (1961).
8. Tiselius, A., and Kabat, E. A., *J. Exp. Med.*, **69**, 119 (1939).
9. Brattsten, I., Colldahl, H., and Laurell, A. H. F., *Acta Allergol.*, **8**, 339 (1955).
10. Stanworth, D. R., *Immunology*, **2**, 384 (1959).
11. Franklin, E. C., and Stanworth, D. R., *J. Exp. Med.*, **114**, 521 (1961).
12. Heremans, J. F., Heremans, M. T., and Schultze, H. E., *Clin. Chim. Acta*, **4**, 96 (1959).
13. Stanworth, D. R., *Intern. Arch. Allergy*, **28**, 71 (1965).
14. Vaerman, J. P., Epstein, W., Fudenberg, H., and Ishizaka, K., *Nature*, **203**, 1046 (1964).
15. Fireman, P., Vannier, W. E., and Goodman, H. C., *J. Exp. Med.*, **117**, 203 (1963).
16. Ishizaka, K., Ishizaka, T., and Hornbrook, M., *J. Allergy*, **34**, 395 (1963).
17. Ishizaka, K., and Ishizaka, T., *J. Allergy*, **37**, 169 (1966).
18. Stanworth, D. R., *Adv. Immunol.*, **3**, 181 (1963).
19. Rowe, D. S., and Fahey, J. L., *J. Exp. Med.*, **121**, 171 (1965).
20. Terr, A. I., and Bentz, J. D., *J. Allergy*, **35**, 206 (1964).
21. Andersen, B. R., and Vannier, W. E., *J. Exp. Med.*, **203**, 117 (1963).
22. Ishizaka, K., and Ishizaka, T., *J. Allergy*, **42**, 330 (1968).
23. Ishizaka, K., Ishizaka, T., and Hornbrook, M. M., *J. Immunol.*, **97**, 35 (1966).
24. Johansson, S. G. O., Bennich, H., and Wide, L., *Immunology*, **14**, 265 (1968).
25. Johansson, S. G. O., and Bennich, H., *Immunology*, **13**, 381 (1967).
26. Bennich, H., Ishizaka, K., Ishizaka, T., and Johansson, S. G. O., *J. Immunol.*, **102**, 826 (1969).
27. Stanworth, D. R., Humphrey, J. H., Bennich, H., and Johansson, S. G. O., *Lancet*, ii, 330 (1967).
28. Bennich, H., Ishizaka, K., Johansson, S. G. O., Rowe, D. S., Stanworth, D. R., and Terry, W. D., *Bull. World Health Org.*, **38**, 151 (1968).
29. Ogawa, M., Kochwa, S., Smith, C., Ishizaka, K., and McIntyre, O. R., *New Engl. J. Med.*, **281**, 1217 (1969).
30. Stanworth, D. R., Humphrey, J. H., Bennich, H., and Johansson, S. G. O., *Lancet*, ii, 17 (1968).
31. Stanworth, D. R., Housley, J., Bennich, H., and Johansson, S. G. O. (in preparation).
32. Stanworth, D. R., Housley, J., Bennich, H., and Johansson, S. G. O., *Immunochemistry*, 7, 321 (1970).
33. Sheard, P., Killingback, P. G., and Blair, A. M. J. N., *Nature*, **216**, 283 (1967).
34. Van Arsdel, P. P., and Sells, C. J., *Science*, **141**, 1190 (1963).
35. Goodfriend, L., and Luhovy, J. I., *Intern. Arch. Allergy*, **33**, 171 (1968).
36. Shore, P. A., Burkhalter, A., and Cohn, U. H., *J. Pharmacol. Exp. Ther.*, **127**, 182 (1959).
37. Schild, H. O., in *Biochemistry of the Acute Allergic Reactions* (edit. by Austen, K. F., and Becker, E. L.), 99 (CIOMS Symp., 1968).
38. Lichenstein, L. M., in *Biochemistry of the Acute Allergic Reactions* (edit. by Austen, K. F., and Becker, E. L.), 153 (CIOMS Symp., 1968).
39. Stanworth, D. R., and Henney, C. S., *Immunology*, **12**, 1267 (1967).
40. Hastie, R., *Clin. Exp. Immunol.* (in the press).
41. Stanworth, D. E., *Immediate Hypersensitivity* (monograph, in preparation).
42. Henney, C. S., and Stanworth, D. R., *Nature*, **210**, 1071 (1966).

遗传性过敏性湿疹）之一是人（或动物）受到寄生虫感染时，这一点可能意义重大。

（张锦彬 翻译；刘京国 审稿）

43. Stanworth, D. R., *Clin. Exp. Immunol.*, **6**, 1 (1970).

44. Levine, B. B., *J. Immunol.*, **94**, 111, 121 (1965).

45. De Weck, A. L., and Schneider, C. H., in *Bayer Symp. on Problems in Allergy*, **1** (1969).

46. Ishizaka, K., and Ishizaka, T., *J. Immunol.*, **100**, 554 (1968).

47. Forsgren, A., and Sjöquist, J., *J. Immunol.*, **97**, 822 (1966).

48. Stanworth, D. R., Matthews, N., and Sjöquist, J. (in preparation).

49. Matthews, N., and Stanworth, D. R., *Proc. Eighteenth Bruges Symp. Protides of Biological Fluids* (in the press).

50. Fernö, O., Högberg, B., and Uvnäs, B., *Acta Pharmacol.*, **17**, 18 (1960).

51. Becker, E. L., and Austen, K. F., *J. Exp. Med.*, **124**, 379 (1966).

52. Lichenstein, L. M., *Proc. Eighth Symp. Collegium Intern. Allergologicum*, Switzerland (1970).

53. Catt, K. J., *Lancet*, i, 763 (1970).

54. Jaques, R., *Intern. Arch. Allergy*, **28**, 221 (1965).

55. Ishizaka, K., and Ishizaka, T., *Clin. Exp. Immunol.*, **6**, 25 (1970).

56. Patterson, R., *Prog. Allergy*, **13**, 332 (1969).

57. Stanworth, D. R., *Clin. Allergy*, **1** (in the press).

58. Vide, L., Bennich, H., and Johansson, S. G. O., *Lancet*, ii, 1105 (1967).

59. Rowe, D. S., *Bull. World Health Org.*, **40**, 613 (1969).

60. Miller, H. R. P., and Jarrett, W. F. H. (in the press).

61. Stewart, D. F., *Austral. J. Agric. Res.*, **4**, 100 (1953).

62. Bennich, H., and Johansson, S. G. O., *Vox Sang.*, **19**, 1 (1970).

Egg Transfer in Domestic Animals

L. E. A. Rowson

Editor's Note

Although the first successful egg transfer experiments were performed in the late nineteenth century, the technique was almost totally disregarded, both as a method of livestock improvement and as a tool for reproductive research, until after the Second World War. Here Tim Rowson describes the revival that subsequently ensued, which he based on his 1970 Hammond Memorial Lecture. Rowson, a pioneer in cattle embryo transfer, points out that egg transfer yields pregnancies in cows, but that the low success rates may be due to the surgical techniques used for their initial retrieval. The subsequent development of non-surgical methods in the mid-1970s was a major milestone towards the widespread application of embryo-transfer technology in cattle.

ALTHOUGH the first successful experiments on egg transfer were carried out by Walter Heape of Cambridge late in the nineteenth century, the technique was almost totally disregarded both as a method of livestock improvement and as a tool for reproductive research until after the Second World War. When one considers the enormous value of the technique in genetics and studies on reproductive physiology, this fact is quite astonishing. Egg transfer is, however, now used as a routine technique for research in many small mammals and during the past ten or fifteen years has been extensively used for studies in the larger domestic species, particularly the sheep, goat, pig and cow. In the polytocous species such as the small mammals and even the pig, it is not normally necessary to induce superovulation in order to provide the requisite number of eggs for experimentation but in most of the domestic species this is an important requirement.

Superovulation

Two chief sources of gonadotrophins have been used. In the small mammals and in some of the earlier work on the domestic species, gonadotrophins of pituitary origin were extensively used and were obtained either from the horse, pig or sheep. Their use in the larger animals was, however, rather expensive and to obtain optimum results it was usually necessary to give repeated injections during the follicular phase of the cycle. The alternative source, gonadotrophins obtained from the serum of mares pregnant between 50 and 80 days (PMSG), is readily obtainable more cheaply and is effective when administered as a single injection. The follicular response to such gonadotrophins was consistently satisfactory but the ovulatory response often left a great deal to be desired. Many animals developed large numbers of follicles of which either only a few or none at all ovulated and this phenomenon seemed to vary with differing batches of PMSG; some

家畜卵移植

罗森

编者按

虽然第一次成功的卵移植实验早在19世纪晚期就已经实现，但是直到第二次世界大战结束之前这项技术几乎被人们完全遗忘了，不论是作为改良家畜的方法，还是作为研究繁殖的工具。蒂姆·罗森在本文中描述了这项技术在第二次世界大战之后的复兴，文章取材于1970年他在哈蒙德纪念讲座上发表的讲话。作为牛胚胎移植的先驱者，罗森指出卵移植可以使母牛怀孕。但成功率很低，这可能与最初用于回收卵子的外科手术有关。随后在20世纪70年代中叶，非外科手术方法的推广成为牛胚胎移植技术得以广泛应用的一个重要里程碑。

虽然第一次成功的卵移植实验早在19世纪晚期就已经由剑桥大学的沃尔特·希普实现，但是直到第二次世界大战结束之前这项技术几乎被人们完全遗忘了，不论是作为改良家畜的方法，还是作为研究繁殖的工具。就遗传学技术和生殖生理学研究的巨大价值而论，这一事实非常令人迷惑不解。虽然在过去的10年或15年中，卵移植一直被广泛地用于研究较大的家畜，特别是绵羊、山羊、猪和牛，但是现在它已成为研究多种小型哺乳动物的常规技术。对于多胎的家畜，例如小型哺乳动物，甚至包括猪，要提供实验所需的卵数一般不需要诱导其超数排卵。但对于大多数家畜来说，这是一个很重要的必备条件。

超数排卵

目前使用过的促性腺激素主要有两种来源。来自马、猪或绵羊的垂体促性腺激素被广泛应用于小型哺乳动物和对家畜进行的一些早期研究中。然而，在研究较大的动物时使用促性腺激素成本就会很高，并且为了得到最佳结果通常需要在卵泡期内反复注射。促性腺激素也可以从怀孕50天～ 80天的母马血清中获得（PMSG），这不仅很容易通过较为便宜的方式得到，而且单次注射也能有效。卵泡对这类促性腺激素的反应一向是令人满意的，但是排卵反应通常还有很大的改善空间。很多动物能产生大量的卵泡，其中只有几个排卵或者全部不排卵，这种现象似乎随着PMSG批次的不同而变化；其中一些不仅可以非常高效地诱发卵泡反应，还可以诱发排卵，而另一些则只能诱发卵泡反应。这一问题至今尚未得到彻底解决，不过现

induced both follicular response and ovulation very efficiently but others induced only the first of these. This problem has not yet been completely solved, but it is beginning to look as though the ratio of the two components of PMSG (that is, the follicle stimulating hormone (FSH) and the luteinizing hormone (LH)) may be important. It is well known that a proportion of both cattle and sheep injected with PMSG do not exhibit oestrus at the expected time or, even if they do, the eggs are in some cases unfertilized. It has also been clear for several years that the injection of such gonadotrophins may cause the premature ovulation of a follicle which is present in the ovary at the time of injection and it seems likely that this effect is related to the LH content of the injected gonadotrophin. In such circumstances either the animal would not exhibit oestrus or the effect of the FSH component would cause sufficient follicular response for oestrus to occur but the progesterone secreted by the newly ovulated follicle might adversely affect fertilization. This effect could be explained in a number of ways, including failure of capacitation or failure of sperm transport to the site of fertilization. It is quite common in these conditions to find an absence of spermatozoa in the zona pellucida of the ovulated eggs.

Once a satisfactory batch of PMSG has been indentified, however, it is very efficient, provided that the animal is not overstimulated. In the cow and sheep, for example, it is inadvisable to stimulate the animal to produce more than about twenty ovulations because stimulation beyond this point seems to upset completely the physiological mechanisms of the reproductive tract and the proportion of unfertilized eggs increases rapidly. It is usual, therefore, to aim at an ovulation rate of about ten eggs per animal in these species. The response also varies from species to species; for example, a dose of 1,200 IU of PMSG given to a medium to large sheep will produce a response similar to a dose of 2,000 IU given to a cow weighing six or seven times as much. In most species it is unnecessary to give exogenous LH to obtain ovulation if normal oestrus is exhibited.

Recovery of Eggs

To obtain optimum results after egg transfer it is necessary to transfer the egg to a site where it would be expected to be in normal physiological conditions; because transfer to the uterus is simpler than transfer to the oviduct, it is usual to attempt to recover the eggs at a time when they have entered or are just about to enter the uterus. The timing of this will vary with the species, but is usually 4 to 5 days from the onset of oestrus in the cow and sheep and rather earlier for the pig. A surgical approach is necessary if maximum recovery of egg is to be achieved. This involves a simple laparotomy; the introduction of a fine cannula into the ovarian end of the oviduct and the injection of the flushing fluid into the ovarian portion of the uterus. The eggs are collected in special cups which can be placed directly under the microscope without further manipulation; the state of cleavage of the recovered eggs can then be identified. In the horse and pig there is a valve-like structure at the utero-tubal junction and it is not always possible to carry out retrograde flushing; in such cases it is usual to clamp off the top portion of the uterus, flush the oviduct towards the clamp and insert the recovery tube in the tip of the uterus, so that the fluid containing the eggs can be milked back into the collecting cup. Although this

在人们开始注意到 PMSG 中两种组成成分（即促卵泡激素（FSH）和促黄体生成激素（LH））的比值可能是一个重要的参数。众所周知，有一定比例注射过 PMSG 的牛和绵羊在预期的时间内并没有发情表现，或者即使有发情表现，卵子在某些情况下也是未受精的。另外，几年前人们就已经知道注射这种促性腺激素可能会导致注射时存在于卵巢中的卵泡过早排卵，并且这种效应似乎很可能与注射的促性腺激素中的 LH 含量有关。在过早排卵的情况下，要么动物未显示发情，要么在 FSH 组分的作用下产生了足以导致发情的卵泡反应，但是如果因此排卵，由新排卵的卵泡分泌的孕酮可能会对受精起反作用。可以从很多方面来解释这一效应，包括精子获能的丧失或者精子未能移动到受精点。在这些情况下通常会发现在排出的卵细胞的透明带中并没有精子存在。

虽然如此，我们一度鉴定出了一批令人满意的 PMSG，在动物没有受到过度刺激的情况下它是非常有效的。例如，对于牛和绵羊来说，刺激它们一次排出多于 20 个卵子是不可取的，因为超过这种程度的刺激似乎会完全打乱其生殖系统的生理机能，并且未受精的卵的比例会迅速增加。因此，通常的目标是使每头牛和每只绵羊的排卵率约为 10 个卵子。对 PMSG 的反应也会随着物种的不同而有所不同；例如，给予大中型绵羊 1,200 IU 剂量的 PMSG 将与给予一头相当于绵羊体重 6 倍或 7 倍的牛 2,000 IU 剂量所产生的反应相似。对于大多数物种而言，如果已经表现出正常的发情，就没有必要再给予外源的 LH 来诱导排卵了。

取　卵

为了在卵移植之后得到最佳结果，有必要将卵移植到一个处于正常生理条件下的部位；因为移植到子宫比移植到输卵管更简单，所以通常在卵子已经进入或即将进入子宫时进行取卵。取卵时间会随物种的不同而有所变化，通常是在牛和绵羊发情开始后的 4 天～5 天，如果是猪则会更早些。要想使卵子的回收率达到最大就必须进行外科手术。为此需要实施一次简单的剖腹手术；将一根细管插入输卵管的卵巢端，并向子宫的卵巢端注射冲洗液。卵子被收集到特殊的杯子里，我们可以直接把这些杯子放在显微镜下观察而不需要进一步的处理；由此可以识别出回收卵的卵裂状态。在马和猪的子宫输卵管交界处有一个类似于阀门的结构，此处并不总能实现反向冲洗；在这种情况下，通常需要夹住子宫的上半部分，同时在子宫顶端插入回收管并朝夹子的方向冲洗输卵管，这样就使含有卵子的液体像挤奶一样回流到回收杯里。虽然这项技术看上去比较原始，但是取卵的效果与直接冲洗法一样好。

technique seems rather crude, egg recovery is as efficient as recovery by the direct flushing method. About 90% of the eggs shed can usually be collected. The most effective flushing medium varies with the species—for the sheep it is homologous serum, for the pig Tyrode plus a little albumen and for the cow TCM 199.

Transfer and Synchronization

The technique of egg transfer is simple—the egg is merely picked up in the medium selected, using a Pasteur pipette attached to a 2 ml. syringe, and the tip of the pipette stabbed through the uterine wall; the egg is then expelled directly into the lumen. A minimum of flushing medium is used, seldom more than 0.1 ml.

The stage of the cycle of the donor and recipient animals must be closely synchronized to obtain optimum pregnancy rates. The degree of variability which can be tolerated is known to be ±2 days in the sheep and ±1 day in the cow; the best results are obtained, however, when oestrus in donor and recipient occurs on the same day (Fig. 1).

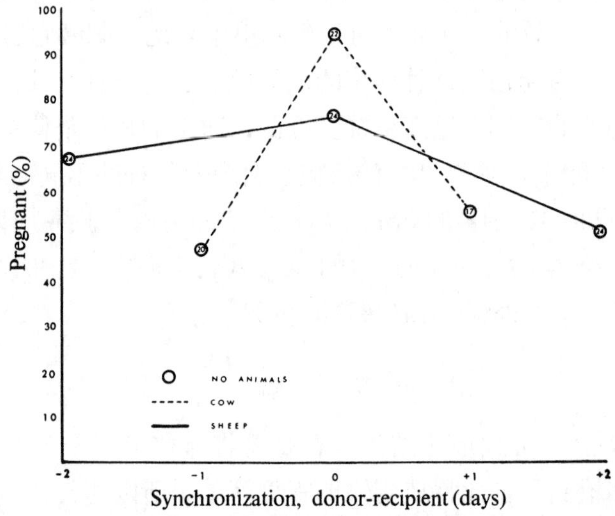

Fig. 1. The effect of varying the degree of synchronization of oestrus on fertility in cattle (- - -) and sheep (——). Numbers of animals are given in the figure.

The results obtained in cases of exact synchronization are remarkably good and are possibly influenced by the fact that any unfertilized, abnormal or retarded eggs are rejected and not transferred after egg recovery from the donor animal.

If the eggs are to be transferred on the day of recovery it is quite satisfactory to store them in the flushing medium in an incubator at 37°C. Sheep pregnancies have been achieved successfully after up to 72 h of storage at 8°C, but the percentage of successful transfers decreases rapidly after 24 h. It has also been shown that the transfer of fertilized eggs of the pig, cow and sheep to the oviduct of the rabbit will result in their continued normal development for longer periods; they can then be recovered and successfully retransferred

通常可以收集到约90%的流出卵子。最有效的冲洗媒介随着物种的不同而有所不同——对于绵羊而言用的是同种血清，猪用的是加了少量蛋白的台罗德氏溶液，牛为组织培养基 199（TCM 199）。

移植和同步性

这种卵移植技术很简单——只需用与 2 ml 注射器相连的巴斯德吸管吸取选定媒介中的卵子，然后将吸管的尖端刺入子宫壁，这样卵子就可以直接排入到内腔。冲洗媒介的使用量越少越好，一般不会超过 0.1 ml。

为了获得最佳妊娠率，供体动物和受体动物的发情期必须保持基本同步。已知的允许误差是：绵羊为 ±2 天，牛为 ±1 天；然而，当供体动物和受体动物在同一天发情时我们得到的结果最好（图 1）。

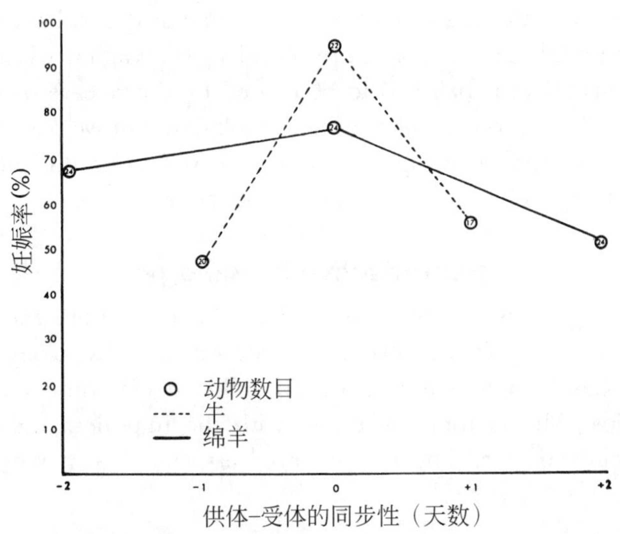

图 1. 发情同步程度的变化对牛（- - -）和绵羊（——）生育率的影响，动物的数目显示于图中。

精确同步时所得的结果非常好，但以下处理可能会对结果有影响，即在从供体动物中回收卵子之后进行移植之前去除了所有未受精的、异常的或发育迟缓的卵子。

如果卵子在回收的当日进行移植，则将它们储存在被置于 37℃恒温箱中的冲洗媒介里就很合适。利用在 8℃下储存达 72 h 的卵子曾使绵羊成功怀孕，但移植的成功率在 24 h 后迅速下降。研究还发现：将猪、牛和绵羊的受精卵移植到兔子的输卵管中会使它们在较长的时间内继续正常生长；随后可将它们回收并成功地再次移植

to suitable recipients. This shows that the requirements for culture of eggs of these species are not very specific and augurs well for *in vitro* culture of such eggs. The use of the rabbit as an incubator would permit the export of large numbers of eggs, for example, of cattle at very little cost, to various parts of the world. The transfer of such eggs after prolonged storage in the rabbit does, however, mean that development of the conceptus will only take place to a certain stage, after which it dies; the membranes are eventually expelled or resorbed. The ultimate solution to the problem of egg storage will lie in deep freezing.

Non-surgical Transfer

Most of the successful work carried out on egg transfer involves a laparotomy and the surgical recovery and transfer of the egg. Access to the uterus of the cow by way of the cervix is quite easy, but attempts to obtain pregnancies after the transfer of eggs in this way have been very disappointing. There are two reasons for this. First, the luteal phase uterus of the cow is very susceptible to infection and the introduction of any organism—very likely when the non-surgical approach is used—will cause pyometritis. Second, the insertion of a catheter, which might be used to introduce the egg, has a stimulating effect on the cervix and uterus so that uterine contractility is enhanced; eggs deposited in this way are frequently expelled and are found in the vagina within a few hours of their deposition. This effect is believed to be caused by the release of oxytocin from the pituitary as a result of the cervical and uterine stimulation, but we have so far been unable to confirm this. We are endeavouring to find an explanation of this phenomenon and to find ways in which the problem of egg expulsion can be overcome.

Reproductive Physiology

The technique of egg transfer has been used in the large domestic animals to study various fundamental and applied problems of reproductive physiology, One of the most interesting aspects has been the use of egg transfer in experiments on utero-ovarian-embryo relationships. The factors concerned with the initiation and formation of the cyclical corpus luteum are well known, but those associated with regression are far less clearly understood.

The uterus is known to be closely involved in the process because hysterectomy in the domestic animal results in persistence of the corpus luteum for prolonged periods and grafting of the removed endometrium between the flank muscles will restore cyclical behaviour and normal corpus luteum regression. The question which arises, therefore, is how the embryo overcomes the normal regression; it would seem from the repeated injection of embryo homogenates into the uterus that it acts in an anti-luteolytic manner. In an attempt to establish at what stage the presence of the embryo within the uterus is essential to prevent corpus luteum regression, eggs have been transferred at later and later stages until it is no longer possible to maintain the corpus luteum as one of pregnancy. In the sheep with a 16 to 17 day cycle the embryo must be present within the uterus by day $12\frac{1}{2}$, otherwise regression cannot be prevented. Similar requirements have not been worked out in detail for other domestic animals, but there are indications that a relatively earlier presence of the embryo within the uterus is necessary in the cow and pig.

到合适的受体中。这说明猪、牛和绵羊的卵子培养条件并不是很特殊，对这些卵子进行体外培养应该会有很好的前景。使用兔子作为恒温箱可以将大量卵子，例如牛卵子，以非常廉价的方式运往世界各地。但是，在用兔子体内长期储存的卵子进行移植时发现：孕体只能发育到某个特定阶段，在这个阶段之后孕体将会死亡；膜最终会被排出或者被再吸收。解决卵子储存问题的最终方法将是低温冷冻。

非手术移植

大多数成功的卵移植实验包括以下两步：先用剖腹和外科手术取卵，然后进行受精卵移植。从子宫颈很容易进入牛的子宫，但是用这种方式进行卵移植之后很难使牛怀孕。原因有两点。第一，牛子宫在黄体期时非常容易被感染，因而引入任何有机体都将导致脓性子宫炎——当使用非手术方式时很容易出现这种情况。第二，为了植入卵子有可能需要插入导管，这会刺激子宫颈和子宫导致宫缩增强。用这种方式放置的卵经常会被排出，在放置后几小时之内就会在阴道中发现这些卵。这种效应被认为是由于子宫颈和子宫受到刺激致使垂体释放催产素造成的，但迄今为止我们还不能证实这一点。我们正努力寻找对这一现象的解释，并在寻找可以克服卵排出问题的办法。

生殖生理学

人们已将卵移植技术应用于大型家畜以研究生殖生理学中的各种基本问题以及应用方面的问题。其中最有趣的一个方面是将卵移植应用于研究子宫 – 卵巢 – 胚胎之间关系的实验。那些影响周期性黄体产生和形成的因素已为人所共知，但是与黄体退化有关的因素却很少有人知晓。

一般认为子宫与此过程密切相关，因为切除家畜的子宫会导致黄体在很长时间内持续存在，并且如果把移去的子宫内膜移植到侧腹肌之间，则周期行为和黄体的正常退化就会得到恢复。因此产生了这样一个问题，即胚胎怎样才能克服正常的衰退；向子宫中反复注入胚胎匀浆似乎可以起到抑制黄体溶解的作用。为了确定妊娠建立的时间，即在什么阶段将胚胎植入子宫才能有效防止黄体退化，我们将卵子分批移植到受体子宫，一直到其不再能维持妊娠黄体为止。绵羊的周期为 16 天 ~ 17 天，因而在 12.5 天之内必须将胚胎置于子宫中，否则无法抑制黄体退化的发生。我们还未详细了解其他家畜的类似条件，但有迹象表明在牛和猪中有必要更早地将胚胎置于子宫内。

There is a remarkable unilateral utero-embryo-ovarian relationship in the cow and the sheep, which has been demonstrated by the use of egg transfer. If, for example, an egg is confined to the uterine horn adjoining the corpus luteum, the pregnancy continues normally; but if a similar egg is confined to the contralateral horn, pregnancy never occurs and the corpus luteum of the non-pregnant side regresses normally. This can be shown to be a unilateral luteolytic effect by removal of the non-pregnant horn adjoining the corpus luteum; the pregnancy will then continue normally. The situation is slightly modified in the case of the pig, for if a similar situation is created by confining embryos to one uterine horn, the lytic effect of the non-pregnant side will eventually result in regression of both sets of corpora lutea. The unilateral nature of these effects would suggest that it is the uterus itself rather than an indirect effect through some other gland which is responsible for lysis. It has been recently shown that radioactive material within the uterine vein is present in higher concentration in the ovarian artery, which runs in close apposition to it, than in other parts of the circulation; this suggests some form of direct transfer or communication between vein and artery although this has not been demonstrated anatomically and would involve a blood flow against the normal pressure differences.

Maturation and Culture

For the study of maturation and *in vitro* fertilization and for the satisfactory culture of eggs, laboratory techniques can only be a guide and it eventually becomes necessary to transfer eggs treated in this way to assess viability and the continuation of normal development. In experiments to assess the fertility of frozen semen it is often convenient to use the two oviducts separately, one as a control, to be injected with normal semen, and the other as a receiver for the experimental sample. Eggs transferred to each oviduct can then be examined and compared. The possibility of using the oviduct connected by fine tubes leading to the exterior as an *in vivo* means of fertilizing follicular eggs, both within and between species, is being investigated.

The demonstration that excellent pregnancy rates after transfer could be achieved with TCM 199 as a storage and flushing medium for cow eggs was followed by an attempt to increase production in this species by the introduction of two eggs into one uterine horn to produce twins. The first experiments were disappointing and only 12.5% produced twins. It was later found that when the two eggs were both deposited in a single uterine horn, migration of one egg did not occur and both implanted on the side; there was consequent competition for nutrients and death of one of the embryos frequently occurred. One egg was deposited in each horn during later experiments and a twinning rate of 73% was obtained. The economic implications of this finding are obvious and a vigorous attempt is being made to solve the problem of the reduced pregnancy rate obtained by non-surgical transfer of eggs so that the twinning technique can be put into practice.

Chimaerism in Cattle

Fusion of the foetal membranes occurs very early in pregnancy in many cattle twins and there is an interchange of haemopoietic and other types of cell between the embryos. The testes cells of the male in mixed sex twins contain many of the XX variety, but it is not

牛和绵羊中存在的子宫－胚胎－卵巢关系很显然具有单侧特性，这已经被卵移植实验所证实。举例来说，如果一个卵子被限制在与黄体相邻的子宫角内，则怀孕可以正常进行；但如果同样的卵子被限制在对侧角，则怀孕永远不会发生且未孕一侧的黄体会正常退化。在切除与黄体相邻的未孕角之后发现这里只有单侧的溶黄体作用；之后怀孕将正常进行。猪的情况略有不同，因为假如同样是将胚胎限制在一个子宫角，则未孕一侧的溶黄体作用将最终导致两部分黄体都退化。上述效应的单侧特性表明，产生溶解现象的原因是子宫本身而非其他一些腺体的间接作用。最近有研究表明：注入子宫静脉中的放射性物质出现在与该静脉紧挨着的并行卵巢动脉中的浓度要高于其出现在血液循环中其他部位的浓度；尽管尚未得到解剖学上的证明，并且这种血液流动可能会使压差不正常，但仍能说明在静脉和动脉之间存在着某种形式的直接迁移或连通。

成熟和培养

为了进行成熟研究和体外受精研究以及更好地培养卵子，实验室技术只能作为一个前导，最终还是需要移植以这种方式处理的卵子来评估它的成活能力以及继续正常发育的能力。在评估冷冻精子受精能力的实验中，一种常用的便利方法是：分别使用两个独立的输卵管，一个作为对照，注入正常精子，另一个注入实验样本。而后对移植到每个输卵管中的卵子进行检查和比较。人们正在研究是否有可能利用与导向体外的细管相连接的输卵管使体内的卵泡卵子受精，不管是在同一物种内部还是在不同物种之间。

在证明用 TCM 199 作为牛卵子的储存和冲洗媒介可以在移植后得到令人满意的妊娠率之后，我们尝试将两个卵子引入到一个子宫角中以产生双胞胎来提高牛的生育率。第一批实验没有得到令人满意的结果，只有 12.5%产生了双胞胎。后来发现：当两个卵子都被置于同一个子宫角时，没有一个卵子会发生迁移，两个卵子都挤在这一端；随即出现两者对养分的竞争，结果常常导致其中一个胚胎的死亡。在后来的实验中，我们将两个卵子分别放到一个子宫角中，从而得到了 73% 的双胎率。这一发现的经济价值是显而易见的，人们正在积极尝试一些方案以解决用非手术方式移植卵子时妊娠率下降的问题，因此双胞胎技术是有可能会被付诸实践的。

牛的嵌合性

在多数双胞胎牛中，胎膜的融合发生在怀孕的最开始阶段，并且造血细胞或其他类型的细胞在两个胚胎之间存在互换现象。在混合性别双胞胎中，雄性的睾丸细

yet established whether germ cells of this type continue through to actual spermatozoa. It seems likely that such a situation would occur in the case of chimaeric male calves. If so, it would mean that a proportion of the offspring of one twin may genetically be by the brother and this could seriously interfere with the accuracy of progeny tests, in which such bulls are used for artificial insemination. The percentage of twins born is not high (usually about 2 to 3%), but from our earlier experiments it was obvious that one of a pair present within a single horn frequently died by day 60 of gestation; as fusion of membranes occurs well before this stage, the fact that a cow has a single calf does not, therefore, mean that it did not start life originally as a twin. Twin ovulations have been estimated to be as high as 11% in the cow by some workers.

Our experiments on twinning open up a simple method of determining whether one twin can sire his brother's offspring. We have been able to transfer the egg of one breed of cattle to one uterine horn and that of a different breed to the other, using breeds which would colour mark their offspring. The use of such twins to inseminate a fairly large number of females should establish with certainty whether the interchanged germ cells of the chimaeras are capable of carrying through spermatogenesis. The use of similar sets of females will also be of interest because one can determine if the mixture of blood types has any effect on milk production and quality by producing twins of widely differing production characteristics such as the Jersey and Friesian. Such chimaeras are also tissue tolerant, so it should also be possible reciprocally to interchange half of the udder and to study milk production at the cellular level with the blood supply remaining constant.

Larger Litters

Transplantation has been used quite extensively to investigate the possibility of increasing the litter size, particularly of the sheep and to some extent of the pig.

Initial experiments involving the transfer of either two or five eggs to recipient sheep indicated that the uterus had a severely limiting effect on the number of offspring which would continue development and it became obvious that the problem was not solely one of availability of sufficient eggs. Lawson has recently investigated this problem in breeds of sheep of normally high, medium and low fertility, by the transfer of five eggs from a common genetic source to each group. At the time of transfer the potential fertility of each group was obtained by counting the number of corpora lutea (indicating the number of eggs shed). This work showed that the high and medium fertility breeds were producing eggs at almost their maximum potential, but that the uterus of the low fertility breed was capable of carrying about twice the maximum number assessed by corpora lutea counts. The last of these is the chief breed of sheep in New Zealand and clearly selection on the basis of high ovulation is necessary if litter sizes are to be increased. Although the litter size of the pig can be increased to some extent, the same uterine limitations apply as described previously for the cow and sheep.

Selection for an overall increase in litter size does, therefore, involve many factors, and in the cow and sheep, for example, there are great variations in cotyledon numbers; these are obviously of importance in multiple births and an examination of their inheritance

胞包含很多 XX 变种，但直到现在仍不清楚这种类型的生殖细胞是否能继续分裂分化到生成实际的精子。也许这样的情况会发生在嵌合体雄性牛仔中。如果事实确实如此，那就意味着在双胞胎之一的后代中有一部分可能在遗传学上应该算是它兄弟的后代，这会严重干扰用后代公牛进行人工授精实验的准确性。双胞胎的出生比例并不高（通常约为 2%~3%），但从我们的早期实验中可以很明显地看到：被置于同一个子宫角的一对卵子中的一个通常会在妊娠 60 天之内死亡；因为膜的融合刚好在这一阶段之前发生，因此，一头母牛只生下一个单胞胎并不意味着它在生命刚开始的时候不是双胞胎。一些研究人员曾估计牛的双胞胎排卵率可高达 11%。

我们的双胞胎实验为确定一个双胞胎是否是它兄弟后代的种畜提供了一个简单的方法。我们已经能够把一个品种的牛的卵子移植到一个子宫角中，而把另一个品种的牛的卵子移植到另一个子宫角，可以用颜色来区别所选用品种的后代。用这种双胞胎对相当数量的母牛进行人工授精就可以明确地知道嵌合体中互换的生殖细胞是否能够实现精子发生。用类似系列的母牛进行实验也同样受到人们的关注，因为人们可以通过制造产奶能力完全不同的双胞胎，如泽西奶牛和弗里斯兰奶牛，来确定血型混合是否会对产奶量和品质有影响。这种嵌合体还具有组织耐受性，所以应该也可以互换一半乳房，并且可以让我们在保持其供血恒定的前提下从细胞水平上研究产奶量。

更多的产仔数

移植技术曾被非常广泛地用于研究增加产仔数的可能性，特别是绵羊和少数几个品种的猪。

最初的实验是将 2 个或者 5 个卵子移植到绵羊受体中，结果表明绵羊子宫对将在其中继续发育的后代的数量有严格的限制。显然，问题不仅仅在于要拥有足够多的卵子。最近劳森采用在正常情况下具有高、中和低生育率的绵羊品种对这一问题进行了研究，方法是把具有共同遗传来源的 5 个卵子分别移植到每个组中。在移植的时候可以通过计算黄体的数量(代表排出卵子的数量)来获得每个组潜在的生育率。这项研究表明：具有中高等生育率的品种排出的卵子数几乎等于它们的最大潜在能力；而低生育率品种的子宫可以容纳的卵子数量是根据黄体数估计出的最大生育能力的两倍。后者是新西兰的主要绵羊品种，很明显，如果要增加产仔数，就必须在高排卵的基础上进行选择。虽然猪的产仔数可以有某种程度的增加，但是猪的子宫也有与前面介绍的牛和绵羊子宫相同的限制作用。

因此，为增加总产仔数而进行的选择确实与很多因素有关。例如，牛和绵羊在绒毛叶数量上的巨大差异；这些在多胎妊娠中显然是很重要的因素，因而有必要检

would be worthwhile. This can only be done at slaughter and would, therefore, require superovulation and removal of fertilized eggs from such animals so that they could be transferred to recipients if a high cotyledon count was found.

Identical Twins

Although identical twins are rarely produced by domestic animals, they are invaluable for research and the possibility of producing them artificially has been enhanced by work on small mammals and the pig. It has been shown that the removal of all but one blastomere of even a four or eight-cell egg can result in normal offspring after the transfer of that egg to a suitable recipient. In principle, therefore, the transfer of removed blastocysts to emptied zonae could lead to the production of identical offspring, but a number of problems have arisen, the chief one being that the blastomere injected into the new zona is very liable to be exuded through the crack made at the time of injection. Attempts are being made to overcome this loss by temporary transfer of such eggs to the rabbit oviduct which coats them with a layer of mucin (Fig. 2); these experiments are still in progress. An alternative approach which seems to be successful in the small mammals is to divide the blastocyst and then to culture the two halves before transfer until they become spherical again. Attempts to use this technique have not yet been made in the domestic animals.

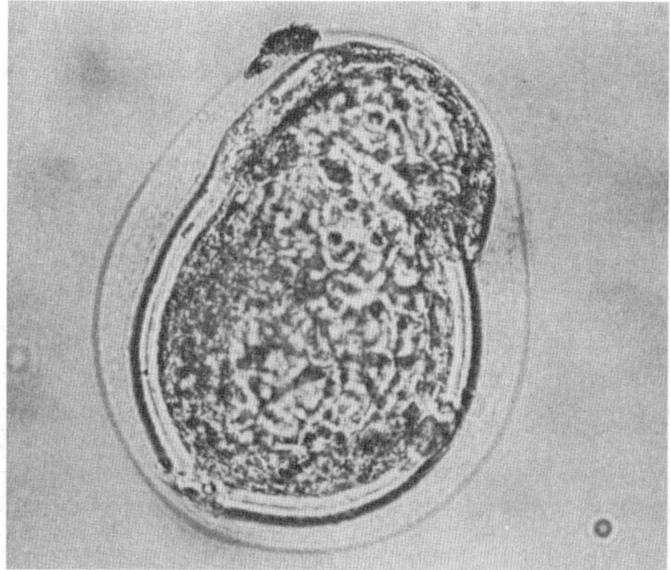

Fig. 2. A sheep blastocyst injected with cells (lower left) when at the eight cell stage and transferred to the rabbit oviduct to become sealed with a mucin coat.

It is clear that the value of egg transfer has only relatively recently been fully realized and there is little doubt that it will be extensively used as a tool for research in reproductive physiology in the future.

(**233**, 379-381; 1971)

L. E. A. Rowson: ARC Unit of Reproductive Physiology and Biochemistry, University of Cambridge, 307 Huntingdon Road, Cambridge.

查它们的遗传特征。这只能在屠宰的时候做到，因此需要超数排卵和从这些动物身上取出受精卵，这样它们才可以被移植到受体中，前提是绒毛叶数量很多。

同卵双胞胎

虽然家畜很少会产下同卵双胞胎，但它们在研究上无比珍贵。通过对小型哺乳动物和猪的研究，人工产生同卵双胞胎的可能性已经有所提高。结果表明：即使在4细胞卵或者8细胞卵的所有卵裂球中只剩下一个没有被去除，也会在将这个卵移植到合适的受体之后产生正常的后代。因此，在原则上，将移除的囊胚移植到空的透明带中是可以产生相同后代的；但会引发很多问题，其中一个主要的问题是，被注入新透明带中的卵裂球很容易从注射时留下的缝隙排出来。现在人们正在尝试利用暂时将这些卵移植到兔子的输卵管中以便使它们的表面裹上一层黏蛋白的方法来克服这种损失（图2）；这些实验仍在进行之中。另一种看似很适合小型哺乳动物的方法是分裂囊胚，然后在移植之前培养分开的两部分，直到它们再次成为球形。目前还没有人尝试过在家畜中使用这项技术。

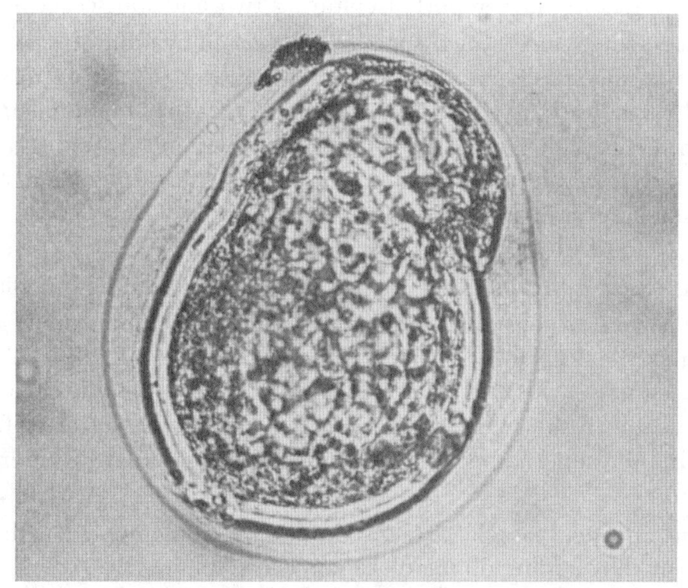

图2. 将处于8细胞阶段的细胞（左下方）注入绵羊囊胚中，并将该囊胚移植到兔子的输卵管中以便裹上一层黏蛋白。

很显然，卵移植的价值直到最近才被人们充分认识到；毫无疑问，在未来它将作为一种研究生殖生理学的工具而得到广泛的应用。

（董培智 翻译；曹文广 审稿）

In vitro Culture of Rabbit Ova from the Single Cell to the Blastocyst Stage

S. Ogawa *et al.*

Editor's Note

Although it was possible at this time to culture fertilized mouse eggs into early embryos, researchers were struggling to replicate their results in other mammalian species and suspected that the tissue culture media played an important role. Here Shyoso Ogawa and colleagues reveal their custom-made synthetic media mix which, they show, supports the *in vitro* development of fertilized rabbit eggs into early embryos. Sixty-five percent of their eggs developed into blastocysts, hollow embryonic structures consisting of around 100 cells, representing a significant increase in the technique's efficiency.

SEVERAL investigators have reported culturing two and four-celled mammalian ova to the blastocyst stage[1-5]. Although successful *in vitro* cultivation of eggs from the single cell to the blastocyst stage has been achieved in the mouse[6-8], few satisfactory methods have been devised for the culture of the ova, at this stage, of other mammalian species, apart from some experiments with human oocytes[9-10]. We now wish to report a higher rate of success with one-celled rabbit ova cultivated *in vitro* to the blastocyst stage, using a nutrient solution composed of several known synthetic media.

Preliminary trials showed that none of the usual chemically defined media were suitable for this work. The best results were obtained with the solution composed of $HamF_{12}$ solution[11] (Nissan 59% v/v); RPMI 1640[12] (Nissan 20% v/v); Eagle's solution[13] (Hanks' solution + Eagle's amino-acids and vitamins, Nissan 20% v/v); sodium caseinate solution (5 g milk casein dissolved slowly in 40–45 ml. of 0.1 N NaOH was centrifuged for 5 min at 3,000 r.p.m., and the supernatant was used as a sodium caseinate solution, 1% v/v).

The medium was adjusted to *p*H 7.5 with $NaHCO_3$, supplemented with 20% v/v calf serum and completed by adding 60 mg of kanamycin sulphate/1.

Ova were flushed with the medium from the oviducts of a rabbit 17–19 h after copulation. In most cases, Earle's flasks (10 ml.) containing 6 ml. of medium were used for cultivation. Two ova were placed in every flask and cultured for up to 4 days at 37°C in a humidified incubator with a constant flow of 5% CO_2 in air. About half of the flasks were stoppered tightly with silicon bungs before being placed in the incubator. After incubation, ova were examined by phase contrast microscopy, and classified according to the final stage of development. The microcinephotographic microscope was also used, together with a specially designed cuvette (a flat slender-rectangular solid type, 3 mm internal depth, with 3–4 ml. volume) filled with the medium. The cleavage and behaviour of the embryonic cells could be recorded through the flat sides without distortion under the inverted

兔卵子从单细胞到囊胚阶段的体外培养

尾川昭三等

编者按

尽管小鼠受精卵已可以被培育成早期胚胎，研究人员仍在尝试将这项研究成果应用到其他哺乳动物中去，并猜测组织培养基在其中发挥了非常重要的作用。尾川昭三和他的同事们在本文中报道了一种特制的合成培养基混合物，他们的实验结果表明这种混合物有助于将受精的兔卵子体外培养至早期胚胎阶段。其中有 65% 的卵子发育成了囊胚。空心的胚胎结构包含了大约 100 个细胞，证明该技术的有效性比以往有了显著的提高。

有几位研究者曾报道过将 2 细胞或 4 细胞的哺乳动物卵子培养至囊胚阶段 [1-5]。尽管已经在小鼠身上成功地实现了从单个卵细胞到囊胚阶段的体外培养 [6-8]，但是除了人类卵母细胞的一些实验以外 [9-10]，人们目前还几乎没有设计出什么令人满意的方法能适用于培养其他哺乳动物的卵子。现在，我们要报道一种能够更成功地实现将单细胞兔卵子体外培养至囊胚阶段的方法，我们使用了一种由若干已知合成培养基组成的营养液。

初步试验显示，所有常用的成分确定的化学培养基都不适用于这项研究。最佳结果是用以下组成的混合溶液得到的：哈姆氏 F_{12} 溶液 [11]（尼桑 59% 体积比）；RPMI 1640[12]（尼桑 20% 体积比）；伊格尔氏溶液 [13]（汉克斯液 + 伊格尔氏氨基酸和维生素，尼桑 20% 体积比）；酪蛋白酸钠溶液（先将 5 克乳酪蛋白缓慢地溶解于 40 毫升 ~ 45 毫升 0.1 摩尔 / 升的 NaOH 中，然后以 3,000 转 / 分的速度离心 5 分钟，得到的上清液就是我们所用的酪蛋白酸钠溶液，1% 体积比）。

用 $NaHCO_3$ 将培养基的 pH 值调至 7.5，添加占总体积 20% 的小牛血清，最后按每升体积 60 毫克的量加入硫酸卡那霉素。

在兔子交配后 17 小时到 19 小时之内，用培养基冲洗其输卵管得到卵子。通常用装入 6 毫升培养基的伊格尔氏培养瓶（容量为 10 毫升）进行培养。每个培养瓶中装有两个卵子，在 37℃下、CO_2 占空气的比例恒为 5% 的加湿培养箱中培养 4 天。在被放入培养箱之前，约有一半培养瓶用硅质的塞子密封。培养结束后，用相差显微镜检测卵子，并根据最终的发育阶段进行分类。我们还使用了带有显微摄影装置的显微镜和一个充满培养基的特制透明容器（一种扁平细长的矩形固体容器，内部深度为 3 毫米，体积为 3 毫升 ~ 4 毫升）。因此，胚细胞在 37℃微型培养容器中的卵

microcinemicroscope kept at 37°C within the microscopic incubator.

Fifty-eight of sixty-seven ova incubated appeared to have undergone normal cleavage to eight cell, morula, or to the blastocyst. Most of them developed to the blastocyst (65%) rather than the morula (24%). Table 1 summaries the results of the studies with either the open or closed system of cultivation.

Table 1. *In vitro* Development of Rabbit Ova from One-celled Stage

Culture system	No. of eggs tested	No. of embryos at final stage of development				
		1~2-cell	8-cell	Morula	Early expanding blastocyst	Expanding blastocyst
Earle's flask						
Open	24	5	4	9	4	2
Closed	27	3	2	3	4	15
Flat cuvette						
Closed	16	1	0	2	4	9

There was a difference in the proportion of ova developing to the blastocyst in the two culture systems; in the open system more embryos remained at the morula stage than in the closed system. There were, however, some cases of the development to the blastocyst stage in the open system. The most advanced development was obtained when we used a closed flask of cuvette without a gas phase. Four consecutive experiments yielded twenty-four expanding blastocysts (56%), eight early expanding blastocysts (19%) and five morulae (12%) from forty-three ova.

Time lapse microcinephotographs (1–8 frames/2 min on 16 mm film under phase contrast) of the embryo which developed from the one-celled stage to the expanding blastocyst (Fig. 1) show a regular pattern of cleavage as the culture passed (two-cell, 20–25 h; eight-cell, 38–45 h; morula, 58–67 h, and early blastocyst, 75–170 h after copulation). This pattern seems to parallel that which can be expected in oviducts and uteri.

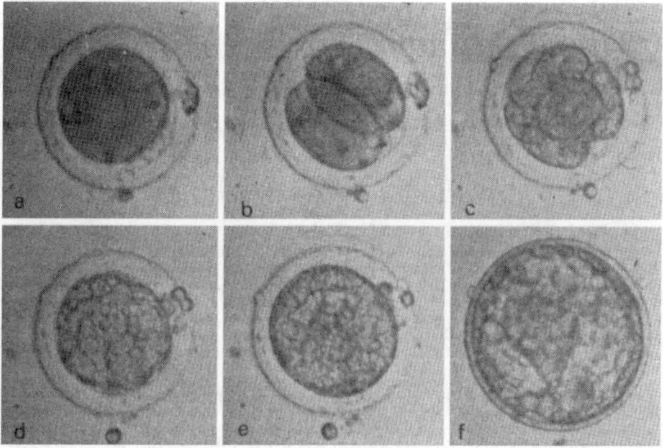

Fig. 1. The development of the rabbit ovum from one-cell to blastocyst stage *in vitro*. From a time-lapse film (16 mm) of an ovum isolated 18 h after mating (× *c.* 200). (*a*) 4.5 h after the start of culture; (*b*) two-cell stage, 6.5 h after (*a*); (*c*) eight-cell stage, 41 h after (*a*); (*d*) late morula to early blastocyst stage, 74 h after (*a*); (*e*) early blastocyst stage, 83.5 h after (*a*); (*f*) expanding blastocyst stage, 122 h after (*a*).

裂和运动状况可以用倒置显微摄像显微镜通过两个扁平端无失真地拍摄下来。

在培养的67个卵子中，有58个经历了正常的卵裂而达到8细胞以至桑葚胚，甚或囊胚的状态。它们中的大多数发育成了囊胚（65%）而不是桑葚胚（24%）。表1概述了在开放或密闭培养体系下得到的研究结果。

表1. 兔卵子从单细胞阶段开始的体外发育

培养体系	检测的卵子数量	处于发育最终阶段的胚胎数量				
		1~2细胞	8细胞	桑葚胚	早期扩张的囊胚	扩张的囊胚
伊格尔氏培养瓶						
开放	24	5	4	9	4	2
密闭	27	3	2	3	4	15
扁平的透明容器						
密闭	16	1	0	2	4	9

在两种培养体系中，发育到囊胚阶段的卵子的比例是不同的：与密闭体系相比，在开放体系中会有更多的胚胎保持在桑葚胚状态。不过，在开放体系中还是有一些胚胎发育到了囊胚阶段。当我们使用一个无气相的密闭小型透明容器作为培养瓶时，所获得的胚胎发育进展最为显著。通过4个连续的实验，我们从43个卵子中培养出了24个扩张的囊胚（56%）、8个早期扩张的囊胚（19%）和5个桑葚胚（12%）。

定时显微摄像记录（每2分钟1张～8张，记录在相差为16毫米的电影胶片上）表明：从单细胞状态发育到扩张囊胚阶段的胚胎（图1）随培养过程显示出常规的卵裂模式（2细胞，交配后20小时～25小时；8细胞，交配后38小时～45小时；桑葚胚，交配后58小时～67小时；早期囊胚，交配后75小时～170小时）。这种模式似乎与我们在输卵管和子宫内分析得到的结果一致。

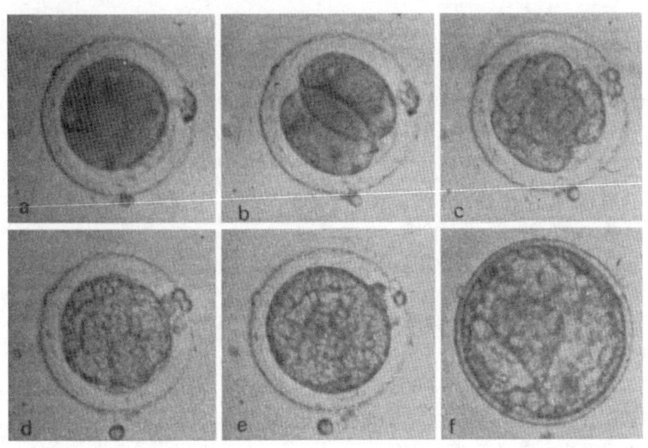

图1. 兔卵子从单细胞到囊胚阶段的体外发育。交配后18小时分离的一个卵子的定时显微胶片（16毫米）（放大约200倍）。(*a*) 培养开始后4.5小时；(*b*) 2细胞阶段，(*a*)之后6.5小时；(*c*)8细胞阶段，(*a*)之后41小时；(*d*) 晚期桑葚胚到早期囊胚阶段，(*a*)之后74小时；(*e*)早期囊胚阶段，(*a*)之后83.5小时；(*f*) 扩张囊胚阶段，(*a*)之后122小时。

We have confirmed by microcinephotography the presence of rhythmical contraction and expansion, like those reported[14-16] in the mouse blastocyst, accompanying cleavage division of the cell mass in the embryo. The movement occurred at the transitional stage from late morula to early blastocyst and increasingly continued throughout the expanding blastocyst stage. The zona seemed to be thinner and elastic, its volume continuing to decrease, while the volume of the blastocoelic cavity continued to increase.

During the expanding stage, many intracellular particles moved up and down sometimes in a circular movement, within the blastocoelic cavity. The cell mass was finally fixed under the marginal expanding membrane of the blastocoelic cavity.

The composition of the culture medium *in vitro* is a critical factor in the cultivation of mammalian ova. Our medium can supply the necessary nutrients and environment to support a high rate of development of one-celled rabbit ova into early blastocysts. Furthermore, because a CO_2 incubator is not needed it seems to be convenient for the study of development of the rabbit ova *in vitro*. We think that this medium could also be used for the culture of the ova of other mammalian species *in vitro*.

This work was supported by a research grant from the Ministry of Education, Japan. We thank Mr. S. Etoh and N. T. Takatsuna for microcinephotography.

(**233**, 422-424; 1971)

Shyoso Ogawa, Kahei Satoh and Hajime Hashimoto: Department of Animal Reproduction, College of Agriculture and Veterinary Medicine, Nihon University, Setagaya-ku, Tokyo.

Received June 2; revised July 5, 1971.

References:

1. Brinster, R. L., *Exp. Cell Res.*, **32**, 205 (1963).
2. Brinster, R. L., *J. Reprod. Fert.*, **10**, 227 (1965).
3. Biggers, J. D., Moore, B. D., and Whittingham, D. G., *Nature*, **206**, 734 (1965).
4. Onuma, H., Maurer, R. R., and Foote, R. H., *J. Reprod. Fert.*, **16**, 491 (1968).
5. Kane, M. T., and Foote, R. H., *Biol. Reprod.*, **2**, 245 (1970).
6. Biggers, J. D., Gwatkin, R. B. L., and Brinster, R. L., *Nature*, **194**, 747 (1962).
7. Whitten, W. K., and Biggers, J. D., *J. Reprod. Fert.*, **17**, 399 (1968).
8. Mukherjee, A. B., and Cohen, M. M., *Nature*, **228**, 472 (1970).
9. Edwards, R. G., Steptoe, P. C., and Purdy, J. M., *Nature*, **227**, 1307 (1970).
10. Steptoe, P. C., Edwards, R. G., and Purdy, J. M., *Nature*, **229**, 132 (1971).
11. Ham, R. G., *Proc. US Nat. Acad. Sci.*, **53**, 288 (1965).
12. Moore, G. G., Gerner, R. E., and Addison, H. F., *J. Amer. Med. Assoc.*, **199**, 519 (1967).
13. Eagle, H., *Science*, **130**, 432 (1959).
14. Kuhl, W., and Friedrich-Freksa, H., *Zool. Anz., Suppl.*, **9**, 187 (1936).
15. Borghese, E., and Cassini, A., in *Cinemicrography in Cell Biology* (edit. by Rose, G. G.), 274 (Academic Press, New York, 1963).
16. Cole, R. J., and Paule, J., *Preimplantation Stages of Pregnancy*, Ciba Foundation Symp. (edit. by Wolstenholme, G. E. W., and O'Conner, M.), 82 (Little Brown, Boston, 1965).

显微摄影方法使我们确信：与那些对小鼠囊胚的报道一样[14-16]，伴随胚胎中内细胞团的卵裂，存在着节律性的收缩和扩张。这种运动发生在从晚期桑葚胚到早期囊胚的过渡阶段，并且在整个扩张囊胚阶段持续增加。透明带似乎会变薄且有弹性，其体积持续减少，与此同时囊胚腔的体积却在不断扩大。

在扩张阶段，囊胚腔中的许多细胞内颗粒有时会沿着圆周轨迹来回移动。最后，内细胞团被固定在囊胚腔边缘扩张膜的下面。

体外培养基的成分是哺乳动物卵子培养的一个关键因素。我们的培养基可以提供必需的营养成分和环境，以保证单细胞兔卵子快速地发育到早期囊胚阶段。此外，因为不需要 CO_2 培养箱，所以对于研究兔卵子的体外发育是很方便的。我们认为这种培养基也能用于体外培养其他哺乳动物的卵子。

这项工作得到了日本教育省提供的科研经费的资助。还要感谢晴代先生和高纲先生在显微摄影方面为我们提供的帮助。

（吴彦 翻译；王晓晨 审稿）

Proposed Mechanism of Force Generation in Striated Muscle

A. F. Huxley and R. M. Simmons

Editor's Note

By the early 1970s, the filamentous sliding nature of muscle fibres was well established. It was known that myosin cross-bridges projected from the thick filaments and interacted with the thin ones, but it was not known how the tension between filaments was generated. Here biologists Andrew Fielding Huxley and Robert M. Simmons from University College London describe a series of experiments measuring the change in tension in striated muscle after a sudden change in length. Cross-bridges, they conclude, are elastic structures with ratchet-like properties: each cross-bridge has three stable positions with progressively lower potential energy. These features enable the cross-bridges to generate the force and movement between muscle filaments.

ONE approach to the elucidation of the kinetics of movement of the "cross-bridges" which are widely assumed to generate the relative force between the thick and thin filaments during contraction of a striated muscle fibre is to record and analyse the transient response of stimulated muscle to a sudden change either of tension or of length. Considerable progress has been made in this way by Podolsky and his colleagues[1-3] who recorded the time course of shortening after a sudden reduction in load. Similar responses have been recorded repeatedly in this laboratory but have not been published because we did not succeed in making the tension change sharply enough to distinguish the component of length change that is truly synchronous with the tension change from that which lags behind the tension change. We therefore turned[4] to the inverse type of experiment, in which the length of the fibre is suddenly altered (by ±0.1–1.5%) and the time course of the resulting tension change is recorded. The results have led to some fairly definite suggestions about the way in which the cross-bridges may actually produce the force between the thick and thin filaments.

All the experiments were carried out on isolated fibres from the semitendinosus muscle of the frog *Rana temporaria*, at 0°–4°C. Length changes, complete in less than 1 ms, were produced by means of a servo system[5], and tension was recorded with a capacitance gauge[6]. Compliance in the apparatus itself is small enough to be completely disregarded; precautions were taken to reduce the compliance in the tendon attachments to a minimum, and in some of the experiments it was eliminated altogether by using the "spot-follower" device[5], which continuously measures the length of a middle segment of the fibre.

横纹肌产生力量的建议机制

赫胥黎，西蒙斯

编者按

人们在20世纪70年代早期就已证实肌丝的滑动特性。那个时候人们就知道肌球蛋白横桥凸出于粗肌丝之上，并能与细肌丝发生相互作用，但尚不知道肌丝之间的张力是如何产生的。在本文中，伦敦大学学院的生物学家安德鲁·菲尔丁·赫胥黎和罗伯特·西蒙斯描述了一系列测量横纹肌在长度突然变化后发生张力改变的实验。他们得出的结论是：横桥是具有棘齿式特性的弹性结构——每个横桥都拥有三个势能逐渐降低的稳定位置。这些特性可以使横桥在肌丝之间产生力和运动。

大家普遍认为，“横桥”会在横纹肌纤维收缩时产生粗肌丝和细肌丝之间的相对推力，有一种方式可以解释横桥运动的动力学，那就是记录并分析受刺激肌肉在张力或长度突然变化时的瞬时反应。波多尔斯基及其同事[1-3]已经在这方面取得了相当大的进展，他们记录了在负载突然减少之后肌纤维收缩的时间进程。我们在实验室中也曾多次观察到类似的反应，但一直没有对外发表，因为我们不能使张力变化的速度快到足以将与张力变化真正同步的长度变化成分和滞后于张力变化的长度变化成分区分开来。因此，我们转而进行[4]相反类型的实验，即在纤维长度突然改变（±0.1%～1.5%）时将由此导致的张力变化的时间进程记录下来。根据这些记录结果，我们得到了一些非常确切的证据，这些证据说明横桥确实可以在粗细肌丝之间产生力。

所有实验都是在0℃～4℃下用林蛙半腱肌的分离纤维进行的。在不足1 ms的时间内通过一个伺服系统[5]产生长度变化，并用电容量表[6]记录张力。装置自身的柔量小到完全可以被忽略；还采取一定的措施使肌腱附着点的柔量减至最小，在其中一些实验中，我们使用了能连续观测纤维中间部分长度的“点跟踪”装置[5]以完全消除柔量的影响。

Responses to Stepwise Length Change

The general time course of tension in these experiments is shown in Fig. 1. This article is concerned only with the changes that occur simultaneously with the length change itself and during the first few milliseconds after it. As has already been briefly reported[4], these changes are of the kind shown in Fig. 2: the tension undergoes a relatively large alteration simultaneously with the step change of length, but recovers quickly towards a level closer to that which existed before the step. The final recovery to the original tension which has been seen by many earlier investigators[7-9] takes place on an altogether slower time scale. The early changes seen in Fig. 2 have only come to light through the improved time resolution of present-day apparatus.

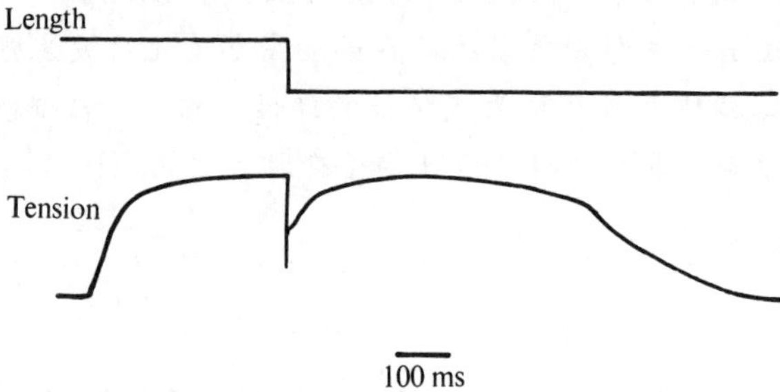

Fig. 1. Isometric tetanus of an isolated muscle fibre (frog, 4°C), with an imposed shortening step. This article is concerned with the tension changes during, and in the first few milliseconds after, the length step; these are shown on a fast time scale in Fig. 2.

The behaviour shown in Fig. 2 suggests the presence of two structural elements in series. One of these would be an elastic element whose length is altered simultaneously with the change of length that is applied to the whole fibre, thus producing the large initial change of tension. The other would be an element with viscous as well as elastic properties, whose length readjusts itself during the period of a few milliseconds immediately after the length change, as a result of the change in tension. As this readjustment proceeds, the imposed length change comes to be shared between the two structures, giving a tension intermediate between the values immediately before and immediately after the length change. At the time of our first note[4] about this quick initial tension recovery, we thought it likely that the recovery took place by sliding, with movement of the cross-bridges, but that the instantaneous elasticity was mostly in the filaments themselves. Since then we have measured these responses in fibres stretched so as to reduce the amount of overlap between thick and thin filaments. We found[10] that responses to a given absolute length change were altered only by a reduction in the scale of the tension changes, which varied in direct proportion to the amount of overlap and therefore also to the number of cross-bridges capable of contributing to tension. On these grounds we now believe that the instantaneous elasticity (or at least the greater part of it) resides in the cross-bridges themselves, as well as the structure responsible for the tension recovery.

对长度阶跃变化的反应

在上述实验中张力的一般时间进程示于图 1。本文中所涉及的仅仅是在长度发生变化的同时以及在随后几个毫秒内的变化。正如我们已经简要报道过[4]的那样，这些变化所属的类型如图 2 所示：在长度发生阶跃变化的同时，张力会出现相当大的改变，但会向接近于发生阶跃变化之前的水平快速恢复。虽然许多早期的研究者[7-9]已注意到张力最终会恢复到初始的状态，但从总体来看会发生在一个较慢的时间尺度上。图 2 所示的早期变化只有用时间分辨率较高的现代装置才能检测到。

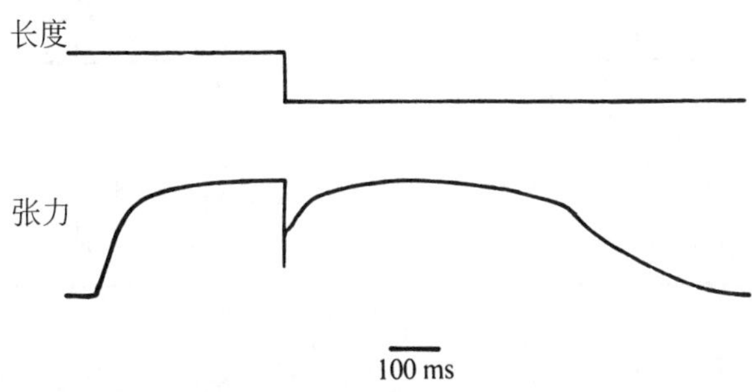

图 1. 一条分离的肌纤维（蛙，4℃）在外加长度阶跃缩短时的等长强直收缩。本文所涉及的是在长度发生变化的同时以及在随后几个毫秒内的张力变化；图 2 所示的就是在快时标上的变化。

图 2 所示的变化说明存在着两个串联的结构成分。其中一个是弹性成分，其长度随着整个纤维的长度变化同步改变，因而张力有很大的初始变化。另一个是同时具有黏性和弹性性质的成分，在长度变化之后的几个毫秒内，这个结构成分的长度会根据张力的变化不断调整。在这种不断调整的过程中，外加的长度变化分摊到两种结构成分上，因而给出的张力值将介于长度变化快要发生之前和长度变化刚好结束之后的数值之间。当我们第一次注意到[4]这种向初始张力快速恢复的现象时，我们认为：恢复很可能是通过滑动进行的，其中伴随着横桥的运动，但瞬时弹性主要发生在肌丝自身之中。此后，我们在伸长量达到可以减少粗细肌丝间交叠量的纤维中观测了上述反应。我们发现[10]：对于给定的绝对长度变化，在反应上的不同仅仅是张力变化尺度有所下降，下降幅度与交叠量成正比，也正比于对张力有贡献的横桥的数量。基于这些结果，我们现在认为：瞬时弹性（或者至少是其中很大的一部分）不但存在于导致张力恢复的结构之中，还存在于横桥自身。

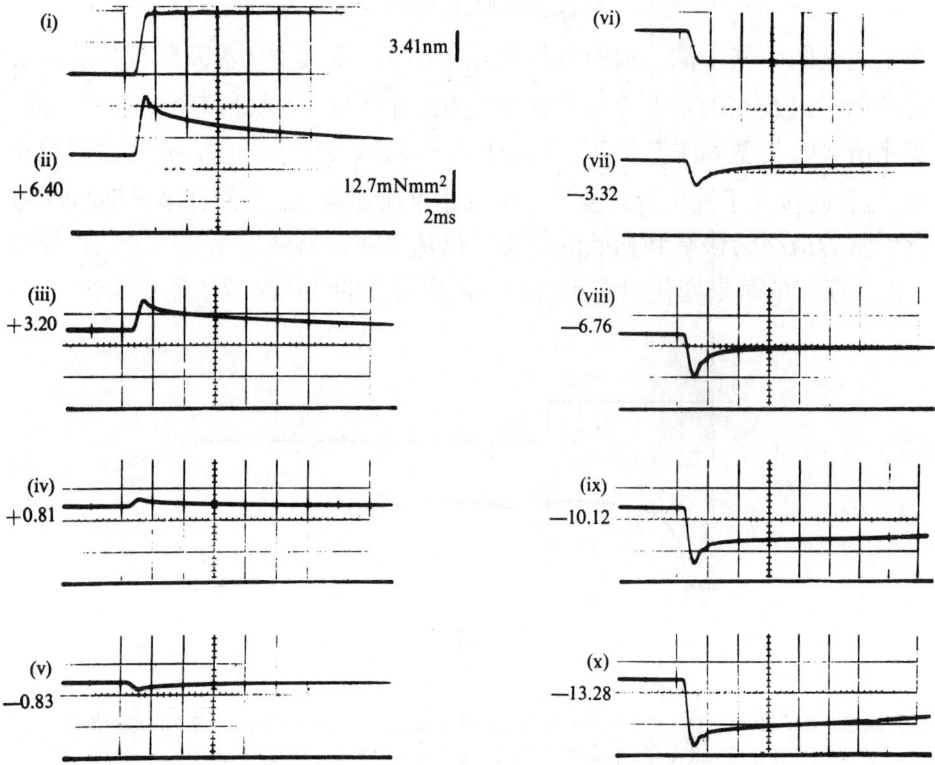

Fig. 2. Transient changes in tension exerted by stimulated muscle fibre when suddenly stretched (ii–iv) or shortened (v, vii–x); same experiment as Fig. 1, which shows the whole of the contraction during which record (ix) was taken. (i) and (vi): records of length change during tension records (ii) and (vii) respectively. The number by each tension record shows the amount of the length change per half-sarcomere, in nm.

The relations between length change and tension are illustrated in Fig. 3. The curve T_1 shows the extreme tension reached during the step change of length. It is somewhat non-linear, becoming stiffer with increasing tension as is commonly found in biological materials. This curve is the best experimental approach that we have to the instantaneous elasticity of the fibre, but it is clear from records such as viii–x in Fig. 2 that the tension drop in the larger shortening steps is cut down because the quick recovery has progressed to an appreciable extent before the length change is complete. The true curve of the instantaneous elasticity is therefore less curved on the shortening side than the curve of T_1 in Fig. 3; it is even possible that it is practically straight, as indicated by the broken line.

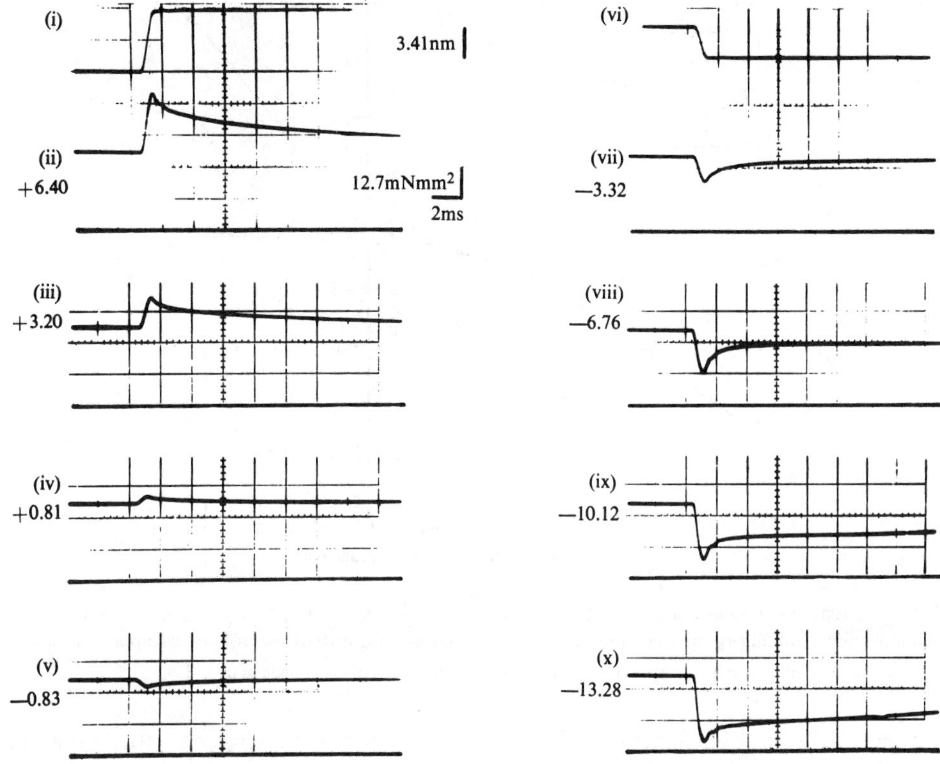

图 2. 受刺激的肌纤维在突然拉伸（ii~iv）或者突然收缩（v，vii~x）时产生的瞬时张力变化；与图 1 对应的实验相同，图 1 显示了记录（ix）中所对应的时间范围内的总长度缩短量。(i）和（vi)：分别为在张力记录 (ii) 和 (vii) 期间的长度变化记录。每幅张力记录图上的数字代表了每个半肌节的长度变化量，单位是 nm。

长度变化与张力之间的关系示于图 3。曲线 T_1 代表在长度发生阶跃变化的过程中所达到的张力极限值。它在某种程度上是非线性的，和大多数生物材料一样，在张力增加时就会变得更硬。这条曲线是我们由纤维瞬时弹性实验得到的最好结果，但是从一些记录，如图 2 的记录 viii~x 中可以清楚地看到，当长度发生较大的阶跃缩短时，张力的下降幅度变小了，因为快速恢复过程在长度变化完成之前就已经发展到了相当的程度。因此，与图 3 中的曲线 T_1 相比，瞬时弹性的真实曲线在阶跃缩短一侧的弯曲程度要小一点；曲线 T_1 在阶跃缩短一侧的部分甚至有可能是直的，如虚线所示。

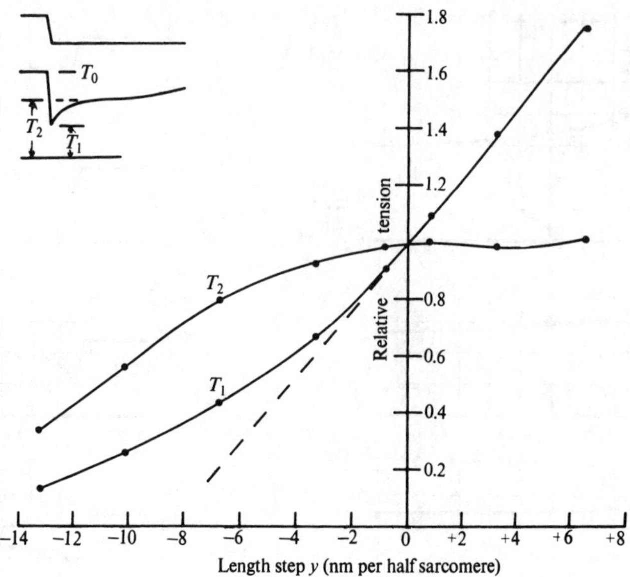

Fig. 3. T_1, Extreme tension reached during step, and T_2, tension approached during quick recovery phase, plotted against amount of sudden stretch (positive) or release (negative). Broken line: extrapolation of the part of the T_1 curve which refers to stretches and small releases. From records in Fig. 2.

In contrast to this straightforward behaviour of instantaneous elasticity, the quick tension recovery is highly nonlinear both in its extent and in its speed. The line T_2 in Fig. 3 shows the level approached at the end of the quick phase of recovery; for moderate amounts of shortening it has the unusual feature of being concave downwards, reflecting the fact that after a small length step the tension returns practically to its previous level (Fig. 2, iv, v). As regards the time course, it is evident from Fig. 2 that the early tension recovery is much more rapid in releases than in stretches, and that its speed varies continuously over a wide range with the size of the length step. The recovery is not exponential, but an estimate of the dominant rate constant can be obtained and is plotted against the size of the stretch or release in Fig. 4; it is roughly fitted by a curve of the form

$$r = \frac{r_0}{2}(1 + \exp -\alpha y) \tag{1}$$

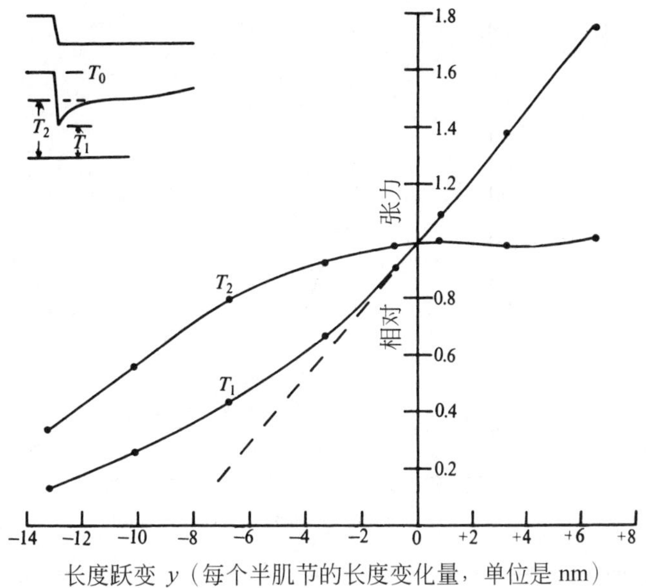

图3. T_1，在长度阶跃变化时达到的张力极限值；T_2，在快速恢复阶段的张力值，横坐标是突然拉伸（正）或释放（负）的量。虚线：从曲线T_1的拉伸和释放量较小部分外推得到的结果。依据是图2中的记录。

与瞬时弹性的这种简单特性相反，快速张力恢复在量上和速度上都是高度非线性的。图3中的曲线T_2显示出了快速恢复阶段末期所达到的水平；当阶跃缩短量适中的时候，它表现出向下凹陷的不寻常特征，这说明在小的长度跃变之后张力几乎可以恢复到原有的水平（图2，iv，v）。至于时间进程，从图2中可以清楚地看到：处于释放状态中的初期张力恢复比处于拉伸状态中的要快很多，并且恢复速度会在很宽的范围内随长度跃变的大小连续变化。恢复并不是指数型的，但可以估计出它的主要速率常数并对拉伸或释放的大小作图，见图4；拟合曲线的表达式大致为：

$$r = \frac{r_0}{2}(1 + \exp - \alpha y) \tag{1}$$

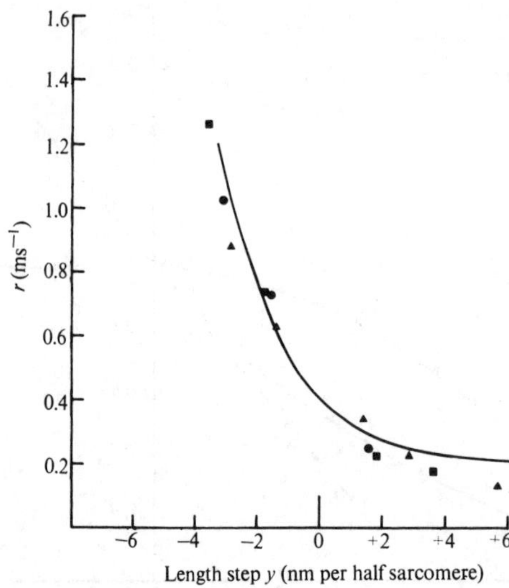

Fig. 4. Rate constant r of quick recovery phase following a length step of magnitude y (positive for stretch). Estimated as $(\ln 3)/t_{1/3}$ where $t_{1/3}$ is the time for recovery from T_1 to $(2T_2+T_1)/3$ (see Fig. 3). From three experiments using the "spot-follower" device; temperature 4°C. The curve is $r = 0.2\,(1+\exp - 0.5\,y)$.

All these features can be given at least a qualitative explanation if we assume that the force on a cross-bridge influences the length changes in that cross-bridge in the way that is assumed by Eyring and others[11,12] in their theory of the visco-elastic behaviour of high molecular weight polymers. The treatment presented in the following paragraphs is meant only to indicate the way in which these features would emerge from such a theory; a more complete treatment will be needed in order to test whether it is fully consistent with the data.

Assumptions

The key assumptions that have to be made are (1) that the movement by which a cross-bridge performs work during the period while it is attached takes place in a small number of steps, from one to the next of a series of stable positions with progressively lower potential energy, and (2) that there is a virtually instantaneous elasticity within each cross-bridge allowing it to shift from one of these stable positions to the next without a simultaneous displacement of the whole thick and thin filaments relative to one another.

These assumptions could be incorporated into some of the mechanisms that have been proposed for the action of the cross-bridges, for example, that of Davies[13], in which each cross-bridge shortens by folding at a number of points, or that of H. E. Huxley[14], in which the head of the myosin molecule (H, Fig. 5) rotates relative to the thin filament, acting as a lever which pulls the thick filament along by a link AB which is also part of the myosin molecule. Our assumptions fit very conveniently on to H. E. Huxley's proposals, and we shall discuss them on this basis in the way illustrated in Fig. 5. The features shown there which are additional to H. E. Huxley's scheme are as follows. (1) The link AB is not inextensible as in H. E. Huxley's proposal but contains the instantaneous elasticity which shows up as curve T_1 (Fig. 3). (2) The head has a small number s of combining sites

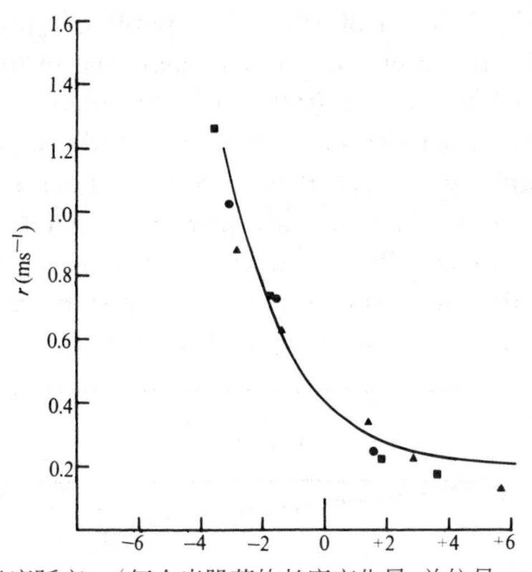

图 4. 紧随幅度为 y（对于拉伸为正值）的长度跃变之后的快速恢复阶段的速率常数 r。估计其表达式为 $(\ln 3)/t_{1/3}$，其中 $t_{1/3}$ 是从 T_1 恢复到 $(2T_2+T_1)/3$ 所需的时间（见图 3）。数据来自使用“点跟踪”装置的三次实验；温度为 4℃。曲线为 $r=0.2(1+\exp-0.5\,y)$。

如果我们假定横桥上的力能够像艾林和其他人[11,12]在高分子聚合物黏弹性行为理论中所设想的那样影响该横桥中的长度变化，那么就至少可以对上述所有特征给出一个定性的解释。在以下段落中所描述的处理方法只是为了说明怎样才能从我们的理论推导出上述特征；若要检验理论与数据是否完全相符，还需要更为全面的考量。

假　定

必须引入的两个关键假定是：(1) 通过少数几个步骤就能产生使横桥在连接时实现其功能的运动，即从一系列势能逐渐降低的稳定位置中的一个转移到下一个。(2) 每个横桥内部都存在着虚拟的瞬时弹性，使其可以从上述稳定位置中的一个转变成另一个，而不会同时出现整条粗肌丝和细肌丝相对于彼此的位移。

这些假定可以与某些曾被用于解释横桥作用的机制相结合。例如：在戴维斯提出的机制中[13]，每个横桥的收缩是通过在许多点的折叠实现的；或者按照赫胥黎的说法[14]，肌球蛋白分子头部（H，图 5）会相对于细肌丝旋转，就像一根杠杆一样沿着同样是肌球蛋白一部分的连接 AB 拉动粗肌丝。我们的假定很容易与赫胥黎的提法相契合，因而我们将在此基础上按照图 5 所描绘的方式来加以讨论。下面列出了一些在赫胥黎理论中没有提到的特征。(1) 连接 AB 并非像赫胥黎所说的那样不能伸展，而是能够表现出如曲线 T_1（图 3）所示的瞬时弹性。(2) 头部有数量为 s 的少量结合点（M_1、M_2 等，图 5），每个结合点都能与细肌丝中肌动蛋白分子上的

(M_1, M_2 and so on, in Fig. 5), each of which is capable of combining reversibly with a corresponding site (A_1, A_2 and so on) on an actin molecule in the thin filament. A single M–A attachment allows variation of θ (rotation in the plane of the diagram of Fig. 5) without hindrance, but no other degree of freedom of the myosin head relative to the actin molecule. (3) The affinity between these myosin and actin sites is smallest for M_1A_1, larger for M_2A_2 and so on. (4) The sites are placed so that the myosin head has (s–1) stable positions, each of which allows two consecutive M and A sites to be attached simultaneously. (5) When the myosin head is in its (s–1)th stable position it can be detached from the thin filament by a process involving the hydrolysis of ATP.

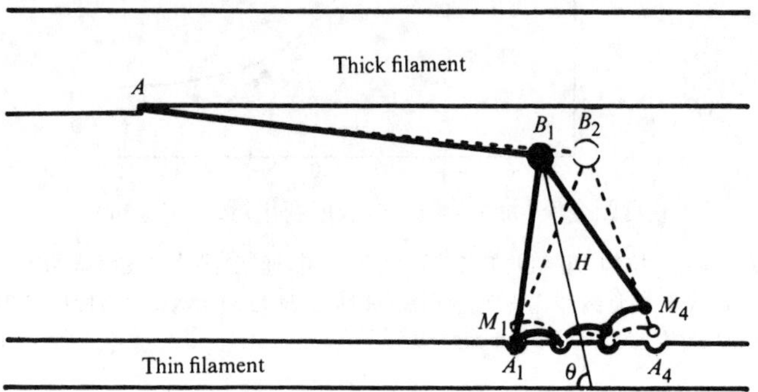

Fig. 5. Diagram showing assumed cross-bridge properties. The myosin head H is connected to the thick filament by a link AB containing the undamped elasticity which shows up as T_1 (Fig. 3) in the whole fibre. Full line shows head in position where M_1A_1 and M_2A_2 attachments are made; broken lines show position where M_2A_2 and M_3A_3 attachments are made.

On this basis the quick tension recovery is due to the tendency for the myosin head to rotate to positions of lower potential energy, while the fact that the recovery occurs at a finite speed is a manifestation of the rate constant for movement of the system from one of the stable positions to the next.

Potential Energy Diagram of a Cross-bridge

Curves i–iv in Fig. 6 show the potential energy diagrams for individual attachment sites (M_1A_1, M_2A_2 and so on) on a single cross-bridge. Each contains a flat-bottomed well extending over the range of myosin head positions where that particular attachment can exist. Curve v is the sum of curves i–iv, and therefore gives the total potential energy of the cross-bridge (in the absence of force in the link AB); it consists of a series of steps, separated by narrow troughs at the positions where two of the links are attached simultaneously. The depth of each trough will depend on the shapes, and on the exact positions, of the sides of the potential wells that contribute to it; it is assumed in Fig. 6 that these are such that the quantities E_1 and E_2 (v, Fig. 6) are the same for each of the troughs.

相应位置（A_1、A_2 等）可逆地结合。单独的一个 M–A 连接可以保证 θ 的变化（在图 5 所示平面内的旋转）不受阻碍，但是肌球蛋白头部相对于肌动蛋白分子没有其他的自由度。(3) 上述肌球蛋白与肌动蛋白结合点之间的亲和性在 M_1A_1 处最小，在 M_2A_2 处要大于 M_1A_1 处，以此类推。(4) 结合点的分布要能确保肌球蛋白头部具有（$s-1$）个稳定位置，每个稳定位置允许两个顺序排列的 M 和 A 结合点同时连接。(5) 当肌球蛋白头部位于它的第（$s-1$）个稳定位置时，它可以借助一种有三磷酸腺苷水解介入的过程与细肌丝分离。

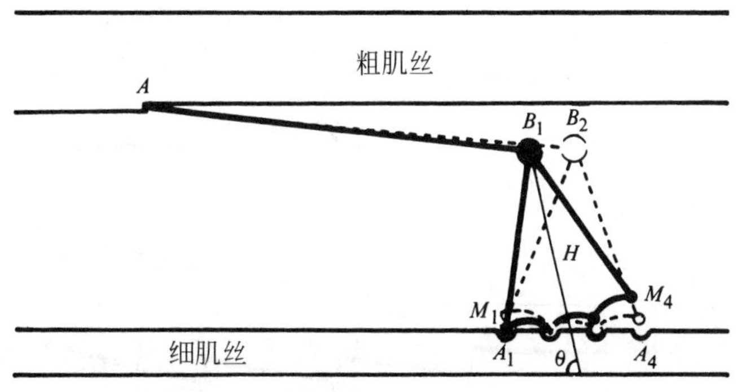

图 5. 假想中的横桥性质示意图。肌球蛋白头部 H 通过连接 AB 与粗肌丝相连，连接 AB 表现出的无阻尼弹性如整个纤维的 T_1 曲线（图 3）所示。实线显示的是当 M_1A_1 和 M_2A_2 发生连接时头部所处的位置；虚线是当 M_2A_2 和 M_3A_3 连接时头部所处的位置。

以此为基础可以认为：出现快速张力恢复的原因是肌球蛋白头部有旋转到势能较低位置的趋势，而以有限速度恢复这一事实是系统从一个稳定位置运动到另一个稳定位置所具有的速率常数的体现。

横桥的势能图

图 6 中的曲线 i~iv 描绘出单一横桥上每对连接点（M_1A_1、M_2A_2 等）的势能图。每条曲线都包含一个平底部分，这个平底部分的延伸范围代表了某个特定连接可以存在于其中的一系列肌球蛋白头部位置。曲线 v 是曲线 i~iv 的叠加，因此给出了横桥（在连接 AB 中不存在力时）的总势能；曲线 v 包含一系列阶跃，阶跃与阶跃之间的分界线是位于同时相连的两个连接处的狭窄凹点。每个凹点的深度取决于导致该凹点的势阱的边缘形状和边缘所在的具体位置；图 6 中假定 E_1 和 E_2 的值（曲线 v，图 6）对于每个凹点都是一样的。

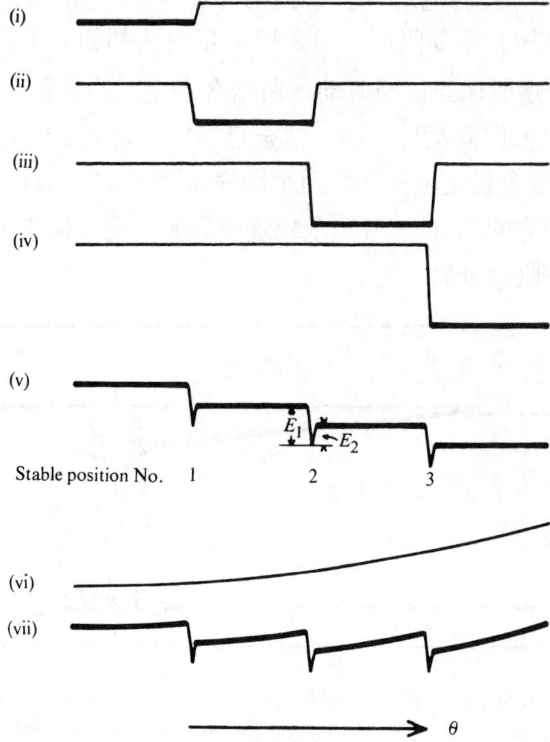

Fig. 6. Potential diagrams relating to the system illustrated in Fig. 5. i–iv, Diagrams for individual attachments M_1A_1, M_2A_2, M_3A_3, M_4A_4 respectively; in each the thick line corresponds to the range of θ within which the corresponding M and A sites are attached; v, sum of i–iv, giving potential energy of a system composed only of a myosin head and a thin filament; vi, potential energy due to stretching the elastic link AB; vii, total potential energy.

The total potential energy of an attached cross-bridge contains also the potential energy of stretching the elastic link AB. The latter term is shown in curve vi, and the total in curve vii.

Responses Expected Theoretically

In the mathematical section at the end of this article, we derive equations describing the response of a system of this kind to a step change of length. The system treated there is simplified by assuming that only two stable positions are available to each attached cross-bridge. The corresponding potential energy diagram is sketched in Fig. 7. This shows that B_2, the activation energy for transfer of a bridge from position 1 to position 2, contains a term W which depends on the force in the link AB, being increased by a stretch and reduced by a release. B_1, however, the activation energy for the reverse transfer, contains no such term and is independent of the force. It is this asymmetry which enables the theory to account for the way in which the rate constant of the quick tension recovery varies with the direction and magnitude of the length step: the theoretical result is expressed in equation (12) which is identical in form with equation (1), already shown (Fig. 4) to represent adequately the experimental data. The theory also leads to equation (16) for

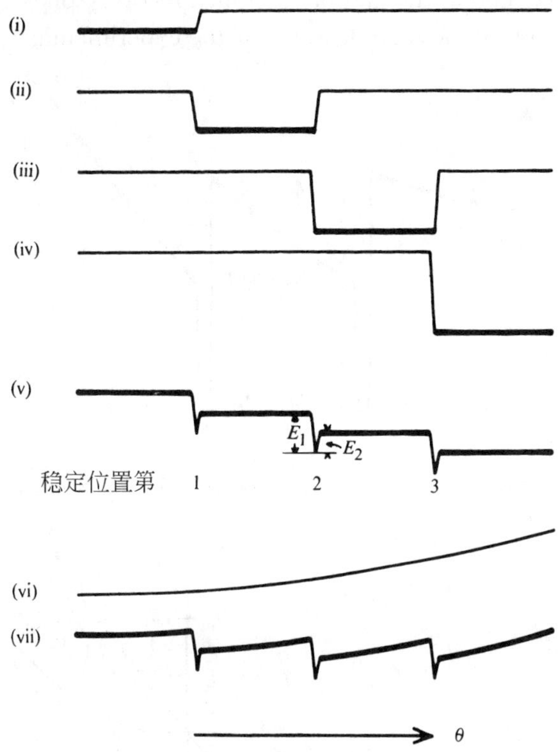

图 6. 与图 5 中所描述的系统相对应的势能图。i~iv 分别为单个连接 M_1A_1、M_2A_2、M_3A_3 和 M_4A_4 的示意图；在每幅图中，粗线代表在对应 M 点和 A 点连接时 θ 的范围；v（i~iv 的叠加）给出一个仅由一个肌球蛋白头部与一条细肌丝组成的系统的势能；vi，由拉伸弹性连接 AB 所产生的势能；vii，总势能。

一个处于连接状态的横桥的总势能还包括拉伸弹性连接 AB 所产生的势能。曲线 vi 代表的是后者，而曲线 vii 则显示了总势能。

从理论上推出的反应

在本文末尾处的数学部分中，我们推导出了一些可以描述这类系统对长度阶跃变化的反应的式子。为了简化要处理的系统，我们假设每个处于连接状态的横桥只有两个可能的稳定位置。对应的势能示意图绘于图 7。图 7 说明一个横桥从位置 1 转移到位置 2 的活化能 B_2 中包含着一个 W 项，该项取决于连接 AB 中的力，在拉伸时增加而在释放时减少。但逆向转移过程的活化能 B_1 却不包含这样的项，所以与该力无关。正是这种不对称性使得我们的理论可以解释快速张力恢复的速率常数随长度跃变方向和大小而变化的形式：理论结果用等式（12）来表示，等式（12）与表达式（1）的形式相同，式（1）已经被实验数据充分证明（图 4）。由我们的理论还可以推出快速恢复阶段末期所能达到的张力水平，即等式（16）；将其绘于图 8 之中，

the tension level approached at the end of the quick recovery phase; this is plotted in Fig. 8 which is seen to reproduce the main features of the experimental T_2 curve in Fig. 3.

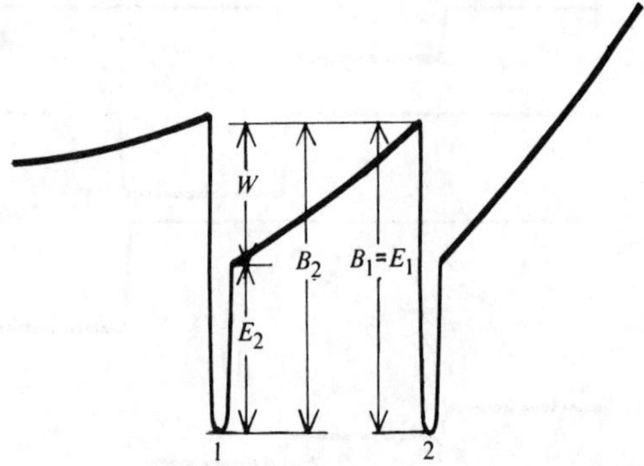

Fig. 7. Potential energy diagram equivalent to Fig. 6 (vii) but referring to the simplified system with only two stable positions.

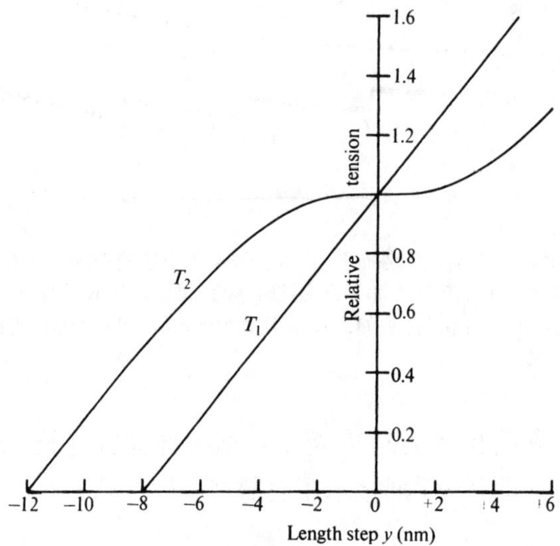

Fig. 8. Curves of T_1 and T_2 calculated from the simplified system, plotted on same scales as the experimental values in Fig. 3. T_2 curve: Equation (16) with $E_1 - E_2 = 4kT$, $1/\alpha = 2$ nm, $h = 8$ nm.

The two striking features of the quick tension recovery are thus accounted for by this theory. The numerical values used in obtaining this degree of agreement are: $E_1 - E_2$, the potential energy difference between the stable positions of attachment of a cross-bridge, is equal to 4 kT; h, the travel of point B between its two stable positions, is 8 nm; K, the stiffness of the link AB, is 2.5×10^{-4} N m^{-1}.

The following considerations show, however, that the assumption that the cross-bridge movement takes place in a single step is probably not correct.

Isometric force and number of bridges. Equations (4) and (7) show that the isometric force

可以看到它也具有在图 3 中由实验得到的曲线 T_2 的主要特征。

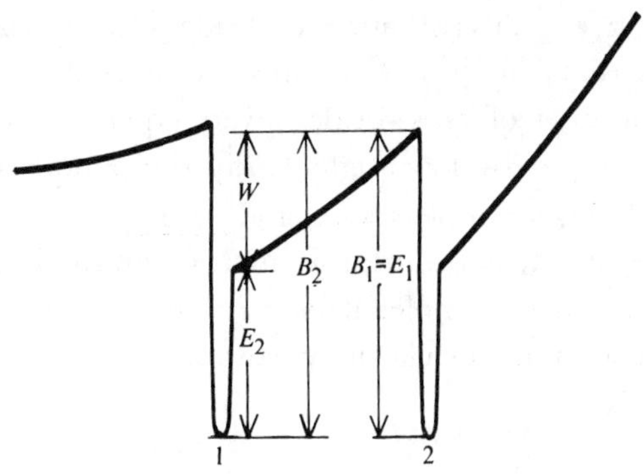

图 7. 与图 6（vii）所示相当的势能图，但所对应的是只有两个稳定位置的简化系统。

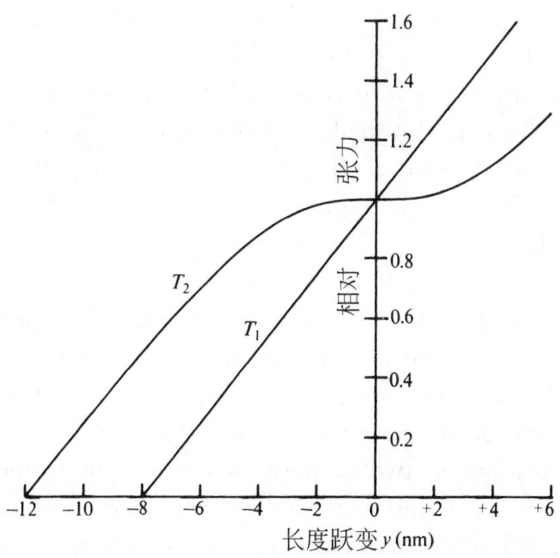

图 8. 由简化系统计算得到的 T_1 曲线和 T_2 曲线，用与图 3 中的实验值相同的标度画出。T_2 曲线：按等式（16）画出，其中 $E_1 - E_2 = 4\,kT$，$1/\alpha = 2$ nm，$h = 8$ nm。

因此，根据这一理论可以解释快速张力恢复的两个显著特征。达到这种吻合程度所要用到的数据是：$E_1–E_2$，一个横桥上两个稳定连接位置之间的势能差，其值等于 $4\,kT$；h，B 点在其两个稳定位置之间的行程，其值为 8 nm；K，连接 AB 的刚度，其值为 2.5×10^{-4} N m^{-1}。

然而，下面的论证过程说明假定横桥的运动仅通过单一步骤完成很可能是不正确的。

等长收缩力与桥的数目。从等式（4）和（7）中可以看出：每个处于连接状态的横

per attached cross-bridge is $(E_1-E_2)/h$; with the values mentioned in the last section this amounts to 2.0×10^{-12} N. To reach a total force of 3×10^5 N m^{-2}, such as real fibres produce, would therefore need 1.5×10^{17} attached bridges m^{-2} in each half-sarcomere. This is about 1.5 times greater than the number of myosin molecules present. The discrepancy would be greater if the number of cross-bridges were equal to the number of projections on the thick filaments detected by low-angle X-ray diffraction since this appears to be about half the number of myosin molecules[15], and greater still if a substantial proportion of the cross-bridges were unattached during isometric contraction. (It is conceivable that each of the two heads on a myosin molecule should be counted separately; this would be appropriate only if the instantaneous elasticity existed within each of the heads, allowing them to shift independently.)

Work per attached cross-bridge. An upper limit to the external work done by the fibre per cross-bridge attachment will be given by the integral in equation (17), between the isometric point $y=0$ and the point ($y=-12.0$ nm) where $\varphi_2=0$. With the adopted values of α and h, this is 3.8 kT. This result is only about half the value which can be calculated as follows from the actual performance of frog muscle. Work can be as much as 40% of (work + initial heat)[16], and the latter quantity is about 11 kcalories/mol of phosphorylcreatine split[17], giving the work term as 4.4 kcalories/mol. This is equal to 7.3 kT per molecule, which will represent the work per cycle of attachment and detachment of a cross-bridge if it is assumed that the cycle is coupled (presumably through ATP) to the hydrolysis of one molecule of phosphorylcreatine.

Probable number of stable positions. The quantitative treatment just presented has thus led to low values for the force and work per cross-bridge. It is just possible that revised values for α and for the number of cross-bridges will resolve these discrepancies, but it seems to us more likely that it will be necessary to assume that the cross-bridge movement takes place not in a single step but in two or perhaps more. This would lead to proportional increases in the calculated force and work, but the expected time course and extent of the quick tension recovery, which are already in good agreement with the experimental results, would not be much altered if each step has the same height E_1-E_2 as has already been assumed and the value of h is reduced so as to keep the same total range of travel. Fig. 5 is drawn for the case where the number of steps is 2 (3 stable positions and 4 points of attachment), which at present seems the most probable number.

Relation to Earlier Theories

The idea of applying Eyring's theory of polymers to muscle is not new. A comprehensive theory of muscle was developed by Polissar[18], in which the shortening of links in actomyosin chains was influenced in this way by the load, but this theory lost its relevance when it became clear that major changes of length take place by sliding, not folding, of the filaments.

The proposal that the sliding movement is generated by the tendency of attachments between the filaments to move through a series of a few positions of progressively lower

桥的等长收缩力为 $(E_1-E_2)/h$；代入最后一部分中提到的数字，其值为 2.0×10^{-12} N。为了使总力和真实纤维中的情况一样达到 3×10^5 N m^{-2}，就需要在每个半肌节中有每平方米 1.5×10^{17} 个连接的横桥。这个值大概是现有肌球蛋白分子数量的 1.5 倍。如果横桥的数目等于用低角度 X 射线衍射法检测到的粗肌丝上的凸起数，那么两者之间的差异还会更大，因为由后者得到的结果似乎只有肌球蛋白分子数量的一半左右 [15]；而如果在等长收缩过程中有相当比例的横桥没有连接，那么这个差值还会再大一些。(有人认为应该分别计算一个肌球蛋白分子的两个头；只有当每个头中都存在瞬时弹性进而可以独立移动时，这样做才是恰当的。)

与每个处于连接状态的横桥相对应的功。每个横桥连接的纤维所做的外功上限可以由等式（17）中的积分给出，积分范围是从等长收缩点 $y = 0$ 到与 $\varphi_2 = 0$ 对应的点 ($y = -12.0$ nm)。代入 α 和 h 的采用值，可以得到 3.8 kT。这一结果只有按照以下方法从青蛙肌肉实际表现计算出来的数值的一半左右。功可以达到（功 + 初始热量）的 40% [16]，（功 + 初始热量）约为 11 kcal/mol，是由磷酸肌酸分解 [17] 得到的，由此计算出功的大小为 4.4 kcal/mol。这相当于每分子 7.3 kT，如果假定连接与解连接的循环与一分子磷酸肌酸的水解偶联（可能是通过三磷酸腺苷），那么这个值就代表了与每个处于单一连接与解连接循环中的横桥相对应的功。

稳定位置的可能数目。由上述定量处理所得到的对应于每个横桥的力值和功值都偏低。虽然通过修正 α 值和横桥数目有可能会解决这个矛盾，但我们认为更有可能的情况是：应该假定横桥的运动不是通过单一步骤完成的，而是通过两步或者更多步完成的。这样会使计算出来的力和功成比例增长，但在每一步都和前述假定一样有同样高度的 $E_1 - E_2$ 且 h 值减少到足以保证总行程大致不变的情况下，已经与实验结果吻合得很好的时间进程和快速张力恢复程度却不会发生很大的变动。图 5 就是按照步骤数为 2 的情况（3 个稳定位置和 4 个连接点）绘制的，也是目前看来最有可能的数目。

与早期理论的关系

将艾林的聚合物理论应用于肌肉算不上新观点。波利萨 [18] 曾提出过一种较为完备的肌肉理论，其中肌动球蛋白链中连接的缩短就是以这种方式受负载影响的。但是当人们逐渐了解到长度的大部分变化是源自肌丝的滑动而非折叠的时候，波利萨理论的价值就不复存在了。

滑动运动的产生是由于肌丝之间的连接具有在一系列势能逐渐递减的若干位置

potential energy was made many years ago by H. H. Weber[19], who also pointed out that in this case the rate constants for shifting from one position to another would be affected by the force on the attachment. He discussed this idea purely on the basis of a translational movement of the thick relative to the thin filament, with a site rigidly fixed to one of the filaments transferring itself from one to the next of the sites on the other filament to which it could be attached. It is difficult to visualize a mechanism of this kind operating over the rather large distances—several nanometers—that the transient responses show to be involved, but the difficulty disappears if there is an elastic structure allowing one of the attachment sites to undergo substantial displacements relative to the filament to which it belongs.

On the kinetic side, the scheme discussed here combines the advantages of the proposals made by A. F. Huxley[20] in 1957 and by Podolsky *et al.*[3]. In each of these schemes the production of force was assumed to occur as an immediate consequence of the formation of a myosin–actin link. In A. F. Huxley's scheme, attachment was assumed to be the main rate-limiting factor in steady shortening, while detachment after the performance of work by a cross-bridge was relatively rapid; with these assumptions it is possible to fit A. V. Hill's relations between load, speed and heat production but not the transient responses discussed here. Podolsky *et al.* reversed the assumptions about the rate constants, making attachment rapid and breakage rate-limiting; in this way they were able to fit many aspects of transient responses but, as they recognized, not the thermal data. With appropriate rate constants for the initial attachment and final detachment of each cross-bridge, the present scheme can probably account for the force–velocity and thermal relationships in the same way as A. F. Huxley's 1957 scheme, while the rapid transfer between stable positions of the myosin head produces transient responses not unlike those which, on the theory of Podolsky *et al.*, result from the rapid formation of new attachments. The present proposal is able to combine the successful aspects of both of these theories because it subdivides the force-generating event into distinct stages, one for attachment and others for stretching the cross-bridge, the latter being much more rapid than the former. This separation also removes another difficulty that both of the earlier theories would very likely have met in a fully quantitative treatment. They assumed that tension is already present in the cross-bridge when it attaches to the thin filament; thermal motion would so seldom bring the cross-bridge to such a large deflexion that it might be impossible to account in that way for the rapidity of contraction that some real muscles achieve.

Mathematical Section

Equations will be derived for the extent and time-course of the quick tension recovery to be expected from a system similar to that shown in Fig. 5 but with only two instead of three stable positions of the myosin head relative to the thin filament. The following additional simplifying assumptions are also made. (1) Actual detachment and re-attachment of cross-bridges are slow enough to be disregarded. (2) Filaments themselves are completely rigid. (3) Filaments undergo no sliding movements except when the total length of the fibre is being altered. (4) The elasticity of the link *AB* obeys Hooke's law. (5) The link *AB* is capable of exerting negative as well as positive tensions. (6) In the

上连续移动的趋势，这个建议是多年以前由韦伯[19]提出的。他还指出，在这种情况下从一个位置移动到另一个位置时的速率常数将会受到连接处的力的影响。他对这一构想的讨论完全是基于粗肌丝相对于细肌丝的平移运动，运动方式是一个牢固地固定于某一条肌丝上的连接点从它能与之连接的另一条肌丝上的一个位置转移到下一个位置。很难想象这种机制会适用于有瞬时反应出现的较大距离（几个纳米），但如果有一种弹性结构能允许其中一个连接点相对于它所在的肌丝发生足量的位移，那么这个难题就会迎刃而解。

从动力学的角度来看，上述理论同时具备赫胥黎 1957 年的提议[20]与波多尔斯基等人的提议[3]所具有的优点。这些理论都假定力的产生是肌球蛋白–肌动蛋白之间形成连接的直接后果。在赫胥黎的理论中，连接被认为是稳态收缩中的主要限速因素，而横桥在产生力之后的解连接过程则要相对快一些；这些假定或许能与希尔所说的负载、速度和放热之间的关系相符合，但无法与本文所讨论的瞬时反应相符合。波多尔斯基等人颠倒了关于速率常数的假设，认为连接是快速的而解连接过程是限速的；由此他们可以解释瞬时反应的很多方面，但不包括热数据，这一点他们也承认。如果知道每个横桥初始连接和最终解连接的速率常数的确切数值，那么目前的理论就有可能可以利用与 1957 年赫胥黎理论相同的方式来解释力–速度与放热的关系，而由肌球蛋白头部在稳定位置之间的快速转移产生瞬时反应的观点也不会与波多尔斯基等人把瞬时反应看作是源自新连接的快速形成的理论有什么不同。目前的提法能够结合这两种理论的成功之处，因为它把产生力的过程细分为不同的阶段，一个是连接阶段，其他几个是横桥的拉伸阶段，后者比前者快很多。这种细分的处理方法还回避了两种早期理论在纯定量处理中很可能会遇到的另一个困难。两种早期理论假定当横桥与细肌丝相接时，张力已经存在于横桥之中；热运动几乎不会使横桥产生如此巨大的偏转，因此用热运动难以解释某些真实肌肉中出现的快速收缩。

数学部分

我们将从与图 5 所示类似但肌球蛋白头部相对于细肌丝只有两个而不是三个稳定位置的系统中导出几个有关快速张力恢复程度和时间进程的表达式。另外还需要做以下假设简化计算。(1) 横桥中实际存在的解连接和重连接过程进行得很慢以至于可以忽略。(2) 肌丝本身是完全刚性的。(3) 只有在纤维总长度发生改变时，肌丝才产生滑动运动。(4) 连接 *AB* 的弹性遵循胡克定律。(5) 连接 *AB* 既能施加正张力，也能施加负张力。(6) 在等长收缩的稳定状态下，每个处于连接状态的横桥在

isometric steady state, every attached cross-bridge spends equal amounts of time in each of the two available positions (this cannot be strictly true in real muscle because the spacings along the thick and thin filaments are not in any simple ratio, so the relative positions of the myosin and actin molecules must vary from one cross-bridge to another). (7) The time taken in transferring from one to the other of the two positions is negligible.

The following notation will be used; n_1: fraction of attached bridges in position 1; n_2 $(=1-n_1)$: fraction of attached bridges in position 2; y: displacement of thick relative to thin filament when fibre is stretched or shortened (zero in isometric state before the applied length change; positive for stretch); y_0: extension of elastic link AB when bridge is midway between positions 1 and 2 (equal to amount of sudden sliding movement needed to bring tension to zero from the isometric state); h: increase in length of AB when bridge shifts from position 1 to position 2; K: stiffness of link AB (assumed to obey Hooke's law); F_1: tension in AB when bridge is in position 1; F_2: tension in AB when bridge is in position 2; φ : time average of F_1 and F_2.

From these definitions,

$$F_1=K(y+y_0-h/2) \text{ and } F_2=K(y+y_0+h/2) \tag{2}$$

and the time average of tension is

$$\begin{aligned}\varphi &= n_1F_1+n_2F_2\\ &= K(y+y_0-h/2+hn_2)\end{aligned} \tag{3}$$

In the isometric state, $y=0$ and $n_2=1/2$, and this equation reduces to the expression for the isometric force per attached cross-bridge

$$\varphi_0=Ky_0 \tag{4}$$

The rate constants k_+ for transfer from position 1 to position 2, and k_- for transfer from 2 to 1, are governed by the energy barriers $B_2(=E_2+W)$ and B_1 $(=E_1)$ respectively (Fig. 7). W is the work done in stretching AB when the bridge transfers from position 1 to position 2, and is given by

$$\begin{aligned}W &= h\frac{F_1+F_2}{2}\\ &= Kh(y+y_0)\end{aligned} \tag{5}$$

from equations (2).

Assuming the ks proportional to $\exp -B/kT$, we have

$$\begin{aligned}k_+ &= k_- \exp (B_1-B_2)/kT\\ &= k_- \exp (E_1-E_2-W)/kT\\ &= k_- \exp (E_1-E_2-Kh(y+y_0))/kT\end{aligned} \tag{6}$$

两个可能位置上的逗留时间都相等（这在真实肌肉中是不可能严格正确的，因为沿着粗肌丝和细肌丝方向的分子排布间距不存在任何简单的比例关系，因此对每个横桥而言，肌球蛋白和肌动蛋白分子的相对位置都不一样）。（7）从两个位置中的一个向另一个转移时所需的时间是可以忽略的。

我们将使用下列符号。n_1：在位置 1 处连接的横桥所占的比例；n_2（$= 1 - n_1$）：在位置 2 处连接的横桥所占的比例；y：当纤维在外力作用下伸长或缩短时粗肌丝相对于细肌丝的位移（在施加长度变化前的等长收缩状态时为 0；对于拉伸为正）；y_0：当横桥处于位置 1 和位置 2 之间时弹性连接 AB 的长度（等于为使张力从等长收缩状态变为 0 所需的突然滑动运动的量）；h：当横桥从位置 1 移向位置 2 时 AB 的长度增量；K：连接 AB（假设其遵循胡克定律）的刚性系数；F_1：当横桥处于位置 1 时 AB 中的张力；F_2：当横桥处于位置 2 时 AB 中的张力；φ：F_1 和 F_2 的时间平均值。

根据上述定义，有：

$$F_1 = K(y + y_0 - h/2) \text{ 和 } F_2 = K(y + y_0 + h/2) \tag{2}$$

而张力的时间平均值为：

$$\begin{aligned}\varphi &= n_1F_1 + n_2F_2 \\ &= K(y + y_0 - h/2 + hn_2)\end{aligned} \tag{3}$$

在等长收缩状态下，有 $y = 0$ 和 $n_2 = 1/2$，由此可以把上式化简为表示每个连接横桥的等长收缩力的式子：

$$\varphi_0 = Ky_0 \tag{4}$$

从位置 1 转移到位置 2 时的速率常数 k_+和从位置 2 转移到位置 1 时的速率常数 k_-分别由能垒 B_2（$= E_2 + W$）和 B_1（$= E_1$）决定（图 7）。W 为横桥从位置 1 转移到位置 2 时纤维拉伸 AB 所做的功，根据式（2）可以化简为：

$$\begin{aligned}W &= h\frac{F_1 + F_2}{2} \\ &= Kh(y + y_0)\end{aligned} \tag{5}$$

假定 k_+ 和 k_- 均与 $\exp -B/kT$ 成比例，我们有：

$$\begin{aligned}k_+ &= k_- \exp(B_1 - B_2)/kT \\ &= k_- \exp(E_1 - E_2 - W)/kT \\ &= k_- \exp(E_1 - E_2 - Kh(y + y_0))/kT\end{aligned} \tag{6}$$

where k_- is constant since B_1 is a fixed quantity independent of the tension in AB.

In the isometric state we have assumed $n_1=n_2$, so $k_+=k_-$; also $y=0$ since y is defined as a length change from the isometric state. It then follows from (6) that

$$E_1-E_2=Khy_0 \tag{7}$$

and (6) becomes

$$k_+=k_- \exp - yKh/kT \tag{8}$$

In the experiments we are considering, y is suddenly altered by imposing a length change on the muscle fibre, and is subsequently held constant. The transfer of myosin heads from one position to the other is then governed by the equation

$$\begin{aligned} \mathrm{d}n_2/\mathrm{d}t &= k_+n_1 - k_-n_2 \\ &= k_+ - (k_+ + k_-)n_2 \end{aligned} \tag{9}$$

This equation represents an exponential approach, with rate constant r given by

$$r = k_+ + k_- \tag{10}$$

towards an equilibrium where

$$n_2=k_+/(k_++k_-) \tag{11}$$

Substituting from (8) into (10) we have

$$r=k_-(1+\exp - yKh/kT) \tag{12}$$

This has the same form as equation (1), and the equations become identical, giving approximate agreement with the experimental results in Fig. 4, if we take

$$Kh=\alpha kT \tag{13}$$

Equation (8) can therefore be written

$$k_+=k_- \exp - \alpha y \tag{14}$$

and (11) becomes

$$n_2 = \frac{1}{2}\left(1- \tanh\frac{\alpha y}{2}\right) \tag{15}$$

其中 k_- 为常数，因为 B_1 是不依赖于 AB 中张力的固定量。

我们曾假定在等长收缩状态下有 $n_1 = n_2$，所以 $k_+ = k_-$；且 $y = 0$，因为 y 被定义为从等长收缩状态开始的长度变化。因此由式（6）可以得到：

$$E_1 - E_2 = Khy_0 \tag{7}$$

而式（6）则变为：

$$k_+ = k_- \exp - yKh/kT \tag{8}$$

在所设想的实验中，我们通过对肌纤维施加一个长度变化而突然改变 y，随后令其保持恒定。因此，利用下式即可确定肌球蛋白头部从一个位置向另一个位置的转移：

$$\begin{aligned} \mathrm{d}n_2/\mathrm{d}t &= k_+ n_1 - k_- n_2 \\ &= k_+ - (k_+ + k_-)\, n_2 \end{aligned} \tag{9}$$

该式呈指数形式，其中由下式给出的速率常数 r

$$r = k_+ + k_- \tag{10}$$

指向的是下面的平衡态：

$$n_2 = k_+ /(k_+ + k_-) \tag{11}$$

将（8）代入（10）后得到：

$$r = k_- (1 + \exp - yKh/kT) \tag{12}$$

上式与式（1）有相同的形式，若要与图 4 中的实验结果大体相符，则两个式子应该相等，如果我们令：

$$Kh = \alpha kT \tag{13}$$

那么就可以将式（8）写作：

$$k_+ = k_- \exp - \alpha y \tag{14}$$

而（11）则变成：

$$n_2 = \frac{1}{2}\left(1 - \tanh\frac{\alpha y}{2}\right) \tag{15}$$

The tension φ_2 at the end of the quick recovery (corresponding to T_2 in the whole fibre) is obtained by combining (3), (13) and (15) to give

$$\varphi_2 = \frac{\alpha kT}{h}\left(y_0 + y - \frac{h}{2}\tanh\frac{\alpha y}{2}\right) \quad (16)$$

The work done during shortening at a low enough speed so that n_2 always has its equilibrium value would be obtained by integrating (16):

$$\int\varphi_2\, dy = kT\left(\frac{\alpha y}{h}\left(y_0 + \frac{y}{2}\right) - \ln\cosh\frac{\alpha y}{2}\right) \quad (17)$$

To match the points in Fig. 4, α is taken as 5×10^8 m^{-1}, the value used for the curve in that figure. (E_1-E_2) is shown by equations (7) and (13) to be equal to $\alpha y_0 kT$. From Fig. 3 (broken line), y_0 is about 8 nm, giving $E_1-E_2=4kT$. h has to be chosen to give the right shape for the curve of T_2 against y (equation 16). A value $4/\alpha$, or 8 nm, is used in Fig. 8; lower values give a less inflected curve and higher values give a curve with a region of negative slope.

Generation of Tension

The tension changes observed in the first few milliseconds after suddenly changing the length of an active muscle fibre suggest thc following mechanism for the generation of tension or shortening by the cross-bridges. Each cross-bridge has three stable positions with progressively lower potential energies, in steps of about 4 times kT, separated by about 4 nm of travel. Transfer from one of these positions to the next is made possible, without simultaneous displacement of the whole filaments through an equally large distance, by the presence of elasticity associated with each individual cross-bridge. The tension generated in this way in the elastic element will show up as such if the muscle length is held constant, or will help to make the filaments slide past each other if shortening is permitted.

A simplified theoretical treatment is given; a more complete treatment will be presented later.

(**233**, 533-538; 1971)

A. F. Huxley and R. M. Simmons: Department of Physiology, University College London, Gower Street, London WC1.

Received August 25, 1971.

References:

1. Podolsky, R. J., *Nature*, **188**, 666 (1960).
2. Civan, M. M., and Podolsky, R. J., *J. Physiol.*, **184**, 511 (1966).
3. Podolsky, R. J., Nolan, A. C., and Zaveler, S. A., *Proc. US Nat. Acad. Sci.*, **64**, 504 (1969).
4. Huxley, A. F., and Simmons, R. M., *J. Physiol.*, **208**, 52*P* (1970).
5. Gordon, A. M., Huxley, A. F., and Julian, F. J., *J. Physiol.*, **184**, 143 (1966).
6. Huxley, A. F., and Simmons, R. M., *J. Physiol.*, **197**, 12*P* (1968).
7. Gasser, H. S., and Hill, A. V., *Proc. Roy. Soc.*, B, **96**, 398 (1924).

在快速恢复结束时的张力 φ_2（相当于整个纤维中的 T_2）可以通过联立（3）、（13）和（15）得到：

$$\varphi_2 = \frac{\alpha kT}{h}\left(y_0 + y - \frac{h}{2}\tanh\frac{\alpha y}{2}\right) \tag{16}$$

在以足够低的速度收缩以使 n_2 始终保持其平衡值的过程中所做的功可以通过积分（16）得到：

$$\int\varphi_2\,\mathrm{d}y = kT\left(\frac{\alpha y}{h}\left(y_0 + \frac{y}{2}\right) - \ln\cosh\frac{\alpha y}{2}\right) \tag{17}$$

为了与图 4 中的点匹配，取 α 为 $5\times10^8\ \mathrm{m}^{-1}$，也是该图中的曲线所用的值。由式（7）和式（13）可以得到（$E_1 - E_2$）等于 $\alpha y_0 kT$。根据图 3（虚线部分），y_0 约为 8 nm，这说明 $E_1 - E_2 = 4kT$。h 的取值必须能够保证以 y 为横坐标的曲线 T_2（式 16）有正确的形状。图 8 中 h 的取值是 $4/\alpha$ 或 8 nm；当 h 取值较低时得到的曲线会更平直一些，而取值较高时在得到的曲线中会出现一段负斜率区。

张力的产生

在突然改变一条活性肌纤维长度之后的最初几个毫秒内所观测到的张力变化意味着由横桥所导致的张力产生或收缩具有如下机制。每个横桥具有三个势能逐渐降低的稳定位置，能量差约为 kT 的 4 倍，之间大约相隔 4 nm。如果假设每个单独的横桥都存在弹性，那么从上述位置中的一个向另一个的转移过程就可以在整条肌丝不发生同样长度的同步位移的前提下实现。以这种方式在弹性成分中产生的张力会表现得如同肌肉长度没有变化一样，或者在允许收缩时有助于肌丝滑过彼此。

本文只给出了简化的理论处理过程；更详细的处理过程将在以后发表。

（王耀杨 翻译；刘京国 审稿）

8. Hill, A. V., *Proc. Roy. Soc.*, B, **141**, 104 (1953).
9. Jewell, B. R., and Wilkie, D. R., *J. Physiol.*, **143**, 515 (1958).
10. Huxley, A. F., and Simmons, R. M., *J. Physiol.*, **218**, 59*P* (1971).
11. Eyring, H., *J. Chem. Phys.*, **4**, 283 (1936).
12. Burte, H., and Halsey, G., *Tex. Res. J.*, 17, 465 (1947).
13. Davies, R. E., *Nature*, **199**, 1068 (1963).
14. Huxley, H. E., *Science*, **164**, 1356 (1969).
15. Huxley, H. E., *J. Mol. Biol.*, 7, 281 (1963).
16. Hill, A. V., *Proc. Roy. Soc.*, B, **127**, 434 (1939).
17. Wilkie, D. R., *J. Physiol.*, **195**, 157 (1968).
18. Polissar, M. J., *Amer. J. Physiol.*, **168**, 766, 782, 793 and 805 (1952).
19. Weber, H. H., *The Motility of Muscle and Cells*, 32 (Harvard University Press, Cambridge, Massachusetts, 1958).
20. Huxley, A. F., *Prog. Biophys.*, 7, 255 (1957).

Molecular Evolution in the Descent of Man

M. Goodman *et al.*

Editor's Note

Studies of human evolution had been about fossils until Morris Goodman's work showing how comparison of amino-acid sequences of proteins could be used to estimate evolutionary relationships and the rate of evolutionary change. This paper pioneered the idea that human beings shared a common ancestry with chimpanzees that excluded other apes, such as the gorilla. However, the view remained that the rate of change had decelerated among higher primates, allowing for a long ancestry of humans through *Australopithecus* and back to *Ramapithecus* more than 10 million years ago. Later work would suggest that the divergence between apes and humans had been in fact more recent—and palaeontology would show that *Ramapithecus* was more closely related to the orang-utan than humans.

IT is likely that the level of molecular organization in higher animals is now more complicated than would have been possible in the first life forms of three billion years or so ago. Even after the emergence of the genetic code and the biochemical apparatus of translation, natural selection would have had to act over long stretches of evolutionary time to produce the sophisticated structures and functions of the multichained proteins now found in living organisms. To begin with, almost any point mutation in a gene coding for a polypeptide chain would have been tolerable; with the evolution of higher levels of molecular organization, however, increasingly stringent functional restraints would have markedly decreased the number of acceptable mutations.

The notion that the level of molecular organization places restraints on how fast genes evolve is implicit in Ingram's hypothesis[1] that the genes for alpha globin have changed more slowly than those for beta globins. The same alpha chain combines with different types of beta chains in the several tetrameric haemoglobins which are synthesized during the life of a mammal. Since there are more restraints on variations of the alpha chains than of the beta chains, fewer mutations would be tolerated at an alpha locus than at a beta locus and alpha genes would not evolve as rapidly as beta genes. In the same way, there may be differences in the rates of evolution of the globin genes at earlier and later evolutionary stages. The original single-chained globin would no doubt have been recognizable by its three-dimensional structure and haem binding site, but it would yet have had to acquire specific sites for subunit combination, for binding diphosphoglycerate, for allosteric modulation of structure to promote the Bohr effect and for combining with the plasma protein haptoglobin. With these functions, natural selection tolerated fewer mutations in the globin genes. From this it follows that the primordial globin genes

人类起源过程中的分子进化

古德曼等

编者按

莫里斯·古德曼的这项工作展示了如何将蛋白质的氨基酸序列比较用于评估进化关系以及进化变化的速度，在此之前，人类进化的研究工作一直都是针对化石展开的。这篇文章率先提出人类和黑猩猩（不包括其他类人猿，例如大猩猩）拥有共同的祖先。虽然如此，在从南方古猿以及追溯到1,000多万年前的腊玛古猿来考虑人类的远祖之后，作者仍然认为在高等灵长类动物中变化的速度已经减慢。而后来的工作表明，类人猿与人类间分化的时间实际上离现在更近，并且古生物学后来证明腊玛古猿更接近于红毛猩猩，而不是人类。

现今高等动物中的分子组织水平很可能比约30亿年前的第一个可能的生命形态更复杂。甚至在遗传密码和翻译的生化装置出现以后，自然选择仍需要很长的进化时间才能形成现在活有机体中多链蛋白的复杂结构和功能。起先，在编码多肽链的基因中，几乎任何点突变都是可接受的；然而，随着分子组织向更高水平的进化，越来越严密的功能性抑制将使可接受突变的数量显著减少。

英格拉姆假说[1]中隐含着分子组织水平会对基因进化速度加以抑制的概念，该假说认为，α球蛋白的相关基因比β球蛋白的相关基因变化得更慢。在哺乳动物生存期间合成的几种四聚体血红蛋白中，相同的α链结合了不同类型的β链。因为比起β链，针对α链变化的抑制会更多一些，所以α基因座比β基因座能容忍的突变少，并且α基因的进化速度也不可能与β基因一样快。在进化的更早和更晚阶段，球蛋白基因的进化速度同样可能存在差异。原始单链球蛋白无疑可以通过它的三维结构及血红素结合位点来识别，但它仍不得不获得一些特异位点以用于结合亚基，用于结合二磷酸甘油酸异构酯、用于结构的变构调整以促进玻尔效应以及用于与血浆蛋白中的触珠蛋白结合。由于这些功能，自然选择只允许球蛋白基因中发生很少的突变。由此得出的结论是，原始球蛋白基因的进化速度要高于后来脊椎动物的球蛋白基因

evolved more rapidly than the later globin genes of vertebrates, consonant with an earlier hypothesis that molecular evolution at a number of gene loci has decelerated, especially in the lineage leading to man[2-4].

This hypothesis, followed by others[5,6], attributes divergent evolution in proteins to selectively neutral mutations but also implies that more mutations were neutral in primitive organisms than in descendant species. Thus biochemical adaptations in human antecedents must have increased with the development of new organs such as the placenta and the cerebral neocortex and the extra functional restraints would then decrease the chances for fresh mutations to accumulate. An immunological mechanism could have further restricted the degree of genetic variability in higher primate populations, for the haemochorial placenta (with its intimate apposition of foetal and maternal blood circulations) and the long gestation period would have increased the risk to the foetus from maternal immunizations to foetal allotypes.

We present evidence in this article that molecular evolution has indeed been slower in higher primates than in other mammals. In particular, from a detailed analysis of the phylogeny of globin genes, we support Ingram's hypothesis that the evolution of alpha globin genes has been slower than that of the beta genes. Available data on proteins also bear on the cladistic relationships of man and other mammals.

The most penetrating view of the molecular changes in individual genes during the descent of man is provided by comparing the amino-acid sequences of human proteins with homologous proteins in other organisms. We have followed the approach of Fitch and Margoliash[7] in constructing gene phylogenetic trees from amino-acid sequence data and for this purpose have used three computer programs. The first program (MMUTD) calculates mutation distances for every pairwise comparison of aligned amino-acid sequences using a matrix based on the genetic code which gives for each amino-acid pair the minimum number of nucleotides that would need to be changed to bring about such a mutation. The second program[6] (UWPGM) produces from a dissimilarity matrix of the mutation distances a dendrogram of the gene species in the set. This dendrogram is an initial approximation to a cladogram, that is a graph which depicts the order of ancestral branching in the gene species set. The UWPGM builds the tree from the smallest to the largest branches in a series of pairwise clustering cycles, each time grouping together the two members of the set (either singleton species or joined species from a previous cycle) with the smallest mutation distance between them. If all the species in the set had evolved at a uniform, or nearly uniform, rate in their descent from a common ancestor, the dendrogram produced by the UWPGM would be the true cladogram[9]. But neither gene species nor animal species necessarily evolve at uniform rates. Thus, with the third computer program (DENDR) we construct alternative dendrograms. The DENDR

进化速度，这与早前的假说一致。早前的假说认为：很多基因座的分子进化已经发生了减速，特别是在进化成人类的这条支系中[2-4]。

这个假说，还有接下来的其他几个[5,6]，将蛋白质中的趋异进化归因于选择性的中性突变，但也暗示与后代物种相比，原始生物中存在更多的中性突变。因此，随着胎盘和大脑新皮层等新器官的发育，人类祖先的生化适应性必定有所增强，但额外的功能性限制接下来会减少新突变积累的机会。在高等灵长类种群中，有一种免疫机制可能会进一步限制基因变异的量，因为血绒膜胎盘（在这里胎儿血液循环和母体血液循环紧密接触）和长怀孕期增加了由母体免疫系统对胎儿同种抗免疫球蛋白的作用给胎儿带来的危险。

我们将在本文中提供证据证明，高等灵长类动物的分子进化速度确实比其他哺乳动物的分子进化速度慢。特别是，从对球蛋白基因系统发生的详尽分析结果来看，我们支持英格拉姆假说——α 球蛋白基因的进化速度比 β 球蛋白基因的进化速度慢。由蛋白质的可用数据还可以得到人类与其他哺乳动物的进化枝关系。

最敏锐地检查人类起源过程中单独基因的分子变化的手段是比较人类蛋白质与其他生物体内同源蛋白质的氨基酸序列。我们按照菲奇和马格利亚什[7]的方法，由氨基酸序列数据来创建基因系统发生树，为了达到这一目的还使用了三个计算机程序。第一个程序（MMUTD）使用基于遗传密码的矩阵来计算由成对比较对齐的氨基酸序列而得到的突变距离，由此得出为了带来这一突变，每一氨基酸对所需变化的最少核苷酸数。第二个程序[6]（UWPGM）由突变距离的相异度矩阵绘制出该系列基因的系统树图。该系统树图是一个进化树的初级近似，图中描绘了该系列基因的祖先分枝的顺序。UWPGM 在一系列配对的聚类循环中建立起从最小到最大分枝的系统树，每次总是把系列中拥有最小突变距离的两个成员（不管是单件种类还是来自前一个循环的联合种类）分在一组。如果系列中所有种类的进化速度在一个共同祖先的后代中保持一致，或接近一致，那 UWPGM 绘制的系统树图就是真实的进化树[9]。但基因类型和动物物种都不一定会按均一的速度进化。因此，我们利用第三个计算机程序（DENDR）建立了若干个备选的系统树图。由 DENDR 程序计算出

program calculates the patristic mutation length* for each pair of adjacent nodes in a dendrogram, sums the patristic lengths through the sequence of nodes between species to calculate a reconstructed mutation distance for each pair of species in the set, and then compares the reconstructed mutation distances to the given mutation distances in the original dissimilarity matrix to calculate a coefficient for the entire dendrogram called "average percentage standard deviation" (APSD)[7]. Among the alternative dendrograms the one with least deviation between original and reconstructed mutation distances (that is, the lowest APSD coefficient) is considered the closest approximation of those tried to a true phylogenetic tree. This approach to finding the best fitting tree has proved capable of capturing the cladistic relationships among gene species even when markedly dissimilar amounts of mutational change characterize their descent from a common ancestor[10].

Globin Phylogenetic Tree

Table 1 is the matrix of mutation values from all possible pairing of sixty-eight metozoan globin chain sequences[11-47]. These chains range in size from 135 to 153 amino-acid residues. They include, apart from carp alpha, chicken alpha, and mammalian alpha and non-alpha haemoglobin chains, a globin of lamprey, myoglobins of horse and sperm whale, and a globin of the insect chironomus. The chicken alpha chain is the first non-mammalian tetrapod haemoglobin chain to be sequenced[17]. It diverges less from human (and chimpanzee) alpha than from any other mammalian alpha (Table 1). This is not due to any special relationship between chicken and man, but, as will be shown, results from the marked conservativism of higher primate alpha chains.

Table 1. Mutation Values from Pairwise Comparisons of Vertebrate Globin Sequences

Full name of globin species given in legend to Fig. 1. The half matrix lists minimum numbers of mutations interrelating pairs of polypeptide chains. These mutational values are converted into percentage mutational divergence

$$\left(\frac{\text{No. of mutations}}{\text{No. of shared amino-acids}} \times 100\right)$$

values, the mutation distances used for constructing the phylogenetic tree in Fig. 1. Alpha llama, sheep foetal beta, and goat A' beta are not completely sequenced. Hence their mutation values may not represent full mutation values.

* The patristic mutation lengths associated with any node, *N*, on a dendrogram are calculated as follows. Any node *N* divides the species of which it is the common ancestor into two sets: set *A* and set *B*. All remaining species under study comprise set *C*. Let D_{ab} be the average of dissimilarity values for pairs of species such that the first species is a member of set *A* and the second species is a member of set *B*, and let D_{ac} and D_{bc} be similarly calculated. The length from set *A* to node *N* is called *a*; the length from set *B* to node *N* is called *b*; and the length from set *C* to node *N* is called *c*. These lengths are calculated by the following formulas:

$$a = \frac{D_{ab} + D_{ac} - D_{bc}}{2}$$

$$b = \frac{D_{ab} + D_{bc} - D_{ac}}{2}$$

$$c = \frac{D_{ac} + D_{bc} - D_{ab}}{2}$$

For the special case that *N* is the apex of the tree (that is, the common ancestor of all species under study): $a = b = \frac{D_{ab}}{2}$ c is undefined.

系统树图中每对相邻节点的父系突变长度 *，沿着两物种间的一系列节点加和父系长度，以得到系列中每对物种的重组突变距离，然后比较重组突变距离与原始相异度矩阵中给出的突变距离，以计算出整个系统树图的系数，这个系数被称为“平均百分比标准偏差”（APSD）[7]。在备选的系统树图中，原始和重组突变距离之间具有最小偏差（即 APSD 系数最低）的那一个系统树图被认为最接近于真实的进化系统发生树。这种寻找最适树图的方法已被证明可以捕捉到在基因类型中存在的进化枝关系，甚至在来自一个共同祖先的后代具有迥然不同的突变变化总量时也是可行的 [10]。

球蛋白系统发生树

表 1 是 68 种后生动物球蛋白链序列的所有可能配对的突变值矩阵 [11-47]。在大小上这些链的范围是 135 个 ~ 153 个氨基酸残基。除包括鲤鱼 α 以外，还包括：鸡 α、哺乳动物 α 和哺乳动物非 α 血红蛋白链，七鳃鳗的某一球蛋白，马和抹香鲸的肌红蛋白，摇蚊属昆虫的某一球蛋白。鸡 α 链是第一条被测序的非哺乳类四足动物血红蛋白链 [17]。它偏离人类（和黑猩猩）α 链的程度要小于偏离其他哺乳类 α 链的程度（表 1）。这并不是因为鸡和人之间存在着某种特殊关系，而是由下面将要讲到的高等灵长目动物 α 链的格外保守造成的。

表 1. 由成对比较脊椎动物球蛋白序列得到的突变值

图 1 两侧的说明文字给出了各类球蛋白的全称。该半矩阵列出了与各多肽链配对相关的最小突变数。这些突变值被转化成百分比形式的突变偏离值：

$$\left(\frac{\text{突变数}}{\text{共享的氨基酸数}}\right) \times 100$$

突变距离被用来构建图 1 中的系统发生树。对美洲驼 α、绵羊胚胎 β 和山羊 A′ β 的测序还不完全。因此它们的突变值也许不能代表完整的突变值。

---

* 计算与系统树图上任一节点 N 相关的父系突变长度的方法如下。任意节点 N 将以 N 为共同祖先的物种分为两组：A 组和 B 组。所有剩下的被研究物种组成 C 组。令 D_{ab} 为配对物种相异值的平均值，所以第一个物种是 A 组中的一个成员，第二个物种是 B 组中的一个成员。以同样的方式计算 D_{ac} 和 D_{bc}。从 A 组到节点 N 的长度称为 a；从 B 组到节点 N 的长度称为 b；从 C 组到节点 N 的长度称为 c。这些长度按下式计算：

$$a = \frac{D_{ab} + D_{ac} - D_{bc}}{2}$$

$$b = \frac{D_{ab} + D_{bc} - D_{ac}}{2}$$

$$c = \frac{D_{ac} + D_{bc} - D_{ab}}{2}$$

作为特例 N 为树顶（即所有被研究物种的共同祖先）：$a = b = \frac{D_{ab}}{2}$，未定义 c。

	INSECT CTT3	INSECT CTT3 SUB	Mb. S. WHALE	Mb. HORSE	GLOBIN LAMPREY	α CARP	α CHICKEN	α RABBIT	α RABBIT SUB	α TREE SHREW	α HORSE SLOW	α HORSE FAST	α HORSE SLOW 24F	α HORSE FAST 24F	α DONKEY	α PIG	α LLAMA	α BOVINE	α GOAT II	α GOAT B	α GOAT A	α SHEEP D	α SHEEP A	α MOUSE NB	α MOUSE C-57	α SIFAKA	α LEMUR	α BUSH BABY	α MULATTA	α FUSCATA	α GORILLA	α CHIMP	α HUMAN	γ HUMAN	β KANGAROO	β KANGAROO -2	β SIFAKA	β LEMUR	β HORSE SLOW	β PIG	β LLAMA	β BOVINE F	β SHEEP F	β BOVINE A	β BOVINE B	β BARBARY C	β SHEEP C	β GOAT C	β SHEEP B	β SHEEP A	β GOAT A	β GOAT A'	β MOUSE SEC	β MOUSE C-57	β MOUSE AKR	β RABBIT	β SP. MONKEY	β TAMARIN	β SQ. MONKEY	δ SQ. MONKEY	δ TAMARIN	δ SP. MONKEY	β MULATTA	β FUSCATA	δ HUMAN	β GORILLA	β CHIMP	β HUMAN
INSECT CTT3	0																																																																			
INSECT CTT3 SUB	2	0																																																																	
Mb. S. WHALE	160	159	0																																																														
Mb. HORSE	164	163	21	0																																																													
GLOBIN LAMPREY	145	145	149	152	0																																																											
α CARP	154	154	150	146	130	0																																																									
α CHICKEN	142	143	155	159	130	101	0																																																							
α RABBIT	145	145	153	153	123	94	54	0																																																					
α RABBIT SUB	143	143	154	154	127	96	53	9	0																																																			
α TREE SHREW	145	145	148	151	125	93	60	37	37	0																																																	
α HORSE SLOW	137	137	153	154	120	94	51	31	33	35	0																																																
α HORSE FAST	138	138	153	155	120	95	52	32	34	36	1	0																																															
α HORSE SLOW 24F	138	138	154	155	120	93	52	32	34	36	1	2	0																																														
α HORSE FAST 24F	139	139	154	156	120	94	53	33	35	37	2	1	1	0																																													
α DONKEY	138	138	154	155	119	92	53	33	35	36	2	3	1	2	0																																												
α PIG	138	138	146	148	119	88	56	28	29	31	21	20	21	20	22	0																																											
α LLAMA	130	130	140	139	107	75	48	28	28	28	21	22	22	23	23	24	0																																										
α BOVINE	141	142	143	146	120	92	50	29	31	29	23	23	24	24	25	18	26	0																																									
α GOAT II	138	139	140	144	116	89	55	30	30	25	24	23	25	24	24	22	27	11	0																																								
α GOAT B	138	139	141	145	116	89	53	31	33	25	21	20	22	21	21	20	27	10	5	0																																							
α GOAT A	137	138	143	146	116	88	52	30	32	25	20	19	21	20	20	19	26	10	4	1	0																																						
α SHEEP D	138	139	145	150	118	95	56	34	36	29	25	25	26	26	26	20	30	14	8	35	5	0																																					
α SHEEP A	139	140	145	149	118	95	56	34	36	28	24	24	25	25	25	19	29	13	7	4	4	1	0																																				
α MOUSE NB	143	144	149	150	119	91	54	34	32	28	28	29	29	30	30	29	26	24	27	27	26	31	30	0																																			
α MOUSE C-57	143	144	148	149	118	88	51	31	29	26	27	28	28	29	29	26	23	21	24	24	23	28	27	3	0																																		
α SIFAKA	148	148	154	156	120	96	57	34	33	33	37	38	38	39	39	32	32	34	34	36	35	38	38	30	27	0																																	
α LEMUR	143	143	152	154	120	90	54	30	29	31	29	30	30	31	30	24	30	28	27	27	26	28	29	28	25	18	0																																
α BUSH BABY	144	144	153	156	118	93	49	32	31	30	27	28	28	29	29	25	26	25	24	24	23	27	27	25	22	25	19	0																															
α MULATTA	141	142	152	154	116	95	46	29	30	29	19	20	20	21	21	21	26	19	23	21	20	25	24	21	20	24	20	14	0																														
α FUSCATA	141	142	152	154	116	95	46	29	30	29	19	20	20	21	21	21	26	19	23	21	20	25	24	21	20	24	20	14	0	0																													
α GORILLA	137	138	146	148	109	89	45	28	29	31	21	22	22	23	23	20	22	20	23	22	21	26	25	22	19	24	18	13	6	6	0																												
α CHIMP	140	141	151	153	113	93	45	28	29	31	22	23	23	24	24	20	23	20	24	23	22	27	26	22	19	23	18	14	5	5	1	0																											
α HUMAN	140	141	151	153	113	93	45	28	29	31	22	23	23	24	24	20	23	20	24	23	22	27	26	22	19	23	18	14	5	5	1	0	0																										
γ HUMAN	165	164	159	157	138	105	109	105	105	108	99	100	99	100	99	98	90	101	102	100	100	102	101	104	101	106	105	106	106	106	100	105	105	0																									
β KANGAROO	166	165	160	155	144	114	114	113	113	116	109	109	109	109	109	105	102	106	107	105	105	109	108	114	111	114	113	115	112	112	108	111	111	60	0																								
β KANGAROO -2	166	165	160	155	144	114	114	113	113	116	109	109	109	109	109	105	102	106	107	105	105	109	108	114	111	114	113	115	112	112	108	111	111	61	1	0																							
β SIFAKA	164	163	161	157	144	110	113	114	115	119	106	107	105	106	105	27	98	109	109	106	105	107	107	115	113	115	111	115	112	112	107	112	112	51	72	72	0																						
β LEMUR	166	165	162	158	141	111	116	111	112	115	104	105	103	104	103	99	97	103	104	102	101	102	101	110	108	114	110	109	110	110	106	110	110	55	71	71	34	0																					
β HORSE SLOW	154	152	155	155	141	104	110	102	104	100	95	96	94	95	94	91	89	97	96	94	93	94	95	97	95	101	99	99	99	99	98	100	100	47	59	58	44	41	0																				
β PIG	152	152	146	143	135	107	107	103	104	101	95	96	94	95	94	90	89	94	94	92	91	93	92	100	98	106	101	102	99	99	98	100	100	50	52	53	43	39	26	0																			
β LLAMA	149	149	155	153	142	107	107	102	102	98	95	95	94	94	94	90	88	92	92	90	89	90	89	96	93	103	99	97	98	98	98	98	98	48	54	53	49	44	29	25	0																		
β BOVINE F	157	156	157	156	144	101	110	106	107	111	103	104	102	103	102	99	90	102	103	103	102	105	105	107	105	108	104	104	105	105	103	105	105	49	66	66	45	47	34	28	34	0																	
β SHEEP F	138	139	141	142	135	89	95	94	95	97	86	87	85	86	85	86	78	89	91	89	88	89	89	93	91	96	92	92	89	89	89	90	90	46	55	55	43	43	38	31	37	12	0																
β BOVINE A	165	164	160	156	143	107	115	107	107	107	100	101	99	100	99	92	94	100	100	98	97	97	96	103	100	105	101	100	101	101	98	100	100	46	54	54	52	45	30	26	25	26	28	0															
β BOVINE B	168	167	161	158	142	107	114	107	107	108	98	99	97	98	97	92	95	100	100	98	97	97	96	104	101	105	102	101	102	102	99	101	101	48	56	56	51	45	31	26	29	27	27	4	0														
β BARBARY C	151	151	152	151	132	98	100	97	98	97	93	94	92	93	92	88	84	94	96	94	93	92	91	96	93	96	94	93	95	95	93	93	93	44	58	58	49	48	38	30	28	29	24	22	24	0													
β SHEEP C	157	156	158	157	137	101	103	98	100	98	92	93	91	92	91	87	83	92	93	91	90	89	88	96	93	96	95	94	95	95	92	93	93	45	60	60	48	48	35	32	24	34	28	23	25	9	0												
β GOAT C	154	154	157	156	132	98	98	95	97	94	89	90	88	89	88	85	81	89	91	89	88	86	85	93	90	93	92	91	92	92	90	90	90	41	57	57	46	45	34	31	25	32	26	23	25	8	1	0											
β SHEEP B	158	158	157	155	142	107	107	103	100	101	96	97	95	96	95	88	91	94	93	93	92	92	91	98	95	100	96	95	99	99	96	97	97	49	53	53	50	45	36	25	32	31	27	14	16	21	22	21	0										
β SHEEP A	157	157	156	154	134	103	105	100	99	97	94	95	93	94	93	87	86	90	89	91	90	89	88	94	91	96	93	92	95	95	92	93	93	50	57	57	50	45	37	30	32	30	28	17	19	19	18	16	7	0									
β GOAT A	158	158	153	151	135	102	104	101	98	98	95	96	94	95	94	88	87	91	90	92	91	90	89	95	92	97	93	92	96	96	93	94	94	48	54	54	51	45	39	30	33	30	27	14	16	18	19	17	7	4	0								
β GOAT A'	146	146	142	141	134	92	96	93	90	88	85	86	84	85	84	81	79	84	83	85	84	84	83	88	85	90	86	85	88	88	85	86	86	45	48	48	48	39	38	30	34	31	28	15	17	19	20	18	8	5	1	0							
β MOUSE SEC	149	148	150	148	133	100	103	99	100	101	94	95	94	95	94	87	87	94	93	90	89	91	90	98	95	102	99	97	98	98	94	96	96	44	58	59	49	47	38	36	39	44	40	37	39	37	39	37	39	39	39	36	0						
β MOUSE C-57	149	148	150	148	133	100	103	99	100	101	94	95	94	95	94	87	87	94	93	90	89	91	90	98	95	102	99	97	98	98	94	96	96	44	58	59	49	47	38	36	39	44	40	37	39	37	39	37	39	39	39	36	0	0					
β MOUSE AKR	153	152	149	147	132	99	102	97	98	100	93	94	93	94	93	88	86	93	92	89	88	90	89	97	94	101	98	96	97	97	93	95	95	45	57	58	49	47	40	38	41	44	41	38	40	39	41	39	40	40	41	38	3	3	0				
β RABBIT	160	159	163	161	143	109	110	104	105	109	101	102	100	101	100	97	95	101	97	95	94	97	96	105	103	110	104	102	102	102	100	103	103	46	51	52	42	39	30	27	25	37	36	30	32	33	37	34	31	32	32	30	32	32	33	0												
β SP. MONKEY	159	158	162	160	147	108	111	108	109	109	101	102	100	101	100	99	96	103	101	99	98	100	99	104	101	109	106	104	104	104	100	103	103	48	54	55	41	42	28	28	22	38	36	27	29	35	31	31	31	32	32	30	34	34	35	18	0											
β TAMARIN	156	155	162	160	143	109	106	104	105	105	97	98	96	97	96	96	93	99	98	96	95	96	95	100	97	105	102	100	100	100	97	99	99	44	53	54	39	39	28	30	25	40	37	30	32	36	33	32	32	34	34	32	35	35	36	16	5	0										
β SQ. MONKEY	157	156	161	159	145	108	112	107	108	109	100	101	99	100	99	97	94	102	101	99	98	100	99	105	102	109	105	104	104	104	99	102	102	47	58	59	40	42	30	30	26	39	37	28	30	35	31	31	32	33	33	31	35	35	36	20	6	7	0									
δ SQ. MONKEY	161	160	161	159	143	108	110	109	110	106	99	100	98	99	98	98	97	102	100	98	97	99	98	104	101	107	105	103	104	104	100	103	103	48	58	59	41	44	33	32	27	38	37	28	28	32	30	28	33	33	33	32	33	33	34	21	11	13	13	0								
δ TAMARIN	161	160	162	160	144	108	110	109	110	106	99	100	98	99	98	98	97	102	100	98	97	99	98	103	100	107	105	103	104	104	100	103	103	49	58	59	41	42	32	32	28	39	38	28	28	34	32	30	33	33	33	32	33	33	34	22	10	12	14	3	0							
δ SP. MONKEY	159	158	161	158	145	106	112	108	109	106	100	101	99	100	99	96	95	101	100	98	97	99	98	103	100	108	104	103	104	104	99	102	102	47	58	59	40	41	30	30	26	38	37	25	27	34	30	30	31	32	32	31	32	32	33	22	8	11	10	5	4	0						
β MULATTA	160	159	161	158	145	106	110	102	103	110	99	100	98	99	98	98	96	102	100	98	97	99	98	108	105	103	99	101	101	101	97	100	100	45	50	51	41	38	33	31	30	38	36	28	30	36	36	34	33	35	31	29	35	35	35	18	12	12	14	17	16	16	0					
β FUSCATA	160	159	161	158	145	106	110	103	104	111	100	101	99	100	99	99	97	103	101	99	98	100	99	107	104	104	100	102	102	102	98	101	101	44	50	51	40	37	32	30	29	37	35	27	29	35	35	33	32	33	30	28	34	34	34	17	11	11	13	16	15	15	1	0				
δ HUMAN	158	157	164	160	144	109	112	106	107	111	100	101	99	100	99	98	97	103	100	98	97	94	98	106	103	107	102	102	103	103	96	100	100	49	57	58	43	43	32	30	27	39	37	28	30	36	34	32	32	34	33	32	37	37	38	21	12	13	14	14	13	12	12	11	0			
β GORILLA	153	153	157	155	137	103	105	95	96	100	93	94	92	93	92	90	89	95	92	90	89	91	90	96	93	97	94	93	93	93	90	91	91	45	49	50	41	38	31	29	26	37	36	23	25	34	33	33	32	34	32	30	31	31	33	16	9	10	11	16	14	13	9	8	12	0		
β CHIMP	152	152	158	156	138	104	107	96	97	101	94	95	93	94	93	91	90	96	93	91	90	92	91	97	94	98	95	94	94	94	91	92	92	46	50	51	40	39	30	28	25	36	35	24	26	33	32	32	31	33	33	31	30	30	32	15	8	9	10	15	13	12	10	9	11	1	0	
β HUMAN	157	156	162	160	144	107	113	101	102	107	98	99	97	98	97	95	94	101	97	95	94	97	96	102	99	103	100	99	98	98	94	97	97	49	53	54	42	40	30	28	25	37	35	25	27	34	33	33	32	33	33	31	31	31	32	16	8	9	10	15	13	12	10	9	11	1	0	0

	昆虫 CTT3	昆虫 CTT3 侧枝	肌红蛋白抹香鲸	肌红蛋白马	球蛋白七鳃鳗	α 鲤鱼	α 鸡	α 兔	α 兔侧枝	α 树鼩	α 马慢	α 马快	α 马慢 24F	α 马快 24F	α 驴	α 猪	α 美洲驼	α 牛	α 山羊 Ⅱ	α 山羊 B	α 山羊 A	α 绵羊 D	α 绵羊 A	α 小鼠 NB	α 小鼠 C-57	α 马达加斯加狐猴	α 狐猴	α 丛猴	α 普通猕猴	α 日本猕猴	α 大猩猩	α 黑猩猩	α 人	γ 人	β 袋鼠	β 袋鼠-2	β 马达加斯加狐猴	β 狐猴	β 马慢	β 猪	β 美洲驼	β 牛 F	β 绵羊 F	β 牛 A	β 牛 B	β 巴巴里绵羊 C	β 绵羊 C	β 山羊 C	β 绵羊 B	β 绵羊 A	β 山羊 A	β 山羊 A'	β 小鼠 SEC	β 小鼠 C-57	β 小鼠 AKR	β 兔	β 蜘蛛猴	β 怪柳猴	β 松鼠猴	δ 松鼠猴	δ 怪柳猴	δ 蜘蛛猴	β 普通猕猴	β 日本猕猴	δ 人	β 大猩猩	β 黑猩猩	β 人
昆虫 CTT3	0
昆虫 CTT3 侧枝	2	0
肌红蛋白抹香鲸	160	159	0
肌红蛋白马	164	163	21	0
球蛋白七鳃鳗	145	145	149	152	0
α 鲤鱼	154	154	150	146	130	0
α 鸡	142	143	155	159	130	101	0
α 兔	145	145	153	153	123	94	54	0
α 兔侧枝	143	143	154	154	127	96	53	9	0
α 树鼩	145	145	148	151	125	93	60	37	37	0
α 马慢	137	137	153	154	120	94	51	31	33	35	0
α 马快	138	138	153	155	120	95	52	32	34	36	1	0
α 马慢 24F	138	138	154	155	120	93	52	32	34	36	1	2	0
α 马快 24F	139	139	154	156	120	94	53	33	35	37	2	1	1	0
α 驴	138	138	154	155	119	92	53	33	35	36	2	3	1	2	0
α 猪	138	138	146	148	119	88	56	28	29	31	21	20	21	20	22	0
α 美洲驼	130	130	140	139	107	75	48	28	28	28	21	22	22	23	23	24	0
α 牛	141	142	143	146	120	92	50	29	31	29	23	23	24	24	25	18	26	0
α 山羊 Ⅱ	138	139	140	144	116	89	55	30	30	25	24	23	25	24	24	22	27	11	0
α 山羊 B	138	139	141	145	116	89	53	31	33	25	21	20	22	21	21	20	27	10	5	0
α 山羊 A	137	138	143	146	116	88	52	30	32	25	20	19	21	20	20	19	26	10	4	1	0
α 绵羊 D	138	139	145	150	118	95	56	34	36	29	25	25	26	26	26	20	30	14	8	35	5	0
α 绵羊 A	139	140	145	149	118	95	56	34	36	28	24	24	25	25	25	19	29	13	7	4	4	1	0
α 小鼠 NB	143	144	149	150	119	91	54	34	32	28	28	29	29	30	30	29	26	24	27	27	26	31	30	0
α 小鼠 C-57	143	144	148	149	118	88	51	31	29	26	27	28	28	29	29	26	23	21	24	24	23	28	27	3	0
α 马达加斯加狐猴	148	148	154	156	120	96	57	34	33	33	37	38	38	39	39	32	32	34	34	36	35	38	38	30	27	0
α 狐猴	143	143	152	154	120	90	54	30	29	31	29	30	30	31	30	24	30	28	27	27	26	28	29	28	25	18	0
α 丛猴	144	144	153	156	118	93	49	32	31	30	27	28	28	29	29	25	26	25	24	24	23	27	27	25	22	25	19	0
α 普通猕猴	141	142	152	154	116	95	46	29	30	29	19	20	20	21	21	21	26	19	23	21	20	25	24	21	20	24	20	14	0
α 日本猕猴	141	142	152	154	116	95	46	29	30	29	19	20	20	21	21	21	26	19	23	21	20	25	24	21	20	24	20	14	0	0
α 大猩猩	137	138	146	148	109	89	45	28	29	31	21	22	22	23	23	20	22	20	23	22	21	26	25	22	19	24	18	13	6	6	0
α 黑猩猩	140	141	151	153	113	93	45	28	29	31	22	23	23	24	24	20	23	20	24	23	22	27	26	22	19	23	18	14	5	5	1	0
α 人	140	141	151	153	113	93	45	28	29	31	22	23	23	24	24	20	23	20	24	23	22	27	26	22	19	23	18	14	5	5	1	0	0
γ 人	165	164	159	157	138	105	109	105	105	108	99	100	99	100	99	98	90	101	102	100	100	102	101	104	101	106	105	106	106	106	100	105	105	0
β 袋鼠	166	165	160	155	144	114	114	113	113	116	109	109	109	109	109	105	102	106	107	105	105	109	108	114	111	114	113	115	112	112	108	111	111	60	0
β 袋鼠-2	166	165	160	155	144	114	114	113	113	116	109	109	109	109	109	105	102	106	107	105	105	109	108	114	111	114	113	115	112	112	108	111	111	61	1	0
β 马达加斯加狐猴	164	163	161	157	144	110	113	114	115	119	106	107	105	106	105	27	98	109	109	106	105	107	107	115	113	115	111	115	112	112	107	112	112	51	72	72	0
β 狐猴	166	165	162	158	141	111	116	111	112	115	104	105	103	104	103	99	97	103	104	102	101	102	101	110	108	114	110	109	110	110	106	110	110	55	71	71	34	0
β 马慢	154	152	155	155	141	104	110	102	104	100	95	96	94	95	94	91	89	97	96	94	93	94	95	97	95	101	99	99	99	99	98	100	100	47	59	58	44	41	0
β 猪	152	152	146	143	135	107	107	103	104	101	95	96	94	95	94	90	89	94	94	92	91	93	92	100	98	106	101	102	99	99	98	100	100	50	52	53	43	39	26	0
β 美洲驼	149	149	155	153	142	107	107	102	102	98	95	95	94	94	94	90	88	92	92	90	89	90	89	96	93	103	99	97	98	98	98	98	98	48	54	53	49	44	29	25	0
β 牛 F	157	156	157	156	144	101	110	106	107	111	103	104	102	103	102	99	90	102	103	103	102	105	105	107	105	108	104	104	105	105	103	105	105	49	66	66	45	47	34	28	34	0
β 绵羊 F	138	139	141	142	135	89	95	94	95	97	86	87	85	86	85	86	78	89	91	89	88	89	89	93	91	96	92	92	89	89	89	90	90	46	55	55	43	43	38	31	37	12	0
β 牛 A	165	164	160	156	143	107	115	107	107	107	100	101	99	100	99	92	94	100	100	98	97	97	96	103	100	105	101	100	101	101	98	100	100	46	54	54	52	45	30	26	25	26	28	0
β 牛 B	168	167	161	158	142	107	114	107	107	108	98	99	97	98	97	92	95	100	100	98	97	97	96	104	101	105	102	101	102	102	99	101	101	48	56	56	51	45	31	26	29	27	27	4	0
β 巴巴里绵羊 C	151	151	152	151	132	98	100	97	98	97	93	94	92	93	92	88	84	94	96	94	93	92	91	96	93	96	94	93	95	95	93	93	93	44	58	58	49	48	38	30	28	29	24	22	24	0
β 绵羊 C	157	156	158	157	137	101	103	98	100	98	92	93	91	92	91	87	83	92	93	91	90	89	88	96	93	96	95	94	95	95	92	93	93	45	60	60	48	48	35	32	24	34	28	23	25	9	0
β 山羊 C	154	154	157	156	132	98	98	95	97	94	89	90	88	89	88	85	81	89	91	89	88	86	85	93	90	93	92	91	92	92	90	90	90	41	57	57	46	45	34	31	25	32	26	23	25	8	1	0
β 绵羊 B	158	158	157	155	142	107	107	103	100	101	96	97	95	96	95	88	91	94	93	93	92	92	91	98	95	100	96	95	99	99	96	97	97	49	53	53	50	45	36	25	32	31	27	14	16	21	22	21	0
β 绵羊 A	157	157	156	154	134	103	105	100	99	97	94	95	93	94	93	87	86	90	89	91	90	89	88	94	91	96	93	92	95	95	92	93	93	50	57	57	50	45	37	30	32	30	28	17	19	19	18	16	7	0
β 山羊 A	158	158	153	151	135	102	104	101	98	98	95	96	94	95	94	88	87	91	90	92	91	90	89	95	92	97	93	92	96	96	93	94	94	48	54	54	51	45	39	30	33	30	27	14	16	18	19	17	7	4	0
β 山羊 A'	146	146	142	141	134	92	96	93	90	88	85	86	84	85	84	81	79	84	83	85	84	84	83	88	85	90	86	85	88	88	85	86	86	45	48	48	48	39	38	30	34	31	28	15	17	19	20	18	8	5	1	0
β 小鼠 SEC	149	148	150	148	133	100	103	99	100	101	94	95	94	95	94	87	87	94	93	90	89	91	90	98	95	102	99	97	98	98	94	96	96	44	58	59	49	47	38	36	39	44	40	37	39	37	39	37	39	39	39	36	0
β 小鼠 C-57	149	148	150	148	133	100	103	99	100	101	94	95	94	95	94	87	87	94	93	90	89	91	90	98	95	102	99	97	98	98	94	96	96	44	58	59	49	47	38	36	39	44	40	37	39	37	39	37	39	39	39	36	0	0
β 小鼠 AKR	153	152	149	147	132	99	102	97	98	100	93	94	93	94	93	88	86	93	92	89	88	90	89	97	94	101	98	96	97	97	93	95	95	45	57	58	49	47	40	38	41	44	41	38	40	39	41	39	40	40	41	38	3	3	0
β 兔	160	159	163	161	143	109	110	104	105	109	101	102	100	101	100	97	95	101	97	95	94	97	96	105	103	110	104	102	102	102	100	103	103	46	51	52	42	39	30	27	25	37	36	30	32	33	37	34	31	32	32	30	32	32	33	0
β 蜘蛛猴	159	158	162	160	147	108	111	108	109	109	101	102	100	101	100	99	96	103	101	99	98	100	99	104	101	109	106	104	104	104	100	103	103	48	54	55	41	42	28	28	22	38	36	27	29	35	31	31	31	32	32	30	34	34	35	18	0
β 怪柳猴	156	155	162	160	143	109	106	104	105	105	97	98	96	97	96	96	93	99	98	96	95	96	95	100	97	105	102	100	100	100	97	99	99	44	53	54	39	39	28	30	25	40	37	30	32	36	33	32	32	34	34	32	35	35	36	16	5	0
β 松鼠猴	157	156	161	159	145	108	112	107	108	109	100	101	99	100	99	97	94	102	101	99	98	100	99	105	102	109	105	104	104	104	99	102	102	47	58	59	40	42	30	30	26	39	37	28	30	35	31	31	32	33	33	31	35	35	36	20	6	7	0
δ 松鼠猴	161	160	161	159	143	108	110	109	110	106	99	100	98	99	98	98	97	102	100	98	97	99	98	104	101	107	105	103	104	104	100	103	103	48	58	59	41	44	33	32	27	38	37	28	28	32	30	28	33	33	33	32	33	33	34	21	11	13	13	0
δ 怪柳猴	161	160	162	160	144	108	110	109	110	106	99	100	98	99	98	98	97	102	100	98	97	99	98	103	100	107	105	103	104	104	100	103	103	49	58	59	41	42	32	32	28	39	38	28	28	34	32	30	33	33	33	32	33	33	34	22	10	12	14	3	0
δ 蜘蛛猴	159	158	161	158	145	106	112	108	109	106	100	101	99	100	99	96	95	101	100	98	97	99	98	103	100	108	104	103	104	104	99	102	102	47	58	59	40	41	30	30	26	38	37	25	27	34	30	30	31	32	32	31	32	32	33	22	8	11	10	5	4	0
β 普通猕猴	160	159	161	158	145	106	110	102	103	110	99	100	98	99	98	98	96	102	100	98	97	99	98	108	105	103	99	101	101	101	97	100	100	45	50	51	41	38	33	31	30	38	36	28	30	36	36	34	33	35	31	29	35	35	35	18	12	12	14	17	16	16	0
β 日本猕猴	160	159	161	158	145	106	110	103	104	111	100	101	99	100	99	99	97	103	101	99	98	100	99	107	104	104	100	102	102	102	98	101	101	44	50	51	40	37	32	30	29	37	35	27	29	35	35	33	32	33	30	28	34	34	34	17	11	11	13	16	15	15	1	0
δ 人	158	157	164	160	144	109	112	106	107	111	100	101	99	100	99	98	97	103	100	98	97	94	98	106	103	107	102	102	103	103	96	100	100	49	57	58	43	43	32	30	27	39	37	28	30	36	34	32	32	34	33	32	37	37	38	21	12	13	14	14	13	12	12	11	0
β 大猩猩	153	153	157	155	137	103	105	95	96	100	93	94	92	93	92	90	89	95	92	90	89	91	90	96	93	97	94	93	93	93	90	91	91	45	49	50	41	38	31	29	26	37	36	23	25	34	33	33	32	34	32	30	31	31	33	16	9	10	11	16	14	13	9	8	12	0
β 黑猩猩	152	152	158	156	138	104	107	96	97	101	94	95	93	94	93	91	90	96	93	91	90	92	91	97	94	98	95	94	94	94	91	92	92	46	50	51	40	39	30	28	25	36	35	24	26	33	32	32	31	33	33	31	30	30	32	15	8	9	10	15	13	12	10	9	11	1	0
β 人	157	156	162	160	144	107	113	101	102	107	98	99	97	98	97	95	94	101	97	95	94	97	96	102	99	103	100	99	98	98	94	97	97	49	53	54	42	40	30	28	25	37	35	25	27	34	33	33	32	33	33	31	31	31	32	16	8	9	10	15	13	12	10	9	11	1	0	0

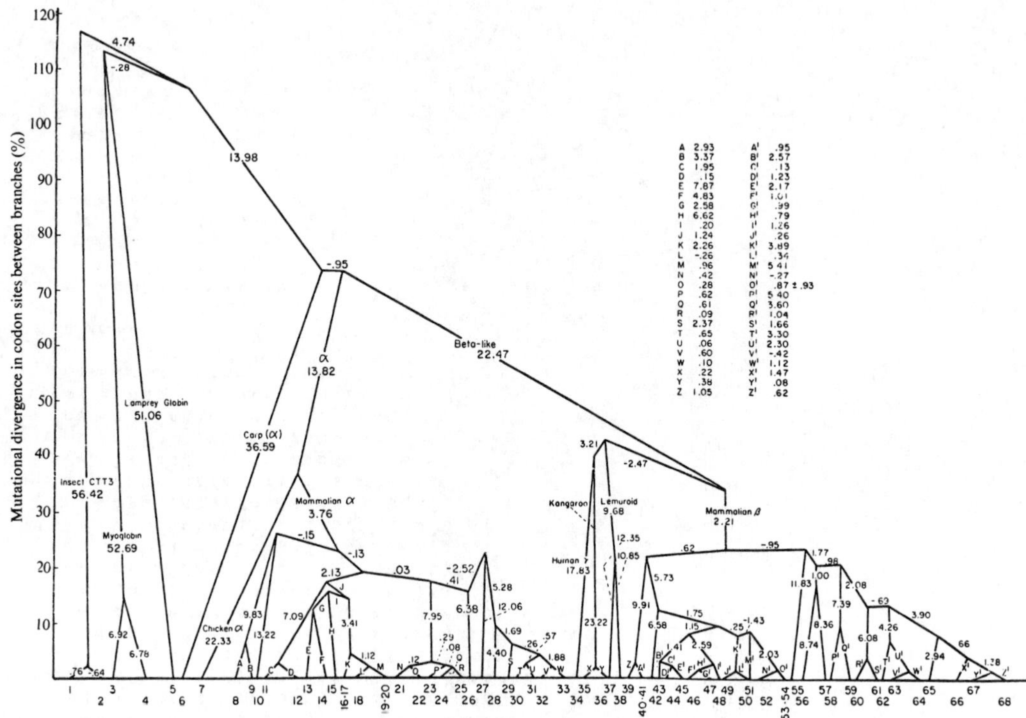

Fig. 1. Globin phylogenetic tree. 1, Insect CTT-3 (*Chironomus thummi*)[11]; 2, insect CTT-3 (sub)[11]—alternative amino-acids (39 Thr, 57 Ile) were used for ambiguous positions; 3, myoglobin sperm whale[13]; 4, myoglobin horse—sequence alignment as given in ref. 14; 5, globin lamprey[15]; 6, alpha carp[16]; 7, alpha chicken[17]; 8, alpha rabbit[18]; 9, alpha rabbit (sub)[18]—alternative amino-acids (29 Leu, 48 Leu, 70 Thr, 76 Val, 80 Ser, 113 His) were used for ambiguous positions; 10, alpha tree shrew[19]; 11, alpha mouse NB[20,21]; 12, alpha mouse C-57Bl[20,21]; 13, alpha sifaka (*Propithecus verreauxi*)[19]; 14, alpha lemur (*Lemur fulvus*)[19]; 15, alpha bush baby (*Galago crassicaudatus*)[19]; 16, alpha *Macaca mulatta*[22]; 17, alpha *Macaca fuscata* (G. Matsuda, unpublished); 18, alpha gorilla[23]; 19, alpha chimpanzee[24]; 20, alpha human[25,26]; 21, alpha donkey[27]; 22, alpha horse slow (24 Phe)[27]; 23, alpha horse fast (24 Phe)[27]; 24, alpha horse fast (24 Tyr)[27]; 25, alpha horse slow (24 Tyr)[28]; 26, alpha pig—sequence alignment as given in ref. 14; 27, alpha llama—sequence alignment as given in ref. 14; 28, alpha bovine[29]; 29, alpha goat II (non-allelic)[30]; 30, alpha goat B (allelic)[30]; 31, alpha goat A[30]; 32, alpha sheep D[31]; 33, alpha sheep A[32,33]; 34, gamma human[34]; 35, beta kangaroo[35]; 36, beta kangaroo-2 (2 Gln)[35]; 37, beta sifaka[19]; 38, beta lemur[19]; 39, beta mouse AKR[36]; 40, beta mouse Sec[36]; 41, beta mouse C-57Bl[36]; 42, beta rabbit[37]; 43, beta squirrel monkey (*Saimiri sciures*)[38]; 44, beta tamarin (*Saguinus nigricollis*)[38]; 45, beta spider monkey (*Ateles geoffroyi*)[38]; 46, delta squirrel monkey[38]; 47, delta tamarin[38]; 48, delta spider monkey[38]; 49, beta *Macaca fuscata*[39]; 50, beta *Macaca mulatta*[22]; 51, delta human[40]; 52, beta gorilla[23]; 53, beta chimpanzee[24]; 54, beta human[25]; 55, beta horse slow[42]; 56, beta pig—sequence alignment as given in ref. 14; 57, beta llama—sequence alignment as given in ref. 14; 58, beta bovine foetal[42]; 59, beta sheep foetal[43]; 60, beta bovine A[44]; 61, beta bovine B[44]; 62, beta sheep barbary C[45]; 63, beta goat C[47]; 64, beta sheep C[46]; 65, beta sheep B[46]; 66, beta sheep A[46]; 67, beta goat A[47]; 68, beta goat A' (allelic)[47].

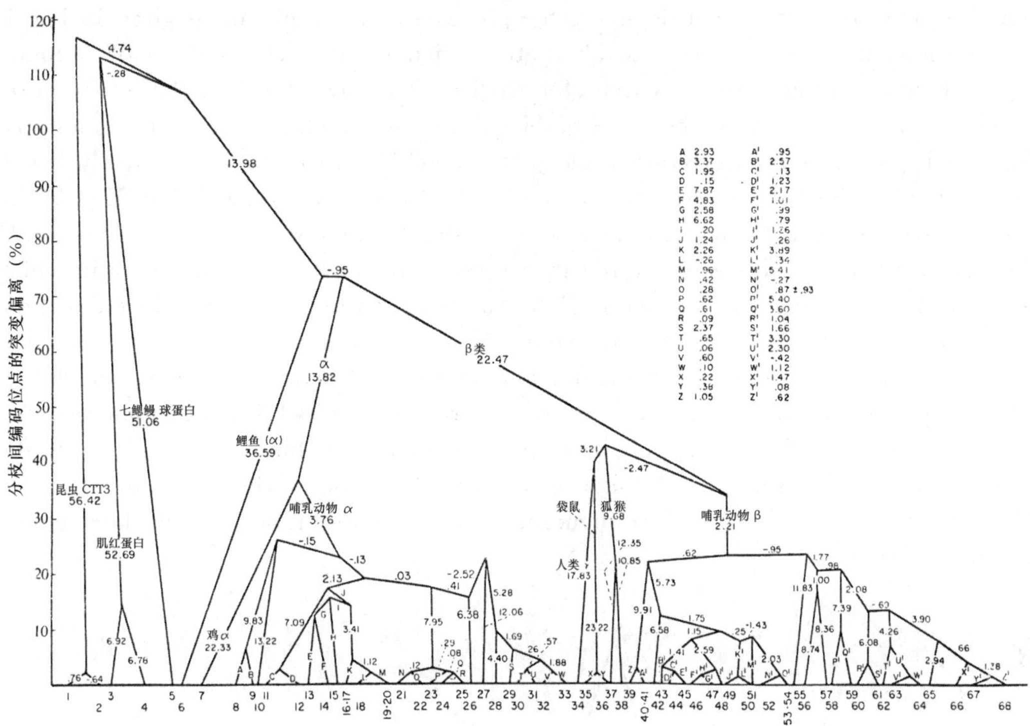

图 1. 球蛋白系统发生树。1，昆虫 CTT-3（吐氏摇蚊）[11]；2，昆虫 CTT-3（侧枝）[11]—— 可替代氨基酸（39 苏氨酸，57 异亮氨酸）被用于发生替代的位置；3，肌红蛋白抹香鲸 [13]；4，肌红蛋白马 —— 序列比对见参考文献 14；5，球蛋白七鳃鳗 [15]；6，α 鲤鱼 [16]；7，α 鸡 [17]；8，α 兔 [18]；9，α 兔（侧枝）[18]—— 可替代氨基酸（29 亮氨酸，48 亮氨酸，70 苏氨酸，76 缬氨酸，80 丝氨酸，113 组氨酸）被用于发生替代的位置；10，α 树鼩 [19]；11，α 小鼠 NB[20,21]；12，α 小鼠 C-57Bl[20,21]；13，α 马达加斯加狐猴（维氏冕狐猴）[19]；14，α 狐猴（褐狐猴）[19]；15，α 丛猴（粗尾丛猴）[19]；16，α 普通猕猴 [22]；17，α 日本猕猴（松田，尚未发表）；18，α 大猩猩 [23]；19，α 黑猩猩 [24]；20，α 人 [25,26]；21，α 驴 [27]；22，α 马慢（24 苯丙氨酸）[27]；23，α 马快（24 苯丙氨酸）[27]；24，α 马快（24 酪氨酸）[27]；25，α 马慢（24 酪氨酸）[28]；26，α 猪 —— 序列比对见参考文献 14；27，α 美洲驼 —— 序列比对见参考文献 14；28，α 牛 [29]；29，α 山羊 II（非等位基因）[30]；30，α 山羊 B（等位基因）[30]；31，α 山羊 A[30]；32，α 绵羊 D[31]；33，α 绵羊 A[32,33]；34，γ 人 [34]；35，β 袋鼠 [35]；36，β 袋鼠–2（2 谷氨酰胺）[35]；37，β 马达加斯加狐猴 [19]；38，β 狐猴 [19]；39，β 小鼠 AKR[36]；40，β 小鼠 Sec[36]；41，β 小鼠 C-57Bl[36]；42，β 兔 [37]；43，β 松鼠猴（松鼠猴）[38]；44，β 柽柳猴（黑须柽柳猴）[38]；45，β 蜘蛛猴（黑掌蛛猴）[38]；46，δ 松鼠猴 [38]；47，δ 柽柳猴 [38]；48，δ 蜘蛛猴 [38]；49，β 日本猕猴 [39]；50，β 普通猕猴 [22]；51，δ 人 [40]；52，β 大猩猩 [23]；53，β 黑猩猩 [24]；54，β 人 [25]；55，β 马慢 [42]；56，β 猪 —— 序列比对见参考文献 14；57，β 美洲驼 —— 序列比对见参考文献 14；58，β 牛胚胎 [42]；59，β 绵羊胚胎 [43]；60，β 牛 A[44]；61，β 牛 B[44]；62，β 巴巴里绵羊 C[45]（译者注：一种北非产的大角羊）；63，β 山羊 C[47]；64，β 绵羊 C[46]；65，β 绵羊 B[46]；66，β 绵羊 A[46]；67，β 山羊 A[47]；68，β 山羊 A'（等位基因）[47]。

The best tree so far obtained depicting the phylogeny of the globins is given in Fig. 1. Fig. 2 shows the best of forty-one alternatives tried for the restricted set of tetrapod alpha chains, and Fig. 3 the best tree of forty-five alternatives tried for the restricted set of mammalian non-alpha or beta-like globin chains. The APSD coefficient was lowered from 15.11 for the alpha tree constructed by the UWPGM to 11.14 for the tree in Fig. 2, and from 11.32 for the beta-like tree by the UWPGM to 8.40 for the tree in Fig. 3. The original tree for the sixty-eight globin species constructed by the UWPGM had an APSD coefficient of 18.52. This was lowered by several points on trying a dendrogram which followed the branching order shown in Figs. 2 and 3, but which otherwise maintained the topology of the UWPGM tree. Shifting (*a*) the carp alpha branch from its UWPGM union with other alphas to a union with the line ancestral to both mammalian beta-like globins and tetrapod alphas, and (*b*) the lamprey globin branch from its UWPGM union with the branch ancestral to all the globin chains in vertebrate haemoglobins over to union with the myoglobin branch, lowered the APSD coefficient to 12.27 and produced the tree shown in Fig. 1. So far eighteen alternative configurations for the major branches of the large globin tree have been tried.

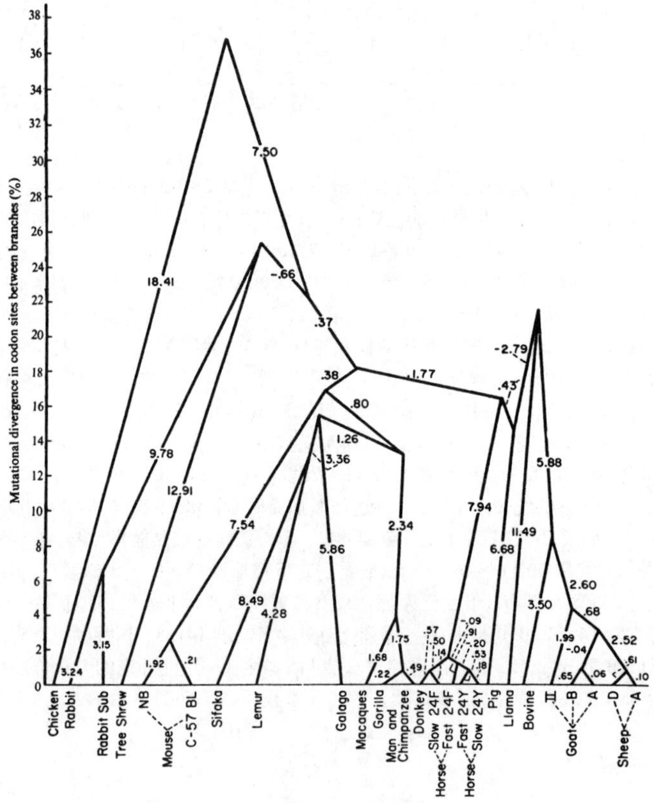

Fig. 2. Chicken and mammalian alpha globin phylogenetic tree.

图1给出了目前所能获得的描述球蛋白系统发生的最佳树图。图2显示的是就四足动物α链的有限系列进行尝试得到的41个备选树图中的最佳树图，而图3是就哺乳动物非α或称β类球蛋白链的有限系列进行尝试得到的45个备选树图中的最佳树图。APSD系数从由UWPGM程序构建的α树图的15.11下降到图2中树图的11.14，从由UWPGM建立的β类树图的11.32下降到图3中树图的8.40。68种球蛋白类型的原始树图是由UWPGM程序构建的，其APSD系数为18.52。在尝试按图2和图3所示的枝序作出系统树图时，这个值会有所下降，但整张图依旧保持着UWPGM树图的拓扑结构。（*a*）将鲤鱼α枝从含有其他α枝的UWPGM联合体置换到含哺乳动物β类球蛋白和四足动物α球蛋白共同祖先的支系联合体中，（*b*）将七鳃鳗球蛋白枝从含脊椎动物血红蛋白中所有球蛋白链的祖先分枝的UWPGM联合体置换到肌红蛋白枝的联合体中，将使APSD系数下降到12.27，建立的树如图1所示。到目前为止已尝试过这一大球蛋白树主要分枝的18种备选结构。

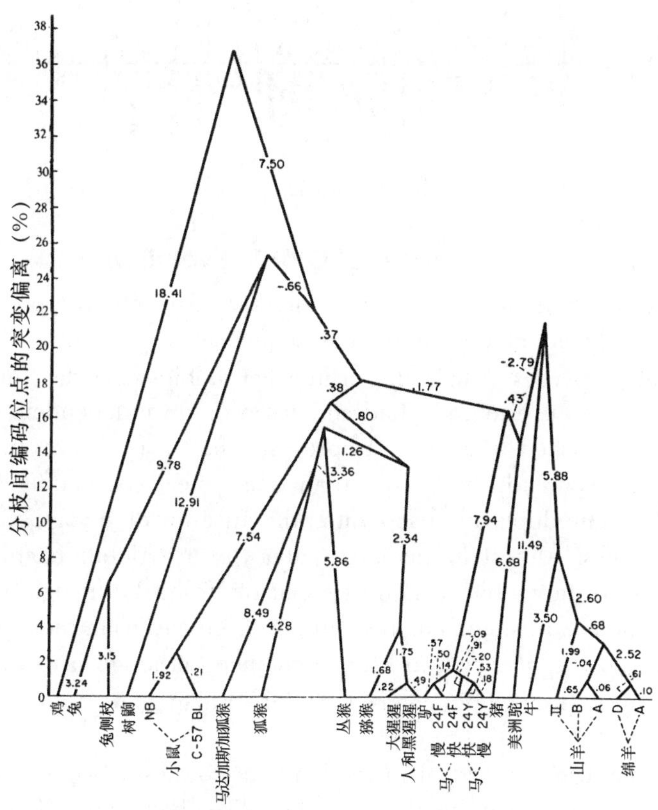

图 2. 鸡和哺乳动物的α球蛋白系统发生树。

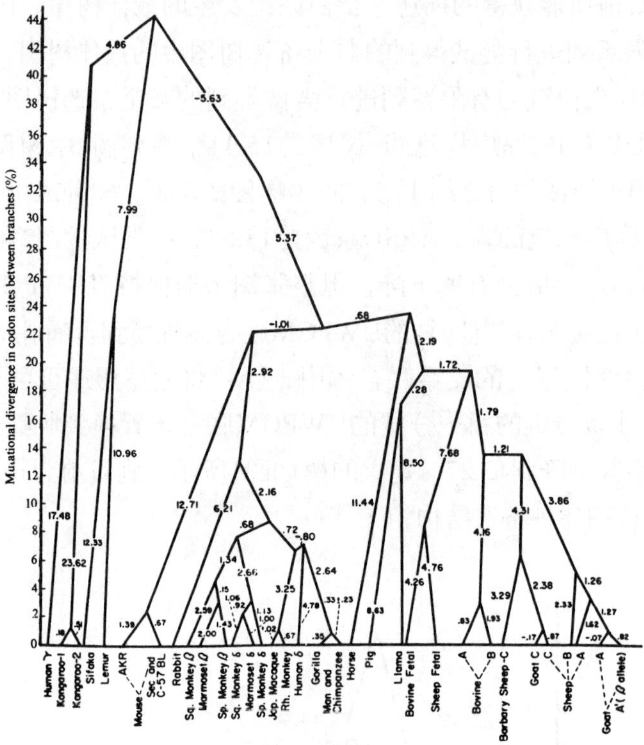

Fig. 3. Mammalian beta-like globin phylogenetic tree.

Course of Globin Evolution

The recognizable homologies between insect haem binding protein and vertebrate myoglobins and haemoglobins in codon sequence and in three-dimensional structure and functional properties[47] place the primordial globin gene deep in the Pre-Cambrian epoch before the protovertebrates had embarked on an independent course of evolution. Fig. 1 reveals that when man and lamprey still had a common ancestor, the primitive vertebrate genome already had more than one gene locus coding for globin chains; it was then that a gene duplication separated the ancestor of myoglobin genes from that of haemoglobin genes. Almost immediately after the myoglobin–haemoglobin duplication, the line leading to the globin in lamprey split off from the myoglobin side as a separate branch. Myoglobins are single chained proteins. Similarly, lamprey globins, including the one which has been sequenced, are single chained, although in certain physico-chemical conditions these lamprey globin chains form larger aggregates[45].

In the early bony fishes ancestral to the land vertebrates a haemoglobin gene duplicated to produce the split between the alpha and beta-like chain genes. This is depicted in Fig. 1 where the tetrapod alpha and beta-like lines separate from each other immediately after their separation from the carp alpha line. Much later in the lineage leading to therian mammals a gamma–beta chain gene duplicated to produce, on the one hand, the first separate ancestor of the genes for the kangaroo non-alpha chain, the human gamma chain (found in the haemoglobin of the human foetus) and the lemuroid non-alpha

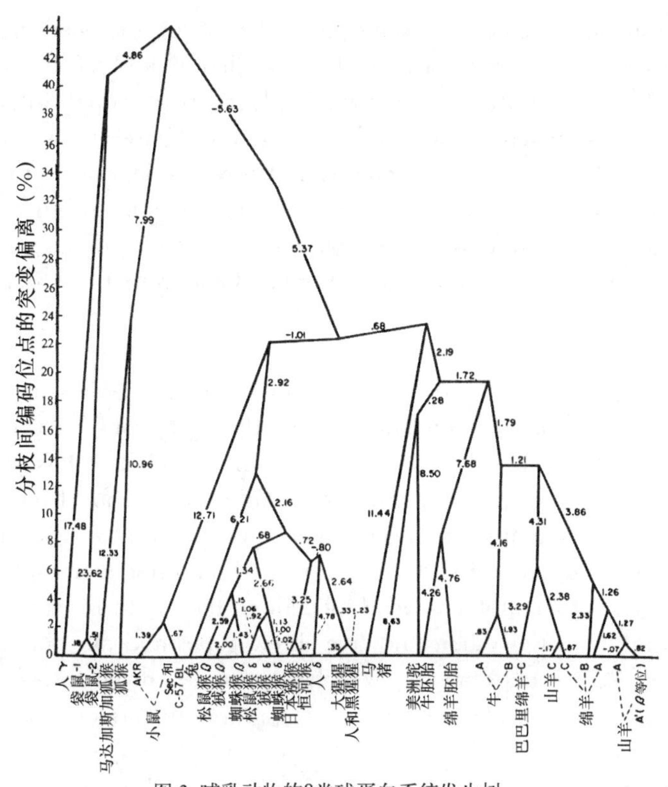

图 3. 哺乳动物的β类球蛋白系统发生树。

球蛋白进化过程

昆虫血红素结合蛋白和脊椎动物肌红蛋白、血红蛋白在编码序列和三维结构上可识别的同源性以及它们的功能特性[47]使原始球蛋白基因深藏在原始脊椎动物开始独立进化过程前的前寒武纪时代。图1表明：在人和七鳃鳗仍然拥有一个共同祖先时，原始脊椎动物的基因组中就已经有不止一个基因座来编码球蛋白链了；这时某一基因的复制使肌红蛋白基因的祖先从血红蛋白基因的祖先中分离出来。几乎紧随肌红蛋白–血红蛋白复制，七鳃鳗的球蛋白很快就从肌红蛋白的一侧分离而成为一条独立的分枝。肌红蛋白是单链蛋白。七鳃鳗球蛋白（包括那条已经测序的）同样也是单链的；但在某些特定的物理和化学条件下这些七鳃鳗球蛋白链会形成更大的聚集体[45]。

在陆地脊椎动物的祖先早期硬骨鱼中，一个血红蛋白基因的复制产生了α链和β类链基因间的分化。如图 1 所示：在从鲤鱼α支系分离后，四足动物α和β类支系很快就相互分开。很久之后，在向兽亚纲哺乳动物进化的支系中，一个γ–β链基因的复制：一方面产生了袋鼠非α链、人类γ链（在人类胚胎的血红蛋白中被发现）和狐猴非α链基因的第一个分离祖先，另一方面产生了哺乳动物β链基因的分离祖先。

chains and, on the other hand, the separate ancestor for the mammalian beta chain genes. The first separate ancestor of the lemuroid (lemur and sifaka) non-alpha globin chains may have been a gamma–beta hybrid produced by unequal homologous crossing over between originally linked gamma and beta genes. There are two residues (13 Ser, 21 Glu) in kangaroo, lemur and sifaka chains and three additional residues in the lemur and sifaka chains (9 Ala, 50 Ser, 52 Ser) which are homologous to human gamma chain. Furthermore, lemur and sifaka chains resemble more human gamma from the N-terminal end than human beta and resemble more human beta from the C-terminal end than human gamma.

At the nodes of Fig. 2 for the two most ancient branching points among mammalian alpha genes, ancestral rabbit and tree shrew genes diverged from each other immediately after their separation from the branch to all other mammalian alphas. Tree shrew alpha actually shows a greater mutation distance from rabbit, and rabbit from tree shrew, than either shows from other mammalian alpha chains (Table 1). This type of pattern, which resembles that of the human gamma, kangaroo and lemuroid non-alpha branch of the beta-like tree, suggests that alpha gene duplications had occurred in the early therian mammals producing ancestral rabbit and tree shrew genes and also the ancestor of other mammalian alphas. Recent duplications of alpha genes are known to occur in present day mammals, for example, in buffalo[50], horse[27], goat[52], macaques[54], mouse[56] and probably rabbit[18].

During eutherian phylogeny, beta chain genes duplicated independently in separate mammalian lines and certain of the descending duplicated genes ultimately coded for new proteins. For example, postnatal human blood contains in addition to its major haemoglobin (A) a minor haemoglobin (A_2) in which a different beta chain, called delta chain[1,57], combines with alpha chain. Similarly, the ceboid monkeys have a minor haemoglobin[38,58] with a beta chain (again called delta) different from the beta chain of the principal ceboid haemoglobin, Fig. 3 depicts in the early Anthropoidea, shortly after the branching apart of ceboids and catarrhine primates, two independent beta gene duplications: one on the catarrhine side coincided with the hominoid-cercopithecoid split to give rise to hominoid delta, and the other on the ceboid side produced the ancestor of ceboid deltas. Independent beta gene duplications also occurred during the evolution of the ruminants. Before the splitting apart of bovines and caprines, a descendant gene from a beta duplication in the early ruminant line began to code for the bovid foetal beta chain. The caprines (but not the bovines) have, like hominoids and ceboids, a second adult haemoglobin[45-47,59] (in this case, called haemoglobin C) normally present in negligible quantities, but synthesized in large amounts in animals under severe anaemic stress. In this second adult caprine haemoglobin, the beta C chain is found rather than the regular beta chain of caprine haemoglobin A. The ancestral *A–C* gene duplicated in the caprine line to give rise to the linked *A* and *C* gene loci shortly after the ancestral separation of bovines and caprines (Fig. 3).

Later in caprine evolution, at some point before the ancestral separation of sheep and goat, a mutation at the A locus produced the first separate ancestor of the present day sheep beta B gene. The period of separate evolution of sheep betas B and A seems to be

狐猴（狐猴和马达加斯加狐猴）非 α 球蛋白链的第一个分离祖先也许是一个由最初相连的 γ 和 β 基因之间的不等同源交叉产生的 γ–β 杂种。在袋鼠、狐猴和马达加斯加狐猴链中有 2 个氨基酸残基（13 丝氨酸，21 谷氨酸）；在与人类 γ 链同源的狐猴和马达加斯加狐猴链中还有 3 个额外的残基(9 丙氨酸,50 丝氨酸,52 丝氨酸)。此外，从 N 端看，狐猴和马达加斯加狐猴链与人类 γ 链的相似程度要高于与人类 β 链的相似程度；但从 C 端看，它们更像人类的 β 链而不是 γ 链。

图 2 中，在哺乳动物 α 基因中的 2 个最原始分枝节点上，兔和树鼩的祖先的基因在它们与所有其他哺乳动物 α 枝分离后很快就各自分开了。从树鼩 α 链到兔 α 链之间显示出的突变距离，实际上比这两者分别到其他任一哺乳动物 α 链的突变距离都要大（表 1）。这种模式很类似 β 类树中人 γ 枝以及袋鼠和狐猴非 α 枝的类型，这说明 α 基因复制发生在早期的兽亚纲哺乳动物中，由此产生了兔和树鼩祖先的基因以及其他哺乳动物 α 的祖先。已知最晚的 α 基因复制发生在现代哺乳动物中，例如水牛 [50]、马 [27]、山羊 [52]、猕猴 [54] 和小鼠 [56]，还可能包括兔 [18]。

在真兽次亚纲系统发生史中，β 链基因在分离的哺乳动物支系上独立复制，并且某些后来复制的基因还能最终编码新蛋白质。比如：初生婴儿的血液中除了包含主要的血红蛋白（A）外，还含有一种少量的血红蛋白（A_2），其中有一个不同质的 β 链（被称为“δ 链”[1,57]）与 α 链相结合。悬猴也拥有一种含量少的血红蛋白 [38,58]，其 β 链（仍被称为 δ）不同于悬猴主要血红蛋白的 β 链。图 3 描绘了在早期的类人猿亚目中，悬猴和狭鼻猴灵长类发生分离后，很快出现了 2 个独立的 β 基因复制：一个位于狭鼻猴那边，正好那时发生人科 – 猕猴科分离从而导致了人科 δ 链的形成；另一个位于悬猴那边，产生了悬猴 δ 链的祖先。独立的 β 基因复制也发生在反刍动物的进化中。牛和羊分离前，由早期反刍动物支系上的一个 β 基因复制的子代基因已经开始编码牛胚胎 β 链。羊（而不是牛）像人科动物和悬猴一样拥有第二种成体血红蛋白 [45-47,59]（这里指血红蛋白 C），正常情况下只显示出可忽略的量，但在严重贫血压力下，动物体内会大量合成。在此第二种成体羊血红蛋白中发现了 β C 链而不是羊血红蛋白 A 中的普通 β 链。在牛和羊的祖先分离后不久，羊支系上的祖先 *A–C* 基因发生复制，产生了连锁的 *A* 和 *C* 的基因座（图 3）。

稍后在羊进化过程中，在绵羊和山羊的祖先发生分离前的某个点上，A 基因座上的一个突变产生了今天绵羊 β B 基因的第一个分离祖先。绵羊 β B 和 β A 分离进化的时期看上去比绵羊和山羊 β A 的进化时间更长。今天绵羊的 β B 与绵羊或山羊

greater than that of sheep and goat beta As. Present day sheep beta B differs from sheep or goat beta A by seven point mutations, whereas sheep beta A differs from goat beta A by four point mutations (Table 1). The sheep A and B-beta results illustrate that typical allelic proteins within a mammalian population can differ by several or more amino-acids. Bovine A and B beta alleles provide another example of such allelic variation; they differ by four point mutations (Table 1). Allelic human haemoglobins, on the other hand, differ from common human haemoglobin A mostly by only single point mutations.

The difference between bovid and human populations with respect to the kind of allelic variation expressed may be because the human foetus is at greater risk from maternal isoimmunizations than the bovid foetus. Bovids, like many mammals, have placentas of the epitheliochorial type[60] which separate the maternal and foetal vascular systems by several layers of tissue and thereby minimize the chance of maternal immune attacks on the foetus. In contrast the haemochorial placenta of the higher primates maximizes the opportunity for such attacks. Thus, the type of allelic beta chains with multiple differences in bovids, if found in man, could give rise to maternal immune reactions and be selected against. The first placentas to evolve in mammals were probably of the epitheliochorial type[60], so that the early mammals would not have been subjected to the immunological mechanism restricting genetic variability which we postulate for higher primate populations.

Rates of Globin Evolution

The patristic distance data, tabulated in Table 2, demonstrate that rates of evolution have varied markedly among vertebrate globin genes. The mammalian alpha genes diverge the least from the common vertebrate globin ancestor and the myoglobin genes; the group of non-alpha kangaroo, lemuroid, and human gamma genes diverge the most; lamprey globin, carp alpha, and chicken alpha each show more mutational change from this common ancestor than does any mammalian alpha (Fig. 1; Table 2, column 4).

Table 2. Percentage Mutational Divergence of Vertebrate Globins during Descent

Globin species	From ancestral eutherian mammal MMUTD†	From ancestral eutherian mammal PML	From ancestral Hb–Mb PML
Human alpha	6.3	6.13	39.34
Chimp alpha*	6.3	6.13	39.34
Gorilla alpha*	7.0	5.86	38.12
Mulatta alpha	7.1	5.57	41.68
Fuscata alpha	7.1	5.57	41.68
Mouse C57 alpha	8.5	8.50	39.75
Mouse NB alpha	10.6	10.21	41.65
Goat A alpha	10.6	8.96	36.26
Lemur alpha*	11.3	10.45	41.46
Bush baby alpha*	11.3	8.67	40.67

的 β A 的区别在于 7 个点突变,而绵羊 β A 与山羊 β A 的区别只有 4 个点突变(表 1)。绵羊 A 和 B β 的结果表明：在一个哺乳动物种群中，典型等位基因蛋白质之间会在几个或更多氨基酸上有所不同。牛 A 和 B β 等位基因为这类等位基因变异提供了另一个例子；它们的区别表现在 4 个点突变（表 1)。另一方面，等位的人血红蛋白与普通的人血红蛋白 A 相比，最多只在一个点突变上有所不同。

牛种群和人种群在表达的等位基因变异的类别上存在不同，这可能是因为人胚胎比牛胚胎更容易受到来自母体的同种免疫的威胁。和很多哺乳动物一样，牛拥有上皮绒膜类型的胎盘 [60]，这使母体和胚胎的血管系统之间有几层组织相隔，因而最大限度地减小了母体免疫系统攻击胚胎的机会。反之，高等灵长类的血绒膜胎盘却使这类攻击的机会最大化。因此，在牛体内可以有存在多方面差异的等位 β 链基因类型，如果在人体内也是这样，就会引起母体免疫反应，从而被母体选择为敌对方。哺乳动物演变的第一个胎盘很可能属于上皮绒膜类型 [60]，因此早期哺乳动物不会因为免疫机制而使基因变异受到限制，而我们假设在高等灵长类动物中是存在这一限制的。

球蛋白进化速度

列于表 2 中的父系长度数据表明，脊椎动物球蛋白基因的进化速度相互之间差异显著。哺乳动物 α 基因与脊椎动物球蛋白基因的共同祖先基因及肌红蛋白基因的偏离最少；袋鼠非 α、狐猴非 α 和人 γ 基因与祖先的偏离最多；七鳃鳗球蛋白、鲤鱼 α 和鸡 α 与任意其他哺乳动物 α 相比，都显示出了和共同祖先之间的更大突变变化（图 1；表 2，第 4 列)。

表 2. 脊椎动物球蛋白在进化过程中的突变变化百分比

球蛋白种类	自原始真兽次亚纲哺乳动物的MMUTD计算值†	自原始真兽次亚纲哺乳动物的父系突变长度	自祖先血红蛋白–肌红蛋白的父系突变长度
人 α	6.3	6.13	39.34
黑猩猩 α*	6.3	6.13	39.34
大猩猩 α*	7.0	5.86	38.12
普通猕猴 α	7.1	5.57	41.68
日本猕猴 α	7.1	5.57	41.68
小鼠 C57 α	8.5	8.50	39.75
小鼠 NB α	10.6	10.21	41.65
山羊 A α	10.6	8.96	36.26
狐猴 α*	11.3	10.45	41.46
丛猴 α*	11.3	8.67	40.67

Continued

Globin species	From ancestral eutherian mammal MMUTD†	From ancestral eutherian mammal PML	From ancestral Hb–Mb PML
Goat B alpha*	11.3	9.55	36.85
Goat II alpha	11.3	13.04	37.74
Bovine alpha*	11.3	9.16	38.08
Horse slow alpha	12.0	9.97	40.08
Horse fast alpha*	12.7	10.32	39.44
Sheep A alpha	12.7	11.56	37.61
Horse slow 24F alpha*	12.7	10.72	38.86
Horse fast 24F alpha*	13.4	10.99	39.37
Donkey alpha*	13.4	11.15	39.00
Sheep D alpha*	13.4	12.07	41.71
Pig alpha*	13.4	9.25	37.30
Rabbit alpha*	15.6	12.36	43.22
Rabbit (sub) alpha*	16.3	12.27	43.66
Sifaka alpha*	17.7	14.66	44.50
Tree shrew alpha*	17.7	12.25	43.68
Llama alpha*	19.1	11.27	40.46
Chicken alpha	—	—	49.18
Human beta	7.5	7.19	48.46
Chimp beta*	7.5	6.73	46.60
Squirrel monkey beta	7.5	8.68	50.94
Gorilla beta*	8.2	6.98	46.39
Fuscata beta*	8.2	8.06	50.21
Spider monkey beta	8.2	7.67	50.67
Tamarin beta	8.8	8.24	49.73
Mulatta beta	8.9	8.71	50.29
Rabbit beta	8.9	8.12	50.64
Human delta*	9.5	8.77	50.04
Tamarin delta	10.2	9.33	51.33
Spider monkey delta	10.2	8.54	50.81
Squirrel monkey delta	10.9	9.39	51.35
Pig beta*	14.2	11.78	48.27
Bovine A beta	16.4	12.47	48.71
Llama beta*	17.7	11.65	47.89
Bovine B beta	17.8	11.37	49.33
Sheep B beta	18.4	13.78	47.76

续表

球蛋白种类	自原始真兽次亚纲哺乳动物的MMUTD计算值†	自原始真兽次亚纲哺乳动物的父系突变长度	自祖先血红蛋白–肌红蛋白的父系突变长度
山羊 B α*	11.3	9.55	36.85
山羊 II α	11.3	13.04	37.74
牛 α*	11.3	9.16	38.08
马慢 α	12.0	9.97	40.08
马快 α*	12.7	10.32	39.44
绵羊 A α	12.7	11.56	37.61
马慢 24F α*	12.7	10.72	38.86
马快 24F α*	13.4	10.99	39.37
驴 α*	13.4	11.15	39.00
绵羊 D α*	13.4	12.07	41.71
猪 α*	13.4	9.25	37.30
兔 α*	15.6	12.36	43.22
兔（侧枝）α*	16.3	12.27	43.66
马达加斯加狐猴 α*	17.7	14.66	44.50
树鼩 α*	17.7	12.25	43.68
美洲驼 α*	19.1	11.27	40.46
鸡 α	—	—	49.18
人 β	7.5	7.19	48.46
黑猩猩 β*	7.5	6.73	46.60
松鼠猴 β	7.5	8.68	50.94
大猩猩 β*	8.2	6.98	46.39
日本猕猴 β*	8.2	8.06	50.21
蜘蛛猴 β	8.2	7.67	50.67
柽柳猴 β	8.8	8.24	49.73
普通猕猴 β	8.9	8.71	50.29
兔 β	8.9	8.12	50.64
人 δ*	9.5	8.77	50.04
柽柳猴 δ	10.2	9.33	51.33
蜘蛛猴 δ	10.2	8.54	50.81
松鼠猴 δ	10.9	9.39	51.35
猪 β*	14.2	11.78	48.27
牛 A β	16.4	12.47	48.71
美洲驼 β*	17.7	11.65	47.89
牛 B β	17.8	11.37	49.33
绵羊 B β	18.4	13.78	47.76

Continued

Globin species	From ancestral eutherian mammal MMUTD†	From ancestral eutherian mammal PML	From ancestral Hb–Mb PML
Horse beta	19.0	12.02	48.59
Goat A beta*	19.1	12.64	46.92
Sheep A beta	19.8	14.33	46.93
Bovine foetal beta	19.8	16.53	52.30
Goat C beta*	20.3	14.11	47.46
Mouse AKR beta*	20.4	13.09	49.29
Mouse Sec and C-57 Beta*	20.7	12.37	49.19
Goat A' beta*	21.2	13.23	47.46
Sheep C beta	21.7	15.15	48.58
Sheep Barb. C beta*	21.7	15.19	48.46
Sheep foetal beta*	24.2	17.03	50.50
Kangaroo-1 beta	—	—	59.68
Kangaroo-2 beta	—	—	59.84
Human gamma	—	—	54.07
Lemur beta*	—	—	53.56
Sifaka beta*	—	—	55.06
Carp alpha	—	—	50.57
Lamprey globin	—	—	50.78
Mb–horse	—	—	59.47
Mb–sperm whale	—	—	59.33
Insect CTT-3	—	—	—
Insect CTT-3 (sub)	—	—	—

* Sequence in part or wholly by composition and homology.

† Values of rabbit alpha, rabbit (sub) alpha, tree shrew alpha, mouse NB alpha and mouse C-57 alpha are directly from ancestral mammalian alpha. Values of prosimian alphas are through ancestral primate alpha node while those of catarrhine alphas are through both ancestral primate alpha node and ancestral catarrhine alpha node. Values of mouse AKR beta, mouse Sec and C-57 betas and rabbit beta are directly from ancestral mammalian beta while those of ungulate betas and primate betas are through ancestral ungulate beta node and ancestral Anthropoidea beta node, respectively.

While the patristic mutation lengths from Fig. 1 reveal that the proportion of codon positions showing mutations varied markedly among the principal branches of vertebrate globins, they do not show any especially marked variation among mammalian alpha genes, or among typical mammalian beta genes. But the patristic mutation lengths for the mammalian alpha portion of the globin tree from Fig. 2 and for the mammalian beta portion from Fig. 3 reveal that rates of evolution have indeed varied markedly both among mammalian alpha genes and among mammalian beta genes (Table 2, column 3). In both sets of mammalian genes, the most slowly evolving lines were those which led to man and

续表

球蛋白种类	自原始真兽次亚纲哺乳动物的MMUTD计算值†	自原始真兽次亚纲哺乳动物的父系突变长度	自祖先血红蛋白–肌红蛋白的父系突变长度
马 β	19.0	12.02	48.59
山羊 A β*	19.1	12.64	46.92
绵羊 A β	19.8	14.33	46.93
牛胚胎 β	19.8	16.53	52.30
山羊 C β*	20.3	14.11	47.46
小鼠 AKR β*	20.4	13.09	49.29
小鼠 Sec 和 C-57 β*	20.7	12.37	49.19
山羊 A' β*	21.2	13.23	47.46
绵羊 C β	21.7	15.15	48.58
巴巴里绵羊 C β*	21.7	15.19	48.46
绵羊胚胎 β*	24.2	17.03	50.50
袋鼠 –1 β	—	—	59.68
袋鼠 –2 β	—	—	59.84
人 γ	—	—	54.07
狐猴 β*	—	—	53.56
马达加斯加狐猴 β*	—	—	55.06
鲤鱼 α	—	—	50.57
七鳃鳗球蛋白	—	—	50.78
肌红蛋白 – 马	—	—	59.47
肌红蛋白 – 抹香鲸	—	—	59.33
昆虫 CTT-3	—	—	—
昆虫 CTT-3（侧枝）	—	—	—

* 序列已部分或全部通过组成和同源性测定。

† 兔 α、兔（侧枝）α、树鼩 α、小鼠 NB α 和小鼠 C-57 α 的值直接从原始哺乳动物 α 算起。原猴亚目 α 的值用到了原始灵长类 α 节点，而狭鼻猴 α 同时用到了原始灵长类 α 节点和原始狭鼻猴 α 节点。小鼠 AKR β、小鼠 Sec β、小鼠 C-57 β 以及兔 β 的值直接从原始哺乳动物 β 算起，而有蹄类 β 和灵长类 β 分别用到了原始有蹄类 β 节点和原始类人猿 β 节点。

虽然图 1 中的父系突变长度揭示出发生突变的编码位置比例在脊椎动物球蛋白各主分枝之间差异很大，但是在哺乳动物 α 基因或在典型哺乳动物 β 基因中，编码位置比例并没有表现出特别显著的变化。然而，图 2 球蛋白树中哺乳动物 α 部分的父系突变长度和图 3 中哺乳动物 β 部分的父系突变长度显示，哺乳动物 α 基因之间和哺乳动物 β 基因之间的进化速度确实有很大的差异（表 2，第 3 列）。在两个系列的哺乳动物基因中，进化最慢的支系是进化成为人类和其他高等灵长类的支系。在这些大的球蛋白系列中（图 1），很早之前就已分离的种类之间的突变距离，例如任

other higher primates. In the large globin series (Fig. 1) the mutation distance between very anciently separated species, such as between any alpha and any beta globin chain, would be grossly underestimated due to uncounted multiple mutations at the same nucleotide sites. Thus, differences in rates of evolution among homologous mammalian genes, which are reflected in the patristic mutation lengths shown in Figs. 2 and 3, are smoothed out in the lengths shown in Fig. 1 where the apportioning of the mutation distances between the more recently separated genes is disproportionately affected by the mutation distance comparisons to the many, very anciently separated globin genes. Kimura[61] concluded that homologous genes evolved at uniform rates because about the same number of amino-acid substitutions separate any mammalian beta from any mammalian alpha and from carp alpha. But he failed to consider that the further in the past the separation of two lineages the larger the number of multiple mutations that must have occurred at the same nucleotide sites.

To detect a larger proportion of the accumulated mutations in a gene lineage, we constructed ancestral sequences (Table 3) for branching points of the globin phylogenetic tree and, using the MMUTD program, determined the mutation distances through the consecutive ancestors to the contemporary polypeptide chains. The mutational change in codon sites from the node for ancestral carp–tetrapod haemoglobin chains to the node for ancestral chicken–mammal alphas and from the latter to the ancestral node for mammalian alpha chains was 23.9 and 9.2% respectively. In the descent of the beta-like line, the mutational change from the ancestral carp–tetrapod node to the ancestral gamma–beta node and from this node to the ancestral node for typical mammalian beta chains was 53.2 and 2.0% respectively. The further mutational change in descent of each eutherian alpha and beta line is given in column 2 of Table 2. These results show that after the alpha–beta globin gene duplication in early vertebrate history, after the ancestral separation 380 million years ago of teleosts and tetrapods, many more mutations accumulated during the descent towards mammals in the beta-like line than in the alpha line. Moreover, during eutherian phylogeny, both alpha and beta chain evolution was much slower (on the average, almost by half) in higher primates than in other mammals. Furthermore, from the ancestral carp–tetrapod node about as much mutational change accumulated up to the ancestral chicken–mammalian alpha branching point as after this point in the descending mammalian alpha line. Yet one has to go back about 280 million years ago to the early reptiles to find the most recent common ancestor of birds and mammals. Thus the evidence suggests that the descending alpha genes evolved at a more rapid rate in earlier less advanced vertebrates than in later higher vertebrates.

一α和任一β球蛋白链之间的距离，可能由于同一核苷酸位点处的未计数多重突变而被严重低估。因此，图2和图3中父系突变长度所反映出的同源哺乳动物基因间进化速度的不同，在图1所示的这个长度中并未体现出来，其中更后期分离的基因间突变距离的分配，不成比例地受到与许多非常远古时就已分离的球蛋白基因间突变距离相比较的影响。木村资生[61]认为：同源基因以一致的速度进化，因为任一哺乳动物β与任一哺乳动物α之间以及任一哺乳动物β与鲤鱼α之间的氨基酸替换数大致相等。但是他没有考虑到以下这一点：两条支系分离的时间越早，在同一核苷酸位点处发生的多重突变数目也就越多。

为了在一个基因支系中检测到更高比例的积累突变，我们构建了球蛋白系统发生树中各分枝点的祖先序列（表3），并用MMUTD程序确定了各祖先基因到当代多肽链的突变距离。从鲤鱼–四足动物祖先血红蛋白链的节点到鸡–哺乳动物祖先α的节点，以及从后者到哺乳动物祖先α链节点之间的编码位点的突变变化分别是23.9%和9.2%。在β类支系的演变中，从鲤鱼–四足动物祖先节点到γ–β祖先节点，以及从该节点到典型哺乳动物β链祖先节点之间的突变变化分别是53.2%和2.0%。表2的第2列给出了每个真兽次亚纲α和β支系在进化过程中的进一步突变变化。这些结果显示：在早期脊椎动物史中，α–β球蛋白基因的复制发生在3.8亿年前硬骨鱼和四足动物的祖先分离后；在这之后，就指向哺乳动物的进化过程而言，积累于β类支系中的突变数量远远高于积累于α支系中的突变数量。此外，在真兽次亚纲系统发生过程中，高等灵长类α和β链进化的速度远比其他哺乳动物慢（平均而言，几乎慢一半）。而且，从鲤鱼–四足动物祖先节点到鸡–哺乳动物祖先α分枝点间所积累的突变变化，大致与哺乳动物α支系在鸡–哺乳动物祖先α分枝点之后的进化过程中所积累的突变变化相等。但是我们必须回到大约2.8亿年前的早期爬行类中来寻找鸟类和哺乳类离现在最近的共同祖先。因此，这些证据表明：世代相传的α基因在早期低等脊椎动物中的进化速度要高于在后期高等脊椎动物中的进化速度。

Table 3. Probable Ancestral Residues for Vertebrate Globin Chains at Different Nodes during Descent

Residue position	11	13	14	15	16	18	19	21	22	23	24	25	27	28	29	30	31	32	34	36	38	40	42	43	44	45	46	48	50	51	52	54
β-Mammal	His	Thr	Ala	Glu	Glu	Ala	Ala	Thr	Ala	Leu	Trp	Gly	Val	Asx	—	—	Val	Asp	Val	Gly	Ala	Gly	Leu	Leu	Val	Val	Tyr	Trp	Gln	Arg	Arg	Glu
γ-β-Mammal	His	Thr	Ala	Glu	Glu	Ala	Ala	Thr	Ala	Leu	Trp	Gly	Val	Asn	—	—	Val	Asp	Val	Gly	Ala	Gly	Leu	Leu	Val	Val	Tyr	Trp	Gln	Arg	Arg	Glu
α-Mammal	—	Ser	Pro	Ala	Asp	Thr	Asn	Lys	Ala	Ala	Trp	Gly	Ile	Gly	Gly	His	Ala	Gly	Tyr	Ala	Ala	Glu	Met	Phe	Leu	Ser	Phe	Thr	Lys	Thr	Tyr	Pro
α-Chicken—Mammal	—	Ser	Ala	Ala	Asp	Ala	Asn	Lys	Ala	Ala	Trp	Ala	Ile	Gly	Gly	His	Ala	Glu	Tyr	Ala	Ala	Glu	Met	Phe	Ile	Gly	Phe	Thr	Lys	Thr	Tyr	Pro
α-β-Carp	—	Ser	Ala	Ala	Asp	Ala	Ala	Lys	Ala	Ala	Trp	Ala	Ile	Gly	Gly	His	Ala	Asp	Tyr	Gly	Ala	Gly	Met	Phe	Ile	Gly	Thr	Thr	Lys	Thr	Thr	Ala

	55	57	61	62	63	64	65	66	68	69	70	71	73	79	81	83	84	85	86	88	90	92	93	94	97	98	99	100	103	106	107	110
β-Mammal	Ser	Gly	Ser	Ala	Asx	Ala	Val	Met	Asn	Pro	Lys	Val	Ala	Leu	Asx	Phe	Ser	Asp	Gly	Asx	Leu	Asn	Leu	Lys	Thr	Phe	Ala	Gln	Glu	Cys	Asp	His
γ-β-Mammal	Ser	Gly	Ser	Ala	Ser	Ala	Val	Met	Asn	Pro	Lys	Val	Ala	Leu	Asx	Phe	Ser	Asp	Gly	Asx	Leu	Asn	Leu	Lys	Thr	Phe	Ala	Gln	Glu	Cys	Asp	His
α-Mammal	His	—	His	—	—	—	—	—	Ser	Ala	Gln	Val	Gly	Ala	Asp	Leu	Thr	Asn	Ala	Gly	Leu	Asp	Leu	Pro	Ala	Leu	Ser	Ala	Asp	Ala	His	Arg
α-Chicken—Mammal	His	—	His	—	—	—	—	—	Ser	Ala	Gln	Val	Gly	Ala	Asp	Leu	Thr	Asn	Ala	Gly	Leu	Asp	Leu	Pro	Ala	Leu	Ser	Ala	Asp	Ala	His	Arg
α-β-Carp	His	Gly	Pro	—	—	—	—	—	Ser	Ala	Gln	Val	Gly	Ile	Asp	Val	Ser	Asp	Ala	Gly	Leu	Asp	Leu	Glu	Ala	Leu	Ala	Ala	Glu	Ala	His	Arg

	114	117	120	121	122	123	124	125	126	128	129	131	132	133	134	135	136	138	140	142	143	144	147	148	149	151	152	153	155	156	159	
β-Mammal	Glu	Arg	Gly	Asn	Val	Leu	Val	Ile	Val	Ala	His	Phe	Gly	Lys	Glu	Phe	Thr	Glx	Gln	Ala	Tyr	Gln	Val	Ala	Gly	Ala	Asn	Ala	Ala	His	His	
γ-β-Mammal	Glu	Arg	Gly	Asn	Val	Leu	Val	Ile	Val	Ala	His	Phe	Gly	Lys	Glu	Phe	Thr	Glx	Gln	Ala	Phe	Gln	Val	Ala	Gly	Ala	Asn	Ala	Ala	His	His	
α-Mammal	Val	Lys	Ser	His	Cys	Leu	Leu	Val	Thr	Ala	Ala	Leu	Pro	Ala	Asp	Phe	Thr	Ala	His	Ser	Leu	Asp	Leu	Ala	Ser	Ser	Thr	Val	Thr	Ser	Arg	
α-Chicken—Mammal	Val	Lys	Ser	His	Cys	Leu	Leu	Val	Thr	Ala	Ser	Leu	Pro	Ala	Glu	Phe	Thr	Glu	His	Ser	Leu	Asp	Leu	Ala	Ala	Ser	Thr	Val	Thr	Ser	Arg	
α-β-Carp	Glu	Lys	Ser	Asn	Ala	Leu	Val	Val	Val	Ala	Ser	Leu	Pro	Ala	Glu	Phe	Thr	Glu	His	Ser	Val	Asp	Leu	Ala	Ala	Ala	Asn	Ala	Thr	Ser	Arg	

Amino-acid residues of vertebrate globins are aligned so as to maximize homology by introducing the necessary gaps (refs. 14 and 15). The residue positions are numbered from 1 to 165 in order to accommodate lamprey globin. In this alignment carp alpha, tetrapod alphas, and mammalian betas begin at position 10 and end at position 159, with a gap at position 96. Carp alpha and tetrapod alphas have in addition gaps at position 11 and at position 62 through 66. Further gaps introduced are: tetrapod alphas at positions 57 and 78; mammalian betas at positions 29, 30, and 78; carp alpha at position 73. Residue positions omitted are the same as chicken[17] when aligned as described. Probable ancestral residues of mammalian alpha and beta chains at branching points within the span of eutherian phylogeny are given in the previous investigation[10]. Ancestral sequences were reconstructed by approximation after the rules described by Fitch and Margoliash[7]. In brief, an ancestral residue is selected if it gives fewer mutations in the lines of descent than any other residue which could be selected and, where more than one residue meets this criterion, distributes the mutations as equally as possible between the lines of descent.

表 3. 在进化过程不同节点上的脊椎动物球蛋白链所可能含有的原始氨基酸残基

残基位置	11	13	14	15	16	18	19	21	22	23	24	25	27	28	29	30	31	32	34	36	38	40	42	43	44	45	46	48	50	51	52	54
β-哺乳动物	组氨酸	苏氨酸	丙氨酸	谷氨酸	谷氨酸	丙氨酸	丙氨酸	苏氨酸	丙氨酸	亮氨酸	色氨酸	甘氨酸	缬氨酸	天冬氨酸或天冬酰胺	–	–	缬氨酸	天冬氨酸	缬氨酸	甘氨酸	丙氨酸	甘氨酸	亮氨酸	亮氨酸	缬氨酸	缬氨酸	酪氨酸	色氨酸	谷氨酰胺	精氨酸	精氨酸	谷氨酸
γ-β-哺乳动物	组氨酸	苏氨酸	丙氨酸	谷氨酸	谷氨酸	丙氨酸	丙氨酸	苏氨酸	丙氨酸	亮氨酸	色氨酸	甘氨酸	缬氨酸	天冬酰胺	–	–	缬氨酸	天冬氨酸	缬氨酸	甘氨酸	丙氨酸	甘氨酸	亮氨酸	亮氨酸	缬氨酸	缬氨酸	酪氨酸	色氨酸	谷氨酰胺	精氨酸	精氨酸	谷氨酸
α-哺乳动物	–	丝氨酸	脯氨酸	丙氨酸	天冬氨酸	苏氨酸	天冬酰胺	赖氨酸	丙氨酸	丙氨酸	色氨酸	甘氨酸	异亮氨酸	甘氨酸	甘氨酸	组氨酸	丙氨酸	甘氨酸	酪氨酸	丙氨酸	丙氨酸	谷氨酸	蛋氨酸	苯丙氨酸	亮氨酸	丝氨酸	苯丙氨酸	苏氨酸	赖氨酸	苏氨酸	酪氨酸	脯氨酸
α-鸡–哺乳动物	–	丝氨酸	丙氨酸	丙氨酸	天冬氨酸	丙氨酸	天冬酰胺	赖氨酸	丙氨酸	丙氨酸	色氨酸	丙氨酸	异亮氨酸	甘氨酸	甘氨酸	组氨酸	丙氨酸	谷氨酸	酪氨酸	丙氨酸	丙氨酸	谷氨酸	蛋氨酸	苯丙氨酸	异亮氨酸	甘氨酸	苯丙氨酸	苏氨酸	赖氨酸	苏氨酸	酪氨酸	脯氨酸
α-β-鲤鱼	–	丝氨酸	丙氨酸	丙氨酸	天冬氨酸	丙氨酸	丙氨酸	赖氨酸	丙氨酸	丙氨酸	色氨酸	丙氨酸	异亮氨酸	甘氨酸	甘氨酸	组氨酸	丙氨酸	天冬氨酸	酪氨酸	甘氨酸	丙氨酸	甘氨酸	蛋氨酸	苯丙氨酸	异亮氨酸	甘氨酸	酪氨酸	苏氨酸	赖氨酸	苏氨酸	苏氨酸	丙氨酸

残基位置	55	57	61	62	63	64	65	66	68	69	70	71	73	79	81	83	84	85	86	88	90	92	93	94	97	98	99	100	103	106	107	110
β-哺乳动物	丝氨酸	甘氨酸	丝氨酸	丙氨酸	天冬氨酸或天冬酰胺	丙氨酸	缬氨酸	蛋氨酸	天冬酰胺	脯氨酸	赖氨酸	缬氨酸	丙氨酸	亮氨酸	天冬氨酸或天冬酰胺	苯丙氨酸	丝氨酸	天冬氨酸	甘氨酸	天冬氨酸或天冬酰胺	亮氨酸	天冬酰胺	亮氨酸	赖氨酸	苏氨酸	苯丙氨酸	丙氨酸	谷氨酰胺	谷氨酸	半胱氨酸	天冬氨酸	组氨酸
γ-β-哺乳动物	丝氨酸	甘氨酸	丝氨酸	丙氨酸	丝氨酸	丙氨酸	缬氨酸	蛋氨酸	天冬酰胺	脯氨酸	赖氨酸	缬氨酸	丙氨酸	亮氨酸	天冬氨酸或天冬酰胺	苯丙氨酸	丝氨酸	天冬氨酸	甘氨酸	天冬氨酸或天冬酰胺	亮氨酸	天冬酰胺	亮氨酸	赖氨酸	苏氨酸	苯丙氨酸	丙氨酸	谷氨酰胺	谷氨酸	半胱氨酸	天冬氨酸	组氨酸
α-哺乳动物	组氨酸	–	组氨酸	–	–	–	–	–	丝氨酸	丙氨酸	谷氨酰胺	缬氨酸	甘氨酸	丙氨酸	天冬氨酸	亮氨酸	苏氨酸	天冬酰胺	丙氨酸	甘氨酸	亮氨酸	天冬氨酸	亮氨酸	脯氨酸	丙氨酸	亮氨酸	丝氨酸	丙氨酸	天冬氨酸	丙氨酸	组氨酸	精氨酸
α-鸡–哺乳动物	组氨酸	–	组氨酸	–	–	–	–	–	丝氨酸	丙氨酸	谷氨酰胺	缬氨酸	甘氨酸	丙氨酸	天冬氨酸	亮氨酸	苏氨酸	天冬酰胺	丙氨酸	甘氨酸	亮氨酸	天冬氨酸	亮氨酸	脯氨酸	丙氨酸	亮氨酸	丝氨酸	丙氨酸	天冬氨酸	丙氨酸	组氨酸	精氨酸
α-β-鲤鱼	组氨酸	甘氨酸	脯氨酸	–	–	–	–	–	丝氨酸	丙氨酸	谷氨酰胺	缬氨酸	甘氨酸	异亮氨酸	天冬氨酸	缬氨酸	丝氨酸	天冬氨酸	丙氨酸	甘氨酸	亮氨酸	天冬氨酸	亮氨酸	谷氨酸	丙氨酸	亮氨酸	丙氨酸	丙氨酸	谷氨酸	丙氨酸	组氨酸	精氨酸

残基位置	114	117	120	121	122	123	124	125	126	128	129	131	132	133	134	135	136	138	140	142	143	144	147	148	149	151	152	153	155	156	159	
β-哺乳动物	谷氨酸	精氨酸	甘氨酸	天冬酰胺	缬氨酸	亮氨酸	缬氨酸	异亮氨酸	缬氨酸	丙氨酸	组氨酸	苯丙氨酸	甘氨酸	赖氨酸	谷氨酸	苯丙氨酸	苏氨酸	谷氨酰胺或谷氨酸	谷氨酰胺	丙氨酸	酪氨酸	谷氨酰胺	缬氨酸	丙氨酸	甘氨酸	丙氨酸	天冬酰胺	丙氨酸	丙氨酸	组氨酸	组氨酸	
γ-β-哺乳动物	谷氨酸	精氨酸	甘氨酸	天冬酰胺	缬氨酸	亮氨酸	缬氨酸	异亮氨酸	缬氨酸	丙氨酸	组氨酸	苯丙氨酸	甘氨酸	赖氨酸	谷氨酸	苯丙氨酸	苏氨酸	谷氨酰胺或谷氨酸	谷氨酰胺	丙氨酸	苯丙氨酸	谷氨酰胺	缬氨酸	丙氨酸	甘氨酸	丙氨酸	天冬酰胺	丙氨酸	丙氨酸	组氨酸	组氨酸	
α-哺乳动物	缬氨酸	赖氨酸	丝氨酸	组氨酸	半胱氨酸	亮氨酸	亮氨酸	缬氨酸	苏氨酸	丙氨酸	丙氨酸	亮氨酸	脯氨酸	丙氨酸	天冬氨酸	苯丙氨酸	苏氨酸	丙氨酸	组氨酸	丝氨酸	亮氨酸	天冬氨酸	亮氨酸	丙氨酸	丝氨酸	丝氨酸	苏氨酸	缬氨酸	苏氨酸	丝氨酸	精氨酸	
α-鸡–哺乳动物	缬氨酸	赖氨酸	丝氨酸	组氨酸	半胱氨酸	亮氨酸	亮氨酸	缬氨酸	苏氨酸	丙氨酸	丝氨酸	亮氨酸	脯氨酸	丙氨酸	谷氨酸	苯丙氨酸	苏氨酸	谷氨酸	组氨酸	丝氨酸	亮氨酸	天冬氨酸	亮氨酸	丙氨酸	丙氨酸	丝氨酸	苏氨酸	缬氨酸	苏氨酸	丝氨酸	精氨酸	
α-β-鲤鱼	谷氨酸	赖氨酸	丝氨酸	天冬酰胺	丙氨酸	亮氨酸	缬氨酸	缬氨酸	缬氨酸	丙氨酸	丝氨酸	亮氨酸	脯氨酸	丙氨酸	谷氨酸	苯丙氨酸	苏氨酸	谷氨酸	组氨酸	丝氨酸	缬氨酸	天冬氨酸	亮氨酸	丙氨酸	丙氨酸	丙氨酸	天冬酰胺	丙氨酸	苏氨酸	丝氨酸	精氨酸	

对齐脊椎动物球蛋白的氨基酸残基的目的是为了通过引入必要的缺口以使同源性最大化（参考文献14和15）。残基位置按1到165编号以适应七鳃鳗球蛋白。在这一对齐过程中，鲤鱼α、四足动物α和哺乳动物β蛋白从位置10开始，结束于位置159，缺口位于位置96。鲤鱼α和四足动物α还在位置11和从位置62到66的地方有缺口。引入的另一些缺口是：四足动物α在位置57和78；哺乳动物β在位置29、30和78；鲤鱼α在位置73。省略的残基位置与用同样方法对齐鸡球蛋白的氨基酸残基时[17]一样。在之前的研究中，我们已经给出了哺乳动物α和β链在真兽次亚纲系统发生跨度内的分枝处可能出现的原始残基[10]。按菲奇和马格利亚什描述的规则[7]以粗略估计值重建祖先序列。简言之，如果一个原始残基能在进化支系中比其他可能被选择的残基带来的突变少，那么它将被选择；如果不止一个残基符合这个标准，那么就在不同的进化支系间尽可能均等地分配这些突变。

The patristic distance data on beta genes show that ceboid delta, bovid foetal beta, and caprine C beta initially evolved more rapidly than the genes from which they were duplicated (Fig. 3), but later in descent their rates of evolution slowed. Note (Table 1) that ceboid deltas actually diverge less from each other than ceboid betas, that sheep and bovine foetal betas diverge less from each other than sheep and bovine adult betas, and that sheep and goat C betas diverge less than sheep and goat A betas. Presumably mutations had accumulated freely at the duplicated loci until certain fortuitous mutations resulted in the emergence of new proteins with useful functions. Then natural selection not only caused these particular advantageous mutant genes to spread rapidly through the populations in which they occurred but also drastically slowed their further evolution.

Following the gamma–beta gene duplication, more mutational change occurred in the descent of human gamma than in most typical eutherian beta chains. Nevertheless amino-acid composition data[62] indicate that slightly less divergence exists between gamma chains of man and baboon than between either beta or alpha chains of these animals. Thus when sequence data become available on gamma chains of different higher primates, the pattern of decelerating evolution will probably be found again. This pattern is presented by the myoglobin genes, for the divergence between horse and sperm whale myoglobins (Fig. 1 and Table 1) is no more than that between any two alpha chains belonging to different eutherian orders; yet the overall mutational change in the descent of the myoglobin branch is larger than in the branch containing the various chains of vertebrate haemoglobin (Table 2, column 4). Ohno[63] describes early vertebrate history as a time when tetraploidization and bursts of gene duplication were occurring in our ancestors; at the later, mammalian stage specialized functions were found for pre-existing redundant genes, but the level of genome organization was such that viable tetraploidization was no longer possible. If mutations accumulate more readily in silent genes than in functional ones, this is equivalent to the hypothesis that molecular evolution decelerated during the descent of man.

In lemurs and other prosimians, foetal haemoglobin seems to be the same as adult haemoglobin[64]. But in the lineage leading to higher primates a specialized function, coding for the non-alpha chain of foetal haemoglobin, was found for the gamma locus. A further advance took place when from certain beta duplications of the early Anthropoidea functioning delta genes and a new haemoglobin type emerged in later Ceboidea and Hominoidea. The new level of organization imposed additional functional restraints, which increased the likelihood that new mutations would be detrimental, thereby slowing the further evolution of the haemoglobin genes. While the alpha and beta chains of man are among the most slowly evolved globins, those of the kangaroo and lemuroids are among the most rapidly evolved. As previously discussed, an additional selective factor, maternal isoimmunizations to foetal antigens, might have also acted to decelerate molecular evolution in the higher primates. The opportunity for such attacks would be virtually absent in kangaroo, and minimal in lemuroid primates which have epitheliochorial type placentas[65] rather than the haemochorial type of higher primates.

关于β基因的父系突变长度数据显示，悬猴δ、牛胚胎β和羊C β在开始时曾比复制为它们的基因进化得更快（图3），但是在后代中它们的进化速度变慢了。注意（表1）：悬猴δ相互之间的偏离实际上比悬猴β相互之间的偏离少，绵羊胚胎和牛胚胎β比绵羊成体和牛成体β之间的偏离少，以及绵羊和山羊C β比绵羊和山羊A β之间的偏离少。假设突变一直在复制基因座处自由积累，直到某些偶然的突变能够产生具有实用功能的新蛋白质。而后自然选择不但使这些特定的有利突变基因在它们发生的种群内部迅速扩散，而且也极大地减缓了它们的进一步进化。

在γ–β基因复制之后，人类γ后代中发生的突变变化超过了大多数典型真兽次亚纲β链中发生的突变变化。虽然如此，氨基酸构成数据[62]显示，人类和狒狒γ链间存在的偏离略少于它们的β链或α链之间的偏离。因此，当可以获得各种灵长类动物的γ链序列数据时，或许将会再次发现减速进化的趋势。这种趋势在肌红蛋白基因中有所体现，因为马和抹香鲸肌红蛋白间的偏离（图1和表1）不会高于真兽次亚纲下不同目动物的任何两种α链间的偏离；而肌红蛋白分枝在进化过程中的全部突变变化要大于含脊椎动物血红蛋白多种链的分枝中的突变变化（表2，第4列）。大野乾[63]将早期脊椎动物史描述为我们祖先体内发生基因复制的四倍体化和爆发的时代；后来，哺乳动物阶段的特异化功能被发现源自早先存在的冗余基因，但基因组组织的等级使得本来可以实现的四倍体化变得不再可能。如果突变在沉默基因中的积累速度比在功能基因中更快，这就等价于假设在人类进化过程中分子进化速度放慢。

在狐猴和其他原猴亚目动物中，胚胎血红蛋白与成体血红蛋白看似一样[64]。但在通向更高等灵长类动物的支系中发现，编码胚胎血红蛋白非α链的特定功能源自γ基因座。当早期类人猿亚目的功能性δ基因发生某些β复制时会导致进化的推进，从而在后来的悬猴总科和人猿总科中出现了一种新的血红蛋白类型。组织的新等级会产生附加的功能性抑制，这就增加了新突变为有害突变的可能性，从而使血红蛋白基因的进一步进化放缓。虽然人类的α和β链处于进化最缓慢的球蛋白行列，但袋鼠和狐猴的相同基因却在进化最快的球蛋白行列中。正如前面讨论的那样，一个附加选择因子，即母体和胚胎抗原不相容引起的同族免疫，可能也会使高等灵长类动物的分子进化减速。在袋鼠中发生这类攻击的可能性几乎不存在；在有上皮绒膜型胎盘而非高等灵长类中的血绒膜型胎盘的狐猴类灵长目动物[65]中，这种攻击的可能性也非常小。

Cladistic Relationships of Man and Other Mammals

The phylogenetic trees constructed from amino-acid sequence data not only reveal differences in relative rates of evolution among homologous genes but also provide evidence on the cladistic relationships of the animal species in which the genes occur. The branching topologies of alpha chain genes and beta chain genes depict a very close relationship, with no detectable mutational divergence, between man and chimpanzee, and between this man–chimpanzee branch and gorilla, in which the divergence is only one point mutation per chain. A beta phylogenetic tree, which includes the partially sequenced gibbon beta chain, joins gibbon to the branch leading to chimpanzee, man and gorilla, and then joins cercopithecoid (macaques) and hominoid branches[10]. Orang-utan haemoglobin chains have not been sequenced. Amino-acid composition data, however, show orang-utan diverging from the branch of African apes and man[66]. In the beta phylogenetic tree (Fig. 3), the ceboids branch away from the line leading to cercopithecoids and hominoids just before these two catarrhine groups branch apart. The phylogenetic tree of alpha genes (Fig. 2) joins the prosimians (galago together with the lemuroids, lemur and sifaka) to the line leading to higher primates, next joins the mouse branch to the base of the primate branch, and then joins the ancestral mouse–primate lineage with that of the ungulates. Both alpha and beta trees join caprines to bovines, llama and pig lines to the bovids, the horse branch with these artiodactyls, and then these ungulates with the lineage descending towards primates. Lemuroids do not produce a globin chain directly homologous to mammalian beta chains, so the phylogenetic tree of beta genes cannot provide cladistic evidence on the lemuroids. Otherwise the topology of the beta tree is similar to the alpha, except that rabbit beta joins primate betas before mouse.

Sequence data have also been obtained on mammalian fibrinopeptides A and B. They contain no more than forty amino-acid residues, all of which, with the exception of one or two invariant sites, have accepted amino-acid substitutions without harm to the function of fibrinogen. Thus these peptide sequences have been able to undergo such extensive divergent evolution that they are capable of depicting cladistic relationships among present day mammals. A matrix of mutation values for thirty-six mammalian combined A and B fibrinopeptide sequences[67-75] is given in Table 4, and the best tree of thirty alternatives tried portraying the phylogeny of these sequences is shown in Fig. 4. The APSD coefficient of this tree is 10.80 compared with 14.63 for the UWPGM tree.

人类和其他哺乳动物的进化枝关系

用氨基酸序列数据建立的系统发生树不仅揭示了同源基因间相对进化速度的不同，同时也为存在这些基因的动物物种的进化枝关系提供了证据。α 链基因和 β 链基因的分枝拓扑结构显示出在人类和黑猩猩间具有非常紧密的关系，未曾探测到两者之间存在突变偏离；人 – 黑猩猩枝同大猩猩枝之间的关系也很密切，每条链上只有一个点突变的偏离。包含部分测序的长臂猿 β 链的 β 系统发生树，将长臂猿连接到通向黑猩猩、人和大猩猩的分枝，然后与猕猴科（猕猴）分枝和人科分枝相连 [10]。目前还没有对红毛猩猩血红蛋白链进行测序。然而氨基酸组成数据显示，红毛猩猩是从非洲类人猿和人类的分枝中分离出来的 [66]。在 β 系统发生树上（图 3），悬猴分枝正好在猕猴科和人科这两个狭鼻猴组分叉前离开了通向这两科的支系。α 基因系统发生树（图 2）将原猴亚目（丛猴和狐猴科动物——狐猴和马达加斯加狐猴）连接到通向高等灵长类的支系上，其次将小鼠分枝结合到灵长类分枝的基部，然后又把小鼠 – 灵长类祖先的支系添加到有蹄类支系。α 树和 β 树都将羊支系连接到牛支系上，将美洲驼和猪支系连接到牛上，将马分枝添加到这些偶蹄类动物，然后把有蹄类添加到向灵长类进化的支系。狐猴科动物并不产生与哺乳动物 β 链直接同源的球蛋白链，所以 β 基因系统发生树不能提供关于狐猴科的进化枝证据。除了兔 β 是在小鼠前加入灵长类 β 之外，β 树的拓扑结构与 α 树类似。

关于哺乳动物血纤维蛋白肽 A 和 B 的序列数据也已得到。它们包含的氨基酸残基数不超过 40 个，在所有残基中，除了 1 或 2 个不变位点以外，其余都已接受了不损害血纤维蛋白原功能的氨基酸替代。因此，这些肽序列已经能够承受住偏离足够大的进化，以至于可以描述现今哺乳动物间的进化枝关系。表 4 给出了一个由 36 种哺乳动物的突变值构成的矩阵，这 36 种动物都兼有 A 和 B 血纤维蛋白肽序列 [67-75]，为描述这些序列系统发生进行尝试而得到的 30 个备选树图中的最佳树图如图 4 所示。该树的 APSD 系数为 10.80，与之相对比，UWPGM 树的 APSD 系数为 14.63。

Table 4. Mutation Values from Pairwise Comparisons of Mammalian Fibrinopeptide A and B Sequences

	Rabbit	Rat	Kulatta	Drill	Vervet	Gibbon-1	Gibbon-2	Man	Chimp	Horse	Mule-1	Zebra-2	Zebra-1	Donkey	Mule-2	Pig	Camel	Llama	Vicuna	Reindeer	Mule Deer	Muntjak	Sika Deer	Red Deer	Am. Elk	Pronghorn	Sheep	Goat	Bovine	Eur. Bison	Cape Buffalo	Water Buffalo	Cat	Dog	Fox	Kangaroo
Rabbit	0																																			
Rat	22	0																																		
Kulatta	20	16	0																																	
Drill	18	15	6	0																																
Vervet	17	15	3	3	0																															
Gibbon-1	16	12	5	6	3	0																														
Gibbon-2	17	11	4	7	4	1	0																													
Man	18	14	5	8	5	4	3	0																												
Chimp	18	14	5	8	5	4	3	0	0																											
Horse	20	19	15	18	16	14	15	19	19	0																										
Mule-1	20	19	15	18	16	14	15	19	19	0	0																									
Zebra-2	22	22	18	19	19	17	18	22	22	8	8	0																								
Zebra-1	22	22	17	19	18	16	17	21	21	7	7	2	0																							
Donkey	24	22	18	19	19	17	18	22	22	6	6	2	2	0																						
Mule-2	24	22	18	19	19	17	18	22	22	6	6	2	2	0	0																					
Pig	24	26	19	22	19	16	17	22	22	15	15	15	14	14	14	0																				
Camel	19	21	15	19	16	12	13	17	17	14	14	14	13	14	14	13	0																			
Llama	20	20	16	20	17	13	14	18	18	15	15	15	14	15	15	12	1	0																		
Vicuna	20	20	16	20	17	13	14	18	18	15	15	15	14	15	15	12	1	0	0																	
Reindeer	21	26	15	20	18	14	15	19	19	17	17	17	16	17	17	17	14	15	15	0																
Mule Deer	21	26	15	20	18	14	15	19	19	16	16	18	17	18	18	18	12	13	13	1	0															
Muntjak	23	24	17	20	20	16	17	21	21	18	18	18	18	18	18	16	14	14	14	4	5	0														
Sika Deer	21	22	17	23	21	17	18	22	22	19	19	19	18	19	19	17	14	14	14	6	7	4	0													
Red Deer	21	23	16	22	20	18	19	21	21	20	20	20	19	20	20	18	15	15	15	7	8	5	1	0												
Am. Elk	21	23	16	22	20	18	19	21	21	20	20	20	19	20	20	18	15	15	15	7	8	5	1	0	0											
Pronghorn	19	31	22	26	25	21	22	26	26	21	21	20	22	22	22	25	20	21	21	14	15	15	16	17	17	0										
Sheep	22	26	14	21	19	14	15	20	20	14	14	16	15	16	16	19	16	17	17	11	9	12	14	15	15	18	0									
Goat	22	26	14	21	19	14	15	20	20	14	14	16	15	16	16	19	16	17	17	11	9	12	14	15	15	18	0	0								
Bovine	26	24	19	20	20	17	18	20	20	20	20	22	22	21	21	18	18	18	18	16	16	13	16	15	15	24	22	22	0							
Eur. Bison	24	22	17	19	17	14	15	17	17	19	19	20	19	19	19	17	15	15	15	13	13	12	13	12	12	22	19	19	3	0						
Cape Buffalo	23	19	12	16	14	9	10	14	14	16	16	15	14	14	14	11	12	12	12	11	11	11	12	13	13	19	16	16	5	3	0					
Water Buffalo	25	23	16	21	19	14	15	19	19	18	18	20	19	19	19	17	14	15	15	13	13	14	15	16	16	20	17	17	7	6	2	0				
Cat	23	22	18	19	18	14	15	18	18	23	23	22	21	22	22	17	19	20	20	21	22	22	21	22	22	29	23	23	23	22	15	21	0			
Dog	25	23	18	26	25	22	23	25	25	22	22	20	19	18	18	23	21	22	22	23	24	25	24	25	25	27	23	23	29	27	18	25	19	0		
Fox	25	23	18	26	25	22	23	25	25	23	23	21	20	19	19	24	21	22	22	22	23	25	24	25	25	27	24	24	29	27	18	25	20	2	0	
Kangaroo	20	30	14	18	18	16	17	21	21	25	25	20	19	20	20	24	22	23	23	24	25	27	27	27	27	27	25	25	29	27	20	26	22	13	13	0

The half matrix lists minimum numbers of mutations interrelating pairs of fibrinopeptides. Numbers of shared amino-acids vary considerably in mammalian fibrinopeptides. For example, human has thirty amino-acid residues and rhesus has twenty-five residues, all the positions of which are shared by human. The mutation value for the human–rhesus pair presented in the table is obtained from the twenty-five shared amino-acid positions. Cape buffalo has thirty-six residues, only twenty-six positions of which are shared by human. Thus the mutation value for the human–Cape buffalo pair is obtained from these twenty-six shared positions. The alignments for most of the fibrinopeptide sequences can be found in ref. 14. The mutation values are converted into % mutational divergence values for constructing the phylogenetic tree in Fig. 4.

表 4. 由成对比较哺乳动物血纤维蛋白肽 A 和 B 序列得到的突变值

	兔	大鼠	普通猕猴	鬼狒	黑长尾猴	长臂猿-1	长臂猿-2	人	黑猩猩	马	骡-1	斑马-2	斑马-1	驴	骡-2	猪	骆驼	美洲驼	骆马	驯鹿	黑尾鹿	鹿	梅花鹿	赤鹿	美洲麋鹿	叉角羚	绵羊	山羊	牛	欧洲野牛	南非水牛	水牛	猫	狗	狐狸	袋鼠
兔	0																																			
大鼠	22	0																																		
普通猕猴	20	16	0																																	
鬼狒	18	15	6	0																																
黑长尾猴	17	15	3	3	0																															
长臂猿-1	16	12	5	6	3	0																														
长臂猿-2	17	11	4	7	4	1	0																													
人	18	14	5	8	5	4	3	0																												
黑猩猩	18	14	5	8	5	4	3	0	0																											
马	20	19	15	18	16	14	15	19	19	0																										
骡-1	20	19	15	18	16	14	15	19	19	0	0																									
斑马-2	22	22	18	19	19	17	18	22	22	8	8	0																								
斑马-1	22	22	17	19	18	16	17	21	21	7	7	2	0																							
驴	24	22	18	19	19	17	18	22	22	6	6	2	2	0																						
骡-2	24	22	18	19	19	17	18	22	22	6	6	2	2	0	0																					
猪	24	26	19	22	19	16	17	22	22	15	15	15	14	14	14	0																				
骆驼	19	21	15	19	16	12	13	17	17	14	14	14	13	14	14	13	0																			
美洲驼	20	20	16	20	17	13	14	18	18	15	15	15	14	15	15	12	1	0																		
骆马	20	20	16	20	17	13	14	18	18	15	15	15	14	15	15	12	1	0	0																	
驯鹿	21	26	15	20	18	14	15	19	19	17	17	17	16	17	17	17	14	15	15	0																
黑尾鹿	21	26	15	20	18	14	15	19	19	16	16	18	17	18	18	18	12	13	13	1	0															
鹿	23	24	17	20	20	16	17	21	21	18	18	18	18	18	18	16	14	14	14	4	5	0														
梅花鹿	21	22	17	23	21	17	18	22	22	19	19	19	18	19	19	17	14	14	14	6	7	4	0													
赤鹿	21	23	16	22	20	18	19	21	21	20	20	20	19	20	20	18	15	15	15	7	8	5	1	0												
美洲麋鹿	21	23	16	22	20	18	19	21	21	20	20	20	19	20	20	18	15	15	15	7	8	5	1	0	0											
叉角羚	19	31	22	26	25	21	22	26	26	21	21	20	22	22	22	25	20	21	21	14	15	15	16	17	17	0										
绵羊	22	26	14	21	19	14	15	20	20	14	14	16	15	16	16	19	16	17	17	11	9	12	14	15	15	18	0									
山羊	22	26	14	21	19	14	15	20	20	14	14	16	15	16	16	19	16	17	17	11	9	12	14	15	15	16	0	0								
牛	26	24	19	20	20	17	18	20	20	20	20	22	22	21	21	18	18	18	18	16	16	13	16	15	15	24	22	22	0							
欧洲野牛	24	22	17	19	17	14	15	17	17	19	19	20	19	19	19	17	15	15	15	13	13	12	13	12	12	22	19	19	3	0						
南非水牛	23	19	12	16	14	9	10	14	14	16	16	15	14	14	14	11	12	12	12	11	11	11	12	13	13	19	16	16	5	3	0					
水牛	25	23	16	21	19	14	15	19	19	18	18	20	19	19	19	17	14	15	15	13	13	14	15	16	16	20	17	17	7	6	2	0				
猫	23	22	18	19	18	14	15	18	18	23	23	22	21	22	22	17	19	20	20	21	22	22	21	22	22	29	23	23	23	22	15	21	0			
狗	25	23	18	26	25	22	23	25	25	22	22	20	19	18	18	23	21	22	22	23	24	25	24	25	25	27	23	23	29	27	18	25	19	0		
狐狸	25	23	18	26	25	22	23	25	25	23	23	21	20	19	19	24	21	22	22	22	23	25	24	25	25	27	24	24	29	27	18	25	20	2	0	
袋鼠	20	30	14	18	18	16	17	21	21	25	25	20	19	20	20	24	22	23	23	24	25	27	27	27	27	27	25	25	29	27	20	26	22	13	13	0

半矩阵列出了与各血纤维蛋白肽链配对相关的最小突变数。在哺乳动物血纤维蛋白肽中，共享氨基酸的数量差别非常大。例如：人有 30 个氨基酸残基，恒河猴有 25 个残基，所有这 25 个位置对于人来说都是共享的。表中出现的人－恒河猴配对突变值就是从这 25 个共享氨基酸位置获得的。南非水牛有 36 个残基，其中只有 26 个位置与人类共享。因此，人－南非水牛配对突变值就是从这 26 个共享位置获得的。大多数血纤维蛋白肽序列的比对结果可在参考文献 14 中找到。为了构建图 4 中的系统发生树，将突变值都转化为百分比形式的突变偏离值。

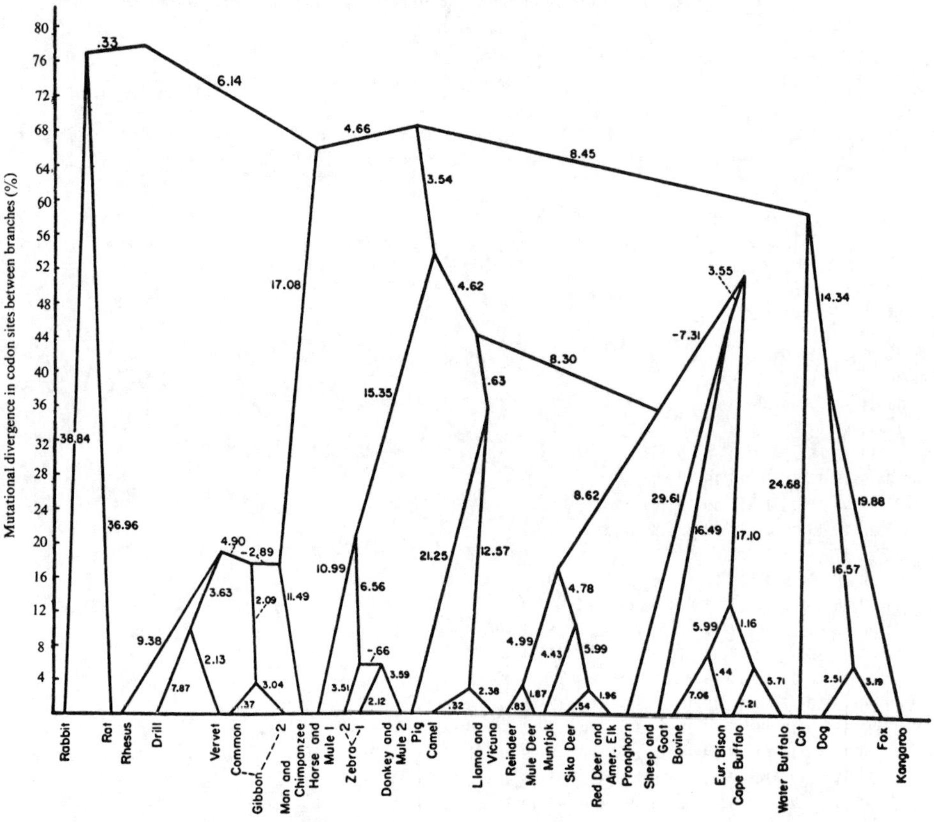

Fig. 4. Mammalian fibrinopeptide phylogenetic tree. 1, Rabbit[67,68]; 2, rat[67]; 3, *Macaca mulatta*[69]; 4, drill (*Mandrillus leucophaeus*)[70]; 5, vervet (*Cercopithecus aethiops*)[69]; 6, gibbon-1 (3 Gly–numbered from right to left)[71]; 7, gibbon-2 (3 Ser)[71]; 8, man[69]; 9, chimpanzee[72]; 10, horse[67,68]; 11, mule-1[67,68]; 12, zebra-2[73]; 13, zebra-1[73]; 14, donkey[67]; 15, mule-2[67,68]; 16, pig[67,68]; 17, camel[74,75]; 18, llama[67,68]; 19, vicuna (*Vicugna vicugna*)[74,75]; 20, reindeer (*Rangifer tarandus*)[74,75]; 21, mule deer (*Odocoileus hemionus hemionus*)[74]; 22, muntjak (*Muntiacus muntjak*)[74,75]; 23, sika deer (*Cervus nippon*)[67,68]; 24, red deer (*Cervus elaphus*)[67,68]; 25, American elk (*Cervus canadensis*)[74,75]; 26, pronghorn (*Antilocapra americana*)[74,75]; 27, sheep[67,68]; 28, goat[67,68]; 29, bovine[67,68]; 30, European bison[67,68]; 31, Cape buffalo[67,68]; 32, water buffalo[67,68]; 33, cat[67,68]; 34, dog[67,68]; 35, fox[67,68]; 36, kangaroo[73].

The branching arrangement among ungulate fibrinopeptides is essentially the same as that for ungulates in the alpha and beta globin phylogenetic trees, except that more ungulate species are represented. Pronghorn and caprines join bovines and then these bovoids join cervids (reindeer and various deers). Yet caprines and cervids on the average show less mutation distance from each other than from bovines (Table 4). Doolittle and Blombäck[76] discussed the possibility that caprines might have a more recent common ancestor with cervids than with bovines, and, indeed, the tree produced in our study by the UWPGM algorithm depicts such an ancestry relationship. Nevertheless, the best arrangements of artiodactyl branches with respect to lowering the APSD coefficient placed the caprines closer to bovines than to cervids in agreement with morphological evidence on ruminant relationships.

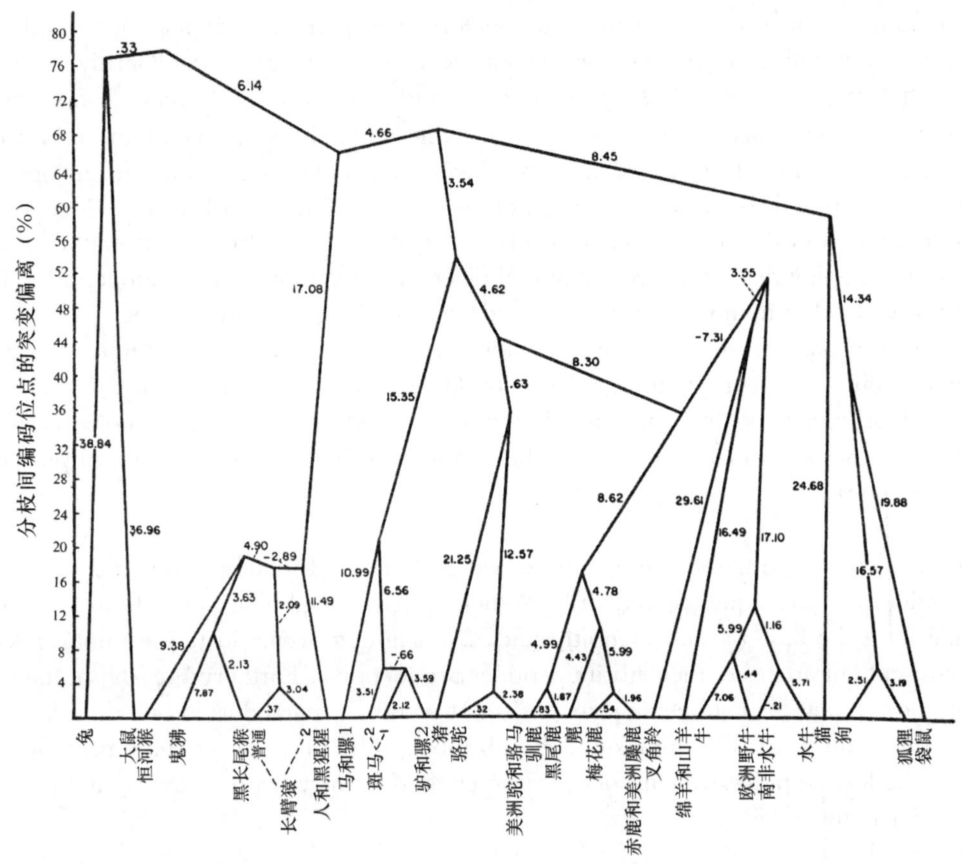

图 4. 哺乳动物血纤维蛋白肽的系统发生树。1，兔[67,68]；2，大鼠[67]；3，普通猕猴[69]；4，鬼狒（黑脸山魈）[70]；5，黑长尾猴（非洲绿猴）[69]；6，长臂猿 –1（3 甘氨酸，编号从右到左）[71]；7，长臂猿 –2（3 丝氨酸）[71]；8，人[69]；9，黑猩猩[72]；10，马[67,68]；11，骡 –1[67,68]；12，斑马 –2[73]；13，斑马 –1[73]；14，驴[67]；15，骡 –2[67,68]；16，猪[67,68]；17，骆驼[74,75]；18，美洲驼[67,68]；19，骆马（小羊驼）[74,75]；20，驯鹿（角鹿）[74,75]；21，黑尾鹿（北美骡鹿）[74]；22，麂（赤麂）[74,75]；23，梅花鹿（梅花鹿）[67,68]；24，赤鹿（马鹿）[67,68]；25，美洲麋鹿（加拿大马鹿）[74,75]；26，叉角羚（北美羚羊）[74,75]；27，绵羊[67,68]；28，山羊[67,68]；29，牛[67,68]；30，欧洲野牛[67,68]；31，南非水牛[67,68]；32，水牛[67,68]；33，猫[67,68]；34，狗[67,68]；35，狐狸[67,68]；36，袋鼠[73]。

除了能代表更多的有蹄类物种以外，有蹄类血纤维蛋白肽间的分枝排列与有蹄类在 α 及 β 球蛋白系统发生树中的分枝排列基本相同。叉角羚与羊连接到牛科，然后这些牛总科动物又连到鹿科（驯鹿和各种鹿）。然而平均而言，羊和鹿相互之间的突变距离比羊和牛的突变距离小（表 4）。杜利特尔和布隆贝克[76]讨论过以下这种可能性，即羊和鹿的共同祖先可能比羊和牛的共同祖先出现得更晚，我们在研究中用 UWPGM 算法构造的树图的确描绘出了这样一种祖先关系。尽管如此，为降低 APSD 系数，偶蹄类分枝的最佳排列使羊和牛之间的位置比羊和鹿之间的位置更接近，这与证明反刍动物间关系的形态学证据一致。

The joining of carnivores with ungulates before they join the Primates fits established ideas on mammalian phylogeny, for the Palaeocene Condylartha and Carnivora (from which the two ungulate orders, Perissodactyla and Artiodactyla, descended) are traced to late Cretaceous palaeoryctoid insectivores[77]. The marked divergence of the rabbit line from all other mammals is also in agreement with fossil data on mammalian phylogeny[77]. Although joining rat to rabbit at the apex of the tree is not in accord with the idea[77] from palaeontological evidence that rodents evolved out of the basal Primates, it may be noted that rat fibrinopeptides show less mutational divergence from primate fibrinopeptides than from those of other mammals (Table 4). Furthermore, the globin phylogenetic trees joined mouse to the base of the Primates. A finding at variance with our general knowledge of mammalian phylogeny is the grouping of kangaroo, a marsupial, with the carnivore branch. Fibrinopeptides from additional marsupials need to be sequenced, however, if we are to construct a phylogenetic tree for the relationship between marsupial and placental fibrinopeptides.

In the descent of primate fibrinopeptides, where hominoids (chimpanzee and man) and cercopithecoids (macaque, vervet, drill) branch apart, the gibbons split off as a distinct lineage from the base of the cercopithecoids. No splitting occurs in the terminal descent of man and chimpanzee: their fibrinopeptides are identical. Furthermore, the mutational divergence among the catarrhine primates is not nearly as marked as in other comparable taxonomic groups. In this connexion, the fibrinopeptides of the catarrhine primates are among the least rapidly evolving ones in the class Mammalia as revealed by the patristic mutation lengths in Fig. 4.

In addition to the amino-acid sequence data on haemoglobin chains and fibrinopeptides, there are partial sequence data on the erythrocyte enzyme carbonic anhydrase B which can be used to depict cladistic relationships among catarrhine primates[78]. Of the 133 homologous residues examined, chimpanzee and man differ at only one site, whereas orang utan differs from chimpanzee at three sites and from man at four sites. These hominoids differ from cercopithecoids (vervet, irus and rhesus macaques, and baboon) at four to six sites. Thus again the very close genetic relationship between man and chimpanzee is highlighted. The findings from the various sets of amino-acid sequence data support the conclusion drawn from extensive comparative immunodiffusion studies[79-82] on primate proteins that in the descent of the hominoids, after the early divergence of gibbon and orang utan lines, there was still a common ancestor for man, chimpanzee, and gorilla. The cladistic relationships depicted by the protein sequences support a classification of the Hominoidea which places *Gorilla* and *Pan* with *Homo* in the family Hominidae[80-82]. This agrees with the view of Darwin[83] that man originated in Africa from the same ancestral stock as chimpanzee and gorilla.

Four to five million year old fossil remains of man's immediate generic ancestor, *Australopithecus*, have recently been found in Pliocene deposits of Ethiopia[84], and fourteen million year old remains of *Ramapithecus*[86] (*Kenyapithecus*[87]), interpreted as a lineal ancestor of *Australopithecus*, date from the terminal Miocene of Kenya. Another Kenya hominoid,

在加入灵长目之前，食肉动物就已经和有蹄类动物相结合的现象符合哺乳动物系统发生学已建立的观点，因为古新世的踝节目和食肉目（有蹄类动物下的两个目——奇蹄目和偶蹄目即来自于此）可追溯到晚白垩纪的古掘猥科食虫动物[77]。兔支系与所有其他哺乳动物的显著分离也符合哺乳动物系统发生的化石证据[77]。虽然大鼠与兔在树顶处结合并不符合由古生物学证据推出的结论[77]，即啮齿类是从灵长类基部逐步进化而来，但可以发现大鼠血纤维蛋白肽同灵长类血纤维蛋白肽之间显示出的突变偏离要小于大鼠与其他哺乳动物血纤维蛋白肽之间的突变偏离（表 4）。而且，球蛋白系统发生树中的小鼠是连接到了灵长类动物的基部。有一个发现与我们对哺乳动物系统发生的常识性认识不符，即将一种有袋类动物——袋鼠结合到食肉动物枝上。虽然如此，如果我们要构建一个关于有袋类和有胎盘类血纤维蛋白肽间关系的系统发生树，还需要对其他一些有袋类动物的血纤维蛋白肽进行测序。

在灵长类血纤维蛋白肽的进化中，长臂猿在人科（黑猩猩和人）和猕猴科（猕猴、黑长尾猴和鬼狒）枝分开的位置从猕猴科基部分离出来，成为一条单独的支系。在人和黑猩猩的演变终点并没有分离发生：它们的血纤维蛋白肽完全相同。而且，狭鼻猴灵长类中的突变偏离远远没有其他类似分类学组群中的那样显著。在这个系统发生树中，狭鼻猴灵长类的血纤维蛋白肽属于哺乳纲中进化速度最慢的群体之一，这一点可以从图 4 所列的父系突变长度中发现。

除了血红蛋白链和血纤维蛋白肽的氨基酸序列数据外，还有红细胞酶碳酸酐酶 B 的部分序列数据，后者可用于描述狭鼻猴灵长类间的进化枝关系[78]。在检测过的 133 个同源残基中，黑猩猩和人只在 1 个位点上有所不同，而红毛猩猩和黑猩猩有 3 个位点不同，和人有 4 个位点不同。人科和猕猴科（黑长尾猴、食蟹猕猴、恒河猴和狒狒）有 4 ~ 6 个位点不同。因此人和黑猩猩在遗传学上的密切关系再次凸现出来。源自多系列氨基酸序列数据中的发现证实了从对灵长类动物蛋白质的大量比较免疫扩散研究[79-82]中得到的结论：在人科动物进化过程中，人、黑猩猩和大猩猩在长臂猿支系与红毛猩猩支系发生先期分离后仍拥有共同祖先。由蛋白质序列给出的进化枝关系为人猿总科的一种分类，即把大猩猩和黑猩猩属连同人属一起放到人科[80-82]提供了证据。这和达尔文的观点[83]一致，达尔文认为：人起源于非洲，其祖先血统与黑猩猩和大猩猩的祖先血统相同。

400 万到 500 万年前的人类直系祖先——南方古猿的化石近日在埃塞俄比亚的上新世堆积物中被发现[84]。南方古猿化石和被认为是南方古猿直系祖先的 1,400 万年前的腊玛古猿[86]（肯尼亚古猿[87]）的化石可追溯到肯尼亚的中新世末期。另一种

an eighteen million year old *Dryopithecus* ape, is considered an ancestor of *Gorilla* and to have even then separated from the line to *Ramapithecus* and *Homo*[88]. The hypothesis that molecular evolution decelerated in higher primates reconciles such palaeontological views with protein findings on the close genetic relationship of the African apes to man, DNA reassociation experiments[89,90] on nonrepeating polynucleotide sequences (the DNA fraction best suited for cladistic comparisons) further demonstrate the close relationship between *Pan* and *Homo*. The DNA findings emphasize the possibility that steadily increasing generation lengths may have been an important parameter[79,89,90] in slowing molecular evolution during the past sixty-five million years in the descent of man.

We thank Mr. Walter Farris for assistance, Miss Joan Bechtold and Mrs. Geraldine Fockler for drawings, and the staff of the Wayne State University Computing Center for help. This work was supported by grants from the US National Science Foundation Systematic Biology and US–Japan Cooperative Science Programs.

(**233**, 604-613; 1971)

Morris Goodman and John Barnabas: Department of Anatomy, Wayne State University School of Medicine, Detroit, Michigan 48207, and Plymouth State Home and Training School, Northville, Michigan 48167.

Genji Matsuda: Department of Biochemistry, Nagasaki University School of Medicine, Nagasaki.

G. William Moore: Institute of Statistics, Biomathematics Program, North Carolina State University, Raleigh, North Carolina 27607.

Received February 17; revised March 18, 1971.

References:

1. Ingram, V. M., *Nature*, **189**, 704 (1961).
2. Goodman, M., *Human Biol.*, **33**, 131 (1961).
3. Goodman, M., in *Classification and Human Evolution*, **203** (Aldine Press, 1963).
4. Goodman, M., in *Protides of Biological Fluids*, **70** (Elsevier, Amsterdam, 1965).
5. Kimura, M., *Nature*, **217**, 624 (1968).
6. King, J. L., and Jukes, T. H., *Science*, **164**, 788 (1969).
7. Fitch, W. M., and Margoliash, E., *Science*, **155**, 279 (1967).
8. Sokal, R., and Sneath, P. H. A., *Principles of Numerical Taxonomy*, **309** (W. H. Freeman, San Francisco, 1963).
9. Moore, G. W., thesis, Univ. North Carolina (1970).
10. Barnabas, J., Goodman, M., and Moore, G. W., *Comp. Physiol. Biochem.*, **39B**, 455 (1971).
11. Braun, V., Crichton, R. R., and Braunitzer, G., *Z. Physiol. Chem.*, **349**, 45 (1968).
12. Braunitzer, G., Buse, G., and Braig, S., *Z. Physiol. Chem.*, **350**, 1477 (1969).
13. Edmundson, A. B., *Nature*, **205**, 883 (1965).
14. Dayhoff, M. O., *Atlas of Protein Sequence and Structure*, **4** (1969).
15. Rudloff, V., Zelenik, M., and Braunitzer, G., *Z. Physiol. Chem.*, **344**, 284 (1966).
16. Hilse, K., and Braunitzer, G., *Z. Physiol. Chem.*, **349**, 433 (1968).
17. Matsuda, G., Takei, H., Wu, K. C., Mizuno, K., and Shiozawa, T., *Eighth Intern. Cong. Biochemistry*, Abstract, 4 (1970).
18. Von Ehrenstein, G., *Cold Spring Harbor Symp. Quant. Biol.*, **31**, 705 (1966).
19. Hill, R. L., unpublished data taken from *Handbook of Biochemistry* (The Chemical Rubber Company, 1968).
20. Rifkin, D. B., Hirsch, D. I., Rifkin, M. R., and Konigsberg, W., *Cold Spring Harbor Symp. Quant. Biol.*, **31**, 715 (1966).
21. Popp, R. A., *J. Mol. Biol.*, **27**, 9 (1967).
22. Matsuda, G., Maita, T., Takei, H., Ota, H., Yamaguchi, M., Miyanchi, T., and Migita, M., *J. Biochem.* (Japan), **64**, 279 (1968).
23. Zuckerkandl, E., and Schroeder, W. A., *Nature*, **192**, 984 (1961).

肯尼亚人科动物——1,800万年前的森林古猿被认为是大猩猩的祖先，而且在那个时候已经从通往腊玛古猿和人属的支系上分离出来[88]。分子进化在高等灵长类中减速的假说使古生物学观点与证明非洲类人猿和人的密切遗传学关系的蛋白质测序结果一致，在非重复多聚核苷酸序列（这种DNA片段最适合进化枝比较）上进行的DNA重组实验[89,90]进一步证明黑猩猩属和人属之间的密切关系。DNA的研究结果强化了如下的可能性：稳步增长的世代长度可能就是证明在过去6,500万年的人类起源过程中分子进化速度减慢的一个重要参数[79,89,90]。

我们要感谢沃尔特·法里斯先生的帮助，感谢琼·贝克托尔德小姐和杰拉尔丁·弗克勒夫人的制图，还要感谢韦恩州立大学计算中心团队的帮助。这项工作得到了来自美国国家科学基金会的系统生物学计划和美日合作科学计划的拨款支持。

（邓铭瑞 翻译；崔巍 审稿）

24. Rifkin, D. B., and Konigsberg, W., *Biochim. Biophys. Acta*, **104**, 457 (1965).

25. Braunitzer, G., Gehring-Mueller, R., Hilschmann, N., Hilse, K., Hobom, G., Rudloff, Y., and Wittman-Liebold, B., *Z. Physiol. Chem.*, **325**, 283 (1961).

26. Schroeder, W. A., Shelton, J. R., Shelton, J. B., and Cormick, J., *Biochemistry*, **2**, 1353 (1963).

27. Kilmartin, J. V., and Clegg, J. B., *Nature*, **213**, 269 (1967).

28. Matsuda, G., Gehring-Mueller, R., and Braunitzer, G., *Biochem. Z.*, **338**, 669 (1963).

29. Schroeder, W. A., Shelton, J. R., Shelton, J. B., Robberson, B., and Babin, D. R., *Arch. Biochem. Biophys.*, **120**, 1 (1967).

30. Huisman, T. H. J., Brandt, G., and Wilson, J. B., *J. Biol. Chem.*, **243**, 3637 (1968).

31. Huisman, T. H. J., Dozy, A. M., Wilson, J. B., Effremov, G. D., and Vaskov, B., *Biochim. Biophys. Acta*, **160**, 467 (1968).

32. Beale, D., Lehman, H., Drury, A., and Tucker, E. M., *Nature*, **209**, 1099 (1966).

33. Wilson, J. B., Brandt, G., and Huisman, T. H. J., *J. Biol. Chem.*, **243**, 3687 (1968).

34. Schroeder, W. A., Shelton, J. R., Shelton, J. B., Cormick, J., and Jones, R. T., *Biochemistry*, **2**, 992 (1963).

35. Air, G. M., and Thomson, E. O. P., *Austral. J. Biol. Sci.*, **22**, 1437 (1969).

36. Rifkin, D. B., Rifkin, M. R., and Konigsberg, W., *Proc. US Nat. Acad. Sci.*, **55**, 586 (1966).

37. Braunitzer, G., Best, J. S., Flamm, U., and Schrank, B., *Z. Physiol. Chem.*, **347**, 207 (1966).

38. Boyer, S. H., Crosby, E. F., Thurmon, T. F., Noyes, A. N., Fuller, G. F., Leslie, S. E., and Sheppard, M. K., *Science*, **116**, 1428 (1969).

39. Matsuda, G., Maita, T., Ota, H., Tachiwaka, I., Tanaka, Y., Araya, A., and Nakashima, Y., *Intern. J. Protein Res.*, **3**, 41 (1971).

40. Ingram, V. M., and Stretton, A. O. W., *Biochim. Biophys. Acta*, **63**, 20 (1962).

41. Smith, D. B., *Canad. J. Biochem.*, **46**, 825 (1968).

42. Babin, D. R., Schroeder, W. A., Shelton, J. R., Shelton, J. B., and Robberson, B., *Biochemistry*, **5**, 1297 (1966).

43. Wilson, J. B., Edwards, W. C., McDaniel, M., Dobbs, M. M., and Huisman, T. H. J., *Arch. Biochem. Biophys.*, **115**, 385 (1966).

44. Schroeder, W. A., Shelton, J. R., Shelton, J. B., Robberson, B., and Babin, D. R., *Arch. Biochem. Biophys.*, **120**, 124 (1967).

45. Huisman, T. H. J., Dasher, G. A., Moretz, W. H., Dozy, A. M., and Wilson, J. B., *Biochem. J.*, **107**, 745 (1968).

46. Boyer, S. H., Hathaway, P., Pascasio, F., Orton, C., Bordley, J., and Naughton, M. A., *Science*, **153**, 1539 (1966).

47. Huisman, T. H. J., Adams, H. R., Dimock, M. O., Edwards, W. E., and Wilson, J. B., *J. Biol. Chem.*, **242**, 2534 (1967).

48. Huber, R., Epp, O., and Formanck, H., *J. Mol. Biol.*, **42**, 59 (1969).

49. Briehl, R. W., *J. Biol. Chem.*, **238**, 2361 (1963).

50. Balani, A. S., and Barnabas, J., *Nature*, **205**, 1019 (1965).

51. Balani, A. S., Ranjekar, P. K., and Barnabas, J., *Comp. Physiol. Biochem.*, **24**, 809 (1968).

52. Huisman, T. H. J., Wilson, J. B., and Adams, H. R., *Arch. Biochem. Biophy.*, **121**, 528 (1967).

53. Ranjekar, P. K., and Barnabas, J., *Indian J. Biochem.*, **6**, 1 (1969).

54. Barnicot, N. A., and Huehns, E. R., *Nature*, **215**, 1485 (1967); Wade, P. T., Skinner, A. F., and Barnicot, N. A., *Protides in Biological Fluids*, 263 (Pergamon, London, 1970).

55. Oliver, E., and Kitchen, H., *Biochim. Biophys. Acta, Res. Commun.*, **31**, 749 (1968).

56. Hilse, K., and Popp, R. A., *Proc. US Nat. Acad. Sci.*, **61**, 930 (1968).

57. Kunkel, H. G., and Wallenius, G., *Science*, **122**, 288 (1955).

58. Kunkel, H. G., Ceppellini, R., Mueller-Eberhard, V., and Wolf, J., *J. Clin. Invest.*, **36**, 1615 (1957).

59. Huisman, T. H. J., Reynolds, C. A., Dozy, A. M., and Wilson, J. B., *J. Biol. Chem.*, **240**, 2455 (1965).

60. Hamilton, W. J., Boyd, J. D., and Mossman, H. W., *Human Embryology* (Williams and Wilkins, Baltimore, 1952).

61. Kimura, M., *Proc. US Nat. Acad. Sci.*, **63**, 1181 (1969).

62. Buettner-Janusch, J., and Buettner-Janusch, V., *Amer. J. Phys. Anthrop.*, **33**, 73 (1970).

63. Ohno, S., *Evolution by Gene Duplication* (Springer, Berlin, 1970).

64. Buettner-Janusch, J., and Hill, R. W., in *Evolving Genes and Proteins*, 167 (Academic Press, London and New York, 1965).

65. LeGros Clark, W. E., *The Antecedents of Man* (Edinburgh University Press, Edinburgh, 1959).

66. Buettner-Janusch, J., Buettner-Janusch, V., and Mason, G. A., *Arch. Biochem. Biophys.*, **133**, 164 (1969).

67. Blombäck, B., Blombäck, M., and Grondahl, N. J., *Acta Chem. Scand.*, **19**, 1789 (1965).

68. Blombäck, B., Blombäck, M., Grondahl, N. J., and Holmberg, E., *Arkiv. Kemi.*, **25**, 411 (1966).

69. Blombäck, B., Blombäck, M., Grondahl, N. J., Guthrie, C., and Hinton, M., *Acta Chem. Scand.*, **19**, 1788 (1965); Blombäck, B., Blombäck, M., and Edman, B., *Biochim. Biophys. Acta*, **115**, 371 (1966).

70. Doolittle, R. F., Glasgow, C., and Mross, G. A., *Biochim. Biophys. Acta*, **175**, 217 (1969).

71. Mross, G. A., Doolittle, R. F., and Roberts, B. F., *Science*, **170**, 468 (1970).

72. Doolittle, R. F., and Mross, G. A., *Nature*, **225**, 643 (1970).

73. Blombäck, B., and Blombäck, M., in *Chemotaxonomy and Serotaxonomy*, 3 (Academic Press, London and New York, 1968).

74. Doolittle, R. F., Schubert, D., and Schwartz, S. A., *Arch. Biochem. Biophys.*, **118**, 456 (1967).

75. Mross, G. A., and Doolittle, R. F., *Arch. Biochem. Biophys.*, **122**, 674 (1967).

76. Doolittle, R. F., and Blombäck, B., *Nature*, **202**, 147 (1964).

77. McKenna, M. C., *Ann. NY Acad. Sci.*, **167**, 217 (1969).

78. Tashian, R. E., and Stroup, S. R., *Biochem. Biophys. Res. Commun.*, **41**, 1457 (1970).

79. Goodman, M., *Human Biol.*, **34**, 104 (1962).

80. Goodman, M., *Ann. NY Acad. Sci.*, **102**, 219 (1962).

81. Goodman, M., *Human Biol.*, **35**, 377 (1963).

82. Goodman, M., *Primates in Med.*, **1**, 10 (1968).

83. Darwin, C., *The Descent of Man and Selection in Relation to Sex* (Appleton, New York, 1871).

84. Patterson, B., Behrensmeyer, A. K., and Sill, W. D., *Nature*, **226**, 918 (1970).

85. Howell, F. C., *Nature*, **223**, 1234 (1969).

86. Simons, E. L., *Nature*, **221**, 448 (1969).

87. Leakey, L. S. B., *Nature*, **213**, 155 (1967).

88. Pilbeam, D. R., *Peabody Mus. Bull. Yale*, **31** (1969).

89. Kohne, D. E., *Quart. Rev. Biophys.*, **3**, 327 (1970).

90. Kohne, D. E., Chison, J. A., and Hoyer, B. H., *Carnegie Inst. Yr Book*, **69**, 488 (1971).

Logarithmic Relationship of DDE Residues to Eggshell Thinning

L. J. Blus *et al.*

Editor's Note

Beginning in the 1960s, there was great concern among environmentalists and naturalists, particularly in the United States, about the effects of small quantities of chemicals such as pesticides in the environment. This paper by Lawrence Blus and colleagues argues that the pesticide called DDE had been responsible for a decline in the population of brown pelicans off the coast of California, producing evidence that the pesticide reduced the thickness of the shell from which young pelicans hatched. The publication of this paper caused some controversy with several scientists producing conflicting evidence, but the eventual outcome of the controversy was that DDE was banned from use in agriculture in the United States. The population of brown pelicans off the coast of California has recovered.

SHELL thinning has been recorded in eggs of the brown pelican (*Pelecanus occidentalis*) which were collected in widely separated nesting colonies[1-3], and it has been reported among numerous other species of wild birds[4-9]. The condition has been used as an indicator of population trend; thinning of 15 to 20% has been associated with declining populations of several species[9].

A cause-and-effect relationship has been established between DDE (1,1-dichloro-2,2-*bis* (*p*-chlorophenyl) ethylene) in the diet and shell thinning; the associated residues in eggs of American kestrels (*Falco sparverius*) fed DDE were comparable with those in eggs of peregrines (*Falco peregrinus*)[10] collected in the field. It is necessary to understand the quantitative relationships between residue content and shell thinning in order to evaluate the problem properly. We have studied this relationship as derived from a study of eggs of the brown pelican and we compare here these results with those reported for other species.

The questions to be explored are (1) whether a concentration-response relationship, paralleling the traditional dose-response relationship, in fact exists between DDE in eggs and the thinning of the shells, and (2) the nature of such a mathematical relationship. The amount of DDE in the egg is taken as an index to the concentration of residues in the female[11-13], the physiological processes of which determine shell thickness. DDE is the only residue considered here, because this residue consistently accounts for a significant percentage of eggshell thinning in these pelicans (our unpublished work).

Eighty eggs of brown pelicans were used in the primary analysis of the problem. Seventy eggs were collected in 1969 from twelve colonies, one in California, two in South Carolina

DDE残留量对数与蛋壳薄化的关系

布卢斯等

编者按

从 20 世纪 60 年代起，尤其是在美国，环境人士和自然主义者们就开始格外关注像杀虫剂这样的小量化学物质对环境产生的影响。这篇由劳伦斯·布卢斯及其同事撰写的文章认为，一种被称为 DDE 的杀虫剂是造成加利福尼亚海岸褐鹈鹕种群数量下降的元凶，并给出证据证明杀虫剂使得孵出小鹈鹕的蛋壳发生了薄化。这篇论文发表后引起了几位科学家之间的争论，他们给出的证据相互矛盾，但争论的最终结果是 DDE 在美国农业生产中被禁用。加利福尼亚州沿岸的褐鹈鹕种群数量得以回升。

对蛋壳薄化的报道已涉及在相距较远的几个褐鹈鹕鸟群聚集地收集的鸟蛋 [1-3]，对于许多其他野生鸟类也有过类似的报道 [4-9]。蛋壳薄化已被用于指示鸟群数量的变化趋势；有几种鸟类的数量曾在薄化达到 15% ~ 20%时出现下降 [9]。

人们已经确定了饲料中 DDE（1,1– 二氯 –2,2– 双 (对 – 氯苯基) 乙烯）含量和蛋壳薄化之间的因果关系；将用 DDE 喂养的美洲隼所生的蛋与从野地里收集的游隼蛋 [10] 相对比，两者所含的相关残留物差不多。为了对这一问题进行正确的评估，我们必须了解残留物含量与蛋壳薄化之间的定量关系。我们通过对褐鹈鹕蛋的一项研究得到了两者之间的关系，并在这里将所得结果与其他鸟类的已知结果进行比较。

需要探讨的问题有：(1）与传统的剂量 – 反应关系相类似的浓度 – 反应关系是否确实可以反映蛋中 DDE 与蛋壳薄化之间的关系，(2）这种数学关系的本质。蛋中的 DDE 含量反映了雌鸟体内的残留物浓度 [11-13]，而雌鸟的生理过程决定了蛋壳的厚度。DDE 是本文中唯一考虑的残留物，因为这种残留物一直被看作是引起鹈鹕蛋蛋壳薄化的主要因素（我们尚未发表的研究结果）。

在对这一问题进行初步分析时我们使用了 80 枚褐鹈鹕蛋。其中有 70 枚蛋是 1969 年从 12 个鸟群聚集地收集的：1 个在加利福尼亚州，2 个在南卡罗来纳州，9

and nine in Florida. Ten more eggs were collected from one of the South Carolina colonies in 1970. Analytical procedures for preparation, clean up, and residue analysis followed those used at the Patuxent Wildlife Research Center[14-16]. The samples were prepared and analysed by Soxhlet extraction and clean up by acetonitrile partitioning and "Florisil" column. The residues were separated and removed in four fractions from a silica gel thin layer plate. Sample fractions were analysed by electron capture chromatography on a column of 3% OV-1 on "Chromosorb W" H.P., or on a column of 3% OV-17 on "Gas Chrom Q". The DDT metabolites were confirmed on a column of 3% XE-60 on "Gas Chrom Q". Some of the samples were treated with alkali to confirm DDT (1,1,1-trichloro-2,2-*bis* (*p*-chlorophenyl) ethane). Each of the three glass columns was 1.8 m long and the outside diameter was 0.64 cm.

The average recovery of the chlorinated insecticides and their metabolites from fortified eagle tissue ranged from 75 to 112%; the recovery value for DDE was 95%[4]. The pelican egg residues were not corrected for recovery. Residues are expressed on a fresh wet weight basis. We found that certain external egg measurements were significantly correlated with the weight of the contents of fresh eggs. The resulting regression equation was used to convert weight of the contents of addled eggs to a fresh wet weight basis (unpublished work of L. F. Stickel, S. N. Wiemeyer and L. J. B.). The statistical analyses were performed after the residues were transformed to logarithms.

The percentage of eggshell thinning was measured by comparison with the pre-1947 mean thickness as computed by Anderson and Hickey[17]. These means were 0.557 mm for Florida and South Carolina eggs and 0.579 mm for California eggs. Eggshell thinning occurred in eleven of the twelve colonies where eggs were collected. In eggs from the nine Florida colonies the mean change in thickness varied from a 13% decrease to a 0.4% increase. There was a 17% decrease in shell thickness of eggs from the two South Carolina colonies, and a 35% decrease in shell thickness of eggs from the California colony.

The thinning of eggshells of the brown pelican proved to be related to the concentrations of DDE in the eggs (Fig. 1). The relationship was essentially linear ($P<0.01$) on a logarithmic residue scale. We used the following regression equation to describe this relationship:

$$\hat{Y} = b_0 - b \log_{10} X$$

where $\hat{Y}$ = % of pre-1947 eggshell thickness; b_0 = expected % of pre-1947 thickness at 1 p.p.m. of DDE; b = the regression coefficient; and $\log_{10} X$ = p.p.m. of DDE on a fresh wet weight basis.

个来自佛罗里达州；还有 10 枚蛋是 1970 年从南卡罗来纳州的一个鸟群聚集地收集的。遵照帕图克森特野生动物研究中心过去所使用的分析步骤[14-16]进行样品的制备、清洗和残留物分析。用索氏萃取法制备和分析样品，用乙腈分配和“硅酸镁载体”柱清洗样品。残留物从硅胶薄层板上分离下来并被分成 4 份。4 份样品采用电子俘获气相色谱法进行分析，分析是在以惠普公司的“硅烷化白色硅藻土”为担体、以 3% OV-1 为固定相的色谱柱，或者以“硅藻土型色谱载体 Q”为担体、以 3% OV-17 为固定相的色谱柱上进行的。DDT 代谢物是用以“硅藻土型色谱载体 Q”为担体、以 3% XE-60 为固定相的色谱柱进行确认的。部分样品用碱处理以确证 DDT (1,1,1– 三氯 –2,2– 双 (对 – 氯苯基) 乙烷)。3 个玻璃色谱柱的长度均为 1.8 m，外径为 0.64 cm。

在处理后的鹰组织中含氯杀虫剂及其代谢物的平均回收率为 75% ~ 112%；DDE 的回收率为 95%[4]。对鹈鹕蛋中的残留物没有进行过回收率的校正。残留物是用鲜湿重来表示的。我们发现对鸟蛋外部的某些测量结果与新鲜鸟蛋内含物的重量显著相关。用得到的回归方程可以将臭蛋内含物的重量转化成以鲜湿重为基准的结果(施蒂克尔、威迈耶和布卢斯尚未发表的研究结果)。统计分析是在将残留量取对数之后进行的。

蛋壳薄化的百分比是与安德森和希基计算的 1947 年以前的蛋壳平均厚度[17]相比较后得到的。佛罗里达州和南卡罗来纳州的鸟蛋蛋壳平均厚度为 0.557 mm，加利福尼亚州为 0.579 mm。在采集鸟蛋的 12 个鸟群聚集地中有 11 个发生了蛋壳薄化。在位于佛罗里达州的 9 个聚集地采集的鸟蛋中，蛋壳厚度的平均变化从下降 13%到增加 0.4%不等。来自南卡罗来纳州 2 个聚集地的鸟蛋蛋壳厚度出现了 17% 的下降，而加利福尼亚州的鸟蛋蛋壳厚度下降了 35%。

可以证明褐鹈鹕蛋蛋壳薄化与蛋中 DDE 浓度之间存在着关联（图 1）。当对残留量取对数时，两者之间的关系基本上是线性的（$P<0.01$）。我们使用以下的回归方程来描述这种关系：

$$\hat{Y} = b_0 - b \log_{10} X$$

式中：$\hat{Y}$= 相对于 1947 年前蛋壳厚度的百分数；b_0 = 当 DDE 含量为 1 ppm 时相对于 1947 年前蛋壳厚度的预期百分数；b = 回归系数；$\log_{10} X$ = 根据鲜湿重计算出来的、以 ppm 为单位的 DDE 含量。

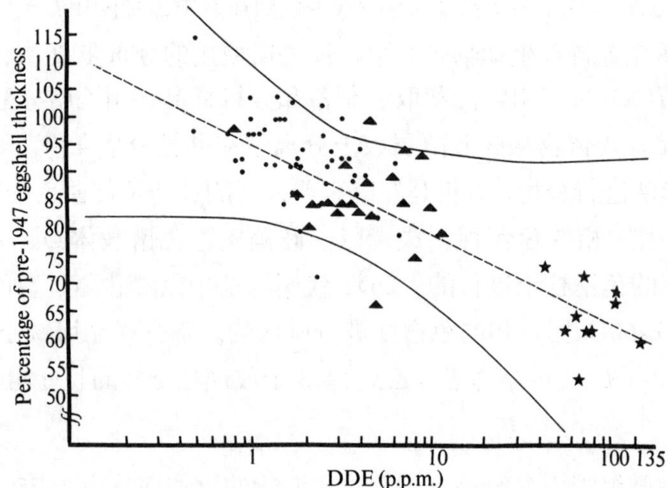

Fig. 1. Association of DDE residues in eighty brown pelican eggs from Florida (●), South Carolina (▲) and California (★) with the % of pre-1947 eggshell thickness. Solid lines represent 95% confidence limits. $\hat{Y} = 95.787 - 15.689 \log_{10} X$; $r = -0.80$ ($P<0.01$).

The significance of the regression indicates that the shell thickness decreases in a predictable manner as the DDE concentration increases. It also means that the percentage change was greater per unit of DDE when the concentration of DDE was lower. In other words, the lower concentrations were more effective. For example, the calculated percentage of thinning per p.p.m. of DDE is 4.2 at 1 p.p.m., 3.0 at 5 p.p.m., 1.9 at 10 p.p.m., and 0.4 at 100 p.p.m.

Analysis of covariance revealed no significant differences ($P>0.05$) among slopes of the three regression lines that were plotted using data from each of the three states from which the eggs were collected. Thus, most of the variation in eggshell thickness (64%; $P<0.01$) may be explained by the common regression used here.

The prediction of a no-effect level is also theoretically possible, but should be approached with caution because of the scarcity of eggshell measurements in eggs with lower residues, and the hazard of making predictions of *X* far removed from the mean[18]. The estimated no-effect level is 0.5 p.p.m., but the validity of this estimate is questionable because of the forementioned reasons. In the eggs from Florida that contained 0.52 p.p.m. of DDE, the thickness (0.64 mm) was slightly less than 0.65 mm, the maximum pre-1947 measurement; thus, the possibility of thinning was not ruled out completely in any of the eighty eggs. Anderson *et al.*[4] indicated an apparent absence of a no-effect level of DDE on the eggshell thickness of the double-crested cormorant (*Phalacrocorax auritus*) and the white pelican (*Pelecanus erythrorhynchos*). The relationship of the logarithm of DDE to eggshell thinning also seems to apply to the prairie falcon (*Falco mexicanus*)[5] and the double-crested cormorant[4].

The relationship between log dose and response is often encountered and is in accord

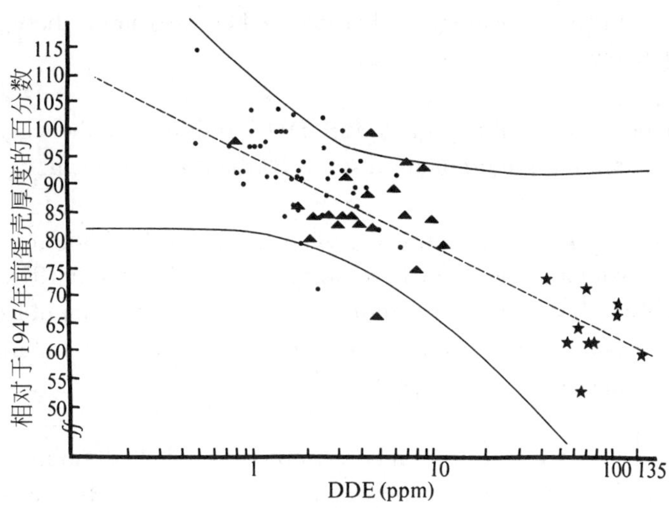

图 1. 在 80 枚来自佛罗里达州（●）、南卡罗来纳州（▲）和加利福尼亚州（★）的褐鹈鹕蛋中，DDE 残留量与相对于 1947 年前蛋壳厚度的百分数之间的关系。实线代表 95 % 的置信区间。$\hat{Y} = 95.787 - 15.689 \log_{10} X$；$r = -0.80$（$P<0.01$）。

回归的显著性表明：蛋壳厚度会以一种可预计的方式随 DDE 浓度的增加而下降。这也意味着当 DDE 浓度较低时，对应于每个单位 DDE 的蛋壳厚度百分数的变化值也会较大。换句话说，浓度越低效果越明显。例如：当 DDE 浓度为 1 ppm 时，对应于每 ppm DDE 的蛋壳薄化百分数的计算值为 4.2；当浓度达到 5 ppm 时为 3.0；10 ppm 时为 1.9；100 ppm 时为 0.4。

由协方差分析结果可知：根据在 3 个州采集的鸟蛋数据绘制而成的 3 条回归线在斜率上并没有显著的差异（$P>0.05$）。因此，采用本文中的公共回归方程或许可以解释大部分的蛋壳厚度变化（64% ；$P<0.01$）。

虽然预测出无作用剂量在理论上是可行的，但在实施时应小心谨慎，因为一方面我们缺少较低残留量下鸟蛋蛋壳厚度的测量数据，另一方面预测远离平均值的 X 是很冒险的 [18]。无作用剂量的估计值是 0.5 ppm；不过，由于前述原因，这个估计值的有效性是有待商榷的。在 DDE 含量为 0.52 ppm 的来自佛罗里达州的鸟蛋中，其厚度（0.64 mm）比 1947 年前的厚度测量结果的最大值 0.65 mm 只略微少一点；因此不能把这 80 枚蛋中任何一枚发生薄化的可能性完全排除掉。安德森等人 [4] 指出：DDE 对双冠鸬鹚和白鹈鹕蛋蛋壳厚度的影响显然不存在无作用剂量。DDE 含量的对数与蛋壳薄化之间的关系似乎也适用于草原隼 [5] 和双冠鸬鹚 [4]。

经常会遇到剂量对数与反应之间相关的情况，这种关系与毒理学和药理学的理

with toxicological and pharmacological theory[19-22]. The association between residue and response is poorly known.

The relationship between shell thinning and DDE residues in the eggs is particularly important in understanding the population status of the brown pelican.

The brown pelican population in South Carolina has been declining for at least 12 yr[1,23] and the number of young fledged per breeding female in 1969 and 1970 was well below the estimated number necessary to maintain a stable population (unpublished work of C. Henry). The brown pelican in California experienced reproductive failure in 1969[2-3], but the Florida population seems stable over the past 3 yr[24-25].

The brown pelican seems to be unusually susceptible to shell thinning, a 15% thinning being associated with DDE residues between 4 and 5 p.p.m. In the declining South Carolina colonies eight of the eleven eggs collected in 1969 and two of ten collected in 1970 contained DDE concentrations greater than 4 p.p.m. Thirteen of the twenty-one eggs collected in the 2 yr exceeded 15% shell thinning. In California, where reproductive failure occurred, a 35% decrease in eggshell thickness was associated with 71 p.p.m. of DDE in the egg.

In contrast, only ten of the forty-nine brown pelican eggs collected in Florida were more than 15% thinner than the pre-1947 average; and residues exceeded 4 p.p.m. in only four eggs. It seems probable, therefore, that some pelicans in Florida are adversely affected by shell thinning although the extent of this thinning is not great enough for obvious effects on the size of the populations.

In contrast to the brown pelican, the herring gull (*Larus argentatus*) showed no thinning when DDE residues were between 4 and 5 p.p.m. and only 11% thinning when residues were near 80 p.p.m[8]. The eggshell thinning response induced by DDE in the brown pelican was similar to that found in the prairie falcon[5]. The double-crested cormorant[4] seemed somewhat more resistant to thinning by DDE than these two species, but shell thinning generally occurred in cormorant eggs that contained less than 5 p.p.m. of DDE and 15% thinning was recorded in one colony when DDE egg residues averaged just under 20 p.p.m.

The concentration-effect relationship involving the logarithm of DDE and eggshell thinning seems to exist and to follow a mathematically similar pattern in different species, but to operate at different levels in different species. A level of DDE in eggs that would result in population collapse among brown pelicans would not be expected to effect an overall population change among herring gulls.

We thank the many individuals who helped us with the collection of eggs.

(**235**, 376-377; 1972)

论是一致的[19-22]。至于残留量和反应之间的关系目前还很少有人认识到。

蛋壳薄化与蛋中DDE残留量之间的关系对于理解褐鹈鹕种群数量的状况尤为重要。

南卡罗来纳州的褐鹈鹕种群数量已经持续下降了至少12年[1,23]，在1969年和1970年间，每只雌鸟繁殖的幼鸟数量远远达不到能够维持种群数量稳定所必需的预估值（亨利尚未发表的研究结果）。加利福尼亚州的褐鹈鹕在1969年出现了繁殖障碍[2-3]，但佛罗里达州褐鹈鹕的种群数量在过去3年中似乎是保持稳定的[24-25]。

褐鹈鹕似乎非常容易受到蛋壳薄化的影响，当DDE残留量为4 ppm ~ 5 ppm时，蛋壳会发生15%的薄化。在种群数量不断下降的南卡罗来纳州，1969年采集的11枚鸟蛋中有8枚、1970年采集的10枚鸟蛋中有2枚所含的DDE浓度超过了4 ppm。在这两年之内收集的21枚鸟蛋中有13枚的蛋壳薄化率超过了15%。在种群出现繁殖障碍的加利福尼亚州，蛋壳厚度下降35%的蛋中DDE残留浓度为71 ppm。

相比之下，在佛罗里达州采集的49枚褐鹈鹕蛋中，只有10枚比1947年前的平均水平变薄了15%以上；而且只有4枚蛋的残留物浓度超过了4 ppm。所以，似乎很可能达到这样的薄化程度还不足以对鸟群数量产生明显的影响，但佛罗里达州的一些鹈鹕的确受到了蛋壳薄化的不利影响。

与褐鹈鹕相对比，当DDE残留量在4 ppm ~ 5 ppm之间时，黑脊鸥并没有出现蛋壳薄化现象；只有在残留量接近80 ppm时才显现出11%的薄化[8]。DDE对褐鹈鹕蛋壳薄化的影响类似于草原隼[5]。与这两个物种相比，双冠鸬鹚[4]似乎对DDE引起的蛋壳薄化有更强的抵抗力，但在DDE浓度小于5 ppm的鸬鹚蛋中往往会发生蛋壳薄化，有人曾报道在一个蛋中DDE平均残留量略小于20 ppm的鸬鹚聚集地中发现蛋壳薄化达到了15%。

在各种不同的物种中似乎都存在着DDE浓度对数与蛋壳薄化之间的浓度–效应关系，并且遵循着相似的数学模式，但在不同物种中有不同程度的体现。导致褐鹈鹕种群数量下滑的蛋中DDE残留水平未必会使黑脊鸥的总体数量发生变化。

在此我们要感谢很多帮助我们采集鸟蛋的人。

（吕静 翻译；周江 审稿）

Lawrence J. Blus, Charles D. Gish, Andre A. Belisle and Richard M. Prouty: Patuxent Wildlife Research Center, Laurel, Maryland 20810.

Received October 1, 1971.

References:

1. Blus, L. J., *Bioscience*, **20**, 867 (1970).
2. Keith, J. O., Woods, jun., L. A., and Hunt, E. G., *Trans. N. Amer. Wildl. Nat. Res. Conf.*, **35**, 56 (1970).
3. Risebrough, R. W., Davis, J., and Anderson, D. W., *Oregon State Univ. Env. Health Sci. Series*, No. 1, 40 (1970).
4. Anderson, D. W., Hickey, J. J., Risebrough, R. W., Hughes, D. F., and Christensen, R. E., *Canad. Fld Nat.*, **83**, 91 (1969).
5. Fyfe, R. W., Campbell, J., Hayson, B., and Hodson, K., *Canad. Fld Nat.*, **83**, 191 (1969).
6. Ratcliffe, D. A., *J. Appl. Ecol.*, **7**, 67 (1970).
7. Ratcliffe, D. A., *Nature*, **215**, 208 (1967).
8. Hickey, J. J., and Anderson, D. W., *Science*, **162**, 271 (1968).
9. Anderson, D. W., and Hickey, J. J., *Proc. Fifteenth Ornithol. Cong.* (in the press).
10. Wiemeyer, S. N., and Porter, R. D., *Nature*, **227**, 737 (1970).
11. Cummings, J. G., Eidelman, M., Turner, V., Reed, D., Zee, K. T., and Cook, R. E., *J. Assoc. Off. Agric. Chem.*, **50**, 418 (1967).
12. Cummings, J. G., Zee, K. T., Turner, V., and Quinn, F., *J. Assoc. Off. Agric. Chem.*, **49**, 354 (1966).
13. Noakes, D. N., and Benfield, C. A., *J. Sci. Food Agric.*, **16**, 393 (1965).
14. Krantz, W. C., Mulhern, B. M., Bagley, G. E., Sprunt, A., IV, Ligas, F. G., and Robertson, jun., W. B., *Pest. Monit. J.*, **4**, 136 (1970).
15. Mulhern, B. M., *J. Chromatog.*, **34**, 556 (1968).
16. Mulhern, B. M., Reichel, W. L., Locke, L. N., Lamont, T. G., Belisle, A., Cromartie, E., Bagley, G. E., and Prouty, R. M., *Pest. Monit. J.*, **4**, 141 (1970).
17. Anderson, D. W., and Hickey, J. J., *Wilson Bull.*, **82**, 14 (1970).
18. Snedecor, G. W., and Cochran, W. G., *Statistical Methods*, sixth ed. (Iowa State University Press, Ames, 1967).
19. Scholz, J., *Nature*, **207**, 870 (1965).
20. Hayes, jun., W. J., *Toxicol. Appl. Pharmacol.*, **11**, 327 (1967).
21. Kinoshita, F. K., Frawley, J. P., and DuBois, K. P., *Toxicol. Appl. Pharmacol.*, **9**, 505 (1966).
22. Ortega, P., Hayes, jun., W. J., Durham, W. F., and Mattson, A., *US Public Health Serv. Publ. Health Monograph*, No. 43, 1 (1956).
23. Beckett, T. A., III, *Chat*, **30**, 93 (1966).
24. Williams, jun., L. E., and Martin, L., *J. Florida Acad. Sci.*, **31**, 130 (1968).
25. Williams, jun., L. E., and Martin, L., *Proc. Twenty-fourth Ann. Conf. SE Assoc. Game Fish Comm.* (1970).

Disagreements on Why Brown Pelican Eggs Are Thin

W. Hazeltine

Editor's Note

After the decline of the American brown pelican population in the 1960s, Lawrence Blus suggested that DDE, a breakdown product of the pesticide DDT, was responsible for the species' thinning eggshells, and the belief became well established. Here William Hazeltine challenges the statistics and methodology of Blus, questioning his use of whole egg analysis when DDE residues are found in the yolk and are metabolised during incubation. Blus's data, he argues, show correlation not cause and effect, and he warns against calls to ban pesticides based on these "faulty" conclusions. "There is no place in science for suppressed conflicting data, or an "end justifies the means" philosophy," he says. The agricultural use of DDT was nonetheless banned in most developed countries in the 1970s and 80s.

I Find four points of disagreement with the proposal of Blus *et al.*[1,2] that DDE is the probable cause of thin eggshells in brown pelicans.

First, his Fig. 1 is based on data points of residue and shell thickness for three separate colonies composed of two subspecies. If this overall correlation is meaningful, each colony should show a similar trend in itself, which is not the case. None of the separate colonies shows a clear trend in itself.

Second, within the lines for 95% confidence limits in Fig. 1, it is possible to have a series of shell measurements parallel to the abscissa. In other words, his 95% confidence limit could include a series of eggs with residues extending from none to a maximum, with no change in shell thickness. This suggests extremely variable supporting values for the regression.

Third, whole egg analysis is inappropriate for incubated eggs because the residues are found in the yolk, and during incubation yolk contents including residues are subject to metabolic changes. I show below that DDE residues are metabolized as brown pelican eggs are incubated. It is not possible to determine the extent of incubation of Blus's eggs from the published literature, but a table was presented at the public hearing on DDT (Exhibit R-74) which shows that Blus's South Carolina and Florida eggs contained embryos; two were even "heard peeping".

Fourth, the use of Pearson's product moment correlation coefficient (r) is not an appropriate method for values which are not bi-variantly normally distributed. Using logarithms for residue levels shows that residues are not normally distributed around a

在褐鹈鹕蛋蛋壳薄化原因上的不同意见

黑兹尔坦

编者按

在20世纪60年代美国褐鹈鹕种群数量下降之后，劳伦斯·布卢斯提出：杀虫剂DDT的降解产物DDE就是造成褐鹈鹕蛋蛋壳薄化的元凶，这一观点得到了大家的普遍认可。在本文中，威廉·黑兹尔坦对布卢斯的统计数据和研究方法提出了质疑，在发现DDE残留存在于蛋黄中并且在孵化时还会发生代谢的情况下，他质疑布卢斯所使用的全蛋分析法。他认为，布卢斯的数据所表现出来的关系并非因果关系，他呼吁不要根据这些"错误"的结论而禁用杀虫剂。他说："科学上绝对不允许隐瞒对立数据，也不能采取'只要目的正当就可以不择手段'的处世态度。"尽管如此，在20世纪七八十年代，大多数发达国家仍然禁止了DDT在农业生产上的使用。

我对布卢斯等人[1,2]提出的关于DDE可能会引起褐鹈鹕蛋蛋壳薄化的观点有四点不同意见。

第一，他文章中的图1是基于三个鸟群聚集地中两个亚种的残留物和蛋壳厚度的数据点绘制而成的。如果这种整体相关性是有意义的，那么每个聚集地本身都应该表现出类似的趋势，但实际情况并非如此。没有哪个单独的聚集地表现出明确的自身趋势。

第二，在图1中的95%置信区间内，可能会有一系列平行于横坐标的蛋壳厚度测定值。换句话说，他的95%置信区间可以包括一系列残留量从零到最大而蛋壳厚度不变的鸟蛋。这说明回归所用的数据带有很大的不确定性。

第三，全蛋分析不适合孵化蛋，因为残留物是在蛋黄中被发现的，在孵化期间蛋黄中的物质（包括残留物）会发生代谢变化。我在下文中将证明DDE残留物在褐鹈鹕蛋的孵化过程中发生了代谢变化。从布卢斯已发表的文章中我们无法确定他所使用的蛋的孵化程度，但从DDT公开听证会上展示的一张表格（展品R-74）中可以看出：布卢斯所使用的南卡罗来纳州鸟蛋和佛罗里达州鸟蛋中含有胚胎；甚至可以从其中的两枚蛋中"听到吱吱声"。

第四，如果不是双变量的正态分布就不适合使用皮尔逊积矩相关系数（r）。对残留物浓度取对数的结果说明：残留量并没有在某个平均值附近呈正态分布。应该

mean value. Non-parametric statistics should be used[3].

To understand that Blus's claims about pelican eggs and residues may not be correct, we need to look at some unpublished analyses provided by G. R. Arnett, director of the California Department of Fish and Game (CDFG). These findings were from eggs collected on Anacapa Island off the California coast. Four of these eggs were part of 25, collected on April 18, 1969 (ref. 4). Mr. Blus's sample of 10 was from the same collection (11 eggs are not accounted for). A sample of May 27, 1969, represents eggs taken when adults were shot off their nests on Anacapa Island by State and Federal researchers[5]. The female with the higher residue of DDE had the thicker shelled egg of the two females collected.

The data of Table 1 show the following points. 1, Lipid levels in the yolk drop nearly ten-fold during incubation (26.9% to 2.8%). 2, Yolk plus embryo weight increases about three-fold (13.6 to 49.7 g) during incubation. 3, Whole egg residues of DDE seem to fit a similar ten-fold level of decrease with incubation (357.5 to 29.4 p.p.m.). 4, Whole egg DDE residues in variously incubated eggs therefore should not be expected to correlate with shell thickness, even if a correlation existed in newly laid eggs.

Table 1. Anacapa Island (California) Brown Pelican Egg Data, supplied by the California Department of Fish and Game

Date collected	Weight yolk (g)	% lipid in yolk	Shell thickness (mm)	Incubated	Residues in μg/g of sample (p.p.m. whole yolk)					Calculated p,p'DDE lipid basis (p.p.m.)
					p,p'DDE	p,p'DDMU	p,p'DDD	p,p'DDT	o,p'DDT	
18/4/69	15.1	21.8	0.42	no	357.5	20.9	8.3	6.9	4.3	1,640
	38.7	6.3	0.38	yes	61.1	4.1	1.6	2.0	1.2	970
	15.3	23.7	0.37	no	204.4	20.3	11.4	10.5	5.0	860
	17.5	22.9	0.37	no	232.1	30.3	17.2	14.7	10.5	1,010
27/5/69	13.6	26.2	0.36	no	178.4	13.0	8.9	7.5	–	680
	16.8	26.9	0.34	no	183.0	13.4	9.1	8.9	–	680
	47.9	4.4	0.40	yes	55.8	2.4	1.1	1.4	–	1,270
	25.3	12.2	0.40	yes	143.4	7.4	3.0	2.9	–	1,172
	49.7	2.8	0.38	yes	29.4	3.3	2.2	1.7	–	1,045

Eggs submitted for analysis by James Keith, US Department of Interior, Bureau of Sport, Fishery and Wildlife.

Dr. E. H. Dustman (personal communication) supplied the data used to establish the points in Fig. 1 of Blus's paper[1]. These data had one obvious typographical error for Anacapa Island, which was corrected. Using Blus's data and those from Table 1, Spearman's rank values were calculated[6] for brown pelican eggs (Table 2).

使用非参数统计方法[3]。

为了说明布卢斯对鹈鹕蛋和残留物的论述有可能是错误的，我们需要回顾一下美国加州渔猎部（CDFG）主任阿内特提供的一些尚未发表的分析结果。该研究使用的鸟蛋采集自加州海岸的阿纳卡帕岛。在全部 25 枚鸟蛋中，有 4 枚是在 1969 年 4 月 18 日采集的（参考文献 4）。布卢斯先生的鸟蛋样品中有 10 枚也是从这个地方采集的（剩余 11 枚鸟蛋未做说明）。其中一个 1969 年 5 月 27 日的样品是由州和联邦研究人员在阿纳卡帕岛将成鸟从巢中打落后取得的数枚鸟蛋中的一枚[5]。从捉到的两只雌鸟的情况来看，DDE 残留量较高的雌鸟会产出蛋壳厚度更厚的蛋。

表 1 中的数据可以说明以下几点：1. 在孵化期间蛋黄中脂类成分的含量降低至约为孵化前含量的 1/10（从 26.9%降低到 2.8%）；2. 在孵化期间蛋黄加胚胎的重量增加了大约 3 倍（从 13.6 g 增加到 49.7 g）；3. 在孵化期间 DDE 全蛋残留似乎降至原来的 1/10（从 357.5 ppm 降低到 29.4 ppm）；4. 因此在孵化程度不同的蛋中，DDE 全蛋残留与蛋壳厚度之间不应该存在相关性，即使相关也只能发生在刚产下的蛋中。

表 1. 由加州渔猎部提供的来自阿纳卡帕岛（加利福尼亚州）的褐鹈鹕蛋数据

采集日期	蛋黄重量 (g)	脂类成分在蛋黄中所占的%	蛋壳厚度 (mm)	是否已孵化	每 g 样品中以 μg 表示的残留物含量（ppm 全蛋黄）					脂类成分中 p,p′DDE 含量的计算值 (ppm)
					p,p′DDE	p,p′DDMU	p,p′DDD	p,p′DDT	o,p′DDT	
1969 年 4 月 18 日	15.1	21.8	0.42	否	357.5	20.9	8.3	6.9	4.3	1,640
	38.7	6.3	0.38	是	61.1	4.1	1.6	2.0	1.2	970
	15.3	23.7	0.37	否	204.4	20.3	11.4	10.5	5.0	860
	17.5	22.9	0.37	否	232.1	30.3	17.2	14.7	10.5	1,010
1969 年 5 月 27 日	13.6	26.2	0.36	否	178.4	13.0	8.9	7.5	–	680
	16.8	26.9	0.34	否	183.0	13.4	9.1	8.9	–	680
	47.9	4.4	0.40	是	55.8	2.4	1.1	1.4	–	1,270
	25.3	12.2	0.40	是	143.4	7.4	3.0	2.9	–	1,172
	49.7	2.8	0.38	是	29.4	3.3	2.2	1.7	–	1,045

用于分析的鸟蛋是由美国内政部渔业和野生动物体育局的詹姆斯 · 基思提供的。

达斯特曼博士（个人交流）提供了用于绘制布卢斯文章[1]中图 1 的数据。在这些数据中，有一个来自阿纳卡帕岛的数据存在明显的印刷错误，现在已经被纠正。由布卢斯的数据和表 1 中的数据可以计算出褐鹈鹕蛋的斯皮尔曼等级系数[6]（表 2）。

Table 2. Spearman's Rank Values for Blus's and Arnett's Data

Whole egg DDE × shell thickness	
Blus's data, South Carolina, n=21,	r_s = −0.0964 NS
Blus's data, Anacapa Is., n=10,	r_s = +0.1454 NS (thickest shells × highest DDE)
	r_s = +0.393 NS (thinnest shells × highest DDE)
CDFG, Anacapa Is., n=9,	r_s = +0.4771 NS
Blus+CDFG, Anacapa Is., n=19,	r_s = −0.0548 NS
Lipid basis DDE × shell thickness	
CDFG, Anacapa Is., n=9,	r_s = +0.9834 highly significant, nearly perfect (as residues increased, shells were thicker)

From these data, I conclude that the statement of Blus *et al.*[1], "A concentration-effect relationship seems to exist between DDE in eggs and shell thinning", is correct, but not in the way Blus *et al.* intended. Whole egg residues of incubated eggs seem to be valueless in determining a cause and effect relationship. The CDFG data (Table 1) show a nearly perfect correlation of lipid DDE residues to shell thickness, and the relationship is positive. Even though thicker shells are associated with higher residues, I do not believe that DDE causes shell thickening in brown pelican eggs. The data appear to completely refute that DDE causes thinning of pelican eggs. If DDE were causally related to shell thinning, it would be correlated in every sample and other factors should be investigated to find the cause of the thin shelled eggs observed in this species.

Blus *et al.* mention peregrine falcon egg residues, suggesting that cause and effect studies were conducted with this species. Arguments for the cause of collapse in peregrine falcon populations in the Arctic may[8] suffer from the same problems as the work of Blus *et al.* Three populations of varying reproductive success are considered together, unidentified, and whole egg DDE residues are used to determine a regression coefficient (r) for residues correlated to eggshell values. There is no mention of stage of embryo development, lipid level or which of the three populations provide the points in the figure showing the regression line. There are further problems of non-uniform methods of egg content preservation used in this study. More important is evidence that the available eggs were as thin in 1952 and 1964 as they were in 1970 when the alleged collapse occurred. Scientific persecution appears to be just as viable a hypothesis as DDE or other causes for problems of reproduction failure in these northern peregrine populations.

In order to check further, the lipid residues given by Risebrough[7] for Anacapa Island eggs were evaluated by Spearman's rank (Table 3).

Table 3. Spearman's Rank Values from Risebrough's Data

Lipid basis residue × shell thickness, Anacapa Is.	
DDE; n=65,	r_s=−0.4318 highly significant
PCB; n=65,	r_s=−0.1361 NS

Risebrough gives r values for DDE as −0.5605 (Sig<0.001) and PCB as −0.2527 (Sig<0.05).

表 2. 由布卢斯的数据和阿内特的数据计算出的斯皮尔曼等级系数

全蛋中的 DDE × 蛋壳厚度	
布卢斯的数据，南卡罗来纳州，n=21，	r_S = –0.0964 无显著性差异
布卢斯的数据，阿纳卡帕岛，n=10，	r_S = +0.1454 无显著性差异（蛋壳厚度最大值 ×DDE 最高值）
	r_S = +0.393 无显著性差异（蛋壳厚度最小值 ×DDE 最高值）
CDFG，阿纳卡帕岛，n=9，	r_S = +0.4771 无显著性差异
布卢斯 + CDFG，阿纳卡帕岛，n=19，	r_S = –0.0548 无显著性差异
脂类成分中的 DDE × 蛋壳厚度	
CDFG，阿纳卡帕岛，n=9，	r_S = +0.9834 非常显著，近乎完美（蛋壳随残留物的增加而变厚）

根据这些数据，我发现布卢斯等人[1]关于“鸟蛋中 DDE 残留与蛋壳薄化之间似乎存在着浓度 – 效应关系”的论断是正确的，但并不是以布卢斯等人预想的方式呈现。用孵化蛋的全蛋残留数据来确定因果关系似乎是没有意义的。CDFG 的数据（表 1）表明脂类成分中 DDE 残留与蛋壳厚度之间存在着一种近乎完美的相关性，而且是正相关。尽管较厚的蛋壳与较高的残留量相关联，但我并不认为 DDE 残留会导致褐鹈鹕蛋的蛋壳变厚。这些数据彻底驳斥了 DDE 残留会使褐鹈鹕蛋蛋壳变薄的观点。如果 DDE 残留与蛋壳薄化之间存在着因果关系，那么在所有样本中都应该存在这样的关系，我们应该调查其他一些因素来找出导致该物种中出现蛋壳薄化的原因。

布卢斯等人提到了游隼蛋中的残留物，说明他们也对这个物种进行过因果关系的研究。某些人提出的有关北极地区游隼种群数量下降原因的论据[8]也存在与布卢斯等人的研究一样的问题。不加区分地同时考虑三个繁殖成功率不同的种群，用全蛋 DDE 残留确定残留量相对于蛋壳厚度的回归系数（r）。没有提到胚胎发育阶段、脂类水平以及在回归线中的点到底来自三个种群中的哪一个。进一步的问题包括没有用同样的方法来保存该项研究中所用的蛋的内容物。更重要的是：在可用于研究的鸟蛋中，1952 年和 1964 年的鸟蛋与 1970 年（也就是所谓发生种群数量下降的年份）的鸟蛋都显示出同样的蛋壳厚度。这种 DDE 假说或者其他一些用于解释北部游隼种群繁殖出现障碍的假说似乎正成为一种现实存在的科学迫害。

为了进行进一步核对，我用斯皮尔曼等级系数评估了由赖斯布拉夫[7]提供的有关阿纳卡帕岛鸟蛋脂类成分中残余物的数据（表 3）。

表 3. 根据赖斯布拉夫的数据得到的斯皮尔曼等级系数

脂类成分中的残留物 × 蛋壳厚度，阿纳卡帕岛	
DDE；n=65，	r_S = –0.4318 非常显著
PCB；n=65，	r_S = –0.1361 无显著性差异

赖斯布拉夫给出的 r 值是：对于 DDE 为 –0.5605（统计显著性 <0.001），对于 PCB（译者注：多氯联苯）为 –0.2527（统计显著性 <0.05）。

The data necessary to arrive at an understanding of pesticide residue levels, rapid residue loss rates in eggs with incubation, metabolism of lipids and other pertinent factors should have been published, along with the original contentions and conclusions. That DDE is the cause of thin brown pelican or peregrine eggs is well established in the popular press and scientific literature, but the underlying data to test the conclusion are just now becoming available and do not support such a conclusion. This is reason for concern about actions to ban pesticides based on these faulty conclusions. There is no place in science for suppressed conflicting data, or an "end justifies the means" philosophy. Scientists can be responsible agents for change only when they consider and present all the available data which bear on their conclusion.

(**239**, 410-411; 1972)

William Hazeltine: 26 Rosita Way, Oroville, California 95965.

Received July 10, 1972.

References:

1. Blus, L. J., Gish, C. D., Belisle, A. A., and Prouty, R. M., *Nature*, **235**, 376 (1972).

2. Blus, L. J., Heath, R. G., Gish, C. D., Belisle, A. A., and Prouty, R. M., *Bio-Science*, **21** (**24**), 1213 (1971).

3. Risebrough, R. J., Cross-examination at Public Hearing on DDT, Washington, DC, transcript pages 8473–8475 (Jan. 7, 1972).

4. Risebrough, R. W., Sibley, F. C., and Kirven, M. N., *American Birds*, **25**(**1**), 8 (1971).

5. Keith, J. O., Woods, L. A., and Hunt, E. G., *Trans. North American Wildlife Conference*, **35**, 56 (1970).

6. Lathrop, R. G., *Introduction to Psychological Research* (Harper and Row, London, 1969).

7. Risebrough, R. W., *6th Berkeley Symposium on Mathematical Statistics and Probability*, MS as exhibit Int. EDF 29 at Public Hearing on DDT, Washington, DC (in the press).

8. Cade, T. J., Lincer, J. L., White, C. M., Roseneau, D. G., and Swarts, L. G., *Science*, **172**, 955 (1971).

在提出新论点和结论时，应该同步发表一些能说明杀虫剂残留水平、孵化蛋中残留物快速损失速率、脂类代谢以及其他相关因素的必要数据。大众媒体和科学文献已经在公众中确立了 DDE 能引起褐鹈鹕蛋或者游隼蛋蛋壳薄化的观念，但是能检验这一结论的基础数据才刚刚发表，并且这些数据并不支持这一结论。这就是我担心人们基于这些错误结论而禁用杀虫剂的理由。科学上绝对不允许隐瞒对立数据，也不能采取“只要目的正当就可以不择手段”的处世态度。只有在考虑并提供所有支持其结论的有效数据的情况下，科学家们才能成为改变世界的可依赖动力。

（吕静 翻译；周江 审稿）

Use of "Whole Egg Residues" in Pesticide/ Eggshell Studies

K. H. Lakhani

Editor's Note

A year earlier, William Hazeltine had criticized the idea that brown pelican eggshell thinning was caused by DDE, on the basis that the underlying studies relied on the inappropriate analysis of whole egg residues. DDE residues, he argued, are found in the yolk, where they are metabolised during incubation. Here K. H. Lakhani criticizes Hazeltine's methodology and argues that the original data linking eggshell thinning with DDE are likely to be "even more significant than suggested by the probability value quoted." The case for the harmful effects of DDT eventually became strong enough to motivate a ban in many countries.

HAZELTINE[1] questions the well established belief that DDE probably causes eggshell thinning in brown pelicans chiefly by finding four points of disagreement with the proposal by Blus *et al.*[2] that a concentration-effect relationship seems to exist between DDE in eggs and shell thickness. In their subsequent reply, Blus *et al.*[3] refute Hazeltine's criticisms. There is a further argument against Hazeltine's third criticism of Blus *et al.*'s conclusions; and the argument has an important bearing on the interpretation of analyses based on "whole egg residues".

Hazeltine expresses the view that the "whole egg residues" are valueless in pesticide/ eggshell studies of incubated eggs on the basis of his assumption that the residues are metabolized during incubation. He provides no data for residues expressed on a whole egg (fresh weight) basis, and gives no acceptable evidence to show that DDE or other residues are metabolized (that is, broken down chemically), as distinct from mobilized. Even if we accept that the residues are indeed metabolized during incubation, this does not necessarily make the use of the whole egg residues inappropriate. If DDE residues are metabolized during incubation, then the whole egg residues for incubated eggs will tend to be low. Bearing in mind the thinning of eggshells by withdrawal of calcium in embryonic development during incubation, it is obvious that in the presence of DDE metabolism during incubation, the use of the whole egg residues for pesticide/eggshell studies would provide a conservative test if the correlation between the eggshell thickness and the levels of residue is asserted to be negative. Blus *et al.*'s "highly significant regression ($P<0.01$)" with a negative slope is then likely to be even more significant than suggested by the probability value quoted.

(**242**, 340-341; 1973)

关于在杀虫剂/蛋壳研究中使用“全蛋残留数据”

拉卡尼

编者按

一年以前，威廉·黑兹尔坦批驳了DDE引起褐鹈鹕蛋蛋壳薄化的观点，他的根据是：基于此的研究建立在了不恰当分析全蛋残留数据的基础之上。他指出：DDE残留物是在蛋黄中被发现的，而蛋黄在孵化期间会发生代谢变化。在本文中，拉卡尼对黑兹尔坦的分析方法提出了批评，他认为证明蛋壳薄化与DDE相关的原始数据很可能“比黑兹尔坦列出的概率值更加显著”。后来，DDT的有害效应案例严重到了促使很多国家颁布禁令的程度。

目前普遍认为DDE可能就是造成褐鹈鹕蛋蛋壳薄化的元凶，但黑兹尔坦[1]对这一观点表示怀疑，他尤其针对布卢斯等人[2]关于鸟蛋中DDE残留物与蛋壳厚度之间存在浓度–效应关系的结论提出了四点不同意见。布卢斯等人[3]在随后的回复中驳斥了黑兹尔坦的批评意见。本文将进一步反驳黑兹尔坦对布卢斯等人的结论提出的第三点批评意见；而这一见解对于解释基于“全蛋残留数据”的分析有着重要的意义。

黑兹尔坦表示，在孵化蛋的杀虫剂/蛋壳研究中使用“全蛋残留数据”是没有意义的，他的根据是残留物在孵化过程中会发生代谢。然而，他既没有提供以一枚全蛋（鲜重）为基础的残留物数据，也没有给出任何令人信服的证据来证明DDE或者其他残留物发生了代谢（即化学分解）而不是迁移。即使我们承认残留物在孵化过程中确实发生了代谢，也不见得不能使用全蛋残留数据。如果DDE残留物在孵化过程中发生了代谢，那么孵化蛋的全蛋残留数值将会非常低。别忘了在孵化期间的胚胎发育阶段会因为钙损失而造成蛋壳薄化，显然当孵化过程中确实存在DDE代谢时，在杀虫剂/蛋壳研究中使用全蛋残留数据就为我们提供了一种保守的方式来验证蛋壳厚度和残留物水平之间是否是负相关。因而布卢斯等人得到的斜率为负的“非常显著的回归结果（$P<0.01$）”很可能比黑兹尔坦列出的概率值更加显著。

（吕静 翻译；周江 审稿）

K. H. Lakhani: Natural Environment Research Council, The Nature Conservancy, 19-20 Belgrave Square, London SW1.

Received November 13; revised December 18, 1972.

References:

1. Hazeltine, W., *Nature*, **239**, 410 (1972).

2. Blus, L. J., Gish, C. D., Belisle, A. A., and Prouty, R. M., *Nature*, **235**, 376 (1972).

3. Blus, L. J., Gish, C. D., Belisle, A. A., and Prouty, R. M., *Nature*, **240**, 164 (1972).

DDE in Eggs and Embryos of Brown Pelicans

I. C. T. Nisbet

Editor's Note

Here Ian Nisbet criticizes William Hazeltine's preceding dismissal of the link between brown pelican eggshell thinning and the DDT metabolite, DDE. The association of eggshell thinning in birds with DDE is, he argues, "one of the best known of all environmental phenomena," and points out that Hazeltine's sample was "too small and homogeneous to demonstrate anything." The link between eggshell thinning and DDE had been shown so many times in both wild and captive species that at least one journal had complained about receiving further "verifications of phenomena already fully demonstrated." He concludes scornfully that Hazeltine's data "add nothing to previous knowledge except for his demonstration that yolks grow larger during incubation."

HAZELTINE[1] claims to have demonstrated that DDE residues are metabolized by developing embryos of brown pelicans (*Pelecanus occidentalis*). This conflicts with experimental work[2] on Japanese quail (*Coturnix coturnix*), in which incubation did not seem to affect total residue levels in eggs, except for some conversion of *p,p'*-DDT to *p,p'*-DDE.

Hazeltine's claim seems to be based on a comparison of DDE levels in the yolks of four incubated and five fresh eggs. As he made clear in the text, the column labelled "Weight yolk" in his Table 1 actually refers to the yolk plus embryo. The apparent 3.2-fold decrease in the mean concentration of DDE in "yolk" simply reflects the 2.6-fold increase in the weight of the yolk + embryo, as the embryos absorbed material from the white. In Hazeltine's sample the whole-egg residues of DDE were slightly smaller in the incubated eggs (mean 2.53 mg) than in the fresh eggs (3.62 mg), but the difference is not statistically significant and in any case this does not demonstrate metabolism of DDE. It is more simply explained by the fact that eggs with higher levels of DDE are broken more frequently during incubation, so that they are less frequently available for collection: this differential breakage has been demonstrated not only in the brown pelican[3], but also in other species[4-7]. For the same reason[3-7] samples of incubated eggs are expected to have thicker shells, on the average, than fresh eggs, as in Hazeltine's sample (means 0.390 mm and 0.372 mm, respectively).

Hazeltine also reported that eggshell thickness was positively correlated with DDE concentration in the lipids of the "yolk". But the latter is not an appropriate measure of the DDE levels in the eggs, because it increases during incubation as the egg lipids are reduced by metabolism[8] (from 3.80 g to 2.26 g in Hazeltine's sample). The most appropriate measures are the whole-egg residue or whole-egg concentration, which

褐鹈鹕蛋和胚胎中的DDE

尼斯比特

编者按

在本文中，伊恩·尼斯比特驳斥了威廉·黑兹尔坦先前提出的关于褐鹈鹕蛋蛋壳薄化与DDT代谢产物——DDE之间并不存在关联的论点。尼斯比特认为鸟类蛋壳薄化与DDE之间的关系是“所有环境现象中最为著名的现象之一”，并且指出黑兹尔坦的样本“范围太小并且均一，根本说明不了什么问题”。蛋壳薄化与DDE之间的关系已经在野生物种和圈养物种中多次被发现，以至于至少有一家杂志开始抱怨说又收到了“已完全得到证实的现象的证据”。在文章最后，他嘲讽黑兹尔坦得到的数据“除了能说明蛋黄在孵化过程中会长大以外，并没有给出任何新的认识”。

黑兹尔坦[1]声称他已经证明DDE残留物在褐鹈鹕的胚胎发育中发生了代谢。这与用日本鹌鹑得到的实验结果[2]并不一致，日本鹌鹑蛋中的残留物水平似乎并没有受到孵化过程的影响，除了有一部分*p,p′*–DDT转化成*p,p′*–DDE以外。

黑兹尔坦的结论似乎是根据对4枚孵化蛋蛋黄和5枚鲜蛋蛋黄中DDE水平的对比结果得到的。正如他在发表的文章中所述：表1中标有“蛋黄重量”的那一列实际上指的是蛋黄加上胚胎的重量。“蛋黄”中DDE平均浓度显著降低至原来的1/3.2，这只能说明蛋黄+胚胎的重量增加至2.6倍，因为胚胎会从蛋白中吸收物质。在黑兹尔坦的样品中，孵化蛋的全蛋DDE残留物（平均值为2.53 mg）比鲜蛋的全蛋DDE残留物（3.62 mg）略微少一些，但是这种差别在统计学上不是特别显著，而且无论如何也不能由此证明DDE发生了代谢。用以下事实来解释会更为简单：DDE残留量较高的鸟蛋在孵化过程中更容易破碎，所以很少能收集到这样的蛋。这种易碎程度有差别的现象不仅存在于褐鹈鹕中[3]，在其他鸟类中也有类似情况[4-7]。由于同样的原因[3-7]，孵化蛋样品的蛋壳平均厚度通常会高于鲜蛋，黑兹尔坦的样品也不例外（平均厚度分别为0.390 mm和0.372 mm）。

黑兹尔坦还指出：蛋壳厚度与“蛋黄”脂类成分中的DDE浓度呈正相关。但是后者不能正确地反映出蛋中的DDE含量，因为在孵化过程中DDE含量会随着蛋中脂类的代谢分解（黑兹尔坦的样品是从3.80 g降到了2.26 g）而逐渐增加[8]。最适合的测量方式是全蛋残留量或者全蛋浓度，这两个参数可以反映雌鸟在产蛋时的

reflect the levels of DDE circulating in the female at the time of laying[9-11]. In Hazeltine's sample, the Spearman rank correlation coefficient r_s between whole-egg residue of DDE and eggshell thickness is +0.244 (not +0.477 as stated in Hazeltine's Table 2). This is not statistically significant, which is not surprising in a small sample with a very small range in eggshell thickness: but the points fall close to the regression lines of Risebrough[12] and Blus *et al.*[13]. As Hazeltine points out (his Table 3), in Risebrough's larger sample the correlation is negative and highly significant (r_s=−0.4318, P<0.01).

Hazeltine also stated that non-parametric statistics should have been used by Blus *et al.*[13] to test the relationship between eggshell thickness and DDE residues, because these variables were "not bi-variantly normally distributed". This would be valid if the goal of the study had been simply to demonstrate association of eggshell thinning with DDE, but Blus *et al.*[13] were attempting to define the dose-response relation. Parametric regression techniques require only that the dependent variable be random and normally distributed about the regression line[14]. This condition was satisfied by the brown pelican eggshell data of both Risebrough[12] and Blus *et al.*[13], and both found a good fit to a logarithmic dose-response relation. A logarithmic relation between eggshell thickness and DDE residues in eggs has also been reported in wild peregrine falcons (*Falco peregrinus*)[4,15] and in experimental mallards (*Anas platyrhynchos*)[16]. The advantage of applying a logarithmic transformation to the DDE variable is that it is then possible to use multiple linear regression techniques to separate the effects of co-existing pollutants. This was done by both Blus *et al.*[17,18] and Risebrough[12], who found independently that most of the variance in eggshell thickness was explained by the log DDE and DDE variables, and no significant fraction by any other chemicals.

Hazeltine also criticized Blus *et al.* for combining data from brown pelicans of two different subspecies. Blus *et al.* corrected for the known difference between the subspecies by using as dependent variable the eggshell thickness expressed as a percentage of the pre-1947 mean for the subspecies. In any case this criticism cannot be applied to Risebrough's data, which were all drawn from Pacific Coast populations[12].

The association of eggshell thinning in birds with *p,p'*-DDE is one of the best known of all environmental phenomena[19]. It has been shown so many times in both wild and captive species[3-7,12,13,16-18] that at least one journal has complained about receiving further "verifications of phenomena already fully demonstrated"[20]. Scientific work in this area is now devoted to elucidating details such as the form of the dose-response relation and the reasons for the known interspecific differences in sensitivity. The brown pelican is one of the species in which the phenomenon has been explored most fully[3,12,13,17,18,21,22]. It is therefore somewhat odd to find another letter[1] discussing whether the association exists, based on a sample too small and homogeneous to demonstrate anything. In fact, Hazeltine's nine eggs fall exactly into the pattern of eggshell thinning known from previous work[12,13,22], and his data add nothing to previous knowledge except for his demonstration that "yolks" grow larger during incubation.

(**242**, 341; 1973)

DDE 循环水平[9-11]。在黑兹尔坦的样品中，全蛋 DDE 残留和蛋壳厚度之间的斯皮尔曼等级相关系数 r_s 为 +0.244（而非黑兹尔坦文章中表 2 所列出的 +0.477）。这在统计学上不具有显著性，在蛋壳厚度范围非常狭小的小样本中出现这样的结果并不令人惊奇；而这些点都落在赖斯布拉夫[12]和布卢斯等人[13]的回归线附近。正如黑兹尔坦所指出的（他文章中的表 3），在赖斯布拉夫采用更大样本进行分析时，相关性呈负相关，并且非常显著（r_s = –0.4318，$P < 0.01$）。

黑兹尔坦还指出：布卢斯等人[13]应该使用非参数统计方法来检测蛋壳厚度与 DDE 残留物之间的关系，因为这些变量不是“双变量的正态分布”。如果研究目的只是为了证明蛋壳薄化与 DDE 有关，那么使用非参数统计应该没有问题，但布卢斯等人[13]的目标是要确定剂量–反应关系。参数回归方法仅需要因变量是随机的且在回归线附近呈正态分布[14]。赖斯布拉夫[12]和布卢斯等人[13]的褐鹈鹕蛋壳数据都可以满足这些条件，并且他们都发现剂量对数与反应之间符合很好的拟合关系。还有人报道在野生游隼[4,15]和实验绿头鸭[16]的蛋壳厚度与 DDE 残留物之间也存在对数关系。对 DDE 变量进行对数变换的好处在于随后可以使用各种多元线性回归技术以排除共存污染物的干扰。布卢斯等人[17,18]和赖斯布拉夫[12]都是这样做的，他们各自独立地发现用 log DDE 和 DDE 变量可以解释大多数蛋壳厚度的变化，而任何其他化学物质都无法作为显著的因子。

黑兹尔坦还批评布卢斯等人混合了来自两种不同褐鹈鹕亚种的数据。布卢斯等人曾通过以下方式校正过两个亚种之间的已知差异——用现在的蛋壳厚度占 1947 年前几个亚种蛋壳厚度平均值的百分数作为因变量。这种批评无论如何不适用于赖斯布拉夫的数据，因为赖斯布拉夫的数据全部来自太平洋沿岸的种群[12]。

鸟蛋蛋壳薄化与 p,p'–DDE 相关是所有环境现象中最为著名的现象之一[19]。这一现象已经在野生物种和圈养物种中多次被发现[3-7,12,13,16-18]，以至于至少有一家杂志开始抱怨说又收到了“已完全得到证实的现象的证据”[20]。目前这一领域的科学研究集中在解释一些细节问题上，如剂量–反应关系的形式以及种间存在敏感度差异的原因。褐鹈鹕是人们对上述现象研究得最全面的物种之一[3,12,13,17,18,21,22]。因此，当我发现有一篇快报[1]还在讨论这种关系是否存在时，确实感到有点奇怪，文章中所依据的样本范围太小并且均一，根本说明不了什么问题。实际上，黑兹尔坦的 9 枚鸟蛋也完全落在前面研究[12,13,22]得到的蛋壳薄化分布之内，他的数据除了能说明“蛋黄”在孵化过程中会长大以外，并没有给出任何新的认识。

（吕静 翻译；周江 审稿）

I. C. T. Nisbet: Massachusetts Audubon Society, Lincoln, Massachusetts 01773, and Environmental Defense Fund, 1712 N Street, Washington DC 20036.

Received November 24, 1972.

References:

1. Hazeltine, W., *Nature*, **239**, 410 (1972).
2. Cooke, A. S., *Pest. Sci.*, **2**, 144 (1971).
3. Keith, J. O., Woods, L. A., jun., and Hunt, E. G., *Proc. 35th N. Amer. Wildl. Conf.*, 56 (1970).
4. Ratcliffe, D. A., *J. Appl. Ecol.*, 7, 67 (1970).
5. Hickey, J. J., and Anderson, D. W., *Science*, **162**, 271 (1968).
6. Longcore, J., Samson, F. B., and Whittendale, T. W., jun., *Bull. Environ. Contam. Toxicol.*, **6**, 485 (1971).
7. Switzer, B., Lewin, V., and Wolfe, F. H., MS presented as exhibit USDA-RBTL-9 at public hearings on DDT (Environmental Protection Agency, Washington DC, 1972).
8. Romanoff, A. L., and Romanoff, A. J., *The Avian Egg* (Wiley, New York, 1949).
9. Cummings, J. G., Eidelman, M., Turner, V., Reed, D., Zee, K. T., and Cook, R. E., *J. Assoc. Off. Agric. Chem.*, **50**, 418 (1967).
10. Cummings, J. G., Zee, K. T., Turner, V., and Quinn, F., *J. Assoc. Off. Agric. Chem.*, **49**, 354 (1966).
11. Noakes, D. N., and Benfield, C. A., *J. Sci. Food Agric.*, **16**, 393 (1965).
12. Risebrough, R. W., *Proc. 6th Berkeley Symposium on Mathematical Statistics and Probability* (in the press).
13. Blus, L. J., Gish, C. D., Belisle, A. A., and Prouty, R. M., *Nature*, **235**, 376 (1972).
14. Steel, R. G. D., and Torrie, J. H., *Principles and Procedures of Statistics*, 187 (McGraw-Hill, New York, 1960).
15. Cade, T. J., Lincer, J. L., White, C. M., Roseneau, D. G., and Swarts, L. G., *Science*, **172**, 955 (1971).
16. Heath, R. G., MS presented as exhibit R-128 at public hearings on DDT (Environmental Protection Agency, Washington DC, 1972).
17. Blus, L. J., Heath, R. G., Gish, C. D., Belisle, A. A., and Prouty, R. M., *BioScience*, **21**, 1213 (1971).
18. Blus, L. J., MSS presented as exhibits R-70 and R-71 at public hearings on DDT (Environmental Protection Agency, Washington DC, 1972).
19. Peakall, D. B., *Bird Study*, **18**, 47 (1971).
20. Behnke, J. A., *BioScience*, **22**, 73 (1972).
21. Schreiber, R. W., and Risebrough, R. W., *Wilson Bull.*, **80**, 119 (1972).
22. Blus, L. J., *BioScience*, **20**, 867 (1970).

Problems Still with Scrapie Agent

Editor's Note

In this leading article, *Nature* drew attention to other diseases with similarities to scrapie, notably Creutzfeld–Jacob disease. The sequel, now widely influential in agriculture, is the recognition in 1986 that a disease in British cattle called bovine spongiform encephalopathy was the cow's equivalent of scrapie. Within five years it had been established that human beings were at risk from eating infected beef, and the European trade in cattle was disrupted for several years afterwards. The cause of scrapie and the other diseases has been linked to the genetics of a protein present in all animals called the prion protein, which is surmised to be capable of existing in more than one form, at least one of which coagulates in nerve cells, leading to severe encephalitis and death.

IN working with the class of agents which cause scrapie, transmissible mink encephalopathy, kuru and Creutzfeld–Jacob disease there have been two recurrent problems. These are the difficulty of distinguishing primary from secondary events in the disease and the need to use whole-animal assays, which last many months or even years, in order to detect and titrate the infective agents.

Scrapie agent has now replicated in a cell culture line for more than 100 tissue culture passages, but without any signs of its presence, so this has not provided any basis for a quicker assay of infectivity. Equally frustrating have been the attempts to develop immune assays specific for infectivity, although Gardiner showed six years ago that scrapie tissues have increased antigenicity. The average concentration of infective units of scrapie in brain is ultimately higher than elsewhere but only reaches about the same order as the number of brain cells, and it has not been possible to purify and concentrate the agent. This is generally regarded as one reason why many workers have failed to detect a specific immune response, for the "antigen" has been a more or less crude preparation of tissues from affected hosts. Such tissues presumably contain many antigens associated with the chronic degenerative changes in the brain and, even with the infective tissues outside the central nervous system, the failure to observe tissue damage may simply reflect the insensitivity of present techniques. On the biochemical side, the inability to differentiate between the primary aspects of agent replication and the secondary tissue changes has produced a decade of studies, the results of which are mostly equivocal.

Within a month, two entirely different methods have been published which may reduce to a few days the time needed for assays in this context. It is therefore opportune to discuss their likely merits in the light of the types of difficulty which have beset work with scrapie. Carp, Licursi, Merz and Merz[1] used both multiple sclerosis and scrapie, and they report

关于瘙痒症病原体的待解决问题

编者按

这篇发表于《自然》杂志上的社论将人们的注意力吸引到了类似于瘙痒症的其他疾病，尤其是克罗伊茨费尔特－雅各布病上。继之而来的是人们在1986年认识到英国牛得的一种被称为牛海绵状脑病（译者注：俗称疯牛病）的疾病就相当于牛瘙痒症，这对当前的农业造成了广泛的影响。人们仅用5年时间就证实人类食用带有病毒的牛肉会有被感染的危险，因而欧洲的牛贸易在此后数年内遭到了重创。瘙痒症及其他疾病的病因被认为与一种存在于所有动物体内的蛋白质的遗传有关，该蛋白质被称为朊蛋白，据推测它能够以多种形式存在，其中至少有一种会凝结在神经细胞中，从而导致严重的脑炎甚至死亡。

在研究引起瘙痒症、传染性水貂脑病、库鲁病和克罗伊茨费尔特－雅各布病这一类疾病的病原体时，总会反复遇到两个难题。一个难题是很难把疾病中的原发事件和继发事件区分开；另一个难题是为了发现和滴定传染性病原体，需要使用整个动物进行分析，这种分析花费的时间长达数月甚至数年。

瘙痒症病原体在一个培养细胞系中复制，目前已经组织培养传代100余次，但是没有任何迹象表明它的存在，所以这种方法不能为感染性的快速检测提供任何依据。尝试开发专门用于检测感染性的免疫测定法也同样遭到了挫折，尽管加德纳在六年前就已经发现瘙痒症组织具有更强的抗原性。脑内瘙痒症感染单位的平均浓度最终会超过体内其他部位，但只能达到和脑细胞数量相近的量级，因而目前还不可能纯化和富集这些病原体。大家普遍认为这就是为什么许多研究者没有检测到特异性免疫应答的一个原因，因为“抗原”一般是受感染宿主组织的粗提物。这些组织被认为含有很多与脑内慢性退行性改变相关的抗原，即使对于中枢神经系统之外的感染性组织，未能观察到组织损伤也只能说明目前技术的灵敏度低。在生物化学方面，人们为解决无法鉴别病原体复制产生的原发因素和继发性组织改变的问题已经研究了十年，但得到的结果大多是模棱两可的。

在一个月之内，研究人员报道了两种完全不同的方法，这两种方法都能使检测瘙痒症病原体所需的时间减少到几天。因此，现在正好可以探讨一下它们对解决瘙痒症研究时所遇到的难题是否会有帮助。卡普、利库尔西、默茨和默茨[1]同时研究

a replicating factor which is first detectable in mice by a reduction in the proportion of circulating polymorphs the day after injection with tissues from cases of multiple sclerosis. A similar effect was found to occur three days after injection with scrapie tissues. With both scrapie and multiple sclerosis tissues the effect persists at least for many months.

The other report is on page 104 of this issue of *Nature*. Field and Shenton have used a lymphocyte sensitization technique which differentiates efficiently between tissues from uninfected mice and those from mice affected by scrapie: the method involves a combined *in vivo* and *in vitro* test which takes about a week. It will be a real advance if both these reports fulfil the expectations which they will generate.

The most exciting possibilities come from the work of Carp and associates, because they may not only provide a rapid assay but may also bridge the gap between scrapie-like diseases and multiple sclerosis. The confidence of their report, with its modest claims for the assay, rests on the finding of a consistent difference between tissues from normal and affected subjects in the ability to produce the polymorph reduction. A number of different tissues were used, which came from several human, sheep and mouse sources. The authors are careful to state that the "virus" in the mouse assay may well be unrelated to the aetiology of multiple sclerosis, and in the case of scrapie they only conclude that it may be related to scrapie agent in some way, even though the assay virus seems to have some properties in common with scrapie agent. This is prudent because they have not yet produced evidence to fulfil the criteria which earlier workers with scrapie found necessary even before it was easy to work with scrapie in laboratory animals[2]. The high thermal stability of scrapie permits serial passage using boiled inocula each time: this eliminates from the inocula conventional viruses which could be producing misleading effects. Even though the polymorph assay may be specific for scrapie, multiple sclerosis and any related diseases, it is important to know if the effect comes from a non-causal secondary virus normally present in such diseases. Another conceivable possibility is that multiple sclerosis and scrapie tissues merely release a latent virus in the mice, which would not necessarily invalidate the assay though it would be less interesting: this alternative is perhaps excluded by the authors' statement that "the characteristics of the multiple sclerosis and scrapie factors that induce the decrease are similar but not identical". Their findings will need independent confirmation and may not prove to be an assay of the respective causal agents, but the prospects are that this approach will not incur problems of dealing with effects merely due to tissue damage.

In the case of the lymphocyte sensitization test it is to be hoped that the claim of Field and Shenton, that it "enables the presence of scrapie to be established within 8 days" rather than several months, will prove correct. As the evidence stands, however, this is too broad a generalization and other kinds of evidence are needed. The problem is to find a way of distinguishing between the agent and tissue damage antigens, and the authors are aware of this difficulty. The scrapie brain and spleen tissues so far tested presumably come from mice at an advanced stage of the disease, and chronic degenerative changes

了多发性硬化症和瘙痒症，他们利用在注射多发性硬化症动物组织一天后小鼠血液循环中多形核白细胞的比例降低首次在小鼠中检测到了复制因子。在注射瘙痒症组织三天后也会出现类似的效应。如果同时注射瘙痒症组织和多发性硬化症组织，则该效应至少能够持续数月。

另一篇报告发表在本期《自然》杂志的第 104 页。菲尔德和申顿用到了能有效鉴别未感染小鼠组织和感染瘙痒症小鼠组织的淋巴细胞致敏技术：该方法采用了体内体外联合实验，耗时大约一周。如果这两篇报告都能达到预期的目标，那么就将出现一项重大的突破。

最令人兴奋的可能性来自卡普及其同事的工作，因为他们有可能不仅提供了一种快速检测的方法，而且还在瘙痒症类疾病与多发性硬化症之间建立了某种联系。这篇对检测过程进行了适度报道的报告的可信性是基于他们发现正常组织与患病组织在导致多形核白细胞减少方面所具有的差异总是保持一致。实验中使用了来自数个人、羊和小鼠的很多种不同组织。作者很谨慎地声明在小鼠中检测到的“病毒”很可能与多发性硬化症的病因无关，而关于瘙痒症，他们只是推测它可能在某些方面与瘙痒症病原体有关，虽然检测到的病毒看似具有一些与瘙痒症病原体相同的性质。这样做是非常谨慎的，因为他们得到的证据还不符合那些早期的瘙痒症研究者在能够轻易用实验动物来研究瘙痒症之前就已经发现的必备标准[2]。瘙痒症病原体的高度耐热性使我们在每次进行连续传代培养时都可以使用煮沸的接种物，这样就可以避免由接种物本身通常会带有的病毒所导致的误导性结果。即使多形核白细胞检测法只对瘙痒症、多发性硬化症以及其他相关疾病有效，了解清楚这种作用是否源自正常存在于这类疾病中的非致病性次要病毒也是很重要的。另一种可能性是多发性硬化症组织和瘙痒症组织在小鼠中仅仅释放了一种潜伏病毒，尽管它没那么引人注目，但也不一定就能证明这种检测法是无效的。作者的一句话也许能排除这种可能，即“引起这种减少的多发性硬化症因子和瘙痒症因子的特征虽然相似但并不相同”。卡普等人的发现还需要独立的证据去证明，而且这一效应可能并不能作为检测各自致病原的方法，但其前景在于不会造成需要排除仅由组织损伤引起的效应的麻烦。

对于淋巴细胞致敏法，菲尔德和申顿声称这一方法“能够在 8 天内诊断出瘙痒症”，而不是几个月。我们希望他们的判断将被证明是正确的。但是，从支持上述论点的论据来看，这只是一个太宽泛的概括，还需要其他类型的证据。问题在于要找到一种鉴别病原体和组织损伤抗原的方法，而且这些作者也知道解决这个问题的难度。据推测目前被检测的瘙痒症脑组织和脾组织都来源于疾病晚期的小鼠，其慢性

will have been present in the brain for many weeks. It cannot be assumed that the spleens of such animals have not been chronically exposed to any changed antigens arising in the brain. If the test is eventually shown only to detect tissue damage as an index of agent concentration it will still be useful because it is "exquisitely responsive to minor changes in antigenic determinant structure" and can evidently detect more subtle degrees of damage than can at present be quickly estimated by conventional histological means. Some of these possible secondary effects could be excluded by using it for a time sequence study of spleen in the four weeks after injection, when the concentration of the agent is rapidly increasing there but long before gross signs of damage appear in the brain, or, better, to compare the new and conventional assays using the tissue culture line developed by Haig and Clarke[3].

But even in these cases the difficulty remains because of the high sensitivity of their technique, which may be detecting antigens from unseen damage. Its value will only be established when it can be shown that there is a close correlation between the titre estimates of the conventional assay and this new one in a large range of circumstances. This range will need to include various severe physical and chemical treatments of tissue preparations, which could alter the antigenic spectrum without removing all the infectivity. If a close correlation cannot be established between the two types of assay, it will be necessary to await the purification of scrapie as an antigen.

It may therefore be premature to use the method for collateral evidence that there is more scrapie agent in neuronal than glial preparations, as is done in the accompanying communication by Narang, Shenton, Giorgi and Field (Page 106). At this stage standard evidence of infectivity titre is also needed for these two cell populations, and this would give further confidence in the use of the proposed assay. Meanwhile it is difficult to assess the significance of their electron micrographs of particles, but the detection of these in serial passages using only boiled inocula would be more impressive than other evidence.

As the lymphocyte sensitization assay takes only a few days it should be possible to deal quite quickly with some of the points raised here, even though, as Field describes elsewhere, the technique involves "a most fickle and unstable instrument which makes great demands upon the patience and endurance of the observer".—From a Correspondent.

(**240**, 71-72; 1972)

References:

1. Carp, R. I., Licursi, P. C., Merz, P. A., and Merz, G. S., *J. Exp. Med.*, **136**, 618 (1972); *Infect. Immunity*, **6**, 370 (1972).

2. Mackay, J. M. K., Smith, W., and Stamp, J. T., *Vet. Rec.*, **72**, 1002 (1960).

3. Haig, D. A., and Clarke, M. C., *Nature*, **234**, 106 (1971).

退行性改变已经在脑内存在了许多周。我们不能认为这些动物的脾脏从未长期暴露在任一产生于脑内的已改变的抗原中。如果最终发现这种方法只能检测作为病原体浓度指标的组织损伤，它仍是有用的，因为它能“对抗原决定结构的微小改变作出灵敏的反应”，而且与目前能够快速估计出损伤度的常用组织学方法相比，这种方法可以明确地检测到更精细的损伤度。采用这种方法对注射后 4 周内的脾进行时间序列研究也许能排除一些可能出现的继发反应，在 4 周之内病原体浓度会快速增加，而脑内损伤的明显症状还远未出现，或者说，与黑格和克拉克[3]发明的利用组织培养系的常规检测法相比，这种新方法更具优势。

然而即便如此，困难仍然存在，因为这项技术的灵敏度很高，有可能检测到不可见损伤产生的抗原。只有在能证明这种新方法的测定值在大多数情况下与用常规分析方法得到的滴度值密切相关时，其价值才能得到认可。上述过程必须包括在组织制备时所进行的各种各样严格的物理、化学处理。这些处理可能在保留感染性的同时改变了抗原谱。如果无法验证这两种检测法之间的高度相关性，那就只有等到瘙痒症组织能够作为一种抗原被纯化出来之后才能证明这种新方法的可用性了。

因此，像纳兰、申顿、乔治和菲尔德在本卷后一篇文章中（第 106 页）所做的那样，用这种方法作为协同证据来证明在神经元样本中的瘙痒症病原体比在胶质细胞样本中的更多还为时尚早。在现阶段，我们还需要这两种细胞群体的感染性滴度的标准证据，有了这个证据，我们才能更自信地使用电子显微镜检测法。同时，虽然很难评估这些微粒的电子显微照片有多大意义，但是在仅用煮沸过的接种物所获得的系列传代细胞中检测到这些微粒本身就会比其他证据的意义更重大。

因为用淋巴细胞致敏法进行检测只需几天，所以有可能很快就能解决这里提出的一些问题，尽管正如菲尔德在其他文章中所指出的：这项技术使用了“最易变化和最不稳定的检测手段，因而非常需要检测者的耐心和毅力”。——来自一位通讯作者。

（毛晨晖 翻译；刘佳佳 审稿）

Rapid Diagnosis of Scrapie in the Mouse

E. J. Field and B. K. Shenton

Editor's Note

Scrapie is an ancient agricultural disease of sheep whose cause was unknown until the 1990s. Here two British scientists describe a means for the rapid diagnosis of the presence of scrapie by the use of infected mice. At the time there was no tangible idea of what the infective agent might be—a virus, or perhaps even an agent lacking nucleic acid? The development of a rapid test for scrapie nevertheless enabled studies of its transmission between animals to be carried out.

RESEARCH into scrapie, an enigma of veterinary medicine which may embody principles of considerable interest for human disease, commonly involves demonstration of agent activity or its titration. A great step forward was made by Chandler[1] when he showed the disease could be produced in mice, and biological titration in these animals is currently widely employed. This takes 6–8 months. From a recent study of multiple sclerosis, with which scrapie has been linked[2], we were led into an immunological study of scrapie itself. We found that brain or spleen from a scrapie mouse when injected into a guinea-pig (adult Hartley) led to the appearance of blood lymphocytes which were more highly sensitized to scrapie brain or spleen than to normal brain or spleen.

In early experiments, guinea-pigs were inoculated intracutaneously with 0.1 ml. of 10^{-1} suspension of scrapie brain or scrapie spleen in sterile saline. Control animals were similarly injected with brain or spleen from normal mice which had been injected some weeks previously with normal brain suspension. At intervals after 5 days, 2–3 ml. of blood was removed by cardiac puncture and lymphocyte sensitization to the scrapie and normal suspensions estimated by the sensitive and highly specific macrophage electrophoretic migration (MEM) method[3]. In principle the method depends on the release by sensitized lymphocyte of a protein which slows migration of normal guinea-pig macrophages in an electric field (macrophage slowing factor (MSF)—which may be identical with macrophage inhibitory factor (MIF)). Our measurements have been carried out in a Zeiss cytopherometer and full experimental details have been given[4].

Guinea-pig blood lymphocytes were prepared by the method of Coulson and Chalmers[5], as modified by Hughes and Caspary[6], and normal guinea-pig macrophages by washing out the peritoneal cavity with heparinized Hanks solution 6–8 days after injection of sterile liquid paraffin. In order to obviate a two-way mixed lymphocyte reaction (at least for the duration of the test) the normal exudate was subjected to 100 rad γ-irradiation[3].

快速诊断小鼠瘙痒症

菲尔德，申顿

编者按

瘙痒症是在绵羊中发现的一种古老的农业病。在20世纪90年代以前，人们一直不了解瘙痒症的病因。在这篇文章中，两位英国科学家描述了一种利用感染的小鼠快速诊断瘙痒症的方法。那时，人们完全不知道其传染性病原体大概是什么样的——到底是一种病毒还是一种没有核酸的病原体。尽管如此，研制出快速检测瘙痒症的方法能使人们对瘙痒症在动物之间传播的研究得以开展下去。

瘙痒症是兽医学界的一个谜，它可能体现了人类疾病的某些关键法则，人们对瘙痒症的研究通常涉及病原体活性或其滴定方法的确定。钱德勒[1]在这方面取得了一项突破性的进展，他发现小鼠会感染这种病，因而人们目前在广泛地使用小鼠进行生物滴定。这类测试需要6到8个月。从一项与瘙痒症有关的疾病[2]——多发性硬化症的最新研究结果中受到启发，我们开始进行瘙痒症本身的免疫学研究。我们发现：患瘙痒症小鼠的脑组织或者脾组织被注射到豚鼠（成年的哈特利种）中后将导致其血液中出现一种淋巴细胞，这种淋巴细胞对瘙痒症脑或脾组织的敏感性要高于其对正常脑或脾组织的敏感性。

在早期实验中，我们给豚鼠皮内接种的是0.1 ml 用无菌生理盐水稀释10倍的瘙痒症脑或脾的悬液。同样也给对照组动物注射了正常小鼠脑或脾的悬液，而这些正常小鼠又在数周前被注射过正常动物的脑悬液。5天后，我们通过心脏穿刺得到2 ml ~ 3 ml血，并用灵敏和高度特异的巨噬细胞电泳迁移（MEM）法估测淋巴细胞对瘙痒症悬液和正常悬液的致敏程度[3]。这种方法的基本原理是已致敏的淋巴细胞会释放出一种能减慢正常豚鼠巨噬细胞在电场中迁移速度的蛋白（巨噬细胞致缓因子（MSF）——可能与巨噬细胞抑制因子（MIF）为同一分子）。我们在测量中使用了蔡司细胞电泳测量装置，全部实验细节都已公布[4]。

豚鼠血淋巴细胞采用由库尔森和查默斯发明[5]并经过休斯和卡斯帕里改进[6]的方法获取。而正常豚鼠的巨噬细胞则是用肝素化的汉克氏平衡盐溶液从注射无菌液态石蜡6天~8天后的腹腔内冲洗出来的。为了消除双向混合淋巴细胞反应（至少是在测试期间），正常动物的渗出液要经过100拉德的γ射线照射[3]。

10^{-1} scrapie brain, scrapie spleen, normal brain and normal spleen suspensions were used as test antigen and for later testing lymphocyte sensitization. 0.5×10^{6} guinea-pig blood lymphocytes were incubated for 90 min at 20°C with 0.1 ml. of a 10^{-1} suspension of scrapie mouse brain or spleen (cleared by spinning at 1,800*g* for 10 min) in the presence of 10^{7} irradiated normal macrophages. Control tubes comprised lymphocytes and macrophages without antigen. The migration time of macrophages in each specimen was measured by timing ten cells (readily identified by their size and paraffin droplet content) in each direction of the potential difference so that a mean (with SD) from twenty readings could be calculated. A full protocol from one specimen is given by Caspary and Field[4]. All measurements were made "blind" and results unscrambled later. If t_c= migration time of control macrophage in the absence of antigen and t_e= migration time when antigen is present; then in general $t_e > t_c$ and $t_e - t_c/t_c\times100$ is a measure of lymphocyte sensitization to the antigen.

The scrapie brain and spleen with which the guinea-pigs were inoculated were titred out by inoculation into groups of 6 mice at dilutions of 10^{-1} through 10^{-7}. The animals were observed for eight months and all clinical scrapie diagnoses were checked histologically.

Guinea-pigs were immunized with either scrapie or normal mouse brain and spleen, 0.1 ml. of a 10^{-1} suspension being inoculated intracutaneously in the dorsum of the right foot (Table 1*a*, *b*). The guinea-pig lymphocyte sensitization to normal brain and normal spleen is always greater when the animal has been injected with scrapie material than with normal (as pointed out by Gardiner[7] with respect to circulating antibody in rabbit experiments). Moreover, the scrapie-normal difference (when scrapie brain or spleen is used as test antigen for the lymphocytes) is greater in the guinea-pig immunized with scrapie brain than with normal brain. The difference appears to be greatest between five and thirteen days. Table 1*b* shows that the same is true when guinea-pigs immunized with scrapie or normal spleen are tested for cellular sensitization to scrapie and normal brain and spleen. The results with spleen are particularly interesting since this organ whilst rich in agent ($LD_{50}=10^{-4.7}$) shows no morphological change[8] so that the test antigen being used might well be the scrapie agent itself (or scrapie-altered but morphologically normal membrane).

Table 1. Demonstration of Scrapie Activity in Mouse Brain and Spleen

	a, Mouse scrapie brain injected guinea-pig 10^{-1}				*b*, Mouse scrapie spleen injected guinea-pig 10^{-1}			
	5 days	13 days	22 days	34 days	7 days	11days	25 days	35 days
EF	12.0	15.3	11.1	4.7	6.6	9.2	11.3	2.2
Normal brain	11.3	12.5	10.3	3.4	3.4	8.9	7.6	0.8
Scrapie brain	16.5	17.9	14.3	5.4	10.1	14.2	12.3	2.6
Brain difference	5.2	5.4	4.0	2.0	6.7	5.3	4.7	1.8
Normal spleen	5.2	6.1	6.1	1.7	11.5	12.2	10.8	1.8
Scrapie spleen	9.6	11.1	10.4	3.5	17.4	17.9	15.3	3.9
Spleen difference	4.4	5.0	4.3	1.8	5.9	5.7	4.7	2.1

我们用稀释10倍的瘙痒症脑、瘙痒症脾、正常脑和正常脾悬液作为测试抗原，随后再用它们检测淋巴细胞的致敏程度。将0.5×10^6个豚鼠血淋巴细胞与0.1 ml稀释10倍的瘙痒症小鼠脑或者脾组织悬液（净化的方法是在相对离心力为1,800 *g*下离心10分钟）以及10^7个经过照射的正常巨噬细胞的混合物在20℃下孵育90分钟。对照管内包含淋巴细胞和巨噬细胞，但没有抗原。每个样本中的巨噬细胞迁移时间都是由测量沿电势差任一方向移动的10个细胞（很容易通过它们的大小和石蜡滴的含量来鉴别）的迁移时间得到的，因而可以计算20个读数的平均值（包括标准差）。卡斯帕里和菲尔德[4]已经针对单独样品给出了完整的实验方案。所有测量均采用“盲法”，对结果的分析随后再进行。如果设t_c=无抗原组对照巨噬细胞的迁移时间，t_e=有抗原组的迁移时间；则一般会有$t_e>t_c$，而淋巴细胞对该抗原的致敏程度可以用$\frac{t_e-t_c}{t_c}\times100$来度量。

用于给豚鼠接种的瘙痒症脑和脾组织的滴度是通过将其稀释10倍~10^7倍后接种到6只为一组的小鼠体内得到的。我们对这些小鼠进行了为期8个月的观察，并用组织学方法检测了瘙痒症的所有临床诊断指标。

我们用患瘙痒症小鼠的脑和脾组织或者正常小鼠的脑和脾组织对豚鼠进行免疫，方法是将0.1 ml稀释10倍的悬液皮内接种到豚鼠右足的背部（表1*a*和1*b*）。相对于注射正常脑组织的豚鼠来说，那些注射瘙痒症脑组织的豚鼠，其淋巴细胞对正常脑和脾的敏感程度通常会更高一些（正如加德纳[7]在用实验研究兔子循环抗体时所指出的）。此外，用瘙痒症脑免疫的豚鼠比用正常脑免疫的豚鼠具有更显著的瘙痒症组织与正常组织之间的差异（当使用瘙痒症脑或者脾检测淋巴细胞的抗原时）。这种差异似乎在第5天~第13天之间最为明显。表1*b*说明当用瘙痒症脾或正常脾免疫的豚鼠被用于检测对瘙痒症脑和脾以及对正常脑和脾的细胞致敏程度时将会得出同样的结果。用脾实验的结果非常有意思，因为这种器官虽然富含病原体（半数致死量$LD_{50}=10^{-4.7}$），却没有出现形态上的改变[8]，所以测试时使用的实验抗原很可能就是瘙痒症病原体本身（或者是虽然被瘙痒症改变但在形态上仍然正常的膜）。

表1. 小鼠脑和脾内的瘙痒症病原体活性数据

	a. 注射稀释10倍的瘙痒症小鼠脑悬液的豚鼠				*b*. 注射稀释10倍的瘙痒症小鼠脾悬液的豚鼠			
	5天	13天	22天	34天	7天	11天	25天	35天
EF	12.0	15.3	11.1	4.7	6.6	9.2	11.3	2.2
正常脑	11.3	12.5	10.3	3.4	3.4	8.9	7.6	0.8
瘙痒症脑	16.5	17.9	14.3	5.4	10.1	14.2	12.3	2.6
脑的差别	5.2	5.4	4.0	2.0	6.7	5.3	4.7	1.8
正常脾	5.2	6.1	6.1	1.7	11.5	12.2	10.8	1.8
瘙痒症脾	9.6	11.1	10.4	3.5	17.4	17.9	15.3	3.9
脾的差别	4.4	5.0	4.3	1.8	5.9	5.7	4.7	2.1

Continued

	Mouse normal brain injected guinea-pig				Mouse normal spleen injected guinea-pig			
	5 days	13 days	22 days	34 days	7 days	11 days	25 days	35 days
EF	12.0	13.5	9.6	3.0	6.2	8.7	8.2	2.4
Normal brain	9.9	13.7	9.4	2.5	4.0	6.2	6.9	1.5
Scrapie brain	12.1	15.5	10.3	3.2	5.7	8.2	8.4	2.4
Brain difference	2.2	1.8	0.9	1.3	1.7	2.0	1.5	0.9
Normal spleen	4.0	5.9	6.2	1.2	13.9	13.9	11.4	2.7
Scrapie spleen	6.2	7.7	7.6	2.2	14.6	15.7	13.3	3.5
Spleen difference	2.2	1.8	1.4	1.0	0.7	1.8	1.9	0.8

Adult Hartley guinea-pigs inoculated intracutaneously in the dorsum of the right foot with 0.1 ml. 10^{-1} crapie brain or spleen (titre 10^{-6}; $10^{-4.7}$ respectively) and lymphocyte sensitization to scrapie and normal brain measured at intervals by the macrophage electrophoresis method of Field and Caspary[3]. Results expressed as percentage slowing (*loc. cit.*).

In the case of inoculation with scrapie brain, however, the difference may perhaps be attributed to morphological changes (especially astrocyte increase) in the inoculated material. Having established the quantitative difference in response to scrapie as opposed to normal tissue we used this method to titre out scrapie activity.

Guinea-pigs were injected with scrapie brain at dilutions of 10^{-1} to 10^{-6} and the lymphocytes examined for sensitization to scrapie brain or spleen and normal brain or spleen at six or seven days and at seventeen or eighteen days. For comparison a similar study was made in guinea-pigs injected with normal brain (Tables 2 and 3). In the scrapie brain sensitized guinea-pig the difference in lymphocyte sensitization to scrapie as compared with normal brain is 2.5% (P<0.01) even at 10^{-6} original inoculum level when the guinea-pig is tested at six days but falls to 2.0% (P =0.1–0.05) by sixteen days. When sensitization to scrapie spleen and normal spleen in these animals is compared, the difference is significant only in guinea-pigs which have received 10^{-4} scrapie brain.

Table 2. Titration of Scrapie Activity from Mouse Brain

	Lymphocytes tested at 6 days					Lymphocytes tested at 16 days		
Antigen	10^{-1}	10^{-2}	10^{-3}	10^{-4}	10^{-6}	10^{-2}	10^{-4}	10^{-6}
	% macrophage slowing							
	Scrapie brain							
Normal mouse brain	11.3	11.0	8.1	5.7	3.7	11.1	5.9	3.9
Scrapie brain	16.5	14.1	11.7	9.6	5.2	14.7	9.9	5.9
Brain difference	5.2	3.1	3.6	3.9	2.5	3.6	4.0	2.0
Normal mouse spleen	5.2	6.1	4.7	3.9	1.7	5.9	4.7	2.9
Scrapie mouse spleen	9.6	10.8	8.1	7.4	3.2	10.1	8.1	4.4
Spleen difference	4.4	4.7	3.4	3.5	1.5	4.2	3.4	1.5
	Normal brain							
Normal mouse brain		8.9		7.1	5.9	9.6	5.5	4.4
Scrapie mouse brain		10.3		7.9	6.7	10.1	6.5	5.2
Brain difference		1.4		0.6	0.8	0.5	1.0	0.8
Normal mouse spleen		5.4		2.5	3.0	6.1	4.4	3.5

续表

	注射正常小鼠脑悬液的豚鼠				注射正常小鼠脾悬液的豚鼠			
	5天	13天	22天	34天	7天	11天	25天	35天
EF	12.0	13.5	9.6	3.0	6.2	8.7	8.2	2.4
正常脑	9.9	13.7	9.4	2.5	4.0	6.2	6.9	1.5
瘙痒症脑	12.1	15.5	10.3	3.2	5.7	8.2	8.4	2.4
脑的差别	2.2	1.8	0.9	1.3	1.7	2.0	1.5	0.9
正常脾	4.0	5.9	6.2	1.2	13.9	13.9	11.4	2.7
瘙痒症脾	6.2	7.7	7.6	2.2	14.6	15.7	13.3	3.5
脾的差别	2.2	1.8	1.4	1.0	0.7	1.8	1.9	0.8

在成年哈特利豚鼠右足背部皮内接种 0.1 ml 稀释 10 倍的瘙痒症脑或者脾组织液（滴度分别为 10^{-6} 和 $10^{-4.7}$），然后每隔一定时间用菲尔德和卡斯帕里发明的巨噬细胞电泳迁移法 [3] 测定淋巴细胞对瘙痒症脑和正常脑的致敏程度。结果以迁移减慢的百分比表示（见上文）。

然而，在接种瘙痒症脑的实验中，其差异或许可以归因于接种物的形态变化（尤其是星形细胞的增加）。在确认对瘙痒症组织和对正常组织的反应存在量上的差异之后，我们就可以使用这种方法来测定瘙痒症病原体的活性了。

给豚鼠注射稀释10倍~10^6倍的瘙痒症脑组织，然后在第6天或第7天以及第17天或第18天检测淋巴细胞对瘙痒症脑或脾以及对正常脑或脾的致敏程度。为了便于对比，我们用注射过正常脑组织的豚鼠进行了类似的实验（表2和表3）。在瘙痒症脑致敏的豚鼠中，即便原始接种物被稀释至10^6倍，到第6天时淋巴细胞对瘙痒症脑和正常脑的致敏程度之差也可以达到2.5%（$P<0.01$），但在第16天时降到了2.0%（$P=0.1\sim0.05$）。当比较豚鼠对瘙痒症脾和正常脾的致敏程度时，只有在注射稀释至10^4倍及以下瘙痒症脑的豚鼠中才会出现明显的差别。

表2. 小鼠脑的瘙痒症病原体活性滴定

	测试第6天的淋巴细胞					测试第16天的淋巴细胞		
抗原	稀释10^1倍	稀释10^2倍	稀释10^3倍	稀释10^4倍	稀释10^6倍	稀释10^2倍	稀释10^4倍	稀释10^6倍
	巨噬细胞迁移减慢的%							
	瘙痒症脑							
正常小鼠脑	11.3	11.0	8.1	5.7	3.7	11.1	5.9	3.9
瘙痒症脑	16.5	14.1	11.7	9.6	5.2	14.7	9.9	5.9
脑的差别	5.2	3.1	3.6	3.9	2.5	3.6	4.0	2.0
正常小鼠脾	5.2	6.1	4.7	3.9	1.7	5.9	4.7	2.9
瘙痒症小鼠脾	9.6	10.8	8.1	7.4	3.2	10.1	8.1	4.4
脾的差别	4.4	4.7	3.4	3.5	1.5	4.2	3.4	1.5
	正常脑							
正常小鼠脑		8.9		7.1	5.9	9.6	5.5	4.4
瘙痒症小鼠脑		10.3		7.9	6.7	10.1	6.5	5.2
脑的差别		1.4		0.6	0.8	0.5	1.0	0.8
正常小鼠脾		5.4		2.5	3.0	6.1	4.4	3.5

Continued

	Normal brain							
Scrapie mouse spleen		6.2		3.8	4.2	7.2	5.7	4.5
Spleen difference		0.8		1.3	1.2	1.1	1.3	1.0

Guinea-pigs inoculated with scrapie mouse brain with a titre of $10^{-5.6}$. 0.1 ml. inoculated at different dilutions.

Table 3. Titration of Scrapie Activity from Mouse Spleen

	Lymphocytes tested at 7 days						Lymphocytes tested at 17 days		
Antigen	10^{-1}	10^{-2}	10^{-3}	10^{-4}	10^{-5}	10^{-6}	10^{-2}	10^{-4}	10^{-6}
	% macrophage slowing								
	Scrapie spleen								
Normal mouse brain	3.4	4.5	2.7	1.0	1.7	0	4.7	1.5	0.7
Scrapie brain	10.1	10.3	7.2	4.4	3.8	1.2	10.1	4.9	1.5
Brain difference	6.7	5.8	4.5	3.4	2.1	1.2	5.4	3.4	0.8
Normal mouse spleen	11.5	12.6	10.4	10.6	8.1	4.5	11.1	10.0	4.4
Scrapie mouse spleen	17.4	17.0	15.3	14.4	10.8	6.3	15.0	13.8	6.2
Spleen difference	5.9	4.4	4.9	3.8	2.7	1.8	3.9	3.8	1.8
	Normal spleen								
Normal mouse brain		4.5		1.5		1.0	6.4	3.0	1.8
Scrapie mouse brain		5.9		2.9		1.2	7.6	4.4	2.9
Brain difference		1.4		1.4		0.2	1.2	1.4	1.1
Normal mouse spleen		11.1		9.1		4.7	10.5	7.9	4.7
Scrapie mouse spleen		12.4		10.2		6.4	11.5	8.9	5.6
Spleen difference		1.3		1.1		1.7	0.9	1.0	0.9

Guinea-pigs inoculated with spleen suspension with a titre of 10^{-5}. 0.1 ml. inoculated at different dilutions.

Animals injected with normal brain showed no significant difference when their lymphocytes were tested with scrapie as opposed to normal brain or spleen (though in general the values with the former were higher).

Guinea-pigs were sensitized by injecting scrapie spleen at 10^{-1} to 10^{-6} and their lymphocytes tested for sensitization to scrapie brain and spleen (Table 3). A significant difference between sensitization to scrapie and normal spleen still exists at 10^{-5} ($P<0.01$) with spleen as antigen but barely with brain (2.1% difference; $P=0.05$), showing apparent scrapie antigenicity in the 10^{-5} dilution.

The titre of the scrapie brain inoculated into the test guinea-pigs was calculated to be $LD_{50}=10^{-5.6}$. We noted that 1/6 mice developed scrapie at 10^{-6} and 1/6 at 10^{-7} and the immunological test suggests that some activity is still present at 10^{-6}. The titre of spleen used for immunizing the guinea-pigs was $LD_{50}=10^{-5}$. The results of *in vivo* titration therefore agree with the *in vitro* immunological assay, and the significance of these results is two-fold. (1) Attempts to show the existence of circulating antibody and/or specific antigen in scrapie disease have not been successful[7,8]. Gardiner did show by inoculation of rabbits with scrapie and non-scrapie spleen material that the latter had increased antigenicity *in*

续表

	正常脑							
瘙痒症小鼠脾		6.2		3.8	4.2	7.2	5.7	4.5
脾的差别		0.8		1.3	1.2	1.1	1.3	1.0

接种滴度为$10^{-5.6}$的瘙痒症小鼠脑组织的豚鼠。对每种不同的稀释度都接种0.1 ml。

表3. 小鼠脾的瘙痒症病原体活性滴定

	测试第7天的淋巴细胞						测试第17天的淋巴细胞		
抗原	稀释10^1倍	稀释10^2倍	稀释10^3倍	稀释10^4倍	稀释10^5倍	稀释10^6倍	稀释10^2倍	稀释10^4倍	稀释10^6倍
	巨噬细胞迁移减慢的%								
	瘙痒症脾								
正常小鼠脑	3.4	4.5	2.7	1.0	1.7	0	4.7	1.5	0.7
瘙痒症脑	10.1	10.3	7.2	4.4	3.8	1.2	10.1	4.9	1.5
脑的差别	6.7	5.8	4.5	3.4	2.1	1.2	5.4	3.4	0.8
正常小鼠脾	11.5	12.6	10.4	10.6	8.1	4.5	11.1	10.0	4.4
瘙痒症小鼠脾	17.4	17.0	15.3	14.4	10.8	6.3	15.0	13.8	6.2
脾的差别	5.9	4.4	4.9	3.8	2.7	1.8	3.9	3.8	1.8
	正常脾								
正常小鼠脑		4.5		1.5		1.0	6.4	3.0	1.8
瘙痒症小鼠脑		5.9		2.9		1.2	7.6	4.4	2.9
脑的差别		1.4		1.4		0.2	1.2	1.4	1.1
正常小鼠脾		11.1		9.1		4.7	10.5	7.9	4.7
瘙痒症小鼠脾		12.4		10.2		6.4	11.5	8.9	5.6
脾的差别		1.3		1.1		1.7	0.9	1.0	0.9

接种滴度为10^{-5}的瘙痒症小鼠脾组织悬液的豚鼠。对每种不同的稀释度都接种0.1 ml。

对于注射正常脑组织的动物，其淋巴细胞对瘙痒症脑或脾的致敏程度与其对正常脑或脾的致敏程度相比并无显著差异（尽管从总体情况来看前者的数值要大一些）。

给豚鼠注射稀释 10 倍～10^6 倍的瘙痒症脾组织使之致敏，然后检测它们的淋巴细胞对瘙痒症脑和脾的致敏程度（表 3）。在用脾作为抗原时，淋巴细胞对瘙痒症脾和正常脾的致敏程度在稀释 10^5 倍时仍存在明显的差异（$P<0.01$），但在用脑作抗原时则只有很小的差异（差异为 2.1%，$P=0.05$），这说明在稀释 10^5 倍时会出现明显的瘙痒症抗原性。

对接种到实验豚鼠上的瘙痒症脑组织的滴度进行计算所得的结果是$LD_{50}=10^{-5.6}$。我们注意到在稀释10^6倍时有1/6的小鼠染上了瘙痒症，在稀释10^7倍时也会有1/6发病，免疫学实验证明在稀释10^6倍时仍存在一些活性。用于免疫豚鼠的脾组织的滴度为$LD_{50}=10^{-5}$。这样，体内滴定结果就与体外免疫学检测结果取得了一致，因而上述结果的重要性是双重的。（1）目前人们在瘙痒症患者体内寻找循环抗体和/或特异性抗原的尝试尚未取得成功[7,8]。加德纳在将瘙痒症和非瘙痒症脾组织接种到兔子中

vivo. The present work extends these very important findings in that it suggests that this is true also of lymphocyte sensitization and that the reactivity is greater against scrapie material (both brain and spleen) than against normal material. This suggests that some specific antigen—perhaps the agent itself, or specifically agent-altered material since the spleen is so potent as a testing tissue—may be active. The macrophage electrophoretic migration (MEM) method is exquisitely responsive to minor changes in antigenic determinant structure and appears to be able to distinguish between normal and scrapie tissue. It is, of course, not possible to decide whether the difference resides in the presence of a specific scrapie agent or whether we are dealing with a structural (antigenic) change induced by the agent. However, if scrapie infected brain is split into a neuronal and glial compartment by the method of Giorgi (unpublished) then, contrary to expectations based on the precocious glial hypertrophy, higher scrapie titre is associated with the neuronal rather than the glial fraction and it is in the former that particles with the size and structural characters postulated for scrapie agent have been found[9]. (2) Lymphocyte sensitization test enables the presence of scrapie to be established and titration carried out within 8 days, an important saving in time during the study of this fascinating condition. Further experiments are in progress involving the use of Freund's adjuvant to boost the responses. A full account of these experiments will be published elsewhere.

(**240**, 104-106; 1972)

E. J. Field and B. K. Shenton: Medical Research Council, Demyelinating Diseases Unit, Newcastle General Hospital, Westgate Road, Newcastle upon Tyne NE4 6BE.

Received July 25, 1972.

References:

1. Chandler, R. L., *Lancet*, ii, 1378 (1961).
2. Field, E. J., *Int. Rev. Exp. Path.*, **8**, 129 (1969).
3. Field, E. J., and Caspary, E. A., *Lancet*, ii, 1337 (1970).
4. Caspary, E. A., and Field, E. J., *Brit. Med. J.*, **2**, 613 (1971).
5. Coulson, A. S., and Chalmers, D. G., *Immunology*, **12**, 417 (1967).
6. Hughes, D., and Caspary, E. A., *Int. Arch. Allergy*, **37**, 506 (1970).
7. Gardiner, A. C., *Res. Vet. Sci.*, 7, 190 (1966).
8. Chandler, R. L., *Vet. Rec.*, **71**, 58 (1959).
9. Narang, H. K., Shenton, B., Giorgi, P. P., and Field, E. J., *Nature*, **240**, 106 (1972).

后确实发现后者在体内的抗原性增加了。我们的研究拓展了这些非常重要的发现，因为它表明淋巴细胞的致敏也是存在的，并且对瘙痒症组织（无论是脑还是脾）的反应性要比对正常组织的强。这意味着某些特异性抗原可能是具有活性的，这些特异性抗原也许就是病原体本身，或者是特异性的病原体改变的物质，因为脾脏是一种非常有效的检测用组织。巨噬细胞电泳迁移（MEM）法对抗原决定结构的微小变化反应很敏感，应该能够辨别出正常组织和瘙痒症组织。当然，我们不可能裁断这种差异能不能归因于某种特定的瘙痒症病原体，也不能说明我们是否检测到了由病原体导致的结构（抗原性）改变。但是，如果利用乔治的方法（尚未发表）将瘙痒症感染的脑分为神经元部分和胶质细胞部分，那么与根据胶质细胞过早肥大得到的预期相反，对应于较高滴度瘙痒症病原体的并不是胶质细胞部分，而是神经元部分，也正是在神经元内人们找到了具有疑似瘙痒症病原体大小和结构特征的颗粒[9]。

（2）我们可以通过淋巴细胞致敏实验来确定瘙痒症病原体的存在，并能在8天内得到滴定的结果，这为研究这个非常吸引人的问题节省了大量的时间。我们正在进行更多的实验，其中包括用弗氏佐剂来加强反应。这些实验的完整细节将在其他论文中详述。

（毛晨晖 翻译；刘佳佳 审稿）

Scrapie Agent and Neurones

H. K. Narang *et al.*

Editor's Note

The search for the infectious agent behind scrapie and similar neurodegenerative disorders had previously focused on a non-neuronal cell type called glia, which appeared swollen in the infected brain. Here Harash Narang and colleagues provide evidence that the scrapie agent primarily targets neurons rather than glia. They see numerous tiny elongated particles inside the neurons of scrapie-infected rats brains which, they suggest, could be the elusive scrapie agent. It is now thought the infectious agent in most transmissible spongiform encephalopathies is a misshapen protein called a prion, and that whilst neurons are the prime target of damage within the brain, glia are also involved.

THE nature of the scrapie agent[1,2] is of considerable interest especially as to whether it is a virus in the strict sense, or some replicating agent devoid of nucleic acid[3,4], or indeed whether scrapie disease represents a transmissible progressive biochemical transformation of cellular membranes[5,6]. Thorough electron microscope studies have been carried out in the disease and, although a number of particles have (not unexpectedly) been described[7,8], their role in scrapie is unconfirmed, especially as none has the special features associated with virions or has the size or other characters currently attributed to the agent.

During careful search in the brain of the scrapie rat granular inclusion bodies were seen in the cytoplasm of apparently normal neurones and their axons in the sites studied; the cortex, Ammon's horn and thalamus. At high magnification these uninteresting neuronal inclusions comprised very large numbers of minute elongated particles about 60 nm long and 20 nm wide, with a dense linear core about 4 nm wide (Figs. 1 and 2). When these particles had once been recognized they were also quickly found in blocks of scrapie rat brain which had been examined years ago.

瘙痒症病原体与神经元

纳兰等

编者按

以前对瘙痒症和类似的神经退行性疾病致病因子的搜寻集中在被称为胶质细胞的非神经元细胞上，因为它在被感染的脑内会出现肿胀现象。在这篇论文中，哈拉什·纳兰及其同事提出证据证明瘙痒症病原体的主要靶标是神经元而非胶质细胞。他们在感染瘙痒症的大鼠脑神经元中发现了数量众多的细长形小颗粒，并指出这些小颗粒可能就是难于发现的瘙痒症病原体。现在人们认为：大多数传染性海绵状脑病的致病因子是一种被称为朊病毒的畸形蛋白质，虽然神经元是脑内受损害的主要靶标，但胶质细胞也会受到牵连。

瘙痒症病原体的本质[1,2]是个非常有趣的问题，尤其在于：它是一种严格意义上的病毒，还是某种没有核酸的可复制病原体[3,4]；或者，瘙痒症是否确实反映了细胞膜的一种可传染的进行性生物化学转化[5,6]。人们利用电子显微镜对该病进行了深入的研究，尽管已经对大量的颗粒进行过（意料之中的）描述[7,8]，但它们在瘙痒症中的作用尚不确定，尤其是因为所有这些颗粒都不具备与病毒相关的特征，其大小或其他特征也与目前所认为的病原体不相符合。

对患瘙痒症大鼠脑内进行仔细搜索时，在皮层、阿蒙氏角和丘脑部位看似正常的神经细胞胞质及其轴突中发现了一些颗粒状包涵体。在高倍镜下，这些不引人注目的神经元包涵体中含有数量众多的长约 60 nm、宽约 20 nm 的细长形颗粒，颗粒中有一约 4 nm 宽的致密线性核（图1和图2）。首次识别出这些颗粒之后，很快在数年前已研究过的患瘙痒症大鼠脑块中再次找到。

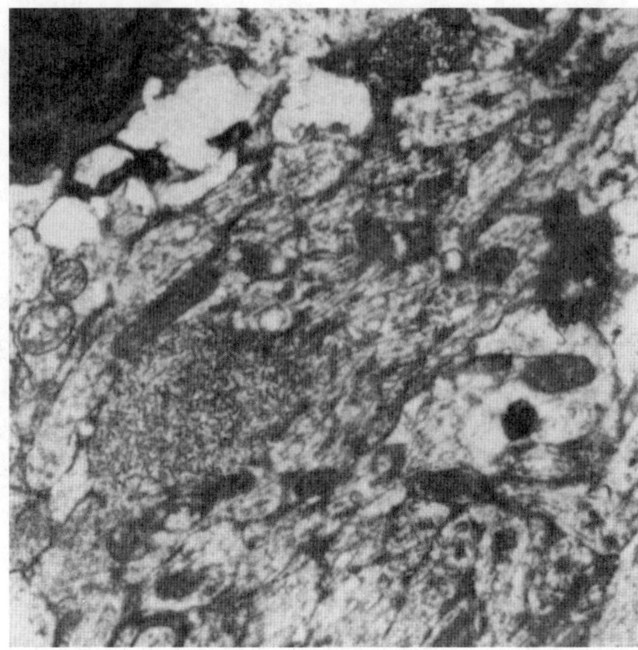

Fig. 1. Scrapie rat thalamus. Axon close to a nerve cell body showing a colony-like accumulation of small elongated particles not bounded by membrane. The neurotubular and other constituents of the axon are normal. ×20,625.

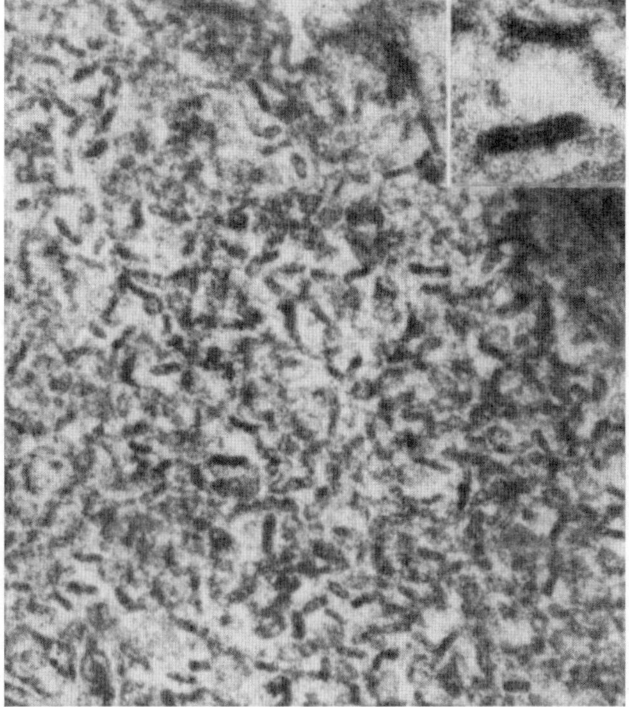

Fig. 2. High power view of colony. Note elongated particles often with dense core (see inset), about 60 nm long and 20 nm wide. The core is about 4 nm thick. Tilting and rotating the preparation confirm the discrete character of the particles. ×72,000; inset, ×208,000.

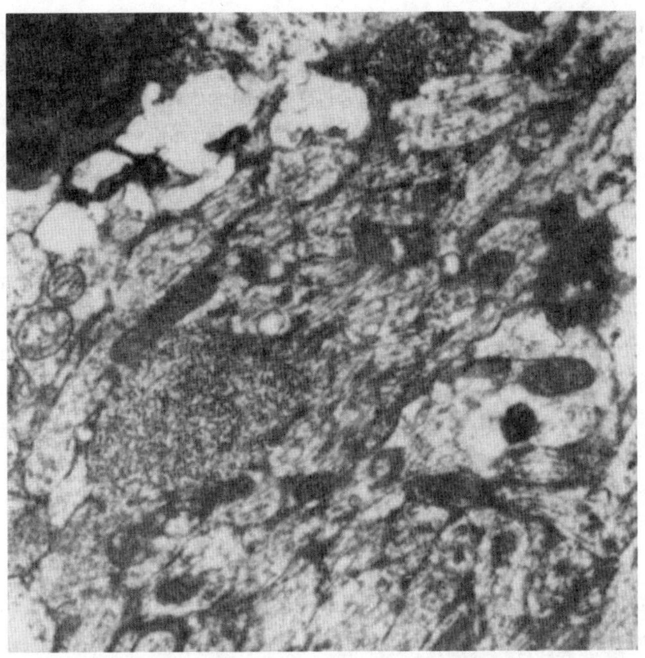

图 1. 患瘙痒症大鼠的丘脑。在靠近神经细胞体的轴突内显示无包膜细长形小颗粒的集落样聚集。轴突的神经管和其他成分都正常。放大 20,625 倍。

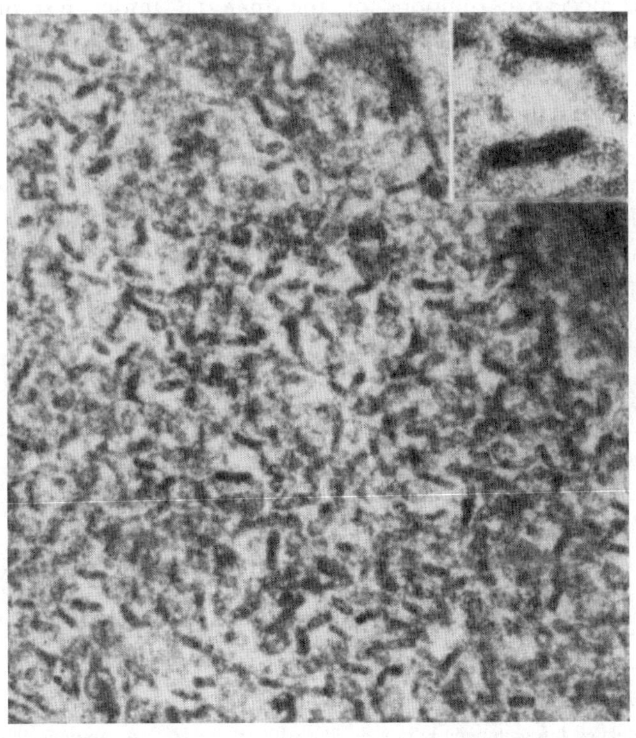

图 2. 集落的高倍图像。注意，长约 60 nm、宽约 20 nm 的细长形颗粒通常含有致密核（见右上角的小图）。核的宽度约为 4 nm。通过倾斜和旋转该样本证明这些颗粒是分离的。放大 72,000 倍；右上角小图，放大 208,000 倍。

A thorough and realistic search of comparable normal rat brain has not revealed similar particles though the vagaries of electron microscope sampling must constantly be borne in mind. The size of the particles is about that deduced from modern infectivity filtration experiments[9]. Moreover the size of central rod accords well with that of the target nucleic acid estimate made by Alper *et al.*[3,4] and by Latarjet *et al.*[10].

If the particles described are indeed scrapie agent, it is surprising to find them in neurones rather than glial cells since there is widespread agreement that the first element in the nervous system to react in the scrapie process is the astroglia[11-13] and the working assumption has been made that colonization of astroglial cells by the agent is responsible for their precocious hypertrophy. Indeed we have tended to concentrate search on glial cells, even though unusual neurological reactions may be detected early in the incubation period[14]. Simultaneously with the electron microscope findings here reported, unexpected collateral evidence has emerged that the agent may in fact proliferate in neurones rather than glial cells.

Preparations of neuronal and glial perikarya have been obtained from scrapie mouse brain using a modification (Giorgi, unpublished work) of the method of Sellinger *et al.*[15]. To these preparations we have applied the macrophage electrophoretic method of titration of scrapie agent[16] in a study of the apportionment of scrapie activity in infected brain as between neuronal and glial compartments. The method briefly consists in immunizing guinea-pigs with successive dilutions of the material under test and measuring the degree of sensitization developed by the lymphocytes to scrapie brain (or spleen) and to normal brain (or spleen) as antigen. To our surprise there was clearly more scrapie activity associated with the neurone enriched fraction than with the glial compartment (Table 1). It might well be that the scrapie agent (or process) affects, or at least reaches greater development within, neurones and that glial changes, even though they are more conspicuous, have the consequential character they are usually assigned in neuropathology. The present work illustrates the synergism of different disciplines in approaching the problem of slow infection.

Table 1. Immunological Titration of Scrapie Activity

	Neuronal compartment				Glial compartment			
Antigen	10^{-1}	10^{-2}	10^{-3}	10^{-4}	10^{-1}	10^{-2}	10^{-3}	10^{-4}
Normal mouse brain	9.1	9.9	8.5	6.7	8.6	7.7	7.1	6.9
Scrapie mouse brain	14.6	13.9	11.3	9.2	11.3	8.1	8.4	8.1
Brain difference	5.5	4.0	2.8	2.5*	2.7	0.4	1.3	1.2
Normal mouse spleen	6.0	5.9	5.7	4.9	6.2	6.2	5.6	4.4
Scrapie mouse spleen	9.7	9.4	9.1	8.2	9.1	8.2	6.9	6.5
Spleen difference	3.7	3.5	3.4	3.3	2.9	2.0	1.3	2.1

* With the degree of scatter in our readings a difference of 2.5% corresponds to $P<0.01$.

Guinea-pigs inoculated intracutaneously in right foot with 10^{-1} to 10^{-4} preparations of neurones and glial cells from mouse scrapie brain (0.1 ml. in each case; equivalent to 300 μg to 0.3 μg protein). Sensitization of the guinea-pig lymphocytes with respect to 10^{-1} scrapie brain and spleen and also to normal brain

在对正常大鼠脑内进行认真而彻底的搜寻之后并没有发现类似的颗粒，不过应该提请注意的是，电子显微镜在取样上具有难以预测性。这些颗粒的大小与由现代传染性筛选实验推断的结果大体一致[9]。此外，中心杆状体的大小与阿尔珀等人[3,4]和拉塔尔热等人[10]估计的目标核酸的尺寸非常符合。

如果所描述的颗粒确实是瘙痒症病原体，那么在神经细胞而不是胶质细胞内发现这些颗粒是很令人惊讶的，因为大家普遍认为在神经系统内对瘙痒症过程作出反应的第一个元素是星形胶质细胞[11-13]，并已假定病原体在星形胶质细胞内的定居是胶质细胞早期肥大的原因。确实，我们一直倾向于集中精力搜索胶质细胞，尽管在潜伏早期就有可能检测到不正常的神经反应[14]。现在又有一个意想不到的证据可以证明本文所报道的电子显微镜结果，即病原体实际上有可能是在神经元内而不是在胶质细胞内增殖的。

利用经改进的塞林杰等人的方法[15]（乔治，尚未发表的研究结果），从患瘙痒症小鼠脑内提取出了神经元和胶质细胞的核周体。在研究瘙痒症活性在感染脑的神经元和胶质细胞间的分配时，我们运用巨噬细胞电泳法对这些提取物进行瘙痒症病原体的滴定[16]。这个方法只需用连续稀释度的待测物免疫豚鼠，并测量淋巴细胞相对于作为抗原的瘙痒症脑（或者脾）和正常脑（或者脾）的致敏程度。令我们惊讶的是，瘙痒症病原体活性在神经元富集区要明显高于胶质细胞区（表1）。瘙痒症病原体（或者过程）可能会影响神经元，或者至少在神经元内有更显著的发展；而胶质细胞的变化，尽管更加明显，却通常是神经病理学中继发的特征。本文可以说明不同学科在解决慢性感染问题中的协同作用。

表 1. 瘙痒症病原体活性的免疫滴定分析

	神经元区				胶质细胞区			
抗原	10^{-1}	10^{-2}	10^{-3}	10^{-4}	10^{-1}	10^{-2}	10^{-3}	10^{-4}
正常小鼠脑	9.1	9.9	8.5	6.7	8.6	7.7	7.1	6.9
瘙痒症小鼠脑	14.6	13.9	11.3	9.2	11.3	8.1	8.4	8.1
脑的差别	5.5	4.0	2.8	2.5*	2.7	0.4	1.3	1.2
正常小鼠脾	6.0	5.9	5.7	4.9	6.2	6.2	5.6	4.4
瘙痒症小鼠脾	9.7	9.4	9.1	8.2	9.1	8.2	6.9	6.5
脾的差别	3.7	3.5	3.4	3.3	2.9	2.0	1.3	2.1

* 读数的离散度为：2.5%的差异，对应于$P<0.01$。

在豚鼠的右足皮内接种浓度为 10^{-1}~10^{-4} 的患瘙痒症小鼠脑内神经元和胶质细胞的提取物（每次 0.1 ml，相当于 300 μg ~ 0.3 μg 蛋白质）。采用巨噬细胞电泳减慢法测量豚鼠淋巴细胞相对于 10^{-1} 浓度的瘙痒症

and spleen is measured by the macrophage electrophoretic slowing method and expressed as percentage slowing of macrophages[17].

(**240**, 106-107; 1972)

H. K. Narang, B. Shenton, P. P. Giorgi, E. J. Field: Medical Research Council, Demyelinating Diseases Unit, Newcastle General Hospital, Westgate Road, Newcastle upon Tyne NE 4 6BE.

Received July 25, 1972.

References:

1. *Nature*, 214, 755 (1967).

2. *Lancet*, ii, 705 (1967).

3. Alper, T., Haig, D. A., and Clarke, M. C., *Biochem. Biophys. Res. Commun.*, **22**, 278 (1966).

4. Alper, T., Cramp, W. A., Haig, D. A., and Clarke, M. C., *Nature*, **214**, 764 (1967).

5. Gibbons, R. A., and Hunter, G. D., *Nature*, **215**, 1041 (1967).

6. Hunter, G. D., in *Proc. VI Int. Congr. Neuropathol.*, 802 (Masson, Paris, 1970).

7. David-Ferreira, J. B., David-Ferreira, K. L., Gibbs, C. J., and Morris, J. A., *Proc. Soc. Exp. Biol. Med.*, **127**, 313 (1968).

8. Bignami, A., and Parry, H. B., *Science*, **171**, 389 (1971).

9. Hunter, G. D., *J. Infect. Dis.*, **125**, 427 (1972).

10. Latarjet, R., Muel, B., Haig, D. A., Clarke, M. C., and Alper, T., *Nature*, **227**, 1341 (1970).

11. Pattison, I. H., in *Slow, Latent and Temperate Virus Infections, NINDB Monograph No. 2*, 249 (1965).

12. Hadlow, W. J., discussion in *Slow, Latent and Temperate Virus Infections, NINDB Monograph No. 2*, 303 (1965).

13. Field, E. J., *Int. Rev. Exp. Pathol.*, **8** (Acad. Press, New York and Canada, 1969).

14. Savage, R. D., and Field, E. J., *Anim. Behav.*, **13**, 443 (1965).

15. Sellinger, O. Z., Azcurra, J. M., Johnson, D. E., Ohlsson, W. G., and Lodin, Z., *Nature New Biology*, **230**, 253 (1971).

16. Field, E. J., and Shenton, B., *Nature*, **240**, 104 (1972).

17. Field, E. J., and Caspary, E. A., *Lancet*, ii, 1337 (1970).

脑和脾以及正常脑和脾的致敏程度，并用减慢的巨噬细胞百分比来表示[17]。

（毛晨晖 翻译；李素霞 审稿）

Calcium Ions and Muscle Contraction

S. Ebashi

Editor's Note

Here biologist Setsuro Ebashi from the University of Tokyo sums up what is known about the role of calcium ions in muscle contraction. In the 1940s, researchers injected calcium ions into single muscle fibres causing their contraction. Then in the 1960s, Ebashi helped demonstrate the involvement of two proteins, troponin and tropomyosin. Troponin, found on the thin actin filament, acts as the calcium receptor protein, while tropomyosin mediates the effect of calcium ions on troponin. But a closer look reveals the system to be more complex, he warns, with calcium sometimes acting as an inhibitor of inhibition. Ebashi was viewed as a pioneer of muscle research and his concept of a calcium receptor protein was later extended to other cell types.

ONE of the characteristics of muscle research in Japan is the close personal contact and willing cooperation that exists among research workers. This pleasant atmosphere was the result of the paternal leadership of Professor H. Kumagai, who in 1955 invited young muscle scientists to join a research group in which enthusiastic and heated discussions on muscle contraction could take place. The problem implicit in the title of this article had also been grappled with under these conditions.

The mechanism of such a complex biological phenomenon as muscle contraction can be examined in two different ways. One is concerned with the underlying elementary process, and the other with the regulatory mechanism through which this process comes into operation as a biological function. The elementary process of muscle contraction involves the interaction of myosin and actin in the presence of ATP (MgATP). On the other hand, the Ca ion plays an indispensable and unparalleled role in turning the interaction of myosin and actin into real function[1,2] and its place in this regulatory mechanism may correspond to that of ATP in the elementary process (Fig. 1).

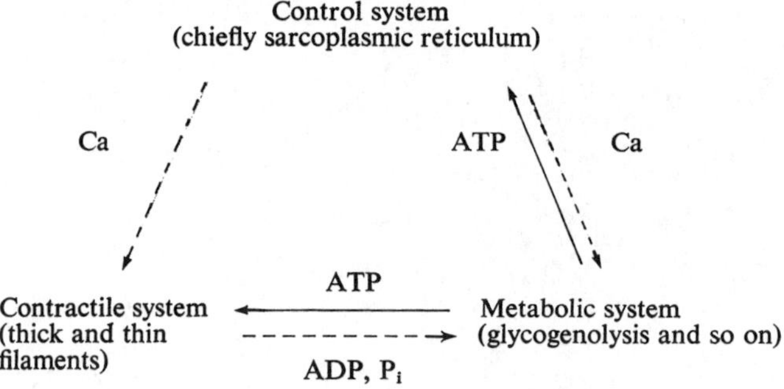

Fig. 1. Schematic representation of the functional organization of muscle.

钙离子与肌肉收缩

江桥节郎

编者按

在本文中，东京大学的生物学家江桥节郎总结了当时人们对钙离子在肌肉收缩中所起作用的认识。在20世纪40年代，研究人员发现，向单肌纤维中注射钙离子会使肌肉产生收缩。随后，江桥节郎在60年代又证明了有两种蛋白质——肌钙蛋白和原肌球蛋白也参与其中。存在于细肌丝（又称肌动蛋白丝）中的肌钙蛋白可以作为钙离子的接受蛋白，而原肌球蛋白的作用则是介导钙离子对肌钙蛋白的影响。但是他提醒大家：进一步的研究发现，该系统其实还要更复杂一些，因为钙离子有时候会表现为抑制收缩作用的抑制剂。江桥节郎被视为肌肉学研究的先驱，后来他提出的钙离子结合蛋白的概念又被拓展到其他细胞类型。

日本肌肉学研究的特色之一是，科研人员之间能保持密切的私人关系并愿意相互合作。这一良好氛围得益于熊谷洋教授所开创的家长式领导体制。在1955年，他曾邀请一批年轻肌肉学家加入一个研究小组中去，在这个小组中大家可以就肌肉收缩问题进行热情洋溢的讨论。本文标题所隐含的问题也曾在这样的讨论中被大家激烈地争辩过。

可以用两种不同的方式来研究像肌肉收缩这么复杂的生物学现象的机理。一种关注的是这个现象背后的基本过程，另一种则关注这一过程赖以发挥生物学功能的调节机制。肌肉收缩的基本过程包括在ATP（MgATP）存在时肌球蛋白与肌动蛋白之间的相互作用。另一方面，钙离子在使肌球蛋白与肌动蛋白间相互作用转变为真正生物学功能[1,2]的过程中发挥着不可或缺的独特作用，而钙离子在肌肉收缩调节机制中的地位可以与ATP在基本过程中的地位相对应（图1）。

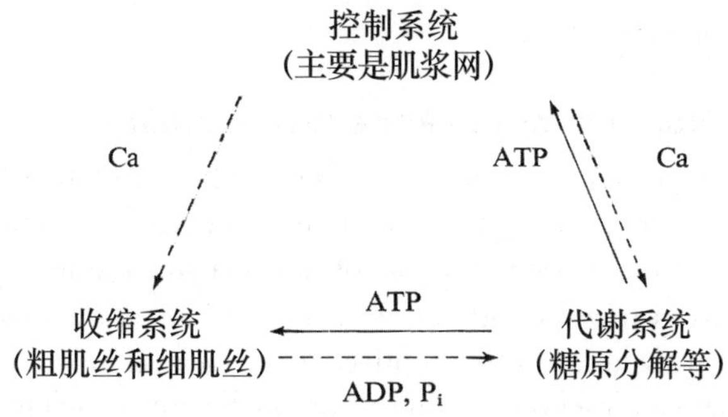

图1. 肌肉中功能性组织的示意图。

The first indication of the crucial role of the Ca ion in the contraction of living muscle came at almost the same time as the establishment of the myosin-actin-ATP system by Szent-Györgyi and his school[3] in 1941–42. Three years after the work of Heilbrunn[4] in 1940, Kamada and Kinosita[5] injected a minute quantity of Ca ions through a micropipette of diameter 2 to 5 μm into a single muscle fibre, which responded by contracting locally in a reversible manner. Unfortunately, this elegant pioneering work was not developed in war-torn Japan; in the meantime, Heilbrunn and Wiercinski[6] presented essentially the same result. These remarkable observations, however, were then ignored for more than ten years by most muscle scientists, probably because the elementary process of muscle contraction, namely the interaction of myosin, actin and ATP, does not require the presence of the Ca ion.

Establishment of the Ca Theory

The wide recognition of the part played by the Ca ion is chiefly based on several findings[1,2,7] reported between 1959 and 1961. First, it turned out that the contractile processes *in vitro* are regulated by quite low concentrations of Ca ions; 10^{-7} M Ca ion keeps the contractile system in a relaxed state, and the presence of 10^{-5} M Ca ion activates it fully. Second, the sarcoplasmic reticulum, the internal membrane system, pumps up Ca ions from the surrounding medium in the presence of ATP. The capacity and affinity of the membrane system for Ca ions are sufficient to keep the contractile system in a relaxed state.

These data allow a fairly detailed picture of the sequence of events during contraction of muscle to be built up. The action potential evoked by the transmitter derived from nerve terminals finally reaches the interior of a muscle fibre through the T-system—an extension of the sarcolemma—and exerts an effect on the terminal cistern—the swollen part of the sarcoplasmic reticulum in close contact with the T-system—to release the Ca ions accumulated there. Ca ions then reach the contractile system by simple diffusion and induce the interaction of thin filaments, chiefly composed of actin, with thick filaments made up of myosin. When the action potential dies away, Ca ions are recaptured from the contractile elements by the whole surface of the sarcoplasmic reticulum and relaxation follows. It is of considerable significance that glycogenolyses, one of the principal energy sources for contraction, is also initiated by the same concentration of Ca ions as that which activates the contractile system[8].

Troponin as the Receptive Protein

At first sight it seems strange that the Ca ion was ignored for so long after the injection experiments mentioned above. An explanation may, however, be found in the report published in 1963 that the regulatory function of the Ca ion requires a corresponding protein system different from myosin and actin[1,2,9]. (It was later shown by Kendrick-Jones *et al.*[10] that, in molluscan muscle, the Ca-receptive site is located in the myosin molecule; thus the regulation mechanism involves an interesting phylogenic problem.)

在1941年~1942年间，人们首次认识到了钙离子在活体肌肉收缩中的重要作用，这一时间与森特–哲尔吉及其追随者[3]建立肌球蛋白–肌动蛋白–ATP系统的时间几乎相同。在海尔布伦[4]1940年发表其研究结果之后三年，镰田和木下[5]利用一根直径为2 μm~5 μm的超微针头向一个单肌纤维中注射了微量的钙离子，从而引发局部的可逆收缩。遗憾的是，这一杰出的先驱性工作因为日本饱受战火摧残而未能继续下去。与此同时，海尔布伦和威尔辛斯基[6]也发表了基本相同的结果。然而，这些不同寻常的观察结果随后被大多数从事肌肉研究的科学家忽视了十多年，这可能是因为肌肉收缩的基本过程，即肌球蛋白、肌动蛋白和ATP的相互作用，并不需要钙离子的存在。

钙离子理论的建立

对钙离子所起作用的广泛重视主要基于1959年到1961年间发表的几项研究成果[1,2,7]。首先，人们发现在体外实验中，收缩过程是被极低浓度的钙离子所调控的；10^{-7} M的钙离子可使收缩系统处于松弛状态，而10^{-5} M的钙离子则可以使之完全被激活。其次，在ATP存在时，肌肉细胞的内膜系统（肌浆网）会将钙离子从周围环境中泵入。内膜系统对钙离子的吸收力和亲和性足以使收缩系统维持在松弛状态。

根据这些数据就可以获得在肌肉收缩过程中所发生的一系列事件的精细图画：从神经末梢传出的信号所引发的动作电位最终通过T管系统（肌纤维膜的延伸结构）到达肌纤维内部，并在终池（肌浆网与T管系统紧密相连的肿胀部分）发挥作用，从而使在那里蓄积的钙离子被释放出来。随后钙离子通过简单扩散到达收缩系统，并引起主要由肌动蛋白构成的细肌丝与主要由肌球蛋白构成的粗肌丝之间的相互作用。当动作电位逐渐消失时，肌浆网的整个表面就会把钙离子从收缩系统中收回，随后将再次出现松弛状态。肌肉收缩的主要能量来源之一——糖原分解也是由与激活收缩系统浓度相同的钙离子所触发的[8]，这一点非常重要。

接受钙离子的肌钙蛋白

乍一看人们似乎会感到奇怪：为什么钙离子在上文提到的注射实验之后那么久还一直被人们忽略？不过，我们也许可以从 1963 年发表的报告中找到对此问题的一种解释：钙离子必须通过相应的蛋白质系统才能发挥调节作用，而这种蛋白质系统并不是肌球蛋白或肌动蛋白 [1,2,9]。（后来，肯德里克－琼斯等人发现 [10]：在软体动物的肌肉中，接受钙离子的位点位于肌球蛋白分子的内部；因此上述调控机制就会涉及一个有趣的系统发生问题。）

This regulatory protein system, composed of tropomyosin and troponin, a new protein, is distributed along the actin filament at intervals of 38 to 39 nm (Fig. 2 and refs. 1 and 2). The function of tropomyosin, a fibrous protein about 40 nm long, was thus clarified some seventeen years after its isolation by Bailey[12] in 1946: it mediates the effect of Ca ions on troponin bound to the actin filament. (The discovery of the new protein system led to the isolation of new myofibrillar proteins like α-actinin, β-actinin and M-protein, which are involved in the structural arrangement of thin and thick filaments; these proteins, together with troponin and tropomyosin, are called "regulatory proteins"[13] by contrast with the chief contractile proteins, namely myosin and actin.)

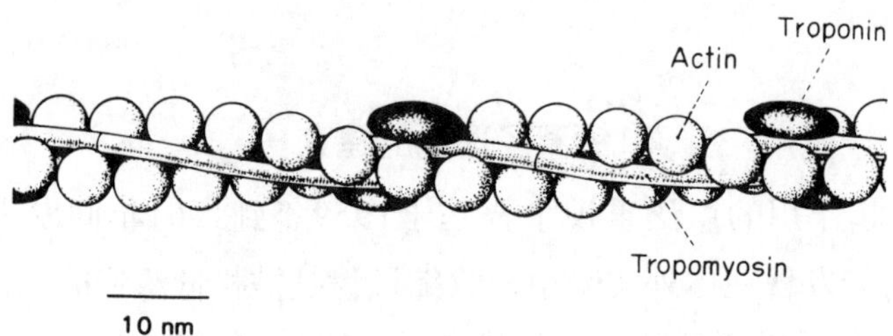

Fig. 2. A model for the fine structure of the thin filament. One tropomyosin molecule combines with one troponin molecule and seven actin molecules. The functional integrity of this morphological unit has recently been noticed and discussed[11]. (Fig. 5 in ref. 2 as modified by I. Ohtsuki.)

It must be emphasized that the Ca ion is not a simple activator of the contractile system. In the absence of Ca ions troponin, in the presence of tropomyosin, exerts an inhibitory effect on actin filaments. When Ca ions are bound to troponin, this inhibitory action does not take place and contraction follows; thus the Ca ion behaves as an inhibitor of inhibition, or a depressor[1,2]. Substantial support for this idea has come from the work of Ishiwata and Fujime[14] in Oosawa's laboratory, which is based on the measurement of the flexibility of the actin-tropomyosin-troponin complex using quasielastic light scattering technique.

Troponin is not a single protein but a complex of three components[15,16]. One component (of molecular weight 22,000 to 23,000) can act with tropomyosin to repress the interaction of actin with myosin. The light Ca-binding component (of molecular weight 17,000 to 18,000) opposes this repressing action; this effect is scarcely influenced by Ca ions. The heavy tropomyosin-binding component (of molecular weight 38,000 to 40,000) inactivates the derepressing action of the light component in the absence of Ca ions, that is in the relaxed state, through its interaction with the light component. It thus seems that the derepressing action of Ca ions is not exerted directly on the inhibitory component, but indirectly through the interaction of the other two components[17].

The subtle structural change that takes place in the thin filament—the troponin-tropomyosin-actin complex—under the influence of Ca ions has become one of the vital problems in muscle science[18]; the answer will undoubtedly provide a crucial clue about the interaction of myosin and actin. In this way, studies on the regulatory protein systems are now closely connected with those on the elementary process.

这种蛋白质调节系统由原肌球蛋白和一种新发现的蛋白——肌钙蛋白组成，分布在肌动蛋白丝中，间距为 38 nm ~ 39 nm（图 2 和参考文献 1、2）。原肌球蛋白是一种长约 40 nm 的纤维状蛋白，虽然贝利[12]早在 1946 年就分离出了这种蛋白，但直到约 17 年以后人们才把它的功能搞清楚：它负责介导钙离子对结合在肌动蛋白丝上的肌钙蛋白的影响。（这一新蛋白质系统的发现引导人们分离出了一些新的肌原纤维蛋白，如 α– 辅肌动蛋白、β– 辅肌动蛋白和 M 蛋白，这些蛋白都与细肌丝和粗肌丝的结构排布有关，它们与肌钙蛋白和原肌球蛋白一起被称为“调节蛋白”[13]，而与主收缩蛋白——肌球蛋白和肌动蛋白是相对的。）

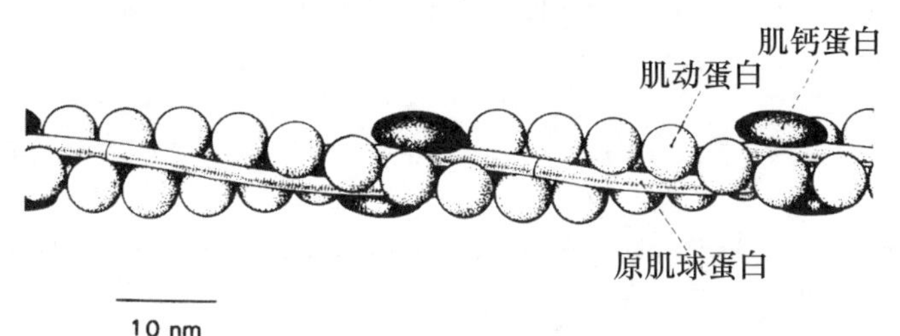

图 2. 细肌丝精细结构的模型。1 个原肌球蛋白分子与 1 个肌钙蛋白分子和 7 个肌动蛋白分子相结合。最近有人已经在关注和讨论这种形态单元的功能完整性[11]。（大月根据参考文献 2 中的图 5 作了修改）

必须着重指出，钙离子并不是一个简单的收缩系统激活剂。在存在原肌球蛋白但不存在钙离子的情况下，肌钙蛋白将发挥抑制肌动蛋白丝的作用。而在钙离子与肌钙蛋白结合后，就不会发生这种抑制作用了，随后出现的将是收缩过程。因而钙离子的表现就如同是抑制收缩作用的一种抑制剂[1,2]。石渡和藤明[14]在大沢实验室中的研究工作为上述想法提供了强有力的支持，他们的这一结论来源于用准弹性光散射技术对肌动蛋白 − 原肌球蛋白 − 肌钙蛋白复合物弹性的测量结果。

肌钙蛋白不是一种单一的蛋白，而是由三种成分组成的复合物[15,16]。其中有一种成分（分子量为22,000~23,000）可以和原肌球蛋白作用从而抑制肌动蛋白与肌球蛋白之间的相互作用。另一种成分是较轻的与钙结合的蛋白（分子量为17,000~18,000），其功能与上述抑制作用相反，这种作用几乎不受钙离子的影响。最后一种成分是较重的与原肌球蛋白结合的蛋白（分子量为38,000~40,000），它可以在没有钙离子存在的情况下抑制较轻成分的去抑制活性，即通过与较轻成分的相互作用使肌肉处于松弛状态。这样看来，钙离子的去抑制作用并不是直接作用于抑制成分之上，而是通过另两种成分之间的相互作用间接实现的[17]。

细肌丝内部，即肌钙蛋白 − 原肌球蛋白 − 肌动蛋白复合物，在钙离子作用下所发生的细微结构改变已成为肌肉学中的重要课题之一[18]。其结果无疑将会为肌球蛋白与肌动蛋白间相互作用的研究提供一个关键的线索。从这个角度讲，对调节蛋白系统的研究是和对基本过程的研究紧密联系在一起的。

Protein System and Living Fibres

Another remarkable contribution of Japanese scientists to the study of muscle is the so-called "skinned fibre", a muscle fibre from which the sarcolemma has been removed, which was developed by Natori[19] in 1949. This is quite a suitable material for investigating the relationship between contractile processes and the concentrations of Ca ions, and also the mechanism by which Ca is released from the sarcoplasmic reticulum. The confirmation of the role of Ca ion with the help of this kind of fibre[20] has certainly bridged the gap between the experiments on injection of Ca ions into muscle fibres and those involving extracts of the protein system. The proposal that a "Ca-induced Ca-release mechanism" is at work in the sarcoplasmic reticulum[2,21,22] is also based on experiments with skinned fibres.

(**240**, 217-218; 1972)

Setsuro Ebashi: Department of Pharmacology, Faculty of Medicine, and Department of Physics, Faculty of Science, University of Tokyo, Tokyo.

References:

1. Ebashi, S., and Endo, M., *Prog. Biophys. Mol. Biol.*, **18**, 123 (1968).
2. Ebashi, S., Endo, M., and Ohtsuki, I., *Quart. Rev. Biophys.*, **2**, 466 (1969).
3. Szent-Györgyi, A., *Studies from Institute of Medical Chemistry, University of Szeged*, 1 (1941-42); **2** (1942).
4. Heilbrunn, L. V., *Physiol. Zool.*, **13**, 88 (1940).
5. Kamada, T., and Kinosita, H., *Jap. J. Zool.*, **10**, 469 (1943).
6. Heilbrunn, L. V., and Wiercinski, F. J., *J. Cellular Comp. Physiol.*, **29**, 15 (1947).
7. Weber, A., *J. Biol. Chem.*, **234**, 2764 (1959).
8. Ozawa, E., Hosoi, K., and Ebashi, S., *J. Biochem.*, **61**, 531 (1967); Ozawa, E., *ibid.*, **71**, 321 (1972).
9. Ebashi, S., *Nature*, **200**, 1010 (1963).
10. Kendrick-Jones, J., Lehman, W., and Szent-Györgyi, A. G., *J. Mol. Biol.*, **54**, 327 (1970).
11. Bremel, R. D., and Weber, A., *Nature New Biology*, **238**, 97 (1972).
12. Bailey, K., *Nature*, **157**, 368 (1946).
13. Maruyama, K., and Ebashi, S., *The Physiology and Biochemistry of Muscle as a Food* (edit. by Briskey, E. J., Cassens, R. G., and Marsh, B. B.), **2**, 373 (Univ. of Wisconsin Press, 1970).
14. Ishiwata, S., and Fujime, S., *J. Mol. Biol.*, **68**, 511 (1972).
15. Hartshorne, D. J., and Mueller, H., *Biochem. Biophys. Res. Commun.*, **31**, 647 (1968).
16. Greazer, M. L., and Gergely, J., *J. Biol. Chem.*, **246**, 4226 (1971).
17. Ebashi, S., in *Thirty-seventh Cold Spring Harbor Symposium on Quantitative Biology* (*The Mechanism of Muscle Contraction*) (in the press).
18. Oosawa, F., Hanson, J., *et al.*, Haselgrove, J. C., and Huxley, H. E., in *Thirty-seventh Cold Spring Harbor Symposium on Quantitative Biology* (*The Mechanism of Muscle Contraction*) (in the press).
19. Natori, R., *J. Physiol. Soc. Japan* (in Japanese), **11**, 14 (1949); Natori, R., *Jikeikai Med. J.*, **1**, 119 (1954).
20. Podolsky, R. J., and Costantin, L. L., *Fed. Proc.*, **23**, 933 (1964).
21. Ford, L. E., and Podolsky, R. J., *Science*, **167**, 58 (1970).
22. Endo, M., Tanaka, M., and Ogawa, Y., *Nature*, **228**, 34 (1970).

蛋白质系统与活体纤维

日本科学家对肌肉学研究的另一个重要贡献是所谓的“脱鞘纤维”——一种去掉了肌纤维膜的肌肉纤维。“脱鞘纤维”是由名取礼二[19]于1949年开发的。这是一种非常适合于研究肌肉收缩过程与钙离子浓度关系，以及研究钙离子从肌浆网中释放的机理的材料。在脱鞘纤维的帮助下，人们明确了钙离子的功能[20]，这无疑可以使我们在向肌纤维中注射钙离子的实验和有关蛋白质系统提取物的实验之间建立起一定的关联。此外，关于“钙离子引发的钙离子释放机制”适用于肌浆网的假说[2,21,22]也是建立在脱鞘纤维的实验基础之上的。

（张锦彬 翻译；刘京国 审稿）

Transmission of Kuru from Man to Rhesus Monkey (*Macaca mulatta*) $8\frac{1}{2}$ Years after Inoculation

D. C. Gajdusek and C. J. Gibbs, jun.

Editor's Note

This paper, which focuses on the transmissible spongiform encephalopathy kuru, highlights the lengthy incubation times that can occur between exposure to the infectious agent and development of symptoms. Daniel Carleton Gajdusek and Clarence J. Gibbs focus on a rhesus monkey inoculated with a suspension of kuru-infected human brain that took 8 years to become ill. At that time, Creutzfeldt–Jakob disease (CJD), known to share many similarities with kuru, had not been shown to cause disease in old-world monkeys. But the researchers warn that none of these animals, inoculated with CJD-infected human brain suspensions, have been observed long enough to confirm this assumption.

NEUROLOGICAL disease has appeared in a female rhesus monkey (*Macaca mulatta*) following an asymptomatic incubation period of 8 years and 5 months after inoculation intracerebrally (i.c.) and intravenously (i.v.) with a 10% suspension of brain tissue from a human kuru patient. Clinical signs were remarkably similar to those observed in patients naturally affected with kuru and in sub-human primates in which the disease has been experimentally induced by inoculation of the kuru virus. Histopathological examination of the brain of this rhesus monkey has confirmed this first successful transmission of kuru to an old-world monkey. Previously the chimpanzee and four species of new-world monkeys (spider, *Ateles*; capuchin, *Cebus*; squirrel, *Saimiri*; and woolly, *Lagothrix*) have been found to be susceptible to kuru.

In August 1963, one chimpanzee (A2), 5 rhesus, 4 cynomolgus and 3 African green monkeys were inoculated i.c. (0.2 ml.) and i.v. (0.3 ml.) with a 10% suspension of brain tissue from kuru patient Enage. Thirty months after inoculation chimpanzee A2 developed experimental kuru and was killed in the terminal stages of disease 4 months after onset. During subsequent months (between 1 month and 62 months after inoculation) 1 rhesus (16L), the 4 cynomolgus and the 3 African green monkeys died of intercurrent infections without signs of neurological disease; histological examinations by light and electron microscopy revealed no pathological lesions in the central nervous system. The remaining 4 rhesus monkeys remained clinically well until January 1972 (101 months after inoculation) when one (11L) was noted to have occasional tremors and locomotor ataxia. The animal climbed with reluctance and caution and became withdrawn and docile. Her hair coat became rough and there was piloerection over the entire body. Neurological signs became progressively worse and one and a half months after onset she developed

接种$8\frac{1}{2}$年后库鲁病从人传染到恒河猴（猕猴）

盖杜谢克，小吉布斯

编者按

这篇文章讨论的重点是一种传染性海绵状脑病——库鲁病，文中特别强调了在接触病原体到发病之间的潜伏期可能会很长。丹尼尔·卡尔顿·盖杜谢克和克拉伦斯·吉布斯重点介绍了一只在接种人类库鲁病患者脑组织悬浮液 8 年后染病的恒河猴。当时，与库鲁病有很多相似之处的克罗伊茨费尔特–雅各布综合征（简称克雅氏病）尚不能使旧大陆猴致病。但研究人员提醒公众：目前人们对接种过克雅氏病患者脑悬浮液的动物的观察时间还不够长，因而不足以证明上述假设。

在雌恒河猴（猕猴）脑内和静脉内接种人类库鲁病患者的 10% 脑组织悬浮液，经过 8 年零 5 个月的无症状潜伏期后，恒河猴出现了神经病学疾病。临床症状与在自然感染库鲁病的患者中和在通过接种库鲁病病毒实验性诱发疾病的亚人类灵长动物中观察到的症状非常相似。用组织病理学方法对这只恒河猴的脑所作的检验已证实，库鲁病首次成功传染给了一只旧大陆猴。先前已发现黑猩猩和 4 种新大陆猴（蜘蛛猴，蛛猴属；僧帽猴，卷尾猴属；松鼠猴，松鼠猴属；绒毛猴，绒毛猴属）对库鲁病易感。

我们于 1963 年 8 月在 1 只黑猩猩（A2）、5 只恒河猴、4 只食蟹猴和 3 只非洲绿猴脑内接种（0.2 ml）和静脉内接种（0.3 ml）库鲁病患者埃纳格的 10% 脑组织悬浮液。接种 30 个月后，黑猩猩 A2 染上了实验性库鲁病并在该病发作后 4 个月进入晚期时死亡。在接种后的若干个月内(从接种后 1 个月到 62 个月)，1 只恒河猴(16L)、4 只食蟹猴和 3 只非洲绿猴死于没有神经病学疾病症状的并发性感染；用光学和电子显微镜进行组织学检查，结果显示在中枢神经系统中并无病理性损伤。其余 4 只恒河猴在临床上一直很健康，直到 1972 年 1 月（接种后 101 个月）时才发现其中有一只（11L）出现了偶尔发抖和运动性共济失调。这只猴子在攀爬时显得非常勉强和小心并且越来越孤僻和温顺。她的毛发变得粗糙并且全身都被立毛所覆盖。神经病学症状逐渐加重，她的四肢和躯体在发作后一个半月均出现了阵挛性痉挛，几乎整

clonic jerks of all four limbs and trunk and almost continuous coarse generalized tremors. Two and a half months after onset, although alert to her surroundings, she lay down on her side and had to be fed by hand. She was killed at this advanced stage of disease. At no time during clinical illness was there any fever nor were significant changes noted in haematological and serum chemistry values.

At necropsy the animal was thin with scant subcutaneous and omental fat. The brain was firm and, on cutting, a blanching of the grey matter was noted. There were no other gross pathological changes.

Preliminary histological examination of the brain by light and electron microscopy revealed extensive neuropathological lesions restricted to the grey matter. There was moderate to severe status spongiosus of the cerebral cortex and basal ganglia, most severe in the deeper layers of the cortical mantle and less extensive in the dentate nucleus of the cerebellum (Fig. 1). In all areas examined there was marked intraneuronal vacuolation, loss of neurones and astroglial proliferation and hypertrophy.

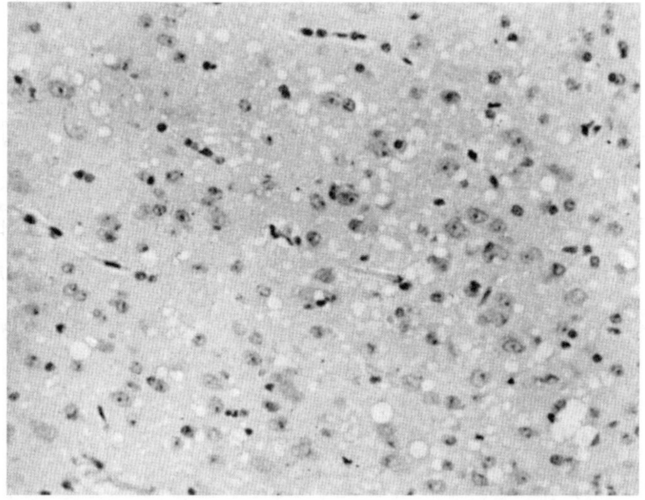

Fig. 1. Status spongiosis in the cerebral cortex of a rhesus monkey (11L) dying with kuru. In the areas of spongiform alteration there is neuronal loss and gliosis. Haematoxylin and eosin stain.

To date Creutzfeldt–Jakob disease, which resembles kuru in both the neuropathology of its cellular lesion and in many properties of its virus, has shown the same species specificity as kuru except for its failure thus far to cause disease in any of the old-world monkeys inoculated. However, none of the animals inoculated with brain suspensions from Creutzfeldt–Jakob disease has yet been observed over asymptomatic incubation periods as long as the 8.5 years required for disease to appear in this kuru-affected rhesus monkey. Rhesus monkeys inoculated with C–J disease virus 42 months ago are still under observation and remain well.

个身体都在不断地发抖。发作后两个半月时，虽然对周围环境依旧警觉，但她侧躺着不动，只能由专人喂食。这只恒河猴死于该病晚期。在临床患病期间，她既没有发过烧，也没有在血液学和血清生化指标上出现可观察到的显著变化。

在尸体剖检时发现，这个动物的皮下和网膜脂肪都很薄。脑很硬实，我们在切片时注意到灰质发生了漂白。没有出现其他的大体病理改变。

我们用光学和电子显微镜对猴子的脑进行了初步的组织学检查，结果显示广泛的神经病理学损伤仅限于灰质内。在大脑皮层和基底神经节处有中度至重度的海绵状状态，最为严重的是皮层外膜的较深层，而在小脑齿状核处则比较少（图 1）。在所有检查区域中都出现了明显的神经细胞内空泡、神经元丢失及星型胶质细胞的增生与肥大。

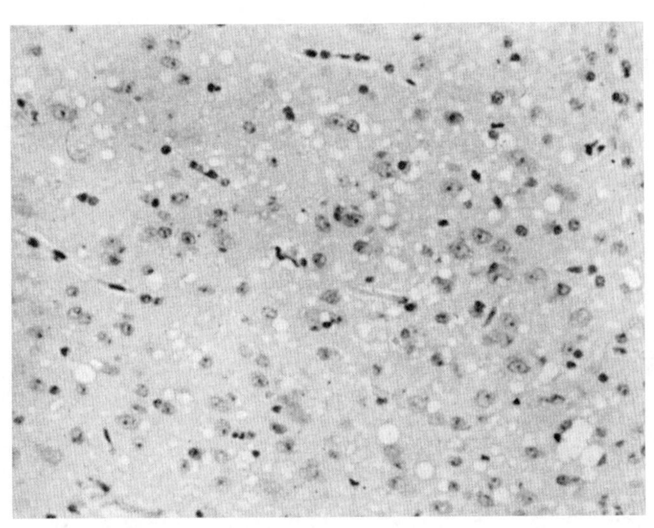

图 1. 一只将死于库鲁病的恒河猴（11L）脑皮层中的海绵状状态。在海绵状改变区域中出现了神经元丢失和神经胶质增生。苏木素和伊红染色。

迄今为止，克罗伊茨费尔特－雅各布综合征在细胞损伤的神经病理学上和病毒的许多特性上都类似于库鲁病。除了尚不能在接种的旧大陆猴中引发疾病外，该病与库鲁病有相同的种属特异性。然而，目前还没有人对接种克雅氏病患者脑悬浮液的动物无症状潜伏期的观察达到 8.5 年，8.5 年是这只感染库鲁病的恒河猴从接种到发病所需的时间。我们正在观察 42 个月前接种克雅氏病病毒的恒河猴，它们现在仍很健康。

It is of interest that the two spongiform encephalopathies of animals have been transmitted recently to old-world monkeys: scrapie to the cynomolgus monkey[1] and mink encephalopathy to the rhesus monkey[2]. In view of the rapid decrease in the incubation period of kuru in the chimpanzee and some new-world monkeys on serial passage, it seems likely that the host range may be altered by serial passage in different hosts, even perhaps on blind passages in hosts not yet demonstrated to be susceptible.

We thank Drs. Peter Rampert and Reid Heffner for electron microscopy.

(**240**, 351; 1972)

D. C. Gajdusek and C. J. Gibbs, jun.: National Institute of Neurological Diseases and Stroke, National Institutes of Health, Bethesda, Maryland.

Received June 26; revised July 31, 1972.

References:

1. Gibbs, jun., C. J., and Gajdusek, D. C., *Nature*, **236**, 73 (1972).
2. Marsh, R. F., Burger, D., Eckroade, R., ZuRhein, G. M., and Hanson, R. P., *J. Infect. Dis.*, **120**, 713 (1969).

值得关注的是，最近有两种动物海绵状脑病传染到了旧大陆猴中：羊瘙痒症传染到食蟹猴中[1]；水貂脑病传染到恒河猴中[2]。鉴于库鲁病在黑猩猩和一些新大陆猴中连续传代的潜伏期迅速缩短，可以猜测在不同寄主内的连续传代也许会改变寄主范围，甚至在寄主中盲传时也有可能是易感的，尽管现在没有证据能证明这一点。

感谢彼得·兰伯特博士和里德·赫夫纳博士为我们提供了电子显微镜。

（李梅 翻译；袁峥 审稿）

Inhibition of Prostaglandin Synthetase in Brain Explains the Antipyretic Activity of Paracetamol (4-Acetamidophenol)

R. J. Flower and J. R. Vane

Editor's Note

A year before this paper, British pharmacologist John Robert Vane proposed a mode of action for aspirin-like drugs, ascribing their therapeutic effects to the suppression of prostaglandin production. Here, Vane and colleague Rod Flower lend support to the theory with the demonstration that paracetamol's fever-easing ability occurs because the enzyme required to make prostaglandin is inhibited in the brain. Prostaglandins occur in most tissues and organs, so the find supported the idea that aspirin-like drugs targeting tissue-specific prostaglandin release could yield a new generation of anti-inflammatory compounds with greater specificity. It also helped explain the previously puzzling observation that paracetamol is inactive against dog spleen synthetase—tissues from different body regions show differing levels of sensitivity to aspirin-like drugs.

INHIBITION of prostaglandin biosynthesis by aspirin-like drugs[1-3] has now been confirmed in several systems[4-7]. The theory[1] that this anti-enzyme action is the basis of the clinical effects of aspirin-like drugs has recently been reviewed[8-11] in detail. One of the few anomalies was that paracetamol (4-acetamidophenol) which has no anti-inflammatory activity, but is analgesic and antipyretic[12], was inactive against dog spleen synthetase (ID_{50}=100 μg ml.$^{-1}$). A possible explanation for this discrepancy is that synthetase systems from different regions of the body show different sensitivities to drugs.

Prostaglandins are themselves pyrogenic[13] and occur in the central nervous system[14]. Feldberg *et al.*[15] found that a prostaglandin-like substance appeared in the cerebrospinal fluid (CSF) at the same time as the fever produced by intravenous injections of pyrogen. Paracetamol produced a prompt defervescence, accompanied by a reduction to normal concentrations of the prostaglandin-like substance in the CSF. It seems clear that paracetamol acts centrally as an antipyretic, and we have, therefore, investigated the effect of this drug on a prostaglandin synthetase system derived from brain.

Rabbit brain was used, but, in case the effects were due to species differences, confirmatory evidence was obtained on dog brain.

New Zealand white rabbits (or mongrel dogs) of either sex were killed by an overdose of pentobarbitone sodium given intraperitoneally or intravenously. The brain was removed as

对大脑前列腺素合成酶的抑制可以解释扑热息痛（4–乙酰氨基酚）的退热作用

弗劳尔，文

编者按

在这篇论文发表的前一年，英国药理学家约翰·罗伯特·文就已提出阿司匹林类药物的一种作用方式是通过抑制前列腺素合成而发挥疗效。在本文中，文及其同事罗德·弗劳尔对上述观点给予了支持，他们指出：扑热息痛之所以具有退热能力是因为大脑中合成前列腺素所需的酶受到了抑制。因为前列腺素存在于大多数组织和器官中，所以该发现支持了以下观点，即通过以特定组织中前列腺素释放为靶向的阿司匹林类药物可能得到一类新的特异性更强的抗炎化合物。这篇文章也有助于我们解释以前发现的一个令人困惑的现象：扑热息痛对狗脾的前列腺素合成酶无效——身体不同部位的组织对阿司匹林类药物的敏感性不同。

目前，已有多个系统的研究[4-7]证实，阿司匹林类药物[1-3]可以抑制前列腺素的生物合成。近来亦有文献[8-11]详细综述了有关理论[1]，即对前列腺素合成酶的拮抗作用是阿司匹林类药物临床功效的基础。也有少数例外情况，其中之一是没有抗炎作用、但有止痛和退热作用的扑热息痛（4–乙酰氨基酚）[12]对狗脾前列腺素合成酶无效（ID_{50}=100 μg/ml）。出现这种不一致的原因可能是身体不同部位的合成酶系统对药物的敏感性不同。

前列腺素本身有致热作用[13]，存在于中枢神经系统中[14]。费尔德伯格等人[15]发现：当通过静脉注射热原导致发热的同时，脑脊液（CSF）中也出现了一种前列腺素样的物质。应用扑热息痛能够迅速退热，并且CSF中前列腺素样物质的浓度会下降至正常水平。显然，扑热息痛是作用于中枢的退热剂，因而我们研究了该药物对大脑前列腺素合成酶系统的影响。

我们应用兔脑进行研究；为了防止其效果存在种属差异，又应用狗脑进行了研究确证。

腹腔或静脉注射过量戊巴比妥钠处死任意性别的新西兰白兔（或杂种狗）。以最快的速度取出脑组织，用剪刀剪碎，再用冷的克雷布斯溶液冲洗以去除残留血液，

quickly as possible, cut into pieces with scissors, washed in cold Krebs solution to remove any residual blood, and homogenized in four times its volume of ice cold phosphate buffer (*p*H 7.4) for 1 min at full speed in a Waring blender. The homogenate was then centrifuged at 100,000*g* for 1 h to separate the soluble cytoplasmic fraction which contains an NAD^+-dependent prostaglandin destroying factor. After the supernatant had been discarded pellets were resuspended in phosphate buffer in the proportion (by volume) of 1:3. This suspension (1 ml.) was added to 1 ml. of the buffer containing 20 μg sodium arachidonate, 100 μg reduced glutathione and 10 μg hydroquinone. Drugs tested as inhibitors were added to the reaction mixtures in varying concentrations. Samples were incubated aerobically for 20 min at 37°C with shaking, after which time the reactions were stopped by heating the tubes in boiling water for 1 min.

Prostaglandin was routinely bioassayed using the rat fundic strip[16], but the identity of the biologically active reaction products was verified by parallel bioassay using the rat fundic strip, rat colon and chick rectum, superfused in series with Krebs solution containing antagonists[17] and by thin layer chromatography (TLC) using the AI and AII systems of Greén and Samuelsson[18]. More than 90% of the biological activity eluted from the TLC plates corresponded to the prostaglandin E_2 standard. The potency of the inhibitors of prostaglandin synthesis was calculated from the curves plotted for % inhibition versus log concentration and expressed as the concentration (μg $ml.^{-1}$) required to produce 50% inhibition (ID_{50}) of the control synthesis. Brain homogenates were freshly prepared for every experiment.

Zero time activity of the incubation mixture was equivalent to 275±54 ng prostaglandin E_2 (mean ±s.e.). After incubation, there was an average increase in prostaglandin content of 180±31 ng $ml.^{-1}$.

Table 1 shows the ID_{50} concentrations of indomethacin, aspirin and paracetamol against the synthetase derived from rabbit brain compared with their ID_{50} values against the dog spleen enzyme which we reported earlier[6]. Aspirin had similar activities in both systems. However, indomethacin was much less active as an inhibitor of prostaglandin production in brain tissue than in spleen and paracetamol was a more potent inhibitor of brain enzyme than of spleen enzyme. In dog brain also, paracetamol had a similar relatively high potency (ID_{50}=12.5 μg $ml.^{-1}$).

Table 1. Inhibition of Prostaglandin Formation by Anti-Inflammatory Drugs

	Dog spleen synthetase ID_{50} μg $ml.^{-1}$	Rabbit brain synthetase ID_{50} μg $ml.^{-1}$
Indomethacin	0.06	1.3
Sodium aspirin	6.6	11.0
4-Acetamidophenol	100.0	14.0

There is a remarkable correlation between the clinical actions of these drugs and their

然后加入 4 倍体积冰块预冷的磷酸盐缓冲液（pH 值 7.4），用韦林氏搅切器以最高速度匀浆 1 min。随后将该匀浆以 100,000g 离心 1 h，分离出含有 NAD^+ 依赖性前列腺素破坏因子的可溶性胞质组分（译者注：NAD 是一种传递电子的辅酶，成分为烟酰胺腺嘌呤二核苷酸）。弃除上清液后，将沉淀用比例为 1:3（体积）的磷酸盐缓冲液再次悬浮。取该悬浮液（1 ml）加入 1 ml 含 20 μg 花生四烯酸钠、100 μg 还原型谷胱甘肽和 10 μg 对苯二酚的缓冲液中。将不同浓度的待测药物加入上述反应混合液。将样品置于 37℃下有氧振荡孵育 20 min，随后将试管放入沸水中加热 1 min 以中止反应。

按照常规用大鼠胃底条 [16] 对前列腺素的生成量进行生物测定。但在对这些生物活性反应产物进行鉴别时采用了大鼠胃底条、大鼠结肠和小鸡直肠平行测定法，用含拮抗剂的克雷布斯溶液对它们进行连续灌流 [17]；还采用了格伦和萨穆埃尔松发明的在 AI 和 AII 系统中进行的薄层色谱法（TLC）[18]。90% 以上从 TLC 平板上洗脱的生物活性物质与前列腺素 E_2 的标准品相符。以抑制百分比对浓度的对数作图来计算前列腺素合成抑制剂的效能，效能的大小用对前列腺素合成对照组产生 50% 抑制（ID_{50}）所需的浓度（μg/ml）来表示。每次实验均使用新鲜制备的脑匀浆。

孵育混合物的零时刻活性相当于 275 ng ± 54 ng 前列腺素 E_2（均值 ± 标准误差）。孵育后，前列腺素含量平均上升了 180 ng/ml ± 31 ng/ml。

表 1 给出了消炎痛、阿司匹林和扑热息痛抑制兔脑前列腺素合成酶的 ID_{50} 浓度，以及与我们以前曾报道过的它们抑制狗脾酶 ID_{50}[6] 的比较。阿司匹林在两种系统中的活性相似。不过作为前列腺素合成的抑制剂，消炎痛在脑组织中的活性要远远低于在脾脏中的活性，而扑热息痛对脑组织中酶的抑制作用则要强于对脾脏中酶的抑制作用。扑热息痛在狗脑中同样具有相对较强的效能（ID_{50}=12.5 μg/ml）。

表 1. 抗炎药物对前列腺素合成的抑制作用

	狗脾合成酶 ID_{50} μg/ml	兔脑合成酶 ID_{50} μg/ml
消炎痛	0.06	1.3
阿司匹林钠	6.6	11.0
4–乙酰氨基酚	100.0	14.0

这些药物的临床疗效与其抑制某一种前列腺素合成酶的效能之间存在着显著的

potency against one or other of the enzymes. All three compounds are antipyretic (a central action) and all have an inhibitory effect on the brain enzyme in concentrations found in the plasma after therapeutic doses[6]. Both as antipyretics and as inhibitors of brain prostaglandin synthetase, the descending order of potency is indomethacin; aspirin; paracetamol[12,13]. Furthermore, indomethacin is about twelve times more potent than aspirin in both tests. The sensitivity of prostaglandin synthetase derived from brain tissue to paracetamol seems to occur in other species also, for Willis *et al.*[11] found that paracetamol inhibited enzymes from mouse and gerbil brain (ID_{50}=20 μg ml.$^{-1}$).

Indomethacin and aspirin are both anti-inflammatory (a peripheral action) and both are active against the spleen enzyme (a peripheral tissue) in therapeutic concentrations. The much higher potency of indomethacin as a synthetase inhibitor both in dog spleen[6] and in bull seminal vesicles[5] is also reflected against inflammation in animal models[5,19] and in man[12]. Paracetamol has no anti-inflammatory activity and is not active against the dog spleen enzyme in therapeutic concentrations.

In addition to providing further support for Vane's theory[1,8], we believe that our results illustrate the mechanism by which the clinical actions of the aspirin-like drugs are determined by the differential sensitivity of the prostaglandin synthetase systems of the "target" tissues. Thus they support the idea[1,8] that a study of prostaglandin synthetase systems from different tissues will lead to aspirin-like drugs with a greater specificity of action.

We thank Dr. J. Pike of the Upjohn Company, Kalamazoo, for his gift of prostaglandins, and the Wellcome Trust and the MRC for grants.

(**240**, 410-411; 1972)

R. J. Flower and J. R. Vane: Department of Pharmacology, Institute of Basic Medical Sciences, Royal College of Surgeons of England, Lincoln's Inn Fields, London WC2A 3PN.

Received September 14, 1972.

References:

1. Vane, J. R., *Nature New Biology*, **231**, 232 (1971).
2. Smith, J. B., and Willis, A. L., *Nature New Biology*, **231**, 235 (1971).
3. Ferreira, S. H., Moncada, S., and Vane, J. R., *Nature New Biology*, **231**, 237 (1971).
4. Smith, W. L., and Lands, W. E. M., *J. Biol. Chem.*, **246**, 6700 (1971).
5. Tomlinson, R. V., Ringold, H. J., Qureshi, M. C., and Forchielli, E., *Biochem. Biophys. Res. Comm.*, **46**, 552 (1972).
6. Flower, R., Gryglewski, R., Herbaczynska-Cedro, K., and Vane, J. R., *Nature New Biology*, **238**, 104 (1972).
7. Sykes, J. A. C., and Maddox, I. S., *Nature New Biology*, **237**, 59 (1972).
8. Vane, J. R., *Hospital Practice*, **7**, 61 (1972).
9. Vane, J. R., in *Inflammation: Mechanisms and Control* (edit. by Lepow, I. H., and Ward, P. A.) (Academic Press, New York, 1972).
10. Vane, J. R., *Proc. Fifth Intern. Congr. Pharmacol.* (Karger, Basle, in the press).
11. Willis, A. L., Davison, P., Ramwell, P. W., Brocklehurst, W. E., and Smith, J. B., in *Prostaglandins in Cellular Biology* (edit. by Ramwell, P. W., and Pharriss, B. B.), 227 (Plenum Press, New York and London, 1972).

相关性。所有这三种化合物均可退热（主要作用），且在使用治疗剂量后的血浆浓度水平下对脑组织酶都有抑制作用[6]。作为退热剂和大脑前列腺素合成酶的抑制剂，以上药物的效能从高到低依次为：消炎痛、阿司匹林、扑热息痛[12,13]。此外，在两组实验中消炎痛的作用都约为阿司匹林的 12 倍。来自脑组织的前列腺素合成酶对扑热息痛的敏感性似乎在其他动物中也存在，因为威利斯等人[11]发现扑热息痛也能抑制小鼠和沙鼠脑中的酶（ID_{50} = 20 μg/ml）。

消炎痛和阿司匹林都有抗炎作用（次要作用），两者在治疗浓度下都具有拮抗脾酶（外周组织）的活性。消炎痛在狗脾[6]和牛精囊[5]中表现为更强的合成酶抑制剂，这种更强的效能同样体现在其对动物模型[5,19]和人体[12]的抗炎作用上。而扑热息痛没有抗炎作用，在治疗浓度下也没有抑制狗脾酶的活性。

除了能为文氏理论[1,8]提供进一步的支持以外，我们认为我们的结果还阐明了一种机制，即阿司匹林类药物的临床功效取决于其对“靶”组织前列腺素合成酶系统的敏感性差异。因此也支持了这样一个观点[1,8]，即对不同组织前列腺素合成酶系统的研究将会促使我们研发出特异性更强的阿司匹林类药物。

感谢美国卡拉马祖市普强制药公司的派克博士为我们提供了前列腺素，还要感谢英国维康信托基金会和英国医学研究理事会为我们提供资金上的援助。

（周志华 翻译；王昕 审稿）

12. Woodbury, D. M., in *The Pharmacological Basis of Therapeutics* (edit. by Goodman, L. S., and Gilman, A.), fourth ed. (Macmillan, New York, 1970).

13. Milton, A. S., and Wendlandt, S., *J. Physiol.*, **207**, 76P (1970).

14. Coceani, F., and Wolfe, L. S., *Canad. J. Physiol. Pharmacol.*, **43**, 445 (1965).

15. Feldberg, W., Gupta, K. P., Milton, A. S., and Wendlandt, S., *Brit. J. Pharmacol.* (in the press).

16. Vane, J. R., *Brit. J. Pharmac. Chemother.*, **12**, 344 (1957).

17. Gilmore, N., Vane, J. R., and Wyllie, J. H., *Nature*, **218**, 1135 (1968).

18. Greén, K., and Samuelsson, B., *J. Lip. Res.*, **5**, 117 (1964).

19. Collier, H. O. J., *Adv. Pharmac. Chemother.*, **7**, 333 (1969).

A Possible Role for Histone in the Synthesis of DNA

H. Weintraub*

Editor's Note

In the 1950s, it was recognized that native DNA found in living cells is intimately associated with protein molecules called histones, but the function of these proteins was not known. This article by Harold Weintraub at the University of Pennsylvania, Philadelphia, suggested that histones regulate the synthesis of DNA. The modern view is that histone molecules are structural elements in the chromosome, and that they serve as a means of binding the double helix of DNA to a platform. It is also now know that chemical modification of the histone molecules can determine which genes in the genome are sued for the production of proteins in the living cell. Weintraub's conclusion, in other words, was only part of the story.

WHEN protein synthesis is inhibited in primitive chick erythroblasts, linear DNA synthesis continues for about 45 min, but at half the control rate[1]. The depression in DNA synthesis occurs within 25 s of the addition of cycloheximide, is readily reversible, and is manifested as a decrease by half in the gross rate of DNA chain elongation. The labelled DNA made under these conditions is covalently bound to previously synthesized DNA and associated with single stranded segments well over 10^7 molecular weight. The effects of cycloheximide on DNA synthesis are due neither to a direct action on the DNA replicase, nor to an effect on cell permeability or nucleotide metabolism mediated by decreased protein synthesis. It is also unlikely that this behaviour is a consequence of the accumulation of an inhibitor in the absence of protein synthesis[1]. All our data indicate that when the synthesis of protein is inhibited, the synthesis of DNA proceeds normally except that the rate of chain elongation is slower. This applies to each replicon in each S phase cell. Here I present evidence that the proteins, termed chain elongation proteins (CEP), involved in this fine control of DNA synthesis are histones.

Cycloheximide Resistance

It was previously shown[1] that low concentrations of cycloheximide or puromycin, although inhibiting bulk protein synthesis by as much as 30%, resulted in no inhibition of DNA synthesis. This could indicate that the synthesis of the CEP was relatively resistant to low concentrations of inhibitor, or that a pool of this protein existed within the cell. The latter possibility was unlikely since higher concentrations of inhibitor lead to a maximal effect on DNA synthesis within 25 s. The dose response curves to cycloheximide for DNA synthesis and the synthesis of various protein fractions are shown in Fig. 1. DNA synthesis was linear

* Present address: MRC Laboratory of Molecular Biology, Hills Road, Cambridge CB2 2QH, England.

组蛋白在DNA合成中的可能角色

温特劳布 *

编者按

在20世纪50年代，人们就认识到存在于活细胞中的天然DNA与被称为组蛋白的蛋白质分子密切相关，但是这类蛋白质的功能尚不为人所知。来自费城宾夕法尼亚大学的哈罗德·温特劳布在本文中指出，组蛋白调控DNA的合成。现代的观点是：组蛋白分子是染色体中的结构组分，DNA双螺旋就是通过它们与基体相连接的。现在人们还知道，通过对组蛋白分子的化学修饰可以确定基因组中有哪些基因参与了活细胞中蛋白质的合成。换句话说，温特劳布的结论只回答了一部分问题。

当小鸡前成红细胞中的蛋白质合成被抑制后，线性DNA的合成仍可以持续45 min左右，但合成率只有对照组的一半[1]。DNA的合成在加入放线菌酮后25 s之内就会被抑制，这种抑制作用很容易得到恢复，并且已经证明此时DNA链的总延伸速率下降了一半。在上述条件下形成的标记DNA与之前合成的DNA以共价键相连，并且还结合到了分子量大大超过10^7的单链DNA片段上。放线菌酮对于DNA合成的抑制作用既不是因为它可以直接抑制DNA复制酶的活性，也不是因为它能通过抑制蛋白质合成来影响细胞的通透性或核苷酸的代谢。这种现象亦不可能是在蛋白质合成停止后由某种抑制性物质的积累导致的结果[1]。我们的所有实验数据均表明：当蛋白质合成受到抑制时，DNA的合成仍可以正常进行，只不过链延伸的速率有所减慢。这适用于所有处于S期的细胞的各个复制子。我将在本文中提出证据证明，这种参与DNA合成过程中此类精细调控的被称作链延伸蛋白（CEP）的蛋白就是组蛋白。

放线菌酮抗性

之前的研究结果表明[1]：虽然低浓度的放线菌酮或嘌呤霉素会使总蛋白合成量减少30%，但它们并不影响DNA的合成。这也许可以说明CEP的合成对于低浓度的抑制剂有相当的抗性，或者在细胞内存在着一定量的CEP库存。后一种可能性不大，因为高浓度抑制剂可以在25 s之内达到对DNA合成的最大抑制。放线菌酮对DNA合成及多种蛋白质合成的量效曲线示于图1。在至少30 min内，DNA合成的抑制

* 现在的地址：英国剑桥医学研究理事会分子生物学实验室，希尔斯路，剑桥 CB2 2QH，英国。

for at least 30 min. The curve for DNA synthesis normalized to the maximum amount of DNA inhibition is also given. Data show that histone synthesis is relatively resistant to cycloheximide[3,4] and that the dose response curve of histones follows the normalized dose response curves for DNA synthesis. SDS-acrylamide gel analysis of the cytoplasmic fraction and the acidic nuclear protein fraction at 1, 5, and 20 μM cycloheximide detected no proteins resistant to cycloheximide using a double-label analysis. Similar experiments showed all histones to be equally affected at each cycloheximide concentration.

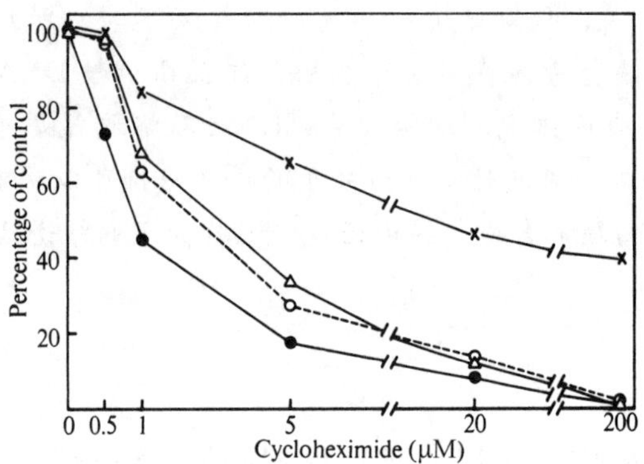

Fig. 1. Resistance of histone synthesis to cycloheximide. Four day old erythroblasts were incubated *in vitro*[1] in L-leucine-4, 5-^{3}H (20 μCi ml.$^{-1}$; 35 Ci mM^{-1}) or ^{3}H-methyl-TdR (20 μCi ml.$^{-1}$; 15 Ci mM^{-1}) for 30 min in the presence and absence of the stated concentrations of cycloheximide. The various protein fractions were then isolated[2] from cells labelled with leucine. Briefly, nuclei were obtained from washed cells using 0.5% "Nonidet" in RSB (0.01 M NaCl; 0.01 M Tris-HCl, *p*H 7.4; 0.003 M $MgCl_2$). Histone (○---○) was extracted twice from washed nuclei using 200 volumes of 0.25 M H_2SO_4. The residual proteins were termed the acidic nuclear proteins (●—●). DNA(×—×), added to the cytoplasmic fraction, sedimented and extracted with acid, failed to show any bound histone-like protein. The incorporation is expressed as a percentage of that incorporated by control cells either by an internally controlled double-label analysis[2] or by direct measurements. Incorporation of ^{3}H-TdR into TCA precipitable material was measured directly[1]. "Normalized DNA" (△—△) represents the amount of inhibition by cycloheximide as a percentage of the maximum amount achieved at 200 μM (68% inhibition). All isotopes were obtained from New England Nuclear Corporation; all analogues and inhibitors, from Sigma Corporation. Radioactive incorporation was measured using a "Beckman LS-200 Liquid Scintillation Counter".

The Effects of Amino-acid Analogues

Compared to other proteins, histones are unusual in that they contain no tryptophan, little tyrosine, but much arginine. It is shown in Table 1 that DNA synthesis is (*a*) resistant to high concentrations of the tryptophan analogues, methyl-tryptophan and beta indole acrylic acid, (*b*) only slightly sensitive to high concentrations of the tyrosine analogues, methyl-tyrosine and iodo-tyrosine, and, (*c*) extremely sensitive to low does of the arginine analogue, canavanine. Also that the tryptophan and tyrosine analogues can interfere with the metabolism of their corresponding amino-acids as equimolar concentrations can markedly depress uptake of labelled amino-acid into TCA precipitable material. Addition of cycloheximide with canavanine results in only a light potentiation. This indicates that

率与放线菌酮的浓度呈线性关系。本文还给出了归一化为最大 DNA 合成抑制量的 DNA 合成曲线。这些数据表明：组蛋白的合成对于放线菌酮具有相当大的抗性[3,4]，并且组蛋白的量效曲线与对 DNA 合成进行归一化后的量效曲线是一致的。在放线菌酮浓度为 1 μM、5 μM 和 20 μM 时对细胞质组分以及酸性核蛋白组分进行十二烷基硫酸钠（SDS）– 丙烯酰胺凝胶电泳分析的结果显示：通过双标记分析可以得到，这些蛋白质中没有哪个蛋白质的合成对放线菌酮有抗性。类似的实验表明：在各种放线菌酮浓度下，所有组蛋白都受到相同程度的影响。

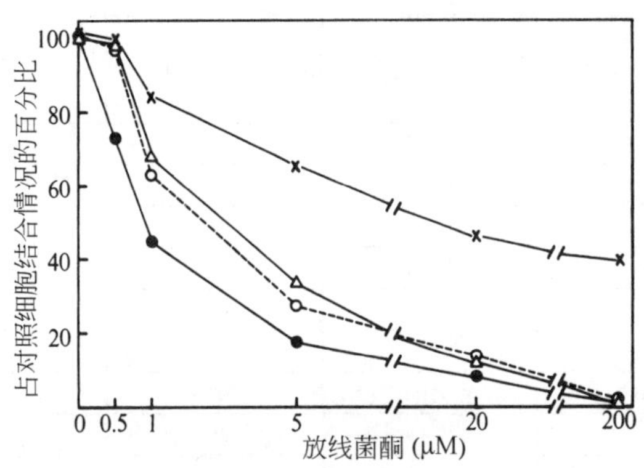

图 1. 组蛋白合成对放线菌酮的抗性。将第 4 天的鸡成红细胞在体外[1]与 L– 亮氨酸 –4, 5–^{3}H（20 μCi/ml, 35 Ci/mM）或 ^{3}H– 甲基胸腺嘧啶脱氧核苷（20 μCi/ml, 15 Ci/mM）一起孵育 30 min，有一组不加入放线菌酮，另外几组中放线菌酮的浓度如图所示。然后从亮氨酸标记的细胞中分离出各种蛋白质组分[2]。简言之就是用含 0.5%“诺乃洗涤剂”的网织红细胞标准缓冲液（0.01 M NaCl；0.01 M Tris-HCl，pH 值 7.4；0.003 M $MgCl_2$）从冲洗后的细胞中分离细胞核组分。用 200 倍体积的 0.25 M H_2SO_4 从冲洗后的细胞核中抽提 2 次以得到组蛋白（○—○）。残留的蛋白质被称为酸性核蛋白（●—●）。将 DNA（×—×）加入细胞质组分中，然后在酸性条件下进行沉淀和抽提，并未发现其结合了类似组蛋白的蛋白。通过内参双标记分析法[2]或直接测定法用占对照细胞掺入情况的百分比来表征掺入率。三氯乙酸（TCA）沉淀物中 ^{3}H– 胸腺嘧啶脱氧核苷的掺入率是通过直接测定得到的[1]。“归一化的DNA”(△—△)代表放线菌酮的抑制率占放线菌酮在浓度为200 μM时所获得的最大抑制率(68%)的百分比。所有放射性同位素均来自新英格兰核公司，所有氨基酸类似物和抑制剂均来自西格马公司。利用“贝克曼公司的 LS-200 液闪计数器”来测量放射性的掺入。

氨基酸类似物的作用

与其他蛋白质相比，组蛋白的特殊性在于它不含色氨酸，只含很少量的酪氨酸，但含有大量的精氨酸。表 1 说明 DNA 的合成：(*a*) 耐受高浓度的色氨酸类似物——甲基 – 色氨酸以及 β– 吲哚丙烯酸；(*b*) 对高浓度的酪氨酸类似物——甲基 – 酪氨酸和碘代酪氨酸只表现出轻度的敏感；(*c*) 对低剂量的精氨酸类似物——刀豆氨酸极度敏感。此外，色氨酸和酪氨酸类似物可以干扰与它们对应的两种氨基酸的代谢，因为当摩尔浓度相等时，在 TCA 沉淀物中这两种放射性标记的氨基酸含量明显下降。同时加入放线菌酮和刀豆氨酸只能起到轻微的强化作用。这表明上述两种药物可能

both drugs are probably affecting the same step in DNA synthesis, made more likely by the observation (not shown) that the kinetics of inhibition by canavanine mimic those of cycloheximide. In canavanine, ^{3}H-leucine incorporation is decreased by 10%. (Fig. 1 shows that a comparable amount of inhibition by cycloheximide has no effect on DNA synthesis.) This might indicate that canavanine affects DNA synthesis by inhibiting the synthesis of a specific class of proteins, rather than producing altered proteins of a specific class.

Table 1. Effect of Amino-acid Analogues on the Synthesis of DNA

Treatment	Amino-acid ratio[4] (histone to nuclear acidic)*	Ratio of analogue to amino-acid†	% Inhibition of ^{3}H-TdR incorporation	Ratio of analogue to amino-acid†	% Inhibition of labelled amino-acid uptake‡
C-^{3}H-tryptophan	0	100:1	0	1:1	22%
3β-Indole acrylic acid	0	100:1	0	1:1	24%
3-Iodo-L-tyrosine	0.09	100:1	22%	1:1	25%
O-C-^{3}H-L-tyrosine	0.09	100:1	24%	1:1	26%
plus cyclo.			56%		
Cyclo. alone			53%		
L-Canavanine	1.83	1:1	62%		
plus cyclo.			68%		
Cyclo. alone			58%		

Cells were treated with the analogues and labelled concurrently with ^{3}H-TdR for 30 min.

* Ratio of amino-acid in histone to that in nuclear acidic protein for the amino-acid corresponding to the given analogue. Incorporation of ^{3}H-TdR was measured as a percentage of an untreated control culture.

† Ratio of analogue to amino-acid in the medium.

‡ A 1:1 ratio of analogue to amino-acid was tested for its effect on the incorporation of labelled amino-acid for 30 min into TCA precipitable material. L-tryptophan-^{3}H(G) (3 Ci mM^{-1}) was at 5 μCi ml.$^{-1}$ and L-Tyrosine-3,5-^{3}H (25 μCi mM^{-1}) was at 10 μCi ml.$^{-1}$. In all cases background was obtained from a sample taken at time zero. To test for effects of the analogues at levels other than protein synthesis, cycloheximide (20 μM) was added alone and together with the analogue and the degree of inhibition monitored.

Actinomycin Resistance

Fig. 2*a* shows the dose response for DNA synthesis and the synthesis of the various protein fractions after a 90 min pre-incubation with actinomycin D[5-7]. Both histone and DNA synthesis are again comparatively resistant to low concentrations of inhibitor, both follow curves compatible with those obtained using cycloheximide. Fig. 2*b* gives the time course for the inhibition of histone, DNA, and RNA synthesis. After a lag of about 30 min, both histone and DNA synthesis decrease to a constant rate. Aside from this lag period the kinetics for both DNA and histone synthesis are similar to those obtained from cultures inhibited with sub-maximal doses of cycloheximide (5 μM). In actinomycin at 2.5 μg ml.$^{-1}$, RNA synthesis is about 18% of controls and constant by 5 min. If actinomycin had been affecting DNA synthesis directly, presumable by binding to the DNA, it might have been expected that DNA and RNA inhibition follow the same time course. This is clearly not the case. Additional experiments have shown that pre-incubation with actinomycin

影响了 DNA 合成中的同一步骤，观察到（未显示）刀豆氨酸的抑制动力学曲线类似于放线菌酮的抑制动力学曲线更加大了这种可能性。当加入刀豆氨酸时，^{3}H– 亮氨酸的掺入率减少了 10%。（图 1 说明：在放线菌酮对蛋白质合成造成相当大的抑制时，DNA 的合成并没有受到影响。）这可能表明刀豆氨酸是通过抑制某种特殊类型蛋白质的合成而不是通过产生变性的特殊类型蛋白质来影响 DNA 的合成的。

表 1. 氨基酸类似物对 DNA 合成的影响

处理方式	氨基酸比例[4]（组蛋白与细胞核酸性蛋白之比）*	氨基酸类似物与氨基酸之比 †	对掺入 ^{3}H– 胸腺嘧啶脱氧核苷的抑制率 %	氨基酸类似物与氨基酸之比 †	对结合放射性标记氨基酸的抑制率 %‡
C–^{3}H– 色氨酸	0	100:1	0	1:1	22%
3β– 吲哚丙烯酸	0	100:1	0	1:1	24%
3– 碘 –L– 酪氨酸	0.09	100:1	22%	1:1	25%
O–C–^{3}H–L– 酪氨酸	0.09	100:1	24%	1:1	26%
加放线菌酮			56%		
只有放线菌酮			53%		
L– 刀豆氨酸	1.83	1:1	62%		
加放线菌酮			68%		
只有放线菌酮			58%		

在用氨基酸类似物处理细胞时，同时使用 ^{3}H– 胸腺嘧啶脱氧核苷标记细胞 30 min。

* 组蛋白中对应于给定类似物的氨基酸与细胞核酸性蛋白质中对应于给定类似物的氨基酸之间的比例。用相当于未经处理的对照培养物的百分比来表示 ^{3}H– 胸腺嘧啶脱氧核苷的掺入率。

† 培养基中氨基酸类似物与相应氨基酸之间的比例。

‡ 使用比例为 1:1 的氨基酸类似物与氨基酸来检验其对标记氨基酸掺入 30 min 后造成的影响。L– 色氨酸 –^{3}H(G) (3 Ci/mM) 的浓度为 5 μCi/ml；L– 酪氨酸 –3,5–^{3}H (25 μCi/mM) 的浓度为 10 μCi/ml。在所有实验中均以零时刻的样品作为本底。为测试氨基酸类似物对除蛋白质合成以外的其他过程的影响，采用单独加入放线菌酮（20 μM）以及将其与氨基酸类似物一起加入的方法并对 DNA 合成的抑制率进行检测。

放线菌素抗性

用放线菌素 D 预孵育 90 min 后，DNA 合成及多种蛋白质合成的量效曲线如图 2*a* 所示[5-7]。组蛋白和 DNA 合成又一次对低浓度抑制剂表现出相当程度的抗性，两者都与用放线菌酮得到的曲线相吻合。对组蛋白、DNA 以及 RNA 合成的抑制随时间变化的曲线如图 2*b* 所示，在滞后大约 30 min 之后，组蛋白和 DNA 合成都降至一个恒定的水平。除了都存在这种滞后期以外，DNA 合成和组蛋白合成的动力学曲线均类似于用次最大量（5 μM）放线菌酮抑制培养物时得到的结果。当放线菌素的浓度为 2.5 μg/ml 时，细胞中 RNA 的合成量约为对照组的 18%，并且在 5 min 后就能达到这种抑制效果。如果放线菌素直接影响 DNA 合成，假如是通过与 DNA 结合，那么其对 DNA 和 RNA 合成的抑制作用就应该遵循相同的时间进程。显然实际情况

(2.5 μg ml.$^{-1}$) fails to add to the inhibition of DNA synthesis caused by cycloheximide. This supports the idea that both actinomycin and cycloheximide are affecting the same protein species. Although it is possible that both drugs are inhibiting the synthesis of an RNA species necessary for chain elongation, it becomes somewhat less likely since in both cases bulk RNA synthesis follows a different time and dose dependence than DNA synthesis.

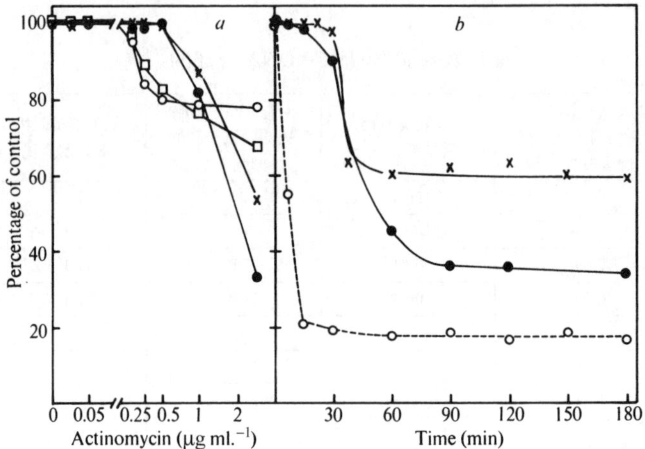

Fig. 2. *a*, Resistance of histone synthesis to low concentrations of actinomycin D. After 90 min incubation in actinomycin D at the stated concentrations, the cells were labelled with ^{3}H-leucine (20 μCi ml.$^{-1}$) for 30 min; the various protein fractions isolated; and the incorporation compared to control cells, either directly, or using double-label methods. ●—●, Histone; ×—×, DNA; □—□, nuclear acidic proteins; ○—○, haemoglobin. *b*, Kinetics of incorporation after 2.5 μg ml.$^{-1}$ actinomycin D. Incorporation of ^{3}H-TdR (10 μg ml.$^{-1}$), ^{3}H-5-uridine (10 μCi ml.$^{-1}$ 20 Ci mM^{-1}), and ^{3}H-leucine (20 μCi ml.$^{-1}$) into histone was measured as a rate percentage of controls after addition of actinomycin D (2.5 μg ml.$^{-1}$). ×—×, ^{3}H-TdR; ●—●, ^{3}H-leucine in histone; ○--○, ^{3}H-U.

Relation to DNA Synthesis

It was previously shown[1] that after release of FUdR blockage of DNA synthesis, there is a lag in the onset of inhibition of DNA synthesis by cycloheximide. The lag period increased with increasing exposure to FUdR, but the length of the lag was extremely short compared to the length of exposure, for example, 10 h in FUdR resulted in a lag time of about 20 min. The lag period was interpreted as indicating the accumulation of a pool of chain elongation protein. The disparity between the exposure to FUdR and the length of the lag could indicate either of two processes; (*a*) chain elongation protein is rapidly synthesized and partially degraded, or (*b*) the synthesis of chain elongation protein is coupled to the synthesis of DNA, but the coupling is leaky.

Synthesis of nuclear histone is sensitive to inhibition by cytosine arabinoside[6,7] as shown in Fig. 3. A similar curve is obtained for FUdR. At high levels of DNA inhibiton, histone synthesis is not completely inhibited; thus histone synthesis is coupled to DNA synthesis in these cells in a leaky fashion. Acrylamide gel analysis has demonstrated that although the synthesis of all histone species is inhibited, the slightly lysine-rich histones are somewhat

并非如此。另有一些实验表明，用放线菌素（2.5 μg/ml）对细胞进行预孵育并不能增加放线菌酮对 DNA 合成的抑制作用。这说明放线菌素和放线菌酮影响的是同一类蛋白质。尽管这两种药物都有可能会抑制链延伸所必需的某一种 RNA 的合成，不过这种情况比它们共同影响 DNA 合成的可能性略低一些，因为这两种抑制剂对大部分 RNA 合成的抑制时间—剂量曲线并不相同。

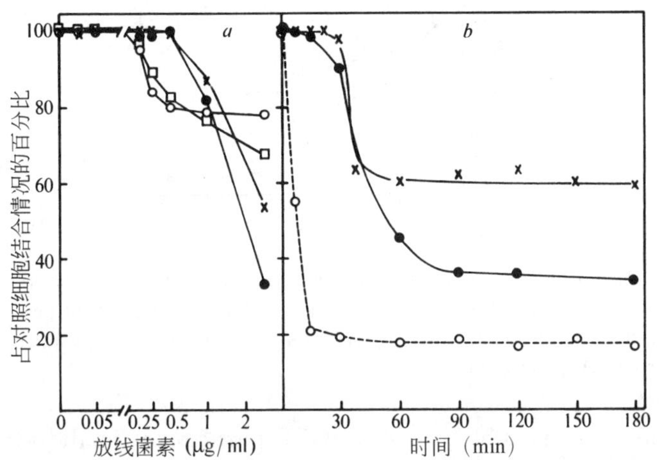

图 2. *a*, 组蛋白合成对低浓度放线菌素 D 的抗性。使用如图所示的各种浓度的放线菌素 D 孵育细胞 90 min 后，再用 ^{3}H– 亮氨酸（20 μCi/ml）标记 30 min；然后分离不同的蛋白质组分，并用直接测定法或双标记法比较实验细胞和对照细胞的掺入率。●—●为组蛋白；×—× 为 DNA；□—□为细胞核酸性蛋白；○—○为血红蛋白。*b*, 加入 2.5 μg/ml 放线菌素 D 之后的掺入动力学曲线。加入放线菌素 D (2.5 μg/ml) 之后，分别用占对照组的百分比来表示 ^{3}H– 胸腺嘧啶脱氧核苷（10 μCi/ml）、^{3}H–5– 尿嘧啶核苷（10 μCi/ml, 20 Ci/mM）和结合组蛋白的 ^{3}H– 亮氨酸（20 μCi/ml）的掺入率。×—× 为 ^{3}H– 胸腺嘧啶脱氧核苷；●—● 为组蛋白中的 ^{3}H– 亮氨酸；○ --- ○为 ^{3}H–5– 尿嘧啶核苷。

与 DNA 合成的关系

此前已证明[1]：在用 5– 氟尿嘧啶脱氧核苷（FUdR）来抑制 DNA 的合成后，放线菌酮对 DNA 合成的抑制作用在起始阶段有一个滞后。滞后期会随着 FUdR 处理时间的增加而延长，但滞后期的长度与处理时间相比是非常短的，例如，用 FUdR 处理 10 h 只会导致 20 min 左右的滞后。对于这个滞后期，人们认为其表明了细胞内存在 CEP 的聚集。FUdR 处理时间与滞后期长度之间的差异很大表明下列两种机制中的一种在起作用：(*a*) 链延伸蛋白被快速合成，且有一部分发生了降解；或（*b*）链延伸蛋白的合成与 DNA 的合成同步，不过这种同步是有缺陷的。

如图 3 所示，细胞核组蛋白的合成对阿糖胞苷的抑制作用很敏感[6,7]。由 FUdR 也可以得到类似的抑制曲线。当 DNA 的合成被显著抑制时，组蛋白的合成并没有被完全抑制；因此，在这些细胞中，组蛋白的合成与 DNA 的合成并不是完全同步的。由丙烯酰胺凝胶分析结果可知：尽管各类组蛋白的合成都受到了抑制，但赖氨

more sensitive[8]. Attempts to detect other proteins, cytoplasmic and nuclear acidic, which follow either of the two criteria listed above have not yet been successful.

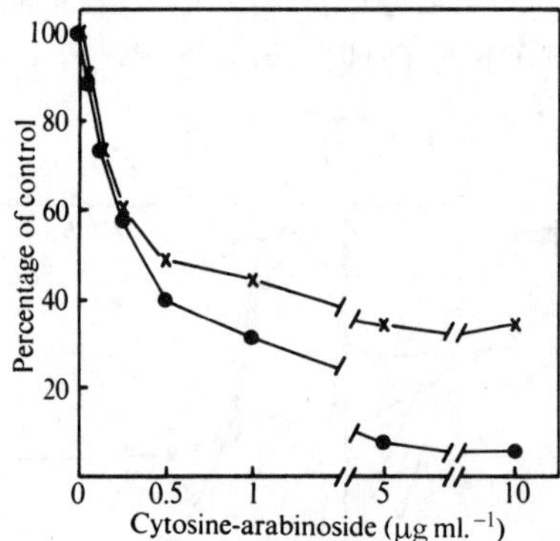

Fig. 3. Leaky coupling of histone synthesis to DNA synthesis. After 90 min in the stated concentrations of cytosine arabinoside, control and treated cells were labelled with ^{3}H-TdR (20 μCi ml.$^{-1}$) and ^{3}H-leucine (25 μCi ml.$^{-1}$) for 30 min. The inhibition of DNA (●—●) and histone (×—×) synthesis as a function of dose was measured as described in Fig. 2.

Given these results, it is possible to ask if CEP participates in the same step of chain elongation as deoxynucleotide addition. Two possibilities can be schematized as follows:

$$\text{(1) XTP + DNA} \xrightarrow{\text{CEP + replicase}} \text{DNA–XMP}$$

$$\text{(2) XTP + DNA + replicase} \longrightarrow \text{XMP–DNA–replicase} \xrightarrow{\text{CEP}} \text{XMP–DNA–CEP + replicase}$$

If TTP is made rate-limiting with FUdR such that the overall rate of precursor incorporation is some 10–30% of controls, then, given the results described in the previous paragraph, the synthesis of the CEP would also be inhibited. If deoxynucleotide addition and the CEP share the same step (1), then an initial inhibition of DNA synthesis should eventually lead to a subsequent inhibition of DNA synthesis because the concentration of two participants (TTP and CEP) of a multi-molecular reaction would then be decreased. Moreover, this process is auto-catalytic and the rate of DNA synthesis should gradually drift toward 50% of the control rate. This is the amount of DNA synthesis not dependent on concurrent protein synthesis. On the other hand, if deoxynucleotide addition and CEP participate in two separate reactions (2), the initial amount of inhibition should be stable, and a drift toward 50% inhibition should not be observed. If CEP is histone, Fig. 3 shows that both steps of the overall reaction depicted in (2) are inhibited to about the same extent. Either one becomes rate limiting and no secondary decreases in the overall reaction rate will occur. Experiments designed to distinguish between these two possibilities favour the second mechanism. With either FUdR or cytosine arabinoside it was found that various low levels of inhibition were stable over at least 3–4 h. These

酸含量略多的组蛋白会更敏感一些[8]。迄今为止试图检测出其他两种蛋白——胞质蛋白和细胞核酸性蛋白是否也符合上述两个标准之一的尝试并没有取得成功。

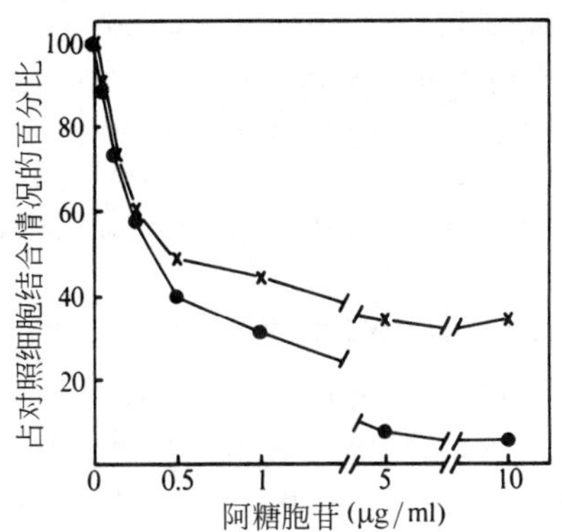

图 3. 组蛋白合成与 DNA 合成的不完全同步性。在用如图所示浓度的阿糖胞苷处理细胞 90 min 之后，将对照细胞和实验细胞用 ^{3}H– 胸腺嘧啶脱氧核苷（20 μCi/ml）和 ^{3}H– 亮氨酸（25 μCi/ml）标记 30 min。采用图 2 中所描述的方法来测量 DNA（●—●）合成以及组蛋白（×—×）合成的抑制率与剂量之间的关系。

从这些结果分析，很可能产生这样的疑问：CEP 是否参与了与脱氧核苷酸插入相同的链延伸步骤。下面用反应式列出了两种可能机制：

$$(1)\ \mathrm{XTP + DNA} \xrightarrow{\mathrm{CEP} + \text{复制酶}} \mathrm{DNA{-}XMP}$$

$$(2)\ \mathrm{XTP + DNA} + \text{复制酶} \longrightarrow \mathrm{XMP{-}DNA{-}}\text{复制酶} \xrightarrow{\mathrm{CEP}} \mathrm{XMP{-}DNA{-}CEP} + \text{复制酶}$$

如果胸腺嘧啶核苷三磷酸（TTP）在 FUdR 的作用下反应速率受限，因而前体掺入的总速率约为对照组的 10% ~ 30%，那么根据上一段中所描述的结果，CEP 的合成也会受到抑制。如果脱氧核苷酸插入和 CEP 是在同一步中的话（反应式 1），那么 DNA 合成一旦受到抑制将最终导致随后的 DNA 合成过程被抑制，因为在多分子反应中两种反应物（TTP 和 CEP）的浓度都会不断下降。此外，这一过程是自动催化的，DNA 的合成率应该会逐渐降低至对照组的 50%。这就是不依赖于同步蛋白质合成的 DNA 合成的量。另一方面，如果脱氧核苷酸插入与 CEP 参与的是两个不同的反应（反应式 2），那么起始抑制率应该是一个恒定值，而且也观察不到抑制率向 50% 漂移。如果 CEP 就是组蛋白，则图 3 表明反应式（2）中所描述的总反应中的两个步骤将受到程度大致相同的抑制。这两个步骤都不是限速步骤，并且总反应速率也不会出现继发性降低。为鉴别这两种可能性而设计的实验倾向于支持第二种机制。不管加入的是 FUdR 还是阿糖胞苷，在至少 3 h ~ 4 h 内，抑制率都维持在较低

experiments measured the incorporation of ^{3}H-deoxyadenosine into alkali stable, TCA precipitable material, a high concentration (10 μg ml.$^{-1}$) of deoxyadenosine in the medium was used to diminish effects from small fluctuations in the internal pool size. The acid soluble pool was monitored for all time points, and the precipitable incorporation adjusted accordingly. Similar conclusions could be derived from analogous experiments using labelled thymidine and deoxycytidine. I conclude that the process of DNA chain elongation *in vivo* is separable into at least two steps and that the step sensitive to decreased levels of CEP is different from that responsible for nucleotide addition.

A Presumptive Replication Complex

Pulse labelled DNA in erythroblasts displays characteristics which are similar to those described in HeLa cells[9,10]. Nascent DNA is preferentially extracted in either phenol or chloroform-isoamyl alcohol and also sediments to the top of a CsCl gradient. This behaviour, which probably indicates the association of nascent DNA with a low density, hydrophobic material, is, for erythroblasts, resistant to former treatment with pronase SPS, high salt, RNAase, or periodate, but sensitive to shear, sonication, alpha-amylase and phospholipase C (Sigma). "STS" gels have demonstrated contaminants in both of these last two enzyme preparations. It would be premature, therefore, to define the complexing agent in terms of either phospholipid or polysaccharide on the basis of these enzyme sentitivities. When pure DNA is incubated with either of these enzymes neither a loss in TCA precipitable counts per min nor a decrease in molecular weight in alkaline sucrose gradients could be detected. Because denatured DNA is preferentially extracted with chloroform[11], and as pulse-labelled DNA might be partially denatured, reconstruction experiments were done in which labelled, denatured DNA was added to nuclei before treatment. These experiments showed that although the labelled, denatured DNA was preferentially extracted with chloroform, it sedimented to a higher density than native DNA on CsCl and that both of these characteristics were unaffectled by shear, alpha-amylase, phospholipase C, or sonication and denaturation. It is therefore unlikely that the behaviour of pulse-labelled DNA in erythroblasts results only from some single-stranded property. Most evidence indicates that pulse-labelled DNA is associated with some nuclear structure which confers certain physical properties on it.

Fig. 4*a* shows the percentage of water soluble counts (counts not extracted with phenol) after a pulse of ^{3}H-TdR (5 min) plotted as a function of the chase time in cold medium. Graphs are given for chases in the presence and absence of cycloheximide. Addition of the drug during the pulse period makes little difference. Care was taken to expose the DNA to the same amount of shear for each point. Better separation of the complex is obtained if, instead of extracting the DNA, it is sedimented to equilibrium on CsCl (4*b*). The kinetics displayed in Fig. 4*a* and *b* are consistent with the model proposed by Friedman and Mueller[9] where the extractability of pulse-labelled DNA is a function of the distance of that DNA from an extractable replication complex. The differing behaviour of cycloheximide treated cells is then explained by a slower rate of DNA chain elongation during the chase period. Recent findings make it unlikely that this complex is associated with the nuclear membrane[11].

水平。利用这些实验可以检测 ^{3}H– 脱氧腺苷掺入到碱性条件下稳定且 TCA 可沉淀的物质中的比例，在培养基中加入高浓度（10 μg/ml）的脱氧腺苷以消除因内部积存量微小波动而造成的影响。我们在各个时间点都监测了酸溶性库存，沉淀中放射性标记的含量会随之变化。由使用放射性标记的胸腺嘧啶脱氧核苷和脱氧胞苷的类似实验也可以得到相近的结论。我认为 DNA 链在体内的延伸过程至少可以分为两个步骤，并且对 CEP 浓度下降敏感的步骤与插入核苷酸的步骤是两个不同的步骤。

一个假定的复制复合物

成红细胞中的脉冲标记 DNA 呈现出与 HeLa 细胞（译者注：是指源自一名美国妇女的子宫颈癌细胞的细胞。这名美国妇女名叫 Henrietta Lacks，于 1951 年死于癌症）中脉冲标记 DNA 相似的性质 [9,10]。新生成的 DNA 更容易被苯酚或者氯仿 – 异戊醇抽提，并且会沉于 CsCl 密度梯度的顶部。这一现象可能说明新生成的 DNA 结合了一种低密度且疏水的物质。在成红细胞中，这一现象对于链霉蛋白酶 SPS、高盐环境、RNA 酶或高碘酸盐的前期处理都不敏感，却对剪切力、超声、α– 淀粉酶和磷脂酶 C（西格马公司）的作用敏感。“STS”凝胶实验已证明在后两种酶制剂中都有污染物存在。因此，根据对这些酶的敏感性，目前还不能确定上述复杂现象到底是由磷脂引起的还是由多糖引起的。当用这两种酶中的任意一种与纯 DNA 一起孵育时，并没有发现 TCA 沉淀计数出现下降，也没有检测到碱性蔗糖梯度中的分子量减小。由于变性的 DNA 更容易被氯仿抽提出来 [11]，而脉冲标记的 DNA 也可能会发生部分变性，因此我们重新设计了实验：在处理细胞之前，将已做好标记的变性 DNA 加入细胞核中。这些实验表明：尽管那些带有标记的变性 DNA 更容易被氯仿抽提出来，但是在 CsCl 密度梯度离心中，它会沉降在密度高于天然 DNA 的位置上，并且这两种性质都不受剪切力、α– 淀粉酶、磷脂酶 C 或者超声和变性处理的影响。因此，成红细胞中脉冲标记 DNA 的特性不太可能仅由一些单链 DNA 的性质所导致。大多数证据表明，脉冲标记的 DNA 是与某种细胞核结构结合在一起的，而赋予它特定物理性质的正是这些核结构。

图 4*a* 描绘了经 ^{3}H– 胸腺嘧啶脱氧核苷脉冲处理（5 min）后水溶性组分（即不能被苯酚抽提的组分）中放射性计数的百分比随着在冷媒中追踪时间的变化而改变的趋势。图中分别给出了存在或不存在放线菌酮时的追踪结果。在脉冲期间加入这种药物对实验结果几乎没有什么影响。小心操作以使 DNA 在每个时间点都受到相同的剪切力。如果用 CsCl 密度梯度离心法来代替抽提 DNA 的方法，就可以更好地分离 DNA 复合物（图 4*b*）。图 4*a* 和 4*b* 中的动力学曲线与弗里德曼和米勒 [9] 提出的模型一致。在他们的模型中，脉冲标记 DNA 的可抽提性随着该 DNA 与某种可抽提复制复合物之间的差距而改变。细胞在放线菌酮作用下所表现出来的不同行为可以用在追踪期间内 DNA 链延伸的速率降低来解释。最近的研究结果表明这一复合物不太可能与细胞核膜结合 [11]。

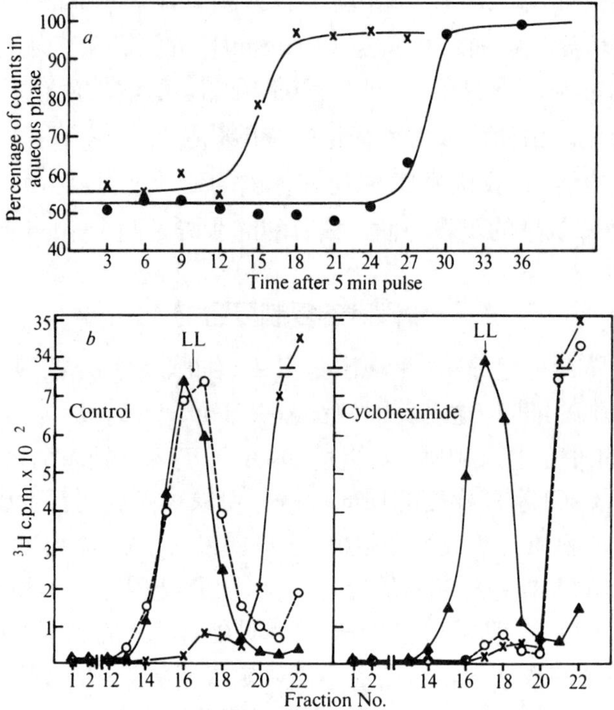

Fig. 4. Chasing of nascent DNA from a presumed replication complex in the presence and absence of cycloheximide (50 μM). Cells were pulsed for 5 min with ^{3}H-TdR (50 μCi mi.$^{-1}$) and chased for increasing periods of time in the presence or absence of cycloheximide. *a*, The isolated nuclei in RSB were treated with SDS to 0.1%. An equal volume of phenol was added and the suspension mixed on a "Vortex" for 30 s. After centrifugation in a bench centrifuge, the water phase was removed and precipitated with TCA (10%) and carrier and then counted. ×—×, Control; ●—●, cycloheximide. *b*, The SDS treated nuclei were sedimented to equilibrium on CsCl[1] after a 30 s agitation on a "Vortex" mixer. Elimination of SDS fails to affect either the extraction or the sedimentation results. LL, Native DNA. ×—×, 5′ pulse; ○--○, 10′ chase; ▲—▲, 40′ chase.

Histones Remove Nascent DNA from the Replication Complex

The effect of added whole calf thymus histone (Sigma) on the extractability of nascent DNA is shown in Fig. 5. Cells were pulsed for 10 min with ^{3}H-TdR and nuclei isolated in RSB these nuclei do not incorporate ^{3}H-TTP. The nuclei were then suspended in increasing concentrations of histone, albumin, or haemoglobin and incubated at 37°C for 30 min. SDS was then added to 0.1% and the DNA extracted with chloroform-isoamyl alcohol. Depending on the experiment, 5 to 50 μg ml.$^{-1}$ of histone could protect over 90% of the nascent DNA from extraction. This protection was inhibited by 2 M NaCl before boiling the histone solution (5 min) and by exogenous DNA; haemoglobin and albumin were ineffective. Treatment of nuclei with histone before SDS solubilization also removes nascent DNA from the complex that sediments to the top of a CsCl gradient (Fig. 5, inset). High histone concentrations consistently inhibit the reaction with the replication complex. When used in high concentrations protamine and cytochrome c have activity in this assay. Both are about 20% as effective as either whole histone or the lysine-rich histones

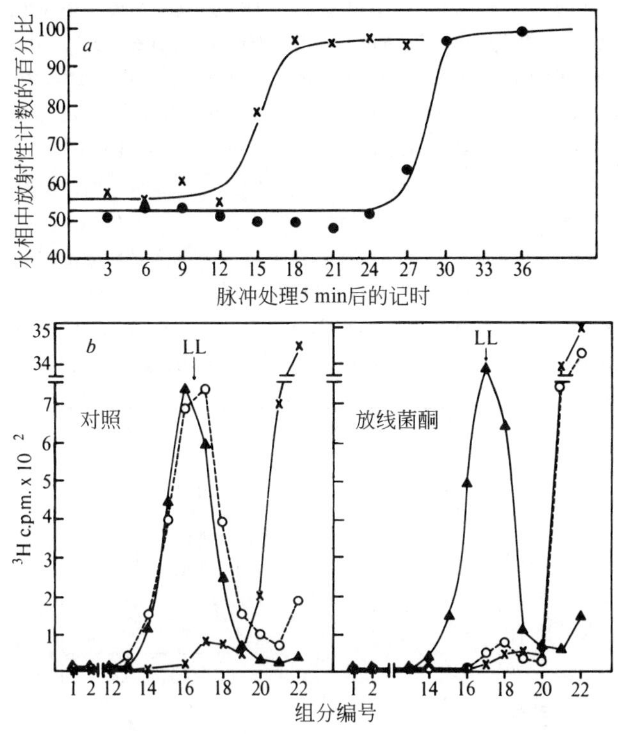

图 4. 分别在放线菌酮存在（50 μM）和不存在时对假定复制复合物中新生 DNA 的追踪结果。用 ^{3}H– 胸腺嘧啶脱氧核苷（50 μCi/ml）脉冲处理细胞 5 min，然后在存在或不存在放线菌酮时追踪放射性标记物随时间推移的变化。*a*, 用网织红细胞标准缓冲液分离细胞核，然后加入 SDS 使其浓度达到 0.1%。加入等体积的苯酚，悬浮液在一台“涡旋”振荡器上混合 30 s。用台式离心机离心后，吸取水相并用 TCA(10%) 及载体进行沉淀并计数。×—× 为对照组，●—●为加入放线菌酮的实验组。*b*, 将经 SDS 处理的细胞核在一台“涡旋”振荡器上振荡 30 s, 然后用 CsCl 密度梯度离心 [1]，去除 SDS 的过程并未影响抽提或沉淀的结果。LL 为天然 DNA，×—× 为脉冲 5 min 的结果；○—○为追踪 10 min 的结果；▲—▲为追踪 40 min 的结果。

组蛋白可以去除复制复合物中新生的 DNA

加入小牛胸腺全组蛋白（西格马公司）后对新生 DNA 的可抽提性产生的效果如图 5 所示。使用 ^{3}H– 胸腺嘧啶脱氧核苷脉冲标记细胞 10 min，随后用网织红细胞标准缓冲液分离细胞核，这些细胞核中不含有 ^{3}H–TTP。随后将细胞核悬浮于浓度逐渐升高的组蛋白、白蛋白或血红蛋白的溶液中，并在 37℃下孵育 30 min。加入 SDS 使其最终浓度达到 0.1%，随后用氯仿 – 异戊醇抽提 DNA。根据不同的实验，加入浓度为 5 μg/ml ~ 50 μg/ml 的组蛋白即可保护抽提物中 90% 以上的新生 DNA。加入 2 M NaCl 然后煮沸组蛋白溶液（5 min）或者加入外源性 DNA 都可以抑制组蛋白的这种保护作用；而加入血红蛋白或白蛋白是无效的。在加入 SDS 增溶之前使用组蛋白处理细胞核也会把新生 DNA 从处于某个 CsCl 密度梯度顶层的复合物中去除（图 5 中的小插图）。高浓度的组蛋白通常会抑制与复制复合物的反应。在上述实验中，高浓度的精蛋白和细胞色素 c 都是有活性的。两者活性都只有低分子浓度全组

when used at low molecular concentrations. Spermine, spermidine, and putrescine at concentrations up to 10^{-3} M fail to dissociate the complex and also fail to relieve the cycloheximide inhibition of DNA synthesis in the intact cell. The effective concentration of histone in these experiments is about the same as that in the cell, although this is a difficult comparison since most cellular histone is bound to DNA. Preliminary results indicate that over 50% of the exogenous histone becomes sequestered in nuclei during the course of experiments similar to those in Fig. 5, and that this histone can protect the nascent DNA from degradation by pancreatic DNAase. Similar experiments to be presented in more detail at a later date have shown that compared to chase DNA, pulse DNA is relatively resistant to DNAase and that this resistance is destroyed by an α amylase preparation that has no effect on the digestion of chase DNA by DNAase.

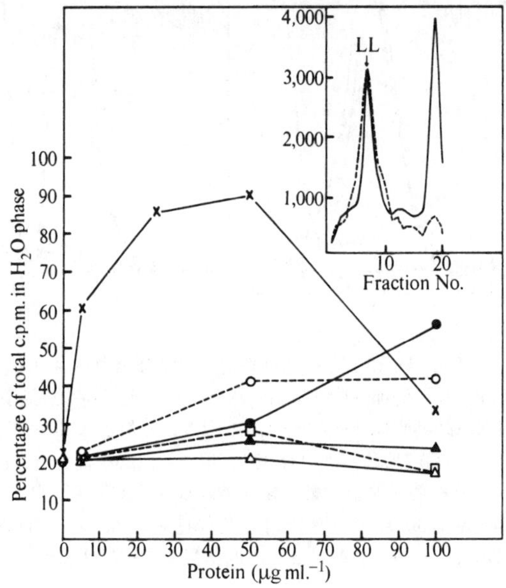

Fig. 5. Removal of nascent DNA from a replication complex by histone. After a 10 min ^{3}H-TdR (25 μCi ml.$^{-1}$) treatment, nuclei were isolated from cells and incubated in RSB at 37°C for 30 min in the stated concentrations of protein under the indicated conditions. Nuclei were then treated with SDS and extracted with chloroform-isoamyl alcohol as described in Fig. 4*a* for phenol extraction. The percentage of the total incorporated counts associated with the water phase was then determined. In representative experiments, all of the extracted counts could be recovered in the non-aqueous phase. Control experiments showed that the extractability of nascent DNA from nuclei lysed in high salt was not dependent on the presence of SDS and that added single-stranded DNA remained at the interphase despite the prior addition of histone. ×—×, Histone; ○--○, heat denatured histone; ●—●, histone and 25 μg DNA; ▲—▲, histone and 2 M salt; △—△, haemoglobin; □--□, BSA. Inset shows the sedimentation behaviour on CsCl of nascent DNA derived from untreated nuclei and nuclei treated with 25 μg ml.$^{-1}$ of histone. By lowering the CsCl concentration compared to that shown in Fig. 4*b*, the complex is seen to move into the gradient. In all experiments, nuclei were at a concentration of about 10^{7} ml.$^{-1}$. SDS gel electrophoresis showed that the histone preparation contained one minor and one major contaminant (%). The preparation also contained no exonuclease activity; endonuclease activity was not monitored. —, Control; ---, histone, LL, native DNA.

I conclude that *in vitro* the proper concentration of whole calf thymus histone can remove

蛋白或富含赖氨酸组蛋白作用时的20%左右。浓度高至 10^{-3} M的精胺、亚精胺和腐胺并不能引起复制复合物的解离，也不能降低放线菌酮对完整细胞中DNA合成的抑制作用。组蛋白在上述实验中发挥作用的有效浓度与它们在细胞中的浓度大致相等，但因为细胞中的组蛋白大多与DNA结合在一起，所以很难进行比较。初步结果显示：在类似于图5的实验中，超过50%的外源性组蛋白会出现在细胞核组分中，而且这类组蛋白可以保护新生DNA不被胰腺DNA酶所降解。近期之内我将更详细报告一些类似实验，它们可以说明脉冲DNA比追踪DNA更能抵抗DNA酶的作用，并且α–淀粉酶会破坏脉冲DNA的这种抵抗能力，但却不影响DNA酶对追踪DNA的降解。

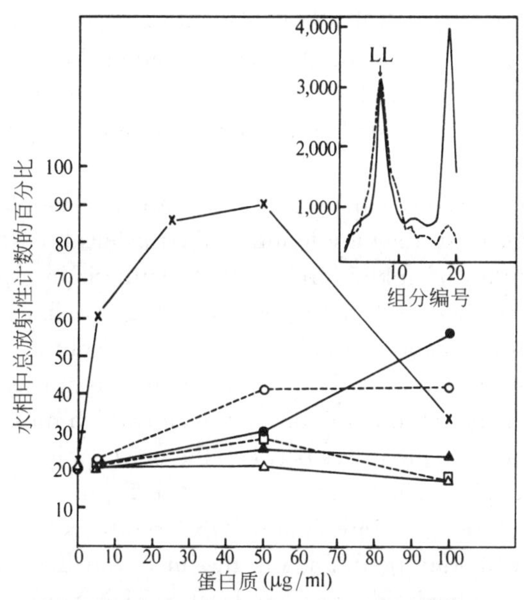

图5. 用组蛋白去除复制复合物中新生的DNA。在用 ^{3}H–胸腺嘧啶脱氧核苷（25 μCi/ml）脉冲标记细胞10 min后，将细胞核从细胞中分离出来，并在37℃及指定条件下用含所示浓度蛋白质的网织红细胞标准缓冲液孵育30 min。反应结束后，用SDS处理细胞核，并用氯仿–异戊醇溶液抽提DNA，采用的方法类似于图4*a*中所述的苯酚抽提法。随后测定水相中总放射性计数所占的百分比。在这些有代表性的实验中，所有抽提物的放射性计数都能在非水相中复原。对照实验表明：在高盐条件下，从细胞核中抽提出来的新生DNA的量与是否存在SDS无关；并且，不管事先是否加入过组蛋白，外源加入的单链DNA都维持在水相和有机相的交界处。×—×为组蛋白，○—○为经加热变性后的组蛋白，●—●为组蛋白及25 μg DNA，▲—▲为组蛋白及2 M NaCl，△—△为血红蛋白，□—□为牛血清白蛋白。图中小插图显示的是从未经处理或经25 μg/ml组蛋白处理的细胞核中抽提出来的新生DNA在CsCl密度梯度离心实验中的沉淀行为。通过使用比图4*b*中更低的CsCl浓度，可以观察到复制复合物溶入了浓度梯度。在所有这些实验中，细胞核的浓度都约为每ml 10^7 个。SDS凝胶电泳实验结果显示，实验中所使用的组蛋白制剂含有一种含量较少和一种含量较多的污染物(%)。这种制剂也不具有核酸外切酶的活性；核酸内切酶的活性未检测。— 为对照组，--- 为添加组蛋白的实验组，LL为天然DNA。

我的结论是：在体外，适当浓度的小牛胸腺全组蛋白可以去除复制复合物中新

nascent DNA from a replication complex. How this activity might relate to DNA synthesis in the intact cell remains to be seen. Other basic proteins have activity in this assay but the histones are much more active at lower molar concentrations. Results with the replication complex are consistent with a previous conclusion that the CEP acts at a step other than nucleotide addition. They are also related to recent data showing that (1) histone stimulates Qβ replicase *in vitro*; (2) bacteria grown in the presence of canavanine accumulate or fail to remove their DNA from complex[13]; and (3) φX capsid proteins are required for φX DNA synthesis[13].

Probable Role of Histones

None of the experiments that I have presented proves that the histones are the DNA chain elongation proteins, although five independent experiments *in vivo* and *in vitro* make this notion likely. Final proof would be the demonstration that the proper concentration histone increases the rate of chain elongation in a cell-free system to that observed *in vivo* and that newly-made histone is located at the growing point of replication in intact cells.

There are several other reasons why the histones are a reasonable candidate for the chain elongation protein. First, they are found in the nucleus bound to the DNA. Second consistent with the rapidity of inhibition of DNA synthesis by cycloheximide, they are usually not found in any detectable free pool. Third, given the preceding observations it follows that the chain elongation protein must move rapidly from its presumed site of synthesis on cytoplasmic ribosomes to the growing point. It is at least questionable whether diffusion can account for the rapidity of this interaction which must involve recognition as well as transport. The electrostatic forces between the newly polymerized DNA phosphates and the basic histones might facilitate this process. Fourth, there are some 3,000 growing points in the average S phase cell[1]. Inhibiton of DNA synthesis by cycloheximide probably occurs within 10 s. The rapidity of inhibition, the linear kinetics after inhibition, and the fact that low levels of cycloheximide give low levels of DNA inhibition indicate that the CEP is probably used stoichiometrically. As our previous experiments showed that all replicons were affected, it follows that at least 20,000 CEPs must be synthesized by the cell per minute in order to sustain the normal rate of replicon growth. These cells make about 40,000 histone molecules per minute[2]; no other protein in the cell, apart from haemoglobin, even approaches this synthetic rate.

The most intriguing feature of the inhibitory effects of cycloheximide on DNA synthesis is that about 50% of the DNA can be replicated. Persistence of this activity can be explained if histone from pre-replicative chromatin acts as a reservoir of histone at the growing point. In conjunction with newly synthesized histone, the reaction of this recycled histone with nascent DNA and the replication complex allows the growing fork to move ahead at the normal rate. In the absence of newly synthesized histone, the rate of chain elongation is slower, but not zero since recycled histone can still be used.

Because all known polymerases have the inherent property to elongate nascent chains, the proposed requirements for histones in this process most probably reflects some higher

生的 DNA。组蛋白的这一功能如何才能与完整细胞中的 DNA 合成相关联还有待于进一步的研究。其他一些碱性蛋白在这个实验中同样具有活性，不过在摩尔浓度更低的条件下，组蛋白比它们的活性高很多。由复制复合物得到的结果与之前认为 CEP 在不同于核苷酸插入的步骤发挥作用的结论一致。这些结果也与最近得到的以下实验数据相吻合：(1) 在体外实验中，组蛋白可以激活 Qβ 复制酶；(2) 在含有刀豆氨酸的环境中生长的细菌会聚集在一起，或者未能使它们的 DNA 与复制复合物分离 [13]；(3) φX DNA 的合成需要 φX 衣壳蛋白的参与 [13]。

组蛋白的可能角色

尽管有 5 个独立的体内和体外实验表明组蛋白有可能就是 DNA 链延伸蛋白，但是我在文中介绍过的所有实验都不能证明这一点。最关键的证据在于：要证明在无细胞系统中，适当浓度的组蛋白可以使 DNA 链延伸的速率增加至细胞内链延伸的速率；并且还需要证明在完整细胞中，新合成的组蛋白分布于 DNA 复制的生长点。

还有其他几条理由可以说明为什么组蛋白是链延伸蛋白的合理候选者。首先，它们位于细胞核中，并且与 DNA 结合。其二，组蛋白很少以游离形式存在，这与放线菌酮能迅速抑制 DNA 的合成相一致。第三，由前面的观测结果可以推出，链延伸蛋白必须能够从它可能被合成的位置——胞质内的核糖体迅速移动到 DNA 合成的生长点。人们至少会对用扩散来解释这种包括识别和运输过程在内的相互作用的快速性表示怀疑。新聚合的 DNA 上磷酸基团与碱性组蛋白之间的静电力可能有利于这种快速移动。第四，在 S 期细胞中，平均会有大概 3,000 个生长点 [1]。放线菌酮对 DNA 合成的抑制很可能在 10 s 之内就能实现。抑制的迅速形成、抑制后得到的线性动力学关系以及低浓度放线菌酮对 DNA 合成的抑制水平也低都说明，CEP 有可能是按量分配的。由于我们之前的实验已经证明所有的复制子都会受到影响，因此为了维持复制子增长的正常速度，细胞每分钟必须合成至少 20,000 个 CEP。这些细胞每分钟大约可以合成 40,000 个组蛋白分子 [2]，除血红蛋白以外，细胞中还没有哪种蛋白能达到这样的合成速度。

放线菌酮对 DNA 合成的抑制作用的最有趣特征是大约 50% 的 DNA 可以被复制。如果复制前染色质中的组蛋白可以源源不断地为复制生长点提供组蛋白，那么这种活性的持续存在就可以得到解释。这种被循环利用的组蛋白与新合成的组蛋白一起与新生 DNA 及复制复合物发生反应，使得复制又能够以正常的速度向前推移。当不存在新合成的组蛋白时，链延伸速率有所减慢，但不会为 0，因为还能使用可循环利用的组蛋白。

由于所有已知的聚合酶都具有延长新生 DNA 链的固有特性，因此 DNA 合成过程对组蛋白的需要很可能反映的是更高级的机制，在这种机制中，新 DNA 的合成

order mechanism by which a synthesis of new DNA is coupled to the generation of a very defined and inviolable chromosomal structure.

I thank H. Holtzer, J. Flaks, and A. Kozinski for their help and criticisms. This work was supported by grants from the National Institutes of Health, the United States Public Health Service, and the Helen Hay Whitney Foundation.

(**240**, 449-453; 1972)

Harold Weintraub: Department of Anatomy, University of Pennsylvania, Philadelphia, Pennsylvania 19104.

Received June 12; revised October 12, 1972.

References:

1. Weintraub, H., and Holtzer, H., *J. Mol. Biol.*, **65**, 13 (1972).
2. Weintraub, H., Campbell, G., and Holtzer, H., *J. Mol. Biol.* (in the press).
3. Spalding, J., Kajiwara, K., and Mueller, G. C., *Proc. US Nat. Acad. Sci.*, **56**, 1535 (1966).
4. Malpoix, P., Zampetti, F., and Fievez, M., *Biochim. Biophys. Acta*, **182**, 214 (1969).
5. Freedman, M. L., Honig, G. R., and Rabinovitz, M., *Exp. Cell Res.*, **44**, 263 (1966).
6. Borun, T. W., Scharff, M. D., and Robbins, E., *Proc. US Nat. Acad. Sci.*, **58**, 1977 (1967).
7. Mueller, G. C., and Kajiwara, K., *Biochim. Biophys. Acta*, **119**, 557 (1966).
8. Sadgopal, S., and Bonner, J., *Biochim. Biophys. Acta*, **186**, 349 (1969).
9. Friedman, D. L., and Mueller, G. C., *Biochim. Biophys. Acta*, **174**, 253 (1969).
10. Pearson, G. D., and Hanawalt, P. C., *J. Mol. Biol.*, **62**, 65 (1971).
11. Fakan, S., Turner, G., Pagana, J., and Hancock, R., *Proc. US Nat. Acad. Sci.*, **69**, 2300 (1972).
12. Kuo, C. H., and August, J. T., *Nature*, **237**, 105 (1972).
13. Schachtele, C. F., Anderson, D. L., and Rogers, P., *J. Mol. Biol.*, **49**, 255 (1970).
14. Iwaya, M., and Denhardt, D. T., *J. Mol. Biol.*, **57**, 159 (1971).

与一种固定不变的染色体结构的生成相关联。

感谢霍尔泽、弗拉克斯和科津斯基对我的帮助及批评。这项工作的资金来自美国国立卫生研究院、美国公共卫生署和海伦·海·惠特尼基金会。

（张锦彬 翻译；孙军 审稿）

Biochemical Evidence for the Bidirectional Replication of DNA in *Escherichia coli*

W. G. McKenna and M. Masters

Editor's Note

Whereas vertebrate DNA is linear and packaged into multiple different chromosomes, the DNA of *Escherichia coli* is circular and housed within a single chromosome. This bacterial DNA replicates sequentially from a fixed point, and here W. G. McKenna and Millicent Masters present biochemical evidence suggesting that replication takes place simultaneously in both directions from the point of origin. The premise was corroborated by David Prescott and Peter Kuempel, who published autoradiographs of bidirectional bacterial DNA replication around the same time. The phenomenon of bidirectional DNA replication is now well accepted, and the process is known to proceed in three stages: initiation, elongation and termination, regulated by a complex and efficient set of catalytically active proteins.

THE chromosome of *Escherichia coli* is a single circular DNA molecule which replicates sequentially[1-3] from a fixed origin[4-14,15]. Although it was initially thought that replication proceeds in a unique direction, Masters and Broda[16] and subsequently Bird *et al.*[17] and Hohlfeld and Vielmetter[18] have presented genetic evidence that replication takes place simultaneously in both directions from the origin. We here present confirmatory biochemical evidence that this is so. Since this manuscript was submitted, Prescott and Kuempel[19] have published autoradiographs of bidirectional replication of DNA.

Bidirectional Replication of DNA

We adapted to *E. coli* the method with which Weintraub[20] demonstrated that DNA replication in developing chick erythroblasts is bidirectional rather than unidirectional. The technique allows one to analyse the way in which the segments of DNA made immediately after initiation of replication are synthesized.

Cells of a thymine-requiring mutant of *E. coli* are allowed to initiate DNA replication synchronously with the thymine analogue bromouracil (BU) present in place of thymine. BU is then rapidly replaced by ^{3}H-thymine and replication allowed to proceed for a much longer period. The DNA is extracted avoiding degradation of the daughter DNA and divided into two aliquots, one of which is subjected to ultraviolet light to degrade the sections of DNA containing BU, as low doses of ultraviolet light lead to specific breakdown of the sugar–phosphate backbone of DNA adjacent to incorporated BU residues[21]. The newly synthesized radioactive DNA strand segments are separated from the parental DNA strands and analysed with respect to molecular weight, in an alkaline sucrose gradient.

大肠杆菌DNA双向复制的生物化学证据

麦克纳，马斯特斯

编者按

脊椎动物的 DNA 都是线性的且被组装到多条不同的染色体中，而大肠杆菌的 DNA 却是环状的且全部位于一条染色体中。大肠杆菌 DNA 的复制从一个固定位点开始连续进行，在本文中麦克纳和米莉森特·马斯特斯提出生物化学证据证明其复制是从起始点开始朝两个方向同时进行的。大约在同一时间，戴维·普雷斯科特和彼得·金佩尔发表了细菌 DNA 双向复制的放射自显影照片，从而证实了上述假设。目前，DNA 双向复制的现象已经被人们广为接受，该过程分三步进行：起始、延伸和终止，这些步骤都受到一套复杂而高效的、有催化活性的蛋白质的调节。

大肠杆菌的染色体是一个单一的环状 DNA 分子，其连续复制[1-3]始于一个固定的起始点[4-14,15]。尽管起初认为其复制是沿单一方向进行的，但马斯特斯、布罗达[16]和随后的伯德等人[17]以及霍尔菲尔德、菲尔梅特[18]提出的遗传学证据表明复制是从起始点开始沿两个方向同时发生的。在本文中，我们用确凿的生物化学证据证明事实确实如此。在这篇文章的手稿提交之后，普雷斯科特和金佩尔[19]随即发表了 DNA 双向复制的放射自显影照片。

DNA 的双向复制

我们将温特劳布[20]证实发育过程中小鸡成红细胞的 DNA 复制是双向而不是单向时所用的方法应用于大肠杆菌。这项技术能帮助我们分析在复制刚刚开始时形成的 DNA 片段是如何被合成的。

我们用胸腺嘧啶类似物——溴尿嘧啶（BU）取代胸腺嘧啶以使需胸腺嘧啶的大肠杆菌突变株细胞进行 DNA 同步复制。随后迅速用 ^{3}H– 胸腺嘧啶代替 BU，并且使复制过程延续更长的时间。在保证子代 DNA 不发生降解的前提下将 DNA 抽提出来并分为两等份，其中一份用紫外线照射以降解 DNA 中含有 BU 的部分，因为小剂量紫外线照射即可导致 DNA 上与掺入的 BU 残基相邻的糖 – 磷酸骨架发生特异性的断裂[21]。将这些新合成的具有放射活性的 DNA 片段从亲代 DNA 链中分离出来并在碱性蔗糖梯度中对分子量进行分析。

As Fig. 1 shows, if replication is unidirectional (*A*), BU will be at one end only of each single-stranded segment. As the duration of BU uptake is small compared with that of ^{3}H-thymine, treatment with ultraviolet will make little difference to the size of segments as compared with the untreated control; both should form homogeneous bands in the same region of the sucrose gradient. By contrast, if replication is bidirectional (Fig. 1*B*), the bromouracil should be located in the middle of each segment, so that treatment with ultraviolet light breaks the segments in two and their molecular weight is approximately halved (see below) as compared with the control. In this case irradiated and unirradiated DNA will appear as separate bands on the sucrose gradient.

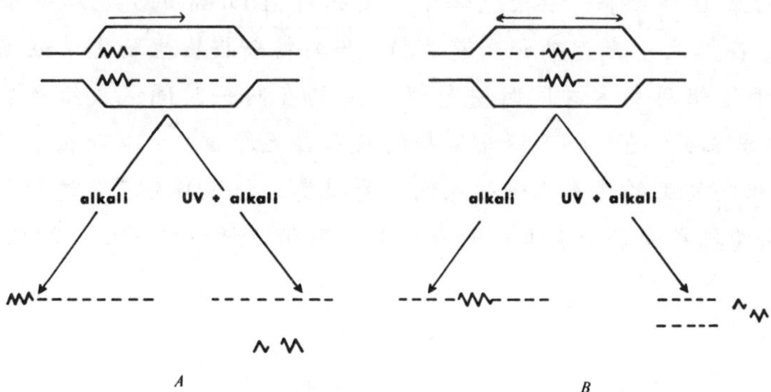

Fig. 1. Diagrammatic representation of the predicted behaviour of unidirectionally (*A*) as opposed to bidirectionally (*B*) replicating origin DNA after the following treatment. Synchronized cells are pulse for 15 s with 5-bromouraci (~~~BU-DNA) and then with ^{3}H-thymine for 100 s (- - -, ^{3}H-DNA; ——, unlabelled DNA). Lysates of cultures are either irradiated with ultraviolet light, or left unirradiated, and centrifuged on alkaline sucrose gradients.

Initiation of DNA replication was synchronized in cells of *E. coli* B/r (Thy$^-$, Leu$^-$, His$^-$, Cys$^-$, Xyl$^-$), growing with a generation time of 40 min on a minimal salts medium, by allowing them to terminate current rounds of DNA replication while preventing the initiation of new ones, by withholding required amino acids. Thymine was then removed from the medium and amino acids returned to allow synthesis of the proteins required for initiation of replication. Replication was then initiated synchronously by adding BU, which, after 15 s, was replaced with ^{3}H-thymine for 100 s. Replication was then stopped, DNA extracted, irradiated if required and centrifuged. Fig. 2 shows the difference in sedimentation behaviour of the irradiated and unirradiated extracts: there is a clear shift in molecular weight of the newly synthesized DNA after exposure to ultraviolet. From the positions of the peaks in a series of experiments we have determined[22] that the ratio of the molecular weights of the untreated to ultraviolet-treated DNA segments is 2.6±0.3. This agrees well with the predicted ratio of 2.32 calculated by assuming that all BU moieties will have been removed from the irradiated DNA segments leaving only the thymine-containing DNA. (The fact that increasing the exposure to ultraviolet twenty-fold leads to no further reduction in molecular weight (data not shown) supports this assumption.) The BU incorporated in this experiment therefore behaves as if it were incorporated into the middle of a DNA fragment, as in Fig. 1*B*, and not as if it were incorporated at its end (Fig. 1*A*).

如图1所示，如果复制是单向的(*A*)，那么BU将只会出现在每个单链片段的一端。因为摄取BU的时间比摄取 ^{3}H–胸腺嘧啶的时间短，所以经过紫外线处理与没有接受过处理的对照物在片段大小上几乎没有什么差异；两者均会在蔗糖梯度的同一区域中形成类似的条带。反之，如果复制是双向的（图1*B*），那么溴尿嘧啶就应该位于每个片段的中间，因而用紫外线处理会使片段一分为二，它们的分子量约为对照物的一半（参见下文）。在这种情况下，照射过和没有照射过的DNA将在蔗糖梯度中表现为两条分离的条带。

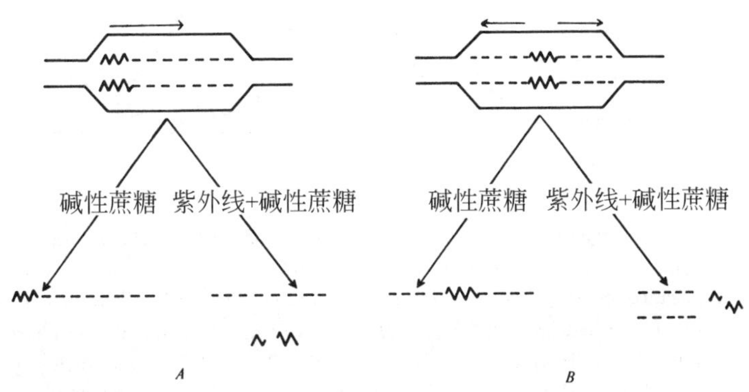

图1. 经过下述处理后DNA在起始点发生单向复制（*A*）与发生双向复制（*B*）的预期结果对比示意图。同步化的细胞先用5–溴尿嘧啶处理15 秒（~~~BU–DNA），接着用 ^{3}H–胸腺嘧啶处理100秒（- - -，^{3}H–DNA；——，未标记的DNA）。培养物的裂解液或者被紫外线照射，或者不被紫外线照射，随后都用碱性蔗糖梯度法离心分离。

在大肠杆菌 B/r（Thy$^-$、Leu$^-$、His$^-$、Cys$^-$、Xyl$^-$）的细胞内进行DNA复制起始同步化处理，该菌在低盐培养基中每40分钟繁殖一代，通过撤除所必需的氨基酸可以使它们结束当前的DNA复制循环，同时防止下一代复制的开始。然后从培养基中移除胸腺嘧啶，重新供给氨基酸以便能够合成启动复制所需的蛋白质。随后加入BU启动同步复制，15秒之后用 ^{3}H–胸腺嘧啶取代BU再处理100秒。复制随即终止，然后提取DNA，根据要求选择是否进行紫外线照射，最后离心。图2显示出接受照射和未接受照射的提取物在沉降行为上的不同：新合成的DNA在接受紫外线照射后出现了分子量上的明显变化。根据一系列实验中的波峰位置，我们断定[22]未经处理的与经紫外线处理的DNA片段的分子量之比为2.6±0.3。这与由假定所有BU部分均从照射过的DNA片段中被移去只剩下含胸腺嘧啶的DNA而计算出来的预期比率2.32非常吻合。(将紫外线的照射时间延长20倍并未使分子量持续下降（数据未给出），这一事实证实了上述假设。）因此，在这个实验中BU的掺入位置好像是一个DNA片段的中间，如图1*B*所示，而不太像是结合在了它的末端（图1*A*)。

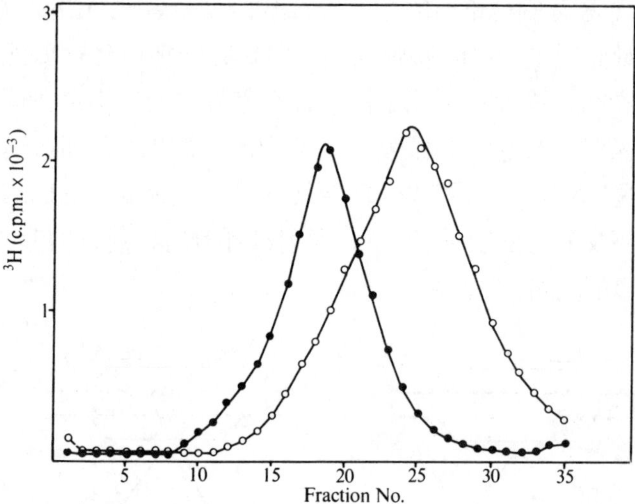

Fig. 2. Centrifugation of irradiated (○—○) and unirradiated (●—●) DNA in which initiation was allowed to take place in bromouracil. 50 ml. cultures at densities of 5×10^8 cells ml.$^{-1}$ were synchronized with regard to DNA replication by 100 min of amino acid starvation followed by thymine starvation for one mass doubling time[25]. The cells were collected on "Millipore 47 mm HA" membrane filters and washed with prewarmed buffer. The cells were pulsed for 15 s on the filter with 5-bromouracil (10 μg ml.$^{-1}$) and washed with prewarmed buffer. Finally the cells were suspended in 10 ml. M9-glucose + aa + thymine (1 μg ml.$^{-1}$) + ^{3}H-thymine (20 μCi μg^{-1}). The cells were vigorously aerated for 100 s. DNA replication was stopped and the cells lysed (Fig. 3). In this case 0.2 ml. samples of the crude lysates were exposed to ultraviolet if required (dose 3×10^5 erg) and layered on 5 ml. linear 5–20% alkaline sucrose gradients (0.1 M NaOH, 0.9 M NaCl) for centrifugation in a "Beckman L2-65B" centrifuge using an SW50.1 rotor at 20,000 r.p.m. for 12 h. Ten drop fractions were collected on "Whatman 3 MM" paper filters, washed in 5% ice cold TCA followed by 80% ethanol. The samples were counted in PPO-POPOP scintillant (Fig. 3).

By using a mouse satellite DNA fraction of known molecular weight as a standard we calculate that the newly synthesized, ^{3}H-labelled single-stranded DNA fragments have a molecular weight of 2×10^7. Assuming that the *E. coli* chromosome has a molecular weight of $1–2\times10^9$ and requires approximately 40 min for its synthesis this is about the size of the piece that we would have expected to have been synthesized during the 2 min period of the experiment. The fact that both the irradiated and unirradiated DNAs are homogeneous and of the size predicted, strongly supports the idea that BU is incorporated into the middle of daughter strands of DNA whose replication is bidirectional, and not into a fragment of DNA arising in one of the ways set out below.

BU Found Only in Daughter DNA

The results obtained above, that BU present during the initiation of replication is incorporated internally into a DNA fragment, could have been generated in ways other than that set out in Fig. 1*B*. According to the model presented there, BU is flanked by ^{3}H-thymine labelled daughter DNA. By contrast, a number of other models predict that ^{3}H-thymine would appear only to one side of the BU; the stretch of DNA on the other

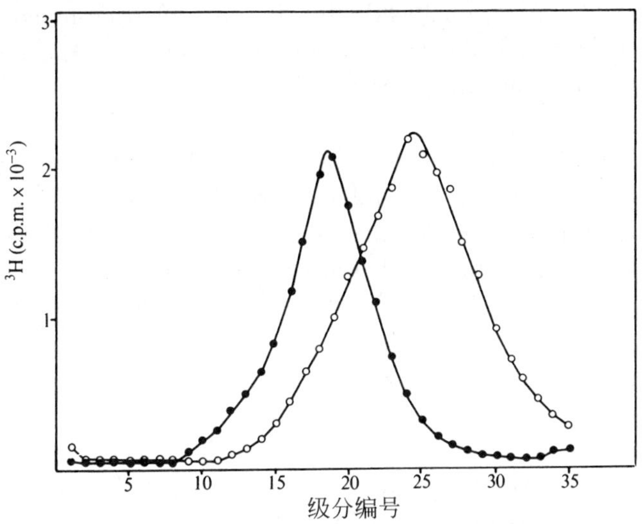

图 2. 复制起始于溴尿嘧啶的照射 DNA（○—○）和未照射 DNA（●—●）的离心结果。对于密度为 5×10^8 个细胞 / 毫升的 50 毫升培养物，通过 100 分钟氨基酸缺乏以及随后长达分裂增殖一倍所需时间的胸腺嘧啶缺乏，来实现 DNA 复制同步化[25]。将这些细胞收集于“密理博 47 毫米 HA”滤膜上，并用预热的缓冲剂洗涤。在滤膜上将细胞用 5– 溴尿嘧啶（10 微克 / 毫升）处理 15 秒，然后用预热的缓冲剂洗涤。最后将这些细胞悬浮于 10 毫升 M9– 葡萄糖 + 氨基酸 + 胸腺嘧啶（1 微克 / 毫升）+ ^{3}H – 胸腺嘧啶（20 微居里 / 微克）中。对这些细胞剧烈通气达 100 秒。随后终止 DNA 的复制并裂解细胞（图 3）。此时，0.2 毫升的粗裂解样品可根据需要选择用紫外线照射（剂量为 3×10^5 尔格）或者不用紫外线照射，然后使其在 5 毫升 5% ~ 20% 线性梯度碱性蔗糖溶液（0.1 摩尔 / 升 NaOH，0.9 摩尔 / 升 NaCl）中分层，方法是在转子为 SW50.1 的“贝克曼 L2-65B”型离心机中以 20,000 转 / 分的转速离心分离 12 小时。10 滴分离出来的级分被收集于“沃特曼 3 MM”滤纸上，先用 5% 冷 TCA 后用 80% 乙醇溶液冲洗。样品用 PPO–POPOP 闪烁体计数（图 3）。

利用分子量已知的小鼠卫星 DNA 组分作为标准，我们计算出了新合成的、^{3}H– 标记的单链 DNA 片段的分子量为 2×10^7。假设大肠杆菌染色体的分子量是 $1\times10^9\sim2\times10^9$，并且假设其合成过程需要 40 分钟左右，则得到的产物大小与我们预期 2 分钟实验中能够合成出来的大小大致相等。照射过的和未照射过的 DNA 都很均一且大小与预期一致的事实充分证实了这一观点，即 BU 被掺入到了双向复制 DNA 的子链中间，而非被掺入到由下文中所列举的任意一种方式形成的 DNA 片段中。

BU 仅在子代 DNA 中出现

上文中得到的结果，即在复制起始阶段存在的 BU 被掺入到了 DNA 片段中间，也可能是由与图 1*B* 所示有所不同的方式产生的。根据上文中提出的模型，^{3}H– 胸腺嘧啶标记的子代 DNA 应位于 BU 的两侧。与之相反，另有很多模型预言 ^{3}H– 胸腺嘧啶只可能出现在 BU 的一侧；在 BU 另一侧的 DNA 链将包含未标记的胸腺嘧啶。然

side of the BU would contain unlabelled thymine. It should be noted, however, that for this to take place a series of coincidences would be required to explain the observed sizes and homogeneity of the fragments described above.

The first way in which this could come about would be if parental DNA were used to prime daughter DNA replication, as in rolling circle[23,24], and some other[25] models of DNA replication. According to these models daughter and parental DNA[16] are covalently linked and so first the BU and then the ^{3}H-thymine would be incorporated in continuation of high molecular weight parental DNA. If this DNA were to fragment during extraction so as to yield pieces of 2×10^7 molecular weight, some of these pieces could be expected to contain BU sequences flanked by parental and daughter DNA which could behave as described above.

If termination of rounds of chromosome replication did not occur during the synchronization treatment, but only after BU and labelled thymine had been added to the medium, linkage between parental and daughter DNA would also be found. Although Stein and Hanawalt[26] and Kuempel (submitted for publication) failed to detect linkage between parental and daughter DNA in *E. coli* 15T$^-$, we thought it essential also to exclude such linkage in B/r under the conditions of our experiment.

Another source of error could be the retention of an internal pool of thymine after termination of rounds of replication or some leakiness in the thymine requirement of the strain used in these experiments. Initiation of synthesis could then occur during the thymine starvation period and the BU, when added, incorporated in continuation of a previously synthesized piece of DNA. If the amount of synthesis occurring before the addition of BU were approximately equal to that which occurs during the later pulse of ^{3}H-thymine, pieces could be generated which would behave as described above, and unidirectional synthesis mistaken for bidirectional.

We therefore excluded these possibilities by performing an experiment which showed that BU used to initiate replication is not added to a previously synthesized piece of DNA. Bacteria were grown for three generations in a medium containing ^{14}C-thymine to totally label parental DNA. The required amino acids were removed and the cells allowed to terminate rounds of replication in ^{14}C-thymine. After termination thymine was removed, amino acids re-added and incubation continued until all cells were ready to initiate replication. ^{3}H-bromouracil was then added to the medium and replication allowed to continue for 100 s. The DNA was then isolated and the strands separated and centrifuged on caesium chloride gradients (Fig. 3). The parental ^{14}C-containing DNA is of light density; daughter DNA which is fully substituted with ^{3}H-BU is of heavy density. Any DNA stretches which contained both parental and daughter DNA, or BU added to a fragment synthesized during thymine starvation from a retained internal pool, would be doubly labelled and of a density intermediate between that of the thymine containing parental and the BU containing daughter DNA. Fragments synthesized by addition of BU to the unlabelled DNA which could be made during thymine starvation by a leaky mutant

而，应该指出的是：如果 ^{3}H– 胸腺嘧啶仅出现在 BU 的一侧，那么就需要用一系列巧合事件来解释上述观察到的分子量大小及 DNA 片段的均一性。

造成这种情况的第一个可能原因是：如果亲代 DNA 被用于引发子代 DNA 的复制，如在滚环 [23,24] 和其他一些 [25]DNA 复制模型中那样。根据这些模型，子代和亲代 DNA[16] 是以共价键相连的，所以 BU 将首先被掺入到高分子量亲代 DNA 连续体中，然后才是 ^{3}H– 胸腺嘧啶。如果这个 DNA 在抽提的过程中断裂，就会产生分子量为 2×10^{7} 的碎片，其中一部分碎片可能会含有两侧分别是亲代和子代 DNA 的 BU 序列，在这种情况下也会出现上文所描述的现象。

如果在同步化处理的时候，染色体复制循环并未终止，而只有当把 BU 和标记的胸腺嘧啶加入培养基之后循环才终止，那么我们也会发现亲代和子代 DNA 之间存在连接。尽管斯坦和哈纳沃特 [26] 以及金佩尔（已提交出版）都未能在大肠杆菌 $15T^{-}$ 中检测到亲代和子代 DNA 之间的连接，但我们认为在我们的实验条件下排除大肠杆菌 B/r 中存在这种连接也是必要的。

另一个造成判断失误的原因可能是：在复制循环终止后仍存在一个内部的胸腺嘧啶代谢池，或者实验中所用菌株的胸腺嘧啶需求存在渗漏。合成的启动可能会发生在胸腺嘧啶缺乏阶段，随后一旦加入 BU，BU 就会掺入到早期合成的 DNA 片段连续体中。如果在加入 BU 以前的合成量近似等于后期与 ^{3}H– 胸腺嘧啶孵育合成的量，那么就会产生上文中所描述的碎片，这时单向合成就会被错误判断成双向合成。

因此，我们要进行一项实验来排除上述可能性，由这个实验可以说明用于启动复制的 BU 并没有掺入到先前合成的 DNA 片段上。细菌在含有 ^{14}C– 胸腺嘧啶的培养基中生长三代就可以将亲代 DNA 全部标记。撤除所需的氨基酸，细胞就会在 ^{14}C– 胸腺嘧啶中终止复制循环。终止之后移除胸腺嘧啶，再次加入氨基酸，继续培养直到所有细胞准备好开始复制。然后，将 ^{3}H– 溴尿嘧啶添加到培养基中，并使复制过程持续 100 秒。随后提取 DNA，解链成单链，并用 CsCl 密度梯度法离心（图 3）。含有 ^{14}C 的亲代 DNA 是低密度的；完全被 ^{3}H–BU 取代的子代 DNA 是高密度的。任何同时包含亲代 DNA 和子代 DNA 的链，或者包含 BU 掺入到在缺少胸腺嘧啶时从内部残留代谢池中合成的 DNA 片段上的链，都将被双重标记，且其密度介于含有胸腺嘧啶的亲代 DNA 和含有 BU 的子代 DNA 的密度之间。在缺乏胸腺嘧啶时渗漏的突变体可以利用重新合成的胸腺嘧啶生成未标记的 DNA，在这种 DNA 上掺入 BU 所合成的片段也具有杂合的密度，但不会被双重标记。

using thymine synthesized *de novo* would also be of hybrid density but would not be doubly labelled.

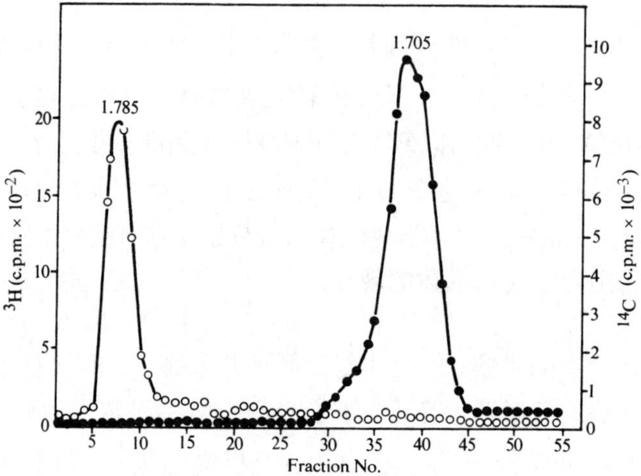

Fig. 3. Separation of daughter and parental DNA on CsCl gradients. *E. coli* B/r (L6) was grown in M9 medium + 0.2% glucose + 10 μg ml.$^{-1}$ thymine, 20 μg ml.$^{-1}$ each of leucine, histidine and cysteine (aa) with rotary shaking at 37°C. The cells were labelled with ^{14}C-thymine (10^{-2} μCi μg^{-1}) for 3 generations and synchronized by 100 min aa starvation followed by thymine starvation for 1 mass doubling time. The cells were washed with prewarmed buffer and resuspended in M9-glucose + aa + 10 μg ml.$^{-1}$ bromouracil + 2 μCi μg^{-1} ^{3}H-bromouracil for 100 s and vigorously aerated. DNA replication was stopped in 50 ml. aliquots (5×10^8 cells ml.$^{-1}$) by pouring the cells with crushed ice into buffer B[28]. After washing twice in 1×SSC the cells were suspended in 2 ml. 0.1×SSC+10^{-3} M EDTA and lysed with 1% Na dodecyl sulphate (SDS)+pronase 1 mg ml.$^{-1}$ (ref. 29). The DNA lysate was extracted twice with an excess volume of 88% phenol—12% meta-cresol—0.1% 8-hydroxy-quinoline followed by gentle shaking in 24:1 mixture of chloroform:octanol to remove the SDS. The DNA was denatured and centrifuged according to the procedure of Stein and Hanawalt[26] except that we used "BDH Analar" grade CsCl and diluted the DNA in 10^{-3} M EDTA. Ten drop fractions were collected into ice cold 5% TCA and counted on "Whatman" GF/C filters in PPO-POPOP scintillant in a "Packard Tricarb" liquid scintillation counter. ○, ^{3}H; ●, ^{14}C.

DNA stretches of intermediate density are clearly identifiable when thymine is replaced by BU in the middle of the replicative cycle of *E. coli*[27]. Fig. 3 gives the data we have obtained from our experiment. No doubly labelled DNA or DNA of intermediate density can be seen. The ^{14}C-DNA bands at 1.705 g cm^{-3} and the ^{3}H-DNA at 1.785 g cm^{-3}. Assuming that the change in density is proportional to the change in molecular weight these densities are what would be expected on the hypothesis that daughter DNA is fully substituted with BU and neither contains, nor is linked to, light DNA containing thymine. It should be noted that this experiment does not eliminate the possibility that daughter DNA is linked to parental DNA for a period of less than 100 s. Because, however, this is shorter than the labelling period used in the experiment reported above, we conclude that linkage between daughter and parental DNA or between daughter DNA synthesized before and after the addition of BU cannot explain the results in Fig. 2.

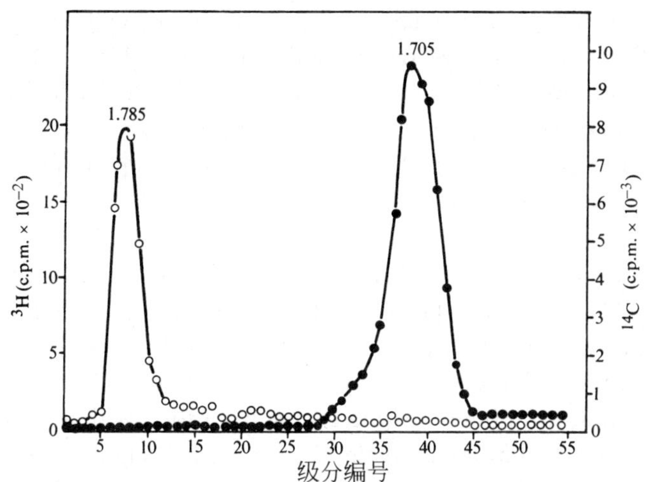

图 3. 用 CsCl 密度梯度离心法分离子代 DNA 和亲代 DNA。大肠杆菌 B/r (L6) 在 M9 培养基 + 0.2% 葡萄糖 +10 微克 / 毫升胸腺嘧啶及浓度均为 20 微克 / 毫升的亮氨酸、组氨酸和半胱氨酸（氨基酸）中生长，37℃旋转振荡培养。细胞用 ^{14}C– 胸腺嘧啶（10^{-2} 微居里 / 微克）标记三代，通过 100 分钟氨基酸缺乏以及随后长达分裂增殖一倍所需时间的胸腺嘧啶缺乏而达到同步化。细胞用预热的缓冲液冲洗，然后重悬于 M9– 葡萄糖 + 氨基酸 +10 微克 / 毫升溴尿嘧啶 +2 微居里 / 微克 ^{3}H– 溴尿嘧啶中 100 秒并剧烈通气。将细胞和碎冰加进缓冲剂 B 中以便终止 50 毫升等分试样（5×10^8 个细胞 / 毫升）中的 DNA 复制 [28]。细胞用 1 × SSC（译者注：氯化钠 – 柠檬酸钠缓冲液）洗涤两次，然后悬浮于 2 毫升 0.1 × SSC + 10^{-3} 摩尔 / 升 EDTA 中，最后在 1% 十二烷基硫酸钠（SDS）+1 毫克 / 毫升链霉蛋白酶中裂解（参考文献 29）。用过量的 88% 苯酚—12% 间甲酚—0.1% 8– 羟基喹啉对 DNA 裂解物进行两次抽提，然后在氯仿 : 辛醇为 24:1 的混合物中轻轻振荡以除去 SDS。利用斯坦和哈纳沃特的方法 [26] 使 DNA 变性并离心，不同之处是我们使用了“英国药品所的分析纯”CsCl，并将 DNA 在 10^{-3} 摩尔 / 升的 EDTA 中稀释。10 滴分离出来的级分被收集到冷的 5% TCA 中，在“沃特曼 GF/C”滤膜上通过“帕卡德公司三汽化器”型液体闪烁计数仪中的 PPO–POPOP 闪烁体进行计数。○，^{3}H；●，^{14}C。

当在大肠杆菌复制周期之中用 BU 取代胸腺嘧啶时，可以清楚地辨认出中间密度的 DNA 链 [27]。图 3 中给出的数据来自我们的实验结果。并未发现双重标记的 DNA 或者中间密度的 DNA。^{14}C–DNA 条带位于 1.705 克 / 立方厘米处，^{3}H–DNA 条带位于 1.785 克 / 立方厘米处。假设密度变化与分子量变化成正比，则上述密度就可以通过以下假设推导出来，即子代 DNA 完全被 BU 置换，其中既不包括含胸腺嘧啶的低密度 DNA，也不会与之相连接。应该指出的是，这个实验并不能排除子代 DNA 在一个少于 100 秒的时间段内与亲代 DNA 相连接的可能性。然而，因为这比上述实验中所用的标记时间更短，所以我们认为：用子代 DNA 与亲代 DNA 之间的连接，或者用在加入 BU 之前和之后合成的子代 DNA 之间的连接，都不能解释图 2 中的结果。

Another possible source of error is that BU continues to be incorporated into DNA for some time after it has been replaced in the medium by ^{3}H-thymine. If BU were simply to compete with thymine for incorporation the result would be stretches of DNA containing both bases which would be expected, on irradiation, to yield many pieces of small size rather than pieces half the size of the initial fragment (Fig. 2). If, however, the BU continued to be incorporated for an appropriate interval to the exclusion of thymine and if replication were unidirectional, pieces could be generated which could be halved in size by the removal of a large number of BU moieties from one end. This could result in our mistaking unidirectional replication for bidirectional.

To eliminate this possibility, initiation was synchronized as described above, replication initiated in BU for 15 s and then the BU replaced with ^{3}H-thymine. Samples were taken at 10 s intervals and the rate of incorporation of the ^{3}H-thymine determined. As can be seen (Fig. 4) the thymine is incorporated linearly for at least 8 min after a lag of no more than 10 s indicating that it is not competing for incorporation with a large internal pool of BU. We therefore conclude that the reduction in molecular weight caused by irradiation of the extracted DNA fragments is not a result of their being composed, to a large extent, of bromouracil.

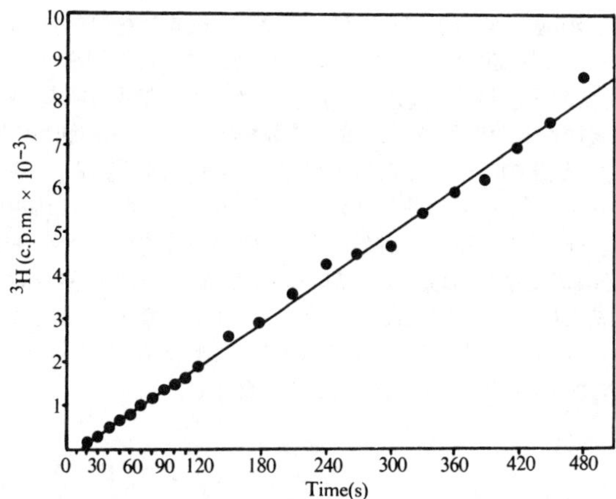

Fig. 4. Thymine incorporation after initiation of replication in BU. Cells were synchronized, labelled with BU and resuspended in ^{3}H-thymine as described in the legend to Fig. 2. Samples are taken every 10 s for 2 min and then every 30 s for another 6 min into ice cold 7% TCA. TCA precipitates are collected on "Millipore" filters and counted as described above.

Rolling Circle Model Excluded

Our experiments demonstrate that when cells initiate replication in BU and are subsequently transferred to a medium containing thymine, the BU behaves as if it were contained in the middle of a DNA fragment. As this fragment does not contain any DNA synthesized before the addition of the BU, the BU must be flanked on both sides by ^{3}H-thymine-containing DNA made after it had been incorporated. This result is consistent only with bidirectional DNA replication of the sort described in Fig. 1*B*, in which each daughter strand elongates in both directions from the origin of replication.

还有一个造成判断失误的原因可能是：在 BU 被培养基中的 ^{3}H– 胸腺嘧啶取代之后，它仍能在一段时间内继续掺入到 DNA 中。如果 BU 与胸腺嘧啶在合成中仅仅是竞争关系，那么结果将是：包含这两种可能碱基的 DNA 链在照射下会产生许多小碎片而不是恰好等于最初片段一半大小的碎片（图 2）。然而，如果 BU 继续被掺入一定的时间以排除胸腺嘧啶，并且如果复制是单向的，那么从一端去除大量的 BU 部分就有可能产生大小减半的碎片。这会使我们将单向复制误解为双向复制。

为了排除这种可能性，如上文所述采用起始同步化处理，在 BU 中启动复制 15 秒后即用 ^{3}H– 胸腺嘧啶取代 BU。每隔 10 秒取一次样，并由此计算 ^{3}H– 胸腺嘧啶的掺入率。可以看出（图 4）：胸腺嘧啶在滞后不超过 10 秒后一直保持线性掺入，时间长达 8 分钟以上，这说明它与某个大的内部 BU 代谢池之间不存在竞争关系。因此，我们可以得出如下结论：由照射提取出的 DNA 片段所产生的分子量下降并不是因为它们含有大量的溴尿嘧啶。

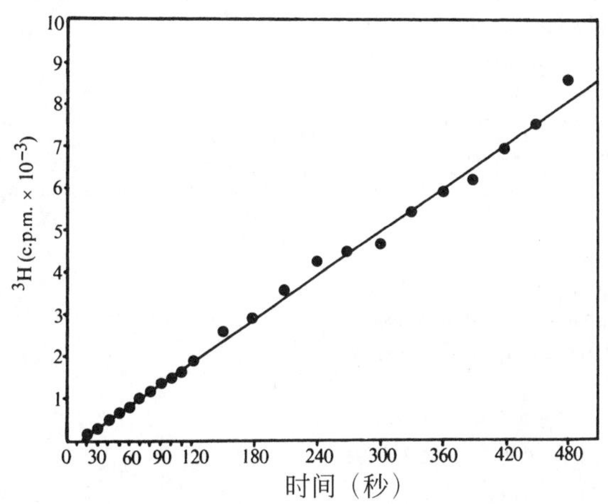

图 4. 在 BU 中启动复制后胸腺嘧啶的掺入情况。按照图 2 注的描述对细胞进行同步化、BU 标记并使其重悬于 ^{3}H– 胸腺嘧啶中。在开始的 2 分钟内，每 10 秒取样一次；随后的 6 分钟即改成每 30 秒取样一次。样品被放入冷的 7% TCA 中。用“密理博”滤膜收集 TCA 的沉淀物，并按照前述方法进行计数。

排除滚环模型

我们的实验证明：如果细胞在 BU 中启动复制，然后再被转移到含胸腺嘧啶的培养基中，那么 BU 的掺入位置就很像是在一个 DNA 片段的中间。因为这个片段并不包含加入 BU 前合成的 DNA，所以 BU 的两侧一定是在它被掺入之后形成的含 ^{3}H– 胸腺嘧啶的 DNA。这一结果只能与图 1*B* 中所描述的双向复制方式相吻合，其中每个子链均从复制起始点开始向两个方向延伸。

The absence of any linkage between parental and daughter DNA excludes rolling circle and related mechanisms, either uni- or bidirectional, from a role in the vegetative replication of the *E. coli* chromosome.

We thank Miss Mora McCallum for instruction in centrifugation techniques and for the gift of ^{32}P mouse satellite DNA, Professor J. M. Mitchison and Dr. W. H. Wain for advice on uptake experiments. Professor W. Hayes, Drs. J. Gross and J. Scaife for help with revision of the manuscript, and Dr. W. D. Donachie for helpful discussion and encouragement.

(**240**, 536-539; 1972)

W. G. McKenna and Millicent Masters: MRC Molecular Genetics Unit, Department of Molecular Biology, University of Edinburgh, Mayfield Road, Edinburgh 9.

Received August 14; revised October 12, 1972.

References:

1. Cairns, J., *J. Mol. Biol.*, **6**, 208 (1963).
2. Meselson, M., and Stahl, F. W., *Proc. US Nat. Acad. Sci.*, **44**, 671 (1958).
3. Lark, K. G., Repko, T., and Hoffman, E. J., *Biochim. Biophys. Acta*, **76**, 9 (1963).
4. Caro, L. C., and Berg, C. M., *J. Mol. Biol.*, **45**, 325 (1969).
5. Berg, C. M., and Caro, L. C., *J. Mol. Biol.*, **29**, 419 (1967).
6. Donachie, W. D., and Masters, M., *Genet. Res.*, **8**, 119 (1966).
7. Donachie, W. D., and Masters, M., in *The Cell Cycle: Gene-Enzyme Interactions* (edit. by Padilla, Whitson and Cameron), 37 (Academic Press, New York and London, 1969).
8. Masters, M., *Proc. US Nat. Acad. Sci.*, **65**, 601 (1970).
9. Helmstetter, C., *J. Bacteriol.*, **95**, 1634 (1968).
10. Pato, M. L., and Glaser, D., *Proc. US Nat. Acad. Sci.*, **60**, 1268 (1968).
11. Abe, M., and Tomizawa, J., *Proc. US Nat. Acad. Sci.*, **58**, 1911 (1967).
12. Cerdá-Olmedo, E., Hanawalt, P. C., and Guerola, N., *J. Mol. Biol.*, **33**, 705 (1968).
13. Wolf, B., Newman, A., and Glaser, D., *J. Mol. Biol.*, **32**, 611 (1968).
14. Ward, C. B., and Glaser, D. A., *Proc. US Nat. Acad. Sci.*, **62**, 881 (1969).
15. Yahara, I., *J. Mol. Biol.*, **57**, 373 (1971).
16. Masters, M., and Broda, P., *Nature New Biology*, **232**, 137 (1971).
17. Bird, R. E., Louarn, J., Martuscelli, J., and Caro, L. G., *J. Mol. Biol.*, **70**, 549 (1972).
18. Hohlfeld, R., and Vielmetter, W. (in the press).
19. Prescott, D. M., and Kuempel, P. L., *Proc. US Nat. Acad. Sci.*, **69**, 2842 (1972).
20. Weintraub, H., *Nature New Biology*, **236**, 195 (1972).
21. Hotz, G., and Wolser, R., *Photochem. Photobiol.*, **12**, 207 (1970).
22. Studier, F. W., *J. Mol. Biol.*, **11**, 373 (1965).
23. Gilbert, W., and Dressler, D., *Cold Spring Harbor Symp. Quant. Biol.*, **33**, 473 (1968).
24. Watson, J. D., in *Molecular Biology of the Gene*, 291 (W. A. Benjamin, Inc., New York, 1970).
25. Yoshikawa, H., *Proc. US Nat. Acad. Sci.*, **58**, 312 (1967).
26. Stein, G. H., and Hanawalt, P. C., *J. Mol. Biol.*, **64**, 393 (1972).
27. Pettijohn, D. E., and Hanawalt, P. C., *J. Mol. Biol.*, **8**, 170 (1964).
28. Oishi, M., *Proc. US Nat. Acad. Sci.*, **60**, 329 (1968).
29. Thomas, jun., C. A., Berns, K., and Kelly, jun., T. J., in *Procedures in Nucleic Acid Research* (edit. by Cantoni and Davies), 535 (Harper and Row, New York, 1966).

因为亲代 DNA 与子代 DNA 之间不存在任何连接，所以滚环以及与滚环相关的其他机制，不管是单向复制还是双向复制，都不能用于解释大肠杆菌染色体的无性复制。

感谢莫拉·麦卡勒姆小姐在离心技术方面为我们提供指导并将 ^{32}P 标记的小鼠卫星 DNA 赠送给我们，感谢米奇森教授及韦恩博士对摄取实验提出了一些建议。感谢海斯教授、格罗斯博士和斯凯夫博士帮助我们修订手稿，感谢多纳基博士和我们进行了有益的讨论而且还鼓励我们。

（董培智 翻译；孙军 审稿）

DNA Replication Sites within Nuclei of Mammalian Cells

J. A. Huberman *et al.*

Editor's Notes

Nearly 20 years after the structure of DNA was published, molecular biologists were still struggling to understand how DNA, the genetic material, is replicated in living cells of higher organisms (eukaryotes). The interest of this article is that it shows how rudimentary were scientists' ideas of that process. Huberman *et al.* studied HeLa cells (a familiar strain of human cells derived from a cancer) during the process of mitosis (cell division). "S" phase is that during which a cell about to divide accumulates the material eventually needed to make two cells. They conclude that DNA is first synthesized near the nuclear membrane but that it later migrates to the whole nucleus.

DNA replication can occur throughout the nucleus and is not restricted to the inner surface of the nuclear membrane.

THE involvement of a membrane site in DNA replication was first suggested by Jacob, Brenner and Cuzin[1] in their "replicon" model for replication of bacterial DNA, but the evidence accumulated since then is inconclusive. In prokaryotic cells, co-sedimentation of DNA replication points and cell membranes has been demonstrated[1-4], and the origin of replication and the membrane found to be associated[5]. A lipid-free replicating DNA–protein complex from *E. coli* has been isolated[6] and it has been reported that only the origin, but not the growing points, of the *E. coli* chromosome is attached to membrane[7].

In cell fractionation experiments with mammalian cells it was also found that replication points, detected by pulse-labelling with ^{3}H-thymidine (TdR), are associated with the nuclear membrane (or some other large, light, hydrophobic cell structure)[8-11,13]. Association between replication points and the nuclear membrane has also been detected by electron microscope autoradiography of thin sections through pulse-labelled, unsynchronized HeLa cells. The label associated with the membrane apparently moved into the nuclear interior after a 1 h chase[13].

On the other hand, most electron microscope autoradiographic experiments suggest that replication can take place throughout the nucleus. For instance, Comings and Kakefuda[14] generally found grains located throughout the nucleus when unsynchronized human amnion cells were pulse-labelled for 5 min or more with ^{3}H-TdR and then

哺乳动物细胞核内的DNA复制位点

休伯曼等

编者按

在DNA结构发表近20年后，分子生物学家们仍然在孜孜不倦地研究作为遗传物质的DNA在高等生物活细胞（真核细胞）中是如何复制的。这篇文章的意义在于，它显示了科学家们对于DNA的复制过程的认识仍然浅显。休伯曼等人研究了海拉(HeLa)细胞（一种源自人体癌细胞的细胞株）的有丝分裂过程。“S”期是指即将分裂的细胞积累分裂成两个细胞所需物质的阶段。他们总结认为DNA首先在核膜附近合成，然后迁移到整个细胞核。

DNA复制发生在整个细胞核内，而不是局限于核膜的内表面。

雅各布、布伦纳和居赞[1]在关于细菌DNA复制的“复制子”模型中首次提出膜位点参与DNA的复制，但是其后这方面所积累的证据并不确凿。在原核细胞中，DNA复制点和细胞膜的共沉淀已经得到证明[1-4]，同时发现，此复制起点能与细胞膜结合[5]。从大肠杆菌中分离得到了不含膜脂的复制的DNA－蛋白复合体[6]，且有报道称仅大肠杆菌染色体上的复制起点与细胞膜相连，生长点与其不相连[7]。

通过^{3}H–胸腺嘧啶核苷酸(TdR)脉冲标记的哺乳动物细胞的分级分离实验发现，复制点与细胞核核膜（或者其他一些大的、轻的、疏水的细胞结构）相结合[8-11,13]。使用薄片电子显微放射自显影法检测脉冲标记非同步的HeLa细胞，同样发现了复制点与细胞核核膜相结合。追踪实验表明与核膜相连的标记1小时后明显转移到细胞核内部[13]。

另一方面，大部分电子显微放射自显影实验表明，复制能在整个细胞核中发生。例如，科明斯和挂札[14]用^{3}H-TdR脉冲标记非同步的人羊膜细胞5分钟或者更长时间，然后将其切片进行放射自显影，研究结果发现颗粒遍布整个细胞核。然而，

sectioned and autoradiographed. When, however, cells that were supposedly synchronized at the beginning of S phase were pulse-labelled for 5 or 10 min, grains were found predominantly over the nuclear membrane. Comings and Kakefuda concluded that initiation of replication takes place on the nuclear membrane, whereas the replication occurs anywhere within the nucleus.

Somewhat different results were obtained by Blondel[15], using KB cells and pulse times of 2 min; Williams and Ockey[16], using Chinese hamster cells and pulse times of 10 min or longer; and Erlandson and de Harven[17], using HeLa cells and pulse times of 15 min. All three groups concluded, in agreement with Comings and Kakefuda[14], that during a large part of the S phase grains are produced over the entire nucleus, but they differed from them in finding a peripheral pattern of grains more frequently at the end of the S phase than at the beginning. A cell fractionation experiment by Kay *et al.*[18] also demonstrated association of late-replicating but not early-replicating DNA with the nuclear membrane.

Higher Resolution Autoradiography

Regardless of the time in S phase at which replication occurs close to the nuclear membrane, one important implication of most of these electron microscope autoradiography experiments is that, in some parts of S phase, at least, replication occurs throughout the nucleus. How can this implication be reconciled with the experiments indicating that all replicating DNA is associated with the nuclear membrane? It is possible that a considerable amount of DNA is synthesized during a pulse as short as 2 min. Huberman and Riggs[19] have shown that the rate of DNA replication in Chinese hamster cells can be as much as 2.5 μm min^{-1}, although most replication seems to occur at rates between 0.5 and 1.2 μm min^{-1}. Thus as much as 5 μm of DNA could be synthesized during a 2 min pulse. If this DNA were synthesized at the nuclear membrane and then stretched out, it could reach the centre of a nucleus of diameter 5–8 μm and give rise to a false impression of the location of sites of DNA synthesis.

This effect could be ruled out by shorter pulses. We calculated that a pulse of 0.5 min would provide sufficient grains, after an exposure of several months, and the resolution (less than 1.25 μm of DNA synthesized) necessary to distinguish between replication solely at the nuclear membrane and replication elsewhere in the nucleus.

In most experiments, we used Chinese hamster (CHO) cells, which can be easily synchronized. The cells were pulse-labelled for 0.5 min with ^{3}H-TdR, then washed, fixed, embedded, sectioned and stained. The sections were placed on grids and autoradiographed by standard techniques[20]. After exposure times of several months, the emulsions were developed and the grids were examined by electron microscopy.

Sites of Replication

Cells from an unsynchronized culture, shown in Fig. 1, illustrate the variety of distribution of grains over the nucleus. The cells in Fig. 1A and B are labelled

如果把同步化为S期初期的细胞脉冲标记5或10分钟，检测到的颗粒则主要分布在核膜上。科明斯和挂札由此推断，复制的起始发生在核膜上，而复制则发生在细胞核内的任何区域。

下面三个小组的研究得到的结果略有不同：布隆德尔[15]用脉冲处理KB细胞（一种口腔癌细胞）2分钟，威廉姆斯和奥克伊[16]用脉冲处理中国仓鼠细胞10分钟或者更长时间，以及厄兰森和德阿尔旺[17]用脉冲处理HeLa细胞15分钟。与科明斯和挂札[14]的结论相一致的是，上述三个研究组都认为在S期的大部分时期，颗粒存在于整个细胞核中。但是与科明斯和挂札的结论不同的是，他们发现与S期初期相比这些颗粒在S期末期出现周边模式的频率更高。凯[18]等人的细胞分级分离实验同样阐明了DNA与核膜的结合主要是发生在复制的后期，而不是复制的初期。

高分辨放射自显影法

不考虑更接近细胞核核膜的复制时期到底出现在S期的哪个阶段，大部分电子显微放射自显影实验得到的一个重要的结论就是，至少在S期的某些阶段内，复制发生在整个细胞核内。这一结论是如何与那些表明正在复制的DNA与核膜相结合的实验保持一致的？在短短2分钟的脉冲时间内合成大量的DNA是可能的。尽管大部分DNA的复制速率在每分钟0.5~1.2 μm，但是休伯曼和里格斯[19]的研究表明，中国仓鼠细胞的DNA复制速率可以达到每分钟2.5 μm。因此在2分钟的时间内可以合成5 μm的DNA。假设这些DNA是在细胞核核膜上合成，然后延伸出去，那么该DNA就可以到达直径为5~8 μm的细胞核的中心，从而造成了对DNA合成位点位置的错误判断。

上述这种影响可以通过缩短脉冲时间来排除。我们计算表明，在曝光几个月以后，0.5分钟的脉冲时间能够提供足够的颗粒，其分辨能力（可分辨小于1.25 μm新合成的DNA）可以将单独发生在细胞核核膜与发生在细胞核的其他区域的复制区分开来。

大部分实验中，我们使用中国仓鼠（CHO）细胞，因为这种细胞容易同步化。将这种细胞用^{3}H-TdR脉冲标记0.5分钟后，进行清洗、固定、包埋、切片以及染色。切片放在铜网上，按照标准技术进行放射自显影[20]。曝光几个月后，显影冲洗，用电子显微镜观察铜网。

复 制 位 点

图1显示了非同步培养的细胞中颗粒在细胞核中的各种分布情况。图1A和图1B的细胞展示的是整个细胞核都被标记的情况（普遍模式），而图1C和图1D的

throughout the nucleus (general pattern), whereas the cells in Fig. 1C and D are labelled predominantly around the nuclear membrane (peripheral pattern). The grains in Fig. 1E are mostly clustered over regions of condensed chromatin, frequently near the membrane. Fig. 1F shows a cell with grains throughout the nucleus but with some concentration around the nuclear membrane.

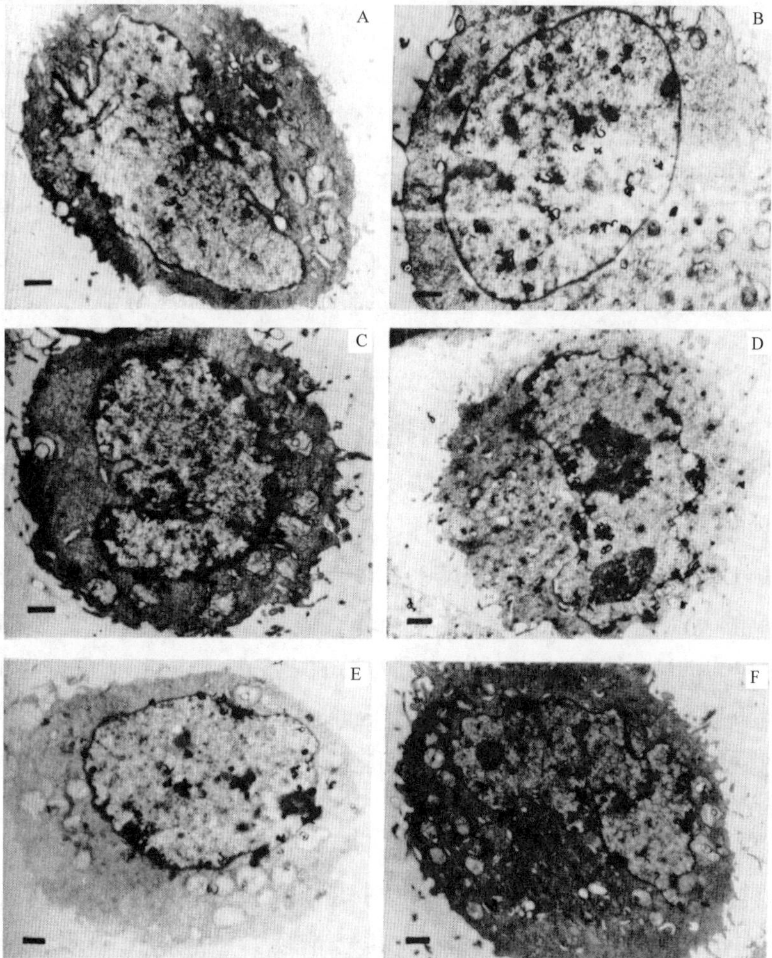

Fig. 1. Autoradiography of unsynchronized CHO cells exposed to ^{3}H-TdR for 0.5 min. A and B, grains distributed over entire nucleus. C and D, grains distributed around nuclear membrane. E, clustered grain distribution. F, mixed grain distribution. CHO cells were grown on Petri plates in Joklik-modified MEM (Grand Island Biological Company) supplemented with 7% foetal calf serum and non-essential amino-acids. Pulse-labelling was performed by first adding 5-fluorouridine deoxyriboside (FUDR; Hoffmann–LaRoche) to a final concentration of 1.6 μg ml.$^{-1}$ to inhibit further biosynthesis of dTTP. After 1 min, ^{3}H-TdR (51 Ci mmol^{-1}; New England Nuclear) was added to 17 μCi ml.$^{-1}$. After 30 s the plates were removed from 37°C, and the medium rapidly sucked off. The plates were then washed with ice-cold isotonic saline containing FUDR at 0.1 μg ml.$^{-1}$ (two changes). Cells were removed from the plates by trypsinization at room temperature in isotonic saline still containing FUDR. The cells were pelleted, then fixed with glutaraldehyde and OsO_4, dehydrated, embedded in epoxy resin[29] and cut into gold-purple sections. The sections were mounted on grids and autoradiographed by the technique of Caro and Van Tubergen[20] except that the acetic acid stop bath was replaced with distilled water. Exposure time was 3.5 months. The bar in each figure represents 1 μm.

细胞展示的是被标记区域主要出现在细胞核核膜周围的情况（周边模式）。图 1E 中颗粒主要集中在浓缩的染色质区域，通常靠近核膜。图 1F 中颗粒遍布细胞的整个细胞核，但是在核膜周围的密度相对较大。

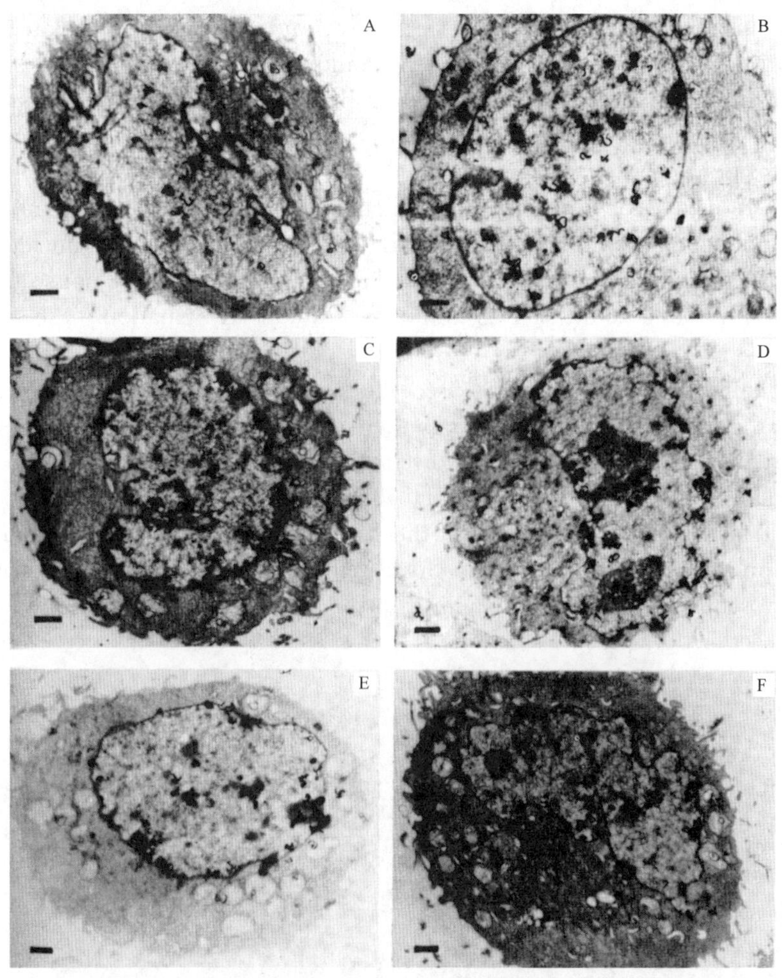

图 1. ^{3}H-TdR 处理非同步 CHO 细胞 0.5 分钟后放射自显影图。图 A 和图 B 展示的是颗粒分布于整个细胞核的情况。图 C 和图 D 展示的是颗粒分布在细胞核核膜周围的情况。图 E 展示的是颗粒的聚集分布情况。图 F 展示的是颗粒的混合分布情况。用改良的乔克利克基础培养基（购自格兰德岛生物公司）在培养皿中培养 CHO 细胞，培养基中添加 7% 的胎牛血清和非必需氨基酸。脉冲标记操作按照以下步骤进行，首先添加 5– 氟尿嘧啶脱氧核苷（FUDR，购自霍夫曼 – 罗氏公司）至终浓度为 1.6 μg · ml^{-1}，以便进一步抑制脱氧胸苷三磷酸 dTTP 的生物合成。1 分钟后，加入 ^{3}H-TdR（51 Ci · mmol^{-1}，购自新英格兰核公司），至终浓度 17 μCi · ml^{-1}。30 秒钟后，将培养皿从 37℃中移走，迅速吸掉培养基。然后用含有 0.1 μg · ml^{-1} FUDR 的冰浴生理盐水冲洗培养板（换液 2 次）。室温下，在含有 FUDR 的生理盐水中进行胰蛋白酶消化，从培养皿中分解细胞。细胞离心后，用戊二醛和四氧化锇固定，脱水，环氧树脂[29]包埋，制成紫金色切片。除了用蒸馏水替代醋酸作为定影液，其他的按照卡罗和范 · 蒂贝根[20]的方法将这些切片放在铜网上进行放射自显影。曝光 3.5 个月。图中标尺代表 1 μm。

Many grains are found in some cells more than 1.25 μm from the nuclear membrane (central grains), suggesting that replication is taking place at sites away from the membrane. To be certain of that conclusion, however, other possible explanations must be ruled out. The central grains cannot be due to background because few or no grains are found over the cytoplasm. They must be due to incorporation into DNA since no grains are found over G1 or G2 cells (see below), and the grains of isolated nuclei can be removed by DNAase. The central grains could also originate from DNA replication occurring on invaginations of the nuclear membrane either just above or just below the section plane. While this might explain some central grains, it certainly cannot explain most of them. Because the cells are sectioned with random orientations, we can get some idea of the frequency of invaginations which might bring nuclear membrane within 1.25 μm of the section plane from the frequency of invaginations in the nuclear periphery. Although some peripheral invaginations are evident in Fig. 1, they are not so frequent that most of the central area could be 1.25 μm or less from an invagination.

In some preliminary experiments we used HeLa cells, which have much more regular nuclei: central grains were also found over their nuclei (Fig. 2). Even serial sections up to 0.5 μm apart through a single cell all showed central grains. The change in pattern from general labelling in early S to peripheral labelling in late S suggests that general labelling is not an artefact of sectioning. Finally, the central grains cannot be accounted for by internal membranes within the nucleus. Such membranes have never been seen by other electron microscopists; apart from invaginations of the peripheral membrane we could see none in our sections.

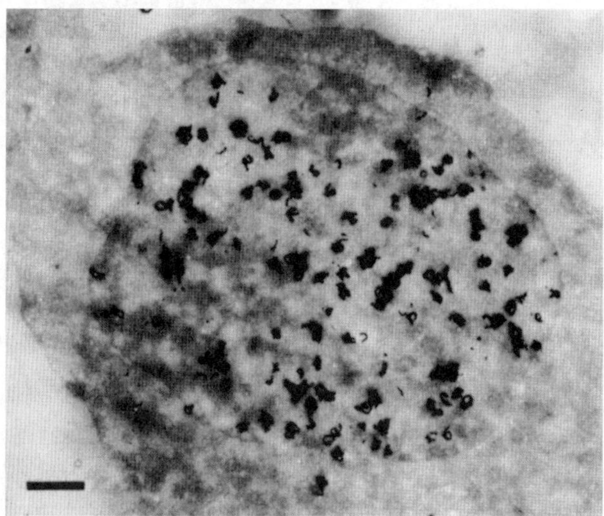

Fig. 2. Autoradiography of a HeLa cell exposed to ^{3}H-TdR for 1.5 min. The cells were grown in suspension culture in Joklik-modified MEM (Grand Island Biological Company) supplemented with 5% horse serum. Pulse-labelling was done by adding ^{3}H-TdR (20 Ci mmol^{-1}, New England Nuclear) to a concentration of 20 μCi ml.$^{-1}$. Exactly 1.5 min after addition of ^{3}H-TdR, the pulse was terminated by addition of an equal volume of ice-cold isotonic saline containing 1% glutaraldehyde. Subsequent processing was as in Fig. 1. Exposure time was 9.5 months. The bar represents 1 μm.

在一些细胞的研究中发现，许多颗粒距离核膜超过 1.25 μm（中间颗粒），这表明复制发生在远离核膜的位点上。但是为了确定这个结论，必须排除其他可能的解释。因为细胞质中几乎没有发现颗粒的存在，所以这些中间颗粒不可能是由于背景造成的。这些颗粒的形成一定是由于放射性标记分子成功结合到 DNA 分子上，因为在细胞的 G1 或者 G2 期也没有发现颗粒的存在（见下文），并且这些从细胞核中分离得到的颗粒可以被 DNA 酶除去。这些中间颗粒也可能是恰好高于或者低于切片平面的核膜内陷位置上 DNA 复制的结果。这或许可以解释一些中间颗粒的存在，但绝不能解释大部分中间颗粒的存在。因为这些细胞的切割方向是随机的，我们可以从细胞核外围内陷的频率获悉能让细胞核核膜在这 1.25 μm 范围内的内陷的概率。尽管图 1 中有一些明显的周边内陷，但是大部分中间区域与一个内陷之间的距离等于或小于 1.25 μm 的概率并不高。

我们使用 HeLa 细胞进行了部分的预实验，在这些实验中，HeLa 细胞具有形状更加规则的细胞核，并且在核周围都出现了中间颗粒（图 2）。甚至在单细胞 0.5 μm 厚的连续切片上都有中间颗粒的存在。从 S 期早期的普遍标记模式到 S 期后期的周边标记模式的变化说明普遍标记并不是由切割造成的假象。最后，细胞核内部的膜结构并不能解释中间颗粒的问题。其他电子显微镜观察者从未观察到这种膜，除了内陷的周边核膜，我们在切片中也没有观察到这种膜的存在。

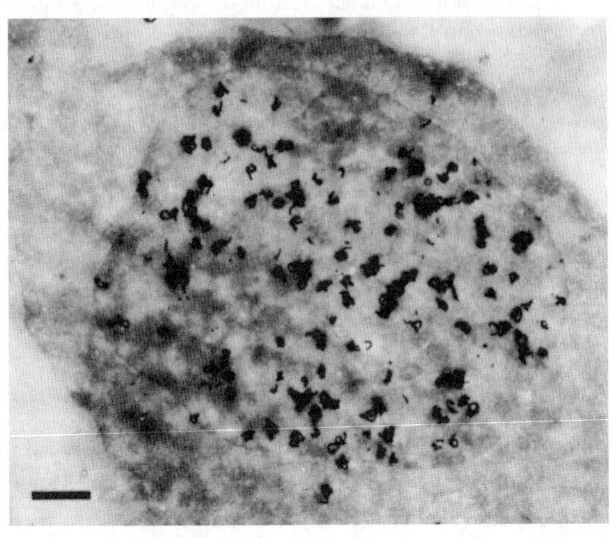

图 2. HeLa 细胞置于 ^{3}H-TdR 1.5 分钟后的放射自显影图。用添加 5% 马血清的改良乔克利克基础培养基（购自格兰德岛生物公司）悬浮培养细胞。加入 ^{3}H-TdR（20 Ci · mmol^{-1}，购自新英格兰核公司），至终浓度为 20 μCi·ml^{-1}，进行脉冲标记。1.5 分钟后，加入等体积冰浴的含 1% 戊二醛的生理盐水终止脉冲反应。随后的步骤同图 1。曝光 9.5 个月。图中标尺代表 1 μm。

To express quantitatively the proportion of cells showing peripheral labelling, we have modified the method of Williams and Ockey[16]. We defined central grains as those further than 1.25 μm from the nearest nuclear membrane, and peripheral grains to be those closer than 1.25 μm to the nuclear membrane. The "central activity" of a cell section was calculated as the ratio of the fraction of central grains : the fraction of nuclear area which was central. Nucleolar areas and grains were excluded because the nucleolus has less DNA than the rest of the nucleus. A histogram showing the frequency of various central activities for an unsynchronized cell population is shown in Fig. 3 (a central activity of 1.0 implies an equal concentration of grains over central and peripheral areas).

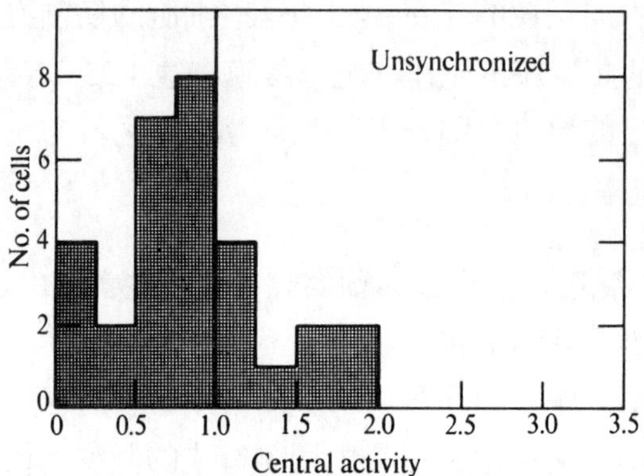

Fig. 3. Distribution of grains over unsynchronized CHO cells exposed to ^{3}H-TdR for 0.5 min. Prints at about ×15,000 magnification were made of autoradiograms of individual randomly chosen cells similar to and including those in Fig. 1. A line was drawn within each nucleus which was at all points 1.25 μm away from the nearest nuclear membrane. This line divided the nucleus into a central area and a peripheral area. Grains were counted over each area, and the size of each area was measured. In these determinations, grains lying over the nucleolus, and nucleolar areas, were ignored because the nucleolus has so much less DNA than the rest of the nucleus. Grains lying outside the nucleus were also ignored. "Central activity" was calculated as the ratio of the fraction of grains which were central to the fraction of area which was central. Thirty nuclei were measured for the histogram.

Only 13% of the cells in Fig. 3 have central activities of less than 0.25, showing that DNA synthesis can go on at sites away from the nuclear membrane. The nuclear membrane is therefore not essential for DNA replication in Chinese hamster cells. But the fact that many cells do show a predominantly peripheral pattern remains to be explained.

Peripheral Replication in Late S Phase

Since previous autoradiographic studies[14-17] had suggested that peripheral labelling occurs only at certain times during the S phase, we decided to try short pulse-labelling of cells synchronized to various times during the cell cycle. We chose "Colcemid" reversal[21] to synchronize the cells. Growing cells were treated with "Colcemid" for a few hours and the cells blocked in mitosis collected by selective trypsinization. These cells (90−100% mitotic) were allowed to continue growing in the absence of "Colcemid". Within 2 h, 98.5% of the cells completed division, and by 18 h after release from mitosis, more than 70% of the cells had divided again. We preferred this synchronization method to methods involving starvation of nucleotides[21], because cells recover excellently and normal DNA replication is not affected.

为了定量标定周边标记细胞的比例，我们对威廉姆斯和奥克伊[16]的方法进行了改进。我们将那些与最邻近的细胞核核膜相距大于 1.25 μm 的颗粒定义为中间颗粒，与核膜相距小于 1.25 μm 的颗粒定义为周边颗粒。一个细胞切片的“中间活性”是以中间颗粒的比例/中间核面积的比例来计算的。这需要排除核仁区域和该区域的颗粒，因为核仁的 DNA 含量比细胞核其他区域要少。图 3 显示了非同步细胞群中不同中间活性细胞的数量柱形图（中间活性 1.0 表示颗粒在中间区域和周边区域分布的密度相同）。

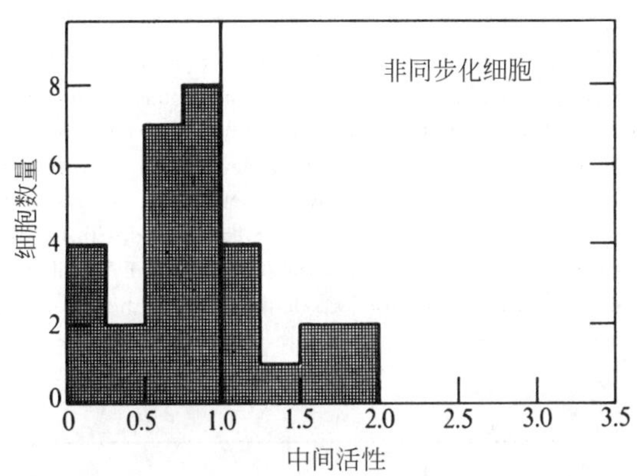

图 3. 非同步 CHO 细胞置于 ^{3}H-TdR 0.5 分钟后颗粒的分布图。随机选择类似图 1 和图 1 中的单个细胞，将其进行放射自显影后以 15,000 倍放大并打印成图。将细胞核内所有与最近邻细胞核膜相距 1.25 μm 的点用线连接起来。这条线将细胞核分成中间区域和周边区域。计算每个区域点的数量，并且测量区域面积的大小。在这种计算中，忽略靠近核仁和核仁区域的颗粒，因为核仁区的 DNA 含量比细胞核其他区域的少得多。位于细胞核外面的颗粒也忽略不计。“中间活性”是计算中间颗粒数量与中间区域面积的比值。该柱形图是对 30 个细胞核进行测定得到的结果。

图 3 中仅有 13% 的细胞中间活性小于 0.25，表明 DNA 合成可以在远离核膜的位点进行。因此在中国仓鼠细胞中，核膜并不是 DNA 复制必需的。但是许多细胞表现出的显著的周边模式仍然需要进一步的研究解释。

S 期后期的周边复制

由于前面的放射自显影研究[14-17]表明周边标记仅发生在 S 期的特定时期，我们决定用短脉冲标记处于细胞周期不同时期的同步化细胞。我们采用“秋水仙酰胺”[21]将细胞同步化。经秋水仙酰胺处理几个小时以后，生长细胞都停滞在有丝分裂期，然后用选择性胰蛋白酶消化收集这些细胞。这些细胞（90%~100% 处于有丝分裂期）在缺乏秋水仙酰胺的条件下可以继续生长。2 小时之内，98.5% 细胞会完成分裂，在完成有丝分裂 18 小时后，70% 以上的细胞会再次分裂。之所以采用这种方法进行同步化而不使用限制核苷酸供应的方法[21]，是因为这样不会影响细胞的复苏和正常的 DNA 复制。

Table 1 and Fig. 4 show the distribution of grains within the nucleus as a function of time after release from mitosis. Peripheral labelling cannot be detected in the early stages of S but becomes predominant in later S.

Table 1. Extent of DNA Synthesis in Synchronized CHO Cells

Hours after release from mitosis	2	4	6	8	10	12	14
Percentage of cells labelled*	26	83	93	89	81	62	48
Average number grains/cell section after 1 month exposure †	<1	6	16	16	24	17	—

* Cells were synchronized as in Fig. 4. At 2 h intervals after release from mitosis, ^{3}H-TdR (20 Ci mmol^{-1}; New England Nuclear) was added to each plate (17 μCi ml.$^{-1}$ final concentration). After 10 min the cells were washed twice with cold isotonic saline, collected by trypsinization at room temperature, allowed to swell in hypotonic medium (2 mM $MgCl_2$, 1 mM EDTA, 10 mM KPO_4, *p*H 7.7) for 10 min, pelleted, fixed with methanol-acetic acid (3:1), spread onto subbed glass slides and allowed to dry. The slides were coated with autoradiographic stripping film ("Kodak AR-10") and exposed for 7 days. After development, the slides were stained with Giemsa stain, then mounted under "Permount". At least 312 cells were scanned to determine the percent of cells labelled at each time point.

† The total number of grains over each nuclear section was determined for the autoradiographs used in Fig. 4. These figures were divided by the exposure time in months. Extra data for 2 h and 8 h after release from mitosis are included here.

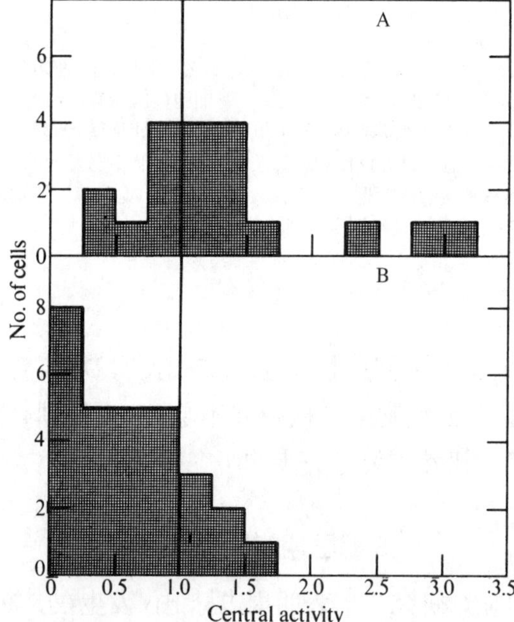

Fig. 4. Distributions of grains over synchronized CHO cells exposed to ^{3}H-TdR for 0.5 min. A, Early S phase (4–6 h after release from mitosis). B, Late S phase (10–12 h after release from mitosis). CHO cells were synchronized by the "Colcemid" reversal method of Stubblefield. Pulse-labelling (performed at 2 h intervals after release from mitosis), preparation for electron microscopy, and autoradiography were performed as in Fig. 1. Exposure time was 2–4 months. Histograms were prepared as in Fig. 3. For A, eight nuclei pulse-labelled at 4 h and eleven nuclei labelled at 6 h after mitotic release were used. For B, thirteen nuclei labelled at 10 h and sixteen nuclei labelled at 12 h were used.

表 1 和图 4 显示的是有丝分裂后细胞核内颗粒随时间变化的关系。虽然在 S 期早期观察不到周边标记，但在 S 期后期则是以周边标记为主。

表 1. 同步化的 CHO 细胞的 DNA 合成情况

有丝分裂后的时间	2	4	6	8	10	12	14
标记细胞的百分比 *	26	83	93	89	81	62	48
曝光 1 个月后平均每个细胞切片的颗粒数量 †	<1	6	16	16	24	17	—

* 细胞的同步化与图 4 相同。有丝分裂后每间隔 2 小时，在每个培养皿中加入 ^{3}H-TdR（20 Ci · mmol^{-1}，购自新英格兰核公司），至终浓度 17 μCi · ml^{-1}。10 分钟后，用冷生理盐水洗细胞两次，室温下用胰蛋白酶消化并收集细胞，在低渗溶液（2 mM $MgCl_2$，1 mM EDTA，10 mM KPO_4，pH 7.7）中溶胀 10 分钟，然后收集沉淀，用甲醇 – 乙酸溶液（3:1）固定，平铺在玻璃片上风干。用放射自显影乳胶片（"柯达 AR-10"）包裹这些玻片，并曝光 7 天。接着，用吉姆萨氏色素染料染色，然后用"Permount"中性树胶封埋。每一个时间点至少要观察 312 个细胞来确定被标记细胞的百分比。

† 对图 4 中用到的放射自显影图片进行了每个细胞核切片上的颗粒的总数的确定。这些图片按照曝光时间进行分类。有丝分裂后 2 小时和 8 小时的数据也包括在内。

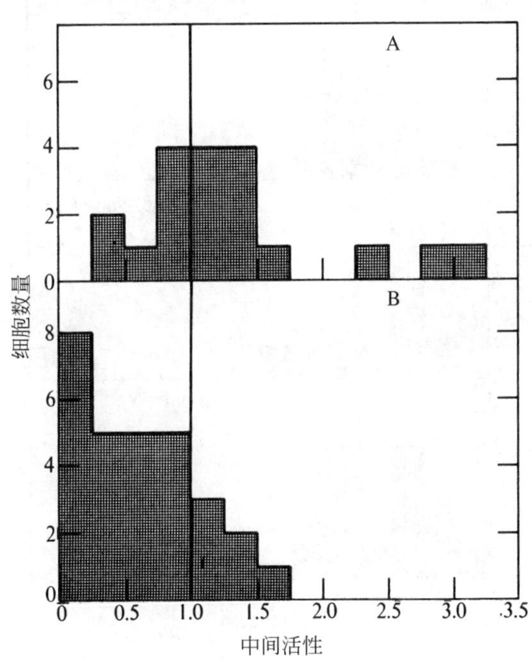

图 4. 置于 ^{3}H-TdR 下 0.5 分钟后的同步化 CHO 细胞中颗粒的分布图。A，S 期早期（有丝分裂后 4~6 小时）。B，S 期后期（有丝分裂后 10~12 小时）。按照斯塔布菲尔德秋水仙酰胺逆转法将 CHO 细胞同步化。脉冲标记(有丝分裂后每隔 2 小时进行)，电镜切片的准备和放射自显影按照图 1 的操作。曝光 2~4 个月。按照图 3 的方法制作柱形图。A 图，有丝分裂后 4 小时统计了 8 个标记的细胞核，有丝分裂后 6 小时统计了 11 个标记的细胞核。B 图，有丝分裂后 10 小时统计了 13 个标记的细胞核，有丝分裂后 12 小时统计了 16 个标记的细胞核。

Stable Location of DNA

If DNA is synthesized at the nuclear membrane (or other localized sites within the nucleus), then the DNA should move toward the membrane for replication and away from the membrane afterwards. The pattern of grains after a long pulse or pulse-chase should then be different from that after a simple short pulse. If, on the other hand, DNA replication can take place anywhere in the nucleus, then DNA movement is unnecessary and the grain pattern after pulse-chase labelling could be identical to that after simple pulse-labelling.

To test these possibilities, we first pulse-labelled synchronized cells for 0.5 min during the first half of the S phase (6 h after release from mitosis), and then chased with non-radioactive thymidine to late S phase (12 h after release from mitosis). Although the cells were examined in late S phase, they showed no higher proportion of peripheral labelling than expected for cells 6 h after release from mitosis (compare Figs. 4 and 5). In addition, unsynchronized cells pulse-labelled for 0.5 min show the same proportion of peripheral and general labelling as unsynchronized cells pulse-labelled for 10 min, or pulse-labelled for 0.5 min and then chased for 6.75 h (Fig. 5). In both of the chase experiments, the cells were more labelled after the chase than before, probably because incorporation of ^{3}H-TdR continues from the cells' internal pools.

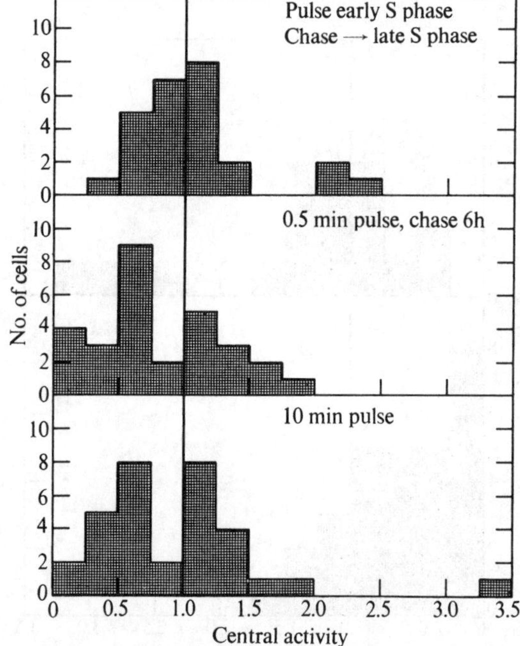

Fig. 5. Distribution of grains over CHO cells after pulse-chase or long pulse-labelling with ^{3}H-TdR. A, Synchronized cells (see Fig. 4) were pulse-labelled 6 h after release from mitosis as in Fig. 1. Then 30 s after addition of label, the medium was sucked off and the plates were washed, then replaced with medium containing 2.5 μg ml.$^{-1}$ of TdR. After 6 h the cells were collected by trypsinization and prepared for electron microscope autoradiography as in Fig. 1. Exposure time was 1.5–4 months. B, Unsynchronized cells were pulse-labelled as in Fig. 1 and chased for 6.75 h as in A. Exposure time was 3.25 months. C, Unsynchronized cells were pulse-labelled as in Fig. 1 except that the ^{3}H-TdR was left in contact with the cells for 10 min instead of 30 s. Exposure time was 1.5–6 months. In all cases, histograms were prepared as in Fig. 3. For A, twenty-four cells were used, whereas for B, twenty-nine cells were used, and for C, thirty-two cells were used.

DNA 稳定的位置

如果 DNA 是在核膜处（或者在细胞核内其他的位置）合成的，那么 DNA 应该首先转移至核膜进行复制，然后再远离核膜。因而长脉冲或者脉冲追踪形成的颗粒模型应该与短脉冲处理所形成的颗粒模型不同。另一方面，如果 DNA 复制可以在细胞核的任意区域进行，那么 DNA 没有必要转移，而且脉冲追踪标记后的颗粒模型应该与简单脉冲标记的颗粒模型一致。

为了验证这些可能性，我们首先脉冲标记处于 S 期的前半期（有丝分裂完成后的 6 小时）的同步化细胞 0.5 分钟，然后用非放射性的胸腺核苷追踪至 S 期的后期（有丝分裂完成后的 12 小时）。尽管对这些细胞的检测是在 S 期后期进行的，但是这些细胞周边标记的比例未如预期，并不比处于 S 期前半期的细胞高（图 4 和图 5 对比）。此外，研究表明脉冲标记非同步化细胞 0.5 分钟与 10 分钟，或者先脉冲标记 0.5 分钟后再追踪 6.75 小时（图 5），其周边标记和普遍标记的比例是相同的。两次追踪实验的结果都显示出有更多的细胞在示踪后被标记，这可能是从细胞内腔持续吸收结合 ^{3}H-TdR 的结果。

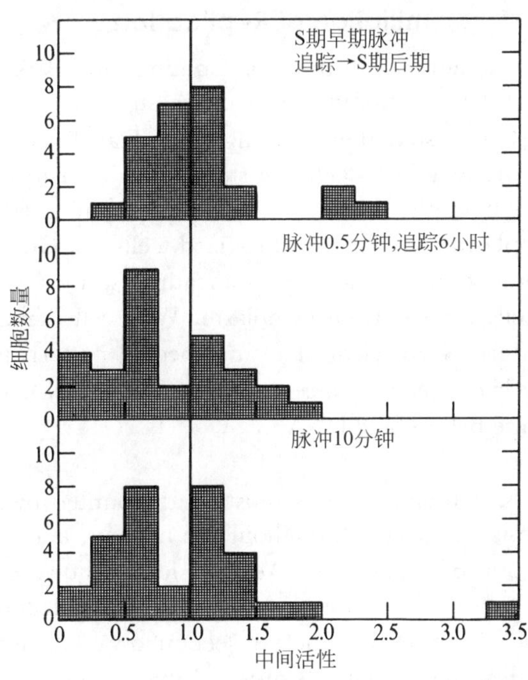

图 5. ^{3}H-TdR 脉冲追踪或者长脉冲标记 CHO 细胞后的颗粒分布图。A，有丝分裂后 6 小时脉冲标记同步化的细胞（同步方法同图 4，标记方法同图 1）。标记 30 秒后，吸掉培养基，冲洗培养皿，然后加入含 2.5 μg · ml^{-1} TdR 的培养基。6 小时后用胰蛋白酶消化收集细胞，按照图 1 的方法准备电子显微放射自显影。曝光时间为 1.5~4 个月。B，按照图 1 的方法脉冲标记非同步化的细胞，并且同图 A 一样追踪 6.75 小时。曝光时间为 3.25 个月。C，除了用 ^{3}H-TdR 处理细胞的时间延长为 10 分钟，而非 30 秒外，其他操作同图 1。曝光时间为 1.5~6 个月。按照图 3 的方法制作柱形图 1。图 A，统计了 24 个细胞；图 B，统计了 29 个细胞；图 C，统计了 32 个细胞。

Thus DNA has a stable location in the nucleus. Because some of the unsynchronized cells must have divided during the chase, our results suggest that DNA bound closely to the membrane in one generation is also bound to the membrane in the succeeding generation.

Because we cannot detect any difference in labelling pattern between a 0.5 min pulse and a 10 min pulse, our results can be compared directly with those of previous investigators[15-17] who have used longer pulse times. Our results agree with most of the earlier results; a general-label pattern is obtained in early S phase and a peripheral one in late S phase[15-17], and chase experiments show that the nuclear position of DNA, once replicated, is stable[15,16].

The increased labelling during chase experiments, together with an apparently lower rate of synthesis in early S phase (Table 1), suggests an explanation for the autoradiographic results of O'Brien *et al.*[13]. In their pulse experiment, generally-labelled cells in early S phase may have been overlooked because they have few grains per nucleus (and a more disperse distribution at that); the increased labelling provided by the chase would have made the generally labelled cells sufficiently obvious to count.

Initiation of Replication

Our results also partially disagree with those of Comings and Kakefuda[14]. They reported that cells synchronized to the beginning of S phase showed solely peripheral labelling. Williams and Ockey[16,22] suggested that Comings and Kakefuda's results might be due to cell damage during the lengthy nucleotide starvation (24 h in the presence of excess thymidine and then 14 h in the presence of amethopterin) used to synchronize their cells. Indeed, the cytoplasms of the synchronized cells in Comings and Kakefuda's experiments contain many large vacuoles (indicating serious cell damage), whereas their unsynchronized cells have normal cytoplasm. Why cell damage should result in a peripheral labelling pattern is not clear. The unexpected pattern may be related to the finding[12] that polyoma virus infection causes mouse satellite DNA, which is normally late replicating[23-25], to replicate before bulk DNA.

Because Comings and Kakefuda's[14] results must be discounted owing to cell damage, it seems that DNA synthesis can occur throughout the nucleus, and the nuclear membrane is not involved in initiation of replication. We have found, however, that in very early S phase (2 and 4 h after release from mitosis), the rate of replication per cell (measured as the average number of grains per nuclear section over labelled nuclei after a fixed exposure time) is much less than in later S phase (Table 1). This suggests that at the true beginning of S phase, the rate might be so low that we cannot detect labelling, let alone determine whether the pattern is peripheral or general.

Euchromatin and Heterochromatin

Heterochromatin has been shown to be replicated later than euchromatin and is condensed during interphase[26,27]. Our sections and those of others[14-17], show a

因此，DNA 在细胞核内有稳定的定位。由于一些非同步化细胞在追踪过程中已经发生分裂，所以我们的结果还揭示了与核膜紧密结合的亲代 DNA 的子代 DNA 也是与核膜紧密结合的。

我们的检测结果显示，0.5 分钟和 10 分钟的脉冲标记模型没有任何差别，所以我们的实验结果可以直接与其他使用长脉冲时间的研究者的结果 [15-17] 进行比较。我们的结果与大多数早期发表的结果一致；普遍标记模型出现在 S 期早期，周边标记模型则出现在 S 期后期 [15-17]，追踪实验还表明，一旦复制开始，DNA 在细胞核的位置是稳定的 [15,16]。

在追踪实验中增加的标记，以及在 S 期早期明显较低的 DNA 合成速率（表 1），可以解释奥布赖恩等人的放射自显影实验结果 [13]。在他们的脉冲实验中，S 期早期标记的细胞通常会由于每个细胞核中颗粒过少（分布过散）而被忽略；而在追踪过程中增加的标记可以提供足够明显的标记的细胞便于统计。

复制的起始

我们的结果也与科明斯和挂札 [14] 的结果有些不同。他们的报道指出，同步至 S 期初期的细胞只表现周边标记。威廉姆斯和奥克伊 [16,22] 认为科明斯和挂札的结果可能是由于为了同步化，而将细胞处于长期缺乏核苷酸（用过量的胸腺核苷酸处理 24 小时，然后用氨甲蝶呤处理 14 小时）的状态，导致细胞损伤。事实上，科明斯和挂札的实验中所用的同步化细胞的细胞质含有许多大的空囊泡（这表明细胞严重损伤），而他们实验中的非同步化细胞的细胞质却正常。细胞损伤导致周边标记模式的原因尚不清楚。这或许与目前发现的由于多瘤病毒感染而导致的鼠卫星 DNA 复制早于主体 DNA 复制 [12] 的情况有些关联，通常情况下卫星 DNA 的复制会相对较晚 [23-25]。

由于出现细胞损伤，科明斯和挂札 [14] 结果的可信度大打折扣，他们的结果让人误以为 DNA 的合成发生在整个细胞核区域，且核膜没有参与复制的起始。然而我们的研究发现，在 S 期较早期（有丝分裂后 2、4 小时），每个细胞的平均复制速率（细胞核经过固定的曝光时间标记以后，每个核切片的平均颗粒数）要远低于 S 期后期（表 1）。这表明，在真正的 S 期初期，复制速率太低以至于检测不到标记，更无从确定是周边模式还是普遍模式。

常染色质和异染色质

已经证明异染色质的复制比常染色质晚，在分裂间期以浓缩的状态存在 [26-27]。我们以及其他研究者的切片观察 [14-17] 表明，浓缩的染色质主要与内核膜相连。因此

preponderance of condensed chromatin attached to the inner nuclear membrane. Thus our results are consistent with the notion[16] that euchromatin is replicated early in S phase and is distributed throughout the nucleus, whereas heterochromatin is replicated in late S phase and is condensed along the inner nuclear membrane and in other discrete areas.

Reinterpretation of Cell Fractionation Experiments

Whereas many of the cell fractionation experiments on the intranuclear location of DNA synthesis[8-11,13] have been interpreted in favour of attachment of replication sites to the nuclear membrane, this is not the only possible interpretation. Equally valid is the possibility that, after cell lysis, special structural properties of the growing points (extensive single strandedness, for instance) might result in their binding more protein, membrane, or other material than bulk DNA. Alternatively, the growing points may actually be attached inside the cell to some material that causes their separation from bulk DNA during lysate fractionation.

As experiments with mammalian[8-11,13] and bacterial[2,3,5] cells used similar techniques and gave the same results, it is possible that the growing point in bacteria may also not be attached to the membrane.

We thank Elaine and Robert Lenk for advice and assistance. This research was supported by grants from the National Science Foundation and the US National Institutes of Health. The electron microscopy was carried out in the Electron Microscope Facility of the Biology Department at MIT.

Note added in proof. Fakan *et al.*[28] have recently obtained results similar to ours in both autoradiographic and cell fractionation experiments with mouse cells.

(**241**, 32-36; 1973)

Joel A. Huberman, Alice Tsai and Robert A. Deich
Departments of Biology, Massachusetts Institute of Technology, Cambridge, Massachusetts 02139

Received May 4; revised December 15, 1972.

References:

1. Jacob, F., Brenner, S., and Cuzin, F., *Cold Spring Harbor Symp. Quant. Biol.*, **28**, 329 (1963).
2. Ganesan, A. T., and Lederberg, J., *Biophys. Biochem. Res. Commun.*, **18**, 824 (1965).
3. Smith, D. W., and Hanawalt, P. C., *Biochim. Biophys. Acta*, **149**, 519 (1967).
4. Tremblay, G. Y., Daniels, M. J., and Schaechter, M., *J. Mol. Biol.*, **40**, 65 (1969).
5. Sueoka, N., and Quinn, W. G., *Cold Spring Harbor Symp. Quant. Biol.*, **33**, 695 (1968).
6. Fuchs, E., and Hanawalt, P. C., *J. Mol. Biol.*, **52**, 301 (1970).
7. Fielding, P., and Fox, C. F., *Biochem. Biophys. Res. Commun.*, **41**, 157 (1970).
8. Friedman, D. F., and Mueller, G. C., *Biochim. Biophys. Acta*, **174**, 253 (1969).
9. Mizuno, N. S., Stoops, C. D., and Sinha, A. A., *Nature New Biology*, **229**, 22 (1971).
10. Hanaoka, F., and Yamada, M., *Biochem. Biophys. Res. Commun.*, **42**, 647 (1971).
11. Pearson, G. D., and Hanawalt, P. C., *J. Mol. Biol.*, **62**, 65 (1971).

我们的结果与以下的想法[16]一致：常染色质的复制发生在S期早期而且分布在细胞核的整个区域，而异染色质的复制发生在S期后期，在内核模和其他分散的区域以浓缩的状态存在。

细胞分级分离实验的重新解释

尽管许多关于DNA在细胞核内合成位点的细胞分级分离实验[8-11,13]已经被诠释为倾向于支持复制位点与细胞核核膜相连的观点，但是这并不是唯一可能的解释。同样合理的一种可能性是，细胞裂解后，由于生长点的某种特殊结构属性（如大多是单链），可能导致其与体内其他大部分的DNA相比能够结合更多的蛋白、膜或者其他物质。当然，也可能生长点确实与细胞内的一些物质相连，造成生长点与体内主体DNA在细胞裂解分离过程中分离。

因为在哺乳动物细胞[8-11,13]和细菌细胞[2,3,5]上开展的实验研究使用了相似的实验技术，并得到相同的结果，所以在细菌体内生长点很有可能也不与膜相连。

特别感谢伊莱恩和罗伯特·伦克提供的建议和帮助。该项目受国家科学基金会和美国国立卫生研究院的资助。电子显微镜实验在麻省理工学院生物系电子显微镜实验室进行。

附加说明：最近福孔等人[28]用小鼠细胞进行自显影和细胞分级分离实验，结果与我们的结果类似。

（吕静 翻译；刘京国 审稿）

12. Smith, B. J., *J. Mol. Biol.*, **47**, 101 (1970).
13. O'Brien, R. L., Sanyal, A. B., and Stanton, R. H., *Exp. Cell Res.*, **70**, 106 (1972).
14. Comings, D. E., and Kakefuda, T., *J. Mol. Biol.*, **33**, 225 (1968).
15. Blondel, B., *Exp. Cell Res.*, **53**, 348 (1968).
16. Williams, C. A., and Ockey, C. H., *Exp. Cell Res.*, **63**, 365 (1970).
17. Erlandson, R. A., and de Harven, E., *J. Cell Sci.*, **8**, 353 (1971).
18. Kay, R. R., Haines, M. E., and Johnston, I. R., *FEBS Lett.*, **16**, 233 (1971).
19. Huberman, J. A., and Riggs, A. D., *J. Mol. Biol.*, **32**, 327 (1968).
20. Caro, L. G., and Van Tubergen, R. P., *J. Cell Biol.*, **15**, 173 (1962).
21. Stubblefield, E., in *Methods in Cell Physiology* (edit. by Prescott, D. M.), **3**, 25 (Academic Press, New York, 1968).
22. Ockey, C. H., *Exp. Cell Res.*, **70**, 203(1972).
23. Tobia, A. M., Schildkraut, C. L., and Maio, J. J., *Biochim. Biophys. Acta*, **246**, 258 (1971).
24. Bostock, C. J., and Prescott, D. M., *Exp. Cell Res.*, **64**, 267 (1971).
25. Flamm, W. G., Bernheim, N. J., and Brubaker, P. E., *Exp. Cell Res.*, **64**, 97(1971).
26. Lima-de-Faria, A., and Javorska, H., *Nature*, **217**, 138 (1968).
27. Brown, S. W., *Science*, **151**, 417 (1966).
28. Fakan, S., Turner, G. N., Pagano, J. S., and Hancock, R., *Proc. US Nat. Acad. Sci.*, **69**, 2300 (1972).
29. Lenk, R., and Penman, S., *J. Cell Biol.*, **49**, 541 (1971).

In vitro Fertilization of Rat Eggs

H. Miyamoto and M. C. Chang

Editor's Note

Chinese American scientist Min Chueh Chang is probably best known for his work developing the oral contraceptive, but here he builds on a wealth of animal *in vitro* fertilization (IVF) research demonstrating IVF of rat eggs. Successful IVF of mammalian eggs had been demonstrated in many other species, and Chang himself had previously achieved IVF in rabbits by implanting black rabbit embryos conceived in the laboratory into a white rabbit. Critically, Chang acknowledged the importance of post-ejaculatory sperm maturation or "capacitation", a phenomenon he had independently co-discovered with reproductive biologist Colin Russell Austin. Chang's research paved the way for IVF of human eggs, culminating in the birth of the world's first "test-tube baby", Louise Brown, in 1978.

AFTER the recognition of "capacitation of sperm" by Austin[1] and Chang[2], successful *in vitro* fertilization of mammalian eggs was achieved in many species[3-17]. Although fertilization of rat eggs *in vitro* has been attempted[1,18-20], incorporation of sperm into the vitellus was observed only after the dissolution of zona pellucida by chymotrypsin[19]. Here we report fertilization of intact rat eggs *in vitro* and the results obtained.

Mature female CD strain rats were kept in a constant temperature room (23–25°C) with artificial light (19.00–7.00 h darkness) and their vaginal smears inspected daily. As ovulation usually occurs at about 2.00 h on the pro-oestrous to oestrous night[21], the eggs were considered to be about 4 to 5 h and 7.5 to 8.5 h after ovulation when pro-oestrous rats were killed at 6.00 to 7.00 and 9.30 to 10.30 h respectively on the next day. Superovulation of mature rats was performed by i. p. injection of 30 IU of pregnant mare's serum (PMS) on the morning of oestrus and 30 IU of human chorionic gonadotrophin (HCG), 52 to 58 h later; the rats were killed 14 to 16 h after injection of HCG. Oviducts were placed under light paraffin oil, equilibrated with 5% CO_2 in air in the presence of a small volume of saline, in a watch glass kept at 37°C and the eggs in cumulus clot were released by dissecting the ampular portion of the oviducts. Sperm for insemination were prepared. The eggs were inseminated with sperm prepared as in Fig. 1.

After incubation for 8 to 12 h, the eggs were removed, mounted on a slide and examined. They were then fixed with neutral formalin and stained for the assessment of fertilization. The eggs which had sperm within their perivitelline space (supplementary sperm) were defined as "penetrated" eggs. When these eggs had either enlarged sperm head(s) or male pronucleus(ei) with fertilizing sperm tail(s) in or on the vitellus, they were considered as undergoing fertilization.

大鼠卵细胞的体外受精

宫本，张明觉

编者按

美籍华裔科学家张明觉最广为人知的工作是开发了口服避孕药，本文中他以大量的动物体外受精实验阐明了大鼠卵细胞的体外受精过程。在很多其他物种中，哺乳动物的卵细胞成功的体外受精已得到阐明。先前张明觉已经亲自完成了兔子的体外受精，他在实验室中将已受孕的黑兔的胚胎移植到一只白兔体内。关键问题是张明觉认识到射精后精子成熟或者称为“精子获能”的重要性，精子获能这种现象是他独立发现的，同时生殖生物学家科林·罗素·奥斯汀也独立发现了相同的现象。张明觉的研究为人类卵细胞的体外受精奠定了基础，最终在 1978 年诞生了世界上第一个“试管婴儿”路易丝·布朗。

奥斯汀[1]和张明觉[2]发现“精子获能”后，科学家已经在很多物种中实现了哺乳动物的体外受精[3-17]。尽管也有人尝试大鼠卵细胞的体外受精[1,18-20]，但只有在透明带被胰凝乳蛋白酶溶解后，才能观察到精子进入卵黄。我们在本文中报道了完整的大鼠卵细胞的体外受精及获得的结果。

成熟的 CD 系雌性大鼠饲养在具有人工照明（19:00~7:00 黑暗）且保持恒定室温（23~25℃）的恒温室中，并且每日进行阴道涂片检查。从发情前期过渡到发情期的夜晚，排卵通常发生在 2:00 左右[21]，因此发情前期的大鼠分别于次日 6:00~7:00 和 9:30~10:30 被处死时，卵细胞被视为处于排卵后大约 4~5 h 和 7.5~8.5 h。通过在发情期早上腹腔注射 30 国际单位的孕马血清（PMS）并在 52~58 h 后注射 30 国际单位的人绒毛膜促性腺激素（HCG），实现成熟大鼠超排卵；注射 HCG 14~16 h 后处死大鼠。在表面皿上盛放少量盐水并用轻质石蜡油覆盖，在含 5%CO_2 的空气中于 37℃进行平衡，再将输卵管置于轻质石蜡油下，然后通过切开输卵管壶腹部释放卵丘细胞块内的卵子。准备好授精用的精子。以图 1 中的方式用准备好的精子对卵子授精。

培养 8~12 h 后，移出卵细胞，将其封装于一个载玻片并进行镜检。然后用中性福尔马林固定并染色，以评估受精状况。在卵周隙空间内存在精子（额外精子）的卵细胞被定义为“被穿透的”卵细胞。若在这些卵细胞的卵黄中或卵黄上，要么有膨大的精子头部，要么存在带有受精精子尾部的雄性原核，便可将其视为正在受精。

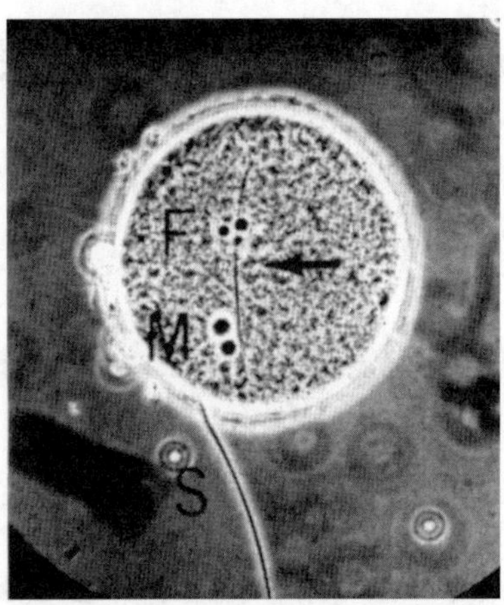

Fig. 1. A rat egg at pronuclear stage, fertilized *in vitro*, showing male pronucleus (M), female pronucleus (F), the fertilizing sperm tail (arrow) and a sperm attached to the zona pellucida (S). Photographed before fixation under a phase-contract microscope. Sperm for insemination were prepared either from the cauda epididymis or from the uterus of a female mated 0.5 to 1 h, or 10 to 11 h, previously. A drop of dense sperm mass was put into a watch glass and covered with 3 ml. of medium (modified Krebs–Ringer bicarbonate solution containing 114.2 mM NaCl, 4.78 mM KCI, 1.71 mM $CaCl_2 \cdot 2H_2O$, 1.19 mM KH_2PO_4, 1.19 mM $MgSO_4 \cdot 7H_2O$, 25.07 mM $NaHCO_3$, 0.55 mM sodium pyruvate, 21.58 mM sodium lactate and 5.55 mM glucose, to which 4 mg $ml.^{-1}$ of crystalline bovine serum albumin, 50 μg $ml.^{-1}$ of streptomycin sulphate and 75 μg $ml.^{-1}$ of penicillin G (potassium salt) were added). The *p*H value of the medium was adjusted to 7.4 to 7.5 by addition of 1 M NaOH and the final solution was filtered through a millipore filter to ensure asepsis. Blood of female rats was collected by heart puncture, allowed to clot and then centrifuged. Serum was sterilized by filtration through a millipore filter and stored under paraffin oil at 2 to 5°C for no longer than five days. The uterine fluid was collected from oestrous rats by means of a syringe needle. Insemination was performed by adding 0.4 ml. of the sperm suspension to the egg clot under the oil, after removing oviducts and débris. In some cases the egg clot was washed three times with the same medium to remove the oviducal fluid. After adding 0.1 or 0.2 ml. of heated rat serum (at 56°C for 40 min), the eggs and sperm were thoroughly mixed and incubated at 37°C in an atmosphere of 5% CO_2 in air. The final concentration of sperm in the suspension was 1,000 to 4,000 sperm mm^{-3}.

Fig.1 shows a rat egg fertilized *in vitro* at the pronuclear stage. Table 1 shows that the eggs recovered from naturally-ovulated rats have a better chance ($P<0.01$) of fertilization *in vitro* than those recovered from superovulated rats (highest penetration rates: 45% compared with 22%). The proportion of penetrated eggs was higher ($P<0.05$) following insemination with sperm recovered from the uterus 10 to 11 h (17–45%) than 0.5 to 1 h after mating (7%) for both the naturally-ovulated and superovulated eggs. The eggs recovered 4 to 5 h after ovulation appeared to have a better chance of penetration (24–45%) than those recovered 7.5 to 8.5 h after ovulation (1–5%). By washing the egg clot before insemination, the proportion of penetration was decreased for both the naturally-ovulated (24% compared with 45%) and superovulated eggs (9–11 compared with 17–22%, not statistically significant), indicating that some beneficial factor may have been removed by washing. Although there was no obvious difference when different proportions of rat serum were added in the medium, sperm penetration was not observed when whole rat serum was used in a few experiments. Sperm penetration, however, was not observed in 52 naturally-ovulated eggs and 229 superovulated eggs when epididymal sperm, which had been pre-incubated for 2 to 7 h in the medium containing rat serum and uterine

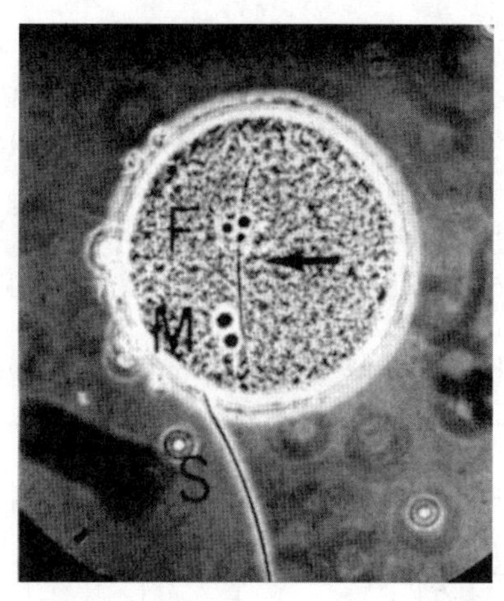

图 1. 一个在体外受精过程中处在原核阶段的大鼠卵细胞，图中显示雄性原核（M），雌性原核（F），受精精子的尾部（箭头）和一个附着在透明带上的精子（S）。样品固定前在相差显微镜下照相。用于授精的精子来源于附睾尾部或已经交配的 0.5~1 h 或 10~11 h 的雌性大鼠子宫。向表面皿上滴加一滴浓稠的精液，其上再覆盖 3 ml 培养基（该培养基是改良 Krebs–Ringer 碳酸氢钠溶液，含有 114.2 mM NaCl，4.78 mM KCl，1.71 mM $CaCl_2 \cdot 2H_2O$，1.19 mM KH_2PO_4，1.19 mM $MgSO_4 \cdot 7H_2O$，25.07 mM $NaHCO_3$，0.55 mM 丙酮酸钠，21.58 mM 乳酸钠和 5.55 mM 葡萄糖，并加入 4 mg·ml^{-1} 结晶牛血清白蛋白，50 μg·ml^{-1} 硫酸链霉素和 75 mg · ml^{-1} 青霉素 G（钾盐））。用 1M NaOH 将培养基 pH 值调至 7.4~7.5 并将终溶液用微孔过滤器过滤以保证无菌。心脏穿刺收集雌性大鼠血液，凝固后离心分离。血清通过微孔过滤器过滤除菌后储存于石蜡油下，在 2~5℃保存不超过 5 天。用注射器针头从发情期雌性大鼠体内收集子宫液。去除输卵管及其碎片后，在石蜡油下，向卵细胞团块加入 0.4 ml 精子悬浮液进行授精。某些情况下，卵细胞团块用相同培养基清洗 3 次以除去输卵管液。加入 0.1 或 0.2 ml 温育处理（56℃，40 min）的大鼠血清后，卵细胞和精子充分混合，并在含 5% CO_2 的空气环境中 37℃培养。悬浊液中精子的终浓度为每毫升 1,000~4,000 个精子。

图 1 显示了一个处于原核阶段的大鼠卵细胞体外受精的情况。表 1 显示从自然排卵的大鼠体内收集到的卵细胞比那些从超排卵大鼠体内收集到的卵细胞有更多机会（$P<0.01$）发生体外受精（最高穿透率为 45%比 22%）。从子宫中分别收集大鼠交配后 0.5~1 h 和 10~11 h 的精子，对卵细胞授精并计算穿透率。无论自然排卵还是超排卵的卵细胞，用 10~11 h 精子授精的卵细胞穿透率（17%~45%）高于（$P<0.05$）用 0.5~1 h 精子授精的卵细胞穿透率 (7%)。排卵后 4~5 h 收集的卵细胞显示其被穿透的可能性（24%~45%）比排卵后 7.5~8.5 h 收集的卵细胞（1%~5%）要高。授精前清洗卵细胞团块，则正常排卵（45% 比 24%）和超排卵（17%~22% 比 9%~11%，无显著性差异）的卵细胞穿透率均下降，说明某些有益的因子可能被冲洗掉了。尽管不同比例的大鼠血清加入培养基后并未观察到明显差异，但在一些实验中，培养基内加入大鼠全血清后未观察到精子的穿透。然而，当使用在含大鼠血清和子宫液的培养基中预先培养了 2~7 h 的附睾精子时，在 52 个自然排卵的卵细胞和 229 个超

fluid, were used, they remained motile for longer than uterine-incubated sperm. This, and the higher proportion of eggs penetrated by sperm recovered from the uterus 10 to 11 h after mating demonstrated clearly that rat sperm do need capacitation in the female tract before they are capable of penetrating the egg. Hamster[6,8], mouse[15], guinea pig[17] and perhaps human[9,10] sperm can be capacitated *in vitro*, but we have shown the difficulty of capacitating rat sperm *in vitro*. Rat sperm is similar to rabbit sperm in this respect.

Table 1. *In vitro* Fertilization of Rat Eggs

Mature female rats	Age of eggs (h after ovulation)	Sperm used	Medium used(medium: serum)	No. of females used	No. of eggs examined	No. of penetrated eggs (%)	No. of eggs undergoing fertilization (%)	No. of polyspermic eggs (%)
Naturally-ovulated	4–5	Uterine 0.5–1 h after mating	4:1	5	54	4 (7)	2 (4)	0
	4–5	Uterine 10–11 h after mating	4:0	5	56	6 (11)	4 (7)	1 (25)
			2:1	10	84	38 (45)	22 (26)	0
	4–5	Uterine 10–11 h after mating	4:1	5	55	25 (45)	25 (45)	9 (36)
	4–5*	Uterine 10–11 h after mating	2:1	5	54	13 (24)	10 (19)	1 (10)
	7.5–8.5	Uterine 10–11 h after mating	2:1	6	78	1 (1)	1 (1)	0
	7.5–8.5	Uterine 10–11 h after mating	4:1	6	74	4 (5)	3 (4)	0
	4–5	Epididymal †	With serum and uterine fluid	5	52	0	0	0
Super-ovulated	1–3	Uterine 0.5–1 h after mating	4:1	12	280	19 (7)	4 (1)	0
	1–3	Uterine 10–11 h after mating	2:1	13	363	62 (17)	30 (8)	1 (3)
	1–3	Uterine 10–11 h after mating	4:1	4	101	22 (22)	13 (13)	1 (8)
	1–3*	Uterine 10–11 h after mating	2:1	5	136	12 (9)	8 (6)	0
	1–3*	Uterine 10–11 h after mating	4:1	4	133	14 (11)	8 (6)	0
	1–3	Epididymal†	With serum and uterine fluid	8	229	0	0	0

* Egg clot was washed three times before insemination.

† Epididymal sperm were incubated in the presence of uterine fluid for 2–7 h before insemination.

Of 126 eggs undergoing fertilization, 114 (90%) were monospermic and 12 (10%) were polyspermic; 12 had an enlarged sperm head and 114 had at least one male pronucleus. Fifty monospermic eggs and 12 polyspermic eggs had 1 to 26 supplementary sperm.

排卵的卵细胞中，均未观察到精子的穿透，它们比子宫培养的精子保持更长时间的活力。这种情况以及交配后 10~11 h 从子宫回收的精子，其穿透卵细胞的比例较高的事实明确说明大鼠精子在能够穿透卵细胞之前的确需要在雌性生殖道中获能。仓鼠 [6,8]，小鼠 [15]，豚鼠 [17]，可能还包括人 [9,10] 的精子能够体外获能，但是我们已经看到了大鼠精子体外获能是很困难的。在这方面大鼠和兔的精子是类似的。

表 1. 大鼠卵细胞的体外受精

成熟雌性大鼠	卵细胞年龄（排卵后时间 (h)）	所用的精子	所用的培养基（培养基:血清）	所用的雌性动物数量	检验的卵细胞数量	穿透的卵细胞数量（%）	经历受精的卵细胞数量（%）	多精受精的卵细胞数量（%）
自然排卵	4~5	交配后 0.5~1 h 的子宫中的精子	4:1	5	54	4（7）	2（4）	0
	4~5	交配后 10~11 h 的子宫中的精子	4:0	5	56	6（11）	4（7）	1（25）
			2:1	10	84	38（45）	22（26）	0
	4~5	交配后 10~11 h 的子宫中的精子	4:1	5	55	25（45）	25（45）	9（36）
	4~5*	交配后 10~11 h 的子宫中的精子	2:1	5	54	13（24）	10（19）	1（10）
	7.5~8.5	交配后 10~11 h 的子宫中的精子	2:1	6	78	1（1）	1（1）	0
	7.5~8.5	交配后 10~11 h 的子宫中的精子	4:1	6	74	4（5）	3（4）	0
	4~5	附睾的精子 †	血清和子宫液	5	52	0	0	0
超排卵	1~3	交配后 0.5~1 h 的子宫中的精子	4:1	12	280	19（7）	4（1）	0
	1~3	交配后 10~11 h 的子宫中的精子	2:1	13	363	62（17）	30（8）	1（3）
	1~3	交配后 10~11 h 的子宫中的精子	4:1	4	101	22（22）	13（13）	1（8）
	1~3*	交配后 10~11 h 的子宫中的精子	2:1	5	136	12（9）	8（6）	0
	1~3*	交配后 10~11 h 的子宫中的精子	4:1	4	133	14（11）	8（6）	0
	1~3	附睾的精子†	血清和子宫液	8	229	0	0	0

* 授精前清洗 3 次卵细胞团块。

† 授精前附睾精子在子宫液存在的情况下孵育 2~7 h。

在 126 个经历受精的卵细胞中，114 个（90%）是单精受精的，而 12 个（10%）是多精受精的；12 个有膨大的精子头部，而 114 个有至少一个雄性原核。50 个单精

Supplementary sperm (1 to 11) were also found in 88 unfertilized eggs, which indicates the occurrence of a vitelline block to sperm penetration as the eggs deteriorated *in vitro*. The incidence of polyspermy of rat eggs fertilized *in vivo* is about 1.2%[22]. The *in vitro* incidence of polyspermy is comparatively high as in the case of hamster[6] and mouse eggs[15]. Although fertilization of rat eggs *in vitro* was reported recently by Bregulla[20], the method used, the photographs published, and the high frequency of cleavage of unfertilized rat eggs *in vivo*[23], make his claim unconvincing.

This work was supported by grants from the US Public Health Service and the Ford Foundation. We thank Mrs Rose Bartke for assistance.

(**241**, 50-52; 1973)

H. Miyamoto and M. C. Chang
Worcester Foundation for Experimental Biology, 222 Maple Avenue, Shrewsbury, Massachusetts 01545

Received June 19; revised August 8, 1972.

References:

1. Austin, C. R., *Aust. J. Sci. Res., Series B, Biol. Sci.*, **4**, 581 (1951).
2. Chang, M. C., *Nature*, **168**, 697 (1951).
3. Thibault, C., Dauzier, L., and Wintenberger, S., *C. R. Soc. Biol.*, **148**, 789 (1954).
4. Chang, M. C., *Nature*, **184**, 466 (1959).
5. Brackett, B. G., and Williams, W. L., *Fertil. Steril.*, **19**, 144 (1968).
6. Yanagimachi, R., and Chang, M. C., *Nature*, **200**, 281 (1963).
7. Barros, C., and Austin, C. R., *J. Exp. Zool.*, **166**, 317 (1967).
8. Yanagimachi, R., *J. Reprod. Fertil.*, **18**, 275 (1969).
9. Edwards, R. G., Bavister, B. D., and Steptoe, P. C., *Nature*, **221**, 632 (1969).
10. Edwards, R. G., Steptoe, P. C., and Purdy, J. M., *Nature*, **227**, 1307 (1970).
11. Whittingham, D. G., *Nature*, **220**, 592 (1968).
12. Iwamatsu, T., and Chang, M. C., *Nature*, **224**, 919 (1969).
13. Cross, P. C., and Brinster, R. L., *Biol. Reprod.*, **3**, 298 (1970).
14. Mukherjee, A. B., and Cohn, M. M., *Nature*, **228**, 472 (1970).
15. Iwamatsu, T., and Chang, M. C., *J. Reprod. Fertil.*, **26**, 197 (1971).
16. Hamner, C. E., Jennings, L. L., and Sojka, N. J., *J. Reprod. Fertil.*, **23**, 477 (1970)
17. Yanagimachi, R., *Anat. Rec.*, **174**, 9 (1972).
18. Long, J. A., *Univ. Calif. Publ. Zool.*, **9**, 105 (1912).
19. Toyoda, Y., and Chang, M. C., *Nature*, **220**, 589 (1968).
20. Bregulla, K., *Arch. Gynäk.*, **207**, 568 (1969).
21. Everett, J. W., *Endocrinology*, **43**, 389 (1948).
22. Austin, C. R., and Braden, A. W. H., *Aust. J. Biol. Sci.*, **6**, 674 (1953).
23. Austin, C. R., *J. Endocrinol.*, **6**, 104 (1949).

受精卵细胞和 12 个多精受精卵细胞有 1~26 个额外精子。在 88 个未受精的卵细胞中也发现了额外精子（1~11 个），这提示由于卵细胞在体外变质，导致卵黄阻断精子的穿透。大鼠卵细胞体内受精时多精受精的发生率约为 1.2%[22]。而相比之下，仓鼠[6]和小鼠[15]卵细胞体外多精受精发生率要高。尽管最近布雷古拉报道了大鼠卵细胞体外受精[20]，但其使用的方法、发表的照片以及体内未受精的大鼠卵细胞高频率的卵裂[23]，使得他的论断没有说服力。

本研究得到了美国公共卫生署和福特基金会的资助。我们感谢罗丝·巴特克女士的帮助。

（吴彦 翻译；曹文广 审稿）

Do Honey Bees Have a Language?

P. H. Wells and A. M. Wenner

Editor's Note

In 1946, German ethologist Karl von Frisch proposed that honey bees communicate by means of an elaborate dance, a method used to guide foragers to food. Bee biologists Patrick H. Wells and Adrian M. Wenner greeted the theory with skepticism, and published data suggesting that bees find food by odour, not language. In 1973, the year von Frisch received a Nobel Prize for his work on insect communication, Wells and Wenner published counterarguments to their critics and additional data that they believed backed up their olfactory theory. Here they conclude that the "language hypothesis has little heuristic value and fails as a predictive tool". Today, further observations have convinced most biologists that von Frisch was in fact correct.

Von Frisch and later adherents of the theory that honey bees communicate by means of an elaborate dance are challenged by controlled experiments which show that their data can be explained in terms of olfactory cues.

WHENEVER new data and interpretations are presented which cannot be reconciled with established beliefs, it is proper and desirable that supporters of the traditional view should examine these data and interpretations with great care, and offer such objections as can be generated. In 1969, on the basis of studies of the recruitment of honey bee foragers to food sources, we compared the predictive powers of the classical "language" hypothesis of von Frisch[1] with those of a simple olfaction model for forager recruitment and concluded that olfaction provides the best interpretation of our data[2].

Investigators who believe that honey bees have a language have since challenged our conclusion[3-5], and two other recent articles[6-7], while not including a discussion of our results, cannot be reconciled with our interpretations. Our purpose here is to examine the principal objections to our interpretations and present additional data.

Procedure and Results

From earlier experiments, we knew that scent fed into a hive on one day influences forager recruitment on the next. We established a hive and a line of three experimental sites, each 200 m from it, in a dry field. The sites were approximately 150 m apart and the prevailing light breeze never blew towards or from the hive. Instead, the wind moved from site 1 to site 3, with site 2 in the middle.

蜜蜂有语言吗？

韦尔斯，温纳

编者按

1946年，德国动物行为学家卡尔·冯·弗里施提出蜜蜂是通过一套精心设计的舞蹈语言来和同伴交流，引导觅食蜜蜂找到食物的。蜜蜂生物学家帕特里克·韦尔斯和阿德里安·温纳对该理论持怀疑态度，并发表数据提示蜜蜂是通过气味觅食的，而并非舞蹈语言。1973年，冯·弗里施因其在昆虫通讯领域的工作而获得诺贝尔奖，韦尔斯和温纳发表了他们对批评者的反驳论点和一些他们认为可以支持嗅觉理论的更多数据。本文中他们得出这样的结论“语言假说没有启发价值，也不能成为一种预测工具”。今天，进一步研究使大多数生物学家相信实际上冯·弗里施是正确的。

冯·弗里施提出蜜蜂通过精心设计的舞蹈语言通讯的理论，他及其该理论的后继支持者受到来自对照实验的挑战，这些对照实验的数据能够用嗅觉信号来解释。

每当新的数据和解释与已建立的理论不一致时，传统观点的支持者应当特别仔细地检查这些数据和解释，并且提供能够提出的反对意见，这样才会令人满意。1969年，在招引觅食蜂到食物源的研究的基础上，我们将冯·弗里施[1]经典的“语言”假说与那些简单的嗅觉模型在招引觅食蜂的预测能力方面进行了比较，得出的结论是，嗅觉为我们的数据提供了最好的解释[2]。

认为蜜蜂有语言的研究者已经质疑过我们的结论了[3-5]，在其他两篇最近的论文[6,7]中，虽然没有包含对我们结果的讨论，但也与我们的解释不一致。在此，我们的目的是检查对我们理论的主要反对意见，并补充更多数据。

步骤和结果

从早期研究中我们知道，前一天注入一个蜂巢的气味会影响第二天对觅食蜂的招引。我们在干燥的野外建一个蜂巢，并且在距它200米的地方设立了排成一行的三个实验站点。那些实验站点间距约150米，而且通常微风不会吹向蜂巢或从蜂巢吹来。相反，风从1号站点吹向3号站点，而2号站点在中间。

During experimental periods of three hours, we fed scented sucrose to ten marked regular foragers at each of sites 1 and 3. Recruited bees were captured and killed. On a following day we fed unscented sucrose at sites 1 and 3 and planted scented sucrose at site 2, which had not been visited by bees from our hive. In this situation, the language hypothesis predicts that recruited foragers would be captured at sites 1 and 3, both visited by successful foragers from the hive. If, however, recruited foragers locate food sources by olfaction, they should arrive at site 2 which was not visited by any bees from the hive.

In four repetitions of this experiment (on days 2, 7, 12 and 16) we captured a total of twenty-five new recruits at site 1, 224 at site 2, and eight at site 3. Thus, newly recruited foragers had located a site in the field not previously visited by them or their hivemates, and had failed to arrive at locations about which they should have been well informed (according to the language hypothesis).

When foragers are experienced at a food site, they are again recruited to it by odour cues rather than by language[8,9] and, when odour cues are rigorously excluded, recruitment of new foragers fails[2,10]. For example, on control days of our experiments reported in 1969 when no scented sucrose was made available (days 4, 9 and 14), arrival of recruits at our feeding stations was drastically curtailed[2]. As a further control, we allowed ten bees to fly to each of four unscented feeders for 3 h (July 25, 1968) for a total of 1,374 round trips. During this period only five recruits from a hive of 60,000 bees found the feeders[2].

Thus, both experienced and inexperienced foragers seem to require olfactory cues to locate a food source in the field. In the absence of compelling evidence to the contrary, it is conservative to suggest that olfaction alone is sufficient to account for recruitment of honey bees to a food source.

Objections to Our Model

In spite of his polemic, Dawkins[3] does offer one substantive objection by arguing that bees are easily distracted from their "intended" (linguistically communicated) goals. In this view, the bees which arrive at site 2 have been distracted from sites 1 and 3, a possibility which deserves serious consideration. But in our experiment, site 2 was downwind from 1 and upwind from 3. Is it likely that recruited foragers could be simultaneously distracted upwind and downwind from their intended goals, or that any animal can be attracted to a downwind odour source? In any case, when all three sites were provided with unscented sucrose (days 9 and 14), recruits failed to arrive at sites about which they were supposedly linguistically informed[2]. We conclude that Dawkins's "distraction" model is not supported by the data.

Esch and Bastian[6] reported interesting new data but did not relate their findings to the literature on forager recruitment. In each of their experiments, a group of bees was trained to feed at a scent-marked site near the hive, and ten of the foragers were marked and confined temporarily in a cage. The feeder was next moved to a new location 200 m from the hive and the number of regular foragers reduced to one marked bee. The ten

在 3 个小时的实验中，我们在 1 号站点和 3 号站点用有香味的蔗糖各喂食 10 只标记后的普通觅食蜂。捕获新招引的蜜蜂并杀死。第二天，我们在 1 号站点和 3 号站点提供无香味的蔗糖，同时在 2 号站点提供有香味的蔗糖，其中 2 号站点还没有蜂巢里的蜜蜂来过。在这种情况下，按照语言假说预测将在 1 号站点和 3 号站点捕获新招引的觅食蜂，因为这两个站点均被来自蜂巢的觅食蜂光顾过。然而，如果蜜蜂觅食是靠嗅觉来定位食物源，它们应该去该蜂巢的蜜蜂从来没有光顾过的 2 号站点。

在 4 次重复实验（第 2、7、12、16 天）中，我们在 1 号站点共捕获 25 只新招引蜂，在 2 号站点共捕获 224 只，在 3 号站点共捕获 8 只。因此，新招引的觅食蜂在野外找到了一个它们及其同伴以前没有去过的站点，却没能到达那个它们本应该已经获知（根据语言假说）的地方。

当觅食蜂在一个食物站点有过经验后，它们会通过气味而不是语言被这个食物站再次吸引 [8,9]，当气味信号被完全排除后，食物站就不能再吸引新觅食蜂了 [2,10]。例如，在 1969 年我们做的对照实验中，当利用没有香味的蔗糖时（第 4、9 和 14 天），到达食物站的新招引蜂的数量大幅度减少了 [2]。作为进一步的对照，我们释放了 10 只蜜蜂飞向 4 个无气味的食物站（1968 年 7 月 25 日），3 小时内共飞行了 1,374 个来回。在此期间，一个蜂巢 60,000 只蜜蜂中只有 5 只发现了食物 [2]。

由此看来，在野外，不管是有经验还是没有经验的觅食蜂似乎都需要用嗅觉信号来确定食物源的位置。由于缺少有力的反面证据，保守地来看，只靠嗅觉就足以解释蜜蜂如何寻找食物源。

对我们模型的反对意见

尽管道金斯 [3] 对我们的观点进行抨击，他的确提供了一个实际的反驳论据，即蜜蜂很容易受到干扰，而偏离“既定”（通过语言交流而确定的）目标站点。在这种观点下，到达 2 号站点的蜜蜂已经不能专心于 1 号站点和 3 号站点，这是一个值得认真考虑的可能原因。但是在我们的实验中，2 号站点在 1 号站点的顺风方向，3 号站点的逆风方向。是否可能是因为新招引的觅食蜂会受到干扰而同时沿着顺风与逆风方向偏离既定的目标站点，或者任何一种动物都会被处于顺风位置的气味源吸引。无论如何，当三个站点都提供没有香味的蔗糖时（第 9 和 14 天），蜜蜂没能到达原以为可根据语言交流而获知其位置信息的站点 [2]。由此我们得出结论，道金斯的“干扰”模型没有得到数据支持。

埃施和巴斯蒂安 [6] 报道了有趣的新数据，但却没有将他们的发现与招引觅食蜂的文献联系起来。在他们的每次实验中，训练一组蜜蜂使其在蜂巢附近具有气味标志的站点处觅食，其中 10 只觅食蜂被标记并暂时被关在笼子中。接着将食物站移动

marked former foragers were then released and observed, to discover whether they would be re-recruited to the scent-marked feeder at the new location, as they should be on the hypothesis of linguistic communication by dance attendance.

Fourteen of seventy experimental bees did attend dances and subsequently found the new feeder location. Nineteen additional marked bees attended dances and flew from the hive without arriving at the feeder. The remaining thirty-seven marked experimental bees had no contact with the dancer and did not fly to the new location of the feeder.

Ten of the fourteen successfully re-recruited bees required between one and nine exploratory flights with intermediate contacts with the dancer. Only four succeeded on the first flight, two of them within 1 min of flight time. Many of the nineteen unsuccessful dance attenders also made several flights from the hive; they attended between five and thirty-one dances (mean, 17.5) and made between one and nine exploratory flights (mean, 3.4), without finding the feeder.

For the fourteen successfully re-recruited bees the mean time in flight between first attending the dancer and locating the feeder was 8.5 min, compared with less than 0.5 min for experienced foragers[11]. Esch and Bastian did not report the durations of flights by unsuccessful dance attenders.

In spite of these negative results, Esch and Bastian consider that their data support the language hypothesis. Because four bees found the feeder on their first flights, it is inferred that, "since they could not have searched the entire experimental area in such a short time...", successful re-recruits must have had prior knowledge (obtained from the dance) of the feeder location.

By focusing attention on the performance of successful bees Esch and Bastian have failed to recognize that most of their data actually contradict the predictions of the language hypothesis. The population of experimental animals must be viewed as a whole. In reality, there was a total of 100 exploratory flights by marked bees after attending the dancer. If recruited bees simply flew out equally in all directions from the hive, like spokes in a wheel, 2% would be headed within 7° of the feeder location, a result which is consistent with the observations of Esch and Bastian.

More generally, the data obtained by Esch and Bastian do not bear out the claim by von Frisch[12]:"...the tail-wagging dance makes known the distance and the compass direction to the goal. ... This description of the location enables the newcomers to fly rapidly and with certainty to the indicated flowers, even when these are kilometres away... an accomplishment on the part of the bees that is without parallel elsewhere in the entire animal kingdom". Nor do they eliminate the alternative possibility, that re-recruited foragers were flying about, using wind and odour, while seeking the scent to which they had been trained.

到离蜂巢 200 米的新位置，常规采食的觅食蜂的数量减少到 1 只。然后释放先前标记的 10 只蜜蜂并观察，以便揭示他们是否能重新被招引到新位置上的具有气味标记的食物站，当通过表演舞蹈实现语言通讯的假设成立时，它们应该能重新被招引。

70 只实验蜜蜂中有 14 只蜜蜂的确参与跳舞，接着找到了食物站的新位置。另外 19 只被标记的蜜蜂跳舞后从蜂巢离开但是没有到达食物站。剩下的 37 只被标记的实验蜜蜂没有与跳舞的蜜蜂接触，也没有飞往位于新位置的食物站。

14 只成功招引的蜜蜂中有 10 只蜜蜂需要依靠与舞蹈者之间的直接接触并进行 1 至 9 次的探索飞行。只有 4 只蜜蜂一次成功，其中的两只飞行不到一分钟的时间。19 只不成功的跳舞的蜜蜂大部分也离开蜂巢飞行了几次，它们参加了 5 到 31 次的跳舞（平均 17.5 次），进行了 1 至 9 次的探索飞行（平均 3.4 次），但是没有找到食物站。

对于 14 只成功招引的蜜蜂，从第一次跳舞到找到食物站的平均时间为 8.5 分钟，相比较而言，有经验的觅食蜂只需不到 0.5 分钟 [11]。埃施和巴斯蒂安没有报道尝试了跳舞却没有成功的蜜蜂的飞行时间。

尽管这些都是否定的结果，埃施和巴斯蒂安仍然认为他们的数据支持语言假说。因为 4 只蜜蜂在第一次飞出就找到食物站，也“因为它们不可能在如此短的时间内就找遍了整个实验区域……”，从而推测成功招引的蜜蜂一定是事先知道了食物站的位置（从舞蹈中获知）。

由于注意力集中在观察成功找到食物站的蜜蜂的行为上，埃施和巴斯蒂安并没有意识到他们大部分的数据实际上与先前的蜜蜂语言假说的预言相矛盾。实验动物群体应该被视为一个整体。实际上，被标记的蜜蜂参与跳舞后共进行了 100 次的探索飞行。如果从蜂箱中飞出的被招引蜜蜂是简单地以均等的机会飞向了各方向，如同车轮上的辐条，则 2% 的蜜蜂会飞向与食物站位置偏离 7° 的范围内，这个结果与埃施和巴斯蒂安的观察相一致。

从更普遍的意义上来说，埃施和巴斯蒂安获取的数据并不能支持冯 · 弗里施的论断 [12]：“……尾部摇摆的舞蹈能够给出目标的距离和方向。……这种对于方位的描述能够使得新来的蜜蜂迅速而准确地找到被指示的花朵，即使这些花儿在数公里外……在整个动物王国里，蜜蜂的这一系列技能是独一无二的。”他们也不能排除下面这种可能性，即重新招引的觅食蜂可能利用风向和气味来寻找它们曾经受到训练的气味。

The article by Gould, Henerey and MacLeod[7] contains many data not available earlier, but some of their data do not fall in line with what one might expect from the dance language hypothesis. As in the study by Esch and Bastian, their experiments were "... designed to examine the behaviour of individual recruits as each attended a dance and subsequently arrived at a feeding station". In their series 1 experiment, marked bees were trained to two feeding stations 120 m distant from and in opposite directions from the hive, and many workers in the hive were given individual marks. During experiments, of 277 potential recruits observed to attend dances, 240 failed to arrive at any station, and only thirty-seven were subsequently captured at the feeding stations. Of these, a third arrived at a station in a direction opposite to and 240 m distant from that "indicated" in the dance manoeuvre. This result contradicts the prediction of the language hypothesis that all thirty-seven of the successful recruits should have arrived at the "correct" station.

These investigators also found that successful recruits spend a considerable amount of time in flight before reaching the food source. The direct line flight time between hive and a feeding station located at 120 m is less than 25 s[11], and recruited bees generally fly from the hive within a minute and a half after leaving a dancing bee (50% leave within 30 s)[13]. The data in Table 4 of Gould *et al.*[7] reveal that the twenty-five bees which did arrive at the "correct" station flew an average of more than thirty times longer than would be necessary to "fly rapidly and with certainty" to the food source. The twelve bees which ended up at the station in the opposite direction averaged only thirty-six times as long as necessary for a direct flight.

A greater number of marked bees arrived at the station regularly visited by the foragers which recruited them than to the one 240 m from it, and some successful bees found the food source quickly (two in less than a minute), so Gould *et al.* concluded that quantitative information was communicated by the dance. The authors thereby focus attention on successful performers as did Esch and Bastian.

In another experiment, these authors fed sucrose of high molarity at one station and very low molarity at the other. Most of the dancing in the hive was by foragers regularly visiting the high molarity food source; and during the experiments most of the recruits captured were at that location. This correlation is accepted by the authors as evidence of linguistic communication.

As Gould *et al.* carefully pointed out, the validity of their interpretations for both of the above experiments depends upon the assumption that perfect odour symmetry existed between the two stations. In experiments of this type even small asymmetries of location odour do influence recruitment[13]. Gould *et al.* located their stations in aromatic vegetation. They got uniformly high recruitment whether or not they added scent to the food, yet there is considerable evidence that recruitment is minimal to an unscented location[2,10]. The high recruitment they obtained indicates that the feeding locations had odours detectable by bees. The unanswerable question is: Did asymmetries of these odours exist between two feeding locations 240 m apart?

古尔德、黑纳雷和麦克劳德的论文[7]有很多早期无法获得的数据，而其中的一部分数据也不符合舞蹈语言交流假说的预测结果。埃施和巴斯蒂安的研究中，他们的实验“……设计的目的是观察单个参与舞蹈并据此到达既定食物站的被招引蜜蜂的行为”。在他们的系列 1 实验中，被标记蜜蜂能被训练到达距离蜂巢 120 米远在相反方向上的两个食物站点，同时给蜂巢中的许多工蜂做了个体标记。实验过程中，在观察到的 277 只参与舞蹈的潜在被招引蜜蜂中，240 只未能到达任何一个食物站点，只有 37 只随后在食物站点被捕获。在这些蜜蜂当中，1/3 到达的是在距离舞蹈动物指定目标相反方向 240 米远的区域。这个结果与语言假说的预测结果（所有 37 只成功招引的蜜蜂应该到达这个“正确”地点）相矛盾。

这些研究者也发现成功招引的新蜂在到达食物源之前花费了相当长时间飞行。蜂巢与食物地点之间 120 米的距离的直线飞行时间不超过 25 秒[11]。招引的蜜蜂通常在接触舞蹈蜂后在一分半钟之内离开蜂巢(其中 50% 的蜜蜂在 30 秒之内离开)[13]。古尔德等人文章中表 4 的数据[7]显示 25 只蜜蜂到达“正确”食物站的时间比“快速而准确地飞行”至食物源的时间平均多花费了 30 倍以上。而 12 只到达相反方向食物站的蜜蜂，其平均飞行时间也仅是直线飞行所需时间的 36 倍。

被招引到先前觅食蜂经常到的食物站点的标记蜜蜂的数量，比被招引至距离该地点 240 米远的食物站点的蜜蜂数量更多，其中一些成功到达的蜜蜂能够迅速发现食物源（有 2 只在一分钟之内）。因此，古尔德等人推断蜜蜂通过舞蹈来传递交流定量信息。同埃施和巴斯蒂安一样，作者将关注点放在了成功的跳舞蜜蜂上面。

在另一实验中，研究者在一个地点为蜜蜂喂食了高摩尔浓度蔗糖，而在另一地点喂食低摩尔浓度蔗糖。蜂巢内大部分的舞蹈都是由经常光顾高摩尔浓度食物源的觅食蜂舞出的；实验期间，大部分新招引到的蜜蜂都是在该位置被捕获的。研究者认为这种相关性是语言交流的佐证。

正如古尔德等人所严肃指出的，上述两个实验解释的正确性建立在两地点之间气味均匀分布的假设基础上。因为这类实验中，站点之间即使少量的气味不对称都能影响对蜜蜂的招引[13]。古尔德等人将地点设置在芳香植物中，无论是否给食物添加香味，他们都能捕捉到很多新的蜜蜂，而另一个有力的证据是，在无香味的地点，招引的蜜蜂数量极少[2,10]。他们所获得的大量新招引蜂提示食物地点存在的气味能被蜜蜂觉察。但是不能解释的问题是：相距 240 米的两个食物地点之间气味是否存在不均匀性？

Mautz[5] also was concerned with the behaviour of individually marked workers which came in contact with a forager dancing in the hive. In his experiments, 32% of marked workers that attended dances found the feeder, and the amount of time the potential recruits attended the dancer was positively correlated with success at finding the food. In addition, the average flight time for successful recruits was more than ten times that expected of experienced foragers and the mean flight time of bees which flew out but failed to find the feeder was twice that of the successful recruits. Mautz noted that one bee found a goal at 400 m after following only five waggle cycles of the dance.

As with Gould *et al.*, and with Esch and Bastian, Mautz assumed that his feeder locations were not recognizable by bees as olfactorily distinctive and that only linguistic information could be communicated more effectively by increased duration of contact with the dancer. He does not discuss the possibility that a recruit which spends more time attending a dancing bee might, by so doing, gain a more accurate impression of the odour characteristics of a specific location. Again we see a supporter of the language hypothesis focusing attention on the successful recruits.

Of the several papers we review, Lindauer's[4] was the most deliberate and direct effort to obtain data contradictory to our findings. The first type of experiment undertaken by Lindauer resembled those of Gould *et al.*[7] in design, and yielded similar data. Forager bees were trained from a hive to two feeding sites, and were then fed high molarity sugar at one and low molarity sugar at the other. Virtually all dancing was by bees visiting the high molarity sugar site, and most successful recruits landed there. Lindauer assumed, as did Gould *et al.* and Mautz, a perfect symmetry (to bees) of environmental odours at his locations. From this he inferred that, "if the recruits followed only the odour signals given by the dancers… they should have appeared in equal numbers at (both) sites". In accordance with his assumption Lindauer attributed the asymmetry in recruit arrivals to linguistic communication. Thus, the untested assumption of station symmetry (to bees) is central to the reasoning of Lindauer as well as to the above authors.

In his second experiment, Lindauer put out three stations in a geometry similar to that used in our experiments[2] and trained bees to stations 1 and 3 but not to 2. Then, with scented sucrose at all three stations (a variation from our design), he collected recruits. Although the stations were deliberately asymmetrical (no foragers at station 2), approximately one-fourth of all captured recruits arrived at station 2. The language hypothesis predicts that the recruits should have travelled only to stations 1 and 3.

In order to reconcile these data with the language hypothesis Lindauer generated *ad hoc* the auxiliary hypothesis that potential recruits alternate their attentions between dancers visiting sites 1 and 3, and subsequently "integrate the directions communicated by both groups of foragers". Although the "integration" model satisfies Lindauer as an explanation of his data, it could not possibly predict the distribution of recruits in our earlier experiments[14,15] or, indeed, in the recent experiments of Gould *et al.*[7].

莫茨[5]同样关注那些有单独标记且在蜂巢中与舞蹈蜂接触的工蜂的行为。在他的实验中，32%参与舞蹈的被标记工蜂找到了食物站，潜在被招引的蜜蜂参与舞蹈的时间与成功找到食物成正相关。另外，成功招引到的蜜蜂的平均飞行时间是有经验的觅食蜂的预计时间的十倍多，而飞出去却不能找到食物站的蜜蜂的平均飞行时间是成功招引到的蜜蜂的两倍。莫茨注意到一只蜜蜂仅跟跳了 5 圈摇摆舞，就找到了 400 米外的食物。

与古尔德等人以及埃施和巴斯蒂安一样，莫茨假设食物站不能被蜜蜂嗅觉分辨识别，那么只有语言信息可以通过增加与舞蹈蜂接触的时间来实现更有效的交流。但他并没有讨论这样的可能性，即招引蜂通过花费更多的时间参与舞蹈，借此可能会获得对一个特定地点的气味特征更为精确的印象。我们再次看到了一个只片面关注成功招引蜂的语言假说的支持者。

在我们所研究的诸多文章中，林道尔的文章[4]是最为深思熟虑的，而且获得了与我们的发现直接相反的数据。林道尔所做的第一个实验在设计上类似于古尔德等人[7]的实验，也得出了相似的数据。蜂巢里的觅食蜂被训练在蜂巢与两个食物站点之间飞行，然后，一处站点喂食高摩尔浓度蔗糖，另一处站点喂食低摩尔浓度蔗糖。事实上，所有的舞蹈都由光顾过高摩尔浓度蔗糖站点的蜜蜂舞出，并且绝大多数招引成功的蜜蜂都降落在那儿。和古尔德等人以及莫茨一样，林道尔假设在他的站点周围环境气味的分布（对于蜜蜂而言）是完全均匀的。由此他推断“如果新招引的蜜蜂仅仅是追寻跳舞的蜜蜂所给的气味信号，那么它们出现在两个站点的数量应该相同”。为了与这一推论相符合，林道尔将招引到的蜜蜂到达的不均等归因于语言交流。因此，就如同以上几位作者一样，（对于蜜蜂而言）未经验证的环境气味均匀性假设对林道尔推理论证至关重要。

在第二个实验中，林道尔设置了 3 个在几何位置上与我们的实验[2]中所设置的相似的站点，并且训练蜜蜂到 1 号站点和 3 号站点而非 2 号站点。然后在 3 个站点都涂上有香味的蔗糖（与我们实验设计的不同之处），他收集了招引来的蜜蜂。尽管这些站点被故意设置得不对称（在 2 号站点没有觅食蜂），但被捕获的招引蜂中有将近 1/4 到达了 2 号站点。而蜜蜂语言假说预测招引到的蜜蜂应该只到达 1 号站点和 3 号站点。

为了用语言假说解释这些数据，林道尔提出特别的辅助性假设——潜在的招引蜂将它们的注意力在访问了 1 号站点和 3 号站点的舞蹈蜂之间转换，随后“综合两组觅食蜂交流的信息所指示的方向”。虽然这个“综合”模型满足了林道尔对于自己数据的解释，但是它不可能解释我们先前实验[14,15]中招引蜂的分布，事实上也解释不了在古尔德等人[7]最近的实验中招引蜂的分布。

Lindauer got good recruitment at his stations 1 and 3 when presenting unscented food at those locations. Interpreted in the light of our data[2,10] these results suggest that Lindauer's feeding sites were in some measure distinctive to bees. This may explain the asymmetry of recruitment he observed when foragers to station 1 were killed and bees visited only a single station. Unfortunately, the precise location and ecology of the sites for these experiments are not given in his paper (except that it was near Frankfurt), nor does he provide dates, wind speed and directions, number and sequence of observations contributing to the total data and other relevant information. These omissions make any interpretation difficult and speculative. Another portion of Lindauer's paper offers some philosophical and rhetorical objections to our paper, but we will consider these later.

Thus all the authors we have discussed[4-7] explicitly or implicitly assumed that the feeding stations they established in the field were in no way distinctive to bees and that the environment can be made symmetrical with respect to odours and other factors. They also assumed that, in order to find the food, a recruit must have prior quantitative information about its location. These assumptions lead the authors to interpret arrival of new workers at sites visited by dancing foragers as definitive evidence of linguistic communication. As pointed out earlier they have focused their attention on successful recruits. Experiments based on these assumptions invariably lead to affirmation of the consequence of the hypothesis (if bees have a language, recruits will reach the food. Some recruits find the food. Therefore, bees have a language). This reasoning is deductively invalid[19].

The experimental designs of these investigators suffer yet another serious disadvantage. It is not logically possible experimentally to establish the validity of the station symmetry assumption (failure to display asymmetry does not prove that none exists). It is quite practical, however, to seek positive evidence of station asymmetry in experiments of this design. One way would be to remove all directional information from dances performed by foragers regularly visiting food sites. If, in spite of this, recruits preferentially arrive at the feeder regularly visited by the foragers which recruited them, it could be concluded that they had done so without use of prior directional information, and that honey bees can exploit subtle environment asymmetries in the recruitment of new foragers to food sources.

We have incorporated the use of "directionless" dances in the design of the experiments reported below.

Conditions and Observations

In the summer of 1970 we did experiments using a single frame observation hive and two feeding stations located 150m from it in approximately opposite directions (west and east of the hive, but slightly north of it). The hive and stations were located at the University of California, Santa Barbara. Both stations were located in dry fields, devoid of green vegetation. The West Station was nearer the sea and upwind; the East Station somewhat closer to the University buildings. Otherwise, the sites appeared to be symmetrical.

当林道尔在 1 号站点和 3 号站点提供没有香味的食物，他在这些站点仍然获得了良好的招引效果。用我们的数据[2,10]来解释，这些结果说明林道尔的喂食地点对于蜜蜂来说具有某种程度的特殊性。这可以解释当去 1 号站点的觅食蜂被杀死以及蜜蜂仅访问单独一个站点时所观察到的招引的不对称性。遗憾的是这些实验中准确的地点和这些地点的生态条件在他的文中都没有交代(除了交代地点挨近法兰克福)，他也没有提供日期、风速、方向、对整体数据有用的观察数量和顺序以及其他有关信息。这些遗漏使得任何解释都很困难且具有投机性。林道尔文章的另一部分也对我们的论文在哲学上和修辞上提出了一些异议，我们将稍后考虑这些。

所有我们讨论的这些作者[4-7]都直接或者含蓄地假设他们在野外建立的这些喂食站点从任何方面对蜜蜂来说都是没有特别之处，并且认为在气味和其他因素方面的环境条件是可以被设置为均匀的。他们也假设，为了发现食物，一只新蜜蜂必须预先掌握足够多的相关地点的量化信息。这些假设导致作者把新的工蜂能够到达跳舞觅食蜂造访的地方解释为语言交流的确切证据。正如早先所指出的一样，他们把注意力放在了招引成功的新蜜蜂上面。因此基于这些假设的实验不可避免的导致了对语言假说的肯定（如果蜜蜂有语言，招引的蜜蜂就可以到达食物源从而发现食物，因此，蜜蜂有语言）。这种演绎推理是没有根据的[19]。

这些研究者的实验设计还有一个严重的缺陷。建立站点对称性假设的有效性在实验逻辑上是不可能的（不能显示不对称性，并不能证明其不存在）。但是，在这个实验设计中寻找地点不对称性的确凿证据却很实用。有一种方式是从频繁访问食物点的觅食蜂所进行的舞蹈中去掉所有与方向相关的信息。尽管如此，如果招引蜂仍倾向于到达招引它们的觅食蜂经常造访的食物站，那么就可以得出结论在这个过程中它们没有使用已有的方向信息，并且蜜蜂可利用细微的环境不对称性来招引觅食蜂至食物源。

我们在下面的实验设计中已经包括了“没有方向性”舞蹈的使用。

条件和观察

在 1970 年的夏天我们利用一个单一架构的观察蜂巢和大致位于相反方向（蜂巢的东、西方向，但略微偏北）相距约 150 米的两个喂食站点做了一些实验。这个蜂巢和站点位于加州大学圣巴巴拉分校。两个喂食站点位于干旱且没有绿色植物的地方。西部的站点靠近海和逆风向。东部的站点稍微靠近大学的建筑。另外，两个站点位于对称的位置。

Ten marked foragers were trained to feed at each station; all successful recruits were either killed or marked as replacements for regular foragers which were then killed. Observations were made at the hive and stations during 3-h periods (08:30–11:30), August 21 to September 4, 1970, and food was provided only during those periods. A westerly breeze prevailed during this period (up to 5 mile h^{-1} on most days) and the usual temperature was 65°–75°F.

A principal objective of the experiments was to examine relationships among food molarity, forager behaviour, and recruitment. Therefore, scent added to the food was held constant. All solutions used were scented with lavender oil (32 $\mu l.l.^{-1}$). Sucrose solutions of 0.8 M, 1.3 M, 1.8 M and 2.3 M were used. On any given day a solution of lower molarity was offered at one station and one of higher molarity at the other. Rotation of molarities between the stations controlled for possible inherent differences in location attractiveness, and on each day the stations acted as controls against each other. The experiments were run blind: the observers were not informed of molarities of sucrose provided at the feeding stations on any given day.

The hive had glass sides and was equipped with a clear plastic atrial chamber inside a diffusely lighted observation room. Hive conditions (well populated, and with little natural forage) ensured that our marked foragers danced in the atrial chamber, rather than on the comb. These dances were on a horizontal surface, and contained no discernible directional information. The straight (waggle) portions of the dance were apparently randomly oriented and highly variable within each dance episode.

Altogether, a total of 1,793 dances by our marked bees and 136 dances by unmarked bees foraging on natural sources were observed during eleven days (33 h) of experimentation. Thus, during our study periods, 93% of all dancing in the hives was by our marked foragers.

Although we were largely successful in eliminating oriented dances from our hive, occasionally a marked forager did enter and perform a dance episode on the vertical surface of the comb. These occurrences will be of great interest to proponents of the language hypothesis, so we will present data on oriented dances before proceeding to other aspects of the study.

There was a total of only 50 apparently oriented dances recorded during 33 h of experimentation; 1.5 such dances h^{-1}, on the average, or approximately 2.8% of all dance episodes recorded for our marked foragers. These occurred equally (24 to 26) for the two stations and did not correlate with recruitment (Table 1).

每一个站点都训练10只标记的觅食蜂在该处觅食。所有成功招引的蜜蜂都被杀死，或者作为被杀死的常规觅食蜂的替代品被标记起来。1970年8月21日到9月4日期间，从8:30到11:30这三个小时内我们对蜂巢和喂食站点进行了观察，并且只有在这段时间才有食物供应。这段时间西风盛行（多数日子风速小于5英里/小时），日常气温为65~75 °F。

这些实验的一个主要目的就是在于探寻：食物摩尔浓度、觅食蜂的行为以及招引新成员这三者之间的联系。因此，添加在食物中的气味是恒定的。所有使用的溶液都加入了有香味的薰衣草油（浓度是32 $\mu l \cdot l^{-1}$）。实验中还用到了浓度分别是0.8M、1.3M、1.8M和2.3M的蔗糖溶液。在同一天里，一个站点使用低浓度的蔗糖溶液，另一个站点则使用高浓度的溶液。在不同站点里轮换使用不同浓度溶液，以控制在站点吸引力方面的内在差异，每天这些站点互相作为对照。实验是以单盲设计：观察者并不知道每天在食物站点提供的蔗糖溶液的浓度。

蜂巢具有玻璃外壳，在散射光照射的观察室中间配置了一个透明的塑料小房间。蜂巢的条件（适宜的种群数量，几乎没有自然食物源）确保了标记过的觅食蜂在小房间里跳舞，而不是在巢脾上。它们所跳的舞蹈是在同一个水平面上的，其中没有任何可识别的方向信息，舞蹈中直线（摇摆）的那部分很显然是随机导向的，而且在每一个舞蹈节拍内的导向都是高度变化着的。

总之，在实验观察的11天中（33小时）总共有1,793次舞蹈是由标记过的蜜蜂跳的，136次舞蹈是没标记过的以自然食物源为食的觅食蜂跳的，因此在我们研究期间蜂巢中93%的舞蹈是由我们标记过的觅食蜂所舞出的。

尽管我们在消除蜂巢中的导向性舞蹈方面取得了巨大的成功，但是我们偶然发现一只被标记了的觅食蜂进入蜂巢并且在垂直于巢脾的平面跳舞，这种现象将会使支持语言假说的人产生极大的兴趣，所以我们将在继续其他方面研究之前展示有关导向性舞蹈的研究数据。

在33个小时的实验中，总共只记录有50次明显的导向性舞蹈。平均大概每小时1.5次这种舞蹈，约为我们记录到的标记的觅食蜂所有舞蹈节拍总数的2.8%。这种舞蹈在两个站点的发生概率均等（24比26），且与招引蜂的数量不相关（表1）。

Table 1. Dances by Unmarked Bees, Oriented Dances, Disoriented Dances, and Recruitment

	First hour	Second hour	Third hour	Total
Dances by unmarked bees	77	38	21	136
Oriented dances by marked bees	23	9	18	50
Disoriented dances by marked bees	337	675	731	1,743
Successful recruits	64	115	143	322

Data are partitioned according to time of occurrence in our experiments. Total data for 33 h of observation are presented.

These oriented dances were not distributed evenly throughout the experimental period, but occurred sporadically. For example, 40% of them occurred on one day, August 30, 1970. Occurrence of a block of oriented dances on a given day did not appear to influence recruitment. A comparison of data for August 30, 1970 (which had the greatest number of on-comb dances during the eleven days of experimentation) with the following (more typical) day is illustrative, and is presented as Table 2.

Table 2. Comparison of Visitation by Marked Foragers, Oriented and Disoriented Dances by Marked Foragers, and Successful Recruitment of New Bees for August 30, 1970 (with many Oriented Dances) and August 31, 1970 (with few)

	Trips by marked foragers		Oriented dances		Disoriented dances		Successful recruits	
	1.3 M	1.8 M	1.3 M	1.8 M	1.3 M	1.8 M	1.3 M	1.8 M
August 30, 1970	319	267	9	19	76	110	20	65
August 31, 1970	304	312	1	2	59	143	19	67
Total	623	579	10	21	135	253	39	132

Table 1 also shows that dancing by unmarked bees foraging on natural food sources declined as the experiment progressed. This is consistent with our findings that insertion of a new food source in the hive-environment system may change the behaviour of bees regularly foraging on an established one[10]. Presumably this alteration is mediated through a change in the available recruit pool.

From our data on oriented dances (Tables 1 and 2) it is not possible to infer that these dances are responsible for the observed recruitment of new foragers to those feeders we had established in the field. There is, however, a positive correlation between the number of disoriented dances (or total dances) and recruitment of new foragers (Table 3).

表 1. 未被标记的蜜蜂的舞蹈，导向性舞蹈，非导向性舞蹈，招引蜂

	第一小时	第二小时	第三小时	合计
未标记蜜蜂的舞蹈	77	38	21	136
标记过的蜜蜂的导向性舞蹈	23	9	18	50
标记过的蜜蜂非导向性舞蹈	337	675	731	1,743
成功的招引蜂	64	115	143	322

根据实验中的发生时间，将数据分组。列出了 33 个小时中观测到的所有数据。

这些导向性舞蹈并没有均匀地分布于整个实验周期中，而是偶发性的。例如它们当中 40% 是在 1970 年 8 月 30 日这一天发生的。这种导向性舞蹈集中发生在某一天并不影响蜜蜂的招引。1970 年 8 月 30 日（这一天在巢脾上进行的舞蹈的数量是这十一天的研究中最多的一次）的数据和接下来（更典型的）一天的数据的对比更具代表性，列于表 2。

表 2. 1970 年 8 月 30 日（发生多次导向性舞蹈）和 1970 年 8 月 31 日（发生少量导向性舞蹈）标记觅食蜂的导向性和非导向性舞蹈及新蜜蜂成功招引的情况的比较

	标记过的觅食蜂的飞行次数		导向性舞蹈数量		非导向性舞蹈数量		成功招引的蜜蜂数量	
	1.3 M	1.8 M	1.3 M	1.8 M	1.3 M	1.8 M	1.3 M	1.8 M
1970 年 8 月 30 日	319	267	9	19	76	110	20	65
1970 年 8 月 31 日	304	312	1	2	59	143	19	67
合计	623	579	10	21	135	253	39	132

表 1 也显示，随着实验的进展没有被标记的以自然资源为食的蜜蜂的舞蹈数量在减少，这和我们的发现是一致的，即我们在蜂巢周围的环境中放入一个新的食物源可能会改变蜜蜂在已建立的食物站点进行常规采食的行为 [10]，推测这种变化可能是通过改变有效食物源来介导的。

从蜜蜂导向性舞蹈的数据（表 1 和表 2）不可能推出以下结论：所观察到的我们在野外建立的食物站能够招引到新觅食蜂的现象应该归因于这种舞蹈。然而非导向性的舞蹈（或者总的舞蹈）的数量与新招引的觅食蜂的数量呈正相关（表 3）。

Table 3. Mean Number of Trips by Marked Foragers; of Disoriented Dances by those Foragers; and of Successful Recruits per 3 h Observation Period for 0.8 M, 1.3 M, 1.8 M and 2.3 M Lavender Scented Sucrose Solutions

Molarity	Forager trips	Disoriented dances	Successful recruits
0.8	308	26	5
1.3	320	86	19
1.8	293	117	42
2.3	231	107	17

Not less than four or more than seven observation periods at each molarity.

The experimental design is not rigorous enough to define the asymmetry in environmental cues used by recruits while searching for and locating a particular feeder in the field, but quantitative directional information does not appear to be a part of that system.

While doing this set of experiments, we also gathered information on Nasanov gland exposures at the food source by bees foraging on various molarities of sucrose. Our previous studies indicate that Nasanov gland exposure by regular foragers fails to attract undisturbed bees but apparently does provide a point of reference for disoriented members of the colony. Gland exposure is apparently a function of the interest potential recruits may have in a food source, rather than food quality *per se*[2,10]. Nasanov gland exposure by foragers visiting established food sources in the field was depressed at the lowest molarity we used (Table 4), a fact which indicates that few recruits were in the field searching for that source.

Table 4. Relationships Among Molarity of Sucrose Solutions Provided, Forager Visitations, and Nasanov Gland Exposures by Marked Foragers

Molarity	Forager trips	Nasanov gland exposures
0.8	308	24
1.3	320	77
1.8	293	85
2.3	231	71

Means for 3 h observation periods are given ($7 \geq$ periods/molarity ≥ 4).

Further Considerations

We have done several experiments which were designed to determine whether honey bees use linguistic communication under defined conditions[2,8,9,14-16]. Results indicated that the language hypothesis has little heuristic value and fails as a predictive tool. The results further suggest that if one wished bees to pollinate a particular crop, it should be

表 3. 标记觅食蜂的平均飞行次数；其非导向性的舞蹈的平均次数；每 3 小时观测周期内蜜蜂被 0.8 M、1.3 M、1.8 M 和 2.3 M 有薰衣草香味的蔗糖成功招引所需平均飞行次数

摩尔浓度	觅食蜂飞行次数	非导向性舞蹈数量	成功招引的蜜蜂数量
0.8	308	26	5
1.3	320	86	19
1.8	293	117	42
2.3	231	107	17

每个摩尔浓度观察期不少于 4 个且不多于 7 个。

这个实验设计对招引蜂在野外寻找和定位特定食物源时所用的环境不对称信号的定义不够严密，但是定量的导向性信息看起来并不是该实验系统的一部分。

当做这一系列的实验时，我们还收集了蜜蜂在不同摩尔浓度的蔗糖上觅食时分泌在食物源上的奈氏腺体分泌物的信息。我们之前的研究提示常规采食的觅食蜂分泌的奈氏腺分泌物并不能吸引未受干扰的蜜蜂，但很明显能给迷失方向的蜂群提供一些参照。觅食蜂腺体分泌物的气味显然是促进潜在招引蜂对食物源，而不是食物本身质量的兴趣[2,10]。觅食蜂造访野外已建立的食物源时奈氏腺分泌物在最低浓度食物站是非常少的（表 4），这个事实说明几乎没有招引蜂在野外寻找到该食物源。

表 4. 蔗糖溶液摩尔浓度、蜜蜂造访次数和标记觅食蜂奈氏腺分泌物之间的关系

摩尔浓度	觅食蜂的飞行次数	奈氏腺分泌物
0.8	308	24
1.3	320	77
1.8	293	85
2.3	231	71

给出了 3 小时观察期内的平均值（7 ≥ 观察期个数 / 摩尔浓度 ≥ 4）。

进一步的思考

我们已经设计了多次实验来确定蜜蜂在限定的条件下是否使用语言进行交流[2,8,9,14-16]。结果表明语言假说没有一点启发价值并且也不能作为预测性工具。结果

useful to regulate the odour carried into the hive by foragers, rather than the angle of dance on the comb.

More generally, we feel that honey bee foraging ecology is regulated by a complex system of hive, environmental and behavioural variables. With sufficient knowledge, it should be possible to manipulate the system and predict the consequences for the foraging ecology of a honey bee colony. We then might say that we understand colony function in terms which are neither teleological nor anthropomorphic; and we would be in a position to attack pragmatically problems of a comparative or practical nature. Accordingly, we have examined interrelationships among Nasanov gland exposure, visits of marked foragers to a food site, dancing in the hive, and recruitment of new foragers, and correlated these with the amount of odour in sucrose solutions provided in the field. In these experiments we have had some success in manipulating the system[10], as have Waller[17] and Friesen[18].

Waller showed that association of odours with food in the field, but not odours alone, could increase bee populations in experimental plots[17]. Friesen examined interrelationships among wind speed and direction, forager flight paths, odour levels and locations, forager visitation and recruitment of new foragers to sites in the field. He concluded that, after recruits leave the hive, their success and distribution depend upon an interaction of field variables affecting the distribution of odours[18]. Thus, evidence from several sources indicates that odour is of great importance in the system of variables which regulate honey bee foraging ecology.

Our comparison of the predictive values of the olfaction and language hypotheses of honey bee forager recruitment was an attempt to determine just how important odour is in this system[2]. To our surprise, our results and interpretations generated a controversy which deeply polarizes interested biologists. Along this line we must stress that the controversy does not emerge from a difference of opinion about the acceptability of various parts of a body of evidence.

The ingenious step and fan experiments of von Frisch have been repeated many times, and our own published data clearly show that when his procedures are closely followed, without insertion of additional controls, distributions of successful recruits may be expected to resemble those obtained by him[14,15]. Similarly, when Lindauer used an experimental geometry modelled after ours, he obtained data consistent with ours[4]; and the data presented in Table 3 of this paper resemble those obtained by Gould *et al.* and by Lindauer when they used similar experimental designs[4,7]. Neither the repeatability of experiments on honey bee behaviour nor the care and accuracy with which workers in this field gather and report data appears to be in question.

Proponents of the language hypothesis argue that the many repetitions of the von Frisch experiments make it highly probable that his interpretation is correct. One must realize, however, that the amount of additional confirmation affected by each new favorable repetition of an experiment becomes smaller as the number of previous repetitions

进一步说明如果人们希望蜜蜂为一种特定作物授粉，那么调控觅食蜂携带到蜂巢的气味应该是有用的，而不是调控在巢脾上跳舞的角度。

大体上说，我们认为，蜜蜂觅食生态是由蜂巢、环境和行为上的变量组成的复杂系统所调控的。有了足够的了解后，就有可能操控这个系统，并预测一个蜂群觅食生态的结果。然后我们可以说，我们明确地理解了蜂群功能，它既没有目的性也不是拟人的，我们将处于务实地解决可比性或实用性方面的问题的位置。因此我们检测了奈氏腺体分泌、标记觅食蜂对食物站的造访、在蜂巢中跳舞以及招引到的新觅食蜂之间的相互关系，并将这些与在野外所提供的蔗糖溶液中芳香物质的含量联系起来。在这些实验中，我们在操控这个系统方面取得了一些成功[10]，正如沃勒[17]和弗里森[18]取得的成功一样。

沃勒指出，野外食物与气味的结合（并不是单独的气味）能够增加实验点蜜蜂的数量[17]。弗里森检测了风速和风向、觅食蜂的飞行路径、气味水平和位置、觅食蜂的造访次数和野外站点新觅食蜂的招引之间的相互关系。他得出的结论是，招引蜂离开蜂巢后，它们的成功招引和分布取决于影响气味分布的各野外变量之间的相互作用[18]。因而，各种来源的证据表明气味在调节蜜蜂觅食生态的变量系统中起着很重要的作用。

嗅觉理论和语言假说在觅食蜂招引的预测价值上的比较是一种用于确定气味在系统中重要性程度的尝试[2]。令我们惊奇的是：我们的研究结果和解释引起了相关生物学家高度两极化的争论。在这方面我们必须强调这种争论的出现并不是源于针对同一证据主体不同方面的可接受性的观点差异。

冯·弗里施的这种独创性的步骤和有趣的实验已经被重复了许多次，我们自己发表的数据也清楚地表明：在没有加入额外对照实验的情况下，严格按照冯·弗里希的步骤操作，成功招引到的蜜蜂的分布与他得到的分布数据类似[14,15]。同样，林道尔用我们的几何实验模型模拟，他得到的数据和我们的一致[4]，本文表 3 所示数据与古尔德等人以及林道尔在用相似实验设计时所获得的数据类似[4,7]。实验中蜜蜂行为上的重复性和该领域工作人员在收集和报道数据时的注意力和精确性似乎都没有问题。

语言假说的支持者辩称：冯·弗里施实验的众多重复使得他的解释极有可能是正确的。然而人们必须认识到，随着重复实验数量的增加，每一个新的圆满重复实验

grows[19]. Thus, even under the most favourable conditions, sheer number of repetitions is never sufficient to render a hypothesis immune to challenge. Furthermore, in the case of the bee language hypothesis, the existence of a considerable body of unfavourable evidence greatly diminishes the weight of even a large body of confirming data[20].

Any hypothesis which is generated as an explanation of certain observed events will, of course, imply their occurrence. The events to be explained will then be taken as supporting evidence for it. If the hypothesis is valid it also may lead to *a priori* predictions of facts and events in conditions different from those leading to its formulation.

It is exactly here that we have difficulty with the language hypothesis of honey bee orientation. Each time we have inserted previously omitted controls or altered the experimental design to create new test implications, the distribution of successful recruits to feeders located in the field is no longer that predicted by the language hypothesis. Similarly, much of the evidence recently obtained by Esch and Bastian, by Mautz, by Gould *et al.* and by Lindauer is contrary to the predictions of that hypothesis[4-7].

The difficulties we had in reconciling our data[14-16] with the language hypothesis led us to propose the alternative olfaction model[21,22]. The olfaction hypothesis allowed a satisfactory explanation of the distribution of successful recruits in those experiments which incorporated new controls[14-16] and could encompass the large body of existing data on honey bee forager recruitment. Furthermore, it had the advantage of simplicity.

Thus, in 1967 we had two hypotheses, olfaction and language, each generated *a posteriori* to explain an existing body of evidence. The language hypothesis had been well articulated by von Frisch twenty years earlier[1], but it had been expanded and modified through the years in an attempt to explain newer results. Olfaction, while new in the sense that it challenged the then currently accepted hypothesis and two decades of thought habits, was essentially a return to the position held by von Frisch and others in the earlier years of this century.

Lindauer not only presented new data but challenged our interpretation on philosophical grounds when he invoked Aristotelian (Darwinian) teleology as an argument in favour of language. He argued that in nature "each morphological structure and behavioural act is associated with a special function. On this basis alone, it would seem unlikely that information contained in the waggle dance of a honey bee is not transmitted to her nest mates"[4]. We have answered his challenge in part by discussing the relevant data in terms of philosophy of science which we consider to be more powerful[19,20].

An inherent weakness of Lindauer's teleological argument can be illustrated by giving one interesting example from another field of biology. Methyl eugenol is extremely attractive to male oriental fruit flies. In one test in Hawaii 1,300 male *Dacus dorsalis* were attracted a half-mile upwind to a muslin screen that had been treated with it. Methyl eugenol is not produced by the female fruit flies nor does it attract them. It is not a component of the

所带来的肯定性却在减弱[19]。因此，即使在最佳的条件下，重复实验的绝对数量也决不足以让一个假说免于挑战。而且，在蜜蜂语言假说的案例中，相当数量的不利证据的存在使大部分已经确认数据的重要性大打折扣[20]。

任何一个源自解释特定观察事件的假说都一定会暗示这些事件的发生。这些应该被解释的事件随后被作为假说的支持证据。如果假说是成立的，则可根据假说推导出对事实和现象的先验预测，即使这些事件发生的条件与假说成立的条件不同。

在这里我们很清楚在蜜蜂导向性的语言假说中存在很多问题。每次我们都会加入以前漏掉的对照实验或者改变实验设计来建立新的实验，但是被野外食物源成功招引到的蜜蜂的分布不再能够通过语言假说来预测。同样地，最近埃施和巴斯蒂安、莫茨、古尔德等人和林道尔获得的一些证据也与假说所预测的相矛盾[4-7]。

在将我们的数据[14-16]与语言假说协调一致时遇到的困难使得我们提出另外的嗅觉模型[21,22]。嗅觉假说可以很好地解释那些包含了新对照的实验中成功招引到的蜜蜂的分布[14-16]，而且可以涵盖觅食蜂招引研究中的大部分现有数据。另外，该假说还具有简单的优点。

因此，在 1967 年我们有两个假说，嗅觉假说和语言假说，二者都能对现有的部分证据进行解释。语言假说已经在二十年前被冯 · 弗里施所阐明[1]，但在近几年人们尝试去解释新结果的过程中，它被不断拓展和完善。嗅觉假说是一种新的假说，对已被接受的观点和 20 年来的思维习惯提出挑战，本质上也回到了 20 世纪初冯 · 弗里施等人与更早之前的观点抗争时所处的位置。

当林道尔引用亚里士多德（达尔文）的目的论作为支持语言假说的论据时，他不仅提出了新的数据，而且从哲学基础上挑战了我们的解释。他认为在自然界中“每种形态结构和行为都是与一种特殊功能相关联的。单单以此为基础，蜜蜂摇摆舞中所含的信息没有传递给它的同伴的现象似乎不可能发生”[4]。我们已经以更有说服力的科学哲学形式，通过对其相关数据的讨论来部分地回答他的质疑[19,20]。

用一个来自生物学其他领域的有趣的例子可以阐明林道尔目的论的一个先天缺点。甲基丁香酚对雄性东方果蝇极具吸引力。在一个实验中，夏威夷的 1,300 只雄性东方果蝇被甲基丁香酚处理过的薄纱屏幕吸引，迎风飞了半英里。甲基丁香酚不由雌性果蝇产生，也不吸引雌性果蝇。它不是该果蝇自然食物中的组成部分，而且

natural food of this fly and probably has no nutritional value. Yet male oriental fruit flies are irresistibly attracted to it and "apparently cannot stop feeding when they have free access to it, and they kill themselves with over indulgence"[23]. Their behaviour under these circumstances certainly cannot be construed as adaptive.

Admittedly, this is an extreme example of non-adaptive behaviour, but it does reveal a weakness in the teleology argument. The mere presence of a characteristic behavioural pattern in an animal cannot be construed as purposeful, adaptive or "associated with a special function".

Lindauer[4] also challenged us on grounds that we have not individually discussed three specific situations investigated earlier by von Frisch and his colleagues. First, at distances quite close to the hive, recruited foragers may be captured at scented dishes other than the one visited by the foragers which recruit them, while at distances of 100 m and beyond "the recruited workers all fly in the direction of the feeding plate". That the latter part of this quoted statement does not hold is well documented by the data of Gould *et al.*[7] as well as by ours[2,14,15], and even by Lindauer's own data[4] unless one is willing to accept his "integration" model. As for the first part of the statement, when the feeding dishes are close to the hive, we agree with von Frisch and Lindauer; the bees seem to be using odour.

A second situation involves disoriented dances performed by foragers on a horizontal surface. In an experiment using a tilted hive, von Frisch got approximately equal recruitment in four directions while regular foragers were fed in only one direction. Later that day, with the hive put upright again, he got preferential recruitment in the direction of the feeder visited by foragers[12]. We prefer a design in which two or more stations visited by bees simultaneously serve as controls against each other, and we set up our experiment accordingly. Our data on the effectiveness of disoriented dances (Table 3) seem to be in disagreement with those of von Frisch. During our 11 days of observation, new recruits preferentially arrived at the sites visited by dancing foragers even though there was no directional component in the dances.

The third situation involves the detour of foragers around an obstacle to a feeder on the other side of it. In one experiment performed by von Frisch, marked foragers were trained to fly around a rocky ridge to a scented feeder. Scent plates were placed on top of the ridge and 50 m laterally to the direct line between hive and feeder. During a 90 min period, three new recruits were captured at the lateral stations, eight on top of the ridge and twenty-three at the feeder. Because the new recruits apparently failed to follow the flight path used by the foragers which recruited them, von Frisch feels that they were linguistically informed of the feeder location[12].

We must argue that the flight paths and duration of searching by the twenty-three successful recruits are not known, and that a failure of bees to use the detour (if indeed they did not use it) does not differentially support either language or olfaction. Since neither hypothesis predicts that recruits will follow the same path as the marked foragers,

可能没有营养价值。然而，雄性东方果蝇却无法抵抗地被它所吸引并且“当它们可以自由接近该物质时，它们显然控制不住贪吃的欲望，以致死于这种过度的嗜好”[23]。在这种情况下，它们的行为当然不能被视为适应性。

不可否认，这是一个极端的非适应性行为的例子，但它确实揭示了目的论的一个缺点。动物单纯的特异行为模式不能被解释为是有目的性的，适应性的或是“与某种特殊功能相关的”。

林道尔[4]还质疑我们，没有对较早前由冯·弗里施和他的同事们所做研究中的三个具体情况进行单独讨论。首先，在食物站相当接近蜂巢时，招引的觅食蜂可能会在有香味的食物站捕获，而不是那些招引它们的觅食蜂所造访的食物站。而在食物站距离蜂巢 100 米或更远的情况下“所有的觅食蜂都向食物源的方向飞去”。后半部分所引述的观点并不成立，这已经被古尔德等人[7]和我们[2,14,15]的数据，甚至林道尔自己的数据[4]很好地证明了，除非有人愿意接受他的“综合”模型。至于第一部分，我们同意冯·弗里施和林道尔的观点：当食物站离蜂巢很近时，蜜蜂似乎是用气味去寻找食物。

第二种情况是觅食蜂在水平面上表演的非导向性舞蹈。在一个采用倾斜蜂巢的实验中，冯·弗里施仅从一个固定方向喂食常规采食的觅食蜂，但他在四个方向招引到的蜜蜂数量大致相等。当天晚些时候，重新将蜂巢放正，他发现被觅食蜂造访过的食物源方向更具招引的优势[12]。我们更倾向于另外一个实验设计，实验中蜜蜂可以造访两个或更多的站点，同时它们相互之间可以作为对照，我们据此设置了实验。我们发现，我们关于非导向性舞蹈有效性的实验数据（见表 3）与冯·弗里施的数据不符。在我们 11 天的观察中，新的招引蜂会优先抵达跳舞的觅食蜂造访过的站点，即使该舞蹈中没有导向性的元素。

第三种情况是蜜蜂绕过障碍找到位于另一边的食物站。冯·弗里施在一个实验中，训练被标记的觅食蜂绕过岩石山脉去寻找有香味的食物站。有香味的食物盘被放置在山顶，且与蜂巢和食物站之间连线横向相距 50 米。在 90 分钟的观察期中，3 只新招引蜂在旁侧站点被捕获，8 只在山顶被捕获，23 只到达了准确的食物站点。基于新招引蜂显然不能按照之前招引它们的觅食蜂的飞行的路线飞行，冯·弗里施认为，蜜蜂通过语言来交流食物站的位置[12]。

我们要说明的是，23 只成功到达的招引蜂的飞行路线和搜寻时间还不清楚，蜜蜂绕道而行的失败（如果事实上它们并没有绕道）在支持语言假说或嗅觉假说方面没有任何差异。因为两个假说都无法预测新招引蜂将遵循标记觅食蜂相同的路径，

the experiment is not crucial and the results do not support or refute either hypothesis.

We believe that experimentation does not "prove" or "disprove" hypotheses, but rather affects their credibility. If the credibility of a hypothesis at any given time is determined by the total body of relevant information available, then the language hypothesis of forager recruitment was very credible in 1946 when von Frisch proposed it. It is less so now, for the body of relevant information is quite different and includes much unfavourable evidence. In fact, it is so much less credible that we no longer can believe that honey bees communicate linguistically.

Do honey bees have a language? That is a question which may never be answered with certainty. It may be more useful to examine assumptions critically, state hypotheses and their consequences with precision, review the evidence objectively, and ask: can we now believe that honey bees have a language? Thus, it appears that the honey bee forager recruitment controversy is not about the nature of evidence but rather about the nature of hypotheses[25]. It is not what investigators observe (the data) but what they believe (infer) that is at the heart of the controversy.

We thank Nelson Dee and Stephanie Niebuhr for technical assistance. The research was supported by the US National Science Foundation. An analysis of earlier events and attitudes leading up to the bee language controversy is available elsewhere[24].

(**241**, 171-175; 1973)

Patrick H. Wells and Adrian M. Wenner

Department of Biology, Occidental College, Los Angeles, California 90041, and Department of Biological Sciences, University of California, Santa Barbara, California 93106

References:

1. Von Frisch, K., *Osterr. Zool. Z.*, **1**, 1 (1946); translation, *Bull. Anim. Behav.*, **5**, 1 (1947).
2. Wenner, A., Wells, P., and Johnson, D., *Science*, **164**, 84 (1969).
3. Dawkins, R., *Science*, **165**, 751 (1969).
4. Lindauer, M., *Amer. Nat.*, **105**, 89 (1971).
5. Mautz, D., *Z. Vergl. Physiol.*, 72, 197 (1971).
6. Esch, H., and Bastian, J. A., Z. *Vergl. Physiol.*, **68**, 175 (1970).
7. Gould, J. L., Henerey, M., and MacLeod, M. C., *Science*, **169**, 544 (1970).
8. Johnson, D. L., and Wenner, A. M., *Anim. Behav.*, **14**, 261 (1966).
9. Johnson, D. L., *Anim. Behav.*, **15**, 487 (1967).
10. Wells, P. H., and Wenner, A. M., *Physiol. Zool.*, **44**, 191 (1971).
11. Wenner, A. M., *J. Apic. Res.*, **2**, 25 (1963).
12. Von Frisch, K., *Tansprache und Orientierung der Bienen* (Springer-Verlag, Berlin, 1965); translation, *The Dance Language and Orientation of Bees* (Harvard University Press, Cambridge, 1967).
13. Johnson, D. L., and Wenner, A. M., *J. Apic. Res.*, **9**, 13 (1970).
14. Johnson, D. L., *Science*, **155**, 844 (1967).
15. Wenner, A. M., *Science*, **155**, 847 (1967).
16. Wenner, A. M., Wells, P. H., and Rohlf, F. J., *Physiol. Zool.*, **40**, 317 (1967).

所以该实验并不是关键性的而且其结果不支持或驳斥任何假说。

我们认为上述实验并不能“证明”或“反驳”这些假说，但是却会影响它们的可信度。如果一个假说的可信度在任何特定时间都是由所能得到的有关信息的总体决定的，那么早在 1946 年冯 · 弗里施提出觅食蜂招引的语言假说时，它就是非常可信的。但现在可信度降低了，因为相关信息的总体情况已经颇为不同，并包括许多不利的证据。事实上，语言假说的可信度太低了，我们不能够再相信蜜蜂是通过语言沟通的。

蜜蜂有语言吗？这个问题可能永远也无法得到确定的回答。批判性地检验假设，精确地陈述假说及其结果，客观审查证据，并自问：我们现在可以认为蜜蜂有语言吗？这样做也许更有用。由此看来，关于觅食蜂招引的争议，并不在于证据本身，而在于假说本身[25]。不是研究者观察到了什么（数据），而是他们相信什么（推论），这才是争议的核心。

我们感谢纳尔逊 · 迪伊和斯蒂芬妮 · 尼布尔提供技术支持。这项研究得到了美国国家科学基金会的资助。对引起蜜蜂语言争议的早期事件和看法在其他文献[24]中也有分析。

（董培智 翻译；张健旭 审稿）

17. Waller, G. D., *J. Apic. Res.*, **9**, 9 (1970).

18. Friesen, L., *Biol. Bull.* (in the press).

19. Hempel, C. G., *Philosophy of Natural Science* (Prentice-Hall, Englewood Cliffs, 1966).

20. Popper, K., in *British Philosophy in Mid-Century* (edit. by Mace, C. A.) (Macmillan, New York, 1957).

21. Wenner, A. M., and Johnson, D. L., *Science*, **158**, 1076 (1967).

22. Wenner, A., and Wells, P., *XXI Int. Apic. Congr. Summ.*, **88** (1967).

23. Steiner, L. F., *J. Econ. Ent.*, **45**, 241 (1952).

24. Wenner, A. M., *The Bee Language Controversy* (Educational Programs Improvement Corporation, Boulder, 1971).

25. Altmann, S. A., *Nature*, **240**, 361 (1972).

T and B Lymphocytes and Immune Responses

M. C. Raff

Editor's note

In this paper, Canadian-born cell biologist Martin C. Raff sums up what was then known about T and B lymphocytes. The two white blood cell types are found in peripheral lymphoid organs where they look identical, and Raff had previously identified a T cell marker, an achievement that won him immediate international recognition. T and B cells have different origins, properties and immunological functions which modulate each other's activities. Their discovery, Raff says, marks a new era in immunology in which powerful research tools and accessible models are likely to light on biological issues as well as disease.

The recognition of two distinct classes of lymphocytes has been a turning point in immunology. Immunological models and tools may help to provide the answers to many biological problems.

IMMUNOLOGY has become an exciting science of its own. Nonetheless, what is being learned about lymphocytes and the immune responses that they mediate has important implications for medicine and other branches of biology. Unfortunately, the private language of immunology has made it difficult for non-immunologists to join in the excitement. This article attempts to review what is known in general terms about the cellular basis of immunity. (For a more detailed review of lymphocytes and their roles in immune responses, see ref. 1.)

Immunology is concerned with the specific responses an animal makes when foreign materials (antigens or immunogens) are introduced into its body. Such immune responses are made by all vertebrates and consist of the production of specific immunoglobulin protein molecules (antibodies) and/or specifically reactive cells, both of which can circulate in the blood and react specifically with antigen. As a result of this reaction, the foreign material may be inactivated (for example, bacterial toxins), killed (for example, infecting organisms or transplanted cells) and/or phagocytosed by cells of the reticuloendothelial system. On the other hand, in some cases, such immune responses may have deleterious effects on the host, such as in hypersensitivity reactions (hayfever and drug allergy, for example), where antigen reacting with antibody fixed to basophils and mast cells causes the release of histamine and other pharmacological mediators of inflammation. In general, immune responses which can be transferred to another animal by means of serum from a sensitized donor (containing antibody) are termed humoral immune (or antibody)

T淋巴细胞、B淋巴细胞与免疫应答

拉夫

编者按

在这篇文章中，出生于加拿大的细胞生物学家马丁·拉夫就T淋巴细胞和B淋巴细胞已有的知识进行了总结。在外周淋巴器官中发现了两种不同类型的白细胞，这两类白细胞在外周淋巴器官中看起来是一样的。拉夫在之前的研究中就已经鉴定了一种T细胞标志物，并因此立即获得国际认可。T细胞和B细胞具有不同的起源、特性和免疫功能，两者在功能上是相互调控的。拉夫说他们的这一发现标志着免疫学的新时代。在这个新时代中，强大的研究工具和可利用的研究模型很有可能为生物学问题和疾病的研究带来曙光。

认识到两类截然不同的淋巴细胞是免疫学研究中一个转折点。免疫学模型和工具将有助于解答多种生物学问题。

虽然免疫学已经发展成了一门独立的振奋人心的学科，但是关于淋巴细胞及其所介导的免疫应答的研究对医学和生物学其他分支学科也具有重要的影响。然而遗憾的是，免疫学独有的语言体系使得非免疫学家很难参与到这个振奋人心的领域中。因此，我们撰写了这篇综述文章，力图对已知的免疫学的细胞学基础进行概括性阐述。（关于淋巴细胞和它们在免疫应答中的作用，更详细的综述见参考文献1。）

免疫学是一门研究动物对进入其体内的外源性物质（抗原或免疫原）所产生的特异性反应的科学。所有的脊椎动物都可以产生这种免疫应答。这个过程包括特异性免疫球蛋白分子（抗体）和（或）具有特异性反应活性的细胞的产生。它们都可以在血液中循环，并与抗原特异地结合。这种反应会使外源性物质失活（如细菌毒素）、被杀死（如受感染的器官或移植的细胞）和（或）被网状内皮系统中的细胞吞噬。另一方面，在某些情况下，这种免疫应答可能会对宿主自身产生有害的作用。例如在超敏反应中（比如花粉过敏和药物过敏），抗原与吸附在嗜碱性粒细胞和肥大细胞上的抗体发生反应，从而引起组织胺和其他药理炎症介质的释放。一般来讲，那些可以通过将致敏供体（携带抗体）的血清转移到另一个动物体而引起的免疫应答称为体液免疫（或抗体）反应，而那些不能通过血清只能通过致敏细胞才能转移的免

responses, whereas those that can be transferred by sensitized cells but not by serum are called cell-mediated immune responses.

While immunochemists were unravelling the structure of antibody in the 1950s and early 1960s, cellular immunologists were demonstrating that lymphocytes are the principal cells involved in immune reactions. The most convincing experiments were those showing that relatively pure populations of rat lymphocytes obtained from the chief lymphatic vessel, the thoracic duct, could transfer both cellular and humoral immunity to irradiated rats, which could not respond immunologically themselves as their lymphocytes had been killed by the radiation (reviewed in ref. 2). In addition, depleting animals of lymphocytes by prolonged drainage of the thoracic duct was found to impair their immune responsiveness[2]. Thus lymphocytes, whose origins and functions had been a mystery for so long, were established as "immunocompetent" cells.

It was soon realized that lymphocytes are not a homogeneous population. Several lines of evidence suggested that there are two distinct types of immunocompetent lymphocytes: one which requires the thymus gland for development and is responsible for cell-mediated immunity and another which develops independently of the thymus and mediates humoral antibody responses. The evidence came from studies in birds, rodents and man in the 1960s. In birds[3,4] and rodents[5] it was found that removing the thymus from an embryo or newborn markedly impaired the cell-mediated immune responses of the animals when they grew up, but had much less effect on humoral immunity. On the other hand, removal at hatching of the bursa of Fabricius[3,4], a cloacal lymphoid organ unique to birds, impaired the bird's ability to make antibody, but had little effect on cell-mediated immunity. Investigations of patients with immunological deficiency diseases also showed that humoral and cell-mediated immunity could be separately affected (reviewed in ref. 6): patients with Bruton-type congenital agammaglobulinaemia could not make antibody and were deficient in lymphoid cells producing antibody, but had normal cell-mediated immunity, whereas children with congenitally hypoplastic thymus glands (for example, Di George's syndrome) had markedly impaired cell-mediated immunity but could make relatively normal amounts of antibody in response to some antigens.

In the past few years the two-lymphocyte model of immunity has been firmly established (at least in birds and mammals), with two "central" lymphoid organs—the bursa, or its mammalian equivalent (still unidentified), and the thymus—producing lymphocytes independently of antigen, and seeding them out to the "peripheral" lymphoid organs (that is, lymph nodes, spleen and gut-associated lymphoid tissues) where they await contact with antigen which will induce them to differentiate into "effector" cells (see later). In the peripheral lymphoid tissues the lymphocytes derived from thymus are referred to as T cells, while those derived from the bursa in birds, or its equivalent in mammals, are called B cells[7].

疫应答则被称为细胞免疫反应。

二十世纪五十年代和六十年代早期，当免疫化学家们正在努力解析抗体结构的时候，细胞免疫学家们正致力于阐明淋巴细胞是参与免疫反应的主要细胞。最具说服力的实验是在大鼠中进行的：研究人员首先利用放射性射线照射杀死了受体大鼠的淋巴细胞从而使其自身无法进行免疫应答（见参考文献2），然后将从大鼠的主淋巴管—胸导管中分离出来的相对纯的淋巴细胞移植到受体大鼠体内，结果发现这样可以使受体大鼠同时获得细胞免疫和体液免疫。此外，研究人员还发现通过胸导管持续引流使动物体内的淋巴细胞耗尽会削弱免疫反应性[2]。至此，长久以来起源和功能一直是个谜的淋巴细胞，终于被确认为“免疫活性”的细胞。

很快人们便认识到，淋巴细胞并不是一个同质的群体。许多证据表明存在两种截然不同的具有免疫活性的淋巴细胞：一种淋巴细胞需在胸腺中发育并与细胞免疫有关；另一种淋巴细胞的发育则与胸腺无关并且介导体液免疫应答。支持这一观点的证据来自二十世纪六十年代在鸟类、啮齿动物以及人类中的研究：在鸟类[3,4]和啮齿动物[5]中，摘除胚胎或新生动物的胸腺，它们长大后细胞免疫应答会显著受损，而对体液免疫影响不大。另一方面，在孵化阶段摘除法氏囊[3,4]（鸟类特有的泄殖腔淋巴器官）会削弱鸟类产生抗体的能力，而几乎不影响细胞免疫。对免疫缺陷病患者的研究也表明，体液免疫和细胞免疫分别受到不同因素的影响（见参考文献6）：由于缺乏产生抗体的淋巴细胞，患有布鲁顿型先天性丙种球蛋白缺乏症的病人不能产生抗体，但是细胞免疫正常；而先天性胸腺发育不全的儿童（如迪乔治综合征患者）则表现为细胞免疫功能显著受损，但对某些抗原产生免疫应答时可以产生相对正常量的抗体。

在过去的几年中，免疫的双淋巴细胞模型（至少在鸟类和哺乳动物中）已经确立。在这个模型中，有两个“中枢”淋巴器官——法氏囊或在哺乳动物中具有的同功能器官（目前尚不明确）和胸腺——产生不依赖于抗原的淋巴细胞，并将它们运送到“外周”淋巴器官（淋巴结、脾脏和肠道相关淋巴组织），淋巴细胞在那里等候与抗原接触从而诱导其进一步分化为“效应”细胞（见下文）。其中，在外周淋巴组织中胸腺来源的淋巴细胞被称为T细胞，而鸟类的法氏囊或哺乳动物相应器官来源的淋巴细胞被称为B细胞[7]。

Phylogeny

Until recently it was thought that specific immune responses were confined to vertebrates. There is now evidence, however, that some invertebrates, such as annelids and tunicates, can reject foreign tissues and that these primitive immunological responses can display specificity and possibly short-term memory[8] (that is, an increased and/or faster response on second exposure to the same antigens). These reactions are mediated by macrophage-like cells (coelomocytes) and possibly by soluble effector molecules having relatively little specificity[8]. As there is no evidence that invertebrates have lymphocytes or immunoglobulins, it seems likely that specific cellular immunity evolved before the appearance of these two principal mediators of vertebrate immunity.

All vertebrates have lymphocytes and probably thymus tissue (at least at some stage in their development) and are capable of producing antibody and cell-mediated immune responses[8]. Lower vertebrates (lampreys and hagfish, for example) have little organized lymphoid tissue and can produce only one class (IgM-like) of antibody. Rudimentary lymph node-like structures are first found in Amphibia which make two classes of antibody. Birds are the first vertebrates in which a clear dichotomy of the lymphoid system has been established, and are unique in having two discrete central lymphoid organs, thymus and bursa, producing T and B lymphocytes respectively. Mammals have abundant and highly organized lymphoid tissues, can elaborate a variety of different classes of antibody (such as IgG, IgM, IgA, IgE, IgD in man) and have distinct T and B lymphocyte populations, although the site of B cell development is still uncertain. It is not known whether vertebrates below birds have separate classes of T and B cells.

Development of T Lymphocytes

In most animals, lymphocytes first appear in the foetal thymus. The thymus anlage is composed of epithelial cells and is derived from the third and fourth pharyngeal pouches. Although in the past it had been suggested that thymus lymphocytes (thymocytes) develop from thymus epithelial cells, experiments in chickens and mice have clearly established that haemopoietic stem cells from foetal yolk sac and liver migrate into the thymus anlage and there proliferate and differentiate into thymus lymphocytes, presumably under the inductive influence of the thymus epithelium[9]. In mice (gestation 20 days) the first stem cells, which seem to be large basophilic blast-like cells, arrive in the thymus around day 11, and the first small lymphocytes are seen by day 15 or 16 of embryonic life[9]. Using radioactive[10], chromosome[5,11] and surface antigenic[9] markers, it has been shown that lymphocytes migrate from thymus to peripheral lymphoid tissues to make up the T lymphocyte population. Although this begins just before birth in mice, most of the seeding occurs in the first week of life[9]. Therefore, if the thymus is removed in the first days of life the mouse will grow up with a marked deficiency of T cells and thus impaired cell-mediated immunity, whereas thymectomy done later in life has much less effect[5]. In adult animals, stem cells from bone marrow migrate to thymus, and thymus lymphocytes continue to seed to the periphery, but these processes take place at a much reduced rate by comparison with the foetus and newborn[5,11].

系统发育

直到最近，人们都一直认为只有脊椎动物才具有特异性免疫反应。然而，现在有证据表明，诸如环节动物和被囊动物等无脊椎动物也可以排斥外源组织，并且这种原始的免疫应答也具有一定的特异性和短期的记忆效应[8]（即在第二次接触到相同抗原时可以产生更强和（或）更快的免疫应答）。这些反应是由巨噬细胞样细胞（体腔细胞）介导的，也可能是由特异性相对较差的可溶性效应分子介导的[8]。由于并没有证据表明无脊椎动物中具有淋巴细胞或免疫球蛋白，因此特异性细胞免疫很可能是在脊椎动物的两种主要免疫细胞出现之前形成的。

所有脊椎动物都有淋巴细胞，也可能有胸腺组织（至少在发育的某个阶段中），并且可以产生抗体、引发细胞免疫应答[8]。低等脊椎动物（如七鳃鳗类和盲鳗类）几乎没有系统化的淋巴组织且只能产生一类抗体（IgM 样抗体）。初级淋巴结样结构是在两栖动物中首次发现的，这种结构可以产生两类抗体。第一种被发现具有明显两类淋巴系统的脊椎动物是鸟类；而且鸟类独一无二地具有两个独立的中枢淋巴器官——胸腺和法氏囊，二者可以分别产生 T 细胞和 B 细胞。尽管 B 细胞的发育位点尚未确定，但可以肯定的是哺乳动物具有大量的、高度系统化的淋巴组织，可以产生各种不同种类的抗体（例如人体可以产生 IgG、IgM、IgA、IgE、IgD 等），并且具有截然不同的 T 细胞和 B 细胞群。我们还不知道比鸟类低等的脊椎动物是否具有独立的 T 细胞和 B 细胞群。

T 淋巴细胞的发育

在大多数动物中，淋巴细胞首先在胎儿胸腺中出现。胸腺原基由上皮细胞组成，来源于第三、第四咽囊。过去人们认为胸腺淋巴细胞（胸腺细胞）是由胸腺上皮细胞发育而来，但是鸡和小鼠的实验清楚表明，胎儿卵黄囊和肝脏中的造血干细胞会迁移到胸腺原基并在那里增殖、分化成胸腺淋巴细胞，这些过程大概是在胸腺上皮诱导作用下进行的[9]。在小鼠（妊娠 20 天）中，在胚胎期第 11 天左右，可以观察到首批类似大嗜碱性母细胞的干细胞到达胸腺；而在胚胎期第 15 天或第 16 天的时候可以检测到首批小淋巴细胞[9]。利用放射性示踪[10]、染色体[5,11]和细胞表面抗原标志物[9]可以清楚地显示淋巴细胞从胸腺迁移到外周淋巴组织并分化为 T 淋巴细胞群。尽管这一过程在小鼠临出生前开始，但是大部分迁移过程发生在其出生后的第一周[9]。因此，如果在小鼠出生的第一天便摘除其胸腺，那么其长大后则表现为明显的 T 细胞缺陷从而损伤细胞免疫；然而如果晚些时候进行胸腺切除，那么影响会小很多[5]。在成年动物中，骨髓来源的干细胞会迁移到胸腺，胸腺淋巴细胞也会持续地向外周组织迁移。但是与胚胎和新生动物相比，成年动物的这一过程要缓慢得多[5,11]。

Most thymus lymphocytes are immunologically incompetent (that is, they cannot respond to antigen) and differ in other ways from peripheral T cells, suggesting that there is another differentiation step from thymocyte to T lymphocyte. Recently it has been demonstrated that there is a small subpopulation (~2 to 5%) of thymus cells, located in the thymus medulla, which is immunologically competent and has most of the properties of peripheral T lymphocytes[9,12,13]. This suggests that the second differentiation step may occur within the thymus and that T cell development may be visualized as stem cell→thymocyte→ "mature" thymus lymphocyte→peripheral T lymphocyte (Fig. 1). This scheme is almost certainly an oversimplification, however, for there is some evidence that cells may leave the thymus at varying stages of maturation, or perhaps as distinct cell lines, giving rise to subpopulations of peripheral T cells with different properties and functions[13]. In addition, the role of putative thymus humoral factors or hormones (thymosin, for example) is still unclear, although there is evidence that they probably do not induce stem cells to differentiate to lymphocytes outside the thymus, but may influence peripheral T cells in some way[14].

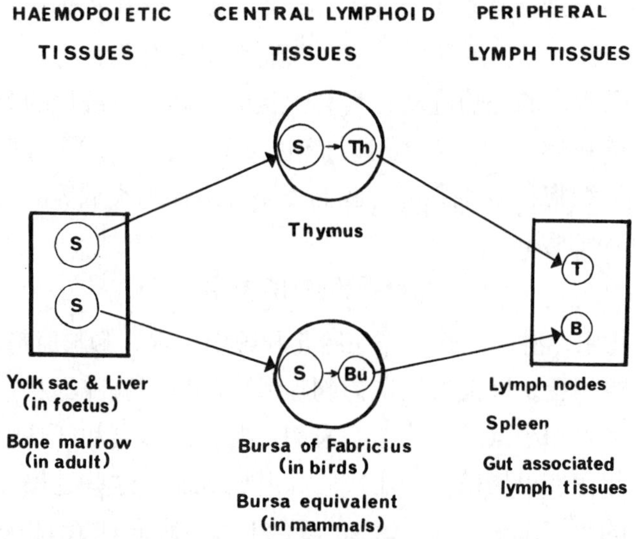

Fig. 1. Diagrammatic (and oversimplified—see text) representation of T and B lymphocyte development showing migration of stem cells (S) to thymus and bursa where they differentiate to thymus (Th) and bursal (Bu) lymphocytes, some of which migrate to the peripheral lymphoid tissues as T and B lymphocytes respectively.

Development of B Lymphocytes

In birds, B cell development is dependent on the bursa of Fabricius which arises as a sac-like evagination of the dorsal wall of the cloaca on day 5. Chromosome marker studies have shown that stem cells (morphologically identical to those seen in the foetal thymus) begin to migrate from yolk sac to the bursa around days 12 to 13 and there differentiate to lymphocytes within 1 or 2 days[9]. By day 14, bursa lymphocytes with IgM

大多数胸腺淋巴细胞没有免疫活性（也就是说它们不能对抗原作出应答），并且在其他方面它们与外周T淋巴细胞也有所不同，这表明从胸腺细胞到T细胞存在另一个分化过程。最近有研究表明，在胸腺髓质中存在胸腺细胞小的亚群（约2%~5%），它们具有免疫活性并且具有外周T淋巴细胞的大部分特征[9,12,13]。这表明第二个分化步骤可能是在胸腺内完成的，可以把T细胞的发育看作如下过程：干细胞→胸腺细胞→“成熟”胸腺淋巴细胞→外周T淋巴细胞（见图1）。但是几乎可以肯定的是，上述模型只是一种过于简化的说明。因为有证据表明，细胞在成熟的各个阶段细胞都可能离开胸腺，或者作为不同的细胞系产生若干具有不同特性和功能的外周T细胞亚群[13]。另外，人们假定的胸腺体液因子或激素（例如胸腺素）的功能仍不清楚。尽管有证据表明在胸腺外它们很可能不能诱导干细胞分化为淋巴细胞，但它们可能以某种方式影响着外周T淋巴细胞[14]。

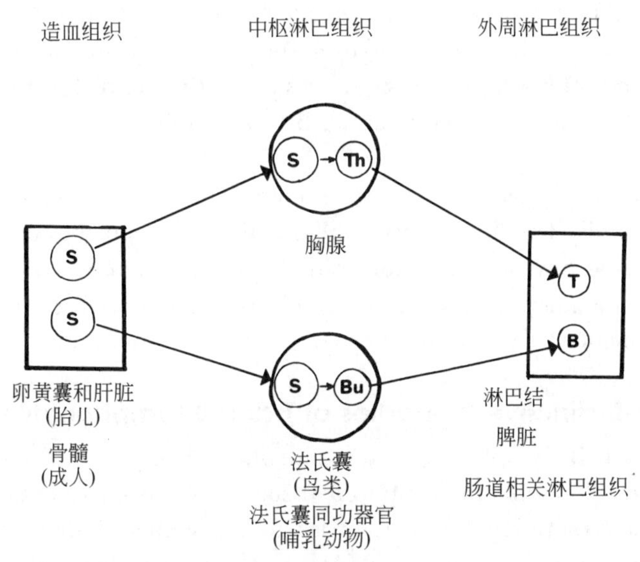

图 1. T淋巴细胞和B淋巴细胞的发育过程示意图（过于简化的——见正文）。干细胞（S）迁移到胸腺和法氏囊组织，并在那里分别分化成胸腺淋巴细胞（Th）和法氏囊淋巴细胞（Bu）。然后这些细胞的一部分迁移到外周淋巴组织中，分别成为T淋巴细胞和B淋巴细胞。

B淋巴细胞的发育

在鸟类中，B细胞的发育依赖于法氏囊。法氏囊是在胚胎发育第五天由泄殖腔背壁外翻所形成的一个囊状结构。染色体标记研究显示，在胚胎发育第12天到第13天时便有干细胞（与胎儿胸腺中检出的干细胞形态一致）从卵黄囊迁移到法氏囊中，并且在1~2天内在那里分化成淋巴细胞[9]。到第14天可以观察到表面有IgM的法氏囊

on their surface can be seen, and bursa lymphocytes bearing IgG are seen a few days later[15]. The migration of bursal lymphocytes to peripheral lymphoid tissues has been demonstrated by isotope labelling experiments. Embryonic bursectomy results in marked depletion of peripheral B lymphocytes and a marked impairment in antibody (that is, immunoglobulin) production[15]. Recently it has been found that injecting anti-μ antibody (that is, specific for the heavy chains of IgM) before hatching, combined with neonatal bursectomy, suppresses later production of IgG as well as IgM[15]. This suggests that even B cells that will eventually produce IgG initially express IgM on their surface, and is strong evidence for an IgM→IgG switch within individual B cells. Whether this switch is driven by antigen, as suggested by experiments in mice[16], or occurs independently of antigen stimulation, as suggested by experiments in chickens[15], is unsettled.

In mammals, it is still not clear where stem cells differentiate to B-type lymphocytes, although it is known not to be in the thymus. It has been suggested that gut-associated lymphoid tissues (like Peyer's patches, tonsils, appendix, and so on) may serve as "bursa-equivalent", but there is little evidence to support this. In rodents, at least, there is increasing evidence that lymphocytes are produced in large numbers in the haemopoietic tissues themselves[17] (that is, liver in embryos and bone marrow in adults) and it seems likely that these tissues not only supply the stem cells for both T and B cell populations but are also the sites where stem cells differentiate to B-type lymphocytes.

It is not clear at what stage stem cells are committed to becoming lymphocytes or to becoming T or B cells. The finding of multipotential haemopoietic stem cells (that is, cells capable of becoming any of the mature blood cell types, lymphoid or myeloid) in early mouse embryonic thymus[18] suggests that commitment may not occur until stem cells enter the microenvironment of the thymus or bursa (or bursa equivalent).

Distinctive Properties of T and B Lymphocytes

As resting T and B lymphocytes are morphologically indistinguishable and are found together in all peripheral lymphoid tissue, it has been essential to find ways of distinguishing and separating them in order to study their individual properties. The demonstration of important surface differences between them has been particularly useful in this regard. Some of these surface differences can be recognized by antibody[19]. For example, the θ alloantigen (defined by alloantibody made in one strain of mouse against thymocytes of another strain) is present on mouse thymocytes and T cells, but absent from B lymphocytes, and this has proved to be a convenient surface marker for T cells in mice[19]. On the other hand, readily demonstrable surface immunoglobulin (Ig) (refs. 20, 21) and the heteroantigen, "mouse-specific B lymphocyte antigen" (MBLA) (ref. 19)—defined by hetero-antibody made in rabbits against mouse B cells—can serve as B cell markers. With antisera reacting specifically with the surface of one or other lymphocyte type, either cell population can be killed in the presence of complement, and thus eliminated from a cell suspension. Alternatively, one can use antibody on digestible solid-phase immunoabsorbents[22], or fluoresceinated antibody and fluorescence-activated electronic cell sorting[23], to purify either type of cell. In addition to surface antigenic

淋巴细胞，几天后可以观察到产生 IgG 的法氏囊淋巴细胞[15]。同位素标记实验证明法氏囊淋巴细胞可以迁移到外周淋巴组织。在胚胎时期摘除法氏囊会导致外周 B 淋巴细胞的显著缺失和抗体（即免疫球蛋白）产量的显著减少[15]。最近人们发现，向孵化前的胚胎注射抗 μ 抗体（特异靶向 IgM 重链）加上在胚胎孵化后摘除其法氏囊，抑制 IgM 的同时，也会抑制随后 IgG 的产生[15]。这表明即使是那些最终产生 IgG 的 B 细胞，最初在其表面也会表达 IgM；同时强有力地证明，单个的 B 细胞中存在 IgM 到 IgG 的转换。但是这种转换是由抗原驱动（如小鼠实验结果显示[16]），还是与抗原刺激无关（如鸡实验结果显示[15]），目前尚不清楚。

在哺乳动物中，人们尚不清楚干细胞是在哪里分化成 B 淋巴细胞的，只知道肯定不是在胸腺中。有人认为哺乳动物的肠道相关淋巴组织（如派伊尔节、扁桃体、阑尾等）可能充当“法氏囊同功结构”，不过尚缺乏证据证明这一点。但是至少在啮齿类动物中，越来越多的证据表明大量的淋巴细胞是在造血组织中（如胚胎的肝脏、成年动物的骨髓）产生的[17]；并且似乎这些组织不仅为 T 细胞群和 B 细胞群提供干细胞，而且也是干细胞分化为 B 淋巴细胞的场所。

目前还不清楚哪个阶段的干细胞会分化为淋巴细胞或成为 T 细胞、B 细胞。研究人员在早期小鼠胚胎胸腺中观察到了多能造血干细胞（即可以分化为任何一种成熟血细胞、淋巴细胞或骨髓细胞的细胞）[18]，这表明干细胞直到进入胸腺或法氏囊（或法氏囊同功结构）微环境以前都不能完成分化。

T 淋巴细胞和 B 淋巴细胞的不同性质

由于处于静息状态的 T 细胞和 B 细胞在形态上无法区分，并且都分布于外周淋巴组织中，因此寻找区别、分离它们的方法对研究其各自的特性来说极其重要。就这一点来说，揭示它们之间重要的表面差异尤为有效。有些表面差异可以通过抗体识别出来[19]。例如 θ 同种异型抗原（由同种抗体识别，用一个种系的小鼠胸腺细胞免疫另一个种系的小鼠得到该抗体）存在于小鼠胸腺细胞和 T 细胞，而不存在于 B 淋巴细胞，因此其被证实为小鼠 T 细胞合适的表面标志物[19]。另一方面，已确实证明的细胞表面免疫球蛋白（Ig）（参考文献 20、21）和“小鼠特异性 B 淋巴细胞抗原”（MBLA）(参考文献 19) 可以作为 B 细胞的标志物，MBLA 是通过兔抗鼠 B 细胞产生的异种抗体识别的异种抗原。在补体存在下，抗血清能与一种或其他类型淋巴细胞表面发生特异反应，杀死相应细胞群，从而将它们从细胞悬液中清除。利用耦联到可消化的固相免疫吸附剂上的抗体[22]，或荧光标记的抗体，通过荧光激发的电子细胞分选技术[23]，就可纯化出任意一种细胞类型。除了上述 T 细胞与 B 细胞的

differences between T and B cells, the latter can bind antibody-antigen-complement complexes by means of surface complement receptors[24], and antibody-antigen complexes by means of receptors for the Fc part of complexed Ig[25]; resting T cells do not have these receptors. The functions of Fc and complement receptors on B cells are unknown, but it has been suggested that they may be important in antigen localization in the lymphoid tissues, in B cell activation by antigen and/or in putative killing by B cells of target cells coated with antibody.

Most T lymphocytes continuously recirculate between blood and lymph, passing out of the blood through specialized post-capillary venules in lymph nodes and Peyer's patches, passing through the substance of the lymphoid tissues and entering the efferent lymph; they then re-enter the bloodstream by way of the thoracic duct[2,5]. Although most B lymphocytes seem not to recirculate, some apparently do, but through different areas of the lymphoid tissues and with a slower transit time than T cells[26]. In the peripheral lymphoid tissues, T and B cells are found in more or less separate areas, the so-called thymus-dependent areas (periarteriolar sheath of spleen, paracortex of lymph nodes, and interfollicular areas of gastrointestinal lymphoid tissues) and thymus-independent areas (lymph follicles and peripheral regions of splenic white pulp, follicles and medulla of lymph nodes and follicles of gastrointestinal lymphoid tissues) respectively[27]. When radiolabelled T or B cells are injected into an animal, they migrate specifically to their respective areas[27]. Although both T and B lymphocyte populations are heterogeneous[1], T cells have a longer generation time[28] on average and are slightly larger[29], more dense[24], less adherent[24] (to various materials such as glass, plastic, nylon, and so on) and more negatively charged than B cells[30]. In addition, T lymphocytes are preferentially depleted by anti-lymphocyte serum[31] (which acts principally on recirculating cells), but in general are less sensitive to cytotoxic drugs (for example, cyclophosphamide[32]), corticosteroids[33] and irradiation[34]. T and B cells also differ in their *in vitro* responses to a variety of "mitogens", such as plant extracts (phytomitogens), bacterial products (like endotoxin) or antibodies to lymphocyte surface antigens, which stimulate a relatively large proportion of T and/or B lymphocytes to divide and differentiate into blast cells. Although pokeweed stimulates both T and B cell proliferation, concanavallin A (Con A), phytohaemag-glutinin (PHA) and lentil stimulate only T cells, and lipopolysaccharides (for example, E. *coli* endotoxin) and anti-Ig sera stimulate only B cells[35]. It is of interest that although soluble Con A and PHA selectively activate T cells, they bind equally well to B cells, and if covalently linked to solid-phase materials they stimulate B cell proliferation[35]. Mitogen stimulation of lymphocytes is being intensively studied as a possible model of lymphocyte activation by specific antigen. These studies have made it clear that there is more to lymphocyte activation than simple binding of ligand to surface receptors.

Antigen Recognition and Specific Lymphocyte Receptors

The central dogma of immunology is the clonal selection hypothesis which suggests that at some time in ontogeny and independently of antigen, individual lymphocytes (or clones of lymphocytes) become committed to responding to one, or a relatively small number of antigens; they express this commitment through antigen-specific receptors

表面抗原差异外，B 细胞还可以通过细胞表面的补体受体与抗体－抗原－补体复合物结合[24]，或通过复杂的免疫球蛋白的 Fc 片段受体与抗原－抗体复合物结合[25]；而静息状态的 T 细胞则不具有这些受体。虽然目前还不清楚 B 细胞上 Fc 片段和补体受体的功能，但有人认为它们可能在淋巴组织中抗原的定位、抗原激活 B 细胞和（或）一般认定的 B 细胞杀死抗体包被的靶细胞的过程中起重要作用。

大多数 T 淋巴细胞都会在血液和淋巴之间反复循环。T 细胞可以通过淋巴结和派伊尔节内特化的毛细血管后微静脉从血液中出来，穿过淋巴组织基质进入输出淋巴管；然后通过胸导管重新回到血液循环系统中[2,5]。尽管大多数 B 淋巴细胞似乎并不参与上述循环过程，但其中有些 B 淋巴细胞确实会通过淋巴组织的不同部位，以比 T 细胞慢的运送速度进行再循环[26]。在外周淋巴组织中，T 细胞和 B 细胞的分布区域基本上是相互分隔的，分别为所谓的胸腺依赖区（脾脏动脉周围鞘、淋巴结副皮质区、胃肠淋巴组织的滤泡间区）和非胸腺依赖区（淋巴滤泡和脾脏白髓的外周区域、淋巴结中的滤泡和髓质以及胃肠淋巴组织滤泡）[27]。将放射性标记的 T 细胞或 B 细胞注入动物体后，它们会特异地迁移到各自的区域[27]。尽管 T 淋巴细胞群和 B 淋巴细胞群都是异质的[1]，但是与 B 细胞相比，T 细胞的平均增代时间更长[28]，并且体积略大[29]，密度更高[24]，粘附性更小[24]（与玻璃、塑料、尼龙等各种材料的粘附性相比），带的负电荷也较多[30]。此外，T 细胞更容易被抗淋巴细胞的血清清除[31]（主要针对反复循环的细胞），但通常对细胞毒性药物（如环磷酰胺[32]）、皮质类固醇[33]和辐射[34]的敏感性较低。另外它们在体外对于各种“有丝分裂原”的反应也是不同的，比如植物提取物（如植物有丝分裂原）、细菌代谢产物（如内毒素）或淋巴细胞表面抗原的抗体，其均可刺激很大一部分 T 和（或）B 淋巴细胞分裂、分化为母细胞。尽管美洲商陆可以同时刺激 T 细胞和 B 细胞增殖，但是伴刀豆球蛋白 A（Con A）、植物凝集素（PHA）和扁豆素只能刺激 T 细胞增殖，而脂多糖（如大肠杆菌内毒素）和抗免疫球蛋白血清则只能刺激 B 细胞增殖[35]。有意思的是，尽管可溶的 Con A 和 PHA 只能选择性地激活 T 细胞，但它们同样可以很好地与 B 细胞结合；而一旦将它们共价耦联到固相支持物上，它们也可以刺激 B 细胞增殖[35]。作为淋巴细胞被特异抗原活化的一种可能的模型，有丝分裂原刺激淋巴细胞活化被广泛研究。这些研究使得人们逐渐认识到淋巴细胞的激活不仅仅是配体与细胞表面受体结合那么简单。

抗原识别与特异性淋巴细胞受体

免疫学的中心法则是克隆选择学说。该学说提出在个体发生的一定时间内，单个淋巴细胞（或淋巴细胞克隆）通过细胞表面抗原特异性受体只对某一种或相对少数的几种抗原起反应，这一过程与抗原无关。于是，一旦抗原进入体内便会筛选出

on their surface. Thus, when an antigen is introduced into the body it selects out those lymphocytes which already have receptors for the antigen on their surface; the interaction of antigen with receptors initiates the activation of the specific cells. There is now an impressive body of evidence supporting the clonal selection hypothesis for both T and B lymphocytes. Thus T and B cells have been shown to bind antigen to their surface[36] (although it has been more difficult to demonstrate T cells binding antigen than B cells) and in general only a small proportion of lymphocytes (~1 in 10^4 to 10^5 in unimmunized animals) bind any one antigen. Furthermore, if lymphocytes are exposed to a highly radioactive antigen, both T and B cell responses to that antigen can be selectively abolished, while responses to other antigens are unaffected[37]. Similarly, B cells capable of responding to a particular antigen specifically adhere to glass beads coated with the antigen and can thus be specifically removed from a cell suspension[38]. Although T cells tend not to adhere under these conditions[38] for reasons that are unclear, T cells responsive to cell surface alloantigens can be selectively removed in cell monolayers bearing the specific alloantigens[39].

In 1900, Ehrlich proposed that cells producing antitoxins (now known to be B cells) had antitoxin molecules as receptors on their surface. The more recent version of the receptor hypothesis suggests that B lymphocytes have antibody molecules (that is, Ig) as receptors for antigen, which, at least in their combining sites, are identical to the antibody which the cell or its progeny will eventually secrete. There is now good evidence for this view, in that B cells have been shown to have Ig molecules on their surface (~10^4 to 10^5 a cell) (refs. 20, 40) and anti-Ig antibody inhibits their ability to bind or respond to antigens (reviewed in ref.1). There is also increasing evidence that the antigen-specificity of receptors and secreted antibody are the same for any one B lymphocyte clone[41,42]. The Ig class of the receptors and that of the ultimately secreted antibody may not, however, always be the same, for B cell precursors of some IgG secretory cells seem to have IgM receptors[15,16]. As different antibody classes (for example, IgG and IgM) seem to be able to share the same specificity (that is different Ig constant regions can be associated with identical Ig variable regions[43]) and IgM→IgG switch within a single clone need not imply a switch in specificity. In mice, at least, there is some evidence that most virgin B cells have IgM receptors (in its 7-8S monomeric form[44]) which may switch class after a primary exposure to antigen[16]. The more fundamental question of how antibody diversity is generated, that is how an animal develops the ability to synthesize such a large number of different Ig molecules (receptors and secreted antibodies) is still being debated. Germ-line theories, which suggest that one is born with a large number of variable region Ig genes, are competing with various somatic theories, which postulate that one is born with few variable region Ig genes and that some somatic process (for example, mutation or recombination) creates a large number.

The chemical nature of receptors on T cells is probably the most controversial issue in cellular immunology at present. The simplest and most logical view, that only antibody can recognize antigen and that all antigen-specific receptors must be Ig, has been challenged by the failure of many investigators to demonstrate Ig directly on the surface of T cells, or to inhibit various T cell responses with anti-Ig sera. Indeed, there is now growing support

表面已有相应受体的淋巴细胞；抗原与受体的相互作用可以激活这些特异性细胞。现在关于T淋巴细胞和B淋巴细胞的克隆选择学说都有了令人信服的证据。因此，人们发现T细胞和B细胞都可以结合抗原至其表面[36]（尽管曾经证明T细胞结合抗原比证明B细胞结合抗原要困难得多），并且通常只有一小部分的淋巴细胞（在未免疫的动物中约为万分之一到十万分之一）能够与任何抗原结合。此外，如果用强放射性抗原作用于淋巴细胞，无论是T细胞反应还是B细胞反应都会被选择性摧毁，而针对其他抗原的免疫反应却不会受到影响[37]。同样，如果把某种特定的抗原耦联到玻璃微珠上，那么可以与这种抗原反应的B细胞会特异性地黏附于玻璃微珠上，从而将它们从细胞悬液中特异地分离出来[38]。尽管出于某种未知原因，在此条件下T细胞通常不能黏附[38]，但用载有特异性同种抗原的单层细胞，可以选择性去除识别这种抗原的T细胞[39]。

早在1900年，埃尔利希便提出那些可以产生抗毒素的细胞（现在知道是B细胞）在其表面有作为受体的抗毒素分子。这种受体学说的最新说法提出，B淋巴细胞表面有作为抗原受体的抗体分子（即免疫球蛋白），这些受体与B细胞或其子细胞最终分泌的抗体至少在结合位点上是相同的。现在有很好的证据支持这一观点：B细胞表面确实有免疫球蛋白分子（每个细胞大约10^4~10^5个）（参考文献20、40），并且这些抗Ig的抗体抑制它们与抗原结合或应答的能力（有关综述见参考文献1）。此外，越来越多的证据表明，对于任何一个B淋巴细胞克隆来说，受体和分泌的抗体具有相同的抗原结合特异性[41,42]。但是，受体与最终分泌的抗体在免疫球蛋白亚型方面可能不总是一样的，比如一些分泌IgG细胞的B细胞前体似乎有IgM受体[15,16]。这些不同亚型的抗体（例如IgG与IgM）可能具有相同的抗原识别特异性（也就是说不同的免疫球蛋白恒定区可以与同一可变区组合[43]），而且单个克隆内IgM到IgG的转换并不意味着特异性的转变。至少在小鼠中有证据表明，大多数未经过抗原激活的B细胞表面都有IgM受体（以7-8S的单体形式存在[44]），而在与抗原初次接触后，这些B细胞则会转换类型[16]。关于抗体多样性是如何产生这一最根本的问题，也就是动物如何产生如此庞大的各种各样的免疫球蛋白分子（包括受体和分泌的抗体），目前仍存有争议。胚系学说认为动物生来就具有大量免疫球蛋白可变区基因；而体细胞突变理论认为动物生来只有很少的免疫球蛋白可变区基因，在随后的一些体细胞过程（例如突变或重组）中产生了大量可变区基因。

T细胞表面受体的化学本质大概是当前细胞免疫学中最富争议性的问题。最简单也最合理的观点是：只有抗体可以识别抗原，而且所有抗原特异性受体必然是免疫球蛋白。但是由于许多研究人员都无法证明T细胞表面有免疫球蛋白的存在，或使用抗免疫球蛋白血清能抑制T细胞与抗原的作用，因此这一观点受到了质疑。事

for the idea that surface components other than classical immunoglobulin may play an important role in T cell recognition of and/or response to at least some antigens. The principal candidates for such T cell "receptors" are the products of the immune response (Ir) genes that are genetically linked to the chief histocompatibility loci[45]. These Ir genes influence T cell responses to a variety of antigens[46]. The exquisite specificity of T cell responses, which resembles very closely the specificity of antibody and B cell recognition[47], taken together with the various (but still controversial) demonstrations of Ig on T cells (reviewed in ref. 1), makes one reluctant, however, to give up the idea that T cells have Ig receptors. It is possible that T cells (and possibly B cells) have at least two "recognition" systems, one involving Ig and another mediated by Ir gene products, the general importance of each varying depending on the antigen, the response and/or the subclass of T cell. The putative non-Ig recognition system could be analogous to the primitive recognition of foreignness seen in invertebrates.

Functions of T and B Cells

When an antigen combines with its corresponding receptors on a T or B lymphocyte, one of at least three things can happen to the lymphocyte: first, it may be stimulated to divide and differentiate to become an effector cell in some type of immune response (that is, it is induced to respond immunologically); second, it may become immunologically tolerant or paralysed, so that it will not be able to respond the next time antigen is given; it is not known if such cells are killed or simply inactivated in some way; third, it may be unaffected by the encounter. In addition, if the animal makes an immune response to the antigen, on subsequent exposure to the same antigen, it will usually give a faster, greater and sometimes qualitatively different response. This altered state of immune reactivity to a specific antigen is called immunological memory. It is likely that memory involves both clonal expansion (that is, division of virgin lymphocytes to give an increased number of cells able to respond on second exposure) and differentiation of virgin cells to memory cells[1], but it is unclear whether memory cells are simply retired effector cells, cells at an earlier stage of differentiation than effector cells, or are derived by differentiation along a separate memory pathway.

The "decision" of an individual lymphocyte on encounter with antigen—whether to "turn-on", "turn-off" or ignore—depends largely on the nature and concentration of the antigen, and upon complex interactions with other lymphocytes and with macrophages. Although most immunogens can stimulate both T and B cell responses, some, particularly those with repeating identical determinants and which are poorly catabolized—the so-called "thymus-independent antigens" (for example, pneumococcal polysaccharide, *E.coli* endotoxin, polyvinylpyrrolidone)—chiefly stimulate B cells (reviewed in ref.1), whereas others preferentially activate T cells[48]. In general, T cells respond to lower concentrations of antigen than do B cells, and although T cells may be paralysed at very low and very high concentrations of antigen (low and high zones of tolerance respectively) B cells seem to be paralysed only at high antigen concentrations[49]. The way in which the antigen-receptor interaction signals a lymphocyte is unknown, although it probably involves allosteric changes and/or redistribution (for example, aggregation into patches or

实上，越来越多的证据表明，至少在T细胞识别和（或）应答某些抗原中，除了传统的免疫球蛋白外，细胞表面组分也可能发挥重要作用。这些T细胞“受体”的优先候选分子很可能是免疫应答（Ir）基因编码的产物，Ir基因在遗传上与主要组织相容性基因座相关联[45]。这些Ir基因会影响T细胞对多种抗原的应答[46]。T细胞应答的精准特异性非常类似于抗体的特异性和B细胞识别的特异性[47]；再加上种种（尽管存有争议）实证表明T细胞上有免疫球蛋白（有关综述见参考文献1），使得人们不愿放弃T细胞表面具有免疫球蛋白受体的观点。也许T细胞（可能也包括B细胞）表面至少具有两套“识别”系统，其中一套包含免疫球蛋白，另一套通过Ir基因编码的产物介导，对于不同的抗原、不同的免疫应答和（或）不同的T细胞亚型来说，这两套识别系统可能具有不同的重要性。此外，这种假定的非免疫球蛋白识别系统可能与无脊椎动物对外源性物质的简单识别类似。

T细胞和B细胞的功能

当抗原与T淋巴细胞或B淋巴细胞表面相应的受体结合时，淋巴细胞至少会发生下列三个事件中的一种：第一种是在某些免疫应答中，淋巴细胞在抗原刺激下会分裂并分化为效应细胞（即被诱导产生免疫应答）；第二种是这些淋巴细胞可能对这种抗原产生耐受或麻痹，以至于下次遇见该抗原时不能发生免疫应答，目前尚不清楚这些淋巴细胞是被杀死了还是仅仅在某种程度上失活了；第三种是淋巴细胞对于抗原的刺激不产生任何反应。此外，如果动物对某种抗原做出过免疫应答，那么当其再次接触此抗原时，通常会产生更快速、更强烈甚至有时会是性质改变了的免疫应答。这种针对同一特定抗原的免疫反应改变了的状态被称为免疫记忆。这种记忆可能既包括克隆扩增（即未经免疫的淋巴细胞通过分裂，使得再次接触同一抗原时，能产生应答的效应细胞数增加），也包括未经免疫的淋巴细胞分化为记忆细胞[1]。但目前尚不清楚记忆细胞只是“退休”的效应细胞——比效应细胞处于更早的分化阶段，还是通过一个独立的记忆通路分化出来的细胞。

当淋巴细胞与抗原接触之后，这个淋巴细胞的“决定”是“激活”、“耐受”还是忽略主要取决于抗原的性质和浓度，还取决于它们与其他淋巴细胞和巨噬细胞的复杂的相互作用。尽管大多数的免疫原都可以激发T细胞和B细胞产生免疫应答，但是有些抗原，尤其是那些具有相同重复的抗原决定簇并且不容易在体内被代谢的抗原——所谓的“胸腺非依赖性抗原”（如肺炎球菌的荚膜多糖、大肠杆菌内毒素、聚乙烯吡咯烷酮）——首要激活B细胞（有关综述见参考文献1），其他抗原则优先激活T细胞[48]。通常来讲，相比B细胞而言，T细胞可以对更低浓度的抗原做出应答，而且尽管过高或过低浓度的抗原（分别为高、低耐受区间）都可以引发T细胞的耐受性，但是B细胞只会对高浓度的抗原产生耐受性[49]。虽然人们并不清楚抗原和受体的相互作用是如何向淋巴细胞传递信号的，但是这个过程可能包含了膜上结合的受体的构

localization over one pole–cap formation[50]) of the membrane-bound receptors.

The most important differences between T and B cells concern their different functions in immune responses. When B cells are activated by antigen they divide and differentiate into blast cells with abundant endoplasmic reticulum, and some go on to become plasma cells. These cells remain in the lymphoid tissues for the most part and secrete large amounts of antibody which circulates in the blood. Individual antibody-secreting cells can be detected by a variety of techniques, the most common being the plaque-forming cell assay, in which anti-erythrocyte antibody released from single B cells lyses erythrocytes in their immediate environment in the presence of complement. Antibodies, in conjunction with various accessory cells (macrophages, mast cells and basophils, for example) and particular serum enzymes (complement components, for example), are responsible for a variety of hypersensitivity reactions and protective immunity against many pathogenic organisms. In addition, antibody serves to regulate the function of both T and B cells, inhibiting their responses by competing with lymphocyte receptors for the antigenic determinants, diverting antigen from the lymphoid tissues or by forming tolerogenic antibody-antigen complexes[51], and enhancing responses by localizing antigen to appropriate lymphoid tissues or perhaps forming immunogenic antibody-antigen complexes. It is also possible (but not established) that B cells themselves play a direct part in transporting antigen (perhaps as antigen-antibody ± complement complexes adhering to Fc or complement receptors on B cells) and/or in killing target cells with coated antibody[52].

When T cells are activated by antigen, they proliferate and differentiate to become blast cells, but they do not develop significant amounts of endoplasmic reticulum and do not become antibody-secreting cells. They do, however, secrete a variety of non-antigen-specific factors ("lymphokines") such as migration inhibition factors (MIF), chemotactic factors, cytotoxic factors and mitogenic factors, at least some of which presumably play a role in cell-mediated immune responses, for which T cells are primarily responsible[53]. The precise chemical nature of these factors, the relationship between them, their significance and mechanisms of action are, however, incompletely understood. Cell-mediated immune responses include delayed hypersensitivity, contact sensitivity, rejection of foreign tissues, graft *versus* host responses (where injected foreign T lymphocytes respond against the antigens of the recipient, often resulting in recipient death) and immunity to various microbes. In all of these responses, T cells enlist the help of macrophages (probably through the secretion of lymphokines). The latter are usually the predominant cells at the site of these reactions[54]. T cells can also be demonstrated to respond to antigen *in vitro* by dividing, secreting lymphokines, killing target cells, or supporting viral replication (reviewed in ref.1). Whether T cells themselves can directly kill target cells, or do so only by activating other cells (such as macrophages) is still controversial, although there is increasing evidence that they can become "killer cells" under some circumstances[55].

Although T cells do not themselves secrete antibody in the usual sense, it is now known that they play an important role in helping B cells to make antibody responses to most immunogens. Thus, in these responses T cells are referred to as "helper" cells, and B

象变化和（或）其在细胞表面的重新分布（例如聚集成斑或定位在一极聚集成帽状[50]）。

T 细胞和 B 细胞最大的区别在于它们在免疫应答中的功能不同。当 B 细胞被抗原激活后，它会分裂并分化为含有丰富内质网的母细胞，且部分母细胞会继续分化为浆细胞。这些细胞大多存在于淋巴组织中，并分泌大量的抗体进入血液循环。有很多方法可以用来检测分泌抗体的细胞，其中最常见的是溶血空斑实验。该实验的原理是在补体存在的情况下，单个 B 细胞分泌的抗红细胞抗体可以使红细胞发生溶血，从而在每个 B 细胞周围形成一个空斑。抗体连同各种辅助细胞（如巨噬细胞、肥大细胞、嗜碱性粒细胞）以及特殊的血清酶（如补体成分）参与了一系列超敏反应以及抵抗多种病原体的保护性免疫。此外，抗体还可以调节 T 细胞和 B 细胞的功能，通过与淋巴细胞表面的受体竞争结合抗原决定簇、将抗原转移出淋巴组织或者通过形成耐受性抗原－抗体复合物[51]，抗体可以抑制淋巴细胞的免疫应答；而通过将抗原定位于适当的淋巴组织或通过形成免疫性的抗体－抗原复合物，抗体可以增强淋巴细胞的免疫应答。B 细胞自身也可能（但不确定）直接参与抗原转运（可能通过黏附在 Fc 或 B 细胞补体受体上的抗原−抗体 ± 补体复合物来实现）和（或）杀死被抗体覆盖的靶细胞[52]。

当 T 细胞被抗原激活后，它们会增殖、分化为母细胞，但不会产生发达的内质网也不会成为可以分泌抗体的细胞。然而，它们会分泌诸如迁移抑制因子（MIF）、趋化因子、细胞毒素、促有丝分裂因子等各种非抗原特异性因子（“淋巴因子”）。至少部分因子可能在 T 细胞起主要作用的细胞免疫应答中发挥作用[53]。但是人们尚不完全了解这些因子的化学本质、它们之间的相互关系、它们的重要性和作用机制。细胞免疫应答包括迟发性超敏反应、接触过敏、对外来组织的排斥、移植物抗宿主反应（移植的外来 T 淋巴细胞对受体抗原的反应，经常会导致受体死亡）以及机体对各种微生物的免疫力。在所有这些免疫应答中，T 细胞要发挥功能都需要巨噬细胞（可能是通过分泌淋巴因子的方式）的参与，并且巨噬细胞通常在反应位点扮演主要的角色[54]。在体外实验中，T 细胞也可以通过分裂、分泌淋巴因子、杀伤靶细胞或者维持病毒复制（有关综述见参考文献 1）等方式对抗原产生应答。尽管越来越多的证据表明，在某些环境下 T 细胞可以成为“杀手细胞”[55]，但是关于 T 细胞自身是否可以直接杀伤靶细胞还是必须通过激活其他细胞（如巨噬细胞）杀伤靶细胞目前尚有争论。

尽管 T 细胞本身一般不分泌抗体，但现在已经知道它们在帮助 B 细胞对免疫原作出应答方面发挥着重要的作用。因此，在这类免疫应答中，T 细胞被称为“辅助”

cells as "antibody-forming precursor" cells. The first direct evidence for such T-B cell cooperation was provided in 1966 by the observation that irradiated mice given both thymus cells and bone marrow cells made a far greater antibody response to sheep erythrocytes (SRBC) than recipients of either thymocytes or bone marrow cells alone[56]. Subsequently it was shown that all of the antibody-secreting cells (that is, those making anti-SRBC antibody) in this type of experiment came from the bone marrow inoculum[57]. Independent studies with chemically defined antigens showed that T-B cell cooperation in antibody responses involved T cells responding to one antigenic determinant on an immunogen and helping B cells to respond to different determinants on the same immunogen[58]. Although it is clear that cooperation is usually mediated by such an "antigen bridge" between T cell and B cell receptors, it is still uncertain whether the bridge is between T and B cells themselves, or between shed T cell receptors (perhaps taken up on the surface of macrophages) and B cells, and whether the bridge serves to "present" antigen to B cells in a particularly immunogenic form (concentrated and multivalent, for example) or to bring B cells close to T cells or a third party cell (such as macrophage) so that a nonspecific, short-range factor (for example, chemical mediator or membrane-membrane interaction) can operate between them (Fig. 2). Although it has been shown that T cells can secrete non-specific factors which can enhance B cell responses[59], their role in normal T-B cell cooperation is still uncertain. There is recent evidence that, in some *in vitro* responses at least, cooperation may involve the release by T cells of antigen-specific IgM-like factors (?receptors) complexed with antigen, which are subsequently taken up on macrophages[60].

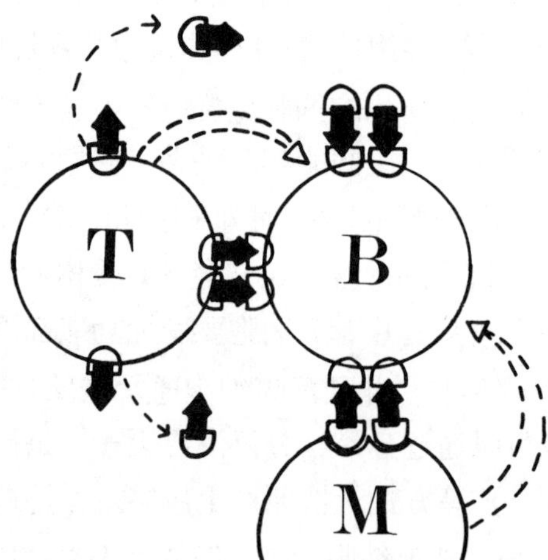

Fig. 2. Possible mechanisms of T-B cell collaboration in humoral antibody responses. The "antigen bridge" (→) between T and B cell receptors could serve to : (i) present antigen to B cells on the surface of T cells or as a matrix of released T cell receptors complexed with antigen, either free or on the surface of third party cells such as macrophages, or (ii)bring B cells together with T cells or a third party cell so that a short-range factor can operate between them.

细胞，而B细胞则被称为“抗体－形成前体”细胞。关于T-B细胞协作的首例直接证据是在1966年发现的。研究人员发现，向经放射照射的小鼠同时移植胸腺细胞和骨髓细胞，其产生抗绵羊红细胞（SRBC）抗体的能力比那些只移植胸腺细胞或骨髓细胞的小鼠产生抗体的能力强的多[56]。后来人们发现，在这类实验中，所有抗体分泌细胞（也就是产生抗SRBC抗体的细胞）都来自移植的骨髓细胞[57]。用化学成分确定的抗原进行的独立研究显示，在抗体免疫应答中，T-B细胞的协作包括T细胞对免疫原上的某一抗原决定簇进行识别，并且帮助B细胞对同一免疫原上不同抗原决定簇产生反应[58]。尽管目前已经清楚这种协作通常是通过T细胞和B细胞受体之间的“抗原桥”来介导的，但人们尚不清楚这个桥是介于T细胞和B细胞之间还是介于脱落的T细胞受体（可能被巨噬细胞摄取后呈递在其表面）与B细胞之间。也不清楚这个桥的作用是以特定的免疫原形式（如浓缩或多价）向B细胞“呈递”抗原，还是把B细胞与T细胞或第三方细胞（如巨噬细胞）拉到一起，以便于非特异性的、小范围内起作用的因素（如化学媒介或细胞膜－细胞膜相互作用）在它们之间发挥作用（图2）。尽管已经证明T细胞可以分泌一些能够加强B细胞免疫应答的非特异性因子[59]，但是人们并不清楚这些因子在正常的T-B细胞协作中发挥着什么样的作用。最近有证据表明，至少在体外免疫应答中，T-B细胞协作可能与T细胞分泌一种具有抗原特异性的IgM样因子（受体？）有关。这种因子可以与抗原结合形成复合物，然后会被巨噬细胞吞噬[60]。

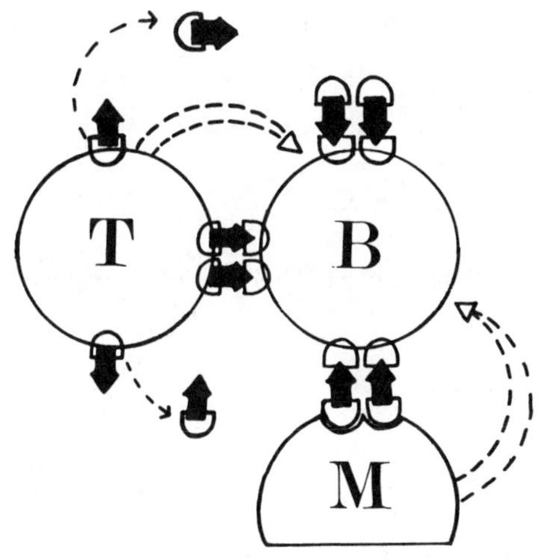

图 2. 体液免疫应答中T-B细胞协作的可能机理。T细胞和B细胞受体之间的“抗原桥”（→），可能发挥如下作用：(i) 在T细胞表面将抗原呈递给B细胞；或者作为释放T细胞受体-抗原复合物的基体，这种复合物可以是游离的，也可以是位于第三方细胞如巨噬细胞表面的。(ii) 把B细胞和T细胞或第三方细胞拉到一起，从而使小范围内起作用的因素在它们之间发挥功能。

There are antigens ("thymus-independent antigens") which seem to be able to stimulate at least some B cell clones to secrete IgM antibody without the help of T cells (reviewed in ref. 1), suggesting that T-B cell collaboration is not always essential for antibody production. Nonetheless, the discovery that T cells cooperate with B cells in humoral immunity has been an important advance and has explained the previous paradox of impaired antibody responses in T cell deficient animals. There is recent indirect evidence that T cells can inhibit B cell activity as well as enhance it[61], and that they can enhance[62] and inhibit[63] the functioning of other T cells. It is not known if these interactions involve antigen bridging between the receptors of the interacting cells. Taken together with the enhancing and inhibiting effects of secreted antibody on both T and B cell functions, a picture is emerging of a highly complex and finely controlled immune system, with each type of cell and response modulating the others.

Way Ahead

With the recognition that there are two distinct classes of lymphocytes with different origins, properties and immunological functions which modulate each other's activities, the door has opened to a new era of immunology. The resulting insight into the functioning of the immune system in health and disease has paved the way for rational attempts to manipulate selectively the different cell types and their various responses for the benefit of patients with infection, autoimmune disease, cancer, immune deficiency states, and organ grafts. And present day immunology provides a number of readily accessible models and powerful tools for studying a variety of biological problems, including differentiation, genetic control, cell interactions, and membrane receptor-ligand interactions.

I am grateful to M. F. Greaves, N. A. Mitchison and J. J. T. Owen for helpful discussion. The bibliography, which is incomplete, is meant only as an arbitrary way into the relevant literature.

(**242**, 19-23; 1973)

Martin C. Raff
Medical Research Council Neuroimmunology Project, Zoology Department, University College, Gower Street, London WC1

References:

1. Greaves, M. F., Owen, J. J. T., and Raff, M. C., *T and B Lymphocytes: Origins, Properties and Roles in Immune Responses* (Excerpta Medica, Amsterdam, in the press).
2. Gowans, J. L., and McGregor, D.D., *Prog. Allergy*, 9,1 (1965).
3. Warner, N. L., Szenberg, A., and Burnet, F. M., *Austral. J. Exp. Biol. Med.*, **40**, 373 (1962).
4. Cooper, M. D., Peterson, R. D. A., South, M. A., and Good, R. A., *J. Exp. Med.*, **123**, 75 (1966).
5. Miller, J. F. A. P., and Osoba, D., *Physiol. Rev.*, **47**, 437 (1967).
6. Good, R. A., Biggars, W. D., and Park, B. H., in *Progress in Immunology* (edit. By Amos, B.), 699 (Academic Press, New York, 1971).
7. Roitt, I. M., Greaves, M. F., Torrigiani, G., Brostoff, J., and Playfair, J. H. L., *Lancet*, ii, 367 (1969).
8. Hildeman, W. H., and Clem, L. W., in *Progress in Immunology* (edit. By Amos, B.), 1305 (Academic Press, New York, 1971).
9. Owen, J. J. T., in *Ontogeny of Acquired Immunity, a Ciba Foundation Symposium*, 35 (Associated Science Publishers, Amsterdam, 1972).

在没有T细胞辅助的情况下，有一些抗原（“胸腺非依赖性抗原”）可以刺激部分B细胞分泌IgM抗体（有关综述见参考文献1）。这表明T-B细胞的协作并不总是抗体产生所必需的。不过发现体液免疫中T细胞与B细胞的协作仍然是免疫学研究中的一个重大进展，它解释了之前提出的缺乏T细胞的动物为什么会出现抗体反应受损这一问题。最近还有间接证据表明T细胞不但可以抑制B细胞的应答，也可以增强B细胞的活性[61]。同时，它们还可以增强[62]或抑制[63]其他T细胞的功能。人们并不清楚这些相互作用是否包括相互作用的细胞之间受体的抗原桥连接。总之，考虑所有这些分泌的抗体对T细胞和B细胞功能的增强或抑制作用，我们可以想象免疫系统是高度复杂且精密调控的，在这个系统中每种细胞和免疫应答都可以调控另一种。

展　望

随着我们的认识，免疫系统中存在着两种不同来源、不同性质、不同免疫功能的淋巴细胞，它们相互调控着彼此的活性，免疫学研究已进入一个新时代。在健康和疾病方面，对免疫系统功能的深入了解使我们有可能通过合理的实验，选择性地控制不同免疫细胞类型及其多样应答，以改善被传染病、自身免疫疾病、癌症、免疫缺陷疾病或器官移植所困扰的患者的生活。另外，现代免疫学也提供了许多可用于研究多种生物学问题（包括分化、遗传控制、细胞间相互作用以及细胞膜受体-配体间相互作用）的有效模型和强大工具。

非常感谢格里夫斯、米奇森和欧文宝贵的建议。此外，参考文献引用可能不全面，仅作为查阅相关文献的一个途径。

（张锦彬 翻译；秦志海 审稿）

10. Weissman, I., *J. Exp. Med.*, **126**, 291 (1967).

11. Davies, A. J. S., *Transplant. Rev.*, **1**, 43 (1969).

12. Blomgren, H., and Andersson, B., *Exp. Cell Res.*, **57**, 185 (1969).

13. Raff, M. C., and Cantor, H., in *Progress in Immunology* (edit. by Amos, B.), 83 (Academic Press, New York, 1971).

14. Stutman, O., Yunis, E. J., and Good, R. A., *J. Exp. Med.*, **130**, 809(1969).

15. Cooper, M. D., Lawton, A. R., and Kincade, P. W., in *Current Problems in Immunobiology* (edit. by Hanna, M. G.), **1**, 33 (Plenum Press, New York, 1972).

16. Pierce, C. W., Solliday, S. M., and Asofsky, R., *J. Exp. Med.*, **135**, 698 (1972).

17. Everett, N. B., and Caffrey, R. W., in *The Lymphocyte in Immunology and Haemopoiesis* (edit. by Yoffey, J. M.), 108 (Arnold, London, 1966).

18. Metcalf, D., and Moore, M. A. S., *Haemopoietic Cells* (North Holland, Amsterdam, 1971).

19. Raff, M. C., *Transplant. Rev.*, **6**, 52 (1971).

20. Raff, M. C., *Immunology*, **19**, 637 (1970).

21. Rabellino, E., Colon, S., Grey, H. M., and Unanue, E. R., *J. Exp. Med.*, **133**, 156 (1971).

22. Schlossman, S. F., and Hudson, L., *J. Immunol.*, **110**, 313 (1973).

23. Hulett, H. R., Bonner, W. A., Barret, J., and Herzenberg, L. A., *Science*, **166**, 747 (1969).

24. Bianco, C., Patrick, R., and Nussenzweig, V., *J. Exp. Med.*, **132**, 702 (1970).

25. Basten, A., Miller, J. F. A. P., Sprent, J., and Pye, J., *J. Exp. Med.*, **135**, 610 (1972).

26. Howard, J. C., *J. Exp. Med.*, **135**, 185 (1972).

27. Parrott, D. M. V., and de Sousa, M. A. B., *Clin. Exp. Immunol.*, **8**, 663 (1971).

28. Sprent, J., and Miller, J. F. A. P., *Eur. J. Immunol.*, **2**, 384 (1972).

29. Howard, J. C., Hunt, S. V., and Gowans, J. L., *J. Exp. Med.*, **135**, 200 (1972).

30. Wioland, M., Sabulovic, D., and Burg, C., *Nature New Biology*, **237**, 275 (1972).

31. Lance, E. M., *Clin. Exp. Med.*, **6**, 789 (1970).

32. Turk, J. L., and Poulter, L. W., *Clin. Exp. Med.*, **10**, 285 (1972).

33. Cohen, J. J., and Claman, H. N., *J. Exp. Med.*, **133**, 1026 (1971).

34. Cunningham, A. J., and Sercarz, E. E., *Eur, J. Immunol.*, **1**, 413 (1972).

35. Greaves, M. F., and Janossy, G., *Transplant. Rev.*, **11** (1972).

36. Roelants, G., *Nature New Biology*, **236**, 252 (1972).

37. Basten, A., Miller, J. F. A. P., Warner, N. L., and Pye, J., *Nature New Biology*, **231**, 104 (1971).

38. Wigzell, H., *Transplant. Rev.*, **5**, 76 (1970).

39. Brondz, B. D., *Transplant. Rev.*, **10**, 112 (1972).

40. Rabellino. E., Colon, S., Grey, H. M., and Unanue, E. R., *J. Exp. Med.*, **133**, 156 (1971).

41. Mäkelä. O., and Cross, A., *Prog. Allergy*, **14**, 154 (1970).

42. Cozenza, H., and Köhler, H., *Proc. US Nat. Acad. Sci.*, **69**, 2701 (1972).

43. Pink, R., Wang, A. -C., and Fudenberg, H. H., *Ann. Rev. Med.*, **22**, 145 (1971).

44. Vitetta, E. S., Baur, S., and S., Uhr, J. W., *J. Exp. Med.*, **134**, 242 (1971).

45. Shevach, E. M., Paul, W. E., and Green, I., *J. Exp. Med.*, **136**, 1207 (1972).

46. Benacerraf, B., and McDevitt, H. O., *Science*, **175**, 273 (1972).

47. Schlossman, S. F., *Transplant Rev.*, **10**, 97 (1972).

48. Alkan, S. S., Williams, F. B., Nitecki, D. E., and Goodman, J. W., *J. Exp. Med.*, **135**, 1228 (1972).

49. Mitchison, N. A., in *Cell Interactions and Receptor Antibodies in Immune Responses* (edit. by Mäkelä, O., Cross, A., and Kosunen, T.), 249 (Academic Press, New York, 1971).

50. Taylor, R. B., Duffus, W. P. H., Raff, M. C., and de Petris, S., *Nature New Biology*, **233**, 225 (1971).

51. Schwartz, R. S., in *Progress in Immunology* (edit. by Amos, B.), 1081 (Academic Press, New York, 1971).

52. MacLennan, I. C. M., *Transplant. Rev.*, **13**, 67 (1972).

53. *Mediators, of Cellular Immunity* (edit. by Lawrence, H. S., and Landy, M.) (Academic Press, New York, 1969).

54. Lubaroff, D. M., and Waksman, B. H., *J. Exp. Med.*, **128**, 1437 (1968).

55. Brunner, K. T., and Cerottini, J. -C., in *Progress in Immunology* (edit. by Amos, B.), 385 (Academic Press, New York, 1971).

56. Claman, H. N., Chaperon, E. A., and Triplett, R. F., *Proc. Soc. Exp. Biol.*, **122**, 1167 (1966).

57. Davies, A. J. S., Leuchars, E., Wallis, V., Marchant, R., and Elliot, E. V., *Transplantation*, **5**, 22 (1967).

58. Mitchison, N. A., Rajewsky, K., and Taylor, R. B., in *Developmental Aspects of Antibody Formation and Structure* (edit. by Sterzl, J., and Riha, I.), 2, 547 (Academia, Prague, 1970).

59. Schimpl, A., and Wecker, E., *Nature New Biology*, **237**, 15 (1972).

60. Feldmann, M., *J. Exp. Med.*, **136**, 737 (1972).

61. Jacobson, E. D., Herzenberg, L. A., Riblet, R., and Hersenberg, L. A., *J. Exp. Med.*, **135**, 1163 (1972).

62. Cantor, H., and Asofsky, R., *J. Exp. Med.*, **135**, 764 (1972).

63. Gershon, R. K., Cohen, P., Hencin, R., and Liebhaber, S. A., *J. Immunol.*, **108**, 586 (1972).

On Estimating Functional Gene Number in Eukaryotes

S. J. O'Brien

Editor's Note

Around 30 years before the Human Genome Project mapped the tens of thousands of protein-coding genes of the human genome, debate over the eukaryotic gene complement was rife. The total number of genes was thought to be far less than the amount of DNA in the haploid genome, leading some to suggest that over 90% of the eukaryotic genome was nonfunctional or "junk". Here geneticist Stephen J. O'Brien questions this assumption, arguing that the evidence for junk DNA is based on the response of the functioning genes to natural selection. Non-coding DNA is now thought to comprise most of the human genome, but the term "junk" is used with caution since functions have been ascribed to some so-called "junk" sequences.

MANY recent studies have been concerned with the construction of biological model systems to describe adequately regulation of gene action during development of eukaryotes[1-5]. The number of genes in mammals and *Drosophila* has been suggested to be 1 to 2 orders of magnitude less than the amount of available DNA per haploid genome could provide[2-7]. Although *Drosophila* and mammalian nuclei contain enough unique DNA to specify for respectively 10^5 and 10^6 genes of 1,000 nucleotide pairs[8,9], it has been argued that a much lower estimate of functional gene number is more reasonable[2-7]. Conversely, these conclusions indicate that more than 90% of the eukaryotic genome may be composed of nonfunctional or noninformational "junk" DNA. Here we demonstrate these estimations have not been fundamentally proven; rather they are based on simplifying assumptions of questionable validity, in some cases contradictory to experimental data.

The perceptive model proposed by Crick[5] provides that the structural genes for proteins are situated generally in the interbands observed in the giant salivary gland chromosomes of *Drosophila*. The chromosome bands, which contain all but a few % of the DNA, are the sites of regulatory elements and presumably large amounts of noninformational DNA. The model thus predicts approximately 5,000 structural genes in *Drosophila*, the approximate number of salivary gland bands which can be observed.

This model is strongly supported by the elegant work of Judd *et al.*[10] who examined 121 lethal and gross morphological point mutations that map in the *zeste* to *white* region of the tip of the X chromosomes in *D. melanogaster*. There are 16 salivary gland chromosome bands or chromomeres in this region corresponding to 16 complementation groups of the morphological or lethal point mutations. In addition a series of overlapping deficiencies

真核生物功能基因数的估计

奥布赖恩

编者按

在人类基因组计划定位了数万个人类基因组中的蛋白质编码基因之前，30 年间关于真核生物的基因数目存在很多争议。基因的总数被认为远少于单倍体基因组中 DNA 的数量，这导致一些人提出 90% 以上的真核生物基因组是没有功能的或者是“垃圾”。在本文中，遗传学家斯蒂芬·奥布赖恩质疑了这个假说，指出垃圾 DNA 实为功能基因对自然选择的反应的证据。现在认为非编码 DNA 占据了人类基因组的大部分，但是“垃圾”一词要谨慎使用，因为已经发现了一些所谓的“垃圾”序列的功能。

近来有许多研究都着眼于构建生物模型系统来充分描述真核生物发育过程中基因行为的调节 [1-5]。就一个单倍体基因组所能够容纳的 DNA 量来说，已发现哺乳动物和果蝇中的基因数目要比其少 1 到 2 个数量级 [2-7]。以 1,000 个核苷酸对构成一个基因来计算，尽管果蝇和哺乳动物细胞核含有足够的非重复 DNA 分别形成 10^5 和 10^6 个基因 [8,9]，但是看来对功能基因的数量更低的估计是更合理的 [2-7]。反过来说，这些结论则提示真核生物基因组超过 90% 的 DNA 可能由非功能性或者不编码信息的“垃圾”DNA 组成。本文中我们的结果显示这些估计都没有得到有力地证明，相反它们都是基于一些简化的假设所获得，而这些假设本身的正确性值得怀疑，其中有些甚至与实验数据相矛盾。

克里克 [5] 提出的模型指出，在果蝇巨大唾液腺染色体中，编码蛋白质的结构基因通常都位于观察到的染色体条带之间。调控元件和大量含有无编码信息的 DNA 序列位于染色体条带上，这些染色体条带包含了绝大多数 DNA。据此该模型预测果蝇大约有 5,000 个结构基因，这也是能够观察到的唾液腺染色体带的大致数量。

该模型得到了贾德等人的出色工作的有力支持 [10]。他们研究了黑腹果蝇位于 X 染色体末端 *zeste* 到 *white* 区域内的 121 个致死性和显著影响形态的点突变。该区域内一共有 16 个唾液腺染色体带或染色粒，与 16 个形态学改变或致死性点突变的互补群互为对应。此外一系列重叠的缺陷支持一个条带一个互补群的关系。外推至整

supports the 1 band : 1 complementation group relationship. Extrapolation over the entire genome gives approximately 5,000 complementation groups or genes to 5,000 chromosome bands. There are also estimates available on the total number of lethal loci in the *Drosophila* genome. By screening for large numbers of lethal chromosomes either in natural populations or following irradiation, it is possible to relate the frequency of allelism to the number of lethal loci by a simple Poisson distribution: and the number of lethals thus measured in *Drosophila* gives a result between 1,000–2,000[11,12].

The problem with extrapolation of the fine structure analysis and the lethal data to the functional gene number is our inability to answer the question: how many genes when mutated are capable of producing a lethal or gross morphological phenotype? The answer is not known specifically but the available data suggest that only a very small percentage of all gene products are critical enough to kill the organism if absent. In *Drosophila* over 30 genes have a known gene product[13], of which there are 14 at which "null" alleles eliminate the protein, its activity, or the RNA product entirely, and of these (Table 1) only the *bobbed* locus has lethal alleles[14]. Most alleles, however, at that locus, which is the structural gene for ribosomal RNA, are viable even at very low levels of rRNA. The other genes, which code for enzymes whose function *a priori* seemed essential for normal metabolism, are in no case lethal when homozygous for completely "null" alleles.

Table 1. Genes in *Drosophila melanogaster* with Known Gene Products and Recovered "Null" Alleles

Locus	Product	Number of "null" alleles	Reference
Est-C	Esterase-C	1	37
Est-6	Esterase-6	1	37
Aph	Alkaline phosphatase	1	38
Acph-1	Acid phosphatase	15	39
rosy	Xanthine dehydrogenase	79	40
Aldox	Aldehyde oxidase	2	41
Zw	Glucose-6-phosphate dehydrogenase	5	*
6-Pgd	6-Phosphogluconate dehydrogenase	1	*
Adh	Alcohol dehydrogenase	14	42,43
Idh	Isocitrate dehydrogenase	2	44
αGpdh-1	α-Glycerophosphate dehydrogenase	4	15
bobbed	Ribosomal RNA	25	45
vermilion	Tryptophan pyrrolase	10	17,46
cinnebar	Kynurenine hydroxylase	3	47

*W. J. Young, personal communication.

个基因组，大约5,000个这种互补群或者基因对应于5,000个染色体带。也有人估计了果蝇基因组中致死基因座的总数。通过筛查自然群体中或者经过辐射处理后的群体中的大量可致死染色体，就有可能通过简单的泊松分布将等位性的频率与致死基因座的数目联系起来，这样得到的果蝇中致死基因座的数目大约是1,000~2,000个[11,12]。

在这种利用精细结构分析和致死性数据外推到功能基因数目过程中存在一个我们无法回答的问题：有多少基因突变后能够产生致死性的或者显著影响形态的表型？这无法明确地回答，但是已有数据提示所有基因产物中只有很少一部分重要到其缺失能够导致生物的死亡。在果蝇中，超过30个基因的基因产物是已知的[13]，其中只有14个在具有“无效”等位基因时能够导致蛋白质及其活性或RNA产物的完全缺失，而这其中（表1）只有*bobbed*基因座具有致死性的等位基因[14]。然而，这个基因座是核糖体RNA（rRNA）的结构基因，其大多数等位基因即使在所产生的rRNA浓度非常低时也能有活力。其他编码酶的基因即使它们对于正常代谢是必需的，在纯合的完全“无效的”等位基因中也没有一个是致死性的。

表1. 具有确定基因产物和可恢复性“无效”等位基因的黑腹果蝇基因

位点	产物	无效等位基因数目	文献
Est-C	酯酶–C	1	37
Est-6	酯酶–6	1	37
Aph	碱性磷酸酶	1	38
Acph-1	酸性磷酸酶	15	39
rosy	黄嘌呤脱氢酶	79	40
Aldox	醛氧化酶	2	41
Zw	葡萄糖–6–磷酸脱氢酶	5	*
6-Pgd	6–磷酸葡萄糖酸脱氢酶	1	*
Adh	乙醇脱氢酶	14	42,43
Idh	异柠檬酸脱氢酶	2	44
αGdph-1	α–甘油磷酸脱氢酶	4	15
bobbed	核糖体 RNA	25	45
vermilion	色氨酸吡咯酶	10	17,46
cinnebar	犬尿氨酸羟化酶	3	47

* 扬，个人交流。

Null alleles at the first eleven loci above were detected by the loss of histochemical stain development on an electrophoretic gel. The sensitivity of this assay detects at least 5% of normal enzyme levels. In several cases (*Acph-1, rosy, Adh, αGpdh-1*), analytical enzyme assays with a sensitivity near 0.1% of wild type enzyme levels also failed to detect trace activity in "null" homozygotes. In the two cases where cross reacting material (CRM) was measured (*Acph-1* and *ry*) it was also negligible.

Null alleles of at least two of the loci were induced in a crossing scheme that would have recovered lethal alleles (*αGpdh-1* and *Acph-1*). A lethal "null" allele would also be detected as an exceptional heterozygote with normal alleles of different eletrophoretic mobilities in those cases of "null" alleles discovered in natural or laboratory populations (*Est-C, Est-6, Aph, Aldox,* and *Idh*).

Five of the fourteen loci have alleles which produce visible recessive phenotypes; *ry*, *cn* and *v* affect eye colour, *bb* affects bristles, and *αGpdh-1* "null" mutations which, although they appear morphologically normal, lack ability to sustain flight. The fraction 5 of 14 should not, however, be taken as an estimation of the fraction of loci at which "null" alleles produce an observable phenotype. This number is probably an overestimate because 4 of the 5 loci in question (all except *αGpdh-1*) were discovered initially as morphological mutations and their gene product was deduced and identified from their visible phenotype.

The eye colour mutations affect enzymes involved in the biosynthesis of eye pigments, and the *bobbed* locus, which shows a syndrome of effects usually associated with protein synthesis, was identified as the gene for rRNA. The phenotype of the *αGpdh-1* "null" mutations might easily have been missed had not the importance of the enzyme in insect flight been known previously[15]. The 11 other loci were identified only as the genes for selected enzymes, and of these none exhibited lethality or any morphological phenotype when "null" alleles were found.

In two cases double "null" mutants of alkaline and acid phosphatase (R. S. MacIntyre, personal communication) and of *Zw* and *6-Pgd* (W. J. Young, personal communication) were constructed and proved viable, fertile, and morphologically normal. Also, in two of the five cases where there is an observable phenotype, *bb* and *αGpdh-1*, there occurs a modification of the phenotype in the afflicted stocks. In the case of *bb* the diminished rDNA cistrons become "magnified" to approach the wild type rRNA levels within a few generations[16]. Flies genetically deficient for α-glycerophosphate dehydrogenase lack the ability to sustain flight due to their disrupted α-glycerophosphate cycle[15], but after 25 generations this phenotype becomes modified and flies recover the ability to fly normally (S. O'Brien, unpublished data). Biological adaptive capacity for physiological compensation for lesions in the structural genes of important functions must be very extensive to protect the fly so efficiently from genetically sensitive loci even in the presumably critical functions.

One might argue that even the smallest cytologically observable mutations in most cases are recessive lethals[10,14,17]. Resolution of such cytology, however, demands that at least 1

上表中前 11 个基因座中的无效等位基因是通过凝胶电泳组织化学染色条带丢失检测出来的。这种方法的敏感性至少能够检测出正常酶水平 5% 的量。在一些例子中（*Acph-1*、*rosy*、*Adh*、α*Gdph-1*），灵敏度接近 0.1% 野生型酶水平的酶分析法也不能在“无效”纯合子中检测到痕量的酶活性。在通过交叉反应物质（CRM）检测的两个例子中（*Acph-1* 和 *ry*），测到的酶活性也是微乎其微。

在杂交实验中，无效等位基因中的至少两个基因座被诱导而恢复了致死性（α*Gdph-1* 和 *Acph-1*）。在自然或实验室群体中发现的“无效”等位基因中（*Est-C*、*Est-6*、*Aph*、*Aldox* 和 *Idh*），也能检测到一种致死性“无效”等位基因，这种等位基因与有着不同电泳迁移率的正常等位基因形成一个异常杂合子。

14 个基因座中有 5 个具有能够产生可见的隐性表型的等位基因，其中 *ry*，*cn* 和 *v* 影响眼睛的颜色，*bb* 影响刚毛，而 α*Gdph-1*“无效”突变使果蝇尽管在形态学上表现正常，但丧失了持续飞行的能力。但是 5/14 这个比例并不能作为产生可见表型的“无效”等位基因在基因座中所占比例的估算值。因为所研究的 5 个基因座中的 4 个（除了 α*Gdph-1*）最初就是作为形态学的突变而被发现的，其基因产物已经从它们的可见表型中推断和鉴定出来，所以这个数值很可能是被高估的。

眼睛颜色相关基因的突变会影响眼色素生物合成过程中的相关酶，而通常显示产生一系列蛋白质合成相关症状的 *bobbed* 基因座已被确定为 rRNA 基因。如果不是以前就清楚 α*Gdph-1* 产生的酶在昆虫飞行中的重要性 [15]，就很容易忽视 α*Gdph-1* 的“无效”突变的表现型。另外 11 个基因座都仅仅是被鉴定为特定酶的基因，并没有发现其“无效”等位基因具有致死性或者导致任何形态表型的改变。

有两个构建了双重“无效”突变体的例子，碱性和酸性磷酸酶突变体（麦金太尔，个人交流）以及 *Zw* 和 *6-Pgd* 突变体（扬，个人交流），它们均被证实能够生存、繁殖并且形态上正常。同时，在 5 个具有可见表型的例子中，*bb* 和 α*Gdph-1* 这两个突变的受累动物出现了表型修饰。在 *bb* 中，减少的 rDNA 顺反子被“放大”，以至在数代内达到野生型 rRNA 的水平 [16]。遗传上缺乏 α–甘油磷酸脱氢酶（α*Gdph-1*）的果蝇由于 α–甘油磷酸循环被破坏而丧失了持续飞行的能力 [15]，但是经过 25 代以后，这种表型发生了变化而且果蝇恢复了正常飞行的能力（奥布赖恩，未发表数据）。对具有重要功能的结构基因损伤进行生理补偿的生物学适应能力一定非常广泛，使得果蝇能相当有效地免受这些遗传敏感的基因座的影响，即使是那些功能被假定很关键的基因座。

有人可能说大多数情况下即便是最小的细胞学可见突变都是隐性致死的 [10,14,17]。

of the 5,000 chromomeres of *Drosophila* polytene chromosomes must be absent to detect a deletion. The precision of the technique then is at the level of 10^6 nucleotides, the average amount per chromomere, enough DNA for 20 genes of average length. I suggest that there could be up to 20 functional genes in each region of which only one might be lethal in its mutant configuration.

A second widely used argument which suggests a minimum of informational DNA in the eukaryote genome (less than 10% of the available DNA) states that the mutational genetic load would be inordinate if mammals used all their DNA to carry and transmit biological information. Ohno[4] states that with a mutation rate of 10^{-5} in mammals containing enough DNA for 3×10^6 genes, if all this DNA were informative, each gamete would contain 30 new mutations, which would produce a genetic load sufficient to have exterminated mammals years ago. Evaluation of these mutational and substitutional load restrictions on functional gene number depends upon the unresolved question of the selective neutrality of gene substitutions, and will be treated from both perspectives.

If one accepts that the majority of gene substitutions and polymorphisms are selectively neutral, then the restrictions imposed by a genetic load on functional gene number become negligible. Neutral gene substitutions certainly cannot contribute to any accumulating substitutional or mutational load which depends upon selective disadvantage for its action. We must therefore estimate whether the number of functional genes are minimal, or rather that most gene substitutions are inconsequential with respect to natural selection. Proponents of selective neutrality feel that most substitutions are neutral, which removes any restrictions on large numbers of functional gene loci.

There have been serious objections raised concerning the role of selective neutrality[18,19]. One of the weakest tenets of this hypothesis is that it is based very heavily on the multiplicative aspect of fitness, which assumes that selection acts independently and in an additive fashion over all loci in a population. That this is not the case has been argued cogently by several authors[20-22]. The main point is that selection acts on the whole organism, not on the genotype at each polymorphic locus in each organism in a population[22]. If multiplicative fitness is an unrealistic assumption, then besides questioning selective neutrality as a major force, it also removes the restrictions imposed by the mutational and substitutional load on the number of functional genes.

There are a number of ways, suggested by myself and others, that a population can escape the rigours of multiplicative fitness, or more specifically, immediate selective consequence. These include diploidy[23], epistasis[20,21], synonymous base substitutions[7], frequency dependent selection[24], linkage disequilibrium[25], and alternative metabolic pathways (Table 1). All these factors, because they can effectively shield new mutations from the rigours of natural selection, even though the mutations may be deleterious in another genetic environment, counter the assumption of multiplicative fitness. If this assumption is removed, so also is the necessity of restrictive genome size in *Drosophila* and mammals.

但是要确定这种细胞学上的改变，所需要的分辨率是至少能检测到果蝇多线染色体中 5,000 个染色粒中的一个发生了缺失。这个技术精确度是在 10^6 个核苷酸水平，差不多是一个染色粒的平均大小，即足以构成 20 个平均长度的基因的 DNA。我认为每个区域可有多达 20 个功能基因，在其突变谱中可能只有一个是致死性突变。

另一个被广泛采用的论点提出了真核生物基因组中信息 DNA 的最小量（小于可获得的 DNA 的 10%），并且认为如果哺乳动物用所有的 DNA 来携带和传递生物学信息的话，那么突变了的遗传负荷就会过度。大野 [4] 认为对含有足够组成 3×10^6 个基因的 DNA 的哺乳动物来说，如果突变率是 10^{-5}，而且所有的 DNA 都是携带信息的，那么每个配子会含有 30 个新的突变，这样的遗传负荷会使得哺乳动物在很多年前就灭绝了。评估这些突变和替换的负荷所限制的功能基因数目取决于基因替换的中性选择这个尚未解决的问题，而且需要从两方面进行考虑。

如果接受大部分基因替换和多态性都是中性选择的，那么遗传负荷对于功能基因数目的限制就变得微不足道了。中性的基因替换当然不会有助于替换负荷或突变负荷的累积，因为这些负荷的作用是由其选择劣势决定的。因此我们必须估计是否有个最少的功能基因数目，或者更确切地说，大部分基因替换不是自然选择的结果。中性选择的支持者们觉得大部分替换都是中性的，这就消除了突变对于大量功能基因座的任何限制。

对于中性选择的作用曾有很多严肃的反对意见 [18,19]。该假设最薄弱的一条是它很大程度上基于对适应的倍增性，即假设选择的作用是独立的并且以加和的方式在群体中所有的基因座发挥作用。数名作者已经中肯地指出事实并非如此 [20-22]。要点在于选择是作用于整个生物体，而不是群体中每个个体的每个多态位点的基因型 [22]。如果倍增性适应是一种不现实的假设，那么除了对于中性选择作为主要作用力的质疑，突变和替换负荷对于功能基因数目的限制也可不再考虑。

包括我本人在内，很多人认为一个种群有多种方法摆脱所谓倍增性适应，或者更确切地说是直接选择的结果。这些包括二倍性 [23]，上位显性 [20,21]，同义碱基替换 [7]，频率依赖的选择性 [24]，连锁不平衡 [25] 和替代代谢通路等（表 1）。由于这些因素可以有效地保护新的突变免于遭受严酷的自然选择，尽管突变在其他遗传环境中可能是有害的，这些因素也可以与倍增性适应的假设相抗衡。如果不考虑这个假设，那么果蝇和哺乳动物中也不必考虑对于基因组大小的限制。

Long sequences (150–300 nucleotides) of polyadenylic acid are generally attached to messenger RNA in eukaryote cells[26-28]. Although post-transcriptional addition of poly A to messenger RNA has been postulated[29-31], the presence of poly T of comparable length in the nuclear DNA suggests transcriptional addition also[32]. RNA–DNA hybridization kinetics show that up to 0.55% of mammalian nuclear DNA anneals with poly A, corresponding to 1.1% poly dA–dT sequences[32]. This suggests a minimum of 5×10^4 poly dA–dT sites. If each of these sequences is transcribed with an adjacent structural gene, the number of functional genes must be greater than 5×10^4 by the addition of post-transcriptionally added poly A messages, plus non-messenger RNA genes, plus all non-transcribed regulatory genes. This number may be considerable.

In the cellular slime mould, *Dictyostelium discoideum*, 28% of the nonrepetitive nuclear genome is represented in the cellular RNA during the 26 h developmental cycle[33]. If only one of the complementary strands of DNA of any gene is transcribed, the estimate represents 56% of the single copy DNA. Because the nonrepetitive genome size of *Dictyostelium* contains approximately 3×10^7 nucleotide pairs[34], there are at least 16,000 to 17,000 RNA transcripts of average gene size (1,000 nucleotides) present over the cell cycle. Similarly, 10% of the mouse single copy sequences are represented in the cellular RNA of brain tissue. This hybridization result implies that a minimum of 300,000 different sequences of 1,000 nucleotides each are present in the mouse brain alone[35]. Results of RNA–DNA annealing experiments with *Drosophila* larval RNA indicate that between 15–20% of the unique nuclear genome is represented in larval RNA (R. Logan, personal communication), which corresponds to 30,000–40,000 RNA gene transcripts of average length. As the mouse and *Drosophila* data include only certain tissues and developmental times respectively, they probably are underestimates of the total unique DNA transcribed by 10–30%, based upon the degree of differences in RNA sequences exhibited at various developmental stages in *Dictyostelium*.

Interpretation of DNA–RNA hybridization experiments as an estimation of functional genes could be argued to be invalid because a large proportion of cellular RNA is the rapidly degraded "heterogeneous nuclear RNA" which never leaves the nucleus for translation[26,28,36]. RNA does not have to be translated to have a function, indeed RNA has a number of functions other than translation. Three points support gene function of such RNA when considered together: first, the actual presence of the gene; second, the transcription of information, and third, the transcription of different non-repetitive sequences at different developmental times and in different tissues[33,35].

The major arguments supporting the contention that much of eukaryotic DNA is neither transcribed nor functional are based essentially on the response of the functioning genes to natural selection. The tremendous amounts of physiological and/or genetic compensatory mechanisms which defer the presumed deleterious effects of mutations make such arguments subject to re-evaluation. Furthermore, the molecular data with the poly A sites and RNA transcript estimates suggest greater amounts of gene action than have been presumed.

真核细胞中的信使 RNA 一般连有很长的（150~300 个核苷酸）多聚腺苷酸序列[26-28]。尽管人们假设多聚腺苷酸添加到信使 RNA 末端发生在转录后[29-31]，细胞核 DNA 中存在类似长度的多聚胸苷酸提示转录时添加也是可能的[32]。RNA–DNA 杂交动力学显示高达 0.55% 的哺乳动物核 DNA 退火后结合多聚腺苷酸，对应于 1.1% 的多聚脱氧腺苷酸 – 脱氧胸苷酸序列[32]。这提示至少有 5×10^4 个多聚脱氧腺苷酸 – 脱氧胸苷酸位点。如果这些序列的每一个都与邻近的结构基因一起转录，再加上转录后加入的多聚腺苷酸序列、非信使 RNA 基因以及所有的非转录性调节基因，那么功能基因数肯定会超过 5×10^4。这个数目可能是相当可观的。

在细胞型黏菌盘基网柄菌阿米巴虫中，28% 的非重复性核基因组在 26 小时的发育周期中表达为细胞 RNA[33]。如果任何基因中只有互补 DNA 链中的一条被转录，那么此估计值代表 56% 的单拷贝 DNA。因为黏菌的非重复性基因组含有将近 3×10^7 个核苷酸对[34]，所以在细胞周期中存在至少 16,000 到 17,000 个平均基因大小（1,000 个核苷酸）的 RNA 转录产物。与之类似，小鼠脑组织中 10% 的单拷贝序列表达为细胞 RNA。这个杂交结果提示单独在小鼠脑组织内部可能存在至少 300,000 个平均拥有 1,000 个核苷酸的不同序列[35]。用果蝇幼虫 RNA 进行的 RNA–DNA 退火实验显示 15%~20% 的特异性核基因组都表达为幼虫 RNA（洛根，个人交流），相当于 30,000~40,000 个平均长度的 RNA 基因转录产物。由于小鼠和果蝇的数据仅仅分别包括了特定的组织和发育阶段，所以根据黏菌不同发育阶段 RNA 序列差异的程度，很可能将转录的特异 DNA 总量低估了 10%~30%。

将DNA–RNA 杂交实验作为对功能基因数量的估计的解释可能会被认为不可靠，因为细胞 RNA 的大部分都是很快降解的“核不均一 RNA”，它们从不离开细胞核去进行蛋白质翻译[26,28,36]。但是 RNA 并不是一定要翻译成蛋白质才具有功能，事实上 RNA 除了翻译之外还有很多功能。综合考虑有三点可以支持 RNA 的基因功能：第一，基因的实际存在；第二，信息的转录；第三，不同发育阶段和不同组织中不同的非重复序列的转录[33,35]。

主流的观点认为大部分的真核 DNA 既不用于转录，也非功能性，支持这一论点的主要依据是功能基因对于自然选择的反应。大量延缓了突变可能产生的有害作用的生理和（或）遗传补偿机制使得我们需要重新评价这个论点。此外，针对多聚腺苷酸位点和 RNA 转录物估算的分子数据显示比预计数量更多的基因活动的存在。

Although it is impossible to measure exactly the number of functional genes in eukaryotes, the acceptance of evidence for these minimum amounts seems a little premature.

Supported by a postdoctoral award from the National Institute of General Medical Science.

I thank Drs. R. J. MacIntyre, W. Sofer, R. C. Getham, J. Bell, and M. Mitchell for criticism and discussion.

(**242**, 52-54; 1973)

S. J. O'Brien

Gerontology Research Center, National Institute of Child Health and Human Development, National Institutes of Health, Baltimore City Hospitals, Baltimore, Maryland 21224

Received August 28; revised December 11, 1972.

References:

1. Tomkins, G. M., Gelehrter, T. D., Granner, D., Martin, D., Samuels, H. H., and Thompson, E. B., *Science*, **166**, 1474 (1969).
2. Britten, R. J., and Davidson, E. H., *Science*, **165**, 349 (1969).
3. Ohno, S., *Nature*, **234**, 134 (1971).
4. Ohno, S., *Devel. Biol.*, **27**, 131 (1972).
5. Crick, F., *Nature*, **234**, 25 (1971).
6. Ohta, T., and Kimura, M., *Nature*, **233**, 118 (1971).
7. Muller, H. J., in *Heritage from Mendel* (edit. by Brink, R. A.), 419 (University of Wisconsin Press, Madison, 1967).
8. Laird, D. C., and McCarthy, B. J., *Genetics*, **63**, 865 (1969).
9. Britten, R. J., and Kohne, D. E., *Science*, **161**, 529 (1968).
10. Judd, B. H., Shen, M. W., and Kaufman, T. C., *Genetics*, **71**, 139 (1972).
11. Wallace, B., *Topics in Population Genetics*, **45** (W. W. Norton and Co., New York, 1968).
12. Herskowitz, I. H., *Amer. Nature.*, **84**, 225 (1950).
13. O'Brien, S. J., and MacIntyre, R. J., *Drosophila Information Service*, **46**, 89 (1971).
14. Lindsley, D., and Grell, E. H., *Genetics Variations of Drosophila melanogaster* (Carnegie Inst. Publ. No. 627, 1967).
15. O'Brien, S. J., and MacIntyre, R. J., *Genetics*, **71**, 127 (1972).
16. Ritossa, F., Malva, C., Boncinelli, E., Graziani, F., and Polito, L., *Proc. US Nat. Acad. Sci.*, **68**, 1580 (1971).
17. Lefevre, G., *Genetics*, **63**, 589 (1969).
18. Richmond, R., *Nature*, **225**, 1025 (1970).
19. Arnheim, N., and Taylor, C. E., *Nature*, **223**, 900 (1969).
20. Sved, J. A., *Amer. Nat.*, **102**, 283 (1968).
21. Smith, J. M., *Nature*, **219**, 1114 (1968).
22. Milkman, R. D., *Genetics*, **55**, 493 (1967).
23. Muller, H. J., *Amer. J. Hum. Genet.*, **2**, 111 (1950).
24. Kojima, K., and Tobari, Y., *Genetics*, **63**, 639 (1969).
25. O'Brien, S. J., and MacIntyre, R. J., *Nature*, **230**, 335 (1971).
26. Edmonds, M., Vaughan, M. H., and Nokatzato, H., *Proc. US Nat. Acad. Sci.*, **68**, 1336 (1971).
27. Lee, Y. S., Mendecki, J., and Brawerman, G., *Proc. US Nat. Acad. Sci.*, **68**, 1331 (1971).
28. Darnell, J. E., Wall, R., and Tushinski, R. J., *Proc. US Nat. Acad. Sci.*, **68**, 1321 (1971).
29. Edmonds, M., and Abrams, R., *J. Biol. Chem.*, **235**, 1142 (1960).
30. Niessing, J., and Sekeris, C. E., *FEBS Lett.*, **22**, 83 (1972).
31. Darnell, J. E., Philipson, L., Wall, R., and Adesnik, M., *Science*, **174**, 507 (1971).

尽管不可能精确测定真核生物功能基因的数量，但接受这些最小数量的证据似乎还有些为时过早。

本研究由国立综合医学研究所的博士后奖金资助。

感谢麦金太尔、索弗、盖塞姆、贝尔、米切尔博士的意见和讨论。

（毛晨晖 翻译；曾长青 审稿）

32. Shenkin, A., and Burdon, R. H., *FEBS Lett.*, **22**, 157 (1972).
33. Firtel, R. A., *J. Mol. Biol.*, **66**, 363 (1972).
34. Firtel, R. A., and Bonner, J., *J. Mol. Biol.*, **66**, 339 (1972).
35. Hahn, W. E., and Laird, C. D., *Science*, **173**, 158 (1971).
36. Soeiro, R., Vaughan, M. H., Warner, J. R., and Darnell, J. E., *J. Cell Biol.*, **39**, 112 (1968).
37. Johnson, F., Wallis, B., and Denniston, C., *Drosophila Information Service*, **41**, 159 (1966).
38. Johnson, F. M., *Drosophila Information Service*, **41**, 157 (1966).
39. Bell, J. B., MacIntyre, R. J., and Olivieri, A., *Biochem. Genet.*, **6**, 205 (1972).
40. Glassman, E., *Fed. Proc.*, **24** (Suppl. 14-15), 1243 (1965).
41. Dickinson, W. J., *Genetics*, **66**, 487 (1970).
42. Grell, E., *Ann. NY Acad. Sci.*, **151**, 441 (1968).
43. Sofer, W., and Hatkoff, M. A., *Genetics*, **72**, 545 (1972).
44. Tobari, Y., and Kojima, K., *Genetics*, **70**, 347 (1972).
45. Ritossa, F. M., Atwood, K. C., and Spiegelman, S., *Genetics*, **54**, 819 (1966).
46. Baglioni, C., *Nature*, **184**, 1084 (1959).
47. Ghosh, D., and Forrest, H. S., *Genetics*, **55**, 423 (1967).

Fusion of Rat and Mouse Morulae and Formation of Chimaeric Blastocysts

G. H. Zeilmaker

Editor's Note

Over ten years earlier, Polish embryologist Andrzej Tarkowski made chimaeric mice by fusing together eggs taken from different mouse strains. Here, Gerard H. Zeilmaker manages to fuse early mouse embryos with early rat embryos to produce chimaeric blastocysts—slightly more-developed embryonic structures containing a mix of mouse and rat cells. The study paved the way for production of viable intraspecies chimaeras, such as the goat-sheep mix or "geep", which has helped answer fundamental questions about development. The fusion of human DNA with bovine and rabbit eggs has since seen early-stage chimeras yield human stem cells without the need for human eggs, a major hurdle in stem cell research. These techniques for creating chimaeras could also help save endangered species.

THE induction of chimaerism by the aggregation of blastomeres of different mouse strains is one of the most remarkable recent contributions of experimental embryology[1,2].

Last year it was pointed out (Dr. A. McLaren, private communication) that the possibility to fuse morulae of different species had not been sufficiently explored[3]. I therefore undertook a series of experiments to investigate whether aggregation of rat and mouse morulae can be induced, using basically the same techniques as those for aggregation of mouse morulae[3,4].

Mouse morulae (C3Hf×Swiss, 8–16 cell stage) were flushed from the utero-tubal region with culture medium on day 2 of pregnancy at 21:00 h, day 0 of plug, 14 h of light, 10 h of darkness, the middle of the dark period at midnight. Rat morulae (R, Amsterdam Wistar, 8–16 cell stage) were isolated by flushing on day 3 at 21:30h. The isolated morulae were stored separately at 37°C under oil and 5% CO_2 in air in pyruvate containing culture medium[5].

大鼠与小鼠桑葚胚的融合以及嵌合囊胚的形成

泽尔马克

编者按

十多年前，波兰胚胎学家安杰伊·塔尔科夫斯基通过融合不同种系的小鼠卵细胞，培育出了嵌合小鼠。在本文中，赫拉德·泽尔马克尝试将小鼠与大鼠的早期胚胎融合得到嵌合囊胚，也就是包含了小鼠细胞和大鼠细胞并且有了些许进一步发育的胚胎结构。这项研究为可育的种内嵌合体的产生，比如山羊和绵羊的杂交体即“山绵羊”，奠定了基础，这帮助人们回答了一些关于发育的基本问题。在人类 DNA 与牛、兔的卵细胞融合实验中观察到，在不需要人类卵细胞的条件下，早期嵌合体就能产生人类干细胞，这在干细胞研究中跨越了一个大的障碍。这些产生嵌合体的技术还有助于挽救濒临灭绝的物种。

通过将不同种系小鼠的囊胚细胞融合诱导产生嵌合体是近期实验胚胎学领域最卓越的成就之一[1,2]。

去年，有人（麦克拉伦博士，私人交流）指出，将不同物种的桑葚胚融合在一起的可能性，这种可能性还没有得到充分的研究[3]。因此我用与融合小鼠桑葚胚基本相同的技术[3,4]进行了一系列的实验，来研究能否诱导大鼠和小鼠桑葚胚的聚集。

在妊娠第 2 天的 21:00，用培养液将小鼠桑葚胚（C3Hf× 瑞士种，8 细胞至 16 细胞阶段）从子宫－输卵管区域洗脱出来，未经历栓塞，见光 14 小时，置于暗处 10 小时，并使黑暗阶段的中期处于午夜。在妊娠第 3 天的 21:30，洗脱出大鼠桑葚胚（R，阿姆斯特丹大鼠，8 细胞至 16 细胞阶段）。分离出来的桑葚胚分别储存在 37℃ 下油封的含丙酮酸的培养基中，培养空间空气中的 CO_2 含量保持在 5%[5]。

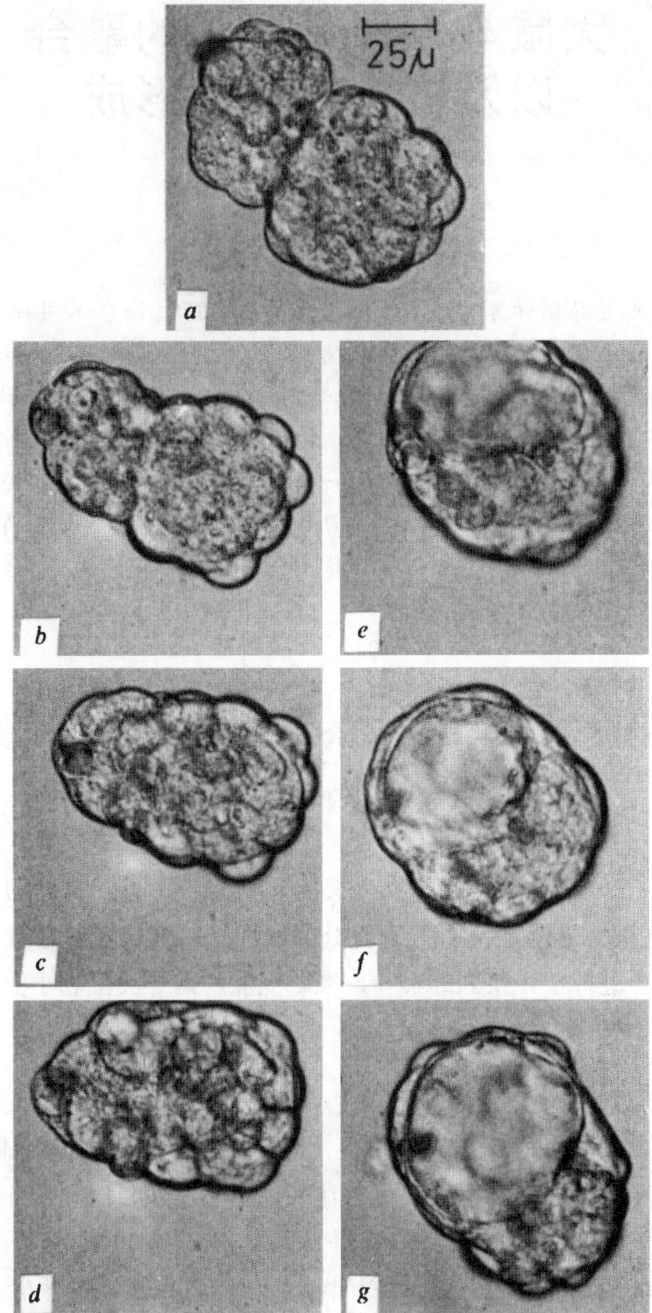

Fig. 1. *a*, Mouse (left) and rat morulae in apposition, shortly after withdrawal of egg holders. Developmental stages after 2 (*b*), 4 (*c*), 6 (*d*), 8 (*e*), 11 (*f*) and 12.5 h (*g*). In *d* the blastocoelic cavity is formed.

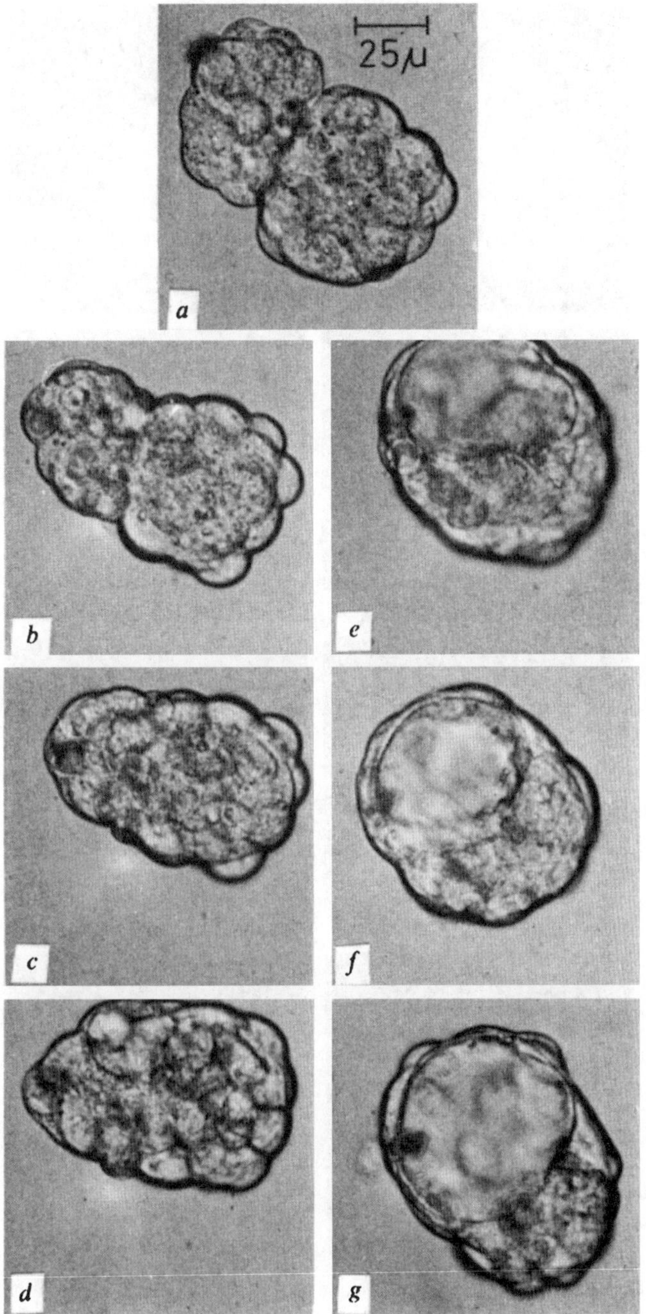

图 1. *a*,撤掉胚胎固定器后不久,小鼠(左侧)和大鼠的桑葚胚紧邻。图中显示经过 2 小时(*b*)、4 小时(*c*)、6 小时 (*d*)、8 小时 (*e*)、11 小时 (*f*) 和 12.5 小时 (*g*) 后的发育阶段。在 *d* 阶段,囊胚腔已经形成。

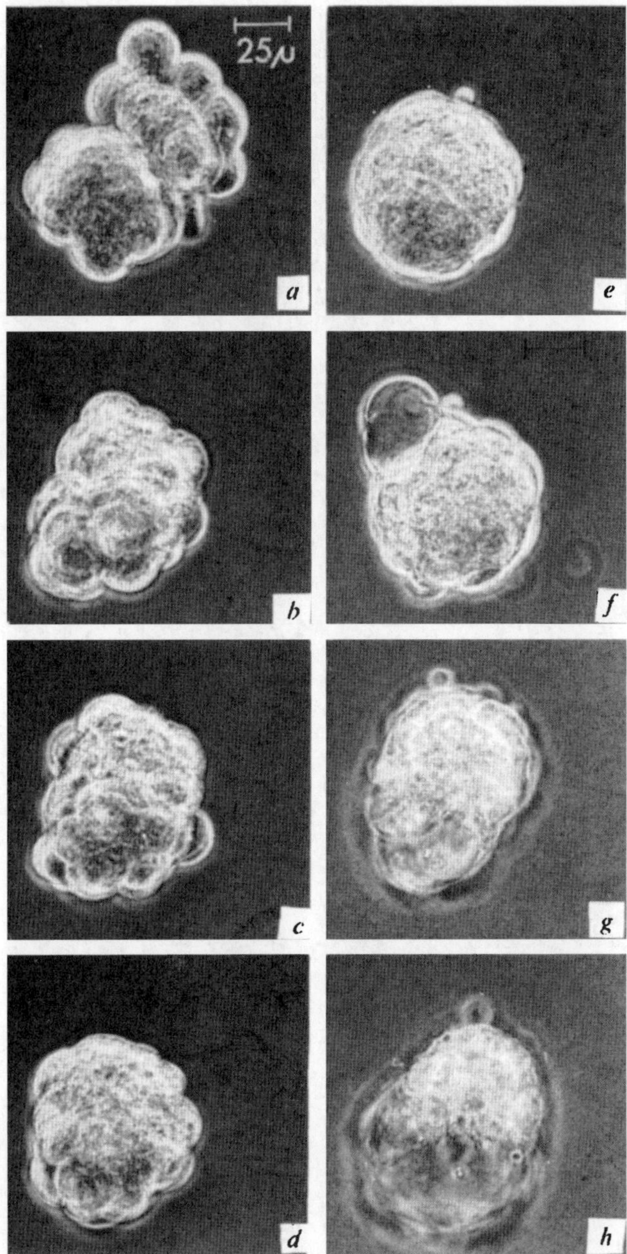

Fig. 2. Mouse (left) and rat (upper right) morulae in the process of fusion. Pictures taken after 30 min (*a*), 2 (*b*), 6 (*c*), 8 (*d*), 12.5 (*e*), 18.3 (*f*), 24 (*g*) and 34 h (*h*). Borderline between rat and mouse morulae still visible in (*e*). First cavity formed in (*f*), final cavity in (*g*).

The zona pellucida was dissolved by incubation of the eggs for 3 to 5 min in 0.5% pronase[6]. Final removal of the zona occurred in culture medium by forcing the eggs through a narrow pipette. In each experiment one mouse and one rat morulae were brought into contact in an oil drop culture at 37°C by means of 2 egg holders with a closed lumen[7] driven by a micromanipulator. During this apposition phase, which lasted

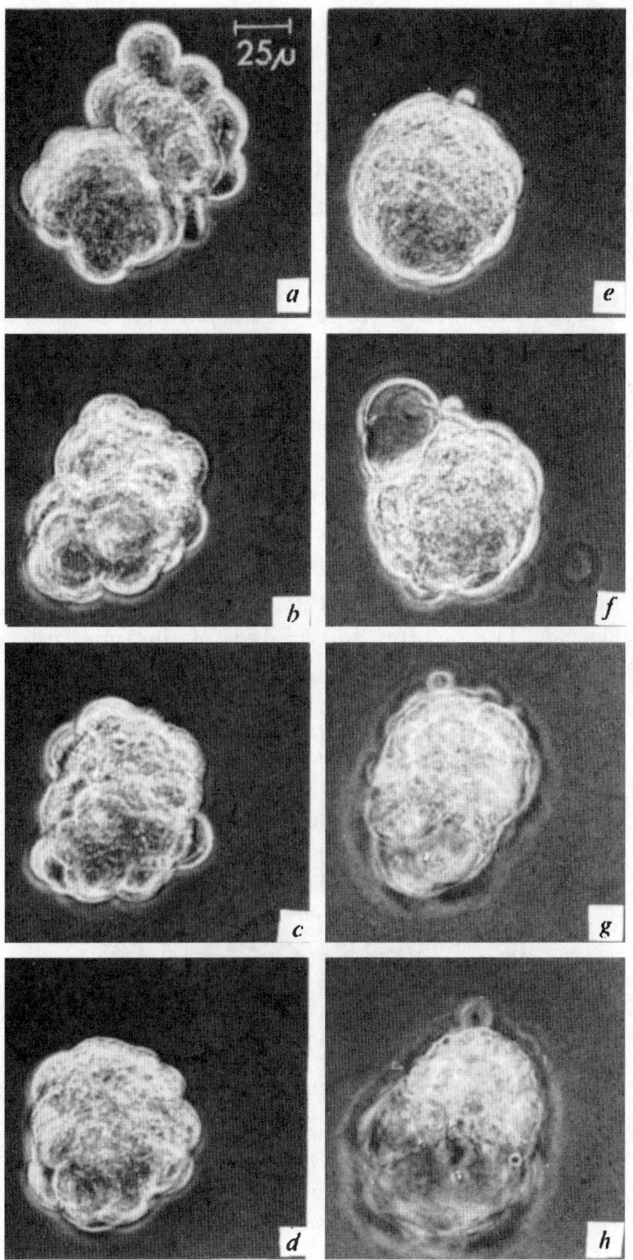

图 2. 融合过程中的小鼠（左侧）和大鼠（右上）桑葚胚。图片采自融合开始 30 分钟（*a*）、2 小时（*b*）、6 小时（*c*）、8 小时（*d*）、12.5 小时（*e*）、18.3 小时（*f*）、24 小时（*g*）和 34 小时（*h*）后。（*e*）阶段仍可见到大小鼠桑葚胚的界线。第一空腔在（*f*）阶段形成，最终的空腔在（*g*）阶段形成。

将卵细胞在 0.5% 的链霉蛋白酶中温育 3 到 5 分钟即可溶解透明带[6]。在培养基中，通过压迫卵细胞穿过狭窄的移液管从而最终去除透明带。在每个实验中，借助两个由显微操纵器驱动的含有封闭腔[7]的卵细胞固定器，一个小鼠的桑葚胚和一个大鼠的桑葚胚在 37℃下的油滴培养中彼此接触。在这个持续 20 到 30 分钟的接触阶

20 to 30 min, the Petri dish was not closed but a humid gas stream was directed over the oil surface in order to maintain *p*H and osmolarity of the medium.

After removal of the egg holders the development of the pair of morulae was followed continuously with a time-lapse movie camera connected to the inverted microscope. Photographs were taken every 30 s with flash light.

A total of sixteen identical experiments were carried out and, in twelve, successful aggregation had occurred in which both aggregate partners showed active proliferation as judged by the temporary bulging of dividing cells at the surface of the large morulae. Sometimes the individual morulae could be recognized for a long time after aggregation by a slight indentation. The way in which the blastocoelic cavity formed varied. In certain cases it appeared where the junction between the two morulae could be seen earlier (Fig. 1*b*–*d*), in others an abortive blastocoelic cavity formed in one of the partners (Fig. 2*f*), which disappeared later. In the same aggregate the lasting cavity formed in the other partner (Fig. 2*g*). In one experiment one of the aggregate partners formed a cavity before the cell mass had become round (Fig. 3*b*); this disappeared and a final one was formed after aggregation (Fig. 3*d*). The pictures and analysis of the time-lapse photographs show that both rat and mouse cells contribute to the formation of a large blastocyst.

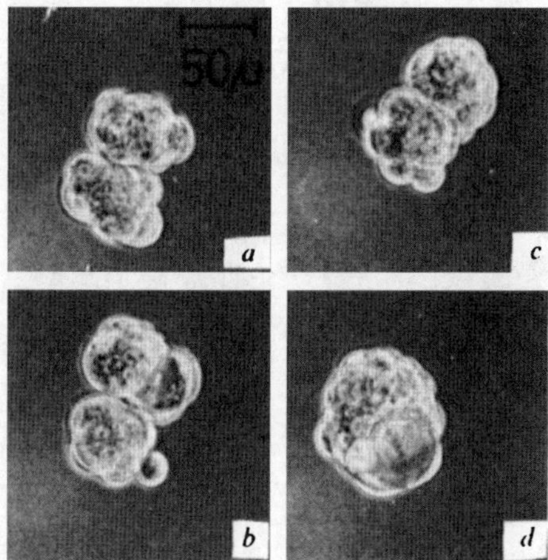

Fig. 3. Fusion of rat and mouse morulae after formation and disappearance of a cavity in the mouse morula. Pictures taken at approximately 30 min (*a*), 6.6 (*b*), 10 (*c*) and 19h (*d*).

A general feature of the chimaeric blastocysts was that the inner cell mass was comparatively large.

Preliminary observations have shown that the developmental stage of the rat eggs is critical for successful aggregation and the use of a micromanipulator was also very helpful.

A copy of the time-lapse movie is available for study.

(**242**, 115-116; 1973)

段，培养皿并不关闭，使潮湿气流直接吹向油表面以维持培养基的 pH 和渗透压。

移除卵细胞固定器以后，用连接在倒置显微镜上的间隔定时摄像机持续追踪这一对桑葚期胚胎的发育过程，每隔 30 秒打开闪光灯进行一次照相。

一共进行了 16 组相同的实验，其中有 12 组发生了成功的胚胎聚集性融合。根据在巨大的桑葚胚表面正在分裂的细胞的暂时性突起可以看出，融合双方都进行着活跃的增殖。有时通过细微的压痕可以鉴定出融合很长时间后的单个桑葚胚。囊胚腔形成的方式是多样化的。在某些情况中，它出现在能够在较早期看到的两个桑葚胚的连接位置（图 1*b*~*d*）；而在其他情况中，发育不完全的囊胚腔是在其中一个胚胎中形成的（图 2*f*），并在之后消失。在同一个融合胚胎中，最终的囊胚腔在另一个参与融合的胚胎内部形成（图 2*g*）。在一组实验中，参与融合的胚胎之一在细胞团变圆之前就形成了腔（图 3*b*），然后消失，最终的腔是在融合之后形成的（图 3*d*）。间隔定时拍摄的图片和分析显示大鼠和小鼠细胞均参与了巨大囊胚的形成。

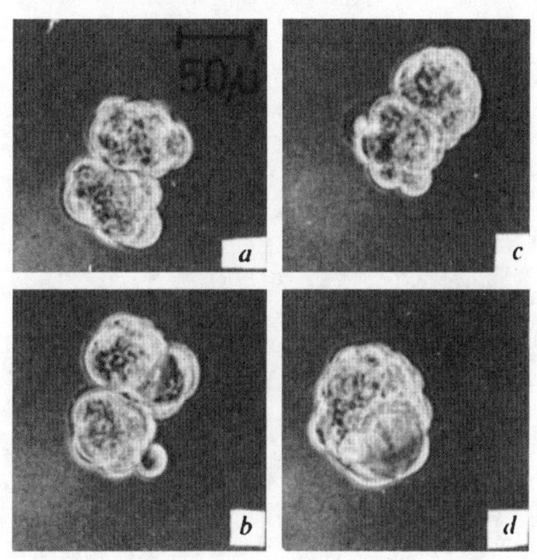

图 3. 在小鼠桑葚胚内形成腔并消失以后，大鼠和小鼠的桑葚胚发生融合。图片拍摄于融合开始大约 30 分钟（*a*）、6.6 小时（*b*）、10 小时（*c*）和 19 小时（*d*）。

嵌合囊胚的总体特征就是内细胞团相对较大。

初步的观察显示大鼠卵细胞的发育阶段对成功的融合至关重要，而且使用显微操作器非常有帮助。

我们可提供间隔定时拍摄所获得的录像拷贝用于研究。

（毛晨晖 翻译；梁前进 审稿）

G. H. Zeilmaker
Department of Endocrinology, Growth and Reproduction, Erasmus University, PO Box 1738, Rotterdam

Received September 29; revised October 20, 1972.

References:

1. Tarkowski, A., *Nature*, **190**, 857 (1961).
2. Mintz, B., *J. Exp. Zool.*, **157**, 273 (1964).
3. Mintz, B., in *Preimplantation Stages of Pregnancy*, CIBA Foundation Symposium (edit. by Wolstenholme, G. E. W., and O'Connor, M.), 207 (Churchill, London, 1965).
4. Tarkowski, A., in *Preimplantation Stages of Pregnancy*, CIBA Foundation Symposium (edit. by Wolstenholme, G. E. W., and O'Connor, M.), 183 (Churchill, London, 1965).
5. Biggers, J., Wittingham, D. G., and Donahue, R. P., *Proc. US Nat. Acad. Sci.*, **58**, 560 (1967).
6. Mintz, B., *Science*, **138**, 594 (1962).
7. Lin, T. P., *Science*, **151**, 333 (1966).

Linkage Analysis in Man by Somatic Cell Genetics

F. H. Ruddle

Editor's Note

Today, linkage analysis, where the location of a gene is worked out relative to a known sequence, is a sophisticated affair involving high-throughput machinery and computers. But in the 1970s researchers had to resort to cell culture methods, mapping genes from somatic cell hybrids. Here geneticist Frank Ruddle describes the state of play. Hybrids combine different genomes within a single cell, and the discovery that chromosomes from parental genomes can be lost or segregated from the hybrid cell, enabled researchers to effectively isolate and identify chromosome fragments and assign genes to chromosomes. The technique correctly assigned many genes to their chromosomes, but was "unacceptably slow". Somatic cell recombination and gene transfer, Ruddle prophetically muses, could provide the way forwards.

Techniques for the study of somatic cell genetics, and particularly those involving the expression of enzyme markers in hybrid cells, have already made possible a large number of gene–chromosome assignments. Genetic and family studies, as well as cellular studies on recombination and gene transfer, promise more and quicker results in the future.

SOMATIC cell hybridization, first demonstrated by Barski *et al.*[1], was an important early step in the formulation of somatic cell genetic systems, allowing as it does the combination of genetically different genomes within a single cell. In a series of investigations Ephrussi and his colleagues showed that hybrid combinations could be obtained between the cells of different species[2], and that chromosomes of one or both parental genomes could be lost or segregated from the hybrid cell[3]. Weiss and Green[4] first demonstrated the practical application of somatic fusion–segregation systems for the purpose of gene mapping in man (see below). Other investigators have contributed useful procedures which enhance the formation of hybrid cells and their enrichment. These developments now allow a completely new cell culture approach to gene mapping in man.

Cultivation of Hybrid Cells

For the formation of hybrid cells the parental cells are mixed together and co-cultivated. Membrane fusion can be enhanced by treatment with inactivated Sendai virus[5,6] or with lysolecithin[7]. Fusion between two parental cells of different origins gives rise to a binucleate heterokaryon. Heterokaryons have a short life expectancy, and following their first mitosis, generally form mononucleated or hybrid daughter cells which contain chromosomes from both parental genomes. In many parental cell combinations, the hybrids have an infinite life expectancy and can be grown into large clonal cell

人类体细胞遗传学连锁分析

拉德尔

编者按

连锁分析是指分析确定某一基因相对于一段已知序列的位置，在今天，这是一项涉及高通量检测体系和计算机数据处理设备的复杂事务。但是在20世纪70年代，研究者们必须借助细胞培养方法，通过体细胞杂交来定位基因。在本文中遗传学家弗兰克·拉德尔描述了这项技术的研究进展。不同的基因组在一个细胞内形成杂交体以及亲本基因组染色体在杂交细胞中可能丢失或者发生分离的发现，都有助于研究者有效地分离、鉴定染色体片段，并将基因定位于染色体上。人们曾利用体细胞杂交技术将许多基因定位于染色体上，但其速度之慢让人难以接受。拉德尔预见性地思忖着，体细胞重组和基因转移或许能够为基因连锁分析提供更加便捷的途径。

体细胞遗传学的研究技术，尤其是涉及杂交细胞中酶标记表达的技术，已经使得对大量的基因–染色体匹配成为可能。遗传学和家族研究，以及基于重组和基因转移的细胞学研究，使我们有可能在未来更快更多地获得成果。

体细胞杂交是由巴尔斯基等人[1]率先证明的，其在构建体细胞遗传学系统的早期迈出了重要的一步，它使得在遗传学上不同的基因组可以在一个单独的细胞中进行组合。埃弗吕西和他的同事们经过一系列的研究发现不同物种的细胞之间可以进行杂交组合[2]，并且来自一个或者两个亲代的染色体组可能会从杂交细胞中丢失或者分离[3]。韦斯和格林[4]首次实现了体细胞融合–分离系统在人类基因作图中的实际应用（见下文）。其他研究者们也提出了一些有用的方法，可以增进杂交细胞的生成及其富集。现在，这些技术的发展使得我们可以得到一种全新的细胞培养方法来进行人类基因作图。

杂交细胞的培养

将亲代细胞混合并共同培养，用以形成杂交细胞。我们可以利用失活的仙台病毒[5,6]或者溶血卵磷脂[7]处理混合细胞，来促进细胞膜的融合。两种不同起源的亲代细胞融合后会产生一种双核的异核体。这些异核体的期望寿命很短，在接下来的第一次有丝分裂中会生成单核的或者含有来自两个亲代基因组的染色体的杂交子细胞。在许多亲代细胞的组合中，杂交细胞具有无限长的期望寿命，并可以生长成为一个巨大的克隆细胞群。在人类–小鼠和人类–中国仓鼠的杂交细胞中，会出现人

populations. In man–mouse and man–Chinese hamster hybrids there is a unilateral loss or segregation of human chromosomes. The segregation of human chromosomes is variable in extent in different clones, and in many instances clones can be obtained which maintain for many generations a partial human chromosome constitution. Thus, it is possible in effect to sample different numbers and combinations of human chromosomes in a series of man–rodent hybrids of independent origin. Each clone represents a partial human karyotype superimposed on an intact mouse or Chinese hamster genome. The experimental isolation of partial human chromosome complements forms the basis of somatic cell linkage analysis.

Enzyme complementation has been used to enrich for hybrids in mixed populations of parental cells. Littlefield has shown that drug resistance mutant cell lines can be useful in this regard[8]. Mutant cell lines can be selected which are deficient in the enzymes hypoxanthine-guanine phosphoribosyltransferase (HGPRT) and thymidine kinase (TK) by exposing cells to the antimetabolites thioguanine and BUdR respectively. HGPRT deficient cells cannot incorporate hypoxanthine, whereas TK deficient cells cannot metabolize thymidine. If *de novo* synthesis of purines and pyrimidines is blocked by the antimetabolite aminopterin, cells become dependent for survival on exogenous hypoxanthine and thymidine. HGPRT and TK deficient cells are thus conditional lethal mutants which are killed by aminopterin irrespective of the availability of hypoxanthine and thymidine. The fusion of HGPRT deficient with TK deficient parental cells yields hybrid cells whose enzyme deficiencies are complemented and which can grow in nonpermissive selection medium containing hypoxanthine, aminopterin, and thymidine (HAT medium). Kusano *et al.* have shown that adenine phosphoribosyltransferase (APRT) deficiency mutations can be used similarly for hybrid cell selection[9]. Conditional lethal mutants other than those based on drug resistance can also be used for hybrid selection. Puck *et al.* have used nutritional auxotrophs with good results[10,11], and it is likely that temperature sensitive mutations can be used in the same way[12]. Moreover, it is possible to make use of conditional mutant established rodent cell lines in combination with diploid human fibroblasts or leucocytes which have low growth potentials *in vitro*[12,13]. One can select against the rodent parent using nonpermissive medium and against the human diploid parent by virtue of its inherently poor growth characteristics.

Conditional lethal cell mutants in rodent cell populations are extremely useful for genetic analysis. In nonpermissive conditions only hybrids which retain the complementing human gene will survive. Thus, if one forms hybrids between HGPRT deficient rodent cells and wild type human cells ($HGPRT^+$), and cultivates them in HAT medium, only those cells which retain the human HGPRT gene will survive. Generally, the intact human *X* chromosome which carries the HGPRT gene is retained in the complemented hybrid. It is possible to conceive of a series of rodent cell lines each of which carry different conditional lethal mutations which are complemented by genes on each of the human autosomes and sex chromosomes. Such a panel of rodent cell lines would be extremely useful in mapping studies because each would produce hybrids in which the segregation of a specific human chromosome would be fixed. Conditional lethal mutants

类染色体单方面缺失或者分离的现象。在不同的克隆中，人类染色体的分离存在着不同程度的变化，在许多情况下，获得的克隆细胞都保持一部分人类染色体组分，并延续多代。因此，在获得的一系列独立来源的人类–啮齿目动物杂交细胞中，很可能抽取到不同数量和组合的人类染色体。每个克隆体都表现出部分人类染色体的核型，这些核型叠加在小鼠或者中国仓鼠的完整的基因组中。因此对部分人类染色体组进行实验分离是进行体细胞基因连锁分析的基础。

酶互补技术已经应用于亲代混合细胞群中富集杂交细胞。利特菲尔德已经证明了耐药性突变细胞系在这方面很有用[8]。我们可以分别通过抗代谢物硫鸟嘌呤和溴脱氧尿嘧啶核苷（BUdR）处理的方法筛选具有次黄嘌呤–鸟嘌呤磷酸核糖转移酶（HGPRT）和胸苷激酶（TK）缺陷的突变细胞系。HGPRT 缺陷的细胞不能利用次黄嘌呤，而 TK 缺陷的细胞不能代谢胸腺嘧啶。如果细胞体内嘌呤和嘧啶的从头合成被抗代谢物氨基蝶呤所抑制，那么细胞就只能依靠外源的次黄嘌呤和胸腺嘧啶来维持生存。因此，HGPRT 和 TK 缺陷的细胞是一种条件致死突变体，在没有外源的次黄嘌呤和胸腺嘧啶时，细胞将被氨基蝶呤杀死。将 HGPRT 缺陷的亲代细胞和 TK 缺陷的亲代细胞进行融合，产生的杂交后代的酶缺陷相互补偿，使其可以在非许可性选择培养基（含有次黄嘌呤、氨基蝶呤和胸腺嘧啶，即 HAT 培养基）中生长。草野等人证明了腺嘌呤磷酸核糖转移酶（APRT）的缺失突变可以应用在类似的杂交后代选择中[9]。除了这些基于耐药性的突变体以外，其他条件致死突变体也可用于杂交后代的筛选。普克等人利用营养突变体获得了不错的结果[10,11]，而温度敏感型突变也很有可能以相同的方式得到应用[12]。而且，我们还可以利用这些条件突变体建立啮齿目动物细胞系，并将其与体外生长势较低的人类二倍体成纤维细胞或白细胞进行组合[12,13]。人们可以利用非许可性培养基从啮齿目动物的亲代细胞筛选杂交后代，也可以利用人类二倍体细胞固有的缓慢生长特点筛选杂交后代。

啮齿目动物细胞群中的条件致死细胞突变体对遗传分析来讲是非常有用的。在非许可性选择条件下，只有那些保留了互补的人类基因的杂交后代才会存活下去。因此，如果用 HGPRT 缺失的啮齿目动物细胞与野生型人类细胞（$HGPRT^+$）生成杂交后代，并在 HAT 培养基中进行培养，那么只有那些保留人类 HGPRT 基因的细胞才会存活。一般来说，带有 HGPRT 基因的完整人类 X 染色体会保留在补偿了这一缺陷基因的杂交后代中。我们可以建立一系列啮齿目动物细胞系：每个细胞带有不同的条件致死突变，这些突变体分别能被不同的人类常染色体和性染色体上的基因所补偿。这样的一组啮齿目动物细胞系在基因作图研究中将起到极大的作用，因为上述每一个细胞系都可以产生固定了某一条发生分离的人类染色体的杂交后代。表

of this type are tabulated in Table 1. It should be pointed out that the drug resistance complementation systems also lend themselves to counter selection. Cells which retain TK, APRT, and HGPRT activity are susceptible to the antimetabolites BUdR, fluoroadenine, and thioguanine, respectively[8-10]. Thus it is possible to use these agents in permissive medium to select against hybrid cells which have retained human chromosomes 17, 16 and *X*.

Table 1. Conditional Drug Resistance and Nutritional Auxotrophic Genetic Markers

Rodent mutation	Rodent parent	Complementing human enzyme	Human linkage unit
HGPRT deficiency	Mouse	$HGPRT^{+}$	X
TK deficiency	Mouse	TK^{+}	17
APRT deficiency	Mouse	$APRT^{+}$	16
Glycine A auxotroph	Chinese hamster	Serine hydroxymethylase	12
Adenine B auxotroph	Chinese hamster	Unknown	4 or 5

Rodent–human Hybrids

In rodent–human hybrids, the human chromosomes are unilaterally segregated, both homologous human and rodent enzymes are expressed and can be identified, and human and mouse chromosomes can be discriminated and accurately identified on an individual basis. These hybrids are therefore particularly suitable for human gene linkage analysis.

The loss of human chromosomes from mouse–human and Chinese hamster–human hybrids is well documented, but the mechanism of loss is poorly understood. Preferential loss of human chromosomes in rodent *X* human hybrids, irrespective of the origins of the parental cell populations, is the rule, and only one possible exception has been reported[15]. A mechanism of loss suggested by Handmaker (personal communication) is that human chromosomes cannot attach efficiently to the hybrid spindle apparatus and thus have higher incidence of loss. Another possibility, which is not necessarily incompatible with this, is a mechanism of segregation based on random non-disjunction of mouse and human chromosomes in combination with the preferential selection of hybrids which possess partial human karyotypes[16]. Nabholz *et al.*[17] have suggested that chromosomes are lost by two temporarily distinct processes. Early loss, possibly during the first several mitotic divisions after fusion, can result in the abrupt loss of a few or many human chromosomes. Late loss is characterized by slow progressive loss in some instances over many cell generations. Nabholz *et al.*[17] have also presented evidence that human chromosomes are segregated non-randomly into hybrid clones. Preliminary results in our laboratory, based on twenty-eight independent hybrid clones, indicate a very low frequency of retention of human chromosome 9 (7%) compared with the overall frequency of human chromosome retention (29%). It has been reported that hybrids with two rodent genomes (2s hybrids) retain more human chromosomes than 1s hybrids[15]. The relationship between rodent and human chromosome number is significant and should be resolved, because it is fundamental to the problem of chromosome segregation.

1 中列出了这种类型的条件致死突变体。但是应该指出的是，自身的抗药性互补系统也会产生抗基因选择的效应。那些保留 TK、APRT 和 HGPRT 活性的细胞分别易受抗代谢物 BUdR、氟腺嘌呤和硫鸟嘌呤的影响 [8-10]。因此，我们可以利用这些突变体在许可性培养基中对那些保留人类的 17 号、16 号染色体和 X 染色体的杂交细胞进行选择。

表 1. 条件抗药性与营养缺陷型遗传标记

啮齿动物突变体	啮齿动物亲代	互补的人类酶	人类连锁遗传单元
HGPRT 缺陷	小鼠	$HGPRT^+$	X 染色体
TK 缺陷	小鼠	TK^+	17 号染色体
APRT 缺陷	小鼠	$APRT^+$	16 号染色体
甘氨酸 A 型营养缺陷	中国仓鼠	丝氨酸羟甲基酶	12 号染色体
腺嘌呤 B 型营养缺陷	中国仓鼠	未知	4 号或 5 号染色体

啮齿目动物 – 人类杂交细胞

在啮齿目动物 – 人类杂交细胞中，人类染色体发生单向分离；人类和啮齿目动物的同源酶表达并且被鉴定；在每个杂交后代上，我们可以清楚地区分并准确地鉴定人类和小鼠的染色体。因此，这些杂交细胞特别适用于人类基因连锁分析。

虽然在小鼠 – 人类和中国仓鼠 – 人类的细胞杂交中，人类染色体的缺失是有据可查的，但我们对缺失的机制还知之甚少。无论亲本细胞群的起源如何，在啮齿目动物 – 人类杂交细胞中，人类染色体更容易缺失是普遍的情况，研究者只报道了一个可能的例外 [15]。汉德梅克提出了一种缺失机制（个人交流），即人类染色体不能有效地附着在杂交后代的纺锤体上，所以才会有较高的缺失发生率。另外一种并不与之相矛盾的可能性是一种分离机制，它基于小鼠和人类染色体的随机不分离以及含有部分人类细胞核型的杂交细胞的偏好性选择 [16]。纳布霍尔茨等人 [17] 认为染色体的缺失是通过临时的两个不同的过程完成的。早期的缺失可能发生在细胞融合后的最先几次有丝分裂过程中，可引起几条或多条人类染色体的突然缺失。后来的缺失经确定是缓慢的逐步缺失，在某些情况下可持续多代细胞。纳布霍尔茨等人 [17] 也已给出证据表明，人类染色体是非随机地分离到杂交克隆中的。我们对 28 个独立的杂交克隆进行了实验，初步的结果显示相比于人类染色体保留的总频率（29%）来说，这些杂交克隆对人类 9 号染色体具有非常低的保留频率（7%）。据报道，拥有两个啮齿目基因组的杂交体（2s 杂交体）要比只有一个啮齿目基因组的杂交体（1s 杂交体）保留更多的人类染色体 [15]。啮齿目动物与人类染色体数目之间的关系很重要，并且应该得到解决，因为这是解析染色体分离机理所需要弄清的基本问题。

The amino-acid constitution of homologous enzymes between man and rodents generally differs to some degree as a result of evolutionary divergence, and it is generally possible to detect these differences by electrophoretic procedures. There is thus a very large potential catalogue of genetic markers in the rodent–human cell hybrid system, limited only by the development of adequate test procedures. A compilation of isozyme procedures has been reported by Ruddle and Nichols[18].

It is important for genetic testing that the enzyme markers be constitutive—that is, they must invariably be expressed if the corresponding structural gene is retained in the hybrid. Facultative markers are defined as those markers which are subject to modulation and which may not be expressed even if the corresponding cistron is present. It is very difficult to define phenotypes as being absolutely constitutive or facultative. Generally speaking, enzymes which are expressed in all cell types *in vivo* and which contribute to vital metabolic pathways are termed constitutive, whereas enzymes which are restricted to one or a few specialized cell types and which do not participate in vital metabolic activities at a cellular level are termed facultative. Good evidence exists for the modulation of certain facultative functions in hybrid combinations between parental cells of different epigenetic types[19]. This necessarily complicates the linkage analysis of such phenotypes. Phenotypes classified as constitutive may under certain conditions be modulated. For example, hybrid clones have been recovered by Ricciuti[20] which possess normal C-7 human chromosomes, but which do not express detectable levels of mannose-phosphate isomerase (MPI) activity. Linkage analysis in other cell hybrids shows a strong correlation between C-7 and MPI. I have concluded that MPI may represent a partially constitutive phenotype. Such phenotypes pose problems for linkage analysis, but they also provide useful material for studies on phenotype modulation.

Cytological Identification of Chromosomes

It is now possible to identify all of the chromosomes of the Chinese hamster, laboratory mouse, and man by cytological procedures. Caspersson and co-workers[21] have shown that quinacrine binds differentially to specific regions of the human chromosomes, and that each chromosome possesses a unique banding pattern. Mouse chromosomes are similarly unique and the banding patterns have now been correlated with known murine linkage groups[22]. Giemsa banding procedures provide results comparable with those of quinacrine[23]. Pardue and Gall have introduced an *in situ* annealing technique which makes possible the localization of highly redundant DNA in mouse and human chromosomes[24]. Purified isotopically labelled redundant DNA is annealed to intact chromosomes which have been pretreated with DNA denaturation agents. The labelled redundant DNA is hybridized to complementary DNA in the chromosome, and its location revealed by autoradiography. Pardue and Gall have shown that the murine redundant DNA (satellite DNA) is restricted to the centromere regions and that denaturation followed by Giemsa staining reveals positively staining, constitutive heterochromatin regions in the chromosomes. Arrighi and Hsu[25] have adapted this method to the analysis of human chromosomes and subsequent studies have shown a

由于进化趋异，人类和啮齿目动物间的同源酶的氨基酸组成在一定程度上普遍存在不同。通常情况下，我们可以通过电泳的方法来检测这些差异。因此，在啮齿目动物–人类细胞的杂交体系中存在着一大类具有潜力的遗传标记，这些遗传标记的发现取决于适当的检测方法的发展。拉德尔和尼科尔斯[18]已经报道了一种同工酶编译方法。

对于遗传学检测来说，很重要的是酶标记是组成型的，也就是说，如果相应的结构基因保留在杂交后代中，那么它们就会得以表达。功能型标记被定义为那些受调节的酶标记，这种酶即使在相应的顺反子存在下，也可能不表达。很难将表型定义为绝对的组成型或功能型。一般来讲，我们把那些可以在体内所有细胞类型中进行表达并且对重要代谢过程有贡献的酶称作组成型的，而那些受限于一个或几个特殊的细胞类型中并且在细胞水平上不参与重要代谢活动的酶被称为功能型的。有证据表明，不同的表观遗传类型的亲代细胞之间的杂交组合体具有某些兼性功能调节[19]。这必定会使得这类表型的连锁分析变得复杂化。那些属于组成型的表型在某种情况下可能也是受调节的。例如，里丘蒂筛选到的具有正常 C-7 人类染色体的杂交克隆[20]，但它并没有表达出达到可检测水平的甘露糖磷酸异构酶（MPI）活性。而在其他细胞杂交中进行的连锁分析表明，C-7 和 MPI 之间高度相关。我认为 MPI 也许代表一种部分组成性的表型。虽然这些表型给连锁分析带来了问题，但它们也为表型调节的研究提供了有价值的资料。

染色体的细胞学鉴定

现在我们可以通过细胞学的方法识别中国仓鼠、实验小鼠和人类的所有染色体。卡斯佩松和他的同事们[21]证明了喹吖因染料可区别性地结合到人类染色体的特定区域，并且每一条人类染色体都有一个独一无二的带型。小鼠的染色体也具有类似的独特性，其带型已经与目前已知的鼠类连锁群相联系[22]。吉姆萨显带方法得到的结果与喹吖因方法得到的类似[23]。帕杜和高尔发明了一种原位退火技术，这使得对小鼠和人类染色体中大量冗余 DNA 进行定位成为可能[24]。将提纯的用同位素标记的冗余 DNA 退火至完整的、已经用 DNA 变性剂进行了预处理的染色体。标记的冗余 DNA 就与染色体中的互补 DNA 进行杂交，其位置可用放射自显影技术显示出来。帕杜和高尔表明鼠类冗余 DNA（卫星 DNA）是被限制在着丝粒区域中的；在变性后的吉姆萨染色结果中出现阳性染料附着，在染色体上属于组成型异染色质区域。阿里吉和徐[25]采用这种方法分析了人类染色体，并在随后的研究中显示出组成型异染色质和冗余 DNA 之间的一致性关系[26]。人类和小鼠的卫星 DNA 是特异的，且无相互反应。因此，可以利用原位杂交技术来区分人类和小鼠的着丝粒区域，这在人类

correspondence between constitutive heterochromatin and redundant DNA[26]. Human and mouse satellite DNA are specific and do not cross react. It is thus possible by hybridization *in situ* to distinguish human and mouse centromeric regions, which has proved useful in the detection of human–mouse chromosome translocations[27] (see below). Several laboratories have reported evidence indicating that the centromeric constitutive heterochromatin in man has different physical properties unique for several of the human chromosomes[28].

Assigning Genes to Chromosomes

Linkage of enzyme phenotypes can be inferred from their concordant segregation. The human chromosomes maintain their integrity for the most part, seldom undergoing rearrangement or deletion, and the concordant segregation of markers thus provides evidence for their location on the same chromosome irrespective of map distance. It is therefore appropriate to employ the term "synteny" coined by Renwick to signify merely localization on the same chromosome. Synteny testing is performed by comparing the segregation pattern of all markers in all pairwise combinations. The synteny test is less biased if performed on clones of independent origin and the detection of valid syntenic relationships is enhanced by using clones derived from separate hybridization experiments, using different hybrid combinations. This generally entails computer analysis because of the number of clones and markers involved.

Individual genes or syntenic genes can be assigned to specific chromosomes by tabulating the human chromosomes in each of the clones and correlating them with the enzyme markers. The concordant presence or absence between a chromosome and a phenotype provides evidence for the assignment of the gene governing a particular phenotype to a specific chromosome. Twenty to thirty metaphases are analysed per clone by means of quinacrine banding, Giemsa banding, or constitutive heterochromatin staining techniques. Identification is enhanced if cells are first photographed by quinacrine fluorescence and then by constitutive heterochromatin staining procedures.

It is frequently possible to strengthen the assignment of a gene to a particular chromosome by correlating the frequency of a particular chromosome within a clone with the intensity of expression of an assigned phenotype(s). Discrepant clones of two classes can occur, however. In the first, presuming a valid assignment, the chromosome cannot be detected but the phenotype is present. We have demonstrated cryptic, rearranged chromosomes in a number of such instances which explain an apparently discordant chromosome/gene relationship[29]. A second type of discrepant clone involves the presence of a specific chromosome, but the absence of its corresponding phenotype(s). Clones of this type are difficult to explain, but could involve subtle rearrangements in chromosome structure, gene mutation, modulation of gene expression, or technical failure in phenotype detection.

It is possible to assign genes to particular regions of chromosomes such as chromosome arms, or band regions as defined by particular staining reactions. This can be accomplished by making use of chromosome rearrangements such as translocations and deletions

–小鼠染色体易位的检测上已经证明是有用的[27]（见下文）。数个实验室已发表证据表明，人类着丝粒组成型异染色质具有异于人类某些染色体才具有的物理性质[28]。

基因与染色体的匹配

酶表型的连锁关系可以从与它们对应的染色体分离中推断出来。人类染色体绝大部分都保持着它们的完整性，很少会发生重排或者缺失，因此在不考虑图谱距离的情况下，与酶标记一致的分离为它们定位在对应的染色体上提供了依据。所以，采用伦威克创造的术语“同线性”来表示仅在同一染色体上的定位，这是很合适的。同线性检测是通过对比所有配对组合中全部遗传标记的分离模式来进行的。如果在独立起源的克隆体上进行检测，则同线性检测的结果的偏差较小。采用独立杂交实验的源于不同杂交组合的克隆体，可以加强有效的同线性关系检测。通常这样的检测需要计算机来进行分析，因为所涉及的克隆体和遗传标记数量庞大。

我们可以对每个克隆体的人类染色体进行列表并将它们与酶标记相联系，以此将单个基因或者同线性基因定位到特定的染色体中。染色体和酶表型同时存在或者缺失为控制将某一特定表型的基因定位到特定染色体中提供了证据。通过喹吖因显带、吉姆萨显带或者组成型异染色质染色技术，我们对每个克隆体的20~30个有丝分裂中期的染色体组型进行了分析。如果先用喹吖因对细胞进行荧光成像，然后再使用组成型异染色质染色技术，识别效果会增强。

通过将一个克隆体中某一特定的染色体频率与指定表型的表达强度联系起来，往往可以强化该基因在特定染色体上的定位。然而，克隆体可以产生两种类型的差异克隆。第一种类型，假设基因分配有效，染色体不能检测到，但其表型存在。我们已经证明了在大量案例[29]中存在隐性的、重排的染色体，能够解释这种染色体与基因之间关系的不一致。第二种类型的差异克隆体是存在特定的染色体，但缺少其相应的表型。这种类型的克隆体很难解释，其可能涉及染色体结构的精细重排、基因突变、基因表达的调节，或者在表型检测上的技术缺陷。

将基因定位到染色体的特定区域中是可能实现的，例如染色体臂或者由特定的染色反应界定的带型区域。利用染色体重排，如在人类亲代细胞群中的染色体易位和缺失，或者发生在杂交细胞中的染色体自发重排，我们可以完成上述的基因定位。

in the human parental cell population or by making use of spontaneous chromosome rearrangements which are generated in the hybrids. Translocations of chromosomes to or between chromosomes *X*, 17, and 16 are useful because these chromosomes possess selectable loci. A number of translocations affecting the same chromosome but with different breakpoints can be used to restrict the localization of genes. For this purpose it will be particularly useful to characterize and store in central repositories all detected human translocations, to serve as a library of rearrangement products for future somatic cell gene mapping. Programs of human mutant cell banking are now being formulated in several countries. An example of regional linkage assignments based on translocations in parental cells is cited below for the *X* chromosome.

It is also possible to make use of spontaneous, sporadic chromosome rearrangements which occur in hybrid cells to fix the location of genes within subregions of particular chromosome. Human chromosomes within hybrids may undergo rearrangement and even translocation to mouse chromosomes[29] and it has been possible to make use of such rearrangements to restrict the localization of the thymidine kinase gene to the long arm of human chromosome 17[29]. The translocation of human chromosome segments to the mouse chromosome set is significant from a genetic point of view because it may serve to restrict the further segregation of human genes involved in the translocation. It is conceivable that treatment of parental cells or hybrids with physical or chemical chromosome breaking agents could be used to induce chromosome rearrangements. Enrichment procedures could be devised to select particular classes of rearrangement products. Such systems could be used to increase the resolution of subregional chromosome gene assignments.

Known Human Linkage Groups

A significant number of syntenic relationships and chromosome assignments has been established by somatic cell genetics. A survey of the current linkage information is presented below for each of the human chromosomes in turn. The results are also summarized in Table 2. For chromosome 1, Van Cong *et al.*[30] have reported a syntenic relationship between phosphoglucomutase-1 (PGM_1) and peptidase C (Pep C) using mouse–human hybrids. This synteny has been confirmed by Ruddle *et al.*[31]. Using Chinese hamster–human hybrids, Westerveld *et al.*[32] have reported a synteny between 6-phosphogluconate dehydrogenase (PGD) and Pep C. Taken together, these findings imply that PGD, Pep C, and PGM_1 are all syntenic. Pep C has been assigned to chromosome 1 using mouse–human cell hybrids[31]. This assignment has been confirmed by the assignment of PGD to chromosome 1 independently by Bootsma *et al.* (personal communication) and Hamerton *et al.*[34] using Chinese hamster–human hybrids. If the findings based on cell hybrids are combined with linkages known from human pedigree analysis, the following additional gene markers can be assigned to chromosome 1: zonular pulverulent cataract, Duffy blood group, auriculo-osteodysplasia, salivary amylase, pancreatic amylase, elliptocytosis and rhesus blood group. For complete literature citations see Ruddle *et al.*[31]

在X、17号和16号染色体或它们之间进行的染色体易位是非常有用的，因为这些染色体具有筛选标记的位点。可以采用一些作用于同一染色体但具有不同断裂位点的染色体易位来限定基因的位置关系。为此目的，将所有检测到的人类染色体易位进行表征分析并集中储存，这对将来建立体细胞基因作图中重排产物的基因库非常有用。目前有几个国家正在酝酿建设人类突变细胞库。下文中将会提到的一个例子是基于亲代细胞X染色体上的染色体易位的区域连锁基因定位的。

我们还能够利用发生在杂交细胞中自发的、零星的染色体重排将基因定位在某个特定染色体的亚区域内。杂交后代中的人类染色体可以发生重排甚至易位至小鼠的染色体中[29]。我们已可以利用这样的重排技术将胸苷激酶基因定位到人类17号染色体的长臂上[29]。从遗传学的观点上来看，人类染色体片段易位至小鼠的染色体组中具有重要的意义，因为它可以用于限制参与染色体易位的人类基因的进一步分离。可以想象的是，用物理的或者化学的染色体断裂剂处理亲代细胞或者杂交细胞可以诱导染色体重排。同时人们设计出富集流程来筛选特定种类的重排产物。这样的体系可以用来提高亚区域内染色体基因定位的分辨率。

已知的人类连锁群

大量的同线性关系和染色体重排已经通过体细胞遗传学建立起来。下面将依次对每条人类染色体目前的连锁资料进行调查。并将调查结果归纳于表2中。对于1号染色体，范康等人[30]对小鼠–人类杂交细胞进行研究，报道了葡糖磷酸变位酶–1（PGM_1）和肽酶C（Pep C）之间的同线性关系。这样的同线性被拉德尔等人[31]所证实。韦斯特费尔德等人[32]利用中国仓鼠–人类杂交细胞得到了6–磷酸葡萄糖酸脱氢酶（PGD）和Pep C之间的同线性关系。结合两个研究结果，提示PGD、Pep C和PGM_1都是同线性关系。利用小鼠–人类杂交细胞，人们已经将Pep C基因定位到1号染色体中[31]。这个定位已经分别被布茨马等人（个人交流）以及哈默顿等人[34]独立地利用中国仓鼠–人类杂交细胞将6–磷酸葡萄糖脱氢酶（PGD）基因定位到1号染色体中所证实。如果将上述的基于杂交细胞的研究结果与从人类谱系分析中得到的已知连锁相结合，那么下列补充的基因标记就可以被定位到1号染色体中：带粉状白内障、达菲血型、耳骨发育异常、唾液淀粉酶、胰淀粉酶、椭圆形红细胞增多症和恒河猴血型。完整的引用文献请参见拉德尔等人的文章[31]。

Table 2. Assignments of Genes to Chromosomes

Chromosome 1	PGM_1, 17190; PGD, 17220; Pep C, 17000
Chromosome 2	IDH, 14770; MOR, 15425
Chromosome 3	—
Chromosome 4–5	Adenine B^+, 10265
Chromosome 6	MOD, 15420; IPO-B, 14745
Chromosome 7	MPI, 15455; PK_3, 17905
Chromosome 8–9	—
Chromosome 10	GOT, 13825
Chromosome 11	LDH-A, 15000; Es-A_4, 13340; KA, 14875
Chromosome 12	LDH-B, 15010; Pep B, 16990; $GlyA^+$ (serinehydroxymethylase ?), 13845
Chromosome 13	—
Chromosome 14	NP, 16405
Chromosome 15	
Chromosome 16	APRT, 10260
Chromosome 17	TK, 18830
Chromosome 18	Pep A, 16980
Chromosome 19	GPI, 23575
Chromosome 20	ADA, 10270
Chromosome 21	IPO-A, 14744; AVP, 10745
Chromosome 22	—
Chromosome X	HGPRT, 30800; PGK, 31180; GPD, 30590; α-Gal, 30150
Chromosome Y	—

These genes were assigned or confirmed by cell hybrid analysis. Each trait is identified by McKusick's human gene catalogue number[59]. IPO-A and B used here agree with the original designation of Brewer[33].

Preliminary results in our laboratory (R. P. Creagan and F. H. R.) suggest that isocitrate dehydrogenase (IDH) and NADP-malate dehydrogenase (MOD) which have been shown to be syntenic[44] can be assigned to chromosome 2. No loci have been assigned to chromosome 3. For chromosomes 4 and 5, Kao and Puck[35] using Chinese hamster–human hybrids have demonstrated a positive association between hamster adenine B auxotrophy and a human B group chromosome when hybrids are propagated on minimal medium. The specific enzyme involved is unknown. Chen *et al.*[36] using mouse–human hybrids have shown that cytoplasmic malate dehydrogenase (MOD) is assignable to chromosome 6 and there is evidence to support a syntenic relationship between indolephenol oxidase, tetrameric form B (IPO-B) and MOD (J. A. Tischfield, R. P. Creagan and F. H. R., unpublished).

表 2. 基因在染色体中的定位情况

1 号染色体	PGM_1，17190；PGD，17220；Pep C，17000
2 号染色体	IDH，14770；MOR，15425
3 号染色体	—
4~5 号染色体	腺嘌呤 B^+，10265
6 号染色体	MOD，15420；IPO-B，14745
7 号染色体	MPI，15455；PK_3，17905
8~9 号染色体	—
10 号染色体	GOT，13825
11 号染色体	LDH-A，15000；Es-A_4，13340；KA，14875
12 号染色体	LDH-B，15010；Pep B，16990；$GlyA^+$（丝氨酸羟甲基化酶？），13845
13 号染色体	—
14 号染色体	NP，16405
15 号染色体	
16 号染色体	APRT，10260
17 号染色体	TK，18830
18 号染色体	Pep A，16980
19 号染色体	GPI，23575
20 号染色体	ADA，10270
21 号染色体	IPO-A，14744；AVP，10745
22 号染色体	—
X 染色体	HGPRT，30800；PGK，31180；GPD，30590；α-Gal，30150
Y 染色体	—

这些基因的定位或确认是通过细胞杂交分析完成的。每一个特征都与麦库西克的人类基因目录编号一致[59]。这里提及的吲哚苯酚氧化酶 A 和 B（IPO-A 和 B）与布鲁尔[33]最初的命名一致。

我们实验室（克里根和拉德尔）初步的研究结果表明具有同线性关系[44]的异柠檬酸脱氢酶（IDH）和 NADP–苹果酸脱氢酶（MOD）基因都在 2 号染色体上。在 3 号染色体上没有可确定的位点。对于 4 号和 5 号染色体，高和普克[35]利用中国仓鼠 – 人类杂交细胞证明了杂交后代在基本培养基中进行繁殖时，仓鼠的腺嘌呤 B 营养缺陷型与人类的 B 族染色体之间呈现正相关。但是目前还不清楚具体参与的特异性酶。陈等人[36]利用小鼠 – 人类杂交细胞证明了胞质苹果酸脱氢酶（MOD）基因可以被定位到 6 号染色体中。也有证据表明四聚体型吲哚苯酚氧化酶 B（IPO-B）和 MOD 之间存在同线性的关系（蒂施菲尔德、克里根和拉德尔，未发表）。

McMorris *et al.*[37] using mouse *X* human hybrids have assigned mannose phosphate isomerase (MPI) to chromosome 7. Shows[38] has reported a syntenic relationship between MPI and the leucocytic form of pyruvate kinase$_3$ (PK$_3$).

There are no assignments to chromosomes 8 or 9. There is evidence from mouse–human hybrids for the assignment of the cytoplasmic form of glutamate oxaloacetate transaminase (GOT) to chromosome 10 (unpublished work of R. P. Creagan, J. A. Tischfield, F. A. McMorris, M. Hirschi, T. R. Chen and F. H. R.)

Boone *et al.*[29] using mouse–human hybrids have assigned lactate dehydrogenase A (LDH-A) to chromosome 11. Shows[39] using mouse–human hybrids has reported a syntenic association between LDH-A and human esterase A$_4$ (EsA$_4$). Van Someren *et al.*[40] have reported a syntenic association between glutamic-pyruvic transaminase (GPT-C) and LDH-A. The possibility exists, however, that their enzyme detection system is recording LDH-A activity. This may also apply to LDH-B and GPT-B (see below). Nabholz *et al.*[17] using mouse–human cells have reported a positive correlation between the segregation of LDH-A or B activity and sensitivity of hybrid cells to anti-human cytotoxic antisera. Puck *et al.*[41] have reported on the segregation of a possibly similar human antigen(s) in Chinese hamster–human hybrid cells, which they have found to be syntenic with LDH-A.

Ruddle and Chen[42] and Chen *et al.*[36] using mouse–human hybrids have demonstrated positive correlation between lactate dehydrogenase B (LDH-B) and chromosome 12. Hamerton *et al.*[34] have confirmed this assignment using Chinese hamster–human hybrids. In mouse–human hybrids a syntenic relationship has been demonstrated between LDH-B and peptidase-B (Pep-B)[14,43]. The Pep-B/LDH-B synteny has been confirmed by Shows[39,44], Van Cong *et al.*[30] and van Someren *et al.*, who have also reported a syntenic association between glutamic-pyruvic transaminase-B (GPT-B) and LDH-B[40]. Jones *et al.*[45] using Chinese hamster–human hybrids have reported a syntenic relationship between LDH-B and the complement to the Chinese hamster glycine auxotrophic mutant A. Serine hydroxymethylase has been implicated as the specific deficiency in glycine A auxotrophy.

No assignments have been made to chromosome 13. For chromosome 14, Ricciuti and Ruddle (ref. 46 and F. Ricciuti and F. H. R., unpublished) using a 14/*X* translocation in a human diploid fibroblastic cell strain (KOP) hybridized to a mouse cell line have demonstrated a segregation of nucleoside phosphorylase (NP) with the *X* linked markers, HGPRT, GPD, and PGK. Somatic cell genetic[46] and family studies[47] have provided evidence for the autosomal linkage of NP. The studies of Ricciuti and Ruddle thereby support the assignment of NP to chromosome 14. Unreported experiments from our laboratory using a 14/22 translocation also confirm the assignment of NP to 14. Hamerton *et al.*[34] have reported results based on Chinese hamster–mouse hybrids which are consistent with the above findings of Ricciuti and Ruddle. No assignments have been made to chromosome 15.

Tischfield and Ruddle (unpublished) using an adenine phosphoribosyltransferase (APRT)

麦克莫里斯等人[37]利用小鼠–人类杂交细胞将甘露糖磷酸异构酶（MPI）基因定位到7号染色体上。肖[38]报道了MPI和丙酮酸激酶3（PK_3）的白细胞型之间存在同线性关系。

没有基因定位于8号或9号染色体上。来自小鼠–人类杂交细胞的证据显示与细胞质型相关的谷氨酸–草酰乙酸转氨酶（GOT）基因可被定位到10号染色体上（克里根、蒂施菲尔德、麦克莫里斯、赫胥、陈和拉德尔，未发表）。

布恩等人[29]利用小鼠–人类杂交细胞将乳酸脱氢酶A（LDH-A）定位到11号染色体上。肖[39]利用小鼠–人类杂交细胞报道了LDH-A和人类酯酶A4（EsA_4）之间的同线性关系。范索梅伦等人[40]报道了谷氨酸–丙酮酸转氨酶C（GPT-C）和LDH-A之间的同线性关系。然而，他们的酶检测系统存在着同时记录LDH-A活性的可能。这样的方法也被应用到LDH-B和GPT-B中（见下文）。纳布霍尔茨等人[17]利用小鼠–人类细胞杂交后代报道了LDH-A或B活性性状分离与杂交细胞抗人类细胞毒性的抗血清敏感性之间的正相关性。普克等人[41]报道了在中国仓鼠–人类杂交细胞中可能类似于人类抗原的分离，并发现其与LDH-A具有同线性关系。

拉德尔和陈[42]以及陈等人[36]利用小鼠–人类杂交细胞证明了乳酸脱氢酶B（LDH-B）和12号染色体之间的正相关性。哈默顿等人[34]利用中国仓鼠–人类杂交细胞证实了这样的基因定位关系。在小鼠–人类杂交细胞中，已经证明LDH-B和肽酶B（Pep-B）之间具有同线性关系[14,43]。肖[39,44]、范康等人[30]和范索梅伦等人证实了LDH-B和Pep-B之间的同线性关系。范索梅伦等人还报道了谷氨酸–丙酮酸转氨酶B（GPT-B）和LDH-B之间的同线性关系[40]。琼斯等人[45]利用中国仓鼠–人类杂交细胞后代报道了LDH-B和中国仓鼠甘氨酸营养缺陷突变体A的补偿物之间的同线性关系。丝氨酸羟甲基化酶可能与甘氨酸A型营养缺陷中的某种缺陷有关。

13号染色体中没有可确定的位点。对于14号染色体，里丘蒂和拉德尔（文献46，以及里丘蒂和拉德尔未发表结果）利用人类二倍体成纤维细胞系（KOP）与小鼠细胞系杂交中的14号染色体和X染色体的易位（14/X易位）证明了核苷磷酸化酶（NP）与X染色体上的连锁标记HGPRT、葡萄糖–6–磷酸脱氢酶（GPD）和磷酸甘油酸激酶（PGK）的分离。体细胞遗传学[46]和家族研究[47]为NP与常染色体连锁提供了证据。因而，里丘蒂和拉德尔的研究结果支持了NP基因被定位到14号染色体的结论。我们实验室利用染色体14/22易位也证明了NP基因位于14号染色体（未发表）。哈默顿等人[34]利用中国仓鼠–人类杂交细胞报道的研究成果与上述里丘蒂和拉德尔的发现是一致的。15号染色体上没有找到可确定的基因。

蒂施菲尔德和拉德尔（未发表）利用腺嘌呤磷酸核糖转移酶（APRT）缺陷的小

deficient mouse cell line hybridized to normal human diploid cells have obtained evidence for the assignment of APRT to chromosome 16. On evidence from family studies, Robson *et al.*[48] have assigned α-haptoglobin to chromosome 16. APRT activity variants have been reported in man, and it would be reasonable to identify kindreds in which α-haptoglobin and APRT variants are jointly expressed to test for linkage between these two markers.

For chromosome 17, Green[49] and Migeon and Miller[50] using mouse–human cell hybrids assigned thymidine kinase (TK) to an E group chromosome. This assignment was based on the earlier findings of Weiss and Green[4] and has since been verified by Boone and Ruddle[51]. Miller *et al.*[52], Ruddle and Chen[53], and Boone *et al.*[29] have now assigned TK specifically to chromosome 17. Boone *et al.*[29], making use of a spontaneously occurring 17 translocation to a mouse chromosome, have provided evidence for the assignment of TK to the long arm of chromosome 17. Kit *et al.*[54] and McDougall *et al.*[55] have demonstrated that adenovirus 12 infection induces host TK activity, and concurrently a secondary constriction in the proximal segment of the long arm of 17. These findings suggest that the TK gene may be located near the adeno-12 induced gap region.

Creagan *et al.* (unpublished) using mouse–human cell hybrids have provided evidence for the assignment of peptidase-A (Pep-A) to chromosome 18.

Glucosephosphate isomerase (GPI) has been assigned to chromosome 19 on the basis of evidence from mouse–human cell hybrids[37]. Hamerton *et al.*[34] have confirmed this assignment using Chinese hamster–human cell hybrids. Linkage studies in the mouse have revealed a loose linkage between GPI and β haemoglobin: it will be of interest to test for a similar linkage relationship in human kindreds.

Boone *et al.*[29] using mouse–human hybrids reported a weak association between cytoplasmic isocitrate dehydrogenase (IDH), cytoplasmic maleate oxidoreductase (MOR) and chromosome 20. More extensive data from mouse–human hybrids have now shown that IDH and MOR cannot be assigned to 20 and using mouse–human hybrids we have obtained evidence of the assignment of tissue-specific adenosine deaminase (ADA) to chromosome 20 (J. A. Tischfield, R. P. Creagan and F. H. R., unpublished). ADA is asyntenic with both IDH and MOR. Family studies have demonstrated linkage between HL-A phosphoglucomutase 3, P blood group, and ADA[56].

On chromosome 21, Tan *et al.*[57], using somatic cell hybrids, have provided evidence for the syntenic association between indolephenoloxidase-A, dimeric (IPO-A) and a genetic factor (AVP) which controls an antiviral response specifically induced by human interferon. The genetic factor may regulate the interferon receptor and/or the antiviral protein. We have also shown that the interferon and AVP loci are asyntenic[57]. These results confirm earlier studies by Cassingena *et al.*[58] on their asyntenic association based on monkey–rat somatic cell hybrids. Tan *et al.*[57] have assigned AVP/IPO-A to chromosome 21. No assignments have yet been made to chromosome 22.

鼠细胞系与正常的人类二倍体细胞进行杂交，得到了将APRT基因定位到16号染色体的证据。在体细胞的家族研究所得到的证据基础上，罗布森等人[48]将α–珠蛋白定位到16号染色体上。人类APRT活性变体已有报道，通过α–珠蛋白和APRT变体的共表达现象来分析检测两个遗传标记间相互连锁的遗传关系是合理的。

对于17号染色体，格林[49]还有米金和米勒[50]利用小鼠–人类杂交细胞将胸腺嘧啶激酶（TK）基因定位到E组染色体中。这个基因定位关系基于韦斯和格林[4]的早期发现，并且已经被布恩和拉德尔[51]所证实。米勒等人[52]、拉德尔和陈[53]以及布恩等人[29]现在已将TK基因特异性地定位到17号染色体上。利用17号染色体对小鼠染色体的自发易位，布恩等人[29]提出证据表明TK基因应该定位在17号染色体的长臂上。基特等人[54]和麦克杜格尔等人[55]已证明腺病毒12感染会诱导宿主的TK活性，同时在最接近17号染色体长臂的区域引起二级收缩。这些发现提示TK基因也许就位于腺–12诱导的缺口区域附近。

克里根等人（未发表）利用小鼠–人类细胞的杂交后代提供了肽酶A（Pep-A）基因可被定位到18号染色体上的证据。

来自小鼠–人类细胞杂交后代的证据表明葡萄糖磷酸异构酶（GPI）基因位于19号染色体上[37]。哈默顿等人[34]后来利用中国仓鼠–人类细胞的杂交证实了这一基因定位关系。在小鼠身上进行的连锁分析提示葡萄糖磷酸异构酶（GPI）与β–血红蛋白之间松散的连锁关系：在人类近亲中检测类似的连锁关系将会非常有趣。

布恩等人[29]利用小鼠–人类细胞的杂交后代报道了细胞质异柠檬酸脱氢酶（IDH）、细胞质马来酸氧化还原酶（MOR）和20号染色体之间存在着弱关联。现在，来自小鼠–人类细胞杂交后代的更广泛的数据表明IDH和MOR基因是不能够被定位到20号染色体上的。我们利用小鼠–人类杂交细胞得到了组织特异性的腺苷脱氨酶（ADA）基因位于20号染色体上的证据（蒂施菲尔德、克里根和拉德尔，未发表）。ADA与IDH和MOR之间均是非同线性的。体细胞家族的研究证明人类白细胞抗原葡糖磷酸变位酶3、P血型和ADA之间存在着连锁关系[56]。

对于21号染色体，谭等人[57]利用体细胞杂交证明了二聚体型吲哚苯酚氧化酶A、（IPO-A）和一种遗传因子（精氨酸加压素，AVP）之间存在同线性关系。这种遗传因子可在人类干扰素的诱导下控制抗病毒反应，可以调节干扰素受体和（或）抗病毒蛋白。我们还证明了干扰素和AVP位点之间是非同线性关系[57]。这些结果证实了卡西季娜等人[58]早期在猴子–大鼠体细胞杂交上得到的干扰素和AVP之间非同线性关系的结论。谭等人[57]已经将AVP / IPO-A基因定位到21号染色体上。在22号染色体上还没有发现基因定位关系。

Glucose-6-phosphate dehydrogenase (GPD), hypoxanthine-guanine phosphoribosyltransferase (HGPRT), and phosphoglycerate kinase (PGK) have all been assigned to the *X* chromosome by segregation analysis in families[59]. Nabholz *et al.*[17] using mouse–human hybrids confirmed the *X* linkage of HGPRT. Meera Kahn *et al.*[60] have demonstrated the *X* linkage of PGK by cell hybrid analysis. Ruddle *et al.*[61] using mouse–human hybrids confirmed the *X* linkage of HGPRT, glucose-6-phosphate dehydrogenase (GPD), and Phosphoglycerate kinase (PGK). Grzeschik *et al.*[62] have recently provided evidence based on Chinese hamster–human hybrids for the assignment of α-galactosidase (α-Gal) to the *X* chromosome. In an earlier report, Grzeschik *et al.*[63] analysed cell hybrids between human KOP cells which possess a 14/*X* translocation (KOP) and mouse and Syrian hamster cells. They observed an infrequent segregation of HGPRT and GPD from PGK, which led them to postulate the assignment of PGK to the long arm, and the possible assignment of HGPRT and GPD to the short arm, although assignment to the long arm was not altogether discounted. Ricciuti and Ruddle (ref. 46 and unpublished results) using the same KOP material hybridized to mouse cells have obtained data which indicate that all three markers are located on the *X* long arm. This suggests that PGK is proximal to the centromere and distant from the other two markers. HGPRT and GPD seem to be close together and distal to the centromere with respect to PGK; preliminary evidence indicates that HGPRT is proximal to GPD. P. Gerald and co-workers (personal communication) have recently studied a human cell with a 19/*X* translocation hybridized to mouse cells. A translocation product composed of the 19 long arm, the proximal half of 19 short arm, and the distal half of the *X* long arm was correlated with GPI, HGPRT, and GPD, but not PGK. This result is consistent with the human *X* linkage map proposed by Ricciuti and Ruddle[46]. It also confirms the assignment of GPI to 19.

No assignments have been made to the Y chromosome.

Possibilities for New Approaches

The development of a detailed human genetic map is certain to provide insight into the evolutionary origins of man and the primates. LDH-A and LDH-B are located on chromosomes 11 and 12 respectively. These chromosomes are similar in size, centromere position, and banding pattern, which is consistent with the occurrence of a primordial polyploid event in the early primate genome as discussed by Comings[64]. Somatic cell genetic analysis should be feasible for representative members of the order primates using rodent–primate hybrids. Linkage data from such hybrids should provide information on the relatedness of these forms, and yield estimates for rates of evolutionary divergence, especially when combined with comparative studies on the chromosome constitutions and amino-acid sequences of proteins in the representative specimens.

It is already obvious that somatic cell genetics has contributed and will in the future contribute important data to human genetics. Moreover, these developments enhance

通过对体细胞杂交细胞系分离的分析发现，葡萄糖–6–磷酸脱氢酶（GPD）、次黄嘌呤 – 鸟嘌呤磷酸核糖转移酶（HGPRT）和磷酸甘油酸激酶（PGK）基因都可以被定位到 X 染色体上[59]。纳布霍尔茨等人[17]利用小鼠 – 人类细胞杂交证实了 X 染色体与 HGPRT 之间的连锁关系。米拉 · 卡恩等人[60]通过对细胞杂交的分析证明了 PGK 与 X 染色体的连锁关系。拉德尔等人[61]利用小鼠 – 人类细胞杂交证实了 X 染色体与 HGPRT、葡萄糖–6–磷酸脱氢酶（GPD）和磷酸甘油酸激酶（PGK）之间的连锁关系。最近，格尔策希克等人[62]利用中国仓鼠 – 人类细胞杂交提供了 α–半乳糖苷酶（α-Gal）定位到 X 染色体上的证据。在早期的报道中，格尔策希克等人[63]分析了人类 KOP 细胞（具有 14/X 的易位）与小鼠和叙利亚仓鼠细胞之间的细胞杂交。他们观察到 HGPRT 和 GPD 与 PGK 的低频分离，这使得他们推测出这样的结论：PGK 基因应该位于染色体长臂，而 HGPRT 和 GPD 基因则在短臂上（尽管其在长臂上的可能性并没有完全排除）。里丘蒂和拉德尔（文献 46 和未发表结果）利用同样的 KOP 细胞与小鼠细胞进行杂交得到的数据显示上述的三种酶标记都位于 X 染色体的长臂上。这提示 PGK 的位点接近着丝粒并远离其他两种酶标记。HGPRT 和 GPD 基因似乎挨得很近，且相对于 PGK 基因而言远离着丝粒；初步证据提示 HGPRT 基因的位置最接近 GPD 基因。近来，杰拉尔德和他的同事们（个人交流）对 19/X 易位的人类细胞与小鼠细胞的杂交进行了研究。杂交后得到的易位产物由 19 号染色体的长臂、19 号染色体短臂近端的一半和 X 染色体长臂远端的一半组成，该产物与 GPI、HGPRT 和 GPD 相关，而与 PGK 无关。该结果与里丘蒂和拉德尔[46]提出的人类 X 染色体连锁图谱是相一致的，还证实了 GPI 基因位于 19 号染色体上。

在 Y 染色体上没有发现基因定位关系。

新方法的契机

建立详细的人类基因图谱无疑使人们能够更加深入地了解人类与灵长类动物的进化起源。LDH-A 和 LDH-B 基因分别位于 11 号和 12 号染色体上。这两条染色体的大小、着丝粒的位置以及带型都是相似的。这与科明斯[64]讨论的在早期灵长类动物基因组中发生过原始多倍体的事实相一致。通过啮齿目 – 灵长类的细胞杂交，对具有代表性的灵长目动物进行体细胞遗传分析应是可行的。这些杂交后代的连锁数据可以提供有关这类关联性的信息，并且可以对进化趋异的速率进行评估，尤其是在结合了对代表性样本中染色体结构和氨基酸序列的比较研究以后。

很显然，体细胞遗传学已经并将继续为人类遗传学提供重要的数据。此外，体细胞遗传学的发展提高了家族和群体遗传研究的重要性和在今后的作用。通过体细胞遗传学中准确的信息，即某些基因对是同线性的，已经绘制的基因间图谱得到完

the significance and future role of family and population genetic studies. Already map distances between genes have been refined or established by the certain knowledge from somatic cell genetics that certain gene pairs are syntenic. We should expect a fruitful interaction and collaboration between practitioners of somatic cell genetics and classical genetics. We must, however, keep clearly in mind that somatic genetic procedures as they now exist are still unacceptably slow and linkage estimates cannot yet be made. If we are soon to develop genetic maps of man comparable to those available for the lower eukaryotes, it will be necessary to develop new procedures. Possibilities for this are to be found in somatic cell recombination and gene transfer.

(**242**, 165-169; 1973)

Frank H. Ruddle
Kline Biology Tower, Yale University, New Haven, Connecticut 06520

References:
1. Barski, G., Sorieul, S., and Cornefert, F., *CR Acad. Sci.*, **251**, 1825 (1960).
2. Ephrussi, B., and Weiss, M. C., *Proc. US Nat. Acad. Sci.*, **53**, 1040 (1965).
3. Ephrussi, B., and Weiss, M. C., *Develop. Biol. Suppl. I.*, 136 (1967).
4. Weiss, M. C., and Green, H., *Proc. US Nat. Acad. Sci.*, **58**, 1104 (1967).
5. Harris, H., and Watkins, J. F., *Nature*, **205**, 640 (1965).
6. Okada, Y., and Murayama, F., *Exp. Cell Res.*, **52**, 34 (1968).
7. Lucy, J. A., *Nature*, **227**, 815 (1970).
8. Littlefield, J. W., *Science*, **145**, 709 (1964).
9. Kusano, T., Long, C., and Green, H., *Proc. US Nat. Acad. Sci.*, **68**, 82 (1971).
10. Puck, T. T., and Kao, F. A., *Proc. US Nat. Acad. Sci.*, **58**, 1227 (1967).
11. Kao, F. T., and Puck, T. T., *Nature*, **228**, 329 (1970).
12. Thompson, L. H., Mankovitz R., Baker, R. M., Tell, J. E., Seminovitch, L., and Whitmore, G. F., *Proc. US Nat. Acad. Sci.*, **66**, 377 (1970).
13. Davidson, R. L., and Ephrussi, B., *Nature*, **205**, 1170 (1965).
14. Santachiara, A. S., Nabholz, M., Miggiano, V., Darlington, A. J., and Bodmer, W., *Nature*, **227**, 248 (1970).
15. Jami, J., Grandchamp, S., and Ephrussi, B., *CR Acad. Sci.*, **272**, 323 (1971).
16. Ruddle, F. H., *Adv. Hum. Genet.*, **3**, 173 (1972).
17. Nabholz, M., Miggiano, V., and Bodmer, W., *Nature*, **223**, 358 (1969).
18. Ruddle, F. H., and Nichols, E., *In Vitro*, **7**, 120 (1971).
19. Davidson, R. L., *In Vitro*, **6**, 411 (1971).
20. Ricciuti, F., thesis, Yale Univ. (1972).
21. Caspersson, T., Zech, L., Johansson, C., and Modest, E., *Chromosoma*, **30**, 215 (1970).
22. Miller, O. J., Miller, D. A., Kouri, R. E., Alderdice, P. W., Dev, V. G., Grewal, M. S., and Hutton, J. J., *Proc. US Nat. Acad. Sci.*, **68**, 1530 (1971).
23. Sumner, A. T., Evans, H. J., and Buckland, R. A., *Nature New Biology*, **232**, 31 (1971).
24. Pardue, M. L., and Gall, J. G., *Science*, **168**, 1356 (1970).
25. Arrighi, F. E., and Hsu, T. C., *Cytogenetics*, **10**, 81 (1971).
26. Chen, T. R., and Ruddle, F. H., *Chromosoma*, **34**, 51 (1971).
27. Ruddle, F. H., *Symp. Intern. Soc. Cell Biol.*, **9**, 233 (1970).
28. Bobrow, M., Madan, K., and Pearson, P. L., *Nature New Biology*, **238**, 122 (1972).
29. Boone, C. M., Chen, T. R., and Ruddle, F. H., *Proc. US Nat. Acad. Sci.*, **69**, 510 (1972).
30. Van Cong, N., Billerdon, C., Picard, J. Y., Feingold, J., and Frizal, J., *CR Acad. Sci.*, **272**, 485 (1971).
31. Ruddle, F. H., Ricciuti, F., McMorris, F. A., Tischfield, J., Creagan, R., Darlington, G., and Chen, T. R., *Science*, **176**, 1429 (1972).
32. Westerveld, A., and Meera Khan, P., *Nature*, **236**, 30 (1972).
33. Brewer, G. J., *Amer. J. Hum. Genet.*, **19**, 674 (1967).
34. Hamerton, J., *Cytogenetics* (in the press).
35. Kao, F. T., and Puck, T. T., *Proc. US Nat. Acad. Sci.*, **69**, 3273 (1972).

善或建立。我们可以期待从事体细胞遗传学和经典遗传学研究的科研人员进行富有成果的互动与协作。但是，我们必须牢记，现在的体外遗传学的研究速度仍然慢得令人难以接受，也不能进行连锁评估。如果我们要尽快作出人类的基因图谱，即类似于那些从低等真核生物上获得的，那么必须要制定新的研究方法。而对这种新的研究方法的发现将寄希望于对体细胞重组和基因转移的研究。

（刘振明 翻译；梁前进 审稿）

36. Chen, T. R., McMorris, F. A., Creagan, R., Ricciuti, F., Tischfield, J., and Ruddle, F. H., *Amer. J. Hum. Genet.* (in the press).

37. McMorris, F. A., Chen, T. R., Ricciuti, F., Tischfield, J., Creagan, R., and Ruddle, F. H., *Science* (in the press).

38. Shows, T. B., Abstr., *Amer. J. Hum. Genet.*, **24**, 13a (1972).

39. Shows, T. B., *Proc. US Nat Acad. Sci.*, **69**, 348 (1972).

40. van Someren, H., Meera Khan, P., Westerveld, A., and Bootsma, R., *Nature New Biology*, **240**, 221 (1972).

41. Puck, T. T., Wuthier, P., Jones, C., and Kao, F., *Proc. US Nat. Acad. Sci.*, **68**, 3102 (1971).

42. Ruddle, F. H., and Chen, T. R., in *Genetics and the Skin, Twenty-first Ann. Symp.: Biol. Skin* (1972).

43. Ruddle, F. H., Chapman, V. M., Chen, T. R., and Kleke, R. J., *Nature*, **227**, 251 (1970).

44. Shows, T., *Biochem. Genet.*, **7**, 193 (1972).

45. Jones, C., Wuthier, P., Kao, F., and Puck, T. T., *J. Cell Physiol.* (in the press).

46. Ruddle, F. H., in *The Use of Long Term Lymphocytes in the Study of Genetic Diseases* (edit. by Bergsma, D., Smith, G., and Bloom, A.) (National Foundation, in the press).

47. Edwards, Y. H., Hopkinson, D. A., and Harris, H., *Ann. Human Genetics*, **34**, 395 (1971).

48. Robson, E. B., Polani, P. E., Dart, S. J., Jacobs, P. A., and Renwick, J. H., *Nature*, **223**, 1163 (1969).

49. Green, H., in *Heterospecific Genome Interaction Wiston Inst. Lymp. Monograph*, No. **9**, 51 (1969).

50. Migeon, B. R., and Miller, O. S., *Science*, **162**, 1005 (1968).

51. Boone, C. M., and Ruddle, F. H., *Biochem. Genet.*, **3**, 119 (1969).

52. Miller, O. J., Alderdice, P. W., Miller, D. A., Breg, W. R., and Migeon, B. R., *Science*, **173**, 244 (1971).

53. Ruddle, F. H., and Chen, T. R., in *Perspectives in Cytogenetics* (edit. by Wright, S. W., and Crandall, B. F.) (Charles C. Thomas, Illinois, 1971).

54. Kit, S., Nakajima, K., and Dubbs, D. R., *J. Virol.*, **5**, 446 (1970).

55. McDougall, J. K., *J. Gen. Virol.*, **12**, 43 (1971).

56. Edwards, J. E., Allen, F. H., Glenn, K. P., Lamm, L. U., and Robson, E. B., *Histocompatibility Testing* (in the press).

57. Tan, Y. H., Tischfield, J., and Ruddle, F. H., *J. Exp. Med.* (in the press).

58. Cassingena, R., Chany, C., Vignal, M., Suarez, H., and Estrade, S., *Proc. US Nat. Acad. Sci.*, **68**, 580 (1971).

59. McKusick, V. A., *Mendelian Inheritance in Man*, third ed. (Johns Hopkins Press, Baltimore, 1971).

60. Meera Khan, P., Westerveld, A., Grzeschik, K. H., Deys, B. F., Garson, O. M., and Siniscalco, M., *Amer. J. Hum. Genet.*, **23**, 614 (1971).

61. Ruddle, F. H., Chapman, V. M., Ricciuti, F., Murnane, M., Klebe, R., and Meera Khan, P., *Nature*, **232**, 69 (1971).

62. Grzeschik, K. H., Romeo, G., Grzeschik, A. M., Banhof, S., Siniscalco, M., van Someren, H., Meera Khan, P., Westerveld, A., and Bootsma, R., *Nature* (in the press).

63. Grzeschik, K. H., Alderdice, P. W., Grzeschik, A., Opitz, J. M., Miller, O. J., and Siniscalco. M., *Proc. US Nat. Acad. Sci.*, **69**, 69 (1972).

64. Comings, D. E., *Nature*, **238**, 455 (1972).

Isolation of the Islets of Langerhans for Transplantation

D. R. Thomas *et al.*

Editor's Note

This paper describes a method for isolating rat islets of Langerhans as a potential transplant source for diabetes. Attempts to treat diabetes by transplanting a whole pancreas had met with limited success, and efforts to isolate the component insulin-producing islets of Langerhans were hampered by methodological problems and low viability. The revised method, by D. R. Thomas and colleagues at Sheffield in England, includes digestion, washing and separation steps and yielded large numbers of viable islets of Langerhans. Islet cell transplantation is currently being assessed as a potential treatment for type I diabetes mellitus, where the immune system destroys insulin-producing beta cells in the islets of Langerhans. Successful transplants have reduced or removed the need for insulin therapy.

PANCREATIC transplantation with the aim of treating diabetes mellitus has so far met with little success. Of 23 patients thus treated and reported to the Transplant Registry in 1971[1], 15 died within 3 months and the longest survived one year. One of the major problems has been to overcome pancreatic exocrine digestion, and pancreatic duct ligation ("Banting pancreas") before transplantation has been performed. It was shown by Dragstedt[2], however, that dogs treated in this way often became diabetic or showed a diabetic glucose tolerance test after several months, probably due to fibrosis and consequent ischaemia of the Islets. Transplantation of the whole gland with its vascular supply is a major undertaking and the problems of thrombosis, leakage and digestion, coupled with immunological rejection, have prevented success so far.

Attempts have been made to isolate the Islets of Langerhans from the pancreas in order to study glucose metabolism. A micro-dissection technique was described by Hellerstrom[3] but this was tedious and produced only small numbers of Islet clumps. Subsequently Moskalewski[4] described a method of collagenase digestion which was successful in the rabbit and improved the yield, and Lacy and Kostianovski[5] modified the method for the rat. There were still difficulties in harvesting the liberated Islets, however, even with methods of separation such as zonal centrifugation and density gradients, and there were also problems with viability of the cells.

We have developed a technique by which large numbers of Islets of Langerhans are prepared consistently from the rat pancreas for purposes of transplantation. Viability was confirmed by transplantation.

Initially, albino outbred rats and subsequently young adult inbred hooded rats (strain PVG/C) were used. Following killing by cervical dislocation, the ventral surface of the

用于移植的朗格汉斯岛（胰岛）的分离

托马斯等

编者按

本文介绍了一种用于分离大鼠朗格汉斯岛（胰岛）的方法，以用作治疗糖尿病的潜在移植来源。通过移植整个胰腺治疗糖尿病的尝试收效甚微，而分离产生胰岛素的胰岛的尝试受到了方法学问题和低生存力的限制。由托马斯和他的同事在英国谢菲尔德所改良的方法，获得了大量可存活的胰岛，这种方法包含消化、洗涤和分离几个步骤。现在胰岛细胞移植正被评估是否可作为一种治疗Ⅰ型糖尿病的潜在方法，在Ⅰ型糖尿病中，免疫系统会破坏产生胰岛素的胰岛β细胞。成功的移植可以减少或消除对胰岛素治疗的依赖。

迄今为止，胰腺移植用于治疗糖尿病成效甚微。1971 年，经过这样治疗并在移植登记处登记备案的 23 例患者中 [1]，有 15 例在 3 个月内死亡，生存时间最长的患者也只有 1 年。一个主要的问题就是患者在移植前需要克服胰腺外分泌消化和胰管结扎（“班廷胰腺”）。但是，德拉格施泰特 [2] 发现，经过这样治疗的狗在数月后会患糖尿病或者表现出糖尿病糖耐量试验的结果，这很可能是由于胰岛的纤维化以及继发的贫血造成的。伴血液供应的全腺体移植是目前的主流，但是伴随免疫排斥发生的血栓、渗漏和消化等问题使得该方法到目前为止也没能取得成功。

人们试图从胰腺中分离出胰岛以研究葡萄糖代谢。赫勒斯特伦 [3] 描述了一种显微解剖技术，但是这种方法非常烦冗，而且只能得到少数的胰岛细胞团。此后，莫斯卡勒夫斯基 [4] 描述了一种胶原酶消化法，该法已经在兔子中取得成功，而且分离出的胰岛数量也得到提高，后来莱西和科斯蒂安诺夫斯基 [5] 在大鼠中对这个方法进行了改进。但是，即便使用了诸如区带离心和密度梯度离心的分离方法，仍然很难获得游离的胰岛，而且这些细胞的生存能力也是一个大问题。

我们建立了一种方法，并用此法从大鼠胰腺中稳定制备了大量用于移植的胰岛，并且通过移植确认了细胞的生存能力。

起初使用的是白化的远交系大鼠，后来使用年轻成年的头部有斑点的顶罩大鼠（PVG/C 系）的近交系。颈椎脱臼法处死大鼠后，刮除腹面毛发，用溶于酒精的氯

animal was shaved, prepared with chlorhexidine in spirit and the abdomen opened with a midline incision. Using a dissecting microscope, magnification × 10, a fine polyethylene catheter ("Intracath"—B. R. Bard, London) was introduced into the common bile duct and secured with a 2-0 linen thread ligature. The lower end of the duct was occluded with an artery forceps just before its entry into the duodenum. The pancreas was distended by injection of 10 ml. Hanks' solution containing bovine albumen (fraction V), 2 mg · ml.$^{-1}$. It was found that after some practice this procedure could be completed within 5 min of the death of the animal. The pancreas was removed, transferred to a glass Petri dish, cut into small pieces with scissors and any excess fat removed. It was then transferred to a tube to which a further 10 ml. of Hanks' solution was added. The pancreatic tissue sank to the bottom and any remaining fat floated on the surface and was readily aspirated and discarded. The prepared tissue was transferred to a small conical flask together with 2 ml. Hanks' solution containing glucose 0.6 mg ml.$^{-1}$ and collagenase type 1 (Sigma Chemicals, London). The stoppered flask was placed in a shaking water bath at 37°C for about 30 min. The exact time for separation was determined by frequent sampling and examination under the dissecting microscope.

The digested pancreas was transferred to a tube and diluted with further cold Hanks' solution containing glucose in the same concentration as before and gently centrifuged for 1 min. The supernatant was discarded and fresh medium added, and the resultant suspension filtered into a Petri dish with a blackened base for examination under the dissecting microscope at × 10 magnification.

Initially the view was obscured by fine fragments of acinar tissue, which were removed by gently agitating the Petri dish and allowing them to become suspended in the Hanks' solution, then aspirated and discarded, the Islets remaining on the bottom of the vessel. They could then be seen clearly (Fig. 1) by the aid of dark ground illumination as yellowish white domes and were picked out with a finely drawn Pasteur pipette. Fig. 2 shows the histological appearance of an isolated pancreatic Islet. Its architecture and cells appear normal.

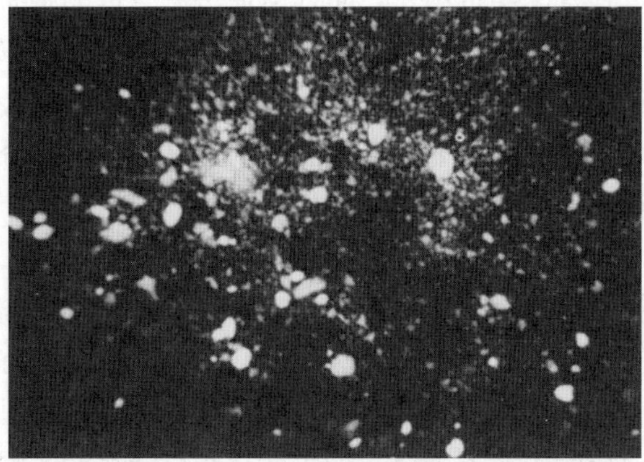

Fig. 1. Islet cell tissue seen after isolation from the pancreas using dark ground illumination. Dissecting microscope, ×5.

己定处理后选取正中线切口开腹。使用立体显微镜放大 10 倍，将一条细小的聚乙烯导管（“Intracath”（译者注：一种留置针）—巴德，伦敦）插入胆总管，并用 2–0 的亚麻缝合线固定。在导管恰好进入十二指肠前，用动脉钳夹闭其下端。注射 10 ml 含有 2 mg · ml^{-1} 牛血清蛋白（第五组分）的汉克氏液（译者注：一种用于细胞培养的平衡盐溶液）使胰腺膨胀。经过一些实践之后我们发现，这些步骤可以在动物死后 5 分钟之内完成。将胰腺摘除并转移到一个玻璃培养皿内，用剪刀将其剪碎并去除所有多余的脂肪。然后将其转移到一个试管内，并加入 10 ml 的汉克氏液。胰腺组织沉到管底，所有残存的脂肪都会漂浮在表面，将它们吸出并丢弃。将制备好的组织转移到一个小的锥形瓶内，该瓶内装有 2 ml 含 0.6 mg · ml^{-1} 葡萄糖和 I 型胶原酶（西格玛化学公司，伦敦）的汉克氏液。把盖好的瓶子置于 37℃水浴摇床内约 30 分钟。分离过程的限速步骤是立体显微镜下的频繁取样和检查。

将消化好的胰腺转移到一个试管内，用含有相同浓度葡萄糖的冷的汉克氏液稀释，温和离心 1 分钟。弃上清液并加入新鲜的培养液（汉克氏液），得到的悬浮液过滤到培养皿内；该培养皿的底为暗色，以便于在放大 10 倍的立体显微镜下进行观察。

起初，许多细小的腺泡组织碎片使得视野很模糊；轻轻地摇动培养皿使其悬浮在汉克氏液中，然后将其吸出并丢弃，而胰岛则存留在容器底部。通过暗场照明能够清晰地看到它们呈黄白色的圆顶状（图 1），用尖端拉长的巴斯德吸管将其移出。图 2 显示了分离出来的胰岛的组织学形态。其结构和细胞表现正常。

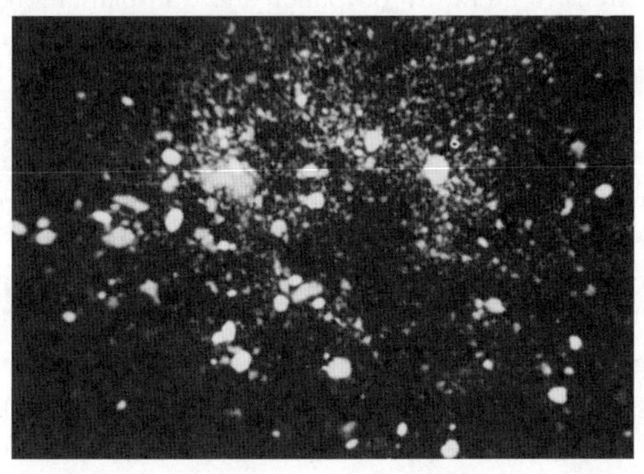

图 1. 通过暗场照明观察到的从胰腺中分离出来的胰岛细胞组织。立体显微镜下放大 5 倍。

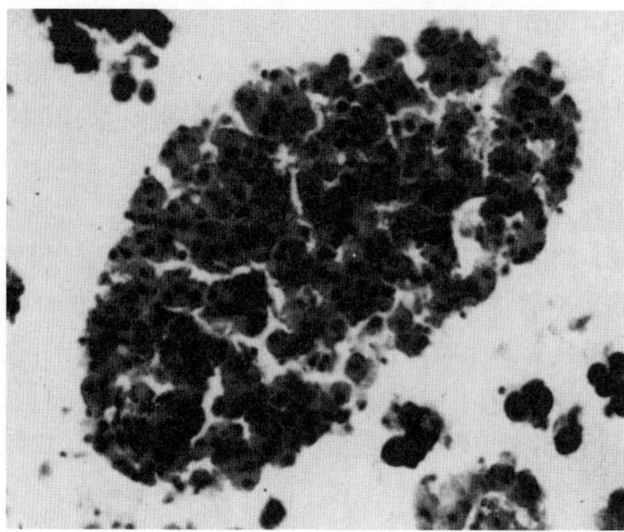

Fig. 2. Isolated pancreatic Islet. Normal architecture and appearance of cells. Haematoxylin and eosin, ×300.

Yields of up to 350 Islets per rat pancreas have been achieved using this method. To obtain larger quantities, rat pancreases have been processed in batches of four.

Viability of the isolated cells was confirmed by transplantation beneath the renal capsule and into the testis of isogeneic rats. The longest period of follow-up was one month, when viable looking Islet cells containing beta cell granules staining with aldehyde-fuchsin were seen. A similar method has been successfully applied in the rabbit.

The relationship of the Islets of Langerhans to diabetes mellitus was established in 1889 by Von Mering and Minkowski[6], and in 1892 Hedon[7] demonstrated that subcutaneous implantation of a small piece of pancreas could delay the appearance of diabetes in an animal that had undergone pancreatectomy. The concept of Islet cell grafting appears to be neglected in the extensive literature on pancreatic transplantation, although one early report suggests that transplanted Islets may modify alloxan diabetes in the rat[8].

We have now established a successful method of isolation of Islets of Langerhans in an animal strain in which inbred immunologically isogeneic lines are available, making possible transplantation studies uncomplicated by rejection problems. These cells can be grafted and survive for appreciable periods of time as shown by the viability studies. Histological and functional studies of Islet cell transplantation will be reported elsewhere.

We acknowledge a grant from the Endowment Fund of the United Sheffield Hospitals and from the Medical Research Council. We thank Dr. Laurence Henry for his help with the histology and Mr A. Tunstill and his staff for photography.

(**242**, 258-260; 1973)

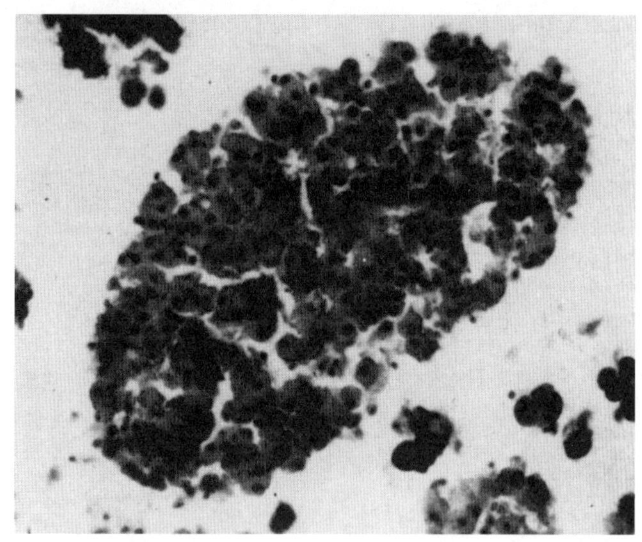

图 2. 分离出的胰岛。正常的结构和细胞外观。苏木精—伊红染色，放大 300 倍。

用这种方法能够在每个大鼠胰腺中得到多达 350 个胰岛。为了获得更多胰岛，大鼠胰腺以四个为一组一起处理。

将分离的胰岛细胞移植到同种大鼠的肾小囊内或者睾丸内以确认其生存能力。最长的随访期是 1 个月，此时可以观察到成活的胰岛细胞含有能被醛复红染色的 β 细胞颗粒。同样的方法在兔子中也获得了成功。

1889 年，冯 · 梅灵和明科夫斯基阐述了胰岛和糖尿病之间的关系[6]。1892 年，埃东[7]证明了移植到皮下的一小块胰腺可以延缓进行胰腺切除术后的动物出现糖尿病症状。在大量有关胰腺移植的文献中，人们都忽略了胰岛细胞移植的想法，尽管有一篇早期报道提出移植的胰岛可能会减轻在大鼠中由四氧嘧啶诱导的糖尿病[8]。

现在，我们已经在近交的免疫同系的动物品系中成功地建立了一种分离胰岛的方法，这使得移植研究摆脱排斥问题的困扰成为可能。正如在对胰岛细胞的生存力研究中所证实的，这些细胞能够被移植并能生存相当长的时间。胰岛细胞移植的组织学和功能方面的研究将另作报道。

我们感谢联合谢菲尔德医院的捐助基金和医学研究理事会的资金。我们感谢劳伦斯·亨利博士帮助进行组织学检查，感谢滕斯蒂尔先生及其职员在成像方面的帮助。

（毛晨晖 翻译；王敏康 审稿）

D. R. Thomas, M. Fox and A. A. Grieve
Royal Hospital, Sheffield

Received October 12, 1972. Requests for reprints to M. F.

References:

1. *Brit. Med. J.*, i, 326 (1972).
2. Dragstedt, L. R., *Ann. Surg.*, **118**, 576 (1943).
3. Hellerstrom, C., *Acta Endocr.*, 45, 122 (1964).
4. Moskalewski, S., *Gen. and Comp. Endocrin.*, **5**, 342 (1965).
5. Lacy, P. E., and Kostianovski, M., *Diabetes*, **16**, 35 (1967).
6. Von Mering, J., and Minkowski, O., in Major, R. M., *Classic Descriptions of Disease*, second ed., 246 (Charles C. Thomas, Springfield, Ill., 1939).
7. Hedon, E., *Compt. Rend. Soc. Biol.*, 44, 678 (1892).
8. Younoszai, R., Sorenson, R. L., and Lindall, jun., A. W., *Diabetes*, 19, Suppl 1, 406 (1970).

Evidence for an Advanced Plio-Pleistocene Hominid from East Rudolf, Kenya

R. E. F. Leakey

Editor's Note

Richard Leakey followed his father Louis' vocation in palaeontology at first very reluctantly, but soon began to make remarkable discoveries of his own. This one, from the eastern shore of Lake Rudolf (now Lake Turkana), was indeed spectacular—a skull and other skeletal material attributed to the genus *Homo*. The face of "1470 man" (after its catalogue number, KNM-ER 1470), with its remarkably human shape and relatively high forehead, was especially emotive. The specimen was later assigned to a new taxon, *Homo rudolfensis*. Notable in this study is Leakey's assertion of the age of this specimen at more than 2.6 million years: this was later found, after much controversy, to be an overestimate of almost a million years.

Four specimens collected last year from East Rudolf are provisionally attributed to the genus *Homo*. One, a cranium KNM-ER 1470, is probably 2.9 million years old.

PRELIMINARY descriptions are presented of four specimens collected from East Rudolf during 1972. Most of the collection recovered during this field season has been reported recently in *Nature*[1]; the specimens described here are sufficiently important to be considered separately and in more detail. The collections of fossil hominids recovered from East Rudolf during earlier field seasons and detailed descriptions of some of these specimens have been published previously[2-5].

The specimens described here are: (1) a cranium, KNM-ER 1470; (2) a right femur, KNM-ER 1472; (3) a proximal fragment of a second right femur, KNM-ER 1475; and (4) an associated left femur, distal and proximal fragments of a left tibia, and a distal left fibula, KNM-ER 1481. They were all recovered from area 131 (see Fig. 1) and from deposits below the KBS Tuff which has been securely dated at 2.6 m.y.[6].

Area 131 consists of approximately 30 km^2 of fluviatile and lacustrine sediments. The sediments are well exposed and show no evidence of significant tectonic disturbance; there is a slight westward dip of less than 3°. Several prominent marker horizons provide reference levels and have permitted physical correlation of stratigraphical units between area 131 and other areas in the East Rudolf locality.

来自肯尼亚鲁道夫湖以东的一个高级上新世-更新世人科动物证据

利基

编者按

理查德·利基最初并不情愿承接他父亲路易斯的职业，从事古生物学的研究，但是很快他就有了自己的重大发现。在鲁道夫湖（即现在的图尔卡纳湖）东畔发现了一件头骨和其他的骨骼材料，这些标本均可被归入人属，这是十分令人震惊的发现。其中，带有显著人类形态和较高前额的“1470号人”（根据其标本号“KNM-ER 1470”命名）的面部尤为令人激动。这个标本后来被归类为一个新种：人属鲁道夫种（*Homo rudolfensis*）。本次研究中值得关注的一点是，利基原本断定这个标本的年代至少为260万年前，但是经过后来的许多争论，人们发现这个年代被高估了约100万年。

去年在鲁道夫湖以东搜集到的四件标本暂时被归入人属。其中的一件颅骨KNM-ER 1470可能有290万年的历史。

本文是对1972年间在鲁道夫湖以东搜集到的四件标本的初步描述。本次野外作业期间搜集到的大部分标本最近已经在《自然》杂志上进行了报道[1]；本文描述的标本特别重要，因此需要单独研究并进行更详细的介绍。以往在鲁道夫湖以东的野外考察中发掘到的人科动物化石标本及对其中部分标本的详细描述也已经发表[2-5]。

本文描述的标本包括：(1) 一件颅骨，KNM-ER 1470；(2) 一件右侧股骨，KNM-ER 1472；(3) 另一件右侧股骨的近端破片，KNM-ER 1475；以及 (4) 相关联的左侧股骨，左侧胫骨的远端和近端破片以及左侧腓骨的远端，KNM-ER 1481。这些标本都是从131区（见图1）的被确定为有260万年历史的KBS凝灰岩以下的沉积物中发掘出来的[6]。

131区由大约30 km^2的河流沉积物和湖沼沉积物组成。这些沉积物都充分暴露，并且没有证据显示发生过重大的地质构造扰动；只有一个向西倾斜的不足3°的小倾角。几个明显的标志性地层可作为参考地层，据此可以将鲁道夫湖以东地点的131区和其他几个区之间的地层单元进行岩性对比。

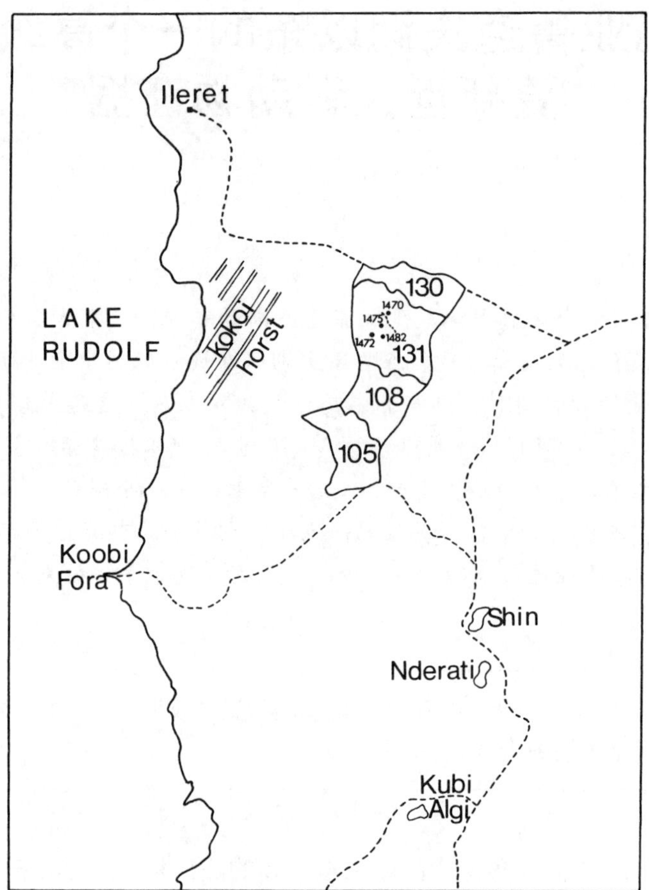

Fig. 1. Map showing sites of discovery of fossil hominids KNM-ER 1470, 1472, 1475 and 1481 in the East Rudolf locality. Succession shown in Fig. 2 was taken from the position indicated by the dotted line.

Several tuffs occur in the vicinity of area 131. The lowest of these is the Tulu-Bor Tuff which is not exposed in the area itself but does outcrop nearby in several stream beds. Above this horizon, in a composite section, there is some 60 m of sediment capped by the prominent KBS Tuff. This latter tuff has been mapped into areas 108 and 105 (also shown in Fig. 1) from where samples have been obtained for K/Ar dates. An account of the geology is given by Vondra and Bowen[7]. A section showing the vertical position of these four hominids in relation to the KBS Tuff is given in Fig. 2.

At present, analysis of samples collected for dating from the KBS Tuff in area 131 has proved inconclusive because of the apparent alteration of the sanidine felspars. This was not seen in the 105/108 samples from the same horizon which provided the date of 2.61 m.y. and there is no reason to suspect the validity of that date (personal communication from J. A. Miller).

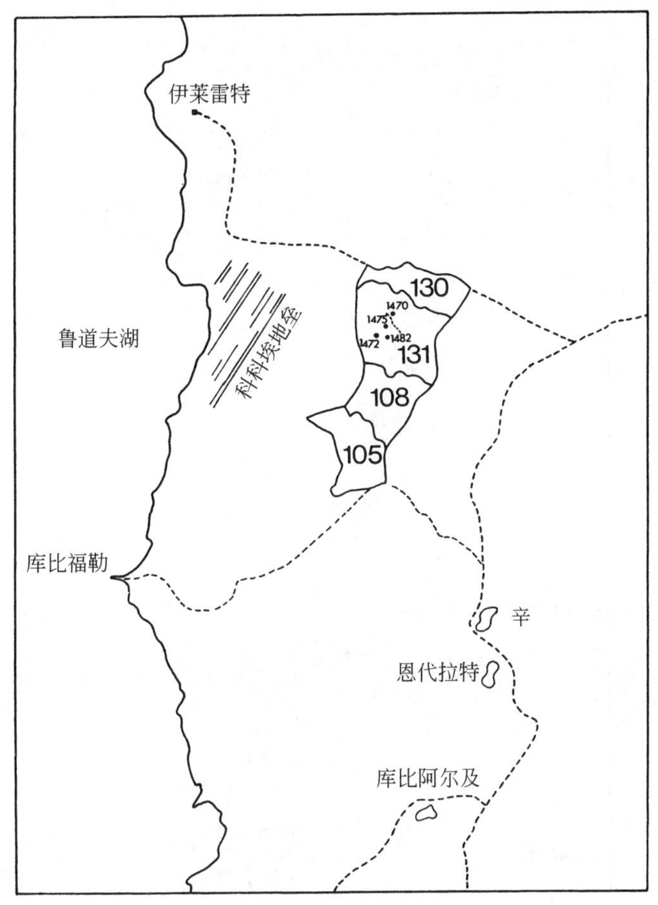

图 1. 此图显示了在鲁道夫湖以东遗址出土人科动物化石 KNM-ER 1470、1472、1475 和 1481 的地点位置。图 2 显示的剖面序列取自本图中虚线所示的位置。

131 区周围存在好几层凝灰岩。其中处于最底层的是 Tulu-Bor 凝灰岩，它本身在该区域并没有暴露出来，仅在几处河床附近有露头。在这层之上，是一个复合剖面，其中有被 KBS 凝灰岩覆盖着的约 60 m 厚的沉积物。这些凝灰岩已经在 108 区和 105 区被定位（见图 1），K/Ar 年代测定的样品就是在这两个区采集的。冯德拉和鲍恩已经对该处的地质情况进行了说明[7]。图 2 给出了垂直方向上这四个人科动物标本的位置与 KBS 凝灰岩关系的剖面示意图。

目前，由于透长石存在明显的变化，所以对从 131 区的 KBS 凝灰岩层采集到的样品进行的年代测定分析结果已不具有决定性。但这种现象并未在 105/108 区同一层位采集到的被确定为 261 万年前的样品中出现，因此，没有理由怀疑 261 万年这一年代的真实性（来自与米勒的个人交流）。

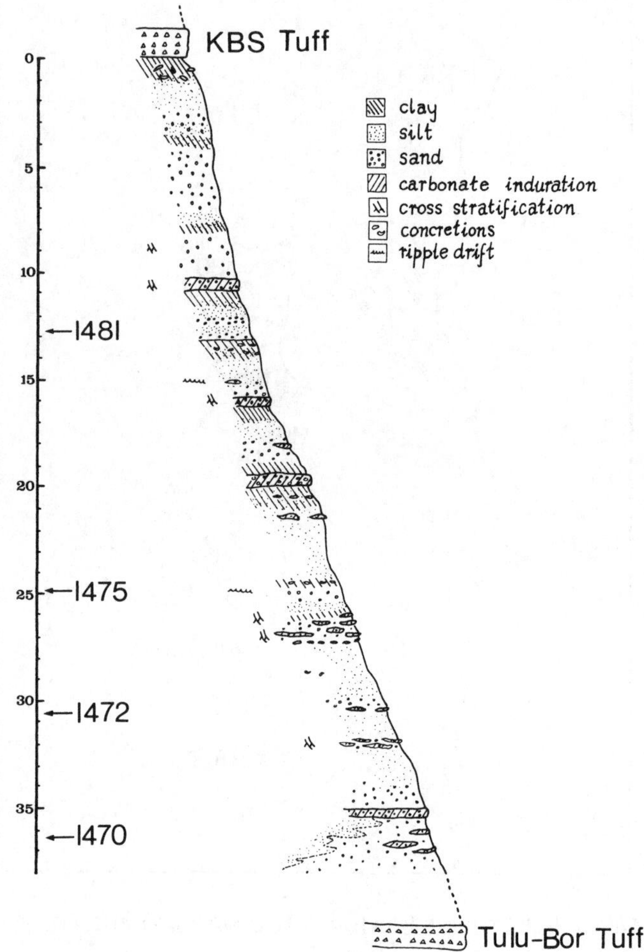

Fig. 2. Stratigraphical succession of the sediments in area 131 and the vertical relationships of the fossil hominids KNM-ER 1470, 1472, 1475 and 1481 to the KBS Tuff. Dotted line shown in Fig. 1 marks the position at which the section was taken.

Detailed palaeomagnetic investigation of the sedimentary units is being undertaken by Dr. A. Brock (University of Nairobi). Systematic sampling closely spaced in the section has identified both the Mammoth and Kaena events in area 105 between the Tulu-Bor and KBS Tuffs, a result which supports the 2.61 m.y. date on the latter. The mapping of several horizons has established a physical correlation between areas 105 and 131. During the 1973 season, the area 131 succession will be sampled in detail in an attempt to confirm this correlation. Available evidence points to a probable date of 2.9 m.y. for the cranium KNM-ER 1470, and between 2.6 and 2.9 m.y. for the other specimens reported here.

Collections of vertebrate fossils recovered from below the KBS Tuff in areas 105, 108 and 131 all show the same stage of evolutionary development and this evidence supports the

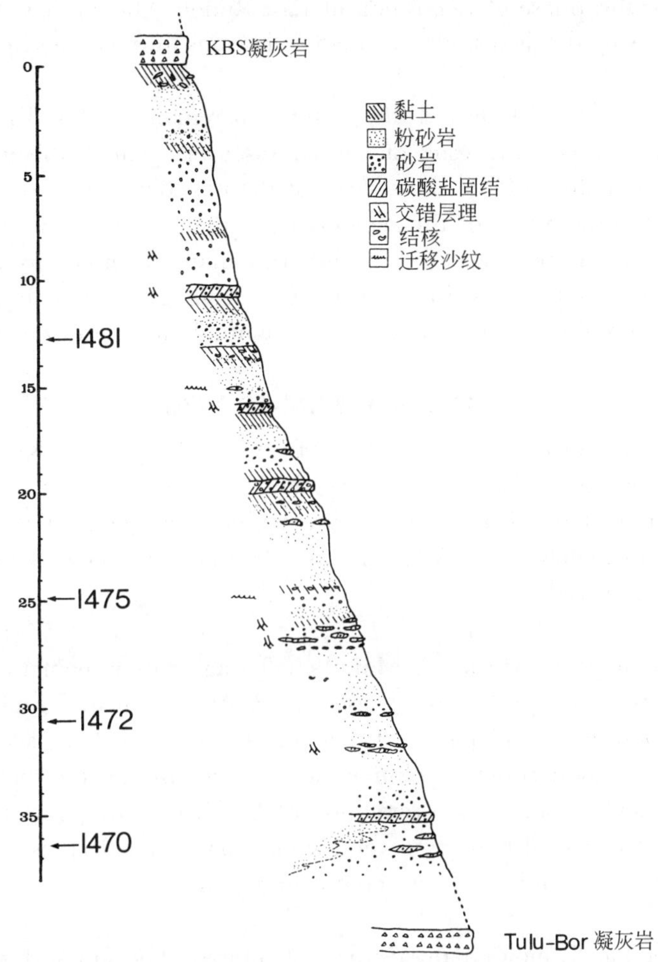

图 2. 此图显示了 131 区沉积物的地层序列以及人科动物化石 KNM-ER 1470、1472、1475 和 1481 与 KBS 凝灰岩在垂直方向上的关系。图 1 中虚线表示该剖面所选择的位置。

布罗克博士（内罗毕大学）正在对沉积物单元进行详细的古地磁学研究。在剖面处进行的小间隔的系统采样工作已经识别出了 105 区的 Tulu-Bor 和 KBS 凝灰岩层间与马莫斯事件和凯纳事件相对应的地层，该结果支持基于 KBS 凝灰岩得到的 261 万年的年代测定结果。对几处地层进行的绘图已经在 105 区和 131 区之间建立了岩性上的相关关系。在 1973 年的挖掘工作中，我们将对 131 区地层序列进行详细采样以期进一步确定这一关系。现有证据表明颅骨 KNM-ER 1470 的年代可能为 290 万年，而本文报道的其他标本的年代则介于 260 万年到 290 万年之间。

从 105、108 和 131 区的 KBS 凝灰岩之下的地层发掘出来的脊椎动物化石都显示出它们处于同样的演化阶段，这一证据支持鲁道夫湖以东沉积物的年代测定结果。

indicated age for this phase of deposition at East Rudolf. Maglio[8] has discussed the fossil assemblages following detailed studies of field collections from various horizons.

The cranium (KNM-ER 1470) and the postcranial remains (KNM-ER 1472, 1475 and 1481) were all recovered as a result of surface discovery. The unrolled condition of the specimens and the nature of the sites rules out the possibility of secondary deposition—there is no doubt in the minds of the geologists that the provenance is as reported. All the specimens are heavily mineralized and the adhering matrix is similar to the matrix seen on other fossils from the same sites. In due course, microscopic examination of thin sections of matrix taken from the site and on the fossils might add further evidence.

Cranium KNM-ER 1470

Cranium KNM-ER 1470 was discovered by Mr. Bernard Ngeneo, a Kenyan, who noticed a large number of bone fragments washing down a steep slope on one side of a gully. Careful examination showed that these fragments included pieces of a hominid cranium. An area of approximately 20 m × 20 m was subsequently screened and more than 150 fragments were recovered.

The skull is not fully reconstructed. Many small fragments remain to be included and it may be some time before the task is completed. At present the cranial vault is almost complete and there are good joins between the pieces. The face is less complete and although there are good contacts joining the maxilla through the face to the calvaria, many pieces are still missing. The orientation of the face is somewhat uncertain because of distortion of the frontal base by several small, matrix filled cracks. The basicranium shows the most damage and is the least complete region.

The cranium (see Fig. 3) shows many features of interest. The supraorbital tori are weakly developed with no continuous supratoral sulcus. The postorbital waisting is moderate and there is no evidence of either marked temporal lines or a temporal keel. The vault is domed with steeply sloping sides and parietal eminences. The glenoid fossae and external auditory meati are positioned well forward by comparison with *Australopithecus*. The occipital area is incomplete but there is no indication of a nuchal crest or other powerful muscle attachments.

In view of the completeness of the calvaria, it has been possible to prepare in modelling clay an endocranial impression which has been used to obtain minimum estimates for the endocranial volume. Six measurements of the endocast by water displacement were made by Dr. A. Walker (University of Nairobi), and gave a mean value of 810 cm^3. Further work on this will be undertaken but it seems certain that a volume of greater than 800 cm^3 for KNM-ER 1470 can be expected.

马利奥[8]在对来自各个不同地层的采集物进行详细研究后，对该化石群进行了讨论。

颅骨（KNM-ER 1470）和颅后骨骼（KNM-ER 1472、1475 和 1481）都是在地表采集的。标本的状况和发现标本的地点的性质排除了二次沉积的可能性——地质学家们都确认关于化石出产层位的报道属实。所有标本的石化程度都很高，附着于其上的围岩与同一遗址的其他化石中看到的围岩都很相似。对从该遗址和化石上取得的围岩进行薄切片显微镜检测已经列入研究计划之中，可能会进一步充实这方面的证据。

颅骨 KNM-ER 1470

颅骨 KNM-ER 1470 由肯尼亚的伯纳德 · 恩格奈奥先生发现，他注意到在一条冲沟一侧的陡坡上有大量骨骼破片被冲下来。经过仔细研究后发现这些破片包括了数块人科动物颅骨。随后对约 20 m × 20 m 的区域进行了筛查，又发现了 150 多块破片。

该头骨没能完全重建出来。还有许多小碎片没有找到，所以重建整个头骨的工作尚需时日才能完成。现在头骨穹隆几乎是完整的，各个碎片之间的连接情况很好。与之相比，面部则稍欠完整，尽管上颌骨从面颅到脑颅之间的部分有着很好的连接，但是还有许多骨片缺失。因为几处小的、填有填充物的裂缝造成额底有些变形，所以脸部的朝向尚未确定。颅底是受损最严重的区域，也是最不完整的部位。

颅骨（见图 3）显示出许多有趣的特征。眶上圆枕并不发达，也没有连续的圆枕上沟。眶后缩狭程度中等，没有显著的颞线或颞龙骨突。颅顶两侧的斜坡和顶骨结节突出组成了颅顶的圆拱。与南方古猿相比，颞骨关节窝和外耳道的位置更靠前。枕区不完整，但没有显示存在项嵴或其他有力的肌肉附着的迹象。

鉴于颅骨的完整性，可以使用黏土制作颅内模，以获得其颅容量的最小估计值。沃克博士（内罗毕大学）通过测量排水量得到了内腔模型的 6 项尺寸，给出了 810 cm^3 的颅容量平均值。对于这方面的进一步研究还会继续，但可以预料到的是，KNM-ER 1470 的颅容量似乎肯定大于 800 cm^3。

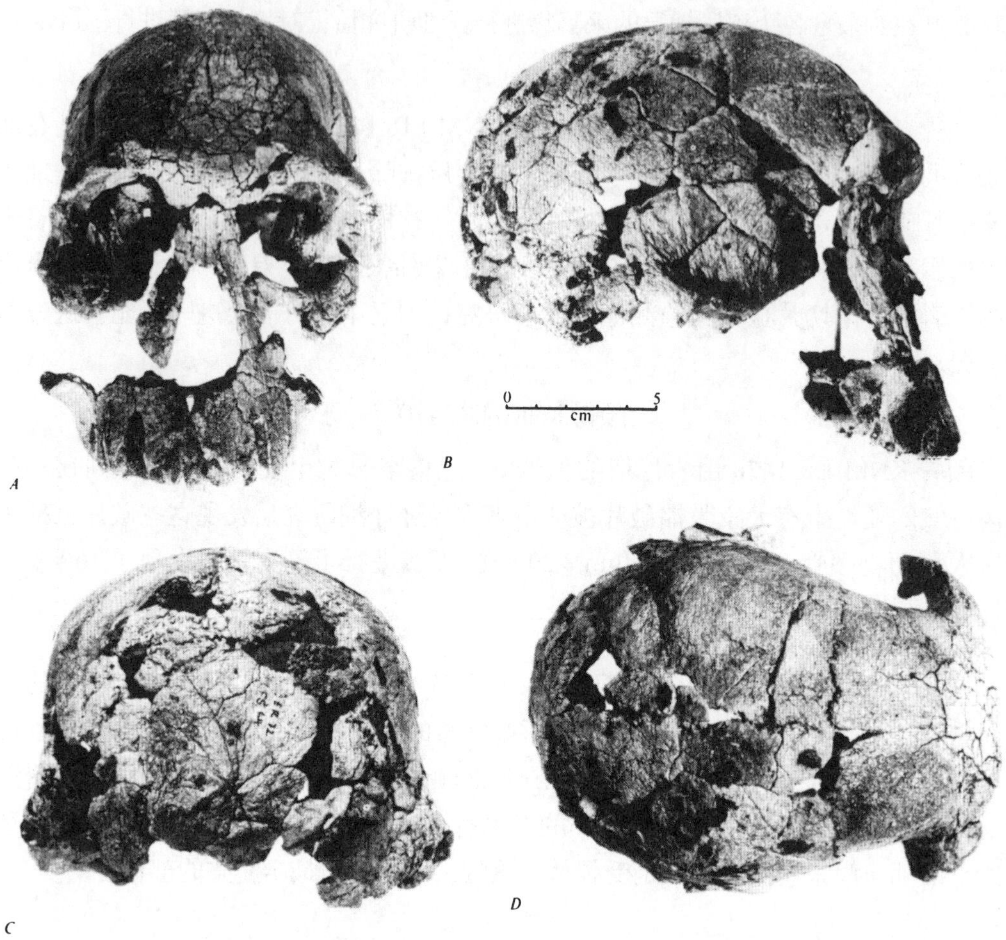

Fig. 3. Cranium KNM-ER 1470. *a*, Facial aspect; *b*, lateral aspect; *c*, posterior aspect; *d*, superior aspect.

The palate is shallow, broad and short with a nearly straight labial border that is reminiscent of the large *Australopithecus*. The great width in relationship to the length of the palate does contrast markedly, however, with known australopithecine material. The molars and premolar crowns are not preserved, but the remaining roots and alveoli suggest some mesiodistal compression. The large alveoli of the anterior teeth suggest the presence of substantial canines and incisors.

Femur KNM-ER 1472

KNM-ER 1472, a right femur, was discovered as a number of fragments by Dr. J. Harris. It shows some features that are also seen in the better preserved left femur, KNM-ER 1481, but other features, such as the apparently very straight shaft and the bony process on the anterior aspect of the greater trochanter, require further evaluation.

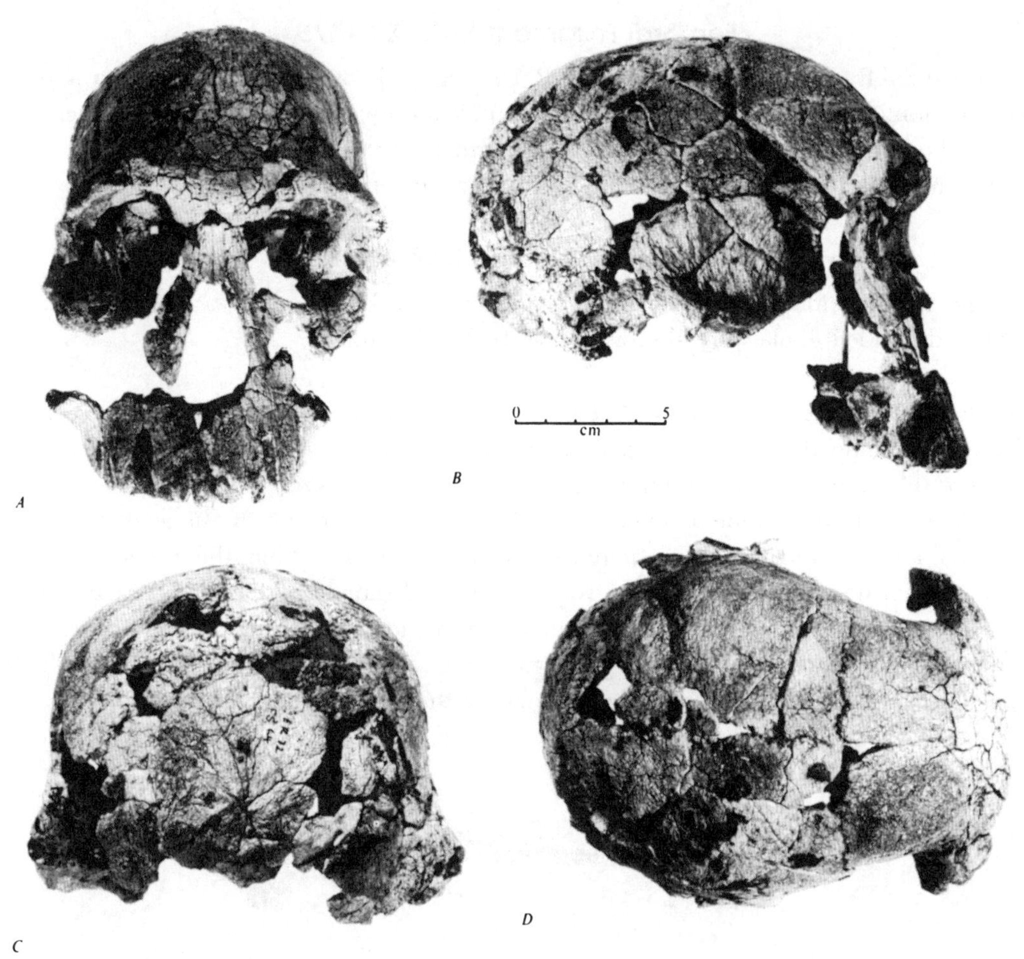

图 3. 颅骨 KNM-ER 1470。*a*，前面观；*b*，侧面观；*c*，后面观；*d*，上面观。

腭浅、宽而短，唇边缘接近笔直，这使人想起了大型南方古猿。然而，腭宽度与长度的比例偏大，这又与已知的南方古猿材料有着明显的不同。臼齿和前臼齿齿冠没有保存下来，但是残存的牙根和齿槽表明其受到了来自近中远侧的挤压。前牙的巨大齿槽表明其具有相当大的犬齿和门齿。

股骨 KNM–ER 1472

KNM-ER 1472 是一件右侧股骨，它被哈里斯博士发现时是一堆破片。该标本具有的某些特征在保存相对较好的左侧股骨 KNM-ER 1481 中也有体现，但是其他特征，例如非常笔直的骨干和大转子前面的骨质突起等，还需要进一步评估。

Femoral Fragment KNM-ER 1475

The proximal fragment of femur, KNM-ER 1475, was discovered by Mr. Kamoya Kimeu. Its condition is such that a final taxonomic identification will be difficult and it is therefore included only tentatively in this report. This fragment shows some features such as a short, more nearly cylindrical neck, which are not seen in the femurs of *Australopithecus*.

Associated Skeleton KNM-ER 1481

A complete left femur, KNM-ER 1481, associated with both ends of a left tibia and the distal end of a left fibula were also discovered by Dr. J. Harris.

The femur (see Fig. 4) is characterized by a very slender shaft with relatively large epiphyses. The head of the femur is large and set on a robust cylindrical neck which takes off from the shaft at a more obtuse angle than in known *Australopithecus* femurs. There is a marked insertion for gluteus maximus and the proximal region of the shaft is slightly flattened antero-posteriorly. The femoro-condylar angle is within the range of *Homo sapiens*. When the femur is compared with a restricted sample of modern African bones, there are marked similarities in those morphological features that are widely considered characteristic of modern *H. sapiens*. The fragments of tibia and fibula also resemble *H. sapiens* and no features call for specific comment at this preliminary stage of study.

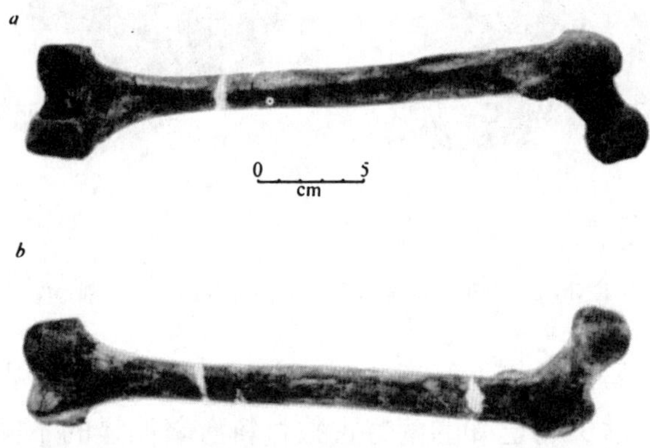

Fig. 4. Left femur KNM-ER 1481. *a*, Posterior aspect; *b*, anterior aspect.

Homo or *Australopithecus*?

The taxonomic status of the material is not absolutely clear, and detailed comparative studies which should help to clarify this problem have yet to be concluded. The endocranial capacity and the morphology of the calvaria of KNM-ER 1470 are characters that suggest inclusion within the genus *Homo*, but the maxilla and facial region are unlike those of any known form of hominid. Only the flat fronted wide palate is suggestive of *Australopithecus*, but its extreme shortening and its shallow nature cannot be matched in existing collections representing this genus. The postcranial elements cannot readily be distinguished from *H. sapiens* if one considers the range of variation known for this species.

股骨破片 KNM–ER 1475

股骨近端的破片，KNM-ER 1475，由卡莫亚·基梅乌先生发现。鉴于该标本的情况，很难对其进行最终的分类学鉴定，因此在本文中只是暂且将其列出。该破片的一些特征，例如短的、接近圆柱形的颈部等特征在南方古猿的股骨中都不曾出现。

相关联的骨骼 KNM–ER 1481

哈里斯博士还发现了一根完整的左侧股骨 KNM-ER 1481，这根股骨与同时发现的左侧胫骨两端以及左侧腓骨远端相关联。

该股骨（见图 4）的特征是具有很细长的骨体以及相对较大的骨骺。股骨头部大，长在一个粗壮的从骨体生出的圆柱状颈上，该颈部与骨体形成的钝角比已知的南方古猿股骨的要大。臀大肌有一个明显的插入，骨体的近端区域在前后方向上有些扁平。股骨髁突与股骨的角度属于智人的范畴。当把该股骨与一个特定的现代非洲骨骼标本进行比较时，普遍认为它们在形态学特征上与现代智人的特点有显著相似性。胫骨和腓骨破片也与智人相似，就目前的初步研究阶段来看，还没有需要特别讨论的特征。

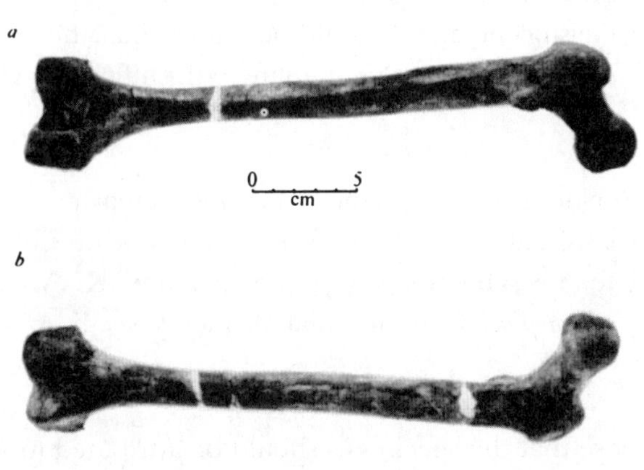

图 4. 左侧股骨 KNM-ER 1481。a，后面观；b，前面观。

是人还是南方古猿？

该材料的分类学位置还不十分清楚，有助于阐明该问题的详细比较学研究尚未得出结论。KNM-ER 1470 的颅容量和颅顶形态学特征表明其属于人属，但是其上颌骨和面部区域又不像任何已知的人科动物。只有前移的宽阔腭表明其可能属于南方古猿，但是它极短且浅的特征却又不能与现存的代表此属的标本相匹配。如果考虑到智人种已知的变异范围，其颅后骨骼结构无法与智人明确地区别开。

The East Rudolf area has provided evidence of the robust, specialized form of *Australopithecus* from levels which span close to 2 m.y. (2.8 m.y.–1.0 m.y.)[1]; throughout this period the morphology of this hominid is distinctive in both cranial and postcranial elements. The cranial capacity of the robust australopithecine from Olduvai Gorge, *A. boisei*, has been estimated for OH 5 to be 530 cm^3 (ref. 9); this is the same value as that estimated by Holloway for the only specimen in South Africa of *A. robustus* which provides clear evidence of cranial capacity[9]. Holloway has also found the mean cranial capacity of six specimens of the small gracile *A. africanus* from South Africa[10] to be 422 cm^3. Thus, to include the 1470 cranium from East Rudolf within the genus *Australopithecus* would require an extraordinary range of variation of endocranial volume for this genus. This seems unacceptable and also other morphological considerations argue strongly against such an attribution.

The Olduvai Gorge has produced evidence of an hominine, *H. habilis*; the estimated endocranial volumes for three specimens referred to this species are 633, 652 and 684 cm^3 (ref. 10). The Olduvai material is only known from deposits that are stratigraphically above a basalt dated at 1.96 m.y. (ref. 11). At present therefore there does not seem to be any compelling reason for attributing to this species the earlier, larger brained, cranium from East Rudolf.

The 1470 cranium is quite distinctive from *H. erectus* which is not certainly known from deposits of equivalent Pleistocene age. It could be argued that the new material represents an early form of *H. erectus*, but at present there is insufficient evidence to justify this assertion.

There is no direct association of the cranial and postcranial parts at present, and until such evidence becomes available, the femora and fragment of tibia and fibula are only provisionally assigned to the same species as the cranium, KNM-ER 1470. Differences from the distinctive *Australopithecus* postcranial elements seem to support this inferred association.

For the present, I propose that the specimens should be attributed to *Homo* sp. indet. rather than remain in total suspense. There does not seem to be any basis for attribution to *Australopithecus* and to consider a new genus would be, in my mind, both unnecessary and self defeating in the endeavor to understand the origins of man.

I should like to congratulate Mr. Ngeneo and Dr. Harris for finding these important discoveries. Dr. Bernard Wood spent many long hours at the site screening for fragments and assisted my wife, Meave, and Dr. Alan Walker in the painstaking reconstruction work. I thank them all. The support of the National Geographic Society, the National Science Foundation, the W. H. Donner Foundation and the National Museum of Kenya is gratefully acknowledged.

(**242**, 447-450; 1973)

鲁道夫湖以东地区将近200万年跨度（距今280万到100万年）的地层中出土了粗壮种南方古猿这一特化类群存在的证据[1]；这整个时期里，这种人科动物在颅骨和颅后骨骼结构的形态方面都是独特的。奥杜威峡谷发现的一种粗壮型南方古猿——南方古猿鲍氏种OH5的颅容量估计为530 cm^3（参考文献9）；这与霍洛韦估计的南非唯一一个能提供明确的颅容量数据的南方古猿粗壮种的数值相同[9]。霍洛韦还计算了在南非[10]发现的小型纤细南方古猿非洲种的六个标本的颅容量，其平均值为422 cm^3。因此，要将鲁道夫湖以东发现的1470号颅骨纳入南方古猿属中，就要求该属的颅容量具有更大的变异范围。这似乎并不能为人们所接受，另外还有其他的形态学因素也强烈反对这种分类。

奥杜威峡谷已经出土了一种人类——能人的化石；估计三个被归入该种的标本的颅腔容量分别为633 cm^3、652 cm^3和684 cm^3（参考文献10）。我们仅仅知道奥杜威标本是从年代为距今196万年的玄武岩之上的地层堆积物中得到的（参考文献11）。因此现在似乎还没有令人信服的理由可以将该标本划分到这种鲁道夫湖以东发现的更早期的、具有较大大脑的颅骨的物种中去。

1470号颅骨与从更新世同期沉积物中采集到的研究不充分的直立人非常不同。可以这样说，这件新标本代表了直立人的一种早期形式，但是现在还没有足够的证据来证实这种说法。

目前这具颅骨和颅后部分没有直接的联系，除非出现相反证据，否则只能暂时将股骨和胫骨、腓骨的破片与颅骨KNM-ER 1470归为同一种。新标本与独特的南方古猿颅后骨骼结构之间存在的差异似乎支持这种推测出的联系。

目前，我提议将该标本归入人属未定种，而非像原来那样完全悬而未决。我认为，将其划分为南方古猿似乎没有任何依据，而将其定义为一种新属既没有必要，而且在理解人类起源这一问题上也只会弄巧成拙。

我想祝贺恩格奈奥先生和哈里斯博士取得了这些重要的发现。伯纳德·伍德博士在遗址处花费了很长时间筛查骨骼破片，并协助我的妻子米芙和艾伦·沃克博士进行艰苦的重建工作。我要感谢他们所有人。对于国家地理学会、国家科学基金会、唐纳基金会和肯尼亚国家博物馆的支持，也表示衷心的感谢。

（刘皓芳 翻译；吴新智 崔娅铭 审稿）

R. E. F. Leakey
National Museums of Kenya, PO Box 40658, Nairobi

Received January 23, 1973.

References:

1. Leakey, R. E. F., *Nature*, **242**, 170 (1973).
2. Leakey, R. E. F., *Nature*, **231**, 241 (1971).
3. Leakey, R. E. F., *Nature*, **237**, 264 (1972).
4. Leakey, R. E. F., Mungai, J. M., and Walker, A. C., *Amer. J. Phys. Anthrop.*, **35**, 175 (1971).
5. Leakey, R. E. F., Mungai, J. M., and Walker, A. C., *Amer. J. Phys. Anthrop.*, **36**, 235 (1972).
6. Fitch, F. J., and Miller, J. A., *Nature*, **226**, 223 (1970).
7. Vondra, C., and Bowen, B., *Nature*, **242**, 391 (1973).
8. Maglio, V. J., *Nature*, **239**, 379 (1972).
9. Tobias, P. V., *The Brain in Hominid Evolution* (Columbia University Press, New York and London, 1971).
10. Holloway, R. L., *Science*, **168**, 966 (1970).
11. Curtis, G. H., and Hay, R. L., in *Calibration of Hominoid Evolution* (edit. by W. W. Bishop and J. A. Miller) (Scottish Academic Press, Edinburgh, 1972).

Practical Application of Acupuncture Analgesia

S. B. Cheng and L. K. Ding

Editor's Note

Around this time, the widespread use of acupuncture analgesia across China was drawing interest from the West. Here Chinese medical doctors S. B. Cheng and L. K. Ding sum up pros and cons for the use of the technique in surgery. The procedure, which is cheaper and carries fewer complications than analgesic drugs, works particularly well in patients with a confident outlook. But it fails to dull the nerves as effectively as drug analgesia, and only really works well in upper-body operations such as thyroidectomies and lung surgery. Given this, the duo suggests that acupuncture analgesia sometimes be considered an alternative but not a replacement to drug therapy.

Acupuncture analgesia can sometimes be considered as an alternative to drugs, but it cannot be considered as a universal replacement. There are certain cases, however, where acupuncture analgesia is better than drugs and this technique needs to be developed further in order to assess its true significance. In this article two Chinese doctors, one of whom has conducted twenty-four operations using acupuncture analgesia, assess the advantages and disadvantages of using the technique in surgery.

SINCE the reported success of the use of acupuncture analgesia in several hundred thousand operations in China, great interest in this type of analgesia has been aroused all over the world. In Japan, America and other nations, medical people have begun to investigate the application of acupuncture analgesia in their daily medical and clinical work. This article is an attempt to give some explanation of the technique in order to assist others who are experimenting with it. It is based on the personal experience of one of us who has used this type of analgesia in twenty-four operations—three mitral stenoses, five lobectomies, six gastrectomies and ten thyroidectomies.

Acupuncture analgesia uses no anaesthetic drugs. Instead it is based on Chinese traditional medical theory, *Ching-lo*, which states that by applying pressure to certain specific points on the body, these points (on the meridians) will become numb. Pressure is applied by using a certain method of needling so that pain is either dulled or removed altogether. Thus, in using this technique during surgical operations, sensation in the area in which the operation is to be performed can be dulled while the patient remains entirely conscious.

At present in China this technique is widely used in both urban and rural hospitals and in mobile medical clinics. It has been used for surgery on the brain, neck, chest, abdomen and limbs. In addition, it has been applied in obstetric operations and in operations on

针刺镇痛的实际应用

郑，丁

编者按

在这个时期，针刺镇痛在中国的广泛应用吸引了西方的关注。本文中中医师郑和丁总结了在手术中使用这项技术的利弊。这种技术与麻醉剂相比成本更低廉，带来的并发症更少，在有信心的患者身上发挥了特别好的功效。但是它并不像麻醉剂那样有效地使神经麻木，而且实际上只能在上半身的手术中发挥好的疗效，如甲状腺切除术和肺部手术。考虑这些，两位医师建议某些时候将针刺镇痛作为药物疗法的另一种选择而不是替代治疗。

针刺镇痛有时候可以作为药物的另一种选择，但是不能完全取代药物的作用。尽管确实在一些病例中针刺镇痛的效果要好于药物，但这门技术需要进一步的发展以评估其真正的重要意义。在这篇文章中，两位中国医生评估了在手术中使用这门技术的优点和缺点，其中一位已经用针刺镇痛施行了 24 例手术。

据报道在中国使用针刺镇痛成功进行了成百上千例手术，这种镇痛方法引起了全世界人们的广泛关注。在日本、美国和其他国家，医学工作人员已经开始研究将针刺镇痛应用到他们的日常医疗和临床工作中。本文试图对这门技术进行一些解释，以便于帮助那些正在研究这门技术的人。这主要来自我们其中一位作者的个人经验，这位作者已经应用这种镇痛方法进行了 24 例手术——3 例二尖瓣狭窄、5 例肺叶切除术、6 例胃切除术和 10 例甲状腺切除术。

针刺镇痛不使用任何麻醉剂，而是基于传统中医学理论——经络，该理论认为将通过向身体上某些特定位点施压，这些点（位于经脉上）会变得麻木。压力的给予是通过一种特定的针刺方法，这样疼痛感就会变得迟钝或者完全消失。因此在手术中使用该技术就可以在患者保持神智完全清醒的状态下，使需要手术的部位的感觉变得迟钝。

目前该技术在中国的城市和乡村医院以及流动诊所均得到广泛使用。它已经被用于脑部、颈部、胸部、腹部和四肢的手术。此外它也被用于产科手术以及耳、鼻

the ears, nose and throat. Acupuncture has attained a definite place in Chinese medical practice and, when it can be used appropriately, has become the first choice of anaesthetic.

The use of acupuncture analgesia depends on the availability of properly trained manpower. A skilled medical team including anaesthetists is necessary. The team should be composed both of those who have expertise in the use of anaesthetic drugs and of those familiar with acupuncture analgesia. Those who have a thorough knowledge of both techniques are obviously the most valuable. Before the operation, one or two members of the team need to visit the patient to find out the patient's mental attitude, his desire to use this kind of analgesia and the nature of his illness. For their part, the team member or members should give a complete explanation of the anaesthesia technique to the patient.

In addition, the team of surgeons must be skilled, alert and capable. The doctors must be those whom the patient trusts, and they must have a thorough understanding of the patient's illness and mental attitude. Before the operation, they must give the patient a complete explanation of the surgical procedure and teach the patient to do the appropriate exercises such as deep breathing. The doctors must work quickly, steadily and accurately and must make the incisions and do suturing in the shortest possible time.

Particular Applications

To illustrate the use of this technique in surgery, we shall describe its application in a thyroidectomy, where the best results have been obtained. Two preliminary steps are necessary. First, the consent of the patient must be obtained. In addition to explaining the procedures to him, it is highly recommended that a demonstration be performed on the patient before his operation. Second, the individuals who administer the analgesia must have a thorough understanding of the procedures. We suggest that the anaesthetists practise acupuncture on their own bodies in order to discover the various methods of using the needle, the different sensations acupuncture produces and the proper depth to which the needle needs to be inserted.

About a half hour before surgery, 50 mg pethidine or other sedative is given to the patient by injection. The acupuncture needles are then inserted at two points on the hand and forearm. The first point is called the *hu ku* and is located between the thumb and the forefinger on each hand. The needle is inserted to a depth of 0.5 inch or until the patient begins to feel sensations of aching, heaviness, fullness and numbness. The other point is the *nei kuan* which is located posteriorly about 2 inches above the wrist. The needle is inserted to a depth of 0.5 to 1 inch, again until similar sensations are felt. If the needles are placed in the proper position and are rotated in a circular manner for about 20 min, an analgesic effect will be induced. The ensuing surgical procedure must be light and fast. When the skin is cut, the technique of "flying knife–rapid cutting" must be used as some pain may be caused when the incision is made.

Acupuncture analgesia does not completely remove pain. For this reason, electric

和喉部的手术。针刺疗法在中医实践中的地位是不可否认的，而且运用恰当时，它势必成为麻醉的首选。

针刺镇痛的使用取决于经过合格训练的专业人员。而且还需要一个包括麻醉科医生在内的训练有素的医疗团队。这个团队的成员应该包括那些对于麻醉剂的使用有丰富经验的人和熟悉针刺镇痛的人。当然能够全面理解两种技术的人是最受欢迎的。手术前，该团队的一或两名成员需要访视患者以了解该患者的精神状态、他对于使用这种镇痛方法的诉求以及他的疾病的状况。他们的任务是向患者完整地解释这种镇痛方法。

此外，外科医生团队也要训练有素、机警并有足够的能力来完成手术。这些医生必须取得患者的信任，他们必须对患者的疾病和精神状态有充分的了解。手术之前他们必须向患者详细解释手术的过程并教会患者作适当的练习，比如深呼吸。医生的工作必须要迅速、稳定和准确，并且必须在尽可能短的时间内完成切口和缝合。

特别应用

为了说明一下该技术在手术中的应用，我们将以其在甲状腺切除术中的应用为例，因为在这种手术中取得的效果最好。手术前的两个预备步骤很重要。首先，需要获得患者的知情同意。除了向他解释整个手术过程，我们强烈建议手术前在该患者身上进行一次演示。其次，施行针刺镇痛的术者必须熟练掌握整个过程。我们建议施针者在自己的身体上练习针刺技术以发现不同的用针方法、针刺产生的不同感受和针刺入的合适深度。

大约手术前半小时，给患者注射 50 mg 哌替啶或者其他镇静剂。然后将针刺入手和前臂上的两个点。第一个点被称为合谷，位于拇指和食指之间。针刺入的深度是 0.5 英寸或者直到患者出现疼痛、沉重、胀和麻木的感觉。另一个点被称为内关，位于腕掌侧正中向上 2 英寸。刺入的深度是 0.5 英寸到 1 英寸，也是直到患者出现类似的感觉为止。如果针刺的位置是准确的并且施行捻转手法大约 20 分钟，就能够产生镇痛效果。接下来的手术过程必须轻柔而且迅速。当切开皮肤时，必须使用“飞刀 – 快切”技术，因为切割时还是可能引起一定程度的疼痛。

针刺镇痛不能完全消除疼痛。出于这个原因，不能用电灼来止血。取而代之的

cauterization must not be used to stop bleeding. Instead, either ligation or pressure must be used. In addition, the patient must be warned that some uncomfortable feeling may be caused when the muscle is pulled around the neck and thyroid gland. The suturing procedure may also cause some pain and must be done as quickly as possible with sharp suture needles. If the level of pain increases, then the needles may be rotated again—or if that fails to bring relief, of course, drug anaesthesia may be resorted to. The whole operation should be performed very quickly to obtain the best results.

Advantages

There are many advantages to the use of acupuncture analgesia. Here we shall discuss three. First, as we have mentioned, acupuncture analgesia, unlike drug anaesthesia, dulls the nerves without causing the patient to lose consciousness; thus, it allows the patient and doctor to cooperate with each other. For example, again in a thyroidectomy, the doctor can talk with the patient at any time, and by listening to the patient's voice can discover if any injury has been caused to the recurrent laryngeal nerve. Another example is found in cases which involve heart and lung surgery. After the opening incision has been made the doctor can direct the patient to do deep abdominal breathing in order to prevent a sudden shift in the mediastinum and to keep the lung inflated so as not to interfere with the operation and ventilation of the patient.

Second, by comparison with the use of anaesthetic drugs, there are fewer physiological and psychological complications. During operations using acupuncture, we have observed that the blood pressure, pulse and breathing remain regular. This is often not the case with drugs. After an operation with acupuncture analgesia, there are no side effects nor evidence of the complications which may follow operations done with drug anaesthesia. Moreover, because there are few physiological reactions, the patient recovers his normal physical and mental state very quickly. In addition, the patient is more psychologically fit after an operation using acupuncture analgesia. According to our observations, the patient has been aware of and understood the entire surgical process as he has been conscious during the operation; therefore, when the operation is finished, he feels the surgery has gone smoothly and has been quite safe. He is very happy with the results. Often, he immediately wants to get off the operating table, walk around and eat.

Third, acupuncture analgesia is more convenient and comparatively cheaper than other known techniques.

Sometimes Inappropriate

One should not assume, however, that acupuncture analgesia is appropriate for all operations or a complete substitute for drug anaesthesia. Again according to our observations, the following factors impose limits on the use of this technique. First, acupuncture analgesia produces different results on different parts of the body. Our past experience has indicated that the results are better if the parts of the body involved are in the chest cavity and above. At present acupuncture analgesia is used in operations

方法是结扎或者压迫。此外，必须预先告知患者当牵拉颈部和甲状腺周围的肌肉时可能出现一些不适的感觉。缝合的过程也会产生一些疼痛，因此也需要用锋利的缝针尽可能快地完成。如果疼痛的程度加重了，需要再次捻转针灸针，如果这样还不能缓解疼痛，就需要加用麻醉剂了。整个手术过程必须非常快地进行以获得最好的效果。

优　点

使用针刺镇痛有很多优点。这里我们只讨论三点。首先，正如我们所提过的，针刺镇痛在保持患者意识的情况下可以使神经麻木，这不同于麻醉剂麻醉。这样便于医生和患者间的合作。再以甲状腺切除术为例，医生能够随时与患者交谈，而通过听患者的声音就能够判断是否损伤了喉返神经。另一个例子就是心肺手术。在打开切口之后，医生能够指导患者进行深度的腹式呼吸以防止胸腔纵膈膜突然摆动。同时还能保持肺处于膨胀状态，这就不会干扰手术以及患者的呼吸。

其次，与使用麻醉剂相比，针刺镇痛的生理和心理并发症更少。利用针刺镇痛手术时，我们发现血压、脉搏和呼吸仍然保持平稳。这通常在药物麻醉时不容易实现。利用针刺镇痛进行手术后,不会出现药物麻醉所导致的副作用或者并发症。此外，由于出现的生理反应较少，患者容易迅速恢复正常的生理和心理状态。而且针刺镇痛手术后的患者心理更加健康。根据我们的研究发现，由于患者在手术中一直保持清醒，他们已经意识到并了解整个手术过程。因此，在手术结束以后他会觉得手术进行得很顺利而且非常安全。他会对结果非常满意。通常他们会立即想从手术床上下来，四处走动和吃饭。

第三，针刺镇痛相对已知的其他技术更加方便而且相对便宜。

有些时候不适用

但是我们不能认为针刺镇痛适于所有的手术，或者可以完全替代药物麻醉。同样根据我们的发现，下面的因素制约了这项技术的应用。首先，针刺镇痛在身体的不同部位产生的效果是不同的。我们过去的经验指出，对胸腔及其以上的身体部位使用针刺镇痛会取得更好的效果。目前，针刺镇痛已经用于头、颈和胸部的手术。正如前面所提及的，在甲状腺切除术中获得最满意的效果，因此我们用这种手术作

involving the head, neck and chest. As mentioned above, the best results have been achieved in thyroidectomies, so this type of surgery is being used for demonstration purposes. Other types of operation in which acupuncture analgesia has achieved good results are lung operations. In surgery on the abdomen and limbs, however, the results have been disappointing.

Second, the surgical cases must be selected carefully. When choosing a patient to undergo surgery with acupuncture analgesia, his illness must be considered. For example, in operations such as gastrectomies for gastric ulcers and pyloric stenosis, only if the operating time is short is it appropriate to use acupuncture analgesia. But for gastrectomies for stomach cancer in which a rather wide and long investigation is necessary, it is not wise to use acupuncture analgesia. For simple chest surgery such as localized pulmonary tuberculosis, acupuncture analgesia can be used. For exploratory surgery which takes time and in operations for widespread adhesions, it is, however, not advisable to use this technique. In conclusion, at the present time, cases should be chosen in which the nature of the disease is not too complicated and in which the operating time is 1 to 2 h.

Third, the emotional state of the patient must be considered. Because the success of the use of acupuncture analgesia depends on the willingness and understanding of the patient, this is a decisive factor. Before it is used, the entire procedure must be explained to the patient and his active cooperation must be enlisted. Thus, we select those who are emotionally stable, who have a high degree of confidence in the advantages of using acupuncture analgesia, and who are able to follow the doctor's requests and carry out his instructions. This kind of person is usually strong, energetic and young. Unsuitable types of patient include children under 10, who cannot cooperate with the doctor, and highly nervous individuals. In China, because thought preparation is thorough and deep, there are many people who have great faith in the doctors and nurses, and therefore many now volunteer for and even request acupuncture analgesia.

Problems

Acupuncture analgesia in clinical use is still in the initial stages of development. Our experience indicates that several problems have arisen in its application.

Acupuncture analgesia does not dull the nerves as completely and effectively as drugs. Thus, it is not suitable for every operation. On some occasions when the first incision is made, when suturing commences and/or when the procedure is long, the pain threshold may be decreased so as to render the analgesic effect relatively ineffective. It is therefore necessary to have drug anaesthesia available in case it must be used to complete the operative procedure.

Acupuncture analgesia does not cause complete relaxation of the muscles. For example, in abdominal surgery, during the operation the stomach and intestines might be disturbed which could result in a nervous reflex tightening the muscles and hence pain to the patient. If this condition does arise, drug anaesthesia may have to be used.

为示范。其他针刺镇痛取得良好效果的就是肺部手术。但是从腹部和四肢的手术取得的效果来看，结果并不令人满意。

其次，手术病例的选择必须非常谨慎。当选择一个患者进行针刺镇痛手术时，必须要考虑他的病情。比如，在进行胃溃疡和幽门梗阻的胃部切除术时，只有手术时间短的患者才适合用针刺镇痛。而胃癌进行胃部切除术时，需要进行非常广泛和长时间的观察，因此并不适合用针刺镇痛。对于简单的胸部手术，比如局部肺结核，可以使用针刺镇痛。但是对于非常耗时的探查性手术和广泛粘连的手术，我们不建议使用这种技术。总之，目前应该选择疾病本身不是很复杂，并且手术时间在 1 小时到 2 小时之间的病例。

第三，必须考虑患者的情绪状态。因为针刺镇痛成功使用取决于患者的意愿和理解，这是一个决定因素。在使用前，必须向患者解释整个过程，而且必须获得其主动配合。因此我们选择那些情绪稳定、高度相信针刺镇痛的优势以及能够听从医生的要求并执行医生指令的患者。这类患者通常比较强壮、精力充沛并且年轻。不合适的患者包括不能和医生合作的 10 岁以下的儿童和高度紧张的患者。在中国，因为思想准备工作做得非常全面和深入，许多患者非常相信医生和护士，因此有很多人自愿甚至主动要求进行针刺镇痛。

问　题

临床使用针刺镇痛仍处在发展的初始阶段。我们的经验表明在其应用过程中出现了一些问题。

针刺镇痛使神经麻木的作用不像药物那么完全和有效。因此它不适用于所有手术。有些情况下，当切开第一刀、开始缝合和（或）手术时间很长时，痛阈可能会下降导致镇痛的效果变得不理想。因此，有必要准备好麻醉剂，在必要的时候使用，以保证整个手术的完成。

针刺镇痛不能让肌肉完全放松。比如，在腹部手术中，胃和肠道受到刺激后可能导致产生神经反射使肌肉紧张，因此造成患者的疼痛。如果确实出现了这种情况就需要使用药物麻醉。

The theoretical investigation of the effects and uses of acupuncture analgesia is still in the early stages. The results of the use of this technique are only being brought to light through continued experience and experiments. Thus, at present, there is a method of systematic investigation to provide a firm basis for our work. We are now attempting to investigate the relationships between drugs and acupuncture analgesia. Although we believe acupuncture analgesia can be used widely, we feel it is misleading to say that it can and will replace the use of drug anaesthesia. At this stage we can only say that in certain cases acupuncture analgesia is better than drugs. Acupuncture analgesia is only a new anaesthetic technique, and in some contexts at least it may prove to be a better kind of anaesthetic skill. We believe that both acupuncture analgesia and drugs should be used in surgical work in order to further our understanding of the use of this new technique. We admit that it is a radical step to say it is possible merely to insert a needle, in the absence of drugs, to relieve pain. But at a time when the medical field is changing rapidly and new ideas are ever present, the importance of this development must not be neglected.

(**242**, 559-560; 1973)

S. B. Cheng and L. K. Ding
Chinese Medical Research Centre, 566-568 Nathan Road, Kowloon, Hong Kong

针刺镇痛的效果和应用的理论研究仍然处于早期阶段。使用该技术的效果只能通过后续的经验和试验才受到人们的关注。因此，目前需要一个系统的研究方法来给我们的工作提供坚实的基础。我们现在正在尝试研究药物麻醉和针刺镇痛之间的关系。尽管我们相信针刺镇痛能够被广泛应用，但是我们认为那种认为它能完全替代药物麻醉的说法是一种误导。现阶段，我们只能说在某些情况下针刺镇痛比药物好。针刺镇痛只是一种新的麻醉技术，而且，至少在某些情况下，它是一种比较好的麻醉方法。我们相信针刺镇痛和药物麻醉都应该在手术中使用以便于加深我们对这项新技术的理解。我们承认仅仅通过刺入一根针而不用药物就可以缓解疼痛是比较激进的。但是在这个医学领域变幻莫测、新想法不断涌现的年代，这种技术的重要性一定不能被忽视。

（毛晨晖 翻译；于天源 审稿）

Resonance Raman Spectroscopy of the Photoreceptor-like Pigment of *Halobacterium halobium*

R. Mendelsohn

Editor's Note

The purple membrane of the salt-loving photosynthetic bacterium *Halobacterium halobium* (and related organisms) contains the protein bacteriorhodopsin, which uses the energy of absorbed light to drive hydrogen ions (protons) through its central channel, creating a gradient in acidity across the membrane that can be tapped for chemical energy. This system has become archetypal in the study of biological energy conversion and proton transfer, as well as attracting technological interest for solar energy. Here Richard Mendelsohn of King's College, London, uses spectroscopy to identify the fundamental process of light absorption by the light-sensitive chemical group in the core of the protein. The details of how photochemical changes in this group lead to the movement of protons are still being worked out.

Use of the resonance Raman technique has shown that the colour of the purple membrane pigment of *H. halobium* arises from an unprotonated Schiff base whose electron density is perturbed by further (electrostatic) interaction with a protein. The chromophore probably consists of a charge transfer complex between retinyllysine and an appropriate side chain of the protein.

RESONANCE Raman spectra have been reported for several molecules of biological interest, including haemoglobin[1,2], cytochrome C[2,3], rubredoxin[4] and several carotenes[5] and retinals[6,7]. The resonant enhancement of certain Raman-active vibrations occurs when molecules are excited by light of a wavelength lying within an electronic transition[8], and the effect provides a structural probe at unusually low solution concentration. Of particular importance to biological systems is the feasibility of *in situ* examination.

One potentially useful area of application for this technique is to photoreceptor pigments. The bathochromic shifts in visual pigments have not yet been satisfactorily explained. The absorption maxima of these pigments, in which there appears to be a Schiff base of 11-*cis* retinal with the ε-amino group of a lysine residue, vary from 430 to 562 nm (ref. 9). Solutions of the Schiff base itself, however, absorb near 360 or 440 nm, depending on whether the nitrogen atom is unprotonated or protonated, respectively. Thus another specific interaction occurs between protein and chromophore which

盐生盐杆菌中类感光色素的共振拉曼光谱

门德尔松

编者按

嗜盐光合细菌盐生盐杆菌（及相关生物）的紫膜中含有细菌视紫红质蛋白，能够利用所吸收光的能量，使氢离子（质子）穿过中央通道，在膜两侧产生酸度梯度，进而被开发为化学能。这个系统已成为研究生物能量转化和质子转移的原始模型，还引发了研究太阳能技术的兴趣。伦敦国王学院的理查德·门德尔松利用光谱确定了蛋白质中心光敏化学基团光吸收的基本过程。这些基团中光化学变化如何导致质子转移的细节仍需进一步阐明。

共振拉曼技术的使用表明，盐生盐杆菌紫膜色素的颜色来自一种未质子化的席夫碱，该碱的电子密度受到蛋白质的远程（静电）相互作用的干扰。发色团可能包含一个由视黄基赖氨酸与蛋白质的适当支链构成的电荷转移复合体。

共振拉曼光谱已被报道应用于多种具有生物学意义的分子，其中包括血红蛋白[1,2]、细胞色素 C[2,3]、红素氧还蛋白[4]，以及几种胡萝卜烯[5]和视黄醛[6,7]。分子在受到能引起电子跃迁的波长的光的激发时，其特征拉曼活性振动就会共振增强[8]，这种效应提供了一种极低浓度下的结构探针。对于生物体系来说尤为重要的是它使原位检测变得可行。

感光色素就是这种技术的潜在应用领域之一。视色素的红移现象目前尚没有令人满意的解释。这些色素的最大吸收峰位于 430 nm 到 562 nm（参考文献 9），显示其中含有由赖氨酸残基中的 ε− 氨基基团和 11− 顺式视黄醛所形成的席夫碱。而席夫碱的溶液，根据氮原子未被质子化与质子化的不同，其对应的吸收峰分别位于 360 nm 和 440 nm 附近。由此，蛋白质与发色基团之间的另一特定相互作用的出现，

perturbs the chromophore absorption band and gives rise to pigment colour[10]. Several theories proposed to explain visual pigment spectra have been reviewed by Abrahamson and Ostroy[11]. The two most important ones are those of: (1) Morton *et al.*[12,13], which assumes that the primary bond is a protonated Schiff base in which the bathochromic shifts are provided by negatively charged groups appropriately positioned adjacent to the polyene chain; (2) the Dartnall formulation[9], which assumes that the primary bond is an unprotonated Schiff base, with secondary shifts caused by an optimally placed pair of charges, producing massive dipoles in the polyene. Little experimental evidence is available with which to test either description.

Because of the anticipated extensive photochemical changes associated with the shining of intense laser radiation on a sample containing rhodopsin, I decided to examine initially the photoreceptor-like pigment of *Halobacterium halobium*. The "purple membrane" fragment containing the pigment offered several advantages for a laser-Raman spectroscopic study. (*a*) The primary structure of the chromophore is retinyllysine, the evidence being much the same as that for rhodopsin[14]. (*b*) Unlike rhodopsin, the pigment does not bleach or undergo photochemical changes on even prolonged exposure to intense (>200 mW) laser light, but simply undergoes a reversible spectral shift from λ_{max}=558 nm to λ_{max}=570 nm on exposure to red or blue light respectively[14]. (*c*) The membrane is easily handled at room temperature in aqueous suspension.

The purple membrane used in the study described here was the gift of Drs D. Oesterhelt and A. E. Blaurock.

Cells of *Halobacterium halobium* were grown and the purple membrane fragments isolated as previously described[14]. Suspensions of membrane used for Raman spectroscopy had an optical density of 3.0 at λ_{max}.

Raman spectra of purple membrane fragments excited with 4880 Å and 5145 Å radiation are shown in Fig. 1. The most intense vibration in each spectrum occurs at 1,531 cm^{-1} and about fifteen other peaks have also been distinguished. The only significant difference in the two spectra is the relative intensity of the 1,568 cm^{-1} vibration compared with that at 1,531 cm^{-1}. The former appears about four times more intense when excited by 4880 Å radiation than when excited by 5145 Å excitation. This observation is explained below.

干扰了发色基团的吸收带，从而产生了色素的颜色[10]。阿伯拉罕森和奥斯特罗伊[11]曾综述了几种解释视色素光谱的理论。其中最重要的两种是：(1) 莫顿等[12,13]提出，假定主价键为质子化的席夫碱，其中多烯链邻近适当位置上的负电子基团导致了红移；(2) 达特诺[9]的说法，假定主价键为未质子化的席夫碱，其中处于最优位置上的电子对引发二次频移，在多烯上生成大量偶极。几乎没有实验证据可用来验证任一理论。

由于对含视紫红质样品进行强激光照射会产生大量可预期的光化学变化，笔者决定首先研究盐生盐杆菌中的类感光色素。包含色素的“紫膜”碎片为激光－拉曼光谱的研究提供了几个有利条件。(*a*) 发色基团的主要结构是视黄基赖氨酸，提供的证据对于视紫红质差不多同样有效[14]。(*b*) 与视紫红质不同，这种色素即使长期暴露在强烈（> 200 mW）激光中也不会脱色或发生光化学变化，当暴露在红光或蓝光中，在两种光之间只会发生可逆的光谱频移，由 λ_{max} = 558 nm 移到 λ_{max} = 570 nm[14]。(*c*) 这种膜在室温下的水悬浮液中很容易处理。

这里描述的本研究中所用的紫膜，为厄斯特黑尔特和布劳罗克两位博士所赠。

培养盐生盐杆菌和分离紫膜碎片的方法如先前所述[14]。用于拉曼光谱检测的膜悬浮液在最大吸收峰 λ_{max} 处光密度为 3.0。

图 1 所示为分别以 4880 Å 和 5145 Å 的射线激发紫膜碎片的拉曼光谱。每张谱图中，都有一个位于 1,531 cm^{-1} 处的最强振动，以及其他约十五个可分辨出的峰。两张谱图中唯一的显著区别，就是位于 1,568 cm^{-1} 处的振动与 1,531 cm^{-1} 处的相比，其相对强度不同。用 4880 Å 的射线激发时，1,568 cm^{-1} 处的峰比用 5145 Å 的射线激发时强约四倍。下面将解释这一观测结果。

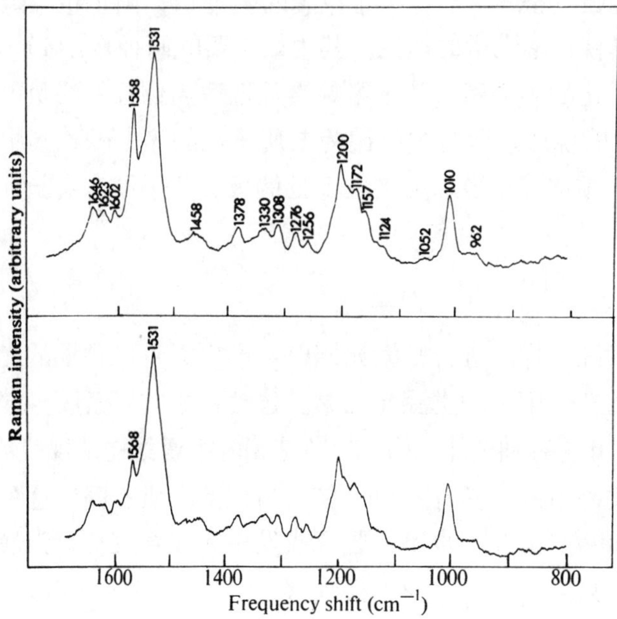

Fig. 1. Raman spectra of an aqueous suspension (5.5×10^{-5} M in the purple protein, *p*H 7.0, unbuffered) of purple membrane fragments obtained with a "Spex" 1401 Raman Spectrophotometer using: (Top) 100 mW of 4880 Å radiation from an Ar^+ laser. Resolution, 7 cm^{-1}; time constant 2 s; scanning rate 25 cm^{-1}min^{-1}; photon counting detection; 1 mm i.d. capillary cell, transverse excitation. (Bottom) 125 mW of 5145 Å radiation from an Ar^+ laser. Conditions as above. Polarization measurements indicated that all observed vibrations had depolarization ratios $0\leq\rho\leq0.3$. Spectra of membrane fragments suspended in 4 N NaCl solution are identical with those shown, as are spectra obtained in 0.05 M phosphate buffer (*p*H 7.0).

Evidence for Resonance Enhancement

Several experiments showed that the observed spectra were resonance-enhanced. (*a*) The protein concentration as measured from the visible spectrum (ε_{max}=54,000 l/mol cm)[14] was 5.5×10^{-5} M or two orders of magnitude less than that required to obtain ordinary protein Raman spectra[15]. (*b*) On bleaching the pigment with 0.1 M cetyltrimethyl-ammonium bromide (*p*H 7.9), a procedure which leaves the Schiff base intact but destroys the purple colour[14], no spectrum could be observed. (*c*) An excitation profile (variation of Raman intensity with excitation wavelength) experiment was carried out as follows. Samples of membrane were prepared containing 0.1 M Na_2SO_4 in which the non-resonant enhanced symmetric stretching vibration of the SO_4^{2-} ion at 983 cm^{-1} was used as an internal standard. The ratios of the intensities of the peaks at 1,531 and 1,568 cm^{-1} to the intensity of the standard were determined as a function of excitation wavelength using six lines available from the argon ion laser, all of which lie on the high frequency side of the pigment absorption band. The results are shown in Fig. 2. It is clear that there is substantial deviation from the $1/\lambda^4$ law for intensity of scattered light, as the observed ratios would not vary with wavelength if this relationship were obeyed. The two vibrations studied appear to be enhanced by (vibronic) coupling to different components of the membrane absorption spectrum. The peak at 1,531 cm^{-1} increases in relative intensity as

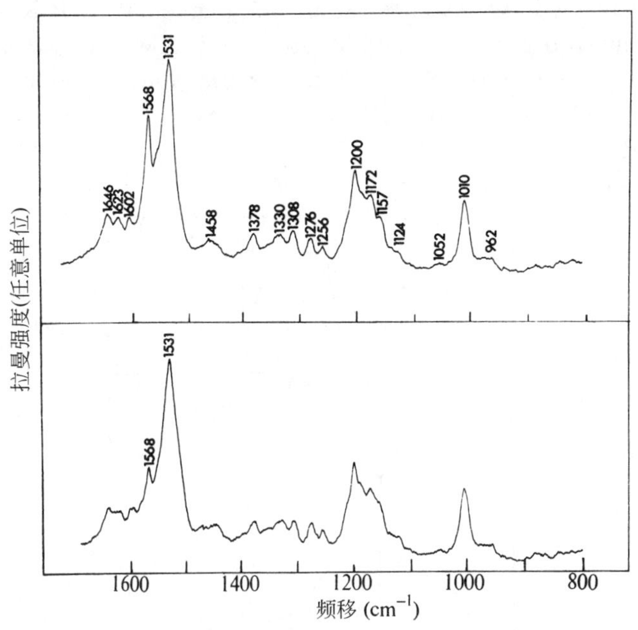

图 1. 使用“Spex”1401 型拉曼光谱仪，得到紫膜碎片水悬浮液（紫色蛋白质浓度为 5.5×10^{-5} M，pH 7.0，不含缓冲剂）的拉曼光谱图：（上图）100 mW Ar^+激光产生的 4880 Å 射线。分辨率 7 cm^{-1}；时间常数 2 s；扫描速率 25 $cm^{-1}\cdot min^{-1}$；光子计数检测；1 mm 内径的毛细管池，横向激发。（下图）125 mW Ar^+激光产生的 5145 Å 射线。其他条件同上。偏振测量指出，所有观测到的振动具有的退偏比值为 $0\leq\rho\leq0.3$。悬浮于 4 N NaCl 溶液中的膜碎片的谱图与所示谱图是一致的，在 0.05 M 磷酸盐缓冲溶液（pH 7.0）中得到同样的谱图。

共振增强的证据

若干实验表明观测到的谱图有共振增强。(*a*) 通过可见光谱测定（$\varepsilon_{max}=54{,}000$ $L\cdot mol^{-1}\cdot cm^{-1}$）[14] 的蛋白质浓度为 5.5×10^{-5} M，或者说，比要获得正常的蛋白质拉曼光谱所需的浓度低两个数量级 [15]。(*b*) 若使用 0.1 M 的溴化十六烷基三甲铵（pH 7.9）使色素脱色，这个过程可以使席夫碱保持完好但破坏紫色 [14]，导致不能观测到光谱。(*c*) 激发曲线（拉曼强度随着激发波长的变化）实验步骤如下。在膜样品中加入 0.1 M 的 Na_2SO_4，以 SO_4^{2-} 离子在 983 cm^{-1} 处的没有共振增强的对称伸缩振动作为内标。测定以位于 1,531 cm^{-1} 处和 1,568 cm^{-1} 处峰强度与内标强度的比值作激发波长的函数；光源是氩离子激光，可以产生六条都位于色素吸收带高频端的谱线。结果如图 2 所示。很明显，确实存在着对于散射光强度的 $1/\lambda^4$ 定律的严重偏离，因为如果遵循该关系的话，所观测到的比值是不会随波长而变化的。所研究的两种振动看来是为膜吸收光谱中不同部分的（电子振动）耦合所增强的。随着激发光波长向着色素的 λ_{max} 方向的增长，位于 1,531 cm^{-1} 处峰的相对强度会增加。由此与主要

the exciting line increases in wavelength toward the pigment λ_{max}. It is therefore coupled to the main pigment absorption. The 1,568 cm^{-1} band, however, is increased in relative intensity as the wavelength is decreased, and it is therefore not coupled to the visible component of the membrane spectrum but to an absorption located at shorter wavelength than the main band.

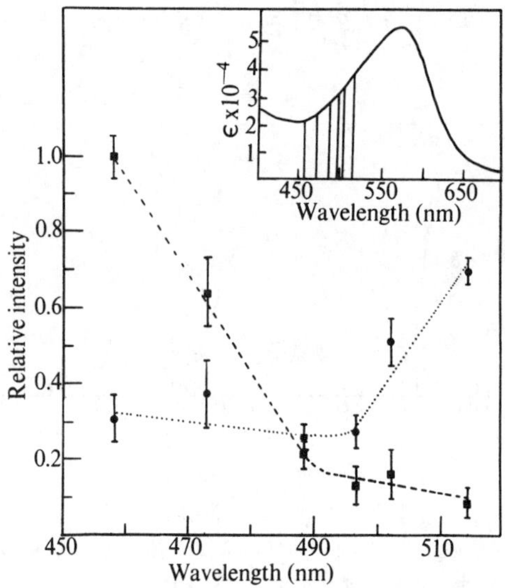

Fig. 2. Excitation profile results. ---, Ratio of the integrated intensities of 1,568 cm^{-1} vibration compared with 983 cm^{-1} vibration of SO_4^{2-} ion used as internal standard. ..., Same ratio comparing the 1,531 cm^{-1} vibration with the standard. ■ and ●, Experimental points, observed at the indicated laser wavelength. Error bars represent standard deviations based on four measurements. Insert, pigment absorption band of purple membrane. Vertical lines indicate positions of laser lines located at 457.9, 472.7, 488.0, 496.5, 501.7 and 514.5 nm. The visible absorption spectrum was recorded on a "Unicam" SP-800A spectrophotometer using cells of 2 mm path length.

All of the other vibrations (except that attributable to solvent O–H bending at 1,646 cm^{-1}) behave in a fashion similar to the 1,531 cm^{-1} vibration and hence arise from coupling to the 560 nm band of the pigment.

Analysis of Spectral Data

Spectral analysis is simplified by the fact that only chromophore vibrations are observed in the resonance Raman effect[1-3]. Protein vibrations are of insufficient intensity to be seen and therefore do not complicate the spectra. In this work the region 1,500–1700 cm^{-1} which contains C=C and C=N stretching vibrations will be considered. The region below 1,500 cm^{-1}, which contains C–C stretching and C–H bending modes, will be discussed in a future publication.

Rimai and coworkers have made a detailed Raman spectroscopic study of various retinals[7] and their Schiff bases[6]. They found that those vibrational modes most strongly enhanced are contributed by C=C and C–C stretching in the conjugated chain, which occur near

色素吸收耦合起来。不过，位于 1,568 cm^{-1} 附近的吸收带相对强度随着激发光波长减少而增加，因此不会与膜光谱的可见部分耦合，而是与位于比主吸收带波长短的吸收部分耦合。

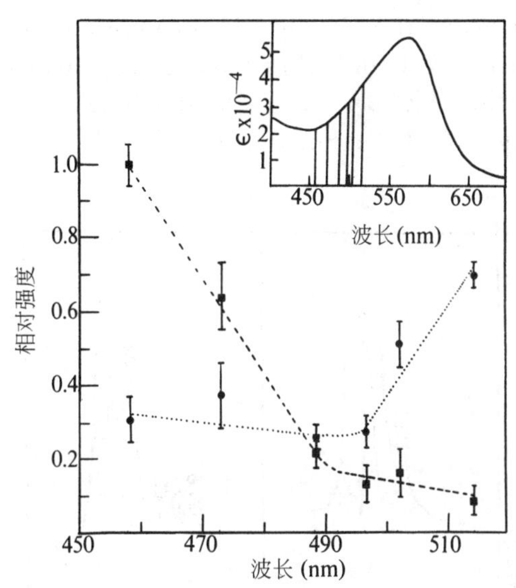

图 2. 激发曲线结果。---，1,568 cm^{-1} 处振动峰强与作为内标的 SO_4^{2-} 在 983 cm^{-1} 处振动峰强的比值。…，1,531 cm^{-1} 处振动峰强与同样的内标峰强的比值。■和●为在上述激光波长条件下观测到的实验值。误差线代表基于四次实验得到的标准偏差。插图为紫膜的色素吸收带。垂直线表示位于 457.9 nm、472.7 nm、488.0 nm、496.6 nm、501.7 nm 和 514.5 nm 的激光束。可见光吸收光谱是由“Unicam”SP-800A 光谱仪使用 2 mm 路径长度的样品池得到的。

所有的其他振动（除了 1,646 cm^{-1} 处归属为溶剂 O—H 的振动外）都具有与 1,531 cm^{-1} 处振动类似的振动方式，因此也是由与色素 560 nm 处吸收带的耦合产生的。

光谱数据分析

在共振拉曼效应中只能观测到发色团的振动这一事实简化了谱图分析 [1-3]。蛋白质振动的强度尚不足以被看到，因而不会使谱图复杂化。在本工作中要考虑的是包含 C══C 与 C══N 伸缩振动的 1,500～1,700 cm^{-1} 区域。低于 1,500 cm^{-1} 的包含 C—C 伸缩振动与 C—H 弯曲振动的区域，将会在以后的发表物中进行讨论。

里毛伊与他的同事们已经对各种视黄醛 [7] 及其席夫碱 [6] 的拉曼光谱进行了详细研究。他们发现，增强最明显的是在共轭链中的 C══C 和 C—C 伸缩振动，分别位

1,570 and 1,200 cm^{-1}, respectively. In addition, they identified C=O and C=N stretching vibrations in the range 1,600–1,670 cm^{-1}.

In Fig. 3*a* and *c*, the 1,500–1,700 cm^{-1} region of the Raman spectra of unprotonated and protonated retinylhexylamine are shown. The C=N frequencies in this model Schiff base are expected to be similar to those of the pigment Schiff base and can be used to decide whether the nitrogen of the latter is protonated. In the unprotonated form (Fig. 3*a*), ν (C=N) occurs as a weak band near 1,623 cm^{-1} while ν (C=C) appears strongly at 1,579 cm^{-1}. In protonated retinylhexylamine (Fig. 3*c*) the C=N stretching frequency is broadened and shifted to 1,645 cm^{-1}, while ν (C=C) is shifted to 1,559 cm^{-1}.

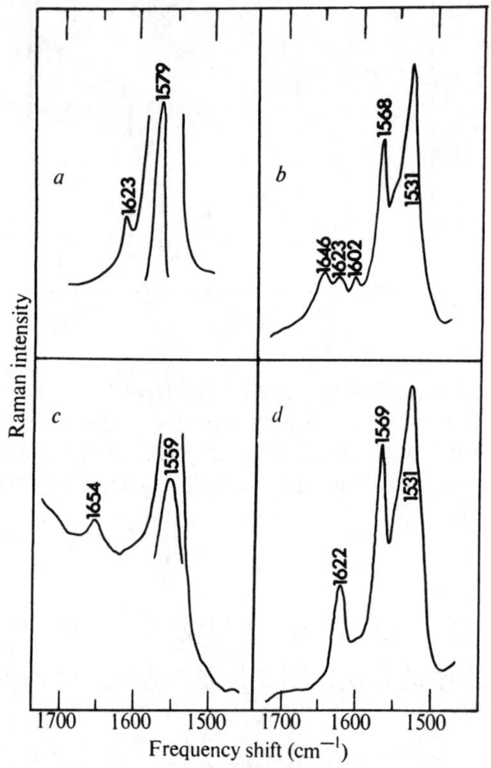

Fig. 3. 1,500–1,700 cm^{-1} region of the Raman spectra of *a*, unprotonated retinylhexylamine (10^{-3} M in C_6H_{14} solution, λ_{max}=360 nm); *b*, purple membrane fragments in H_2O suspension (see Fig. 1 (top), caption); *c*, protonated retinylhexylamine (10^{-4} M in acidified EtOH solution, λ_{max}=445 nm); *d*, purple membrane fragments in D_2O suspension (conditions as in Fig. 1 (top) except for solvent). All spectra recorded using 4880 Å radiation. The difference in concentration in *a* and *c* reflects the increased resonant enhancement as the retinylhexylamine absorption band is shifted into the visible region. There is little or no change in the Raman frequencies of this Schiff base as a function of solvent. The best spectra of unprotonated and protonated retinylhexylamine were obtained in the solvent indicated.

In Fig. 3*b*, the 1,500–1,700 cm^{-1} range of the purple membrane spectrum is shown. This region is complicated by the presence of a weak solvent OH bending vibration at

于大约 1,570 cm^{-1} 和 1,200 cm^{-1} 处。另外，他们还在 1,600~1,670 cm^{-1} 区域中指认出 C═O 和 C═N 的伸缩振动。

在图 3*a* 和 3*c* 中所示分别为未质子化的和质子化的视黄基己胺拉曼光谱图的 1,500~1,700 cm^{-1} 区域。在这种形式的席夫碱中，预期 C═N 的振动频率会与视色素席夫碱中的相似，从而可用于确定后者中的氮原子是否被质子化。在未质子化的形式中（图 3*a*），*v*（C═N）为位于 1,623 cm^{-1} 附近的一个弱吸收带，而位于 1,579 cm^{-1} 处的 *v*（C═C）看起来很强。在质子化了的视黄基己胺中（图 3*c*），C═N 伸缩振动峰增宽并移向 1,645 cm^{-1} 处，而 *v*（C═C）则移向 1,559 cm^{-1} 处。

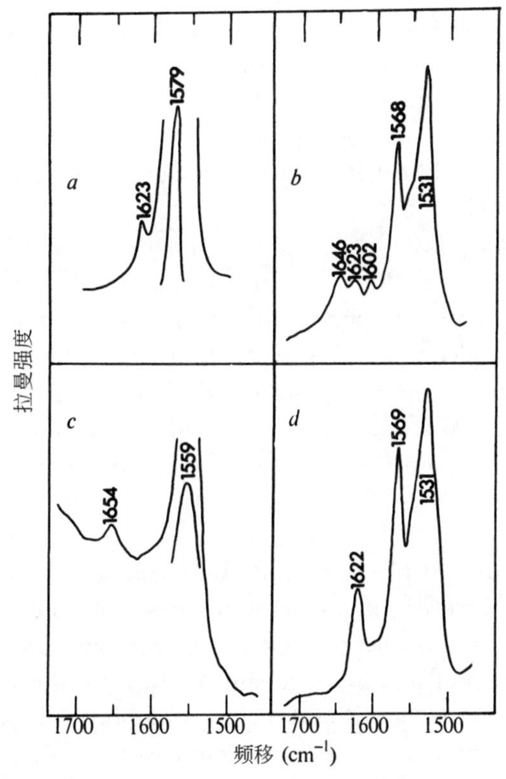

图 3. 拉曼光谱图的 1,500~1,700 cm^{-1} 区域，其中 *a*，未质子化的视黄基己胺（浓度 10^{-3} M 的 C_6H_{14} 溶液，λ_{max} = 360 nm）；*b*，紫膜碎片在水中的悬浮液（见图 1 的上图，图注）；*c*，质子化的视黄基己胺（浓度 10^{-4} M 的酸化乙醇溶液，λ_{max} = 445 nm）；*d*，紫膜碎片在 D_2O 中的悬浮液（除溶剂外，条件与图 1 的上图相同）。所有谱图都是在 4880 Å 光照射下记录的。在 *a* 和 *c* 之间的浓度差别反映出，共振增强在视黄基己胺的吸收带向可见光区域移动时变大。这种席夫碱在不同溶剂下其拉曼频率改变很少甚至没有变化。在已提到的溶剂中获得了未质子化和质子化的视黄基己胺的最佳谱图。

图 3*b* 中所示为紫膜光谱的 1,500~1,700 cm^{-1} 区域。这个区域由于有 1,646 cm^{-1} 处的溶剂 OH 产生的微弱的弯曲振动出现而复杂化了，不过位于 1,531 cm^{-1}、1,568 cm^{-1}、

1,646 cm^{-1}, although distinct peaks are seen at 1,531, 1,568, 1,602, and 1,623 cm^{-1}. The peak at 1,568 cm^{-1} is assigned to ν (C=C) of the Schiff base. The excitation profile results described above indicate that this vibration arises from retinyllysine other than that in the pigment. Whether this free Schiff base is significant to the structure of the pigment is not clear. On suspension of the purple membrane in D_2O, the 1,646 cm^{-1} band disappears as expected, while the vibration near 1,623 cm^{-1} becomes more prominent (Fig. 3*d*) than in the H_2O preparation. This band, assigned to C=N stretch of the Schiff base, appears at the same frequency as in unprotonated retinylhexylamine and strongly suggests that the Schiff base in the pigment itself is unprotonated.

The strongest peak in the purple membrane Raman spectrum at 1,531 cm^{-1} has no counterpart in the spectra of the model Schiff bases, all of which have intense C=C stretching vibrations above 1,550 cm^{-1}. The intensity of the 1,531 cm^{-1} band implies strong vibronic coupling to the pigment absorption and it is suggested that, by its magnitude and position, the vibration is still due to retinal C=C stretch. The frequency decrease indicates that the π electron density of the conjugated system has been perturbed and electrons removed from the C=C bonds. Such a process would reduce the C=C stretching force constants and lower ν (C=C).

The above observations indicate that pigment colour arises from an unprotonated Schiff base whose π electron density is perturbed by further (electrostatic) interaction with the protein. Additional evidence for this description was obtained when it was noticed that the purple membrane λ_{max} is highly solvent dependent (C. W. F. McClare, personal communication). Addition of a small amount of chloroform reversibly shifts λ_{max} of the pigment to 500 nm and the main Raman frequency to 1,520 cm^{-1}. The chloroform seems to penetrate to the chromophore and thereby perturbs its electronic arrangement.

The exact geometry of the purple membrane pigment must await a three-dimensional structure determination; however, one model which this study points to is that of a charge transfer complex[16] between retinyllysine and an appropriate side chain of the protein. Such interactions can account for the large shift in λ_{max} (from 360 nm to ~560 nm) of the chromophore as well as the perturbation of chromophore vibration frequencies upon complex formation. In addition λ_{max} in such complexes are quite solvent dependent[16], as observed in the present case. That such a mechanism is feasible has been shown by Ishigami *et al.*[17], who observed charge transfer between tryptophan and 9-*cis* retinal in acidified methanol solution (λ_{max} of complex =520 nm). Furthermore, I have observed charge transfer between indole and *trans* retinal (λ_{max}=625 nm) in the same solvent. It is therefore quite conceivable that an appropriately positioned tryptophan residue on the protein interacts with the π electrons of the Schiff base and produces the purple pigment colour.

A Raman spectroscopic study of rhodopsin and the intermediates present in its low temperature bleaching sequence would yield valuable information as to the nature of the Schiff base-protein interactions that determine pigment colour. The work described here

1,602 cm^{-1} 和 1,623 cm^{-1} 处的峰还是明显可见。位于 1,568 cm^{-1} 处的峰归属于席夫碱的 v（C══C）。前面描述过的激发曲线实验结果指出，这种振动来自视黄基赖氨酸，而不是色素中的振动。尚不清楚这种游离的席夫碱对于色素结构是否有重要意义。至于紫膜在 D_2O 中的悬浮液，1,646 cm^{-1} 处的吸收带如同预期那样地消失了，而 1,623 cm^{-1} 附近的振动变得比水溶液中的更显著（图 3*d*）。这一归属于席夫碱中 C══N 伸缩振动的吸收带，显示与未质子化的视黄基己胺中振动频率相同，从而强烈暗示着色素中的席夫碱是未质子化的。

紫膜拉曼光谱中位于 1,531 cm^{-1} 处的最强峰在模型席夫碱的谱图中找不到对应峰，后者在 1,550 cm^{-1} 以上有很强的 C══C 伸缩振动峰。1,531 cm^{-1} 吸收带的强度暗示着对于色素吸收存在强烈的电子振动耦合，并且指出，从其数值和位置来看，该振动仍是源于视黄基中的 C══C 伸缩。频率的降低意味着共轭系统中 π 电子密度受到扰动以及电子远离了 C══C 键。这一过程将会减小 C══C 伸缩的力常数并降低 v（C══C）。

上述观测指出，色素颜色来自一种未质子化的席夫碱，该碱的 π 电子密度受到其与蛋白质的远程（静电力）相互作用干扰。当注意到紫膜的 λ_{max} 是高度依赖于溶剂时（麦克莱尔，个人交流）就得到了支持这一陈述的另一证据。加入少量氯仿能够使色素的 λ_{max} 可逆地移向 500 nm，而主拉曼频率移至 1,520 cm^{-1} 处。氯仿似乎渗透进了发色团并以此干扰其电子排布。

紫膜色素的具体的几何形状还得等待其三维结构的确定；但是，本研究所针对的模型是一个由视黄基赖氨酸与蛋白质的适当支链构成的电荷转移复合体 [16]。这种相互作用可以解释发色团 λ_{max} 的巨大位移（从 360 nm 到约 560 nm），以及复合体形成后对发色团振动频率的干扰。此外，就目前所观测到的来说，这种复合体中的 λ_{max} 是非常依赖于溶剂的 [16]。石上等 [17] 已经说明了这一机制的可能性，他在酸化的甲醇溶液中观测到色氨酸与 9–顺式视黄醛的电荷转移（该复合体的 λ_{max} = 520 nm）。而且，笔者曾观测到吲哚与反式视黄醛（λ_{max} = 625 nm）在相同溶剂中的电荷转移。因此，极有可能是处于蛋白质适当位置上的色氨酸残基与席夫碱的 π 电子相互作用并产生了紫色素的颜色。

对视紫红质及其在低温脱色过程中出现的中间体的拉曼光谱研究，能获得关于决定色素颜色的席夫碱 – 蛋白质相互作用本质的宝贵信息。本工作描述了这种研究

illustrates the feasibility of such studies and the power of the resonance Raman technique in providing a structural probe for the photoreceptor pigments.

I thank Drs A. E. Blaurock, W. R. Lieb, and C. W. F. McClare for detailed discussions. This work was supported by the National Research Council of Canada.

(**243**, 22-24; 1973)

R. Mendelsohn
Biophysics Department, King's College, 26-29 Drury Lane, London WC2B 5RL

Received February 2, 1973.

References:

1. Spiro, T. G., and Strekas, T. C., *Biochim. Biophys. Acta,* **263**, 830 (1972).
2. Spiro, T. G., and Strekas, T. C., *Proc. US Nat. Acad. Sci.,* **69**, 2622 (1972).
3. Spiro, T. G., and Strekas, T. C., *Biochim. Biophys. Acta,* **278**, 188 (1972).
4. Long, T. V., Loehr, T. M., Allkins, J. R., and Lovenberg, W., *J. Amer. Chem. Soc.,* **93**, 1809 (1971).
5. Rimai, L., Kilponen, R. G., and Gill, D., *J. Amer. Chem. Soc.,* **92**, 3824 (1970).
6. Heyde, M. E., Gill, D., Kilponen, R. G., and Rimai, L., *J, Amer. Chem. Soc.,* **93**, 6776 (1971).
7. Rimai, L., Gill, D., and Parsons, J. L., *J. Amer. Chem. Soc.,* **93**, 1353 (1971).
8. Behringer, J., in *Raman Spectroscopy* (edit. by Szymanski, H. A.) (Plenum Press, New York, 1967).
9. Dartnall, H. J. A., and Lythgoe, J. N., *Vision Res.,* **5**, 81 (1964).
10. Hubbard, R., *Nature,* **221**, 432 (1969).
11. Abrahamson, E. W., and Ostroy, S. E., *Prog. Biophys. Mol. Biol.,* **17**, 181 (1967).
12. Morton, R. A., and Pitt, G. A. J., *Biochem. J.,* **59**, 128 (1955).
13. Kropf, A., and Hubbard, R., *Ann. NY Acad. Sci.,* **74**, 266 (1958).
14. Oesterhelt, D., and Stoeckenius, W., *Nature New Biology,* **233**, 149 (1971).
15. Lord, R. C., and Mendelsohn, R., *J. Amer. Chem. Soc.,* **94**, 2133 (1972).
16. McGlynn, S. P., *Rad. Res. Suppl.,* **2**, 300 (1960).
17. Ishigami, M., Mieda, Y., and Mishima, K., *Biochim. Biophys. Acta,* **112**, 372 (1966).

的可行性，以及共振拉曼技术为感光色素提供结构探针的能力。

我要感谢布劳罗克、利布、麦克莱尔这几位博士所做的详细讨论。本研究由加拿大国家研究委员会赞助。

（王耀杨 翻译；李芝芬 审稿）

Double Helix at Atomic Resolution

J. M. Rosenberg *et al.*

Editor's Note

By this time, the double-helix structure of DNA proposed by Francis Crick and James Watson was not in doubt. All the same, no one had carried out X-ray crystallography at sufficiently high resolution to see the positions of the atoms directly. That is what Alexander Rich, a postdoctoral student of Linus Pauling, and his colleagues report here—not for DNA itself, but for a double-helical form of a small fragment of RNA. This offered the first direct view of a Watson–Crick base pair, and confirmed its structure beautifully. Nadrian Seeman, one of Rich's coauthors, went on to pioneer the use of DNA as a construction material for molecular nanotechnology.

The sodium salt of the dinucleoside phosphate adenosyl–3′, 5′–uridine phosphate crystallizes in the form of a right-handed antiparallel double helix with Watson–Crick hydrogen bonding between uracil and adenine. A sodium ion is located in the minor groove of the helix complexed to both uracil rings.

MANY important functions in molecular biology are determined by antiparallel double helical nucleic acids with complementary base pairing between the two strands, as first described by Watson and Crick[1]. Although considerable work has been carried out on the structures of double helical nucleic acids[2-8], the fine details of their molecular architecture have never been available at atomic resolution. The X-ray diffraction patterns shown by fibres of DNA and RNA characteristically die out at resolution greater than about 3 Å. Because of this, there has been considerable discussion[9-13] about the actual nature of the detailed stereochemistry and hydrogen bonding in double helical DNA. One way to obtain additional information about these atomic details is to crystallize oligonucleotide fragments of known sequence which may form double helices in the crystal lattice. Here we report the single crystal X-ray analysis of the sodium salt of adenosyl–3′, 5′–uridine phosphate (ApU) which forms a right-handed antiparallel double helix in which the ribose phosphate backbones are held together by Watson–Crick hydrogen bonding between the adenine and uracil residues. This is the first crystal structure in which the atomic details of double helical nucleic acids can be visualized. In addition, this is the first single crystal structure showing Watson–Crick base pairing between adenine and uracil.

Experiments

The sodium salt of ApU (Miles) was mixed stoichiometrically with bromotertiary butyl amine in a 40% solution of 2–methyl–2,4–pentanediol (MPD). A small volume of the solution was placed in equilibrium with a large reservoir of 60% MPD in a closed

原子分辨率的双螺旋

罗森堡等

编者按

直到现在，由弗朗西斯·克里克和詹姆斯·沃森提出的DNA双螺旋结构仍无可置疑。尽管如此，利用X射线晶体学的方法获得高分辨率结构以便直接观察原子位置的研究还没有人进行过。莱纳斯·鲍林的博士后亚历山大·里奇以及他的同事在本文中报道了一小段RNA（而不是DNA）的双螺旋结构。此项工作使人们第一次直接观察到沃森–克里克碱基配对，并且对其结构的正确性进行了精彩的证实。里奇的一位合作者纳德里安·西曼继而开创性地将DNA作为构建材料应用到了分子纳米技术中。

二核苷磷酸腺苷–3′, 5′–尿苷磷酸的钠盐以右手反向平行双螺旋的形式结晶，尿嘧啶和腺嘌呤之间形成沃森–克里克氢键。一个钠离子位于双螺旋的小沟，与两侧的两个尿嘧啶环结合。

分子生物学中的许多重要功能皆由两条链之间形成碱基互补配对的反向平行双螺旋核酸决定，沃森和克里克首次对其进行了描述[1]。尽管人们对双螺旋核酸的结构已经开展过大量的研究[2-8]，却还从未获得过精细的原子分辨率的分子结构。由DNA和RNA纤维获得的X射线衍射图样的分辨率无法超过3 Å。基于这一现实，人们对双螺旋DNA中精细的立体化学结构及氢键形成的真实本质开展了诸多讨论[9-13]。将已知序列的寡聚核苷酸片段进行结晶是获得更多DNA的原子结构细节的一种途径，它们可能在晶格中形成双螺旋结构。本文报道了由腺苷–3′,5′–尿苷磷酸（ApU）的钠盐所形成的右手反向平行双螺旋的单晶体X射线分析结果，其中的核糖磷酸骨架通过在腺嘌呤和尿嘧啶残基之间形成沃森–克里克氢键而结合在一起。这是第一个可以看到双螺旋核酸原子细节的晶体结构。此外，这也是第一个显示腺嘌呤和尿嘧啶之间的确能形成沃森–克里克碱基配对的单晶体结构。

实　验

在40%的2–甲基–2,4–戊二醇（MPD）溶液中，将ApU（迈尔斯公司生产）的钠盐与三溴丁胺按化学反应计量比混合。在一个密闭容器中，将少量该混合物溶液置于含60% MPD的大体积溶液中在4℃进行平衡。静置两周后，开始出现具有清

container which was stored at 4°C. After standing for 2 weeks, small prismatic crystals with well-defined faces began to appear. Crystal growth continued slowly for several months and yielded crystals suitable for X-ray analysis. The amine was put into the solution in the hope that it might become the cation of the structure; however, subsequent analysis revealed that the sodium salt had crystallized rather than the bromoamine. A crystal measuring 0.2 mm × 0.15 mm × 0.05 mm was mounted on the tip of a glass fibre for X-ray analysis. The crystal was found to be monoclinic, space group $P2_1$, with cell dimensions a = 18.025 Å, b = 17.501 Å, c = 9.677 Å, β = 99.45°. The crystal density measured in a density gradient was 1.53. In order to obtain a calculated density near the observed density, it was necessary to assume four molecules of Na^+ApU^- and twenty-two water molecules in the unit cell. This surprisingly high degree of hydration was subsequently shown to be a low estimate as solution of the structure revealed twenty-four water molecules in the unit cell. Three-dimensional X-ray diffraction intensity data were collected out to a resolution of 0.8 Å, with a "Picker" FACS-1 diffractometer in an "Omega" step scan mode, using Nickel filtered CuKα radiation. The data were collected at 8°C and were corrected for Lorentz and polarization effects; no absorption correction has been applied because of the low mass absorption coefficient (18.0).

Solution of the Structure

Although there was no crystallographic two-fold axis in the lattice, the presence of a peak nearly 40% the height of the origin on the Harker section of the Patterson function suggested a non-crystallographic two-fold axis in the structure. From this we inferred the double helical nature of ApU. This peak was assumed to contain all vectors from each atom of one ApU molecule to the corresponding atom of the other independent molecule, symmetry related. From previous work with the protonated dinucleoside phosphate uridyl–3′,5′–adenosine phosphate[14] (UpA), it was known that resolution difference Patterson techniques were effective in discriminating vectors arising from two second-row atoms in a large structure. (Resolution difference Patterson techniques are those in which two Patterson or superposition functions, for example, multiple minimum functions or N-atom symmetry minimum functions, are compared. These two Pattersons are (1) the standard F^2 Patterson, calculated using all the diffraction data, and (2) a Patterson calculated only from the higher-order reflexions whose Fs should contain a larger contribution from the heavier atoms sought as they are relatively denser near the atomic centres. The origins of both Pattersons are normalized to the same value. Peaks arising from heavy atom–heavy atom vectors ought to be relatively more prominent than overlapping light atom–light atom vectors in the second map. In this work, the first Patterson contained all the 1 Å data, and the second function contained the data in the shell between 1.5 Å and 1 Å.)

UpA contained seventy-seven first-row atoms. Thus, we hoped to locate the phosphorus atoms by using this vector as the basis vector of a Resolution Difference 2-Atom Symmetry Minimum Function[15,16] (RDSMF(2)). The RDSMF(2) initially indicated the wrong location for the phosphorus atoms, however, which was obvious when Fourier refinement procedures failed to reveal the structure. An (E^2–1) Patterson (E is the

晰表面的细小棱状晶体。晶体持续缓慢生长数月，最终形成适用于X射线分析的晶体。之所以在溶液中加入胺，是希望其成为结构中的阳离子；不过，之后的分析表明，形成晶体的是钠盐而非溴胺。一个大小为0.2 mm × 0.15 mm × 0.05 mm的晶体置于一玻璃纤维的尖端进行X射线分析。分析显示，晶体为单斜晶系，空间群为$P2_1$，晶胞大小为：a = 18.025 Å，b = 17.501 Å，c = 9.677 Å，β = 99.45°。在密度梯度中测得晶体密度为1.53。为了使计算出的密度值接近观测到的密度，我们必须假设在一个晶胞中含有4个分子的Na^+ApU^-和22个分子的水。这种令人惊讶的高度水合现象被后来的结果证明仍然是被低估的，因为结构分析结果表明，一个晶胞含有24个分子的水。利用“皮克”式FACS-1衍射仪、采取“欧米茄”逐步扫描模式、以镍过滤的CuKα做光源，我们采集到了分辨率达到0.8 Å的三维X射线衍射强度数据。数据是在8℃采集的并校正了洛伦兹和偏振效应；由于质量吸收系数比较低（18.0），所以没有进行吸收校正。

结构解析

尽管晶格中没有晶体学二次轴，在帕特森函数的哈克截面上出现了约为初始高度40%的峰值，这提示在该结构中存在非晶体学二次轴。由此我们推断出了ApU的双螺旋本质。我们认为这一峰值包含了一个ApU分子中的每一原子到另一独立分子中相应原子的所有矢量，而且这些矢量是相关对称的。根据以前针对质子化尿苷–3′,5′–腺苷磷酸[14]（UpA）这种二核苷磷酸的研究，人们已经知道，在辨别一个大的结构中源自两个第二排（编者注：应指元素周期表第三周期）原子的矢量时，分辨率差异帕特森技术是有效的。（分辨率差异帕特森技术是指将两种帕特森或叠加函数（如多个最小值函数或N原子对称最小函数）进行比较。这两种帕特森函数是：（1）标准F^2帕特森函数，采用全部衍射数据计算所得；（2）仅利用较高次序的反射数据计算出的帕特森函数，其Fs会包含更多所收集到的较重原子的权重，因为它们的原子中心附近相对更加致密。两种帕特森函数的初始值被归一化为相同的值。在第二张图上，源自重原子–重原子矢量的峰值应当比重叠的轻原子–轻原子向量相对更加明显。在本研究中，第一个帕特森函数包含了所有分辨率为1 Å的数据，而第二个函数包含了1.5 Å到1 Å之间的数据）

UpA这种二核苷磷酸含有77个第一排（编者注：应指元素周期表第二周期）原子。因此，我们希望使用这一矢量作为“分辨率差异2–原子对称最小函数”[15,16]（RDSMF(2)）的基矢量来定位磷原子。然而，RDSMF(2)最初对磷原子的定位是错误的。当傅里叶优化程序无法揭示该结构时，这一点就变得显而易见了。因此我们

quasinormalized structure factor) was therefore calculated and the map and its Fourier coefficients were corrected for non-negativity and minimal bond lengths according to the procedure proposed by Karle and Hauptman[17]: (1) all negative points in the Patterson map and all points within a 0.9 Å radius of the origin were zeroed; (2) the map was Fourier transformed and all resultant (E^2–1) coefficients less than −1 were raised to −1; (3) a new Patterson map was calculated from these coefficients, and the process was iterated until convergence was obtained at forty cycles. From the final amplitudes, a set of thermally sharpened *F*s were generated, and used to calculate Patterson functions. The RDSMF(2) calculated from these Pattersons revealed the proper phosphorus locations. Approximate phosphate orientations were derived from Patterson superpositions. Fourteen cycles of Fourier refinement using the corrected *E*s as Fourier amplitudes revealed the two molecules of ApU, the sodium ions and four water molecules. The other water molecules were located in a series of difference syntheses. The structure has been refined using isotropic thermal parameters and only the 1 Å observed data by full matrix least squares. The current *R* factor is 0.091. (The discrepancy factor, *R*, is defined as $R = \sum ||F_o|-|F_c|| / \sum |F_o|$, where $|F_o|$ and $|F_c|$ are the amplitudes of the observed and calculated structure factors, respectively.) Because this is one of the larger non-centrosymmetric biological crystal structures solved without isomorphous replacement, we will expand on the details of the solution in a later publication.

Structure

In the analysis of this structure, there are two striking features. The non-crystallographic pseudo two-fold axis mentioned above rotates one ApU molecule into the other, so that the structure forms a segment of a right-handed antiparallel double helix in which the bases are hydrogen bonded to each other in the Watson–Crick manner, as shown in Figs. 1 and 2. This is the familiar hydrogen bonding between adenine and uracil which is believed to occur quite generally in double helical nucleic acids. Nonetheless, this type of hydrogen bonding had never been seen previously in single crystal X-ray analysis. It should be noted that the two-fold axis to which we are referring lies half-way between the base planes, rather than in those planes. Axes in the planes are the ones usually discussed in the literature, but the periodic nature of double helices generates symmetry elements at both locations.

计算了一种（E^2–1）帕特森函数（E 是准归一化的结构因子），并根据卡勒和豪普特曼[17]提出的流程，将其图谱和傅里叶系数校正为非负值和最小键长：(1）帕特森图谱上的所有负点及距原点 0.9 Å 半径以内所有的点都被归零；(2）对该图谱进行傅里叶变换，产生的所有（E^2–1）系数小于 –1 的都被提到 –1；(3）根据这些系数计算出一个新的帕特森图谱，重复该过程，直到第 40 轮得到收敛解为止。从最后的振幅中，产生了一组热锐化的 Fs，并用于计算帕特森函数。从这些帕特森函数中计算得到的 RDSMF(2) 揭示了正确的磷原子位置。磷原子的大致取向是通过帕特森叠加得到的。使用校正过的 Es 作为傅里叶振幅，进行 14 次傅里叶优化后，两分子 ApU、钠离子和 4 个水分子的位置得以确定。其他水分子是通过一系列的差值合成而被定位的。使用各向同性热参数和仅 1 Å 的观察数据，通过全矩阵最小二乘法，我们对该结构进行了优化。目前得到的 R 因子为 0.091。(偏离因子 R 被定义为：$R = \sum ||F_o| - |F_c|| / \sum |F_o|$，其中 $|F_o|$ 和 $|F_c|$ 分别是观测到的和计算得到的结构因子的振幅。）因为这是没有用同晶置换解析的、较大的非中心对称生物晶体结构之一，我们将在以后发表的文章中交代解析方法的细节。

结　构

经过分析，我们发现该结构有两个显著的特征。上文提到的非晶体学伪二次轴使一个 ApU 分子旋转到另一个 ApU 上方，因而该结构形成了一个右手反向平行双螺旋片段，其中的碱基以沃森 – 克里克方式相互形成氢键，如图 1 和 2 所示。这种存在于腺嘌呤和尿嘧啶之间的氢键为人们所熟知，它被认为相当普遍地出现在双螺旋的核酸分子中。尽管如此，这种氢键之前从未在单晶体的 X 射线分析中被观察到。应当注意的是，我们所说的二重轴位于两个碱基平面的中间线上，而非位于每一个碱基平面之中。位于碱基平面内的那个轴在文献中经常被讨论到，但是双螺旋的周期性本质使得在这两个位置上都能产生出对称元素。

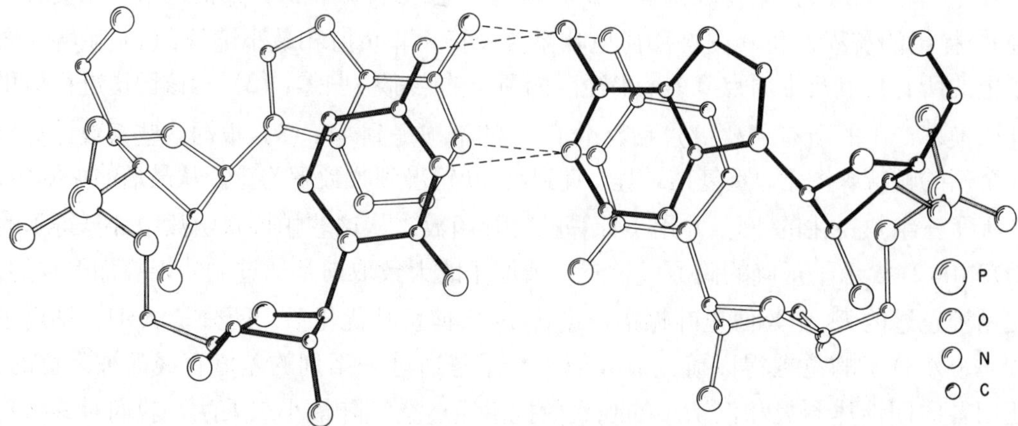

Fig. 1. View of the crystal structure perpendicular to the base planes showing hydrogen bonding (dashed lines) as well the base stacking interactions. The darkest portions of the figure are those nearest the viewer. The rotational relationship is easily seen by comparing the front (black) and rear (white) glycosidic bonds.

The other prominent feature was that the crystal was heavily hydrated, so that the ApU molecules are surrounded by large numbers of water molecules. This observation, coupled with the non-crystallographic nature of the two-fold axis, leads us to believe that the double helical nature of ApU is a function of the molecules themselves, rather than crystal packing forces.

The separation between the base pairs is 3.4 Å and, as can be seen in Fig. 1, there is a considerable degree of stacking in the structure. The NH...O hydrogen bonds between the N6 amino group of adenine and the O4 carbonyl oxygen of uracil (see Fig. 3 for the numbering scheme) have bond lengths of 2.95 Å and 2.91 Å. The NH...N hydrogen bonds between uracil N3 and adenine N1 have lengths of 2.82 Å and 2.86 Å. The important distances across the base pairs between the two ribose carbon atoms C1′ of the glycosidic linkage, are 10.50 Å and 10.53 Å. Similar distances to these have been obtained by analysis of single crystals of intermolecular complexes of purine and pyrimidine derivatives[18] but this is the first time that these distances have been observed in a molecule which forms a fragment of a double helix.

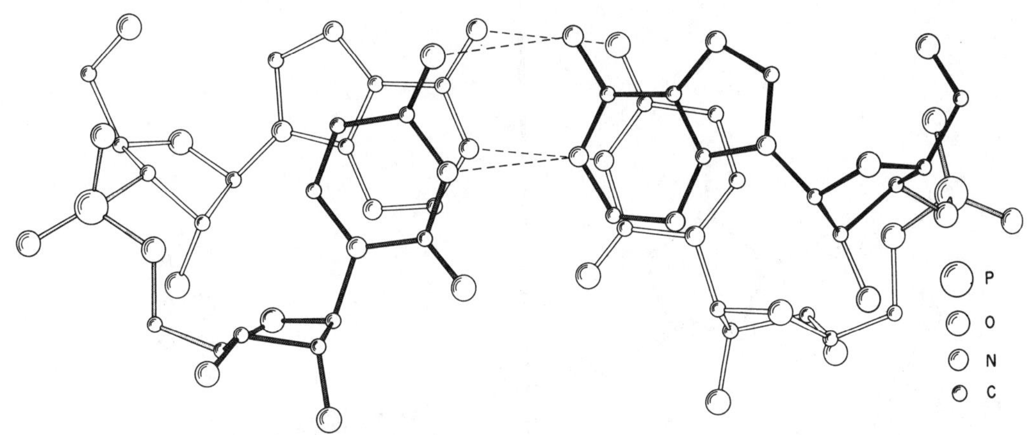

图 1. 从垂直于碱基平面的方向观察晶体结构可以显示氢键（虚线）以及碱基堆积相互作用。图中颜色最深的部分距离观察者最近。通过比较前面的（黑色）和后面的（白色）糖苷键，可以很容易看出碱基之间的旋转关系。

另一个显著的特征是该晶体被高度水合，使得 ApU 分子被大量水分子包围。这一结果结合非晶体学二次轴的性质使我们相信，ApU 的双螺旋本质源于分子本身的作用，而非晶体学堆积所致。

碱基对之间的距离是 3.4 Å，如图 1 所示，该结构中有相当程度的堆积。在腺嘌呤的 N6 氨基与尿嘧啶的 O4 羰基氧（碱基上各个原子的编号规则见图 3）之间的 NH⋯O 氢键的键长分别为 2.95 Å 和 2.91 Å。尿嘧啶 N3 和腺嘌呤 N1 之间的 NH⋯N 氢键长分别为 2.82 Å 和 2.86 Å。另外一对重要的距离是横跨碱基对、参与糖苷键形成的两个核糖碳原子 C1′之间的，分别是 10.50 Å 和 10.53 Å。（译者注：这里的两个距离应该是指图 3 中的上面那对碱基之间的和下面那对碱基之间的相关原子之间距离）人们曾经通过分析嘌呤和嘧啶衍生物的分子间复合物的单晶体，得到过类似的距离 [18]；但在形成了双螺旋片段的分子中测得这些距离，还是第一次。

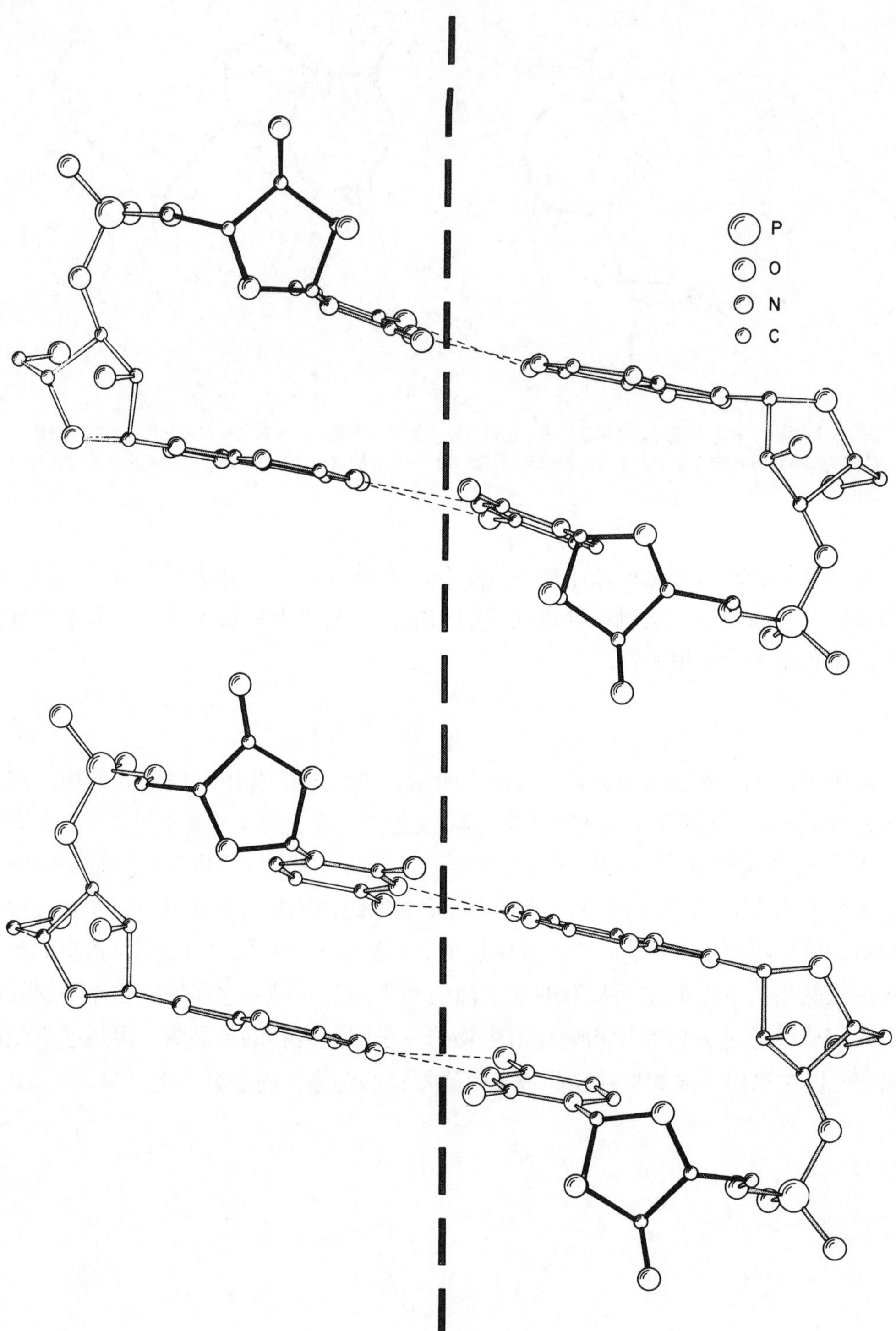

Fig. 2. Comparison of the structures of ApU (upper) and RNA 11 (lower). This view is approximately perpendicular to the helix axis which is indicated by the vertical dashed line.

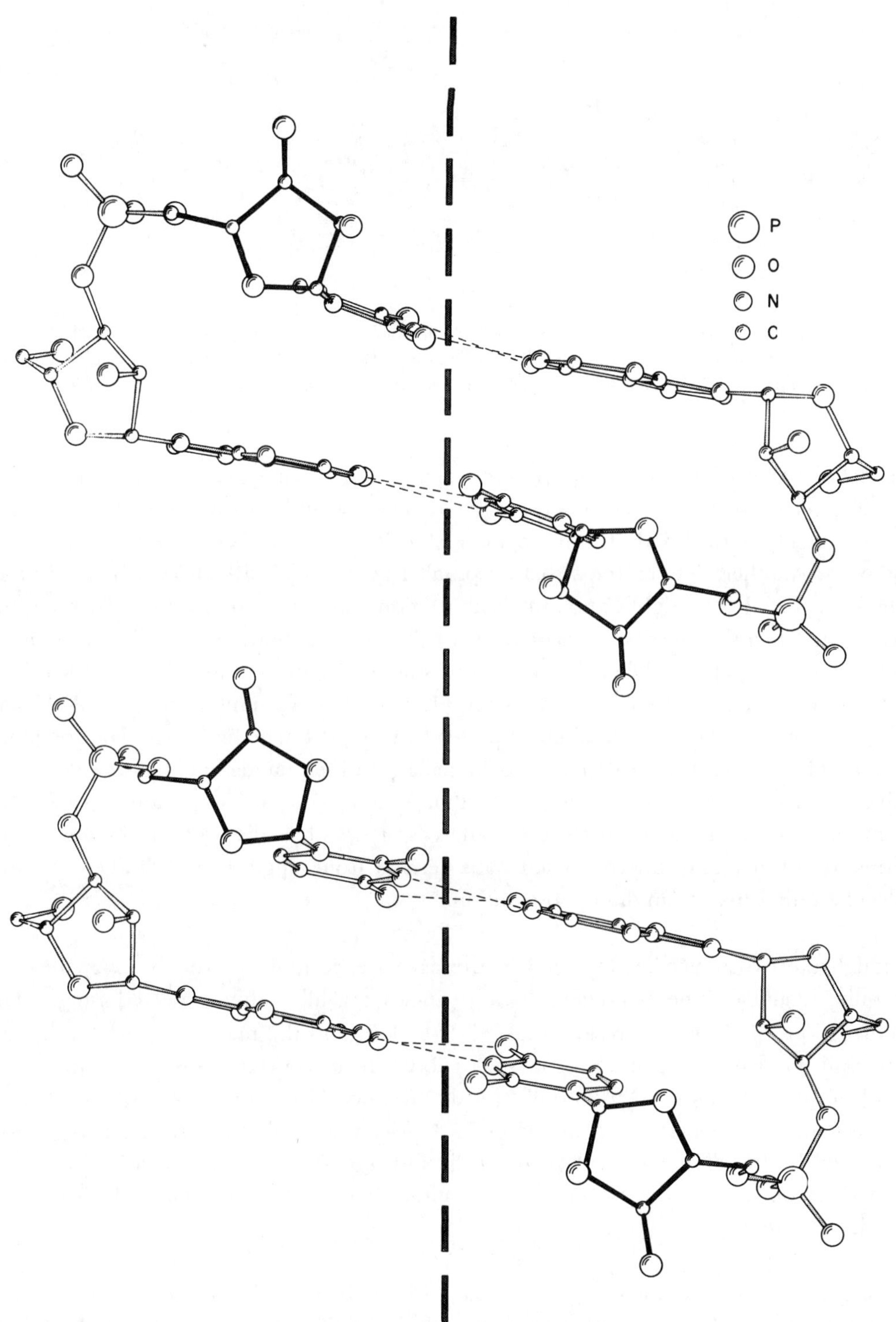

图 2. ApU（上方）和 RNA 11（下方）的结构比较。观察角度大致垂直于竖直虚线所示的螺旋轴。

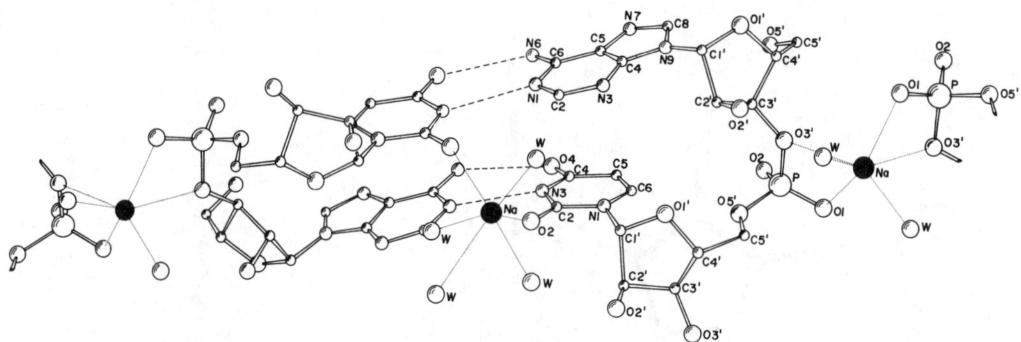

Fig. 3. Perspective view of the structure as seen from the minor groove showing the numbering scheme and the coordination of the sodium ions. Both sodium ions (●) have distorted octahedral coordination involving water molecules and other oxygens. The central sodium ion is 2.36 Å from the uracil O2 atoms.

The helical form of the molecules can be seen in Fig. 1 in which one is looking in a direction perpendicular to the base pair plane. There is clearly a rotational relationship between the glycosidic bonds (C1′ to adenine N9 or C1′ to uracil N1) of the front and rear base pairs. Another view of the structure parallel to the stacked hydrogen bonded bases is shown in Fig. 2. The right-handed helical rotation can readily be seen by observing the orientations of adjoining ribose residues on the ribose phosphate chain. This structure is similar to that which has been deduced from studying double-stranded viral RNA. The naturally occurring material is called RNA 11. Fig. 2 shows similar views of ApU and RNA 11, where the dashed vertical line represents the approximate helix axis. The continued extension of the ApU nucleotide pairs would generate a right-handed double-stranded helix similar to RNA 11. It is important to note that although the helical parameters of RNA 11 can be derived directly from an X-ray analysis of the fibre, the positions of the atomic centres are of necessity inferred from data derived from single crystal studies of small molecules rather than from direct observation.

Although the two molecules of ApU in the asymmetric unit are very similar, they are not quite identical. The bond lengths and angles are all within expected range. The conformations of the ribose residues are all 3′-endo[19], and the nucleosides are orientated in the anticonformation[20], with torsion angles about the glycosidic bonds[19] of 2° and 7° for the adenosine residues, and 29° and 30° for the uridines. Comparison with the protonated UpA structure[21,22] suggests that discrepancies between adenine and uracil torsion angles are caused by the differences between the 3′ and 5′ ends of the molecule as the same pattern was obtained with respect to the 3′ and 5′ ends in that molecule which had the opposite base sequence.

In the protonated structure UpA, it was noted that there were short distances between adenosine C8 and O5′, as well as uridine C6 and O5′ (ref. 14). At that time, it was suggested that this might be a result of an intramolecular attraction which, if of a general nature, would help stabilize the anticonformation of the nucleotide. Accordingly, the

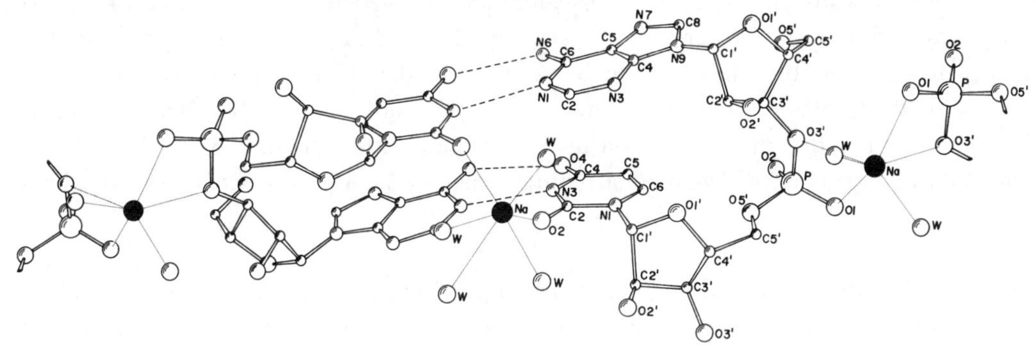

图 3. 从（DNA 双螺旋的）小沟上观察到的透视结构图，显示了核苷酸中原子的编号和钠离子形成的配位结构。两个钠离子（●）均参与形成扭曲的八面体配位，参与成键的原子还有水分子和其他氧原子。位于中心的那个钠离子距离尿嘧啶的 O2 原子的距离是 2.36 Å。

从图 1 中可以看到这些分子的螺旋形态，在此图中，读者的观察方向与碱基对平面垂直。图中前方和后方的碱基对的糖苷键（C1′ 到腺嘌呤 N9 或 C1′ 到尿嘧啶 N1）之间存在明显的旋转关系。图 2 则显示了从平行于以氢键连接的堆积碱基的方向观察到的结构。通过观察核糖磷酸链上的相邻核糖残基的取向，我们可以很容易地看出该分子的右手螺旋结构。这一结构与研究双链病毒 RNA 推导出的结构相似。这种天然形成的分子被称为 RNA 11。图 2 显示了 ApU 和 RNA 11 之间的类似结构，其中垂直的虚线表示近似的螺旋轴。ApU 核苷酸对的持续延伸将会形成一个类似于 RNA 11 的右手双螺旋。值得注意的是，尽管 RNA 11 的螺旋参数可以从 RNA 纤维的 X 射线分析中直接得到，但是其中那些原子中心的位置必须从小分子的单晶体研究数据中得到，而无法直接观察到。

尽管在不对称单元中的两分子 ApU 非常相似，但它们并非完全相同。它们的键长和键角均在预期范围内。核糖残基的构象均为 3′– 内向 [19]，而核苷基团的取向则为反向构象 [20]，其中糖苷键 [19] 对于腺苷残基的扭转角为 2° 和 7°，对于尿嘧啶则为 29° 和 30°。与质子化的 UpA 结构 [21,22] 进行比较表明，腺嘌呤和尿嘧啶在扭转角上的差异是由于分子的 3′ 和 5′ 端之间的差异而引起的；因为在具有相反碱基序列的分子中也观察到了类似的 3′ 和 5′ 端的差异。

在 UpA 的质子化结构中，人们注意到了腺苷的 C8 和 O5′ 之间以及尿嘧啶的 C6 和 O5′ 之间的距离都很短（文献 14）。当时有人提出，这可能是分子内吸引的结果。如果这种吸引力普遍存在的话，它有助于核苷酸反向构象的稳定。因此，人们怀着

related distances in this structure were examined with great interest. Three of the four independent distances were greater than or equal to 3.3 Å; thus, there was no evidence of any interaction. But the distance between one adenine C8 and its ribose O5′ was 3.07 Å. As this O5′ is not covalently linked to a phosphorus atom, the phenomenon may not be relevant to the nucleotides in a helical polynucleotide chain. Nevertheless, for non-helical and 3′ terminal nucleic acid structures, this interaction may yet be seen to play an important role.

The individual double helical fragments are separated in the crystal lattice by a spacing of 3.4 Å, with considerable overlapping of the adenine residues. Besides this intermolecular base stacking, the other important interaction in the direction perpendicular to the base pairs involves the proximity of O1′ of both adenosine riboses with uracil rings. A similar type of interaction was found to be important in the crystal structure of UpA[21,22].

The crystal structure can be visualized as a rod-like entity. The central portions of these rods contain the stacked base pairs. These are flanked by the ribose phosphate backbones, both of which in turn are surrounded by the solvent structure. The large amount of water allows the crystal structure to assume a conformation which is minimally perturbed by lattice interactions. Thus, the double helical structure which we observe here bears a marked resemblance to those structures proposed for solvated polynucleotides. It should be noted in passing, however, that the water is held rather firmly, as the intensity data were collected from the crystal while it was mounted in air.

The ribose phosphate chains are packed together with the phosphate groups facing each other. In the layer between the ribose phosphate moieties are sodium ions which are found in two distinct sites: one, complexed between phosphate groups, and the other, surprisingly, bound to the uracil residues. The sodium ions and their ligands are shown in Fig. 3. The sodium ion in the centre is located on the pseudo two-fold axis in the minor groove of the helix with octahedral coordination which includes the free O2 atoms of the uracils; the remaining ligands are water molecules. As shown in Fig. 3, the other sodium ion rests on an intermolecular pseudo-dyad axis, also exhibiting octahedral coordination, which includes two adenosyl O3′ atoms and two phosphate oxygen atoms.

The stability of the sodium ion position between two phosphate groups is readily understood on electrostatic grounds. What is not so obvious is why the second sodium ion is complexed to the two uracils in the minor groove of the helix. Five angstroms away from this site there is another position between two phosphate groups which could comfortably accommodate the sodium ion and is occupied only by solvent. It is possible that the structure is stabilized considerably by the sodium ion in the minor groove site. Should this prove to be of more general occurrence in other crystals containing A–U sequences, it is possible that this type of coordination may be of importance in polynucleotide structure and function.

浓厚的兴趣对该结构中的相关距离进行了研究。在 4 种独立距离中，有 3 种大于或等于 3.3 Å；因此，没有证据表明发生了任何相互作用。但腺嘌呤的 C8 与其核糖 O5′ 的距离是 3.07 Å，由于这个 O5′ 与磷原子并非共价相连，这一现象可能与核苷酸是否位于螺旋状的多聚核苷酸链上无关。尽管如此，对于非螺旋的和有 3′ 端的核酸结构而言，这种相互作用可能仍然起着重要作用。

每一个双螺旋片段都以 3.4 Å 的距离分布在晶格上，其腺嘌呤残基之间存在相当程度的重叠。除了这种分子之间的碱基堆积外，另一种重要的相互作用存在于碱基对的垂直方向上，由两个腺苷的核糖中的 O1′ 与两个尿嘧啶环之间的相互接近而产生。人们发现在 UpA 晶体中有一种类似的相互作用也十分重要 [21,22]。

晶体结构可以被看作是一种类似杆状的实体。这些杆状实体的中心部分为堆积的碱基对，两侧为核糖磷酸骨架，而这两条核糖磷酸骨架又被溶剂结构所包围。大量的水的存在使得晶体结构呈现出一种构象，该构象可以使晶格之间的扰动降到最低。因此，我们这里所观察到的双螺旋结构与那些溶剂化的多聚核苷酸的结构之间存在显著的相似性。不过，应当顺便提出的是，核酸分子相当牢固地结合了水，因为晶体的衍射强度数据是将其置于空气中时采集的。

核糖磷酸链组装在一起，磷酸基团两两相对。在两个核糖磷酸的中间层内，钠离子存在于两个位点：其一是络合在磷酸基团之间，另一个则是出乎意料地与尿嘧啶残基络合。钠离子及其配体见图 3。钠离子位于双螺旋小沟的伪二次轴上，以其为中心，可以与包括尿嘧啶的自由 O2 原子和水分子形成八面体配位 。如图 3 所示，另一个钠离子位于分子内的伪二次对称轴上，同样以八面体配位形式存在，参与配位的包括两个腺苷的 O3′ 原子和两个磷酸基团上的氧原子。

从静电学角度就能很容易地理解钠离子可以稳定存在于两个磷酸基团之间。然而，我们不清楚的是为何第二个钠离子与螺旋小沟的两个尿嘧啶络合。距离这一位点 5Å 处，在两个磷酸基团之间有另一位点，完全可以容纳钠离子，但是该位点仅被溶剂分子所占据。该结构的稳定性很大程度上可能是通过位于小沟上的钠离子而实现的。如果能证明在其他包含 A–U 序列的晶体中这一现象也常见的话，这种类型的配位可能对多聚核苷酸的结构和功能起重要的作用。

As noted above, one of the most significant features of the organization of the crystal structure of ApU is that the two independent molecules in the asymmetric unit have very similar but not quite identical conformations. Thus, they are related by a pseudo two-fold rotation axis while in an antiparallel double helical polynucleotide there is presumed to be a true two-fold axis which relates the backbones of the two antiparallel chains. We do not completely understand why the crystal structure of ApU did not adopt a true two-fold axis. In this regard, however, we recently discovered that the closely related dinucleoside phosphate, guanosyl-3′, 5′-cytidyl phosphate does indeed form a right-handed antiparallel double helix in the crystalline state in which this two-fold axis is crystallographic (unpublished results of R. O. Day, N. C. Seeman, J. M. R., and A. R.). This recent observation lends support to our earlier emphasis on the importance of the pseudo two-fold axis in the present crystal structure.

Among the significant facts which we learn by analysing double helical fragments are the important detailed parameters determining the conformation of the polynucleotide chain, especially those parts dealing with the geometry of the phosphate group. The structure of the ribose phosphate chain is of central importance in understanding the physical properties and behavior of polynucleotide chains[23]. This information should prove valuable in interpreting the details of the molecular structure in polynucleotide double helices, as well as in the more complex forms of RNA such as those observed in tRNA[24].

This research was supported by grants from the National Institutes of Health, the National Science Foundation and the American Cancer Society. N. C. S. is a postdoctorate fellow of the Damon Runyon Foundation, J. M. R. is a predoctoral trainee of the National Institutes of Health, F. L. S. is a postdoctoral fellow of the American Cancer Society, H. B. N. was an NIH postdoctoral fellow.

We thank Bob Rosenstein, Roberta Ogilvie Day, Don Hatfield, Sung-Hou Kim and Gary Quigley for useful discussions and encouragement, and John Genova and Tim O'Meara for technical assistance.

(**243**, 150-154; 1973)

John M. Rosenberg, Nadrian C. Seeman, Jung Ja Park Kim, F. L. Suddath, Hugh B. Nicholas and Alexander Rich
Department of Biology, Massachusetts Institute of Technology, Cambridge, Massachusetts 02139

Received January 3, 1973.

References:

1. Watson, J. D., and Crick, F. H. C., *Nature*, **171**, 737 (1953).

2. Fuller, W., Wilkins, M. H. F., Wilson, H. R., and Hamilton, L. D., *J. Mol. Biol.*, **12**, 60 (1965).

3. Langridge, R., and Gomatos, P. J., *Science*, **141**, 694 (1963).

4. Tomita, K., and Richa, A., *Nature*, **201**, 1160 (1964).

5. Arnott, S., Dover, S. D., and Wonacott, A. J., *Acta Cryst.*, B, **25**, 2192 (1969).

6. Rich, A., Davies, D. R., Crick, F. H. C., and Watson, J. D., *J. Mol. Biol.*, **3**, 71 (1961).

正如以上所言，ApU 晶体结构组织的最显著的特征是，位于不对称单元中的两个独立分子之间具有非常相似但不完全相同的构象。因此，尽管在这里它们是以一条伪二重旋转轴相关联，但当处于一条反向平行的双螺旋多聚核苷酸中时，应该有一条真正的二重轴把两条反向平行链的骨架关联起来。我们还没有完全理解为何 ApU 的晶体结构没有采用真正的二重轴。不过，关于这一点，我们最近发现与之密切相关的二核苷磷酸，鸟苷–3′,5′–胞苷磷酸在晶体状态下确实形成了右手反向平行双螺旋，且形成的是晶体学二次轴（戴、西曼、罗森堡和里奇的未发表成果）。此前我们强调了伪二次轴对该晶体结构的重要性，这一近期结果支持了我们的结论。

我们通过分析双螺旋片段所得到的重要结果有，确定多聚核苷链构象的重要详细参数，特别是关于磷酸基团几何结构的部分。核糖磷酸链的结构对于理解多聚核苷酸链的物理性质和行为都至关重要[23]。对于解释多聚核苷酸双螺旋以及形式更为复杂的 RNA 中的（如在 tRNA 中观察到的）分子结构细节，这些信息将会非常有价值[24]。

此研究得到了国立卫生研究院、国家科学基金会和美国癌症协会的经费资助。西曼是戴蒙·鲁尼恩基金会的博士后学者，罗森堡是国立卫生研究院的博士前实习生，苏达茨是美国癌症协会的博士后学者，尼古拉斯是国立卫生研究院的博士后学者。

在此我们感谢鲍勃 · 罗森斯坦、罗伯塔 · 奥格尔维 · 戴、唐 · 哈特菲尔德、金圣浩和加里 · 奎格利的有益讨论和鼓励，以及约翰 · 吉诺瓦和蒂姆 · 奥马拉的技术支持。

（周志华 翻译；吕增益 审稿）

7. Arnott, S., Hukins, D. W. L., and Dover, S. D., *Biochem. Biophys. Res. Comm.*, **48**, 1392 (1972).

8. Arnott, S., and Hukins, D. W. L., *Biochem. Biophys. Res. Comm.*, 47, 1504 (1972).

9. Donahue, J., *Science*, **165**, 1091 (1969).

10. Wilkins, M. H. F., Arnott, S., Marvin, D. A., and Hamilton, L. D., *Science*, **167**, 1693 (1970).

11. Crick, F. H. C., *Science*, **167**, 1694 (1970).

12. Arnott, S., *Science*, **167**, 1694 (1970).

13. Donahue, J., *Science*, **167**, 1700 (1970).

14. Seeman, N. C., Sussman, J. L., Berman, H. M., and Kim, S.-H., *Nature New Biology*, **233**, 90 (1971).

15. Corfield, P. W. R., and Rosenstein, R. D., *Trans. Amer. Cryst. Assoc.*, **2**, 17 (1966).

16. Seeman, N. C., thesis, Univ. Pittsburgh (1970).

17. Karle, J., and Hauptman, H., *Acta Cryst.*, **17**, 392 (1964).

18. Voet, D., and Rich, A., *Prog. Nucl. Acid. Res. Mol. Biol.*, **10**, 183 (1970).

19. Sunderalingam, M., *Biopolymers*, 7, 821 (1969).

20. Donahue, J., and Trueblood, K. N., *J. Mol. Biol.*, **2**, 363 (1960).

21. Sussman, J. L., Seeman, N. C., Kim, S.-H., and Berman, H. M., *J. Mol. Biol.*, **66**, 403 (1972).

22. Rubin, J., Brennan, T., and Sundaralingam, M., *Biochemistry*, **11**, 3112 (1972).

23. Kim, S. -H., Berman, H. M., Seeman, N. C., and Newton, M. D., *Acta Cryst.* (in the press).

24. Kim, S. H., Quigley, G. J., Suddath, F. L., MacPherson, A., Sneden, D., Kim, J. J. P., Weinzierl, J., and Rich, A., *Science*, **179**, 285 (1973).

Eukaryotes–prokaryotes Divergence Estimated by 5S Ribosomal RNA Sequences

M. Kimura and T. Ohta

Editor's Note

This brave stab at estimating one of the earliest but most crucial events in evolution marks an early attempt at using comparative molecular data to calibrate evolutionary history. Motoo Kimura and Tomoko Ohta compared sequences from 5S ribosomal RNA—one of the most "conserved" of all genetic sequences, meaning that it differs little across species—to estimate the divergence date between eukaryotes (such as humans and yeast) and prokaryotes (here two bacterial species). The estimate of two billion years would not surprise anyone today, but it caused some surprise then by implying that yeast is more closely related to humans than to bacteria.

DATING the principal events in the history of life on the Earth is an interesting subject in evolutionary studies. Here we estimate the time of divergence of the eukaryotes and the prokaryotes through comparative studies of 5S ribosomal RNA sequences, coupled with those of cytochrome *c*. By prokaryotes we mean primitive forms having no true nucleus (bacteria and blue-green algae), while by eukaryotes we mean higher nucleated organisms such as plants and animals including yeasts and fungi. The principle we use in our estimation is that the rate of nucleotide substitutions in the course of evolution is constant per year per site for each informational macromolecule as long as the structure and function of the molecule remain unaltered.

To estimate the evolutionary distances (number of mutant substitutions) among 5S rRNA sequences, we made the alignment shown in Fig.1, using published data[1–3] on human, yeast and bacterial (*Escherichia coli* and *Pseudomonas fluorescens*) sequences. To arrive at this alignment, previous attempts[1,3] involving two or three sequences were helpful. The alignment is made in such a way that the number of matches between sequences is maximized while keeping the gaps inserted as few as possible. It involves a trial and error process, shifting various regions, and counting the number of nucleotides by which the two sequences agree with each other, followed by calculation of probability that this or better agreement occurs by chance. Figure 1 shows clearly (as was noted already by others) the marked conservative nature of this molecule as shown by the fact that only a small number of gaps need be inserted to obtain homology. The observed differences between sequences in terms of the fraction of different sites are given in Table 1. The mutational distance, that is, the average number of nucleotide substitutions per site, was estimated using the formula

$$K=-\frac{3}{4}\ln(1-(\frac{4}{3}\lambda))$$

通过5S核糖体RNA序列估计真核生物-原核生物分化

木村资生，太田朋子

编者按

对进化中最早但是最关键事件的估计，这种大胆的尝试标志着应用比较的分子数据来标定进化历史的早期努力。5S 核糖体 RNA（5S rRNA）序列是最保守的遗传序列之一，其在物种之间差异很小。通过比较该序列，木村资生和太田朋子对真核生物（例如人类和酵母）与原核生物（本文中以两种细菌为例）之间的差异进行了估算。如今，对“二十亿年”的估计恐怕不会令人惊讶，但是研究显示酵母与人类的关系比其与细菌的关系更近却让大家感到意外。

确定地球生命史中重要事件的发生年代是进化研究中一个有趣的课题。在此我们通过对 5S rRNA 序列的比较研究，外加对细胞色素 *c* 的研究，估计了真核生物和原核生物分离的时间。对于原核生物，我们是指没有真正细胞核的简单构造（如细菌和蓝－绿藻），而对于真核生物，我们是指高等有细胞核的生物，如植物和动物，包括酵母和真菌。我们在估计中使用的原则是，只要分子的结构和功能保持不变，那么对于每个信息大分子而言，在进化过程中，每年每个位点的核苷酸替代率是恒定的。

为了估计 5S rRNA 序列的进化距离（突变替代的数量），我们用已发表的人类、酵母和细菌（大肠杆菌和荧光假单胞菌）的数据 [1-3] 进行了序列比对，如图 1 所示。为了获得这个比对，之前涉及两到三个序列比对的尝试 [1,3] 是有帮助的。这个比对是按照将两个序列间匹配的数量最大化同时又使插入的间隙数尽可能少的原则建立的。这是一个试错的过程，移动各个区域，记录两个序列之间一致的核苷酸数量，接下来计算当前结果的概率或随机出现的更好的一致性的概率。图 1 明确显示了（正如其他人已经注意到的）该分子显著的保守性，事实上只需插入少量间隙就能获得同源性。表 1 以差异位点所占序列比例的形式给出了序列之间的差异。突变距离，也就是每个位点的核苷酸替代的平均数量，通过公式

$$K=-\frac{3}{4}\ln(1-(\frac{4}{3}\lambda))$$

where λ is the fraction of sites by which two homologous sequences differ. The formula was derived under the assumptions that in the course of evolution nucleotide substitutions occur spatially at random and with uniform probabilities and that each of the four bases (A, C, G, U) mutates to any of the remaining three with equal probability. (For details see ref. 4.) The equivalent formula has been previously derived by Jukes and Cantor[5]. We should also note that Dayhoff's[3] empirical relationship tabulated in her Table 11–3 is practically equivalent to this formula especially for the purpose of comparing different K values.

```
Human          -G-UCUACGGCC-AUACCACCCUGAACGCGCCCGAUCUCGUCUGAU-CUCGGAAGCUAAGCAG
Yeast          -G-GUUGCGGCC-AUACCAUCUAGAAAGCACCGUUCUCCGUCCGAUAACCUGUAGUUAAGCUG
E. coli        UGCCUGGCGGCC-GUAGCGCGGUGGUCCCACCUGACCCCAUGCCGAACUCAGAAGUGAAACGC
P. fluorescens UGUUCUUUGACGAGUAGUGGCAUUGGAACACCUGAUCCCAUCCCGAACUCAGAGGUGAAACGA

               GGUCGGGCCUG-GUUAGUACUUGGAUGGGAGACCGCCUGGGAAUACCGGGUGCUGUAG-GCUU
               GUAAGAGCCUGACCGAGUAGUGUAGUGGGUGACCAUACGCGAAACCUAGGUGCUGCA--AUCU
               CGUAGCGCC---GAUGGUAGUGUG--GGGUCUCCCCAUGCGAGAGUAGGGAACUGCCAGGCAU
               UGCAUCGCC---GAUGGUAGUGUG--GGGUUUCCCCAUGUCAAGAUCUCG-ACCAUAGAGCAU
```

Fig. 1. Alignment of 5S rRNA sequences.

Table 1. Fraction of Different Sites between 5S rRNA Sequences

	Yeast	*E. coli*	*P. fluorescens*
Human	0.395	0.457	0.478
Yeast		0.474	0.565
E. coli			0.319

The average mutational distance between the eukaryotes (man, yeast) and the prokaryotes (*E. coli*, *P. fluorescens*) turned out to be

$$K_{eu\text{–}pro}=0.817\pm0.158$$

where the standard error was obtained from four observations (comparisons). On the other hand, the corresponding quantity between the human and yeast was

$$K_{h\text{–}y}=0.561\pm0.095$$

where the error is a theoretical one computed by equation (5) in ref. 4. It may be interesting to note here that yeast is more closely related to man than to the bacteria, supporting the thesis (compare ref. 6) that the division between the eukaryotes and prokaryotes is more basic than divisions within eukaryotes. The remoteness of the eukaryotes–prokaryotes divergence relative to the human–yeast divergence can be estimated by the ratio $K_{eu\text{–}pro}/K_{h\text{–}y}$ which is approximately 1.5. This is much lower than the corresponding estimate of McLaughlin and Dayhoff[7] who obtained the ratio 2.6 using data on cytochrome c, c_2 and tRNA sequences. It is also lower than the corresponding estimate, ~2, obtained by Jukes (1969) (quoted in ref.7). Hoping to resolve this discrepancy we calculated the evolutionary

来估计，其中 λ 是两个同源序列相异位点的比例。这个公式基于如下假设得出：进化过程中核苷酸替代在空间上以相同概率随机发生，并且四种碱基（腺嘌呤、胞嘧啶、鸟嘌呤、尿嘧啶）中的每一种都能以相等的概率突变成其他三种中的任何一种（详细信息请看参考文献 4）。之前朱克斯和康托尔就已推导出此当量公式[5]。我们也应当注意到，戴霍夫[3]在她的表 11-3 中列出的经验关系实际上和这个公式等价，尤其是当用于比较不同 K 值时。

```
人类        - G- UCUACGGCC- AUACCACCCUGAACGCGCCCGAUCUCGUCUGAU- CUCGGAAGCUAAGCAG
酵母        - G- GUUGCGGCC- AUACCAUCUAGAAAGCACCGUUCUCCGUCCGAUAACCUGUAGUUAAGCUG
大肠杆菌    UGCCUGGCGGCC- GUAGCGCGGUGGUCCCACCUGACCCCAUGCCGAACUCAGAAGUGAAACGC
荧光假单胞菌 UGUUCUUUGACGAGUAGUGGCAUUGGAACACCUGAUCCCAUCCCGAACUCAGAGGUGAAACGA

            GGUCGGGCCUG- GUUAGUACUUGGAUGGGAGACCGCCUGGGAAUACCGGGUGCUGUAG- GCUU
            GUAAGAGCCUGACCGAGUAGUGUAGUGGGUGACCAUACGCGAAACCUAGGUGCUGCA- - AUCU
            CGUAGCGCC- - - GAUGGUAGUGUG- - GGGUCUCCCCAUGCGAGAGUAGGGAACUGCCAGGCAU
            UGCAUCGCC- - - GAUGGUAGUGUG- - GGGUUUCCCCAUGUCAAGAUCUCG- ACCAUAGAGCAU
```

图 1. 5S rRNA 序列的比对图

表 1. 5S rRNA 序列间差异位点的比例

	酵母	大肠杆菌	荧光假单胞菌
人类	0.395	0.457	0.478
酵母		0.474	0.565
大肠杆菌			0.319

真核生物（人、酵母）和原核生物（大肠杆菌、荧光假单胞菌）间的平均突变距离就是

$$K_{真核-原核}=0.817\pm0.158$$

其中，标准差由四组观察结果（比较）获得。另一方面，人和酵母菌之间相对应的值为

$$K_{人-酵母}=0.561\pm0.095$$

其中，误差是由参考文献 4 中的方程（5）计算出的一个理论值。此处有个有趣的现象值得注意，相比与细菌的关系，酵母和人的关系更近一些，这支持了真核生物和原核生物之间的差异比真核生物内的差异更为根本这一论点（对照参考文献 6）。真核生物 – 原核生物之间的差异相对于人 – 酵母之间差异的遥远程度可以通过比值 $K_{真核-原核}/K_{人-酵母}$来估计，大约是 1.5。这比麦克劳克林和戴霍夫相应的估计[7]低很多，他们用细胞色素 c、c_2 和转运 RNA（tRNA）序列数据得出的比值为 2.6。这也比朱克斯（1969）（在参考文献 7 中引用）得出来的约等于 2 的估计值要低。我们用麦克

distances using the same data[3] on tRNA as McLaughlin and Dayhoff[7] but restricting our treatment only to paired regions. The reason for doing this is that it seems as if there is no excess of highly conserved regions (as inferred from our statistical analysis of the frequency distribution of the number of evolutionary changes per site in the alignment of Fig. 1), so that evolutionary change appears to be uniform over the entire sequence and in this respect 5S rRNA might be more similar to the paired than unpaired regions of tRNA. The evolutionary distances turned out to be $K_1=0.836\pm0.136$ for the eukaryotes–prokaryotes divergence but $K_2=0.420\pm0.029$ for the average of rat–yeast ($tRNA^{Ser}$) and wheat–yeast ($tRNA^{Phe}$) divergences. It may be seen that although K_1 is comparable to $K_{eu\text{–}pro}$, K_2 is clearly lower than $K_{h\text{–}y}$, so that the ratio $K_1/K_2=1.99$ is still considerably higher than $K_{eu\text{–}pro}/K_{h\text{–}y}=1.46$. It is possible that the difference is due to sampling error, with the true value lying somewhere between these two. At any rate we should take these estimates as tentative (including the problem that might arise because of the multiplicity of ribosomal genes). There is some reason to believe, however, that our estimate of 1.5 is consistent with the fossil records as explained below.

From comparative studies of cytochrome *c* sequences among eukaryotes, we can estimate the remoteness of the human–yeast divergence relative to the mammal–fish divergence. This allows us to estimate the absolute time of the human–yeast divergence as it is known from classical palaeontological studies that the common ancestor of the fish and the mammals goes back to some 400 m.y. (compare refs 8 and 9). From a number of comparisons involving various species of fish and mammals (data taken from ref. 3), we obtained the results that the mutational distance (in terms of the number of amino acid substitutions) between mammals and yeast is about three times that between mammals and the fish. This puts the time of the human-yeast divergence back to about 1.2×10^9 yr. This agrees with Dickerson[10] who obtained 1,200±75 m.y. as the estimated date of the branch point for animals/plants/protists. We should note here that we avoided using the cytochrome c_2 sequence of the bacterium *Rhodospirillum rubrum* to estimate the prokaryotes–eukaryotes divergence, because there seems to be some difference in function between cytochromes *c* and c_2 (compare ref. 7).

Multiplying 1.2×10^9 yr by the ratio $K_{eu\text{–}pro}/K_{h\text{–}y}\approx1.5$, we arrive at the result that the divergence between the eukaryotes and prokaryotes goes back to some 1.8×10^9 yr. Recent studies on Precambrian fossils (compare ref.11) suggest that the eukaryotes evolved from prokaryotes at some point between the Bitter Springs formation (10^9 yr old) and the Gunflint formation (2×10^9 yr old). With additional relevant data forthcoming (for example, ref. 12) we hope that the studies of molecular evolution will soon supply a more accurate date.

We tentatively conclude that the eukaryotes diverged from prokaryotes nearly 2×10^9 yr ago, thus opening up the way toward "higher organisms".

We thank Dr. K. Miura and Mr. H. Komiya for calling our attention to the relevant literature on 5S rRNA sequences and Drs. T. Maruyama, S. Takemura and S. Kondo for

劳克林和戴霍夫[7]使用过的同一组 tRNA 数据[3]计算了进化距离，但是我们的处理只限定在配对区域，希望可以解决这个差异。这样做的原因是看起来好像没有额外的高度保守区（正如通过比对图 1 中每个位点的进化改变数的频度分布进行的统计分析推测出的一样），所以在整个序列上进化改变似乎是均衡的，并且在这方面 5S rRNA 和 tRNA 的未配对区相比可能和配对区更相似。对于真核生物 – 原核生物之间的差异，进化距离为 K_1=0.836±0.136，而对大鼠 – 酵母（色氨酸 tRNA）和小麦 – 酵母（苯丙氨酸 tRNA）之间差异的平均值，进化距离为 K_2=0.420±0.029。可以看出，尽管 K_1 和 $K_{真核-原核}$相当，K_2 却明显比 $K_{人-酵母}$小，所以比值 K_1/K_2=1.99 仍然比 $K_{真核-原核}/K_{人-酵母}$=1.46 高很多。这个差异有可能是采样误差造成的，真实值可能介于二者之间。至少我们应该将这些作为初步估计值（可能存在由核糖体基因多样性而产生的问题）。然而有理由相信，我们估计的 1.5 是同下面阐述的化石数据相符的。

通过真核生物间细胞色素 c 序列的比较研究，我们能够估计人 – 酵母之间的差异相对于哺乳动物 – 鱼差异的遥远程度。经典的古生物学研究表明鱼和哺乳动物的共同祖先出现在 4 亿年前（对照参考文献 8 和 9），这使我们能估计出人 – 酵母菌趋异的绝对时间。通过对各种鱼类物种和哺乳类物种的大量比较（数据来自参考文献 3），我们得到了哺乳动物和酵母之间的突变距离（依据氨基酸替代的数量）是哺乳动物和鱼之间突变距离的三倍的结果。这将人和酵母分离的时间推到了 12 亿年前。这和迪克森[10]得到的动物、植物和原生生物的分支点是 1,200±75 百万年的结论相符。在此我们应该注明，我们没有使用细菌中的红螺菌的细胞色素 c_2 来估计原核生物 – 真核生物的分离，因为细胞色素 c 和细胞色素 c_2 在功能上似乎有些不同（对照参考文献 7）。

用 1.2×10^9 年乘以比例 $K_{真核-原核}/K_{人-酵母}\approx 1.5$，我们得出真核生物和原核生物的分离发生在大约 1.8×10^9 年前的结论。最近对前寒武纪化石的研究（对照参考文献 11）表明，在苦泉地层（年龄为 10^9 年）和加拿大冈弗林特地层（年龄为 2×10^9 年）之间的某个时候，真核生物从原核生物进化而来。结合将来更多的相关数据（例如参考文献 12），我们希望不久后分子进化的研究将提供一个更加准确的日期。

我们暂且推断大约 2×10^9 年前真核生物从原核生物分离，从此走上了通向“高等生物”的道路。

感谢三浦博士和小宫先生引起我们对 5S rRNA 序列相关文章的关注，感谢丸山

helpful discussions.

(*Nature New Biology*, **243**, 199–200; 1973)

Motoo Kimura and Tomoko Ohta
National Institute of Genetics, Shizuoka-ken 411, Mishima

Received December 29, 1972; revised March 5, 1973.

References:

1. DuBuy, B., and Weissman, S. M., *J. Biol. Chem.*, **246**, 747 (1971).
2. Hindley, J., and Page, S. M., *FEBS Lett.*, **26**, 157 (1972).
3. Dayhoff, M. O., *Atlas of Protein Sequence and Structure 1972* (National Biomedical Research Foundation, Washington, DC, 1972).
4. Kimura, M., and Ohta, T., *J. Mol. Evol.*, **2**, 87 (1972).
5. Jukes, T. H., and Cantor, C. R., in *Mammalian Protein Metabolism* (edit. by Munro, H. N.), 21 (Academic Press, New York, 1969).
6. Margulis, L., *Origin of Eukaryotic Cells* (Yale Univ. Press, New Haven and London, 1970).
7. McLaughlin, P. J., and Dayhoff, M. O., *Science*, **168**, 1469 (1970).
8. Romer, A. S., *The Procession of Life* (Weidenfeld and Nicolson, London, 1968).
9. McAlester, A. L., *The History of Life* (Prentice-Hall, Englewood Cliffs, 1968).
10. Dickerson, R. E., *J. Mol. Evol.*, **1**, 26 (1971).
11. Barghoorn, E. S., *Sci. Amer.*, **224**, 30 (1971).
12. Brownlee, G. G., Cartwright, E., McShane, T., and Williamson, R., *FEBS Lett.*, **25**, 8 (1972).

博士、武村博士和绀户博士的宝贵讨论。

（邓铭瑞 翻译；陈新文 陈继征 审稿）

Isolation and Genetic Localization of Three φX174 Promoter Regions

C. Chen *et al.*

Editor's Note

Four years before the φX174 bacteriophage became the first DNA-based organism to have its genome completely sequenced, Cheng-Yien Chen and colleagues had used restriction enzymes (which cut DNA at sequence-specific locations) to produce a genetic map of this model molecular biology system. Importantly, they managed to isolate and localize three promoter regions (DNA segments that facilitate the transcription of a particular gene) by clever use of a "protective" RNA polymerase protein that let them digest away the exposed, unwanted nucleic acid. Three decades later the φX174 bacteriophage was to court attention again, when researchers reported that they had synthetically assembled its genome from scratch.

SPECIFIC sequences of nucleic acids such as ribosome binding sites[1] (W. Gilbert, cited in ref. 2) and portions of the promoter region[3,4] have been isolated by protecting those sequences with the relevant protein or organelle and digesting away the exposed nucleic acid. We have isolated specific sequences from the promoter regions in bacteriophage φX174 replicative form (RF) DNA using RNA polymerase as the protecting protein. We wish to describe that isolation and the procedures used to localize these RNA-polymerase-protected sequences within the φX174 genome.

We have previously described the use of restriction enzymes to cleave φX174 DNA into specific fragments[5,6] and the genetic assay used to order these fragments with respect to the φX174 recombination map[7]. The φX genome is separated into 11–14 specific pieces using either of the *Haemophilus* restriction enzymes, endonuclease R[5,8] or endonuclease Z[6]. Our approach has been to determine which restriction fragments contain sequences protected by RNA polymerase. As most of the restriction fragments have been ordered with respect to the φX174 genetic map (manuscript in preparation), this localizes the protected sequences within the map as well.

Two procedures were used to identify DNA fragments bearing sequences protected by RNA polymerase. In the first, *Escherichia coli* RNA polymerase was bound to ^{3}H-labelled φX174 RFI (covalently closed circular DNA). After extensive digestion with pancreatic DNase, the fraction of the DNA protected by RNA polymerase was isolated. The protected ^{3}H-DNA was then hybridized to purified ^{32}P-labelled restriction fragments of φX174 RF immobilized on nitrocellulose filters. Retention of ^{3}H counts indicated the presence of homologous sequences within that particular specific restriction fragment.

三个φX 174启动子区的分离和基因定位

陈等

编者按

在φX174噬菌体成为首个被完整测序的基于DNA的生物有机体的四年前，陈成印（音译）和他的同事们就已经使用限制性内切酶（在序列特异性位点切开DNA）得到了这个模式分子生物学系统的遗传图谱。重要的是，他们巧妙地运用一种“保护性”RNA聚合酶蛋白，通过消化掉暴露的、不需要的核酸，分离和定位了三个启动子区域（帮助特定基因进行转录的DNA片段）。三十年后，当研究者们报道了他们从头合成组装了φX174噬菌体的基因组时，它再次引起了人们的关注。

通过相关蛋白或细胞器保护DNA序列并消化掉暴露的核酸，若干特异核酸序列已经被确定，例如核糖体结合位点[1]（吉尔伯特，引自参考文献2）和部分启动子区域[3,4]。我们已经利用RNA聚合酶作为保护蛋白，从噬菌体φX174复制型（RF）DNA的启动子区域中分离到了特定序列。我们想叙述一下这一分离的过程，以及在φX174基因组中定位这些RNA聚合酶保护序列的步骤。

之前，我们已经对采用限制性内切酶将φX174 DNA切割成特定片段[5,6]，及根据φX174重组图谱将这些片段进行排序的遗传分析[7]进行过描述。使用嗜血杆菌限制性内切酶、核酸内切酶R[5,8]或者核酸内切酶Z[6]均可将φX基因组切割成11~14个特定的片段。我们的方法可以确定哪个限制性片段含有被RNA聚合酶所保护的序列。由于绝大多数限制性片段已经根据φX174基因图谱进行了排序（论文撰写中），因此我们的方法也能同样在图谱中定位被（RNA聚合酶）保护的片段。

鉴定含有RNA聚合酶保护序列的DNA片段需要两个过程。首先，让大肠杆菌RNA聚合酶与^{3}H标记的φX174 RFI（共价闭合环状DNA）结合。经过胰DNA酶的充分消化后，RNA聚合酶保护的DNA片段就被分离出来了。受保护的^{3}H–DNA随后与固定在硝酸纤维素滤膜上的^{32}P标记的纯化φX174 RF限制性片段进行杂交。留在滤膜上的DNA的^{3}H计数表明与特定的限制性片段的同源序列的存在。

About 1.9% of φX174 RFI is protected from DNase digestion by RNA polymerase. The DNA–RNA polymerase complex formed is quite stable because an excess of cold RF during digestion does not change the fraction of labelled DNA protected. The sites which can be protected by RNA polymerase are saturated, as increasing the ratio of RNA polymerase to DNA beyond 4 μg polymerase to 0.02 μg DNA (our standard conditions) does not increase the amount of DNA protected.

Assuming the average chain length of a protected site is thirty-five base pairs long[3], the resistant counts represent approximately three to four such sites per genome. The protected ^{3}H-labelled DNA sequences were eluted from a nitrocellulose filter with 0.2% SDS and hybridized to a set of filters loaded with purified restriction fragments of φX174 RF produced by endonuclease R and to a separate set of filters containing the fragments produced by endonuclease Z. The RNA polymerase protected DNA hybridized to filters bearing endonuclease R fragments r2, r4 and r6 and to endonuclease Z fragments z1,z2 and z3 (Table 1).

Table 1. Protection of DNA

Fragments on filter from band	% ^{3}H bound*	Fragments on filter from band	% ^{3}H bound†
r1	0.74	z1	6.0
r2	6.2	z2	3.7
r3	0.97	z3	3.4
r4	4.0	z4	0.7
r5	0.77	z5	0.6
r6	2.4	z6	0.9
r7	0.93	z7	0.6
r8	0.92	z8	0.9
r9	0.80	z9	0.2
RFII	18.6	RFII	18.6

* A blank filter background of 0.39% has been subtracted.
† A blank filter background of 0.35% has been subtracted.
Hybridization of RNA-polymerase-protected ^{3}H-labelled φX174 RF DNA to specific ^{32}P-labelled, endonuclease R and Z fragments. The preparation of ^{32}P-RFI, endonuclease R and Z digestions, electrophoresis and autoradiography have been described previously[5,6]. The ^{32}P endo Z and endo R fragments were eluted from the bands of dry gel in 2×SSC at 65°C for 24 h (N. Axelrod, personal communication). Equimolar amounts of the ^{32}P restriction fragments were immobilized on the membrane filters (25 mm) (Schleicher and Schuell, B6). Each fragment was made up to 20 ml at a final concentration of 0.1×SSC. After heating at 100°C for 5 min and quickly quenching in the ice bath, each fragment was immobilized on a B6 filter according to the method of Raskas and Green[18]. The ^{3}H-RFI-DNA-RNA polymerase complexes were produced as follows. The reaction mixture contained 0.02 μg of ^{3}H-RFI-DNA (about 75,000 c.p.m.) (prepared as in ref. 6), 20 μg of RNA polymerase and 0.1 mM each of ATP and GTP in 0.5 ml of buffer A (8 mM $MgCl_2$, 50 mM KCl, 20 mM Tris, *p*H 7.9, 0.1 mM dithiothreitol). The RNA polymerase had been purified through a glycerol gradient centrifugation as in the method of Burgess[16]. Acrylamide gels had verified that sigma factor was present in our preparation. After 5 min at 37°C, 10 μg of unlabelled RFI-DNA, followed by 200 μg of pancreatic DNase, was added and the digestion was continued for 40 min. The mixture was chilled, diluted with

受RNA聚合酶保护而免于被DNA酶消化的φX174RFI大约有1.9%。形成的DNA–RNA聚合酶复合物非常稳定，因为消化过程中过量的冷RF也改变不了被标记的受保护DNA的比例。受RNA聚合酶保护的位点已经达到饱和，将RNA聚合酶与DNA的比例增加到超过4 μg聚合酶比0.02 μg DNA（我们的标准条件）也不能增加受保护DNA的数量。

假设受保护位点的平均链长是35个碱基对[3]，那么每个基因组中这样的抗消化位点的数量大约为3到4个。用0.2% SDS将^{3}H标记的受保护DNA序列从硝酸纤维素膜上洗脱下来，然后将其分别与含有纯化的由核酸内切酶R制备的φX174RF限制性片段的滤膜及负载了由核酸内切酶Z制备的片段的滤膜进行杂交。RNA聚合酶保护的DNA与负载了核酸内切酶R片段r2、r4和r6以及核酸内切酶Z片段z1、z2和z3的滤膜进行杂交（表1）。

表1. DNA的保护

滤膜上来自条带的片段	^{3}H结合比例%*	滤膜上来自条带的片段	^{3}H结合比例%†
r1	0.74	z1	6.0
r2	6.2	z2	3.7
r3	0.97	z3	3.4
r4	4.0	z4	0.7
r5	0.77	z5	0.6
r6	2.4	z6	0.9
r7	0.93	z7	0.6
r8	0.92	z8	0.9
r9	0.80	z9	0.2
RFII	18.6	RFII	18.6

* 减去了0.39%的空白滤膜本底。

† 减去了0.35%的空白滤膜本底。

RNA聚合酶保护的^{3}H标记的φX174RF DNA与特异性^{32}P标记的核酸内切酶R和Z片段的杂交。^{32}P–RFI的制备、内切酶R和Z的消化、电泳以及放射自显影的方法如前所述[5,6]。在2×SSC 65℃条件下，经过24小时的洗脱，从干燥胶条带中获得^{32}P标记的内切酶R和Z的片段（阿克塞尔罗德，个人交流）。在滤膜（25 mm）上固定等摩尔的^{32}P限制性片段（施莱克尔和许尔，B6）。每个片段溶解于终浓度0.1×SSC的缓冲液中，最终体积20 ml。100℃加热5分钟后在冰浴中快速冷却，然后根据拉斯卡斯和格林的方法[18]将每个片段固定在B6滤膜上。^{3}H–RFI–DNA–RNA聚合酶复合物的制备方法如下：反应混合物是0.5 ml缓冲液A（8 mM $MgCl_2$，50 mM KCl，20 mM Tris，pH7.9，0.1 mM二硫苏糖醇）中含有0.02 μg的^{3}H–RFI–DNA（大约75,000 c.p.m.）（制备方法按照文献6），20 μg的RNA聚合酶，ATP和GTP各0.1 mM的。RNA聚合酶按照伯吉斯的方法用甘油梯度离心纯化[16]。丙烯酰胺凝胶的实验结果证实，在我们的制备物中存在西格玛因子。经过37℃下5分钟，加入10 μg没有标记的RFI–DNA，随后加入200 μg胰DNA酶并继续消化40分钟。反应混合物经过冷的缓冲液A冷却稀释，其中的复合物被收集到硝酸纤维素膜上（密理博，HA）。用0.2%的SDS将受保护的序列从滤膜上洗脱下来。在

cold buffer A and the complexes were collected on a nitrocellulose filter ("Millipore", HA). The protected sequences were freed from the filter by treatment with 0.2% SDS. An aliquot (430 c.p.m.) of the protected sequences was heated at 100°C for 5 min and quickly quenched in ice and hybridized to each filter containing ^{32}P endo Z and endo R fragments. The filters were then counted with a Packard Tri-Carb scintillation counter in a toluene-based scintillation fluid. The percentage of the ^{3}H bound was calculated as the fraction of input counts bound minus the percentage of counts bound to a blank filter. The counts were determined from 100 min counts and were reproducible in four different experiments.

The second procedure used to identify DNA fragments bearing RNA-polymerase-protected sequences was to bind RNA polymerase to ^{32}P-labelled restriction fragments of φX174 RF and then to pass the reaction mixture through a nitrocellulose filter which should retain only the RNA polymerase-^{32}P-labelled DNA fragment complex[9]. The filtrate was then electrophoresed in 3% polyacrylamide gels to determine which DNA fragments were missing. The gel was then fractionated and counted to obtain quantitative data on the recovery of each restriction fragment. The recoveries of these fragments compared with fragments which were not reacted with RNA polymerase showed that fragments r2, r4 and r6 were retained by the filter (Fig. 1). A similar experiment with restriction fragments produced by endonuclease Z showed that fragments z1, z2 and z3 were retained by filtration.

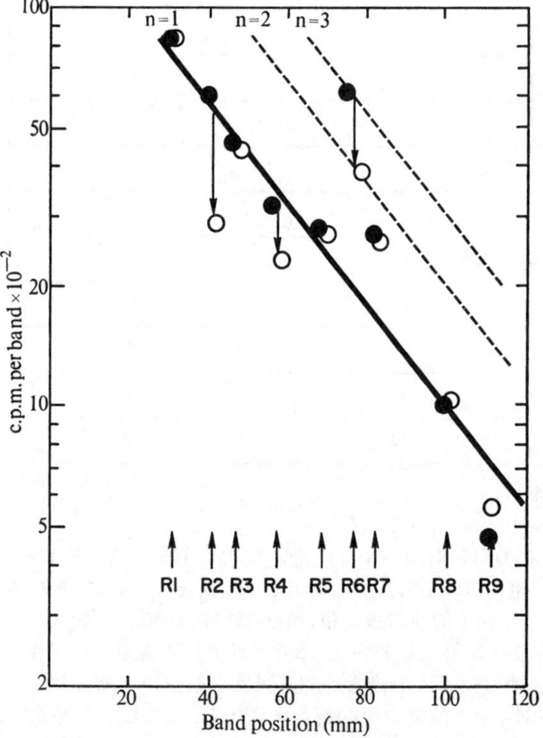

Fig. 1. Quantitative determination of fragment recovery with and without RNA polymerase binding before nitrocellulose filtration. A sample of 200 μl of ^{32}P-φX RF DNA (2.6×10^5 c.p.m., 0.08 μg) was digested with endonuclease R[5]. RNA polymerase (32 μg) was then mixed with 100 μl of the endo R limit digest of the RF to give a final volume of 200 μl in buffer A (8 mM $MgCl_2$, 50 mM KCl, 20 mM Tris, *p*H 7.9, 0.1 mM dithiothreitol) and 0.1 mM each of ATP and GTP. After 5 min at 37°C the reaction mixture

100℃下将一部分（430 c.p.m.）受保护序列加热5分钟，并迅速在冰上冷却，将其分别与含有 ^{32}P标记的内切酶Z和内切酶R片段的滤膜杂交。所得滤膜随后在甲苯为基础的闪烁液中用Packard Tri–Carb型闪烁计数器计数。投入的计数比例减去空白滤膜的计数比例就是 ^{3}H的结合比例。所有数值经由100分钟计数确定，并且在四个不同的实验中具有可重复性。

鉴定含有RNA聚合酶保护序列的DNA片段的第二步是将RNA聚合酶结合到 ^{32}P标记的φX174RF限制性片段上，将反应混合物通过硝酸纤维素滤膜后仅会留下RNA聚合酶–^{32}P标记DNA片段的复合物[9]。将滤过物在3%的聚丙烯酰胺凝胶中进行电泳，以确定哪些DNA片段丢失了。随后将凝胶分成几部分并计数，获得回收后每个限制性片段的定量数据。通过将这些回收后的片段与没有和RNA聚合酶反应的片段进行比较，得知被滤膜拦截的片段为r2、r4和r6（图1）。用核酸内切酶Z制备得到的限制性片段的类似实验显示，片段z1、z2和z3被滤膜拦截。

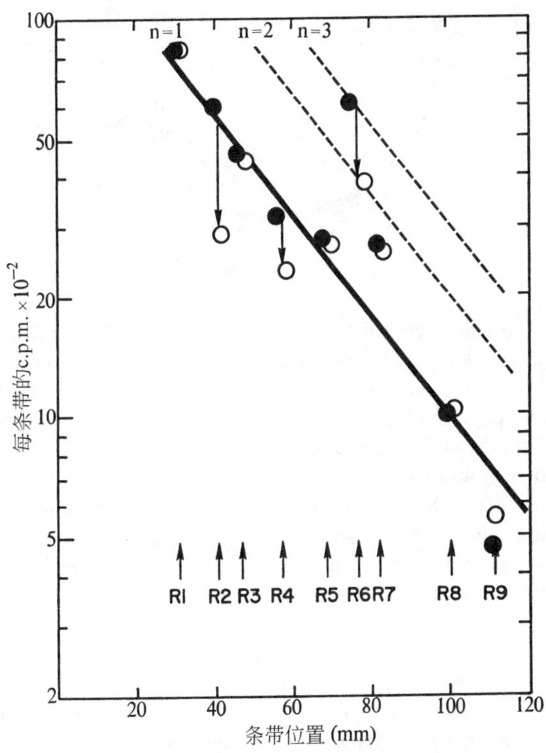

图1. 在硝酸纤维素膜过滤前使用和不用RNA聚合酶的条件下，回收各片段的定量计数。200 μl的 ^{32}P–φX174 RF DNA样本（2.6×10^{5} c.p.m.，0.08 μg）用核酸内切酶R进行消化[5]。RNA聚合酶（32 μg）随后与100μl RF的内切酶R限制性消化产物混合并加入含0.1 mM ATP和GTP的缓冲液A（8 mM $MgCl_2$，50 mM KCl，20 mM Tris，pH 7.9，0.1 mM二硫苏糖醇），使总体积为200 μl。37℃作用5分钟后，将反应混合物用硝酸纤维素膜过滤（密理博，HA）。滤过物约150 μl，进行电泳[5]。剩余的100 μl RF的内

was passed through a nitrocellulose filter ("Millipore", HA). The filtrate containing approximately 150 μl was subjected to electrophoresis[5]. The remaining 100 μl of the endo R limit digest of RF was treated as above except that RNA polymerase was omitted. After electrophoresis the gels were dried on a filter paper, cut into 1 mm segments and counted with a scintillation counter. The tracking dye migrated 142 mm. The integrated counts in the various bands were plotted (log scale) against mobilities (band position in mm). Band R6 has been shown to contain three unresolved fragments of about the same size[5] and so has three times (n=3) the number of counts as those bands containing a single fragment (n=1). Band R7 as those bands containing a single fragment (n=1). Band R7 contains two unresolved fragments (n=2). ●, No RNA-P; ○, plus RNA-P.

The data from these experiments are consistent with the restriction fragment map for φX174. That is, each r fragment containing sequences protected by RNA polymerase overlaps with a z fragment also shown to contain such a sequence (Fig. 2).

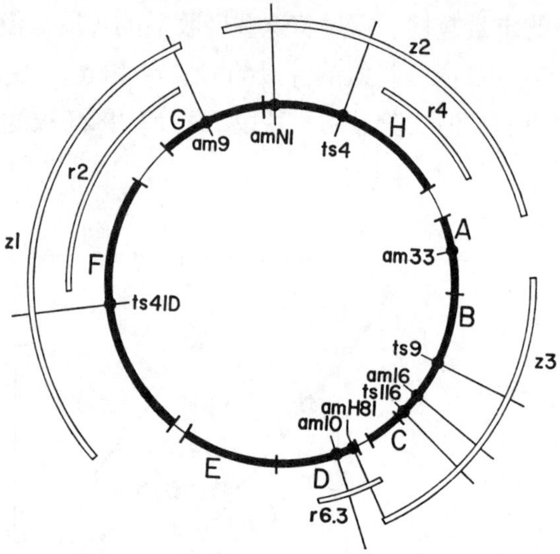

Fig. 2. Locations on the genetic map of φX174 of restriction fragments containing sequences protected by RNA polymerase. The order of eight φX174 genes (*A* through *H*) has been determined by genetic recombination[15]. The gene sizes shown on this map were calculated from published estimates of the molecular weights of the corresponding gene products[19-21]. (The assignments of sizes for genes *A*, *B*, *C* and *E* are uncertain; however, the exact sizes of these genes do not affect the general picture presented here.) As the estimated gene sizes add up to 90% of the total genome size (5,500 nucleotide pairs), the remaining 10% has been arbitrarily distributed as intergenic spacers. Fragments z1, z2 and z3 have been mapped by the genetic assay for DNA fragments[5-7,22] and contain the indicated genetic sites. Fragment r2 is homologous to z1 by hybridization, but does not contain any of the genetic sites shown. A different endonuclease R fragment (r1) contains the site F*ts*41D and extends counterclockwise beyond the end of z1. Therefore, r2 must lie between the sites F*ts*41D and G*am*9 as shown. Fragment r4 is homologous to z2 by hybridization. Also, r4 is adjacent to a fragment (r3) containing the site *am*33 in gene *A*. So r4 must be located in the region indicated. (We know that r4 and r3 are adjacent because electrophoresis of an endonuclease R digest of phage S13 RF gives the normal φX174 pattern of fragments except that r4, r3 and r5 (gene *B*) are missing and are replaced by a single large fragment.) Band R6 has been shown to contain three fragments[5] (named r6.1, r6.2 and r6.3) which have been partially resolved on prolonged electrophoresis. Fragments r6.1 and r6.2 contain genetic sites within cistrons G and H; therefore, fragment r6.3 (containing the indicated sites in cistron D) must be the component of band R6 which shares a polymerase protected sequence with fragment z3.

切酶 R 的限制性消化产物也按如上步骤处理，只是不加入 RNA 聚合酶。电泳后，将凝胶放在一张滤纸上干燥，切成 1 mm 的片段后用闪烁计数器计数。示踪染料迁移了 142 mm。将不同条带上计数值的整数（取对数值）对迁移率（条带位置，单位 mm）作图。条带 R6 含有三种大小几乎相同的不能分解的片段[5]，因此它们的计数是那些都含有单一的片段（n=1）条带的三倍（n=3）。条带 R7 含有两个不能分解的片段（n=2）。●没有 RNA 聚合酶；○加入 RNA 聚合酶。

这些实验获得的数据与 φX174 的限制性片段图谱一致。也就是说，每个含有 RNA 聚合酶保护序列的 r 片段与同样显示含有这些序列的 z 片段有重叠（图 2）。

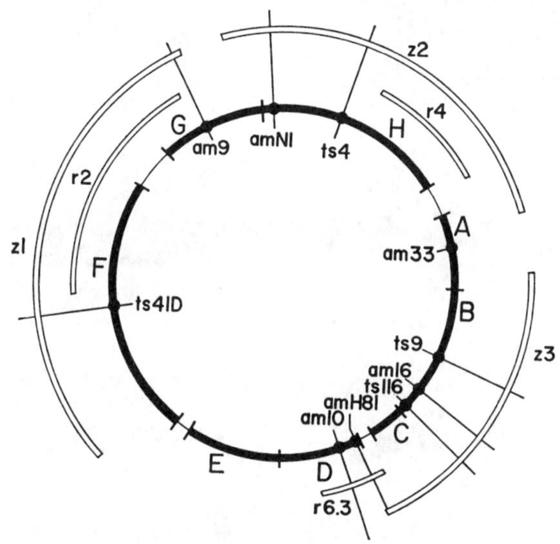

图 2. φX174 遗传图谱上含 RNA 聚合酶保护序列的限制性片段的定位。应用基因重组技术已经确定了 8 个 φX174 基因的顺序（*A* 到 *H*）[15]。本图中显示的基因大小是根据已发表的相应基因产物的分子量的估计值计算出来的[19-21]。（基因 *A*，*B*，*C* 和 *E* 的大小尚未确定；但是，这些基因的准确大小不会影响这里显示的总体图像。）由于估计的基因大小占到了整个基因组大小的 90%（5,500 个核苷酸对），剩余的 10% 是随意分布于基因之间的间隔。片段 z1，z2 和 z3 已经通过 DNA 片段的遗传分析定位于图上[5-7,22]，并含有所显示的基因位点。杂交显示片段 r2 与 z1 同源，但是不含有任何显示的基因位点。另一个不同的核酸内切酶 R 片段（r1）含有位点 F*ts*41D，并逆时针延伸超出 z1 的末端。因此，r2 肯定位于如图所示的 F*ts*41D 和 G*am*9 位点之间。片段 r4 通过杂交定位与 z2 同源。而且，r4 与基因 A 中含有位点 *am*33 的片段（r3）相邻。因此，r4 肯定位于图中所示的区域。（我们知道 r4 和 r3 相邻，因为除了 r4，r3 和 r5（基因 *B*）丢失并被一个大的单片段替代以外，噬菌体 S13 RF 的核酸内切酶 R 消化产物的电泳显示正常的 φX174 片段分布类型。）条带 R6 含有在延时电泳中部分显示的三个片段[5]（称为 r6.1，r6.2 和 r6.3）。片段 r6.1 和 r6.2 含有位于顺反子 G 和 H 内的基因位点；因此，片段 r6.3（含有图中顺反子 D 中所示的位点）肯定是条带 R6 的组成部分，并且与片段 z3 共享一个聚合酶保护序列。

The RNA polymerase DNA fragment complexes were also visualized in the electron microscope. RNA polymerase was reacted with purified restriction fragments r2 and z1 and stained by the Kleinschmidt procedure for microscopy. When r2 was mixed with RNA polymerase the structures shown in Fig. 3*a*, *b*, and *c* were seen. Polymerase was seen bound only at the end of r2, although not every r2 fragment had an enzyme molecule associated with it. On the other hand, when RNA polymerase was mixed with fragment z1 all the binding seen was internal, about one-third of the way from the end (Fig. 3*d*, *e* and *f*). These fragments have not been placed precisely enough with respect to the recombination map to be sure which end of r2 contains the sequences protected by RNA polymerase.

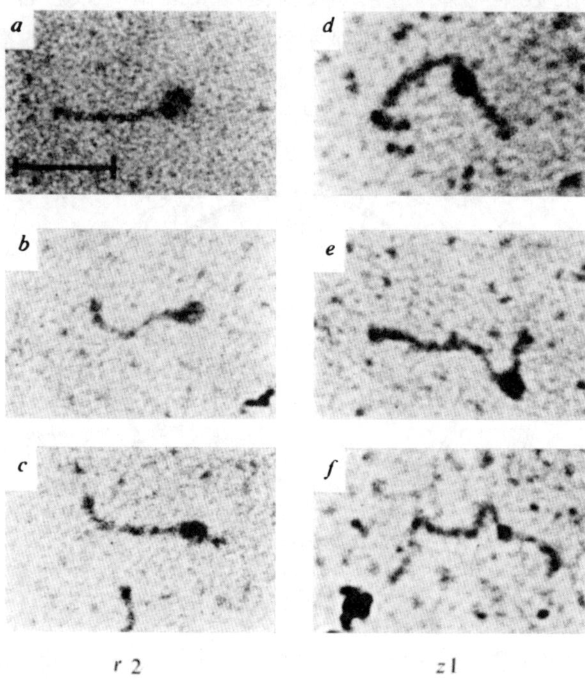

Fig. 3. Electron microscopic pictures of RNA polymerase-restriction fragment mixtures. RNA polymerase was bound to the fragments r2 (*a*, *b*, *c*) and z1 (*d*, *e*, *f*) as described in Fig.1. The RNA polymerase-DNA fragments were prepared for microscopy according to Kleinschmidt's procedure, as modified by Davis and Davidson[17]. Grids containing DNA–RNA polymerase complexes were stained with uranyl acetate and then photographed at a magnification of 20,000 with an AEI model EM6B electron microscope. The bar represents 0.2 μm.

It is clear that RNA polymerase when reacted with DNA in this way protects specific sequences from nuclease digestion. It seems likely to us that these sequences contain the start sites[10] within the promoter regions. RNA polymerase in the presence of ATP and GTP presumably becomes associated with the DNA at entry sites[10] and then drifts to the start site where it is prevented from synthesizing mRNA by the absence of all four triphosphates. It has been shown that the sequences from bacteriophage fd protected in a similar fashion by RNA polymerase contain sequences which specify the initial nucleotides in the *in vivo* messages[3]. Rüger[11] showed that similar sites from T4 phage DNA still serve as a template for RNA polymerase. There is no clear evidence, however, that every sequence protected in this fashion by RNA polymerase contains a start site.

也可在电子显微镜下对 RNA 聚合酶–DNA 片段复合物进行观察。将 RNA 聚合酶与纯化的限制性片段 r2 和 z1 进行反应，并用克莱因施密特方法染色后于显微镜下观察。当 r2 与 RNA 聚合酶混合时，看到了图 3*a*、*b* 和 *c* 所示的结构。聚合酶仅仅结合到 r2 的末端，尽管不是每个 r2 片段都有结合的酶分子。另一方面，当 RNA 聚合酶与片段 z1 混合时，所有可见的结合都位于中间，距末端大约 1/3 处(图 3*d*、*e*、*f*)。对根据重组图谱确定 r2 的哪一端含有 RNA 聚合酶保护序列而言，这些片段的定位还不够精确。

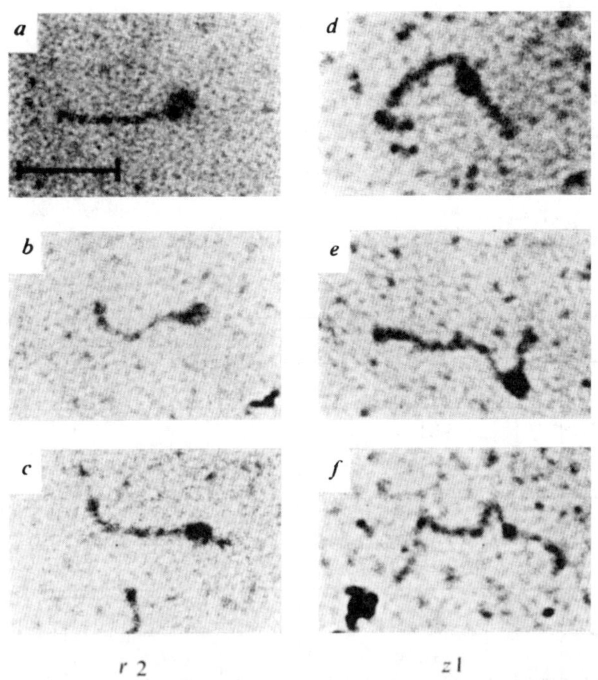

图 3. RNA 聚合酶 – 限制性片段复合物的电子显微镜照片。正如图 1 所示，RNA 聚合酶同片段 r2（*a*、*b*、*c*）和 z1（*d*、*e*、*f*）结合。用于显微观察的 RNA 聚合酶 – DNA 片段按照戴维斯和戴维森改良的克莱因施密特法制备[17]。用醋酸双氧铀对含有 DNA–RNA 聚合酶复合物的样品载网进行染色，然后用 AEI 型 EM6B 电子显微镜在放大 20000 倍时照相。图例标尺表示 0.2 μm。

很清楚的是，当 RNA 聚合酶以这种方式与 DNA 反应时，可以保护特定的序列免受核酸酶消化。在我们看来，这些序列在启动子区可能含有起始位点[10]。在 ATP 和 GTP 存在时，RNA 聚合酶可能在进入位点开始与 DNA 相互作用[10]，然后漂移到起始位点，在那里因为没有四种三磷酸盐而无法合成 mRNA。研究显示，被类似 RNA 聚合酶方式保护的、来自噬菌体 fd 的序列含有在体内信息传递中起到起始核苷酸作用的序列[3]。吕格尔[11]指出 T4 噬菌体 DNA 的类似位点也能作为 RNA 聚合酶的模板。但是，没有明确的证据表明每个被 RNA 聚合酶以这种方式保护的序列都含有起始位点。

Previous studies of the *in vivo* and *in vitro* transcription products of φX174[12,13] and fd[14] have suggested the existence of multiple sites for message initiation and termination. These sites, however, were not ordered with respect to the genetic map. Studies on φX174 *in vivo* translation[15] led to the conclusion that synthesis was initiated with the cistron A and D products. We have now mapped three φX174 sequences which are protected by RNA polymerase. These sequences lie within restriction fragments which span the beginnings of cistrons A, D and G (Fig. 2). However, our data do not prove that these protected sequences are in fact located at the beginnings of the cistrons.

It is interesting that sequences can be isolated from *E. coli* DNA by protection with *E. coli* RNA polymerase which will hybridize to these same restriction fragments from φX174 RF DNA (Table 2) which we have shown to contain the φX174 sequences protected by RNA polymerase. Apparently *E.* coli DNA contains some promoter regions which have start sites similar to those found in φX174 RF DNA.

Table 2. Hybridization of ^{3}H-labelled *E. coli* DNA Sequences Protected by RNA Polymerase to Specific ^{32}P-labelled Endonuclease R and Z Fragments

Fragments on Filter	% ^{3}H bound*	Fragments on filter	% ^{3}H bound†
rl	0.58	zl	1.44
r2	1.4	z2	1.15
r3	0.47	z3	1.11
r4	1.1	z4	0.45
r5	0.54	z5	0.50
r6	1.2	z6	0.56
r7	0.028	z7	0.46
r8	0.6	z8	0.58
r9	0.58	z9	0.38
RFII	6.7	RFII	6.7

* A blank filter background of 0.21% has been subtracted.

† A blank filter background of 0.18% has been subtracted. See legend to Table 1 for the procedures used.

The hybridization data in the two tables indicate a different degree of hybridization with the various restriction fragments. These are reproducible differences. This suggests that the φX174 promoter regions are not necessarily identical. It may be possible to use these techniques to divide promoter regions into several classes. The overlap between r6.3 and z3 containing an RNA-polymerase-protected sequence is less than 100 nucleotide pairs long. This overlap fragment has been isolated from φ174 RF cleaved simultaneously by endonucleases R and Z and has been shown to contain a genetic site (D*am*H81) near the beginning of cistron D (J. H. Middleton, unpublished results).

之前有关 φX174[12,13] 和 fd[14] 的体内和体外转录产物的研究提示存在多个信息起始和终止的位点。但是，这些位点并不是按照遗传图谱排列的。φX174 体内翻译的研究 [15] 得出结论，合成起始于顺反子 A 和 D 的产物。我们目前已经定位了 3 个受 RNA 聚合酶保护的 φX174 序列。这些序列位于跨越顺反子 A，D 和 G 起点的限制性片段内（图 2）。但是，我们的数据还不能证明这些受保护的序列实际就是位于这些顺反子的起点处。

有意思的是，用大肠杆菌 RNA 聚合酶进行保护也能从大肠杆菌的 DNA 中分离出相应序列，这些序列能够与已经证实的同样受 RNA 聚合酶保护的 φX174 序列的 φX174 RF DNA（表 2）限制性片段进行杂交。显然大肠杆菌 DNA 的某些启动子区含有与 φX174 RF DNA 类似的起始位点。

表 2. 受 RNA 聚合酶保护的 ^{3}H 标记大肠杆菌 DNA 序列与特异 ^{32}P 标记的核酸内切酶 R 和 Z 片段的杂交

滤膜上的片段	^{3}H 结合比例 %*	滤膜上的片段	^{3}H 结合比例 %†
r1	0.58	z1	1.44
r2	1.4	z2	1.15
r3	0.47	z3	1.11
r4	1.1	z4	0.45
r5	0.54	z5	0.50
r6	1.2	z6	0.56
r7	0.028	z7	0.46
r8	0.6	z8	0.58
r9	0.58	z9	0.38
RFII	6.7	RFII	6.7

* 减去了 0.21% 的空白滤膜本底。
† 减去了 0.18% 的空白滤膜本底。使用的方法见表 1 注。

两张表格中的杂交数据表明，不同的限制性片段存在不同程度的杂交。这些差异都是可重复的。这提示 φX174 的启动子区域不是完全相同的。用这些技术有可能将启动子区分成不同种类。含有受 RNA 聚合酶保护序列的 r6.3 和 z3 之间的重叠少于 100 个核苷酸对。这个重叠的片段已经从同时被核酸内切酶 R 和 Z 消化的 φX174RF 中分离出来了，并且经证实在靠近顺反子 D 的起始处含有一个遗传位点（D*am*H81）（米德尔顿，未发表的结果）。

We thank June H. Middleton for stocks of endonuclease R and endonuclease Z. This work was supported by US Public Health Service grants from the National Institute of Allergy and Infectious Diseases.

(*Nature New Biology*, **243**, 233–236; 1973)

Cheng-Yien Chen, Clyde A. Hutchison, III and Marshall Hall Edgell
Department of Bacteriology and Immunology and Curriculum in Genetics, School of Medicine, University of North Carolina, Chapel Hill, North Carolina 27514

Received December 14, 1972; revised March 7, 1973.

References:

1. Steitz, J. A., *Nature*, **224**, 957 (1969).
2. von Hippel, P. H., and McGhee, J. P., *Ann. Rev. Biochem.*, **41**, 231 (1972).
3. Okamoto, T., Sugiura, M., and Takanami, M., *Nature New Biology*, **237**, 108 (1972).
4. Heyden, B., Nüsslein, C., and Schaller, H., *Nature New Biology*, **240**, 9 (1972).
5. Edgell, M. H., Hutchison, C. A., III, and Sclair, M., *J. Virol.*, **9**, 574 (1972).
6. Middleton, J. H., Edgell, M. H., and Hutchison, C. A., III, *J. Virol.*, **10**, 42 (1972).
7. Hutchison, C. A., III, and Edgell, M. H., *J. Virol.*, **8**, 181 (1970).
8. Smith, H. O., and Wilcox, K. W., *J. Mol. Biol.*, **51**, 379 (1970).
9. Jones, O. W., and Berg, P., *J. Mol. Biol.*, **22**, 199 (1966)
10. Blattner, F. R., Dahlberg, J. E., Boetliger, J. K., Fiandt, M., and Szybalski, W., *Nature New Biology*, **237**, 232 (1972).
11. Rüger, W., *Biochim. Biophys. Acta*, **238**, 202 (1971).
12. Sedat, J. W., and Sinsheimer, R. L., *Cold Spring Harbor Symp. Quant. Biol.*, **35**, 163 (1970).
13. Hayashi, Y., and Hayashi, M., *Cold Spring Harbor Symp. Quant. Biol.*, **35**, 171 (1970).
14. Okamoto, T., Sugiura, M., and Takanami, M., *J. Mol. Biol.*, **45**, 101 (1969).
15. Benbow, R. M., Hutchison, C. A., III, Fabricant, J. D., and Sinsheimer, R. L., *J. Virol.*, 7, 549 (1971).
16. Burgess, R. R., *J. Biochem.*, **244**, 6160 (1969).
17. Davis, R. W., and Davidson, N., *Proc. US Nat. Acad. Sci.*, **60**, 243 (1968).
18. Raskas, H. J., and Green, M., in *Methods in Virology* (edit. by Maramorosch, K., and Koprowsky, H.), **5**, 247 (1971).
19. Burgess, A. B., and Denhardt, D. T., *J. Mol. Biol.*, **44**, 377 (1969).
20. Godson, G. N., *J. Mol. Biol.*, **57**, 541 (1971).
21. Benbow, R. M., Mayol, R. F., Picchi, J. C., and Sinsheimer, R. L., *J. Virol.*, **10**, 99 (1972).
22. Hutchison, C. A., III, Middleton, J. H., and Edgell, M. H., *Biophys. J.*, **12** (abstracts), 31a (1972).

我们感谢米德尔顿供给核酸内切酶R和Z。本工作得到了美国国家过敏与传染病研究所的公共卫生服务资金的资助。

（毛晨晖 翻译；曾长青 审稿）

Depression of Freezing Point by Glycoproteins from an Antarctic Fish

R. E. Feeney and R. Hofmann

Editor's Note

Fish living in polar waters risk having their blood and cell fluids freeze. In the late 1960s it was found that their blood contains "antifreeze" proteins that reduce the freezing point. One possibility was that these act like other solutes which suppress freezing, such as salt or sugar, simply by their physical presence and not their chemical nature—a so-called colligative property. But here Robert Feeney and R. Hofmann in Zürich, Switzerland, show that this cannot be so for antifreeze proteins from two types of Antarctic fishes. They conclude that these proteins must be exerting more specific effects, perhaps by altering the way ice crystals grow. Modern research now focuses on how antifreeze proteins' surface structure affects ice nucleation.

SCHOLANDER and colleagues[1-3] observed that the blood sera of some polar fishes contain a substance of high molecular weight which lowers the freezing point. The general properties and structures of a family of several glycoproteins with this characteristic have recently been described in Antarctic fishes[4-14]. These "antifreeze" glycoproteins (AFGP) consist of repeating units of the triglycopeptide Ala-Ala-Thr-*o*-disaccharide. Three active glycoproteins have been characterized; these differ only in polymer length, with molecular weights ranging from 10,500 to 21,000 g mol^{-1} as determined by ultracentrifugation, light scattering, or osmotic pressure[6,11]. According to the freezing point depression, however, the apparent molecular weight is only 20 g mol^{-1}, a value equivalent to >500 times the depression calculated from the molecular weights.

The blood sera of two species of Antarctic fish, *Trematomus borchgrevinki* and *Dissostichus mawsoni*, freeze at approximately −2.0°C, slightly below the temperature of the ice-salt water mixture in Antarctica[5]. Approximately one-third of the depression of the freezing point is caused by the AFGP, the remainder by dialysable substances of low molecular weight[5]. We have examined the rates of development of ice crystals and possible equilibria between ice crystals and aqueous phases, using differential thermal analysis (DTA) and direct microscopic observations of freezing and melting, respectively. The AFGP was purified from *T. borchgrevinki* serum as described previously[6,10].

The DTA experiments were carried out in aluminium vessels in a Mettler vacuum thermal analyser. Controls of water or solutions of 1% chicken ovomucoid[15] (a glycoprotein containing approximately 25% carbohydrate) in water and the solutions containing 1% AFGP froze at the same temperatures and had similar freezing and melting curves. Freezing of small volumes (10 λ) of such solutions usually occurred between −15 and

一种南极鱼类的糖蛋白具有降低凝固点的作用

菲尼，霍夫曼

编者按

生活在极地海域中的鱼类有血液和细胞内液体凝固的危险。在 20 世纪 60 年代后期，人们发现它们的血液中含有一种可以降低血液凝固点的“抗冻”蛋白。有可能像盐或糖这类溶质一样，仅依靠其物理特性（即所谓的依数特性）而非化学性质来阻碍结冰。但是本文中的瑞士苏黎世的罗伯特·菲尼和霍夫曼发现的两种南极鱼类的抗冻蛋白的作用机理却并非如此。他们认为这类蛋白或许是通过改变冰晶的生长方式而发挥其独特的抗冻作用。当代的研究主要集中在抗冻蛋白的表面结构如何影响冰晶成核。

朔兰德和他的同事们[1-3]发现某些极地鱼类的血清中含有一种可以降低凝固点的大分子量物质。最近报道了南极鱼类中具有这种性质的这一家族的几种糖蛋白的结构和一般性质[4-14]。这些“抗冻”糖蛋白（AFGP）是由三糖肽丙氨酸－丙氨酸－苏氨酸－邻－二糖的重复单位组成。已经对三种活性糖蛋白做了鉴定，通过超速离心、光散射或者渗透压的方法测定出它们之间唯一的区别就是聚合长度不同，它们的分子量在 10,500 g/mol 到 21,000 g/mol 范围内[6,11]。然而根据凝固点下降的程度，其表观分子量仅为 20 g/mol，这个数值相当于比实际分子量下降了 500 多倍。

博氏肩孔南极鱼和鳞头犬牙南极鱼这两种南极鱼的血清凝固点大约为 –2.0℃，略低于南极冰盐水混合物的温度[5]。凝固点的降低，其中将近 1/3 是由抗冻糖蛋白引起的，其余的是由一些可透析的小分子量物质引起的[5]。我们已经通过示差热分析 (DTA) 以及显微镜直接观察凝固和融化状态的方法分别研究了冰晶的形成速率以及冰晶和水相之间可能存在的平衡关系。从博氏肩孔南极鱼血清中提纯的抗冻糖蛋白在先前的文章已有描述[6,10]。

示差热分析实验是在梅特勒真空热分析仪的铝管中进行的。将对照水或者含有 1% 蛋清粘蛋白[15]（一种含有大约 25% 碳水化合物的糖蛋白）的水溶液和含有 1% 抗冻糖蛋白的水溶液在相同的温度下冻结，并得到相似的冻融曲线。通常情况下，小体积（10λ）的这种溶液在 –15℃至 –20℃时发生凝固，这取决于水的纯度以及实

–20°C depending on the purity of the water and the experimental rates of lowering the temperature. The freezing temperatures in these conditions are usually considered to be related to the rates of nucleation and not to their freezing temperatures in equilibrium with ice. These experiments did not, therefore, indicate that the AFGP functions by inhibiting nucleation, that is, by inhibiting the initial formation of points of crystallization.

Direct observations of melting and freezing in volumes of approximately 50 λ at closely controlled temperatures were made with a Zeiss Universal polarization microscope (magnification 90) equipped with a photographic camera and a television screen monitor. The microscope slides were double-celled, with a compartment on each side of a centre divider. The whole slide was cooled by a slow flow of cold nitrogen gas and the temperature was maintained by heat supplied to the centre of the slide by a heating element in the divider. The thermocouple was Pt-PtRh, positioned in the centre of the slide. The apparatus was calibrated by determining the melting and freezing temperatures of pure water. A series of eight experiments was done and each experiment had five to twenty separate melting and thawing trials. Results of a typical experiment are summarized in Table 1.

Table 1. Freezing and Melting of Water and Solution of Antifreeze Glycoproteins in Water

Temperature adjustments °C phase	Observed changes	
	In water containing ice crystals	In water solution of 1% antifreeze glycoproteins containing ice crystals
0.0 holding	Melt and freeze	Crystals melt
–0.1 lowering	Frozen	Crystals do not melt, liquid does not freeze
–0.7 lowering	Frozen	
–0.8 holding	Frozen	Crystals grow, new crystals form until all solution frozen
–0.7 raising	Frozen	All frozen, no melting
–0.1 raising	Frozen	
0.0 holding	Melt and freeze	Melt

Water and antifreeze glycoprotein solution was initially frozen at –3°C and then allowed to melt at +0.1°C until 5–10% of solution remained as ice crystals. The temperature was then adjusted to 0.0°C and periodically lowered and then raised as indicated. The times at each temperature intermediate between freezing and melting were 5–10 min. All observations were made microscopically as described in the text.

In one series of trials, solutions of egg white ovomucoid were used in the control well of the microscope slide. The AFGP and the ovomucoid were both tested as 1% solutions in water. In all trials, both the melting and freezing point of the ovomucoid solution were –0.02±0.01°C. The melting point of an AFGP frozen solution or of ice crystals in AFGP solution was –0.01°C, and the freezing point of the solution containing ice crystals was –0.80°C. In another series of trials, a solution of AFGP containing a few crystals of ice froze at –0.78°C and melted at 0.00°C. After freezing, the sample was melted at +1°C

验时温度降低的速率。一般认为这种情况下的凝固温度与成核速率有关，而与冰水均衡态的凝固温度无关。因此，这些实验并没有揭示出抗冻糖蛋白对成核的抑制作用，也就是对最初结晶点形成的抑制。

通过配备照相机和电视显示屏的蔡司全能偏光显微镜（放大倍数为 90 倍），我们可以在严格控温的条件下直接观察到大约 50 λ 的体积中的融化与凝固过程。显微镜载片是一个双室的结构，在中心分隔物的两侧分别有一个小室。整个载片是通过缓慢施加的冷氮气流进行冷却的，并通过载片分隔物上带有的加热元件对载片中心部位进行加热以维持温度。载片中心处的热电偶为 Pt–PtRh。此仪器是通过测量纯水融化温度和凝固温度来进行校准的。现在已经完成了一个系列的 8 个实验，每个实验都有 5 到 20 个独立的融化和融解试验。表 1 中总结了其中一个典型实验的结果。

表 1. 水和抗冻糖蛋白水溶液的凝固和融化

温度调节 ℃状态	观察到的变化	
	含有冰晶的水	含有冰晶的 1% 抗冻糖蛋白水溶液
0.0 维持	融化和凝固	晶体融化
–0.1 降低	冻结	晶体不融化，液体不凝固
–0.7 降低	冻结	
–0.8 维持	冻结	晶体生长，新的晶体开始形成，直至所有液体全部冻结
–0.7 升高	冻结	全部冻结，没有融化
–0.1 升高	冻结	
0.0 维持	融化和凝固	融化

水和抗冻糖蛋白溶液事先在 –3℃冻结，然后在 +0.1℃下开始融化直至 5%~10% 的溶液以冰晶形式存在为止。接着将温度调整到 0.0℃并如表所示周期性地降低温度之后再升高。每个融化和冻结温度之间的时间间隔为 5~10 min。所有的结果都是通过文中所述的显微观察得到的。

在一系列的试验中，显微镜载片的对照室中使用的是蛋清粘蛋白溶液。检测的抗冻糖蛋白和蛋清粘蛋白都是浓度为 1% 的水溶液。在所有的试验中，蛋清粘蛋白溶液的熔点和凝固点都是 -0.02 ± 0.01℃。抗冻糖蛋白冷冻溶液或者抗冻糖蛋白溶液中冰晶的熔点为 –0.01℃，而含有冰晶的溶液的凝固点为 –0.80℃。在另外一系列试验中，含有少量冰晶的抗冻糖蛋白溶液在 –0.78℃凝固，在 0.00℃融化。在凝固后，样品在 +1℃时开始融化，直至仅仅剩余一点冰晶（大约 0.1% 的溶液含有晶体）。然

until only a small bundle of crystals remained (approximately 0.1% of the solution had crystals). The sample was then adjusted to −0.60°C and held at this temperature for 300 min. No growth or melting of the crystals occurred. Similar observations were made when the FPDG solutions containing ice crystals were maintained at a temperature slightly less than the melting point (−0.10°C) or slightly more than the freezing point (−0.70°C).

From our experiments we conclude that: (1) Ice formed in a solution of AFGP seems to be normal ice—that is, it melts at 0°C. (2) Freezing and melting of AFGP solutions occur at rates similar to those at which water freezes and melts when equivalent amounts of heat are applied or removed at the respective melting or freezing temperatures. Thus there was no evidence indicating a comparatively rapid development of crystals in AFGP solutions as described by Hargens[16]. (3) It is not possible to prove a mechanism involving nucleation from the DTA experiments on kinetic effects. There are, therefore, no unusual "supercooling effects" as are commonly found in solutions in which initiation of freezing is very slow in the absence of crystals. The data indicate that there is no significant kinetic effect involved in the overall freezing mechanism.

The pronounced hysteresis and absence of equilibrium between the melting and freezing of solutions of AFGP are consistent with a mechanism which is not based on colligative properties. Models for the mechanism could include those postulating effects on either the structure or growth of ice crystals or the structure of water. If a model concerning ice is correct, the mechanism would most likely involve the development of ice crystals[10,17] after nucleation. If the model concerning the structure of water is correct, the mechanism could involve either the structuring of water itself or some intermediate state. The fact that ice crystals in AFGP solutions have normal melting points lends credence to the latter model.

We thank Dr Hansa Ahrends for supervising the DTA experiments and for his advice and suggestions, the Mettler Corp. for assistance, and Dr Richard Criddle for discussions and suggestions regarding the mechanism of action. The investigation was supported by funds from the Eidgenössische Technische Hochschule. R. E. F. was on sabbatical leave from the University of California, Davis.

(**243**, 357-359; 1973)

R. E. Feeney and R. Hofmann

Laboratorium für Festkörperphysik, Eidgenössische Technische Hochschule, Zürich, Hönggerberg, CH-8049 Zürich

Received January 23, 1972.

References:

1. Scholander, P. F., Flagg, W., Walters, V., and Irving, L., *Physiol. Zool.*, **26**, 67 (1953).

2. Scholander, P. F., van Dam, L., Kanwisher, J. W., Hammel, H. T., and Gordon, M. S., *J. Cell Comp. Physiol.*, **49**, 5 (1957).

3. Gordon, M. S., Amdur, B. H., and Scholander, P. F., *Biol. Bull.*, **122**, 52 (1962).

后把样品溶液的温度调整到 –0.60℃，并保持在这个温度 300 min，结果没有发生新的冰晶的生长或冰晶的融化。当含有冰晶的凝固点降低糖蛋白（FPDG）溶液保持在一个略低于其熔点（–0.10℃）或者略高于其凝固点（–0.70℃）的温度时，我们发现了类似的现象。

从这些实验中我们得到如下的结论：(1) 抗冻糖蛋白溶液中形成的冰晶看来是正常的冰，也就是说，其熔点为 0℃。(2) 在各自的融化或凝固温度时给予或去除等量的热量，抗冻糖蛋白溶液和水的凝固及融化发生的速度相近。因此，并没有证据显示哈根斯所报道的抗冻糖蛋白溶液中结晶速度相对较快 [16]。(3) 通过示差热分析实验从动力学效应方面来证明冰晶的成核机制是不可能的。因此，并不存在非同寻常的“过冷效应”。“过冷效应”通常出现在没有晶体时其凝固起始的过程非常缓慢的溶液中，数据显示在整个凝固机理中并没有显著的动力学效应。

抗冻糖蛋白溶液的融化和凝固之间具有明显的滞后并且缺乏平衡，这与其发生的不依据于依数性特征的机理一致。此机理的模型应该包括那些对冰晶的结构或生长，或者对水的结构所提出的假设效应。如果关于冰的模型是正确的，那么这个机理就有可能会涉及成核后冰晶的生长过程 [10,17]。如果关于水的结构的模型是正确的，那么这个机理可能会涉及水本身的结构，或者某种中间状态。抗冻糖蛋白溶液中的冰晶具有正常的熔点，基于这一事实，我们倾向于第二种模型。

我们感谢汉萨 · 阿伦茨博士对示差热分析实验的指导以及他提出的意见和建议，感谢梅特勒公司的协助，感谢理查德 · 克里德尔博士在作用机理方面所做的讨论和给出的建议。该项研究是菲尼在加州大学戴维斯分校休学术年假期间，受到苏黎世联邦理工学院的基金资助完成的。

（刘振明 翻译；黄晓航 审稿）

4. De Vries, A. L., and Wohlschlag, D. E., *Science*, **163**, 1073 (1969).

5. Komatsu, S. K., thesis, Univ. California, Davis (1969).

6. DeVries, A. L., Komatsu, S. K., and Feeney, R. E., *J. Biol. Chem.*, **245**, 2901 (1970).

7. Komatsu, S. K., DeVries, A. L., and Feeney, R. E., *J. Biol. Chem.*, **245**, 2909 (1970).

8. DeVries, A. L., Vandenheede, J., and Feeney, R. E., *J. Biol. Chem.*, **246**, 305(1971).

9. Shier, W. T., Lin, Y., and DeVries, A. L., *Biochem. Biophys. Acta*, **263**, 406 (1972).

10. Vandenheede, J. R., Ahmed, A. I., and Feeney, R. E., *J. Biol. Chem.*, **247**, 7885 (1972).

11. Feeney, R. E., Vandenheede, J., and Osuga, D. T., *Naturwissenschaften*, **59**, 22 (1972).

12. DeVries, A. L., in *Fish Physiology* (edit. by Hoar and Randall), **6**, 157 (Academic Press, New York, 1969).

13. Chuba, J. V., Kuhns, W. J., Nigrelli, R. F., Vandenheede, J., Osuga, D. T., and Feeney, R. E., *Nature*, **242**, 342 (1973).

14. DeVries, A. L., *Science*, **172**, 1152 (1971).

15. Feeney, R. E., and Allison, R. G., *Evolutionary Biochemistry of Proteins* (Wiley-Interscience, New York, 1969).

16. Hargens, A. R., *Science*, **176**, 184 (1972).

17. Scholander, P. F., and Maggert, J. E., *Cryobiology*, **8**, 371 (1971).

Muscular Contraction and Cell Motility

H. E. Huxley

Editor's Note

One of the early triumphs of molecular biology was the working out of the mechanism of muscular contraction. Much of the work was done by Hugh Huxley at the Medical Research Council Laboratory of Molecular Biology at Cambridge. This paper describes Huxley's views on muscular contraction (which are still regarded as essentially correct) and also his preliminary views on a mechanism by which cells are able to move themselves relative to others.

This article is based on a lecture given at the Thirteenth International Congress of Cell Biology, University of Sussex, September 3–8, 1972.

IT has become apparent during the past few years that close similarities exist in a number of instances between some of the proteins directly involved in contraction in striated muscle, and proteins present in certain non-muscle cells in which movement occurs. The question therefore arises whether all these systems share a common basic mechanism, and if so, whether the current picture of the contracting mechanism in striated muscle, which is a fairly detailed one in some respects, can cast any light on these other motile mechanisms, which are less well understood.

Here I first describe certain aspects of the muscle mechanism, especially structural ones, which seem to me particularly relevant to more general questions of motility. Next, I review a number of recent studies which demonstrate in a very decisive way that the similarities between the proteins concerned are ones that relate to the most basic properties and interactions used in the muscle mechanism. Finally, I point out that these considerations suggest a definite mechanism for certain kinds of cell motility. This general type of mechanism ("active shearing") has been suggested by others before[1-3] (though not always in very explicit terms) but it does not seem to have gained general acceptance.

According to the sliding filament model of muscle contraction striated muscles consist of overlapping arrays of actin and myosin filaments which can slide past each other when the muscle changes length. The individual filaments, and the arrays which they form because of their in-register arrangement, remain virtually constant in length. The active sliding force between the filaments is developed by cross-bridges on the thick myosin-containing filaments. These represent the biologically active ends of individual myosin molecules, which can attach to, and exert a longitudinal force on, the thin actin-containing filaments alongside. A cross-bridge is believed to act in a cyclical manner, pulling an actin filament

肌肉收缩与细胞运动

赫胥黎

编者按

分子生物学早期的重大成就之一是阐明了肌肉收缩的机理。其中许多工作都是由剑桥大学医学研究理事会分子生物学实验室的休·赫胥黎教授完成的。这篇论文阐述了赫胥黎有关肌肉收缩的观点（目前为止仍被认为是基本正确的），也包含了他对于细胞相对运动的一些初步的想法。

本文是基于我在“第十三届国际细胞生物学大会”（1972年9月3日到8日，萨塞克斯大学）上的报告完成的。

在过去的几年中，人们发现在许多情况下，横纹肌细胞中与收缩直接相关的蛋白质与发生运动的某些非肌细胞中的蛋白质之间存在着惊人的相似性。我们因此提出疑问，这些系统是否都有着共同的基本机制？如果是的话，目前在某些方面对于横纹肌收缩机理已经相当详尽的研究能否为那些了解相对较少的其他运动机制的研究带来一些启示呢？

在这篇文章中，我首先对肌肉机制的某些方面进行了介绍，尤其是肌肉的结构——我认为这与运动的普遍性问题特别相关。接着，我回顾了近来的很多研究，结果都非常明确地证明，人们所关注的蛋白质之间的相似性与肌肉运动机制中最基本的性质和相互作用有关。最后，我认为这些观点提示了特定种类细胞运动的明确机制。虽然这个普遍性的机制（“主动剪切”）在以前也曾被其他人提到过[1–3]（尽管有时表述地不够明确），但似乎尚未得到广泛认可。

根据肌肉收缩的肌丝滑动模型，横纹肌由一系列互相重叠的肌动蛋白纤维（细肌丝）和肌球蛋白纤维（粗肌丝）排列组成，当肌肉长度改变时这些纤维可以发生相对滑动。不过每条纤维以及由它们有序排布形成的阵列事实上保持着恒定的长度。这种纤维间的主动滑动力来自含肌球蛋白的粗肌丝上的横桥。横桥代表了个体肌球蛋白分子的生物活性末端，它可以结合在附近的细肌丝上并且对其施加一个纵向力。横桥被认为是以循环往复的方式发挥功能的，其每次循环可以将肌动蛋白纤维向A along towards the centre of an A-band for a distance which is probably of the order of 50 to 100 Å, then releasing and reattaching to the actin filament at another point, initially further away from the centre of the A-band and going through the cycle again. A continuous movement of the actin filament is thus produced by the asynchronous action of all the cross-bridges acting upon it from the myosin filaments alongside.

The mechanism requires that the molecules of myosin and actin are assembled in their respective filaments with the appropriate structural polarity. As the cross-bridges have to move so as to draw the actin filaments towards the centre of the A-band, they must be oriented so that they pull in one direction in one half A-band and in the opposite direction in the other half. Thus all the myosin molecules must be oriented in one sense in one half of the length of each thick filament, and in the opposite sense in the other half. This is indeed found to be the case in practice[4]. Also myosin molecules are able to assemble *in vitro* into filaments with this important and characteristic reversal of polarity half way along their length. A similar requirement applies to the actin filaments. Highly specific interactions between an actin monomer and a myosin cross-bridge require that the interacting groups on the two molecules always have the appropriate mutual orientation. Thus actin filaments which interact with opposite ends of a myosin filament must contain actin monomers with opposite polarity. In practice, it is found that all the actin monomers along a given actin filament have the same polarity, and that the polarity reverses at the Z-lines. Thus the actin filaments are attached on either side of the Z-lines with opposite polarity[4].

This means that the direction of the relative force experienced by an actin filament when it interacts with appropriately oriented myosin filaments is specified by the structural polarity of the actin filament itself. Consequently, if an actin filament were attached to some other cellular structure with the same polarity as actin filaments are attached to Z-lines, then the attachment point would experience a force pulling it in the direction of the actin filament. Again, as the force exerted on an actin filament must always be in the same direction, such a filament (or group of similarity polarized filaments) might be maintained in motion over significant distances by interaction with myosin filaments[4], whose orientation could be selected and perhaps enforced by the actin. These considerations are obviously relevant to mechanisms for cell motility and will be discussed again later.

A second characteristic of the sliding filament mechanism as it appears to work in practice is that the relative force between myosin and actin is developed as the result either of an active change in the angle of attachment of the head of the myosin molecules (the S_1 subunit) to the actin filament (Fig. 1), or of an active change in the shape of the S_1 subunit, the attachment to actin remaining rigidly fixed[5]. That is, the relative sliding force appears not to be generated by changes elsewhere in the myosin molecule or filament, not, for example, by an active change in the orientation of the S_1 head generated at the link to the S_2 part of the molecule. The arguments supporting this view, which I believe are very powerful are somewhat involved, and the original papers should be consulted[4,6,7]. In essence, they derive mainly from X-ray diffraction and electron microscope evidence that

带的中心拉动约 50~100 Å 的距离，然后与肌动蛋白纤维脱离，并重新结合到距离 A 带中心更远的另一个位点上，然后重复上述循环。通过肌球蛋白纤维上的所有横桥的异步运动，便可以产生肌动蛋白纤维的连续滑动。

上述机制要求肌动蛋白与肌球蛋白分子在装配的时候以合适的结构极性形成各自的纤维。由于横桥需要通过移动把肌动蛋白纤维向 A 带中心拉动，因此横桥排布必然是具有方向性的，这样才可以使 A 带中一半的横桥向一个方向拉动，而另一半则向相反方向拉动。由此可以推断出，粗肌丝上的肌球蛋白分子肯定是一半沿着一个方向排布，另一半则沿着相反方向排布。事实的确如此[4]。实验证明肌球蛋白分子可以在体外沿其轴方向组装成这种重要而且颇具特点的逆转极性纤维。同样的，肌动蛋白纤维也具有极性特征。肌动蛋白分子单体与肌球蛋白横桥之间高度特异性的相互作用要求这两种分子上相互作用的基团总是具有适当的极性取向。因此，如果两条肌球蛋白纤维具有方向相反的末端，那么与之相互作用的肌动蛋白纤维必然含有极性相反的肌动蛋白单体。事实上，实验表明在一条给定的肌动蛋白纤维上所有的肌动蛋白单体都具有相同的极性。不过，在 Z 线的位置，肌动蛋白纤维的极性会调转过来。因此，肌肉中肌动蛋白纤维是以相反的极性结合在 Z 线两侧的[4]。

这就意味着当肌动蛋白纤维与适当取向的肌球蛋白纤维相互作用时，产生的相对作用力的方向是一定的，且由肌动蛋白纤维本身的结构极性所决定。因此，如同肌动蛋白纤维结合到 Z 线，当肌动蛋白纤维结合到与之相同极性的其他细胞结构时，该连接位点将会承受一个与肌动蛋白纤维极性取向相同的拉力。同样地，由于这些施加在肌动蛋白纤维上的拉力必须总是朝着同一个方向，因此这样一条纤维（或一组具有相同极性的纤维）就可以通过与肌球蛋白纤维的相互作用而持续滑动较远的距离[4]。这种滑动的方向当然也是由肌动蛋白所选择或是强制决定的。上述这些观点与细胞运动机制显著相关并将在后文中再次讨论。

在实际研究中发现的肌丝滑动机制的第二个问题是滑动动力的来源。实验表明肌球蛋白与肌动蛋白的连接保持固定不变，两者间的相对滑动力来源于肌球蛋白分子头部（S_1 亚基）与肌动蛋白纤维结合角度的主动变化（图 1），或者来源于 S_1 亚基形状的主动变化[5]。也就是说，相对滑动力不是通过肌球蛋白分子本身或肌球蛋白纤维上其他部位的改变而产生的，例如，不是通过改变 S_1 与 S_2 的连接处使 S_1 头部取向产生主动变化。我认为一些支持上述观点的论据很有说服力，在此处或多或少会提到，具体内容可参考它们的原始文献[4,6,7]。其实，这些结论主要来源于 X 射线衍射和电子显微镜观察到的结果，即肌球蛋白分子的头部与粗肌丝主链结合的部

the attachment of the heads of the myosin molecules to the backbone of the thick filament is very flexible, whereas their attachment to the actin filaments (in the rigor configuration) is a very rigid one.

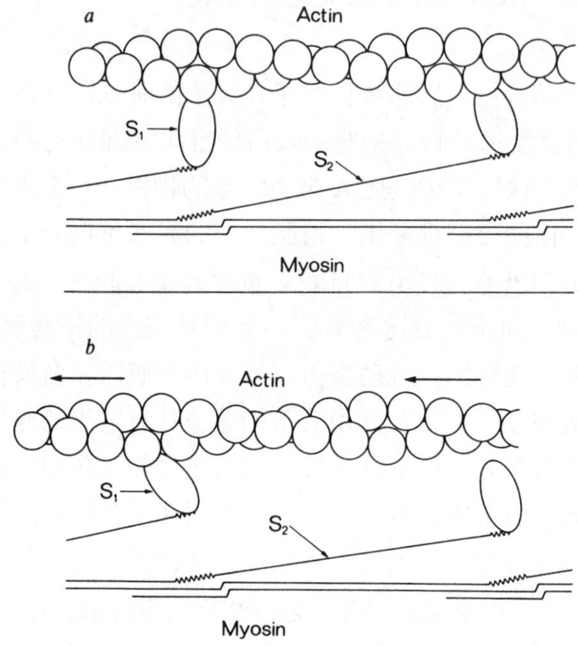

Fig. 1. Active change in angle of attachment of cross-bridges (S_1 subunits) to actin filaments could produce relative sliding movement between filaments maintained at constant lateral separation (for small changes in muscle length) by long range force balance. Bridges can act asynchronously as subunit and helical periodicities differ in the actin and myosin filaments. Only one of the two S_1 subunits in each myosin molecule is shown. *a*, Left hand bridge has just attached; other bridge is already partly tilted. *b*, Left hand bridge has just come to the end of its working stroke; other bridge has already detached, and will probably not be able to attach to this actin filament again until further sliding brings helically arranged sites on actin into favourable orientation.

If this picture is correct, it follows that the basic contractile mechanism is represented by the interacting complex of thin filament and attached myosin head, and that the exact mode of assembly of myosin molecules into filaments, provided their appropriate polarity is maintained, may be capable of some variation. Even amongst striated muscles, variations are observed in non-vertebrate species, and it is perfectly possible that myosin may be organized in quite different forms in other situations whilst the same myosin head–actin filament interaction is still maintained.

A third feature of the sliding filament mechanism in striated muscle is concerned with the exact mode of attachment of the myosin head to the actin filaments. At present, no means are available to arrest the working stroke of the cross-bridge, or to prevent subsequent dissociation in the presence of ATP. In the absence of ATP, however, the myosin cross-bridges remain attached to the actin filament and the muscle is then in rigor. In an *in vitro* system, actin filaments can be "decorated" in the absence of ATP, by the

位具有很好的柔性，而其与肌动蛋白纤维结合的部分（在精确的构型中）则是一个刚性结构。

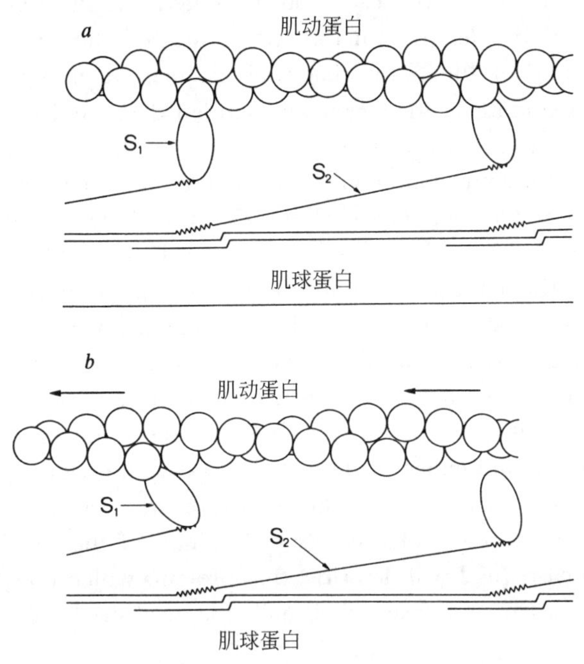

图 1. 横桥（S_1 亚基）与肌动蛋白纤维结合角度的主动变化使肌丝间（在肌肉长度发生微小变化的情况下，长程力的平衡使肌丝间保持恒定的间隔）发生相对滑动。由于肌动蛋白纤维与肌球蛋白纤维的亚基及螺旋周期不同，横桥可以异步方式发挥作用。图中只显示了每个肌球蛋白中两个 S_1 亚单位中的一个。(*a*) 左边的横桥处于刚与肌动蛋白结合的状态；另一个横桥则已经处于部分倾斜的状态。(*b*) 左边的横桥已经接近其摆动过程的末尾；另一个横桥则处于刚从肌动蛋白纤维上解离的状态，并且如果没有进一步的滑动使肌动蛋白上螺旋形排列的位点呈现一个适当的取向，那么这个横桥可能就不会再与该肌动蛋白纤维结合了。

如果这幅图所描述的机制是正确的话，那么由此我们可以得出肌肉收缩的基本原理就是通过细肌丝与结合在它上面的肌球蛋白头部形成具有相互作用的复合物来实现的。同样的，我们还可以得出，如果要维持纤维适当的极性，肌球蛋白分子组装成肌丝纤维的精确模型可能存在某些可变之处。即使都是横纹肌，不同的无脊椎动物物种之间肌球蛋白的组装模式也有所不同，并且很有可能在某些情况下，肌球蛋白可通过一种完全不同的形式组装而仍保持原有的肌球蛋白头部–肌动蛋白的相互作用。

横纹肌肌丝滑动模型的第三个特点是肌球蛋白头部与肌动蛋白纤维之间精确的结合方式。目前，还没有一种办法可以捕捉到横桥的摆动过程，也不能在 ATP 存在的情况下阻止随后横桥与细肌丝的解离。然而，当没有 ATP 存在时，肌球蛋白的横桥会保持与肌动蛋白纤维结合的状态，之后肌肉会处于僵直状态。在一个体外实验系统中，在没有 ATP 的情况下，肌动蛋白纤维可以被肌球蛋白分子、重酶解肌球蛋

attachment of molecules of myosin or of heavy meromyosin (HMM) or of subfragment 1 (S_1). Suggestion of an "angled" configuration of the attached myosin heads was given by the visual appearance of "arrowheads" on decorated thin filaments, but the first good evidence was provided by electron microscope studies of thin sections of glycerinated insect flight muscle, and of the X-ray diagrams given by it in rigor and in the relaxed state[8]. More detailed studies of the negatively stained decorated filaments by the three-dimensional image reconstruction technique[9] have shown that not only are the myosin S_1 subunits (which appear as somewhat elongated structure of dimensions very approximately 150 Å×40 Å×30 Å) tilted at an angle of about 45° to the long axis of the actin filaments, but they are slewed round in a characteristic way, that is rotated by about 45° in a plane parallel to the axis of the actin filament and perpendicular to the plane in which they are tilting. It is this combination of tilting and slewing, together with the distinctive shape of the myosin heads, which gives rise to the very characteristic "arrowhead" appearance (Fig. 2), which would not arise from a straightforward "tilted" attachment. We do not at present know the functional significance of this peculiar feature of the attachment (the tilt itself is, of course, what we should expect to find at the end of the working stroke if the initial attachment was perpendicular, and is in the right direction); but it is diagnostic of a very specific structural interaction and of a very specific shape of the attached molecule. It has been found in many cases, as I will describe, that proteins which may be involved in non-muscular motility can form complexes of virtually identical structure.

A final feature of the contraction mechanism in striated muscle is concerned with regulation of activity, that is with switching contractile activity on and off. This is effected by preventing the attachment of the myosin heads to actin in a relaxed muscle, and allowing it to take place when the muscle is activated. Attachment to actin is a necessary part of the biochemical cycle in which ATP is split and force developed by the actin–myosin system. In its absence, ATP splitting, by the myosin alone, takes place very slowly and no tension is developed by the sliding filament system. Indeed, the filaments will slide past each other passively under the action of relatively small external forces.

白（HMM）或亚片段 1（S_1）的结合所“修饰”。修饰后的肌动蛋白纤维呈现出“箭头”的形状，表明肌球蛋白头部是以一定角度结合的。首次证明这个结论的有力证据来自对甘油处理过的昆虫飞行肌超薄切片的电子显微镜研究，以及飞行肌紧张和放松状态时的 X 射线照片[8]。更细致的研究来自利用三维图像重构技术对修饰后的肌动蛋白纤维负染样品的观察[9]。其结果显示，肌球蛋白 S_1 亚基（看起来是一个细长的结构，三维尺寸约为 150 Å × 40 Å × 30 Å）一方面以与肌动蛋白纤维纵轴方向呈 45 度的角度与其结合，另一方面又以一种特殊的方式缠绕，即其在平行于肌动蛋白横轴的平面内，垂直于倾斜面旋转 45 度角。正是这种倾斜加缠绕式的排列，加之肌球蛋白头部独特的形状，使得整个复合物呈非常独特的“箭头”外观（图 2），这种外观不可能由直接的倾斜连接所导致。目前我们尚不清楚这种奇特的结合方式在功能上的意义（当然，如果肌球蛋白与肌动蛋白开始结合的状态是垂直的，并且其方向也是正确的，那么这种在摆动过程结束时产生的倾斜状态很可能是我们所期望的），不过此类结合方式可以用来鉴定特定的结构相互作用和结合分子的特定形状。正如我将要描述的那样，许多例子表明，和非肌细胞的运动有关的蛋白质也可以形成与上述结构几乎完全相同的复合物。

横纹肌收缩机制的最后一个特点与其活性的调节有关，即收缩活性开启与关闭的切换。这是通过阻止肌球蛋白的头部与肌动蛋白结合从而使肌肉处于放松状态，以及通过允许两者结合从而使肌肉处于兴奋状态来实现。肌球蛋白与肌动蛋白结合是生化循环的一个必要步骤，在这个循环里，ATP 水解，肌动蛋白 – 肌球蛋白系统相互作用产生动力。在肌动蛋白不存在，仅肌球蛋白存在的条件下，ATP 的水解过程变得非常缓慢，不可能通过肌丝滑动系统产生张力。实际上，在较小的外力作用下，这些纤维之间也会被动地发生相对滑动。

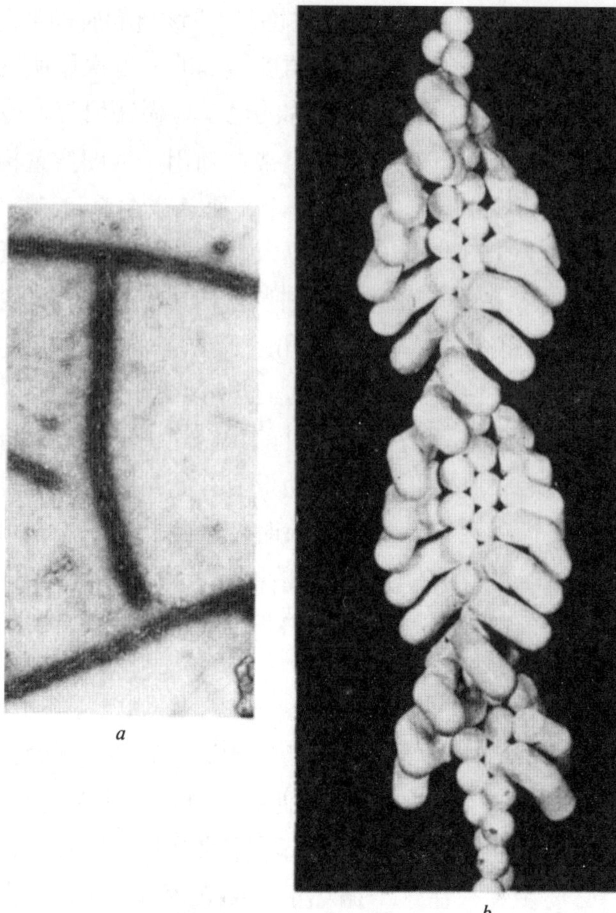

Fig. 2. *a*. Electron micrograph of negatively stained preparation of actin filaments "decorated" with myosin subfragment 1, showing well developed "arrowhead" formations (×138,600). *b*, Simplified model of "decorated" actin filament, based on 3-D reconstruction results; the S_1 subunits are attached to the central core of helically arranged actin subunits in a characteristically tilted and slewed configuration, and it is this that leads to the appearance of arrowheads.

In a relaxed muscle, the concentration of free calcium is kept at a very low value (probably less than 10^{-7} to 10^{-8} M) by the action of the calcium pump in the sarcoplasmic reticulum. Upon activation, calcium is released so that the level of free calcium rises (probably to 10^{-6} to 10^{-5} M) and attachment of myosin to actin can take place. In vertebrate striated muscle, this change is effected by the tropomyosin–troponin system in the actin-containing filaments[10], which prevents attachment of myosin in the absence of free calcium, but allows it to take place in its presence, possibly by a steric blocking mechanism dependent on changes in position of tropomyosin. In molluscan muscles, however (including the striated muscle of *Pecten*), regulation is effected by changes in the myosin molecules to which calcium ions become bound upon activation[11]. The distribution of these two types of regulation, or combinations of them, between species is a subject of very active current research[12]. Nevertheless, all systems so far investigated share the common feature that they are operated by changes in the concentration of free calcium ions over the critical range of 10^{-7} to 10^{-5} M.

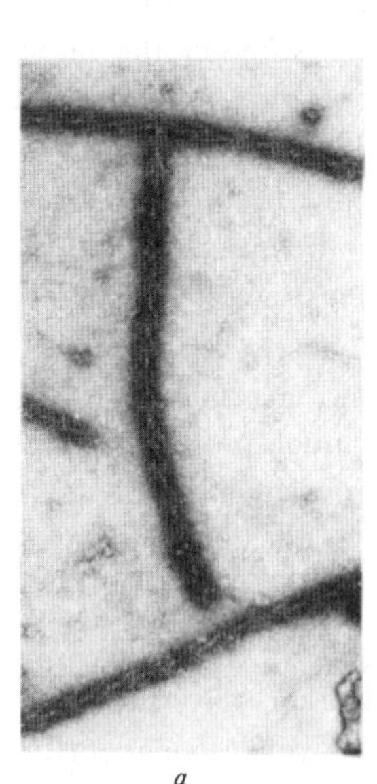

a

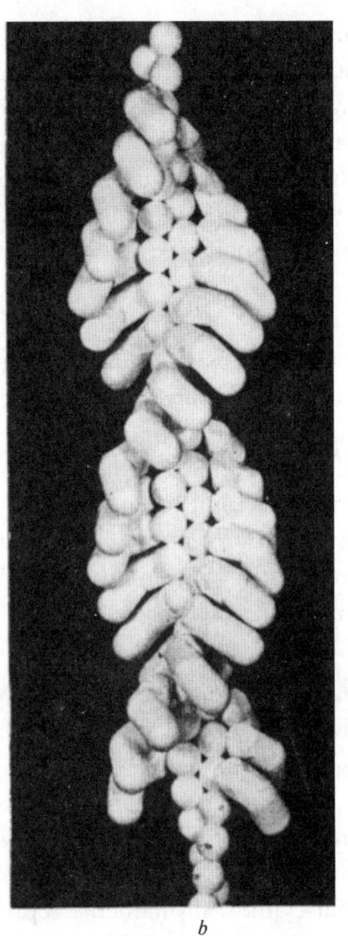

b

图 2.（*a*）经负染处理的肌动蛋白纤维被肌球蛋白S_1亚基“修饰”后的电子显微照片。图中可以非常清楚地观察到“箭头”形状（放大138,600倍）。（*b*）根据三维图像重构结果制作的被“修饰”肌动蛋白纤维的简化模型。肌动蛋白分子呈螺旋状排列，S_1亚基以一种特别的倾斜加缠绕的方式与其中心核结合。这种独特的结合方式导致了“箭头”形状的形成。

在松弛的肌肉中，由于肌质网中钙泵的作用，游离钙离子的浓度保持在一个非常低的水平（可能低于 10^{-7}~10^{-8} M）。一旦被激活，钙离子从肌质网中释放，胞内游离钙离子的浓度升高（可能升至 10^{-6}~10^{-5} M），此时肌球蛋白可以与肌动蛋白结合。在脊椎动物的横纹肌中，这种变化由肌动蛋白纤维中的原肌球蛋白–肌钙蛋白系统产生 [10]。当没有游离钙离子时，肌球蛋白与肌动蛋白的结合被抑制；而当游离钙离子存在时，可能由于原肌球蛋白的位置发生变化，引起空间位阻的改变，从而允许肌球蛋白与肌动蛋白结合。然而在软体动物的肌肉中（包括螺类的横纹肌），钙离子的调节功能是通过激活后直接与肌球蛋白结合令其发生变化来实现的 [11]。关于不同物种中这两种调节方式的分工或者协同的问题是当前非常活跃的研究领域之一 [12]。然而，在目前已经研究过的物种中，所有的系统都具有一个共同特点，即游离钙离子浓度变化的临界范围是 10^{-5}~10^{-7} M。

If one is primarily interested in the molecular mechanism of contraction, then striated muscle is the best system to work with, because of its high degree of order, because of the high concentration and relatively large amounts of the contractile proteins in it, and because of its great robustness and stability, which allow it to be manipulated with ease and fixed for electron microscope examination in a state close to its native one by relatively crude techniques. We must now consider, however, whether the same basic mechanism appears anywhere else in nature, even, in the first instance, in as closely related a system as smooth muscle.

There is no difficulty here with the kind of smooth muscles found, for example, in the adductor and retractors of bivalves; these have been shown to contain abundant thick and thin filaments[13], and the reason that striations are absent is the simple one, suggested by Jean Hanson and myself[14], that the filaments are not arranged in register. It is vertebrate smooth muscle which has until recently presented much more difficulty, as for many years only thin filaments could be seen by electron microscopy and only actin reflexions picked up by X-ray diffraction, in spite of the fact that both actin and myosin were obviously present from biochemical observations.

In the past few years, however, several groups of workers have shown that the form of the myosin component in vertebrate smooth muscles is very dependent on the physical and chemical environment of the muscle; and Lowy *et al.*[15,16] have found that with appropriate conditions before and during fixation, prominent and plentiful ribbon shaped structures can be seen amongst the actin filaments. They have also shown that there is good X-ray evidence for this form of myosin aggregate in comparable specimens before fixation. Some workers believe that cylindrical filaments represent the more natural form of the myosin assembly in smooth muscle, but I think everyone now agrees that myosin is present in smooth muscle in the form of large aggregates and that this probably represents its form when involved in contraction.

The actin component and the regulatory protein system appear to be virtually the same as in vertebrate striated muscle, and the main difference detectable in the purified proteins from smooth muscle lies in the solubility properties of the myosin, that is, its inability to form aggregates with itself at physiological ionic strength except in special conditions. We do not know the functional significance of this, but it is undoubtedly responsible for the difficulties which have occurred in detecting the myosin component by X-ray diffraction and electron microscopy. However, I do not think that there are any serious doubts now that contraction takes place in smooth muscle by the same basic molecular mechanism as in striated muscle and by an essentially similar sliding filament process, although the overall structural organization may be somewhat different.

Non-muscular Systems

Before I consider how some of the other motile systems might function mechanically, it will be helpful to review briefly some of the more recent findings on their protein

对于主要对收缩的分子机制感兴趣的人们来说，横纹肌是最理想的研究系统。因为它具有高度有序的结构、含有浓度高且相对大量的与收缩系统有关的蛋白质以及高度的稳健性和稳定性，便于人们对它进行各种操作，可以通过相对粗糙的技术固定并保持其接近天然的状态以用于电子显微镜的观察。然而，现在我们必须要考虑横纹肌的这种基本机制是否同样出现在自然界中的其他系统，与横纹肌紧密相关的平滑肌是我们要考察的第一个系统。

我们很容易便可以找出一些平滑肌的例子，比如双壳类动物的内收肌和牵缩肌。这些肌肉中也含有丰富的粗肌丝和细肌丝[13]，却没有横纹。在琼·汉森和我[14]看来，缺乏横纹的原因很简单，主要由于这些肌丝没有像在横纹肌中那样高度有序地排列。最近的研究表明，研究脊椎动物平滑肌的难度大得多。多年来，虽然生化实验的结果表明其中显然同时含有肌动蛋白和肌球蛋白，但是人们通过电子显微镜只能观察到其中的细肌丝，通过 X 射线衍射实验也只能检测到肌动蛋白的反射信号。

然而，最近几年，多个研究组的研究结果表明，脊椎动物平滑肌中的肌球蛋白的存在形式高度依赖于肌肉所处的物理和化学环境。洛伊等人[15,16]发现在固定前和固定过程中保持适当的条件，便可以观察到肌动蛋白纤维中分布着大量明显的带状结构。通过 X 射线衍射实验，他们证明了这种固定后所观察到的肌球蛋白的聚集方式与固定前的样品是一致的。过去一些研究人员认为圆柱形纤维更接近于平滑肌中肌球蛋白组装的天然形式，但是我想现在每个人都会认同平滑肌中肌球蛋白以大的聚集体的形式存在的观点。并且，这种大的聚集体可能代表着肌球蛋白参与平滑肌收缩时的状态。

平滑肌中的肌动蛋白组分以及调节蛋白系统几乎与脊椎动物横纹肌中是一样的，其主要不同是可监测到从平滑肌中纯化出的肌球蛋白的溶解度的改变，即在生理离子强度下，肌球蛋白不能自发形成聚集体，只有在特殊条件下才能自发形成。我们不知道这种性质在功能方面的意义，但是毫无疑问，肌球蛋白的这种性质使得人们很难用 X 射线衍射和电子显微镜的方法对其进行观察。虽然平滑肌与横纹肌的总体结构的组织方式有些许不同，但是我认为目前没有任何可靠的理由去怀疑，在平滑肌中发生的收缩与在横纹肌中发生的收缩具有相同的基本分子机制并经过了相似的肌丝滑动过程。

非肌系统

在考虑其他的运动体系可能的作用机理之前，很有必要首先简单回顾一下最近对这些体系的蛋白质组分的研究结果。这样做的目的并非评判许多先行者的研究，

components. This account cannot do justice to many of the pioneers, but very often technical advances have now made it possible to characterize these proteins much more fully and relate them to the proteins from muscle in a decisive way; and this latter evidence is easier to summarize in a short article.

First, consider the amoeba *Acanthamoeba castellanii*, which has the capacity for typical amoeboid movement. Weihing and Korn[17-19] have extracted an actin-like protein from this organism, which is virtually identical to muscle actin in every respect that can be investigated. It forms the same double helical filaments when examined by negative staining in the electron microscope, and these give exactly the same arrowhead structures when combined with muscle myosin, showing that the myosin-binding sites are oriented in a closely similar way[20]. The amoeba actin can activate the ATPase activity of muscle myosin, possibly to a lesser extent than muscle actin can, and this can be regulated by the tropomyosin–troponin system of rabbit muscle[21]. When the molecule is cleaved by cyanogen bromide, three large peptides can be isolated which are almost identical in amino acid composition to the three corresponding cyanogen bromide peptides from muscle actin[19]. As in muscle actin, one of these peptides contains the rare amino acid 3-methylhistidine. Additionally, amoeba actin contains one residue of ϵ-N-dimethyllysine and small amounts of ϵ-n-monomethyllysine.

Next consider the slime mould *Physarum polycephalum*, which shows very active cytoplasmic streaming. It has been shown by Hatano *et al.*[22-24] that an actin-like protein can be extracted from this organism, which shows the same double helical structure as muscle actin in the electron microscope. As in muscle actin, the monomeric form has one mol of bound ATP which is dephosphorylated to ADP when the actin polymerizes in the presence of salts. Adelman and Taylor[25] have shown that slime mould actin can activate the ATPase activity of muscle myosin, and Nachmias *et al.*[26] have shown that the actin can combine with rabbit muscle myosin subfragment 1 to give arrowheads very closely similar to those characteristic of muscle actin. Again, this shows a very specific structural relationship in the complex.

Next consider blood platelets, which undergo a kind of contraction during clot retraction. Several groups of workers[27-30] have shown that the protein known as thrombosthenin A is closely similar to muscle actin in its structure, in its ability to form arrowhead complexes with muscle myosin (usually HMM), in its ability to activate muscle myosin ATPase, in its ability to allow its activating activity to be regulated by the tropomyosin–troponin system, and in its content of 3-methylhistidine.

Again, consider nerve cells actively growing and extending in tissue culture. Fine and Bray[31] have shown that no less than 20% of all the "soluble" protein in developing chick neurones behaves like actin on an SDS gel. This protein forms the characteristic arrowheads with HMM and shows remarkable chemical similarity to muscle actin on two-dimensional electrophoresis diagrams; of fourteen methionine labelled peptides, ten coincided with ones from muscle actin, showing close homology though not necessarily

而是因为技术的进步使得我们现在可以对这些蛋白质的特征有更清楚的认识，并且以一种更明确的方式将它们与肌肉系统中的蛋白质联系起来，而且后者的证据更易于在短的文章中进行总结。

首先来了解一下可以进行典型的阿米巴运动的变形虫卡氏棘阿米巴。魏和科恩[17–19]从这种生物中分离出了一种肌动蛋白样的蛋白质，并发现其与肌肉中的肌动蛋白在研究过的各个方面都几乎完全相同。利用电子显微镜观察负染后的样品可以看到其形成了同样的双螺旋纤维，并且当其与肌肉肌球蛋白结合后也可以形成与肌肉系统完全相同的“箭头”形状，表明该肌动蛋白上肌球蛋白结合位点的结合方式也与肌肉中高度相似[20]。阿米巴肌动蛋白还可以激活肌肉肌球蛋白的 ATP 酶的活性（效果比肌肉肌动蛋白稍弱），这个过程同样可以被兔肌肉中的原肌球蛋白 – 肌钙蛋白系统所调节[21]。阿米巴肌动蛋白可以被溴化氰切割成三个大的肽段，它们与使用同样方法切割肌肉肌动蛋白所得到的多肽在氨基酸组成上几乎完全一致[19]。就像肌肉肌动蛋白一样，其中一条多肽含有稀有氨基酸残基 3– 甲基组氨酸。此外，阿米巴肌动蛋白中还含有一个 ϵ–*N*– 二甲基赖氨酸残基和少量 ϵ–*N*– 单甲基赖氨酸。

接下来我们来了解一下黏菌中的多头绒泡菌，它具有非常活跃的胞质环流。波多野等人[22–24]证明可以从这种生物中分离出一种肌动蛋白样的蛋白质，使用电子显微镜也可以观察到其具有与肌肉肌动蛋白相同的双螺旋结构。与肌肉肌动蛋白一样，这种蛋白质的单体上结合一摩尔的 ATP 分子，并且在一定盐浓度下单体发生多聚化，ATP 便会脱去磷酸水解为 ADP。阿德尔曼和泰勒[25]证明了这种黏菌肌动蛋白同样可以激活肌肉肌球蛋白的 ATP 酶活性，而纳赫米尔斯等人[26]则证明这种肌动蛋白可以与兔肌球蛋白 S1 亚基结合并形成“箭头”结构，这与肌肉中肌动蛋白的特征非常相似。这些结果再一次表明这些运动相关的蛋白质复合物具有高度特异的结构相关性。

接下来我们再了解一下可以在血液凝结过程中收缩的血小板。多个研究小组[27–30]的结果表明血栓收缩蛋白 A 与肌肉中的肌动蛋白在结构上具有很高的相似性——其可以与肌肉肌球蛋白（一般为重酶解肌球蛋白）形成“箭头”状复合物，也可以激活肌肉肌球蛋白 ATP 酶的活性，并且这种活性可以被原肌球蛋白 – 肌钙蛋白系统所调节，此外血栓收缩蛋白 A 也含有 3– 甲基组氨酸。

我们再来看看在组织培养液中活跃生长并向外延伸的神经细胞。法恩和布雷[31]证明在正在发育的鸡神经元中，至少有 20%的“可溶性”蛋白质在 SDS – 聚丙烯酰胺凝胶中的电泳行为与肌动蛋白很相似。这种蛋白质可以与重酶解肌球蛋白结合形成典型的“箭头”结构，并且在双向电泳图谱中显示出与肌肉肌动蛋白具有非常高的化学相似性——在十四个甲硫氨酸标记的肽链中有十个与肌肉肌动蛋白一致，显

complete identity.

Further examples could be given, but I think cases mentioned are sufficient to show decisively that in a number of widely different cells exhibiting different forms of motile activity, a protein closely similar to muscle actin by rather strict criteria is present in significant amounts. And in many cases, the protein can be identified in the cell *in situ*, as bundles of filaments of the appropriate diameter, which can be "decorated" by HMM, and which are in a position where they might be associated with movement[32-34].

In addition to non-muscle actins, myosin-like proteins have also been found in many of these same cells possessing motile activity, and in several instances the protein in question has been identified by rather strict criteria.

In the slime mould *Physarum polycephalum*, Hatano *et al.*[35-37] have isolated a protein which has similar ATPase and actin-combining activities to muscle myosin, the same very characteristic form (being a long rod-shaped molecule with a globular region at one end), and a similar sedimentation constant to muscle myosin (~6.05). Interestingly, the protein differs from striated muscle myosin in being soluble at physiological ionic strength. Adelman and Taylor[25] showed that its molecular weight was close to that of muscle myosin, about 460,000. Nachmias[38,39] showed that the slime mould myosin would attach in the characteristic arrowhead configuration both to slime mould actin and to muscle actin, and also[40] that in the presence of millimolar concentrations of calcium the myosin would assemble itself into rather short but essentially similar bipolar filamentous aggregates like those of muscle myosin.

In the case of blood platelets, Bettex-Gallard *et al.*[41] and Booyse *et al.*[42] have shown that the other component of thrombosthenin is a myosin-like protein, having calcium and magnesium-activated ATPase activity and a molecular weight of approximately 540,000. Adelstein *et al.*[28] have shown that this molecule contains large polypeptide chains of chain weight 200,000, just like muscle myosin, and chains of 16 to 18,000 molecular weight, analogous to the light chains of muscle myosin. The platelet myosin will form excellent arrowhead structures with actin, and will also assemble itself into the characteristic bipolar aggregates at low ionic strength.

In the case of the amoeba *Acanthamoeba*, however, Pollard and Korn[43] have found that, although a myosin like component can be identified in the sense of having ATPase activity and actin-binding ability, this protein is rather different from muscle myosin in other respects, having a much lower molecular weight (~200,000) and lacking the ability to form filaments. This particular species of amoeba may, of course, be a special case, and it should be remembered that myosin-like filaments have been seen in other amoeba (for example, in *Chaos chaos*[44,45] and in *Proteus*[46,47]).

Thus there are a number of instances where it has been demonstrated quite decisively that cells showing various forms of motility contain proteins which behave in many

示其与肌动蛋白具有很高的同源性，不过完全的一致性是不必要的。

类似的例子还可以举出很多，不过我想上面提到的这些例子已经有力地证明了在各种具有不同运动能力的细胞中，都在严格标准下存在相当数量的与肌肉肌动蛋白高度相似的蛋白质。在许多例子中，人们发现这种蛋白质能在细胞中被原位鉴定出，它们是具有适当直径的束状纤维，可以被重酶解肌球蛋白“修饰”，且这些蛋白质通常可能分布在与运动相关的位置[32–34]。

除了肌动蛋白，在许多具有运动能力的非肌细胞中还发现了肌球蛋白样的蛋白质，并且在一些例子中，研究人员是按照相当严格的标准进行鉴定的。

波多野等人[35–37]从多头绒泡菌（一种黏菌）中分离出了一种蛋白质，其具有类似肌肉肌球蛋白的 ATP 酶活性并可与肌动蛋白结合。这种蛋白质与肌肉肌球蛋白一样具有典型的形态（长棒型分子，一侧末端有一个球形头部），且沉降常数（~6.05）也相似。有意思的是，与横纹肌中的肌球蛋白不同，这种蛋白质在其生理离子强度的溶液中，处于溶解状态。阿德尔曼和泰勒[25]证明该蛋白的分子量约为 460,000，与肌肉肌球蛋白的分子量近似。纳赫米尔斯[38,39]证明这种黏菌肌球蛋白可以附着在黏菌肌动蛋白和肌肉肌动蛋白上形成典型的“箭头”结构，而且在毫摩尔浓度的钙离子存在时，这种肌球蛋白可以自组装成本质上与肌肉肌球蛋白形成的聚合物类似的双极性纤维状短聚集体[40]。

贝泰–加拉尔等人[41]以及博伊兹等人[42]发现血小板血栓收缩蛋白中的另一种组分为肌球蛋白样的蛋白质。该蛋白质具有可被钙离子和镁离子激活的 ATP 酶活性，其分子量大约为 540,000。阿德尔斯坦等人[28]证明这个分子与肌肉肌球蛋白一样，含有分子量为 200,000 的多肽链和分子量 16,000~18,000 的类似于肌肉肌球蛋白轻链的小肽链。血小板肌球蛋白可以和肌动蛋白结合形成完美的“箭头”结构，也可以在低离子强度的溶液中自组装成特征性的双极性聚集体。

然而，在对变形虫卡氏棘阿米巴的研究中，波拉德和科恩[43]发现尽管鉴定出一个肌球蛋白样的成分，它具有 ATP 酶的活性和与肌动蛋白结合的能力，但是该蛋白其他方面的特征却与肌肉肌球蛋白大为不同，其分子量只有约 200,000，远低于肌肉肌球蛋白的分子量，并且这种蛋白不能够形成纤维。当然，这种变形虫可能是变形虫的一个特例，值得注意的是，在其他的一些变形虫中已经观察到肌球蛋白样的纤维（例如在多核变形虫[44,45]和变形杆菌[46,47]中）。

综上所述，众多的例子都很确切的证明，具有不同运动形式的细胞却都含有某些在许多关键方面与肌肉中的肌球蛋白和肌动蛋白性质极其相似的蛋白质。涉及以

crucial respects exactly like actin and myosin from muscle. The resemblance is particularly significant in those properties concerned with the activating effect of actin on the ATPase activity of the myosin and with the precise structural form of the actin-myosin complex, as these are so directly linked to the contraction mechanism in striated muscle. It is difficult to believe that such close homology would exist unless the same basic mechanism was involved in each case.

Indeed, one can take this argument a stage further and suggest that the actin filament-myosin head interaction was developed as a motile mechanism very early in evolution, and that cells have been using it ever since, with a high degree of conservation of the essential protein interactions involved. As multicellular organisms developed, certain cells specialized in producing more extensive and more powerful movements, and the contractile proteins in them became organized into the large structures required to integrate the smaller scale motile processes into directed forces. But the same basic molecular mechanism was retained.

In striated muscle, it has been shown that the mechanism depends on the development of a relative shearing force between filaments of actin and myosin. Because so many of the underlying structural and biochemical features of this interaction are shared by the more primitive systems, it seems highly probable, to say the least, that these systems must also operate by an active shearing mechanism, in which sliding forces are developed between polarized actin filaments and some form of myosin assembly. For example, in cases where cytoplasmic streaming is taking place next to a stationary cortical gel layer, one might suppose that actin filaments, attached to the inner cell surface and lying approximately parallel to it, with appropriate structural polarity, generate an active shearing force by their interaction with assemblies of myosin molecules in the more fluid cytoplasm, which is therefore propelled along, carrying with it other cell organelles. If the myosin assemblies, which could be quite small, contained two sets of myosin molecules with opposite polarity (either in the form of bipolar filaments, or perhaps as "face-polar" sheets as described in smooth muscle by Lowy *et al.*[48] and Small and Squire[49]), then the second set could interact with actin filaments of appropriate polarity in solution, and propel those along too. Alternatively, the location of the actin and myosin assemblies could be interchanged.

Cell organelles could be moved more directly by means of attached actin filaments, which could interact with myosin in large or even very small assemblies and propel themselves along in a manner which I suggested some years ago[4], or by the interactions described above.

Where a cell, for instance in tissue culture, is moving over a substratum, it is clear that one part must be anchored to the support and that another part of the cell must move relative to that attachment site and form new attachment sites of its own. As attachment sites on and between cells often show filaments trailing back from them, it is natural to suggest that those filaments represent one component of an active shearing system, say, actin, and that the other components are in the cytoplasm and therefore can flow forward over

下两方面的相似是非常重要的：肌动蛋白对肌球蛋白 ATP 酶活性的激活，以及肌动蛋白－肌球蛋白复合物的精确结构，因为这两点与横纹肌的收缩机制直接相关。我们很难相信如果不是基于某种共同的机制，这些高度同源的分子如何会在每个例子中都存在。

其实，可以将上述讨论继续推进一个阶段，认为在进化早期肌动蛋白纤维－肌球蛋白头部的相互作用已经发展成为了一种运动机制，细胞自此一直使用这种机制，并使这种基本的蛋白质相互作用在进化中高度保守。随着多细胞生物的出现，某些细胞特化为能够产生更大范围更有力运动的特定细胞，其细胞内与收缩有关的蛋白质也组织成更大的结构以便把小尺度的运动整合成为定向的力量。但是，在这个过程中，同一种基本分子机制被保留了下来。

人们已经证明，在横纹肌中这种机制取决于肌动蛋白和肌球蛋白纤维之间的相对剪切力的产生。因为在这种相互作用中如此多的潜在结构和生化特性都为更多原初系统所共有，至少可以说，很有可能的是这些系统也通过一种主动剪切机制来运行，在这种机制中，在极化的肌动蛋白纤维和以某种形式组装的肌球蛋白之间产生了滑动力。例如，当胞质环流发生在稳定的皮质凝胶层附近时，我们可以假设具有适当结构极性的、附着在细胞内表面并与之大致平行的肌动蛋白纤维，可以通过与肌球蛋白聚合体的相互作用在流动性更高的细胞质中产生一个主动的剪切力，从而驱使并携带其他细胞器进行环流。当肌球蛋白聚合体（可以非常小）包含两组具有相反极性的肌球蛋白分子时（可能是以双极性纤维的形式存在，也可能以洛伊等人[48]和斯莫尔、斯夸尔[49]描述的在平滑肌中的“面极性”层的形式存在），第二组肌球蛋白也可以与细胞质中具有适当极性的肌动蛋白纤维相互作用，推动胞质环流。在上述机制中，肌动蛋白和肌球蛋白聚合体的位置可以相互轮流交换。

细胞器可以更直接地通过连接在它上面的肌动蛋白纤维实现移动。这些肌动蛋白纤维可以与或大或小的肌球蛋白聚集体相互作用，从而通过某种方式驱动它们移动（我在几年前便提出了这种方式[4]），或者通过上文所述的相互作用方式运动。

当细胞在某种基质上运动时（例如在组织培养物中），可以肯定的是，细胞中的其中一部分必须锚定在基质支持物上，而另一部分必定对这些附着位点做相对移动，并形成自己新的附着位点。由于在细胞上和细胞间的附着位点上经常可以观察到来自它们的拖尾的纤维，我们很自然的认为这些纤维是剪切系统中的某种组分，比如肌动蛋白，而另一个组分（此处指肌球蛋白，译者注）则因存在于细胞质中而得以

the attached filaments. The filament attachment sites, perhaps analogous to fragments of Z-lines, would pass through the membrane and anchor to the substratum. As the front part of the cell was pushed forward by the internal pressure generated by the cytoplasmic stream, fresh attachment sites for actin filaments might be laid down at the leading edge. As actin filaments polymerized onto these (with appropriate polarity) they would be pulled back by shearing forces developed with the cytoplasm. This might give rise to the appearance of "ruffling" and backwards flow of the unattached membrane of moving cells. When the attachment sites became anchored to the substratum, however, the cell as a whole could then move forward as before over the attachment point (Fig. 3). (See also Bray, *Nature New Biology*, in the press.)

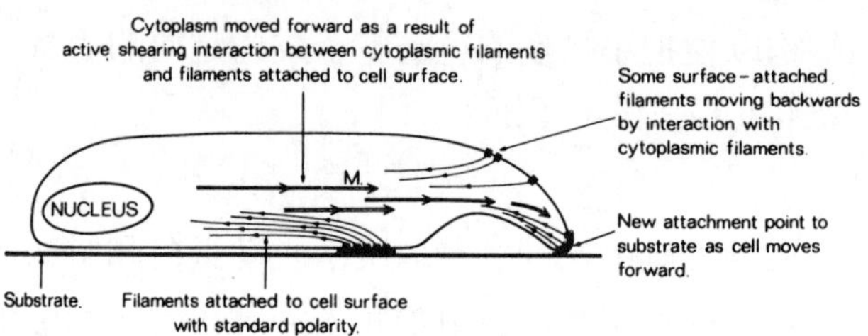

Fig. 3. Diagrammatic representation of a mechanism by which active shearing forces developed between two sets of filaments could produce cytoplasmic streaming and cell movement.

The character of encounters with other cells would depend on whether the external attachment sites on the two interacting cells could attach to each other, or only to a third, freely movable site on the opposing membrane. In the latter case, the attachment sites between cells would be drawn to their edge, sustained overlap could not occur, and indeed the entire actin complement of that region of the cell might become tied up in such junctions, leading to "contact inhibition"[50]. In the former, then, overgrowth could occur. Whilst these latter suggestions are entirely speculative, the force of the earlier arguments may make them worth considering and testing.

Another problem, about which there is very little experimental evidence at present, concerns the regulation mechanisms used in non-muscular motile systems. Because activity in muscle is controlled by small changes in calcium concentration, it would be reasonable to look for similar changes, and for the necessary calcium sequestering structures, in other motile systems. Evidence for the presence of elements of the troponin-tropomyosin system in slime mould[51] and in platelets[52] has recently been described, and evidence for the presence of calcium-sequestering structures has been described in *Spirostomum*[53] and *Physarum*[54].

Finally, I should mention that there are now numerous other instances where actin-like filaments have been implicated in cell movement, though the evidence does not have the decisive character of the examples I have given earlier. In many of these, particularly in the cases of various types of morphogenetic movement, biochemical identification of

向前流动越过附着的纤维。这些纤维的附着位点，可能与肌节中的Z线相似，会穿过细胞膜并锚定在基质上。当细胞的前端被胞质环流产生的内部压力推动向前移动时，一些新的肌动蛋白纤维附着位点也相应地在细胞前沿形成。当肌动蛋白纤维以适当的极性聚合到附着位点时，它们也会被胞质环流产生的剪切力拉向后方。这可能引发"波缘运动"，并使未与基质附着的细胞膜逆向流动。然而，一旦附着位点锚定到了基质上，细胞便可以作为一个整体向附着位点移动了（图3）。（同样可参考布雷即将发表在《自然新生物》上的论文。）

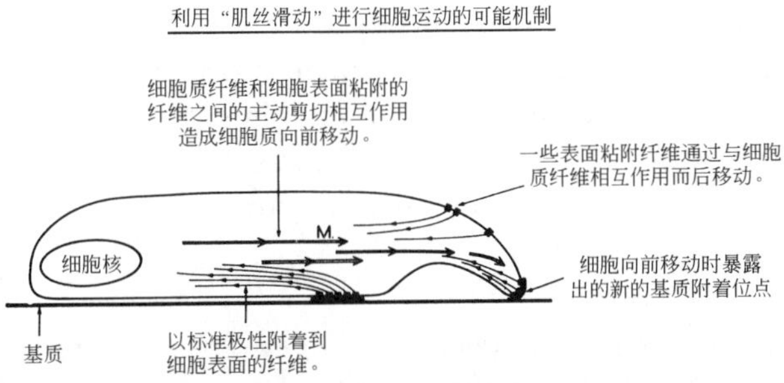

图3. 两组纤维间产生的主动剪切力引起胞质环流和细胞运动的机制。

当细胞在运动中与其他细胞相遇时，其特征取决于两个相遇细胞的外部附着点是彼此直接结合还是与对方膜上可自由移动的第三附着位点结合。在后一种情况下，细胞间的附着位点会被拖到细胞的边缘，而不会发生持续互相重叠的情况，实际上细胞该区域内的肌动蛋白补充因接触而停止，从而导致"接触抑制"[50]。而在前一种情况下，细胞则可能出现过度生长。虽然后面这些想法完全来自推断，但之前的讨论说明这些推断是值得认真考虑和进一步验证的。

另外一个问题是，目前对于非肌运动系统中的调节机理仍缺乏实验证据。由于肌肉活动是通过钙离子浓度的细微变化来调节的，所以我们理应在其他的运动系统中寻找类似的变化和螯合钙离子的必需结构。最近有人报道黏菌[51]和血小板[52]中含有原肌球蛋白–肌钙蛋白系统的组分，并且在旋口虫[53]和绒泡菌[54]中也发现了螯合钙离子的结构存在的证据。

最后，我想指出的是，尽管并不能像在我以前给出的例子一样具有明确的特征，但是现在有大量其他的证据表明肌动蛋白样纤维与细胞运动关系密切。在许多例子中，尤其是在各种形态发生运动中，要想用生化手段鉴定肌动蛋白是非常困难的。

actin will be difficult. There are, however, several instances where actin has been identified by arrowhead formation with HMM[55]. This is a very specific test for an actin-like protein (though not necessarily for its location in polymerized form unless actin-like filaments can be seen in the same position in untreated material). For example, Perry *et al.*[56] showed that a ring of filaments, just below the surface of the cleavage furrow at the newt egg, would bind HMM to give arrowhead complexes. This ring of filaments has been described by several workers on jellyfish eggs[57,58], and also on the eggs of a polychaete worm[58] and on the ciliate *Nassula*[59] as a contractile structure which, by decreasing its diameter, is responsible for at least a large part of the cleavage process. Two opposing sets of actin filament, linked by myosin, could have this property.

Tilney and Mooseker[60] have shown that filaments located in the microvilli forming the brush border of epithelial cells of chicken intestine also give arrowheads with HMM, behave like actin on SDS gels and, especially interestingly, appear clearly to be anchored to the cell membrane; thus they may aid transport by some kind of pumping action.

In summary then, there seems to be good evidence that several types of cell movement are brought about by molecular mechanisms which use proteins very similar to actin and myosin in muscle. A strong presumption exists, therefore, that active sliding or shearing mechanisms are involved. This possibility needs to be explored in each case, especially by structural and mechanical studies on the detailed processes involved.

(**243**, 445-449; 1973)

H. E. Huxley
MRC Laboratory of Molecular Biology, Hills Road, Cambridge CB2 2QH

References:

1. Jarosch, R., *Biochim. Biophys. Acta,* **25**, 204 (1957).
2. Jarosch, R., *Protoplasma*, 7, 478 (1956).
3. Kamiya, N., and Kuroda, K., *Bot. Mag.* (*Tokyo*), **69**, 544 (1956).
4. Huxley, H. E., *J. Mol. Biol.*, **37**, 507 (1963).
5. Huxley, H. E., *Science*, **164**, 1356 (1969).
6. Huxley, H. E., and Brown. W., *J. Mol. Biol.*, **30**, 383 (1967).
7. Huxley, H. E., *J. Mol. Biol.*, **37**, 507 (1968).
8. Reedy, M. K., Holmes, K. C., and Tregear, R. T., *Nature*, **207**, 1276 (1965).
9. Moore, P. B., Huxley, H. E., and DeRosier, D., *J. Mol. Biol.*, **50**, 279 (1970).
10. Ebashi, S., and Endo, M., *Prog. Biophys. Mol. Biol.*, **18**, 123 (1968).
11. Kendrick-Jones, J., Lehman, W., and Szent-Gyorgyi, A. G., *J. Mol. Biol.*, **54**, 313 (1970).
12. Lehman, W., Kendrick-Jones, J., and Szent-Gyorgyi, A. G., *Cold Spring Harbor Symp. Quant. Biol.*, **37**, 319 (1972).
13. Lowy, J., and Hanson, J., *Physiol. Rev.*, **42**, Suppl. 5, 34 (1962).
14. Hanson, J., and Huxley, H. E., *Symp. Soc. Exp. Biol.*, **9**, 228 (1955).
15. Lowy, J., and Small, J. V., *Nature*, **227**, 46 (1970).
16. Lowy, J., Poulsen, F., and Vibert, P., *Nature*, **225**, 1053 (1970).
17. Weihing, R., and Korn, E. D., *Biochem. Biophys. Res. Comm.*, **35**, 906 (1969).
18. Weihing, R., and Korn, E. D., *Biochemistry*, **10**, 590 (1971).

不过，在一些实例中，肌动蛋白通过与重酶解肌球蛋白形成箭头状结构而被鉴定出来[55]。对于肌动蛋白样的蛋白质来说，这是一个非常特异性的检验（检测它在聚合体中的位置不是必要的，除非能够在未经处理的材料的相同位置观察到肌动蛋白样纤维）。例如，佩里等人[56]证明，在蝾螈卵裂沟表面下的环状纤维可以与重酶解肌球蛋白结合并形成箭头状复合物。在水母卵[57,58]、多毛虫卵[58]和蓝色纤毛虫卵[59]的研究中，该环状纤维被认为是收缩性结构，通过缩小直径在卵裂过程中发挥主要作用。由肌球蛋白连接的两组极性相反的肌动蛋白纤维可能具有这种性质。

蒂尔尼和穆斯科尔[60]证明了鸡肠上皮刷状边缘的微绒毛中的纤维也可以与重酶解肌球蛋白结合形成箭头状结构，且其在 SDS– 聚丙烯酰胺凝胶中的特征与肌动蛋白的特征相似。尤其有意思的是，该纤维明显已锚定在细胞膜上。因此，它可能通过某种泵的机制帮助物质转运。

总之，似乎有很好的证据表明不同类型的细胞运动都是通过类似于肌肉中肌动蛋白 – 肌球蛋白系统的分子机制来产生的。因此，我们可以作出一个有力的假设，主动滑动和剪切机制也包含在细胞运动的分子机制中。这一假设还有待在各个实验中进一步验证，尤其需要对涉及的相关精细过程进行结构以及机理方面的研究。

（张锦彬 翻译；周筠梅 审稿）

19. Weihing, R., and Korn, E. D., *Biochemistry*, **11**, 1538 (1972).
20. Pollard, T. D., Shelton, E., Weihing, R., and Korn, E. D., *J. Mol. Biol.*, **50**, 91 (1970).
21. Eisenberg. E., and Weihing, R. R., *Nature*, **228**, 1092 (1970).
22. Hatano, S., and Oosawa, F., *Biochim. Biophys. Acta*, **154**, 507 (1966).
23. Hatano, S., and Oosawa, F., *J. Cell. Physiol.*, **68**, 197 (1966).
24. Hatano, S., Totsuka, T., and Oosawa, F., *Biochim. Biophys. Acta*, **140**, 109 (1967).
25. Adelman, M. R., and Taylor, E. W., *Biochemistry*, **8**, 4964 (1969).
26. Nachmias, V. T., Huxley, H. E., and Kessler, D., *J. Mol. Biol.*, **50**, 83 (1970).
27. Bettex-Gallard, M., and Lüscher, E. F., *Adv. in Protein Chem.*, **20**, 1 (1965).
28. Adelstein, R. S., Pollard, T. D., and Kuehl, W. M., *Proc. US Nat. Acad. Sci.*, **68**, 2703 (1971).
29. Zucker-Franklin, D., and Grasky, G., *J. Clin. Invest.*, **51**, 49 (1972).
30. Adelstein, R. S., and Conti, M. A., *Cold Spring Harbor Symp. Quant. Biol.*, **37** (in the press).
31. Fine, R. E., and Bray, D., *Nature*, **234**, 115 (1971).
32. Wessells, N. K., Spooner, B. S., Ash, J. F., Bradley, M. O., Ludena, M. A., Taylor, E. L., Wrenn, J. T., and Yamada, K. M., *Science*, **171**, 135 (1971).
33. Spooner, B. S., Yamada, K. M., and Wessells, N. K., *J. Cell Biol.*, **49**, 595 (1971).
34. Goldman, R. D., and Knipe, D. M., *Cold Spring Harbor Symp. Quant. Biol.*, **37**, 523 (1972).
35. Hatano, S., and Tazawa, M., *Biochim. Biophys. Acta*, **154**, 507 (1968).
36. Hatano, S., and Ohnuma, J., *Biochim. Biophys. Acta*, **205**, 110 (1970).
37. Hatano, S., and Takahashi, K., *J. Mechanochem. Cell. Motility*, **1**, 7 (1971).
38. Nachmias, V. T., and Ingram, W. C., *Science*, **170**, 743 (1970).
39. Nachmias, V. T., *J. Cell Biol.*, **52**, 648 (1972).
40. Nachmias, V. T., *Proc. US Nat. Acad. Sci.*, **69**, 2011 (1972).
41. Bettex-Gallard, M., Portzehl, H., and Lüscher, E. F., *Nature*, **193**, 777 (1962).
42. Booyse, F. M., Hoveke, T. P., Zschocke, D., and Rafelson, M. E., *J. Biol. Chem.*, **246**, 4291 (1971).
43. Pollard, T. D., and Korn, E. D., *Cold Spring Harbor Symp. Quant. Biol.*, **37** (in the press).
44. Nachmias, V. T., *J. Cell Biol.*, **23**, 183 (1964).
45. Nachmias, V. T., *J. Cell Biol.*, **38**, 40 (1968).
46. Wolpert, L., Thompson, C. M., and O'Neill, C. H., in *Primitive Motile Systems*, **143** (Academic Press, NY, 1964).
47. Pollard, T. D., and Ito, S., *J. Cell Biol.*, **46**, 267 (1970).
48. Small, J. V., Lowy, J., and Squire, J. M., *Proc. First Europ. Biophys. Congr.*, *Baden*, EX1/5 (1971).
49. Small, J. V., and Squire, J. M., *J. Mol. Biol.*, **67**, 117 (1972).
50. Abercrombie, M., *Exp. Cell Res.*, suppl., **8**, 188 (1961).
51. Tanaka, H., and Hatano, S., *Biochim. Biophys. Acta*, **257**, 445 (1972).
52. Cohen, I., and Cohen, C., *J. Mol. Biol.*, **68**, 383 (1972).
53. Ettienne, E. M., *J. Gen. Physiol.*, **56**, 168 (1970).
54. Braatz, R., and Komnick, H., *Cytobiologie*, **2**, 457 (1970).
55. Ishikawa, H., Bischoff, R., and Holtzer, H., *J. Cell Biol.*, **43**, 312 (1969).
56. Perry, M. M., John, H. A., and Thomas, N. S. T., *Exp. Cell Res.*, **65**, 249 (1971).
57. Schroeder, T. E., *Exp. Cell Res.*, **53**, 272 (1968).
58. Szollosi, D., *J. Cell Biol.*, **44**, 192 (1970).
59. Tucker, J. B., *J. Cell Sci.*, **8**, 557 (1971).
60. Tilney, L. G., and Mooseker, M., *Proc. US Nat. Acad. Sci.*, **68**, 2611 (1971).

Effect of Lithium on Brain Dopamine

E. Friedman and S. Gershon

Editor's Note

Lithium's anti-manic properties were first noted in 1949, when Australian medical officer John Cade reported its calming effects on guinea pigs and humans. 21 years later it was licensed for the treatment of bipolar disorder in the United States, and although it remains in widespread use, its mode of action remains unclear. Here Eitan Friedman and Samuel Gershon describe how treating rat brain slices with chronic lithium inhibits dopamine production. The finding correlates with the observation that chronic treatment is needed to lessen manic behavior, and suggests that dopamine metabolism is impaired in patients with bipolar disorder. Today, lithium's effects on dopamine are accepted, but other potential key players in the process have also emerged.

THE efficacy of lithium ion in the therapy of mania is now established[1,2], and much speculation about its mode of action has involved its effects on brain catecholamine metabolism. Treatment with lithium ion increases the turnover rate of whole brain noradrenaline[2-4]; increases intraneuronal while decreasing extraneuronal metabolism of noradrenaline[3,4]; increases uptake of noradrenaline by synaptosomes[5], and reduces the rate of release of ^{3}H-noradrenaline caused by electrical stimulation of striatal slices[6]. The influence of lithium ion on brain dopamine, however, is less fully documented. Here we describe the effect of chronic lithium chloride treatment on dopamine synthesis in striatal brain slices.

Male albino Sprague-Dawley rats (180 to 200 g) were housed four per cage in a regime of 12 h light and 12 h darkness and fed lab chow and water *ad lib.* Lithium chloride, dissolved in distilled water (0.7% w/v) or an equal volume of isotonic sodium chloride, was administered intraperitoneally (i.p.) as acute or chronic daily injections. Animals were killed 60 min after the last injection, the brains were quickly removed, rinsed in ice-cold Krebs–Henseleit physiological solution and the striatum was dissected out as described by Glowinski and Iversen[7]. Striatal slices (0.4 mm thick) were prepared using the slicing guide described by McIlwain[8] and placed in flasks containing 2 ml. of oxygenated cold physiological solution. Incubations were carried out at 37°C in an atmosphere of 95% O_2-5% CO_2. After a 10 min preincubation, 3,5-^{3}H-tyrosine (New England Nuclear Corp.) was added to each flask to give a final concentration of 8.15×10^{-6} M tyrosine and the incubations were continued for a further 45 min. The tissue and incubation media were rapidly separated by filtration and the tissue slices were washed twice with cold physiological solution. The tissue was homogenized in 0.4 N perchloric acid and the media and washings were acidified to give a final 0.4 normality with perchloric acid. The samples were centrifuged and cold tyrosine and dopamine were added to the supernatant as carriers. Labelled and endogenous tyrosine and dopamine were isolated on "Dowex" 50-4X columns (K^{+}

锂对于大脑中多巴胺的作用

弗里德曼，格申

编者按

锂的抗躁狂性质在1949年被首次提出，当时澳大利亚卫生官员约翰·凯德报道了它对豚鼠和人体的镇静作用。21年后，它在美国获得批准用于治疗双相情感障碍，尽管一直被广泛使用，其作用机制仍然不清楚。在本文中，埃坦·弗里德曼和塞缪尔·格申描述了如何用慢性锂处理大鼠脑片抑制多巴胺的产生。减轻躁狂行为需长期治疗，本文中的发现与这一观察结果相关，并提示双相情感障碍患者的多巴胺代谢受损。今天，人们不但认可了锂对多巴胺的影响，还揭示出了此过程中其他潜在关键参与者。

现在锂离子治疗狂躁症的疗效得到了确认[1,2]，很多关于其作用机制的推测都与它对大脑儿茶酚胺代谢的影响有关。用锂离子治疗能够增加全脑去甲肾上腺素的转换率[2–4]，增加神经细胞内去甲肾上腺素代谢的同时降低神经细胞外去甲肾上腺素的代谢[3,4]，增加突触小体对去甲肾上腺素的摄取[5]，并降低纹状体切片因电刺激引起的 ^{3}H–去甲肾上腺素释放的速率[6]。然而，关于锂离子对大脑多巴胺的影响，却较少被详细记录。我们在本文中描述长期氯化锂治疗对纹状体脑切片中多巴胺合成的影响。

白化的雄性SD大鼠（180 g到200 g），每4只一笼，光照周期为12 h光照，12 h黑暗，自由摄取食物水分。氯化锂溶解于蒸馏水（0.7% w/v）或者等体积的等渗氯化钠溶液中，急性腹腔注射或每日慢性腹腔注射。最后一次注射60分钟后处死动物，快速取出脑部，在以冰冷却的克雷布斯–亨泽莱特生理溶液中冲洗，按照格洛温斯基和艾弗森所描述的方法[7]解剖分离出纹状体。用麦基尔韦恩描述的方法[8]制备纹状体切片（厚度：0.4 mm），并置于盛有2 ml充氧冷生理溶液的烧瓶中。在含95% O_2–5% CO_2的气体中37℃进行孵育。在10分钟预培养后，向每个烧瓶中加入3,5–^{3}H–酪氨酸（新英格兰核公司），使酪氨酸终浓度达到 8.15×10^{-6} M，再继续孵育45分钟。通过过滤快速分离组织与培养液，用冷的生理溶液清洗组织切片两次。在0.4 N的高氯酸中将组织匀浆，并用高氯酸酸化培养液和洗液，使终浓度为0.4 N。离心样品，并向上清液中加入冷的酪氨酸和多巴胺作为载体。被标记的内源性酪氨酸和多巴胺用“Dowex”50–4X色谱柱（K^{+}型）进行分离，按照内夫等人[9]所描述的方法用氧化铝进行纯化。利用液体闪烁光谱方法对酪氨酸和多巴胺样品进行放射

form) followed by purification with alumina as described by Neff *et al.*[9]. Aliquots of the tyrosine and dopamine were taken for radioactive determination by liquid scintillation spectroscopy. Endogenous amines were determined fluorometrically by methods described previously[10,11]. Fluorescence was read in the Aminco Bowman spectrofluorometer. Plasma lithium concentrations were determined by atomic absorption spectrometry.

Acute injections of lithium chloride in doses of 2 to 4 mequiv kg^{-1} did not affect endogenous dopamine nor the synthesis of ^3H-dopamine in striatal slices (Table 1). Daily administration of lithium chloride, 1 and 2 mequiv kg^{-1} for 14 days, inhibited striatal dopamine synthesis by 34 and 62% respectively (Table 1). These doses caused a slight but insignificant decrease in endogenous dopamine. There was no alteration in tyrosine specific activity. Serum lithium concentrations of 0.54 to 1.7 mequiv $1.^{-1}$ resulted from the chronic treatment; these levels are within the range associated with lithium therapy. The *in vitro* addition of 4 to 18 mequiv $1.^{-1}$ of lithium ion, or the replacement of similar amounts of sodium by lithium in the media incubating striatal slices obtained from saline-treated rats, did not affect dopamine synthesis (Table 2).

Table 1. Effect of LiCl on Dopamine Synthesis in Striatal Slices

Treatment		*N**	Dose (mequiv kg^{-1})	Plasma Li^+ (mequiv $1.^{-1}$)	Dopamine (μg g^{-1})	^3H-Dopamine (c.p.m. mg^{-1})
Acute	NaCl	16	—	—	4.21±0.08	1,063±21
	LiCl	10	2	1.65	4.34±0.05	1,108±28
	LiCl	10	4	2.84	4.28±0.06	1,162±31
Chronic	NaCl	8			4.02±0.07	1,160±44
	LiCl	8	1	0.64	3.89±0.07	761±25†
	LiCl	8	2	1.52	3.82±0.05	445±24‡

* Number of animals; each provided duplicate striatal samples.
† $P<0.005$.
‡ $P<0.001$.

Table 2. Effect of *in vitro* LiCl or Replacement of Na by Li ion on Striatal Dopamine Synthesis

Concentration of Li^+ (mequiv $1.^{-1}$)	*N**	^3H-Dopamine (c.p.m. mg^{-1})
Added to control media	6	948±24
4	4	1,028±34
6	4	1,006±29
Replacement of Na	6	
4	4	959±28
6	4	1,078±31
18	4	992±32

*Number of animals; each provided duplicate striatal samples.

Thus the inhibition of dopamine synthesis by lithium ion seems to require chronic

性测定。按照以前描述的荧光方法测定内源性胺类[10,11]。用 Aminco-Bowman 荧光分光计读取荧光信号。用原子吸收光谱方法测定血浆锂浓度。

急性注射剂量为 2~4 mEq/kg 的氯化锂，既不影响纹状体切片中内源性多巴胺，也不影响纹状体切片中 ^{3}H– 多巴胺的合成（表 1)。每天给予氯化锂 1~2 mEq/kg，连续 14 天，对纹状体多巴胺合成的抑制分别为 34%和 62%（表 1)。这些剂量使得内源性多巴胺略有下降但无统计学意义。酪氨酸的比活性并无改变。长期治疗造成血清锂浓度在 0.54~1.7 mEq/L 范围；这个浓度水平在锂治疗相关浓度范围之内。在培养液中加入 4~18 mEq/L 锂离子，或者是用等量的锂取代培养液中的钠，孵育经生理盐水处理的大鼠纹状体切片，不会影响多巴胺的合成（表 2)。

表 1. LiCl 对纹状体切片中多巴胺合成的影响

处理方法		N*	剂量 (mEq/kg)	血浆 Li^+ (mEq/L)	多巴胺 (μg/g)	^{3}H– 多巴胺 (c.p.m. mg^{-1})
急性	NaCl	16	—	—	4.21 ± 0.08	1,063 ± 21
	LiCl	10	2	1.65	4.34 ± 0.05	1,108 ± 28
	LiCl	10	4	2.84	4.28 ± 0.06	1,162 ± 31
慢性	NaCl	8			4.02 ± 0.07	1,160 ± 44
	LiCl	8	1	0.64	3.89 ± 0.07	761 ± 25†
	LiCl	8	2	1.52	3.82 ± 0.05	445 ± 24‡

* 动物数量，每只提供两份相同的纹状体样品。

† $P < 0.005$

‡ $P < 0.001$

表 2. LiCl 或以锂离子取代钠离子对离体纹状体多巴胺合成的影响

Li^+的浓度 (mEq/L)	N*	^{3}H– 多巴胺 (c.p.m. mg^{-1})
加入对照培养液	6	948 ± 24
4	4	1,028 ± 34
6	4	1,006 ± 29
取代 Na	6	
4	4	959 ± 28
6	4	1,078 ± 31
18	4	992 ± 32

* 动物数量，每只提供两份相同的纹状体样品。

treatment. High lithium concentrations produced by either an acute injection or by *in vitro* addition of lithium ion are insufficient to inhibit synthesis. These conclusions correlate with the clinical observations that lithium's anti-manic action requires 1 to 2 weeks of daily medication. Our results agree with the observed decrease of dopamine excretion in manic patients undergoing lithium treatment[12]. Although the clinical results may reflect the effect of lithium on the peripheral disposition of dopamine, the data reported here suggest that central dopamine metabolism is altered in the same direction by lithium treatment.

Corrodi *et al.*[14] also found a slight but significant decrease in whole brain dopamine turnover in rats treated with lithium for 3 weeks, judged on the basis of the rate of dopamine decline observed after injection of α-methyl-*p*-tyrosine. They have further found, using the histochemical fluorescence method, an increase in tubero-infundibular dopamine turnover. On the other hand, Ho *et al.*[15] found no alteration in dopamine turnover rates in four brain regions after 4 weeks of treatment with lithium. They used the synthesis inhibition method for estimating dopamine turnover. Although differences in methods and treatment time may have contributed to the discrepancy in these results, the fact that different brain regions were studied may be more significant, especially as we have been concerned only with the corpus striatum.

Messiha *et al.* reported increased urinary levels of dopamine during manic states[12], while others have described the induction of hypomanic and manic behavior by L-dopa in depressed patients with prior episodes of mania[13]. Thus, the inhibition of striatal dopamine synthesis may play some part in the psychomotor action of lithium in mania, or this element may exert some more fundamental action on the disorder itself.

(**243**, 520-521; 1973)

Eitan Friedman and Samuel Gershon
Neuropsychopharmacology Research Unit, New York University Medical Center, 550 First Avenue, New York, New York 10016

Received January 16; revised March 9, 1973.

References:

1. Schou, M., *J. Psychiat. Res.*, **6**, 67 (1968).
2. Gershon, S., *Ann. Rev. Med.*, **23**, 439 (1972).
3. Stern, D. N., Fieve, R. R., Neff, N. H., and Costa, E., *Psychopharmacologia*, **14**, 315 (1969).
4. Schildkraut, J. J., Logue, M. A., and Dodge, G. A., *Psychopharmacologia*, **14**, 135 (1969).
5. Dolburn, R., Goodwin, F., Bunney, W., and Davis, J., *Nature*, **215**, 1395 (1967).
6. Katz, R. I., Chase, T. N., and Kopin, I. J., *Science*, **169**, 466 (1968).
7. Glowinski, J., and Iversen, L. L., *J. Neurochem.*, **13**, 655 (1966).
8. McIlwain, H., *Biochem. J.*, **78**, 213 (1961).
9. Neff, N. H., Spano, P. F., Groppetti, A., Wang, C. T., and Costa, E., *J. Pharmacol. Exp. Ther.*, **170**, 701 (1971).
10. Laverty, R., and Taylor, K. M., *Anal. Biochem.*, **22**, 269 (1968).
11. Udenfriend, S., *Molecular Biology Series* (edit. by Kaplan, N. O., and Scheraga, H. A.), 129 (Academic Press, New York, 1962).
12. Messiha, F., Agallianos, D., and Clower, C., *Nature*, **225**, 868 (1970).
13. Murphy, D. L., Brodie, H. K. H., Goodwin, F. K., and Bunney, jun., W. E., *Nature*, **229**, 135 (1971).
14. Corrodi, H., Fuxe, K., and Schou, M., *Life Sci.*, **8**, 643 (1969).
15. Ho, A. K. S., Loh, H. H., Craves, F., Hitzemann, R. J., and Gershon, S., *Europ. J. Pharmacol.*, **10**, 72 (1970).

由此看来，通过锂离子对多巴胺合成进行抑制似乎需要长期治疗。由急性注射或者体外加入锂离子所产生的高锂浓度，不足以抑制其合成。这些结论与临床观察相关：锂离子的抗躁狂作用，需要每天服药一至两周才能实现。我们的结果与所观察到的接受锂治疗的躁狂症患者多巴胺分泌减少一致[12]。尽管临床结果可能反映的是锂对多巴胺的外周处置效果，但这里报道的数据表明锂治疗使得中枢多巴胺代谢以相同的方向发生变化。

科罗迪等人[14]也发现，接受锂治疗3周的大鼠全脑多巴胺转换出现轻微但有统计学意义的下降；这是基于在注射 α–甲基–p–酪氨酸后观察到多巴胺下降速率得出的判断。运用组织化学荧光方法，他们还进一步发现下丘脑漏斗节结部位多巴胺转换率的增加。另一方面，何等人[15]发现，经过4周锂治疗后，4个脑区中的多巴胺转换率没有发生变化。他们使用合成抑制方法来估算多巴胺转换。尽管检测方法和治疗时间的不同对这些结果的差异可能有影响，实际上对不同脑区的研究可能更有意义，特别是一直以来我们只关心纹状体。

梅西亚等人报道了躁狂状态下尿液中多巴胺水平增加[12]，而其他人描述了对躁狂发作前的抑郁患者用L–多巴治疗可诱发轻躁狂和躁狂行为[13]。由此看来，纹状体多巴胺合成的抑制可能在锂对躁狂症的心理运动作用中扮演某种角色，或者这种元素对障碍本身可能产生更为基础的作用。

（王耀杨 翻译；李素霞 审稿）

L-Glutamic Acid Decarboxylase in Parkinson's Disease: Effect of L-Dopa Therapy

K. G. Lloyd and O. Hornykiewicz

Editor's Note

Parkinson's disease is caused by a malfunction of the pathways in the brain leading to the synthesis of a substance called dopamine, which occurs in two optically active forms: laevo and dextro. The former, known as L-dopa, is used as a means of treating Parkinsonism and appears to have a beneficial effect on the development of symptoms of the disease. This paper is an attempt to unravel what happens when L-dopa is administered to patients. That it is inconclusive is perhaps not surprising, given that the problem is still unresolved.

IN Parkinson's disease severe decreases occur in striatal L-dopa decarboxylase, putaminal synaptosomal uptake of dopamine, and dopamine and homovanillic acid concentrations in the nigro-striato-pallidal complex[1-4]. In addition, L-glutamic decarboxylase (GAD) activity is decreased in the caudate nucleus[5]. The experiments described here were performed to further evaluate this finding and to determine the effect (if any) of L-dopa therapy.

Human or rat brains were dissected and stored on dry ice as previously described[6]. Human material was obtained at autopsy from either control (no known neurological disorder) or Parkinsonian patients. Nine Parkinsonian patients received L-dopa until they died; two others did not receive L-dopa therapy. The patient groups were well matched with respect to age (means: control=68 yr; Parkinsonian=67 yr), sex (controls: ten males and three females; Parkinsonian: nine males and two females) and interval between death and freezing of the tissue (4–21 h for the controls, and 4–17 h for the Parkinsonian patients). None of these parameters had any apparent influence on GAD activity[4].

For the animal experiments, male albino Wistar rats (250–275 g) were fed an aqueous suspension of L-dopa by way of a stomach tube on the following regimen: days 1–37, 50 mg per rat per day; days 38–88, 100 mg per rat per day; days 89–109, 2×100 mg per rat per day. Controls were fed the suspension fluid (water) according to the same schedule. Animals were killed at days 37, 88 and 109. The striata were immediately removed and frozen.

GAD activity was estimated as described for L-dopa decarboxylase[6]. In brief, brain tissue was homogenized in ice-cold isotonic dextrose and incubated at *p*H 7.0 (37°C, 60 min) in the presence of pyridoxal phosphate (0.6 mM) and L-glutamic acid (2.5 mM containing 0.4 μCi DL-glutamic acid-1-^{14}C). The carbon dioxide formed was trapped

帕金森氏症中的L–谷氨酸脱羧酶：L–多巴的治疗效果

劳埃德，霍尼基维茨

编者按

帕金森氏症是由于大脑中被称作多巴胺的一种物质的合成通路出现功能障碍所致，多巴胺以两种光学活性形式存在——左旋和右旋。前者，就是我们所说的L–多巴，用于帕金森氏症的治疗，并且对疾病症状的发展产生有益的影响。这篇文章试图阐明患者服用L–多巴后的反应。由于该问题仍未解决，此处未得出结论也就不足为奇了。

在帕金森氏症中，纹状体L–多巴脱羧酶，壳核突触小体对多巴胺的摄取以及黑质–纹状体–苍白球复合体的多巴胺和高香草酸浓度均严重减少[1-4]。此外，尾状核中的L–谷氨酸脱羧酶（GAD）的活性也降低[5]。此实验的目的就是要进一步评估这个研究发现，并确定L–多巴的治疗效果（如果有效的话）。

按照以前描述的方法，将人或大鼠的脑取出并保存于干冰中[6]。人体材料来自对照（不患有已知的神经性疾病）或者帕金森氏症患者的尸体。9位帕金森氏症患者在死亡前均一直接受L–多巴治疗，另外2例没有接受L–多巴治疗。两组患者在年龄（平均年龄：对照患者 = 68岁，帕金森氏症患者 = 67岁），性别（对照患者：10男3女，帕金森氏症患者：9男2女）以及从死亡到组织冷冻之间的时间间隔（对照患者为4~21小时，帕金森氏症患者为4~17小时）等方面都匹配得非常好。上述参数对GAD活性都没有任何明显的影响[4]。

对于动物实验，雄性白化Wistar大鼠（250~275 g）通过胃管喂食L–多巴的水性悬浊液，具体方案如下：第1~37天，每天每鼠50 mg；第38~88天，每天每鼠100 mg；第89~109天，每天每鼠2 × 100 mg。根据同样的方案，用空白悬浊液（水）喂养对照组。分别在第37，88和109天时处死动物，立即将纹状体取出并进行冷冻。

按照以前描述过的测定L–多巴脱羧酶的方法来评估GAD的活性[6]。简言之，将脑组织在冰冷的等渗葡萄糖溶液中匀浆，在含有磷酸吡哆醛（0.6mM）和L–谷氨酸（2.5 mM，含有0.4 μCi的DL–谷氨酸–1–^{14}C），pH 7.0的溶液中进行孵育（37℃，

in hyamine hydroxide and assayed for radioactivity. Blanks contained *p*-bromo-*m*-hydroxybenzyloxyamine, a potent inactivator of pyridoxal phosphate[7].

The GAD activity in the striatum and substantia nigra of the non-dopa treated Parkinsonian patients was less than 50% of the control mean (Table 1), in agreement with a previous report[1]. In other brain regions GAD activity was within the range of the control values. In the brains of patients who received L-dopa for 8 months or less, GAD activity was significantly lower than in the controls (Table 1). In the caudate nucleus, putamen, substantia nigra, and globus pallidus of patients who received L-dopa continuously for 1 yr or longer, however, GAD activity was within the range of the controls, being significantly greater than that of patients treated for a shorter time. The GAD activity in other brain regions was only slightly higher in patients with the more prolonged L-dopa treatment.

Table 1. Activity of L-Glutamic Acid Decarboxylase in Discrete Brain Regions of Control Patients and Patients with Parkinson's Disease

Caudate nucleus	Putamen	Pallidum externum	Substantia nigra	Temporal cortex
A, Control patients				
1,318±185 (13)	1,243±220 (13)	1,106±306 (13)	1,273±297 (13)	1,024±154 (13)
B, Patients with Parkinson's disease 1. No L-dopa therapy				
641 (2)	583 (2)	776 (2)	526 (2)	663 (2)
2. L-Dopa therapy for 8 months or less (average dose: 3 g day^{-1})				
339±53 (4)*	249±24 (4)*	504±50 (4)	300±102 (4)†	292±101 (3)*
3. L-Dopa therapy for 1 yr or longer (average dose: 4 g day^{-1})				
1,172±173 (5) ‡	887±95 (5)§	1,210±109 (5)§	1,210±110 (4)§	407±67*

Patients on L-dopa received the drug for different periods in daily oral doses ranging from 2 to 6 g. Enzyme activity (mean±s.e.m.) is expressed as nmol carbon dioxide produced per 100 mg of protein per 2 h. (Number of cases in parentheses.) Brain tissue was homogenized in ice-cold isotonic dextrose and incubated for 20 min (37°C) in the presence of phosphate buffer (*p*H 7.0) and 0.6 mM pyridoxal phosphate. L-Glutamic acid was then added (final concentration of 2.5 mM containing 0.4 μCi DL-glutamic acid-^{14}COOH) and after 2 h the reaction was terminated by addition of acid. The evolved carbon dioxide was trapped in hyamine hydroxide and then assayed for radioactivity.

* Significantly different from group *A* ($P<0.01$).

† Significantly different from group *A* ($P<0.001$).

‡ Significantly different from group *B*, 2 ($P<0.05$).

§ Significantly different from group *B*, 2 ($P<0.001$).

In the rats treated chronically with oral L-dopa, after administration of L-dopa for 37 and 38 days the striatal GAD activity was not different from control (Table 2). After 109 days, however, striatal GAD activity was significantly greater than in either those rats which had received lower doses of L-dopa for a shorter period, or the controls. Thus, it was possible to replicate, in controlled laboratory conditions, the increase in striatal GAD activity observed in Parkinsonian patients chronically treated with L-dopa.

60 分钟)。用氢氧化甲基苄氧乙胺吸收生成的二氧化碳，并检测其放射性。空白中含有 *p*–溴–*m*–羟基苯甲氧基羟胺，是磷酸吡哆醛的强力灭活剂 [7]。

在未接受多巴治疗的帕金森氏症患者的纹状体和黑质中，GAD 活性低于对照平均值的 50%（参见表 1)，与之前的报道一致 [1]。在其他脑区中，GAD 活性处于对照组的范围之内。在接受 L–多巴治疗等于或少于 8 个月的患者脑中，GAD 活性明显低于对照组（参见表 1)。对于连续接受 L–多巴治疗 1 年及以上的患者，其尾状核、壳核、黑质和苍白球中 GAD 活性处于对照组的活性范围之内，显著高于接受治疗时间较短的患者。接受更长时间 L–多巴治疗的患者其他脑区 GAD 活性只是略高一点而已。

表 1. 对照患者与帕金森氏症患者不同脑区中 L– 谷氨酸脱羧酶的活性

尾状核	壳核	苍白球外层	黑质	颞叶皮质
A，对照患者				
1,318 ± 185(13)	1,243 ± 220(13)	1,106 ± 306(13)	1,273 ± 297(13)	1,024 ± 154(13)
B，帕金森氏症患者 1．未接受 L–多巴治疗				
641(2)	583(2)	776(2)	526(2)	663(2)
2．L–多巴治疗等于或不足八个月（平均剂量：每天 3 g）				
339 ± 53(4)*	249 ± 24(4)*	504 ± 50(4)	300 ± 102(4)†	292 ± 101(3)*
3．L–多巴治疗一年或更久（平均剂量：每天 4 g）				
1,172 ± 173(5) ‡	887 ± 95(5) §	1,210 ± 109(5) §	1,210 ± 110(4) §	407 ± 67*

不同治疗周期的患者的 L–多巴日口服剂量为 2~6 g。酶活性（平均值 ± 标准差）以每 2 小时 100 mg 蛋白质产生的二氧化碳纳摩尔数表示（括号中为案例数）。脑组织在冰冷的等渗葡萄糖溶液中匀浆，在含有磷酸盐缓冲液（pH 7.0）和 0.6 mM 磷酸吡哆醛的溶液中孵育 20 分钟（37℃)。之后加入 L– 谷氨酸（终浓度为 2.5 mM，含 0.4 μCi 的 DL–谷氨酸 –^{14}COOH〔译注：即具有放射活性的消旋谷氨酸进行示踪，引入放射性原子为 ^{14}C，取代位置是羧基碳原子，放射性强度为 0.4 μCi〕)，2 小时后加入酸终止反应。用氢氧化甲基苄氧乙胺吸收生成的二氧化碳，并检测其放射活性。

* 与 A 组结果相比，差异有统计学意义（$P<0.01$）

† 与 A 组结果相比，差异有统计学意义（$P<0.001$）

‡ 与 B，2 组结果相比，差异有统计学意义（$P<0.05$）

§ 与 B，2 组结果相比，差异有统计学意义（$P<0.001$）

在接受 37 和 38 天 L–多巴治疗后，长期口服 L–多巴的大鼠纹状体 GAD 活性与对照组没有差异（表 2)。但是，经过 109 天治疗后的大鼠纹状体 GAD 活性显著高于那些接受较短时间较低剂量 L–多巴治疗的大鼠或者对照组。由此看来，在可控制的实验室条件下，长期接受 L–多巴治疗的帕金森氏症患者纹状体 GAD 活性增加是可重复的。

Table 2. Effect of Chronic Oral L-Dopa Administration on the GAD Activity of the Rat Striatum

Duration of L-dopa administration (days)	L-Glutamic acid decarboxylase activity		
	Mean±s.e.m.	No. of animals	% of control
Control	7,399±950	7	100.0
37 days	7,360±853	7	99.5
88 days	7,012	2	94.8
109 days	11,310±839	5	152.8*†

Rats were fed an L-dopa suspension (on the regimen described in detail in the text) by means of a metal stomach tube at 1600 h (days 1–88) or at 0800 and 2000 h (days 89–109). Enzymic activity expressed as nmol carbon dioxide produced per 100 mg of protein per 2 h. Age-matched sham-treated control animals were killed with each experimental group and GAD values did not differ between groups; hence all control values were combined. Assay conditions as in Table 1.

* Significantly different from controls, $P<0.002$.

† Significantly different from 37+88 day groups, $P<0.005$.

The GAD activity of the substantia nigra and globus pallidus, but not striatum, in Parkinsonism was reported to be subnormal compared with the frontal cortex[8,9]. Absolute GAD activity was, however, decreased throughout the brain compared with controls. Moreover, there was no indication of whether or not L-dopa had been administered. We find that in Parkinson's disease there is a definite decrease in the activity of striatal GAD and that its amelioration by L-dopa therapy is dependent upon the daily dose of L-dopa and the duration of its administration. This implies that the striatal GAD-L-dopa relationship depends on the concentration of dopamine present. In fact, in rats the increase in striatal GAD induced by L-dopa was dose-dependent. In addition, the L-dopa-treated patients with highest GAD activities were treated for longer (1 year or longer) and received, as a group, a higher dose of L-dopa (Table 1). It has also been shown that the striatal dopamine concentration of chronically L-dopa treated patients is markedly higher than in non-dopa treated patients[5,12,13]. Correspondingly, the more chronically treated group of Parkinsonian patients had an average concentration of striatal dopamine (caudate nucleus: 1.6 $\mu g\ g^{-1}$) distinctly higher than in the group treated for 8 months or less (caudate nucleus: 0.7 $\mu g\ g^{-1}$).

What is the mechanism of the L-dopa-induced increase in GAD activity? Although large diencephalic lesions in experimental animals decrease GAD activity in the substantia nigra and globus pallidus[8,10], no such lesions are apparent in Parkinson's disease[11]. Thus the decrease in GAD activity in Parkinson's disease may be secondary to the severe degeneration of the nigrostriatal dopamine pathway. These GAD-containing neurones could be under a continuous "trophic" influence of the dopaminergic system; degeneration of the dopaminergic pathway might then result in a biochemical "atrophy" of the GAD neurones. Replenishment of striatal dopamine by L-dopa[4,14] possibly reverses the biochemical atrophy of the GAD-containing neurones, a process that may well require prolonged administration of L-dopa.

表 2. 长期口服 L–多巴对大鼠纹状体 GAD 活性的影响

给予 L– 多巴的持续时间（天）	L– 谷氨酸脱羧酶活性		
	平均值 ± 标准误	动物数量	与对照的百分比
对照	7,399±950	7	100.0
37 天	7,360±853	7	99.5
88 天	7,012	2	94.8
109 天	11,310±839	5	152.8*†

通过金属胃管给大鼠喂以 L–多巴悬浊液（文中已具体描述了施用方案），给药时间为 16:00 点（第 1~88 天）或者 08:00 点与 20:00 点（第 89~109 天）。酶活性以每 2 小时内 100 mg 蛋白质所产生的二氧化碳纳摩尔数表示。处死与每个实验组年龄匹配的假饲对照动物后，发现各组间 GAD 值没有差异；因此将全部对照组数据进行合并。分析条件如表 1。

* 与对照组相比，差异具有统计学意义，$P<0.002$。

† 与 37＋88 天实验组相比，差异具有统计学意义，$P<0.005$。

据报道，帕金森氏症患者的黑质和苍白球的 GAD 活性低于额叶皮质，而纹状体的结果并非如此 [8,9]。不过，与对照组相比，整个脑中的绝对 GAD 活性下降。此外，也没有迹象表明患者是否服用过 L–多巴。我们发现，在帕金森氏症中纹状体 GAD 活性有明确降低，并且 L–多巴治疗效果取决于每天使用的剂量与持续时间。这意味着，纹状体 GAD 与 L–多巴的关系取决于多巴胺的实际浓度。实际上，由 L–多巴引起的大鼠纹状体 GAD 增加是剂量依赖性的。另外，在接受 L–多巴治疗的患者中，GAD 活性最高的患者（作为一个组）接受的治疗时间较长（1 年或更久），而且服用 L–多巴的剂量也较大（参见表 1）。还可以看出，接受长期 L–多巴治疗的患者纹状体多巴胺浓度显著高于未接受多巴治疗的患者 [5,12,13]。相应的，接受治疗时间更长的帕金森氏症患者组纹状体多巴胺的平均浓度（尾状核：1.6 μg/g）明显高于接受 8 个月或更短时间治疗的患者组（尾状核：0.7 μg/g）。

由 L–多巴诱导 GAD 活性增加的机制是什么？尽管实验动物的间脑大面积损伤会降低黑质和苍白球的 GAD 活性 [8,10]，但帕金森氏症却没有出现上述损伤 [11]。因此，帕金森氏症的 GAD 活性降低，可能是继发于黑质纹状体多巴胺通路严重退化。这些含有 GAD 的神经元可能受多巴胺能系统的持续“营养”影响；多巴胺能通路的退化可能导致 GAD 神经元的生化性“萎缩”。通过 L–多巴补充纹状体多巴胺 [4,14]，有可能逆转含 GAD 神经元的生化性萎缩，这一过程很可能需要长期服用 L–多巴。

Dopamine might act by removing a repression of GAD synthesis. This is an attractive but possibly incomplete explanation because of the long latency before striatal GAD activity increases. Relatively high concentrations of striatal dopamine may be needed, a process requiring a considerable time. This is supported by the lengthy period of L-dopa administration required to increase striatal GAD activity in rat brain.

An alternative hypothesis is an enhanced synthesis of GAD apoenzyme in compensation for decreased cofactor availability as a result of Schiff base formation with L-dopa[15,16]. Such an *in vivo* formation of Schiff base would not be reflected *in vitro* where an optimal concentration of pyridoxal phosphate is added. Signs of vitamin B_6 deficiency are not, however, apparent during chronic L-dopa therapy, making this a rather unlikely hypothesis.

The increase in striatal GAD activity during chronic L-dopa therapy may have an important clinical correlate as it conspicuously parallels the ameliorative effect of L-dopa on Parkinsonian tremor (which may take from several weeks to months to develop[17]). So far the anti-tremor effect of L-dopa has been without any known neurochemical correlate; our observations suggest that a GABA-containing neurone system may be involved. In this context, a new GABA-like compound has recently been reported to exert a specific anti-tremor effect in Parkinsonian patients[18].

The *p*-bromo-*m*-hydroxybenzyloxyamine was supplied by Dr J. M. Smith jun., of Lederle Laboratories, Pearl River, New York. This work was supported by the Clarke Institute of Psychiatry and Eaton Laboratories, Norwich, New York.

(**243**, 521-523; 1973)

K. G. Lloyd and O. Hornykiewicz

Department of Psychopharmacology, Clarke Institute of Psychiatry, and Department of Pharmacology, University of Toronto, Toronto, Ontario

Received February 14; revised March 18, 1973.

References:

1. Hornykiewicz, O., in *Handbook of Neurochemistry* (edit. by Lajtha, A.), 7, 465 (Plenum, New York, 1972).
2. Lloyd, K. G., and Hornykiewicz, O., *Science*, **170**, 1212 (1970).
3. Lloyd, K. G., and Hornykiewicz, O., *Fifth International Congress on Pharmacology* (Abstract) (1972).
4. Lloyd, K. G., thesis, Univ. Toronto (1972).
5. Bernheimer, H., and Hornykiewicz, O., *Arch. Exp. Path.*, **243**, 295 (1962).
6. Lloyd, K. G., and Hornykiewicz, O., *J. Neurochem.*, **19**, 1549(1972).
7. Perkinson, E. N., and DaVanzo, J. P., *Biochem. Pharmacol.*, **17**, 2498 (1968).
8. McGeer, E. G., McGeer, P. L., Wada, J. A., and Jung, E., *Brain Res.*, **32**, 425 (1971).
9. McGeer, P. L., McGeer, E. G., and Wada, J. A., *Neurology*, **21**, 1000 (1971).
10. Kim, J. S., Bak, I. J., Hassler, R., and Okada, Y., *Exp. Brain Res.*, **14**, 95 (1971).
11. Greenfield, J. G., *Neuropathology*, 530 (Arnold, London, 1958).
12. Davidson, L., Lloyd, K. G., Dankova, J., and Hornykiewicz, O., *Experientia*, **27**, 1048 (1971).

多巴胺可能通过解除GAD合成抑制起作用。虽然这个解释很有吸引力，但考虑到纹状体GAD活性增加前有很长的潜伏期，这个解释可能还是不完整。纹状体多巴胺可能需要达到相当高的浓度，而这个过程需要相当长的时间。大鼠脑纹状体GAD活性增加需要长期服用L–多巴的事实支持这一观点。

另外一种假说是，增强GAD脱辅基酶的合成弥补了由于L–多巴合成席夫碱而导致可用的辅因子减少[15,16]。这种在“体内”形成的席夫碱的过程，在体外实验中即使加入最适浓度的磷酸吡哆醛后也不能反映出来。然而，L–多巴长期治疗并未导致维生素B_6缺乏的体征，使得这一假说不可能成立。

在L–多巴长期治疗期间，纹状体GAD活性增加可能与临床有重要的关联，因为它与L–多巴明显改善帕金森氏症患者颤抖呈现显著的平行关系（这可能要花上几个星期至几个月的时间[17]）。目前为止，L–多巴的抗震颤作用与任何已知的神经化学物质都不相关；我们的观察表明这一过程可能会涉及含γ–氨基丁酸（GABA）的神经元系统。基于这一说法，最近报道了一种新的GABA样化合物，其对帕金森氏症患者具有特异性的抗震颤作用[18]。

p–溴–*m*–羟基苯甲氧基羟胺由纽约珀尔里弗莱德利实验室的小史密斯博士所提供。这项研究由纽约诺威奇克拉克精神病研究所和伊顿实验室提供支持。

（王耀杨 翻译；李素霞 审稿）

13. Rinne, U. K., Sonninen, V., and Hyyppa, M., *Life Sci.*, **10**, (I) 549 (1971).

14. Lloyd, K. G., and Hornykiewicz, O., in *Treatment of Parkinsonism* (edit. by Calne, D. B.) (Raven Press, New York, 1973).

15. Schott, H. F., and Clark, W. G., *J. Biol. Chem.*, **196**, 449 (1952).

16. Kurtz, D. S., and Kanfer, J. N., *J. Neurochem.*, **18**, 2235 (1971).

17. Barbeau, A., *J. Canad. Med. Assoc.*, **101**, 791 (1969).

18. Curci, P., and Prandi, G., *Rev. Farmacol. Terap.*, **111**, 197 (1972).

Neutral Mutations

B. Charlesworth

Editor's Note

Mutations are changes in the structure of a gene brought about by the replacement of one nucleotide by another. Motoo Kimura, a Japanese geneticist, argued strongly in a book with his student Tomoko Ohta that most mutations that occur in living things have no effect on the fitness of the organism and so are neutral in the process of natural selection. In mid 1970s this was a controversial idea, but experience has shown it to be valid in many circumstances.

*THEORETICAL Aspects of Population Genetics**. This is an excellent book, which should be read by everyone interested in population genetics. The authors are two of the world's outstanding theoretical geneticists, and this book is essentially an account of their contributions to the field over the past five years or so. It is admirably clear and concise; the non-mathematically minded are catered for by the relegation of most of the mathematical proofs to an appendix. In chapter 1, the problem of the fixation of mutant genes is treated; this is used in chapter 2 in discussing the interpretation of observations on the rate of evolution as measured from protein sequence data. Later chapters deal with the concept of effective population size, the theory of genetic load, two-locus problems, the maintenance of variability in populations, and the adaptive significance of sex.

Kimura and Ohta believe that most nucleotide substitutions in evolution are due to the random fixation of selectively neutral alleles, and that protein polymorphisms revealed by electrophoresis merely represent a transient phase of this molecular evolution. Most of the mathematical results described in this book have been developed in order to show that this theory can account for the known facts of protein variation and evolution. Kimura and Ohta do a very good job of presenting this view, and only the most dyed-in-the-wool pan-selectionist can fail to be impressed. Nevertheless, there are some facts which are hard to fit into a neutral scheme of things, notably Prakash and Lewontin's discovery of non-random associations between inversions and protein variants in *Drosophila pseudoobscura.* It is a pity that these observations are not discussed in this book.

Many criticisms have, of course, been levelled against the authors' arguments for neutral mutations. In this book, most of them are dealt with quite convincingly. There does, however, seem to be an inconsistency which is hard to overcome. Kimura and Ohta show that, on the neutral mutation theory, the rate of amino acid substitution in evolution is equal to the rate of origin by mutation of new alleles affecting protein structure. If the data from proteins which have been sequenced are taken as representative, the average

---

* By M. Kimura and T. Ohta. Pp. ix+219. (Princeton University: Princeton, New Jersey, 1971.) $12.50.

中性突变

查尔斯沃思

编者按

突变是一种核苷酸被另一种核苷酸置换而造成的基因结构的改变。日本遗传学者木村资生与他的学生太田朋子在书中强有力地论证了：生物中的绝大多数突变都不会对生物体的适应度造成影响，这些突变在自然选择过程中是中性的。尽管在20世纪70年代中期这是一个很有争议的观点，但是实验显示它在多数情况下都是正确的。

《群体遗传学理论》* 是一本极好的图书，每一个对群体遗传学感兴趣的读者都应该阅读它。这本书的作者是世界上两位杰出的理论遗传学家，这本书主要讲述了他们过去五年里在这个领域所做出的贡献。本书内容非常清晰、精炼，大多数的数学证据都降低了难度，使数学手段成为附属工具，从而迎合了非数学思维模式。第1章处理了突变基因的固定的问题；第2章对进化速率观测结果的阐释进行讨论，这用到了第1章的内容，其中进化速率是根据蛋白质序列数据测得的。接下来的章节阐述了有效群体大小的概念、遗传负荷理论、双基因座问题、群体多样性的维护以及性别的适应意义。

木村和太田认为进化中的大多数核苷酸置换是由于可选择的中性等位基因的随机固定，通过电泳显示出来的蛋白质的多态现象仅仅描绘了这个分子进化的一个瞬间相。为了说明这个理论可用来解释蛋白质的变异和进化这些已知的事实，书中描述的大部分数学结果都有所改进。木村和太田非常好地提出了这个观点，只有极少数顽固的泛自由选择论者才能够不为所动。然而，还存在一些不符合中性进化理论框架的事实，特别是普拉卡什和列万廷发现的拟暗果蝇中倒位和蛋白质变异之间存在的非随机关系。令人遗憾的是这些观察结果并没有在这本书中被讨论。

许多批判不可避免地将矛头指向了作者的中性突变论点。在这本书中，大多数对立观点都被令人信服地处理了。然而，似乎还是存在一个很难克服的矛盾。木村和太田表示，中性突变理论认为，在进化中氨基酸被取代的速率与新的等位基因突变影响蛋白质结构的速率相等。如果将蛋白质的序列信息认为是具有代表性的话，

* 作者是木村资生和太田朋子。ix+219页。（普林斯顿大学：普林斯顿，新泽西，1971年。）12.50美元。

rate of mutation to neutral alleles per locus per generation can be computed. Kimura and Ohta find that this rate is less than one-tenth the order of magnitude of the mutation rates per locus measured experimentally in higher organisms. These mutation rates are based almost exclusively on rates of mutation to deleterious alleles. They therefore conclude that "this suggests that the neutral mutations constitute a rather small fraction of the total mutations".

Now the rate of neutral mutation for the whole genome can be estimated by multiplying the rate per nucleotide site by the total number of sites in the genome, calculated from the DNA content of sperm. For man, Kimura and Ohta estimate that between sixty and seventy-five neutral mutations occur per genome per generation. They argue that this means that "nucleotide substitution has an appreciable effect on fitness in only a small fraction of DNA sites", otherwise "the mutational load must be unbearably high for human populations".

This is a puzzling contradiction, which is difficult to resolve without serious damage to some other parts of the case for neutral mutations. For example, if one assumes that much of the DNA is functionless, the argument that most amino acid substitutions cannot be adaptive (because there would otherwise be too high a substitutional load) loses its force. Perhaps Kimura and Ohta would argue that observed mutation rates are biased in favour of loci which are known to have mutated and which therefore must have higher than average mutation rates. This can be tested by measuring mutation rates for electrophoretically detectable loci. Some data of this sort have been published by Mukai and by Kojima, and these agree quite well with usual mutation rates.

I would emphasize again that this is an important and well-written book. The question of the possible selective neutrality of most protein variation is one of the most interesting in contemporary biology. Kimura and Ohta have greatly refined the mathematical tools needed for tackling this problem, and have made us all much more critical in our attitude towards evidence in favour of selection. This book is a valuable account of their work.

(**243**, 551-552; 1973)

我们就可以计算出每代每个基因座中性等位基因的平均突变率。木村和太田发现这个速率比通过实验在高等生物上测量出的每个基因座的突变率的十分之一还低。这些突变率几乎是完全以有害等位基因的突变率为基础的。他们因此得出结论“这些结果证明中性突变在全部突变中只占相当小的一部分”。

现在，可以根据精子 DNA 含量计算基因组核苷酸位点总数，再通过每个核苷酸位点的突变率与基因组核苷酸位点总数的乘积，估算出整个基因组的中性突变率。根据木村和太田的估计，人的每个基因组每代会出现 60 到 75 个中性突变。他们认为这意味着“核苷酸置换仅在小部分的 DNA 位点上才能显著地影响适应度”，否则“人类种群的突变负荷就会高得不可忍受”。

要想解决这个令人迷惑的矛盾而不严重破坏中性突变理论的其他方面是十分困难的。例如，如果假设大多数 DNA 都是没有功能的，那么关于大多数氨基酸置换不能被适应（因为那将出现过高的替代负荷）的论点就失去了它的说服力。也许木村和太田会认为观察到的突变率都偏好于一些已知突变的位点，因此它们本身就具有高于平均水平的突变率。这可以通过检测可被电泳检测到的基因座的突变率来测定。这方面的一些相关数据向井和小岛已经发表，并且与通常的突变率相当吻合。

我想再次强调一下，这是一本非常重要且写得很好的书。多数蛋白的可选择的中性突变是当代生物学界最有意思的问题之一。木村和太田很好的优化了解决这个难题所需要的数学工具，并且使我们所有人对支持选择理论的证据产生了更为批判的态度。这本书是对他们工作的一个有价值的报道。

（姜薇 翻译；梁前进 审稿）

Further Evidence of Lower Pleistocene Hominids from East Rudolf, North Kenya, 1973

R. E. F. Leakey

Editor's Note

After six years of work at East Rudolf, Richard Leakey and his team had found evidence for two kinds of hominid. The first was a robust australopithecine found throughout the deposits and showing relatively little change. The second was a primitive but more evolutionarily variable form of *Homo*, perhaps exemplified by the spectacular skull "1470" Leakey described in 1973. This latest report suggested a hominid of a third kind, with *Homo*-like dentition but a smaller cranium, perhaps similar to the "gracile" australopithecines from Sterkfontein. The increasing wealth of hominid remains, the disputes over attributions to *Homo habilis* and australopithecines, together with the possibility of marked sexual dimorphism, only served to deepen the mysteries of human origin.

Twenty new hominid specimens were recovered from the East Rudolf area in 1973. New evidence suggests the presence of at least three hominid lineages in the Plio-Pleistocene of East Africa.

THIS is a report of the 1973 field season at East Rudolf, Kenya, where the East Rudolf Research Project (formerly Expedition) has now concluded its sixth year of operations. Eighty-seven specimens of fossil hominid were collected[1] from the area during 1968–72; a further twenty specimens were recovered between June and September 1973 from the Upper, Lower and Ileret Members of the Koobi Fora Formation[2]. Exploration to the south of Koobi Fora was begun in 1972 and continued in 1973. No hominids have yet been found in the limited exposures of the Kubi Algi Formation. A notice of two specimens—KNM-ER 1510 and 1590—that were previously[1] mentioned only by number, is included in this report. The 1973 hominids are not here attributed to genera as there are still no clear generic diagnoses available for fossil hominids. With a few exceptions, previous attributions for the East Rudolf hominid collection remain satisfactory.

Archaeological investigation during 1973 was extended under the direction of G. Ll. Isaac, with J. C. W. Harris who conducted major excavations at several sites in the Upper Member of areas 130 and 131. Limited excavation, but extensive prospecting, in the Lower Member produced sufficient results to support further searching for artefacts below the KBS Tuff.

1973年在肯尼亚北部鲁道夫湖以东下更新统发现更多人科动物证据

利基

编者按

理查德·利基及其研究小组在鲁道夫湖以东经过六年的研究工作，发现了两种人科动物的证据。其一是一种粗壮型南方古猿，他在整个沉积序列中都出现并且变化相对较小。其二是一种原始的但具备更多进步特征的人属类型，也许利基1973年所描述的引人注目的“1470”头骨就是其典型代表。这篇最新报道提出了第三种人科动物，他具有与人属相似的齿系，但颅骨更小，可能与在斯泰克方丹发现的“纤细型”南方古猿相似。随着人科动物化石的增加，将其归入能人还是南方古猿尚存争议，还有可能存在的明显性双形，这些都只能使人类起源的疑团更加扑朔迷离。

1973年在鲁道夫湖以东地区发掘出了20件新的人科动物标本。新证据表明在东非地区的上新世–更新世时期至少存在三个人科动物支系。

本文是有关1973年在肯尼亚鲁道夫湖以东野外挖掘的报告，鲁道夫湖以东研究项目组（前身是探险队）已经结束了在那里第六个年头的工作。在1968~1972年间，已经从该地区采集到了87件人科动物化石标本[1]；另外还有20件标本是在1973年6~9月之间从库比福勒组的上段、下段以及伊莱雷特段[2]发掘出来的。对库比福勒南部地区的调查始于1972年，并且在1973年持续进行。目前为止，还没有在库比阿尔及组有限出露的地层中发现过人科动物化石。本次报告对KNM-ER 1510和1590这两件标本也做了简报，而在此之前[1]，只提到过这两件标本的数字编号。由于现在尚无明确的对人科动物化石进行属一级划分的鉴定标准，所以在本文中，并未将1973年发现的人科动物鉴定到属。除个别例外，先前对鲁道夫湖以东采集到的人科动物标本的归属仍然适用。

在艾萨克的指导下，1973年的考古调查工作得到拓展，哈里斯重点在130和131区域的上段的几个遗址进行了发掘。尽管在下段进行的发掘工作较少，但很有前景，发掘结果充分表明在KBS凝灰岩下有进一步寻找人工制品的必要。

During the palaeontological survey, which was supervised by J. M. Harris, all identifiable fragments from certain horizons were collected; new species were recorded and some primate remains were recovered during a limited survey of the Kubi Algi Formation. A detailed account of the East Rudolf fauna will be presented upon conclusion of current studies, but there are clear indications that at times the palaeoenvironment differed from that of the lower part of the Shungura Formation of the Omo Valley in Ethiopia.

In the geological studies, emphasis was placed on microstratigraphy and palaeo-environmental reconstruction. B. Bowen supervised a study of the Lower and Upper Members of areas 130 and 131 which included confirming the stratigraphic relationships of the cranium KNM-ER 1470. The complete section of the Koobi Fora Formation exposed in area 102 was studied by a group from Dartmouth College, New Hampshire, under G. Johnson. A. K. Behrensmeyer completed a preliminary geological investigation of the hominid sites, noting depositional environments and possible association of fauna; further studies are planned.

I. Findlater extended mapping of tuffaceous horizons to the south of Koobi Fora and collected samples for isotope dating. A series of dates has been obtained from material collected during 1972 (unpublished work at Miller, Findlater, Fitch and Watkins). Palaeomagnetic studies complement those of 1972 and there are sufficient data for internal correlations to be made[3].

Hominid Collection

Specimen KNM-ER 1590, reported previously[1], consists of dental and cranial fragments which were collected from area 12, some meters below the KBS Tuff. Both parietals, fragments of frontal and other pieces of cranial vault, the left deciduous c and dm^2, the left and right unerupted C, P^3 and P^4, and the erupted left and right M^1 and left M^2 were recovered. Although the cranium is immature, it was large with a cranial capacity as great as that determined for KNM-ER 1470. The parietals may show some deformation but, in any event, they suggest that the cranium was wide with a sagittal keel.

KNM-ER 1510, also reported previously[1], includes cranial and mandibular fragments. The specimen is poorly mineralised and further geological investigation at the site indicates a Holocene rather than an early Pleistocene provenance as originally thought.

The 1973 hominids and their stratigraphical positions are listed in Table 1. Specimens from area 123 are rare, and their stratigraphical position relative to the Upper and Lower Members of the Koobi Fora Formation needs clarification.

在哈里斯指导下的古生物调查中，收集了所有在特定地层中的可鉴定的化石残片；在库比阿尔及组地层中发现了新物种，并且发现了一些灵长类化石。对于鲁道夫湖以东动物群的详细报告将在本项研究的结论部分中给出，但是有明显迹象表明当时的古环境与埃塞俄比亚奥莫河谷的上古拉组下部是不同的。

在地质学研究方面，重点关注了微观地层学和古环境的重建。鲍恩指导了对130区和131区的上下段地层的研究，包括对产出KNM-ER 1470头骨的地层关系的确认。由约翰逊带领，来自新罕布什尔州达特茅斯学院的研究小组对102区出露完好的库比福勒组的地层剖面进行了研究。贝伦斯迈耶对人科动物化石地点做了初步地质调查，并记录了沉积环境和可能伴生的动物群；进一步的研究已做好了安排。

芬勒特将对凝灰岩层的绘图工作扩展到了库比福勒以南，并且采集了同位素测年样品。从1972年间采集到的材料中已经获得了一系列年代数据（米勒、芬勒特、菲奇和沃特金斯未发表的工作）。古地磁学研究补充了1972年的研究数据，现在已经有足够多数据来相互印证[3]。

人科动物标本

此前报道过的KNM-ER 1590号标本[1]包括牙齿和颅骨残片是采自12区的KBS凝灰岩之下数米处。发现的材料包括如下解剖部位：两侧顶骨、额骨及头盖骨残片，左侧乳犬齿和第二上乳臼齿（dm^2），未萌出的左、右C，P^3及P^4，已萌出的左右M^1及左M^2。尽管颅骨并未发育成熟，但是其颅容量却与已知的KNM-ER 1470号标本的一样大。虽然其顶骨可能有些变形，但无论如何，它们都表明了这是个很宽且具有矢状嵴的颅骨。

KNM-ER 1510号标本此前也报道过[1]，该标本包括颅骨和下颌骨残片。这个标本石化程度很浅，后来对遗址的地质调查结果表明，含该标本的地层时代属于全新世，而非最初认为的早更新世。

表1中列出了1973年发现的人科动物标本及其地层位置。从123区出土的标本很少，其与库比福勒组的上、下段地层的相对位置关系还需要进一步澄清。

Table 1. 1973 hominid collection from East Rudolf

KNM-ER NO.	Specimen	Area	Member
1800	Cranial fragments	130	Lower
1801	Left mandible, P_4, M_1, M_3	131	Lower
1802	Left mandible, P_4-M_2 and right P_3-M_2	131	Lower
1803	Right mandible fragment	131	Lower
1804	Right maxilla, P^3-M^2	104	Upper
1805	Cranium and mandible	130	Upper
1806	Mandible	130	Upper
1807	Right femur shaft	103	Upper
1808	Associated skeletal and cranial fragments	103	Upper
1809	Right femur shaft	127	Lower
1810	Proximal left tibia	123	?Lower
1811	Left mandible fragment	123	?Lower
1812	Right mandible fragment and left I_1 and M_1	123	Lower
1813	Cranium	123	?Lower
1814	Maxillary fragments	127	Upper
1815	Right talus	1	Upper
1816	Immature fragmented mandible	6A	Upper
1817	Right mandible	1	Upper
1818	I^1	6A	Upper
1819	M_3	3	Upper
1820	Left mandible with M_1	103	Upper

A well preserved mandible (Fig. 1), KNM-ER 1802, was discovered by J. Harris *in situ* below the KBS Tuff in area 131. The dentition is only slightly worn, and fragments of both M_3 crowns suggest that death occurred before full eruption. The canines and incisors are represented by roots and by alveoli filled with matrix. The mandible shows some interesting features—moulding of the mandibular body, absence of a strong post-incisive planum, the development of a slight inferior mandibular torus and the distinct eversion of the mandibular body when viewed from below.

表 1. 1973 年从鲁道夫湖以东发掘的人科动物标本

KNM–ER NO.	标本	区	段
1800	颅骨残片	130	下
1801	左下颌骨，P_4，M_1，M_3	131	下
1802	左下颌骨，P_4-M_2 及右 P_3-M_2	131	下
1803	右下颌骨残片	131	下
1804	右上颌骨，P^3-M^2	104	上
1805	颅骨和下颌骨	130	上
1806	下颌骨	130	上
1807	右股骨骨干	103	上
1808	关联的体骨和颅骨残片	103	上
1809	右股骨骨干	127	下
1810	左胫骨近端	123	? 下
1811	左下颌骨残片	123	? 下
1812	右下颌骨残片、左 I_1 和 M_1	123	下
1813	颅骨	123	? 下
1814	上颌骨残片	127	上
1815	右距骨	1	上
1816	未发育完全的残破下颌骨	6A	上
1817	右下颌骨	1	上
1818	I^1	6A	上
1819	M_3	3	上
1820	左下颌骨带 M_1	103	上

哈里斯在 131 区 KBS 凝灰岩之下就地发现了一件保存很好的下颌骨 KNM-ER 1802（图 1）。其齿系只有轻微磨损，两侧的 M_3 牙冠残片表明该个体是在牙齿未完全萌出之前就已死亡。犬齿和门齿只保留着牙根和充满沉积物的齿槽。该下颌骨显示出了一些很有趣的特征——下颌体的形状，不存在明显的门齿后平面，轻微发育的下颌圆枕，以及从下面看时下颌体呈独特的外翻形式。

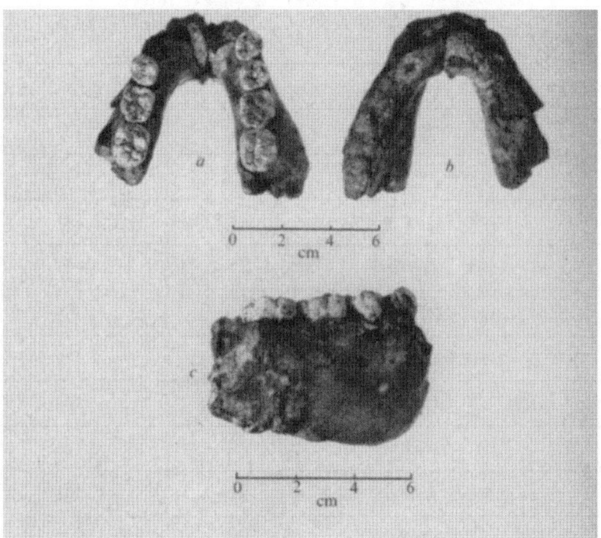

Fig. 1. Mandible, KNM-ER 1802. *a*, Superior view; *b*, inferior view; *c*, right lateral view.

A weathered mandible, KNM-ER 1801, bears some resemblances to KNM-ER 1802, but its worn dentition and loss of surface bone prevent direct comparisons. The relative proportions of the molars and premolars may have been exaggerated by interstitial wear.

A crushed maxillary fragment, KNM-ER 1804, with P^3-M^2 preserved was discovered by R. Holloway. The teeth are complete but worn.

A skull (cranium with associated mandible), KNM-ER 1805, was discovered by P. Abell *in situ* in the BBS Tuff complex in area 130. The specimen is heavily encrusted with a hard matrix and will require careful preparation before its morphology is revealed. Comments here are thus preliminary. The cranium is in pieces which fit together. After preparation, it should be possible to determine the endocranial capacity; at present, a volume of 600–700 cm^3 is suggested. The supraorbital region, much of the face and the greater part of the basicranium have not been preserved. The postorbital region is preserved and the minimum breadth is approximately 90 mm. No distinct temporal lines cross the frontal area although they can be discerned and are still apart at the bregma. There are distinct parasagittal crests. The nuchal attachments are very distinctive and protrude to form a wide bony shelf. The palate is intact; all the teeth are preserved except for the right P^4 and the left I^1. The mandible, small and distinctly robust, is represented by both sides of the body but, except for the right M_2 and M_3, the tooth crowns arc missing. The ascending rami are not preserved. The right M_3 and M_2 are well worn but small. The upper dentition shows wear on all the teeth, including M^3.

A large mandible, KNM-ER 1806, was discovered by Meave Leakey at the same site and horizon as was KNM-ER 1805. There are no tooth crowns preserved and the ascending

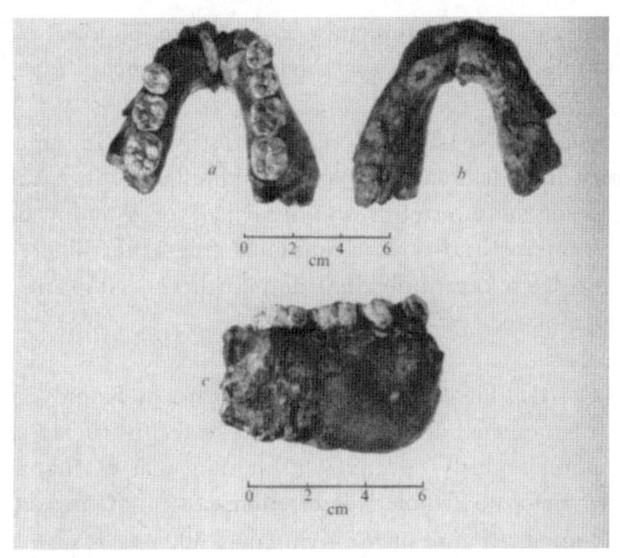

图 1. 下颌骨，KNM-ER 1802。*a*，上面视；*b*，下面视；*c*，右侧面视。

KNM-ER 1801 号标本是一个遭受风化的下颌骨，与 KNM-ER 1802 号标本有些许相像之处，但其严重磨耗的牙系及骨骼表面的破损使得我们无法进行直接比较。臼齿和前臼齿列的相对比例可能被齿间隙磨损放大了。

KNM-ER 1804 是霍洛韦发现的一个压碎了的、保存有 P^3-M^2 的上颌骨残片。其齿列完整但磨耗严重。

KNM-ER 1805 是一个头骨（带有相连的下颌骨及颅骨），由埃布尔在 130 区的 BBS 混杂凝灰岩中就地发现。该标本外面包裹着厚厚的坚硬围岩，要想揭示其形态特征，还需要进行认真细致的修理。因此，此处所做讨论只是初步认识。该颅骨虽支离破碎，但可以拼接到一起。经过修理之后，应该可以确定其颅容量；目前，可以认为其颅容量为 600~700 cm^3。眶上区域、面部大部分以及颅底绝大部分都没有保存下来。眶后区域保存下来了，其最小宽度约为 90 mm。尽管颞线依稀可辨，但在额区表现不明显，并且在前囟处仍然是分开的。有明显的副矢状嵴。项肌附着区明显而突出，形成了一个宽的骨质架。腭骨完好；除了右 P^4 和左 I^1，其余牙齿都保存完好。下颌骨小且非常粗壮，左、右下颌体都保存下来了，除右 M_2 和 M_3 之外，其余牙齿齿冠都缺失了。下颌上升支没有保存下来。右 M_3 和 M_2 深度磨耗，并且很小。上齿系的所有牙齿都有磨耗，M^3 也不例外。

KNM-ER 1806 是一个巨大的下颌骨，由米芙·利基在发现 KNM-ER 1805 的同一地点和同一地层发现。该下颌骨没有任何齿冠保存下来，上升支也丢失了，其他

rami are missing in this otherwise complete specimen. The mandible is typical of the large East Rudolf hominid that I have previously attributed to *Australopithecus*.

A fragmented specimen, KNM-ER 1808, was discovered in area 103 by Kamoya Kimeu. The specimen includes maxillary and mandibular teeth, cranial and mandibular fragments, a fragment of atlas vertebra, the distal half of a femur lacking the condyles, a large segment of humerus and other postcranial fragments. There is little doubt that the various pieces are from one individual and further sieving and excavation will be undertaken in the hope of recovering more material.

A cranium, KNM-ER 1813 (Fig. 2), was discovered *in situ* by Kamoya Kimeu in area 123. The specimen was fragmented but has been partially reconstructed. Plastic deformation is evident. The cranium is partly covered with a thin coat of matrix and considerable preparation is needed before a detailed description can be attempted. The endo-cranial volume is likely to be small; a figure of approximately 500 cm^3 is suggested on the basis of comparative external measurements. Other interesting features include the curvature of the frontals, a postglabella sulcus and the small dentition. The maxilla has well preserved teeth, P^3-M^3, on the left side, but on the right side only the tooth roots and the complete crown of M^3 remain. Both canines and lateral incisors are present but the central incisors seem to have been lost before fossilisation. Both sides of the maxilla fit together to give the form of the dental arcade. The right maxillary fragment includes the malar region and connects with the lateral margin of the right orbit.

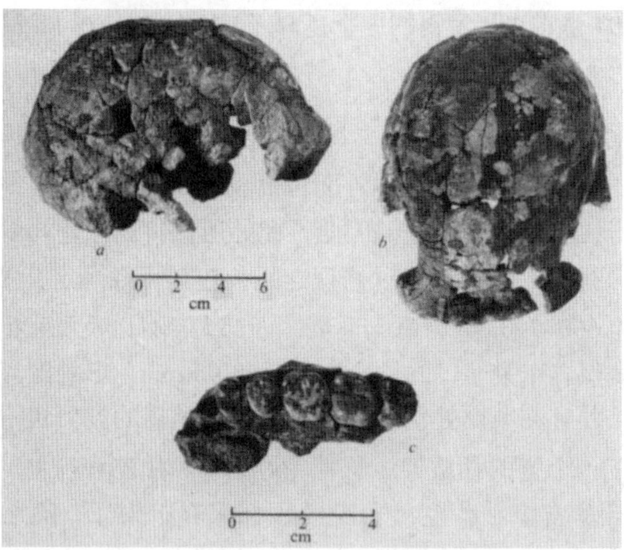

Fig. 2. Cranium, KNM-ER 1813. *a*, Right lateral view; *b*, superior view; *c*, occlusal view of left side of palate.

Other specimens recovered during 1973 are listed in Table 1 and will be described in detail after studies are completed.

方面都很完整。该下颌骨具有我此前归入南方古猿的鲁道夫湖以东出土的巨大人科动物的典型特征。

KNM-ER 1808 是一个破碎的标本，由卡莫亚·基梅乌在 103 区发现。该标本包括上、下颌牙齿、颅骨和下颌骨残片，寰椎残片，缺少股骨髁的股骨远端、一大段肱骨和其他颅后骨残片，几乎可以肯定这些残片均来自同一个个体。为了得到更多材料，我们将进行进一步的筛洗和挖掘工作。

颅骨 KNM-ER 1813（图 2）由卡莫亚·基梅乌在 123 区就地发现。虽然该标本都裂成了碎片，但是已经被部分地修复好了。塑性变形明显。该颅骨的一部分被薄层沉积物所覆盖，在对其进行详细描述之前，需要大量的修理工作。其颅容量可能比较小；根据其外部尺寸的比较，可以估计出其数值约为 500 cm^3。其他有趣的特征包括：额骨的曲率、眉间后槽和小的牙齿。上颌骨上有保存很好的左 P^3-M^3，但是右齿列只有牙根和 M^3 齿冠残存。犬齿和侧门齿都在，但是中门齿似乎在石化之前就已经丢失了。上颌骨的两侧可以拼接到一起，从而可看出齿弓的形状。右上颌骨残片包括颧骨部及与右眼眶侧边相连的部分。

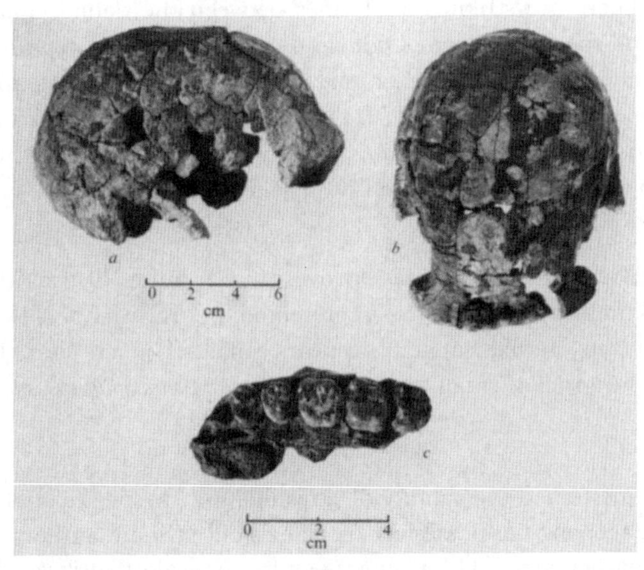

图 2. 颅骨，KNM-ER 1813。*a*，右侧面视；*b*，上面视；*c*，左上颌嚼面视。

1973 年间挖掘的其他标本都在表 1 中列出，并将在相关研究完成以后再做详细描述。

Significance of the 1973 Collection

In previous reports[1,4-7], the East Rudolf hominids were assigned to *Australopithecus*, *Homo* or indeterminate (the last category included both very fragmentary specimens and those of uncertain taxonomic rank).

The East Rudolf specimens that have been attributed to *Australopithecus* span a period of time from 3 million years to just over 1 million years with apparently little morphological change. This form is likely to be the same species as *A. boisei*[8]; it also shows similarities with *A. robustus* from southern Africa. A Pliocene origin is suggested for this specialised group.

Specimens attributed to *Homo* have been recovered from deposits covering a similar time span, but these show greater morphological variability. Those recovered from the Ileret Member seem to differ from those recovered from the Lower Member of the Koobi Fora Formation. The suggestion that a large brained, fully bipedal hominid was living at East Rudolf 3 million years ago was put forward after the 1972 discoveries[7]. This point of view is supported by the cranial fragments, KNM-ER 1590, also from below the KBS Tuff, and this specimen is provisionally attributed to *Homo*.

The 1973 collection from East Rudolf raises many questions. The new mandible, KNM-ER 1802, could be considered as belonging to the same genus and species as KNM-ER 1470 and 1590. There are striking similarities between the dental characters of KNM-ER 1802 and some specimens from Olduvai Gorge such as the type mandible of *Homo habilis*, OH 7. Although the suggested cranial capacity for *H. habilis* is appreciably smaller than that determined for KNM-ER 1470, the discrepancy may be due to the fragmentary material upon which the former estimates were made. I consider that the evidence for a 'small brained' form of *Homo* during the Lower Pleistocene is tenuous.

The cranium, KNM-ER 1813, may prove to be quite distinct from the robust australopithecines and from *Homo*, as represented by KNM-ER 1470. The dentition is 'hominine', yet the cranial capacity appears small. The cranium has some of the features seen in the gracile, small brained, hominid *Australopithecus africanus* Dart, from Sterkfontein.

I have previously questioned the validity of a distinct gracile species of *Australopithecus*[6], but this new evidence reopens the possibility of its existence. Some authors[9,10] have suggested that *Homo habilis*, particularly OH 24, shows features typical of *Australopithecus africanus*. My suggestion here, that *H. habilis* may have affinities with KNM-ER 1470 and 1590, refers only to OH 7 and OH 16. Features of the calvarium of OH 24 show similarities with KNM-ER 1813. The size and morphology of the teeth of the two specimens are alike and the cranial capacities may also be comparable[11,12].

The skull KNM-ER 1805 is undoubtedly important, but its interpretation is enigmatic.

1973 年所采集标本的意义

在之前的报道中[1,4-7]，鲁道夫湖以东的人科动物化石被归入到了南方古猿、人属或分类未定（后者既包括破碎的标本，也有分类位置不明的标本）。

鲁道夫湖以东发现的标本被归入南方古猿，其分布时间在距今300万到100万年，期间几乎未曾发生显著形态变化。这类南方古猿与南方古猿鲍氏种[8]相同；同时也显示了与南非的南方古猿粗壮种的相似性。有人提出这一特定类群起源于上新世。

出土人属化石的地层时代与上述时间段相似，但是这些标本的形态变异较大。从伊莱雷特段地层出土的标本似乎与从库比福勒组下段地层出土的标本有所不同。在 1972 年的发现[7]之后，有人提出 300 万年前在鲁道夫湖以东生活着一种脑量较大并且能完全直立行走的人科动物。这种观点也得到同样是从 KBS 凝灰岩层下出土的 KNM-ER 1590 号颅骨残片的支持，该标本暂时被归入人属。

1973 年在鲁道夫湖以东采集的标本引发了更多问题。可以认为新发现的下颌骨 KNM-ER 1802 与 KNM-ER 1470 和 1590 属于同一属种。KNM-ER 1802 和奥杜威峡谷出土的一些标本，例如能人种的模式标本 OH 7，在牙齿特征方面有着惊人的相似之处。尽管先前提出能人颅容量要明显小于 KNM-ER 1470 的脑量，这一偏差可能是由于前者颅容量的估算是基于残破化石材料的缘故。我认为，那些有关早更新世期间的人属具有“小容量大脑”的证据非常缺乏说服力。

颅骨 KNM-ER 1813 可以证明其与粗壮型的南方古猿及 KNM-ER 1470 号标本所代表的人属都非常不同。该标本的齿系具有“人亚科”的特征，但是其颅容量却显得很小。该颅骨所具有的某些特征与发现于斯泰克方丹的南方古猿非洲种相似，后者是一种纤细且脑量较小的人科动物。

我以前质疑过这种独特的南方古猿纤细种的有效性[6]，但是这个新证据重新提供了其存在的可能性。有些作者[9,10]曾提出，能人，尤其是 OH 24，显示出了典型的南方古猿非洲种的特征。我的观点是，能人可能与 KNM-ER 1470 和 1590 具有亲缘关系，这里所说的能人只是指 OH 7 和 OH 16。OH 24 的颅骨与 KNM-ER 1813 具有相似性。这两个标本的牙齿尺寸和形态特征都很相像，颅容量可能也相当[11,12]。

头骨 KNM-ER 1805 无疑很重要，但是对于它的解释还是一个谜。其相对较大

Its relatively large cranium bears sagittal and nuchal crests but has small teeth; this combination is in contrast to all the specimens previously recovered from East Rudolf.

In any consideration of the affinities of the East Rudolf hominids, the question of sexual dimorphism must not be overlooked. There does seem to be evidence for quite marked sexual dimorphism in one hominid group as demonstrated by the East Rudolf crania, KNM-ER 406 and 732[5]. Unfortunately both crania lack teeth so that the dental characteristics of the alleged female are far from clear.

The possibility of more than two contemporary hominid lineages in the Plio-Pleistocene of East Africa may now have to be recognised, whereas previously one, or at most, two forms were assumed. The attribution of isolated teeth may thus become even more difficult than it is now. Postcranial identifications likewise may be difficult, although the proximal femoral material continues to suggest a morphological dichotomy.

I suggest the following as a basis of nomenclature for Plio-Pleistocene hominids. One genus would include much of the material currently referred to *Australopithecus robustus* and *A. boisei*. A second genus would incorporate many of the gracile specimens from Sterkfontein presently referred to *A. africanus*, perhaps certain specimens from East Rudolf including KNM-ER 1813, and possibly some from Olduvai such as OH 24. A third genus, *Homo*, would incorporate specimens such as KNM-ER 1470 and 1590 from East Rudolf and possibly OH 7 and OH 16 from Olduvai. Some material from South Africa might also be considered within this last category together with later specimens from Olduvai and East Rudolf. The unusual mandible KNM-ER 1482[1], together with the specimen from the Omo area referred to *Paraustralopithecus*[13] and certain other specimens from Omo which are contemporary with the three groups just mentioned, could be considered a fourth form—a remnant of an earlier population that disappeared during the early Pleistocene. All these forms may be traced back well beyond the Plio-Pleistocene boundary.

These remarks are necessarily speculative. A more detailed review of hominid systematics is being prepared in collaboration with B. A. Wood. The wealth of data now available presents a new era in the study of early man. The complexities of dealing with the enlarged sample are challenging, and isolated studies on specific specimens must be replaced by exhaustive studies on all the fossil hominid evidence.

I should like to express appreciation for the financial backing provided by the National Geographic Society, the National Science Foundation, the W. H. Donner Foundation and others. The support and encouragement of the National Museums of Kenya and the Kenya Government made the research possible. Members of the East Rudolf Research Project are too numerous to thank individually but all play a part in a successful field season and are thanked along with those who made important discoveries. I would also

的颅骨具有矢状嵴和项嵴，但是牙齿却很小；这种组合与以前在鲁道夫湖以东发现的所有标本都形成了鲜明对比。

无论从哪方面考虑鲁道夫湖以东人科动物的亲缘关系，其性双形都是一个不容忽视的问题。似乎确能证明有一组人科动物，如鲁道夫湖以东发现的颅骨 KNM-ER 406 和 732 所展示的那样的[5]，具有相当明显的性双形现象。遗憾的是，这两具颅骨都缺失牙齿，所以对所推断的女性个体的牙齿特征仍很不清楚。

现在必须承认，在上新世–更新世时期，东非地区同时生活着两种以上的人科动物，但此前，只推断有一种，最多两种。因此现在要将这些单个牙齿进行归类就变得更加困难了。尽管股骨近端总能表明形态上的歧异性，但对头后骨骼的鉴定也同样困难。

如下是我提出的有关上新世–更新世人科动物命名的基本框架。第一个属，包括目前被归入南方古猿粗壮种和南方古猿鲍氏种的大部分标本。第二个属，包含大量发现于斯泰克方丹的现被归入南方古猿非洲种的纤细类型的标本，也许还包括鲁道夫湖以东发现的一些标本，例如 KNM-ER 1813，甚至还有在奥杜威发现的某些标本，例如 OH 24。第三个属就是人属，该属包含鲁道夫湖以东出土的标本 KNM-ER 1470 和 1590，可能还有奥杜威出土的 OH 7 和 OH 16；也可以考虑将南非发现的一些标本以及后来从奥杜威和鲁道夫湖以东发现的标本一起都归入第三个属中。不寻常的下颌骨 KNM-ER 1482[1] 以及从奥莫地区发现的被归入傍人[13] 的标本及某些其他标本与上述三个属同时代，可以考虑将它们一起归入第四种类型——消亡于早更新世的早期类群的孑遗分子。所有这些类型都可以追溯到上新世–更新世界限之前。

以上论述都是有必要深思的。我正在与伍德共同准备一份更详细的有关人科动物系统分类的综述。现在拥有的大量数据为研究早期人类开辟了新纪元。处理这些新增加标本的复杂程度很具有挑战性，必须停止对特定标本进行孤立的研究，而代之以对所有人科动物化石证据进行彻底详尽的研究。

我想对国家地理学会、国家科学基金会、唐纳基金会和其他组织提供的经济资助表示感谢。感谢肯尼亚国家博物馆和肯尼亚政府提供的支持与鼓励使我们的研究得以顺利进行。由于鲁道夫湖以东研究项目组的成员太多，无法一一致谢，但是我得说所有人都在成功的野外发掘工作中起了相当重要的作用，并同时感谢获得重要

express thanks to my wife Meave who, as always, provided invaluable assistance both at the museum and in the field.

(**248**, 653-656; 1974)

R. E. F. Leakey
National Museums of Kenya, PO Box 40658, Nairobi

Received January 9, 1974.

References:

1. Leakey, R. E. F., *Nature*, **242**, 170 (1973).
2. Bowen, B. E., and Vondra, C. F., *Nature*, **242**, 391 (1973).
3. Brock, A., and Isaac, G. Ll., *Nature*, **247**, 344 (1974).
4. Leakey, R. E. F., *Nature*, **226**, 223 (1970).
5. Leakey, R. E. F., *Nature*, **231**, 241 (1971).
6. Leakey, R. E. F., *Nature*, **237**, 264 (1972).
7. Leakey, R. E. F., *Nature*, **242**, 447 (1973).
8. Tobias, P. V., *Olduvai Gorge*, **2** (Cambridge University Press, 1967).
9. Robinson, J. T., *Nature*, **205**, 121 (1965).
10. Anon., *Nature*, **232**, 294 (1971).
11. Leakey, M. D., Clarke, R. J., and Leakey, L. S. B., *Nature*, **232**, 308 (1971).
12. *Nature*, **239**, 469 (1972).
13. Arambourg, C., and Coppens, Y., *C. r. hebd. Séanc. Acad. Sci., Paris*, **265**, 589 (1967).

发现的人员。我还要感谢我的妻子米芙一如既往地在博物馆和野外工作方面所提供的宝贵帮助。

（刘皓芳 翻译；同号文 审稿）

Kinky Helix

F. H. C. Crick and A. Klug

Editor's Note

DNA in eukaryotes is tightly packaged in chromatin, in which it is wound around disk-shaped proteins called histones. This implies that the elegant double helix of Watson and Crick must be severely distorted. Here Crick, together with biochemist Aaron Klug, suggests that this distortion might involve sharp kinks in the chain, rather than smooth bends. Their argument is purely one of chemical plausibility, and it turned out to be incorrect in detail, although DNA is quite sharply bent in some situations in the cell. But the precise structure of chromatin remains unclear, although it seems apparent that this structure and its regulation are central to the way genes are activated.

DNA in chromatin is highly folded. Is it kinked? And does it kink in other situations?

CHROMATIN is the name given to chromosomal material extracted from the nuclei of cells of higher organisms. It consists mainly of DNA and a set of small rather basic proteins called histones. Other proteins and RNA are present in lesser amounts (see for example ref. 1). Early X-ray work (for review see ref. 2) suggested that there was a structure in chromatin which repeated at intervals of about 100 Å. More recent work using nucleases[3,4] has shown that the DNA in chromatin exists in some regular fold which repeats every 200 base pairs, the best value currently being 205±15 base pairs[5].

The most cogent model for chromatin has been put forward by Kornberg[6] who suggested that the basic structure consists of a string of beads each containing two each of the four major histones, each bead being associated with about 200 base pairs of DNA. Linear arrangements of beads (in a partly extended form) were first seen in the electron microscope by Olins and Olins[7] and called by them *v*-bodies. The exact diameter of a bead in the wet state is rather uncertain but it is probably in the region of 100 Å. Kornberg's model suggested that DNA, when associated with histone, is folded to about one-seventh of its length. This is the value deduced by Griffith[8] from electron micrographs of the mini-chromosome of the virus SV40. A similar value has been obtained by Oudet *et al.*[9] from measurements on adenovirus 2. Other compact models have been proposed by van Holde *et al.*[10] and Baldwin *et al.*[11].

Thus the DNA in chromatin, even at this first level of structure, must be folded considerably since its length is contracted to about one-seventh. Moreover, the basic repeat of 200 base pairs (which is 680 Å long in the B form of DNA) must be folded into a fairly limited space having the dimensions of about 100 Å^3 (ref. 6).

有扭结的螺旋

克里克，克卢格

编者按

真核生物的DNA在染色质中是紧密包装的，围绕在被称为组蛋白的盘状蛋白质周围。这意味着沃森和克里克提出的优美的双螺旋结构会受到严重扭曲。在本文中，克里克与生物化学家阿龙·克卢格一起提出，这种扭曲可能涉及DNA链的急剧扭结，而不是平滑的弯曲。他们的论证只是在化学上说得通，结果证明在细节上是错误的，虽然在某些情况下，细胞内的DNA会发生非常急剧的弯曲。迄今为止，染色质的精确结构尚不为人所知，但显而易见的是，该结构及其调控作用对基因的激活极其重要。

染色质中的DNA是高度折叠的。它是有扭结的吗？在其他条件下它会发生扭结吗？

染色质是指从高等生物的细胞核中提取的染色体物质。它主要由DNA和被称为组蛋白的一组碱性小蛋白组成，其他蛋白质和RNA少量存在（实例见参考文献1）。早期的X射线研究（综述见参考文献2）表明：在染色质中存在一种重复出现的结构，重复间隔大约为100 Å。近期用核酸酶做的研究[3,4]已经表明：染色质中的DNA是规则折叠的，每200个碱基对为一个重复单位，目前的最佳值为205±15个碱基对[5]。

科恩伯格[6]提出的染色质模型是最有说服力的。他认为其基本结构是一串珠粒，每颗珠粒含有四种主要组蛋白中的两个分子，每粒珠子与长度约为200个碱基对的DNA相连接。奥林斯和奥林斯[7]最先用电子显微镜观察到珠粒的线性排列（呈部分延伸状），并把它命名为ν小体（译者注：核小体）。在潮湿状态下，一个珠粒的准确直径尚未确定，不过可能是在100 Å的范围内。科恩伯格的模型表明，当与组蛋白结合时，DNA折叠成其自身长度的七分之一。这个数值是由格里菲思[8]根据病毒SV40的微小染色体的电子显微照片推断出来的。乌代等人[9]对腺病毒2进行的测量也获得了相似的数值。范霍尔德等人[10]和鲍德温等人[11]还提出了其他的紧凑模型。

因此染色质中的DNA即便是在其结构的第一层次上，也必定是高度折叠的，因为其长度被压缩到了原来的约七分之一。而且，含200个碱基对的基本重复单位（在B型DNA中的长度为680 Å）一定是被折叠到一个相当有限的空间内，其尺寸约为100 $Å^3$（参考文献6）。

We have found it very difficult to estimate just how much energy is required to bend DNA "smoothly" to a small radius of curvature, say 30–50 Å, bearing in mind that these numbers are not many times greater than the diameter of the DNA double helix, which is about 20 Å, and that bending a helix destroys its symmetry. We have formed the impression that the energy might be rather high. We therefore asked ourselves whether the folded DNA may consist of relatively straight stretches joined by large kinks. This paper describes a certain type of kink which can be built rather nicely and has interesting properties.

The Stereochemistry of a Kink

No doubt other types of kink could be built, but we have concentrated on one special type which we consider to be rather plausible. We have assumed that all the base pairs of the double helix are left intact (so that no energy is lost by unpairing them), that the straight parts of the DNA on each side of the kink remain in the normal B form, but that at the kink one base pair is completely unstacked from the adjacent one. Thus at each kink the energy of stacking of one base pair on another is lost. Naturally all bond distances and angles (including dihedral angles) have to be stereochemically acceptable.

We find that, given these assumptions, one can convincingly build a neat kink, having a large angle of kink, in one way only; or, more strictly, in a family of ways all very similar to each other. The double helix is bent towards the side of the minor groove. This can be seen in the photograph of one such model shown in Fig. 1.

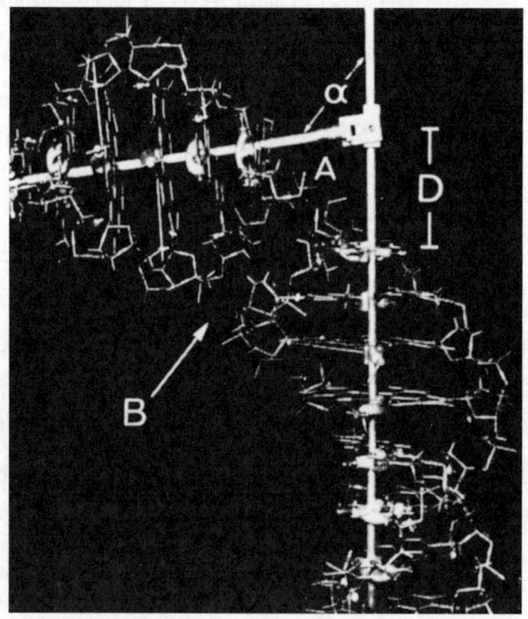

Fig. 1. General view of a model of a kink, taken from the side. For this model $d = 0$, $\alpha = 98°$, $D = 8$ Å and $\theta = 23°$ (see text). The two short lengths of backbone, connecting the two stretches of straight helix, can be seen at A. The region of van der Waals contacts between backbones, which limit the kink angle α, is near B.

我们发现很难估计究竟需要多少能量才能将 DNA“平滑地”弯曲成一个很小的曲率半径，例如 30~50 Å。要知道，这些数值并非比 DNA 双螺旋的直径（大约 20 Å）大许多倍，并且这样弯曲螺旋会破坏它的对称性。我们已经形成了一种想法：这需要的能量也许会相当高。因此我们曾经自问：这种折叠的 DNA 是否可能由相对笔直的舒展部分通过很大的扭结连接形成。本文描述了某一特定类型的扭结，这种扭结恰好可以完美构建，并具有令人感兴趣的性质。

一个扭结的立体化学

尽管其他类型的扭结无疑也可以构建形成，但我们只关注一个特殊的类型，因为我们认为它比较合理。我们假设：DNA 双螺旋中所有的碱基对都保持完好无损(即没有未配对碱基的能量损失)，在扭结每一侧 DNA 的笔直部分都保持正常的 B 型，但是在扭结处，一个碱基对与相邻碱基对完全不是相互堆积的。因此，在每个扭结处，就损失了把一个碱基对堆积到另一个碱基对上的能量。当然，所有的键长和键角（包括二面角）都必须是立体化学上能接受的。

我们发现，根据这些假设，可以令人信服地构建出一个具有很大扭结角度的纯扭结，但是只能以一种方式做到；或者更严格地说，可以用一类彼此非常相似的方式做到。其双螺旋弯向小沟的一侧，我们可以在图 1 所示的照片中看到一个这样的模型。

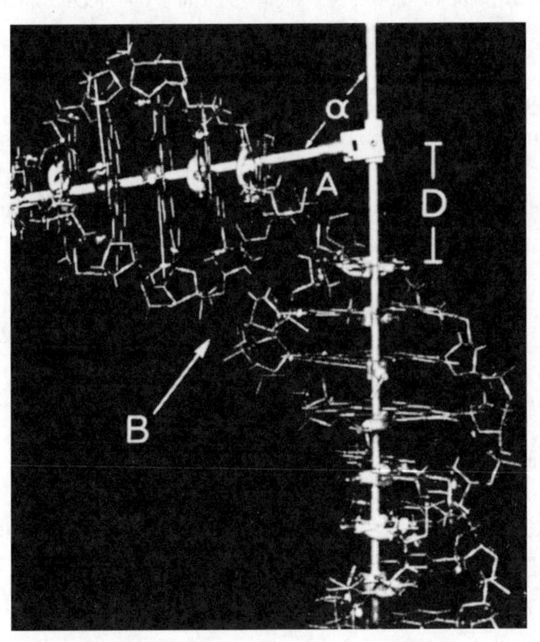

图 1. 从侧面观察的一个扭结模型的整体图。此模型的 $d = 0$，$\alpha = 98°$，$D = 8$ Å，$\theta = 23°$（见正文）。连接两段直螺旋的两段比较短的骨架在 A 处可见。骨架之间的范德华作用区域使扭结角度 α 受到限制，在 B 处可见。

The structure can be built with an approximation to a dyad axis passing through the kink, though we cannot see any strong reason why such symmetry is essential in chromatin. A partial view looking along the pseudo-dyad is shown in Fig. 2.

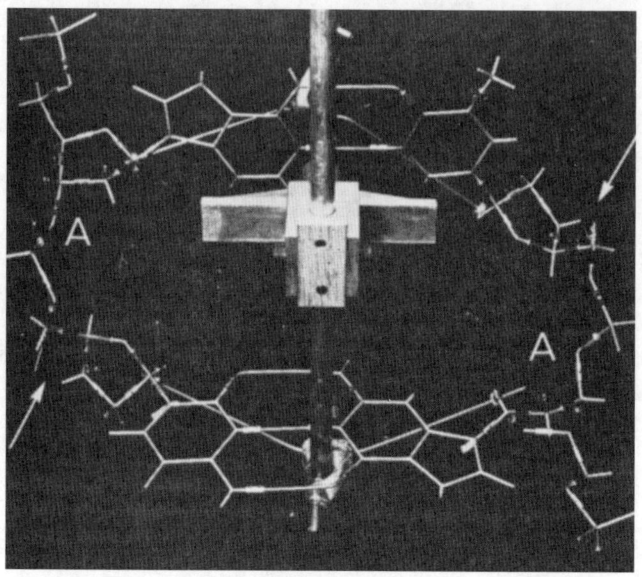

Fig. 2. View of part of the model of Fig. 1 taken approximately looking down the pseudo-dyad. It shows two base pairs, one on either side of the kink. The rest of the model has been blanked out for easier viewing. The two arrows point to the C_4'–C_5' bonds at which the chain conformation is changed by kinking—see Fig. 3. The letters A correspond to the region marked A in Fig. 1.

To our surprise the configuration of the backbone at the kink can be made similar to that of the normal backbone of the B form except that the conformation at the C_4'–C_5' bond in the sugar is rotated 120° about this bond, going from one of the possible staggered configurations to another one as shown in Fig. 3. (We have arbitrarily kept the same pucker of the sugar ring as is found in the B form of DNA.)

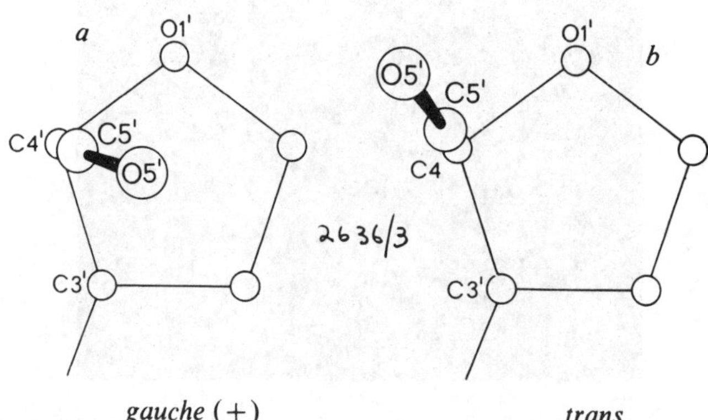

Fig. 3. Diagrams of the deoxyribose ring showing the approximate conformation at the C_4'–C_5' bond (*a*) in the normal straight B form of DNA, and (*b*) in the proposed kink; the two sugars affected are marked in Fig. 2.

这种结构可以用一个二重轴通过扭结来近似构建，虽然我们不知道为什么在染色质中这样的对称性是必不可少的。沿假二重轴方向看的局部图示于图 2。

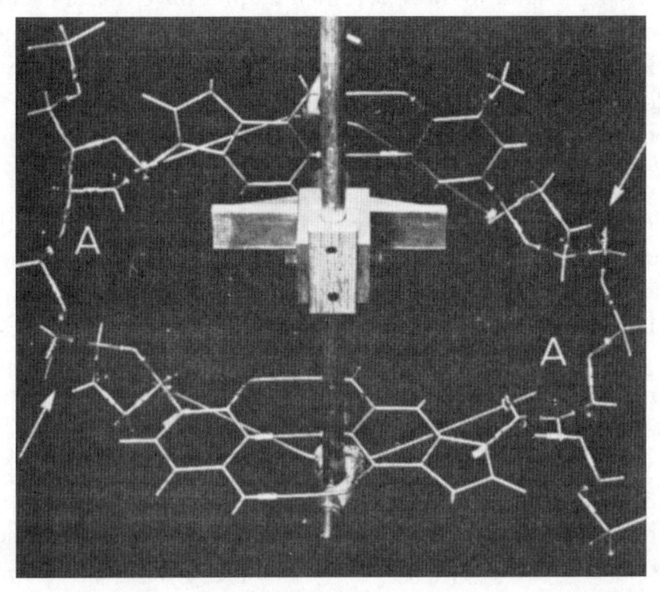

图 2. 从近似俯视假二重轴的角度拍摄到的图 1 模型的局部视图。图中显示了两个碱基对，在扭结的两侧各有一个碱基对。为了便于观察，模型的其他部分被屏蔽掉了。两个箭头指向 $C_4'-C_5'$ 键，在这里扭结使链的构象发生变化（见图 3）。两个字母 A 与图 1 中标记的 A 区相对应。

令我们惊奇的是，扭结处的骨架构型可以做得与 B 型的正常骨架相似，只不过糖中 $C_4'-C_5'$ 键的构象绕此键旋转了 120°，使其从一种可能的交错构型转变为另外一种，如图 3 所示。（我们故意保存了与 B 型 DNA 中存在的相同的糖环折叠。）

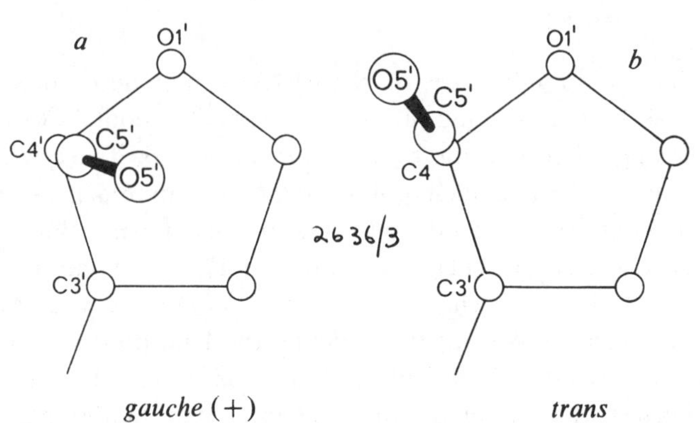

图 3. 脱氧核糖环示意图，显示在 $C_4'-C_5'$ 糖键处的近似构象：(*a*) 在正常的直线 B 型 DNA 中和 (*b*) 在假设的扭结中；受影响的两个糖分子在图 2 中已经标明。

The structure of the backbone at the kink is not one of the preferred configurations that have been listed[12,13] since these involve a weak CH···O close contact between the hydrogen attached to either C_6 of a pyrimidine or C_8 of a purine and the O_5' of the sugar, and by its very nature the type of kink we have assumed cannot have this at every position. In our model this contact is absent for one base of each of the base pairs immediately adjacent to the kink. As explained above, however, the backbone configuration is sufficiently close to that of the B form of DNA (except for the torsion angle about C_4'–C_5') that we feel it is acceptable stereochemically.

The nature of the base pairs on either side of the kink is immaterial to the model though presumably the energy required to unstack these base pairs will depend to some extent on their composition. We have found it difficult to estimate the free energy involved. It is probably a few kilocalories. It is obviously desirable that this figure should be determined as accurately as possible since the ease of making kinks depends on just how big it is. Another important question is how much a DNA double helix can be bent before it kinks. If we denote the mean curvature by κ (where κ equals the reciprocal of the radius of curvature) then we would expect the energy of deformation of a uniformly bent helix, per unit length, to increase at least as fast as κ^2. For a kinked helix, on the other hand, this energy increases only as κ. In this case κ is the mean "curvature" of the segmented double helix which is proportional to the number of kinks per unit length. Thus, as is intuitively obvious, at small κ the double helix will bend, while at large κ it will kink. The value we should like to know is the radius of curvature at which it changes from bending to kinking.

There is probably an appreciable activation energy to the process of making a kink since the C_4'–C_5' bond must pass through the eclipsed configuration. For this reason we consider kinks of this type with a kink angle of about half the full 100° to be unlikely.

Common Features of the Family

A number of very similar structures can be built along these lines and it is not obvious which is to be preferred. They all have certain features in common.

(1) The axes of the two straight parts of the DNA do not necessarily intersect exactly, but may be separated by only a small distance, d, typically about 1 Å or less. Note that d has a sign. (2) The angle between the two axes, α (projected, if necessary, on to a plane perpendicular to the line joining their points of nearest approach) is easily made more than 90° but approaches 100° with difficulty. The model shown in Fig. 1 has $\alpha = 98°$. At the maximum angle (for any particular model) the backbone of one straight part starts to touch the backbone of the other chain of the other straight part. This contact, marked B in Fig. 1 has a (local) dyad axis. (3) If we define the kink point as the point where the local helix axes, on either side of the kink, intersect (or, if they do not intersect, then the midpoint of the shortest straight line between the axes) then the distance, D, from the kink point to the plane of the nearest base pair is appreciable and typically in the region of 7–8 Å.

扭结处的骨架结构并不是已经列举的优选构型[12,13]之一，因为这些优选的构型包含一个弱的 CH···O 紧密接触，这种紧密接触在嘧啶 C_6 或者嘌呤 C_8 上的氢与糖环上的 O_5' 之间形成。而我们所设想的扭结类型，正是由于其本身性质，不可能在每个位置上都有这样的紧密接触。在我们的模型里，在紧邻扭结的每一个碱基对中，有一个碱基里不存在这种紧密接触。然而，正如上面所解释的那样，扭结的骨架构型与 B 型 DNA 的构型十分相似（除了绕 $C_4'-C_5'$ 键的扭角外），因此我们认为其在立体化学上是合理的。

扭结两侧的碱基对的性质对模型并不重要，尽管从理论上推测将碱基去堆积化所需的能量在一定程度上会依赖于碱基的组成。我们发现很难估计这个过程所涉及的自由能，也许只是几个千卡。显然最好是可以尽可能精确地测定自由能，因为扭结形成的容易程度恰恰取决于其所需能量的大小。另外一个重要的问题是：一个 DNA 双螺旋在其发生扭结前，能够弯曲的程度是多少。如果我们用 κ 代表平均曲率（这里的 κ 等于曲率半径的倒数），那么我们可以推测，一个均匀弯曲的双螺旋的形变能量，在单位长度上至少会以 κ^2 的速度增加。而另一方面，对于扭结的双螺旋而言，这个能量仅以 κ 的速度增加。在这种情况下，κ 是指分段的双螺旋的平均“曲率”，它与单位长度上的扭结数成比例。因此，在直观上非常明显的是，当 κ 值小的时候，双螺旋会弯曲；而当 κ 值大的时候，双螺旋会形成扭结。我们想要知道从弯曲变成扭结时的曲率半径。

在发生扭结的过程中，可能需要数量可观的活化能，因为 $C_4'-C_5'$ 键必须经过重叠构型。正因如此，我们认为：这种类型的扭结不太可能具有约为完整 100° 一半的扭结角度。

扭结家族的共同特征

一些非常类似的结构都可以按着这些规则进行构建，而其中究竟哪一个结构更可取，这一点还不是很明显。它们都拥有某些共同特征。

（1）DNA 两个直线部分的轴不一定准确地相交，而或许只由一个很小的距离 d 分隔开，典型的间距为 1Å 或更小。请注意，d 是有符号的。（2）两轴之间的角度 α（必要时，可以将两个轴投影到一个平面上，该平面垂直于两轴之间最靠近点间的连接线）可以很容易地做到大于 90°，但很难达到 100°。图 1 中所示的模型的 α 值为 98°。在角度最大的时候（对于任何特殊的模型来说都是如此），一条链直线部分的骨架就开始接触到另一条链直线部分的骨架。这种接触标示于图 1 中的 B 点，它具有一个（局部的）二重轴。（3）如果我们把扭结位点定义为扭结两侧的两个局部双螺旋两轴间的交点（或者，如果它们不相交，则把扭结位点定义为两个螺旋轴之间最短直线的中点），那么从扭结位点到最近碱基对所在平面的距离 D 就是可以估算的，其典型数值在 7~8 Å 这个区间内。

To introduce our fourth point we must first consider the relationship between three successive straight portions; that is, two kinks in succession. We assume that every kink is exactly the same. The structure formed will depend on the precise number of base pairs between the kinks. (It should be remembered that the B form of DNA has an exact repeat after ten pairs.) For example, if there are ten (or a multiple of ten) base pairs between two kinks, the structure will bend round into, very roughly, three sides of a square. If there are five base pairs between kinks (or an odd multiple of five) then the structure will approximate to a zig-zag. In short the dihedral angle for three successive straight portions will depend on the exact number of base pairs between the two adjacent kinks.

We can now state our fourth point. (4) The kink imparts a small negative twist to the DNA. This is most easily grasped by imagining that the kink is made in two steps; first, the two base pairs to be unstacked are unstacked in the axial direction without kinking the backbone—this reduces the twist of 36° between these residues to about 10–20°—and second, that this extended structure is then kinked. The result is that if successive kinks are made at intervals of $10n$ base pairs (where n is an integer) the DNA instead of folding back to form a "circle" follows instead a left-handed helix, though naturally a kinked helix made of straight segments (see Fig. 4).

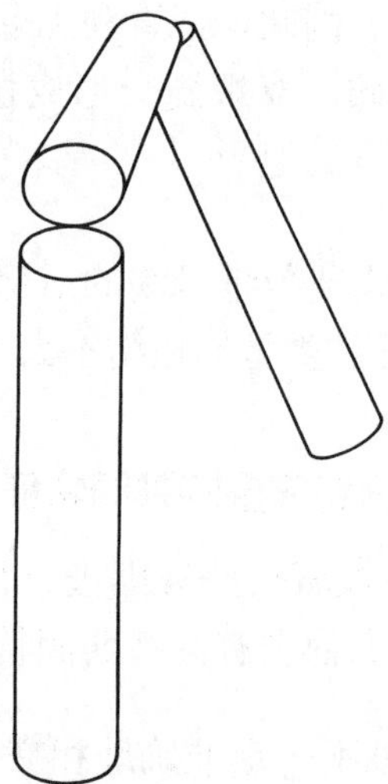

Fig. 4. Each cylinder represents a length of straight double helix. When there are ten base pairs (or an integer multiple of ten base pairs) in the middle stretch the kink has a left-handed configuration, as shown. The dihedral angle used in this paper is zero for the *cis* conformation. Its sign agrees with the usual convention that positive values less than 180° correspond to a right-handed configuration.

The exact dihedral angle associated with three successive straight stretches depends somewhat on the precise details of the kink but is typically about $(m.36°–\theta)$ where m is the number of base pairs between the two kinks and θ, the dihedral correction angle, is not far from 15–20°. A very small rotational deformation of the straight portions could, however, alter this figure a little so the exact value in chromatin (if it does indeed consist of kinked

为了介绍我们的第四点，我们首先要考虑三段连续直线部分之间的关系；也就是说，两个连续的扭结。我们假定每个扭结都是完全相同的。所形成的结构依赖于扭结之间碱基对的精确数目。（应该记住的是，B 型 DNA 在 10 个碱基对以后有完全相同的重复片段）。例如，如果在两个扭结之间存在 10（或者 10 的倍数）个碱基对，那么这个结构会粗略的弯曲环绕成正方形的三个边。如果在两个扭结之间存在 5（或者 5 的奇数倍数）个碱基对，那么这个结构大体上近似于一个锯齿的形状。简言之，三段连续直线部分的二面角会依赖于两个相邻扭结之间的碱基对的准确数目。

现在我们可以陈述我们的第四点了。（4）扭结带给 DNA 一点负面的小扭曲。如果设想扭结是通过两个步骤形成的，就可以很容易地理解这一点：首先，即将被拆开的两个碱基对沿轴线方向拆开，骨架并不发生扭结——这使得这些残基之间的扭曲角度从 36° 减少到大约 10°~20°。第二步，这种在空间上延伸了的结构接着就发生扭结。其结果是，如果连续的扭结在 10*n*（*n* 为整数）个碱基对的间隔上形成，那么 DNA 就不反方向折叠形成一个“环”，而形成左手螺旋，虽然在自然情况下，扭结的螺旋是由直线的片段形成的（见图 4）。

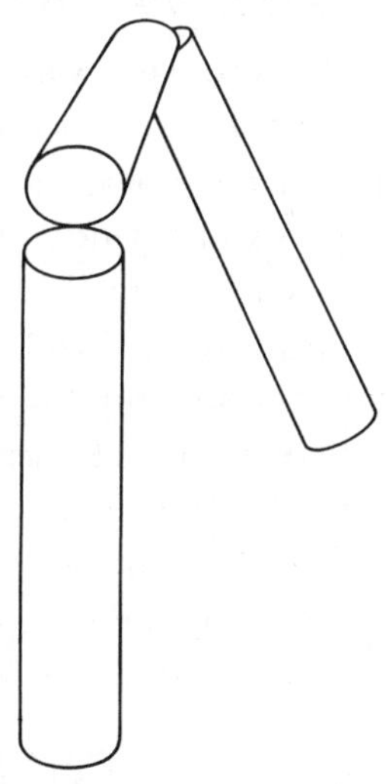

图 4. 每一个圆柱体代表一段直线形双螺旋。当中间呈直线舒展的部分含有 10 个碱基对（或 10 个碱基对的整数倍）时，扭结具有左手构型，如图所示。本文中，为顺式构象所用的二面角为 0。其符号与通常的惯例一致，小于 180° 的正值对应右手构型。

与三段连续直线舒展的片段相关的二面角的准确数值在一定程度上取决于扭结的精确细节，但其典型的数值约为（$m \times 36° - \theta$），m 是两个扭结之间的碱基对数目，θ 是二面角的修正值，大约为 15°~20°。然而，这些直线部分发生的一些微小的旋转性变形，会使得此结构发生少许的改变，因此二面角在染色质中的精确数值（如

helices) will probably be imposed by the histones.

(5) For any model there is a smallest number of base pairs between two adjacent kinks. For models of this family this number is usually three. In particular, the model illustrated in Fig. 1, for which $\alpha = 98°$, can be built with three base pairs between two kinks but not with only two base pairs there. This is probably true for all models of this type for which α is greater than 90°. A model with three base pairs between two kinks exposes these base pairs rather effectively.

It is easy to see that six parameters are needed to describe the relationship between any two (equal) stretches of straight double helix. If these stretches are related by a dyad axis through the kink point then only four parameters are required. These can conveniently be taken to be the four used above: d, α, D and θ.

Another family of models can be made with the kink on the side of the major groove, but such structures have a smaller angle of kink and seem rather awkward to build. We have not explored them further. A rather different type of kink, in which a base pair is undone, has been suggested by Gourévitch *et al.*[14]

The Occurrence of Kinks

The idea that the fold of DNA in chromatin was based on a unit of 10 base pairs was originally suggested to us by experimental evidence discovered by our colleague, Dr. Markus Noll. Noll[15] has shown that the digestion of native chromatin with the nuclease DNase I produces nicks in the DNA which tend to be spaced multiples of 10 bases apart. This suggests that the DNA is folded in a highly regular way and is probably mainly on the outside of the structure[6,15]. More recent work by Noll and Kornberg (unpublished) using micrococcal nuclease points to a structural repeat at intervals of 20 base pairs. Thus a rather neat model for the most compact (wet) form of chromatin can be made in which the DNA is kinked through about 95–100° every 20 base pairs, giving a shallow kinked helix having 10 straight stretches of DNA in each 100 Å repeat. The middle stretches of this repeat would be largely protected by the histones of the bead, the flanking stretches less so. Whether this very simple model is basically correct remains to be seen.

Obviously we should ask whether DNA is kinked in other situations. One interesting possibility is that when the *lac* repressor binds to the operator site on the DNA the double helix becomes kinked. It has been shown by Wang, Barkley and Bourgeois[16] that this binding unwinds the helix by a small angle, either about 40°, or, more likely, about 90° (the value depending on the amount of unwinding assumed to be produced by the standard agent, ethidium bromide). As they point out, this is too small to allow the formation of a Gierer-type loop[17]. It is, however, just what one would expect from a small number of kinks since each kink of the kind we have described unwinds the double helix by about 15° to 25°. For example, an attractive zig-zag model can be imagined with four kinks, each spaced about five base pairs apart. This model places the two sequences related by a dyad,

果染色质确实是由发生了扭结的螺旋组成的话）很有可能是由组蛋白决定的。

（5）对于任何一个模型而言，相邻两个扭结之间总存在着最小数目的碱基对。对于这类模型而言，其最小数目通常是 3 个。特别是在图 1 中所示的模型（其 α 值为 98°），可以在两个扭结之间用 3 个碱基对进行构建，但仅用 2 个碱基对就不行。这个规律可能适用于这一类 α 值大于 90° 的所有模型。在两个扭结之间有 3 个碱基对的模型显示了碱基对的高效性。

我们可以很容易发现，在描述任意两条（相同的）直线双螺旋片段之间的关系时，需要有六个参数。如果这些片段由通过扭结位点的二重轴相联系，那么只需四个参数就可以。这些参数可以方便地采用上面已经采用过的四个参数：d，α，D 和 θ。

另一类模型可以由在大沟一侧发生的扭结来构建。但这类结构的扭结角度比较小，似乎难于构建。我们没有进一步对其进行研究。古雷维奇等人[14]提出了一种非常不同的扭结类型，其中的一个碱基对是解开的。

扭结的发生

染色质中的 DNA 折叠是基于一个 10 个碱基对的单元发生的，这个观点最早是由我们的同事马库斯·诺尔博士根据他所发现的实验证据向我们提出的。诺尔[15]发现，用核酸酶 DNase I 对天然的染色质进行消化，使 DNA 上产生了一些切口，这些切口之间的间隔趋近于为 10 的倍数个碱基。这提示 DNA 是以一种高度规则的方式折叠的，并且切割可能主要发生在染色质结构的外侧[6,15]。最近由诺尔和科恩伯格利用微球菌核酸酶进行的研究（尚未发表）表明，染色质结构的重复是以 20 个碱基对为间隔。由此可以为染色质最紧密的（湿的）形式构建一个比较简洁的模型，其中 DNA 每隔 20 个碱基对发生大约 95° ~ 100° 的扭结，形成一个浅的有扭结的螺旋，在每隔 100 Å 的重复间距上，有 10 个 DNA 的直线形片段。这个重复结构的中间片段大部分受到珠粒的组蛋白的保护，而两侧的段落受到较少保护。这个非常简单的模型是否基本正确，还有待证实。

显而易见，我们应该提出这样的问题：DNA 在其他情况下是否也发生扭结。一个有趣的可能性是：当乳糖阻遏子结合在 DNA 上的操纵子位点时，双螺旋就发生扭结。王、巴克利和布儒瓦等[16]证明：这种结合使螺旋以一个小的角度解旋，要么大约为 40°，更可能的是大约 90°（这个数值依赖于解旋的程度，通常假定解旋是由标准试剂溴化乙啶引起的）。正如他们指出的那样，这个数值太小，不容许吉勒型环[17]的形成。但这恰好是人们从少数扭结中期望看到的情况，因为在我们描述的那种类型中，每个扭结只能使双螺旋以大约 15°~25° 的角解旋。例如，可以设想一个有四个扭结的漂亮锯齿状模型，每个扭结之间相隔大约 5 个碱基对。该模型将两条序列通过二重轴相联系，每条序列具有六个连续的碱基对，分别安插在第一个和最后一个扭结

each of six consecutive base pairs, on either side of the first and last kinks (see Fig. 5). In this position, being near a kink, they are more exposed than they would be in a stretch of unkinked DNA.

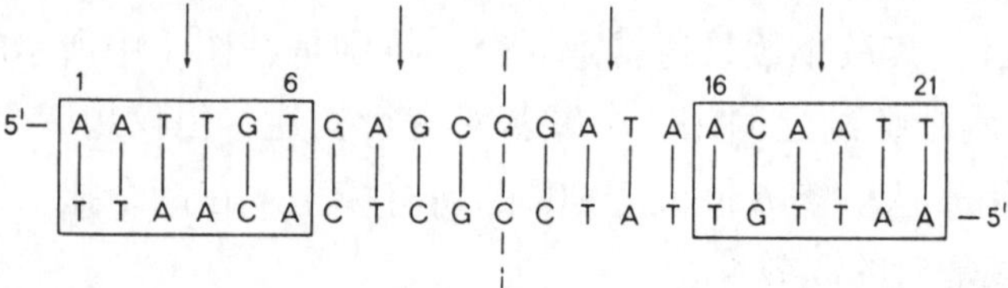

Fig. 5. The minimal base sequence of the *lac* operon, taken from Gilbert and Maxam[18]. The dotted line marks the pseudodyad in the base sequence. The two sets of consecutive base pairs, related by the dyad, are boxed. The arrows show one choice of positions where kinks might occur.

In essence, kinking may be a way of partly exposing a small group of base pairs without too great an expenditure of energy. The exposed side of each of these base pairs is that normally in the major groove. The kink has the effect of displacing one of the phosphate-sugar backbones which normally make up the two sides of this groove. The specific pattern of hydrogen bonding sites in the major groove is thus made more accessible for a few base pairs on either side of a kink. A kink may therefore turn out to be a preferred configuration of DNA when it is interacting specifically with a protein.

Kinks may be suspected in all cases where double-stranded DNA has been shown to adopt a more compact state than the normal double helix. Obvious examples are the folded chromosomes of *Escherichia coli*[19,20] (and no doubt other prokaryotes), the folded DNA in viruses, the ψ phase of naked DNA discovered by Lerman[21] and the shortened form of DNA in alcoholic solutions as described by Lang[22].

One should also ask whether kinks occur spontaneously, as a result of thermal motion, in double-stranded DNA in solution. The frequency at which this occurs clearly depends on the free energy difference involved. If this were, say, about 4 kcalorie then there should be one kink in about 800 base pairs which could be appreciable. Such kinks would occur mainly between A–T pairs. If the free energy were as high as 6 kcalorie this would produce one kink in about every 22,000 base pairs, which would be more difficult to detect.

At the present we have no compelling evidence which shows that DNA in chromatin is kinked rather than bent nor that kinks exist in DNA in other contexts. Nevertheless our model seems to us sufficiently attractive to be worth presenting now for consideration by other workers in the field. Kinks, if they occur, have at least two possible advantages. It has always been a puzzle how to construct hierarchies of helices in a neat way, since bending an existing helix necessarily distorts its regular structure. This distortion becomes more acute as the basic helix is coiled at higher and higher levels. A kink allows such deformations to be local rather than diffuse and makes it easier to build hierarchical models which are neat stereochemically. The other advantage is that, at a kink, several base pairs may be more easily available for specific interaction with a protein. If kinks in

的两侧（参见图 5）。在这个位置上，由于靠近扭结，它们比在一段无扭结的 DNA 上更加的暴露。

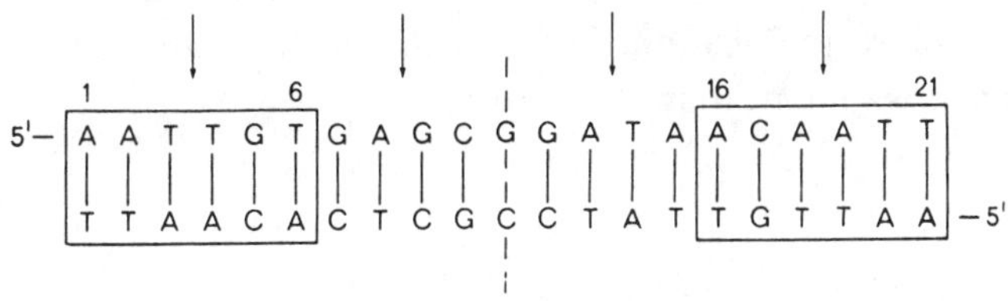

图 5. 乳糖操纵子的最小碱基序列，来自吉尔伯特和马克萨姆的文章[18]。虚线标记的是碱基序列中的假二重轴。方框标记由二重轴联系的两组连续的碱基对。箭头表示扭结可能发生的位置。

从本质上说，扭结也许是一种不用消耗太多能量就可以暴露一小组碱基对的方式。正常情况下这些碱基对被暴露的一侧位于大沟中。大沟的两侧通常是由磷酸–糖骨架构成的，扭结使其中一个骨架发生移位。于是，大沟中氢键位点的特殊分布由此对扭结两侧的一些碱基对变得更易接近。因此，当 DNA 与一个蛋白发生特异性的相互作用时，扭结也许就成了 DNA 的一种优选构型。

当双链 DNA 被证实采用了比正常双螺旋更为紧凑的形态时，扭结都有可能存在。明显的例子有：大肠杆菌（无疑还有其他原核生物）的折叠染色体[19,20]，病毒中的折叠了的 DNA，列尔曼[21]发现的裸露 DNA 的 ψ 相，以及兰[22]所描述的酒精溶液中 DNA 的缩短形式。

人们应该还会提出这样的问题：在双链 DNA 的溶液中，扭结是否会由于热运动而自发地产生。这种扭结发生的频率明显依赖于所涉及自由能的差别。比方说，如果该自由能大约是 4,000 卡路里，那么可以估计，在大约 800 个碱基对上应该有一个扭结。这样的扭结主要发生在 A–T 碱基对之间。如果自由能高达 6,000 卡路里，那么大约每 22,000 个碱基对上会产生一个扭结，这将更加难以察觉。

目前我们还没有令人信服的证据表明染色质中的 DNA 是扭结而不是弯曲，以及在其他环境下 DNA 中存在扭结。但不管怎样，我们的模型对我们来说具有足够的吸引力，值得现在呈现出来，供本领域内其他研究者参考。扭结如果发生的话，至少有两个可能的优点。一直以来让人困惑的是：如何以简洁的方式构建不同层次的螺旋结构，因为使已经存在的螺旋弯曲，必然导致其规则结构的扭曲。随着基本的螺旋在越来越高的层次上卷曲，这种扭曲变得越来越剧烈。扭结允许这样的扭曲在局部发生，而不是扩散开来；并且使构建立体化学上简洁合理的各层次模型变得更容易。另外一个优点是：在扭结处，有几个碱基对可能会更为容易地与蛋白质发

DNA exist they will surely prove to be important.

We thank our colleagues Drs. R. D. Kornberg, M. Noll and J. O. Thomas for communicating their results to us before publication. We also thank them and our other colleagues for many useful discussions on chromatin structure.

(**255**, 530-533; 1975)

F. H. C. Crick and A. Klug
Medical Research Council Laboratory of Molecular Biology, Hills Road, Cambridge, UK

Received April 25; accepted May 6, 1975.

References:

1. *Histones and Nucleohistones* (edit. by Philips, D. M. P.) (Plenum, London and New York, 1971).
2. Pardon, J. F., Richards, B. M., and Cotter, R. I., *Cold Spring Harb. Symp. Quant. Biol.*, **38**, 75-81 (1974).
3. Hewish, D. R., and Burgoyne, L. A., *Biochem. Biophys. Res. Commun.*, **52**, 504-510 (1973).
4. Burgoyne, L. A., Hewish, D. R., and Mobbs, J., *Biochem. J.*, **143**, 67-72 (1974).
5. Noll, M., *Nature*, **251**, 249-251 (1974).
6. Kornberg, R. D., *Science*, **184**, 868-871 (1974).
7. Olins, D. E., and Olins, A. L., *Science*, **183**, 330-332 (1974).
8. Griffith, J., *Science*, **187**, 1202-1203 (1975).
9. Oudet, P., Gross-Bellard. M., and Chambon, P., *Cell*, **4**, 281-299 (1975).
10. Van Holde, K. E., Sahasrabuddhe, B., and Shaw, R., *Nucleic Acid Res.*, **1**, 1579-1586 (1974).
11. Baldwin, J. P., Boseley, P. G., Bradbury, E. M., and Ibel, K., *Nature*, **253**, 245-249 (1975).
12. Arnott, S., and Hukins, D. W. L., *Nature*, **224**, 886-888 (1969).
13. Sundaralingam, M., *Biopolymers*, **7**, 821-869 (1969).
14. Gourévitch, M., *et al.*, *Biochemie*, **56**, 967-985 (1974).
15. Noll, M., *Nucleic Acid Res.*, **1**, 1573-1578 (1974).
16. Wang, J. C., Barkley, M. D., and Bourgeois, S., *Nature*, **251**, 247-249 (1974).
17. Gierer, A., *Nature*, **212**, 1480-1481 (1966).
18. Gilbert, W. and Maxam, A., *Proc. Natl. Acad. Sci. U.S.A.*, **70**, 3581-3584 (1973).
19. Pettijohn, D. E., and Hecht, R., *Cold Spring Harb. Symp. Quant. Biol.*, **38**, 31-41 (1974).
20. Worcel, A., Burgi, E., Robinton, J., and Carlson, C. L., *Cold Spring Harb. Symp. Quant. Biol.*, **38**, 43-51 (1974).
21. Lerman, L., *Cold Spring Harb. Symp. Quant. Biol.*, **38**, 59-73 (1974).
22. Lang, D., *J. Molec. Biol.*, **78**, 247-254 (1973).

生特异性相互作用。如果 DNA 中存在扭结，它们一定会被证明是很重要的。

我们感谢我们的同事科恩伯格、诺尔和托马斯博士在论文发表之前与我们交流他们的研究成果。我们也感谢他们以及我们的其他同事关于染色质结构的许多有益讨论。

（刘振明 翻译；顾孝诚 审稿）

Continuous Cultures of Fused Cells Secreting Antibody of Predefined Specificity

G. Köhler and C. Milstein

Editor's Note

Antibodies are the chemical agents (with well-known but complex molecular structure) that help vertebrate animals to defend themselves against infection and other foreign agents. In principle, antibodies against agents the body has not previously encountered would be invaluable in protecting the lives of human beings. This paper by César Milstein and Georges Köhler describes a technique for making antibodies specific against arbitrary protein structures, so providing a means of defence against unknown or as yet non-existent infectious agents. Such general-purpose antibodies (now known as monoclonal antibodies) are now widely used in research and in the practice of medicine. Milstein and Köhler shared the Nobel Prize in Physiology or Medicine (with Niels K. Jerne) in 1984.

THE manufacture of predefined specific antibodies by means of permanent tissue culture cell lines is of general interest. There are at present a considerable number of permanent cultures of myeloma cells[1,2] and screening procedures have been used to reveal antibody activity in some of them. This, however, is not a satisfactory source of monoclonal antibodies of predefined specificity. We describe here the derivation of a number of tissue culture cell lines which secrete anti-sheep red blood cell (SRBC) antibodies. The cell lines are made by fusion of a mouse myeloma and mouse spleen cells from an immunised donor. To understand the expression and interactions of the Ig chains from the parental lines, fusion experiments between two known mouse myeloma lines were carried out.

Each immunoglobulin chain results from the integrated expression of one of several *V* and *C* genes coding respectively for its variable and constant sections. Each cell expresses only one of the two possible alleles (allelic exclusion; reviewed in ref. 3). When two antibody-producing cells are fused, the products of both parental lines are expressed[4,5], and although the light and heavy chains of both parental lines are randomly joined, no evidence of scrambling of *V* and *C* sections is observed[4]. These results, obtained in an heterologous system involving cells of rat and mouse origin, have now been confirmed by fusing two myeloma cells of the same mouse strain, and provide the background for the derivation and understanding of antibody-secreting hybrid lines in which one of the parental cells is an antibody-producing spleen cell.

可分泌特异性抗体的融合细胞的连续培养

克勒，米尔斯坦

编者按

抗体是一种众所周知但分子结构复杂的化学物质，它帮助脊椎动物抵御感染及其他异物。原则上讲，抗体能够对抗机体先前没有遇到过的物质，对保护人类的生命极具价值。塞萨尔·米尔斯坦和乔治斯·克勒的这篇文章描述了一种制备针对任意蛋白结构的特异性抗体的技术，因此也提供了一种抵御未知的或迄今为止尚不存在的传染源的手段。这种通用抗体（现称为单克隆抗体）如今已被广泛应用于研究和医疗实践。米尔斯坦和克勒（与尼尔斯·杰尼）因此在 1984 年共同获得诺贝尔生理学暨医学奖。

通过对细胞系进行持续的组织培养来生产符合人们预期的特异性抗体的方法引起了人们的广泛关注。现在已经建立起了数量可观的可以永久培养的骨髓瘤细胞系[1,2]，并且还建立了多种筛选流程用于检测其中是否存在抗体活性。然而，这种方法无法提供令人满意的符合人们预期的特异性单克隆抗体。本文中，我们描述了一些可以分泌抗绵羊红细胞（SRBC）抗体的组织培养细胞系的制备过程。这些细胞系是通过融合小鼠骨髓瘤细胞和小鼠脾细胞（来自经过免疫的小鼠）的方法建立起来的。为了了解亲本细胞系中免疫球蛋白（Ig）链的表达和相互作用关系，我们使用了两个已知的小鼠骨髓瘤细胞系进行融合实验。

免疫球蛋白的每条链都是通过一个 *V* 基因（编码可变区）与一个 *C* 基因（编码恒定区）经过整合之后表达的。每个细胞只能表达两个可能的等位基因中的一个（等位基因排斥，详见参考文献 3 中所做的综述）。当两个都能够分泌抗体的细胞融合到一起之后，则两个亲本细胞系各自编码的抗体都会得到表达[4,5]。不过，虽然两个亲本细胞系中的轻链和重链都是随机组合的，但是在融合细胞中并没有发现扰乱 *V* 片段和 *C* 片段之间连接的现象[4]。这些实验结果最初来自由大鼠细胞和小鼠细胞组成的异源系统，现在已经在来源于同一个小鼠品系的两个骨髓瘤融合细胞系中得到证实。这些结果也为了解分泌抗体的杂交系的来源提供了基础，因为该杂交系亲本细胞之一是可分泌抗体的脾细胞。

Two myeloma cell lines of BALB/c origin were used. PlBul is resistant to 5-bromo-2'-deoxyuridine[4], does not grow in selective medium (HAT, ref. 6) and secretes a myeloma protein, Adj PC5, which is an IgG2A(κ), (ref. 1). Synthesis is not balanced and free light chains are also secreted. The second cell line, P3-X63Ag8, prepared from P3 cells[2], is resistant to 20 μg ml^{-1} 8-azaguanine and does not grow in HAT medium. The protein secreted (MOPC 21) is an IgG1(κ) which has been fully sequenced[7,8]. Equal numbers of cells from each parental line were fused using inactivated Sendai virus[9] and samples containing 2×10^5 cells were grown in selective medium in separate dishes. Four out of ten dishes showed growth in selective medium and these were taken as independent hybrid lines, probably derived from single fusion events. The karyotype of the hybrid cells after 5 months in culture was just under the sum of the two parental lines (Table 1). Figure 1 shows the isoelectric focusing[10] (IEF) pattern of the secreted products of different lines. The hybrid cells (samples *c–h* in Fig. 1) give a much more complex pattern than either parent (*a* and *b*) or a mixture of the parental lines (*m*). The important feature of the new pattern is the presence of extra bands (Fig. 1, arrows). These new bands, however, do not seem to be the result of differences in primary structure; this is indicated by the IEF pattern of the products after reduction to separate the heavy and light chains (Fig. 1*B*). The IEF pattern of chains of the hybrid clones (Fig. 1*B*, *g*) is equivalent to the sum of the IEF pattern (*a* and *b*) of chains of the parental clones with no evidence of extra products. We conclude that, as previously shown with interspecies hybrids[4,5], new Ig molecules are produced as a result of mixed association between heavy and light chains from the two parents. This process is intracellular as a mixed cell population does not give rise to such hybrid molecules (compare *m* and *g*, Fig. 1*A*). The individual cells must therefore be able to express both isotypes. This result shows that in hybrid cells the expression of one isotype and idiotype does not exclude the expression of another: both heavy chain isotypes (γ1 and γ2a) and both V_H and both V_L regions (idiotypes) are expressed. There are no allotypic markers for the C_κ region to provide direct proof for the expression of both parental C_κ regions. But this is indicated by the phenotypic link between the *V* and *C* regions.

Table 1. Number of chromosomes in parental and hybrid cell lines

Cell line	Number of chromosomes per cell	Mean
P3-X67Ag8	66,65,65,65,65	65
P1Bul	Ref. 4	55
Mouse spleen cells	–	40
Hy-B(P1-P3)	112,110,104,104,102	106
Sp-1/7-2	93,90,89,89,87	90
Sp-2/3-3	97,98,96,96,94,88	95

在实验中我们使用了两个 BALB/c 小鼠来源的骨髓瘤细胞系。P1Bul 细胞具有 5−溴−2'−脱氧尿嘧啶抗性 [4]，在选择性培养基（HAT，参考文献 6）中不生长，并且可以分泌一种骨髓瘤蛋白——Adj PC5，属于 IgG2A（κ）亚型（参考文献 1）。这种蛋白的轻、重链合成并不是均衡的，也可以检测到分泌的游离免疫球蛋白轻链。另一个细胞系 P3-X63Ag8 来自 P3 细胞 [2]，可以耐受 20 μg · ml^{-1} 的 8−氮鸟嘌呤，并且不能在 HAT 培养基中生长。这个细胞分泌的蛋白（MOPC 21）属于 IgG1（κ）亚型，并且已经测出其全序列 [7,8]。利用灭活的仙台病毒 [9] 将相同数量的亲本细胞融合后，再将 2×10^5 个融合后的细胞分别接种在含有选择性培养基的不同培养皿中。结果含选择性培养基的 10 个培养皿中有 4 个培养皿上有融合细胞生长，这些被认为是独立的杂交细胞系，它们可能来自单一的细胞融合。经过 5 个月的培养后，这些杂交细胞的染色体组型被证明是其两个亲本细胞系相加的结果（表 1）。图 1 显示的是不同细胞系分泌产物的等电聚焦 [10]（IEF）图谱。其中，杂交细胞（图 1 中样品 *c~h*）的图谱比其两个亲本（*a* 和 *b*）或者两个亲本的混合样品（*m*）都要更加复杂。这种新图谱的一个重要特点就是产生了一些额外的条带（图 1 中箭头所示）。然而，我们发现这些新的条带似乎并不是因为蛋白质一级结构的不同而导致的；这一点可通过用还原剂处理样品（以便分开抗体的重链和轻链）后的还原产物的等电聚焦图谱看出（图 1*B*）。杂交细胞克隆（图 1*B*，*g*）的肽链的等电聚焦图谱与其两个亲本细胞克隆的肽链（*a* 和 *b*）的等电聚焦图谱之和相同，并且没有多余的产物。由此我们得出结论，与之前在不同物种来源的细胞间进行的融合实验一样 [4,5]，新的免疫球蛋白分子是两个亲本的重链和轻链相互组合的产物。由于单纯地将两种细胞混合并不能产生这种杂交分子（比较图 1*A* 中 *m* 和 *g*），因此这一过程应该发生在细胞内。相应地，每个融合细胞都应该能够表达这两种同种型。这一结果表明，在杂交细胞中一种同种型或独特型的表达并不会排斥另一种的表达——两种重链同种型（γ1 和 γ2a）以及重链可变区（V_H）和轻链可变区（V_L）（独特型）都有表达。由于缺乏链恒定区（$C_κ$）的同种异型标记，因此我们无法提供直接的证据证明两种亲本的 $C_κ$ 区都有表达。不过由于 *V* 区和 *C* 区是连接在一起的，因此可以间接证明这一结论。

表 1. 亲本细胞系和杂交细胞系中染色体的数量

细胞系	每个细胞的染色体数量	平均值
P3-X67Ag8	66, 65, 65, 65, 65	65
P1Bu1	参考文献4	55
小鼠脾细胞	—	40
Hy-B (P1-P3)	112，110，104，104，102	106
Sp-1/7-2	93，90，89，89，87	90
Sp-2/3-3	97，98，96，96，94，88	95

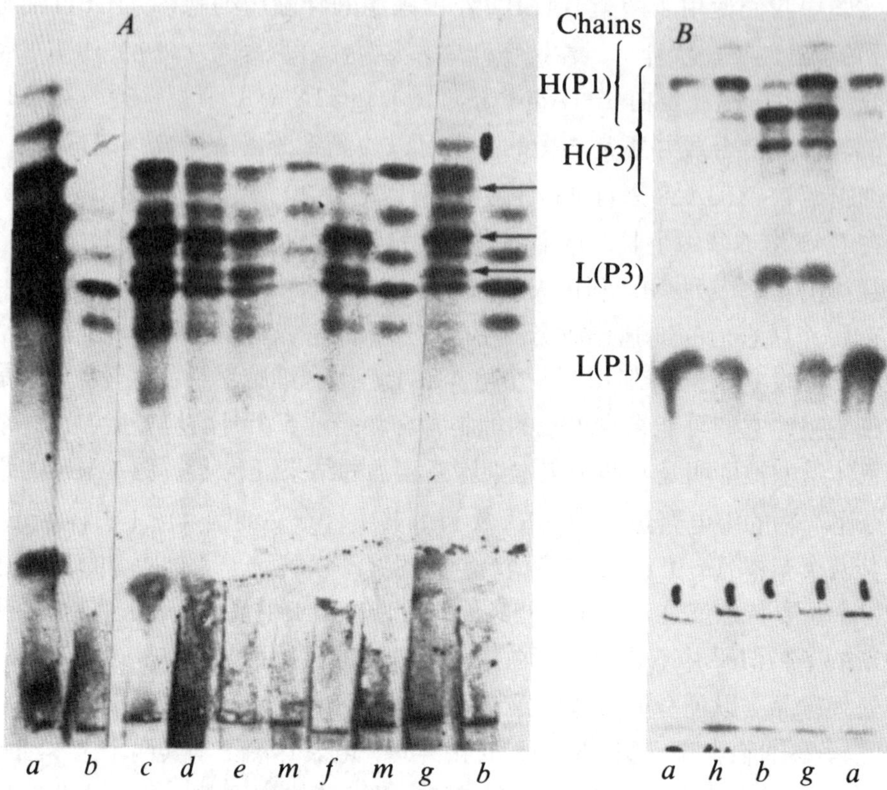

Fig. 1. Autoradiograph of labelled components secreted by the parental and hybrid cell lines analysed by IEF before (*A*) and after reduction (*B*). Cells were incubated in the presence of ^{14}C-lysine[14] and the supernatant applied on polyacrylamide slabs. *A*, *p*H range 6.0 (bottom) to 8.0 (top) in 4 M urea. *B*, *p*H range 5.0 (bottom) to 9.0 (top) in 6 M urea; the supernatant was incubated for 20 min at 37°C in the presence of 8 M urea, 1.5 M mercaptoethanol and 0.1 M potassium phosphate *p*H 8.0 before being applied to the right slab. Supernatants from parental cell lines in: *a*, P1Bu1; *b*, P3-X67Ag8; and *m*, mixture of equal number of P1Bul and P3-X67Ag8 cells. Supernatants from two independently derived hybrid lines are shown: *c–f*, four subclones from Hy-3; *g* and *h*, two subclones from Hy-B. Fusion was carried out[4,9] using 10^6 cells of each parental line and 4,000 haemagglutination units inactivated Sendai virus (Searle). Cells were divided into ten equal samples and grown separately in selective medium (HAT medium, ref. 6). Medium was changed every 3 d. Successful hybrid lines were obtained in four of the cultures, and all gave similar IEF patterns. Hy-B and Hy-3 were further cloned in soft agar[14]. L, Light; H, heavy.

Figure 1*A* shows that clones derived from different hybridisation experiments and from subclones of one line are indistinguishable. This has also been observed in other experiments (data not shown). Variants were, however, found in a survey of 100 subclones. The difference is often associated with changes in the ratios of the different chains and occasionally with the total disappearance of one or other of the chains. Such events are best visualised on IEF analysis of the separated chains (for example, Fig. 1*h*, in which the heavy chain of P3 is no longer observed). The important point that no new chains are detected by IEF complements a previous study[4] of a rat–mouse hybrid line in which scrambling of *V* and *C* regions from the light chains of rat and mouse was not observed.

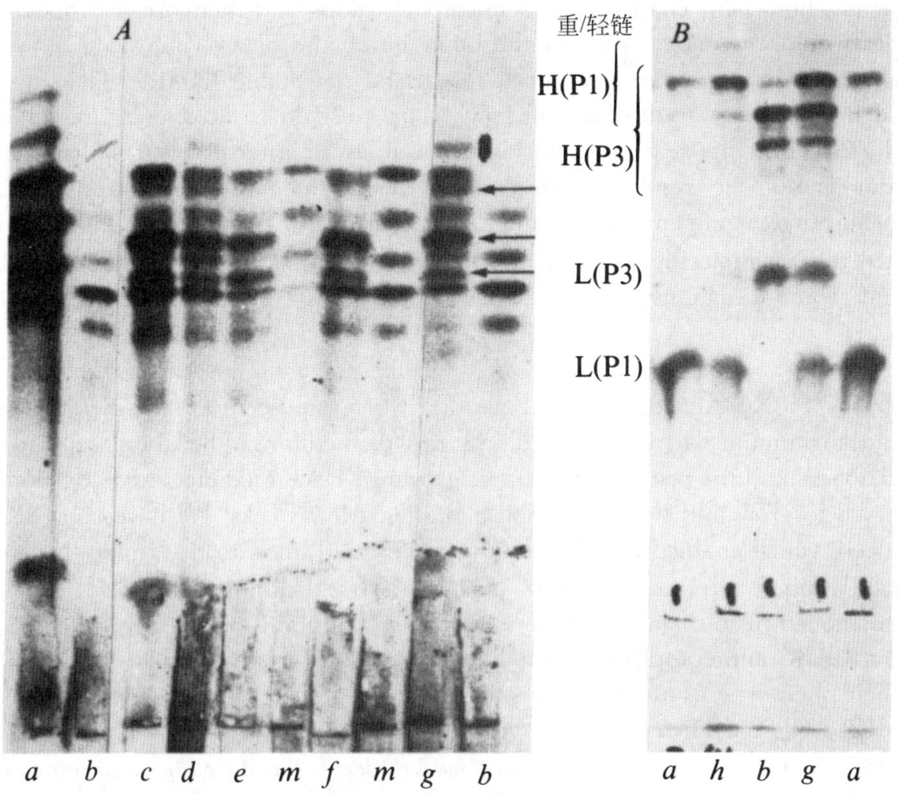

图 1. 放射性标记的亲本细胞系和杂交细胞系分泌产物在被还原前（*A*）和还原后（*B*）经过等电聚焦电泳分析后的放射自显影图像。细胞均培养在含有 ^{14}C– 赖氨酸[14] 的培养基中，细胞培养上清被加到聚丙烯酰胺胶上。*A*，pH 范围为 6.0（底部）到 8.0（顶部），尿素浓度为 4 M。*B*，pH 范围为 5.0（底部）到 9.0（顶部），尿素浓度为 6 M。细胞培养上清先在含有 8 M 尿素、1.5 M 巯基乙醇的 0.1 M 磷酸钾缓冲液（pH 8.0）中 37℃ 孵育 20 分钟，再加到胶上。亲本细胞系上清为：*a*，P1Bul；*b*，P3-X67Ag8；*m*，P1Bul 和 P3-X67Ag8 细胞等量混合物。分别得到的两个杂交系上清为：*c*~*f*，Hy-3 的 4 个亚克隆；*g* 和 *h*，Hy-B 的 2 个亚克隆。各用 10^6 个细胞的亲本细胞以及 4,000 血凝反应单位的灭活仙台病毒（瑟尔公司）进行细胞融合实验[4,9]。融合完成后，细胞被分为均等的 10 份并分别接种到选择性培养基（HAT 培养基，参考文献 6）中。培养基每 3 天换 1 次。最终在 4 个培养皿中获得了成功融合的杂交系，并且它们具有相似的等电聚焦图谱。Hy-B 和 Hy-3 则是在软琼脂中进一步克隆得到[14]。L 代表轻链，H 代表重链。

图 1*A* 显示出来源于不同杂交实验的克隆以及同一个细胞系的不同亚克隆是无法区别的。这在其他实验中也曾经被观察到（数据未显示）。然而，通过对 100 个亚克隆进行分析，我们发现了一些变异体。这种区别通常与不同肽链的比例变化有关，偶尔会发现有一种或另一种肽链完全消失。这种情况在等电聚焦实验分析单个肽链时得到了最好的体现（例如，图 1*h* 中，P3 的重链消失了）。在等电聚焦中观察不到新的条带这点很重要，这印证了早先在大鼠 – 小鼠杂交细胞系中所得到的类似的结论[4]，在这个细胞系中观察不到来自大鼠和小鼠轻链的 *V* 区和 *C* 区发生错误连接。在这个研究中，两个亲本的轻链具有相同的 $C_κ$区，因此无法检测融合细胞中 V_L–C_L 是否错误

In this study, both light chains have identical C_κ regions and therefore scrambled V_L–C_L molecules would be undetected. On the other hand, the heavy chains are of different subclasses and we expect scrambled V_H–C_H to be detectable by IEF. They were not observed in the clones studied and if they occur must do so at a lower frequency. We conclude that in syngeneic cell hybrids (as well as in interspecies cell hybrids) *V–C* integration is not the result of cytoplasmic events. Integration as a result of DNA translocation or rearrangement during transcription is also suggested by the presence of integrated mRNA molecules[11] and by the existence of defective heavy chains in which a deletion of *V* and *C* sections seems to take place in already committed cells[12].

The cell line P3-X63Ag8 described above dies when exposed to HAT medium. Spleen cells from an immunised mouse also die in growth medium. When both cells are fused by Sendai virus and the resulting mixture is grown in HAT medium, surviving clones can be observed to grow and become established after a few weeks. We have used SRBC as immunogen, which enabled us, after culturing the fused lines, to determine the presence of specific antibody-producing cells by a plaque assay technique[13] (Fig. 2*a*). The hybrid cells were cloned in soft agar[14] and clones producing antibody were easily detected by an overlay of SRBC and complement (Fig. 2*b*). Individual clones were isolated and shown to retain their phenotype as almost all the clones of the derived purified line are capable of lysing SRBC (Fig. 2*c*). The clones were visible to the naked eye (for example, Fig. 2*d*). Both direct and indirect plaque assays[13] have been used to detect specific clones and representative clones of both types have been characterised and studied.

The derived lines (Sp hybrids) are hybrid cell lines for the following reasons. They grow in selective medium. Their karyotype after 4 months in culture (Table 1) is a little smaller than the sum of the two parental lines but more than twice the chromosome number of normal BALB/c cells, indicating that the lines are not the result of fusion between spleen cells. In addition the lines contain a metacentric chromosome also present in the parental P3-X67Ag8. Finally, the secreted immunoglobulins contain MOPC 21 protein in addition to new, unknown components. The latter presumably represent the chains derived from the specific anti-SRBC antibody. Figure 3*A* shows the IEF pattern of the material secreted by two such Sp hybrid clones. The IEF bands derived from the parental P3 line are visible in the pattern of the hybrid cells, although obscured by the presence of a number of new bands. The pattern is very complex, but the complexity of hybrids of this type is likely to result from the random recombination of chains (see above, Fig. 1). Indeed, IEF patterns of the reduced material secreted by the spleen-P3 hybrid clones gave a simpler pattern of Ig chains. The heavy and light chains of the P3 parental line became prominent, and new bands were apparent.

连接。另一方面，两个亲本的重链属于不同的亚型，因此我们预测可以通过等电聚焦实验来确定是否发生了 V_H–C_H 的错误连接。在我们研究的克隆中，并没有检测到这种错误连接的发生,即使错误连接确实发生,其概率也是非常低的。我们的结论是,在同源性细胞杂交实验中（不同物种间的细胞杂交也是一样的），*V*–*C* 的整合并不是发生在细胞质中的事件的结果,而是在转录的过程中 DNA 发生易位或者重排导致的。整合 mRNA 分子的发现 [11]，以及在已经定型的细胞中 *V* 片段或 *C* 片段缺失的重链（存在功能缺陷）的发现 [12] 都支持了这一结论。

上述 P3-X63Ag8 细胞系在 HAT 培养基中不能存活。从免疫后的小鼠中获得的脾细胞也会在生长培养基中死去。当这两种细胞在仙台病毒的作用下发生融合后则可以在 HAT 培养基中生长，几周后观察到存活下来的克隆生长并固定下来。我们使用绵羊红细胞作为免疫原刺激小鼠，这使我们可以在获得融合细胞之后，用空斑实验技术 [13] 来检测是否存在可以产生特异性抗体的细胞（图 2*a*）。我们将杂交细胞在软琼脂中进行了克隆 [14]，并且通过覆盖绵羊红细胞和补体便可以很容易地挑出那些可以产生抗体的克隆（图 2*b*）。单个克隆被分离出来之后仍然能保持其表型，几乎所有的来源于纯化细胞系的克隆都具有裂解绵羊红细胞的能力（图 2*c*）。这些克隆是肉眼可见的（例如图 2*d*）。经过直接和间接空斑实验 [13] 来检测特定的克隆，这两种方法检测出的代表性克隆被挑选出来用于进一步的研究。

以下原因证明上述筛选出来的细胞系（Sp 杂交体）是杂交细胞系。首先它们可以在选择性培养基中生长。其次，培养 4 个月后，它们的核型比两个亲本细胞系之和稍小，但比正常的 BALB/c 小鼠细胞的两倍要多。这表明杂交细胞不是由两个脾细胞之间的融合产生的。此外，杂交细胞中含有一个中着丝粒染色体，这也存在于其亲本细胞 P3-X67Ag8 中。最后，杂交细胞分泌的免疫球蛋白中除了含有 MOPC 21 蛋白外，还含有新的未知组分。后者可能表明杂交细胞分泌的抗体来自亲本细胞中的特异性抗绵羊红细胞抗体。图 3*A* 显示了两种 Sp 杂交细胞克隆分泌物质的等电聚焦图谱。在杂交细胞中，来自亲本 P3 细胞系的条带依然可见，尽管一些新产生的条带使其变得模糊。等电聚焦的图谱非常复杂，不过杂交细胞中的条带的复杂情况似乎来源于亲本肽链的随机重组（见上文，图 1）。确实，脾细胞 P3 杂交细胞克隆分泌的样品被还原后，免疫球蛋白链的图谱更简单一些。亲本 P3 细胞系的轻链和重链成为主要的条带，而其他新产生的条带也变得更明显了。

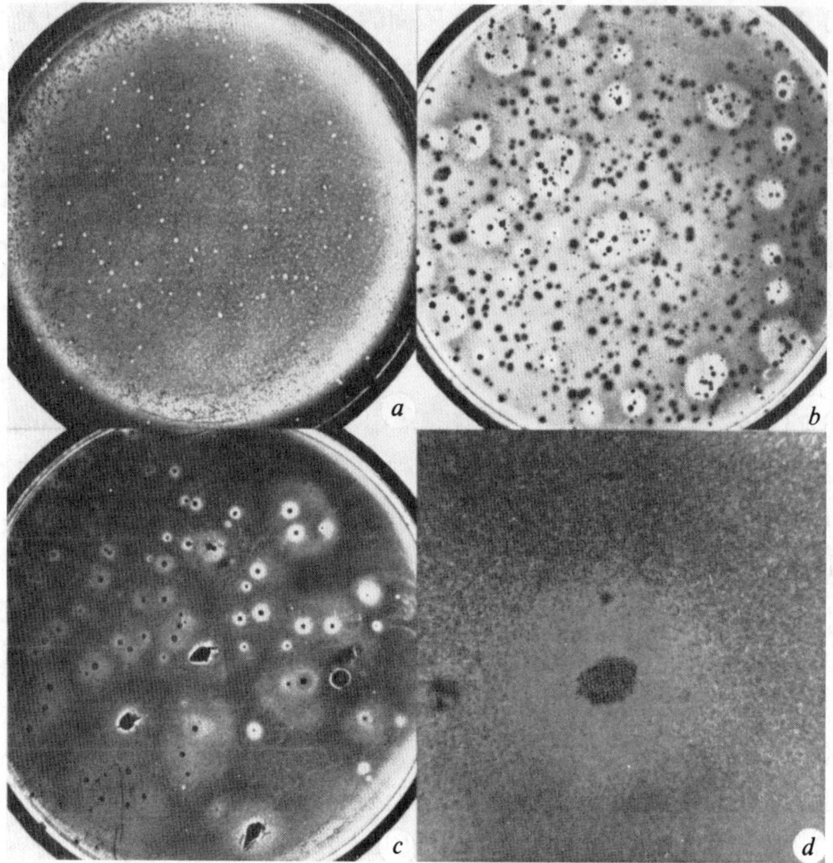

Fig. 2. Isolation of an anti-SRBC antibody-secreting cell clone. Activity was revealed by a halo of haemolysed SRBC. Direct plaques given by: *a*, 6,000 hybrid cells Sp-1; *b*, clones grown in soft agar from an inoculum of 2,000 Sp-1 cells; *c*, recloning of one of the positive clones Sp-1/7; *d*, higher magnification of a positive clone. Myeloma cells (10^7 P3-X67Ag8) were fused to 10^8 spleen cells from an immunised BALB/c mouse. Mice were immunised by intraperitoneal injection of 0.2 ml packed SRBC diluted 1:10, boosted after 1 month and the spleens collected 4 d later. After fusion, cells (Sp-1) were grown for 8 d in HAT medium, changed at 1–3 d intervals. Cells were then grown in Dulbecco modified Eagle's medium, supplemented for 2 weeks with hypoxanthine and thymidine. Forty days after fusion the presence of anti-SRBC activity was revealed as shown in *a*. The ratio of plaque forming cells/total number of hybrid cells was 1/30. This hybrid cell population was cloned in soft agar (50% cloning efficiency). A modified plaque assay was used to reveal positive clones shown in *b*–*d* as follows. When cell clones had reached a suitable size, they were overlaid in sterile conditions with 2 ml 0.6% agarose in phosphate-buffered saline containing 25 μl packed SRBC and 0.2 ml fresh guinea pig serum (absorbed with SRBC) as source of complement. *b*, Taken after overnight incubation at 37°C. The ratio of positive/total number of clones was 1/33. A suitable positive clone was picked out and grown in suspension. This clone was called Sp-1/7, and was recloned as shown in *c*; over 90% of the clones gave positive lysis. A second experiment in which 10^6 P3-X67Ag8 cells were fused with 10^8 spleen cells was the source of a clone giving rise to indirect plaques (clone Sp-2/3-3). Indirect plaques were produced by the addition of 1:20 sheep anti-MOPC 21 antibody to the agarose overlay.

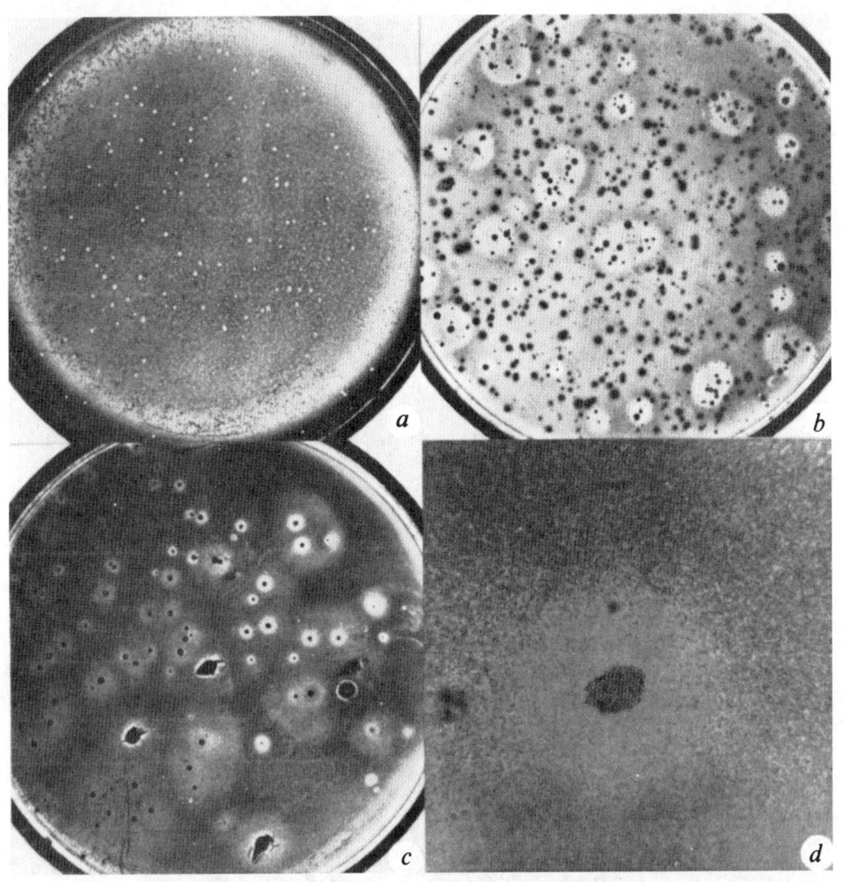

图 2. 可分泌抗绵羊红细胞抗体的细胞克隆的分离。在细胞克隆周围由溶解的红细胞形成的圆环大小代表了其所分泌的抗体的活性大小。直接空斑实验：*a*，6,000 个 Sp-1 杂交细胞；*b*，转接在软琼脂中的 2,000 个 Sp-1 细胞形成的克隆；*c*，对其中一个阳性克隆 Sp-1/7 进行再次克隆；*d*，阳性克隆的高倍放大图像。10^7 个骨髓瘤细胞（P3-X67Ag8）与 10^8 个从免疫后的 BALB/c 小鼠中获得的脾细胞进行融合。首次免疫向小鼠腹膜内注射 0.2 ml 绵羊红细胞（1∶10 稀释），然后在一个月后再加强免疫一次，并于 4 天后收集脾细胞。经过融合后，将细胞（Sp-1）接种在 HAT 培养基中培养 8 天，期间每 1~3 天换液一次。然后将细胞在 DMEM 培养基中培养，并在接下来的两周内补充添加次黄嘌呤和胸腺嘧啶。40 天后，使用如图 *a* 所示的方法检测抗绵羊红细胞抗体的活性。空斑形成细胞数与总杂交细胞数之比为 1/30。这些杂交细胞群体被接种到软琼脂中进行克隆（克隆效率为 50%）。一种改良的空斑实验被用于阳性克隆的选择，如图 *b*~*d* 所示。当细胞克隆长到合适的大小时，在无菌环境下加入 2ml 含 0.6% 琼脂的磷酸盐缓冲液中（其中含有 25 μl 绵羊红细胞和 0.2 ml 新鲜豚鼠血清作为补体成分的来源）铺平板。*b*，37℃孵育过夜后的图像。阳性克隆占总克隆的 1/33。挑选出来一个合适的阳性克隆并悬浮培养。将该克隆命名为 Sp-1/7，如图 *c* 所示再次克隆。超过 90% 的克隆都具有裂解绵羊红细胞的能力。另一个实验使用了 10^6 个 P3-X67Ag8 细胞和 10^8 个脾细胞进行融合，所获得的融合细胞被用于间接空斑实验（克隆命名为 Sp-2/3-3）。在间接空斑实验中，1∶20 稀释的绵羊抗 MOPC 21 细胞抗体被加到融合细胞克隆上层。

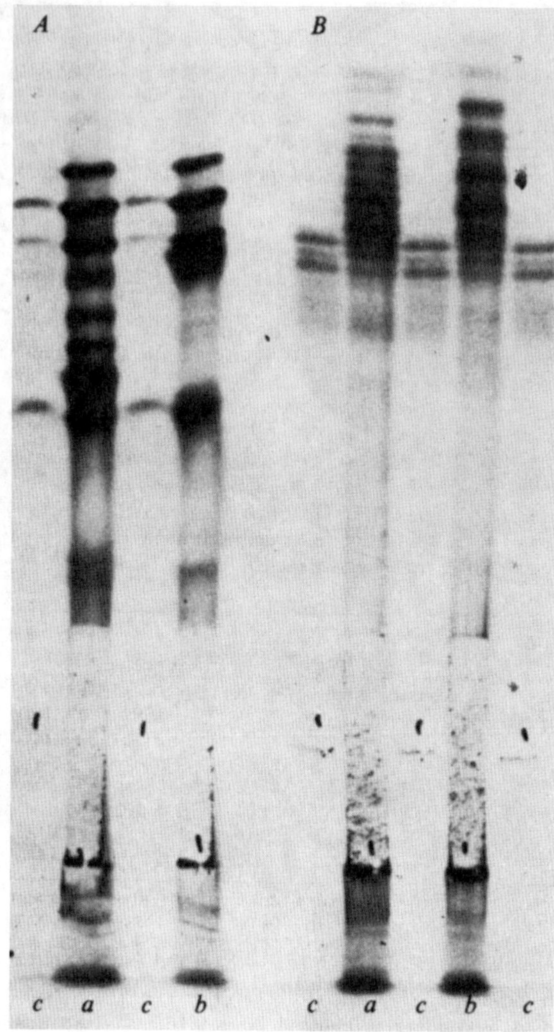

Fig. 3. Autoradiograph of labelled components secreted by anti-SRBC specific hybrid lines. Fractionation before (*B*) and after (*A*) reduction was by IEF. *p*H gradient was 5.0 (bottom) to 9.0 (top) in the presence of 6 M urea. Other conditions as in Fig. 1. Supernatants from: *a*, hybrid clone Sp-1/7-2; *b*, hybrid clone Sp-2/3-3; *c*, myeloma line P3-X67Ag8.

The hybrid Sp-1 gave direct plaques and this suggested that it produces an IgM antibody. This is confirmed in Fig. 4 which shows the inhibition of SRBC lysis by a specific anti-IgM antibody. IEF techniques usually do not reveal 19S IgM molecules. IgM is therefore unlikely to be present in the unreduced sample *a* (Fig. 3*B*) but μ chains should contribute to the pattern obtained after reduction (sample *a*, Fig. 3*A*).

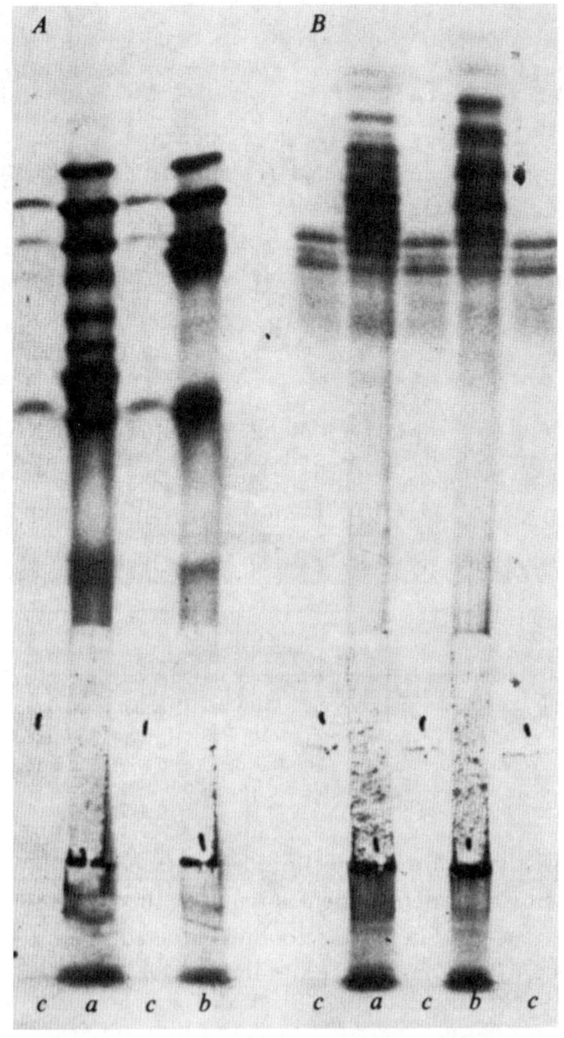

图 3. 抗绵羊红细胞杂交细胞系分泌物的放射自显影图片。使用等电聚焦分析还原前（*B*）和还原后（*A*）的样品。pH 梯度从 5.0（底部）到 9.0（顶部），尿素浓度为 6 M。其他条件与图 1 中相同。*a*，Sp-1/7-2 细胞培养上清；*b*，Sp-2/3-3 细胞培养上清；*c*，骨髓瘤细胞 P3-X67Ag8 细胞培养上清。

杂交细胞 Sp-1 可以直接形成空斑，这表明其可以产生 IgM 抗体。图 4 也进一步证明了这一点：特异性抗 IgM 的抗体可以抑制绵羊红细胞的裂解。等电聚焦实验通常不能分辨 19S 的 IgM 分子。因此，在非还原的样品 *a* 中（图 3*B*）IgM 应该没有显现出来，不过在还原之后 μ 链应该会显示在图中（样品 *a*，图 3*A*）。

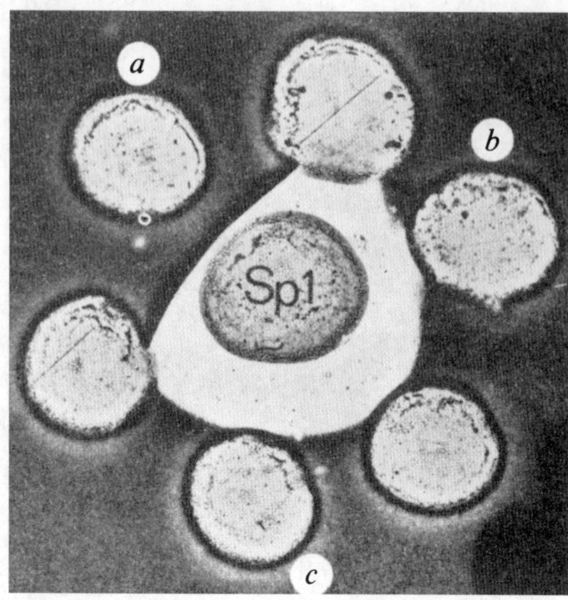

Fig. 4. Inhibition of haemolysis by antibody secreted by hybrid clone Sp-1/7-2. The reaction was in a 9-cm Petri dish with a layer of 5 ml 0.6% agarose in phosphate-buffered saline containing 1/80 (v/v) SRBC. Centre well contains 2.5 μl 20 times concentrated culture medium of clone Sp-1/7-2 and 2.5 μl mouse serum. *a*, Sheep specific anti-mouse macroglobulin (MOPC 104E, Dr. Feinstein); *b*, sheep anti-MOPC 21 (P3) IgGl absorbed with Adj PC-5; *c*, sheep anti-Adj PC-5 (IgG2a) absorbed with MOPC 21. After overnight incubation at room temperature the plate was developed with guinea pig serum diluted 1:10 in Dulbecco's medium without serum.

The above results show that cell fusion techniques are a powerful tool to produce specific antibody directed against a predetermined antigen. It further shows that it is possible to isolate hybrid lines producing different antibodies directed against the same antigen and carrying different effector functions (direct and indirect plaque).

The uncloned population of P3-spleen hybrid cells seems quite heterogeneous. Using suitable detection procedures it should be possible to isolate tissue culture cell lines making different classes of antibody. To facilitate our studies we have used a myeloma parental line which itself produced an Ig. Variants in which one of the parental chains is no longer expressed seem fairly common in the case of P1-P3 hybrids (Fig. 1*h*). Therefore selection of lines in which only the specific antibody chains are expressed seems reasonably simple. Alternatively, non-producing variants of myeloma lines could be used for fusion.

We used SRBC as antigen. Three different fusion experiments were successful in producing a large number of antibody-producing cells. Three weeks after the initial fusion, 33/1,086 clones (3%) were positive by the direct plaque assay. The cloning efficiency in the experiment was 50%. In another experiment, however, the proportion of positive clones was considerably lower (about 0.2%). In a third experiment the hybrid population was studied by limiting dilution analysis. From 157 independent hybrids, as many as 15 had anti-SRBC activity. The proportion of positive over negative clones is remarkably

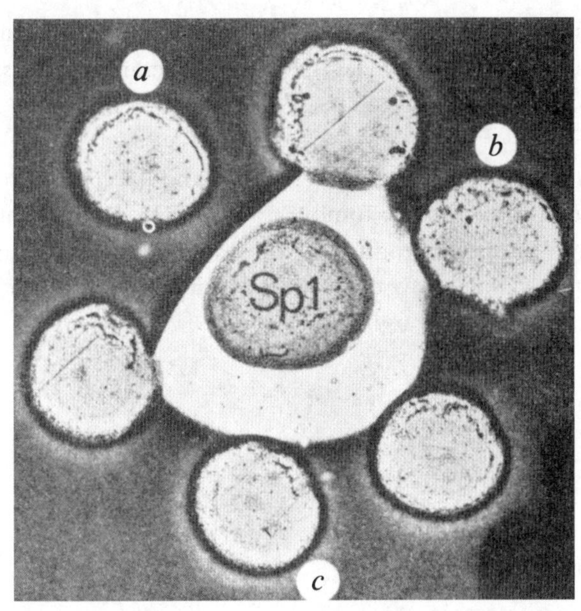

图 4. 杂交细胞克隆 Sp-1/7-2 分泌的抗体抑制红细胞溶解。反应在 9 cm 培养皿中进行。在 5 ml 0.6% 的琼脂糖（溶解在 PBS 中）中按 1/80（体积比）加入了绵羊红细胞。中间的孔中含有 2.5 μl 20 倍浓缩的 Sp-1/7-2 培养上清和 2.5 μl 小鼠血清。*a*，绵羊抗小鼠巨球蛋白（MOPC 104E，范斯坦博士）；*b*，绵羊抗 MOPC 21（P3）IgG1 与 Adj PC-5 结合；*c*，绵羊抗 Adj PC-5（IgG2a）与 MOPC 21 结合。室温下过夜培养后，加入豚鼠血清（用无血清的杜尔贝科培养基稀释 10 倍）。

上述结果表明，细胞融合技术是一种制备针对某种特定抗原的特异性抗体的强有力的工具。并进一步表明这种技术可能分离出能够产生针对同一抗原的具有不同效应功能的抗体的杂交细胞系（直接空斑和间接空斑）。

未克隆的 P3–脾细胞杂交细胞似乎具有异质性。使用合适的检测方法应该可以将产生不同种类抗体的组织培养细胞系分离出来。为了方便研究，我们使用了一种自身可以分泌免疫球蛋白的骨髓瘤细胞作为亲本细胞。在 P1–P3 杂交中，变异体某一亲本链不再表达的现象似乎相当常见（图 1*h*）。因此，筛选那些只表达某种特异性抗体链的细胞系似乎会更简单一些。或者，那些不产生变异体的骨髓瘤细胞也可以用于融合实验。

在实验中我们使用绵羊红细胞作为抗原。三个不同的融合实验都成功地产生了大量抗体生成细胞。第一个实验发现，在最初融合之后三周，33/1086 克隆（3%）在直接空斑实验中呈阳性。这个实验的克隆效率为 50%。然而，在另一个实验中，阳性克隆率则相当低（约为 0.2%）。在第三个实验中，我们使用了有限稀释法对杂交细胞群体进行了分析。在 157 个独立的克隆中，有 15 个克隆具有抗绵羊红细胞的活性。阳性与阴性克隆之比非常高。这可能是因为免疫过后的脾细胞大大增强了杂

high. It is possible that spleen cells which have been triggered during immunisation are particularly successful in giving rise to viable hybrids. It remains to be seen whether similar results can be obtained using other antigens.

The cells used in this study are all of BALB/c origin and the hybrid clones can be injected into BALB/c mice to produce solid tumours and serum having anti-SRBC activity. It is possible to hybridise antibody-producing cells from different origins[4,5]. Such cells can be grown *in vitro* in massive cultures to provide specific antibody. Such cultures could be valuable for medical and industrial use.

(**256**, 495-497; 1975)

G. Köhler & C. Milstein
MRC Laboratory of Molecular Biology, Hills Road, Cambridge CB2 2QH, UK

Received May 14; accepted June 26, 1975.

References:

1. Potter, M., *Physiol. Rev.*, **52**, 631-719 (1972).
2. Horibata, K., and Harris, A. W., *Expl Cell Res.*, **60**, 61-70 (1970).
3. Milstein, C., and Munro, A. J., in *Defence and Recognition* (edit. by Porter, R. R.), 199-228 (MTP Int. Rev. Sci., Butterworth, London, 1973).
4. Cotton, R. G. H., and Milstein, C., *Nature*, **244**, 42-43 (1973).
5. Schwaber, J., and Cohen, E. P., *Proc. Natl. Acad. Sci. U.S.A.*, **71**, 2203-2207 (1974).
6. Littlefield, J. W., *Science*, **145**, 709 (1964).
7. Svasti, J., and Milstein, C., *Biochem. J.*, **128**, 427-444 (1972).
8. Milstein, C., Adetugbo, K., Cowan, N. J., and Secher, D. S., *Progress in Immunology*, II, **1** (edit. by Brent, L., and Holborow, J.), 157-168 (North-Holland, Amsterdam, 1974).
9. Harris, H., and Watkins, J. F., *Nature*, **205**, 640-646 (1965).
10. Awdeh, A. L., Williamson, A. R., and Askonas, B. A., *Nature*, **219**, 66-67 (1968).
11. Milstein, C., Brownlee, G. G., Cartwright, E. M., Jarvis, J. M., and Proudfoot, N. J., *Nature*, **252**, 354-359 (1974).
12. Frangione, B., and Milstein, C., *Nature*, **244**, 597-599 (1969).
13. Jerne, N. K., and Nordin, A. A., *Science*, **140**, 405 (1963).
14. Cotton, R. G. H., Secher, D. S., and Milstein, C., *Eur. J. Immun.*, **3**, 135-140 (1973).

交细胞的存活能力。不过，使用其他抗原是否也能得到类似的结果仍有待进一步的研究。

本研究所用到的所有细胞都来自 BALB/c 小鼠。将融合后产生的杂交克隆注射到 BALB/c 小鼠体内能够诱导实体瘤并产生抗绵羊红细胞活性的抗体血清。此外，也可以使用不同种属来源的抗体生成细胞进行融合实验 [4,5]。体外大规模培养这些细胞可以用于生产特异性抗体，而特异性抗体在医学和工业领域具有非常高的应用价值。

（张锦彬 翻译；胡卓伟 审稿）

Plio-Pleistocene Hominid Discoveries in Hadar, Ethiopia

D. C. Johanson and M. Taieb

Editor's Note

After just three field seasons exploring the Afar Triangle of Ethiopia, Maurice Taieb and colleague Donald Johanson could present AL 288–1, a partial skeleton of a very primitive form that lived more than three million years ago. At first, this skeleton was assigned, if tentatively, to *Australopithecus africanus*, because of resemblances to fossils from Sterkfontein, and other Hadar specimens were thought to be more like *Homo*. Eventually, the Hadar material was ascribed to a new taxon, *Australopithecus afarensis*. The significance of this find was that even the most primitive hominids walked as upright as modern humans. This alone guaranteed the status of AL 288–1 as a truly important fossil. But it was its nickname that made it iconic: "Lucy".

The International Afar Research Expedition has now recovered remains of twelve hominid individuals from geological deposits estimated to be ~3.0 Myr in Hadar, Ethiopia. A partial skeleton represents the most complete hominid known from this period. The collection suggests that *Homo* and *Australopithecus* coexisted as early as 3.0 Myr ago.

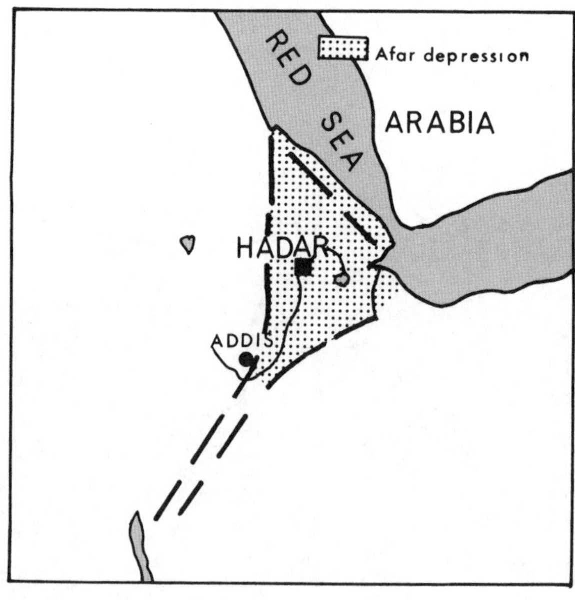

Fig. 1. Location of Hadar.

在埃塞俄比亚哈达尔发现的上新世–更新世时期的人科动物

约翰森，塔伊布

编者按

在埃塞俄比亚阿法尔三角区经过了仅仅三个野外工作季的发掘，莫里斯·塔伊布和同事唐纳德·约翰森就为大家展现了一个生活在超过 300 万年前的、形态非常原始的一具人类的不完整的骨架 AL 288–1。起初，由于认为它与斯泰克方丹的化石类似，因此被暂时归入南方古猿非洲种，而其他哈达尔的标本则被认为更像人属。最终，这具哈达尔的材料被归入一个新的分类单元——南方古猿阿法种。这个发现的意义在于，即使是最原始的人科动物，他们也跟现代人一样直立行走。单是这一点就足以确保 AL 288–1 作为一块真正重要的化石的地位。但是，使它成为标志性化石的却是它的外号——“露西”。

国际阿法尔研究探险队现已修复了 12 个人科动物个体的化石，这些化石均发现在埃塞俄比亚哈达尔的一处估计距今约 300 万年的地层堆积物中。其中一具不完整的骨架代表了现在已知的这一时期最完整的人科动物。该标本提示人属和南方古猿早在 300 万年前就同时存在着。

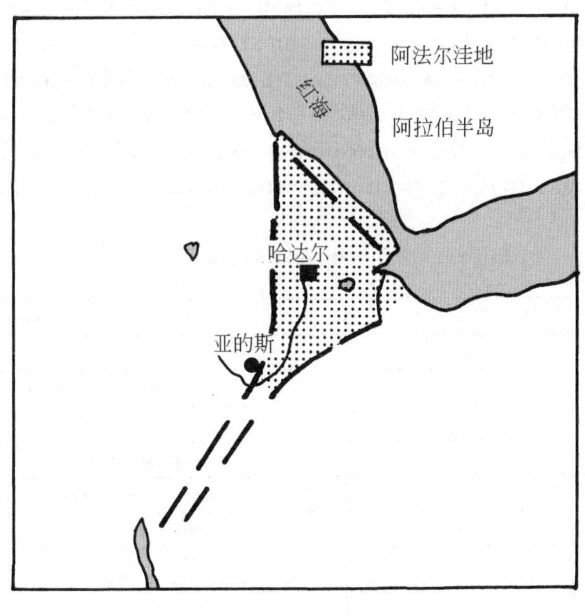

图 1. 哈达尔的位置

FOLLOWING a short reconnaissance expedition in 1972 (ref. 1) to the central Afar, one of us (M. T.) organised the International Afar Research Expedition. We have now codirected two field campaigns (September–December, 1973 and 1974 (refs 2 and 3)) in the area known as Hadar (see Fig. 1).

Although a number of other excellent fossil sites are known from the central Afar[4], Hadar was selected for intensive exploration for a number of reasons: first, the fossil vertebrate assemblage suggested a considerable antiquity (~3–4 Myr); second, the area is extremely rich in splendidly preserved fossils, deposited in low energy environments; third, the deposits are heavily dissected and characterised by clear marker-horizons which are laterally continuous; and fourth, the thick series of lake sediments contains a number of volcanic horizons available for absolute radiometric age determinations.

The Hadar site is located in the Afar depression in the west central Afar sedimentary basin[5] at 11°N and 40°30′E. It now encompasses ~42 km^2, but extension of study into adjacent, previously unexplored regions will substantially increase the area.

Preliminary palaeontological investigations, particularly of the suids and the elephants, have suggested a biostratigraphic correlation of the Hadar Formation with the Usno and the lower portion of the Shungura Formations[5,6]. Two radiometric K–Ar age determinations for a basalt support this with an age estimate of 3.0±0.2 Myr (ref. 5).

All hominid fossils (Table 1), and most of the vertebrate material, were surface finds, although some small scale excavation and sieving operations were undertaken. A painted marker was placed at each palaeontological locality, each located on a map and the geological section carefully examined to determine the horizon yielding fossil material (see Fig. 2). At each hominid locality, a topographical map and detailed stratigraphical sections were drawn. Samples of matrix adhering to fossil hominid specimens are undergoing mineralogical and sedimentological analyses for comparison with sediment samples taken at their collection points. Their completeness and lack of an ataxic abrasion[7] virtually precludes the possibility that the specimens were moved over great distances.

Table 1. Fossil hominid specimens from Hadar

Afar locality (AL)	Date of discovery	Description
	1973	
128–1	30 October	Left proximal femur fragment
129–1a	30 October	Right distal femur
129–1b	30 October	Right proximal tibia
129–1c	30 October	Right proximal femur fragment
166–9	11 December	Left temporal fragment
	1974	
188–1	16 October	Right mandibular corpus; M_2–M_3

1972 年，在一次赴阿法尔中部的短期勘察探险（参考文献 1）之后，我们的一名成员（塔伊布）便组织了国际阿法尔研究探险队。现在我们共同指导了两项野外活动（1973 年和 1974 年的 9~12 月（参考文献 2 和 3）），这两项活动都是在著名的哈达尔地区（见图 1）进行的。

尽管有许多其他极佳的化石遗址都是在阿法尔中部地区发现的[4]，但是选在哈达尔开展集中的发掘工作是有很多原因的：第一，脊椎动物化石组合显示其时代相当古老（距今约 300 万到 400 万年）；第二，这里富含沉积在低能环境中并且保存得极好的化石；第三，这些沉积物被剧烈地切割，并在其侧面显示出连续的标志性地层；第四，厚厚的湖泊沉积物序列包含大量的可以用来测定绝对放射性年代的火山灰层。

哈达尔遗址的地理坐标为北纬 11° 东经 40° 30′，位于阿法尔沉积盆地中心区西部的阿法尔洼地[5]。该遗址现在包含约 42 km^2 的地域，不过对相邻的、以前未开发的地区进行的拓展研究实质上增加了其范围。

初步的古生物学研究，尤其是对猪类和大象的研究，已经提示了哈达尔组与乌斯诺组和上古拉组的下部之间存在生物地层相关性[5,6]。对玄武岩进行了两次放射性 K–Ar 年代测定，估计其年代为距今 3.0±0.2 百万年，这进一步支持了该结论（参考文献 5）。

尽管进行了一些小范围的挖掘和筛查工作，但是所有的人科动物化石（表 1）和大部分脊椎动物材料都是在地表发现的。在每个古生物化石点都放置了一个涂色的标记，然后在地图上一一定位，并且对地质剖面进行了认真的调查以确定产出化石的地层（见图 2）。在每个人科动物化石点，都绘制了地形图和详细的地层剖面图。对人科动物化石表面黏附的围岩样本进行了矿物学和沉积学的分析，并将其与化石采集地的沉积物样本分析结果进行了对比。事实上，它们的完整性和未出现（与周围物质的）无序摩擦所产生的痕迹[7]排除了这些标本远距离搬运的可能性。

表 1. 来自哈达尔的人科动物化石标本

阿法尔地点（AL）	发现日期	说明
	1973 年	
128–1	10 月 30 日	左股骨近端破片
129–1a	10 月 30 日	右股骨远端
129–1b	10 月 30 日	右胫骨近端
129–1c	10 月 30 日	右股骨近端破片
166–9	12 月 11 日	左颞骨破片
	1974 年	
188–1	10 月 16 日	右下颌骨体；M_2~M_3

Continued

Afar locality (AL)	Date of discovery	Description
198–1	18 October	Left mandibular corpus; C–P_4, dm, M_1–M_2
198–17a, b	5 December	Left I^1 and I^2
198–18	5 December	Right I_2
199–1	17 October	Right maxilla; C–M^3
200–1a, b	17 October	Complete maxilla; 16 teeth. Right M_1
211–1	20 October	Right proximal femur fragment
228–1	27 October	Diaphysis of right femur
241–14	22 December	Left lower molar
266–1	16 November	Mandibular corpus; left P_3–M_1, right P_3–M_3
277–1	19 November	Left mandibular corpus; C–M_2
288–1	24 November	Partial skeleton: occipital and parietal fragments; mandibular corpus with left P_3, M_3, right P_3–M_3, mandibular condyles; right scapula fragment; right humerus; proximal and distal left humerus; proximal and distal right and left ulnae; proximal and distal right radius; distal left radius; left capitate; 2 phalanges; 6 thoracic vertebrae and fragments; 1 lumbar vertebra; sacrum; left innominate; left femur; proximal and distal right tibia; right talus right distal fibula; numerous rib fragments.

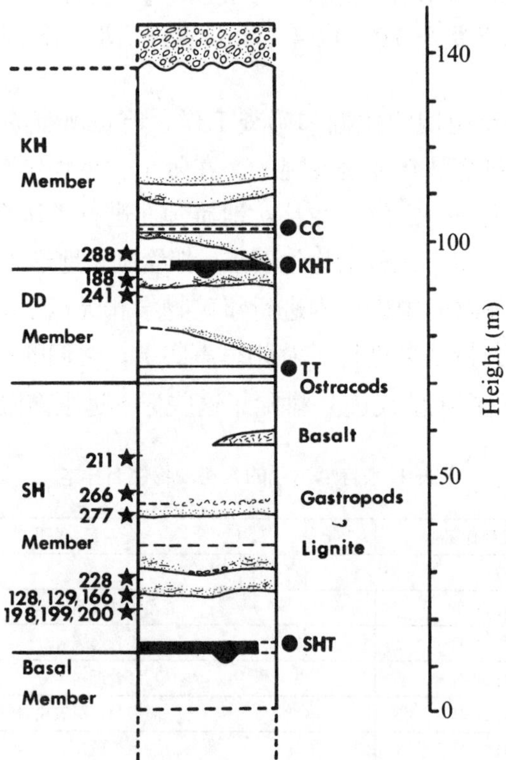

Fig.2. Stratigraphic positions of Hominid Horizons. ★ , Hominid levels; ● , marker beds.

续表

阿法尔地点（AL）	发现日期	说明
198–1	10 月 18 日	左下颌骨体：C~P_4、dm、M_1~M_2
198–17a、b	12 月 5 日	左 I^1 和 I^2
198–18	12 月 5 日	右 I_2
199–1	10 月 17 日	右上颌骨：C~M^3
200–1a、b	10 月 17 日	完整上颌骨：16 颗牙齿。右 M_1
211–1	10 月 20 日	右股骨近端破片
228–1	10 月 27 日	右股骨骨干
241–14	12 月 22 日	左下臼齿
266–1	11 月 16 日	下颌骨体：左 P_3~M_1、右 P_3~M_3
277–1	11 月 19 日	左下颌骨体：C~M_2
288–1	11 月 24 日	不完整的骨架：枕骨和顶骨破片；含左 P_3、M_3、右 P_3~M_3、两侧下颌髁突的下颌骨体；右肩胛骨破片；右肱骨；左肱骨近端和远端；左右尺骨近端和远端；右桡骨近端和远端；左桡骨远端；左头状骨；2 块指骨；6 块胸椎骨和破片；1 块腰椎骨；骶骨；左髋骨；左股骨；右胫骨近端和远端；右距骨；右腓骨远端；许多肋骨破片。

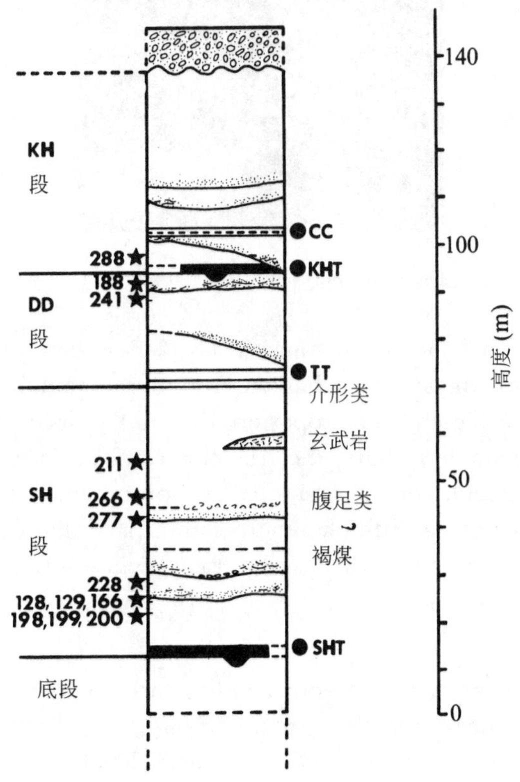

图 2. 人科动物层的地层位置。★：人科动物层；●：标志性层位。

1973 Hominid Discoveries

The first discovery of fossil hominid remains in the Hadar region occurred on October 30, 1973, when four associated leg bone fragments (AL 128, 129) were collected from a mudstone horizon. These consist of fragmentary right and left proximal femora associated with a right proximal tibia and distal femur. Their proximity to one another, as well as the morphological and size similarities of the proximal femora, strongly suggest that they represent a single individual. This provides a complete knee which will greatly enhance our understanding of the biomechanics of this important joint in early hominids (see Fig. 3).

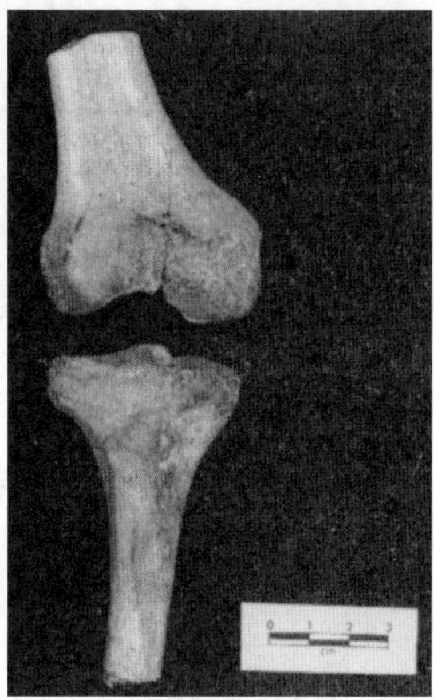

Fig. 3. Anterior view of AL 129–1a and 1c

Both proximal femora lack heads and necks. The left is best preserved with the shaft broken ~ 38.0 mm below the lesser trochanter. From the remaining portions of the femora it is clear that the neck is flattened anteroposteriorly and that its cross section is oval. There is an indication of trochanteric flare, the trochanteric fossa is well marked, a quadrate tubercle is clearly discernible, muscle markings are prominent on the greater trochanter, the spiral line is pronounced and the lesser trochanter is visible from the anterior aspect. Both fragments, particularly the right, exhibit some degree of predepositional crushing, which is suggestive of carnivore activity.

The distal femur is undistorted, and both condyles are complete and intact. It is small, with the lateral condyle measuring 39.0 mm anteroposteriorly and 18.9 mm mediolaterally. This fragment demonstrates a number of anatomical details which are intimately related to bipedal locomotion[8,9]: the bicondylar angle is rather high, the lateral lip of the patellar groove is raised and the lateral condyle flattened and elongated.

1973 年人科动物的发现

1973 年 10 月 30 日在哈达尔地区首次发现了人科动物化石，当时从一处泥岩层采集到了四个相关联的腿骨残段（AL 128 和 AL 129）。这些标本包括残断的左右股骨近端以及与其相关的右侧胫骨近端及股骨远端。它们彼此都很靠近，且股骨近端的形态和尺寸都很相似，这强有力地说明它们来自同一个体。该标本提供了一个完整的膝关节，这将大大增加我们对早期人科动物的这一重要关节的生物力学认识（见图 3）。

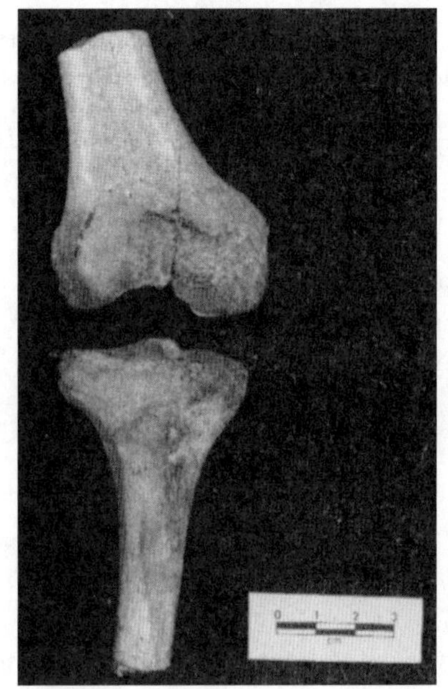

图 3. AL 129–1a 和 1c 的前面观。

两块股骨近端都缺少头部和颈部。左股骨近端保存最好，在小转子下约 38.0 mm 的骨干处有断裂。从股骨的其余部分看，股骨颈在前后方向上很明显是扁平的，其横截面呈椭圆形。有结节外展（编者注：现代人体解剖学中无法查到此解剖部位的对应译法。早期人科动物由于行走姿势不同可能存在大结节边缘外展的现象，故暂译如此）的迹象，转子窝很明显，可以很清楚地辨别出方形结节，肌肉纹理在大转子处非常明显，螺旋线突出，从前面可以看到小转子。两块残段，尤其是右侧残段，呈现出在沉积前就有一定程度破碎的现象，提示食肉动物的活动痕迹。

股骨远端未变形，两个髁突都是完整无损的。这块股骨远端很小，经测量，外侧髁前后方向上有 39.0 mm 长，内外方向上有 18.9 mm 长。该残段展示了许多与双足步行密切相关的解剖学细节特征[8,9]：两个髁突所成角度非常大，髌骨沟的侧唇升高，外侧髁扁平且细长。

The associated proximal tibia is intact except for some slight abrasion around the periphery of the superior articular surface. It also is small, the total preserved articular surface measuring 50.7 mm mediolaterally and 33.0 mm anteroposteriorly. The tibial tuberosity is pronounced and is limited proximally by the transverse groove; the soleal line is distinct on the posterior surface of the shaft and the interosseous membrane attachment is indicated by a roughened line on the medial aspect of the shaft; the head is slightly retroflexed with only minor tibial torsion; and the intercondyloid eminence is prominent with well developed intercondyloid fossae.

A heavily eroded temporal fragment (AL 166-9) constitutes the only other hominid discovery from the 1973 field season. The specimen derives from a sandstone horizon a few metres stratigraphically above the level of the postcranial material. The temporal has been extensively eroded, with most of the petrous portion broken away, and extensive pneumatisation is exposed in the mastoid and zygomatic regions. The mandibular fossa is broad and flat, bordered by a postglenoid process, but open anteriorly with only a slight articular tubercle. There is a prominent entoglenoid process on the medial wall which is broken exposing a large pneumatisation. The mastoid is large and globular. The external auditory meatus is oval in section with a thick tympanic plate.

1974 Hominid Discoveries

Our sample of the Hominidae has been greatly augmented with additional field work. The 1974 investigations yielded remains of 10 additional individuals represented by dental, cranial and postcranial elements, including a remarkably intact partial skeleton.

Stratigraphically just below the 1973 hominid discoveries, one complete (AL 200-1) and one half maxilla (AL 199-1) (see Fig. 4) were recovered by Ato Alemayehu Asfaw within 15 m of one another and from the same mudstone horizon.

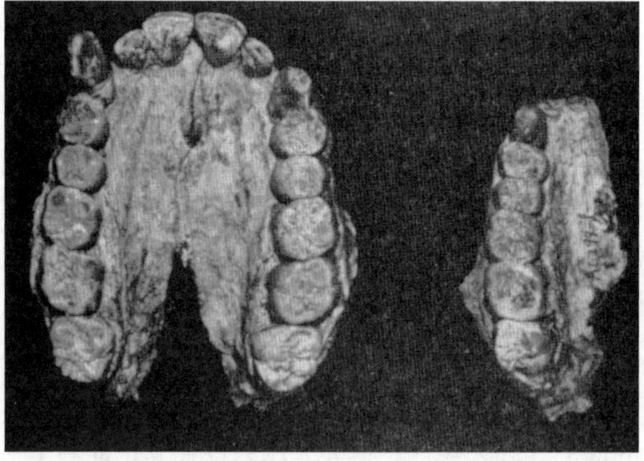

Fig. 4. Occlusal views of AL 200-1a and AL 199-1

The AL 200 specimen is an undistorted palate with full dentition and an associated right

相关的胫骨近端除了在上关节面的边缘有轻微磨损外，其余部分都是完整的。这块胫骨近端也很小，经测量，保存下来的整个关节面在内外方向上长 50.7 mm，前后方向长 33.0 mm。胫骨粗隆突出，其近端以横股沟为界；比目鱼肌线在骨干的后表面上很明显，骨间膜的附着面可以从骨干中部粗糙的线迹看出来；由于胫骨有轻微的扭曲，其头部略显翻转；髁间隆起突出，髁间窝很发达。

一块严重磨损的颞骨破片（AL 166–9）是 1973 年野外工作中仅有的人科动物的其他发现。该标本来源于颅后化石产出层之上数米处的砂岩层。颞骨磨损的范围很大，岩部的大部分都断裂了，大量的气腔暴露在乳突区和颧突区。下颌窝宽阔平坦，后边是关节后突，但是前面仅有一细小的关节结节。在内侧壁有一个凸出的关节内突，内侧壁断裂，暴露出一个大的气腔。乳突大且呈球状。外耳道剖面呈椭圆形，具有很厚的鼓板。

1974 年人科动物的发现

接下来的野外工作大大增加了我们的人科动物标本的数量。1974 年的调查发现了另外 10 个个体的化石，包括齿系、颅骨和颅后等部分，还包括一具较完整的局部骨架。

就在 1973 年发现人科动物的地层之下，阿托·阿莱马耶胡·阿斯富发现并复原了一块完整的上颌骨（AL 200–1）和另外半块上颌骨（AL 199–1）（见图 4），它们之间相距近 15 m，且处于同一泥岩层。

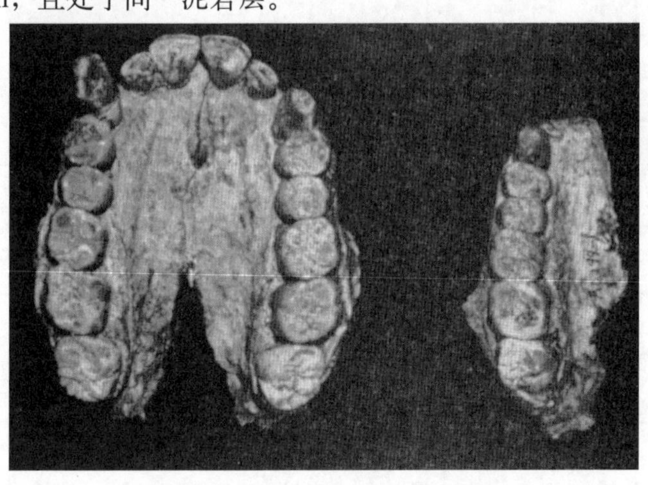

图 4. AL 200–1a 和 AL 199–1 的咬合面观。

AL 200 号标本是一个没有变形的带有整套齿系的上颌和一个相关联的右 M_1。

M_1. The general pattern of dental wear is interesting, in that the incisors exhibit extensive "ribbon like" wear and dentine exposure relative to the postcanine teeth. In addition, the canines and the P^3s have suffered antemortem enamel chippage on their buccal surfaces during the specimen's lifetime. The tooth rows of AL 200 are subparallel and the arch is relatively long. The anterior portion of the arch is broad to accommodate the large central incisors and marked diastemata occur between the lateral incisors and the canines. The palate is shallow, becoming somewhat deeper posteriorly.

In lateral view this specimen exhibits pronounced alveolar prognathism. The zygomatic root is situated above the first molar, the maxillary sinuses are large and the lower nasal margin is guttered.

The right half maxilla is smaller in size and is less complete, with C–M^3, as well as both incisor sockets and the I^2 root. The dentition is smaller than that of AL 200 (Table 2), but the similarity in morphological detail is striking. Although the incisors are lacking, it is apparent from the root sockets that the anterior portion of the arch was broad and somewhat squared-off as in AL 200. The specimen is broken close to the midline exposing the incisive canal. The palate is shallow and the greater palatine foramen is present. The inferior nasal margin is not sharp and in lateral view alveolar prognathism is well developed. The maxillary sinus is large and the zygomatic root is situated above the distal portion of the first molar.

Table 2. Dental measurements of AL 200–1a and AL 199–1 (mm)

	Left		Right	
AL 200–1a	Mesiodistal	Buccolingual	Mesiodistal	Buccolingual
I^1	10.8	8.3	10.9	8.5
I^2	7.4	7.1	7.3	7.0
C	9.4	10.9	9.4	11.0
P^3	9.0	12.2	8.9	12.2
P^4	8.5	12.2	8.5	12.1
M^1	11.8	13.1	11.8	13.2
M^2	13.8	14.8	13.7	15.0
M^3	14.2	15.0	14.3	15.0
AL 199–1				
C			8.7	9.3
P^3			7.3	11.2
P^4			7.1	11.2
M^1			10.1	12.0
M^2			11.7	13.5
M^3			11.3	(12.7)

牙齿磨耗的总体模式很有趣，因为门齿呈现出大量的“带状”磨损而且其牙本质像颊齿一样暴露出来。此外，犬齿和 P^3 在颊面上存在着珐琅质剥落，这些剥落发生在个体生前。AL 200 的齿列接近平行，牙弓相对较长。牙弓的前部很宽可容下大的中门齿，侧门齿与犬齿之间有明显的齿隙。腭很浅，后部稍微变深。

从侧面观看，该样本显示出明显的牙槽前突。颧弓根部位于第一臼齿之上，上颌窦大，鼻腔的下沿呈沟状。

半块右侧上颌骨尺寸更小，也更不完整，保留有 C~M^3，还有两个门齿槽和一个 I^2 的牙根。其齿列比 AL 200 的要小（表 2），但是形态学细节上的相似性很明显。尽管缺少门齿，但从其牙根槽上可以明显看出牙弓的前部很宽，与 AL 200 的形状相似。该标本在靠近中线处断裂，暴露出门齿管。腭很浅，存在较大的腭孔。下鼻缘不清晰，从侧面看，牙槽前突很发达。上颌窦很大，颧弓根部位于第一臼齿远中部分的上方。

表 2. AL 200–1a 和 AL 199–1 的牙齿测量尺寸（mm）

	左		右	
AL 200–1a	近中远中方向	颊舌方向	近中远中方向	颊舌方向
I^1	10.8	8.3	10.9	8.5
I^2	7.4	7.1	7.3	7.0
C	9.4	10.9	9.4	11.0
P^3	9.0	12.2	8.9	12.2
P^4	8.5	12.2	8.5	12.1
M^1	11.8	13.1	11.8	13.2
M^2	13.8	14.8	13.7	15.0
M^3	14.2	15.0	14.3	15.0
AL 199–1				
C			8.7	9.3
P^3			7.3	11.2
P^4			7.1	11.2
M^1			10.1	12.0
M^2			11.7	13.5
M^3			11.3	(12.7)

A partial mandible (AL 266–1) was located in a horizon of sandy clay, ~10 m stratigraphically below the basalt flow[5]. This specimen includes the complete right corpus with P_3–M_3, the symphysis with canine and incisor roots, and a portion of the left corpus with P_3–M_1 (Table 3). The dentition is moderately worn with the third molar having just come into occlusion and showing only minor wear facets. The anterior portion of the dental arch is rounded and the dental rows are straight and slightly divergent posteriorly. The post-incisive planum is only moderately developed and the corpus in the region of M_1 is 21.7 mm thick and 31.0 mm deep.

Table 3 Dental measurements of AL 266–1 (mm)

	Left		Right	
AL 266–1	Mesiodistal	Buccolingual	Mesiodistal	Buccolingual
P_3	9.1	10.1	9.2	10.1
P_4	8.9	11.0	9.4	10.8
M_1	12.1	11.9	12.1	12.0
M_2			13.3	14.0
M_3			15.3	13.7

A right mandibular fragment (AL 188–1) containing M_2–M_3 and roots of P_3–M_1, was collected from a sandstone horizon immediately below KHT. The molars are heavily worn. The mandible exhibits some wind abrasion, but it is possible to estimate a thickness of 20.7 mm and a depth of ~ 32.0 mm in the region of M_1.

AL 277–1, a fragmentary left mandible containing C–M_2 and sockets for I_1–I_2, is derived from a sandstone horizon. It is broken just to the right of the midline and distal to M_2. The occlusal wear is heavy with the canine worn flat. Anteriorly the fragment gives the impression of being deep, measuring ~41.0 mm just distal to P_3. A prominent post-incisive planum is present, as well as a slight inferior mandibular torus. The inferior surface of the symphysis is flattened and has distinct mental spines. The corpus exhibits slight eversion.

An interesting left half mandible (AL 198–1) was recovered from a mudstone horizon, stratigraphically equivalent to AL 199 and AL 200. The specimen is broken near the midline with the I_1 socket and the root for I_2 preserved. The C–P_4 and M_1–M_2 are present, with a retained deciduous molar located distal to the P_4. The deciduous nature of this tooth is confirmed by a radiograph demonstrating a widely divergent root system. Substantial occlusal wear is evident on P_4–M_2. the corpus is broken just distal to M_2, exposing a portion of the mesial root of M_3 immediately superior to the mandibular canal. The mandible is lightly built, being 16.0 mm thick and 31.9 mm deep at M_1. It exhibits some modelling and the inferior margin is thin and not bulbous.

不完整的下颌骨（AL 266–1）是在砂质黏土层中发现的，在地层上位于玄武岩熔体层下约 10 m 处[5]。该标本包括带有 P_3~M_3 的完整右下颌骨体、带有犬齿根和门齿根的联合部，以及带有 P_3~M_1 的部分左侧主体（表 3）。齿列中度磨损，第三臼齿刚刚达到咬合面，仅显示出微小的磨损面。牙弓前部呈圆形，齿列直，只是在后部稍微有些分散。门齿后平面中等发育，M_1 处的下颌体厚 21.7 mm，深 31.0 mm。

表 3. AL 266–1 的牙齿测量尺寸（mm）

	左		右	
AL 266–1	近中远中方向	颊舌方向	近中远中方向	颊舌方向
P_3	9.1	10.1	9.2	10.1
P_4	8.9	11.0	9.4	10.8
M_1	12.1	11.9	12.1	12.0
M_2			13.3	14.0
M_3			15.3	13.7

带有 M_2~M_3 和 P_3~M_1 齿根的右下颌骨破片（AL 188–1）是从紧挨着 KHT 下部的砂岩层中采集到的。其臼齿磨损严重。下颌骨显示出某种程度的风化，但是可以估计出在 M_1 处的厚度为 20.7 mm，深度约为 32.0 mm。

AL 277–1 是一块带有 C~M_2 和 I_1~I_2 牙槽的残破不全的左下颌骨，它们是从砂岩层挖掘出来的。在中线右侧及 M_2 之后发生了断裂。咬合面磨损严重， 犬齿已经磨平。前部断裂让人感觉下颌骨深，P_3 远端深约 41.0 mm。突出的门齿后平面保存下来，还有一个明显的下颌圆枕。联合部的下表面变平，具有明显的颏棘。下颌骨显示出轻微的外翻现象。

在泥岩中发现并复原了一块有趣的左半部分下颌骨（AL 198–1），其位置在地层上与 AL 199 和 AL 200 相当。该标本在接近中线处断裂，I_1 牙槽和 I_2 牙根保存下来。C~P_4 和 M_1~M_2 也保存下来，还有一个位于 P_4 之后的乳臼齿。这颗牙的乳牙性质得到了 X 光照片的证实，显示其牙根分叉。大量的咬合磨损在 P_4~M_2 上表现得很明显。下颌骨体在 M_2 之后断裂，暴露出一部分紧挨在下颌管之上的 M_3 的近中牙根。下颌骨较纤细，在 M_1 处有 16.0 mm 厚、31.9 mm 深。它显示出一定的形态，下缘薄而不鼓。

During screening operations, three additional teeth were located at this locality: a right I_2 (AL 198–18), a left I^2 (AL 198–17a) and a left I^1 (AL 198–17b). The interproximal facets on the upper incisors match perfectly, suggesting that they are from the same individual. Because of the close association of these teeth with the AL 198 mandible, we tentatively assign them to the same individual.

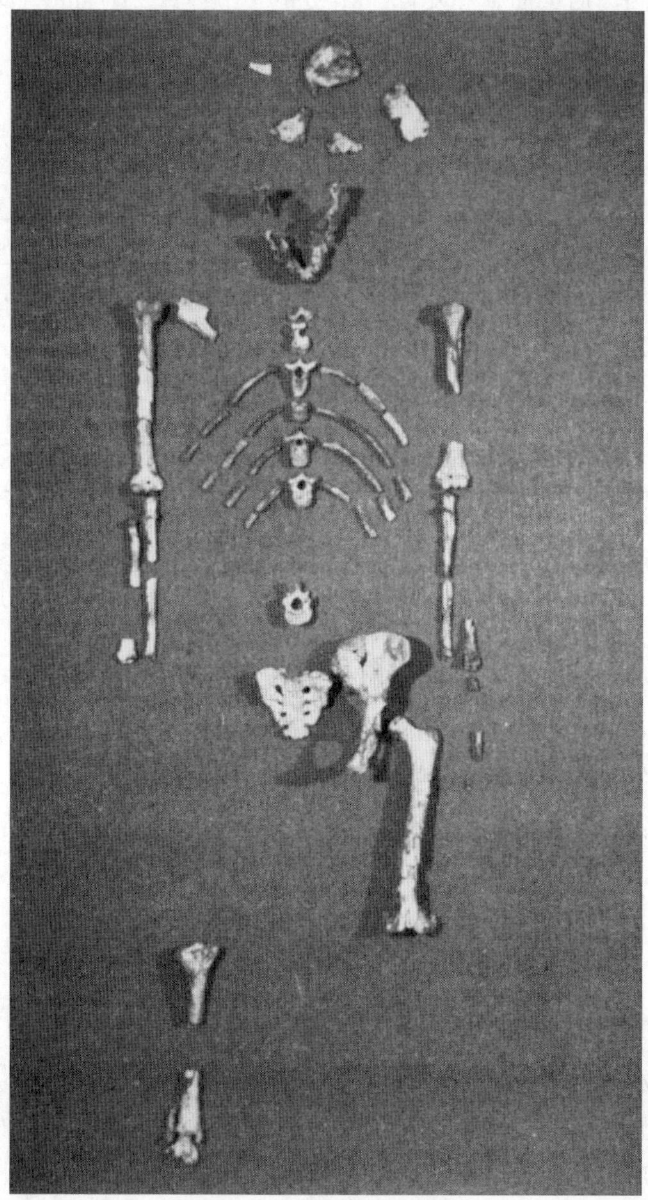

Fig . 5. Partial skeleton (AL 288–1) from Hadar

Stratigraphically just below the basalt flow a fragmentary right proximal femur (AL 211–1) was collected. The specimen is somewhat abraded with the head and neck missing and

筛选工作期间，在该地点发现了另外三颗牙齿，包括：一颗右 I_2（AL 198–18）、一颗左 I^2（AL 198–17a）和一颗左 I^1（AL 198–17b）。上门齿的牙间面很匹配，说明它们来自同一个体。由于这些牙与下颌骨 AL 198 出于同一层，所以我们暂时将它们划分到同一个体。

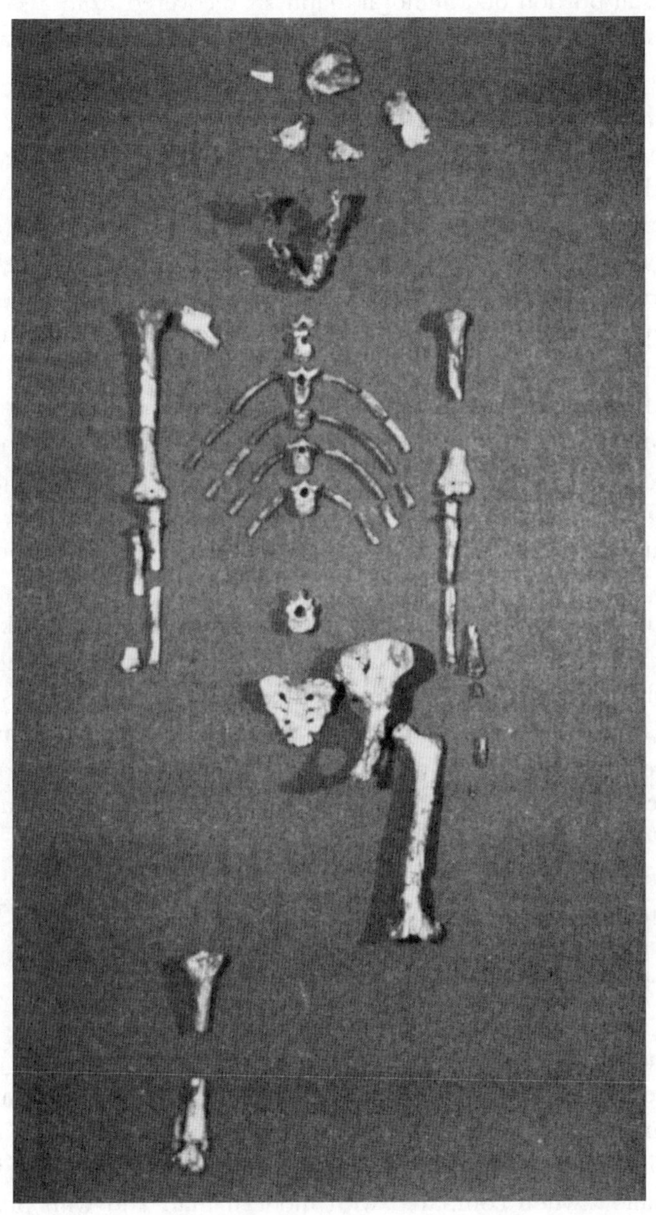

图 5. 来自哈达尔的不完整的骨架（AL 288–1）

就在地层上位于玄武岩熔体之下的位置，采集到了一块残破不全的右侧股骨近端（AL 211–1）。该标本轻微磨损，其头部和颈部缺失，大转子大部分被保存下来。

most of the greater trochanter. The neck is very flattened, no trochanteric flare is present, the intertrochanteric line is weakly expressed, the spiral line is well developed and the lesser trochanter is not visible from the anterior aspect. The specimen is quite large and fairly robust.

AL 228–1 is a distal portion of a femoral diaphysis recovered from a sandstone. It is not large, but exhibits strong development of the linea aspera. It is smaller than AL 129–1A.

The Partial Skeleton

The discovery on November 24 of a partial skeleton (AL 288–1; see Fig. 5) eroding from sand represents the most outstanding hominid specimen collected during the 1974 field season. The stratigraphic horizon yielding the skeleton is situated just above the KHT, which has not yet been dated. Fossil preservation at this locality is excellent, remains of delicate items such as crocodile and turtle eggs and crab claws being found. It is obvious that this discovery provides us with a unique opportunity for reconstructing the anatomy of an early hominid in far more detail than has been previously possible. Extensive descriptive and comparative studies are projected for the AL 288 partial skeleton and will provide us with details of stature, limb proportions, articulations and biomechanical aspects. Three weeks were devoted to intensive collecting and screening to ensure the recovery of all bone fragments from the site. Laboratory preparation and analysis has only just begun, and in this report it is possible to mention only a few salient points.

The mandible is not heavily built; it is 30.0 mm deep and 19.0 mm thick in the region of M_1. The M_3s are fully erupted and occlusal wear facets are just appearing. The symphysis is intact with a slight post-incisive planum. Although the incisor crowns are absent, it is apparent that this region was quite small. The remaining dentition is small and not very worn. The P_3s are interesting with a sloping buccal surface and almost no development of a metaconid. The form of the dental arch as well as the body of the mandible is distinctly V-shaped.

The cranium is not sufficiently complete to estimate cranial capacity. The cranial bones are thin, exhibit no sutures (internally or externally) and no marked development of nuchal or temporal musculature is present.

The left innominate is complete, although it is somewhat distorted in the pubic region and particularly in the area of sacral articulation. In size the specimen resembles Sterkfontein (Sts) 14; the ilium, however, gives the appearance of being higher, and the anterior border is relatively straight. A strongly developed anterior inferior spine is apparent. The acetabulum is shallow when compared with modern man and with Sts 14. The sciatic notch is broad, the subpubic angle obtuse and the pubis exhibits a pronounced ventral arc, all of which suggests the skeleton belonged to a female. When viewed from the superior aspect, the base of the sacrum is divided into thirds, with the diameter of the sacral body equal to each of the alae. This again suggests that AL 288 was female.

颈部非常扁平，没有结节外展的迹象，转子间线的痕迹也不明显，螺旋线发达，从前面看不到小转子。这个标本十分大且相当粗壮。

AL 288–1 是股骨骨干的远端部分，是从砂岩中发现并复原的。该标本并不大，但是表现出粗线很发达的迹象。它比 AL 129–1A 要小。

不完整的骨架

于 11 月 24 日发现的不完整的骨架（AL 288–1，见图 5）受到过沙的侵蚀，它是 1974 年野外考察工作中所采集到的最不同凡响的人科动物标本。发现该骨架的地层就位于 KHT 之上，其年代尚未确定。该发现的化石保存状况非常好，一些易破碎的化石，例如鳄鱼、乌龟蛋以及蟹螯都有发现。很明显，这一发现给我们提供了一个独一无二的机会，让我们能够比之前更详细地对早期人科动物的解剖学结构进行重建。已经计划对不完整的骨架 AL 288 进行大量的描述性和比较性研究，这些研究将为我们提供关于早期人科动物的身材、肢骨比例、关节和生物力学方面的详细信息。我们用了三个星期对该遗址进行深入的搜索和筛选，以确保可以复原所有的骨骼碎片。实验室的准备和分析工作才刚刚开始，在本次报告中，只能提一下其中几个要点。

其下颌骨不是很结实；在 M_1 处有 30.0 mm 深，19.0 mm 厚。两侧 M_3 完全萌出，咬合处磨损面刚刚出现。联合部完整，门齿后平面单薄。尽管门齿齿冠缺失，但是很明显该区域非常小。其余齿系小且磨损不严重。P_3 很有趣，有一个倾斜的颊面而且下后尖几乎不发育。牙弓以及下颌骨体的形状都呈明显的 V 形。

颅骨不太完整，不足以估计出颅容量。颅骨很薄，没有任何缝线（内部或外部），没有明显的颈肌或颞肌发育的迹象。

尽管左侧髋骨在耻骨区，尤其是骶骨关节区域有点变形，但总体较完整。在尺寸上，该标本与斯泰克方丹（Sts）14 号接近，但是髂骨看起来更高，并且其前缘相对直一些。很明显有一个非常发达的前下棘。与现代人或 Sts 14 相比，其髋臼较浅。其坐骨切迹很宽，耻骨下角呈钝角，耻骨有突出的腹侧弧，所有这些都表明该骨架属于一名女性。从上面看，骶骨基底被分成了三部分，骶骨体的直径与每个骨翼都是相等的。这再次表明 AL 288 是一名女性的骨架。

A complete left femur is associated with the innominate. But the distal portion is badly crushed. Its total length has been estimated at 280 mm but may be slightly revised when the distal end is reconstructed. The femur has minimal trochanteric flare, the neck is anteroposteriorly flattened, an intertrochanteric line is present and the lesser trochanter is not visible from the anterior aspect.

The proximal right tibia is nearly identical in size and morphology to AL 129–1b. The distal tibia is associated, and articulates with a talus and a distal fibula.

The right humerus is complete with some crushing of the proximal end. Its total length is estimated at ~ 235 mm giving a value of 83.9 for the humeral–femoral index. The distal end possesses a marked ridge separating the capitulum and trochlea.

Significance of the Hadar Hominids

Detailed studies of the Hadar hominids have just recently been initiated and definitive interpretations are not yet possible. Because of the geological antiquity, and in some instances the completeness of the specimens, however, we proffer a number of preliminary impressions concerning their phyletic affinities and resemblances to other specimens.

The partial skeleton exhibits a number of similarities to the Sterkfontein sample. Specifically, the size of the pelvis and to some degree its morphology are reminiscent of Sts 14. The shallow acetabulum and relatively high ilium, however, demonstrate certain divergences from the Sterkfontein specimen and possibly reflect a somewhat more primitive status for the AL 288 specimen. The associated V-shaped mandible is noteworthy. Leakey[10] has drawn attention to mandibular shape and has suggested that KNM-ER 1482 (Kenya National Museum, East Rudolph), and Omo 18 should be considered primitive because of their V-shaped contours. Should this prove to be a diagnostic character, it is possible that AL 288 retains more primitive features than *Australopithecus africanus*, recognized from Sterkfontein.

Previously the 1973 postcranial material has not been assigned to a taxon. It is now clear that it should probably be included in the same category as the AL 288 specimen because of the striking similarity of the proximal tibial fragments in size and morphology as well as the preserved femoral fragments. This is important because: there is now evidence of at least two individuals of a very small hominid in Hadar, and the AL 128 and 129 specimens are situated stratigraphically 80 m below the partial skeleton (Fig. 2).

The presence of another taxon is suggested by two other specimens in the Hadar hominid collection. AL 211–1 resembles very closely Olduvai Hominid (OH) 20 (ref. 11) as well as the two proximal femora from Swartkrans (SK 82 and SK 97). Similarities are not only in size but also in morphological detail; stout shafts, flattened necks, a lack of trochanteric flare and posteriorly facing lesser trochanters not visible in anterior view.

与髋骨相连的一块完整的左股骨，其远端粉碎严重。其总长度估计为 280 mm，但是重建远端时可能对其进行了些微的修正。该股骨具有最小的结节外展，颈部在前后方向上是扁平的，转子间线存在，从前面看不到小转子。

右侧胫骨近端在尺寸和形态上几乎与 AL 129–1b 完全相同。胫骨远端与一块距骨和一块腓骨远端以关节相连。

右肱骨完整，近端有些粉碎。其总长度估计在 235 mm 左右，说明其肱骨–股骨指数值为 83.9。其远端具有明显的嵴，该嵴将肱骨小头和滑车分离开。

哈达尔人科动物的重要性

最近刚刚启动了对哈达尔人科动物的详细研究，现在还不可能对其进行确定的解释。然而，由于地质年代古老以及有些标本较完整，对于它们的系统亲缘关系及它们与其他标本的相似性，我们提出了许多初步想法。

该不完整的骨架显示出许多与斯泰克方丹标本的相似性。确切地说，骨盆的尺寸以及在某种程度上的形态与 Sts14 颇为相似。但是，较浅的髋臼和相对高的髂骨表明其与斯泰克方丹标本有一定差异，并且 AL 288 标本可能反映了一种更为原始的状态。属于同一个体的 V 形下颌骨值得我们注意。利基 [10] 引起了我们对下颌骨形状的注意，并且提出 KNM-ER 1482（肯尼亚国家博物馆，鲁道夫湖以东）和 Omo 18 应该被归为原始人类标本，因为他们也具有 V 形的下颌。如果证明这是一种具有判别性的特征，那么 AL 288 有可能保留了比南方古猿非洲种更加原始的特征，这些特征正是从斯泰克方丹的标本中辨认出来的。

在此之前，1973 年得到的颅后材料没有被划入到任何分类单元中。现在很明显，它们很可能被划入与 AL 288 标本相同的分类单元中去，因为他们的胫骨近端的尺寸和形态以及保存下来的股骨残段都具有显著的相似性。这很重要，因为现在有证据证明在哈达尔至少有两个个体非常小的人科动物，AL 128 和 AL 129 标本在地层学上位于发现不完整的骨架的地层之下 80 m 处（图 2）。

哈达尔人科动物中的另外两个标本暗示着另一个分类单元的存在。AL 211–1 与奥杜威人科动物（OH）20（参考文献 11）以及斯瓦特克朗斯的两块股骨近端（SK 82 和 SK 97）非常相似。他们不仅在尺寸上具有相似性，在形态学细节方面也具有相似性；例如，强壮的骨干、扁平的颈部、缺少结节外展、从前面看不到朝后的小转子等等。

The temporal fragment (AL 166–9) resembles material from Swartkrans and East Rudolf which has also been assigned to a large *Australopithecus* pattern, particularly in the large bulbous mastoid process and heavy pneumatisation, as well as the broad temporal shelf. It is, however, less typical of this pattern in the broad, flat mandibular fossa and absence of a strong articular tubercle.

The close association of AL 199–1 and AL 200–1, as well as their stratigraphic equivalence, is important because of the high probability that they sample the same taxon. Except for size differences, these two specimens are remarkably similar, and suggest variation within a single taxon. The complete maxilla has large canines, broad central incisors and large posterior dentition. These characters as well as other details suggest resemblances with some *Homo erectus* material, particularly *Pithecanthropus* IV. This dental pattern is also seen in KNM-ER 1590 from East Rudolf, a specimen with large cranial capacity assigned to the genus Homo[10,12]. It must be recognized that other aspects of the AL 200 maxilla are "primitive", such as the guttered nasal margin and the alveolar prognathism.

The AL 266–1 and AL 277–1 mandibles resemble, in details of the dentition and mandibular morphology, other specimens assigned to *Homo*, such as OH 7 (ref. 13) and KNM-ER 1802 (ref. 10).

On the basis of the present hominid collection from Hadar it is tentatively suggested that some specimens show affinities with *A. robustus*, some with *A. africanus* (*sensu stricto*), and others with fossils previously referred to *Homo*.

The understanding of Plio-Pleistocene hominid remains is undergoing intensive revision and re-evaluation. With continued collection of specimens from the three million year-old time range in Hadar and elsewhere, and with additional detailed studies and comparisons, our attempts to interpret the earliest hominids should become more clear.

Special gratitude is expressed to Y. Coppens for his assistance in preparation of this manuscript. Owen Lovejoy also provided helpful comments.

(**260**, 293-297; 1976)

D. C. Johanson* and M. Taieb†

*Cleveland Museum of Natural History, Department of Anthropology, Case Western Reserve University, Cleveland, Ohio 44106

†Laboratoire de Géologie du Quaternaire, CNRS, Meudon-Bellevue, France

Received September 25, 1975; accepted February 9, 1976.

颞骨破片（AL 166–9）与斯瓦特克朗斯标本和已经被归属到一种大型南方古猿的鲁道夫湖以东标本相像，尤其是大的球状乳突、大量的气腔以及宽阔的颞架。但是，宽阔、平坦的下颌窝不是典型特征，且缺少强壮的关节结节。

AL 199–1 与 AL 200–1 关联密切且地层相当，这一点很重要，因为它们来自同一分类单元的可能性极大。除了尺寸上的差异，这两个标本非常相似，这表明它们的差异属于同一分类单元内的差异。完整的上颌骨具有大的犬齿、宽的中门齿和大的后齿列。这些特征及其他细节都表明它们与一些直立人材料，尤其是与猿人属 IV 具有相像之处。在鲁道夫湖以东的 KNM-ER 1590 中也出现了这种牙齿模式，后者具有大的颅容量，被归属到人属中[10,12]。不得不承认，上颌骨 AL 200 的其他方面是“原始的”，例如开沟的鼻缘和牙槽前突。

在齿列和下颌骨的形态学细节方面，下颌骨 AL 266–1 和 AL 277–1 与其他归属到人属的标本相像，例如 OH 7（参考文献 13）和 KNM-ER 1802（参考文献 10）。

根据现在已从哈达尔采集到的人科动物标本，暂且可以认为有些标本与南方古猿粗壮种有亲缘关系，有些与南方古猿非洲种有亲缘关系，其余的则与之前称为人属的化石有亲缘关系。

对上新世–更新世时期人科动物遗存的认识正处于进一步修正与重新评定的时期。随着对哈达尔和其他地方的属于三百万年这一时段的标本的持续收集，以及其他详细的研究和比较，我们尝试阐明最早人科动物的前景将会更加明朗。

特别感谢科庞在准备本文原稿时所提供的帮助。欧文·洛夫乔伊也提供了有益的意见。

（刘皓芳 翻译；赵凌霞 审稿）

References:

1. Taieb, M., Coppens, Y., Johanson, D. C., and Kalb, J., *C. r. hebd. Séanc. Acad. Sci., Paris*, **275D**, 819-822 (1972).

2. Taieb, M., Johanson, D. C., Coppens, Y., Bonnefille, R., and Kalb, J., *C. r. hebd. Séanc. Acad. Sci., Paris*, **279D**, 735-738 (1974)

3. Taieb, M., Johanson, D. C., and Coppens. Y., *C. r. hebd. Séanc. Acad. Sci., Paris*, **281D**, 1297-1300 (1975).

4. Taieb, M., thesis, Université de Paris VI (1974).

5. Taieb, M., Johanson, D. C., Coppens, Y., Aronson, J. L., *Nature*, **260**, 289-293 (1976).

6. Johanson, D. C., Taieb, M., and Coppens, Y., in *African Hominidae of the Plio-Pleistocene* (edit. by Jolly, C. L.,) (Duckworth, London, in the press).

7. Clark, J., Beerbower, J. R., and Kietzke, K. K., *Fieldiana, Geol. Mem.*, **5**, 1-158 (1957).

8. Kern, H. M., and Straus, W. L., *Am. J. Phys. Anthrop.*, **7(1)**, 53-66 (1949).

9. Heiple, K. C., and Lovejoy, C. O., *Am. J. Phys. Anthrop.*, **35(1)**, 75-84 (1971).

10. Leakey, R. E. F., *Nature*, **248**, 653-656 (1974).

11. Day, M. H., *Nature*, **221**, 230-233 (1969).

12. Leakey, R. E. F., *Nature*, **242**, 170-173 (1973).

13. Leakey, L. S. B., Tobias, P. V., and Napier, J. R., *Nature*, **202**, 7-9 (1964).

Single-channel Currents Recorded from Membrane of Denervated Frog Muscle Fibres

E. Neher and B. Sakmann

Editor's Note

The modern view is that nerve cells communicate with the outside world by means of channels on a molecular scale spanning the outer membrane of the cells, using the channels to import or export particular chemicals such as the ions of sodium or potassium. Typically nerve cells have a great variety of channels specific for different chemicals in their exterior membranes. The interest of this paper is that it describes a technique for isolating individual channels by means of exceedingly fine glass pipettes which is now known as the "patch-clamp" technique. The clamp is a means of maintaining a fixed voltage across the membrane of the cells being studied. The authors Bert Sakmann and Erwin Neher were awarded the Nobel Prize for Medicine in 1991.

THE ionic channel associated with the acetylcholine (ACh) receptor at the neuromuscular junction of skeletal muscle fibres is probably the best described channel in biological membranes. Nevertheless, the properties of individual channels are still unknown, as previous studies were concerned with average population properties. Macroscopic conductance fluctuations occurring in the presence of ACh were analysed to provide estimates for single channel conductance and mean open times[1-3]. The values obtained, however, depended on assumptions about the shape of the elementary conductance contribution—for example, that the elementary contribution is a square pulse-like event[2]. Clearly, it would be of great interest to refine techniques of conductance measurement in order to resolve discrete changes in conductance which are expected to occur when single channels open or close. This has not been possible so far because of excessive extraneous background noise. We report on a more sensitive method of conductance measurement, which, in appropriate conditions, reveals discrete changes in conductance that show many of the features that have been postulated for single ionic channels.

The key to the high resolution in the present experiments lies in limiting the membrane area from which current is measured to a small patch, and thereby decreasing background membrane noise. This is achieved by applying closely the tip of a glass pipette, 3–5 μm in diameter, on to the muscle surface, thus isolating electrically a small patch of membrane (Fig. 1). This method has been applied previously in various modifications and mostly with larger pipette tips to muscle[4], molluscan neurones[5,6], and squid axon[7]. The pipette, which has fire-polished edges, is filled with Ringer's solution and contains the cholinergic agonist at micromolar concentrations. Its interior is connected to the input of a virtual-ground circuit, which clamps the potential inside the pipette to ground and at the same time

切除神经的青蛙肌纤维膜上的单通道电流

内尔，萨克曼

编者按

现代观点认为，神经细胞可以借助横跨细胞膜的分子通道输入或输出特定的化学物质，如钠离子或钾离子等，来实现与外部世界的通讯联络。典型的神经细胞外膜上有很多针对不同化学物质的特异通道。本文的亮点在于，描述了一种借助玻璃微电极分离单独通道的技术，即现在所知的“膜片钳”技术。膜片钳是维持被研究细胞跨细胞膜固定电压的一种方法。本文作者贝尔特·萨克曼和埃尔温·内尔获得了 1991 年诺贝尔医学奖。

在骨骼肌纤维的神经肌肉连接处，与乙酰胆碱（ACh）受体相关的离子通道是生物膜中研究的最为深入的通道。然而，单离子通道的性质仍然不清楚，以前的研究工作所涉及的都是群体通道的平均性质。通过对 ACh 存在时所发生的宏观电导率的波动进行分析来评估单通道电导率和平均开放时间 [1-3]。但是所得到的数值取决于基础电导产生波的形状，例如是否为矩形脉冲波 [2]。显然，为了区分在单通道开放或关闭时期预计会出现的电导率的不连续变化，提高电导率的测量技术就非常重要。由于大量外来的背景噪音，至今这种精细技术还没有实现。本文将报道一种用于测定电导率更为灵敏的方法。在适合条件下，这种方法中所出现的电导率的不连续变化显示了被假定为单离子通道的许多特征。

本实验中高分辨率技术的关键在于通过将测定电流的膜的面积限制到很小的小块，从而降低膜的背景噪音。为达到该目的，将直径 3~5 微米的玻璃吸管尖端紧靠在肌肉表面上，从而使一小片膜绝缘（图 1）。以前，这种方法经过不同的改进，多采用较大的尖管，用于肌肉 [4]、软体动物神经 [5,6] 和枪乌贼神经轴突 [7] 的记录。将边缘经火抛光的尖管内充满林格溶液，其内含有微摩尔级浓度的类胆碱激动剂。管内部接有有效接地的输入电路，这是通过夹持管内壁使其电势接地来实现的。与此同时，测定流经管内即流经被管口罩住的小片膜的电流。用两个常规的微电极夹将肌细胞

measures current flowing through the pipette, that is, through the patch of membrane covered by the pipette opening. The interior of the muscle fibre is clamped locally to a fixed value by a conventional two-microelectrode clamp[8]. Thus, voltage-clamp conditions are secured across the patch of membrane under investigation. Since current densities involved are very small, a simple virtual ground inside the pipette is preferable to more complicated arrangements for stabilizing potential described previously[6].

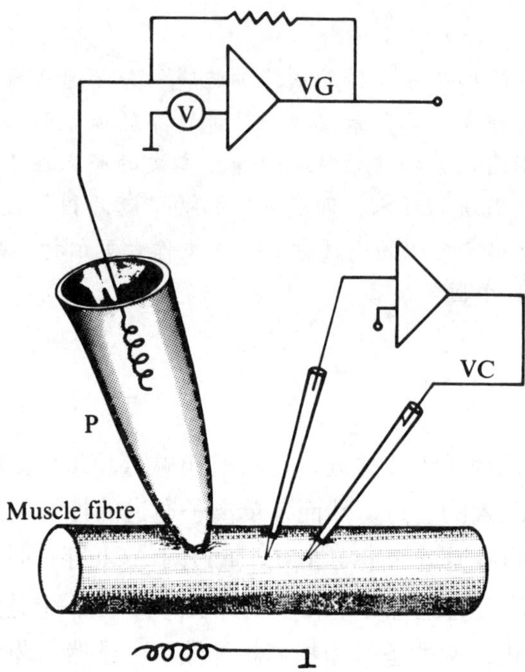

Fig. 1. Schematic circuit diagram for current recoding from a patch of membrane with an extracellular pipette. VC, Standard two-microelectrode voltage clamp circuit to set locally the membrane potential of the fibre to a fixed value. P, Pipette, fire polished, with 3–5 μm diameter opening, containing Ringer's solution and agonist at concentrations between 2×10^{-7} and 6×10^{-5} M. d. c. resistance of the pipette: 2–5 MΩ. The pipette tip applied closely on to the muscle fibre within 200 μm of the intracellular clamp electrodes. VG, Virtual ground circuit, using a Function Modules Model 380K operational amplifier and a 500 MΩ feedback resistor to measure membrane current. The amplifier is mounted together with a shielded pipette holder on a motor-driven micromanipulator. V, Bucking potential and test signal for balancing of pipette leakage and measuring pipette resistance.

The dominant source of background noise in these measurements was the leakage shunt under the pipette rim between membrane and glass. It was constantly monitored by measuring the electrical conductance between pipette interior and bath. Discrete conductance changes could be resolved only when the conductance between pipette interior and bath decreased by a factor of four or more after contact between pipette and membrane. To minimize the leakage conductance, the muscle was treated with collagenase and protease[9]. This enzyme treatment digested connective tissue and the basement membrane, thereby enabling closer contact between glass and membrane. At the same time, however, it made the membrane fragile and more sensitive to damage by the approaching pipette. It did not, however, change the ACh sensitivity of the fibre or alter

（纤维）局部固定[8]。这样，研究中玻璃微管口下方的小片膜就被完全绝缘。由于流过的电流量很小，因此将管内进行简单有效的接地要优于以前所描述的用于稳定电压而进行的各种更复杂的方式[6]。

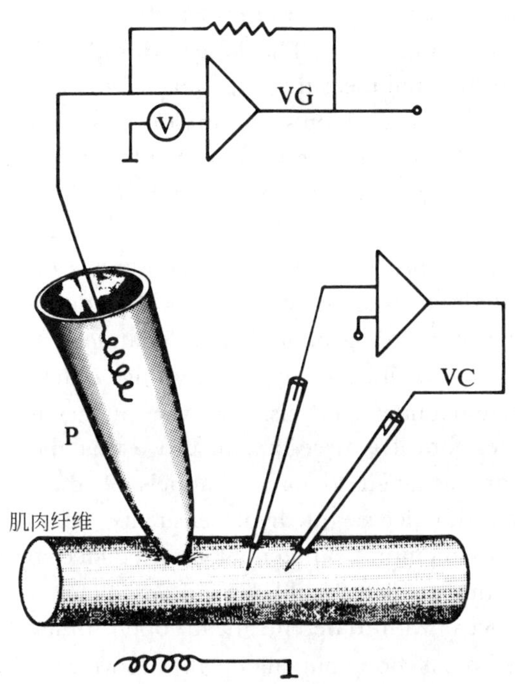

图 1. 用一个胞外玻璃管记录小片膜电流的电路设计图。VC，用于使局部肌细胞纤维膜电势达到一固定值的标准双微电极电压夹钳电路。 P，管口经火抛光的玻璃微管，直径 3~5 μm，装有林格溶液和浓度介于 $2\times10^{-7}\sim6\times10^{-5}$ M 之间的激动剂。管的直流电阻为 2~5 MΩ。管尖紧靠肌纤维细胞，距细胞内夹钳电极 200 μm 以内。VG，有效的接地电路，通过功能模块模型 380K 运算放大器和一个 500 MΩ 的反馈电阻器测定膜电流。放大器与电驱动马达推进的显微操纵器和固定肌细胞夹持器装在一起。V，对抗电势和检测用于平衡移液管泄漏和测定移液管电阻的信号。

在这些测定中，背景噪音的主要来源是介于膜和玻璃之间的管边缘下出现的泄漏。通过借助于测定管内和浴池之间的电导率而进行不断的监控。只有当管和膜接触后管内和室间的电导率降低到 1/4 或更多时，才能分辨出不连续的电导率变化。为了将泄漏电导率降到最低，肌肉都经过了胶原酶和蛋白酶的处理[9]，结缔组织和基膜中的胶原得以消化，从而使玻璃和膜之间可以更加紧密的接触。然而，与此同时，这样的处理也使得膜变得易碎，且对插进的管造成的破坏更敏感。但是，这并没有改变纤维对 ACh 的敏感性，也没有改变 ACh 诱导的电导率出现波动的特性（内

the properties of ACh-induced conductance fluctuations (E.N. and B.S., unpublished).

All experiments were carried out on the extrasynaptic region of denervated hypersensitive muscle fibres. The uniform ACh sensitivity found over most of the surface of these fibres greatly enhanced the probability of the occurrence of agonist-induced conductance changes at the membrane patch under investigation. Extrasynaptic ACh channels of denervated muscle fibres have mean open times which are about three to five times longer than those of endplate channels[1,10-12]. The longer duration facilitated the detection of conductance changes. Additional measures were taken which are known to either increase the size of the elementary current pulse or prolong its duration: the membrane was hyperpolarized up to −120 mV; suberyldicholine (SubCh) was used as an agonist in most of the experiments; the preparation was cooled to 6–8 °C.

Figure 2 shows a current recording taken in the conditions outlined above. Current can be seen to switch repeatedly between different levels. The discrete changes are interpreted as the result of opening and closing of individual channels. This interpretation is based on the very close similarity to single-channel recordings obtained in artificial membrane systems[13]. The preparation under study is, however, subject to a number of additional sources of artefact. Therefore it is necessary to prove that the recorded events do show the properties which are assigned to ionic channels of the cholinergic system. These are: a correlation with the degree of hypersensitivity of the muscle membrane; an amplitude dependent on membrane potential as predicted by noise analysis; a mean length or channel open time, which should depend on voltage in a characteristic manner[2]; pharmacological specificity with different mean open times for different cholinergic agonists[14,15]. The experiments bore out all of the above-mentioned points as outlined below.

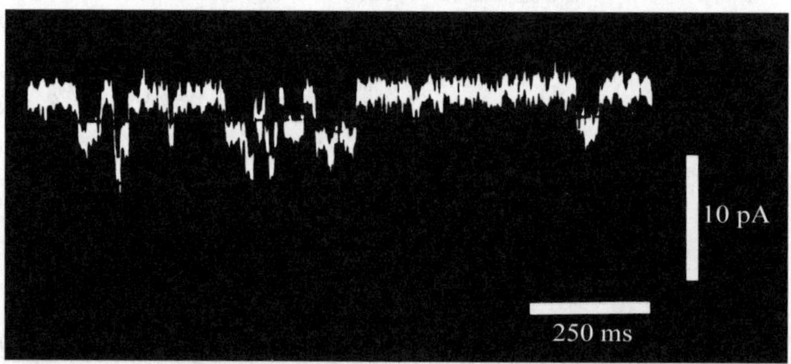

Fig. 2. Oscilloscope recording of current through a patch of membrane of approximately 10 μm^2. Downward deflection of the trace represents inward current. The pipette contained 2×10^{-7} M SubCh in Ringer's solution. The experiment was carried out with a denervated hypersensitive frog cutaneus pectoris (*Rana pipiens*) muscle in normal frog Ringer's solution. The record was filtered at a bandwidth of 200 Hz. Membrane potential: −120 mV. Temperature: 8 °C.

The frequency of occurrence of single blips depended on the sensitivity of the patch under investigation. A plot of the number of current pulses per second against the iontophoretically measured sensitivity of the membrane region, determined either

尔和萨克曼未发表数据)。

所有实验都在去神经后的超敏肌纤维的突触外区域进行。在这些区域都表现出对 ACh 的一致敏感性，这大大提高了所研究的膜片上能发生激动剂诱导的电导率变化的可能性。突触外的去神经肌肉纤维 ACh 通道的平均开放时间比终板通道的开放时期长约 3~5 倍 [1,10-12]。较长的持续开放时间有助于检测电导率的变化。同时还做了能增加基础电流脉冲大小，或者可以延长电流脉冲时间的实验检测：膜被超极化到 −120 mV；在大多数实验中使用的激动剂是环庚基二胆碱（SubCh)；样本被冷却至 6~8℃。

图 2 所示为在上述条件下记录的电流。电流在不同水平间重复转换。由于单通道的开和关，电流出现了不连续的变化。这种解释基于其与在人工膜系统中得到的单通道记录的高度相似性 [13]。然而，本研究的制备物受到许多附加的人工因素的影响。因此,很有必要证实所记录的结果确实说明了类胆碱功能系统的离子通道的性质。这些性质是：与肌肉膜超敏感性程度的相互关系；依赖于噪音分析所预测的膜电势的幅度；以典型方式依赖于电压的通道平均长度或通道平均开放时间 [2]；对不同类胆碱激动剂的不同平均开放时间的药理学特异性 [14,15]。正如下述，实验完全证明了上述所有提到的要点。

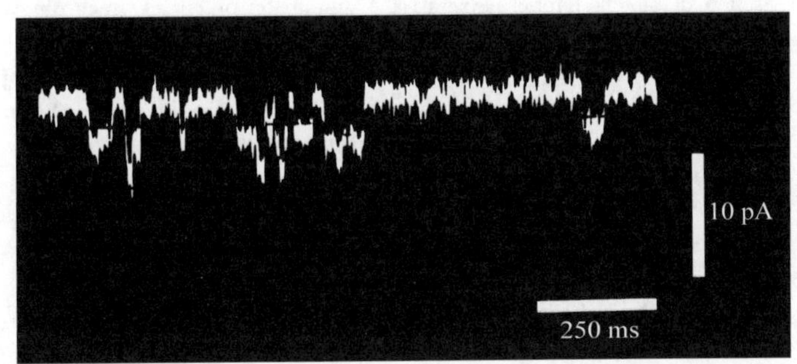

图 2. 示波器对通过约 10 μm^2 的小片膜的电流的记录结果。向下的微小偏差代表的是内部电流。 管内的林格溶液含有 2×10^{-7} M 的 SubCh。本实验采用的是切除了神经的超敏感美洲豹蛙（*Rana pipiens*）肌肉细胞，实验在普通的蛙林格溶液中进行。记录经 200 Hz 的带宽过滤。膜电势 −120 mV，温度 8 ℃。

单个信号的发生频率取决于所研究的小片膜的敏感性。将每秒电流脉冲数对所测定的膜区域离子渗透敏感性作图，在玻璃微管夹住前后立即测定。结果显示，两

immediately before or after the pipette experiment, revealed a distinct correlation between both quantities with a correlation coefficient of 0.91 for a linear regression (Fig. 3*b*). Student's *t* test assigned a significance better than 0.1% to the relationship.

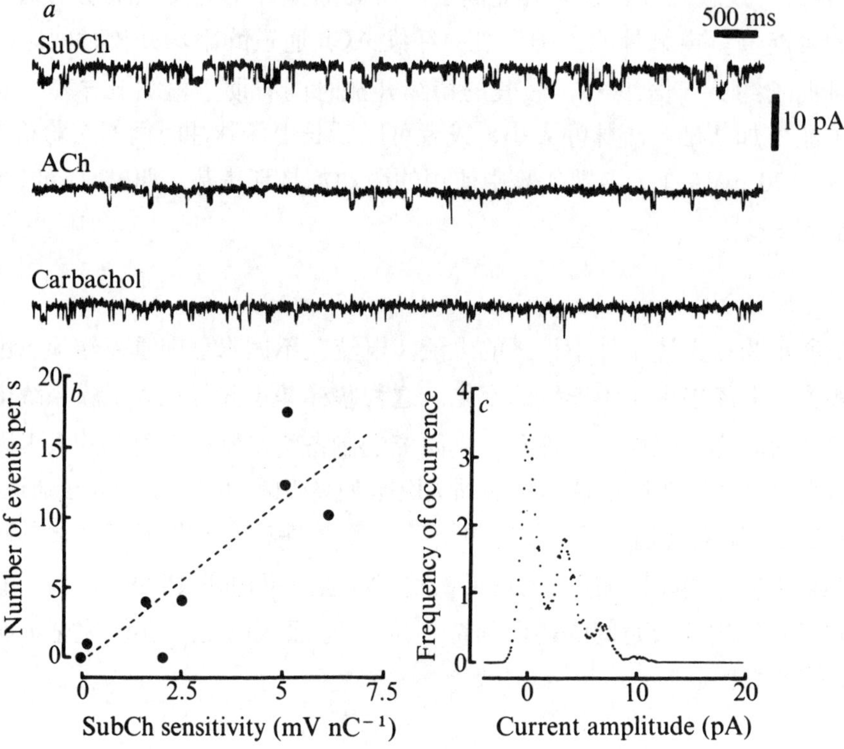

Fig. 3. Characterisation of single-channel currents. *a*, Comparison of current recordings obtained with different cholinergic agonists at concentrations of 2×10^{-7} M(SubCh), 2×10^{-6} M(ACh), and 6×10^{-5} M (carbachol). Downward deflection of the trace represents inward current. Three different experiments. Pen records replayed from analogue tape at a bandwidth of 100 Hz. All experiments at −120 mV membrane potential; 8°C. *b*, Number of current blips per second is plotted against iontophoretic sensitivity of the membrane region under investigation. Sensitivity was determined by 100-ms iontophoretic pulses delivered from a pipette filled with 1 M SubCh (40 MΩ pipette resistance, 10 nA bucking current, sensitivity measured at resting potential). Pooled data from eight experiments. Broken line represents linear regression. *c*, Amplitude histogram of membrane current. Current traces were digitalised and baselines fitted by eye to data records, each 4 s in length. Frequency of occurrence of deviations from the baseline is shown in arbitrary units. Histograms were calculated on a PDP-11 computer; 2–8 histograms were averaged to obtain curves like the one shown. 8°C; −120 mV membrane potential.

To estimate the size of the current pulses amplitude histograms were calculated from the current recordings (Fig. 3*c*). The histograms show a prominent peak of gaussian shape around zero deviation, the width of which is a measure of the high frequency background noise of the current trace. Multiple, equally spaced peaks at larger deviations represent the probabilities that either one, two, or three channels are open simultaneously. The peak separation gives the amplitude of the single-channel contribution, which was 3.4 pA for the histogram shown in Fig. 3*c*.

者的数量都呈现出相关系数为 0.91 的线性回归（图 3*b*），具有明显的相关性。对这种关系来说，*t* 检验给予的显著性要好于 0.1%。

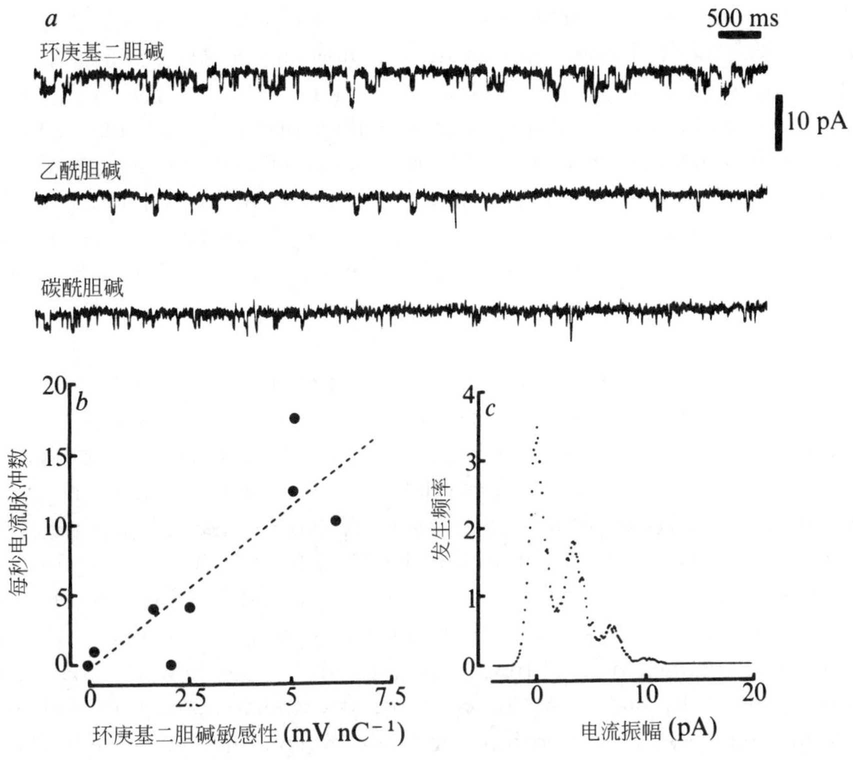

图 3. 单通道电流的特征。*a*，对浓度为 2×10^{-7} M 的环庚基二胆碱、2×10^{-6} M 的 ACh 和 6×10^{-5} M 的碳酰胆碱等所记录的电流进行比较。向下的微量偏差是内部电流。有 3 个不同的实验。记录类似于 100 Hz 的带宽下的重复值。所有实验的膜电势为 −120 mV，实验温度 8℃。*b*，将屏上显示的每秒钟的电流脉冲数对研究膜区的离子电渗敏感性作图。通过充满 1 M SubCh（管电阻 40 ΩM，屏蔽电流 10 nA，在静电势下测定敏感性）的微管中传递过来的 100 ms 的离子电渗脉冲测定敏感性。合并 8 个实验的数据。虚线代表线性回归。*c*，膜电流振幅图。将微量电流数字化，并通过对记录数据的观察设定基线，每次的时间长度为 4 s。由基线发生偏差的频率以任意单位显示。经过 PDP-11 计算机计算并作直方图，如图所示的曲线为对 2~8 张图平均化的结果，温度 8℃，电势 −120 mV。

为了估计电流脉冲的大小，通过计算记录的电流值给出了幅度图（图 3*c*）。该图显示，在零偏差处有一个高斯形状的显著峰，其宽度是在高频率背景噪音下测定的痕量电流值。其后较大偏差处的多个等距离的峰代表了一个、两个或三个通道同时开放的可能性。分开的峰显示了单通道的幅度，如图 3*c* 所示，其值大约为 3.4 pA。

This was obtained from an experiment at −120 mV membrane potential. A similar histogram from the same muscle fibre obtained at −80 mV yields a current pulse amplitude of 2.2 pA. These two values extrapolate to an equilibrium potential of −7 mV. Channel conductance is estimated as 28 pmhos in this case. It scattered from fibre to fibre with a mean value of 22.4±0.3 pmhos (mean±s.e.; number of determinations = 27). This value is somewhat lower than the one derived from noise analysis at normal endplates, which is 28.6 pmhos for SubCh[15]. Higher order peaks in the histograms are not merely scaled images of the zero order peak. They tend to be smeared out due to non-uniformity of current pulse amplitudes. We cannot decide at present whether this is a real feature of the channels or a measurement artefact. Such an effect could arise if not all of the channels are located ideally in the central region of the pipette opening. Current contributions from peripherally located channels would only partially be picked up by the pipette. This source of error is also likely to lead to an underestimate of channel size if the pipette seal is not optimal.

Temporal analysis of the current records was carried out partly by measurement of individual channel length and averaging 40–50 measurements, and partly by calculation of the power spectrum of the current recordings. In the latter case, the cutoff frequency f_0 of the Lorentzian spectrum yielded an estimate of mean channel open time τ (or pulse duration) through the relationship $\tau = 1/(2\pi f_0)$. Values of mean open times obtained by the two methods were consistent within ±30%. For SubCh as an agonist and at a temperature of 8°C it was 45±3 ms (n=11) at −120 mV and 28±3 ms (n=14) at −80 mV. These values are approximately three times longer than the corresponding mean open times of endplate channels derived from noise analysis[15]. Note, however, that lengthening of channel durations by factors of three to five at extrajunctional sites with respect to endplate values has been measured independently by conventional noise analysis[12]. The voltage dependence of the values given above corresponds to an e-fold change per 80 mV, which is within the range of published values[2].

Channel open times were different when different cholinergic agonists were used (Fig. 3*a*). For −120 mV and 8°C, mean channel open time was 45±3 ms (n=11) for SubCh, 26±5 ms (n=4) for ACh, and 11±2 ms (n=3) for carbachol. This sequence reflects the well known relationship between the open times of channels induced by these drugs at normal endplates[14,15] and at extrasynaptic membrane of hypersensitive fibres[12].

The results obtained so far, especially the pharmacological specificity, lead us to conclude that the observed conductance changes are indeed recordings of single-channel currents. They are consistent with the conclusions drawn from statistical analysis of endplate current fluctuations, and show that current contributions of individual channels are of the form of square pulses. In addition, analysis of areas under the peaks of histograms like Fig. 3*c* indicates that in our experimental conditions opening of individual channels is statistically independent, since the probabilities of zero, one, or two channels being open simultaneously follow—within the limits of experimental resolution—a Poisson distribution.

该图（图 3*c*）是在 − 120 mV 膜电势下的实验中所得。在 − 80 mV 下，同样肌肉纤维所得的类似电流脉冲幅度为 2.2 pA。将两个值进行外推，可以得到平衡电势为 − 7 mV。在这种情况下，估计通道电导率为 28 pmhos。不同纤维间的电导率平均值为 22.4 ± 0.3 pmhos（平均值 ± 标准误差；测定数 = 27）。该值稍微低于通常情况下终板噪音分析所得的数值，对 SubCh 该值是 28.6 pmhos[15]。图中更高级别的峰并不仅仅是零级次峰的比例图。由于电流脉冲振幅的不一致性，此类峰趋于扩散。我们目前并不能确定这种情况是通道的真正特征，还是由于测量的人工因素造成的假象。如果不能将所有的通道都理想地定位在管口的中心区域，就可能会引起这种效应。周边通道对于电流的贡献只可能被记录电极捕获一部分。这种误差也可能源于管的密封性没有达到最优而造成的对通道大小的低估。

对于记录的电流值的时间分析，一部分来自对单个通道长度的测定，其值为 40~50 次测量结果的平均值，另一部分来源于对记录的电流的能谱分析。在后种情况下，可以通过洛伦兹谱的截止频率 f_0 估算平均开放时间 τ（或脉冲时长），其关系式为 $\tau = 1/(2\pi f_0)$。两种方法所得的平均开放时间值在 ± 30% 的误差范围以内是一致的。用 SubCh 作为激动剂，温度为 8℃时， − 120 mV 下，开放时间为 45 ± 3 ms（n=11）；− 80 mV 下，开放时间是 28 ± 3 ms（n=14）。这些值大约是由噪音分析所得到的终板通道平均开放时间的 3 倍 [15]。但是值得注意的是，已经通过常规的噪音分析法独立地测定了关于终板值，在额外的连接点处，通道开放期被延长了 3~5 倍 [12]。上述所给值的电压依赖性相当于每 80 mV 发生 e 倍的变化，这在已发表数据的范围之内 [2]。

当使用不同的类胆碱激动剂时（图 3*a*），通道的开放时间是不同的。在 − 120 mV，8℃下，SubCh 作用下的通道平均开放时间为 45 ± 3 ms（n=11）；ACh 作用下的通道平均开放时间为 26 ± 5 ms（n=4）；而碳酰胆碱作用下的通道开放时间则为 11 ± 2 ms（n=3）。这种顺序反映了由这些药物在常规终板 [14,15] 上诱导产生的通道开放时间与超敏感纤维的突触外膜处 [12] 之间的典型关系。

目前为止，所得的结果，特别是药物特异性所得的结果，使我们推断所观察到的电导率变化实际上是记录到的单通道电流。它们与终板电流波动的统计分析所得的结论一致，并且显示了个别通道以矩形脉冲的形式形成电流。此外，对图 3*c* 的峰下区域的分析表明，在我们的实验条件下，个别通道的开放在统计上是独立的。因为在实验分辨率的限制下，零个、一个或两个通道同时开放的概率遵循泊松分布。

Recordings of single-channel currents finally resolves the third level of quantification in the process of neuromuscular transmission after the discovery of endplate currents and miniature endplate currents. It should facilitate discrimination between factors influencing the properties of single channels and agents creating or modifying different populations of channels.

We thank J. H. Steinbach for help with some experiments. Supported by a USPHS grant to Dr. C. F. Stevens, and a stipend of the Max-Planck-Gesellschaft.

(**260**, 799-802; 1976)

Erwin Neher* and Bert Sakmann†
*Yale University School of Medicine, Department of Physiology, New Haven, Connecticutt 06510
†Max-Planck-Institut für Biophysikalische Chemie, 3400 Göttingen, Am Fassberg, West Germany

Received January 26; accepted March 1, 1976.

References:

1. Katz, B., and Miledi, R., *J. Physiol., Lond.*, **224**, 665-699 (1972).
2. Anderson, C. R., and Stevens, C. F., *J. Physiol., Lond.*, **235**, 655-691 (1973).
3. Ben Haim, D., Dreyer, F., and Peper, K., *Pflügers Arch. ges. Physiol.*, **355**, 19-26 (1975).
4. Strickholm, A., *J. Gen. Physiol.*, **44**, 1073-1087 (1961).
5. Frank, K., and Tauc, L., in *The Cellular Function of Membrane Transport* (edit by Hoffman, J.) (Prentice Hall, Englewood Cliffs, New Jersey, 1963).
6. Neher, E., and Lux, H. D., *Pflügers Arch. ges. Physiol.*, **311**, 272-277 (1969).
7. Fishman, H. M., *Proc. Natl. Acad. Sci. U.S.A.*, **70**, 876-879 (1973).
8. Takeuchi, A., and Takeuchi, N., *J. Neurophysiol.*, **22**, 395-411 (1959).
9. Betz, W., and Sakmann, B., *J. Physiol., Lond.*, **230**, 673-688 (1973).
10. Neher, E., and Sakmann, B., *Pflügers Arch. ges. Physiol.*, **355**, R63 (1975).
11. Dreyer, F., Walther, Ch., and Peper, K., *Pflügers Arch. ges. Physiol.*, **359**, R71 (1975).
12. Neher, E., and Sakmann, B., *J. Physiol., Lond.* (in the press).
13. Hladky, S. B., and Haydon, D. A., *Nature*, **225**, 451-453 (1970).
14. Katz, B., and Miledi, R., *J. Physiol., Lond.*, **230**, 707-717 (1973).
15. Colquhoun, D., Dionne, V. E., Steinbach, J. H., and Stevens, C. F., *Nature*, 253, 204-206 (1975).

在发现终板电流和微型终板电流之后，单通道电流的记录最终实现了对神经肌肉传导过程中第三层次的定量分析。这应该有利于辨别影响单通道性质的因子和引起或修饰不同通道群体的化学试剂。

我们感谢斯坦巴克对实验的帮助。本研究受到史蒂文斯博士得到的美国公共卫生署基金的支持，并得到了马克斯·普朗克学会的基金资助。

（荆玉祥 翻译；曾少举 审稿）

Simple Mathematical Models with Very Complicated Dynamics

R. M. May

Editor's Note

Theories in population biology, economics and physics often involve a relatively simple class of equations called difference equations. Here Robert May, a pioneer in the dynamics of nonlinear systems, reviews recent findings showing that systems described by difference equations can exhibit rich dynamical behaviour. May focuses on a simple model of population dynamics. It exhibits not only "fixed points", reflecting a population that settles into an unchanging equilibrium, but an infinite number of oscillating population states and, beyond a threshold value of one key parameter, chaotic fluctuations. May implores that these lessons be taught in introductory mathematics courses. Over the next two decades this is indeed what happened, as chaos became recognized as a ubiquitous phenomenon in nature.

First-order difference equations arise in many contexts in the biological, economic and social sciences. Such equations, even though simple and deterministic, can exhibit a surprising array of dynamical behaviour, from stable points, to a bifurcating hierarchy of stable cycles, to apparently random fluctuations. There are consequently many fascinating problems, some concerned with delicate mathematical aspects of the fine structure of the trajectories, and some concerned with the practical implications and applications. This is an interpretive review of them.

THERE are many situations, in many disciplines, which can be described, at least to a crude first approximation, by a simple first-order difference equation. Studies of the dynamical properties of such models usually consist of finding constant equilibrium solutions, and then conducting a linearised analysis to determine their stability with respect to small disturbances: explicitly nonlinear dynamical features are usually not considered.

Recent studies have, however, shown that the very simplest nonlinear difference equations can possess an extraordinarily rich spectrum of dynamical behaviour, from stable points, through cascades of stable cycles, to a regime in which the behaviour (although fully deterministic) is in many respects "chaotic" , or indistinguishable from the sample function of a random process.

This review article has several aims.

First, although the main features of these nonlinear phenomena have been discovered

具有极复杂动力学行为的简单数学模型

梅

编者按

种群生物学、经济学和物理学中的理论经常会涉及一类相对简单的方程——差分方程。在本文中，非线性系统动力学的先驱者罗伯特·梅对近期的一些发现进行了整理归纳，这些发现表明由差分方程所描述的系统可以呈现丰富的动力学行为。梅关注种群动力学的一个简单模型。该模型不仅展示了反映种群处在不变平衡态上的“不动点”，还显示了无限多的振荡种群态和超过一个关键参数阈值的混沌波动。梅恳请在数学导论课中讲授这些课程。在之后的20年里，这果然发生了，就如混沌已被公认为是自然界的普遍现象一样。

在生物学、经济学和社会科学的很多情境中，都会出现一阶差分方程。这些方程，即使是简单的和确定的，也能呈现出一系列惊人的动力学行为，从稳定点到稳定循环的分岔谱系，再到明显的随机波动。由此引来了很多有趣的问题，其中一些涉及关于轨道精细结构的精巧的数学内容，而另外一些则涉及实际含义与应用。本文是对于这些内容的解释性综述。

在很多学科中，至少在粗略的一级近似下，有很多情况可以用一个简单的一阶差分方程来描述。对于这些模型的动力学性质的研究，通常包括寻找常数平衡解，继而导出线性化的分析，以确定它们在微小扰动下的稳定性：明显的非线性动力学特征通常是不加以考虑的。

但是，最近的研究表明，最简单的非线性差分方程也能够具有一系列异常丰富的动力学行为，从稳定点经由各级稳定循环到达这样一种状态：其中的行为（尽管是完全确定的）在很多方面是“混沌的”，或无法与一个随机过程的样本函数相区别。

这篇综述性文章有以下几个目的：

第一，尽管这些非线性现象的主要特征已被发现并被其他的多位学者独立地再

and independently rediscovered by several people, I know of no source where all the main results are collected together. I have therefore tried to give such a synoptic account. This is done in a brief and descriptive way, and includes some new material: the detailed mathematical proofs are to be found in the technical literature, to which signposts are given.

Second, I indicate some of the interesting mathematical questions which do not seem to be fully resolved. Some of these problems are of a practical kind, to do with providing a probabilistic description for trajectories which seem random, even though their underlying structure is deterministic. Other problems are of intrinsic mathematical interest, and treat such things as the pathology of the bifurcation structure, or the truly random behaviour, that can arise when the nonlinear function $F(X)$ of equation (1) is not analytical. One aim here is to stimulate research on these questions, particularly on the empirical questions which relate to processing data.

Third, consideration is given to some fields where these notions may find practical application. Such applications range from the abstractly metaphorical (where, for example, the transition from a stable point to "chaos" serves as a metaphor for the onset of turbulence in a fluid), to models for the dynamic behaviour of biological populations (where one can seek to use field or laboratory data to estimate the values of the parameters in the difference equation).

Fourth, there is a very brief review of the literature pertaining to the way this spectrum of behaviour—stable points, stable cycles, chaos—can arise in second or higher order difference equations (that is, two or more dimensions; two or more interacting species), where the onset of chaos usually requires less severe nonlinearities. Differential equations are also surveyed in this light; it seems that a three-dimensional system of first-order ordinary differential equations is required for the manifestation of chaotic behaviour.

The review ends with an evangelical plea for the introduction of these difference equations into elementary mathematics courses, so that students' intuition may be enriched by seeing the wild things that simple nonlinear equations can do.

First-order Difference Equations

One of the simplest systems an ecologist can study is a seasonally breeding population in which generations do not overlap[1-4]. Many natural populations, particularly among temperate zone insects (including many economically important crop and orchard pests), are of this kind. In this situation, the observational data will usually consist of information about the maximum, or the average, or the total population in each generation. The theoretician seeks to understand how the magnitude of the population in generation $t+1$, X_{t+1}, is related to the magnitude of the population in the preceding generation t, X_t: such a relationship may be expressed in the general form

$$X_{t+1} = F(X_t) \tag{1}$$

发现，但是，就我所知，还没有一份资料将所有的主要成果汇集起来。因此，我试图采取简略的和描述性的方式给出这样一种概要性说明。这一说明中包括一些新材料，详细的数学证明可以在技术性文献中找到（已用记号标注）。

第二，我要指出一些有趣的数学问题，它们似乎还没有得到完全的解决。其中有一些问题属于实际问题，对看似随机的运动轨道提供一个概率性的描述，即使其根本结构是确定性的。另外一些问题则属于固有的数学趣味，因而把它们看作分岔结构的反常状态，或者是真正的随机行为，当方程（1）中的非线性函数 $F(X)$ 不是解析的时候就会出现。本文的一个目的是激发对于上述问题的研究，特别是对于与数据处理有关的经验问题的研究。

第三，我对上述观点可以找到实际应用的某些领域进行了考察。这些应用包括从抽象的隐喻（如用从一个稳定点到“混沌”的转变来类比流体中湍流的出现）到生物学种群的动力学行为模型（人们可以设法利用野外数据或实验数据来估计差分方程中的参数值）。

第四，本文还包括了对上述一系列动力学行为——稳定点、稳定循环、混沌——可以在二阶或更高阶差分方程（也就是说，二维或更高维，两个或者多个相互作用的物种）中出现的文献的一个极为简略的综述。在这些情况下，混沌的出现通常不需要那么严格的非线性性质。为此，我还对微分方程进行了探讨。看起来，处理混沌行为需要的是一个三维的一阶常微分方程组。

这篇综述性文章以一个热忱的呼吁作为结束，即将这些差分方程引入到初等数学课程中，以便让学生看到简单的非线性方程会呈现复杂的问题，这样使他们的直觉变得更丰富。

一阶差分方程

生态学者所能研究的最简单的系统之一，就是各世代没有交叠的季节性繁殖种群[1-4]。很多自然种群，特别是那些温带地区的昆虫（包括很多种重要经济农作物和果园的害虫），都属于这一类型。在这种情况下，观测到的数据通常会由以下信息组成：每一世代种群的最大值、平均值或者是总数。理论研究工作者试图理解第 $t+1$ 世代的种群数量 X_{t+1} 与其前一世代，即第 t 世代的种群数量 X_t 之间有怎样的关系。这种关系可以表达为如下一般形式：

$$X_{t+1} = F(X_t) \tag{1}$$

The function $F(X)$ will usually be what a biologist calls "density dependent", and a mathematician calls nonlinear; equation (1) is then a first-order, nonlinear difference equation.

Although I shall henceforth adopt the habit of referring to the variable X as "the population", there are countless situations outside population biology where the basic equation (1), applies. There are other examples in biology, as, for example in genetics[5,6] (where the equation describes the change in gene frequency in time) or in epidemiology[7] (with X the fraction of the population infected at time t). Examples in economics include models for the relationship between commodity quantity and price[8], for the theory of business cycles[9], and for the temporal sequences generated by various other economic quantities[10]. The general equation (1) also is germane to the social sciences[11], where it arises, for example, in theories of learning (where X may be the number of bits of information that can be remembered after an interval t), or in the propagation of rumours in variously structured societies (where X is the number of people to have heard the rumour after time t). The imaginative reader will be able to invent other contexts for equation (1).

In many of these contexts, and for biological populations in particular, there is a tendency for the variable X to increase from one generation to the next when it is small, and for it to decrease when it is large. That is, the nonlinear function $F(X)$ often has the following properties: $F(0) = 0$; $F(X)$ increases monotonically as X increases through the range $0<X<A$ (with $F(X)$ attaining its maximum value at $X=A$); and $F(X)$ decreases monotonically as X increases beyond $X=A$. Moreover, $F(X)$ will usually contain one or more parameters which "tune" the severity of this nonlinear behaviour; parameters which tune the steepness of the hump in the $F(X)$ curve. These parameters will typically have some biological or economic or sociological significance.

A specific example is afforded by the equation[1,4,12-23]

$$N_{t+1} = N_t\,(a-bN_t) \qquad (2)$$

This is sometimes called the "logistic" difference equation. In the limit $b=0$, it describes a population growing purely exponentially (for $a>1$); for $b\neq 0$, the quadratic nonlinearity produces a growth curve with a hump, the steepness of which is tuned by the parameter a. By writing $X=bN/a$, the equation may be brought into canonical form[1,4,12-23]

$$X_{t+1} = aX_t\,(1-X_t) \qquad (3)$$

In this form, which is illustrated in Fig. 1, it is arguably the simplest nonlinear difference equation. I shall use equation (3) for most of the numerical examples and illustrations in this article. Although attractive to mathematicians by virtue of its extreme simplicity, in practical applications equation (3) has the disadvantage that it requires X to remain on the interval $0<X<1$; if X ever exceeds unity, subsequent iterations diverge towards $-\infty$

这里的函数 $F(X)$ 通常被生物学家称为“密度制约的”，而数学家则会称之为非线性的；于是是方程（1）就是一个一阶非线性差分方程。

尽管此后我将习惯于将变量 X 称为“种群数量”，但在种群生物学之外还是存在不可胜数的可以应用基本方程（1）的情形。生物学中还有其他一些实例，如在遗传学中 [5,6]（这时该方程描述了基因频率随时间的变化），或者是在流行病学中 [7]（其中 X 是 t 时刻被感染群体所占的比例）。经济学中的例子包括商品数量与价格之间的关系模型 [8]，商业周期理论模型 [9] 以及由各种其他经济学量所产生的时间序列的模型 [10]。方程（1）的一般形式也与社会科学 [11] 有着密切的关系，例如，它会出现在学习理论中（其中 X 可以是一个时间间隔 t 之后能够记住的信息量），或者出现于在各种结构化社会团体内流言传播的问题中（此时 X 是在时间 t 之后听到过流言的人数）。富于想象力的读者还可以为方程（1）设计出其他情境。

在上述多种情境中，特别是对于生物学种群问题，变量 X 有这样一种趋势，当种群很小时，从一个世代到下一个世代逐渐增加，而当种群很大时，则减小。换言之，非线性函数 $F(X)$ 具有以下性质：$F(0) = 0$；当 $0<X<A$ 时，$F(X)$ 随着 X 增加而单调增加（$F(X)$ 在 $X = A$ 处取得最大值）；而在超过 $X = A$ 之后，$F(X)$ 随着 X 的增加而单调减小。此外，$F(X)$ 通常还会包含一个或多个参数，这些参数或可“调节”这种非线性行为的强度，或可调节 $F(X)$ 曲线峰的陡度，它们均有其典型的生物学、经济学或社会学意义。

下列方程提供了一个特别的实例 [1,4,12-23]：

$$N_{t+1} = N_t\,(a-bN_t) \tag{2}$$

这个方程有时被称为“逻辑斯谛”差分方程。在 $b = 0$ 的极限情况下，它描述了纯指数形式的种群增长（其中 $a > 1$）；当 $b \neq 0$ 时，二次非线性性质产生一条有一个峰的生长曲线，其陡度由参数 a 所调节。记 $X = bN/a$，可以将该方程转化为如下规范形式 [1,4,12~23]：

$$X_{t+1} = aX_t\,(1-X_t) \tag{3}$$

如图 1 所示，在这种形式下，可以说明它是最简单的非线性差分方程。在本文的大多数数值实例和说明中，我都会用到方程（3）。尽管对于数学家来说，它那极端简单性的优点很有吸引力，但在实际应用中，方程（3）却存在一点不足，即它要求 X 始终保持在 $0<X<1$ 范围内，如果 X 超过 1，随后的迭代将会发散到 $-\infty$（这意味着种群

(which means the population becomes extinct). Furthermore, $F(X)$ in equation (3) attains a maximum value of $a/4$ (at $X=\frac{1}{2}$); the equation therefore possesses non-trivial dynamical behaviour only if $a<4$. On the other hand, all trajectories are attracted to $X=0$ if $a<1$. Thus for non-trivial dynamical behaviour we require $1<a<4$; failing this, the population becomes extinct.

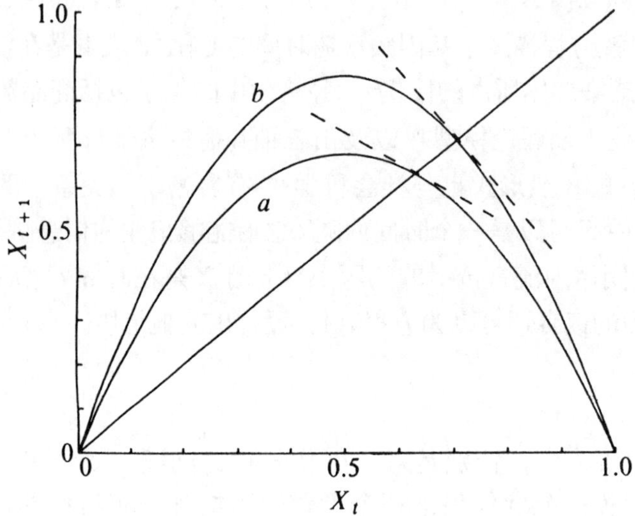

Fig. 1. A typical form for the relationship between X_{t+1} and X_t described by equation (1). The curves are for equation (3), with $a=2.707$ (*a*); and $a=3.414$ (*b*). The dashed lines indicate the slope at the "fixed points" where $F(X)$ intersects the 45° line: for the case *a* this slope is less steep than −45° and the fixed point is stable; for *b* the slope is steeper than −45°, and the point is unstable.

Another example, with a more secure provenance in the biological literature[1,23-27], is the equation

$$X_{t+1} = X_t \exp[r(1-X_t)] \tag{4}$$

This again describes a population with a propensity to simple exponential growth at low densities, and a tendency to decrease at high densities. The steepness of this nonlinear behaviour is tuned by the parameter r. The model is plausible for a single species population which is regulated by an epidemic disease at high density[28]. The function $F(X)$ of equation (4) is slightly more complicated than that of equation (3), but has the compensating advantage that local stability implies global stability[1] for all $X>0$.

The forms (3) and (4) by no means exhaust the list of single-humped functions $F(X)$ for equation (1) which can be culled from the ecological literature. A fairly full such catalogue is given, complete with references, by May and Oster[1]. Other similar mathematical functions are given by Metropolis *et al.*[16]. Yet other forms for $F(X)$ are discussed under the heading of "mathematical curiosities" below.

将会灭绝)。此外，方程(3)中的 $F(X)$ 取得最大值为 $a/4$(在 $X=\frac{1}{2}$ 处)，因此，只在 $a<4$ 时，该方程才具有非平凡的动力学行为。另一方面，若 $a<1$，则所有轨道都被吸引到 $X=0$。于是，为了得到非平凡的动力学行为，我们要求 $1<a<4$，若不满足这一点，种群将会灭绝。

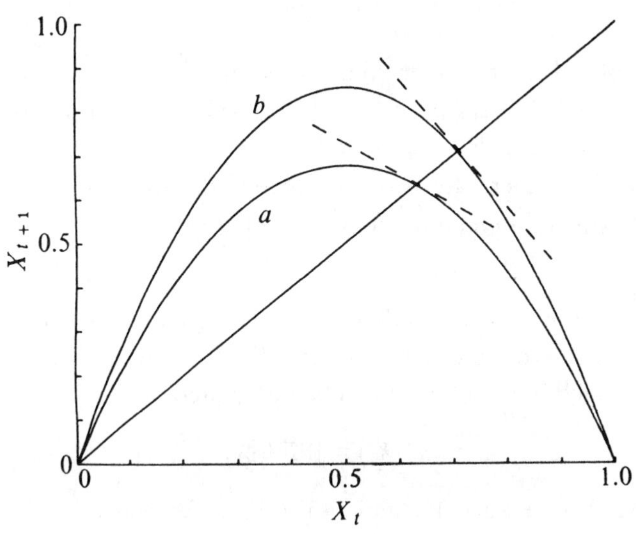

图 1. 方程(1)所描述的 X_{t+1} 与 X_t 之间关系的典型形式。两条曲线对应方程(3)，其中(a)，$a=2.707$；(b)，$a=3.414$。虚线所示为 $F(X)$ 与 45° 线的相交处"不动点"的斜率：对于情况 a，其倾斜程度不足 −45°，因而不动点是稳定的；对于情况 b，其倾斜程度超过 −45°，因而该点是不稳定的。

另一个在生物学文献中有更可靠出处[1,23-27]的实例是方程

$$X_{t+1}=X_t\exp[r(1-X_t)] \tag{4}$$

它所描述的也是这样一个种群，即在低密度时具有简单指数增长的倾向，而在高密度时则具有减小的趋势，其非线性行为的陡度是由参数 r 来调节的。对于单一物种种群在高密度时受一种流行病影响的情况，该模型看起来是合理的[28]。方程(4)中的函数 $F(X)$ 比方程(3)中的稍微复杂一些，但作为补偿，其优势在于，对于所有 $X>0$ 的范围内，局域稳定即意味着全局稳定[1]。

(3)和(4)两种形式绝不是能够从生态学文献中精选出来的形如方程(1)的单峰函数 $F(X)$ 的全部可能。梅和奥斯特[1]给出了这样一个相当完整的目录，并且都附有参考文献。梅特罗波利斯等人[16]给出了其他类似的数学函数。至于 $F(X)$ 的其他形式，将会在"数学方面的好奇心"这一标题下探讨。

Dynamic Properties of Equation (1)

Possible constant, equilibrium values (or "fixed points") of X in equation (1) may be found algebraically by putting $X_{t+1} = X_t = X^*$, and solving the resulting equation

$$X^*=F(X^*) \tag{5}$$

An equivalent graphical method is to find the points where the curve $F(X)$ that maps X_t into X_{t+1} intersects the 45° line, $X_{t+1} = X_t$, which corresponds to the ideal nirvana of zero population growth; see Fig. 1. For the single-hump curves discussed above, and exemplified by equations (3) and (4), there are two such points: the trivial solution X=0, and a non-trivial solution X^* (which for equation (3) is $X^*=1-[1/a]$).

The next question concerns the stability of the equilibrium point X^*. This can be seen[24,25,19-21,1,4] to depend on the slope of the $F(X)$ curve at X^*. This slope, which is illustrated by the dashed lines in Fig. 1, can be designated

$$\lambda^{(1)}(X^*) = [\mathrm{d}F/\mathrm{d}X]_{X=X^*} \tag{6}$$

So long as this slope lies between 45° and −45° (that is, $\lambda^{(1)}$ between +1 and −1), making an acute angle with the 45° ZPG line, the equilibrium point X^* will be at least locally stable, attracting all trajectories in its neighbourhood. In equation (3), for example, this slope is $\lambda^{(1)}=2-a$: the equilibrium point is therefore stable, and attracts all trajectories originating in the interval $0<X<1$, if and only if $1<a<3$.

As the relevant parameters are tuned so that the curve $F(X)$ becomes more and more steeply humped, this stability-determining slope at X^* may eventually steepen beyond −45°(that is, $\lambda^{(1)}<-1$), whereupon the equilibrium point X^* is no longer stable.

What happens next? What happens, for example, for $a>3$ in equation (3)?

To answer this question, it is helpful to look at the map which relates the populations at successive intervals 2 generations apart; that is, to look at the function which relates X_{t+2} to X_t. This second iterate of equation (1) can be written

$$X_{t+2} = F[F(X_t)] \tag{7}$$

or, introducing an obvious piece of notation,

$$X_{t+2} = F^{(2)}(X_t) \tag{8}$$

The map so derived from equation (3) is illustrated in Figs 2 and 3.

方程（1）的动力学性质

利用代数方法，令 $X_{t+1} = X_t = X^*$，即可求出方程（1）中 X 的可能的常数解，即平衡值（或“不动点”），该解可由下列方程给出：

$$X^* = F(X^*) \tag{5}$$

一种等价的图像解法是寻找将 X_t 映射到 X_{t+1} 的曲线 $F(X)$ 与 45° 线 $X_{t+1} = X_t$ 的交点，它对应于种群零增长的理想情况（图 1）。对于前面所讨论的单峰曲线，以方程（3）和（4）为例，这样的点有两个：平凡解 $X = 0$ 和一个非平凡解 X^*（对于方程（3），$X^* = 1-[1/a]$）。

下一个问题与平衡点 X^* 的稳定性有关。可以看到 [24,25,19-21,1,4] 这取决于 $F(X)$ 曲线在 X^* 点处的斜率。如同图 1 中虚线所示，这一斜率可以由下式确定为：

$$\lambda^{(1)}(X^*) = [\mathrm{d}F/\mathrm{d}X]_{X = X^*} \tag{6}$$

只要这个倾斜度位于 45° 到 –45° 之间（也就是说，$\lambda^{(1)}$ 介于 +1 和 −1 之间），与 45°零增长线构成一个锐角，那么平衡点 X^* 至少是局部稳定的，将吸引邻域内的全部轨道。例如，在方程（3）中，这个斜率是 $\lambda^{(1)} = 2 - a$：当且仅当 $1<a<3$ 时，平衡点是稳定的，因而吸引所有源于区间 $0<X<1$ 内的轨道。

由于相关参数的调节使得曲线 $F(X)$ 变得越来越陡峭，决定稳定性的 X^* 点处的倾斜度最终可能超过 −45°（也就是说，$\lambda^{(1)}< -1$），于是平衡点 X^* 不再是稳定的。

接下来会发生什么？例如，当方程（3）中的 $a > 3$ 时会发生什么？

要回答这个问题，考察一下相继间隔 2 世代的种群间关系的映射将会是有益的。也就是说，看一看联系 X_{t+2} 与 X_t 的函数。方程（1）的第二次迭代可以写作

$$X_{t+2} = F[F(X_t)] \tag{7}$$

或者引入一个含义明显的记号，

$$X_{t+2} = F^{(2)}(X_t) \tag{8}$$

用这种方式从方程（3）得出的映射如图 2 和图 3 所示。

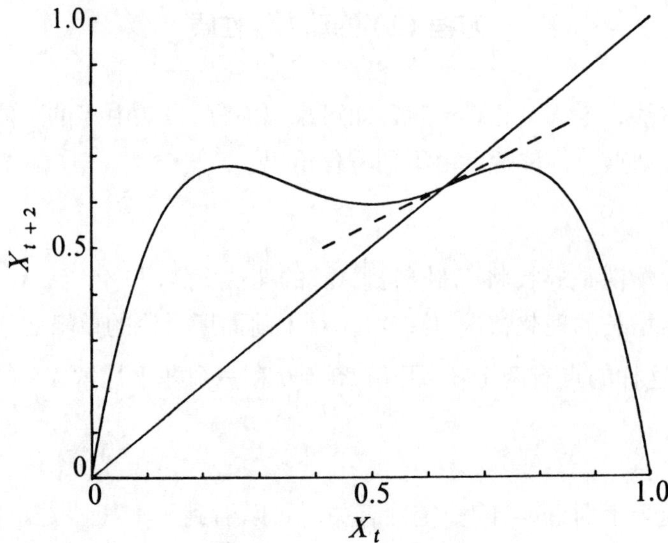

Fig. 2. The map relating X_{t+2} to X_t, obtained by two iterations of equation (3). This figure is for the case (*a*) of Fig.1, a=2.707: the basic fixed point is stable, and it is the only point at which $F^{(2)}(X)$ intersects the 45° line (where its slope, shown by the dashed line, is less steep than 45°).

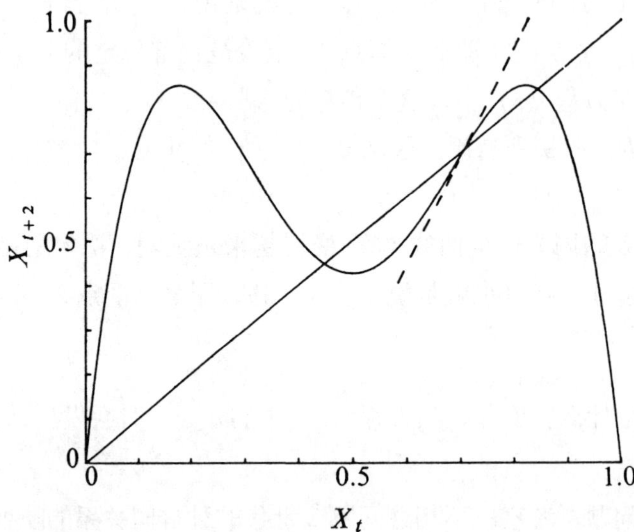

Fig. 3. As for Fig. 2, except that here a=3.414, as in Fig. 1*b*. The basic fixed point is now unstable: the slope of $F^{(2)}(X)$ at this point steepens beyond 45°, leading to the appearance of two new solutions of period 2.

Population values which recur every second generation (that is, fixed points with period 2) may now be written as X^*_2, and found either algebraically from

$$X^*_2 = F^{(2)}(X^*_2) \tag{9}$$

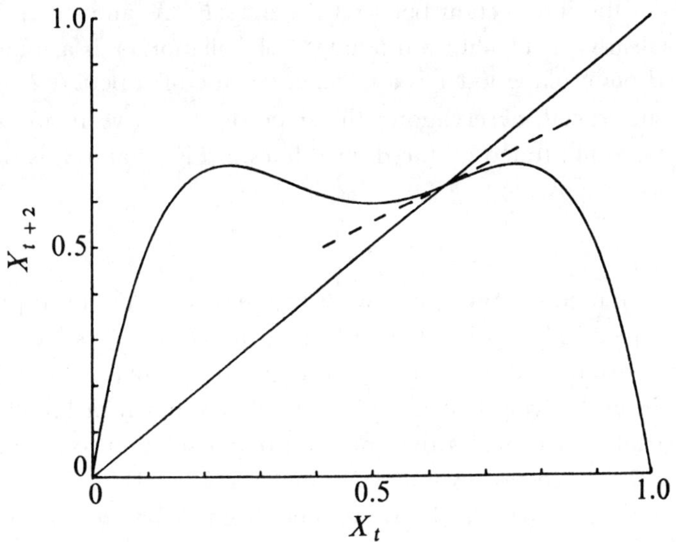

图 2. 联系 X_{t+2} 与 X_t 的映射，通过两次迭代方程（3）而得到。本图对应于图 1 中的情况（a），a = 2.707：基本不动点是稳定的，并且它是 $F^{(2)}(X)$ 与 45° 线的唯一交点（如图中虚线所示，其倾斜度不足 45°）。

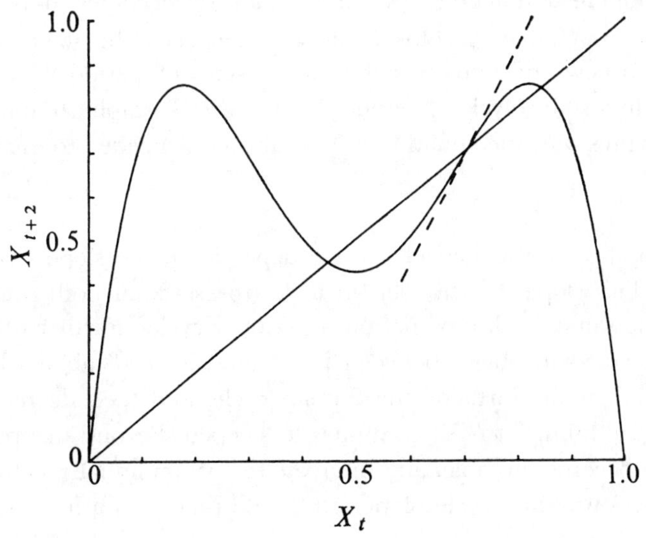

图 3. 与图 2 类似，除了 a = 3.414（这与图 1b 中一样）。基本不动点现在是不稳定的：$F^{(2)}(X)$ 在这点处的倾斜度超过了 45°，导致周期为 2 的两个新解出现。

现在可以将在每隔一世代都会重现的种群数值（即周期为 2 的不动点）记为 X^*_2，接下来，或者用代数方法从

$$X^*_2 = F^{(2)}(X^*_2) \tag{9}$$

or graphically from the intersection between the map $F^{(2)}(X)$ and the 45° line, as shown in Figs 2 and 3. Clearly the equilibrium point X^* of equation (5) is a solution of equation (9); the basic fixed point of period 1 is a degenerate case of a period 2 solution. We now make a simple, but crucial, observation[1]: the slope of the curve $F^{(2)}(X)$ at the point X^*, defined as $\lambda^{(2)}(X^*)$ and illustrated by the dashed lines in Figs 2 and 3, is the square of the corresponding slope of $F(X)$

$$\lambda^{(2)}(X^*)=[\lambda^{(1)}(X^*)]^2 \tag{10}$$

This fact can now be used to make plain what happens when the fixed point X^* becomes unstable. If the slope of $F(X)$ is less than −45°(that is, $|\lambda^{(1)}|<1$), as illustrated by curve *a* in Fig. 1, then X^* is stable. Also, from equation (10), this implies $0<\lambda^{(2)}<1$ corresponding to the slope of $F^{(2)}$ at X^* lying between 0° and 45°, as shown in Fig. 2. As long as the fixed point X^* is stable, it provides the only non-trivial solution to equation (9). On the other hand, when $\lambda^{(1)}$ steepens beyond −45° (that is, $|\lambda^{(1)}|>1$), as illustrated by curve b in Fig 1, X^* becomes unstable. At the same time, from equation (10) this implies $\lambda^{(2)}>1$, corresponding to the slope of $F^{(2)}$ at X^* steepening beyond 45°, as shown in Fig. 3. As this happens, the curve $F^{(2)}(X)$ must develop a "loop", and two new fixed points of period 2 appear, as illustrated in Fig. 3.

In short, as the nonlinear function $F(X)$ in equation (1) becomes more steeply humped, the basic fixed point X^* may become unstable. At exactly the stage when this occurs, there are born two new and initially stable fixed points of period 2, between which the system alternates in a stable cycle of period 2. The sort of graphical analysis indicated by Figs 1, 2 and 3, along with the equation (10), is all that is needed to establish this generic result[1,4].

As before, the stability of this period 2 cycle depends on the slope of the curve $F^{(2)}(X)$ at the 2 points. (This slope is easily shown to be the same at both points[1,20], and more generally to be the same at all k points on a period k cycle.) Furthermore, as is clear by imagining the intermediate stages between Figs 2 and 3, this stability-determining slope has the value $\lambda=+1$ at the birth of the 2-point cycle, and then decreases through zero towards $\lambda=-1$ as the hump in $F(X)$ continues to steepen. Beyond this point the period 2 points will in turn become unstable, and bifurcate to give an initially stable cycle of period 4. This in turn gives way to a cycle of period 8, and thence to a hierarchy of bifurcating stable cycles of periods 16, 32, 64,..., 2^n. In each case, the way in which a stable cycle of period k becomes unstable, simultaneously bifurcating to produce a new and initially stable cycle of period $2k$, is basically similar to the process just adumbrated for $k=1$. A more full and rigorous account of the material covered so far is in ref. 1.

This "very beautiful bifurcation phenomenon"[22] is depicted in Fig. 4, for the example equation (3). It cannot be too strongly emphasised that the process is generic to most functions $F(X)$ with a hump of tunable steepness. Metropolis *et al.*[16] refer to this hierarchy of cycles of periods 2^n as the harmonics of the fixed point X^*.

求出，或者用图像方法找出映射 $F^{(2)}(X)$ 与 45° 线的交点，如图 2 和图 3 所示。很明显，方程（5）的平衡点 X^* 是方程（9）的一个解；周期为 1 的基本不动点是周期为 2 的解的一种简并情况。现在我们来做一番简单却极为重要的考察[1]：曲线 $F^{(2)}(X)$ 在点 X^* 处的斜率定义为 $\lambda^{(2)}(X^*)$，并用虚线表示于图 2 和图 3 中，它是相应的 $F(X)$ 的斜率的平方：

$$\lambda^{(2)}(X^*) = [\lambda^{(1)}(X^*)]^2 \tag{10}$$

现在我们能够用这一事实来解释当不动点 X^* 变得不稳定时所发生的事情。如果 $F(X)$ 的倾斜度小于 −45°（即 $|\lambda^{(1)}| < 1$），正如图 1 中曲线 a 所示，那么 X^* 是稳定的。同样，根据方程（10），这就意味着 $0 < \lambda^{(2)} < 1$ 对应于 $F^{(2)}$ 在 X^* 处的倾斜度介于 0°～45°，如图 2 所示。只要不动点 X^* 稳定，它就能为方程（9）提供唯一的非平凡解。另一方面，当 $\lambda^{(1)}$ 倾斜度超过 − 45°（即 $|\lambda^{(1)}| > 1$）时，正如图 1 中曲线 b 所示，X^* 将变得不稳定。同时，根据方程（10）可知，这意味着 $\lambda^{(2)} > 1$，对应的 $F^{(2)}$ 在 X^* 处的倾角超过 45°，如图 3 所示。在这种情况下，曲线 $F^{(2)}(X)$ 必定会形成一个"循环"，并且出现两个周期为 2 的新不动点，如图 3 所示。

总而言之，随着方程（1）中的非线性函数 $F(X)$ 逐渐变陡峭，基本不动点 X^* 逐渐变成不稳定的。而在这个阶段中，会产生两个最初稳定且周期为 2 的新不动点，在这两点之间，系统在周期为 2 的稳定循环中振荡。图 1、2 和 3 所表示的图像分析方法，以及方程（10）正是要确定这个一般性结果所需要的一切[1,4]。

如前所述，这个周期为 2 的循环的稳定性取决于曲线 $F^{(2)}(X)$ 在两点处的斜率。（两点处的斜率很明显是相同的[1,20]，更一般地，对于周期为 k 的循环，所有 k 个点处的斜率都是相同的。）不仅如此，通过想象图 2 与图 3 之间的过渡阶段可以知道，这个确定稳定性的斜率在 2 点循环刚产生时的值是 $\lambda = +1$，继而随着 $F(X)$ 的峰逐渐变陡，经由零逐渐减少到 $\lambda = -1$。超过该点之后，周期为 2 的各点将会依次变得不稳定，并且产生分岔而形成周期为 4 的初始稳定循环。照此方式，依次可以产生周期为 8 的循环，进而产生一个周期为 16, 32, 64, ⋯, 2^n 的稳定循环的分岔谱系。在各种情况下，周期为 k 的稳定循环逐渐变得不稳定，同时产生分岔而形成周期为 $2k$ 的新初始稳定循环，其方式基本上与对 $k = 1$ 的情况所概括的过程相类似。目前，参考文献 1 包含了更为完整和严格的材料。

针对示例方程（3），图 4 给出了这种"极为美丽的分岔现象"[22]。无论怎样着力强调下列事实都不过分：这一过程对于绝大多数具有一个陡度可调的单峰函数 $F(X)$ 来说是普遍适用的。梅特罗波利斯等[16] 将这种周期为 2^n 的循环谱系称为不动点 X^* 的谐振。

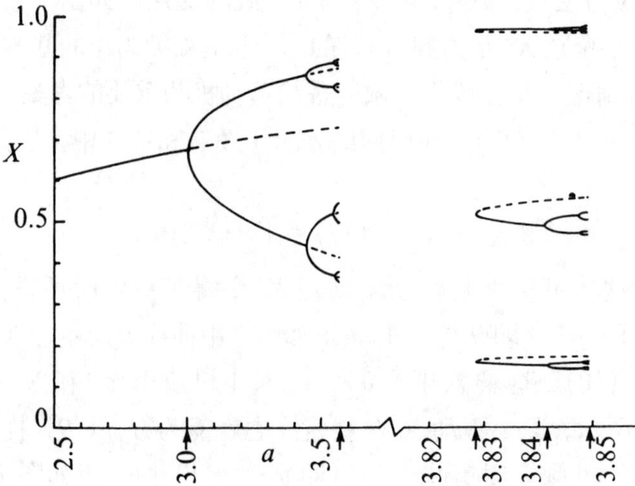

Fig. 4. This figure illustrates some of the stable (——) and unstable (- - - -) fixed points of various periods that can arise by bifurcation processes in equation (1) in general, and equation (3) in particular. To the left, the basic stable fixed point becomes unstable and gives rise by a succession of pitchfork bifurcations to stable harmonics of period 2^n; none of these cycles is stable beyond a = 3.5700. To the right, the two period 3 cycles appear by tangent bifurcation: one is initially unstable; the other is initially stable, but becomes unstable and gives way to stable harmonics of period 3×2^n, which have a point of accumulation at a = 3.8495. Note the change in scale on the a axis, needed to put both examples on the same figure. There are infinitely many other such windows, based on cycles of higher periods.

Although this process produces an infinite sequence of cycles with periods 2^n ($n\rightarrow\infty$), the "window" of parameter values wherein any one cycle is stable progressively diminishes, so that the entire process is a convergent one, being bounded above by some critical parameter value. (This is true for most, but not all, functions $F(X)$: see equation (17) below.) This critical parameter value is a point of accumulation of period 2^n cycles. For equation (3) it is denoted a_c: a_c =3.5700...

Beyond this point of accumulation (for example, for $a>a_c$ in equation (3)) there are an infinite number of fixed points with different periodicities, and an infinite number of different periodic cycles. There are also an uncountable number of initial points X_0 which give totally aperiodic (although bounded) trajectories; no matter how long the time series generated by $F(X)$ is run out, the pattern never repeats. These facts may be established by a variety of methods[1,4,20,22,29]. Such a situation, where an infinite number of different orbits can occur, has been christened "chaotic" by Li and Yorke[20].

As the parameter increases beyond the critical value, at first all these cycles have even periods, with X_t alternating up and down between values above, and values below, the fixed point X^*. Although these cycles may in fact be very complicated (having a non-degenerate period of, say, 5,726 points before repeating), they will seem to the casual observer to be rather like a somewhat "noisy" cycle of period 2. As the parameter value continues to increase, there comes a stage (at a=3.6786.. for equation (3)) at which the first odd period

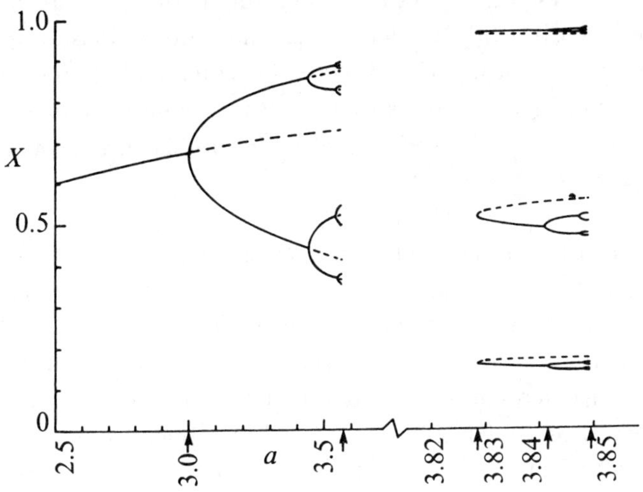

图 4. 此图显示出若干不同周期的稳定的（——）和不稳定的（- - - -）不动点，在一般情况下，它们可以由方程（1）的分岔过程产生，此图是由方程（3）产生。在左边，基本的稳定不动点变得不稳定，从而产生一连串的叉状分岔，形成周期为 2^n 的稳定谐振；超过 $a = 3.5700$ 后，这些循环都不稳定。在右边，通过切分岔出现两个周期为 3 的循环：一个最初就是不稳定的，另一个最初是稳定的，但是逐渐变得不稳定并且让位于周期 3×2^n 的稳定谐振，这些周期以 $a = 3.8495$ 作为累加点。注意 a 轴上有标度的变化，这是为了将两个例子置于同一图中。这样的窗口还有无限多个，它们源于周期更高的其他循环。

尽管该过程产生出一个具有周期 2^n（$n \to \infty$）的无限循环序列，任一循环的参数范围的“窗口”却是稳定地逐渐减小，致使整个过程是收敛的，其上边界由某个临界参数值所确定（这对形如后面的方程（17）的绝大多数函数 $F(X)$ 都成立，但并非全部）。这一临界参数值是周期为 2^n 的各循环累加点。对于方程（3），将它记为 a_c，$a_c = 3.5700\cdots$。

超过这个累加点后（如在方程（3）中就是 $a > a_c$），有无限多个具有不同周期性的不动点和无限多个不同的周期性循环。另外还有无数个初始点 X_0，它们产生完全非周期性的（尽管是有界的）轨道。无论 $F(X)$ 所产生的时间序列持续多久，其模式都绝不重复。这些事实可以通过若干种方法加以确定 [1,4,20,22,29]。对于这种能够出现无限多个不同轨道的情形，李和约克 [20] 称之为“混沌的”。

当参数增加到超过临界值后，开始时全部循环都具有偶数周期，X_t 在不动点 X^* 上方和下方的数值间振荡。尽管这些循环实际上可能是非常复杂的（如每次重复一个 5,726 个点的非简并循环），但对于不认真的观测者来说，它们看起来在某种程度上更像是周期为 2 的“噪声”循环。随着参数值的持续增加，一个新阶段来临（对于方程（3），位于 $a = 3.6786\cdots$ 处），在这里出现了第一个周期为奇数的循环。开

cycle appears. At first these odd cycles have very long periods, but as the parameter value continues to increase cycles with smaller and smaller odd periods are picked up, until at last the three-point cycle appears (at a=3.8284.. for equation (3)). Beyond this point, there are cycles with every integer period, as well as an uncountable number of asymptotically aperiodic trajectories: Li and Yorke[20] entitle their original proof of this result "Period Three Implies Chaos".

The term "chaos" evokes an image of dynamical trajectories which are indistinguishable from some stochastic process. Numerical simulations[12,15,21,23,25] of the dynamics of equation (3), (4) and other similar equations tend to confirm this impression. But, for smooth and "sensible" functions $F(X)$ such as in equations (3) and (4), the underlying mathematical fact is that for any specified parameter value there is one unique cycle that is stable, and that attracts essentially all initial points[22,29] (see ref. 4, appendix A, for a simple and lucid exposition). That is, there is one cycle that "owns" almost all initial points; the remaining infinite number of other cycles, along with the asymptotically aperiodic trajectories, own a set of points which, although uncountable, have measure zero.

As is made clear by Tables 3 and 4 below, any one particular stable cycle is likely to occupy an extraordinarily narrow window of parameter values. This fact, coupled with the long time it is likely to take for transients associated with the initial conditions to damp out, means that in practice the unique cycle is unlikely to be unmasked, and that a stochastic description of the dynamics is likely to be appropriate, in spite of the underlying deterministic structure. This point is pursued further under the heading "practical applications", below.

The main messages of this section are summarised in Table 1, which sets out the various domains of dynamical behaviour of the equations (3) and (4) as functions of the parameters, a and r respectively, that determine the severity of the nonlinear response. These properties can be understood qualitatively in a graphical way, and are generic to any well behaved $F(X)$ in equation (1).

We now proceed to a more detailed discussion of the mathematical structure of the chaotic regime for analytical functions, and then to the practical problems alluded to above and to a consideration of the behavioural peculiarities exhibited by non-analytical functions (such as those in the two right hand columns of Table 1).

始时这些奇循环具有极长的周期，但是随着参数值的持续增加，具有越来越小的奇数周期的循环逐渐凸现出来，直到最后出现三点循环（对于方程（3），位于 $a = 3.8284\cdots$处）。超过这一点，就会出现具有所有整数周期的循环，以及无数个渐近的非周期性轨道：李和约克[20]将他们对这一结果给出的初始证明称为“周期三意味着混沌”。

“混沌”这个词使人们产生了动力学轨道与某些随机过程不可区分的印象。对方程（3）、（4）以及其他类似方程的动力学的数值模拟[12,16,21,23,25]也倾向于支持这种印象。但是，对于诸如方程（3）和（4）中那样光滑而“正常的”函数 $F(X)$，根本性的数学事实是对于任意特定的参数值，只存在唯一的一个循环是稳定的，它实际上吸引了所有的初始点[22,29]（一份简单而清晰的说明参见文献 4，附录 A）。也就是说，有一个循环“拥有”几乎所有的初始点，剩下的其他无限多个循环，以及那些渐近非周期性轨道，拥有一个尽管不可数但却测度为零的点集。

如同表 3 和表 4 所示，任意一个特定的稳定循环都应该占据一个极其狭窄的参数值窗口。这一事实结合与初始条件有关的暂态效果需长时间才能消除，意味着那个唯一循环实际上并不倾向于显露出来，因而对动力学的随机描述看起来是合适的，即使它是基于一个确定性的结构。后文在“实际应用”标题下将对这一点进行进一步说明（译注：实际上后面并没有一个叫作“实际应用”的标题，从具体内容来看，原作者所指的可能是“实际问题”一节）。

表 1 对本节主要信息进行了总结，它将方程（3）和（4）的动力学行为的各种情况分别视为参数 a 和 r 的函数，它们决定了非线性响应的强度。这些性质可以通过图像方法而定性地理解，并且对于方程（1）中任何行为良好的函数 $F(X)$ 都是普遍适用的。

现在我们要对解析函数就混沌区域中的数学结构进行更为详细的探讨，继而对上面涉及的实际问题加以讨论，并对非解析函数（如表1右侧两列中的函数）所呈现出的行为特性加以考察。

Table 1. Summary of the way various "single-hump" functions $F(X)$, from equation (1), behave in the chaotic region, distinguishing the dynamical properties which are generic from those which are not

The function $F(X)$ of equation (1)	$aX(1-X)$	$X\exp[r(1-X)]$	aX; if $X<\frac{1}{2}$ $a(1-X)$; if $X>\frac{1}{2}$	λX; if $X<1$ λX^{1-b}; if $X>1$
Tunable parameter	a	r	a	b
Fixed point becomes unstable	3.0000	2.0000	1.0000*	2.0000
"Chaotic" region begins [point of accumulation of cycles of period 2^n]	3.5700	2.6924	1.0000	2.0000
First odd-period cycle appears	3.6786	2.8332	1.4142	2.6180
Cycle with period 3 appears [and therefore every integer period present]	3.8284	3.1024	1.6180	3.0000
"Chaotic" region ends	4.0000†	∞‡	2.000†	∞‡
Are there stable cycles in the chaotic region?	Yes	Yes	No	No

* Below this a value, $X=0$ is stable.

† All solutions are attracted to $-\infty$ for a values beyond this.

‡ In practice, as r or b becomes large enough, X will eventually be carried so low as to be effectively zero, thus producing extinction in models of biological populations.

Fine Structure of the Chaotic Regime

We have seen how the original fixed point X^* bifurcates to give harmonics of period 2^n. But how do new cycles of period k arise?

The general process is illustrated in Fig. 5, which shows how period 3 cycles originate. By an obvious extension of the notation introduced in equation (8), populations three generations apart are related by

$$X_{t+3} = F^{(3)}(X_t) \tag{11}$$

If the hump in $F(X)$ is sufficiently steep, the threefold iteration will produce a function $F^{(3)}(X)$ with 4 humps, as shown in Fig. 5 for the $F(X)$ of equation (3). At first (for $a<3.8284..$ in equation 3) the 45° line intersects this curve only at the single point X^* (and at $X=0$), as shown by the solid curve in Fig. 5. As the hump in $F(X)$ steepens, the hills and valleys in $F^{(3)}(X)$ become more pronounced, until simultaneously the first two valleys sink and the final hill rises to touch the 45° line, and then to intercept it at 6 new points, as shown by the dashed curve in Fig. 5. These 6 points divide into two distinct three-point cycles. As can be made plausible by imagining the intermediate stages in Fig. 5, it can be shown that the stability-determining slope of $F^{(3)}(X)$ at three of these points has a common value, which is $\lambda^{(3)}=+1$ at their birth, and thereafter steepens beyond $+1$; this period 3 cycle is never stable. The slope of $F^{(3)}(X)$ at the other three points begins at $\lambda^{(3)}=+1$, and

表 1. 对方程（1）中各种“单峰”函数 $F(X)$ 在混沌区域的行为方式的总结，区分出一般性的动力学性质与非一般性的动力学性质

方程（1）中的函数$F(X)$	$aX(1-X)$	$X\exp[r(1-X)]$	若$X<\frac{1}{2}$, aX 若$X>\frac{1}{2}$, $a(1-X)$	若$X<1$, λX 若$X>1$, λX^{1-b}
可调参数	a	r	a	b
不动点变得不稳定	3.0000	2.0000	1.0000*	2.0000
“混沌”区域开始［周期 2^n 循环的累加点］	3.5700	2.6924	1.0000	2.0000
第一个奇数周期循环出现	3.6786	2.8332	1.4142	2.6180
周期为 3 的循环出现［于是任意整数周期存在］	3.8284	3.1024	1.6180	3.0000
“混沌”区域结束	4.0000†	∞‡	2.000†	∞‡
混沌区域中是否有稳定循环?	是	是	否	否

* 在这个 a 值以下，$X=0$ 是稳定的。

† 当 a 值超过此值后，所有的解都被吸引到 $-\infty$。

‡ 事实上，当 r 或 b 变得足够大时，X 将会变得非常小以致等价于 0，从而导致生物学种群模型中的灭绝。

混沌区域的精细结构

我们已经看到初始不动点 X^* 是如何分岔而产生周期为 2^n 的谐振的。但是，周期为 k 的新循环是如何产生的呢?

图 5 中显示了一般性的过程，说明了周期为 3 的循环是如何产生的。通过对方程（8）中所用记号的一个明显推广，相隔三个世代的种群通过下列公式联系起来：

$$X_{t+3}=F^{(3)}(X_t) \tag{11}$$

如果 $F(X)$ 中的峰足够陡峭，三重迭代就会产生一个有 4 个峰的函数 $F^{(3)}(X)$，正如图 5 中所示的方程（3）中 $F(X)$ 所对应的情况。最初（对于方程（3）即为 $a<3.8284\cdots$ 时），45° 线与这条曲线只在唯一的 X^* 处（以及 $X=0$ 处）相交，如同图 5 中用实线所表示的曲线。随着 $F(X)$ 中的峰变陡，$F^{(3)}(X)$ 中的峰和谷变得越来越显著，直到前两个谷下降以及最后一个峰升高至同时触及 45° 线，从而与该线交于 6 个新的点，如图 5 中虚线所示。这 6 个点分成 2 个不同的三点循环。通过想象图 5 中的过渡阶段即可明白，在上述点中的 3 个点处，$F^{(3)}(X)$ 具有相同的决定稳定性的斜率值，即刚产生时的 $\lambda^{(3)}=+1$，之后变得陡峭超过 +1；这个周期为 3 的循环是不稳定的。$F^{(3)}(X)$ 在其他 3 个点处的斜率由 $\lambda^{(3)}=+1$ 开始，随后便向 0 减少，最终结果是一个周期为

then decreases towards zero, resulting in a stable cycle of period 3. As $F(X)$ continues to steepen, the slope $\lambda^{(3)}$ for this initially stable three-point cycle decreases beyond -1; the cycle becomes unstable, and gives rise by the bifurcation process discussed in the previous section to stable cycles of period 6, 12, 24, ..., 3×2^n. This birth of a stable and unstable pair of period 3 cycles, and the subsequent harmonics which arise as the initially stable cycle becomes unstable, are illustrated to the right of Fig. 4.

There are, therefore, two basic kinds of bifurcation processes[1,4] for first order difference equations. Truly new cycles of period k arise in pairs (one stable, one unstable) as the hills and valleys of higher iterates of $F(X)$ move, respectively, up and down to intercept the 45° line, as typified by Fig. 5. Such cycles are born at the moment when the hills and valleys become tangent to the 45° line, and the initial slope of the curve $F^{(k)}$ at the points is thus $\lambda^{(k)} = +1$: this type of bifurcation may be called[1,4] a tangent bifurcation or a $\lambda = +1$ bifurcation. Conversely, an originally stable cycle of period k may become unstable as $F(X)$ steepens. This happens when the slope of $F^{(k)}$ at these period k points steepens beyond $\lambda^{(k)} = -1$, whereupon a new and initially stable cycle of period $2k$ is born in the way typified by Figs 2 and 3. This type of bifurcation may be called a pitchfork bifurcation (borrowing an image from the left hand side of Fig. 4) or a $\lambda = -1$ bifurcation[1,4].

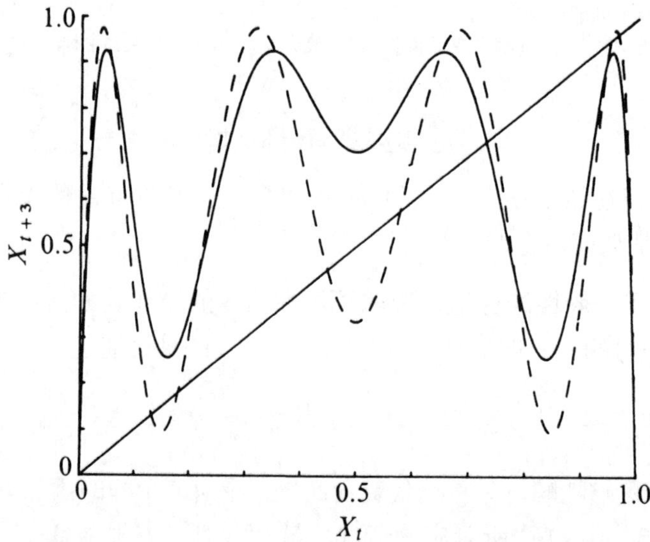

Fig. 5. The relationship between X_{t+3} and X_t, obtained by three iterations of equation (3). The solid curve is for $a = 3.7$, and only intersects the 45° line once. As a increases, the hills and valleys become more pronounced. The dashed curve is for $a=3.9$, and six new period 3 points have appeared (arranged as two cycles, each of period 3).

3 的稳定循环。随着 $F(X)$ 持续倾斜，这一初始稳定三点循环的斜率 $\lambda^{(3)}$ 下降至低于 -1；循环变得不稳定，并且通过前一节中所讨论的分岔过程而形成具有周期 6, 12, 24, ⋯, 3×2^n 的稳定循环。这对周期为 3 的稳定和不稳定循环的产生以及随后由初始稳定循环变得不稳定而导致的谐振的产生，如图 4 的右方所示。

因此，对于一阶差分方程而言，存在着两个基本类型的分岔过程 [1,4]。随着 $F(X)$ 较高次迭代的峰和谷分别上下运动而与 45° 线相交，周期为 k 的真正的新循环成对产生（一个稳定，一个不稳定），正如图 5 所代表的。这些循环产生于峰和谷逐渐变成与 45° 线相切的时刻，因而 $F^{(k)}$ 曲线在那些点处的初始斜率就是 $\lambda^{(k)}=+1$：这种类型的分岔可以称为 [1,4] 切分岔或者 $\lambda=+1$ 分岔。反过来，一个周期为 k 的初始稳定循环也可以随着 $F(X)$ 变陡峭而变得不稳定。这发生在周期为 k 的各点处的 $F^{(k)}$ 斜率倾斜到超过 $\lambda^{(k)}=-1$ 时，于是一个新的周期为 $2k$ 的初始稳定循环就产生了，图 2 和图 3 是此方式特征的典型描述。这种类型的分岔可以称为叉形分岔（借用图 4 左边的图像来说明），或称为 $\lambda=-1$ 分岔 [1,4]。

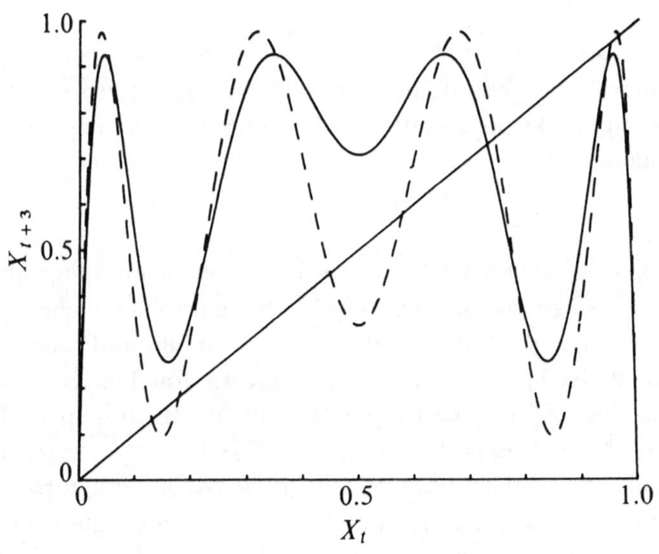

图 5. 通过方程（3）的三次迭代而得到 X_{t+3} 与 X_t 之间的关系。用实线画出的曲线对应于 $a=3.7$，并且只与 45° 线相交一次。随着 a 增大，峰和谷变得更为显著。用虚线画出的曲线对应于 $a=3.9$，出现了 6 个新的周期为 3 的点（分布于 2 个周期为 3 的循环）。

Putting all this together, we conclude that as the parameters in $F(X)$ are varied the fundamental, stable dynamical units are cycles of basic period k, which arise by tangent bifurcation, along with their associated cascade of harmonics of periods $k2^n$, which arise by pitchfork bifurcation. On this basis, the constant equilibrium solution X^* and the subsequent hierarchy of stable cycles of periods 2^n is merely a special case, albeit a conspicuously important one (namely $k = 1$), of a general phenomenon. In addition, remember[1,4,22,29] that for sensible, analytical functions (such as, for example, those in equations (3) and (4)) there is a unique stable cycle for each value of the parameter in $F(X)$. The entire range of parameter values ($1<a<4$ in equation (3), $0<r$ in equation (4)) may thus be regarded as made up of infinitely many windows of parameter values—some large, some unimaginably small—each corresponding to a single one of these basic dynamical units. Tables 3 and 4, below, illustrate this notion. These windows are divided from each other by points (the points of accumulation of the harmonics of period $k2^n$) at which the system is truly chaotic, with no attractive cycle: although there are infinitely many such special parameter values, they have measure zero on the interval of all values.

How are these various cycles arranged along the interval of relevant parameter values? This question has to my knowledge been answered independently by at least 6 groups of people, who have seen the problem in the context of combinatorial theory[18,30], numerical analysis[13,14], population biology[1], and dynamical systems theory[22,31] (broadly defined).

A simple-minded approach (which has the advantage of requiring little technical apparatus, and the disadvantage of being rather clumsy) consists of first answering the question, how many period k points can there be? That is, how many distinct solutions can there be to the equation

$$X^*_k = F^{(k)}(X^*_k)? \tag{12}$$

If the function $F(X)$ is sufficiently steeply humped, as it will be once the parameter values are sufficiently large, each successive iteration doubles the number of humps, so that $F^{(k)}(X)$ has 2^{k-1} humps. For large enough parameter values, all these hills and valleys will intersect the 45° line, producing 2^k fixed points of period k. These are listed for $k \leqslant 12$ in the top row of Table 2. Such a list includes degenerate points of period k, whose period is a submultiple of k; in particular, the two period 1 points ($X=0$ and X^*) are degenerate solutions of equation (12) for all k. By working from left to right across Table 2, these degenerate points can be subtracted out, to leave the total number of non-degenerate points of basic period k, as listed in the second row of Table 2. More sophisticated ways of arriving at this result are given elsewhere[13,14,16,22,30,31].

综上所述，我们得出结论：随着 $F(X)$ 中参数的改变，基本的稳定动力学单元是周期为 k 的循环，由切分岔而产生，以及与其相伴随的周期为 $k2^n$ 的谐振级联，这一谐振由叉形分岔而产生。基于此，恒定的平衡解 X^* 与随之而来的周期 2^n 的稳定循环谱系仅仅是一般现象中的一种特殊情况，当然，很明显，它是一种重要的特殊情况（即 $k = 1$）。此外要记得[1,4,22,29]，对于合理的、解析的函数（如方程（3）和（4）中的那些函数）来说，对 $F(X)$ 中的每个参数值，存在唯一的稳定循环。因此，可以将参数值可取的整个范围（方程（3）中的 $1 < a < 4$，方程（4）中的 $0 < r$）看作由无限多个参数值窗口组成——一些窗口较大，一些窗口则小到难以想象——每个窗口都恰好对应于一个上述基本动力学单元。下文中的表 3 和表 4 对这一情况进行了说明。这些窗口被各个点（周期为 $k2^n$ 的谐振的累加点）分隔开，在这些点处，系统处于真正的混沌状态，没有任何有吸引能力的循环：尽管有无限多个这样的特殊参数值，它们在全部可取值的区间上的测度却是零。

这些各式各样的循环在相关参数值的区间中是如何分布的呢？就我所知，至少有 6 组人曾经独立地回答过这个问题，他们曾经在组合论[18,30]、数值分析[13,14]、种群生物学[1]以及（广义的）动力系统理论[22,31]等不同背景下审视过这个问题。

一种简单的方法（优点是不需要技术性工具，缺点是过于粗略）是，首先要回答下面的问题：可以有多少个周期为 k 的点？也就是说，方程

$$X^*_k = F^{(k)}(X^*_k) \tag{12}$$

可以有多少个不同的解？如果函数 $F(X)$ 具有充分陡峭的峰，正如当参数值充分大时它将会呈现出的那样，每次相继的迭代都会使峰的数目加倍，因此，$F^{(k)}(X)$ 有 2^{k-1} 个峰。对于足够大的参数值，所有这些峰和谷都会与 45° 线相交，产生 2^k 个周期为 k 的不动点。它们被列于表 2 首行中 $k \leqslant 12$ 的范围内。表 2 中包含周期为 k 的简并点，其周期是 k 的一个因数；特别地，两个周期为 1 的点（$X = 0$ 和 X^*）对于所有的 k 来说都是方程（12）的简并解。从左到右横贯表 2 进行处理，可以将这些简并点全部去掉，保留基本周期为 k 的非简并点的总数，列于表 2 的第二行中。其他一些地方[13,14,16,22,30,31]给出了获得这一结果的更为准确的方法。

Table 2. Catalogue of the number of periodic points, and of the various cycles (with periods $k = 1$ up to 12), arising from equation (1) with a single-humped function $F(X)$

k	1	2	3	4	5	6	7	8	9	10	11	12
Possible total number of points with period k	2	4	8	16	32	64	128	256	512	1,024	2,048	4,096
Possible total number of points with non-degenerate period k	2	2	6	12	30	54	126	240	504	990	2,046	4,020
Total number of cycles of period k, including those which are degenerate and/or harmonics and/or never locally stable	2	3	4	6	8	14	20	36	60	108	188	352
Total number of non-degenerate cycles (including harmonics and unstable cycles)	2	1	2	3	6	9	18	30	56	99	186	335
Total number of non-degenerate, stable cycles (including harmonics)	1	1	1	2	3	5	9	16	28	51	93	170
Total number of non-degenerate, stable cycles whose basic period is k (that is, excluding harmonics)	1	–	1	1	3	4	9	14	28	48	93	165

For example, there eventually are $2^6=64$ points with period 6. These include the two points of period 1, the period 2 "harmonic" cycle, and the stable and unstable pair of triplets of points with period 3, for a total of 10 points whose basic period is a submultiple of 6; this leaves 54 points whose basic period is 6.

The 2^k period k points are arranged into various cycles of period k, or submultiples thereof, which appear in succession by either tangent or pitchfork bifurcation as the parameters in $F(X)$ are varied. The third row in Table 2 catalogues the total number of distinct cycles of period k which so appear. In the fourth row[14], the degenerate cycles are subtracted out, to give the total number of non-degenerate cycles of period k: these numbers must equal those of the second row divided by k. This fourth row includes the (stable) harmonics which arise by pitchfork bifurcation, and the pairs of stable–unstable cycles arising by tangent bifurcation. By subtracting out the cycles which are unstable from birth, the total number of possible stable cycles is given in row five; these figures can also be obtained by less pedestrian methods[13,16,30]. Finally we may subtract out the stable cycles which arise by pitchfork bifurcation, as harmonics of some simpler cycle, to arrive at the final row in Table 2, which lists the number of stable cycles whose basic period is k.

Returning to the example of period 6, we have already noted the five degenerate cycles whose periods are submultiples of 6. The remaining 54 points are parcelled out into one cycle of period 6 which arises as the harmonic of the only stable three-point cycle, and four distinct pairs of period 6 cycles (that is, four initially stable ones and four unstable ones) which arise by successive tangent bifurcations. Thus, reading from the foot of the column for period 6 in Table 2, we get the numbers 4, 5, 9, 14.

Using various labelling tricks, or techniques from combinatorial theory, it is also possible to give a generic list of the order in which the various cycles appear[1,13,16,22]. For example, the basic stable cycles of periods 3, 5, 6 (of which there are respectively 1, 3, 4) must appear in

表 2. 周期点数目，以及由具有一个单峰函数 $F(X)$ 的方程（1）所产生的各种循环（周期 k=1，2，…，12）的列表

k	1	2	3	4	5	6	7	8	9	10	11	12
具有周期 k 的点的可能总数	2	4	8	16	32	64	128	256	512	1,024	2,048	4,096
具有非简并周期 k 的点的可能总数	2	2	6	12	30	54	126	240	504	990	2,046	4,020
周期为 k 的循环的总数，包括简并的和（或）谐振的和（或）没有任何局部是稳定的情况	2	3	4	6	8	14	20	36	60	108	188	352
非简并循环的总数（包括谐振和不稳定循环）	2	1	2	3	6	9	18	30	56	99	186	335
非简并的稳定循环的总数（包括谐振）	1	1	1	2	3	5	9	16	28	51	93	170
基本周期为 k 的非简并稳定循环的总数（即不包括谐振）	1	–	1	1	3	4	9	14	28	48	93	165

例如，最终会有 $2^6 = 64$ 个周期为 6 的点，其中包括两个周期为 1 的点、周期为 2 的“谐振”循环，以及一对周期为 3 的三点组（一个稳定和一个不稳定），总共有 10 个以 6 的一个因数为基本周期的点，还剩下 54 个基本周期为 6 的点。

周期为 k 的 2^k 个点分布于周期为 k 或其因数的各个循环之中，后者随着 $F(X)$ 中参数的变化通过切分岔或者叉形分岔而相继出现。表 2 中的第三行列出了所出现的周期为 k 的不同循环总数。在第四行中 [14]，简并循环被去掉，给出了周期为 k 的非简并循环总数：这些数必然等于第二行中的那些数除以 k 所得的商。第四行中包括由叉形分岔而产生的（稳定）谐振，以及由切分岔所产生的稳定–不稳定循环对。通过除去从一出现就不稳定的那些循环，第五行中给出了可能的稳定循环的总数；利用不太平庸的方法也可以得到这些数字 [13,16,30]。最后，我们可以去除由叉形分岔所产生的那些稳定循环，如某些较简单循环的谐振，而得到表 2 中的最后一行，该行列出了基本周期为 k 的稳定循环的数目。

回到周期为 6 的那个例子，我们已经注意到有以 6 的因数为周期的 5 个简并循环。其余 54 个点被分配到周期为 6 的循环的一个稳定的三点循环谐振及由相继的切分岔所产生 4 个不同的周期为 6 的循环对（即 4 个初始的稳定循环与 4 个不稳定循环）之中。于是从表 2 中周期为 6 的那一列的底部读起，我们得到了数据 4、5、9 和 14。

利用各种标记技巧，或者是来自组合论的技术，还有可能给出关于各种循环出现顺序的一个一般性列表 [1,13,16,22]。例如，周期为 3，5，6 的基本稳定循环（分别有 1 个、3 个和 4 个）必定以 6，5，3，5，6，6，5，6 的顺序出现，对照表 3 和表 4。

the order 6, 5, 3, 5, 6, 6, 5, 6: compare Tables 3 and 4. Metropolis *et al.*[16] give the explicit such generic list for all cycles of period $k \leqslant 11$.

Table 3. A catalogue of the stable cycles (with basic periods up to 6) for the equation

$$X_{t+1} = aX_t(1-X_t)$$

Period of basic cycle	*a* value at which:		Subsequent cascade of "harmonics" with period $k2^n$ all become unstable	Width of the range of *a* values over which the basic cycle, or one of its harmonics, is attractive
	Basic cycle first appears	Basic cycle becomes unstable		
1	1.0000	3.0000	3.5700	2.5700
3	3.8284	3.8415	3.8495	0.0211
4	3.9601	3.9608	3.9612	0.0011
5(*a*)	3.7382	3.7411	3.7430	0.0048
5(*b*)	3.9056	3.9061	3.9065	0.0009
5(*c*)	3.99026	3.99030	3.99032	0.00006
6(*a*)	3.6265	3.6304	3.6327	0.0062
6(*b*)	3.937516	3.937596	3.937649	0.000133
6(*c*)	3.977760	3.977784	3.977800	0.000040
6(*d*)	3.997583	3.997585	3.997586	0.000003

Table 4. Catalogue of the stable cycles (with basic periods up to 6) for the equation

$$X_{t+1} = X_t \exp[r(1-X_t)]$$

Period of basic cycle	*r* value at which:		Subsequent cascade of "harmonics" with period $k2^n$ all become unstable	Width of the range of *r* values over with the basic cycle, or one of its harmonics, is attractive
	Basic cycle first appears	Basic cycle becomes unstable		
1	0.0000	2.0000	2.6924	2.6924
3	3.1024	3.1596	3.1957	0.0933
4	3.5855	3.6043	3.6153	0.0298
5(*a*)	2.9161	2.9222	2.9256	0.0095
5(*b*)	3.3632	3.3664	3.3682	0.0050
5(*c*)	3.9206	3.9295	3.9347	0.0141
6(*a*)	2.7714	2.7761	2.7789	0.0075
6(*b*)	3.4558	3.4563	3.4567	0.0009
6(*c*)	3.7736	3.7745	3.7750	0.0014
6(*d*)	4.1797	4.1848	4.1880	0.0083

梅特罗波利斯等[16]对于所有周期 $k \leqslant 11$ 的循环给出了这样一个清晰的一般性列表。

表 3. 对应于方程 $X_{t+1} = aX_t(1-X_t)$ 的稳定循环的目录（基本周期直到 6 为止）

基本循环的周期	满足下列条件的 a 值		随后出现的具有周期 $k2^n$ 的“谐振”级联全都变得不稳定	使基本循环或它的某一个谐振成为吸引子的 a 值范围的宽度
	基本循环首次出现	基本循环变得不稳定		
1	1.0000	3.0000	3.5700	2.5700
3	3.8284	3.8415	3.8495	0.0211
4	3.9601	3.9608	3.9612	0.0011
5（a）	3.7382	3.7411	3.7430	0.0048
5（b）	3.9056	3.9061	3.9065	0.0009
5（c）	3.99026	3.99030	3.99032	0.00006
6（a）	3.6265	3.6304	3.6327	0.0062
6（b）	3.937516	3.937596	3.937649	0.000133
6（c）	3.977760	3.977784	3.977800	0.000040
6（d）	3.997583	3.997585	3.997586	0.000003

表 4. 对应于方程 $X_{t+1} = X_t \exp[r(1-X_t)]$ 的稳定循环的目录（基本周期到 6 为止）

基本循环的周期	满足下列条件的 r 值		随后出现的具有周期 $k2^n$ 的“谐振”级联全都变得不稳定	使基本循环或它的某一个谐振成为吸引子的 r 值范围的宽度
	基本循环首次出现	基本循环变得不稳定		
1	0.0000	2.0000	2.6924	2.6924
3	3.1024	3.1596	3.1957	0.0933
4	3.5855	3.6043	3.6153	0.0298
5（a）	2.9161	2.9222	2.9256	0.0095
5（b）	3.3632	3.3664	3.3682	0.0050
5（c）	3.9206	3.9295	3.9347	0.0141
6（a）	2.7714	2.7761	2.7789	0.0075
6（b）	3.4558	3.4563	3.4567	0.0009
6（c）	3.7736	3.7745	3.7750	0.0014
6（d）	4.1797	4.1848	4.1880	0.0083

As a corollary it follows that, given the most recent cycle to appear, it is possible (at least in principle) to catalogue all the cycles which have appeared up to this point. An especially elegant way of doing this is given by Smale and Williams[22], who show, for example, that when the stable cycle of period 3 first originates, the total number of other points with periods k, N_k, which have appeared by this stage satisfy the Fibonacci series, N_k=2, 4, 5, 8, 12, 19, 30, 48, 77, 124, 200, 323 for k=1, 2, ..., 12: this is to be contrasted with the total number of points of period k which will eventually appear (the top row of Table 2) as $F(X)$ continues to steepen.

Such catalogues of the total number of fixed points, and of their order of appearance, are relatively easy to construct. For any particular function $F(X)$, the numerical task of finding the windows of parameter values wherein any one cycle or its harmonics is stable is, in contrast, relatively tedious and inelegant. Before giving such results, two critical parameter values of special significance should be mentioned.

Hoppensteadt and Hyman[21] have given a simple graphical method for locating the parameter value in the chaotic regime at which the first odd period cycle appears. Their analytic recipe is as follows. Let α be the parameter which tunes the steepness of $F(X)$ (for example, $\alpha = a$ for equation (3), $\alpha = r$ for equation (4)), $X^*(\alpha)$ be the fixed point of period 1 (the nontrivial solution of equation (5)), and $X_{max}(\alpha)$ the maximum value attainable from iterations of equation (1) (that is, the value of $F(X)$ at its hump or stationary point). The first odd period cycle appears for that value of α which satisfies[21,31]

$$X^*(\alpha) = F^{(2)}(X_{max}(\alpha)) \tag{13}$$

As mentioned above, another critical value is that where the period 3 cycle first appears. This parameter value may be found numerically from the solutions of the third iterate of equation (1): for equation (3) it is[14] $a=1+\sqrt{8}$.

Myrberg[13] (for all $k\leqslant 10$) and Metropolis *et al.*[16]. (for all $k\leqslant 7$) have given numerical information about the stable cycles in equation (3). They do not give the windows of parameter values, but only the single value at which a given cycle is maximally stable; that is, the value of a for which the stability-determining slope of $F^{(k)}(X)$ is zero, $\lambda^{(k)} = 0$. Since the slope of the k-times iterated map $F^{(k)}$ at any point on a period k cycle is simply equal to the product of the slopes of $F(X)$ at each of the points X^*_k on this cycle[1,8,20], the requirement $\lambda^{(k)} = 0$ implies that $X = \mathrm{A}$ (the stationary point of $F(X)$, where $\lambda^{(1)} = 0$) is one of the periodic points in question, which considerably simplifies the numerical calculations.

For each basic cycle of period k (as catalogued in the last row of Table 2), it is more interesting to know the parameter values at which: (1) the cycle first appears (by tangent bifurcation); (2) the basic cycle becomes unstable (giving rise by successive pitchfork bifurcations to a cascade of harmonics of periods $k2^n$); (3) all the harmonics become unstable (the point of accumulation of the period $k2^n$ cycles). Tables 3 and 4 extend the work of May and Oster[1], to give this numerical information for equations (3) and (4), respectively. (The points of accumulation are not ground out mindlessly, but are calculated

作为推论还可以知道，若给定最近出现的循环，就有可能（至少在原则上）列出到这一点为止所有已经出现过的循环。斯梅尔和威廉斯[22]给出了做到这一点的一种特别巧妙的方法。例如，他们指出，当周期为 3 的稳定循环最初产生时，在这一阶段出现的具有周期 k 的其他各点的总数，N_k，满足斐波那契序列关系，即当 $k = 1$，2，…，12 时，$N_k = 2$，4，5，8，12，19，30，48，77，124，200，323，这与随着 $F(X)$ 持续变陡峭而最终将出现的周期为 k 的点的总数（表 2 的顶行）是不同的。

这种不动点的总数及其出现顺序的目录都是相对容易构造的。与此相反，对于任一特定的函数 $F(X)$，寻找使得任一循环或其谐振在其中稳定的参数值窗口的数值工作则是相对乏味而笨拙的。在给出这些结果之前，应该先谈谈两个具有特殊重要性的临界参数值。

霍彭施泰特和海曼[21]给出了一种简单的图像方法，能够确定第一个奇数周期循环在混沌区域中出现的参数值的位置。他们的分析方法如下：设 α 是调节 $F(X)$ 陡峭程度的参数（如在方程（3）中 $\alpha = a$，在方程（4）中 $\alpha = r$），$X^*(\alpha)$ 是周期为 1 的不动点（方程（5）的非平凡解），而 $X_{max}(\alpha)$ 是通过方程（1）迭代所得到的最大值（也就是说，$F(X)$ 位于其峰值或平稳点处）。第一个奇数周期的循环当 α 值满足以下方程时出现[21,31]：

$$X^*(\alpha) = F^{(2)}(X_{max}(\alpha)) \tag{13}$$

如同前面已经提到的，另一个临界值是周期为 3 的循环首次出现的位置。这个参数值可以从方程（1）三次迭代的解中通过数值方法得到。对于方程（3），它就是[14] $a = 1+\sqrt{8}$。

迈尔伯格[13]（对于全部 $k \leqslant 10$）和梅特罗波利斯等[16]（对于全部 $k \leqslant 7$）给出了关于方程（3）中稳定循环的数值信息。他们没有给出参数值窗口，而是给出了使得一个给定循环最大程度稳定的单独值。也就是说，使得确定稳定性的 $F^{(k)}(X)$ 斜率为 0（$\lambda^{(k)} = 0$）的 a 的值。由于 k 次迭代映射 $F^{(k)}$ 的斜率在周期为 k 的循环上的任意一点处都恰好等于在该循环上每个点 X^*_k 处 $F(X)$ 斜率的乘积[1,8,20]，对 $\lambda^{(k)} = 0$ 的要求意味着 $X = \mathrm{A}$（即 $F(X)$ 的平稳点，其中 $\lambda^{(1)} = 0$）是我们所讨论的周期点之一，这就在相当程度上简化了数值计算。

对于每个周期为 k 的基本循环（如同表 2 最后一行中所列出的那样），更为有趣的是了解以下几种情况时的参数值：（1）循环首次出现（通过切分岔）；（2）基本循环变得不稳定（通过相继的叉形分岔导致周期为 $k2^n$ 的谐振级联的出现）；（3）所有的谐振都变得不稳定（周期为 $k2^n$ 的循环的累加点）。表 3 和表 4 中拓展了梅和奥斯特[1]的工作，从而分别给出了对应于方程（3）和（4）的数值信息。（累加点并不是不假思索地算出的，而是通过一种快速收敛迭代过程计算出来的，参见文献 1 中的

by a rapidly convergent iterative procedure, see ref. 1, appendix A.) Some of these results have also been obtained by Gumowski and Mira[32].

Practical Problems

Referring to the paradigmatic example of equation (3), we can now see that the parameter interval $1<a<4$ is made up of a one-dimensional mosaic of infinitely many windows of a-values, in each of which a unique cycle of period k, or one of its harmonics, attracts essentially all initial points. Of these windows, that for $1<a<3.5700$ .. corresponding to $k = 1$ and its harmonics is by far the widest and most conspicuous. Beyond the first point of accumulation, it can be seen from Table 3 that these windows are narrow, even for cycles of quite low periods, and the windows rapidly become very tiny as k increases.

As a result, there develops a dichotomy between the underlying mathematical behaviour (which is exactly determinable) and the "commonsense" conclusions that one would draw from numerical simulations. If the parameter a is held constant at one value in the chaotic region, and equation (3) iterated for an arbitrarily large number of generations, a density plot of the observed values of X_t on the interval 0 to 1 will settle into k equal spikes (more precisely, delta functions) corresponding to the k points on the stable cycle appropriate to this a-value. But for most a-values this cycle will have a fairly large period, and moreover it will typically take many thousands of generations before the transients associated with the initial conditions are damped out: thus the density plot produced by numerical simulations usually looks like a sample of points taken from some continuous distribution.

An especially interesting set of numerical computations are due to Hoppensteadt (personal communication) who has combined many iterations to produce a density plot of X_t for each one of a sequence of a-values, gradually increasing from 3.5700 .. to 4. These results are displayed as a movie. As can be expected from Table 3, some of the more conspicuous cycles do show up as sets of delta functions: the 3-cycle and its first few harmonics; the first 5-cycle; the first 6-cycle. But for most values of a the density plot looks like the sample function of a random process. This is particularly true in the neighbourhood of the a-value where the first odd cycle appears (a=3.6786 ..), and again in the neighbourhood of a=4: this is not surprising, because each of these locations is a point of accumulation of points of accumulation. Despite the underlying discontinuous changes in the periodicities of the stable cycles, the observed density pattern tends to vary smoothly. For example, as a increases toward the value at which the 3-cycle appears, the density plot tends to concentrate around three points, and it smoothly diffuses away from these three points after the 3-cycle and all its harmonics become unstable.

I think the most interesting mathematical problem lies in designing a way to construct some approximate and "effectively continuous" density spectrum, despite the fact that the exact density function is determinable and is always a set of delta functions. Perhaps such techniques have already been developed in ergodic theory[33] (which lies at the foundations of statistical mechanics), as for example in the use of "coarse-grained observers". I do not know.

附录 A。）古莫夫斯基和米拉[32]也得到了部分上述结果。

实际问题

参考方程（3）的典型实例，现在我们可以看到，参数区间 $1 < a < 4$ 是由无限多个 a 值窗口的一维马赛克结构所组成的，每个窗口中都有唯一一个周期为 k 的循环，或者是它的一个谐振，实质上吸引了所有的初始点。在这些窗口中，对于 $1 < a < 3.5700\cdots$，即对应于 $k = 1$，其谐振的窗口是目前为止最宽的也是最显著的。超过第一个累加点后，从表 3 中可以看到，这些窗口很狭窄，甚至对于极低周期的循环来说也是如此，并且随着 k 的增加，窗口迅速变得很小。

因此，我们在基础数学行为（它完全是明确的）与人们从数值模拟中得出的“常识”结论之间建立了分界。如果参数 a 在混沌区域中保持为某一恒定值，并且将方程（3）对于任意大的代数进行迭代，X_t 在 0 到 1 区间上的观测值的密度图将会有 k 个相同的峰值（更确切地说，是 δ 函数），对应于适合该 a 值的稳定循环上的 k 个点。但是对于大多数 a 值而言，这个循环会具有一个相当大的周期。不仅如此，它通常还会历经数以千计个迭代才能使与初始条件有关的暂态消除。因此，通过数值模拟而产生的密度图通常看起来像是来自某些连续分布的点的样本。

有一组特别值得关注的数值计算应归功于霍彭施泰特（个人交流），他将很多迭代组合起来，对从 3.5700…逐渐增加到 4 的一系列 a 值中的每一个值，产生一个 X_t 的密度图，这些结果如同电影般呈现出来。如同从表 3 中可以预期的那样，某些更为显著的循环实际上显示为一系列的 δ 函数：3–循环及其前几个谐振；第一个 5–循环；第一个 6–循环。但是对于大多数 a 值，密度图看起来就像一个随机过程的样本函数。在第一个奇数循环出现处的 a 值（$a = 3.6786\cdots$）的邻域中，这一点尤为真实，在 $a = 4$ 的邻域中也是如此：这并不令人惊讶，因为在这些位置中，每个都是累加点。尽管稳定循环的周期性质的实质是不连续变化，但是观测到的密度图倾向于光滑变化。例如，随着 a 向出现 3–循环的数值方向增加，密度图倾向于在三个点附近聚集，并且在 3–循环之后从这三个点处光滑地扩散开，并且其所有的谐振也变得不稳定。

我认为最有趣的数学问题莫过于设计一种方法来构造某种近似而又“有效连续的”密度谱，尽管精确的密度函数是确定性的，并且总是一系列 δ 函数。可能在各态历经理论[33]中已经有了这样一种技术（建立在统计力学的基础上），如利用“粗粒化观测者”的实例。我并不了解。

Such an effectively stochastic description of the dynamical properties of equation (4) for large r has been provided[28], albeit by tactical tricks peculiar to that equation rather than by any general method. As r increases beyond about 3, the trajectories generated by this equation are, to an increasingly good approximation, almost periodic with period $(1/r)\exp(r-1)$.

The opinion I am airing in this section is that although the exquisite fine structure of the chaotic regime is mathematically fascinating, it is irrelevant for most practical purposes. What seems called for is some effectively stochastic description of the deterministic dynamics. Whereas the various statements about the different cycles and their order of appearance can be made in generic fashion, such stochastic description of the actual dynamics will be quite different for different $F(X)$: witness the difference between the behaviour of equation (4), which for large r is almost periodic "outbreaks" spaced many generations apart, versus the behaviour of equation (3), which for $a \rightarrow 4$ is not very different from a series of Bernoulli coin flips.

Mathematical Curiosities

As discussed above, the essential reason for the existence of a succession of stable cycles throughout the "chaotic" regime is that as each new pair of cycles is born by tangent bifurcation (see Fig. 5), one of them is at first stable, by virtue of the way the smoothly rounded hills and valleys intercept the 45° line. For analytical functions $F(X)$, the only parameter values for which the density plot or "invariant measure" is continuous and truly ergodic are at the points of accumulation of harmonics, which divide one stable cycle from the next. Such exceptional parameter values have found applications, for example, in the use of equation (3) with $a = 4$ as a random number generator[34,35]: it has a continuous density function proportional to $[X(1-X)]^{-\frac{1}{2}}$ in the interval $0<X<1$.

Non-analytical functions $F(X)$ in which the hump is in fact a spike provide an interesting special case. Here we may imagine spikey hills and valleys moving to intercept the 45° line in Fig. 5, and it may be that both the cycles born by tangent bifurcation are unstable from the outset (one having $\lambda^{(k)}>1$, the other $\lambda^{(k)}<-1$), for all $k>1$. There are then no stable cycles in the chaotic regime, which is therefore literally chaotic with a continuous and truly ergodic density distribution function.

One simple example is provided by

$$X_{t+1} = aX_t; \text{ if } X_t < \frac{1}{2} \qquad (14)$$
$$X_{t+1} = a(1-X_t); \text{ if } X_t > \frac{1}{2}$$

defined on the interval $0<X<1$. For $0<a<1$, all trajectories are attracted to $X=0$; for $1<a<2$, there are infinitely many periodic orbits, along with an uncountable number of aperiodic trajectories, none of which are locally stable. The first odd period cycle appears at $a=\sqrt{2}$, and all integer periods are represented beyond $a=(1+\sqrt{5})/2$. Kac[36] has given a careful discussion of the case $a=2$. Another example, this time with an extensive biological

当 r 很大时，已经能对方程（4）的动力学性质提供一种有效的随机描述 [28]，不过所使用的处理技巧仅限于这个方程而不是任意的一般方法。随着 r 的增加，大约超过 3 时，由该方程所产生的轨道，在越来越好的近似意义上，几乎是周期性的，其周期为 $(1/r)\exp(r-1)$。

我在本节中要发表的观点是，尽管混沌区域中新颖的精细结构在数学上是令人着迷的，但它与大多数实际用途没有关系。看来下一步要做的是对确定性动力学系统给出一种有效的随机描述。虽然可以用一般性的方式来对不同循环及其出现顺序进行各种叙述，对于不同的 $F(X)$，实际动力学给出的这种随机描述将会大不相同：可以对比方程（4）与方程（3）的行为之间的差异作为证据。前者对于大的 r 几乎是周期性的"爆发"，将很多个代分隔开；后者当 $a \to 4$ 时与伯努利抛币序列没有什么区别。

数学方面的好奇心

如同前面已讨论的，贯穿混沌区域的稳定循环序列存在的本质的原因是，当每次通过切分岔而新生成一对循环时（见图 5），其中一个开始时是稳定的，由光滑的峰和谷与 45° 线相交而产生。对于解析函数 $F(X)$，仅有的能够使密度图或"不变测度"为连续并且真正遍历的参数值都位于谐振的累加点处，这些累加点将一个稳定循环与后一个循环分隔开。这些例外的参数值已经找到了用途，如当 $a = 4$ 时，使用方程（3）作为随机数生成器 [34,35]：在区间 $0 < X < 1$ 中，它具有与 $[X(1-X)]^{-\frac{1}{2}}$ 成正比的连续密度函数。

其峰实际上是一个尖峰的非解析函数 $F(X)$，提供了一种有趣的特殊情况。这里我们可以想象，尖锐的峰和谷移向 45° 线而与其相交，如图 5 所示，因而情况有可能是，对于所有的 $k > 1$，通过切分岔而产生的两个循环（一个 $\lambda^{(k)} > 1$，另一个 $\lambda^{(k)} < -1$）从一开始就都是不稳定的。于是在混沌区域中不存在稳定循环，这与混沌的字面意思相符，它具有连续和真正遍历的密度分布函数。

下面提供一个简单的实例：

$$
\begin{aligned}
&\text{若 } X_t < \frac{1}{2}, \quad X_{t+1} = aX_t \\
&\text{若 } X_t > \frac{1}{2}, \quad X_{t+1} = a(1-X_t)
\end{aligned}
\tag{14}
$$

定义在区间 $0 < X < 1$ 上。当 $0 < a < 1$ 时，所有轨道都被吸引到 $X = 0$；当 $1 < a < 2$ 时，有无限多个周期性轨道，以及无数个非周期轨道，其中没有一个是局部稳定的。第一个奇数周期循环出现在 $a = \sqrt{2}$ 处，而所有的整数周期循环都位于超过 $a = (1+\sqrt{5})/2$ 的地方。卡克 [36] 曾对于 $a = 2$ 的情况进行了细致的讨论。另一个实

pedigree[1-3], is the equation

$$X_{t+1} = \lambda X_t; \text{ if } X_t < 1 \qquad (15)$$
$$X_{t+1} = \lambda X_t^{1-b}; \text{ if } X_t > 1$$

If $\lambda>1$ this possesses a globally stable equilibrium point for $b<2$. For b>2 there is again true chaos, with no stable cycles; the first odd cycle appears at $b=(3+\sqrt{5})/2$, and all integer periods are present beyond b=3. The dynamical properties of equations (14) and (15) are summarised to the right of Table 2.

The absence of analyticity is a necessary, but not a sufficient, condition for truly random behaviour[31]. Consider, for example,

$$X_{t+1} = (a/2)X_t; \text{ if } X_t < \frac{1}{2} \qquad (16)$$
$$X_{t+1} = aX_t(1-X_t); \text{ if } X_t > \frac{1}{2}$$

This is the parabola of equation (3) and Fig. 1, but with the left hand half of $F(X)$ flattened into a straight line. This equation does possess windows of a values, each with its own stable cycle, as described generically above. The stability-determining slopes $\lambda^{(k)}$ vary, however, discontinuously with the parameter a, and the widths of the simpler stable regions are narrower than for equation (3): the fixed point becomes unstable at $a=3$; the point of accumulation of the subsequent harmonics is at a=3.27..; the first odd cycle appears at a=3.44..; the 3-point cycle at a=3.67.. (compare the first column in Table 1).

These eccentricities of behaviour manifested by non-analytical functions may be of interest for exploring formal questions in ergodic theory. I think, however, that they have no relevance to models in the biological and social sciences, where functions such as $F(X)$ should be analytical. This view is elaborated elsewhere[37].

As a final curiosity, consider the equation

$$X_{t+1} = \lambda X_t [1+X_t]^{-\beta} \qquad (17)$$

This has been used to fit a considerable amount of data on insect populations[38,39]. Its stability behaviour, as a function of the two parameters λ and β, is illustrated in Fig. 6. Notice that for $\lambda<7.39$ .. there is a globally stable equilibrium point for all β; for 7.39 .. $<\lambda<$12.50 .. this fixed point becomes unstable for sufficiently large β, bifurcating to a stable 2-point cycle which is the solution for all larger β; as λ increases through the range 12.50 ..$<\lambda<$14.77 .. various other harmonics of period 2^n appear in turn. The hierarchy of bifurcating cycles of period 2^n is thus truncated, and the point of accumulation and subsequent regime of chaos in not achieved (even for arbitrarily large β) until $\lambda>$14.77…

例具有广阔的生物学背景[1-3]，即方程

$$若\ X_t < 1,\quad X_{t+1} = \lambda X_t \qquad (15)$$
$$若\ X_t > 1,\quad X_{t+1} = \lambda X_t^{1-b}$$

若 $\lambda > 1$，则当 $b < 2$ 时，具有全局稳定的平衡点。当 $b > 2$ 时，再次产生没有任何稳定循环的真正的混沌：第一个奇数周期循环出现在 $b = (3+\sqrt{5})/2$ 处，而所有的整数周期循环则位于超过 $b = 3$ 的地方。表 2 右方对方程（14）和（15）的动力学性质进行了概括。

不具有解析性是一个真正的随机行为的必要非充分条件[31]。例如，考虑

$$若\ X_t < \frac{1}{2},\quad X_{t+1} = (a/2)X_t \qquad (16)$$
$$若\ X_t > \frac{1}{2},\quad X_{t+1} = aX_t(1-X_t)$$

这是方程（3）和图 1 中的抛物线，但是 $F(X)$ 的左半段变平而成为一条直线。如同前面的一般性描述，该方程确实具有 a 值的窗口，并且每个窗口都有自己的稳定循环。但是，决定稳定性的斜率 $\lambda^{(k)}$ 随着参数 a 而不连续地变化，因而较简单稳定区域的宽度就变得比方程（3）所对应的更狭窄。在 $a = 3$ 处，不动点变得不稳定；伴随谐振的累加点位于 $a = 3.27\cdots$ 处；第一个奇数周期循环出现在 $a = 3.44\cdots$ 处；3 点循环位于 $a = 3.67\cdots$ 处（对比表 1 的第一列）。

非解析函数表现出的种种行为上的异常，对于探索各态历经理论中的常规问题来说可能是有意义的。不过，我认为，它们与生物学和社会科学中的各种模型没有什么关系，在那些领域中，函数 $F(X)$ 应该是解析的。这一看法在别的地方[37]已有详述。

作为最后一个关注点，考虑方程

$$X_{t+1} = \lambda X_t [1+X_t]^{-\beta} \qquad (17)$$

这一方程曾用来拟合相当大量的昆虫种群数据[38,39]。作为两个参数 λ 和 β 的函数，它的稳定性行为如图 6 所示。注意到当 $\lambda < 7.39\cdots$ 时，对所有的 β 都存在全局稳定的平衡点；当 $7.39\cdots < \lambda < 12.50\cdots$ 时，如果 β 足够大，这个不动点将变得不再稳定，而是分岔成为一个稳定的 2 点循环，它是对于所有较大 β 的解。随着 λ 增大到 $12.50\cdots < \lambda < 14.77\cdots$，各种其他的周期为 2^n 的谐振依次出现。于是周期为 2^n 的分岔循环谱系被截断，而累加点与随后的混沌区域要到 $\lambda > 14.77\cdots$ 时才能得到（即使对于任意大的 β）。

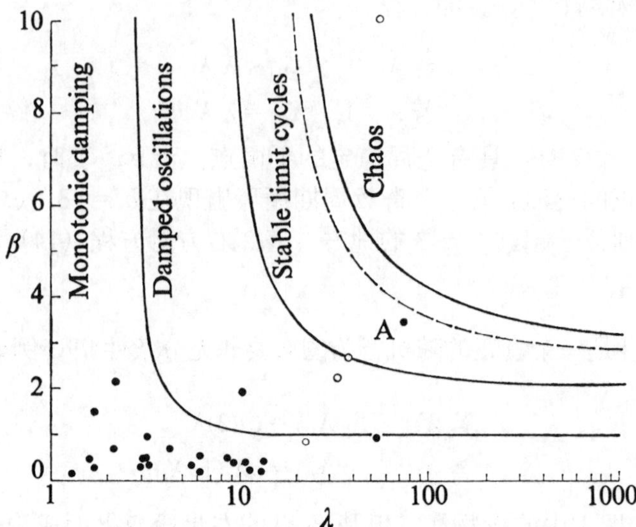

Fig. 6. The solid lines demarcate the stability domains for the density dependence parameter, β, and the population growth rate, λ, in equation (17); the dashed line shows where 2-point cycles give way to higher cycles of period 2^n. The solid circles come from analyses of life table data on field populations, and the open circles from laboratory populations (from ref. 3, after ref. 39).

Applications

The fact that the simple and deterministic equation (1) can possess dynamical trajectories which look like some sort of random noise has disturbing practical implications. It means, for example, that apparently erratic fluctuations in the census data for an animal population need not necessarily betoken either the vagaries of an unpredictable environment or sampling errors: they may simply derive from a rigidly deterministic population growth relationship such as equation (1). This point is discussed more fully and carefully elsewhere[1].

Alternatively, it may be observed that in the chaotic regime arbitrarily close initial conditions can lead to trajectories which, after a sufficiently long time, diverge widely. This means that, even if we have a simple model in which all the parameters are determined exactly, long term prediction is nevertheless impossible. In a meteorological context, Lorenz[15] has called this general phenomenon the "butterfly effect": even if the atmosphere could be described by a deterministic model in which all parameters were known, the fluttering of a butterfly's wings could alter the initial conditions, and thus (in the chaotic regime) alter the long term prediction.

Fluid turbulence provides a classic example where, as a parameter (the Reynolds number) is tuned in a set of deterministic equations (the Navier–Stokes equations), the motion can undergo an abrupt transition from some stable configuration (for example, laminar flow) into an apparently stochastic, chaotic regime. Various models, based on the Navier–Stokes differential equations, have been proposed as mathematical metaphors for this

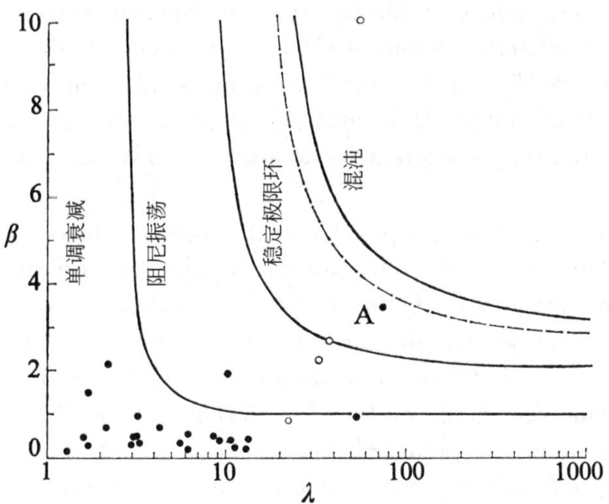

图 6. 实线划分了方程（17）中依赖于密度的参数 β 和种群增长速率 λ 构成的参数平面中的稳定区域；虚线显示了 2 点循环变成周期较高的 2^n 循环的位置所在。实心圆来自对野外种群生命统计表中数据的分析，而空心圆则来自实验室中的种群（引自参考文献 3 和参考文献 39）。

应　用

简单而确定的方程（1）能够具有看起来像某种随机噪声一样的动力学轨道，这一事实具有令人困扰的实际意义。例如，从一个动物种群的普查数据中所具有的看似无规律的波动，不一定能断定它就是不可预测的环境异常或者采样误差造成的：它可能单纯地源于诸如方程（1）那样具有严格确定性的种群增长关系。我在其他地方对这一点进行了更为完整和细致的讨论[1]。

换言之，可以看到，混沌区域中任意相近的初始条件经过充分长的时间都能会进入相互偏离很远的轨道。这意味着，即使我们有一个简单模型，其中所有的参数都是精确确定的，但长期预测也是不可能的。在气象学研究背景下，洛伦茨[15]将这种普遍现象称为“蝴蝶效应”：即使可以用一个所有参数均为已知的确定模型来描述大气，一只蝴蝶翅膀的震颤就可能改变初始条件，从而（在混沌区域中）改变长期预测结果。

流体的湍流提供了一个经典实例，将一系列确定方程（纳维－斯托克斯方程组）中的一个参数（雷诺数）加以调节，流体运动就能够从某种稳定状态（如层流）发生突变而进入一个显然随机的、混沌的区域。作为对这一过程的数学隐喻，已经提出了很多种基于纳维－斯托克斯微分方程组的模型[15,40,41]。在最近一份关于湍流理论

process[15,40,41]. In a recent review of the theory of turbulence, Martin[42] has observed that the one-dimensional difference equation (1) may be useful in this context. Compared with the earlier models[15,40,41], it has the disadvantage of being even more abstractly metaphorical, and the advantage of having a spectrum of dynamical behaviour which is more richly complicated yet more amenable to analytical investigation.

A more down-to-earth application is possible in the use of equation (1) to fit data[1,2,3,38,39,43] on biological populations with discrete, non-overlapping generations, as is the case for many temperate zone arthropods. Figure 6 shows the parameter values λ and β that are estimated[39] for 24 natural populations and 4 laboratory populations when equation (17) is fitted to the available data. The figure also shows the theoretical stability domains: a stable point; its stable harmonics (stable cycles of period 2^n); chaos. The natural populations tend to have stable equilibrium point behaviour. The laboratory populations tend to show oscillatory or chaotic behaviour; their behaviour may be exaggeratedly nonlinear because of the absence, in a laboratory setting, of many natural mortality factors. It is perhaps suggestive that the most oscillatory natural population (labelled *A* in Fig. 6) is the Colorado potato beetle, whose present relationship with its host plant lacks an evolutionary pedigree. These remarks are only tentative, and must be treated with caution for several reasons. Two of the main caveats are that there are technical difficulties in selecting and reducing the data, and that there are no single species populations in the natural world: to obtain a one-dimensional difference equation by replacing a population's interactions with its biological and physical environment by passive parameters (such as λ and β) may do great violence to the reality.

Some of the many other areas where these ideas have found applications were alluded to in the second section, above[5-11]. One aim of this review article is to provoke applications in yet other fields.

Related Phenomena in Higher Dimensions

Pairs of coupled, first-order difference equations (equivalent to a single second-order equation) have been investigated in several contexts[4,44-46], particularly in the study of temperate zone arthropod prey—predator systems[2-4,23,47]. In these two-dimensional systems, the complications in the dynamical behaviour are further compounded by such facts as: (1) even for analytical functions, there can be truly chaotic behaviour (as for equations (14) and (15)), corresponding to so-called "strange attractors"; and (2) two or more different stable states (for example, a stable point and a stable cycle of period 3) can occur together for the same parameter values[4]. In addition, the manifestation of these phenomena usually requires less severe nonlinearities (less steeply humped $F(X)$) than for the one-dimensional case.

Similar systems of first-order ordinary differential equations, or two coupled first-order differential equations, have much simpler dynamical behaviour, made up of stable and unstable points and limit cycles[48]. This is basically because in continuous two-dimensional

的综述中，马丁[42]注意到一维差分方程（1）对于这个问题可能是有帮助的。与早先的模型[15,40,41]相比，它的不足在于其数学隐喻意义更为抽象，而其优势则是提供了一份动力学行为谱系，虽然更为复杂，但更适合于分析研究。

一个更为实际的应用是，有可能利用方程（1）来处理具有离散和非重叠世代的生物学种群数据[1,2,3,38,39,43]，如很多种温带节肢动物符合这种情况。图6显示了在用方程（17）处理可获得的数据时，对24个自然种群和4个实验室种群所估计出的参数λ和β[39]的值。图6中还显示了理论稳定性区域：一个稳定点；它的稳定谐振（周期为2^n的稳定循环）；混沌。自然种群倾向于具有稳定平衡点行为。实验室种群倾向于显示振荡的或混沌的行为；由于在实验室条件下，很多自然死亡因素的缺失，它们的行为可能会表现为夸张的非线性。这可能暗示了，最具振荡性质的自然种群（图6中标记为A）是科罗拉多马铃薯甲虫，它与寄主植物的当前关系缺乏进化谱系。这些评述仅仅是假设性的，出于几方面原因的考虑，必须慎重地对待它们。有两个主要的提醒，即在选择和简化数据时存在技术性困难，且自然界不存在单一物种的种群：为了获得一个一维差分方程而将一个种群与其生物学环境和物理学环境之间的相互作用替换为被动参数（例如λ和β），这可能会严重违背现实情况。

上述观念还有很多其他的用武之地，上面第二节中也间接谈到了一些[5-11]。这篇综述文章的一个目的就是促进其在更多其他领域中的应用。

高维情况中的相关现象

很多对耦合的一阶差分方程（等价于一个二阶方程）已在若干背景中得到了研究[4,44-46]，特别是在对于温带节肢动物的猎物–捕食者系统的研究中[2-4,23,47]。在这些二维系统中，动力学行为的复杂性中还进一步添加了如下影响因素：(1) 即使对于解析函数，也可能存在真正的混沌行为（如方程（14）和（15）），对应于所谓的“奇异吸引子”；(2) 在同一组参数值下，可以出现两个或更多个不同的稳定状态（如一个稳定点和一个周期为3的稳定循环）[4]。此外，与一维情况相比，这些现象的出现通常只需要不那么显著的非线性性质（峰不是那么陡的$F(X)$）。

类似的一阶常微分方程组，或者两个耦合的一阶微分方程组成的系统，具有更简单的动力学行为，包括稳定的和不稳定的点以及极限环[48]。这主要是因为在连续二维系统中，内部和外部的闭曲线是可区分的；动力学轨道不可能彼此交叉。当我

systems the inside and outside of closed curves can be distinguished; dynamic trajectories cannot cross each other. The situation becomes qualitatively more complicated, and in many ways analogous to first-order difference equations, when one moves to systems of three or more coupled, first-order ordinary differential equations (that is, three-dimensional systems of ordinary differential equations). Scanlon (personal communication) has argued that chaotic behaviour and "strange attractors", that is solutions which are neither points nor periodic orbits[48], are typical of such systems. Some well studied examples arise in models for reaction–diffusion systems in chemistry and biology[49], and in the models of Lorenz[15] (three dimensions) and Ruelle and Takens[40] (four dimensions) referred to above. The analysis of these systems is, by virtue of their higher dimensionality, much less transparent than for equation (1).

An explicit and rather surprising example of a system which has recently been studied from this viewpoint is the ordinary differential equations used in ecology to describe competing species. For one or two species these systems are very tame: dynamic trajectories will converge on some stable equilibrium point (which may represent coexistence, or one or both species becoming extinct). As Smale[50] has recently shown, however, for 3 or more species these general equations can, in a certain reasonable and well-defined sense, be compatible with any dynamical behaviour. Smale's[50] discussion is generic and abstract: a specific study of the very peculiar dynamics which can be exhibited by the familiar Lotka–Volterra equations once there are 3 competitors is given by May and Leonard[51].

Conclusion

In spite of the practical problems which remain to be solved, the ideas developed in this review have obvious applications in many areas.

The most important applications, however, may be pedagogical.

The elegant body of mathematical theory pertaining to linear systems (Fourier analysis, orthogonal functions, and so on), and its successful application to many fundamentally linear problems in the physical sciences, tends to dominate even moderately advanced University courses in mathematics and theoretical physics. The mathematical intuition so developed ill equips the student to confront the bizarre behaviour exhibited by the simplest of discrete nonlinear systems, such as equation (3). Yet such nonlinear systems are surely the rule, not the exception, outside the physical sciences.

I would therefore urge that people be introduced to, say, equation (3) early in their mathematical education. This equation can be studied phenomenologically by iterating it on a calculator, or even by hand. Its study does not involve as much conceptual sophistication as does elementary calculus. Such study would greatly enrich the student's intuition about nonlinear systems.

们转向三个或更多个耦合的一阶常微分方程所构成的系统（即三维常微分方程组系统）时，情况在定性的意义上变得更为复杂了，而且在很多方面类似于一阶差分方程。斯坎伦（个人交流）主张，混沌行为和“奇异吸引子”（即既不是点也不是周期轨道的那些解[48]）对于上述系统是典型的。一些获得充分研究的例子都来自针对化学和生物学中的反应－扩散系统的模型[49],以及前面涉及的洛伦茨[15]的模型(三维)、吕埃勒和塔肯斯[40]的模型（四维）。这些系统的高维性质使得对于它们的分析比对方程（1）的分析显得更不明确。

最近有一个从这一视角对某一系统进行研究的清晰且相当令人感到吃惊的例子，即生态学中用来描述竞争物种的常微分方程组。对于一个或两个物种，这些系统是极为乏味的：动力学轨道在某些稳定平衡点上收敛（这可以表示为共存，或者是一个或两个物种趋于灭绝）。不过，正如斯梅尔[50]最近所指出的那样，从某种合理的并且恰当界定的意义上来讲，对于3个或更多个物种，这些一般性的方程能够与任何动力学行为相一致。斯梅尔[50]的讨论是一般性的和抽象的：梅和伦纳德[51]曾详细研究过在3个竞争者的情况下由我们所熟知的洛特卡－沃尔泰拉方程所呈现出来的不寻常的动力学行为。

结　论

尽管还有些实际问题有待于去解决，但本综述中所谈到的观念仍可以在很多领域中具有明显的应用。

不过，最重要的应用可能还是在教学方面。

数学理论中与线性系统有关的精巧内容（傅里叶分析、正交函数等）以及它们在物理科学中很多基本线性问题上的成功应用，倾向于使它们在大学高等数学和理论物理学课程中占有相当重要的地位。以这种方式训练而建立数学直觉的学生，对于诸如方程（3）那样最简单的离散、非线性系统所呈现出的奇特行为会感到不理解。但是这些非线性系统在物理科学之外确实是常见的，而不是例外。

因此，我强烈呼吁在数学教育的早期向所有人介绍诸如方程（3）那样的内容。这个方程可以通过计算器，甚至手算的方式，迭代进行唯象研究。对它的研究不像初等微积分那样涉及很多概念意义上的细微精妙。这样的学习将会大大丰富学生对于非线性系统的直觉。

Not only in research, but also in the everyday world of politics and economics, we would all be better off if more people realised that simple nonlinear systems do not necessarily possess simple dynamical properties.

I have received much help from F. C. Hoppensteadt, H. E. Huppert, A. I. Mees, C. J. Preston, S. Smale, J. A. Yorke, and particularly from G. F. Oster. This work was supported in part by the NSF.

(**261**, 459-467; 1976)

Robert M. May

King's College Research Centre, Cambridge CB2 1ST; on leave from Biology Department, Princeton University, Princeton 08540

References:

1. May, R. M., and Oster, G. F., *Am. Nat.*, **110** (in the press).
2. Varley, G. C., Gradwell, G. R., and Hassell, M. P., *Insect Population Ecology* (Blackwell, Oxford, 1973).
3. May, R. M. (ed.), *Theoretical Ecology: Principles and Applications* (Blackwell, Oxford, 1976).
4. Guckenheimer, J., Oster, G. F., and Ipaktchi, A., *Theor. Pop. Biol.* (in the press).
5. Oster, G. F., Ipaklchi, A., and Rocklin, I., *Theor, Pop. Biol.* (in the press).
6. Asmussen, M. A., and Feldman, M. W., *J. theor. Biol.* (in the press).
7. Hoppensteadt, F. C., *Mathematical Theories of Populations; Demographics, Genetics and Epidemics* (SIAM, Philadelphia, 1975).
8. Samuelson, P. A., *Foundations of Economic Analysis* (Harvard University Press, Cambridge, Massachusetts, 1947).
9. Goodwin, R. E., *Econometrica*, **19**, 1-17 (1951).
10. Baumol, W. J., *Economic Dynamics*, 3rd ed. (Macmillan, New York, 1970).
11. See, for example, Kemeny, J., and Snell, J. L., *Mathematical Models in the Social Sciences* (MIT Press, Cambridge, Massachusetts, 1972).
12. Chaundy, T. W., and Phillips, E., *Q. J. Math. Oxford*, 7, 74-80 (1936).
13. Myrberg, P. J., *Ann. Akad. Sc. Fennicae, A*, I, No. 336/3 (1963).
14. Myrberg, P. J., *Ann. Akad. Sc. Fennicae, A*, I, No. 259 (1958).
15. Lorenz, E. N., *J. Atmos. Sci.*, **20**, 130-141 (1963); *Tellus*, **16**, 1-11 (1964).
16. Metropolis, N., Stein, M. L., and Stein, P. R., *J. Combinatorial Theory*, **15**(A), 25-44 (1973).
17. Maynard Smith, J., *Mathematical Ideas in Biology* (Cambridge University Press, Cambridge, 1968).
18. Krebs, C. J., *Ecology* (Harper and Row, New York, 1972).
19. May, R. M., *Am. Nat.*, **107**, 46-57 (1972).
20. Li, T-Y., and Yorke, J. A., *Am. Math. Monthly*, **82**, 985-992 (1975).
21. Hoppensteadt, F. C., and Hyman, J. M. (Courant Institute, New York University: preprint, 1975).
22. Smale, S., and Williams, R. (Department of Mathematics, Berkeley: preprint, 1976).
23. May, R. M., *Science*, **186**, 645-647 (1974).
24. Moran, P. A. P., *Biometrics*, **6**,250-258 (1950).
25. Ricker, W. E., *J. Fish. Res. Bd. Can.*, **11**, 559-623 (1954).
26. Cook, L. M., *Nature*, **207**, 316 (1965).
27. Macfadyen, A., *Animal Ecology: Aims and Methods* (Pitman, London, 1963).
28. May, R. M., *J. theor. Biol.*, **51**, 511-524 (1975).
29. Guckenheimer, J., *Proc. AMS Symposia in Pure Math., XIV*, 95-124 (1970).
30. Gilbert, E. N., and Riordan, J., *Illinois J. Math.*, **5**, 657-667 (1961).
31. Preston, C. J. (King's College, Cambridge: preprint, 1976).
32. Gumowski, I., and Mira, C., *C. r. hebd, Séanc. Acad. Sci., Paris*, **281a**, 45-48 (1975); **282a**, 219-222 (1976).
33. Layzer, D., *Sci. Am.*, **233**(6), 56-69(1975).
34. Ulam, S. M., *Proc. Int. Congr. Math. 1950, Cambridge, Mass.;Vol. II* . pp.264-273 (AMS, Providence R. I., 1950).
35. Ulam, S. M., and von Neumann, J., *Bull. Am. Math. Soc.* (abstr.), **53**, 1120 (1947).

不仅在研究中，而且在政治和经济领域的日常生活中，如果有更多的人认识到简单的非线性系统并不一定具有简单的动力学性质，那么我们就会生活得更加舒适。

霍彭施泰特、于佩尔、米斯、普雷斯顿、斯梅尔、约克，特别是奥斯特，给了我很大的帮助。本研究部分得到了美国国家科学基金会的支持。

（王耀杨 翻译；李典谟 审稿）

36. Kac, M., *Ann. Math.*, 47,33-49 (1946).

37. May, R. M., *Science,* **181**, 1074 (1973).

38. Hassell, M. P., *J. Anim. Ecol.,* **44**, 283-296 (1974).

39. Hassell, M. P., Lawton, J. H., and May, R. M., *J. Anim. Ecol.* (in the press).

40. Ruelle, D., and Takens, F., *Comm. math. Phys.*,**20**, 167-192 (1971).

41. Landau, L. D., and Lifshitz, E. M., *Fluid Mechanics* (Pergamon, London, 1959).

42. Martin, P. C., *Proc, Int. Conf. On Statistical Physics, 1975, Budapest* (Hungarian Acad. Sci., Budapest, in the press).

43. Southwood, T. R. E., in *Insects, Science and Society* (edit. by Pimentel, D.), 151-199 (Academic, New York, 1975).

44. Metropolis, N., Stein, M. L., and Stein, P. R., *Numer. Math.,* **10**, 1-19 (1967).

45. Gumowski, I., and Mira, C., *Automatica,* **5**, 303-317 (1969).

46. Stein, P. R., and Ulam, S. M., *Rosprawy Mat.,* **39**, 1-66 (1964).

47. Beddington, J, R., Free, C. A., and Lawton, J, H., *Nature,* **255**, 58-60 (1975).

48. Hirsch, M, W., and Smale, S., *Differential Equations, Dynamical Systems and Linear Algebra* (Academic, New York, 1974).

49. Kolata, G. B., *Science,* **189**, 984-985 (1975).

50. Smale, S. (Department of Mathematics, Berkeley: preprint, 1976).

51. May, R. M., and Leonard, W. J., *SIAMJ. Appl. Math.,* **29**, 243-253 (1975).

Australopithecus, Homo erectus and the Single Species Hypothesis

R. E. F. Leakey and A. C. Walker

Editor's Note

Human fossils found in Africa and representing the past three million years or so of human evolution appear to have existed in two distinguishable forms, one of which is heavy boned or "robust" and the other of which has more delicate bones and is called "gracile". This brief paper is a plea by Richard Leakey (Louis Leakey's son) and a Harvard colleague that palaeontologists should acknowledge the coexistence of two species of human beings over the past three million years. The robust species is assumed now to have disappeared.

AN enormous wealth of early hominid remains has been discovered over the past few years by expeditions within eastern Africa. Evidence has been presented for the existence over a considerable period of time of at least two contemporaneous hominid species[1]. Some of this evidence is compelling, but some less so for a variety of reasons such as the lack of association, fragmentary specimens, geological uncertainties, equivocal anatomical differences and suchlike. Many of these new specimens are of great antiquity and have led to suggestions that an early form of the genus *Homo* was contemporary with at least one species of *Australopithecus.* The evidence presented here deals not with the earlier stages of human evolution, but with the unequivocal occurrence of *H. erectus* from the Koobi Fora Formation, east of Lake Turkana (formerly Lake Rudolf).

Among the variety of hypotheses put forward to accommodate the evidence in an evolutionary framework, the most explicit and directly simple is the single species hypothesis[2]. This hypothesis rests on the assumption that dependence upon tools was the primary hominid adaptation that enabled expansion into open country environments. In the clearest exposition of the hypothesis[3], basic hominid characteristics, including bipedal locomotion, reduced canines and delayed physical maturity are seen to have come about in response to more effective and greater dependence on culture. This assumption of a basic hominid cultural adaptation allied with the principle of competitive exclusion[4,5] leads to the conclusion that two or more hominid species would be extremely unlikely to exist sympatrically. Here we present decisive evidence that shows the existence of two contemporaneous hominid species in the Koobi Fora area.

The adult cranium KNM-ER 3733 (ref. 6) was found *in situ* in the upper member of the Koobi Fora Formation[7]. The sediments lie stratigraphically between the KBS and the Koobi Fora/BBS tuff complexes. The cranium consists of a complete calvaria and a great deal of the facial skeleton, including the nasal and zygomatic bones. The nearly

南方古猿、直立人及单物种假说

利基，沃克

编者按

非洲发现的人类化石，代表了过去300万年左右的人类演化，他们似乎存在着两种显然不同的类型：其中一种的骨头粗壮厚重或称“粗壮型”，另一种的骨头纤细易碎，被称为“纤细型”。在这篇短文中，理查德·利基（路易斯·利基的儿子）与他的一个哈佛的同事认为，古生物学家应该承认过去的300万年中同时存在两种人类。其中的粗壮型现在认为已经消失了。

过去几年中，在东非进行的野外调查发现了很多早期人科动物的化石。有证据证明在相当长的一段时间内，在这里同时存在至少两种早期人科动物[1]。这类证据有些让人深信，但有些就不那么令人信服，原因各不相同，比如缺乏相关性，标本不完整，地质年代的不确定性，解剖学上的形态差异模棱两可，等等诸如此类。这些新标本中的很多都很古老，他们暗示人属的早期形式至少与一种南方古猿同时代。这里给出的证据讨论的不是早期人类的演化，而是真正出现在图尔卡纳湖（原名鲁道夫湖）以东库比福勒组的直立人。

在演化框架内针对化石证据提出的各种假说中，单物种假说最为明确和简单[2]。这个假说依赖于如下假设，早期人科动物面对环境作出的第一个适应性的调整就是对工具的依赖，这种调整使得早期人科动物能够向开阔环境扩散。该假说的一个很好的阐释[3]是，对文化更多更有效的依赖促使早期人科动物的一些基本特征发生了改变，包括直立行走、减小的犬齿尺寸以及生理成熟的推迟。这个基本的早期人科动物文化适应的假设与竞争排他原理[4,5]联系在一起，推导出以下结论，两种或更多种人科动物不可能有重叠的生活分布区。而在这里我们将提供确切证据表明在库比福勒地区同时生活着两个人科动物物种。

在库比福勒组上段的地层中就地发现了一个成年颅骨，编号KNM-ER 3733(参考文献6)[7]。地层位置上，沉积物位于KBS与库比福勒/BBS凝灰岩之间。颅骨包括齐全的头盖骨与大量面部骨骼（包括鼻骨与颧骨）。几乎完整的前部牙齿齿槽，前

complete alveoli of the anterior teeth and examples of the premolars and molars are preserved. The third molars were lost before fossilisation. Preparation has been limited so far to the external cranial surfaces and part of the brain case. An endocranial capacity is unknown, but by comparison is likely to be of the order of 800–900 ml. In all its features the cranium is strikingly like that of *H. erectus* from Peking[8]. Such orthodox anthropometric comparisons that can be made at present fall well within the range of the Peking specimens.

The cranium is large (glabella to inion/opisthocranion is 183 mm) with large projecting supraorbital tori and little postorbital constriction (minimum postorbital breadth 91 mm). There is a marked postglabellar sulcus and the frontal squama rises steeply from behind it to reach vertex at bregma. The skull decreases in height from bregma and the occipital bears a pronounced torus where the occipital and nuchal planes are sharply angled. The greatest breadth is low at the angular torus (biauricular breadth 132 mm). There are strong temporal lines that are 60 mm apart at their nearest point. As far as can be judged at present, the vault bones are thick (about 10.0 mm in mid-parietal). The temporal fossae are small. The facial skeleton is partly preserved and shows deep and wide zygomatic portions, longitudinally concave and laterally convex projecting nasals and wide, low piriform aperture. The incisive alveolar plane is short and wide and the large incisive alveoli are set almost in a straight line. There are strong canine juga. The palate is high and roughly square in outline. The facial skeleton is flexed under the calvaria and in the preliminary reconstruction is set at about the same angle as that reconstructed for a female *H. erectus* by Weidenreich[8]. The prosthion to nasion length is 87 mm. This is the best preserved single *H. erectus* cranium known, the facial skeleton being more complete and less distorted than that of "Pithecanthropus VIII" (ref. 9).

Figures 1–3 show KNM-ER 3733 together with KNM-ER 406, another cranium discovered *in situ* in the Upper Member of the Koobi Fora Formation. This latter specimen has been described[10] in some detail and is clearly that of a robust *Australopithecus.* Other specimens of robust *Australopithecus* have been found in deposits of both the Upper and Lower members of the Koobi Fora Formation. Two *in situ* mandibles that must have been associated with this type of cranium are the most massive representatives of this species. KNM-ER 729 (ref. 1) was excavated in 1971 from the base of the Middle Tuff and KNM-ER 3230 (ref. 6) from the upper part of the BBS complex. Other, more incomplete, specimens of robust *Australopithecus* mandibles have come from the higher levels of the Formation between the Karari/ Chari and Koobi Fora/BBS tuffs. The radioisotopic dating of these tuffs, in spite of a continuing controversy over one of them[11], is not an issue here. The contemporaneity of *Homo erectus* and a robust *Australopithecus* is now clearly established over the period during which the Upper Member of the Koobi Fora Formation was deposited. Using the time scale given by Fitch *et al.*[12], this would be from earlier than about 1.3 to earlier than 1.6 Myr ago.

臼齿和臼齿的样本也都保存了下来。第三臼齿在石化前就失去了。目前的修理准备工作还仅限于颅骨的外表面和部分大脑。颅容量尚属未知，但通过对比推测大概在800~900 毫升之间。该颅骨的所有形态特征与周口店直立人[8]表现出惊人的相似。目前能从 KNM-ER 3733 上得到的传统的测量数据正好在周口店直立人标本变化范围之内。

该颅骨很大（从眉间点到枕外隆凸点 / 颅后点为 183 毫米），其眶上圆枕大而凸出，眶后缩狭不明显（最小眶后宽 91 毫米）。眉脊上沟明显，从其后，额骨倾斜并在前囟点达到最高。头骨的高度从前囟处开始降低，枕外圆枕发育，在此处，枕平面和项平面的过渡部分急剧转折。颅骨最大宽处位于角圆枕处（耳点间宽为 132 毫米）。颞线粗重，两侧颞线最近距离 60 毫米。就目前的判断来看，颅顶厚（顶骨中央位置厚大约 10 毫米）。颞窝小。部分被保存下来的面部骨骼显示：颧骨深而宽；突出的鼻的鼻骨纵向呈凹形；横向呈凸形；梨状孔宽而低。门齿齿槽平面短而宽，中央门齿齿槽几乎在一条直线上。有明显隆起的犬齿齿槽轭。上颌高，轮廓大体呈方形。颅顶以下面部骨骼变形，初步重构时角度大约与魏登瑞[8]重构的女性直立人的相同。上齿槽中点到鼻根点长 87 毫米。这是已知的保存得最好的单个直立人颅骨，与“爪哇直立猿人Ⅷ”相比，其面部骨骼更齐全，变形更小（参考文献 9）。

图 1~3 显示的是 KNM-ER 3733 与另外一个就地发现在库比福勒组的上段地层中的颅骨（KNM-ER 406）的对比情况。对后一标本进行过较为详细的描述[10]，很清楚它是属于粗壮型南方古猿的。在库比福勒组上下段的沉积物中都发现过其他的粗壮型南方古猿标本。就地出土的两件下颌骨很可能与这类颅骨相关联，组成这个种最有力的代表。KNM-ER 729（参考文献 1）是在 1971 年从中段凝灰岩的底部挖掘出来的，而 KNM-ER 3230（参考文献 6）是从 BBS 层上部挖掘出来的。其他不完整的粗壮型南方古猿下颌骨标本，来自卡拉里 / 沙里与库比福勒/BBS 凝灰岩之间的较高层位。对这些凝灰岩的放射性同位素测年结果不再存在问题，尽管对其中一个凝灰岩层的测年结果一直争论不断[11]。现在可以肯定的是，在库比福勒组的上段沉积的这个时期内，直立人与粗壮型南方古猿同时存在。参考菲奇[12]等给出的地质年代表，该时期应该是从大约 130 万年前到 160 万年以前。

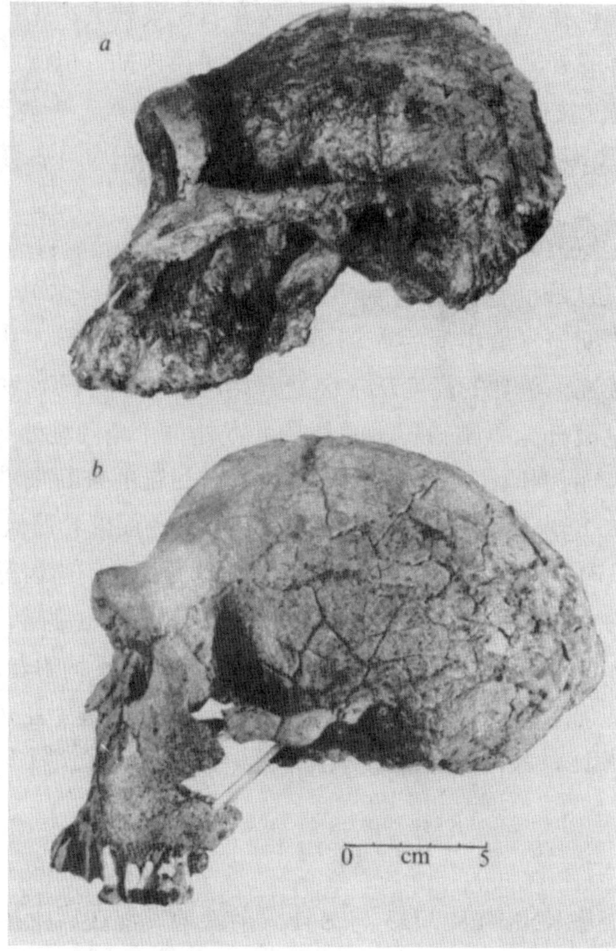

Fig. 1. Lateral aspect of KNM-ER 406 (*a*) and KNM-ER 3733 (*b*).

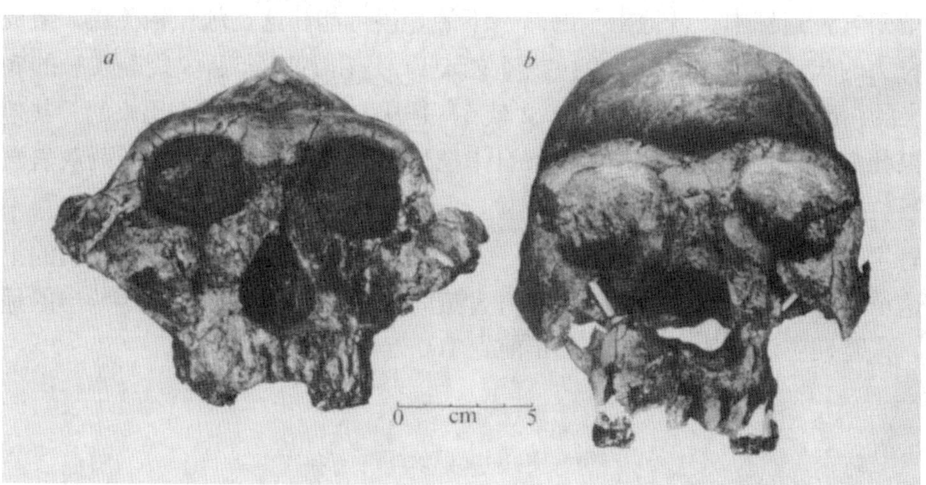

Fig. 2. Frontal aspect of KNM-ER 406 (*a*) and KNM-ER 3733 (*b*).

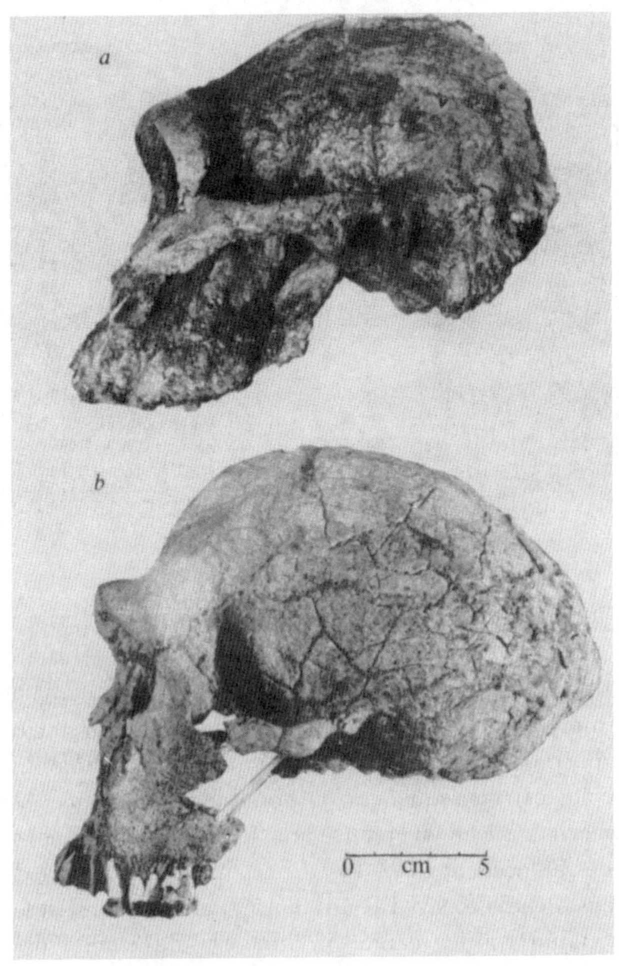

图 1. KNM-ER 406 (*a*) 与 KNM-ER 3733 (*b*) 的侧面观

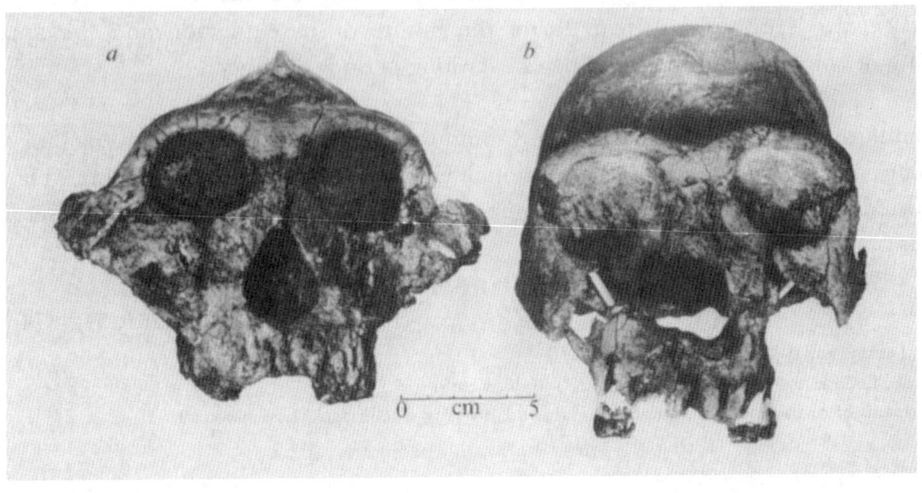

图 2. KNM-ER 406 (*a*) 与 KNM-ER 3733 (*b*) 的正面观

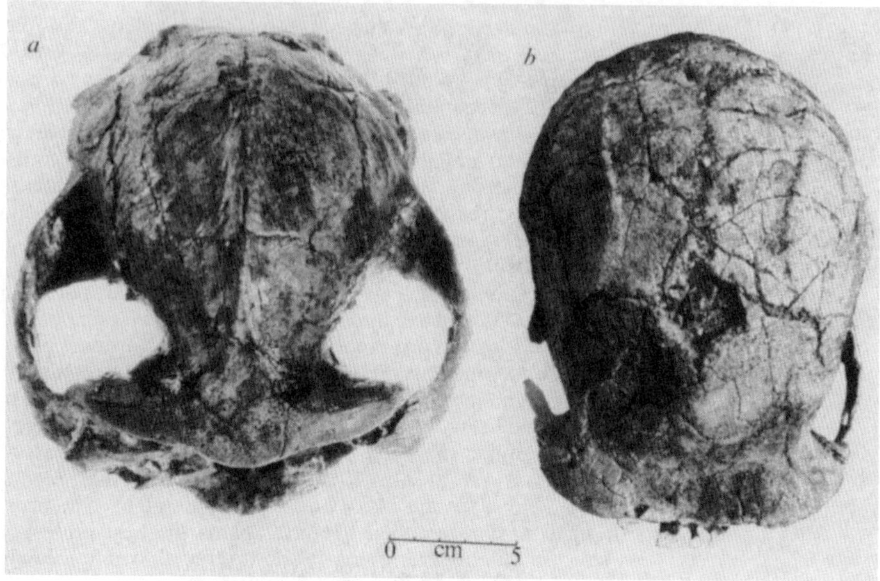

Fig 3. Superior aspect of KNM-ER 406(*a*) and KNM-ER 3733(*b*).

The new data show that the simplest hypothesis concerning early human evolution is incorrect and that more complex models must be devised. The single species hypothesis has served a useful purpose in focusing attention on variability among the early hominids and also on the ecological consequences of hominid adaptations. Alternative concepts, especially those concerning niche divergence and sympatry, should now be formulated. We think that populations antecedent to *H. erectus* ones have been sampled in the Koobi Fora Formation (specimens include KNM-ER 1470, 1590 and 3732). The clear demonstration of at least two hominid species earlier in time should enable us to reconsider our approaches to the problems of earlier hominid evolution. We also think that there is no good evidence for the presence of *Australopithecus* outside Africa, and the finding of an apparently advanced *H. erectus* cranium at Koobi Fora provokes issues such as why *Australopithecus* is only an African form, the nature of its extinction and the apparent stability of some hominid morphologies over long periods of time.

We thank the Museum Trustees of Kenya and the National Museums of Kenya for access to material and facilities and the National Geographic Society for support. The NSF supported A.C.W. We also thank colleagues who helped with suggestions and observations on this paper.

(**261**, 572-574; 1976)

R. E. F. Leakey[*] and Alan C. Walker[†]

[*]National Museums of Kenya, P. O. Box 40658, Nairobi, Kenya

[†]Departments of Anatomy and Anthropology, Harvard University, Cambridge, Massachusetts 02138

Received February 10; accepted May 4, 1976.

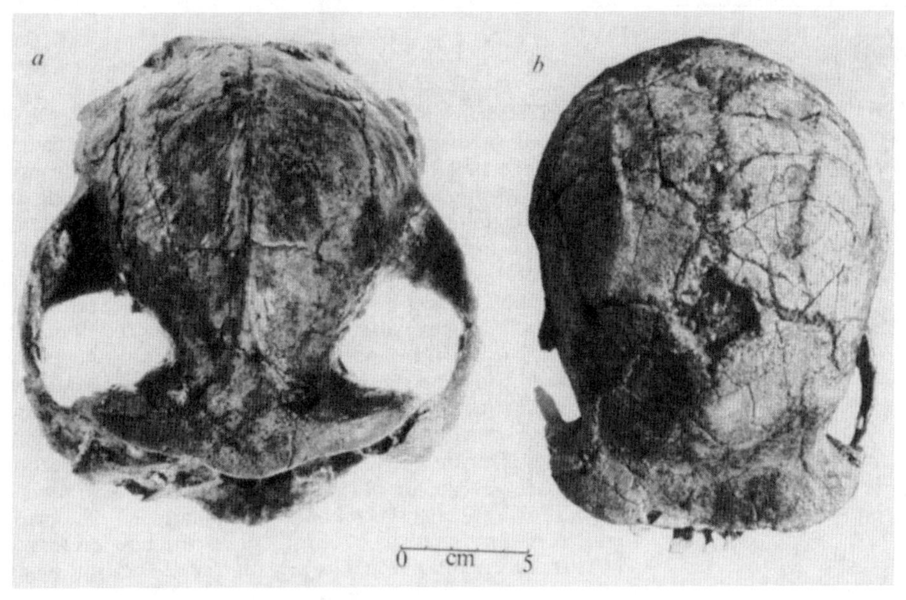

图 3. KNM-ER 406 (*a*) 与 KNM-ER 3733 (*b*) 的顶面

新的资料显示关于早期人类演化的最简单假说是不正确的，需设计出更为复杂的模型。单物种假说有益于把注意力集中在早期人科动物的变异以及早期人类适应性的生态影响上。现在应该阐明一些其他的可供选择的概念了，特别是有关小生境分化与分布区重叠。我们认为已经在库比福勒组采到过比直立人更早的样本（标本包括 KNM-ER 1470、1590 与 3732）。对较早时期内同时存在至少两种早期人科动物的证明应使我们能重新考虑对早期人科动物演化问题的解决方法。我们也认为没有好的证据证明非洲之外存在南方古猿，然而在库比福勒发现了明显更进步的直立人的颅骨引出了诸如此类的问题：为什么南方古猿只在非洲存在？为什么南方古猿会灭绝？以及为什么在相当长的时期内某些早期人科动物形态保持稳定不变？

我们感谢肯尼亚博物馆理事会与肯尼亚国家博物馆为我们提供研究资料和便利。感谢国家地理学会的支持。沃克获得了国家科学基金会的支持。还要感谢我们的同事，他们为本文提供了观测数据并给出了宝贵的建议。

（田晓阳 翻译；刘武 邢松 审稿）

References:

1. Leakey, R. E. F., *Nature,* **231**, 241-245 (1971); **237**, 264-269 (1972); **242**, 447-450 (1973); **248**, 653-656 (1974).
2. Brace, C. L., *The Stages of Human Evolution* (Prentice Hall, Engelwood Cliffs, 1967).
3. Wolpoff, M. H., *Man,* **6**, 601-614 (1971).
4. Gauss, G. F., *The Struggle for Existence* (Williams & Wilkins, Baltimore, 1934).
5. Mayr, E., *Cold Spring Harb. Symp. and Quant. Biol.,* **15**, 108-118 (1950).
6. Leakey, R. E. F., *Nature,* **261**, 574-576 (1976).
7. Bowen, B. E., and Vondra, C. F., *Nature,* **242**, 391-393 (1973).
8. Weidenreich, F., *Palaeont. Sinica,* **10**, 1-291 (1943).
9. Sartono, S., *Koninkl. Ned. Akad. Wet.,* **74**, 185-194 (1971).
10. Leakey, R. E. F., Mungai, J. M., and Walker, A. C., *Am. J. Phys. Anthrop.,* **35**, 175-186 (1971).
11. Curtis, G. H., Drake, Cerling, T., and Hampel, *Nature,* **258**, 395-398 (1975).

12, Fitch, F. J., Findlater, I. C., Watkins, R. T., and Miller, J. A., *Nature,* **251**, 213-215 (1974).

Fossil Hominids from the Laetolil Beds

M. D. Leakey *et al.*

Editor's Note

The Laetolil Beds are deposits of volcanic rocks and ash-falls 30 miles south of Olduvai. Fossils had been gathered there since the 1930s—the Leakeys had collected there on three occasions—but the absolute ages of the fossils could not be estimated due to a lack of a suitable technique. Mary Leakey and colleagues returned there, and armed with the then-new potassium-argon dating method, dated the fossiliferous deposits to more than three and a half million years old. The latest collection included thirteen hominid fossils, all jaws and teeth, which Leakey and colleagues suggested belonged to the genus *Homo*. It was only later that they were reassigned to *Australopithecus afarensis*, the species to which the Hadar skeleton "Lucy" belonged.

Remains of 13 early hominids have been found in the Laetolil Beds in northern Tanzania, 30 miles south of Olduvai Gorge. Potassium–argon dating of the fossiliferous deposits gives an upper limit averaging 3.59 Myr and a lower limit of 3.77 Myr. An extensive mammalian fauna is associated. The fossils occur in the upper 30 m of ash-fall and aeolian tuffs whose total measured thickness is 130 m.

THE fossil-bearing deposits referred to variously as the Laetolil Beds, Garusi or Vogel River Series lie in the southern Serengeti Plains, in northern Tanzania, 20–30 miles from the camp site at Olduvai Gorge.

Fossils have been collected from the area on several occasions, the largest collection being made by L. Kohl-Larsen in 1938–39, who also found a small fragment of hominid maxilla which was named *Meganthropus africanus* by Weinert[1]. L. S. B. Leakey and M. D. L. visited the area in 1935 and in 1959, while a day trip was made in 1964 in company with R. L. H. The faunal material recovered on these occasions was all collected before the advent of isotopic dating and the age of the fossils remained uncertain until potassium–argon dating was carried out on samples of biotite obtained during the 1975 field season. In 1974, however, lava flows which unconformably overlie the Laetolil Beds had been dated by G. H. C. at 2.4 Myr.

We now report evidence that fossiliferous deposits of several different ages exist in the Laetolil area and that specimens found on the surface are not necessarily derived from the same beds. There are, however, noticeable variations in the colour and physical condition of the surface fossils which provide indications of their origin.

在莱托利尔层发现的人科动物化石

利基等

编者按

莱托利尔层是指在奥杜威南部30英里的火山岩和火山灰堆积物。那里的化石采集始于20世纪30年代——利基夫妇曾在那采集过三次——但由于缺乏合适的技术，一直无法确定化石的绝对年龄。玛丽·利基及其同事重返此地，配备了当时新出现的钾氩断代法，测得该含化石的堆积物的年代已超过距今350万年。其最新采集到的标本包括13件人科动物化石，全部是颌骨和牙齿，利基及其同事们认为他们均属于人属。不久以后，他们就被重新厘定为南方古猿阿法种，这个种包括哈达尔骨架“露西”。

在坦桑尼亚北部的莱托利尔层发现了13个早期人科动物的化石，该地在奥杜威峡谷南部30英里处。使用钾氩断代法对含化石的堆积物进行测年，得到其上限年代平均值为距今359万年，下限年代为距今377万年。伴生的哺乳动物群种类很丰富。这些化石出自总厚度为130米的火山灰和风积凝灰岩层最上部30米。

该含有化石的堆积物被不同地称作莱托利尔层、加鲁西或沃格尔河统，它位于坦桑尼亚北部的南塞伦盖蒂平原上，距离奥杜威峡谷营地遗址有20~30英里远。

人们已经数次在该处采集到了化石，采集量最大的一次是科尔–拉森在1938~1939年间所进行的，他还发现了一小块人科动物上颌骨破片，韦纳特将其命名为非洲魁人[1]。路易斯·利基和玛丽·利基于1935年和1959年参观了此地，1964年又与海一起在此逗留了一天。这几次挖掘出的动物化石是在同位素测年法出现之前采集的，所以动物群的年代并不确定。直到用钾氩断代法测得在1975年野外季得到的黑云母样品的年代时，这些化石的年代才得以确定。但在1974年，柯蒂斯就已经确定了不整合地覆盖在莱托利尔层之上的熔岩流的年代为距今240万年。

现在我们报道了莱托利尔地区存在不同年代的含化石的堆积物的证据，而且我们认为地表发现的标本未必源自相同的层位。然而从地表采集到的化石在颜色和物理特征上明显的差异可以暗示它们的来源层位。

Discrepancies in the fauna were noted by Dietrich[2] and Maglio[3] who both postulated faunal assemblages of two different ages. In view of this, it was proposed for a time that the name Laetolil should be abandoned in favour of the more generalised term Vogel River Series, based on the colloquial German name for the Garusi river, which abounds in bird life. As the early fossiliferous deposits here referred to as the Laetolil Beds are not confined to the Garusi valley, this change of name seems unnecessary. Furthermore, the name Laetolil embraces a larger area, because it is the anglicised version of the Masai name (laetoli) for *Haemanthus*, a red lily that is abundant in the locality. The Laetolil Beds, *sensu stricto*, form a discrete unit, distinguishable from later deposits, and M. D. L. considers that the original name proposed by Kent in 1941[4] should be retained for this part of the sequence.

The relationship of the Laetolil to the Olduvai Beds had been under discussion for some years, but in 1969 R. L. H. noted that they underlay bed I at the Kelogi inselberg in the Side Gorge and established that they antedated the Olduvai sequence. This has been further confirmed when tuffs correlatable with bed I as well as an earlier, fossil-bearing series of tuffs were found to lie unconformably on the Laetolil Beds.

Interest in the area was renewed in 1974 after the discovery by George Dove of fossil equid and bovid teeth in the bed of the Gagjingero river, which drains into Lake Masek at the head of Olduvai Gorge. These fossils were found to be eroding from relatively recent deposits, probably the beds named Ngaloba by Kent[4]. Exposures of the Laetolil Beds, not hitherto seen by M. D. L., were found to the east of the Gagjingero river, at the headwaters of the Garusi river and of the Olduvai Side Gorge (referred to as Marambu by Kohl-Larsen[5]). Several fossils, including a hominid premolar, were found at these localities and subsequent visits yielded further hominid remains.

The possibility of establishing the age of the hominid fossils from the Laetolil Beds by radiometric dating and of clarifying the discrepancies in the faunal material led to a 2-month field season during July and August 1975. Samples from the fossiliferous horizons, collected by R. L. H., have now been dated. On the basis of these results the hominid remains and associated fauna can be bracketed in time between 3.59 and 3.77 Myr.

No trace of stone tools or even of utilised bone or stone was observed in the material from the Laetolil Beds, although handaxes and other artefacts occur in conglomerates which are present in certain areas and which are unconformable to the Laetolil Beds.

Stratigraphy of Laetolil Area

The bulk of the faunal remains and all of the hominid remains were found within an area of about 30 km^2 at the northern margin of the Eyasie plateau and in the divide between the Olduvai and Eyasie drainage systems (Fig.1).

迪特里希[2]和马利奥[3]注意到了动物群的差异，他们都主张存在两种不同年代的动物群。鉴于这一点，曾经有人提议基于加鲁西河的口语化的德语名字，应该弃用莱托利尔这一名称而代之以更广义的名称——沃格尔河统，该河流存在着大量鸟类。由于这里提到的称为莱托利尔层的早期含化石的堆积物并不局限于加鲁西河谷，这种名字的变化似乎并不必要。另外，莱托利尔这个名字涵盖更广泛的区域，因为它是一种在当地很繁盛的红色百合花——网球花属的马赛语名称（莱托里）的英语变体。严格来讲，莱托利尔层形成了一个独立的单元，与后期的堆积物是有区别的，玛丽·利基认为应该保留肯特在 1941 年[4]提出的最初的名字作为这部分地层层序的名字。

莱托利尔层和奥杜威层的层序关系已经讨论了数年，但是 1969 年海提出莱托利尔层下伏于侧峡谷的克罗吉岛山的 I 层之下，因此确定了它们早于奥杜威层序。当 I 层与凝灰岩可以对比，以及发现更早的含化石的凝灰岩系不整合地覆盖在莱托利尔层之上时，这一点就得到了进一步的证实。

1974 年，当乔治·达夫在伽津格罗河的河床里发现了马科动物和牛科动物的牙齿后，人们对这一区域的兴趣被再度点燃。伽津格罗河在奥杜威峡谷的源头处流入马赛科湖。这些化石是由于相对更新的堆积物遭遇侵蚀而出露的，这些堆积物很可能是肯特命名的恩加洛巴层[4]。玛丽·利基迄今还没看到过莱托利尔层的出露，出露地层位于伽津格罗河的东部，在加鲁西河和奥杜威侧峡谷（科尔 – 拉森称之为马拉姆布[5]）的源头处。在这些地方发现了好几种化石，其中包括一个人科动物的前臼齿，随后的探查又发现了其他的人科动物化石材料。

为了能够通过放射性测年法确定在莱托利尔层发现的人科动物化石的年代并且澄清动物群的差异，在 1975 年 7~8 月间开展了一次为期两个月的野外工作。海从含化石层采集的样品已经完成了测年。基于这些结果，可以将人科动物化石和伴生动物群的年代圈定在距今 359 万年到 377 万年之间。

尽管在某些区域存在的不整合地覆盖在莱托利尔层之上的砾岩中出现了手斧和其他人工制品，但是在莱托利尔层出土的标本中没有看到石器甚至是被使用过的骨骼或石头的痕迹。

莱托利尔地区的地层情况

大部分动物化石和所有人科动物化石都是在埃亚西高原北部边缘以及奥杜威和埃亚西水系的分水岭（图 1）之间的约 30 平方公里的区域内发现的。

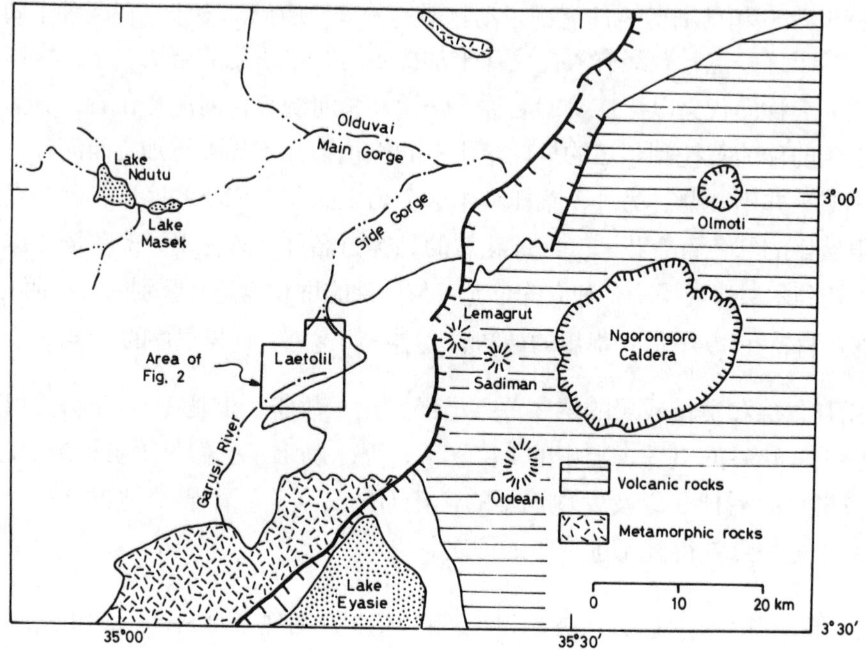

Fig. 1. Map of the southern Serengeti and volcanic highlands.

Kent studied this area as a member of L. S. B. Leakey's 1934–35 expedition, and his short paper is the only published description of the stratigraphy. He recognised three main subdivisions of the stratigraphic sequence, which overlies the metamorphic complex of Precambrian age. The lower unit he named the Laetolil Beds and the upper the Ngaloba Beds. The middle unit consists of olivine-rich lava flows and agglomerate, which are much closer in age to the Laetolil Beds than to the Ngaloba Beds. Kent briefly noted the local occurrence of tuffs younger than the lavas and older than the Ngaloba Beds. He described the Laetolil Beds as subaerially deposited tuffs, and he gave 30 m as an aggregate thickness in the vicinity of Laetolil. The Ngaloba Beds he described as tuffaceous clays, and he gave a thickness of about 5 m at the type locality. Pickering mapped this area as part of a 1:125,000 quarter degree sheet[6] and extended the known occurrence of the Laetolil Beds. He also recognised that at least some of the tuffs are of nephelinite composition.

This picture was modified considerably by stratigraphic work in 1974 and 1975. The Laetolil Beds proved to be far thicker and more extensive than previously recognised. The thickest section, 130 m, was measured in a valley in the northern part of the area (geological localities *A* to *C*, Fig. 2). The base is not exposed and the full thickness of the Laetolil Beds here is unknown. Sections 100–120 m thick and representing only part of the Laetolil Beds were measured about 10 km south-east of Laetolil. The Laetolil Beds are 15–20 m thick at a distance of 25–30 km to the south-west and 10–15 m thick at Lakes Masek and Ndutu, 30 km to the north-west. The Laetolil Beds are tuffaceous sediments, dominantly of nephelinite composition.

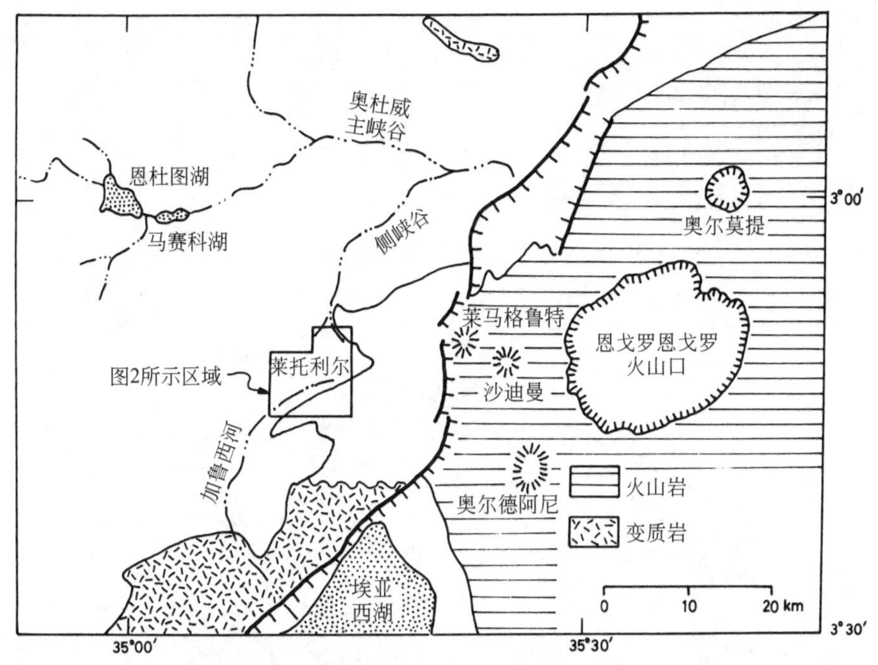

图 1. 南塞伦盖蒂和火山高地的地图。

肯特作为路易斯 · 利基 1934~1935 年间考察队的成员对该区域进行过研究，他写了一篇简短的文章，是唯一一篇已经发表的有关当地地层情况的描述。他把覆盖在前寒武纪变质杂岩之上的地层按层序划分出三个主要亚层。他将下层命名为莱托利尔层，上层命名为恩加洛巴层。中间的单元包括富含橄榄石的熔岩流和集块岩，它们在年代上更接近于莱托利尔层。肯特简要地说明了出现在局部区域的凝灰岩，它在年代上比熔岩要晚而比恩加洛巴层要早。他将莱托利尔层描述为靠近地面堆积的凝灰岩，并认为它在莱托利尔附近区域的总厚度为 30 米。他将恩加洛巴层描述为凝灰质黏土岩，估计在模式产地的厚度约为 5 米。皮克林将该区域作为 1:125,000 四分之一度表的一部分进行了绘图 [6]，并扩展了已知的莱托利尔层，并且他认为其中至少有些凝灰岩是霞石岩成分的。

1974 年和 1975 年，通过地层学研究工作，这幅图得到了相当大的修改。事实证明莱托利尔层远远比之前认为的要厚，面积也更广。最厚的部分有 130 米，是在该区域北部的一条河谷处测量到的（图 2 中的地质点 *A* 到 *C*）。基底没有暴露出来，此处莱托利尔层的全部厚度还是未知的。测量了在莱托利尔东南约 10 千米处的一个仅代表局部莱托利尔层的剖面，其厚度为 100~120 米。莱托利尔层在西南 25~30 千米处的厚度为 15~20 米，在西北 30 千米的马赛科湖和恩杜图湖处的厚度为 10~15 米。莱托利尔层是凝灰质沉积物，主要成分是霞岩。

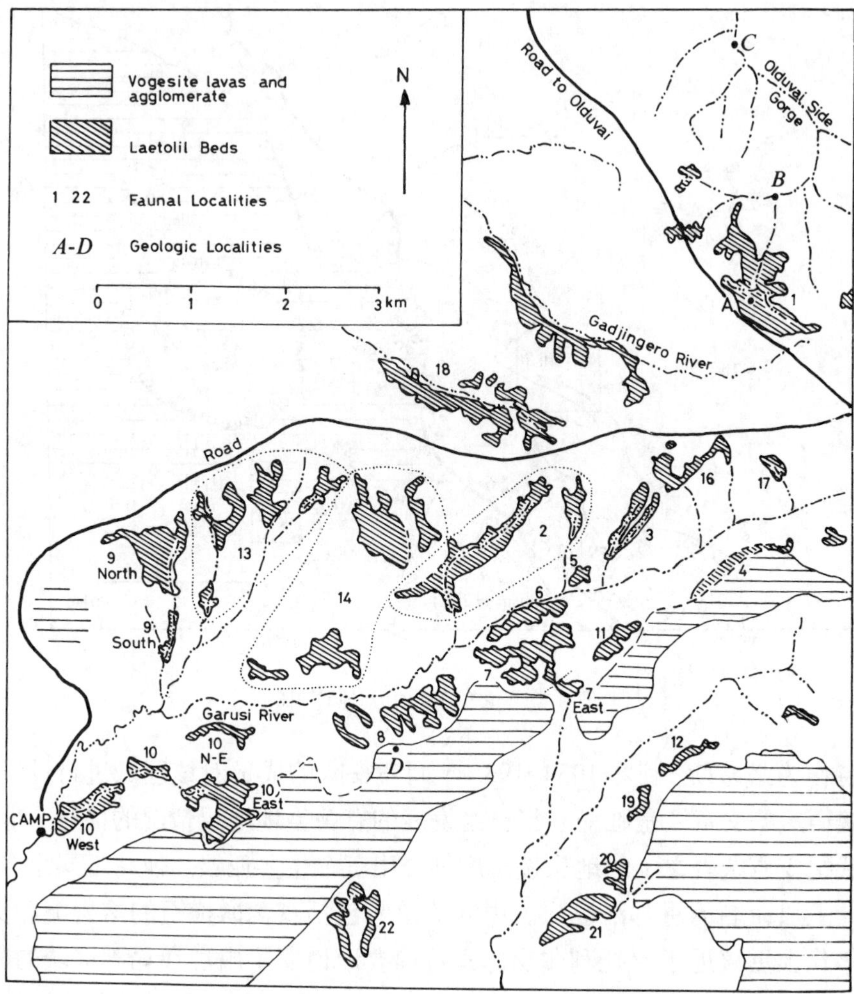

Fig. 2. Map of the Laetolil area showing the fossil beds.

The lavas and agglomerates noted by Kent overlie an irregular surface deeply eroded into the Laetolil Beds. The lavas were erupted from numerous small vents to the south, southwest and west of Lemagrut volcano. Although designated as nephelinite by Kent, the flows proved to be vogesite, a highly mafic lava with interstitial alkali feldspar and phlogopitic biotite. The lavas have reversed polarity as determined by field measurements (personal communication from A. Cox). A sample of lava from geological locality *D* (Fig. 2) gave K–Ar dates of 2.38±0.5 Myr and 2.43±0.7 Myr (Table 1).

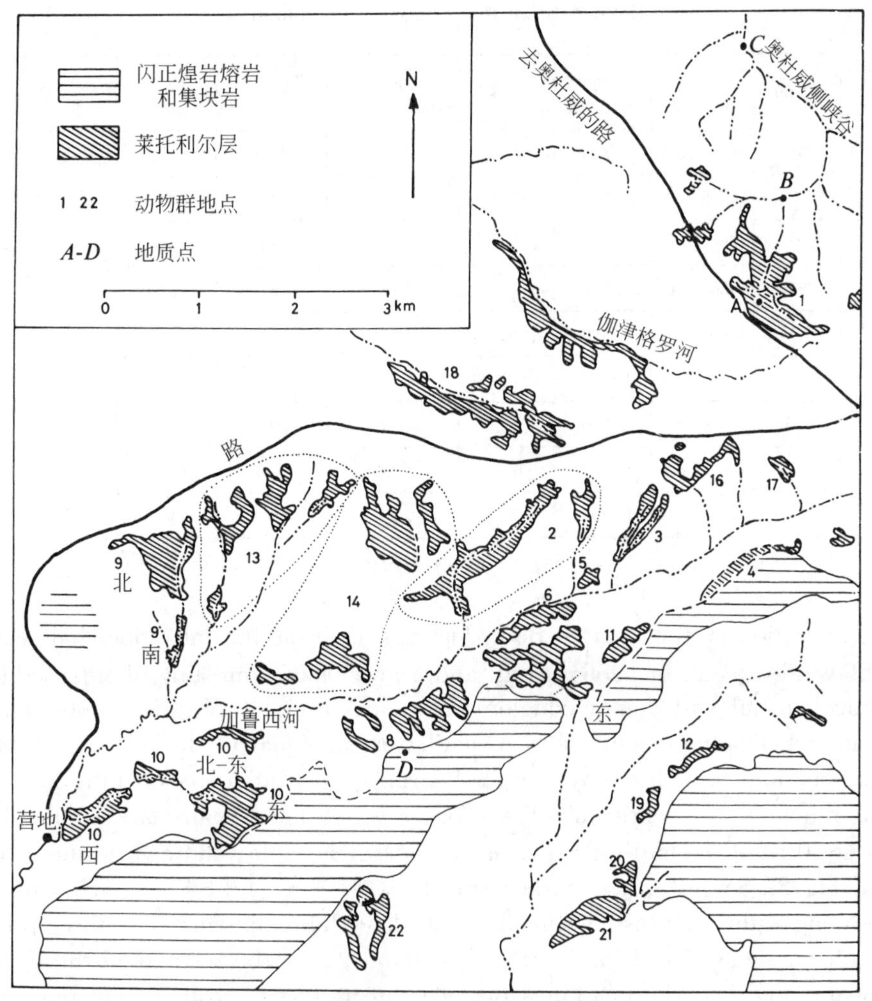

图 2. 标明含化石地层的莱托利尔地区的地图。

肯特指出，熔岩和集块岩覆盖在深入侵蚀到莱托利尔层内部的不规则表面之上。熔岩是从莱马格鲁特火山的南侧、西南侧和西侧的许多小出口喷发出来的。尽管肯特将其指定为霞岩，但是事实证明熔岩流成分是闪正煌岩，这是一种夹有碱性长石和金云母质黑云母的高镁铁质的熔岩。现场测量（与考克斯的个人交流）确定，这种熔岩具有反极性。从地质点 *D*（图 2）取到的一份熔岩样品以钾氩法测定其年代为距今 238 万 ±50 万年和 243 万 ±70 万年（表 1）。

Table 1. K–Ar dates from the Laetolil area

Sample	KA no.	Dated material	Sample weight (g)	% K	mol g^{-1} ^{40}Ar radiation×10^{-11}	% ^{40}Ar atmosphere	Age yr (Myr)	Remarks
Vogesite lava, Location *D*	2835	Whole rock	10.06588	1.013	0.419	74.0	2.38±.05	Unconformably overlying Laetolil Beds
	2837R	Whole rock	11.27408	1.013	0.427	84.4	2.43±.08	
	2929	Whole rock	1.98629	6.96±.03	4.118	77.5	3.41±.08	Treated with dilute HCl
Tuff *c*, location *A*	2977	Whole rock	2.544	6.98±.04	4.462	78.5	3.68±.09	Treated with warm dilute acetic acid
	2979	Whole rock	1.94298	6.95±.04	4.454	79.1	3.69±.10	
Xenolithic horizon, location *A*	2930	Whole rock	3.07730	7.58±.02	4.771	50.8	3.62±.09	Treated with dilute HCl
	2930R	Whole rock	1.83909	7.58±.02	4.733	39.7	3.59±.05	
Ash-fall tuff, location *B*	2932	Whole rock	1.05978	6.00±.1	3.996	75.0	3.82±.16	Crushed, hand picked, treated with dilute HCl
	2938	Whole rock	1.42092	6.49±.04	4.182	78.8	3.71±.12	

$^{40}K/K=1.18\times10^{-4}$; $^{40}K_{\lambda}=5.480\times10^{-10}$ yr^{-1}; $^{40}K_{\lambda\beta}=4.905\times4.905\times10^{-10}$ yr^{-1}; $^{40}K_{\lambda\varepsilon}=0.575\times10^{-10}$ yr^{-1}.

The 130-m section of the Laetolil Beds (Fig. 3) is divisible into an upper half consisting largely of wind-worked, or aeolian, tuff[7] and a lower half consisting of interbedded ash-fall and aeolian tuff with minor conglomerate and breccia. Tephra in the lower half are nephelinite, whereas nephelinite and melilitite are subequal in the upper half. Between these two divisions is a distinctive biotite-bearing coarse lithic-crystal tuff 60 cm thick. Hominid remains and nearly all of the other vertebrate remains are confined to the uppermost 30 m of aeolian tuffs beneath a widespread pale yellow vitric tuff 8 m thick (tuff *d* of Fig. 3). Several other marker tuffs, between 1 m and 30 cm thick, can be used for correlating within the fossiliferous 30-m thickness of sediments. The three prominent marker tuffs, designated *a*, *b*, and *c* (Fig. 3, column 2), can be recognised throughout the area shown in Fig. 2. Widespread horizons of ijolite and lava (mostly nephelinite) xenoliths are at several levels in the fossiliferous part of the section and assist in correlating. Biotite is common in some of the ijolite.

Additional fossiliferous deposits, of mineralogical affinity to the Laetolil Beds and 10–15 m thick, lie between tuff *d* and the bed I (?) tuffs at several places. They are locally separated from tuff *d* and the underlying aeolian tuffs of the Laetolil Beds by an erosional surface with a relief of 8 m. These sediments comprise water-worked tuffs, aeolian tuffs, clay-pellet aggregate of aeolian origin and limestone. The tephra are of phonolite and nephelinite composition. It is not yet clear whether or not these sediments should be regarded as part of the Laetolil Beds.

Beds I and II are represented by sedimentary deposits that lie stratigraphically between the lava and the Ngaloba Beds. The bed II deposits locally contain fossils and artefacts.

表 1. 莱托利尔地区的钾氩年代测定

样品	KA 编号	年代测定材料	样品质量（克）	% K	摩尔 / 克 ^{40}Ar 放射性 $\times 10^{-11}$	% ^{40}Ar 大气	年代（万年）	备注
闪正煌岩熔岩，*D* 地点	2835	全岩	10.06588	1.013	0.419	74.0	238 ± 5	不整合地覆盖在莱托利层之上
	2837R	全岩	11.27408	1.013	0.427	84.4	243 ± 8	
	2929	全岩	1.98629	6.96 ± .03	4.118	77.5	341 ± 8	用稀盐酸处理
凝灰岩 *c*，*A* 地点	2977	全岩	2.544	6.98 ± .04	4.462	78.5	368 ± 9	用温稀醋酸处理
	2979	全岩	1.94298	6.95 ± .04	4.454	79.1	369 ± 10	
捕虏体岩石地层，*A* 地点	2930	全岩	3.07730	7.58 ± .02	4.771	50.8	362 ± 9	用稀盐酸处理
	2930R	全岩	1.83909	7.58 ± .02	4.733	39.7	359 ± 5	
火山灰凝灰岩，*B* 地点	2932	全岩	1.05978	6.00 ± .1	3.996	75.0	382 ± 16	压碎并手工挑选，用稀盐酸处理
	2938	全岩	1.42092	6.49 ± .04	4.182	78.8	371 ± 12	

$^{40}K/K=1.18\times10^{-4}$；$^{40}K_{\lambda}=5.480\times10^{-10}$/ 年；$^{40}K_{\lambda\beta}=4.905\times4.905\times10^{-10}$/ 年；$^{40}K_{\lambda\varepsilon}=0.575\times10^{-10}$/ 年。

莱托利尔层 130 米厚的剖面（图 3）可分成上、下两部分，上半部分由大量风加工过的或者说是风成的凝灰岩 [7] 构成，下半部分则由火山灰和含有少量砾岩和火山角砾岩的风成凝灰岩互层而成。在下半部分的火山喷发碎屑是霞岩，而在上半部分霞岩和黄长岩几乎是等量的。这两部分之间是一层独特的含黑云母的粗粒岩屑晶屑凝灰岩，约 60 厘米厚。人科动物化石和几乎所有其他脊椎动物化石都局限在最上面 30 米的风成凝灰岩内，并位于 8 米厚的广泛分布的浅黄色玻屑凝灰岩（图 3 中的凝灰岩 *d*）之下。其余几种厚度在 1 米至 30 厘米之间的标志性凝灰岩可以作为 30 米厚的含化石沉积物的对比。标为 *a*、*b* 和 *c* 的三种突出的标志性凝灰岩（图 3，第 2 栏），可以通过图 2 所示的整个地区的地图辨认出来。广泛分布的霓霞岩和熔岩（大部分是霞岩）捕虏体层处于该剖面的含化石部分的不同层面上，有利于相互对照。在某些霓霞岩中，黑云母是很常见的。

其他与莱托利尔层具有矿物学相似性的 10~15 米厚的含化石堆积物在好几处地点都存在于凝灰岩 *d* 和 I 层（?）的凝灰岩之间。它们就地被一个起伏幅度达 8 米的侵蚀面所分隔而与凝灰岩 *d* 和下伏的莱托利尔层风成凝灰岩分开。这些沉积物由水成凝灰岩、风成凝灰岩、风成来源的黏土颗粒聚集物和石灰岩构成。火山喷发碎屑的成分是响岩和霞岩。目前还没搞清楚这些沉积物是否应该被看作是莱托利尔层的一部分。

I 层和 II 层以在地层上位于熔岩和恩加洛巴层之间的沉积物为代表。II 层堆积物的局部含有化石和人工制品。

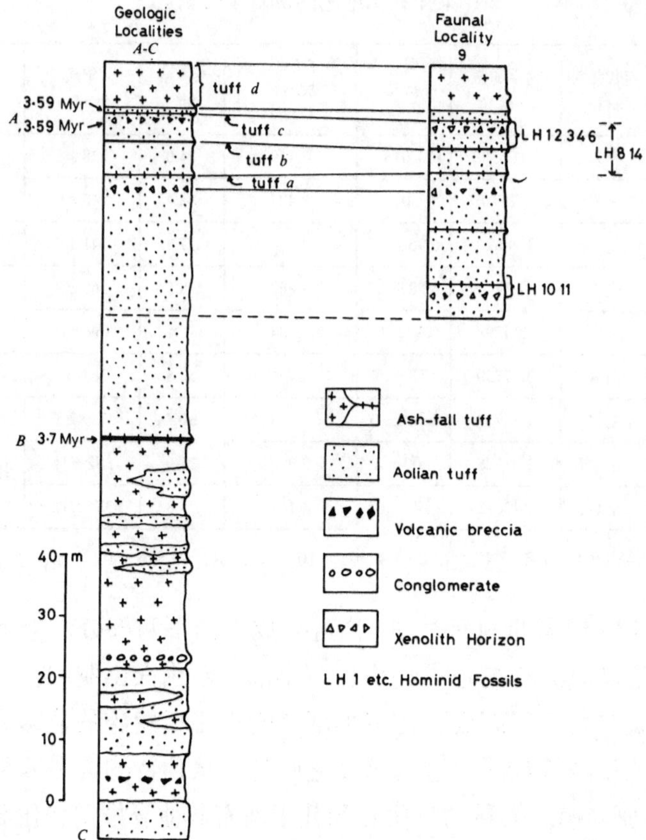

Fig. 3. Stratigraphic column of the Laetolil Beds showing the positions of the dated tuffs and hominid fossils.

Sadiman volcano, about 15 km east of the fossiliferous exposures, seems to have been the eruptive source of the Laetolil Beds. The one K–Ar date, of 3.73 Myr, obtained from Sadiman lava, fits with the K–Ar dates on the Laetolil Beds presented here. This date was published previously as K–Ar 2238 (ref. 8), where it was incorrectly assigned to Ngorongoro because of a mistake in listing the sample numbers.

Potassium–argon Dating

The vogesite lava is composed of approximately 85–90% olivine and augite. The remainder is principally anorthoclase in the groundmass together with a very small amount of phlogopitic biotite. These two minerals proved too fine-grained and sparse for effective separation, and whole-rock samples were used for dating.

In addition to the vogesite lava unconformably overlying the Laetolil Beds, three of the tuffaceous layers within the Laetolil Beds have been dated by the conventional, total degassing K–Ar method (Table 1), and one of these tuffaceous layers has also been dated by the $^{40}Ar/^{39}Ar$ method, using incremental heating.

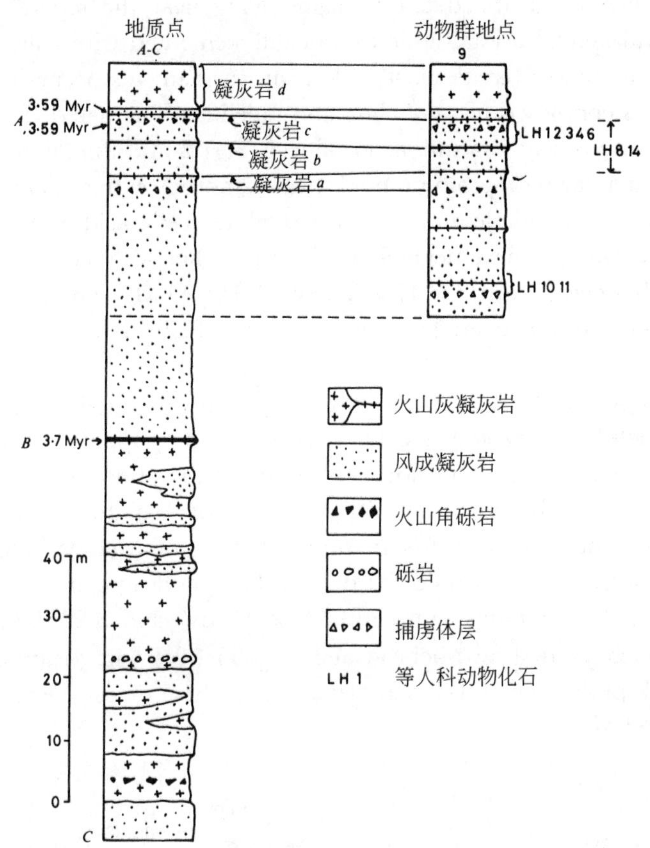

图 3. 显示已确定年代的凝灰岩和人科动物化石位置的莱托利尔层地层柱状剖面。

萨迪曼火山位于含化石的出露地层的东部约 15 千米处，看上去似乎是莱托利尔层的喷发源。其中距今 373 万年这一钾氩年代数据是根据萨迪曼熔岩测出来的，其与本文记述的莱托利尔层的钾氩年代是相符的。这一测年结果曾作为钾氩 2238 发表（参考文献 8），当时由于记录标本编号出现差错而错误地将其归属到了恩戈罗恩戈罗中。

钾氩测年

闪正煌岩熔岩的成分有大约 85%~90% 是由橄榄石和辉石构成。其余成分主要是基质中的歪长石和非常少量的金云母质黑云母。由于这两种矿物非常细小而分散，确实难以有效地分离，所以用完整的岩石样品进行年代测定。

除了不整合地覆盖在莱托利尔层之上的闪正煌岩熔岩外，莱托利尔层中的三个凝灰质地层也都使用传统的、全除气钾氩法（表 1）进行了年代测定，其中一个凝灰质层还使用了阶段加热的 $^{40}Ar/^{39}Ar$ 法进行了年代测定。

Tuff *c* is the uppermost of the dated horizons, lying near the top of the fossiliferous deposits. It is a widespread crystal-lithic air-fall tuff cemented with calcite and generally 10–15 cm thick. Abundant biotite crystals 1–2 cm in diameter occur in the upper part of the tuff, which is composed of nepheline and milelitite. The dated crystals were hand-picked from two outcrops of the tuff at locality *A* (Figs 2 and 3). The cementing calcite adhering to and interleaving the biotite books was removed by treatment with dilute HCl for a few minutes on one sample and with warm dilute acetic acid on two other samples. This treatment was found to have negligible deleterious effects on biotite standard samples. The three dates obtained for tuff *c* (3.41±.08 Myr, 3.68±.09 Myr, and 3.69±.10 Myr) have about equal precision so were averaged to give a date of 3.59 Myr.

Two conventional K–Ar dates and one $^{40}Ar/^{39}Ar$ were obtained from a single biotite crystal from a xenolithic horizon at locality *A* (Figs 2 and 3) approximately 1–2 m below the youngest tuff dated (tuff *c*). This horizon lies within the upper part of the fossiliferous beds and is distinguished by its ejecta of ijolite xenoliths together with nephelinite lava xenoliths. Although the biotite occurs in some of the ijolite clasts, a large single free crystal picked from the tuff was used for dating. The good agreement between the two conventional K–Ar dates for this sample (3.62±0.09 Myr and 3.59±0.05 Myr) and the $^{40}Ar/^{39}Ar$ date which yielded an isochron age of 3.55 Myr (Fig. 4) indicates that initial excess ^{40}Ar is not a problem with this sample, and that these dates give an average age for this horizon of 3.59 Myr.

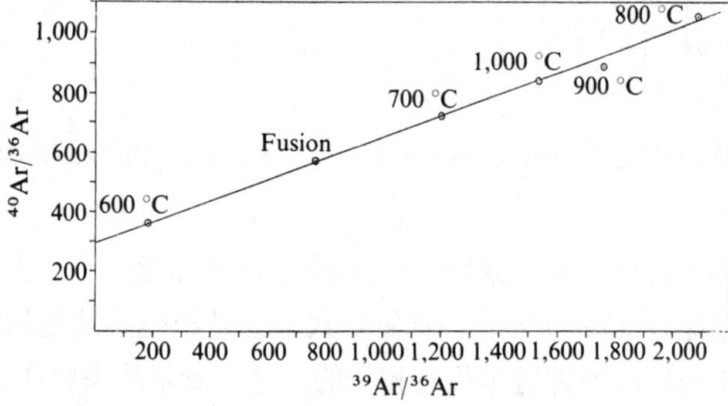

Fig. 4. $^{40}Ar/^{39}Ar$ incremental heating of biotite from xenolithic horizon, locality *C*. Isochron age, 3.55 Myr; $[^{40}Ar/^{39}Ar]_0=294$; $^{40}Ar/^{39}Ar=0.35318$; $J=0.005204$.

Two dates were obtained from a biotite-bearing crystal-lithic tuff 60 cm thick lying approximately 50 m below tuff *c* near the middle of the thickest section of the Laetolil Beds at location *B* (Fig. 2). The tuff is of nephelinite composition, containing abundant augite and altered nepheline and 2–3% of biotite, some crystals of which are as much as 1 cm in diameter. Calcite cements the crystals and fragments together. Biotite was separated by crushing, screening and hand picking and was cleaned of calcite with dilute HCl. Two dates, 3.82±0.16 Myr and 3.71±0.12 Myr, average 3.77 and give a lower limit to the age of the hominid remains (which occur below but close to the dated xenolithic horizon higher in the section) in the 30 m section at locality *A* (Fig. 2) whose base projects approximately 20 m above this tuff at locality *B* (Fig. 3).

凝灰岩 c 位于已经确定年代的地层的最上层，接近含化石堆积物的顶部。该层是广泛分布的由钙质胶结的晶屑岩屑风成凝灰岩，通常有 10~15 厘米厚。大量直径 1~2 厘米的黑云母晶体出现在该凝灰岩的上部，其由霞岩和黄长岩（编者注：英文原文中为 milelitite，可能为 melilitite 的错写，应为黄长岩）构成。从 A 地点的两处凝灰岩露头处手工挑选一些晶体用于年代测定（图 2 和图 3）。对于附着在黑云母上并与书页状黑云母交错在一起的钙质胶结物，其中一个样品用稀盐酸处理几分钟后除去，另外两个标本则用温的稀醋酸除去。这种处理方法对于黑云母标准样品具有极小的负面作用。得到的关于凝灰岩 c 的三个年代（距今 341 万 ±8 万年、368 万 ±9 万年和 369 万 ±10 万年）具有几乎一致的精确性，所以可以取其平均值，即年代为距今 359 万年。

从位于年代最年轻的凝灰岩（凝灰岩 c）之下大约 1~2 米处的 A 地点（图 2 和图 3）的捕虏体岩石层中得到了一块黑云母晶体，两个通过传统的钾氩法测定的年代和一个 $^{40}Ar/^{39}Ar$ 年代都是根据该黑云母晶体测定的。该地层处于含化石层的上部，以其霓霞岩捕虏体及霞岩熔岩捕虏体喷出物为特征。尽管在某些霓霞岩碎屑中也出现了黑云母，但是研究人员从凝灰岩中挑选了一块大的游离的晶体来进行年代测定。对该样品用传统的钾氩法测定的两个年代（距今 362 万 ±9 万年和 359 万 ±5 万年）和 $^{40}Ar/^{39}Ar$ 年代（该方法得到了距今 355 万年的等时线年龄）（图 4）之间具有很好的一致性，这种一致性表明最初过量的 ^{40}Ar 对于这个样品并不构成影响，根据这些年代得出这个地层的平均年龄为距今 359 万年。

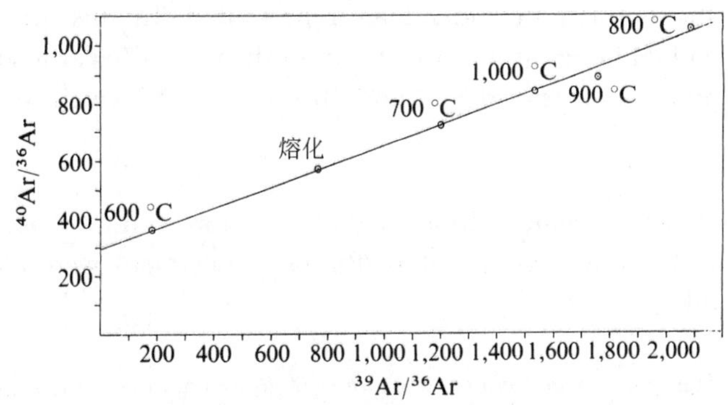

图 4. 从 C 地点的捕虏体岩石层得到的黑云母阶段加热的 $^{40}Ar/^{39}Ar$ 法年代。等时线年龄：距今 355 万年；$[^{40}Ar/^{39}Ar]_0$=294；$^{40}Ar/^{39}Ar$=0.35318；J=0.005204。

从 60 厘米厚的含晶屑岩屑黑云母凝灰岩得到了两个年代，该凝灰岩位于凝灰岩 c 之下约 50 米处，接近 B 地点莱托利尔层最厚剖面的中间（图 2）。该凝灰岩的成分是霞岩，含有丰富的辉石和蚀变的霞石、2%~3% 的黑云母以及直径几乎为 1 厘米的一些晶体。方解石把晶屑和碎屑胶结在一起。通过粉碎、筛选、手工挑选将黑云母分离出来，然后用稀盐酸清洗掉方解石。距今 382 万 ±16 万年和 371 万 ±12 万年这两个年代的平均值是距今 377 万年，给出了在 A 地点的 30 米厚剖面处（图 2）发现的人科动物化石的年代下限（该化石是在下层发现的，但是其与该剖面较高位置的已确定年代的捕虏体岩石地层接近），其基部突出于 B 地点处的这种凝灰岩之上约 20 米处（图 3）。

Fauna

The fossiliferous deposits in the Laetolil area have been subdivided, for purposes of collecting, into 26 localities. These subdivisions are based on existing topographic features such as grassy ridges, lines of trees, stream channels and so on. This has provided a means of dividing the fossiliferous area into units of restricted size, but does not relate to the former topography.

Eighteen of these localities are in the Garusi valley, one in the valley at the head of the Olduvai Side Gorge, one in the Gadjingero valley, five in a valley to the south of the Garusi river and one in an isolated position to the west (Fig. 2).

Identifiable fossils noted on the exposures were either collected and registered or listed on the sites. Specimens from the Laetolil Beds were distinctively cream coloured or white and sometimes chalky in texture, but the surface material also included brown, grey or black specimens, often rolled. These have been excluded from the material under review, together with fossils which have adhering matrix clearly dissimilar from the tuffs of the Laetolil Beds. Among these are all remains of *Hippopotamus, Equus, Theropithecus, Phacochoerus* and *Tragelaphus*, formerly included in the Laetolil fauna, with the exception of *Hippopotamus.*

The fossil material from the Laetolil Beds is dispersed and fragmentary and it is not possible to assess the number of individuals represented. In this article, "numbers of fossils" refer to individual bones and teeth, except in the case of clear association, confined almost entirely to remains of *Serengetilagus* and *Pedetes*, some of which were associated and even articulated.

Table 2 shows the mean percentage frequency of the more common vertebrate groups at several of the richer localities. A total of 6,288 fossil specimens were identified from the localities considered here.

Table 2. Mean percentage frequency of bones of more common vertebrate groups

Group	Mean percentage	Range	Nos of localities
Bovidae	43.0	29.5–57.9	18
Lagomorpha	14.4	5.6–24.0	18
Giraffidae	11.2	6.8–23.2	18
Rhinocerotidae	9.7	5.7–17.3	18
Equidae	4.4	1.7–7.1	18
Suidae	3.6	1.0–7.1	18
Proboscidea	3.4	1.0–5.1	18
Rodentia	3.3	0.9–7.2	17
Carnivora	3.1	0.8–8.2	17

动 物 群

为了采集标本，已将莱托利尔地区的含化石堆积物细分成了 26 个地点。这些分区是基于现存的地形学特征如草嵴、树线、河道等划分的。这提供了一种将含化石区域划分成一定大小单元的方法，但是与先前的地形学研究并无关联。

这些地点中有十八个在加鲁西河河谷，一个位于奥杜威侧峡谷的源头河谷处，一个在伽津格罗河河谷，五个在加鲁西河南部的一个河谷，还有一个在西部的一个孤立的位置（图 2）。

在露头看到的可鉴定化石或者被采集并记录，或者在遗址上标注出来。莱托利尔层的化石都是特殊的乳白色或者白色，并且有时呈白垩的质地，但是从地表采集到的材料也有褐色、灰色或黑色的标本，这些颜色经常是包在标本外面的。本项研究的标本排除了地表采集的标本，同时排除了那些附着有与莱托利尔层凝灰岩明显不同的围岩的化石。这些标本都是河马属、马属、狮尾狒属、疣猪属和林羚属的化石，除了河马属以外，其他原本就包括在莱托利尔动物群中。

莱托利尔层中的化石标本分散而残破不全，因而不可能估计出其代表的个体数量。本文中“化石编号”是指单个的骨骼和牙齿，除非是相互之间关系明确的化石标本，例如塞伦盖蒂兔属和跳兔属化石几乎完全在同一处发现，其中有些骨骼就在一起，甚至是有关节连接的。

表 2 给出了在几处化石较多的地点中较常见的脊椎动物群出现的平均百分频率。在本文研究的地点中总共鉴定出了 6,288 件化石标本。

表 2. 较常见的脊椎动物群的骨骼的平均百分比频率

动物群	平均百分比	范围	所在地点的数量
牛科	43.0	29.5~57.9	18
兔形目	14.4	5.6~24.0	18
长颈鹿科	11.2	6.8~23.2	18
犀科	9.7	5.7~17.3	18
马科	4.4	1.7~7.1	18
猪科	3.6	1.0~7.1	18
长鼻目	3.4	1.0~5.1	18
啮齿目	3.3	0.9~7.2	17
食肉目	3.1	0.8~8.2	17

Reptiles are represented by snake vertebrae at three localities and by tortoises at all localities. The latter have an average frequency of 2.2% and include several giant specimens. Avian remains occur widely and at several localities birds' eggs were completely preserved. There is one example of a shattered clutch of at least eight eggs, rather smaller in size than eggs of domestic fowl. Primates were found at 15 localities, and in one area they constitute 3.8% of the fauna. Both cercopithecines and colobines are present (M. G. Leakey, personal communication).

Rodents are fairly well represented, although not abundant. Of the specimens identified by J. J. Jaeger, the most common are *Pedetes*, *Saccostomus* and *Hystrix*. The carnivore fauna is characterised by a high percentage of viverrids, constituting 32% of all the carnivore specimens. Large carnivores are represented by hyaenids, of which there are several genera, by felids and a machairodont.

Proboscidea include *Deinotherium* and *Loxodonta* sp. (M. Beden, personal communication). There is no evidence that the equid material (other than that derived from later deposits) includes any genus except *Hipparion*. The suids consist only of two genera, *Potamachoerus* and *Notochoerus* (J. Harris, personal communication). The presence of *Ancylotherium* and *Orycteropus*, noted in previous collections, is confirmed and the existence of two rhinocerotids has been established by the discovery of skulls of both *Ceratotherium* and *Diceros*, although only the former was listed previously.

Among the giraffids, *Sivatherium* and a small form of giraffe are equally common. *Giraffe jumae* is also present but is much less well represented. The bovid fauna is chiefly characterised by the very high percentage of *Madoqua* (dikdik). In 18 localities dikdik range from 1.5 to 37.7% of all bovid specimens, with a mean percentage of 15.1%.

The 1975 field season, although mainly confined to surface collection, has established that previous collecting had sampled faunas of several time periods. It is now possible to exclude some genera from the published lists of fauna from the Laetolil Beds[2,9], such as *Theropithecus*, *Tragelaphus*, *Equus* and *Phacochoerus.*

Fossil Hominids

Thirteen new fossil hominid specimens were recovered from the Laetolil site during 1974 and 1975. The remains include a maxilla, mandibles and teeth. This sample displays a complex of characters seemingly demonstrative of phylogenetic affinity to the genus *Homo*, but also features some primitive traits concordant with its great age.

The hominid specimens are listed in Table 3. Provisional stratigraphic correlation and dating have placed the hominid remains as shown in Fig. 3. With the exception of Laetolil hominids (LH) 7 and 8, all specimens retain the matrix characteristic of the Laetolil Beds from which they have been weathered or excavated. There is no reason to doubt that all the specimens derive from the Laetolil Beds as reported.

在其中三处地点发现的爬行动物主要是蛇椎骨，在所有地点都发现了龟甲。后者出现的平均频率为2.2%，包括几个巨大的标本。鸟类化石出现的范围很广，在几处地点，鸟蛋被完整地保存下来了。有一窝蛋至少包括八个个体，但是都已经破碎了，这些蛋的尺寸都比家禽的蛋要小。在15处地点发现了灵长类；有一个地区，它们占到了整个动物群的3.8%。猕猴亚科和疣猴亚科都存在（米芙·利基，个人交流）。

啮齿类化石广泛存在，但总体数量并不多。在耶格鉴定出的标本中，最常见的是跳兔属、南非囊鼠属和豪猪属。食肉类动物以高比例的灵猫类为特征，它们占所有食肉动物标本的32%。大型食肉动物的主要代表是鬣狗类、猫类和一种剑齿虎，其中鬣狗类有好几个属。

长鼻目包括恐象属和非洲象未定种（贝登，个人交流）。没有证据表明除了三趾马之外还有别的马类化石（除了从较晚的堆积物中发现过外）。猪科动物只包括两个属：河猪属和南方猪属（哈里斯，个人交流）。之前的挖掘中曾发现过爪脚兽属和土豚属的存在，此次再次得到确认。白犀和黑犀头骨的发现证实存在两种犀科动物，但是之前只发现过白犀。

在长颈鹿科动物中，西洼兽属和一种小型的长颈鹿都很常见。朱玛长颈鹿也存在，但是保存状况不太好。牛科动物群以极高比例的犬羚属为特征。在18处地点中，犬羚出现的频率占所有牛科标本出现频率的1.5%~37.7%，平均百分比为15.1%。

在1975年的野外工作中，尽管主要局限在地表标本采集方面，但是确认了之前的采集工作所获得的动物群标本属于几个不同的时期。现在可以将一些属从已发表的莱托利尔层的动物群名单中剔除出去了[2,9]，例如狮尾狒属、林羚属、马属和疣猪属。

人科动物化石

1974年到1975年间在莱托利尔遗址挖掘出了13件新的人科动物化石标本。这些化石包括一个上颌骨、一些下颌骨和牙齿。该标本既显示出了一系列似乎证实其在系统发育上与人属具有亲缘关系的特征，但是也具有与其古老年代一致的一些原始特征。

表3列出了这些人科动物标本。根据暂行的地层对比和年代测定确定出的这些人科化石层位如图3所示。除了莱托利尔人科动物（LH）7和8外，所有标本都保留着莱托利尔层特有的围岩，这些标本是从莱托利尔层中被风化出来或挖掘出来的。正如报道的一样，所有这些标本都源自莱托利尔层，这点毫无疑问。

Table 3. Hominid remains recovered in 1974–75

Laetolil hominid (LH)	Locality	Specimen consists of	Discovered by
1	1	RP^4 fragment	M. Muoka
2	3*1	Immature mandibular corpus with deciduous and permanent teeth	M. Muluila
3 (a–t)	7*	Isolated deciduous and permanent teeth, upper and lower	M. Muoka
4	7	Adult mandibular corpus with dentition	M. Muluila
5	8	Adult maxillary row: I^2 to M^1	M. Muluila
6 (a–e)	7*†	Isolated deciduous and permanent teeth, upper	M. Muoka
7	5	RM^1 or ^{2} fragment	M. Muoka
8	11	RM^2, RM^3	E. Kandindi
10	10W	Fragment left mandibular corpus with broken roots	E. Kandindi
11	10W	LM^1 or ^{2}	E. Kandindi
12	5	LM^2 or ^{3} fragment	E. Kandindi
13	8	Fragment right mandibular corpus with broken roots	M. Jackes
14	19	Isolated permanent teeth, lower	E. Kandindi

LH 9 was not valid.

* *In situ.*

† LH 3 and 6 associated in mixed state.

The Laetolil hominid sample consists of teeth and mandibles. Important features of these specimens are described here, followed by a brief preliminary discussion regarding the phylogenetic status of the fossils.

Remains of deciduous and permanent dentitions have been recovered. The dentitions consist of two maxillary and four mandibular partial tooth rows. Compared with the rest of the East African Pliocene/early Pleistocene hominid sample, the Laetolil anterior teeth are large and the postcanine teeth of small to moderate size.

Deciduous Dentition

Canines (LH 2, 3) Single upper and lower deciduous canines are known. The lower is a slightly projecting, sharp conical tooth in its damaged state. It is smaller but its overall morphology is similar to its permanent counterpart.

First molars (LH 2, 3) The upper first deciduous molar displays spatial dominance of the protocone and a well-marked mesiobuccal accessory cusp defined by a strong anterior fovea. The lower first deciduous molar is molarised, with four or five main cusps depending on hypoconulid expression. There is a spatially dominant protoconid with a large flat buccal face, a lingually facing anterior fovea, and an inferiorly projecting mesiobuccal enamel line. Vertical dominance of the protoconid and metaconid is marked in lateral view.

表 3. 1974~1975 年间发掘的人科动物化石

莱托利尔人科动物（LH）	地点	标本组成	发现者
1	1	右 P^4 碎块	穆奥卡
2	3*[1]	保存有乳齿和恒齿的未成年下颌体	穆卢伊拉
3（a~t）	7*	单独的上下乳齿和上下恒齿	穆奥卡
4	7	保存有齿系的成年下颌体	穆卢伊拉
5	8	成年上颌齿列：I^2 到 M^1	穆卢伊拉
6（a~e）	7*†	单独的上乳齿和上恒齿	穆奥卡
7	5	右 M^1 或右 M^2 碎块	穆奥卡
8	11	右 M^2、右 M^3	坎丁迪
10	10W	保存有残破齿根的左下颌体碎块	坎丁迪
11	10W	左 M^1 或左 M^2	坎丁迪
12	5	左 M^2 或左 M^3 碎块	坎丁迪
13	8	保存有残破齿根的右下颌体碎块	杰基斯
14	19	单独的下恒齿	坎丁迪

LH 9 无效。
* 在原地。
† LH 3 和 6 以伴生状态出土。

莱托利尔人科动物化石标本包括牙齿和下颌骨。本文描述了这些标本的重要特征，并且对这些化石的系统发育位置进行了简单的初步探讨。

发掘中也采集到乳齿系和恒齿系标本。齿系包括两个上颌骨和四个下颌骨的部分齿列。与其余的东非上新世 / 早更新世时期的人科动物标本相比，莱托利尔标本的前牙较大，而颊齿则为小到中等尺寸。

乳 齿 系

犬齿（LH 2、3）：这是已知的唯一的上乳犬齿和下乳犬齿。下乳犬齿稍微有些突出，尖锐的圆锥齿处于受损状态。虽然该牙比恒齿小，但是其总体形态与恒齿很相似。

第一臼齿（LH 2、3）：第一上乳臼齿的上原尖很大，一个由发达的前窝限定的近中颊侧附尖明显。第一下乳臼齿臼齿化，依下次小尖的发育状况而有四或五个主尖。下原尖很大，具有大而平的颊面；前窝朝向舌侧；近中颊侧釉质线向下突出。从外侧视下原尖和下后尖明显垂直。

Second molars (LH 2, 3, 6) Upper and lower deciduous second molars take the basic form of the analogous permanent first molars but are smaller in overall size. The dM_2 of LH 2 is the only lower molar in the Laetolil sample bearing any indication of a tuberculum sextum.

Permanent Dentition

Incisors (LH 2, 3, 5, 6, 14) The single upper central incisor is very large and bears pronounced lingual relief. The upper lateral incisors are smaller, with variable lingual relief. The lower incisors are narrow and very tall, with minimal relief.

Canines (LH 2, 3, 4, 5, 6, 14) The incompletely-developed upper canine LH 3 is large, stout and pointed, bearing pronounced lingual relief. LH 6 is slightly smaller, with less lingual relief and a tall, pointed crown. LH 5 bears an elongate dentine exposure on its distal occlusal edge but does not project beyond the occlusal row in its worn state. The lower canines are similar to the uppers in their great size, height and lingual relief.

Premolars (LH 1, 2, 3, 4, 5, 6, 14) The upper premolars tend to be buccolingually elongate and bicuspid, with the lingual cusp placed mesially and an indented mesial crown face. The two lower third premolars each have a dominant buccal cusp with mesial and distal occlusal ridges, and a weak lingual cusp placed mesial of the major crown axis. The long axis of the oval crown crosses the dental arcade contour from mesiobuccal to distolingual, donating a "skewed" occlusal profile to the tooth. The lower fourth premolars are fairly square in shape with major buccal and lingual cusps and moderate talonids.

Molars (LH 2, 3, 4, 5, 6, 7, 8, 11, 12, 14) The upper molars are of moderate size, their relative proportions unknown. They show a basic four-cusp pattern with spatial dominance of the protocone and typical expression of a pit-like Carabelli feature. The lower molars display progressive size increase from first to third in LH 4. They have a fairly square occlusal outline with hypoconulid apressed anteriorly between hypoconid and entoconid and no trace of a tuberculum sextum. The Y-5 pattern of primary fissuration is constant.

Mandibles

Juvenile mandible (LH 2) This specimen (Figs 5 and 6) is incompletely fused at the midline, and the developing crowns of the permanent canines and premolars are exposed in the broken corpus. The first molars have just reached the occlusal plane. Only the posterior aspect of the symphysis shows fairly intact contours, with a concave post-incisive planum and incipient superior transverse torus. The genioglossal fossa is obscured by midline breakage. Associated distortion has artificially increased the bimolar distances.

第二臼齿（LH 2、3、6）：第二上、下乳臼齿与第一恒臼齿具有相似的基本形式，但是总尺寸要小一些。LH 2 的 dM_2 是莱托利尔标本中唯一一枚有第六小尖迹象的下臼齿。

恒 齿 系

门齿（LH 2、3、5、6、14）：唯一的上中门齿非常大，而且具有明显的舌面突起。上侧门齿较小，舌面突起因不同个体而异。下门齿狭窄而高，突起很小。

犬齿（LH 2、3、4、5、6、14）：发育不充分的上犬齿 LH 3 很大、结实而尖锐，具有明显的舌面突起。LH 6 稍微小一点，具有较少的舌面突起和一个高而尖的牙冠。LH 5 在远中侧咬合边缘具有一个延长的牙本质暴露区，但是没有在磨损的状态下突出于咬合列之外。下犬齿与上犬齿在大尺寸、高度和舌面突起方面都很相似。

前臼齿（LH 1、2、3、4、5、6、14）：上前臼齿趋于沿颊舌向延伸并呈双尖型，舌侧主尖位于近中侧以及一个锯齿状的近中牙冠面。两枚第三下前臼齿中的每枚都有一个突出的具有近中和远中咬合嵴的颊侧齿尖，还有一个位于主牙冠轴线中间的比较小的舌侧齿尖。椭圆形齿冠的长轴从近中颊侧到远中舌侧穿过齿弓的轮廓，为牙齿提供了一个“斜的”咬合剖面。第四下前臼齿形状上非常接近正方形，具有颊侧主尖和舌侧主尖以及中等大小的下跟座。

臼齿（LH 2、3、4、5、6、7、8、11、12、14）：上臼齿的大小适中，它们的相对比例还不清楚。这些牙齿显示出一种基本的四尖型，该模式中原尖很大，表现出坑状卡拉贝利的典型特征。LH 4的下臼齿表现出从第一颗到第三颗逐渐增大的特点。它们具有方形的咬合轮廓，其下次小尖在下次尖和下内尖之间向前紧贴，没有第六小尖的迹象。嚼面的主凹谷都是 Y-5 型。

下 颌 骨

未成年下颌骨（LH 2）：该标本（图 5 和图 6）在中线处没有完全融合，发育中的恒犬齿和恒前臼齿牙冠已在断裂的下颌体萌出。第一臼齿刚刚长到咬合面。只有联合部位后面的轮廓是比较完整的，有一个凹进去的后门齿平面和刚开始发育的上横圆枕。颏窝由于中线破裂而模糊不清了。由此引起的扭曲变形不自然地增加了左右臼齿间的距离。

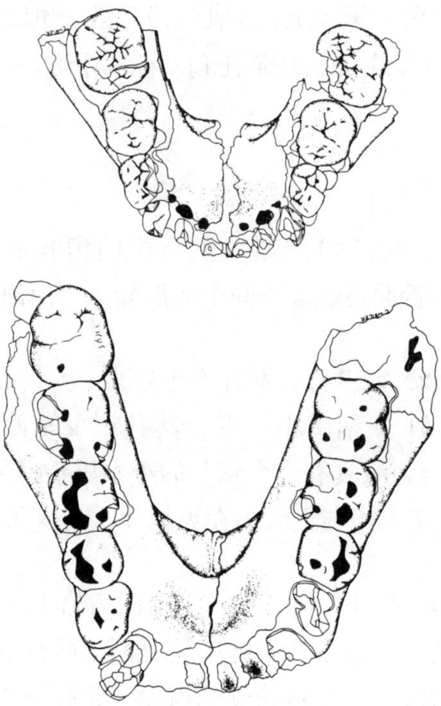

Fig. 5. Occlusal views of juvenile and adult mandibles (LH 2 and 4).

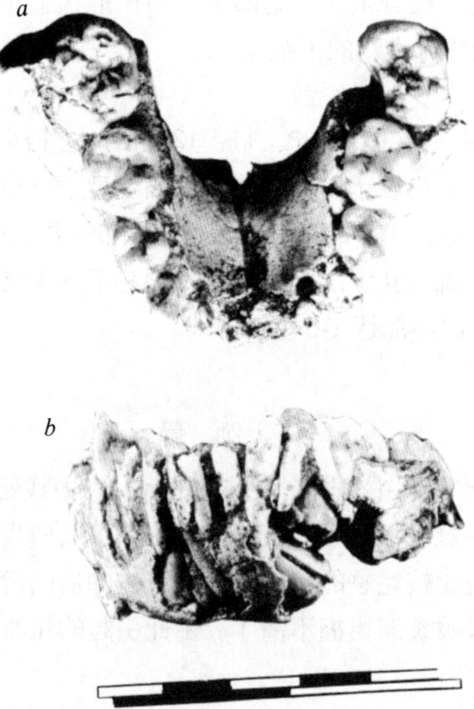

Fig. 6. Juvenile mandible from the Laetolil Beds (LH 2). *a*, Occlusal view; except for the first molars all the teeth are deciduous. *b*, Front view showing the central incisors, canine and P_3 in the bone.

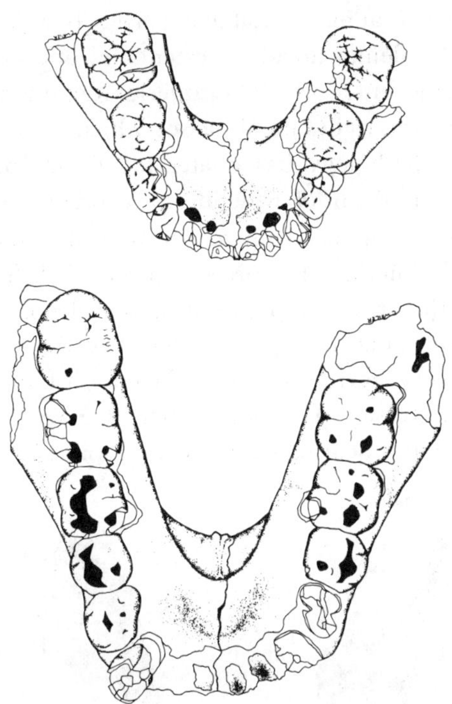

图 5. 未成年和成年下颌骨的嚼面视（LH 2 和 4）。

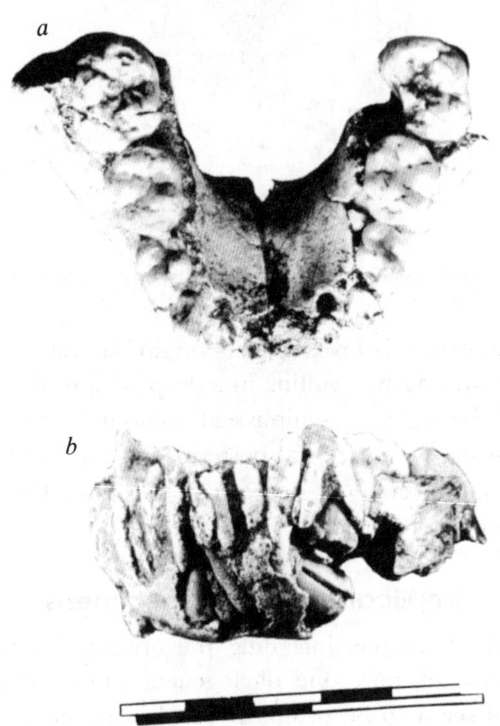

图 6. 莱托利尔层出土的未成年下颌骨(LH 2)。*a*,嚼面视;除了第一臼齿外,所有牙齿都是乳齿。*b*,正面视,展示了骨骼上的中门齿、犬齿和 P_3。

Adult mandible (LH 4) The adult mandibular corpus is well preserved, with the rami missing (Figs 5 and 7). The dental arcade is essentially undistorted, and presents fairly straight sides which converge anteriorly. The anterior dentition has suffered *post mortem* damage and loss, except for the right lateral incisor which seems to have been lost in life. Largely resorptive alveolar pathology has obliterated its alveolus and has affected the adjacent teeth. There is development of wide interproximal facets for the canine teeth but no C/P_3 contact facet. This combines with observation of extensive wear on the buccal P_3 face to suggest that C/P_3 interlock has prevented mesial drift from eliminating the C/P_3 diastema. Judging from the preserved posterior incisor alveoli, these teeth were set in an evenly rounded arcade, projecting moderately anterior to the bicanine axis. The internal mandibular contour is a very narrow parabola in contrast to the wider basal contour which displays great lateral eversion posteriorly. There are weak to moderate superior and inferior transverse tori, the latter bearing strong mental spines.

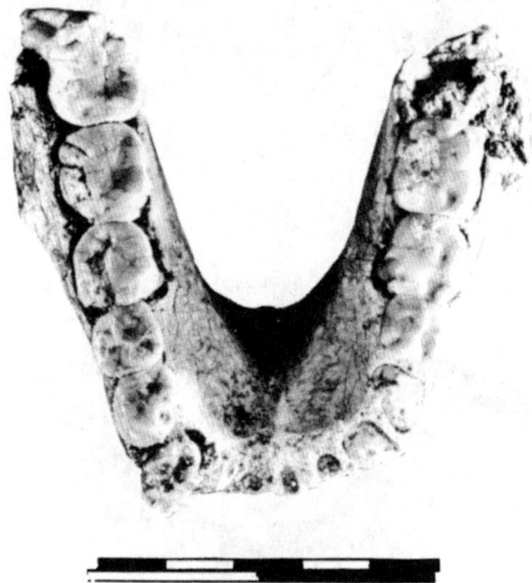

Fig. 7. Adult mandible from the Laetolil Beds (LH 2), occlusal view.

The anterior root of the ramus is broken at its origin, lateral to M_2. Occlusal and basal margins diverge strongly anteriorly, resulting in a deep symphysis. The symphysis is angled sharply posteriorly and the anterior symphyseal contour is rounded and bulbous. The lateral aspects of the corpus have very flat posterior portions and distinctive hollowing in the areas above the mental foramina at the P_3 to P_4 position. The corpus is tall and fairly narrow, especially in its anterior portion.

Implications of the Specimens

The Laetolil fossil hominid sample, including the original Garusi maxillary fragment[5], seems to be representative of only one phylogenetic entity or lineage. The variations observed in the material seem to be primarily size-based and stem from individual and sexual factors.

成年下颌骨（LH 4）：成年下颌体的保存状况很好，下颌支缺损（图 5 和图 7）。齿弓基本没有变形，两侧的齿列很直并且向前集中。除了似乎在生前就已经丢失了的右侧门齿之外，前齿系都是在死后受损和丢失的。很大程度的再吸收性齿槽病使得其齿槽的痕迹已经不清楚了，并且影响到了相邻的牙齿。犬齿发育了宽的邻接面，但没有 C/P_3 接触面。这与观察到 P_3 的颊侧面的较大的磨损共同提示了 C/P_3 咬合阻止了 P_3 的近中向迁移使 C/P_3 间的齿隙消失。从保存下来的后门齿齿槽来判断，这些牙齿是位于曲度均匀的圆形齿弓上的，向前适度突出到左右犬齿间的轴线上。与下颌骨后端显示出明显向外侧外翻的稍宽的基底轮廓相反，下颌骨内轮廓呈现出一条非常窄的抛物线形状。存在从微弱到中度不等的上、下横圆枕，后者具有强壮的颏棘。

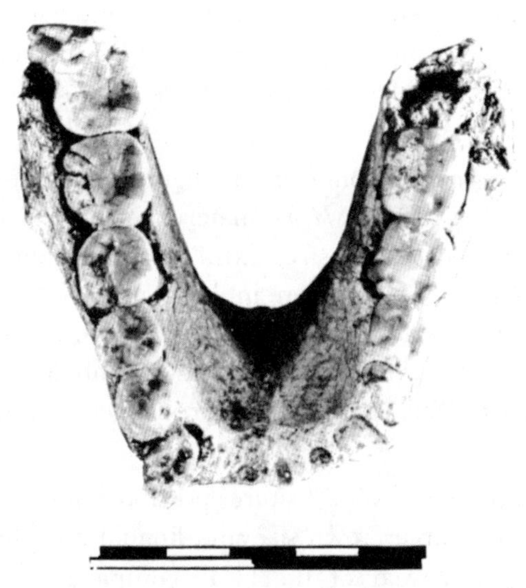

图 7. 莱托利尔层出土的成年下颌骨（LH 2），嚼面视。

下颌支的前根部在前端断裂，侧向延续到 M_2。嚼面和基底缘向前分离很大，使下颌联合部很深。故联合部位后面的角度很尖锐，前面的联合部轮廓圆而呈球根状。下颌体的外侧面具有很平的后部，在 P_3 到 P_4 位置处的颏孔之上的区域具有特殊的中空。下颌体高而相当窄，特别是在前部。

标本的启示

包括最初的加鲁西上颌骨破块 [5] 在内的莱托利尔人科动物化石标本似乎只代表了一个支系，或者叫一个谱系。观察到的标本上的形态差异似乎主要是大小方面的，并且这些差异是由于个体因素和性别因素引起的。

The deciduous teeth, particularly the lower deciduous first molars, display remarkable similarity to hominid specimens from South Africa (Taung; STS 24)[10] as well as to individuals tentatively assigned to *Homo* in East Africa (KNM ER 820, 1507)[11,12]. They depart strongly from the pattern of molarisation displayed in the South African "robust" specimens (TM 1601; SK 61, 64)[10] as well as from East African specimens generally assigned to the same hominid lineage (KNM ER 1477)[13].

The Laetolil permanent canines and incisors are relatively and absolutely large and bear a great deal of lingual relief. They ally themselves similarly to earlier hominid specimens such as STS 3, 50, 51, 52; MLD 11; OH 7, 16; KNM ER 803, 1590 (refs 10, 15, 18–21), as well as to the younger African and Asian specimens usually assigned to *Homo erectus*. These features set the Laetolil specimens apart from the sample including SK 23, 48, 876 and so on; Peninj; KNM CH 1; OH 5, 38; KNM ER 729, 1171 (refs 10, 15, 18–21). The Laetolil permanent premolars show none of the molarisation seen in the latter specimens and bear particularly strong resemblances to South and East African material (STS 51, 52, 55; OH 7, 16, 24; KNM ER 808, 992) (refs 10–12, 14).

The Laetolil permanent molars are consistent in aligning with the South African "gracile" australopithecus and the East African *Homo* material in both size and morphology. The molars do not display the increased size, extra cusps or bulging, expanded, "puffy" development of the individual cusps seen in high frequency among South and East African "robust" forms (SK 6, 13, 48, 52; TM 1517; Peninj; KNM CH 1; KNM ER 729, 801, 802) (refs 10, 17–20). The adult mandible has resemblances to certain East African specimens such as KNM ER 1802, with similar corpus section and basal eversion.

The Laetolil fossil hominids have several features possibly consistent with their radiometric age. These traits include the large crown size and lingual morphology of the permanent canines; the morphology and wear of the C/P_3 complex; the buccolingually elongate upper premolars; the overall square occlusal aspect of the lower permanent molars; the low symphyseal angle; the bulbous anterior symphysis; the relatively straight posterior tooth rows; the low placement of the mental foramina, and the distinctive lateral corpus contours including small, superiorly placed extramolar sulci.

Preliminary assessment indicates strong resemblance between the Laetolil hominids and later radiometrically-dated specimens assigned to the genus *Homo* in East Africa. Such assessment suggests placement of the Laetolil specimens among the earliest firmly dated members of this genus. It should come as no surprise that the earlier members of the genus *Homo* display an increasing frequency of features generally interpreted as "primitive" or "pongid like", which indicate derivation from as yet largely hypothetical ancestors.

Much of the recently discovered comparable fossil hominid material from the Hadar region of Ethiopia shows strong similarity to the Laetolil specimens[23], and further collection combined with detailed comparative analysis of material from both localities

乳齿，尤其是第一下乳臼齿，表现出了与南非的人科动物标本(汤恩;STS 24)[10]以及暂时被归入人属的在东非发现的那些个体（KNM ER 820、1507）[11,12]具有非常显著的相似性。它们与南非“粗壮型”标本（TM 1601；SK 61、64）[10]以及被归入同一人科谱系的东非标本（KNM ER 1477）[13]所表现出来的臼齿化模式非常不同。

莱托利尔的恒犬齿和门齿标本的相对大小和绝对大小都很大，具有大量的舌面突起。它们与早期的人科动物标本在亲缘关系上很相似，例如STS 3、50、51、52；MLD 11；OH 7、16；KNM ER 803、1590（参考文献10、15、18~21），与通常被划分为直立人的幼体非洲标本和亚洲标本在亲缘关系上也很相似。这些特征使得莱托利尔标本与如下标本区别开来：SK 23、48、876等；佩宁伊；KNM CH 1；OH 5、38；KNM ER 729、1171（参考文献10、15、18~21）。莱托利尔恒前臼齿完全没有在后来的标本中所见到的臼齿化现象，并且与南非和东非标本（STS 51、52、55；OH 7、16、24；KNM ER 808、992）（参考文献10~12、14）非常相似。

莱托利尔恒臼齿与南非“纤细型”南方古猿和东非人属标本在大小和形态学上都具有一致性。这些臼齿没有表现出在南非和东非“粗壮型”南方古猿（SK 6、13、48、52；TM 1517；佩宁伊；KNM CH 1；KNM ER 729、801、802）（参考文献10、17~20）中高频出现的尺寸增大、出现额外的牙尖或突起、单个牙尖扩展的、“肿胀的”发育情况等。成年下颌骨与某些东非标本（如KNM ER 1802）具有相像之处，它们具有相似的下颌体部分和基底外翻。

莱托利尔人科动物化石具有的几个特征可能与它们的放射性年代一致。这些特点包括：恒犬齿的大齿冠尺寸和舌侧形态；C/P_3复合体的形态和磨损情况；颊舌向延长的上前臼齿；下恒臼齿大体呈方形的咬合面；小的联合部角度；球根状的联合部前部；相对直的后齿列；颏孔的低位，以及独特的下颌体外侧面轮廓，包括小型的、位置靠上的外侧臼齿槽。

初期评估暗示莱托利尔人科动物和经放射性测年法确定时代较晚的属于人属的东非标本非常相似。这一估计表明了莱托利尔标本在已确定年代的早期人属成员中的地位。早期的人属成员表现出某些特征的出现频率逐渐增加的趋势，这些特征通常被认为是“原始的”或者“似猩猩科的”，这一现象并不足为奇，这表明了它们很可能是从迄今为止在很大程度上还是假定的祖先进化而来的。

最近在埃塞俄比亚的哈达尔地区发现的很多类似的人科动物化石标本显示出与莱托利尔标本极大的相似性[23]，对这两处遗址进行进一步的标本采集并对这些标本进行详细的比较分析，这对于进一步理解人类起源是很关键的。莱托利尔标本丰富

is essential for the further understanding of human origins. The Laetolil collection adds to the developing phylogenetic perspective of the early Hominidae and emphasises the need for taxonomic schemes reflective of and consistent with the evolutionary processes involved in the origin and radiation of this family.

We thank the following for facilities, financial support, permission to examine originals and for discussion: the United Republic of Tanzania, the trustees of the National Museums of Kenya, the NSF Graduate Fellowship Program, the Scott Turner Fund, the Rackham Dissertation Fund, the National Geographic Society, G. Brent Dalrymple, C. K. Brain, T. Gray, D. C. Johanson, K. Kimeu, P. Leakey, R. E. F. Leakey, P. V. Tobias, E. Vrba, A. Walker, C. Weiler and B. Wood.

(**262**, 460-466; 1976)

M. D. Leakey*, R. L. Hay*, G. H. Curtis*, R. E. Drake*, M. K. Jackes* and T.D. White†
*Olduvai Gorge, PO Box 30239, Nairobi, Kenya.
†Department of Anthropology, University of Michigan, Ann Arbor, Michigan 48104.

Received March 30; accepted May 26, 1976.

References:

1. Weinert, H., *Z. Morph. Anthrop.*, **42**, 138-148 (1950).
2. Dietrich, W. O., *Palaeontographica*, **94A**, 43-133 (1942).
3. Maglio, V. J., *Breviora*, 336 (1969).
4. Kent, P. E., *Geol. Mag., Lond.*, **78**, 173-184 (1941).
5. Kohl-Larsen, L., *Auf des Spuren des Vormenschen*, **2**, 379-381 (1943).
6. Pickering, R., *Endulen, Quarter Degree Sheet 52* (Geological Survey in Tanzania, 1964).
7. Hay, R. L., *Bull. Geol. Soc. Am.*, 1281-1286 (1963).
8. Hay, R. L., *Geology of the Olduvai Gorge* (University of California Press, 1976).
9. Hopwood, A. T., in *Olduvai Gorge* (edit. by Leakey, L. S. B.), (Cambridge University Press, 1951).
10. Robinson, J. T., *Transvaal Museum Mem.*, 9, 1 (1956).
11. Leakey, R. E. F., and Wood, B. A., *Am. J. Phys. Anthrop.*, **39**, 355 (1973).
12. Leakey, R. E. F., and Wood, B. A., *Am. J. Phys. Anthrop.*, **39**, 355 (1974).
13. Leakey, R. E. F., *Nature*, **242**, 170 (1972).
14. Leakey, L. S. B., *Nature*, **188**, 1050 (1960).
15. Leakey, L. S. B., and Leakey, M. D., *Nature*, **202**, 3 (1964).
16. Day, M. H., and Leakey, R. E. F., *Am. J. Phys. Anthrop.*, **41**, 367 (1974).
17. Leakey, R. E. F., *Nature*, **248**, 653 (1974).
18. Carney, J., Hill, A., Miller, J., and Walker, A., *Nature*, **230**, 509 (1971).
19. Leakey, R. E. F., Mungai, J. M., and Walker, A. C., *Am. J. Phys. Anthrop.*, **36**, 235 (1972).
20. Leakey, R. E. F., and Walker, A. C., *Am. J. Phys. Anthrop.*, **39**, 205 (1973).
21. Tobias, P. V., *Olduvai Gorge*, **2** (Cambridge University Press, 1967).
22. Leakey, M. D., Clarke, R. J., and Leakey, L. S. B., *Nature*, **232**, 308 (1971).
23. Johanson, D. C., and Taieb, M., *Nature*, 260, 293-297 (1976).

了对早期人科动物的系统演化的认识，强调了分类学框架需要反映涉及人科动物起源和辐射进化的过程并与此过程保持一致。

我们感谢以下团体和个人为我们提供设备、经济支持，允许我们对原始标本进行研究，以及与我们进行讨论：坦桑尼亚联合共和国、肯尼亚国家博物馆的理事们、国家科学基金会研究生奖学金项目、斯科特·特纳基金、拉克姆论文基金、国家地理学会、布伦特·达尔林普尔、布雷恩、格雷、约翰森、基梅乌、菲利普·利基、理查德·利基、托拜厄斯、弗尔巴、沃克、韦勒和伍德。

（刘皓芳 翻译；董为 审稿）

Nucleotide Sequence of Bacteriophage ΦX174 DNA

F. Sanger *et al.*

Editor's Note

By 1977, English biochemist Frederick Sanger had developed the "dideoxy" method for sequencing DNA. Here he applies it to sequence the genome of the ΦX174 bacteriophage, making it the first fully-sequenced DNA-based genome. The single-stranded, circular genome contains just over 5,000 nucleotides. Sanger's group went on to sequence human mitochondrial DNA, and developed the whole-genome shotgun method to decode the genome of bacteriophage lambda, an important virus for molecular biology research. His work laid the foundations for all genome sequencing projects, which continue to shed light on the processes of health, disease, development and evolution. It also earned Sanger his second Nobel Prize in Chemistry. The first, in 1958, was for his work on the structure of proteins.

A DNA sequence for the genome of bacteriophage ΦX174 of approximately 5,375 nucleotides has been determined using the rapid and simple "plus and minus" method. The sequence identifies many of the features responsible for the production of the proteins of the nine known genes of the organism, including initiation and termination sites for the proteins and RNAs. Two pairs of genes are coded by the same region of DNA using different reading frames.

THE genome of bacteriophage ΦX174 is a single-stranded, circular DNA of approximately 5,400 nucleotides coding for nine known proteins. The order of these genes, as determined by genetic techniques[2-4], is *A-B-C-D-E-J-F-G-H*. Genes *F*, *G* and *H* code for structural proteins of the virus capsid, and gene *J* (as defined by sequence work) codes for a small basic protein that is also part of the virion. Gene *A* is required for double-stranded DNA replication and single-strand synthesis. Genes *B*, *C* and *D* are involved in the production of viral single-stranded DNA: however, the exact function of these gene products is not clear as they may either be involved directly in DNA synthesis or be required for DNA packaging, which is coupled with single-strand production. Gene *E* is responsible for lysis of the host.

The first nucleotide sequences established in ΦX were pyrimidine tracts[5-7] obtained by the Burton and Petersen[8] depurination procedure. The longer tracts could be obtained pure and sequences of up to 10 nucleotides were obtained. More recently Chadwell[9] has improved the hydrazinolysis method to obtain the longer purine tracts. These results are included in the sequence given in Fig. 1.

噬菌体 ΦX174 的 DNA 核苷酸序列

桑格等

编者按

至 1977 年，英国生物化学家弗雷德里克·桑格已经开发出了用于 DNA 测序的“双脱氧”法。本文中，他应用这种方法对 ΦX174 噬菌体的基因组进行了测序，使其成为第一个被完全测序的 DNA 基因组。这个单链环状基因组只包含 5,000 多个核苷酸。紧接着桑格的研究小组对人类线粒体 DNA 进行了测序，还开发了全基因组鸟枪法破译了分子生物学研究中一个重要病毒——λ 噬菌体的基因组。桑格的工作为所有的基因组测序项目奠定了基础，这些项目又进一步为健康、疾病、发育和进化的研究进展提供了线索。这些工作使桑格赢得了他的第二个诺贝尔化学奖，早在 1958 年，在蛋白质结构方面的工作使桑格第一次获得诺贝尔化学奖。

应用快速、简便的“加减测序法”，我们确定了大约含有 5,375 个核苷酸的 ΦX174 噬菌体基因组的 DNA 序列。从这一序列发现了这个生物的 9 个已知基因产生蛋白质的许多特征，包括蛋白质和 RNA 的起始位点与终止位点。有两对基因由同一段 DNA 编码，只是具有不同的阅读框。

噬菌体 ΦX174 的基因组是一个单链环状 DNA，由将近 5,400 个核苷酸组成，编码 9 种已知蛋白。通过遗传学技术确定的这些基因的排列顺序 [2-4] 为 *A–B–C–D–E–J–F–G–H*。基因 *F*、*G* 和 *H* 编码病毒衣壳的结构蛋白。基因 *J*（由测序工作确定）编码一个小的碱性蛋白，这个蛋白也是病毒粒子的一部分。基因 *A* 负责双链 DNA 的复制和单链 DNA 的合成。基因 *B*、*C* 和 *D* 与病毒单链 DNA 的合成有关；然而，这些基因产物的确切功能还不是很明确，因为它们要么与 DNA 合成直接相关，要么是与单链 DNA 合成相偶联的 DNA 包装所需要的。基因 *E* 负责宿主的裂解。

在 ΦX 中第一个被测定的序列是嘧啶区域的核苷酸序列 [5-7]，它是通过伯顿和彼得森 [8] 的脱嘌呤方法完成的。我们能够获得长达 10 个核苷酸的纯化片段。最近，查德韦尔 [9] 改进了肼解方法以获得更长的嘌呤区域。在图 1 中给出了上述这些区域的序列。

Nucleotide Sequence of Bacteriophage ΦX174 DNA

```
P1/1                            R5/7b                                  F6/9  T7/8                                                                                   T8/9
GAGTTTTATC GCTTCCATGA CGCAGAAGTT AACACTTTCG GATATTTCTG ATGAGTCGAA AAATTATCTT GATAAAGCAG GAATTACTAC TGCTTGTTTA CGAATTAAAT CGAAGTGGAC
        10         20         30         40         50         60         70         80         90        100        110        120
                                                 End B ↑

                                       A5/18
                     T9/10        H8b/4    T10/4  A18/6                                             F9/13
TGCTGGCGGA AAATGAGAAA ATTCGACCTA TCCTTGCGCA GCTCGAGAAG CTCTTACTTT GCGACCTTTC GCCATCAACT AACGATTCTG TCAAAAACTG ACGCGTTGGA TGAGGAGAAG
       130        140        150        160        170        180        190        200        210        220        230        240
    End A ↑

                                                  F13/17                               F17/16a   R7b/6c                                        F16a/16b
TGGCTTAATA TGCTTGGCAC GTTCGTCAAG GACTGGTTTA GATATGAGTC ACATTTTGTT CATGGTAGAG ATTCTCTTGT TGACATTTTA AAAGAGCGTG GATTACTATC TGAGTCCGAT
       250        260        270        280        290        300        310        320        330        340        350        360
                                                                                                                               mRNA start ↑

                                  F16b/1                                         Z3/7       A6/1
GCTGTTCAAC CACTAATAGG TAAGAAATCA TGAGTCAAGT TACTGAACAA TCCGTACGTT TCCAGACCGC TTTGGCCTCT ATTAAGCTCA TTCAGGCTTC TGCCGTTTTG GATTTAACCG
       370        380        390        400        410        420        430        440        450        460        470        480
                     D start ↑

M1/7     T4/5
AAGATGATTT CGATTTTCTG ACGAGTAACA AAGTTTGGAT TGCTACTGAC CGCTCTCGTG CTCGTCGCTG CGTTGAGGCT TGCGTTTATG GTACGCTGGA CTTTGTAGGA TACCCTCGCT
       490        500        510        520        530        540        550        560        570        580        590        600
                                                                                  E start ↑

                                                       R6c/7a       Z7/5                  H4/13
TTCCTGCTCC TGTTGAGTTT ATTGCTGCCG TCATTGCTTA TTATGTTCAT CCCGTCAACA TTCAAACGGC CTGTCTCATC ATGGAAGGCG CTGAATTTAC GGAAAACATT ATTAATGGCG
       610        620        630        640        650        660        670        680        690        700        710        720

T5/3    Y1/3                                        H13/11                                 M7/3
TCGAGCGTCC GGTTAAAGCC GCTGAATTGT TCGCGTTTAC CTTGCGTGTA CGCGCAGGAA ACACTGACGT TCTTACTGAC GCAGAAGAAA ACGTGCGTCA AAAATTACGT GCGGAAGGAG
       730        740        750        760        770        780        790        800        810        820        830        840
                                                                                                                                  End E ↑

                                 H11/14                                                               H14/12               R7a/6b
TGATGTAATG TCTAAAGGTA AAAAACGTTC TGGCGCTCGC CCTGGTCGTC CGCAGCCGTT GCGAGGTACT AAAGGCAAGC GTAAAGGCGC TCGTCTTTGG TATGTAGGTG GTCAACAATT
       850        860        870        880        890        900        910        920        930        940        950        960
End D ↑ ↑ J start

                Z5/8                                        H12/10
TTAATTGCAG GGGCTTCGGC CCCTTACTTG AGGATAAATT ATGTCTAATA TTCAAACTGG CGCCGAGCGT ATGCCGCATG ACCTTTCCCA TCTTGGCTTC CTTGCTGGTC AGATTGGTCG
       970        980        990       1000 ↑     1010       1020       1030       1040       1050       1060       1070       1080
↑ End J                           F start

                         Y3/2                 F1/14b  T3/1                    H10/7                          Z8/4       F14b/2
TCTTATTACC ATTTCAACTA CTCCGGTTAT CGCTGGCGAC TCCTTCGAGA TGGACGCCGT TGGCGCTCTC CGTCTTTCTC CATTGCGTCG TGGCCTTGCT ATTGACTCTA CTGTAGACAT
      1090       1100       1110       1120       1130       1140       1150       1160       1170       1180       1190       1200

                                  Q1/3c                                                                        R6b/1
TTTTACTTTT TATGTCCCTC ATCGTCACGT TTATGGTGAA CAGTGGATTA AGTTCATGAA GGATGGTGTT AATGCCACTC CTCTCCCGAC TGTTAACCAA ACTACTGGTT ATATTGACCA
      1210       1220       1230       1240       1250       1260       1270       1280       1290       1300       1310       1320

                                                                                                               H7/5
TGCCGCTTTT CTTGGCACGA TTAACCCTGA TACCAATAAA ATCCCTAAGC ATTTGTTTCA GGGTTATTTG AATATCTATAG CGTATTTTAA AGCGCCGTGG ---ATGCCTG ACCGTACCGA
      1330       1340       1350       1360       1370       1380       1390       1400       1410       1420       1430       1440

           A1/12c                                                                                                 A12c/13
GGCTAACCCT AATGAGCTTA ATCAAGATGA TGCTCGTTAT GGTTTCCGTT GCTGCCATCT CAAAAACATT TGGACTGCTC CGCTTCCTCC TGAGACTGAG CTTTCTCGCC AAATGACGAC
      1450       1460       1470       1480       1490       1500       1510       1520       1530       1540       1550       1560

                                  A13/2                                                                                 M3/4
TTCTACCACA TCTATTGACA TTATGGGTCT GCAAGCTGCT TATGCTAATT TGCATACTGA CCAAGAACGT GATTACTTCA TGCAGCGTTA CCATGA-GTT ATTTCTTCAT TTGGAGGTAA
      1570       1580       1590       1600       1610       1620       1630       1640       1650       1660       1670       1680

                                  H5/9a                                                                       Z4/1
AACCTCATAT GACGCTGACA ACCGTCCTTT ACTTGTCATG CGCTCTAATC TCTGGGCATC TGGCTATGAT GTTGATGGAA CTGACCAAAC GTCGTTAGGC CAGTTTTCTG GTCGTGTTCA
      1690       1700       1710       1720       1730       1740       1750       1760       1770       1780       1790       1800

                                                                         H9a/8a                                     F2/11
ACAGACCTAT AAACATTCTG TGCCGCGTTT CTTTGTTCCT GAGCATGGCA CTATGTTTAC TCTTGCGCTG GTTCGTTTTC CGCCTACTGC GACTAAAGAG ATTCAGTACC TTAACGCTAA
      1810       1820       1830       1840       1850       1860       1870       1880       1890       1900       1910       1920

                                                                                                          Q3c/6  F11/7
AGGTGCTTTG ACTTATACCG ATATTGCTGG CGACCCTGTT TTGTATGGCA ACTTGCCGCC GCGTGAAATT TCTATGAAGG ATGTTTTCCG TTCTGGTGAT TCGTCTAAGA AGTTTAAGAT
      1930       1940       1950       1960       1970       1980       1990       2000       2010       2020       2030       2040

                          H8a/6                         Q6/5                                                      Q5/3b
TGCTGAGGGT CAGTGGTATC GTTATGCGCC TTCGTATGTT TCTCCTGCTT ATCACCTTCT TGAAGGCTTC CCATTCATTC AGGAACCGCC TTCTGGTGAT TTGCAAGAAC GCGTACTTAT
      2050       2060       2070       2080       2090       2100       2110       2120       2130       2140       2150       2160

                                                                                                              F7/5b
TCGCAACCAT GATTATGACC AGTGTTTCAG TCGTTCAGTT GTTGCAGTGG ATAGTCTTAC CTCATGTGAC GTTTATCGCA ATCTGCCGAC CACTCGCGAT TCAATCATGA CTTCGTGATA
      2170       2180       2190       2200       2210       2220       2230       2240       2250       2260       2270  ↑   2280
                                                                                                                          End F
```

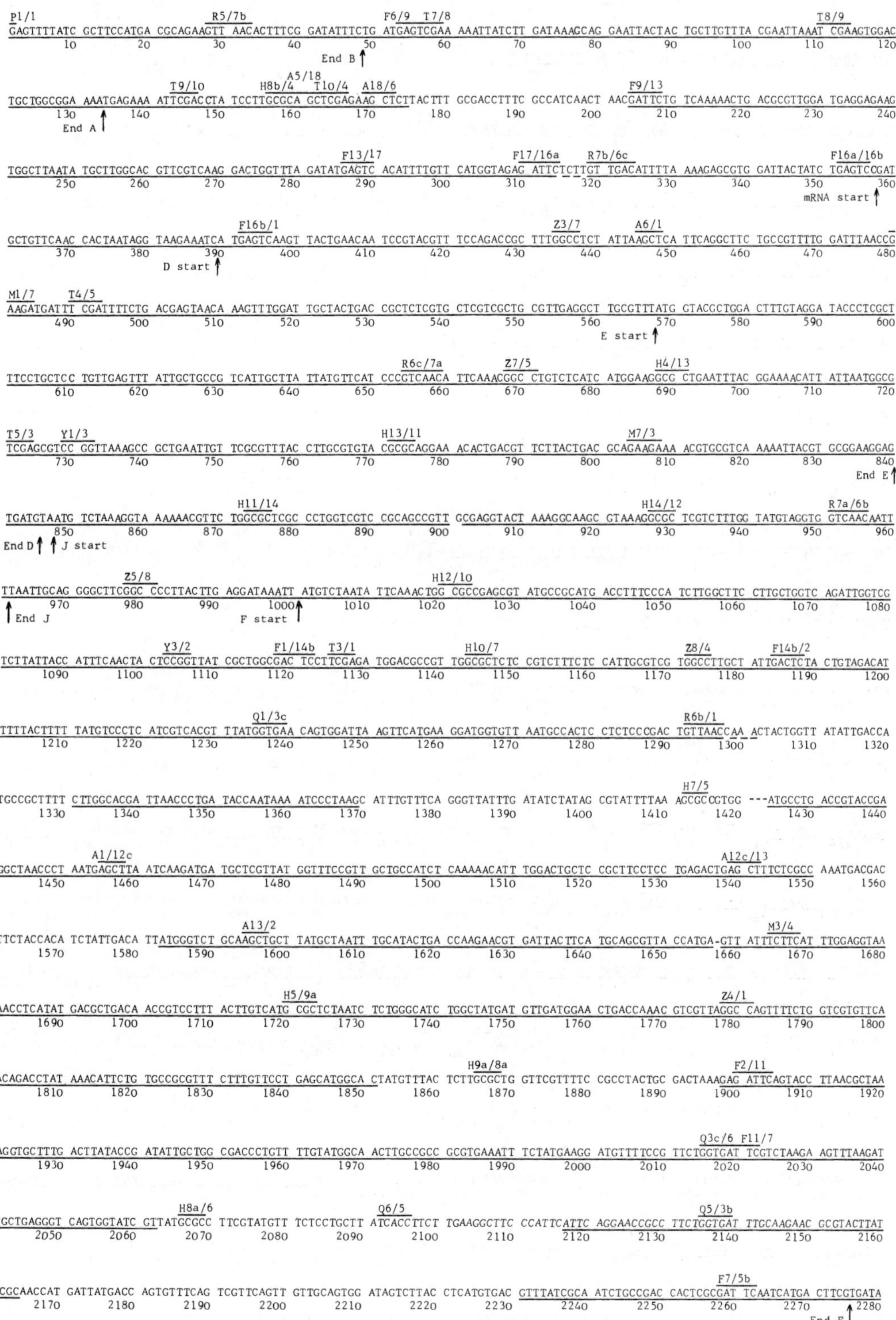

R1/9 H6/3
AAAGATTGAG TGTGAGGTTA TAACCGAAGC GGTAAAAATT TTAATTTTTG CCGCTGAGGG GTTGACCAAG CGAAGCGCGG TAGGTTTTCT GCTTAGGAGT TTAATCATGT TTCAGACTTT
2290 2300 2310 2320 2330 2340 2350 2360 2370 2380 2390 2400
G start ↑

A2/16 A16/15a M4/10 A15a/3 R9/10
TATTTCTCGC CACAATTCAA ACTTTTTTTC TGATAAGCTG GTTCTCACTT CTGTTACTCC AGCTTCTTCG GCACCTGTTT TACAGACACC TAAAGCTACA TCGTCAACGT TATATTTTGA
2410 2420 2430 2440 2450 2460 2470 2480 2490 2500 2510 2520

M10/9 R10/2
TAGTTTGACG GTTAATGCTG GTAATGGTGG TTTTCTTCAT TGCATTCAGA TGGATACATC TGTCAACGCC GCTAATCAGG TTGTTTCAGT TGGTGCTGAT ATTGCTTTTG ATGCCGACCC
2530 2540 2550 2560 2570 2580 2590 2600 2610 2620 2630 2640

F5b/8 M9/2
TAAATTTTTT GCCTGTTTGG TTCGCTTTGA GTCTTCTTCG GTTCCGACTA CCCTCCCGAC TGCCTATGAT GTTTATCCTT TGGATGGTCG CCATGATGGT GGTTATTATA CCGTCAAGGA
2650 2660 2670 2680 2690 2700 2710 2720 2730 2740 2750 2760

Y2/5 F8/4
CTGTGTGACT ATTGACGTCC TTCCCCGTAC GCCGGGCAAT AACGTCTACG TTGGTTTCAT GGTTTGGTCT AACTTTACCG CTACTAAATG CCGCGGATTG GTTTCGCTGA ATCAGGTTAT
2770 2780 2790 2800 2810 2820 2830 2840 2850 2860 2870 2880

Q3b/4 H3/2
TAAAGAGATT ATTTGTCTCC AGCCACTTAA GTGAGGTGAT TTATGTTTGG TGCTATTGCT GGCGGTATTG CTTCTGCTCT TGCTGGTGGC GCCATGTCTA AATTGTTTGG AGGCGGTCAA
2890 2900 2910 2920 2930 2940 2950 2960 2970 2980 2990 3000
End G ↑ H start ↑

Y5/4 Q4/7 Q7/2
AAAGCCGCCT CCGGTGGCAT TCAAGGTGAT GTGCTTGCTA CCGATAACAA TACTGTAGGC ATGGGTGATG CTGGTATTAA ATCTGCCATT CAAGGCTCTA ATGTTCCTAA CCCTGATGAG
3010 3020 3030 3040 3050 3060 3070 3080 3090 3100 3110 3120

Z1/2 A3/9
GCCGCCCCTA GTTTTGTTTC GTGTGCTATT GCTAAAGCTG GTAAAGGACT TCTTGAAGGT ACGTTGCAGG CTGGCACTTC TGCCGTTTCT GATAAGTTGC TTGATTTGGT TGGACTTGGT
3130 3140 3150 3160 3170 3180 3190 3200 3210 3220 3230 3240

A9/12d R2/6a
GGCAAGTCTG CCGCTGATAA AGGAAAGGAT ACTCGTGATT ATCTTGCTGC TGCATTTCCT GAGCTTAATG CTTGGGAGCG TGCTGGTGCT GATGCTTCCT CTGCTGGTAT GGTTGACGCC
3250 3260 3270 3280 3290 3300 3310 3320 3330 3340 3350 3360

Y4/1 F4/14a A12d/7c F14a/12
GGATTTGAGA ATCAAAAAGA GCTTACTAAA ATGCAACTGG ACAATCAGAA AGAGATTGCC GAGATGCAAA ATGAGACTCA AAAAGAGATT GCTGGCATTC AGTCGGCGAC TTCACGCCAG
3370 3380 3390 3400 3410 3420 3430 3440 3450 3460 3470 3480

F12/10
AATACGAAAG ACCAGGTATA TGCACAAAAT GAGATGCTTG CTTATTC-AC AGAAGGAGTC TACTGCTGCG TTGCGTCTAT TATGGAAAAC ACCAATCTTT CCAAGCAACA GCAGGTTTCC
3490 3500 3510 3520 3530 3540 3550 3560 3570 3580 3590 3600

H2/9b A7c/8 F10/15 R6a/4
GAGATTATGC GCCAAATGCT TACTCAAGCT CAAACGGCTG GTCAGTATTT TACCAATGAC CAAATCAAAG AAATGACTCG CAAGGTTAGT GCTGAGGTTG ACTTAGTTCA TCAGCAAACG
3610 3620 3630 3640 3650 3660 3670 3680 3690 3700 3710 3720

F15/5c M2/5 H9b/1 A8/14
CAGAATCAGC GGTATGGCTC TTCTCATATT GGCGCTACTG CAAAGGATAT TTCTAATGTC GTCACTGATG CTGCTTCTGG TGTGGTTGAT ATTTTTCATG GTATTGATAA AGCTGTTGCC
3730 3740 3750 3760 3770 3780 3790 3800 3810 3820 3830 3840

A14/7b
GATACTTGGA ACAATTTCTG GAAAGACGGT AAAGCTGATG GTATTGGCTC TAATTTGTCT AGGAAATAAC CGTCAGGATT GACACCCTCC CAATTGTATG TTTTCATGCC TCCAAATCTT
3850 3860 3870 3880 3890 3900 3910 3920 3930 3940 3950 3960
End H ↑ mRNA start ↑

T1/6
GGAGGCTTTT TTATGGTTCG TTCTTATTAC CCTTCTGAAT GTCACGCTGA TTATTTTGAC TTTGAGCGTA TCGAGGCTCT TAAACCTGCT ATTGAGGCTT GTGGCATTTC TACTCTTTCT
3970 3980 3990 4000 4010 4020 4030 4040 4050 4060 4070 4080
A start ↑

A7b/7a M5/8 F5c/3 T6/2 Q2/3a R4/3 Z2/6b
CAATCCCCAA TGCTTGGCTT CCATAAGCAG ATGGATAACC GCATCAAGCT CTTGGAAGAG ATTCTGTCTT TTCGTATGCA GGGCGTTGAG TTCGATAATG GTGATATGTA TGTTGACGGC
4090 4100 4110 4120 4130 4140 4150 4160 4170 4180 4190 4200

CATAAGGCTG CTTCTGACGT TCGTGATGAG TTTGTATCTG TTACTGAGAA GTTAATGGAT GAATTGGCAC AATGCTACAA TGTGCTCCCC CAACTTGATA TTAATAACAC TATAGACCAC
4210 4220 4230 4240 4250 4260 4270 4280 4290 4300 4310 4320

M8/6 A7a/4
CGCCCCGAAG GGGACGAAAA ATGGTTTTTA GAGAACGAGA AGACGGTTAC GCAGTTTTGC AAGCTGGCTG CTGAACGCCC TCTTAAGGAT ATTCGCGATG AGTATAATTA CCCCAAAAAG
4330 4340 4350 4360 4370 4380 4390 4400 4410 4420 4430 4440

Z6b/6a
AAAGGTATTA AGGATGAGTG TTCAAGATTG CTGGAGGCCT CCACTAAGAT ATCGCGTAGA GGCTTTGCTA TTCAGCGTTT GATGAATGCA ATGCGACAGG CTCATGCTGA TGGTTGGTTT
4450 4460 4470 4480 4490 4500 4510 4520 4530 4540 4550 4560

ATCGTTTTTG ACACTCTCAC GTTGGCTGAC GACCGATTAG AGGCGTTTTA TGATAATCCC AATGCTTTGC GTGACTATTT TCGTGATATT GGTCGTATGG TTCTTGCTGC CGAGGGTCGC
4570 4580 4590 4600 4610 4620 4630 4640 4650 4660 4670 4680

```
                                                                   R1/9                 H6/3
AAAGATTGAG TGTGAGGTTA TAACCGAAGC GGTAAAAATT TTAATTTTTG CCGCTGAGGG GTTGACCAAG CGAAGCGCGG TAGGTTTTCT GCTTAGGAGT TTAATCATGT TTCAGACTTT
      2290       2300       2310       2320       2330       2340       2350       2360       2370       2380       2390       2400
                                                                                                                 G start ↑

                                  A2/16                   A16/15a M4/10                        A15a/3     R9/10
TATTTCTCGC CACAATTCAA ACTTTTTTTC TGATAAGCTG GTTCTCACTT CTGTTACTCC AGCTTCTTCG GCACCTGTTT TACAGACACC TAAAGCTACA TCGTCAACGT TATATTTTGA
      2410       2420       2430       2440       2450       2460       2470       2480       2490       2500       2510       2520

                                  M10/9                            R10/2
TAGTTTGACG GTTAATGCTG GTAATGGTGG TTTTCTTCAT TGCATTCAGA TGGATACATC TGTCAACGCC GCTAATCAGG TTGTTTCAGT TGGTGCTGAT ATTGCTTTTG ATGCCGACCC
      2530       2540       2550       2560       2570       2580       2590       2600       2610       2620       2630       2640

                            F5b/8  M9/2
TAAATTTTTT GCCTGTTTGG TTCGCTTTGA GTCTTCTTCG GTTCCGACTA CCCTCCCGAC TGCCTATGAT GTTTATCCTT TGGATGGTCG CCATGATGGT GGTTATTATA CCGTCAAGGA
      2650       2660       2670       2680       2690       2700       2710       2720       2730       2740       2750       2760

                                  Y2/5                                                                                 F8/4
CTGTGTGACT ATTGACGTCC TTCCCCGTAC GCCGGGCAAT AACGTCTACG TTGGTTTCAT GGTTTGGTCT AACTTTACCG CTACTAAATG CCGCGGATTG GTTTCGCTGA ATCAGGTTAT
      2770       2780       2790       2800       2810       2820       2830       2840       2850       2860       2870       2880

                                  Q3b/4                                                       H3/2
TAAAGAGATT ATTTGTCTCC AGCCACTTAA GTGAGGTGAT TTATGTTTGG TGCTATTGCT GGCGGTATTG CTTCTGCTCT TGCTGGTGGC GCCATGTCTA AATTGTTTGG AGGCGGTCAA
      2890       2900       2910       2920       2930       2940       2950       2960       2970       2980       2990       3000
                       End G ↑  H start ↑

         Y5/4        Q4/7                         Q7/2
AAAGCCGCCT CCGGTGGCAT TCAAGGTGAT GTGCTTGCTA CCGATAACAA TACTGTAGGC ATGGGTGATG CTGGTATTAA ATCTGCCATT CAAGGCTCTA ATGTTCCTAA CCCTGATGAG
      3010       3020       3030       3040       3050       3060       3070       3080       3090       3100       3110       3120

Z1/2                              A3/9
GCCGCCCCTA GTTTTGTTTC GTGTGCTATT GCTAAAGCTG GTAAAGGACT TCTTGAAGGT ACGTTGCAGG CTGGCACTTC TGCCGTTTCT GATAAGTTGC TTGATTTGGT TGGACTTGGT
      3130       3140       3150       3160       3170       3180       3190       3200       3210       3220       3230       3240

                                                            A9/12d                                                R2/6a
GGCAAGTCTG CCGCTGATAA AGGAAAGGAT ACTCGTGATT ATCTTGCTGC TGCATTTCCT GAGCTTAATG CTTGGGAGCG TGCTGGTGCT GATGCTTCCT CTGCTGGTAT GGTTGACGCC
      3250       3260       3270       3280       3290       3300       3310       3320       3330       3340       3350       3360

Y4/1  F4/14a     A12d/7c                                              F14a/12
GGATTTGAGA ATCAAAAAGA GCTTACTAAA ATGCAACTGG ACAATCAGAA AGAGATTGCC GAGATGCAAA ATGAGACTCA AAAAGAGATT GCTGGCATTC AGTCGGCGAC TTCACGCCAG
      3370       3380       3390       3400       3410       3420       3430       3440       3450       3460       3470       3480

                                                       F12/10
AATACGAAAG ACCAGGTATA TGCACAAAAT GAGATGCTTG CTTATTC-AC AGAAGGAGTC TACTGCTGCG TTGCGTCTAT TATGGAAAAC ACCAATCTTT CCAAGCAACA GCAGGTTTCC
      3490       3500       3510       3520       3530       3540       3550       3560       3570       3580       3590       3600

   H2/9b                 A7c/8                                          F10/15                 R6a/4
GAGATTATGC GCCAAATGCT TACTCAAGCT CAAACGGCTG GTCAGTATTT TACCAATGAC CAAATCAAAG AAATGACTCG CAAGGTTAGT GCTGAGGTTG ACTTAGTTCA TCAGCAAACG
      3610       3620       3630       3640       3650       3660       3670       3680       3690       3700       3710       3720

F15/5c          M2/5     H9b/1                                                                                         A8/14
CAGAATCAGC GGTATGGCTC TTCTCATATT GGCGCTACTG CAAAGGATAT TTCTAATGTC GTCACTGATG CTGCTTCTGG TGTGGTTGAT ATTTTTCATG GTATTGATAA AGCTGTTGCC
      3730       3740       3750       3760       3770       3780       3790       3800       3810       3820       3830       3840

                                  A14/7b
GATACTTGGA ACAATTTCTG GAAAGACGGT AAAGCTGATG GTATTGGCTC TAATTTGTCT AGGAAATAAC CGTCAGGATT GACACCCTCC CAATTGTATG TTTTCATGCC TCCAAATCTT
      3850       3860       3870       3880       3890       3900       3910       3920       3930       3940       3950       3960
                                                                  End H ↑                                      mRNA start ↑

                                                                              T1/6
GGAGGCTTTT TTATGGTTCG TTCTTATTAC CCTTCTGAAT GTCACGCTGA TTATTTTGAC TTTGAGCGTA TCGAGGCTCT TAAACCTGCT ATTGAGGCTT GTGGCATTTC TACTCTTTCT
      3970       3980       3990       4000       4010       4020       4030       4040       4050       4060       4070       4080
         ↑ A start

                                             A7b/7a    M5/8  F5c/3                             T6/2      Q2/3a          R4/3  Z2/6b
CAATCCCCAA TGCTTGGCTT CCATAAGCAG ATGGATAACC GCATCAAGCT CTTGGAAGAG ATTCTGTCTT TTCGTATGCA GGGCGTTGAG TTCGATAATG GTGATATGTA TGTTGACGGC
      4090       4100       4110       4120       4130       4140       4150       4160       4170       4180       4190       4200

CATAAGGCTG CTTCTGACGT TCGTGATGAG TTTGTATCTG TTACTGAGAA GTTAATGGAT GAATTGGCAC AATGCTACAA TGTGCTCCCC CAACTTGATA TTAATAACAC TATAGACCAC
      4210       4220       4230       4240       4250       4260       4270       4280       4290       4300       4310       4320

                                        M8/6                  A7a/4
CGCCCCGAAG GGGACGAAAA ATGGTTTTTA GAGAACGAGA AGACGGTTAC GCAGTTTTGC AAGCTGGCTG CTGAACGCCC TCTTAAGGAT ATTCGCGATG AGTATAATTA CCCCAAAAAG
      4330       4340       4350       4360       4370       4380       4390       4400       4410       4420       4430       4440

                                  Z6b/6a
AAAGGTATTA AGGATGAGTG TTCAAGATTG CTGGAGGCCT CCACTAAGAT ATCGCGTAGA GGCTTTGCTA TTCAGCGTTT GATGAATGCA ATGCGACAGG CTCATGCTGA TGGTTGGTTT
      4450       4460       4470       4480       4490       4500       4510       4520       4530       4540       4550       4560

ATCGTTTTTG ACACTCTCAC GTTGGCTGAC GACCGATTAG AGGCGTTTTA TGATAATCCC AATGCTTTGC GTGACTATTT TCGTGATATT GGTCGTATGG TTCTTGCTGC CGAGGGTCGC
      4570       4580       4590       4600       4610       4620       4630       4640       4650       4660       4670       4680
```

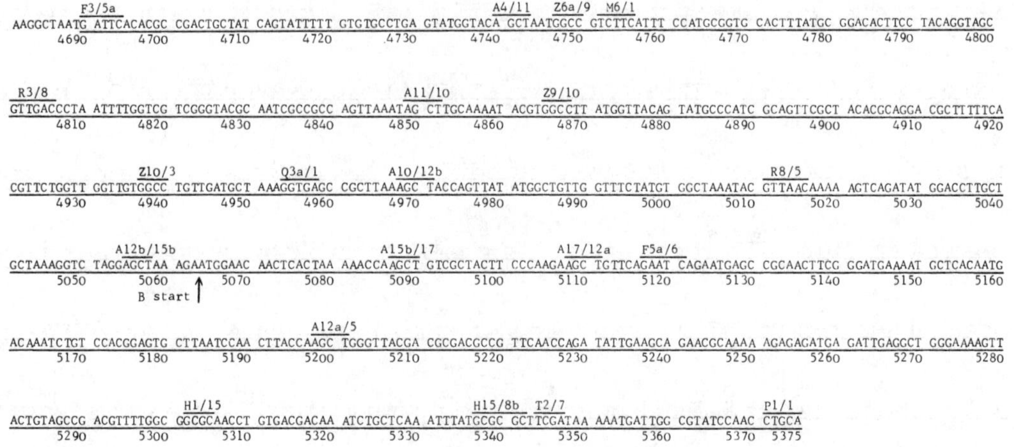

Fig. 1. A provisional nucleotide sequence for the DNA of bacteriophage ΦX174 *am*3 *cs*70. Solid underlining indicates sequences that are fully confirmed; sequences with no underlining probably do not contain more than one mistake per 50 residues. Broken underlining indicates more uncertain sequences. Restriction enzyme recognition sites are indicated (for key to single letter enzyme code see legend to Fig. 2), as are mRNA starts and protein initiation and termination sites. Nucleotides 4,127 to 4,201 have been independently sequenced by van Mansfield *et al.*[58]. The *am*3 codon is at position 587.

More extensive ΦX sequences were obtained using partial degradation techniques, particularly with endonuclease IV (refs 10 and 11). Ziff *et al.*[12,13] used this enzyme in conditions of partial hydrolysis to obtain fragments 50–200 nucleotides long which were purified as ^{32}P-labelled material by electrophoresis on polyacrylamide gels. The fragments came from the same region of the genome and the sequence of a 48-nucleotide long fragment (band 6, positions 1,047–1,094) was determined using mainly further degradation with endonuclease IV and partial exonuclease digestions.

Another 50-nucleotide long fragment was obtained by Robertson *et al.*[14] as a ribosome binding site. The viral (or plus) strand DNA of ΦX has the same sequence as the mRNA and, in certain conditions, will bind ribosomes so that a protected fragment can be isolated and sequenced. Only one major site was found. By comparison with the amino acid sequence data it was found that this ribosome binding site sequence coded for the initiation of the gene *G* protein[15] (positions 2,362–2,413).

At this stage sequencing techniques using primed synthesis with DNA polymerase were being developed[16] and Schott[17] synthesised a decanucleotide with a sequence complementary to part of the ribosome binding site. This was used to prime into the intercistronic region between the *F* and *G* genes, using DNA polymerase and ^{32}P-labelled triphosphates[18]. The ribo-substitution technique[16] facilitated the sequence determination of the labelled DNA produced. This decanucleotide-primed system was also used to develop the plus and minus method[1]. Suitable synthetic primers are, however, difficult to prepare and as DNA fragments generated by restriction enzymes are more readily available these have been used for most of the work reported here.

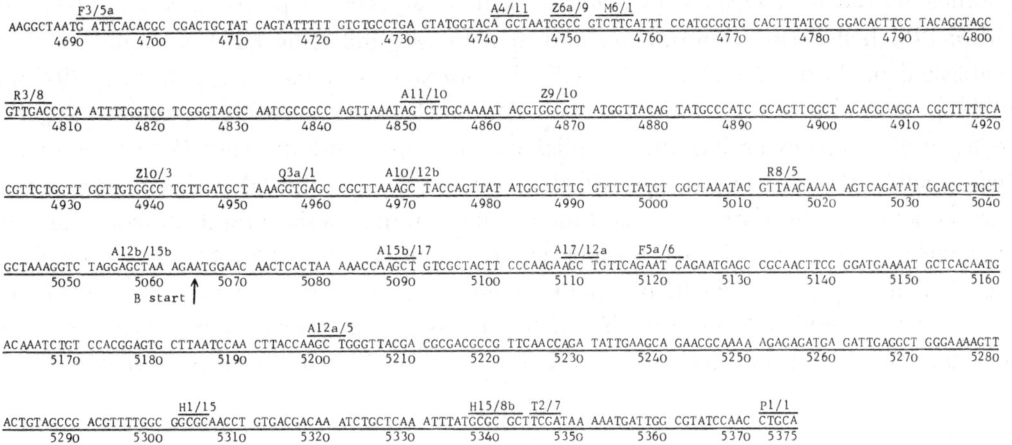

图 1. 噬菌体 ΦX174 *am*3 *cs*70 DNA 的暂定核苷酸序列。下面划实线的序列表示已经被完全确定；下面未划线的序列表示每 50 个核苷酸残基包含不超过 1 个错误碱基。下面划虚线的表示不确定的序列。限制性内切酶识别位点（酶编码的单字母缩写含义见图 2 注），mRNA（信使核糖核酸）转录起始位点，蛋白质翻译起始位点和终止位点均已标明。核苷酸 4,127 到 4,201 已经由范曼斯菲尔德等人[58] 独立完成测序。*am*3 密码子在第 587 位。

利用部分降解技术，尤其是利用核酸内切酶 IV（参考文献 10 和 11），人们测定了更多的 ΦX 序列。齐夫等 [12,13] 利用核酸内切酶 IV 对核酸序列进行部分水解，通过聚丙烯酰胺凝胶电泳的方法纯化获得了 ^{32}P 标记的 50~200 个核苷酸长度的序列。这些片段来自基因组的相同区域，其中片段长度为 48 个核苷酸的序列（条带 6，1,047~1,094 区域）主要是利用核酸内切酶 IV 的进一步降解和外切酶的部分消化来测定的。

罗伯逊等[14] 发现另外一个 50 个核苷酸长度的片段是核糖体结合位点。ΦX 的病毒（正）链 DNA 与其 mRNA 具有相同的序列，并且在某些条件下能与核糖体结合，因此这段受保护的序列才得以分离和测序。用这一方法只发现了一个主要的核糖体结合位点。与氨基酸序列数据比较发现，这段核糖体结合位点的序列（2,362~2,413 区域）编码基因 *G* 蛋白的起始位点 [15]。

在这一时期，利用 DNA 聚合酶进行引物合成的测序技术得到发展 [16]，肖特[17] 合成了一段 10 个核苷酸的序列，该序列与部分核糖体结合位点的序列互补。用这一序列作为引物，利用 DNA 聚合酶和 ^{32}P 标记的三磷酸盐 [18] 对基因 *F* 和 *G* 之间的顺反子间区进行合成。再用核苷酸置换技术 [16] 对带标记的 DNA 产物进行序列鉴定。这种用 10 个核苷酸作为引物的系统也被用于发展加减测序法 [1]。虽然合成完全相配的引物很不容易，但是利用限制性内切酶产生 DNA 片段却容易，因此这个方法在本文报道的大部分工作中都有应用。

Another approach to DNA sequencing is to make an RNA copy using RNA polymerase with α-^{32}P-labelled ribotriphosphates and then to determine the RNA sequence by more established methods. Blackburn[19,20] used this approach on intact single-stranded ΦX and on fragments obtained by digestion with endonuclease IV or with restriction enzymes. Sedat *et al.*[21] were extending their studies on the larger endonuclease IV fragments and their results, taken in conjunction with the transcription of the DNA fragments[20], amino acid sequence of the F protein[22], and the plus and minus method results, made it possible to deduce a sequence of 281 nucleotides (positions 1,016–1,296, Fig. 1) within the *F* gene[23]. Transcription of *Hind*II fragment 10, amino acid sequence data in the *G* gene, and the plus and minus method using *Hind*II fragments 2 and 10 as primers, gave a sequence of 195 nucleotides (positions 2,387–2,582, Fig. 1) at the N terminus of gene *G* (ref. 24).

The "Plus and Minus" Method

Further work on the ΦX sequence has been done using chiefly the plus and minus method primed with restriction fragments. Figure 2 shows the various restriction enzymes used and the fragment maps for each (refs 25–30 and C.A.H., submitted for publication, and N.L.B., C.A.H. and M.S., submitted for publication).

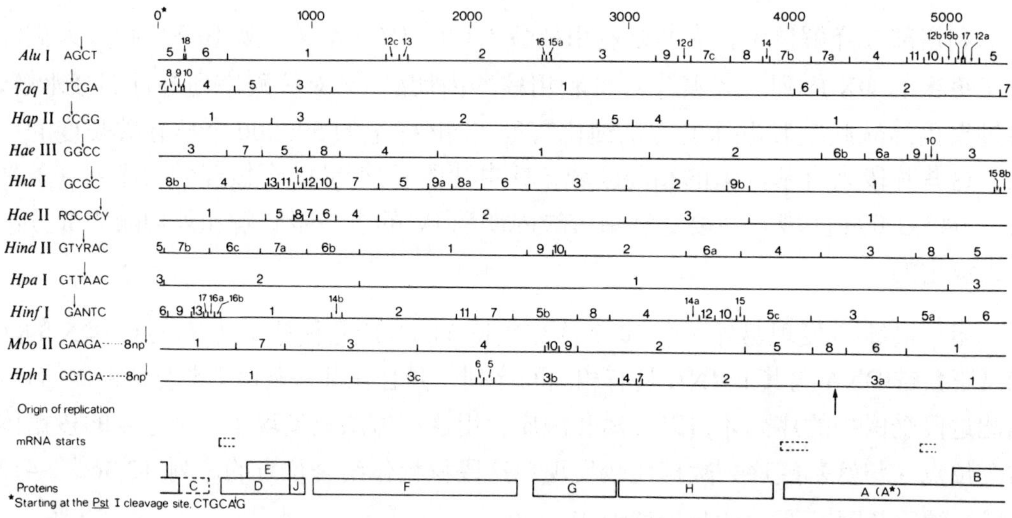

Fig. 2. Fragment maps of restriction enzymes used in the sequence analysis of ΦX174 *am*3 RFI DNA. Fragment maps of ΦX174 have been prepared for *Hind*II (R), *Hae*III (Z) and *Hpa*I+II by Lee and Sinsheimer[25], *Hin*HI and *Hap*II (Y) by Hayashi and Hayashi[26], and for *Alu*I (A) by Vereijken *et al.*[27] and for *Pst*I (P) by Brown and Smith[30]. B.G.B., G.M.A., C.A.H. and D. Jaffe prepared the *Hinf*I (F) map, C.A.H. the *Hph*I (Q) map, and Jeppesen *et al.*[28] the *Hha*I (H), *Alu*I, *Hae*II and *Hap*II maps by using a rapid method depending on priming with DNA polymerase. A rapid two-dimensional hybridisation technique has been developed by C.A.H. (submitted for publication) and recently used for mapping *Mbo*II (M) (N.L.B., C.A.H., and M.S., submitted for publication) and *Taq*I (T)[29]. *Hha*I and *Hinf*I maps have also been prepared by Baas *et al.*[52].

另一种 DNA 测序的方法是利用 RNA 聚合酶和 α–^{32}P 标记的三磷酸核苷酸合成一段 RNA 拷贝，然后利用更加成熟的方法测定 RNA 序列。布莱克本 [19,20] 利用这种方法对完整的 ΦX 单链 DNA 以及用核酸内切酶 IV 或限制性内切酶消化所得的片段进行了测序。赛达特 [21] 等将他们的研究拓展到更长的核酸内切酶 IV 酶切片段，他们的研究结果与 DNA 片段的转录 [20]、F 蛋白的氨基酸序列 [22] 以及加减测序法的结果一起，使得在 *F* 基因中 [23] 推导出一段 281 个核苷酸的序列（1,016~1,269 区域，图 1）成为可能。进而根据 *Hind*II 酶切片段 10 的转录和 *G* 基因的氨基酸序列数据，以及利用 *Hind*II 片段 2 和 10 作为引物的加减法测序，又推断出 *G* 基因 N 末端的 195 个核苷酸的序列（2,387~2,582 区域，图 1）（参考文献 24）。

"加减法"测序

我们用限制性酶切片段作为引物，首先主要使用加减法测序对 ΦX 序列进行了进一步的研究。图 2 显示了所用的各种限制性内切酶和每一片段的图谱（参考文献 25~30，哈奇森，已投稿，以及布朗、哈奇森和史密斯，已投稿）。

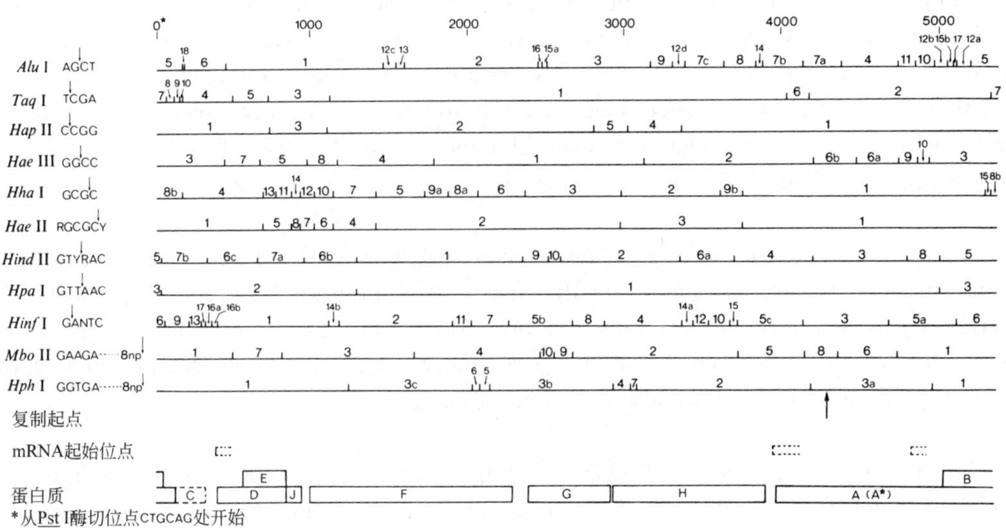

图 2. 在 ΦX174 *am*3 RFI DNA 序列分析中用到的限制性内切酶的片段图谱。李和辛西默[25] 完成了 *Hind*II (R)、*Hae*III (Z)、*Hpa*I+II 的图谱测定，林昌树和林玛丽 [26] 完成了 *Hin*HI 和 *Hap*II (Y) 的图谱测定，费赖伊肯等[27] 完成了 *Alu*I (A) 的图谱测定，布朗和史密斯[30] 完成了 *Pst*I (P) 的图谱测定。通过使用基于引物和 DNA 聚合酶的快速扩增方法，巴雷尔、艾尔、哈奇森和贾菲完成了 *Hinf*I (F) 的图谱，哈奇森完成了 *Hph*I (Q) 的图谱，杰普森等[28] 完成了 *Hha*I (H)、*Alu*I、*Hae*II 和 *Hap*II 的图谱。哈奇森发展了一种快速的二维杂交技术（已投稿），并于最近用于 *Mbo*II (M)（布朗、哈奇森和史密斯，已投稿）和 *Taq*I (T)[29] 的图谱测定。巴斯等[52] 完成了 *Hha*I 和 *Hinf*I 的图谱测定。

Figure 1 shows the combined results of the sequence work to date. The sequence is numbered from the single cleavage site of the restriction enzyme *Pst*I. As with other methods of sequencing nucleic acids, the plus and minus technique used by itself cannot be regarded as a completely reliable system and occasional errors may occur. Such errors and uncertainties can only be eliminated by more laborious experiments and, although much of the sequence has been so confirmed, it would probably be a long time before the complete sequence could be established. We are not certain that there is any scientific justification for establishing every detail and, as it is felt that the results may be useful to other workers, it has been decided to publish the sequence in its present form.

As template we have used both the viral (plus) and complementary (minus) strands of ΦX. Usually it is possible to determine a sequence with a single primer starting at about 15–100 nucleotides from the appropriate restriction enzyme site. In a particularly good experiment the sequence can be read out to 150–200 nucleotides but the results may become less reliable. Most sequences have been derived by priming on both strands; this allows more confidence than when only one strand could be used.

A useful method for confirming runs of the same nucleotide is depurination of ^{32}P-labelled small restriction enzyme fragments or of products of the DNA polymerase priming experiments (ref. 31 and N.L.B. and M.S., in preparation). The most satisfactory way of confirming the DNA sequences is through amino acid sequence data. As the methods used are entirely unrelated, the results of the two approaches complement each other very well and therefore complete sequences can usually be deduced from incomplete data obtained by each method. The complete sequence of genes *G* (ref. 32), *D* (ref. 33), *J* (ref. 33 and Freymeyer, unpublished) and most of *F* have been obtained in this way.

Many of the sequences in Fig. 1 have been amply confirmed and are regarded as established: these are indicated in the figure by underlining. Some sequences are considered to be reasonably accurate and probably contain no more than one mistake in every 50 nucleotides. Sequences that are particularly uncertain—either because of lack of data or conflicting results—are also indicated in Fig. 1.

In considering the sequence of ΦX174 as a functional unit it is convenient to begin in the region between the *H* and *A* genes and to continue around the DNA in the direction of transcription and translation.

A Promoter and Terminator

Sinsheimer *et al.*[34,35] and Axelrod[36] have determined the sequences of the 5′ end of three ΦX *in vitro* mRNA species and have located them on the restriction map. These sequences have been identified on the DNA sequence and one of them (AAATCTTGG) is found only at position 3,954 at which an *in vivo* unstable mRNA start has been located[37]. The sequence to the left of this has some characteristics of typical *E. coli* promoters[38] in that five out of the "ideal" TATPuATPu residues are present. Nearby, to the right of this mRNA initiation,

图 1 显示了到目前为止序列研究的综合结果。序列自限制性内切酶 *Pst*I 的单一切割位点开始编号。与其他测序方法测定的核苷酸序列相比，加减法测序技术自身不能视为一个可以完全可信的系统，可能偶尔会有错误发生。只有通过更多实验才能消除这些错误和不确定因素。尽管很多序列已经用这样的方法完成测序，但完整序列的全部完成可能还需要很长的时间。我们不很确定继续完成每一个细节的科学理由，但是觉得这些结果会对其他研究人员有所帮助，因此决定以目前的形式公布这个序列。

我们把 ΦX 的病毒（正）链和互补（负）链都用作模板。通常情况下，从合适的限制性内切酶切位点开始大概 15~100 核苷酸处，一个单一引物就可以测定一段序列。在特别好的实验中，能够读出 150~200 个核苷酸，但是这个结果的可信度比较低。这里对大部分序列通过引物进行了双链测定，这比仅仅测定单链具有更高的可信度。

确定一个核苷酸是否正确的一个有效方法是对 ^{32}P 标记的小的限制性内切酶切片段或者对 DNA 聚合酶扩增的产物进行脱嘌呤（参考文献 31 以及布朗和史密斯准备发表的结果）。验证 DNA 序列最可信的方法是通过氨基酸序列数据分析。由于运用的方法完全不相关，两种方法的结果彼此之间可以很好地互补，因此我们可以通过每种方法得到的不完整核苷酸数据来推断完整的序列。通过这种方法我们已经获得了 *G* 基因（参考文献 32）、*D* 基因（参考文献 33）和 *J* 基因（参考文献 33 以及弗雷迈耶未发表的结果）的全部序列，以及 *F* 基因的大部分序列。

我们已经充分证实了图 1 中的许多序列：这些在图中用下划线来标明。有些序列相当准确，可能每 50 个核苷酸中不会多于一个错误。特别不确定的序列是因为缺少数据或者是结果相互冲突，这在图 1 中也有标明。

考虑到 ΦX174 序列是一个功能单元，为方便起见以下将从基因 *H* 和基因 *A* 之间的区域开始，沿 DNA 转录和翻译方向进行介绍。

*A*启动子和终止子

辛西默等 [34,35] 和阿克塞尔罗德 [36] 已经在体外确定了 3 种 ΦX mRNA 的 5′ 端序列，并将它们定位在限制性酶切图谱上。这些序列已经在 DNA 序列上得到鉴定，其中一个序列（AAATCTTGG）被发现位于 3,954 位点处，是一个体内不稳定的 mRNA 起始处 [37]。这个序列左侧的序列具有大肠杆菌启动子的一些典型的特征 [38]，

however, is the sequence TTTTTTA which is similar to sequences found at the 3′ ends of a number of mRNAs (see ref. 39) and seems a likely signal for mRNA termination. The presence of a rho-independent termination site in this approximate position has been suggested[36,37], but the relative positions of the initiating and putative termination signals is rather surprising since the terminator for one mRNA would be expected to precede the initiator for the next. One possibility is that the T_6A might be acting as an "attenuator" involved in the control of mRNA production in a similar manner to that suggested for the tryptophan operon by Bertrand *et al.*[40]. If indeed it were acting as a transcription terminator one would expect a small RNA of 20 nucleotides to be produced, but no such product has yet been detected. Recent work, however, (Rosenberg, unpublished and ref. 41) indicates that termination may require the presence of a base-paired loop structure before the termination site. From the DNA sequence such a loop is probably present before the T_6A sequence, but in mRNA starting from the initiation site at position 3,954 this loop is not formed (Fig. 3). Therefore mRNA that had started at an earlier promoter and extended through the *H* gene would be expected to terminate here, whereas mRNA newly initiated at position 3,954 would not. This could be a way in which the phage has economised on the use of DNA—by having the ends of the two mRNAs overlapping.

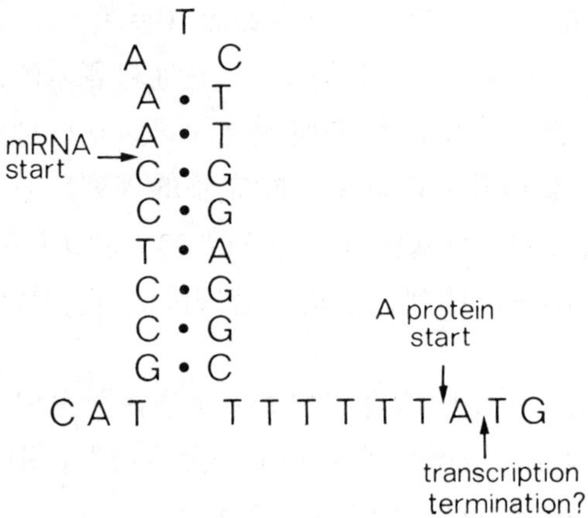

Fig. 3. Potential secondary structure at the *A* mRNA start.

The A Protein

Where the amino acid sequence is available there is no problem in relating the DNA sequence to its coding properties, but it is more difficult to do so in the absence of such data, as is the case for the A protein. One way of identifying the reading phase of the DNA is from the distribution of nonsense codons. Over a sufficiently long sequence that is known to be coding for a single protein there is usually one phase that contains no nonsense codons, and this is identified as the reading phase. This requires completely accurate determination of the DNA sequences however: omission of a single nucleotide may give completely erroneous results. Another approach is possible in the case of ΦX. The results with the *F* and *G* genes[23,24,32] showed an unexpectedly high frequency of

包含5个“理想的”TATPuATPu残基。然而这个mRNA起始处的右侧序列是TTTTTTA，这与出现在许多mRNA 的3′端的序列相似（见参考文献39），似乎是mRNA终止的信号。有人提出不依赖于rho因子的终止位点就在这个位点附近[36,37]，但起始位点和推测的终止信号的相对位置令人惊讶，因为一般认为一个mRNA的终止子应该在下一个启动子之前。一种可能是T_6A充当了“衰减子”来控制mRNA的生成，其调控模式与伯特兰等人[40]提出的色氨酸操纵子相似。如果它确实充当转录终止子，我们可以预计将产生一段20个核苷酸的小RNA，但目前为止还没有检测到这样的产物。然而最近的研究（罗森堡，未发表的结果及参考文献41）表明终止的发生可能需要在终止位点前存在一个碱基配对的环结构。在DNA序列中这样一个环可能出现在T_6A序列的前面，但在从3,954处起始的mRNA中并没有形成这个环（图3）。因此预计由更靠前的一个启动子启动并延伸通过*H*基因的mRNA可能终止于此，而新起始于3,954位点的mRNA并不在此终止。这可能是噬菌体节约使用DNA的一种方式——使两个mRNA的末端重叠。

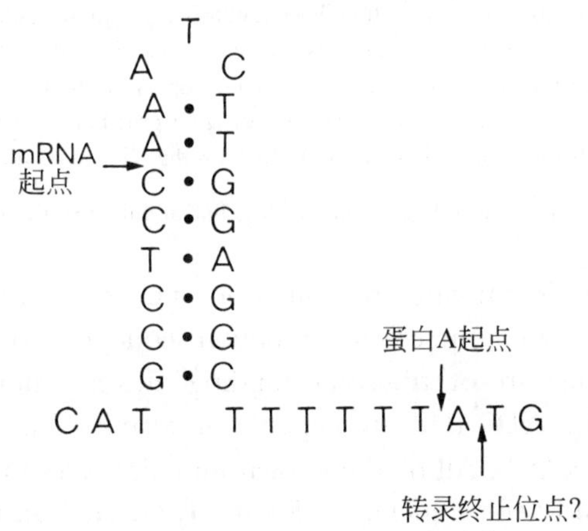

图3. *A*基因mRNA起点可能的二级结构

A 蛋白

如果已知氨基酸序列就很容易推断出DNA序列的编码特性，但是像蛋白A这样缺少这些数据时这样做就很困难。鉴别DNA阅读框的一个方法是依据无义密码子的分布。对于一段足够长并已知能够编码单一蛋白的序列，通常情况下会有一种读法不含有无义密码子，这就可以断定为是阅读框。这需要对DNA序列进行非常精确的测定：缺失一个核苷酸就会导致完全错误的结果。在ΦX中还有一种方法也是可行的。*F*基因和*G*基因的研究结果[23,24,32]显示以T结尾的密码子的频率出乎意料的高。因此编码区的一个倾向就是每个密码子的第三位核苷酸是T，这样就可以

codons ending in T. Therefore in a coding region there is a tendency for every third nucleotide to be a T and it is then possible to define the reading phase. Figure 4 illustrates how this characteristic was used to help determine the reading phase for the A protein and to identify its initiation codon at position 3,973. In a similar way the distribution of Ts may be used to identify errors in the DNA sequence, provided that such errors occur only infrequently.

		Ts in codon position		
		(1)	(2)	(3)
3,910	C C G . T C A . G G A . T T G . A C A . C C C . T C C . C A A . T T G . T A T .	5	2	1
		3	4	3*
3,940	G T T . T T C . A T G . C C T . C C A . A A T . C T T . G G A . G G C . T T T .	2	5	5
		3	5	7*
3,970	T T T . <u>A T G</u> . G T T . C G T . T C T . T A T . T A C . C C T . T C T . G A A .	5	3	7
		5	0	7*
4,000	T G T . C A C . G C T . G A T . T A T . T T T . G A C . T T T . G A G . C G T .	4	2	7

Fig. 4. Identification of the initiation codon for the A protein. Sequences of 30 nucleotides in the region in which the initiation was expected were written down and arbitrarily marked off in triplets. The number of T residues in the first position in each triplet was then counted and listed, and similarly the number of Ts in the second and third positions. The marked preference for Ts in position 3 in the last two lines, as compared with the first two lines, suggests that they are coding for protein and that the triplets are correctly marked off. The most likely initiation codon for the A protein is the ATG in position 3,973.

*These figures refer to the last five codons of the previous line and the first five of the next line.

A different approach to identifying the initiation site and reading phase in a coding sequence is by looking for a characteristic "initiation sequence". Shine and Dalgarno have shown that a common feature of ribosome binding sites is a number of nucleotides (at least three) preceding the ATG that are capable of forming base pairs with a sequence at the 3′ end of 16S rRNA[42,43]. All of the known initiation sites in ΦX174 that have been identified by direct amino acid sequencing (for the F, G, H, J and D proteins) satisfy this criterion (see Table 2) and the fact that the sequence preceding the ATG in position 3,973 also has this characteristic supports its identification as the initiation site for the A protein.

If, as has been suggested[37], some mRNA from the previous promoter does extend beyond the hairpin structure, initiation of A protein synthesis may be controlled by the inclusion of the region complementary to the 16S rRNA in the hairpin loop. This could explain the presence of two types of mRNA covering the *A* cistron, as suggested by Hayashi *et al.*[37]—one unstable and active and the other stable but inactive. The former would be initiated at the A promoter, the latter at an earlier promoter and result from "read-through" at the terminator. The postulated reading frame for the A protein was confirmed by sequencing amber mutants mapping in the N-terminal region of gene *A*. *am*86 proved to be a C → T change at position 4,108 and *am*33 a C → T change at position 4,372. These both result in formation of an amber codon (TAG) in the same reading frame as the proposed

确定阅读框。图 4 阐释了如何利用这个特点来鉴别 A 蛋白的阅读框并确定其起始密码子在 3,973 位点处。与这一方式相似，如果连续是 T 核苷酸的频率非常低的话，多个 T 的连续分布可以用来判定 DNA 序列测定中的错误。

位置	序列	密码子中T的位置 (1)	(2)	(3)
3,910	CCG.TCA.GGA.TTG.ACA.CCC.TCC.CAA.TTG.TAT.	5	2	1
		3	4	3*
3,940	GTT.TTC.ATG.CCT.CCA.AAT.CTT.GGA.GGC.TTT.	2	5	5
		3	5	7*
3,970	TTT.ATG.GTT.CGT.TCT.TAT.TAC.CCT.TCT.GAA.	5	3	7
		5	0	7*
4,000	TGT.CAC.GCT.GAT.TAT.TTT.GAC.TTT.GAG.CGT.	4	2	7

图 4. 蛋白A起始密码子的确定。将预计含有起始位点的30个核苷酸序列按照三联体任意划分。将在每个三联体中第一点为T的数量记录下来，再同样将第二点和第三点出现T的数量也记录下来。将最后两行的第三点优先为T的和前两行比较，表明它们编码蛋白并且三联体划分正确，A蛋白最可能的起始位点是3,973处的ATG。

*这些数字参考前一行后5个密码子和后面一行的前5个密码子。

鉴别编码序列的起始位点和阅读框的另一种方法是寻找“起始序列”的特征。夏因和达尔加诺研究发现核糖体结合位点的一个共同特征就是 ATG 前面存在若干核苷酸(至少有 3 个)可以与 16S rRNA 3′ 端的一段序列形成碱基配对 [42,43]。在 ΦX174 中，通过直接氨基酸序列分析（如蛋白 F、G、H、J 和 D）确定的所有已知起始位点都满足这一特征（见表 2)，事实上在 3,973 位点处的 ATG 之前的序列也具有这一特征，证明了此处正是蛋白 A 的起始位点。

按前面所提及的 [37]，如果来自前一启动子的一些 mRNA 确实延伸并超越了发夹结构，那么发夹环中与 16S rRNA 互补的序列就可以调控蛋白 A 的合成。这样就可以解释林昌树等 [37] 提出的观点：*A* 顺反子中出现两种 mRNA——一个不稳定但有活性，另一个稳定但没有活性。前者可以在 A 启动子处被激活，后者则在靠前一个启动子处被激活，并把终止子“读过去”了。A 蛋白的推测阅读框通过测定基因 *A* 的 N 末端区域的琥珀型突变体而获得确认。*am*86 突变体的 4,108 处 C 变成 T，*am*33 突变体的 4,372 处 C 变成 T。这两者都导致了在前面提到的起始密码子 ATG 的同一个阅读框中形成了一个琥珀（终止）密码子（TAG），阅读框序列延续到终止密码

initiating ATG and the sequence continues to the termination codon at position 133. The A protein, which is the largest coded by ΦX174, is thus 512 amino acids long with a molecular weight of 56,000, in good agreement with SDS gel estimations (see refs 4 and 44). The A* protein, with a molecular weight of about 35,000, is believed to result from an internal translational start in the *A* gene, in the same reading phase[45]. From consideration of possible ribosome binding sequences[42,43] the ATG in position 4,657 seems to be the most likely initiation site for the A* protein.

The Origin of Replication

The origin of ΦX viral strand DNA synthesis has been located in gene *A*, in restriction fragment Z6b (ref. 46). This origin, while coding for part of the A protein, probably corresponds to the position of the plus strand nick made by the same protein[44]. Gaps in this region that are found in replicating double-stranded (RF) DNAs are probably related to the position of the nick. Eisenberg *et al.*[47] have investigated such gaps by depurination analysis and identified, in particular, the product C_6T. The sequence CTC_5 is found in position 4,285 (Fig. 1) and the location of the origin in this region agrees precisely with the results of Baas *et al.*[46]. It is not possible at present to identify the actual position of the origin nick. The region shows no apparent secondary structure or symmetrical sequences, although there is an AT-rich region (4,298–4,307) between two GC-rich regions which might be of significance. Such a region is found near the origin of replication of SV40 DNA (ref. 48).

Table 1. ΦX174 coding capacity

Gene	Protein molecular weight from SDS gels*	Number of nucleotides (Fig. 1)	Protein molecular weight from sequence information
A	55,000–67,000	1,536	56,000
(*A**)	35,000		
B	19,000–25,000	(360)†	13,845‡
C	7,000		
D	14,500	456	16,811‡
E	10,000–17,500	(273)†	9,940
J	5,000	114	4,097‡
F	48,000	1,275	46,400
G	19,000	525	19,053‡
H	37,000	984	35,800
Non-coding and *C*		485	
Total		5,375	

* See ref. 4.

† Values in parenthesis are overlapping sequences and therefore not included in the addition to obtain the total length of DNA.

‡ These values are calculated from the amino acid sequence (in the case of B deduced from the nucleotide sequence). The others are derived using the formula

$$\text{Protein molecular weight} = \frac{\text{No. of nucleotides}}{3\times0.00915}$$

子所在的 133 处。由此推算 A 蛋白是 ΦX174 编码的最大蛋白，由 512 个氨基酸组成，分子量为 56,000，这通过聚丙烯酰胺凝胶电泳得到了很好的验证（见参考文献 4 和参考文献 44）。我们认为分子量为 35,000 的 A* 蛋白是从基因 *A* 内部的一个起点所进行的同一阅读框的翻译所致[45]。考虑到可能的核糖体结合位点[42,43]，4,657 处的 ATG 最可能是 A* 蛋白的起始位点。

复制起点

ΦX 病毒链 DNA 的复制起点被定位在基因 *A* 中限制性片段 Z6b 处（参考文献 46）。当编码 A 蛋白时，这个起点可能就是同一蛋白所形成的正链 DNA 的单链切口的位置[44]。在这一区域发现的正在复制的双链 DNA 的缺口可能与单链切口的位置有关。艾森伯格等[47] 通过脱嘌呤分析，特别是在 C_6T 产物中鉴定了这个缺口。在 4,285 处发现 CTC_5 序列（图 1），这一区域的起始位点与巴斯等[46] 的结果完全一致。目前还不可能确定单链切口起点的实际位置。尽管在两个富含 GC 的区域之间的一段富含 AT 区域 (4,298~4,307 处) 可能有一定意义，但是在这段区域内不存在明显的二级结构或对称序列。这样的区域在 SV40 DNA 复制的起点附近也被发现过（参考文献 48）。

表1. ΦX174编码容量

基因	根据聚丙烯酰胺凝胶电泳获得的蛋白分子量*	核苷酸数量（图1）	从序列信息获得的蛋白分子量
A	55,000~67,000	1,536	56,000
(*A**)	35,000		
B	19,000~25,000	(360) †	13,845‡
C	7,000		
D	14,500	456	16,811‡
E	10,000~17,500	(273) †	9,940
J	5,000	114	4,097‡
F	48,000	1,275	46,400
G	19,000	525	19,053‡
H	37,000	984	35,800
不编码和*C*		485	
总计		5,375	

* 见参考文献4。

† 括号中的值是重叠序列，因此计算 DNA 总长的时候没有包含在内。

‡ 这些值从氨基酸序列计算而来（在蛋白 B 中这些值从核苷酸序列推断而来）。其他的用以下公式得出

$$蛋白质分子量 = \frac{核苷酸数量}{3\times0.00915}$$

B Promoter

The second of the mRNA 5′ sequences (AUCGC)[34] has been mapped in restriction fragment R8 (Fig. 2), which starts about 300 nucleotides on from the proposed A* initiation. The sequence ATCGC is found at positions 4,832 and 4,888 in Fig. 1. The only way we can choose between them at the moment is that the second is preceded by the sequence TACAGTA (position 4,877), which is more akin to sequences found in known promoters[38] than are sequences preceding the other possible mRNA start. Irrespective of which of these sequences is used, the mRNA has a long "leader" sequence (232 or 176 nucleotides) before the next proposed initiation codon (gene *B*).

The B Protein

From a study of the ribonuclease T_1 digestion products of the ribosome binding sites of ΦX mRNAs[49], it was possible to identify an initiating ATG in position 5,064. From the genetic map[2,3], this would be expected to be gene *B* but, as discussed above, the A protein coding sequence extends right through this region, past the *Pst* site at residue 1 in Fig. 1, and terminates at residue 133. The initiating codon contained in the ribosome-protected sequence is, however, out of phase with the A protein reading frame. The proposed B protein coding sequence is one nucleotide to the left of the A protein phase, and continues until a termination codon occurs at position 49. Therefore the B protein coding sequence is totally contained within the *A* gene. These reading frames have been confirmed by sequencing mutants in genes *A* and *B* (*am*16, N.L.B. and M.S., in preparation; *am*18, *am*35, *ts*116 (ref. 50)). Since the B protein has not been purified no protein sequence data is available. The complete amino acid sequence can be predicted from the DNA sequence however. The protein is 120 amino acids long with a molecular weight of 13,845 (including the N-terminal Met). The molecular weight estimates of the B protein obtained by SDS-gel electrophoresis are mostly greater than this (see review, ref. 4), but the electrophoretic mobility varied with gel concentration and cross linker. Such anomalous behaviour suggests that there may be, for instance, carbohydrate attached to the B protein.

The C Protein

The next known gene product, protein C, maps between genes *B* and *D*. Examination of the DNA sequence in this region indicates that the most probable initiating ATG overlaps the termination codon, TGA, of gene *A* in the sequence ATGA at position 134. A possible termination codon for gene *C* could then be at position 391, although the sequence and phasing is not yet confirmed through this region. There is another possible protein initiation codon (position 51, overlapping the B protein termination codon) which would result in a slightly shorter gene product terminating at nucleotide 219. For the C protein, however, we favour the "A terminator" start, since only this reading frame contains a CAA sequence, which by a C → T alteration could give the ochre 6 mutant. Ochre 6 is a gene *C* mutant produced by the decay of ^{3}H-cytosine[51] and has been mapped in fragments A6 and F9 (ref. 52); that is, between nucleotides 170 and 205 (Fig. 1).

B 启动子

第二个 mRNA 5′ 序列（AUCGC）[34] 绘制在限制性片段 R8 中（图 2），它起始于之前所提及的 A* 起始位点后约 300 个核苷酸处。序列 ATCGC 在图 1 中的 4,832 位点和 4,888 位点被发现过。目前在它们之间选择的唯一方法是第二种序列位于序列 TACAGTA（4,877 位点）之后，与其他可能的 mRNA 起点之前的序列相比，序列 TACAGTA 与已知启动子中发现的序列更为相似 [38]。不管采用这些序列中的哪一个，在下一个推测的起始密码子（基因 *B*）之前 mRNA 都有一段长“引导”序列（232 个或 176 个核苷酸）。

B 蛋白

通过对 ΦX mRNA[49] 核糖体结合位点的核糖核酸酶 T_1 消化产物的研究，有可能鉴定出位于 5,064 处的一个起始密码子 ATG。从遗传图谱 [2,3] 可以预测该处是基因 *B*。但如上讨论，A 蛋白编码序列向右延伸穿过这个区域，经过图 1 中位于残基 1 的 *Pst* 位点，并终止于第 133 位残基。然而，这个包含核糖体保护序列的起始密码子不在 A 蛋白阅读框内。推算的 B 蛋白编码序列向 A 蛋白序列左侧移动了一个核苷酸，并延伸到位点 49 处出现终止密码子。因此，B 蛋白编码序列完全包含在 *A* 基因内。这些阅读框通过测定基因 *A* 和基因 *B* 内的突变体（*am*16，布朗和史密斯，准备发表中；*am*18、*am*35、*ts*116（参考文献 50））已经得到确认。因为 B 蛋白还未纯化，所以没有蛋白序列数据。然而可以从 DNA 序列预测完整的氨基酸序列。预计该蛋白长为 120 个氨基酸，分子量为 13,845（包括 N 末端甲硫氨酸）。通过聚丙烯酰胺凝胶电泳得出的 B 蛋白分子量的估计值远大于这个数值（综述见参考文献 4），但电泳迁移率可随凝胶浓度和交联剂变化。这些异常现象说明可能存在与 B 蛋白黏附的物质，例如多糖。

C 蛋白

C 蛋白是下一个已知的基因产物，定位于基因 *B* 和基因 *D* 之间。这一区域的 DNA 序列分析表明，最可能的起始密码子 ATG 与基因 *A* 的终止密码子 TGA（位于位点为 134 的 ATGA 序列中）重叠。虽然序列和阅读框在这个区域尚未获得确定，基因 *C* 的终止密码子位点可能为 391。还有另一个可能的蛋白起始密码子（位点 51，与 B 蛋白终止密码子重叠），它将导致产生稍短的基因产物，终止在核苷酸 219。然而对于 C 蛋白，我们更倾向于“A 终止子”起始，因为只有这个阅读框包含一个 CAA 序列，这个序列通过 C → T 变化可以产生赭石 6 突变体。赭石 6 是基因 *C* 的一个突变体，通过 ^3H– 胞嘧啶 [51] 衰变产生，并已经绘制在片段 A6 和 F9 中（参考文献 52）；也就是说，在核苷酸 170 和 205 之间（图 1）。

Sequence following the *D* Promoter

The mRNA 5′ sequence which maps before the *D* gene (GAUGC)[34] is found at position 358 in Fig. 1. The sequence preceding the messenger start has only four of the TATPuATPu nucleotides[38]. Thirty-two nucleotides after the mRNA initiation is the ATG (position 390) that initiates D protein synthesis. The amino acid sequence of the D protein has been determined almost completely, and nucleotide and amino acid sequences can be correlated to the termination codon at position 846 (ref. 33). The D protein, which is involved in capsid assembly, is 151 amino acids in length, with a molecular weight of 16,811. The *D* termination codon overlaps the initiation codon for gene *J* in the sequence TAATG. A similar structure has also been found by Platt and Yanofsky[53] in the tryptophan operon. The DNA sequence following this initiation codon matches the amino acid sequence of the small basic protein (37 amino acids) of the virion determined by D. Freymeyer, P. R. Shank, T. Vanaman, C.A.H. and M. H. Edgell (personal communication). Benbow *et al.*[2,3] suggested that the mutation *am*6 was located in a gene *J*, coding for the small protein of the virion, and mapping immediately before gene *F*. Although marker rescue experiments indicate that *am*6 is not in this region[54], the DNA sequence shows that there is a gene coding for the virion protein and we have defined this as gene *J* (ref. 33). Since the *J* initiation codon overlaps the *D* termination codon we had to look elsewhere for gene *E*, which genetic mapping[2,3] had placed between them. Amber mutants in gene *E* (*am*3, *am*27, *am*34 and *am*N11) were located by the marker rescue technique and sequenced. All were found to be within the *D* coding sequence, with the mutant amber codons one nucleotide to the right of the *D* reading frame[33]. Thus the *E* coding sequence is completely contained within the *D* coding region but in a different reading frame. The proposed initiation and termination codons for the E protein are at nucleotides 568 and 840, respectively[33], giving a protein 91 amino acids in length with a molecular weight of about 9,900 (including the N-terminal methionine).

The F Protein

Following the *J* gene is an intercistronic region of 39 nucleotides before initiation of the F protein. There is no known function of this apparently untranslated sequence, although the presence of a hairpin structure (positions 969–984) suggests that it could be the site of the *in vivo* messenger termination signal[37] mapped in the region. The F protein is initiated by the ATG at position 1,001. This is the capsid component of the virion, and almost all the amino acid sequence is known[22,24]. There are regions in this gene where the DNA sequence is not completely established, but the protein is about 424 amino acids in length, giving a molecular weight of $\simeq$46,300.

The G Protein Region

The termination signal for the F protein (position 2,276) is followed by an unusually long untranslated sequence of 111 nucleotides until the G protein initiation codon[31]. This region contains a looped structure which was postulated to have some functional role, as yet unknown, in the single-stranded DNA or the mRNA.

*D*启动子之后的序列

在图 1 的 358 位点处是 *D* 基因（GAUGC）[34] 之前的 mRNA 5′ 序列。该序列位于这个信使起点之前，只有 TATPuATPu 核苷酸 [38] 中的 4 个。mRNA 起始之后第 32 个核苷酸是 ATG（390 位点），它起始 D 蛋白的合成。D 蛋白的氨基酸序列基本已经被完全确定，核苷酸和氨基酸序列都与位于位点 846 处的终止密码子相关（参考文献 33）。D 蛋白参与病毒壳体组装，长度为 151 个氨基酸，分子量为 16,811。*D* 的终止密码子与基因 *J* 的起始密码子在序列 TAATG 内发生重叠。普拉特和亚诺夫斯基 [53] 在色氨酸操纵子中也曾发现过相似结构。这个起始密码子后面的序列与弗雷迈耶、尚克、瓦纳曼、哈奇森和埃杰尔（个人交流）确定的病毒小分子碱性蛋白（37 个氨基酸）的氨基酸序列匹配。本博等人 [2,3] 认为 *am*6 突变位于编码病毒小蛋白的基因 *J* 中，在图上位于基因 *F* 之前并与之紧密相邻。虽然标记获救实验表明 *am*6 并不在此区域内 [54]，但 DNA 序列显示这里存在一个编码病毒蛋白的基因，并且我们确定这就是基因 *J*（参考文献 33）。因为 *J* 的起始密码子与 *D* 的终止密码子重叠，我们不得不在别的区域寻找基因 *E*，遗传图谱显示 *E* 在二者之间。通过标记获救技术和测序，我们对基因 *E* 的琥珀突变体（*am*3、*am*27、*am*34 和 *am*N11）进行了定位。发现它们都位于 *D* 编码序列内，这些突变体的琥珀密码子相对于 *D* 阅读框全部整体右移一个核苷酸 [33]。因此，*E* 编码序列完全包含在 *D* 编码区域内，但是阅读框不同。推测 E 蛋白的起始密码子和终止密码子分别在核苷酸 568 和 840 处 [33]，编码蛋白长度为 91 个氨基酸，分子量为 9,900（包含 N 末端甲硫氨酸）。

F 蛋白

J 基因之后与 F 蛋白的起始位点之前存在一个 39 个核苷酸构成的顺反子之间的区域。这个看似不翻译的序列的功能未知，但存在一个发夹结构（969~984 区域），提示其可能是图谱上位于此区域内的体内信使终止信号的位点 [37]。F 蛋白起始于 1,001 位点的 ATG 密码子，是病毒衣壳的组分，并且几乎所有氨基酸序列都是已知的 [22,24]。这个基因中有些区域的 DNA 序列尚未完全完成，但其编码的蛋白长度大约为 424 个氨基酸，分子量约为 46,300。

G蛋白区域

F 蛋白的终止信号（2,276 位点）之后存在一段长达 111 个核苷酸的非翻译序列，直到 G 蛋白起始密码子 [31]。这个区域包含一个环状结构，推测其在单链 DNA 或 mRNA 中有一些未知功能。

Initiation of the G protein at position 2,387 is followed by a sequence of 425 nucleotides until termination at position 2,912, giving a spike protein of molecular weight 19,053. The nucleotide and amino acid sequences of this gene and product are known[24,32].

The H Protein

The initiation codon for the H protein (position 2,923) was identified first on the basis of the distribution of T nucleotides between the three reading phases, and later confirmed by amino acid sequence analysis. Amino acid sequence data on the H protein is minimal but the five peptide sequences known do correspond to the amino acid sequence, deduced from the DNA sequence by using the high frequency of third position T to help in assigning a reading frame to any given region. The DNA sequence is not entirely confirmed but it is possible to write a reasonably accurate amino acid sequence for the H protein. The protein terminates at nucleotide 3,907, in agreement with carboxypeptidase results, giving a spike protein of molecular weight ≃ 35,600 (326 amino acids). The amino acid sequence at the N terminus seems to be particularly rich in hydrophobic residues, which is consistent with its suggested function as the "pilot" protein that reacts with the bacterial membrane[55,56]. After H protein termination there are 66 nucleotides before initiation of the A protein at position 3,973.

Coding Capacity of the ΦX174 Genome

The most striking feature of the ΦX DNA sequence is the way in which the various functions of the genome are compressed within the 5,375 nucleotides. Since the identification of ΦX gene products[2,4] it has been clear that proteins of the accepted molecular weights could not be separately coded on the available length of DNA. However, with the presence of two pairs of overlapping genes (*B* within *A* (ref. 50), *E* within *D* (ref. 33)) the genome has more coding capacity than had been originally supposed on the assumption that each gene was physically separate. Table 1 summarises the molecular weights of the known ΦX-coded proteins. There are other potential initiation sites for polypeptide synthesis (for example, in genes *A*, *F*, *G* and *H*) and further genetic work may clarify whether there are in fact other ΦX genes as yet unidentified.

Initiation of Protein Synthesis

Table 2 lists the protein initiation sequences for genes *A*, *B*, *D*, *E*, *J*, *F*, *G* and *H*. It can be noted that there are no extra precursor sequences in proteins D, J, F, G or H at either the N or C terminus. There seems to be no relationship between the degree of complementarity to the 16S rRNA and the amount of protein synthesised, and we see no other features in the sequence that could explain different efficiencies of translation except where genes overlap.

G 蛋白起始位点 2,387 后是一段 425 个核苷酸的序列，直到终止位点 2,912，产生一个分子量为 19,053 的刺突蛋白。这个基因的核苷酸和氨基酸产物的序列是已知的[24,32]。

H 蛋 白

H 蛋白的起始密码子（位点 2,923）最初是在 3 个阅读框之间 T 核苷酸分布的基础上鉴定出来的，随后通过氨基酸序列分析得到确认。关于 H 蛋白的氨基酸序列数据极少，根据密码子第三位的高频 T 碱基分布有助于鉴定任何给定区域的阅读框，有 5 个已知肽段序列与其氨基酸序列对应。虽然 DNA 序列尚未完全确定，但已可能写出 H 蛋白的比较准确的氨基酸序列。这个蛋白终止于核苷酸 3,907 位点，与羧肽酶结果一致，产生一个分子量约为 35,600（326 个氨基酸）的刺突蛋白。N 末端的氨基酸序列好像特别富含疏水残基，这与其作为“引导”蛋白与细菌膜相互作用的假定功能一致[55,56]。H 蛋白终止之后与 A 蛋白 3,973 起始位点之前之间是一段 66 个核苷酸的序列。

ΦX174基因组的编码容量

ΦX174 DNA 序列最吸引人的特征是基因组的各种功能压缩在 5,375 个核苷酸中。自从鉴定了 ΦX174 的基因产物[2,4]，就明白在它的 DNA 长度内是不可能分别编码出这些分子量已被确认的蛋白的。然而，由于存在两对重叠基因（*B* 在 *A* 内（参考文献 50），*E* 在 *D* 内（参考文献 33）），这与最初认为的每个基因在位置上是分离的这一假设相比，这样的基因组具有更大的编码容量。表 1 总结了已知的 ΦX174 编码蛋白的分子量。多肽合成还存在其他的可能起始位点（例如，在基因 *A*、*F*、*G* 和 *H* 中），进一步的遗传研究可能会阐明是否真的存在尚未鉴定出的其他 ΦX 基因。

蛋白合成的起始

表 2 列出了基因 *A*、*B*、*D*、*E*、*J*、*F*、*G* 和 *H* 的蛋白起始序列。可以看出在蛋白 D、J、F、G 或 H 的 N 末端或 C 末端都没有额外的前体序列。16S rRNA 的互补程度和该蛋白的合成量之间似乎没有关系，并且除基因重叠区域外，我们在这些序列中没有发现其他特征可以解释翻译效率的不同。

Table 2. Initiation sequences of ΦX174 coded proteins

Gene	Sequence
D	C-C-A-C-T-A-A-T-A-G-G-T-A-A-G-A-A-A-T-C-A-T-G-A-G-T-C-A-A-G-T-T-A-C-T
	Ser Gln Val Thr
E	C-T-G-C-G-T-T-G-A-G-G-C-T-T-G-C-G-T-T-T-A-T-G-G-T-A-C-G-C-T-G-G-A-C-T
J	C-G-T-G-C-G-G-A-A-G-G-A-G-T-G-A-T-G-T-A-A-T-G-T-C-T-A-A-A-G-G-T-A-A-A
	Ser Lys Gly Lys
F	C-C-C-T-T-A-C-T-T-G-A-G-G-A-T-A-A-A-T-T-A-T-G-T-C-T-A-A-T-A-T-T-C-A-A
	Ser Asn Ile Gln
G	T-T-C-T-G-C-T-T-A-G-G-A-G-T-T-T-A-A-T-C-A-T-G-T-T-T-C-A-G-A-C-T-T-T-T
	Met Phe Gln Thr Phe
H	C-C-A-C-T-T-A-A-G-T-G-A-G-G-T-G-A-T-T-T-A-T-G-T-T-T-G-G-T-G-C-T-A-T-T
	Met Phe Gly Ala Ile
A	C-A-A-A-T-C-T-T-G-G-A-G-G-C-T-T-T-T-T-T-A-T-G-G-T-T-C-G-T-T-C-T-T-A-T
B	A-A-A-G-G-T-C-T-A-G-G-A-G-C-T-A-A-A-G-A-A-T-G-G-A-A-C-A-A-C-T-C-A-C-T
16S RNA 3′ end	HO A-U-U-C-C-U-C-C-A-C-U-A-G

Where the protein start has been independently confirmed by protein sequencing data the amino acid sequences are indicated. The other initiation regions were identified as described in the text. Sequences complementary to the 3′ end of 16S rRNA (refs 42, 43) are boxed; broken lines indicate further complementarity if some nucleotides are looped out or not matched. Ribosome binding to mRNA has been demonstrated in these regions for genes *J*, *F*, *G* and *B* (ref. 49).

Transcription of ΦX174

The sequences preceding known mRNA starts[34-36] are shown in Table 3. Other studies on promoter sequences[38] have suggested certain features that they may have in common. Although some of these features are present in the sequences preceding the ΦX transcription initiations others are not, and at present it is difficult to suggest what signal on the DNA determines a promoter site or the efficiency with which it initiates RNA synthesis. It is interesting to note that a polymerase binding site found by Chen *et al.*[57], but not associated with any *in vitro* or *in vivo* mRNA starts, mapped near the region where there is the sequence TATGATG characteristic of promoters[38] (positions 2,705–2,711).

表 2. ΦX174编码蛋白质的起始序列

基因	序列	氨基酸序列
D	C-C-A-C-T-A-A-T-A-G-G-T-A-A-G-A-A-A-T-C-A-T-G-A-G-T-C-A-A-G-T-T-A-C-T	Ser Gln Val Thr
E	C-T-G-C-G-T-T-G-A-G-G-C-T-T-G-C-G-T-T-T-A-T-G-G-T-A-C-G-C-T-G-G-A-C-T	
J	C-G-T-G-C-G-G-A-A-G-G-A-G-T-G-A-T-G-T-A-A-T-G-T-C-T-A-A-A-G-G-T-A-A-A	Ser Lys Gly Lys
F	C-C-C-T-T-A-C-T-T-G-A-G-G-A-T-A-A-A-T-T-A-T-G-T-C-T-A-A-T-A-T-T-C-A-A	Ser Asn Ile Gln
G	T-T-C-T-G-C-T-T-A-G-G-A-G-T-T-T-A-A-T-C-A-T-G-T-T-T-C-A-G-A-C-T-T-T-T	Met Phe Gln Thr Phe
H	C-C-A-C-T-T-A-A-G-T-G-A-G-G-T-G-A-T-T-T-A-T-G-T-T-T-G-G-T-G-C-T-A-T-T	Met Phe Gly Ala Ile
A	C-A-A-A-T-C-T-T-G-G-A-G-G-C-T-T-T-T-T-T-A-T-G-G-T-T-C-G-T-T-C-T-T-A-T	
B	A-A-A-G-G-T-C-T-A-G-G-A-G-C-T-A-A-A-G-A-A-T-G-G-A-A-C-A-A-C-T-C-A-C-T	
16S RNA 3′末端	HO A-U-U-C-C-U-C-C-A-C-U-A-G	

由蛋白质序列数据独立确定出的蛋白质起始位点，表中标明了氨基酸序列。其他起始区域按正文描述的方法鉴定。与16S rRNA 3′末端互补的序列（参考文献42，参考文献43）加了方框；虚线方框表示如果一些核苷酸成环或未配对而引起的进一步的互补。基因 *J*、*F*、*G* 和 *B* 的 mRNA 与核糖体结合的区域已经得到阐明（参考文献49）。

ΦX174的转录

表3中显示了已知mRNA起始位点之前的序列[34-36]。对启动子序列的其他研究[38]表明，这些序列可能具有某些相同的特征。尽管在ΦX转录起始之前的序列中只出现了其中一些特征而其他的没有，并且目前很难确定DNA上哪些信号决定着一个启动子位点或这一位点起始RNA合成的效率。值得注意的是，陈等人[57]发现的一种聚合酶结合位点被定位在启动子TATGATG序列[38]（2,705~2,711区域）附近，而与任何体外或体内mRNA合成起始无关。

Table 3. Promoter sequences in ΦX174

Promoter	Sequence	Position	
A promoter	A-G-G-A-T-T-G-A-C-A-C-C-C-T-C-C-C-A-A-T-T-G-T-A-T-G-T[T-T-T-C-A-T-G]C-C-T-C-C-A-A-A-T-C-T - - -	3954	18 nucleotides to A protein start
D promoter	(R7b/R6c) G-T-T-G-A-C-A-T-T-T-T-A-A-A-A-G-A-G-C-G-T-G-G-A-T-T-A-C[T-A-T-C-T-G-A]G-T-C-C-G-A-T-G-C-T	358	32 nucleotides to D protein start
B promoter?	C-A-G-G-T-A-G-C-(R3/R8) G-T-T-G-A-C-C-C-T-A-A-T-T-T-G-G-T-C-G[T-C-G-G-G-T-A]C-G-C-A-A-T-C-G-C-C	4832	232 nucleotides to B protein start
B promoter?	A-G-C-T-T-G-C-A-A-A-A-T-A-C-G-T-G-G-C-C-T-T-A-T-G-G-T[T-A-C-A-G-T-A]T-G-C-C-C-A-T-C-G-C-A	4888	176 nucleotides to B protein start

mRNA initiation sequences[34-36] are underlined. Boxed regions indicate sequences that may correspond to the TATPuATPu sequence found in other promoters[38], taking into account the distance from the mRNA starts.

The Use of Codons in ΦX174

Table 4 shows the codons used in regions where the nucleotide sequence is fully confirmed. It is clear that the pattern established by early observations on non-random use of codons[23,24] is continued now that more information is available. In particular, the preference for T at the third position of the codon is marked throughout the genome, as shown in Table 4. In regions of overlapping genes, one of the pair tends to continue the "third T" trend (*D* and *B*), thus excluding the other (*E* and *A*). This may give some indication of the order in which overlapping genes evolved[33,50]. Another interesting feature is the very low occurrence of codons starting AG, particularly in non-overlapping regions. The base composition of the sequence of ΦX174 DNA shown in Fig. 1 is: A, 23.9%; C, 21.5%; G, 23.3% and T, 31.2%. This is in good agreement with previously determined values (see ref. 4).

Table 4. Codons used in ΦX174

Phe	TTT	39	Ser	TCT	35	Tyr	TAT	36	Cys	TGT	12
	TTC	26		TCC	9		TAC	15		TGC	10
Leu	TTA	19		TCA	16	Ter	TAA	3	Ter	TGA	5
	TTG	26		TCG	14		TAG	0	Trp	TGG	16
Leu	CTT	36	Pro	CCT	34	His	CAT	16	Arg	CGT	40
	CTC	15		CCC	6		CAC	7		CGC	29
	CTA	3		CCA	6	Gln	CAA	27		CGA	4
	CTG	24		CCG	21		CAG	34		CGG	8
Ile	ATT	45	Thr	ACT	40	Asn	AAT	37	Ser	AGT	9
	ATC	12		ACC	18		AAC	25		AGC	5
	ATA	2		ACA	13	Lys	AAA	47	Arg	AGA	6
Met	ATG	42		ACG	19		AAG	31		AGG	1
Val	GTT	53	Ala	GCT	64	Asp	GAT	44	Gly	GGT	38
	GTC	14		GCC	17		GAC	35		GGC	28
	GTA	10		GCA	12	Glu	GAA	27		GGA	13
	GTG	11		GCG	12		GAG	34		GGG	3

The totals are derived from sequences in Fig. 1 which are fully confirmed, that is, 377 codons in gene *A*, 120 in gene *B*, 152 in gene *D*,

表 3. ΦX174 的启动子序列

启动子	序列	位置	说明
A 启动子	A-G-G-A-T-T-G-A-C-A-C-C-C-T-C-C-C-A-A-T-T-G-T-A-T-G-T[T-T-T-C-A-T-G]C-C-T-C-C-A-A-A-T-C-T — — —	3954	距离蛋白A起始位点18个核苷酸
D 启动子	(R7b/R6c) G-T-T-G-A-C-A-T-T-T-T-A-A-A-A-G-A-G-C-G-T-G-G-A-T-T-A-C[T-A-T-C-T-G-A]G-T-C-C-G-A-T-G-C-T	358	距离蛋白D起始位点32个核苷酸
B 启动子？	C-A-G-G-T-A-G-C-G-T-T-G-A-C (R3/R8) -C-C-T-A-A-T-T-T-T-G-G-T-C-G[T-C-G-G-G-T-A]C-G-C-A-A-T-C-G-C-C	4832	距离蛋白B起始位点232个核苷酸
B 启动子？	A-G-C-T-T-G-C-A-A-A-A-T-A-C-G-T-G-G-C-C-T-T-A-T-G-G-T[T-A-C-A-G-T-A]T-G-C-C-C-A-T-C-G-C-A	4888	距离蛋白B起始位点176个核苷酸

mRNA 起始序列[34-36] 加了下划线。从 mRNA 起点的距离上考虑，加方框的区域表示可能是与在其他启动子中发现的 TATPuATPu 序列相对应的序列[38]。

ΦX 174中密码子的使用

表 4 显示了在核苷酸序列已被完全确定的区域中所使用的密码子。很明显，早期对非随机使用密码子的观察结果建立的模式[23,24]不断发展而产生了更多可利用信息。特别是，整个基因组密码子第三位碱基都明显地偏好 T，见表 4 所示。在重叠基因区域，一对基因倾向于延续“第三位 T”的趋势（*D* 和 *B*），因而排除了与另一对的重合（*E* 和 *A*）。这可能给我们一些关于重叠基因进化顺序的提示[33,50]。另一个有趣的特征是密码起始于 AG 的发生率很低，特别是在非重叠区域。图 1 显示了 ΦX174 DNA 序列的碱基组成是：A，23.9%；C，21.5%；G，23.3%；T，31.2%。这与以前确定的数值很相符（见参考文献 4）。

表 4. ΦX174中使用的密码子

Phe	TTT	39	Ser	TCT	35	Tyr	TAT	36	Cys	TGT	12
	TTC	26		TCC	9		TAC	15		TGC	10
Leu	TTA	19		TCA	16	Ter	TAA	3	Ter	TGA	5
	TTG	26		TCG	14		TAG	0	Trp	TGG	16
Leu	CTT	36	Pro	CCT	34	His	CAT	16	Arg	CGT	40
	CTC	15		CCC	6		CAC	7		CGC	29
	CTA	3		CCA	6	Gln	CAA	27		CGA	4
	CTG	24		CCG	21		CAG	34		CGG	8
Ile	ATT	45	Thr	ACT	40	Asn	AAT	37	Ser	AGT	9
	ATC	12		ACC	18		AAC	25		AGC	5
	ATA	2		ACA	13	Lys	AAA	47	Arg	AGA	6
Met	ATG	42		ACG	19		AAG	31		AGG	1
Val	GTT	53	Ala	GCT	64	Asp	GAT	44	Gly	GGT	38
	GTC	14		GCC	17		AGC	35		GGC	28
	GTA	10		GCA	12	Glu	GAA	27		GGA	13
	GTG	11		GCG	12		GAG	34		GGG	3

列出的所有信息都是从图 1 中完全确定的序列中获得，即基因 *A* 中的 377 个密码子，基因 *B* 中的 120 个密码子，基因 *D* 中的 152 个

91 in gene *E*, 38 in gene *J*, 344 in gene *F*, 175 in gene *G* and 49 in gene *H*. Out of a total of 1,346 codons 42.9% terminate in T. The percentages in the different genes are: *A*, 37.1 (non-overlapping region 47.1; overlapping region 15.8); *B*, 34.2; *D*, 42.1; *E*, 14.3; *J*, 47.4; *F*, 52.0; *G*, 54.3; *H*, 49.0. The initiating ATG is included in all cases.

We thank D. McCallum and R. Staden for carrying out the computer data storage and analysis of the sequence.

Note added in proof: J. E. Sims and D. Dressler (personal communication) have independently determined the sequence in positions 263–375 and 4,801–4,940. Their results agree with those given in Fig. 1. They have also identified the "B" mRNA start as being at position 4,888.

(**265**, 687-695; 1977)

F. Sanger, G. M. Air*, B. G. Barrell, N. L. Brown†, A. R. Coulson, J. C. Fiddes, C. A. Hutchison III‡, P. M. Slocombe§ & M. Smith¶

MRC Laboratory of Molecular Biology, Hills Road, Cambridge CB2 2QH, UK

Present addresses: *John Curtin School of Medical Research, Microbiology Department, Canberra City ACT 2601, Australia

†Department of Biochemistry, University of Bristol, Bristol BS8 1TD, UK

‡Department of Bacteriology and Immunology, University of North Carolina, Chapel Hill, North Carolina 27514

§Max-Planck-Institut für Molekulare Genetik, 1 Berlin 33, FRG

¶Department of Biochemistry, University of British Columbia, Vancouver BC, Canada V6T 1W5

Received November 30; accepted December 24 1976.

References:

1. Sanger, F. & Coulson, A. R. *J. Molec. Biol.* **94**, 441-448 (1975).
2. Benbow, R. M., Hutchison, C. A. III, Fabricant, J. D. & Sinsheimer, R. L. *J. Virol.* 7, 549-558 (1971).
3. Benbow, R. M., Zuccarelli, A. J., Davis, G. C. & Sinshiemer, R. L. *J. Virol.* **13**, 898-907 (1974).
4. Denhardt, D. T. *CRC Crit. Rev. Microbiol.* **4**, 161-222 (1975).
5. Hall, J. B. & Sinsheimer, R. L. *J. Molec. Biol.* **6**, 115-127 (1963).
6. Ling, V. *Proc. Natl. Acad. Sci. U.S.A.* **69**, 742-746 (1972).
7. Harbers, B., Delaney, A. D., Harbers, K. & Spencer, J. H. *Biochemistry* **15**, 407-414 (1976).
8. Burton, K. & Petersen, G. B. *Biochem. J.* **75**, 17-27 (1960).
9. Chadwell, H. A. Thesis, University of Cambridge (1974).
10. Sadowski, P. D. & Bakyta, I. *J. Biol. Chem.* **247**, 405-412 (1972).
11. Ling, V. *FEBS Lett.* **19**, 50-54 (1971).
12. Ziff, E. B., Sedat, J. W. & Galibert, F. *Nature New Biol.* **241**, 34-37 (1973).
13. Galibert, F., Sedat, J. W. & Ziff, E. B. *J. Molec. Biol.* **87**, 377-407 (1974).
14. Robertson, H. D., Barrell, B. G., Weith, H. L. & Donelson, J. E. *Nature New Biol.* **241**, 38-40 (1973).
15. Air, G. M. & Bridgen, J. *Nature New Biol.* **241**, 40-41 (1973).
16. Sanger, F., Donelson, J. E., Coulson, A. R., Kössel, H. & Fischer, D. *Proc. Natl. Acad. Sci. U.S.A.* **70**, 1209-1213 (1973).
17. Schott, H. *Makromolek. Chem.* **175**, 1683-1693 (1974).
18. Donelson, J. E., Barrell, B. G., Weith, H. L., Kössel, H. & Schott, H. *Eur. J. Biochem.* **58**, 383-395 (1975).
19. Blackburn, E. H. *J. Molec. Biol.* **93**, 367-374 (1975).
20. Blackburn, E. H. *J. Molec. Biol.* **107**, 417-432 (1976).
21. Sedat, J. W., Ziff, E. B. & Galibert, F. *J. Molec. Biol.* **107**, 391-416 (1976).
22. Air, G. M. *J. Molec. Biol.* **107**, 433-444 (1976).
23. Air, G. M. *et al. J. Molec. Biol.* **107**, 445-458 (1976).

密码子，基因 *E* 中的 91 个密码子，基因 *J* 中的 38 个密码子，基因 *F* 中的 344 个密码子，基因 *G* 中的 175 个密码子，基因 *H* 中的 49 个密码子。共 1,346 个密码子，其中有 42.9% 的末位为碱基 T。在不同基因中的百分比为：*A*，37.1（非重叠区域 47.1；重叠区域 15.8）；*B*，34.2；*D*，42.1；*E*，14.3；*J*，47.4；*F*，52.0；*G*，54.3；*H*，49.0。所有例子中都包含起始密码子 ATG。

我们感谢麦卡勒姆和施塔登进行的计算机数据存储和序列分析。

附加说明：西姆斯和德雷斯勒（个人交流）已经独立确定了 263~375 区域和 4,801~4,940 区域的序列。他们的结果与图 1 给出的一致。他们也鉴定了“B” mRNA 起始于位点 4,888 处。

（郑建全 李梅 翻译；曾长青 审稿）

24. Air, G. M., Blackburn, E. H., Sanger, F. & Coulson, A. R. *J. Molec. Biol.* **96**, 703-719 (1975).
25. Lee, A. S. & Sinsheimer, R. L. *Proc. Natl. Acad. Sci. U.S.A.* **71**, 2882-2886 (1974).
26. Hayashi, M. N. & Hayashi, M. *J. Virol.* **14**, 1142-1152 (1974).
27. Vereijken, J. M., van Mansfeld, A. D. M., Baas, P. D. & Jansz, H. S. *Virology* **68**, 221-233 (1975).
28. Jeppesen, P. G. N., Sanders, L. & Slocombe, P. M. *Nucl. Acids Res.* **3**, 1323-1339 (1976).
29. Sato, S., Hutchison, C. A. III & Harris, J. I. *Proc. Natl. Acad Sci. U.S.A.* (in the press).
30. Brown, N. L. & Smith, M. *FEBS Lett.* **65**, 284-287 (1976).
31. Fiddes, J. C. *J. Molec. Biol.* **107**, 1-24 (1976).
32. Air, G. M., Sanger, F. & Coulson, A. R. *J. Molec. Biol.* **108**, 519-533 (1976).
33. Barrell, B. G., Air, G. M. & Hutchison, C. A. III *Nature* **264**, 34-41 (1976).
34. Smith, L. H. & Sinsheimer, R. L. *J. Molec. Biol.* **103**, 699-735 (1976).
35. Grohmann, K., Smith, L. H. & Sinsheimer, R. L. *Biochemistry* **14**, 1951-1955 (1975).
36. Axelrod, N. *J. Molec. Biol.* **108**, 753-779 (1976).
37. Hayashi, M., Fujimura, F. K. & Hayashi, M. *Proc. Natl. Acad. Sci, U.S.A.* **73**, 3519-3523 (1976).
38. Pribnow, D. *Proc. Natl. Acad. Sci. U.S.A.* **72**, 784-788 (1975).
39. Rosenberg, M., de Crombrugghe, B & Musso, R. *Proc. Natl. Acad. Sci. U.S.A.* **73**, 717-721 (1976).
40. Bertrand, K. *et al. Science* **189**, 22-26 (1975).
41. Sugimoto, K., Sugisaki, H., Okamoto, T. & Takanami, M. *J. Molec. Biol.* (in the press).
42. Shine, J. & Dalgarno, L. *Proc. Natl. Acad. Sci. U.S.A.* **71**, 1342-1346 (1974).
43. Steitz, J. A. & Jakes, K. *Proc. Natl. Acad. Sci. U.S.A.* **72**, 4734-4738 (1975).
44. Henry, T. J. & Knippers, R. *Proc. Natl. Acad. Sci, U.S.A.* **71**, 1549-1553 (1974).
45. Linney, E. & Hayashi, M. *Nature* **249**, 345-348 (1974).
46. Baas, P. D., Jansz, H. S. & Sinsheimer, R. L. *J. Molec. Biol.* **102**, 633-656 (1976).
47. Eisenberg, S., Harbers, B., Hours, C. & Denhardt, D. T. *J. Molec. Biol.* **99**, 107-123 (1975).
48. Subramanian, K. N., Dhar, R. & Weissman, S. M. *J. Biol. Chem.* (in the press).
49. Ravetch, J. V., Model, P. & Robertson, H. D. *Nature* **265**, 698-702 (1977).
50. Smith, M. *et al.* (submitted to Nature).
51. Funk, F. & Sinsheimer, R. L. *J. Virol.* **6**, 12-19 (1970).
52. Baas, P. D., van Heusden, G. P. H., Vereijken, J. M., Weisbeek, P. J. & Jansz, H. S. *Nucl. Acids Res.* **3**, 1947-1960 (1976).
53. Platt, T. & Yanofsky, C. *Proc. Natl. Acad. Sci, U.S.A.* **72**, 2399-2403 (1975).
54. Weisbeek, P. J., Vereijken, J. M., Baas, P. D., Jansz, H. S. & Van Arkel, G. A. *Virology* 72, 61-71 (1976).
55. Jazwinski, S. M., Lindberg, A. A. & Kornberg, A. *Virology* **66**, 283-293 (1975).
56. Kornberg, A. *DNA Synthesis* (W. H. Freeman, San Francisco, 1974).
57. Chen, C. Y., Hutchison, C. A. III & Edgell, M. H. *Nature New Biol.* **243**, 233-236 (1973).
58. van Mansfeld, A. D. M., Vereijken, J. M. & Jansz, H. S. *Nucl. Acids Res.* **3**, 2827-2843 (1976).